Rifts and Passive Margins
Structural Architecture, Thermal Regimes, and Petroleum Systems

Rifts and passive margins are extremely important for the petroleum industry, as they are areas of high sedimentation and can contain significant oil and gas resources. This book provides a comprehensive understanding of rifts and passive margins as a whole. It synthesizes in one volume the existing information devoted to specific aspects of these vitally important hydrocarbon habitats. This collection of state-of-the-art information on the topic facilitates the better use of this knowledge to assess the risks of exploring and operating in these settings and the development of systematic and predictive hydrocarbon screening tools. This book will be invaluable for a broad range of readers, from advanced geology students and researchers to exploration geoscientists to exploration managers exploring for and developing hydrocarbon resources in analogous settings.

Michal Nemčok is a research professor at the Energy & Geoscience Institute at the University of Utah and a visiting professor at the Energy and Geoscience Institute Laboratory at the Geological Institute SAV, Bratislava, Slovak Republic. He has performed structural and hydrocarbon evaluations in numerous rift settings, such as the Pannonian Basin system in the Czech Republic and Slovakia; the Basin and Range province of the western United States; the Bristol Channel Basin in the United Kingdom; the Central Basin in the Democratic Republic of Congo; and the Salton Sea region in California. He has also worked in many passive margin settings, such as West Australia, West India, East India, the eastern United States and Canada, the African margins of the Central Atlantic, the Equatorial Atlantic, the Gulf of Mexico, the Guyana-Suriname Basin, Gabon and Cameroon, the Black Sea region, and southeastern Brazil. The results of most of these studies have been presented in thirty-seven articles published in refereed scientific journals and in three edited books. He is the coauthor of *Thrustbelts: Structural Architecture, Thermal Regimes and Petroleum Systems* (2005, Cambridge University Press).

Rifts and Passive Margins

Structural Architecture, Thermal Regimes, and Petroleum Systems

MICHAL NEMČOK

ENERGY & GEOSCIENCE INSTITUTE
University of Utah
and
Energy and Geoscience Institute Laboratory at the Geological Institute SAV

CAMBRIDGE
UNIVERSITY PRESS

CAMBRIDGE
UNIVERSITY PRESS

University Printing House, Cambridge CB2 8BS, United Kingdom

Cambridge University Press is part of the University of Cambridge.

It furthers the University's mission by disseminating knowledge in the pursuit of education, learning, and research at the highest international levels of excellence.

www.cambridge.org
Information on this title: www.cambridge.org/9781107025837

First published 2016
Reprinted 2017

Printed in the United Kingdom by Print on Demand World Wide

A catalog record for this publication is available from the British Library.

Library of Congress Cataloging in Publication Data
Nemčok, Michal.
Rifts and passive margins : structural architecture, thermal regimes, and petroleum systems / Michal Nemčok, University of Utah.
Includes bibliographical references and index.
ISBN 978-1-107-02583-7 (hardback)
QE606.N44 2016
551.8′72–dc23 2015023014

ISBN 978-1-107-02583-7 Hardback

To Ján, Eva, and Bruce

Contents

Acknowledgments

I wish to thank all those who contributed to the progress of this book, including the work on the earlier report (Energy and Geoscience Institute Report No. 01-00059-5000-50501401). The project called "Hydrocarbon exploration models in deep-water rift and passive margin settings," which took place during 2006–2008, was funded by Gaz de France, Nexen, Reliance, and RWE, which are gratefully acknowledged. Ivana Nemčoková, Marína Matejová, Štefánia Sliacka, and Ray Levey helped with organizing the subsequent work on the book. Detailed discussions with Bruce R. Rosendahl, Andreas Henk, Sudipta T. Sinha, Mainak Choudhuri, Clay Jones, Charles J. Stuart, Joseph N. Moore, Tom Powell, and Mitch Stark improved parts of the book focused on rift-drift transition, finite-element modeling, continental break-up, hot spot effects, fluid flow models, depositional systems at passive margins, and geothermal reservoirs and their controlling factors, respectively. Júlia Kotulová, Paul Sikora, Mauricio Parra, and Ján Král' provided friendly reviews of Chapter 15 and Vitrinite reflectance and Rock-Eval pyrolysis text in Chapter 4, Chronostratigraphy text in Chapter 4, Low-temperature thermochronology text in Chapter 4, and Low-temperature thermochronology text in Chapter 4, respectively. Ivana Nemčoková made the language editing and figure formatting for the entire manuscript. Štefánia Sliacka assisted with permissions and book production processes. Doug Jensen kindly helped with book cover design prior to manuscript submission. Gabi Lach kindly provided a cover photograph. Geological photographs and figures were provided by Robert S. Tye, Gary J. Hampson, Gianreto Manatschal, Sudipta T. Sinha, Achyuta Ayan Misra, Mainak Choudhuri, and Atul Patidar. Ivana Nemčoková, Marína Matejová, Štefánia Sliacka, Matej Molčan, Samuel Rybár, Lucia Ledvényiová, Michal Jánošík, Ondrej Pelech, Patrícia Ekkertová, Andrea Bartošová, Tomáš Klučiar, Ján Nemčok, Jakub Bobrovský, Soňa Tomaškovičová, Marián Bošanský, Viera Šimonová, Eva Komanická, Ivana Perichtová, Jakub Múčka, and Martin Mal'a helped with figure drafting. Marína Matejová and Štefánia Sliacka assisted with references.

Introduction

Over the past decade, a large amount of exploration data has been generated at rift and passive margin settings indicating that some of the basic exploration rules for these settings that control present-day exploration efforts are based on flawed assumptions.

For example, the oil industry views ocean basin as an outboard of interesting areas, understanding oceanic environment as one type of basin. Its underlying oceanic crust is understood as cooler than continental crust and as not having sufficiently thick overburden. As a result, source rocks should be immature and petroleum systems should be missing.

However, geologists have discovered a number of hydrocarbon fields on oceanic crust in the Gulf of Mexico and offshore Nigeria, and there is at least one well, far in front of the continental margin, DSDP well 368, which penetrated organic-rich shale, which is not immature. Furthermore, current academic research indicates that there are several types of oceanic crust and each of them has very different thermodynamic history. Some of them are warmer than "classical oceanic crust."

For example, the oil industry routinely uses several standard source rock maturation/hydrocarbon expulsion timing software packages, and this modeling usually utilizes heat conduction algorithms and omits the heat transfer by fluids for the time being. The reason is that the heat conduction is understood as a dominant type of the heat transfer. The heat transfer by fluids is considered insignificant in current exploration settings and, as such, ignored.

However, there are exploration data from areas such as the Canary Islands or the Espírito Santo Basin, where exploring companies noted a significant impact of geothermal fluids on source rock maturation. Other passive margins start to indicate cementation problems with otherwise good reservoir rocks in discharge areas of geothermal fluids. There are also evidences from the rift settings, such as the Rhine Graben, about fluid flow systems reducing and increasing the organic matter maturation level in recharge and discharge areas of the geothermal fluid flow systems. Furthermore, there are present-day geothermal areas in rift and pull-apart settings where the geothermal industry uses significant thermal anomalies for commercial electricity production. The research the geothermal industry has done indicates that some of the thermal anomalies can have life span, which should be long enough to modify maturation history considerably. As documented by heat flow maps of the Salton Sea region, these anomalies can express themselves even on the elevated background heat flow regime.

We also need to appreciate that widely spaced regional 2-D seismic grids can provide a misleading picture of the regional context needed for successful hydrocarbon exploration. Dense 3-D seismic-based programs, in contrast, focus on the prospect level and, consequently, do not provide sufficient regional context about the depositional section under investigation. What is required are exploration models that bridge the regional and prospect level scales to improve understanding of the elements and processes contributing to basin and petroleum systems development, and the role of their controlling factors.

Using the words of Bruce Rosendahl:

> the initial geological problem to be solved is understanding basin development through the transformation of disorganized rifting to organized seafloor spreading. Thus, we must look at the evolution from indecisive continental rift basins and imperfect accommodation zones (Tanganyika region) to pseudo-volcanic lineaments and transfer faults (Turkana region) to gradual, tortuous, and probably sub-aerial spreading (Afar region) to opening (Red Sea region), if located on the successful limb of a triple junction to discern how these environments affect petroleum system elements and processes.
>
> Of course, all of the petroliferous margins are much more complex, more segmented, and more controlled by deep thermodynamics than one tends to envision. This is especially true on continental shelf edges. Perhaps the best way to imagine these exploration terrains is as a bunch of small, fault-connected basins divided by accommodation zones and transfer faults that hoped to evolve into transform faults, and did so rather imperfectly, or not at all.
>
> As a result, one can imagine a series of basins with varying thermal regimes strung along the continental margins, usually oriented at two different oblique directions to the original margin. The kitchens may generate hydrocarbons but they may well end up on the dividing (and overlying) structures. Successful exploration in deep water margins, thus, seems to require a capability of model-driven interpretation of the imaging data, focusing on basin development,

an understanding of controlling factors to predict thermal history, and the implications of both of these factors on the presence or absence of essential petroleum system elements and processes.

Motivated by the aforementioned examples and discussion, this book tries to provide a comprehensive understanding of rifts and passive margins as a whole. The aim is to synthesize existing information devoted to specific aspects of these most important hydrocarbon habitats. This book assembles this information in one volume, in a manner that permits the use of this knowledge to assess the risks of exploring and operating in these settings.

The plan for this book originated with a three-year-long project called "Hydrocarbon exploration models in deep-water rift and passive margin settings," which took place during 2006–8, which summarized various aspects of exploration in rift and passive margin settings provided by large and diverse literature, and which addressed gaps in knowledge.

This synthesis is completed from results of personal, long-term research on rifts and passive margins, numeric validations of various concepts, and extensive tables documenting various factors influencing structural styles, thermal regimes, and petroleum systems, as well as rates of geologic processes. This book should have value for a broad range of readers, spanning from geology students, to exploration geologists, to exploration managers searching for the hydrocarbons in analogous settings. This book's strategy is to develop a comprehensive appreciation of the factors controlling structural architecture, basin development, and fluid flow in rifted margin settings. Synthesis of this knowledge is then used to tackle what is perhaps the most enigmatic aspect of these environments – the factors controlling thermal regimes and temperature peculiarities of rift and passive margin settings. The final phase of this book is the discussion of various aspects of petroleum systems based on the results of the previous structural and thermal phases. Special attention is given to the oceanic basin – the large frontier region, almost untouched by exploration so far.

As a result, this book is divided into three blocks.

The first block accumulates and synthesizes the structural and fluid flow knowledge necessary for fact-based and forward-thinking exploration plans in the rift and passive margin settings. The text begins with a basic description of structural styles, delineation of extension directions, which is necessary for the proper layout of geophysical surveys; includes determination of crust types, which are necessary for definition of different petroleum system candidates; determination of structural segmenting features, which define the sizes and geometry of a potential petroleum system; and determination of structural timing events, which is necessary for specification of source/reservoir/trap/seal temporal relations. This block is also focused on the mechanics of rifting and its transition to drifting and the controlling role of lithospheric composition and its variations on rifting and breakup. The text tries to contribute to the knowledge of why certain passive margin segments are so prolific and why others are (or should be considered) dry. Ultimately, this means a true understanding of the continuum that exists between pure rift systems and pure pull-apart basins, and how the continuum applies to petroleum systems. The structural block concludes with discussion of the role of preexisting anisotropy on rifting and breakup, and occurrences and controlling factors of associated fluid flow systems.

The second block develops the thermal foundation for successful (and unsuccessful) petroleum system determinations. The text consists of evaluation of critical factors influencing the thermal history of rocks in rift and passive margin settings, including the effects of pre-rift heat flow regime, lithostratigraphy, deposition, structural framework, erosion, deformation, and movement of geothermal fluids.

The third block develops knowledge of source rock distribution, source maturation history, reservoir distribution, seal quality, potential migration scenarios, trap candidates, and preservation issues. The text evaluates the effects of movement of geothermal fluids on diagenetic history of reservoirs and seals, and maturation history of source rocks. Knowledge of reservoir distribution works with prediction tools designed to assess continental margin vertical motion histories and the controlling factors on sediment distribution through breaches of the marginal uplifts.

1 Basic description of structural styles in rift and passive margin settings, including extension directions and key structural elements

In order to place the basins discussed in this book into plate settings, we use the respective categories from the classification of Bally (1982) (Figure 1.1):

1. Basins located on the rigid lithosphere, not associated with formation of megasutures
   1.1. basins related to oceanic crust formation
      1.1.1. rifts
      1.1.2. basins associated with oceanic transform faults
      1.1.3. oceanic abyssal plains
      1.1.4. passive margins
         1.1.4.1. passive margins overlying earlier rift systems
         1.1.4.2. passive margins overlying earlier transform systems
         1.1.4.3. passive margins overlying earlier back-arc basins
   1.2. basins located on pre-Mesozoic continental lithosphere
      1.2.1. cratonic basins
      1.2.2. basins located on earlier rifts
      1.2.3. basins located on earlier back-arc basins
2. Episutural basins located and mostly contained in compressional megasutures
   2.1. basins related to B-subduction zone
      2.1.1. circum-Pacific back-arc basins
         2.1.1.1. back-arc basins floored by oceanic crust
         2.1.1.2. back-arc basins floored by continental or transitional crust
   2.2. basins related to A-subduction
      2.2.1. basins located on continental crust (Pannonian-type basins)
      2.2.2. basins located on transitional or oceanic crust (western Mediterranean basins)
   2.3. basins related to episutural megashear systems
      2.3.1. Great basin–type basin
      2.3.2. California-type basin

Successful exploration in these basins requires a detailed knowledge of their architecture, which is the goal of this chapter.

Rift classifications

Passive versus active rifts

Based on the behavior of the asthenosphere, the rifting is driven by passive and active rifting components (Sengör and Burke, 1978; Figure 1.2), which are understood as rather end-members of the natural rift types driven by the mixture of these components (see, e.g., Huismans, 1999; Davis and Kusznir, 2002; Manatschal, 2004; Huismans and Beaumont, 2005; Simon and Podladchikov, 2006).

Passive rifting is characterized by passive asthenospheric upwelling in response to overlying layers separation controlled by regional tectonic extension (Keen, 1985; Ruppel, 1995; Kincaid et al., 1996). In this case, the lithosphere is thinned only in response to extension. The passive asthenospheric upwelling drives many secondary processes such as decompression-driven melting, crustal/lithospheric magma underplating, eruption of continental flood basalts, the onset of secondary convection, and development of large thermal gradients between extended and unextended regions (e.g., McKenzie and Bickle, 1988; Ruppel, 1995; Huismans et al., 2001).

Active rifting is characterized by active asthenospheric upwelling in response to buoyancy instability, which drives the rifting. This upwelling exhibits broad spatial wave lengths in the order of 200–500 km in proportion to asthenospheric thickness (Ruppel,

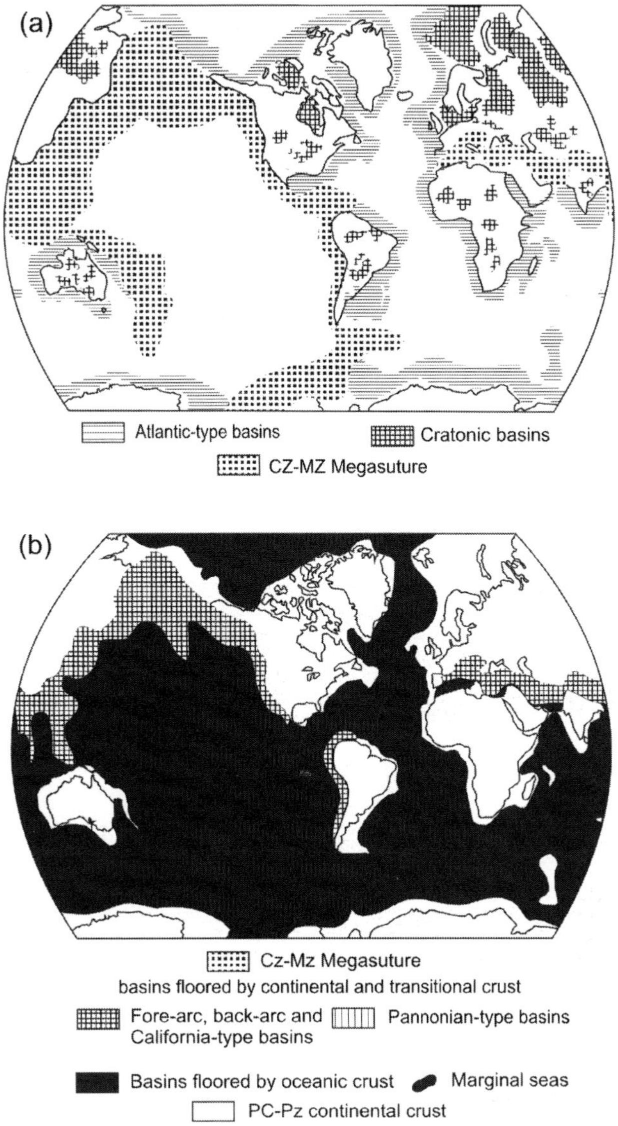

Figure 1.1. Extensional basin classification (Bally, 1982).

1995; Barnouin-Jha et al., 1997). In this case, tensional stresses at the base of the lithosphere are driven by the ascending convecting material (Turcotte and Emerman, 1983); the lithosphere is thermally thinned by heating and adsorption into the asthenosphere, in addition to necking in response to extension, and hence, the volume of asthenosphere rising into the lithosphere exceeds the volume of lithosphere displaced laterally by extension (e.g., Olsen and Morgan, 1995).

Narrow rift, wide rift, and core complex rift modes

The rift styles can be divided into narrow rift, wide rift, and core complex rift modes (Brun and Choukroune, 1983; England, 1983; Buck, 1991; Hopper and Buck, 1996; Brun, 1999 and references therein), reflecting how the lithospheric rheology, lithospheric thickness, existence of mechanic anisotropies, rate of extension, direction of extension, and presence of magma affect their development.

The narrow rift is characterized by the thinning of both crust and lithospheric mantle occupying a narrow zone of up to 100–150 km wide. This style has been attributed to local weakening factors such as thermal lithospheric thinning, local strain weakening affecting the strong layers of the lithospheric multilayer, or local magmatism (England, 1983; Kusznir and Park, 1987; Buck, 1991, 2004). It is characterized by the extension being localized in the weakest and thinnest region. As a consequence, narrow rifts have large lateral gradients in crustal thickness and topography (Corti et al., 2003). The localized lithospheric thinning in narrow rifts is associated with elevated heat flow within rift depressions in comparison to adjacent rift shoulders and cratonic blocks (e.g., Bonatti, 1985; Ruppel, 1995).

Examples of narrow rifts come from the Bajkal Rift (Artemjev and Artyushkov, 1971), the East African Rift system (Rosendahl et al., 1986; Rosendahl, 1987), the Recôncavo-Tucano Rift (Destro et al., 2003a, 2003b), the Gulf of Suez–Red Sea Rift (Bonatti, 1985; Steckler et al., 1988; Patton et al., 1994), and the Rhine Graben (Brun et al., 1992).

The wide rift is typical for its thinning of both crust and lithospheric mantle in a zone that is wider than the lithospheric thickness. Its distributed deformation style is characterized by a large number of separated basins extending across a region more than 1,000 km wide (Brun and Choukroune, 1983). Wide rifts are characterized by high extensional strain, which is not uniformly distributed over the extended region as, for example, in the Basin and Range province (e.g., Bennett et al., 1998; Thatcher et al., 1998). They can contain relatively small lateral gradients in crustal thickness (Shevenell, 2005a) and topography (USGS, 1996). Wide rifts also contain the flat base of the crust, which is attributed to magmatic underplating (e.g., Gans, 1987; MacCready et al., 1997) as contributing to smoothing Moho undulations known from narrow rift terrains (see Brun et al., 1992 and references therein).

This style has been understood as controlled by:

a) a local increase of integrated strength caused by the replacement of the crustal material by stronger material of the lithospheric mantle and accompanying lithospheric cooling driving subsequent extension into weaker unstretched

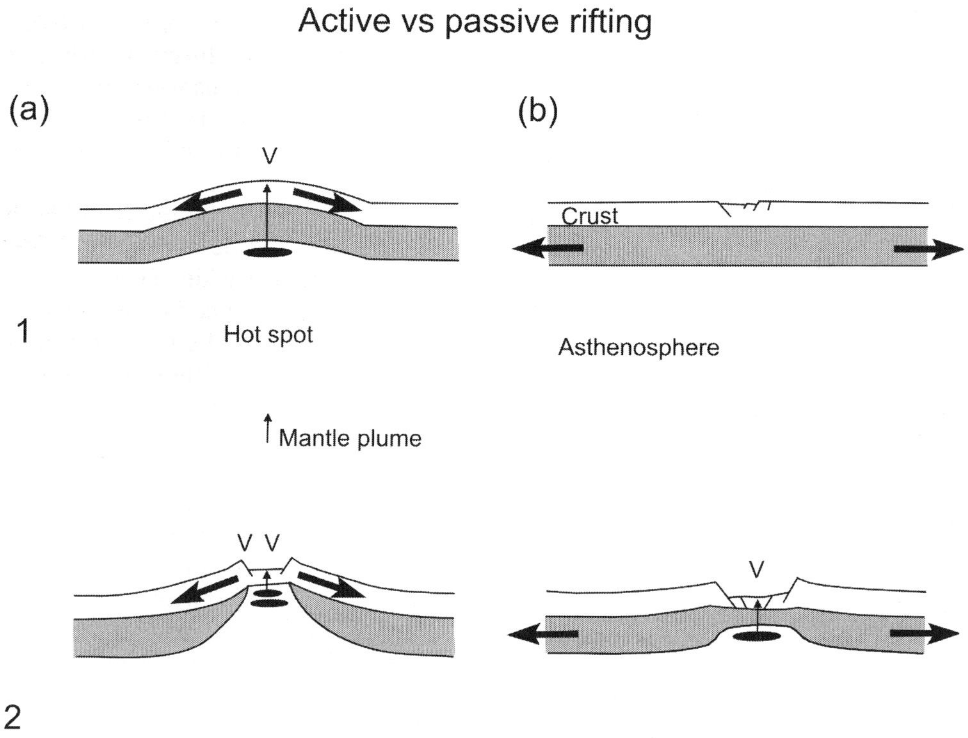

Figure 1.2. Passive versus active rifting (Corti et al., 2003).

lithospheric regions (England, 1983; Houseman and England, 1986);

b) lower crustal flow in response to pressure gradients caused by interaction of regional tectonic, local flexural, and gravity forces, resulting in delocalization of deformation (Buck, 1991; Buck et al., 1999); and

c) the degree of decoupling in lithospheric multilayers containing a viscous lower crustal layer, where the existence of localized versus distributed deformation modes depends on the coupling of the layers and the viscosity of the coupling layer (Huismans et al., 2005; Schueller et al., 2005).

Examples of wide rifts come from the Basin and Range province of the western United States (e.g., Hamilton, 1987), the Pannonian Basin system (e.g., Tari et al., 1992), the Aegean region (Jackson, 1994), and the Tyrrhenian Sea (Faccenna et al., 1996).

The core complex rift mode is characterized by the localized extension of the upper crust accompanied by the broadly distributed extension of the lower crust. This mode is attributed to a rapid lower or even middle crustal flow removing the crustal thickness variations that would otherwise control the wide rift style (Buck, 1991; Martínez-Martínez et al., 2004). In core complex structures, high-grade metamorphic rocks originating in the middle-lower crust are exposed at the surface, exhumed by shallow-angle normal faults, uplift, and erosion. They are typically separated by a detachment carrying low-grade rocks. The crust of the core complex areas has a similar thickness to the crust of surrounding less-extended areas (Corti et al., 2003). A close association of core complexes with wide rifts led Brun (1999) to describe them as local anomalies within wide rifts, developed above heterogeneities in underlying ductile lower crust, which are weak enough to localize stretching. Consequent local enhanced brittle crustal thinning becomes compensated by the upwarping and exhumation of the deep crustal levels. As Martínez-Martínez and colleagues (2004) documented, regions that underwent orogenic thickening and that are characterized by the brittle-to-ductile crustal thickness ratio of 1:3 are especially prone to development of core complexes (e.g., Coney, 1980; Crittenden et al., 1980; Lister and Davis, 1989). However, core complexes have also been described in different extensional settings such as the East African Rift system (e.g., Talbot and Ghebreab, 1997; Morley 1999; Ghebreab and Talbot, 2000), the Tertiary rift basins of the Thai Peninsula (Morley et al., 2001 and references therein), or ultra-slow and slow seafloor-spreading ridges (e.g., Cann et al., 1997; Tucholke et al., 1998; Escartín et al., 2003).

Examples of core complexes come from the Basin and Range province (Wernicke et al., 1988; Hurlow et al., 1991; Wdowinski and Axen, 1992), the Hellenic Arc (Dinter and Royden, 1993, Fassoulas et al., 1994), and the Pannonian Basin system (Tari et al., 1992).

Rift system components

The largest architectural component of the rift terrains is a rift system. An example comes from the East African Rift system (Figure 1.3a; Rosendahl, 1987). The rift systems are generally agreed as evolving in a more-or-less continuous manner without an interruption of a different tectonic process.

The rift system can be divided into rift branches (Figure 1.3b; Rosendahl, 1987) such as the Western and Eastern rift branches of the East African Rift system. They coincide with marked differences in character, whether it is the character of accompanying magmatism, detailed development timing, or opening mechanisms.

Rift branches are formed by rift zones (Figure 1.3c; Rosendahl, 1987). The average rift zone width and length in the northern South Atlantic region are 40 and 500 km, respectively. The rift zone boundaries are represented by offsets, bends, and any other major trend changes in the rift branch. A natural example of a rift zone is the Tanganyika Rift Zone of East Africa. The zones can contain several discrete basins separated by accommodation zones (Bosworth, 1985; Rosendahl et al., 1986), transfer zones (Morley et al., 1990), or interference accommodation zones (Versfelt and Rosendahl, 1989).

Each rift zone propagates until it becomes arrested because of the lack of critical stress available for further propagation. Depending on the reason for the arrest of neighbor rift zones, there are various geometries of the accommodation zones between them. As the rifting evolves, each such distinct offset of the rift branch becomes a potential nucleus for the future transform. Natural examples include transforms such as the dextral Rukwa transform connecting the adjacent Tanganyika and Malawi Rift Zones in East Africa (Chorowicz and Mukonki, 1980; Rosendahl, 1987); the sinistral Sergipe-Alagoas transform linked to the South Gabon and Recôncavo-Tucano-Jatobá Rift Zones in West Africa–Brazil (Rosendahl et al., 2005); the dextral transform connecting the adjacent Cauvery and Krishna-Godavari Rift Zones in India; and the sinistral Gettysburg-Tarfaya transform in the Central Atlantic region (Nemčok et al., 2005b).

Discrete basins of the rift zone are called rift units (Rosendahl, 1987). The average width and length in the northern South Atlantic region are 40 and 120 km, respectively. The ideal rift unit can be envisaged as an isolated half-graben bounded by a spoon-shaped normal fault (Figure 1.4; Rosendahl, 1987). The fault is characterized by the dip-slip and oblique displacements at its center and ends, respectively. A supporting systematic study of slip-vector normal faults comes from normal fault terrains in the Gulfs of Corinth and Evia, Greece, and normal fault terrains in the Central and Southern Apennines, focusing on basin-bounding faults (Roberts, 2007). The common characteristics of these faults are complex slip distributions, involving

(a)

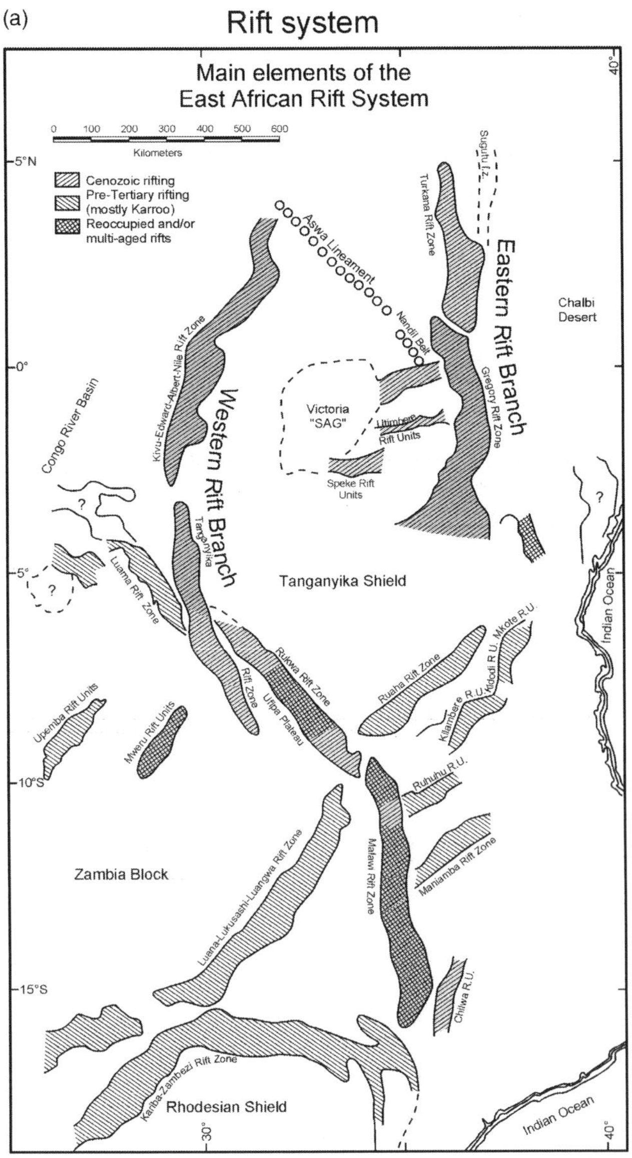

Figure 1.3. (a) Rift system example (Rosendahl, 1987). (b) Rift branch example (Rosendahl, 1987). (c) Rift zone example (Rosendahl, 1987).

converging patterns of slip toward the hanging walls (Figure 1.5). All faults can be characterized by dip-slip vectors dominating in central portions, where the overall displacement is largest, and oblique slip vectors typical for lateral propagation tips, where the displacement converges to zero. Extensive fieldwork documents that these converging slip patterns are the result of single-phase tectonic events, which is in accordance with the results of earthquake research in the Gulf of Corinth (e.g., Jackson et al., 1982). Striae and corrugation populations in various studies of faults indicate that the total displacement is a result of repeated earthquakes, each of them having a converging slip pattern (Roberts and Ganas, 2000). This observation implies

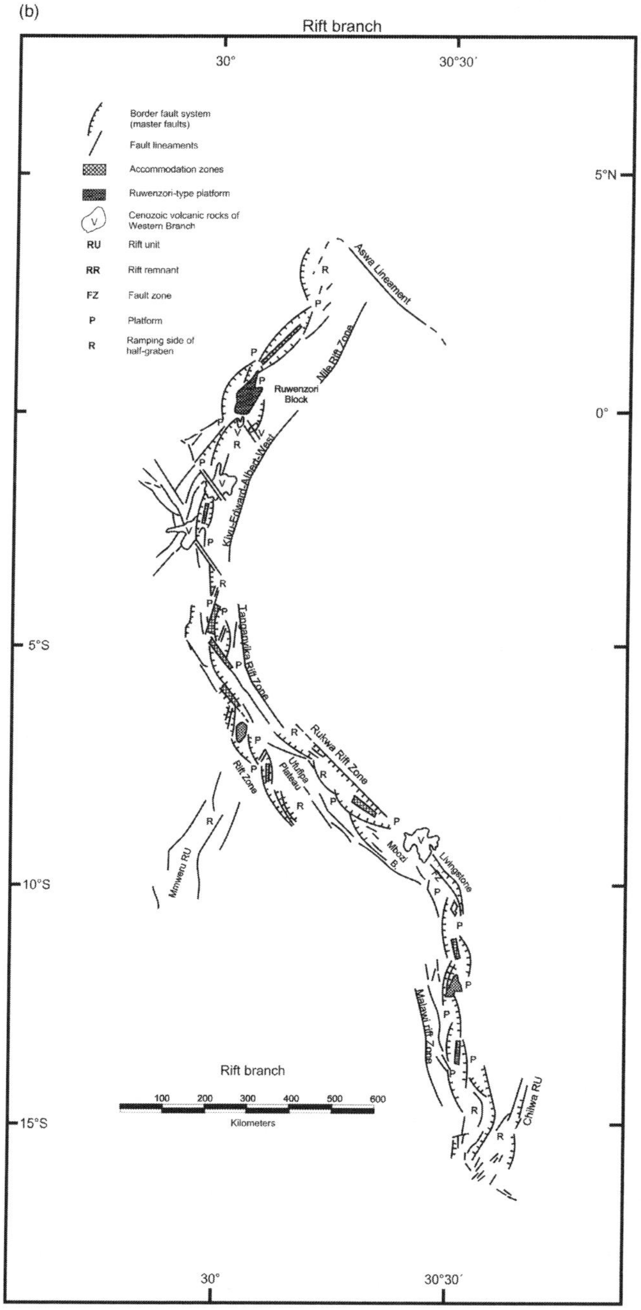

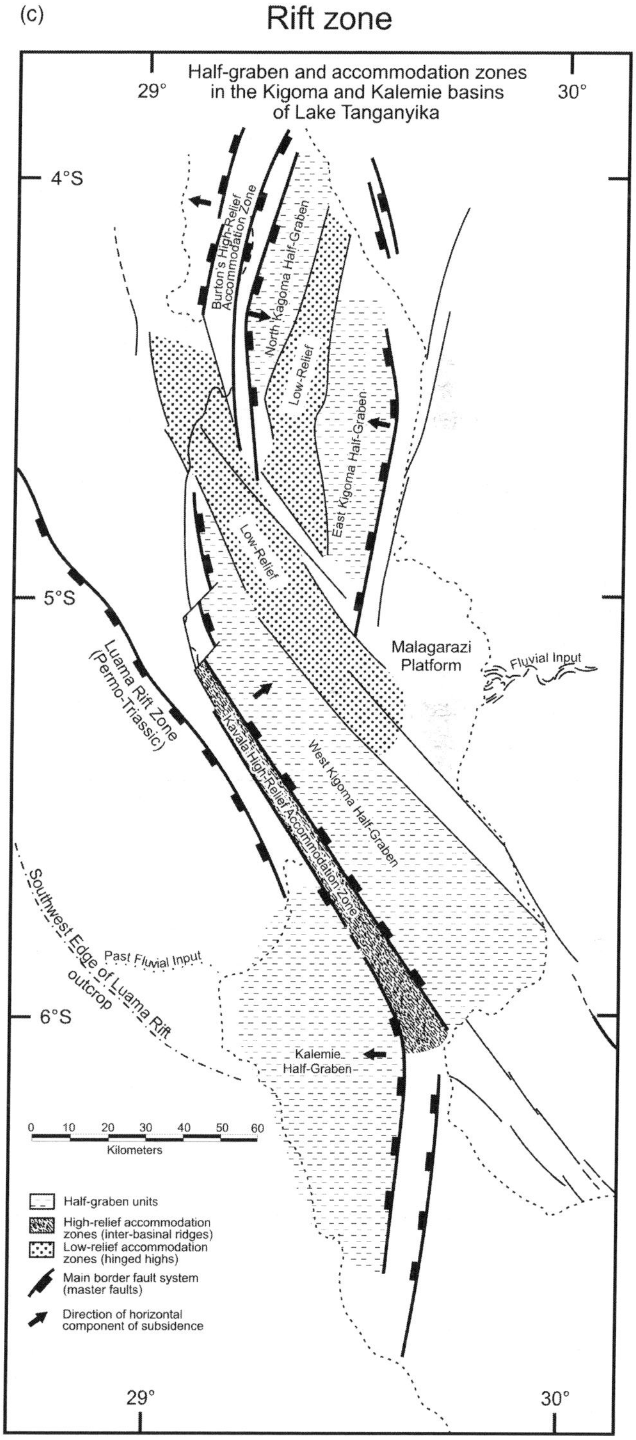

Figure 1.3 (*cont.*)

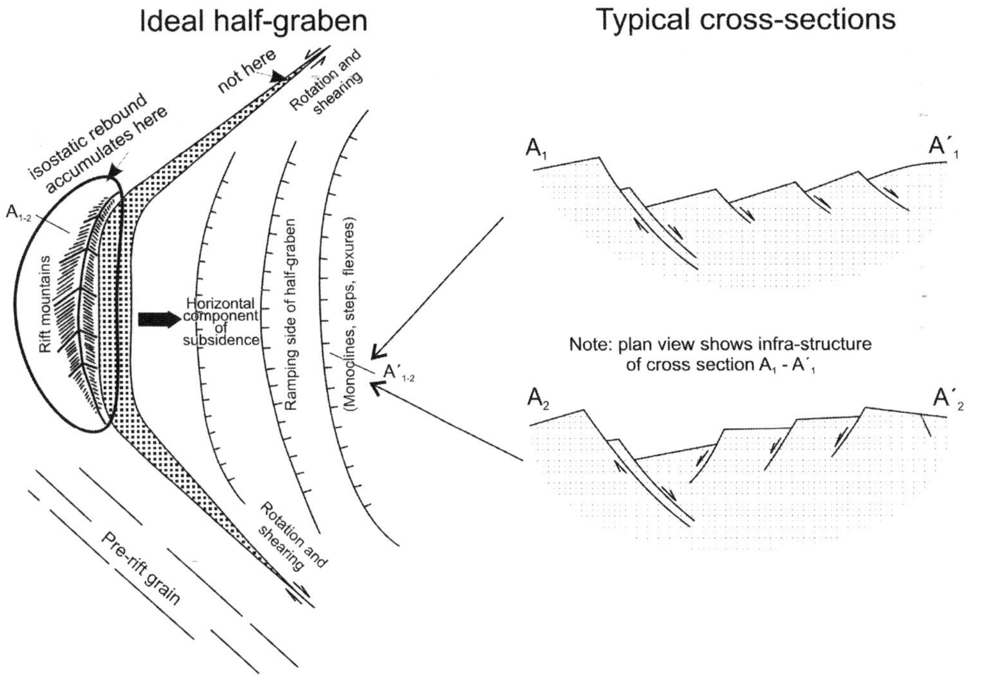

Figure 1.4. Sketch of the ideal rift unit (Rosendahl, 1987).

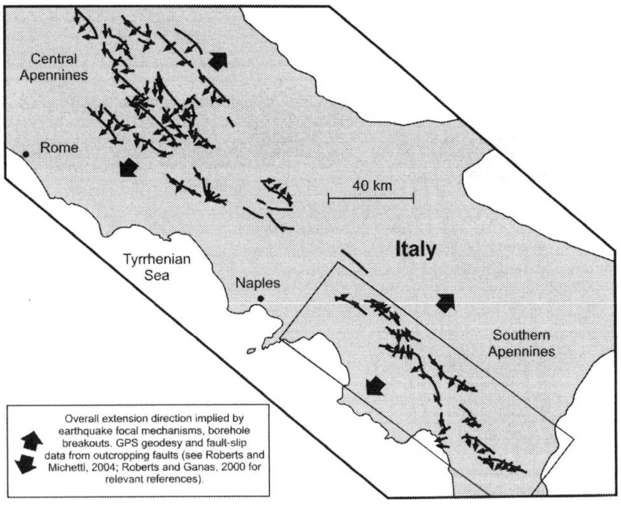

Figure 1.5. Examples of normal faults with converging patterns of slip toward the hanging walls (Roberts, 2007).

spatial and temporal stress field complexity, which causes slip in the direction of shear stress resolved in studies of spoon-shaped faults.

The rift unit footwall undergoes the largest uplift next to the fault with dip-slip displacement. Natural examples come from the Western branch of the East African Rift system (Rosendahl, 1987). The rift unit hanging wall undergoes the largest subsidence next to the fault with dip-slip displacement. Natural examples come from the Strathspey-Brent-Statfjord half-graben, North Sea (McLeod et al., 2002). The ideal spoon-shaped fault case does not occur very frequently in nature because of the influence of preexisting anisotropies and rift unit linkages on the geometry of the boundary faults. The spacing of the boundary faults varies between 2 and 15 km (Angelier and Colletta, 1983; Rosendahl, 1987). The spacing of the internal synthetic and antithetic faults is usually denser.

The rift units are composed of rift blocks (Rosendahl, 1987). This smallest unit of rift architecture has the average width and length of 10 and 40–100 in the East African Rift system (Rosendahl, 1987). The rift blocks can form structural highs or lows, but they can be frequently tilted and rotated, changing their topographic expression along their trend, as can be seen at the North Gabon passive margin.

Linkages of rift architecture elements

Half-grabens forming the rift zone can be linked in various ways, which can be grouped into linkages of half-grabens with opposing and similar polarities (Figure 1.6; Rosendahl, 1987). Both categories can be further subdivided into overlapping and non-overlapping ones.

If the opposing overlapping half-grabens develop roughly coevally, the space problem in the linkage area results in the development of a positive structure known

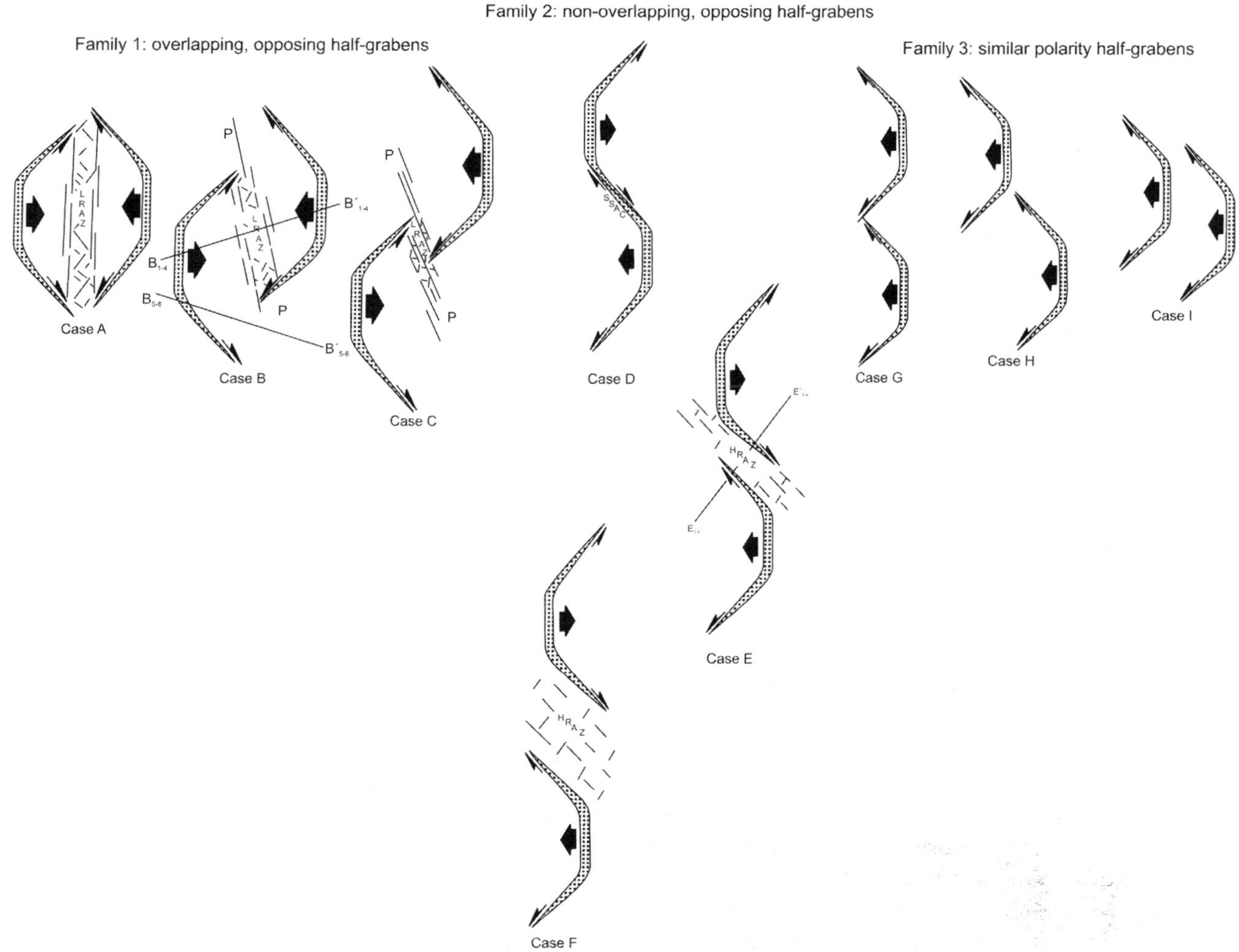

Figure 1.6. Half-graben linkages grouped into families (Rosendahl, 1987). Opposing geometries create either low-relief or high-relief accommodation zones (LRAZ or HRAZ), depending on the extent of overlap. Strike-slip accommodation zones (SSAC) may be the intermediate case.

as a low-relief accommodation zone (Figure 1.6, cases A–C; Rosendahl, 1987; McClay et al., 2002). The subsidence of these hinged highs pales in comparison with that at depocenters near graben-bounding faults (Rosendahl, 1987). Seismic data from Lakes Tanganyika and Malawi (Rosendahl, 1987) and outcrop data from western Tasmania (Noll and Hall, 2006); the Gulf of Corinth, Greece (Jackson et al., 1982; Roberts, 2007); and the Apennines, Italy (Roberts, 2007) indicate that even in the case of roughly coeval grabens, neighbor faults move individually and not exactly simultaneously. Being used in the breakup of the upper brittle crust, the low-relief accommodation zones are also widely recognized on passive margins. Examples come from the Sergipe-Alagoas and Barreirinhas Basins of Brazil (de Azevedo, 1991; Mohriak et al., 1998) and the offshore regions of Nova Scotia and Morocco (Nemčok et al., 2005b).

If the opposing non-overlapping half-grabens are active at roughly the same time, there is no space problem to be solved in the linkage area. This area tends to be "left behind" as a relatively unsubsided structure called a high-relief accommodation zone (Figure 1.6, cases D–F; Reynolds, 1984; Burgess et al., 1988; Rosendahl, 1987). In cross-sections, the high-relief accommodation zones are represented by prominent structural highs (Rosendahl, 1987). All high-relief accommodation zones in the Western branch of the East African Rift system are apparently located along preexisting anisotropies. Switches in half-graben polarities in the Gulf of Suez–Red Sea Rift system occur where rift-bounding faults impinge on or are influenced by preexisting basement structures such as the Precambrian shear zones.

Being used in the breakup of the upper brittle crust during continental breakup (Rosendahl, 1987), the high-relief accommodation zones are also widely recognized on passive margins. Examples come from offshore Benin, Ghana Ridge, Guinea Plateau; offshore Côte d'Ivoire; the Acaraú, Mundaú, Para-Maranhão, and Piauí Basins of the Brazilian Equatorial Atlantic margin (Nemčok et al., 2012a); and offshore Nova Scotia (Nemčok et al., 2005b).

Accommodation zones between opposing graben-bounding faults contain en echelon horst-graben faulting (Moustafa, 2002) or linking transfer faults (e.g., Bally, 1982; Rosendahl, 1987).

If the half-grabens of the same polarity are active at roughly the same time, there are space problems to solve in the linkage area. A large overlap area is affected by tilt, which is driven by uplift of the footwall in relationship to the first bounding fault and subsidence of the hanging wall in relationship to the second bounding fault. Areas characterized by a small overlap develop either direct linkages of the neighbor boundary faults (Noll and Hall, 2006) or relay ramps in the overlap between them (Figure 1.7; McLeod et al., 2002).

Figure 1.7. Small-scale relay ramp, Arches National Park, Utah.

Detailed fracture/fault patterns in accommodation zones between graben-bounding faults of the same polarity are (Moustafa, 2002):

a) relay ramp
b) linking transfer fault
c) en echelon step faulting
d) tilted fault blocks
e) zigzag fault array.

A well-constrained example of the spatial and temporal variation of activity of normal faults with the same polarity and located above the same detachment fault comes from the Corinth-Patras Rift, Greece (Sorel, 2000). The active low-angle, north-dipping detachment fault is indicated by micro-earthquakes occurring on a 12°–20° plane dipping to the north. The timing of the faults above detachment was derived from ages of sediments in respective half-grabens, which are the oldest in the south and become younger northward in jump-like fashion (Sorel et al., 1997; Sorel, 2000; Figure 1.8). The whole described extensional history took place during the last 1 Ma (Sorel, 2000). During this time, the total distance traveled by the rolling hinge of the system was about 30 km.

The northerly shift of the normal fault activity above the detachment fault was accompanied with rotation of the detachment fault into shallower dip geometry in the soon-to-be non-active portion of the system, which expanded from south to north (Sorel, 2000). The rotation caused the higher-friction geometry, which eventually led to stalling of the detachment fault, which progressively expanded to the north.

The lack of deeper seismicity north of the Gulf of Corinth along the detachment fault probably indicates its ductile character at this depth. Earthquakes occur only along the southern portion of the detachment fault, where the lower dip angle results in higher friction

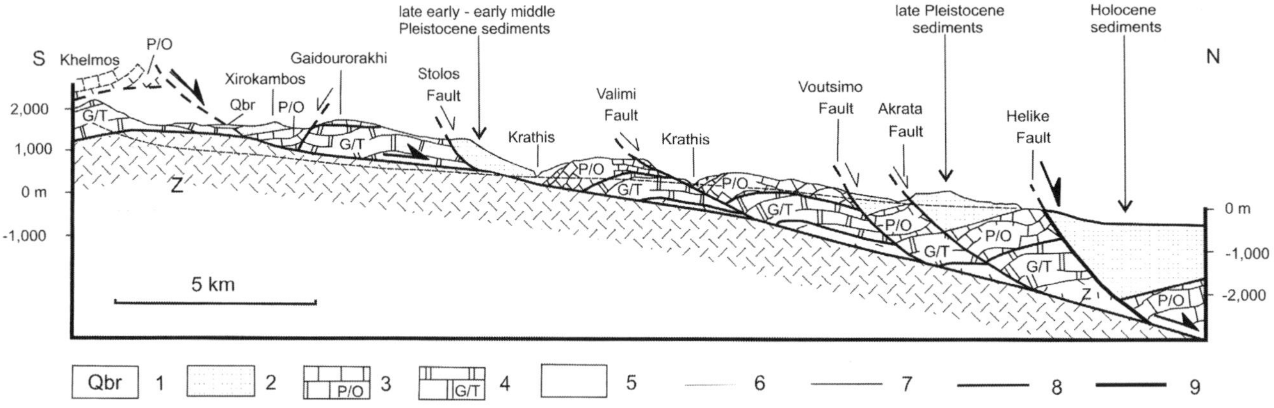

Figure 1.8. Extensional basin system in the Gulf of Patras (Sorel, 2000).

along the fault, and along active normal faults above the detachment fault.

Noll and Hall (2006) studied normal fault growth and linkage rates, especially in comparison to deposition rates, in western Tasmania. The studied array is a series of usually overlapping normal faults having variable polarity, and dips about 60–70°. Their development and further activity could be determined in detail thanks to the existence of four syn-rift sediment packages recording fluvial deposition outpacing the tectonic subsidence, ensuring that thickness variations are highly responsive to fault movements (see McLeod et al., 2002).

The described fault growth and linkage events can be summarized as (Noll and Hall, 2006):

1) initiation and isolated fault growth,
2) followed by limited fault growth interaction,
3) followed by renewed fault initiation, and
4) finally followed by a rift climax.

The hard linkage of rift units and zones is provided by the strike-slip transfer faults. Their function is to accommodate differences in strain amount or structural styles along the strike of the extensional system or the direction of extension (e.g., Gibbs, 1984; Lister et al., 1986; Bosworth, 1995; McClay and Khalil, 1998).

Major transfer faults separating half-grabens with opposing detachment asymmetries are called accommodation zones (Bosworth, 1985). Transfer faults can have a character of pure strike-slip faults when they are parallel to extension direction or a character of oblique slip with transtensional or transpressional character depending on the relationship between their geometry and controlling stress field (Gibbs, 1984; Lister et al., 1986). Transpressional transfer faults are relatively rare. The example is the Jeremboabo sinistral transpressional fault in the failed Recôncavo-Tucano Rift of northeastern Brazil (Destro et al., 2003a).

The preexisting structural grain strongly influences the locations of the syn-rift transfer faults. This can be seen in the Rukwa transfer zone of the East African Rift system, initiated along the initial sinistral shear zone during the Ubendian orogeny and reactivated during Permo-Triassic rifting, which is located between the Tanganyika and Malawi Rift Zones (Versfelt and Rosendahl, 1989).

Transfer faults can be represented by single faults, such as those from the Gulf of Suez–Red Sea Rift region (Younes and McClay, 2002), or they can be represented by wide zones of strike-slip faults with pull-apart basins among them, such as those of the Gettysburg-Tarfaya transfer zone on the U.S. and Moroccan Central Atlantic margins (Nemčok et al., 2005b and references therein). A well-documented case of the transfer fault is the Garlock Fault in southern California, which functions as an intracontinental transform between the northern, westward-extended Basin and Range province, and the southern, non-distended Mojave Desert (Sylvester, 1988).

Pull-apart basins

The remaining category of basins to be described in rift systems is pull-apart basins, which are associated with transfer faults and fault zones. Pull-apart basins are extensional depressions formed at releasing bends (Carey, 1958; Burchfiel and Stewart, 1966), extensional bridges (Petit, 1987), extensional oversteps (Christie-Blick and Biddle, 1985), or releasing junctions (Crowell, 1974) of strike-slip fault systems. Their basin geometry is represented by a rhomb-shaped graben, a rhomb-shaped half-graben, a series of coalescent basins (e.g., Schubert, 1980; Aydin and Nur, 1985), or a more complex combination of fault blocks (Howell et al., 1980; Royden, 1985; Nemčok et al., 2005b).

Analog material modeling results indicate that the geometry in map view is controlled by factors such as depth to the basement, spacing between parallel or en echelon fault strands, and size of an overstep (Hempton and Neher, 1986; Mandl, 1988; Richard et al., 1995) in accordance with interpretations made from natural examples (Sylvester, 1988). They also indicate that internal geometries and basin symmetries are controlled by displacement along the boundary faults (Faugère et al., 1986) and by relative displacement rates along them (Rahe et al., 1998).

Based on their depth of detachment, pull-apart basins can be separated into thick- and thin-skin pull-apart basins (Royden, 1985). Examples of the former come from the Gulf of Paria in Venezuela and Trinidad (Sims et al., 1999) and the Salton Trough pull-apart basins in the United States and Mexico (e.g., Elders et al., 1972; Herzig et al., 1988; Lonsdale, 1991). Examples of the latter come from Death Valley (Burchfiel et al., 1987) and the Ridge Basin in California (e.g., May et al., 1993). In certain cases the depth of detachment within the basin can vary, making a portion of it a thin-skin and a portion of it a thick-skin pull-apart basin as it is in the northern and southern portions of the Vienna Basin (Lankreijer, 1998).

Based on the depth and lithology of detachment, the detachments of pull-apart basins can have brittle or ductile deformation character. Thick-skin types of pull-apart basins with detachment depths below the brittle-ductile transition and pull-apart basins detached along evaporites such as the Gulf of Elat (Ten Brink and

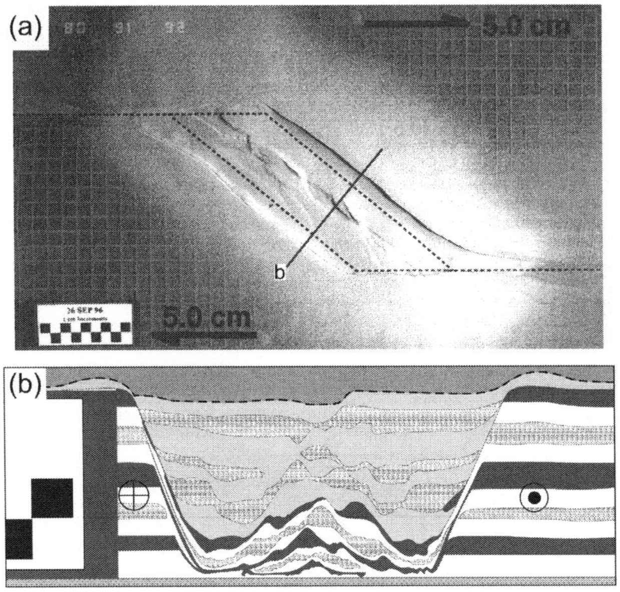

Figure 1.9. Scenario of the pull-apart basin detached along a brittle detachment (Sims et al., 1999).

Ben-Avraham, 1989) form the ductile-detachment case and thin-skin ones form the brittle-detachment case.

A series of experiments by Rahe and colleagues (1998) and Sims and colleagues (1999) focused on the effect of the detachment character on the pull-apart structural architecture.

The model of the pull-apart basin above the brittle detachment (Rahe et al., 1998) records the pull-apart initiation by normal faults, which define the basin edges. Their propagation is closely followed by strike-slip faults parallel to the main bounding strike-slip faults and strike-slip faults with Riedel shear orientation. Riedel shears further propagate until they form cross-basin strike-slip faults, which terminate at basin-bounding normal faults. The pull-apart basin reaches its maturity when a through-going Riedel shear develops, extending between both basin tips at the ends of the main bounding strike-slip faults. The subsidence in the basin, which is located above the brittle detachment horizon, initiates earlier than the subsidence in basins developed above ductile detachments. It is because there is no ductile layer, which would take a certain share of the extensional deformation before the extension in the brittle rock section developed faults controlling the basin subsidence. Unlike in the basin above ductile detachment, the within-basin high in the center of the basin is minimal (Figure 1.9; Sims et al., 1999). The pull-apart basin of this detachment scenario is represented by a single basin with dominant normal faults controlling the basin geometry and subsidence.

The pull-apart basin development above the ductile detachment, regardless of its thickness, initiates with Riedel shears at acute angles to the main boundary strike-slip faults, which try to link them (Figures 1.10, 1.11; Sims et al., 1999). Riedel shear propagation is closely followed by propagation of both conjugate Riedel and antithetic Riedel shears. All mentioned shears develop a complex of strike-slip duplexes. Later, strike-slip faults parallel to the bounding strike-slip faults develop. The basin subsidence begins with the development of normal faults and the initiation of a localized dip-slip component of displacement on Riedel and antithetic Riedel shears. Dip-slip displacement varies along the strike of the pull-apart basin, controlling asymmetric sub-basins with a flip-flop switch along the basin strike. The pull-apart basin of this detachment scenario is represented by a complex of isolated or coalescing sub-basins having strike-slip fault–dominated geometries.

The thickness of the ductile detachment horizon exerts an important control on the basin geometry. The pull-apart basin above a thick ductile detachment horizon is relatively thin (Figure 1.11). Its development can be followed by the development of neighbor basins divided from the initial pull-apart basin by between-basin highs, which do not usually contain syn-rift sediments because they form areas "left behind" as relatively unsubsided structures characterized by high relief.

The pull-apart basin developed above a thin, ductile detachment horizon develops the basin asymmetry, which is controlled by within-basin, strike-slip transfer zones. These transfer zones define within-basin highs. They separate the deeper opposing ends of the elongate internal basins. These highs are covered by syn-rift sediments as they undergo certain subsidence.

A good chance for investigation of the role of detachment character in the pull-apart basin geometry comes from passive margins where a short pull-apart segment of the margin lies between long rift segments of the margin. Here, significant stretching and post-breakup erosion bring once deep-seated detachments into relatively shallow depths. Nice natural examples come from the Equatorial Atlantic margins (Nemčok et al., 2012a).

The seismic image of planar detachment, which never made it to continental breakup, comes from the Ridge Basin, California (Figure 1.12; May et al., 1993). It shows that the hinge area had little or no sediment accommodation. Subsidence and accommodation were greatest to the bounding fault. The total displacement of the sediment fill along the bounding fault causes the oldest stratigraphies to have the steepest dips, while the youngest ones are subhorizontal. The image documents

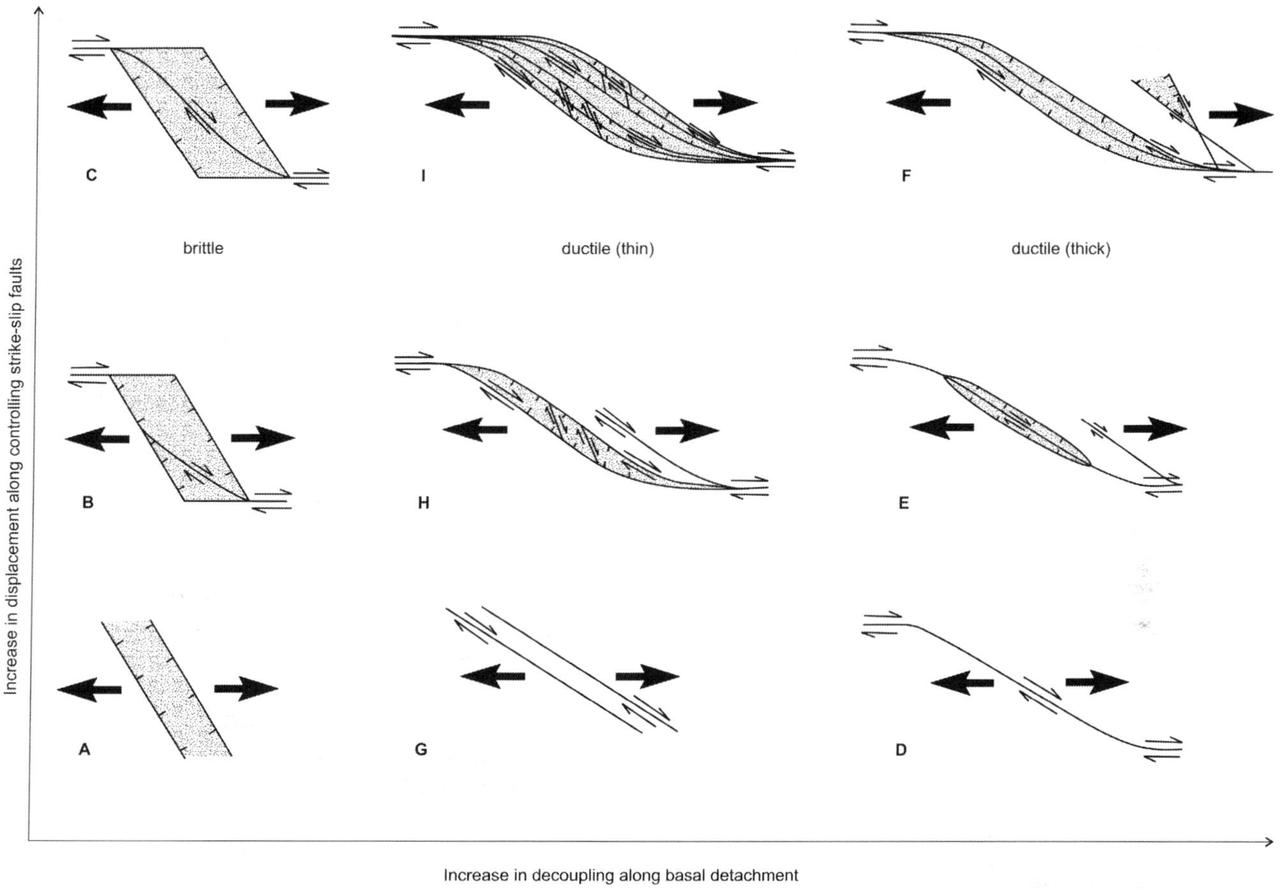

Figure 1.10. Models of the pull-apart basin detached along a brittle, thin ductile and thick ductile detachments (Sims et al., 1999).

an enormous extension indicated by the stratigraphic thickness to vertical thickness ratio approaching 3:5 (May et al., 1993).

The structural architecture of the Benin passive margin allows scientists to study the pull-apart terrain that made it to continental breakup (Nemčok et al., 2012a; Figure 1.13). There are two sets of pull-apart basins divided by the boundary between a bit thicker and a very thin continental crust. Those formed on the thicker crust have a clear distinction between normal and strike-slip faults controlling their rhomb-shaped geometries. Those formed on the thin crust lack a clear distinction between normal and strike-slip faults. They are narrow and controlled by single Riedel shears, which are dextral strike-slip faults with a normal fault component.

According to the analog material modeling discussed earlier, the pull-apart basins that developed in the continental crust during the early stages of stretching – when the crust was thicker and cooler – were detached along brittle detachments. The pull-apart basins that developed in the continental crust during the mature stages of stretching – when the crust was thinner and warmer – were detached along ductile detachments.

As Nemčok and colleagues (2012b) documented, these two crustal zones behaved differently during the early drift characterized by slight rotation of the divergence direction. The thin and still somewhat warmer transitional crust was weak enough to undergo transpressional faulting and folding, while the thicker and cooler continental crust did not experience any significant transpressional deformation.

Differing subsidence patterns of rift and pull-apart basins

Scientists have described how subsidence curves can indicate different basin types (see Allen and Allen, 1990). The example of the curve for the narrow and roughly orthogonal rift comes from the Roer Valley Graben, southeastern Netherlands (Zijerveld et al., 1992). Its subsidence curves show two major regional rifting events. The Late Permian–Early Triassic event is characterized by a subsidence (lasting 10 million years)

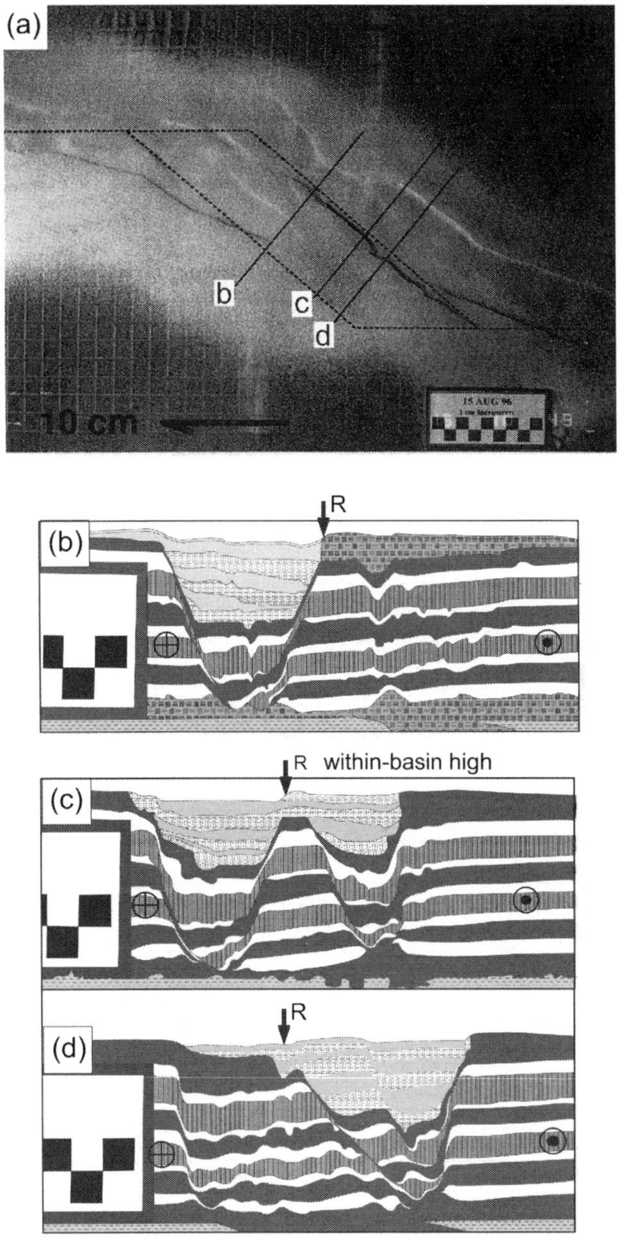

Figure 1.11. Scenario of the pull-apart basin detached along a thick ductile detachment (Sims et al., 1999).

reaching values of roughly 400–900 m and followed by a period of a long-lasting (several tens of million years), smoother, thermally controlled, post-rift subsidence of about 200–250 m (Figure 1.14).

The later history of the East Slovakian and Styrian Basins of the Pannonian Basin system provides examples from the roughly orthogonal wide rift with tendency to form core complexes (see Tari et al., 1992) (Figure 1.15; Lankreijer, 1998; Figure 1.16; Ebner and Sachsenhofer, 1995).

The East Slovakian Basin has a complex history, including a pull-apart basin during Karpatian-Badenian and a rift basin during Sarmatian (Nemčok et al., 2006b and references therein) (Figure 1.15; Lankreijer, 1998). The Sarmatian rift event took place roughly 2.6 Ma. The northwestern part of the basin shows only its early Sarmatian portion, which is minimal. It is the south-central and southeastern parts of the basin that underwent a full Sarmatian rift event. More southerly areas of the basin recorded progressively larger syn-rift subsidence. While the south-central areas show maximally 200–250 m of subsidence, the southeastern areas show a subsidence of 300–500 m. The importance of the rift event correlates also with the significance of the thermally controlled subsidence in these areas. While the thermally controlled subsidence in the south-central areas lasting during Pannonian-Pontian was less than 70 m, the thermally controlled subsidence in the southeastern areas is recorded by a sediment thickness of about 250 m.

Separation of the wide rift event from the Styrian Basin is difficult. Its early, Ottnangian-Karpatian, pull-apart history was followed by Badenian-Sarmatian wide rift history and thermally controlled subsidence history. However, a substantial Karpatian-Badenian magmatism raised the heat flow regimes of the basin to extremely elevated values during late pull-apart and early rift development (Ebner and Sachsenhofer, 1995). Wells Pichla 1 and Übersbach 1 from the southwestern and east-central parts of the basin in Figure 1.16 serve as examples of the wells in areas affected and unaffected by the magmatism, respectively. Subsidence curves provide a relatively clear distinction of the pull-apart history from the subsequent rift event. While the approximately 1.8 Ma-long pull-apart event accumulated 1.2 to 1.6 km of sediment, the approximately 5.2 Ma-long rift event accumulated only about 0.4–1 km of sediment. The area without the influence of extra heat from active magmatism has a subsidence history that can be divided into slower Badenian and accelerated Sarmatian subsidence. The area with extra heat from the active magmatism does not have such a prominent change in subsidence history and accumulates only 0.4 km in comparison with 1 km in the other area. The area with earlier magmatism has relatively subordinate post-rift Pannonian subsidence, which accumulated only about a 50 m-thick sediment section. The area without earlier magmatism has more pronounced and longer (Pannonian-Pontian) thermally controlled subsidence, which accumulated an almost 500 m-thick sediment section.

Examples from the Vienna Basin come as examples from the pull-apart basin (Figure 1.17; Lankreijer, 1998). The Vienna Basin evolution is tied to the

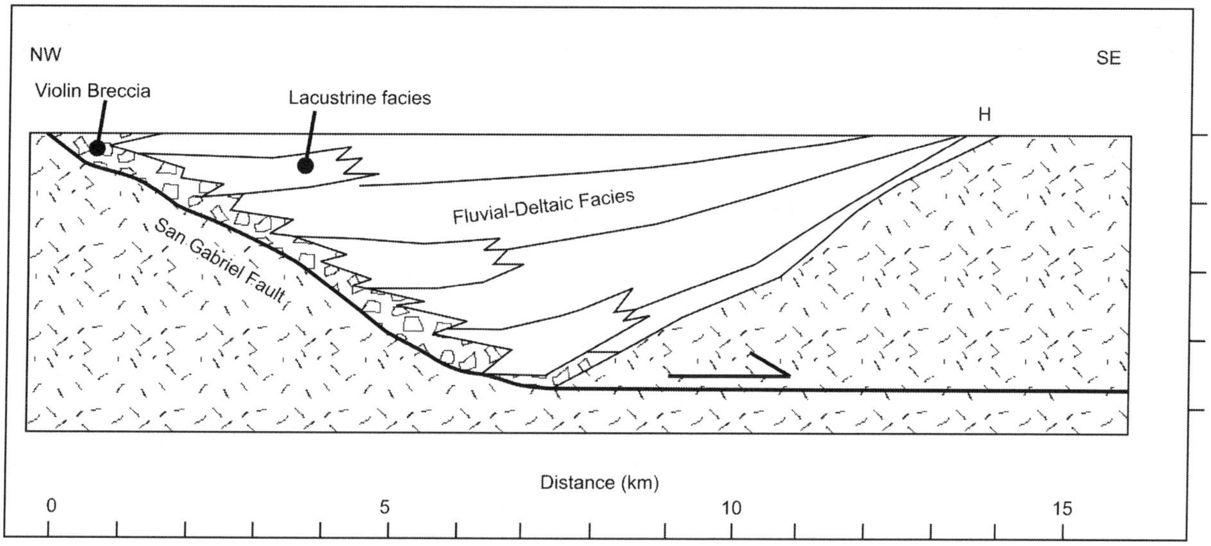

Figure 1.12. Sediment accommodation profile of the Ridge Basin (May et al., 1993). H indicates the hanging-wall hinge as an area of little or no sediment accommodation. Subsidence and accommodation were greatest adjacent to the San Gabriel Fault, where the Violin Breccia was deposited into basinal, marine, lacustrine, and fluvial-deltaic environments.

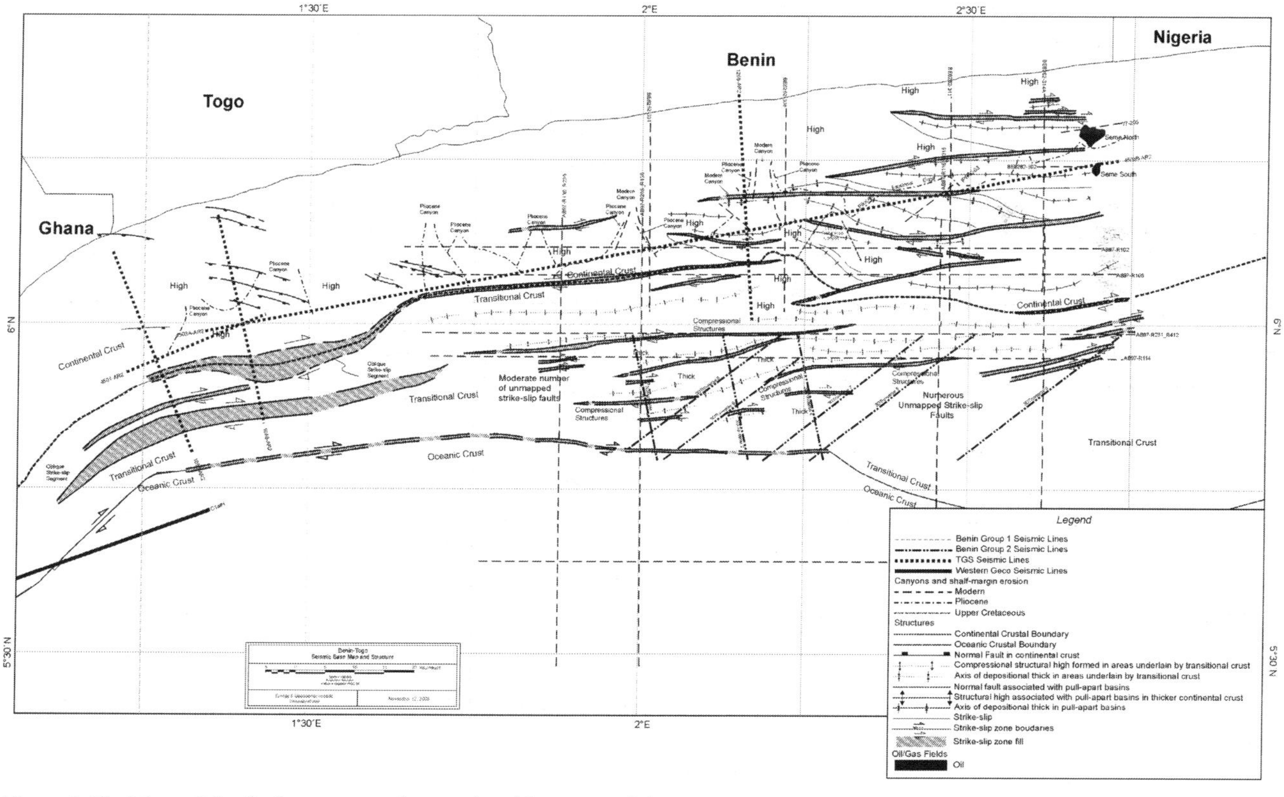

Figure 1.13. Map of the fault pattern and crustal architecture, offshore Togo and Benin (Nemčok et al., 2012b).

disintegration of the internal portion of the East Alpine–West Carpathian orogen (Nemčok et al., 2006b and references therein). Using inclined pre-existing thrust detachments for its detachments, the Austrian, more internal, central and southern portion of the basin is a thick-skin, pull-apart basin while the Czech-Slovakian, more external, northeastern portion has a thin-skin character (Lankreijer, 1998).

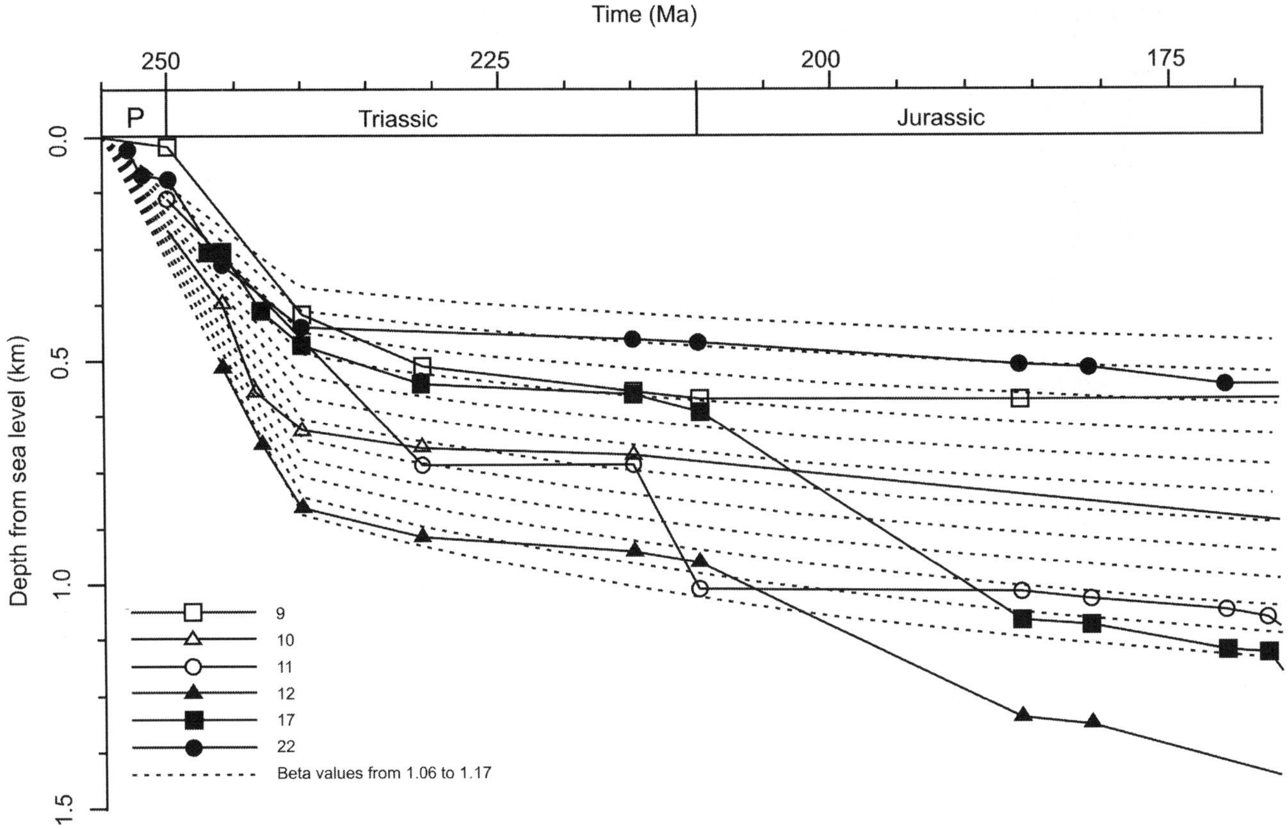

Figure 1.14. Tectonic subsidence of selected wells during and after the Late Permian–Early Triassic rifting event (255–240 Ma) compared to synthetic subsidence curves constructed with a model for one-layer lithospheric stretching with various stretching factors (Zijerveld et al., 1992).

Figure 1.17 shows several sets of the subsidence curves from the Slovak and Austrian portions of the basin. To compare the subsidence curves first to the subsidence curves from the narrow rift (see Figure 1.14), the pull-apart subsidence curves:

1) have extremely short subsidence events; and
2) do not have a significant thermally controlled post-rift subsidence, unless the pull-apart basin is large and thick-skin type.

The longest periods of the pull-apart-controlled subsidence were ten times shorter than the rift subsidence in the previously described narrow rift example.

The global averages of the heat flow anomalies associated with active rifts and pull-apart basins listed by Allen and Allen (1990) imply that rifts should contain a clear thermally controlled post-rift subsidence as part of their subsidence history, while thin-skin pull-aparts might not show it at all.

The main difference between subsidence patterns of narrow rifts, wide rifts, and pull-apart basins seems to be controlled by the activity timing of the controlling fault systems and associated thermodynamics.

Numerous examples of the narrow rift basins (Ksiazkiewicz, 1962; Villemin et al., 1986; Zijerveld et al., 1992; Hendrie et al., 1994; Nemčok and Gayer, 1996) document 10–120 Ma-long subsidence histories, which can be either continuous or separated by periods of tectonic quiescence. Post-rift subsidence histories usually contain reasonably well-defined thermally controlled subsidence and flank and rift locations remain relatively spatially stable throughout the history.

Examples of the wide rift basins (Ebner and Sachsenhofer, 1995; Constenius, 1996; Lankreijer, 1998; Sorel, 2000) document 0.1–9 Ma-long subsidence histories, which can be either continuous or separated by periods of tectonic quiescence. Post-rift histories usually contain reasonably well-defined thermally controlled subsidence. Flank and rift locations vary in time and space.

Examples of the pull-apart basins (Hunt and Mabey, 1966; Graham, 1976; Kerr et al., 1979; Norris and Carter, 1980; Ben-Avraham et al., 1981; Crowell and Link, 1982; Moore and Curray, 1982; Lankreijer, 1998) document 0.25–10 Ma-long subsidence histories, which can be continuous or separated by periods of tectonic quiescence

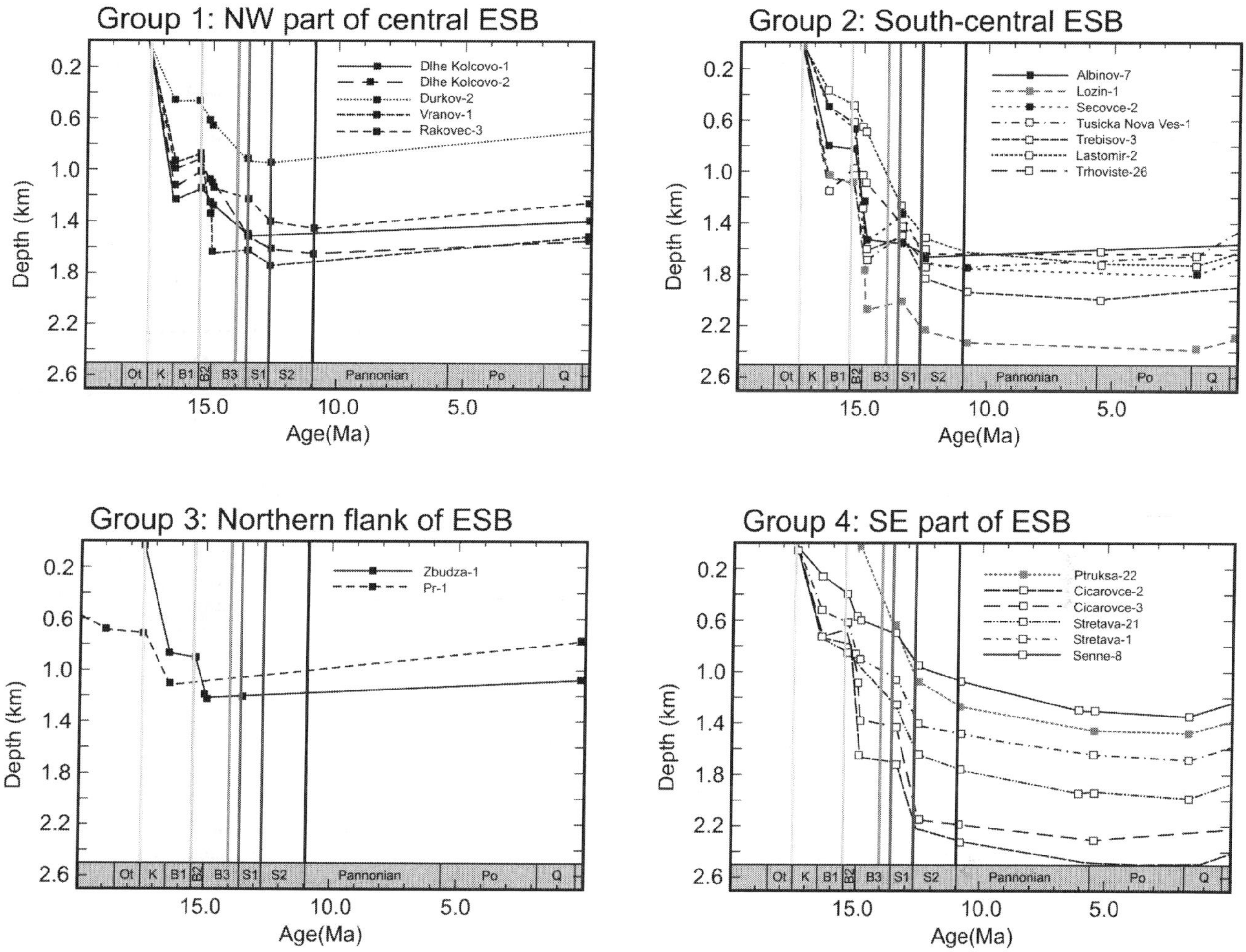

Figure 1.15. Subsidence curves from various areas in the East Slovakian Basin (Lankreijer, 1998).

or periods of contraction. Short intervals of subsidence are represented by 1–2 km-thick stratigraphic sections. Post-rift histories do not usually contain any thermally controlled subsidence. It is most likely depressed because of the "shallow" detachment of the basin in contrast to the rift basin associated with extension of the whole lithosphere. Many basins experience subsequent uplift and erosion (e.g., Moser and Frisch, 1996). Flexural effects at pull-apart margins are reduced during rifting events (Christie-Blick and Biddle, 1985). Subsidence curves along various wells in the same pull-apart basin vary dramatically, in contrast to rift basin subsidence records, because of the fact that a pull-apart basin can alternately experience both extension and shortening on a time scale of thousands to millions of years, through variations in the motion of adjacent crustal blocks (e.g., Miall, 1985; Nilsen and McLaughlin, 1985; Steel et al., 1985). The directions of extension and shortening also tend to vary temporally in the known pull-apart basins (Christie-Blick and Biddle, 1985).

At the end of this subchapter, it is interesting to compare the rates of ocean opening along its segments having the upper crustal breakup controlled by strike-slip fault zones with pull-apart basins with rates of opening along segments controlled by orthogonal rifts, taking a map of oceanic crustal ages from Muller and colleagues (1997).

The Central Atlantic segment between the Guinea Plateau and the southernmost tip of Portugal, dominated by rift breakup control, is about 3,480 km long. Its opening took place during 165–170 Ma, having a rate of 0.08 my^{-1}.

The Southern Atlantic segment between central offshore Namibia and central offshore Cameroon, dominated by rift breakup control, is about 3,560 km long. Its opening lasted from 130–135 Ma to 115–120 Ma, having a rate of 0.24 my^{-1}.

The Equatorial Atlantic segment between central offshore Cameroon and the Guinea Plateau, dominated by strike-slip breakup control, is about 3,090 km long. Its opening lasted about 5 Ma, having a rate of 0.62 my^{-1}.

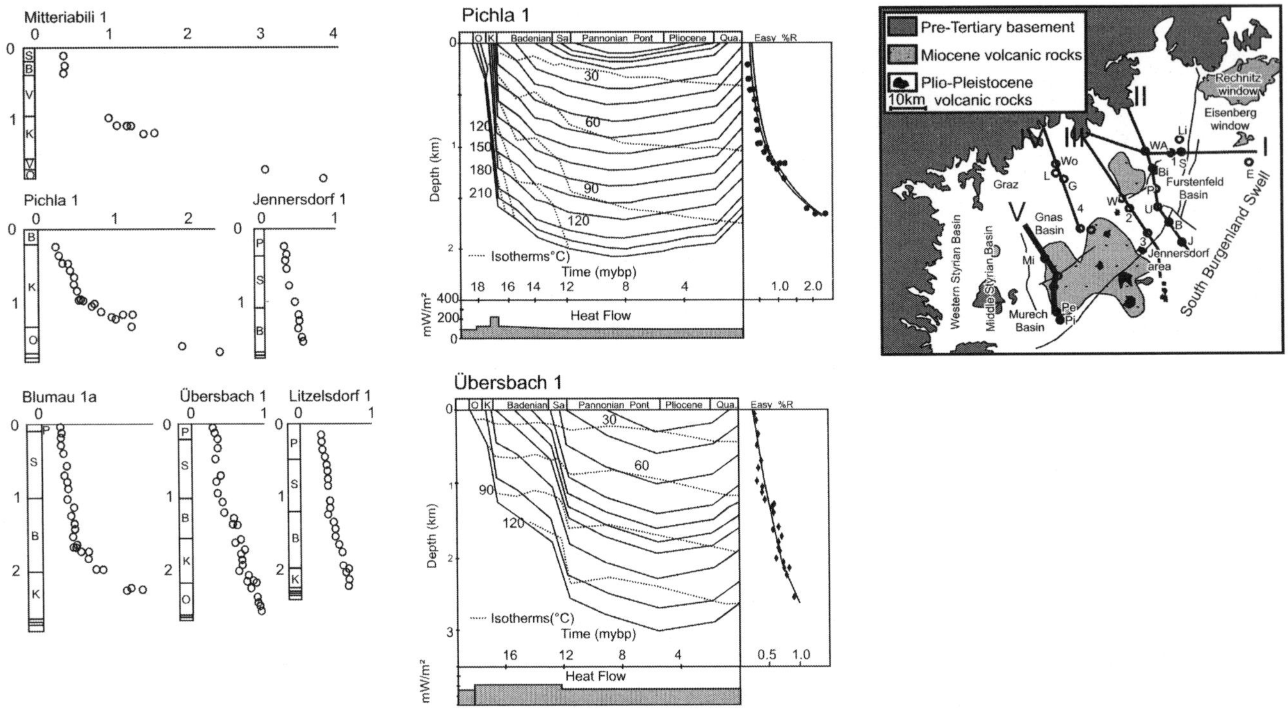

Figure 1.16. Subsidence curves from various areas in the Styrian Basin (Ebner and Sachsenhofer, 1995).

Normal and detachment fault anatomies

Normal faults can be either planar or spoon-shaped (Figure 1.18).

A nice example of the planar normal faults comes from the Simplon Pass in the Western Alps (Bally et al., 1981). Faults are developed in a horizon of very competent dolomite sandwiched by incompetent layers of calcareous material that fills the space between the rotated domino-like fault blocks.

The spoon-shaped faults, which have listric geometry in vertical cross-sections, dominate over planar ones. They can be further subdivided into growth listric normal faults (Figure 1.19) and those that show no growth. While the deformed rocks in the hanging wall of the latter do not indicate a displacement increasing downward by the cumulative effect of summed-up slip events, which is the largest in the oldest stratigraphy, the hanging wall sediments of the former do.

Listric normal faults can be also divided into those developed in gravity glides (e.g., Liro and Coen, 1995; Rowan et al., 1999; Marton et al., 2000) and those that detach deep in the basement (e.g., Brun et al., 1992; Morley et al., 1992; Tari et al., 1992).

Reflection and refraction data indicate that normal faults detach along some mechanical discontinuity. The example comes from offshore East India, which contains a transparent and interpreted brittle domain deformed by listric normal faults detaching roughly at the top of the highly reflective and interpreted ductile domain. The deep reflection seismic profiles from offshore Gabon document that the depth of the detachment fault can range from the top of the lower crust to Moho discontinuity. This data set also documents that the detachments of the early normal faults are located higher than detachments of the later normal faults, which developed in more stretched crust.

Normal fault detachments in regions with preexisting orogenic belts frequently utilize mechanical discontinuity along preexisting thrust detachments/décollements for their location (Bally et al., 1966; Horváth, 1993; Constenius, 1996; Stuart et al., 2011; Figure 1.20), where the extension followed the orogenesis immediately or with short time delay, and offshore East India, Equatorial Atlantic passive margins and northern South Atlantic passive margins, where the extension followed the orogenesis after very long time delay, which is longer than the time necessary for the thermal re-equilibration of the lithosphere.

One of the typical boundaries that structural geologists interpret as detachment level for extensional faults is a brittle–ductile boundary. Its location for crustal rocks is temperature dependent. The reason for using brittle–ductile transition for interpretation of subhorizontal detachments is that in truly plastic materials,

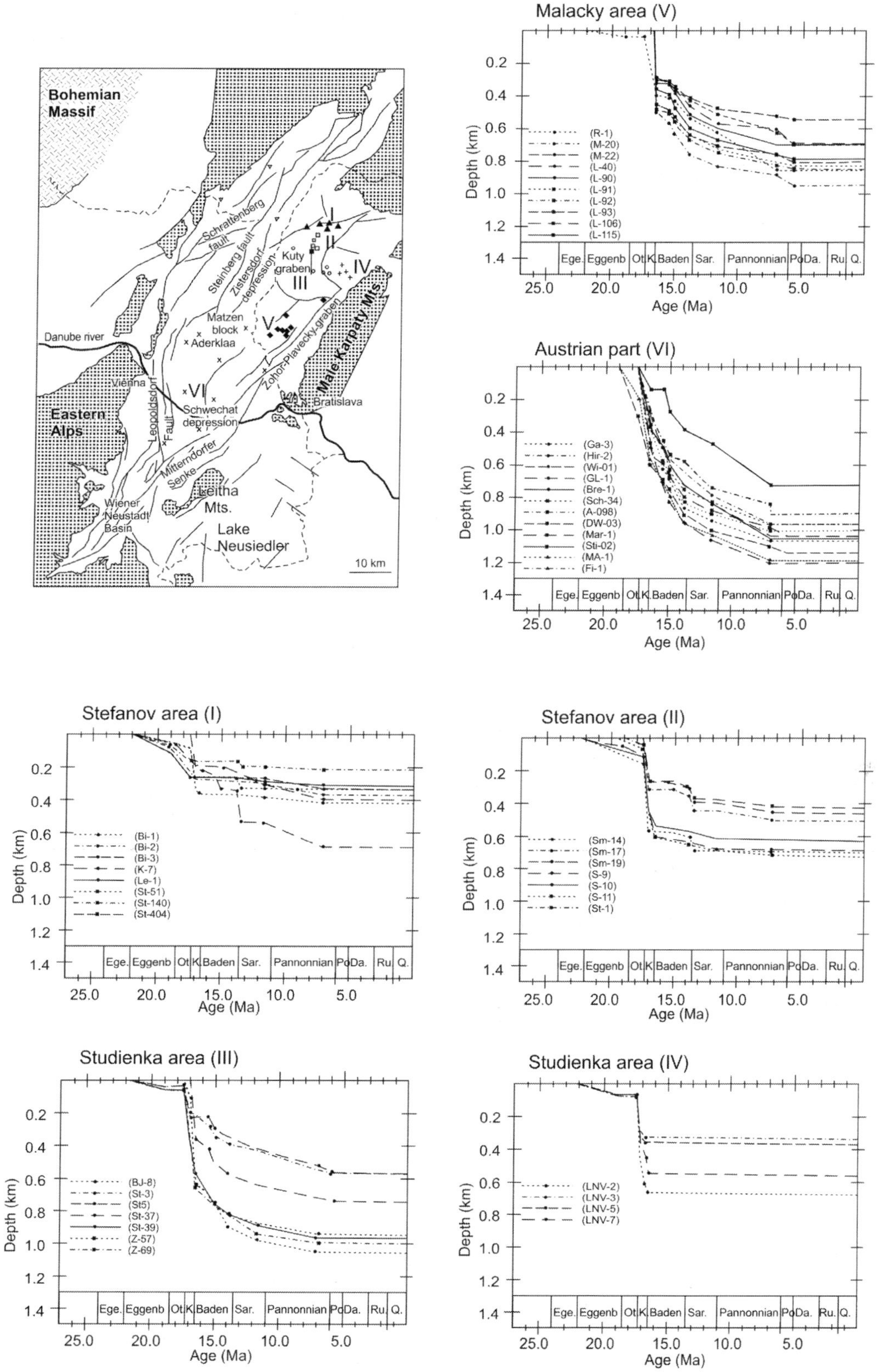

Figure 1.17. Subsidence curves from various areas of the Vienna Basin pull-apart (Lankreijer, 1998).

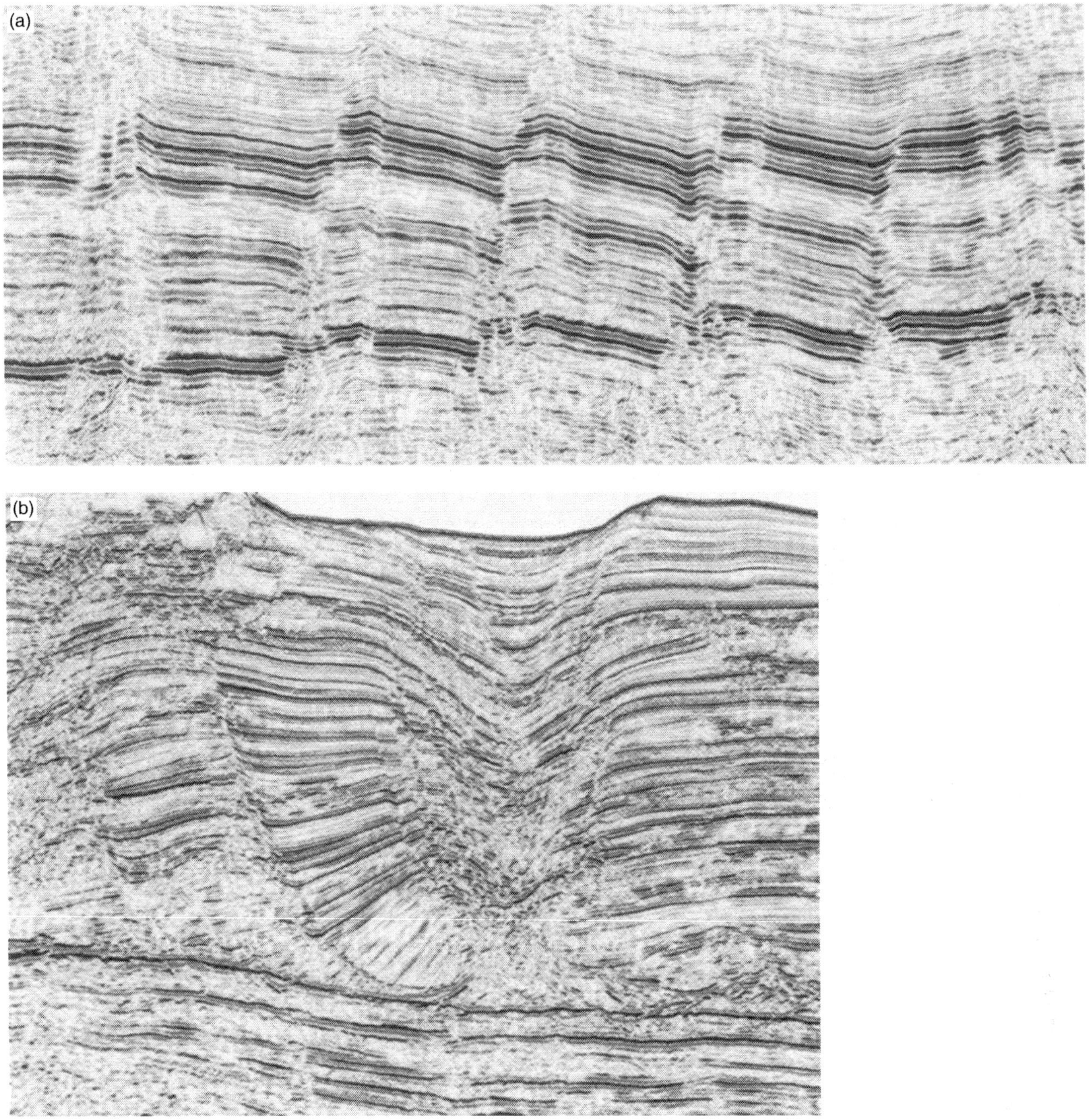

Figure 1.18. (a) Example of the reflection seismic image of the planar normal fault system, Lebanon (TGS-NOPEC, 2003). (b) Example of the reflection seismic image of the listric normal fault system, France (TGS-NOPEC, 2003).

faults initiate as parallel to the slip vectors corresponding to extensional flow (e.g., Bruhn and Choukroune, 1983). However, because real rock may not be truly completely plastic, the interpretation can be used only qualitatively.

The best areas to find the data on the brittle–ductile transition are high-temperature geothermal fields. In The Geysers and the Imperial Valley, California (Gilpin and Lee, 1978; Majer and McEvilly, 1979; Sibson, 1982), shallow bottoming of the earthquakes allows scientists to interpret the shallow brittle–ductile transition. Geothermal deep wells drilled to depths where temperatures exceeded 370–400°C have encountered a host rock with very little permeability and pore fluid pressure significantly higher than hydrostatic in zones very close to the brittle–ductile transition for respective rheologies

Figure 1.19. Listric growth normal fault, indicating syn-sedimentary activity in the Triassic Mercia mudstone, Watchet, southern Bristol Channel (photo courtesy of Rod Gayer). Four anhydrite marker horizons document progressively larger throw with age.

(e.g., Cappetti et al., 1985; Ferrara et al., 1985; Fournier, 1991). Studies of epithermal ore deposits have provided the evidence for the narrow transition from an overlying environment with hydrostatically pressured fluids to the underlying almost lithostatically pressured regime (e.g., Hedenquist et al., 1998; Fournier, 1999).

The studies of orogenic belts have demonstrated that there is a whole system of discontinuities in the "ductile crustal domain" underneath the upper brittle domain, which is without any ductile deformation, and there is usually a progressive transition between the two domains (Brun and Choukroune, 1983).

As the stretching of the crust continues, the pressure-temperature conditions spatially change in time. Regions undergoing progressive warming could find themselves developing detachment faults at progressively shallower levels. On the contrary, regions undergoing progressive cooling develop their detachments at progressively deeper levels. Natural examples come from offshore Gabon where ultra-deep reflection

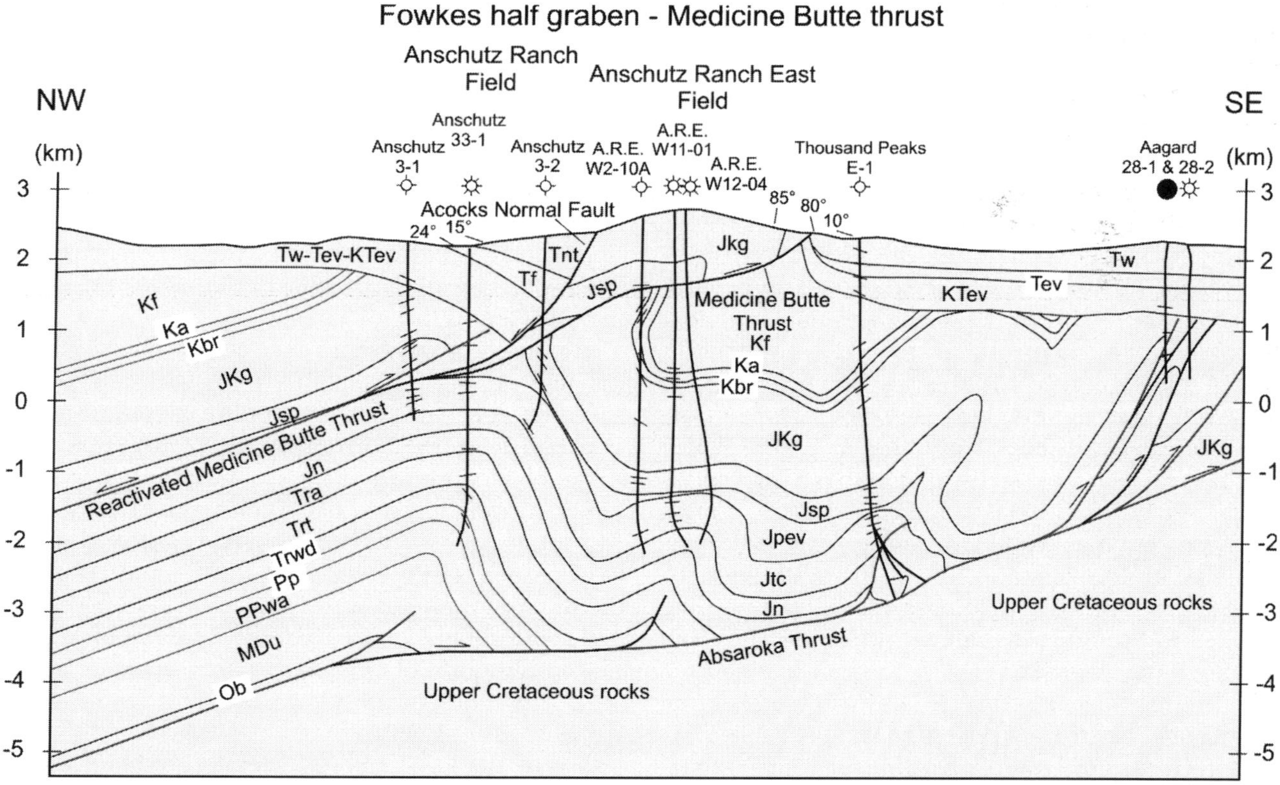

Figure 1.20. Geologic cross-section through the Anschutz Ranch and Anschutz Ranch East fields, Wyoming thrustbelt (Constenius, 1996). Figure shows the Acocks normal fault, the reactivated Medicine Butte thrust, and the Fowkes half-graben formed during the orogenic collapse that took place during 49–20 Ma, about 5–7 Ma after the end of orogenic shortening. Ob-Jtc are formations ranging from the Ordovician Big Horn dolomite to the Jurassic Twin Creek limestone, Jpev is evaporites of the Jurassic Preuss Formation, Jsp-Kf are formations ranging from the Jurassic Stump–Preuss Formations to the Cretaceous Frontier Formation, Kev is the Maastrichtian Evanston Formation, Tev is the Paleocene Evanston Formation, Tw is the upper Paleocene–early Eocene Wasatch Formation.

Table 1.1. Normal fault lengths, including averages, from several extensional provinces.

Region	Fault length			Source of data
	Minimum (km)	Average (km)	Maximum (km)	
Gulf of Corinth, Greece	0.6	1.4	5.0	Jackson et al. (1982)
Eastern Great Salt Lake Valley, Utah	1.1	3.2	8.4	Bryant (1990)
Wasatch Fault near Salt Lake City, Utah	2.6	7.3	10.0	Bryant (1990)
Death Valley, California	8.0	12.3	21.0	Stewart (1983)
Central Apennines, Italy	5.0	17.3	30.0	Roberts (2007)
Western Tasmania, Australia	7.5	17.3	35.0	Noll and Hall (2006)
Gulfs of Corinth and Evia, Greece	12.5	25.4	32.5	Roberts (2007)
Southern Apennines, Italy	10.7	30.4	52.0	Roberts (2007)

seismic lines indicate the existence of the initial shallower detachment system and the subsequent deeper one as the stretching proceeded. Similar observations have been pointed out from a combination of the outcrop and seismic studies by Manatschal (2004 and references therein).

Detachment faults of gravity glide systems are also characterized by their location along a mechanical discontinuity such as the evaporite or shale zone separating two more competent rock sections (Demercian et al., 1993; Hudec and Jackson, 2006; Rowan et al., 1999).

Lithological discontinuities in multilayers used for location of normal fault detachments can be also found in outcrops, such as the Lower Jurassic calcilutite/mudstone multilayer of the Bristol Channel, United Kingdom. Here the outcrops are deformed by meso-scale normal faults detached along horizons of the calcareous mudstones (Nemčok et al., 1995). While each normal fault is a result of one or several slip events, which happen when differential stress is great enough to trigger rupturation, the detachment fault that links a number of normal faults contains structures indicating crack-and-seal deformation mechanisms, implying more steady deformation in comparison with normal faulting events above it.

The size of spoon-shaped normal faults above detachment varies from region to region (Table 1.1). A geologic map of Death Valley (Stewart, 1983) shows about six normal faults having an average length of 12.3 km. Minimum and maximum lengths are 8 and 21 km, respectively. The average length of the normal fault segment from the seismoactive Wasatch normal fault in the vicinity of Salt Lake City is 7.3 km. Minimum and maximum lengths are 2.6 and 10 km, respectively. This indicates that a set of factors controls normal fault propagation and arrest, including the deforming rheology, stress field, or existence of preexisting anisotropy.

The presence of the "ideal" listric trajectories is most likely in gravity glides in very young sediments, which do not possess too many other constraining factors. The fact that forced shear drag (e.g., Hamblin, 1965) and collapsed hanging walls are known to occur (e.g., Doust and Omatsola, 1990) suggest numerous factors controlling the fault geometry instead of just a simple, geometry-based, energy-balance argument. These factors include angle of internal friction of deformed rocks, existence of preexisting anisotropies, and dilatancy effects.

Regardless of the overall geometry of the main normal fault, it typically comprises segmented arrays of linked, en echelon faults that control overall displacement distributions and resulting hanging wall accommodation patterns (e.g., Peacock and Sanderson, 1991; Anders and Schlische, 1994; Gawthorpe et al., 2003). The array is characterized by complex displacement distributions, which result from spatial and temporal variations in fault-slip activity, which is induced by the stress perturbations associated with fault rupture events and interactions of segments within evolving array (e.g., Segall and Pollard, 1980; Cowie and Scholz, 1992; Willemse et al., 1996).

One of the good natural laboratories for the studies of normal faults and their detachments is the Basin and Range province in the southwestern United States (Figure 1.21; Hamilton, 1987). Normal faults are represented by dominating listric and subordinate planar normal faults. Most of the listric faults progressively curve into the detachments. However, several outcrops show major gently dipping faults, which rise for tens of

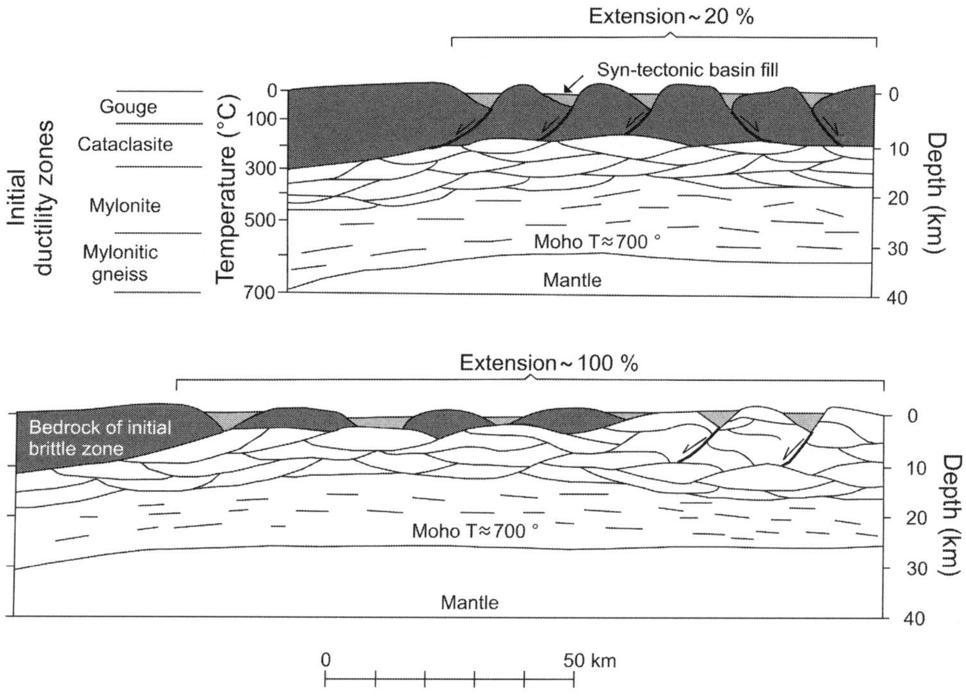

Figure 1.21. Upper and lower crustal architecture in the Basin and Range province (Hamilton, 1987).

kilometers laterally through upper crustal rocks and curve to steeper dips only at shallow depth.

Linkage style of normal faults with their detachment fault can be observed in the Black Mountains, California, where a large, active, normal fault limited to hanging wall rocks ends truncated downward against Badwater, Copper Canyon, and Mormon Point turtlebacks representing the middle crustal lenses (Hamilton, 1987). Each of the three turtleback detachment faults is now active as a low-angle normal fault. Footwall Neogene volcanic rocks, hanging wall Quaternary slide breccias, and the late Quaternary Artist's Drive Fault are all truncated against the gently dipping Badwater turtleback.

The detachment faults define the tops of great lenses (Figure 1.21), which are separated by gently dipping, anastomosing ductile faults and retain pre-faulting fabrics in their interiors. Detachment faults are represented by irregular undulating surfaces or many of them outcrop as broad domes (e.g., Davis, 1980; Davis et al., 1980; Rehrig and Reynolds, 1980). Domes can be symmetrical or irregular. Their typical dimensions are 10–20 by 20–40 km (Hamilton, 1987). Some are elongate in the direction of slip of overlying plates, being formed as large corrugations or mullions (see Wright, 1974; John, 1984). It is interesting to compare these mullions with mullion structures found at ultra-slow and slow seafloor-spreading ridges (e.g., Cann et al., 1997; Tucholke et al., 1998; MacLeod et al., 2002; Escartín et al., 2003) because both settings have rather similar extensional rates (see Macdonald, 1982; Hamilton, 1987; Bown and White, 1994). Other domes are elongate at high angles to the direction of slip of overlying plates, being formed by interactions between diversely shaped middle crustal lenses as they slid apart during the overall extension (Hamilton, 1987).

As the lenses slide apart, the area of their composite top, the detachment faults, is increased and the middle crust is extended by discontinuous ductile flow. This increase of the detachment level area reacts to the increase of the extensional province area. The lower crust flattens pervasively (Hamilton, 1987).

This situation is different from the Rhine Graben situation, where reflection seismic profiles indicate offset of the Moho (Figure 1.22a; Brun et al., 1992), and the graben represents a narrow rift (Brun et al., 1992).

The northern of their two profiles shows significant reduction in thickness of the reflective interval at the lower crust level as the profile leaves both platform areas and enters the basin (Figure 1.22a; Brun et al., 1992). Refraction data indicate Moho on the east platform side of the graben at a depth of about 30 km, then upwarped to about 25 km underneath the graben axis and then down-dropped to about 32 km on the west platform side of the graben. The lowermost

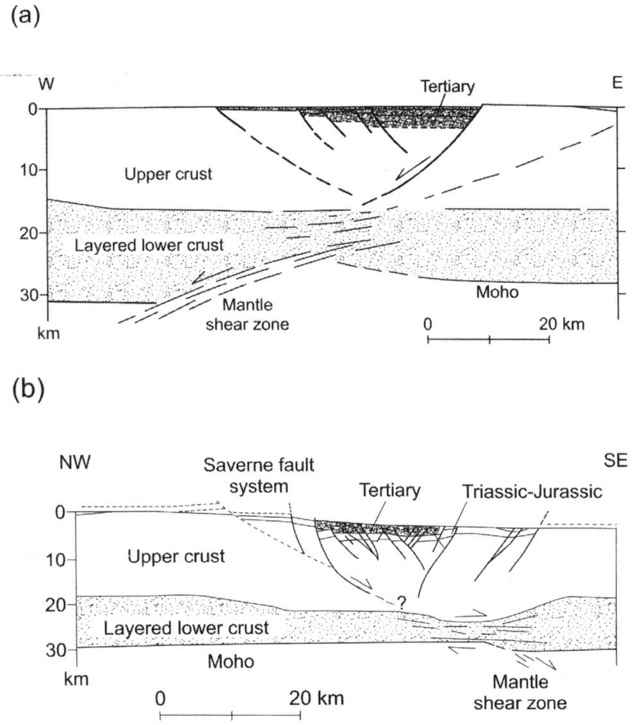

Figure 1.22. (a) Northern reflection seismic profile through the Rhine Graben (Brun et al., 1992). (b) Southern reflection seismic profile through the Rhine Graben (Brun et al., 1992).

reflectors roughly coincide with Moho determined from refraction data.

The southern of their two profiles is somewhat similar, having clearly transparent and reflective upper and lower crusts in both platform sides of the graben (Figure 1.22b; Brun et al., 1992). Moho reflectors underneath both platform sides are exceptionally bright. The lower crust underneath the graben remains at least a bit reflective and its thickness is reduced. The graben asymmetry is opposite to the asymmetry shown by the northern profile. There is no prominent upwarping of Moho underneath the graben, only a Moho offset of about 2.5 km to the east of the basin axis.

Both lines show shallow 10–20° dipping reflectors, which cut Moho in areas of Moho offsets, interpreted as shear zones with localized plastic deformation in the uppermost mantle (Brun et al., 1992). They also show listric upper crustal faults, which seem to detach at middle crustal levels at relatively gently dipping reflectors, interpreted as preexisting Variscan orogenic detachments.

Local variations in temperature, strain rate, and lithology are reflected in variations in initial depth and configuration of the ductile faults that evolve into detachment faults, as demonstrated in the Basin and Range province (Hamilton, 1987).

The detachment faults are represented by brittle gouge in their core, typically a few decimeters to a few meters thick, but ranging from a centimeter to tens of meters. Many gouge zones grade upward into decreasingly brecciated upper-plate rocks. Breccia zones can be up to 100 m thick. Immediate hanging walls of the detachment faults frequently show extensional structures.

Immediate footwalls contain various types of middle crustal rocks such as granites, gneisses, migmatites, or metasedimentary or meta-igneous rocks of any grade from lowest greenschist to highest amphibolite. Gouge zones of detachment faults are commonly bounded sharply against a shear-polished slickenside on a hard, chloritic microbreccia. This cataclasite is typically a few meters thick. Rocks beneath the microbreccia typically show a record of the ductile shear. They are represented by laminated mylonites in the case of quartz-feldspatic rocks and foliated greenschist rocks in the case of metasediments. These ductile rocks typically fade downward after 20 or even 100 meters, defining carapaces semi-parallel to the overlying detachment fault. Mylonitic gneisses are widespread in the most-uplifted footwalls. The mylonitic carapaces and progressively lower-temperature cataclasites are developed in some cases on Tertiary mid-crust granites, barely older than the detachment faults that cut them (e.g., Snoke et al., 1984; Reynolds, 1985).

Beneath the detachment faults, the footwall rocks largely retain pre-detachment fabrics. They contain syn-extensional ductile shear occurring in discrete zones. Such zones divide sub-detachment lithologies into lenses.

Offsets along middle crustal detachments can be as large as 20–60 km (Hamilton, 1987). Stratigraphic omissions are enormous. Detachment faults juxtapose rocks having petrological evidence for middle crustal crystallization with unmetamorphosed supra-crustal rocks.

The ages and intensities of detachment faulting over broad terrains vary complexly. Hanging walls and footwalls of detachment faults display random variations in their initial crustal depth and variable slip vectors. This indicates that detachment faults were never planar but rather undulating surfaces separating middle crustal lenses, which did not deform too much internally and extended by sliding apart in response to the overall extension (Hamilton, 1987).

As extension progressed and the crust thinned, parts of the system rose toward the surface, the ductile deformation boundaries rose as extension, and shear heating possibly outpaced thermal conduction (Davis, 1980, 1983; Rehrig and Reynolds, 1980; Snoke et al., 1984; Reynolds, 1985; Hamilton, 1987). The local extension

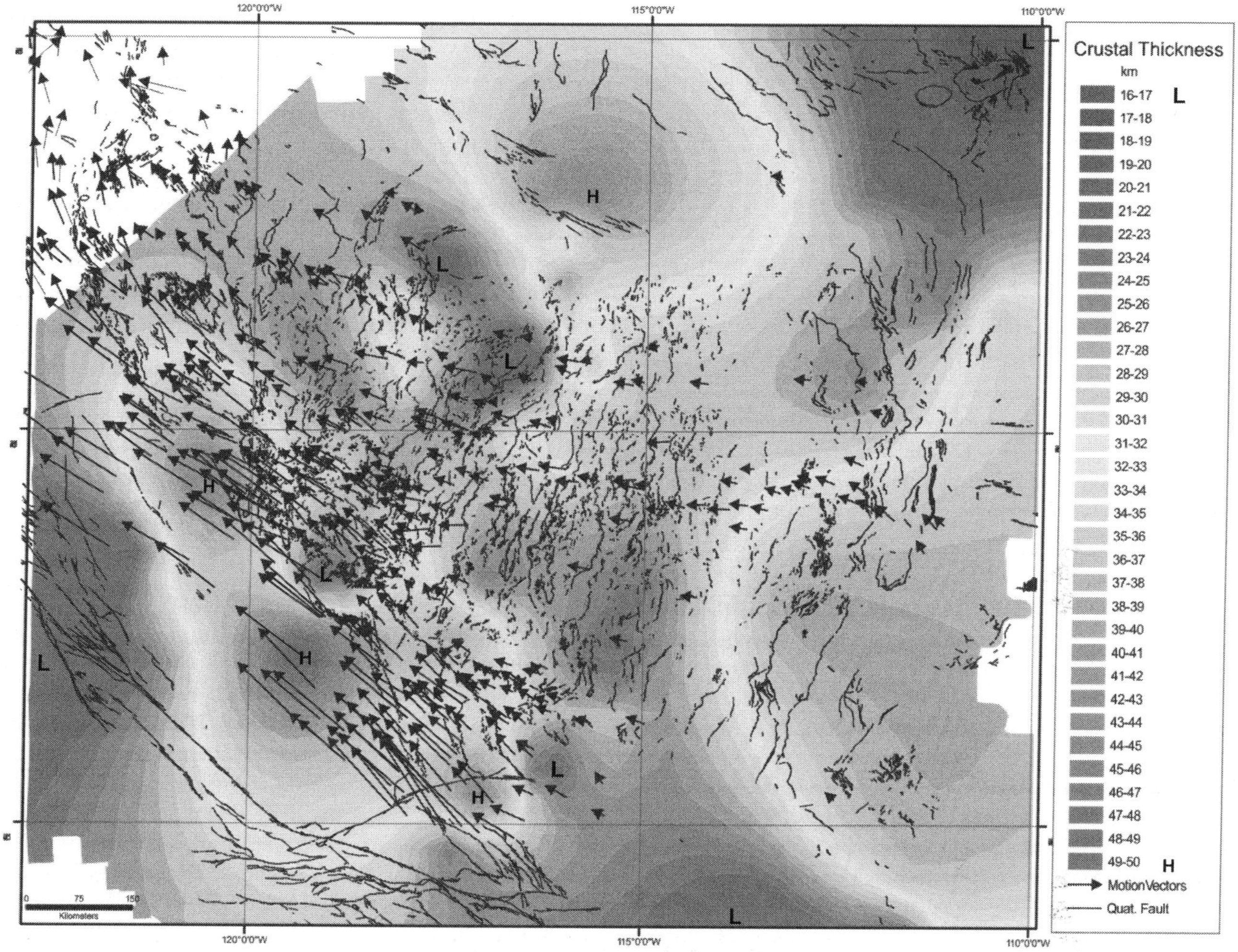

Figure 1.23. Present-day distribution of GPS-based motion vectors in the Basin and Range province drawn on top of the fault pattern active during Quaternary and present-day crustal thickness map (Hammond and Thatcher, 2005; Shevenell, 2005a; USGS, 2006; Kreemer et al., 2003). H and L indicate occurrences of crustal thickness maxima and minima.

rates were spatially and temporally variable. It can be shown by the present-day distribution of GPS-based motion vectors and shear strain (Figure 1.23; Shevenell, 2005b; Kreemer, 2007). The reasons for variable extension rates, apart from variations in regional tectonic stresses, include migrations of the upper crustal stretching controlled by gravity gradients provided that lower crust has relatively low viscosity, as indicated by modeling (Hopper and Buck, 1998).

Superimposed structures and fabrics indicate progressions from deformation under deep, hot conditions to deformation under shallow, cold conditions. The attenuating crust is partially rebuilt by magmatism, and also possibly by phase changes, so the crust is thinned less than we would expect. A support for the basalt contribution to the overall crustal thickness can come from an interpretation of a "basaltic pillow" underneath the Great Basin–Colorado Plateau transition area, based on a crustal layer along the base of the crust having seismic velocities of 7.4–7.5 kms^{-1} (Smith et al., 1989). Although this interpretation was highly debatable, the region with interpreted basaltic layer correlates with heat flow anomaly (Bodell and Chapman, 1982; Powell, 1997) and deep electrical conductivity zone (Porath, 1971), which could represent the zone of most active basaltic underplating and crustal intrusion or fluid exsolution into the lower crust or the mantle melt zone (Wannamaker and Hohman, 1991). In fact, the lower crustal conductivity layer under the actively extending eastern Great Basin from magnetotelluric survey is much stronger than that of the relatively quiescent central Great Basin of northeastern Nevada (Wannamaker et al., 1997).

Other natural cases with magmatic underplating come from the Koka and Sabure Basins in Ethiopia, supported by variations in seismogenic and/or elastic layer, which clearly thins in rift segments characterized by magmatism (Ebinger and Casey, 2001) or the Turkana Rift in northern Kenya, where the location of the anomalously low seismic velocity layer underneath the rift indicates the presence of melt layer (Green et al., 1991; Achauer et al., 1994; Mechie et al., 1994).

As the result of continuing extension, uplift, and erosion, the majority of ranges in the Basin and Range province represent a mature stage of the structural architecture development (compare to the relatively immature stage of the Corinth-Patras Rift in this chapter; Sorel, 2000; Figure 1.8). They are either tilted hanging wall panels of resistant rocks isolated from one another atop detachment faults or outcropping sub-detachment rock complexes (Hamilton, 1987). Hanging wall panels have typical width and length of 10–25 km and 70–150 km, respectively. Gaps between them are mostly filled by syn-rift sediments. A direct contact of the overlying young sediments with initially middle crustal detachment fault is described from the Death Valley region (Hamilton, 1987).

Factors controlling listric normal fault geometry

The way a rock package deforms by normal faulting is controlled by the faulting instability envelope. The typical model for an extended rock section includes a sheet composed of strong (competent) elastic/frictional plastic material, separated from a rigid basement by a thin layer of weaker (incompetent) elastic/frictional plastic material (Mandl, 1988). The stress buildup prior to the normal fault propagation and movement results from horizontal displacement of the back end of the sheet.

The stress transfer is utilized by straining. This straining consists of elastic straining, followed by plastic hardening and later by plastic softening (e.g., Geiser, 1974; Wojtal and Mitra, 1986; Mitra, 1987). Each episode of the detachment fault rupturation, accompanied by elastic unloading, is triggered by preceding stress buildup (Yielding et al., 1981; Rockwell et al., 1988; Philip et al., 1992). Straining related to each stress cycle comprising buildup and release, which controls the episodic normal fault propagation and displacement, leaves its own related shear softening record (e.g., Scholz, 1968; Kranz and Scholz, 1977; Lockner et al., 1992), which is ductile or brittle.

After a certain increment of back-end displacement of the extending sheet, there will be a zone inside the extending sheet that is affected by plastic softening. Such a shear-softening zone creates conditions for spontaneous fault formation by localizing the shear deformation. Each new buildup of the critical stress results in new fault propagation and also in the reactivation of the already existing fault that is the result of previous critical stress releases. When the detachment layer behaves as a viscous material, the stress transmission in the overlying competent sheet is significantly enhanced (Mandl, 1988). The shear resistance of the viscous layer, for example, evaporite, is strain dependent, and the stress transfer can be described by a stress-diffusion model (Bott and Dean, 1973). This model describes how the force applied to the rear causes elastic deformation of the rear part of the competent rock sheet and the onset of viscous flow in the underlying detachment layer. It also describes how the "stress front" advances from the stressed part of the competent sheet into its unstressed part. The time, T, until the elastic sheet stresses to 90 percent of its constant end stress is:

$$T = \eta/E^*(L^2/H^*h), \qquad (1\text{-}1)$$

where h is the thickness and η the viscosity of the viscous detachment layer. L is the length, E is the Young's modulus, and H is the thickness of the elastic sheet. Despite the significant simplifications involved in this concept of the moving "stress front," it serves as a good tool in understanding the stress transfer.

A beautiful natural example of the moving stress front in the elastic sheet resting on the viscous detachment layer comes from the Canyonlands, Utah (Trudgill and Cartwright, 1994; Trudgill, 2002). The region consists of a graben system formed in a 460 m-thick competent layer of Pennsylvanian–Lower Permian sandstones and limestones gliding on the inclined 300 m-thick Pennsylvanian evaporitic layer toward the Colorado River in the northwest, which started after the Pliocene-Pleistocene incision of the river into the stratigraphic section dipping at 2–4° to the northwest. The grabens were progressively developed from northwest to southeast during the past 500,000 years.

To understand why the listric normal faults dominate over planar ones, we need to discuss the factors controlling the listric normal fault geometry, which include:

1) nonplastic processes;
2) 3-D shear strength anisotropy;
3) shear strength variation among sedimentary bodies;
4) dilatancy;
5) pore fluid pressure; and
6) stress changes in the vicinity of the propagating fault.

The nonplastic processes, such as compaction solution, take place in the non-lithified sediment, which undergoes rapid mechanical and chemical compaction with depth. For example, the mud-supported shelf sediment undergoes the porosity reduction of 20–38 percent during the first 100 m of burial (Moore, 1989). Nonplastic processes can increase the angle between σ_1 stress and fault plane (Mandl, 1988), which is in accordance with increasing depth of the listric normal fault with depth. For example, a 12° dip reduction of an initial 60° normal fault takes place with 35 percent of sediment compaction. In extensional settings, this effect can be further enhanced by the layer extension added on top of burial-driven compaction.

The shear strength anisotropy is a directional dependence of the shear strength in rocks (Mandl, 1988). Such anisotropy may be associated with anisotropy in rock fabric, either preexisting or induced by inelastic deformation processes during the extensional stress development cycle. The detailed description of the effect of planar anisotropies on fault geometry is in Chapter 6.

The shear strength variation among sedimentary bodies cut by a propagating fault is probably the least likely factor in listric shape control because it relates to changes in angles of internal friction of deformed rocks, which may result in listric fault shape only in the case when more competent rocks overlie less competent ones. Nice seismic examples of the shear strength variation control over the fault trajectory in cross-section are shown in reflection seismic images from the North Sea (Stewart and Reeds, 2003). The effect of shear strength variation can be increased by stress perturbations at interfaces between materials of distinctly different rheologies. The stress perturbation is especially pronounced by a shearing action of a ductile underflow in viscous layer underneath brittle layers. The underflow typically occurs in gravity glides where shale or salt material is squeezed out from the autochthonous horizon into marginal shale or salt structures (Mandl, 1988).

In dilatant rocks, the angle α between the initiating faults and the maximum principal compressive stress axis, σ_1, should be smaller than the angle $\alpha = \pm (45° - \phi/2)$ where ϕ is the angle of internal friction. It is because the Griffith's cracks are generally oriented parallel to the σ_1 direction and the fractured rock may dilate more easily in the σ_3 direction. The dilatancy effect is largest near the surface, under low confining pressure, while it is suppressed by high confining pressure at larger depths. The propagating fault forms a hybrid between tensile fracture and shear fracture. Its geometry is characterized by the steepest portion of the listric geometry in the shallow subsurface.

A natural example comes from the Schramen ore vein in the Banská Štiavnica, Slovakia, formed in the stratovolcanic rock complex (Nemčok and Lexa, 1990). Its outcrops in the mines at deep levels contain striated slickensides with normal fault displacement and a dip of about 60°, while the vein at shallow levels contains idiomorphic crystals grown into the open space of the subvertical tensile fracture.

The increased pore fluid pressure effect on the fault's listricity is similar to a contrast in angles of internal friction of individual multilayer horizons because the pore fluid pressure controls the porosity, which controls the angle of internal friction in siliciclastic sediments with grains in matrix. If the contrasting layers are not horizontal and at least partially decoupled, further fault trajectory changes are introduced by shear stress on bedding planes (Mandl, 1988). One more aspect of the increased pore fluid pressure factor, in connection with shear strength anisotropy, already mentioned in the text on the dilatancy factor, is the shift of the stress conditions toward dilatational normal faulting.

If we think about the stress field in the area around the normal fault propagation tip as about the subcritical elastic state, the superposition of the homogeneous stress field with maximum principal compressional stress σ_1 oriented at angle of $45° - \phi/2$ to the existing fault plane should not cause any curvature in a map view. However, the stress field is not homogeneous because of the stress changes on the hanging wall side of the normal fault by shear softening of the fault zone material and the synchronous reduction of the shear resistance of the active fault plane. The reduction of the fault plane shear strength is balanced by an increase of the fault-parallel normal stress in the adjacent hanging wall, in order to maintain static equilibrium on the fault (Mandl, 1988). Triggered differential straining produces extra shear stress on the downthrown side of the fault in the vicinity of the fault hinge, which migrates outward. The additional shear stress disturbs the symmetry of the maximum shear stresses, in favor of the shear stresses on the hanging wall side. Therefore, plastic limit state will be preferentially achieved on the hanging wall side of the downward and laterally outward propagating normal fault, and the fault will have a tendency to propagate along a spoon-shaped geometry concave toward the hanging wall, following a minimum energy solution for the normal fault propagation.

One more stress change is controlled by the strike-parallel extension of hanging wall strata driven by much larger subsidence of the hanging wall block at the most mature area of the propagating normal fault versus no subsidence of the hanging wall block at lateral propagation tips. Where developed, it reduces the intermediate principal compressional stress σ_2 oriented parallel to the strike of the normal fault. The local stress regime at each point along the strike of the normal fault is modified by respective bending stresses of

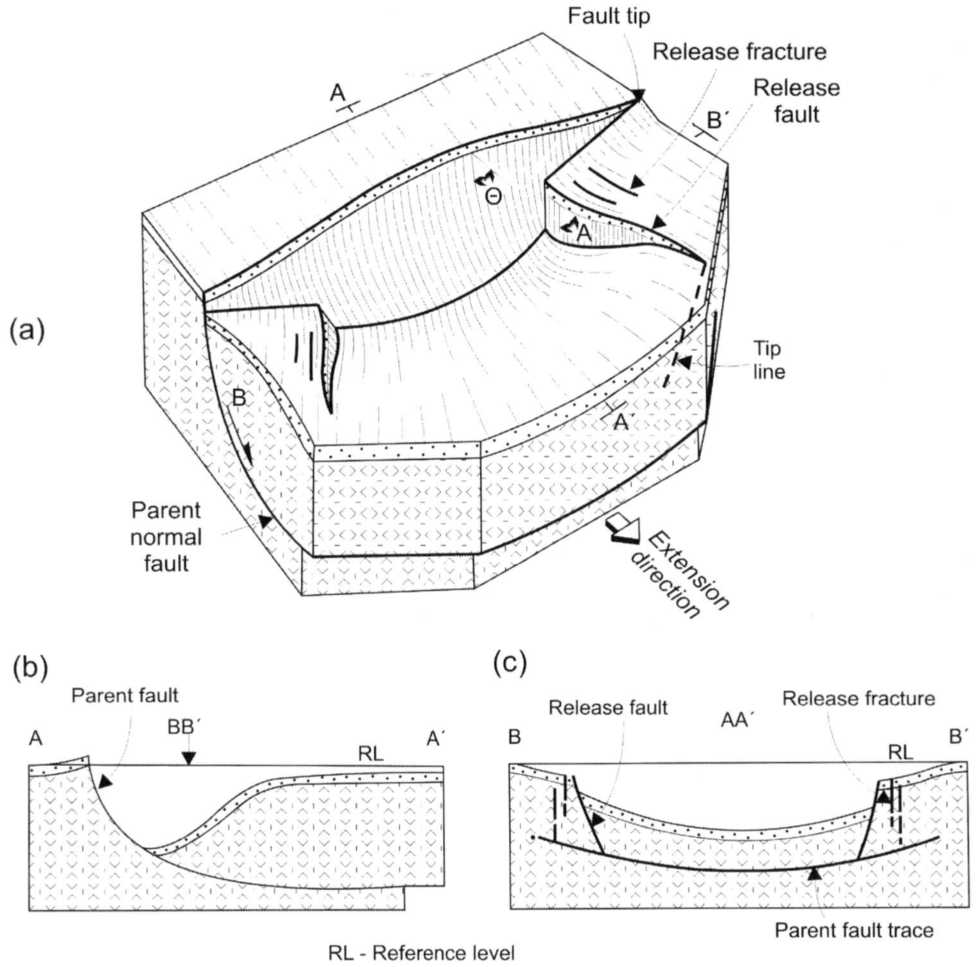

Figure 1.24. Displacement variation along the strike of the normal fault in adjacent footwall and hanging wall (Destro et al., 2003b). Note that a pitch of striation vectors in release faults corresponds to dip of the main normal fault and the release faults die out along the detachment of the main fault.

the hanging wall block caused by variation of displacement along the controlling normal fault.

Other potential stress change controls include syn-propagational deposition, the degree of coupling with the basement, distribution of the detachment horizon, and superimposed stresses.

The stress perturbations inside the hanging wall of the listric normal fault can introduce accommodation faults (Destro, 1995, cited in Destro et al., 2003b). A typical example is the release fault, which develops as a result of varying throws along the strike of a listric parent normal fault (Figure 1.24; Destro et al., 2003b). It does not cut the parent fault. They can form on the strike-slip ramps or at the parent fault tip, or be distributed along the strike of the parent fault. In general, release faults have normal fault displacement. Because of dynamic linkage with their parent fault, they can undergo some component of strike-slip displacement (Destro, 1995, cited in Destro et al., 2003b). Natural examples are provided by the South and North Mata-Catu Faults from the Recôncavo Rift Zone (Destro et al., 2003b).

Factors controlling normal fault spacing

Provided a certain host rheology, thickness, and pore fluid pressure, the fault spacing is controlled by the shear softening, which localizes the shear deformation and makes active fault plane a weakness.

Its role is demonstrated in a finite-element model consisting of the horizontal layer of homogeneous isotropic elastic/frictional plastic material lying on the extendable base and stresses are built by small increments of block movement. After several small strain increments, most of the area is still in elastic strain state; some zones of the plastic hardening are developed and few areas of plastic softening appear. Later the first plastic softening zone develops and its surrounding drops back into elastic strain state. This plastic zone develops from surface to base after several strain increments. After it fully develops by reaching the base of the deformed layer, the second plastic softening zone starts to develop.

It turns out that the extent of the rock volume around the first fault where differential stress dropped to elastic stress state represents the region where the subsequent fault cannot develop. This mechanism controls the spacing of the normal faults, which develop in the deformed layer at the end of the finite-element simulation. The more brittle the rock layer, the greater the shear strength reduction along the propagated fault and the more significant the accompanying stress unloading will be.

The energy-balance control of the fault spacing development are in the competition between the stress unloading by shear softening along existing faults and along fracture clusters controlling their future propagation and stress reloading by continuous extensional straining. During this competition, the width of the area affected by the elastic state increases with (Mandl, 1988):

1) the amount of the total shear strength reduction along the progressively developing fault, and
2) the rate of the plastic softening with respect to the shear strain imposed by the continuous extensional straining of the deforming rock layer (shear strain drop per shear strain increment).

In nature, the fault location is further influenced by rock layer inhomogeneity and anisotropy, which would have a tendency to influence the location of the embryonic normal faults. However, as the extensional straining continues, the normal fault pattern will become more regularly spaced. Therefore, it is possible to find long, continuous outcrops such as those in the Bristol Channel, which contain embryonic normal faults, which remained "frozen" in their initial development stages because they found themselves located in subsequently expanded areas with elastic state. The opposite case of the infill normal faults between initially too widely spaced faults is also possible.

Sensitivity analysis of the extending layer thickness's effect on extensional fault spacing in the finite-element model (Witlox, 1986), on tensile fracture spacing in the outcrop in the Bristol Channel (Nemčok et al., 1995), and on normal fault spacing in analog material model (Allemand and Brun, 1991) indicates linear dependence. A linear dependence was also observed for the relationship of the fault length and thickness of the elastic and/or seismogenic layer in the Afar region (Hayward and Ebinger, 1996).

Architectural styles of normal fault zones

Fault zone development is controlled by numerous factors, the host rock rheology and the amount of the displacement being the most important ones. Depending on the type of fault core, including:

1) single-slip surface,
2) unconsolidated clay-rich gouge zone,
3) brecciated and geochemically altered zones, and
4) highly indurated, cataclasite zone

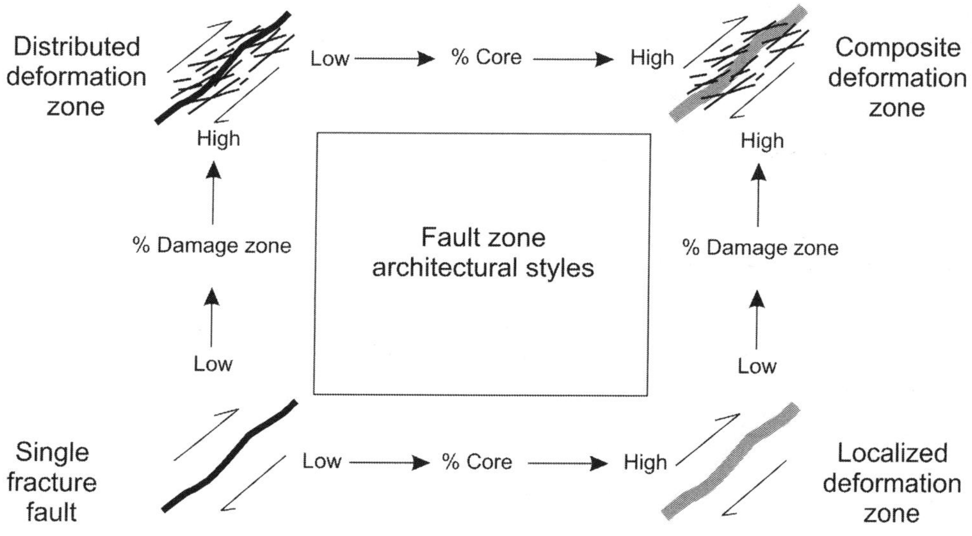

Figure 1.25. Architectural styles of the fault zone (Caine, 1999).

and damage zone, including:

1) small faults,
2) veins,
3) fractures,
4) cleavage, and
5) folds,

and their combination, the fault zones can be divided into (Chester and Logan, 1986; Smith et al., 1990; Caine, 1999; Figure 1.25):

1) single-fracture fault,
2) distributed deformation zone,
3) composite deformation zone, and
4) localized deformation zone.

Factors controlling the fault zone include the host rock lithology, the fault scale, the fault type, the deformation chemistry, the pressure-temperature history, the component percentage, and anisotropy. Natural examples of the single-fracture faults are low displacement faults in brittle rocks, examples of distributed deformation zones come from fault-fracture meshes or small displacement strike-slip faults in dolomite, examples of composite deformation zones come from large displacement normal and strike-slip faults, such as the Stillwater normal fault, Nevada, and examples of localized deformation zones come from rare cases of strike-slip faults such as the San Gabriel Fault, California (Caine, 1999).

2 Mechanics of rifting and transition to drift phases

Introduction

The transformation from disorganized rifting to organized seafloor spreading is one of the fundamental geological problems. Stated in more detail, it's the evolution from indecisive continental rift basins and imperfect accommodation zones, such as those in the Tanganyika region, Tanzania, Burundi, the Democratic Republic of Congo, and Zambia, to pseudo-volcanic lineaments and transfer faults, such as those in the Turkana area, Kenya, and Ethiopia, to gradual, tortuous, and probably subaerial spreading, such as takes place in the Afar region, Ethiopia, to perhaps successful opening as in the Red Sea, if located on the good limb of a triple junction (Rosendahl, 2005, pers. comm.).

All of the various stages of this transition host substantial petroleum deposits. The deepest of them require rather rigorous exploration because their imaging is difficult due to the fact that these petroliferous regions are much more complex, more segmented, and more controlled by thermodynamics at deep levels than we tend to anticipate. This is especially true on either side of the continental shelf edges at passive margins (Rosendahl, 2005, pers. comm.).

Perhaps the best way of imagining the start of the development of these terrains is as a group of small, fault-connected basins divided by accommodation zones and transfer faults that hope to evolve into transform faults, and do so rather imperfectly, or not at all. Successful exploration in these terrains requires deep imaging of the basement and deep crustal structures at scales that can roughly define the location of separate petroleum systems. The deeper the exploration proceeds, the more we need to know about the controlling thermodynamics of the deep structural elements (Rosendahl, 2005, pers. comm.).

The goal of this chapter is to have a discussion about the thermodynamics and the mechanisms of rifting and its transition to drifting, drawing from a combination of numerical and analog model studies with case studies.

Stretching of the lithospheric multilayer

Numerous successful applications of the uniform stretching model (McKenzie, 1978) to initial-stage rift terrains indicate that most of them formed by mechanisms dominated by passive rifting of the lithosphere, discussed in Chapter 1. This model describes an instantaneous lithosphere thinning in uniform fashion, which produces thermal and gravitational instabilities. The equilibration of the gravitational instability takes place instantaneously. This produces the syn-rift subsidence. The equilibration of the thermal instability lasts for a long period of time and produces the post-rift subsidence, which varies with time.

The driving forces of stretching are deviatoric stresses developed in stress cycles by block/plate movements, deviatoric stresses developed over upwelling asthenosphere convection systems, and frictional forces along boundaries of individual lithospheric layers between themselves and with the asthenosphere. Furthermore, partial melting driven by adiabatic decompression of the rising lower lithosphere contributes to lithospheric thinning (McKenzie and Bickle, 1988).

Many studies of the more mature-stage rift terrains, however, required various modifications of the uniform stretching model to explain the observed data. This indicates that active rifting plays a progressively more important role in advanced rifting stages (e.g., Huismans and Beaumont, 2005). Studies of passive margins, structures that went beyond advanced rifting development, document that there are practically no passive margins developed by homogeneous stretching.

The most common difficulties with the application of homogeneous stretching model to rift and passive margin terrains are:

1) The existence of relatively thin syn-rift sediments but thick post-rift sediments (e.g., Royden and Keen, 1980 – eastern Canada margin; Sclater et al., 1980 – intra-Carpathian basins; Beaumont et al., 1982 – Nova Scotia margin; Spadini et al.,

1997 – Black Sea; Walker et al., 1997 – Vøring margin), which is the opposite situation to the one resulting from the homogeneous stretching prediction.

2) The basin uplift during the widespread extension (e.g., Huismans, 1999 – Pannonian Basin system; Ren et al., 2003 – Vøring margin). This can be explained only with thin crust, which contradicts the crustal thickness observations. Uplift usually occurs before the onset of extension or after some initial span of extension.
3) The fact that many basins have accelerated subsidence rates during the post-rift stage (Middleton, 1980 – theoretical study; Hamdani et al., 1994 – Williston Basin), which is at odds with subsidence decrease with cooling predicted by the homogeneous stretching approach.
4) The problem that the homogeneous stretching model (McKenzie, 1978) seems to work only for intracontinental rifts and proximal margins (see Manatschal, 2004), if it does, and it typically breaks down for distal margins because they are deformed by simple shear yet they have no grabens with fill.
5) The mantle lithospheric thinning, which is stronger than crustal thinning, occurs in the late syn-rift to early post-rift evolution in numerous cases (Royden and Keen, 1980; Beaumont et al., 1982).
6) The post-rift evolution of many basins does not show the quiet and smooth subsidence pattern predicted by the model of the thermal relaxation following the syn-rift (Cloetingh, 1986; Ziegler, 1994 – Pannonian Basin system).
7) The existence of the second extension phase and its associated subsidence (Huismans, 1999).
8) The occurrence of alkaline volcanic activity (Huismans, 1999).
9) Strong differential thinning and the coeval occurrence of extension and compression are observed during late syn-rift to post-rift events.

As a response to the inability of the homogeneous stretching model to explain various observed events in the advanced rifts and passive margins, numerous improvements of the homogeneous stretching approach have been suggested, including:

1) The depth-dependent stretching (Royden and Keen, 1980 – eastern Canada margin; Beaumont et al., 1982 – Nova Scotia margin; Roberts et al., 1997 – Norwegian margin; Driscoll and Karner, 1998 and Baxter et al., 1999 – Australian rifted continental margin; Davis and Kusznir, 2004 – Goban Spur, United Kingdom, Galicia Bank, Iberian margin, and Vøring margin, Norway).
2) The active rifting component (e.g., Huismans et al., 2001 – theoretical study).
3) The effect of different rheology development histories of various portions of the lithospheric multilayer (Faugère and Brun, 1984; Davy, 1986; Allemand and Brun, 1991; Davy and Cobbold, 1991; Tron and Brun, 1991; Brun et al., 1994; Brun and Beslier, 1996; Brun, 1999; Mart and Dauteuil, 2000; Corti et al., 2003; Huismans and Beaumont, 2005, 2008, 2011).
4) The interaction between lithospheric rheology and erosion (e.g., Burov and Poliakov, 2001 – theoretical study).
5) The effect of mineral phase transitions on rifting (e.g., O'Connell and Wasserburg, 1972; Podladchikov et al., 1994; Yamasaki and Nakada, 1997; Artyushkov et al., 2000; Simon and Podladchikov, 2006 – theoretical studies). Referenced studies indicate that mantle composition changes during stretching, introducing melting by various mechanisms. The most important material in phase transitions affecting rifting is plagioclase. An example of the particularly important effect of the mineral phase transitions on rifting is the re-fertilization of the mantle, which can lead to strong density reduction, that is, buoyancy-driven uplift of the lithospheric mantle, which has a strong effect on the continental breakup style.

Current research in magma-poor margins and some more magma-rich margins indicates that, instead of the single-stage lithospheric breakup predicted by the uniform stretching model (McKenzie, 1978), the lithospheric multilayer undergoes breakup in several stages (Manatschal, 2004; Huismans and Beaumont, 2005, 2008, 2011; Lavier and Manatschal, 2006) where synchronous crustal and lithospheric mantle breakups are just one of the possible scenarios. The end-member scenarios are represented by the first one with crustal breakup followed by mantle lithosphere breakup and the second one with lithospheric mantle breakup followed by crustal breakup (Huismans and Beaumont, 2011). The development of the magma-rich margin is favored by (Richards et al., 1989; Campbell and Griffiths, 1990; Anderson et al., 1992; King and Anderson, 1995, 1998; Smith and Lewis, 1999; van Wijk et al., 2001; Geoffroy,

2005; Nemčok et al., 2012; own data on the Central Atlantic):

1) hotter mantle, most likely undergoing small-scale convection;
2) high extension rate; and
3) mantle enriched and re-melted by a hot spot.

The different breakup timing of different layers of the lithospheric multilayer, which can be represented by a different number of layers typically from two to four (e.g., Faugére and Brun, 1984; Davy, 1986; Allemand and Brun, 1991; Davy and Cobbold, 1991; Tron and Brun, 1991; Brun et al., 1994; Brun and Beslier, 1996; Brun, 1999; Mart and Dauteuil, 2000; Corti et al., 2003), as discussed in Chapter 5, is the consequence of their different necking length scales (Huismans and Beaumont, 2011). The key in the behavior of the breaking up lithospheric multilayer will be the viscosity contrasts between individual layers, as they do not only influence the stress and strain histories of respective layers but also the histories of their neighbor layers. For example, when the layer behaves as a viscous material, the stress transmission is significantly enhanced in comparison to the elastic/frictional plastic layer (e.g., Mandl, 1988). Furthermore, strong and weak neighbor layers interact mechanically. The strong layer undergoes stress loading and straining, which depend on those in the weak layer. The key in the behavior of the breaking up lithospheric multilayer is, thus, whether rheology of certain layers allows them to stretch but avoid flow while others flow following pressure gradients.

In the numerical simulations discussed in this chapter, the upper part of the lithospheric multilayer can be considered elastic on the time scale of hundreds of millions of years (Ranalli, 1995). The elasticity allows the lithosphere to have a tectonic event record from its earlier development history. It controls the stress transfer in strong layers, which in turn controls the strain location and timing in weak layers.

The mechanical behavior of individual layers in the elastic domain in two dimensions is described by the generalized Hooke's law, relating strains, ε, to stresses, σ, via Young's modulus, E, and Poisson's ratio, ν (e.g., Turcotte and Schubert, 2002) under plane strain conditions ($\varepsilon_2 = 0$) according to:

$$\varepsilon_1 = \frac{(1+\upsilon)}{E}[\sigma_1(1-\upsilon) - \upsilon\sigma_3] \quad (2\text{-}1)$$

$$\varepsilon_3 = \frac{(1+\upsilon)}{E}[\sigma_3(1-\upsilon) - \upsilon\sigma_1] \quad (2\text{-}2)$$

The frictional-plastic deformation in lithospheric layers is specified by a pressure-dependent Drucker-Prager yield criterion. Expressed in terms of cohesion, C, and the angle of internal friction, ϕ, the Drucker-Prager yield criterion can be written as:

$$\sqrt{\frac{(\sigma_1-\sigma_2)^2+(\sigma_2-\sigma_3)^2+(\sigma_1-\sigma_3)^2}{6}} = A + B(\sigma_1+\sigma_2+\sigma_3) \quad (2\text{-}3)$$

with

$$A = \frac{6\,C\cos\phi}{\sqrt{3}(3-\sin\phi)} \quad (2\text{-}4)$$

and

$$B = \frac{2\sin\phi}{\sqrt{3}(3-\sin\phi)} \quad (2\text{-}5)$$

The viscous flow in lithospheric layers is specified by temperature-dependent power-law rheologies based on laboratory analyses (see Goetze and Evans, 1979; Carter and Tsenn, 1987; Ranalli and Murphy, 1987) and the viscosity, η, is controlled by function (Huismans and Beaumont, 2008):

$$\eta = A^{-1/n}(\dot{I}'_2)^{(1-n)/2n}\exp[(Q+Vp)/nRT], \quad (2\text{-}6)$$

where A is the scaling factor, $\dot{I}'_2$ is the second invariant of the deviatoric strain rate tensor ($1/2\dot{\varepsilon}'_{ij}\dot{\varepsilon}'_{ij}$), n is the power law exponent, Q is the activation energy, V is the activation volume, which expresses the viscosity dependence on pressure, p. R is the universal gas constant, and T is the absolute temperature. It needs to be emphasized at this point that the most important problem in defining the creep behavior of rocks prescribed to the lithospheric layers is the justification of extrapolating empirically derived flow laws from the experimental conditions to strain rates and temperatures representative of geological processes (see Carter and Tsenn, 1987).

Because the viscosity is temperature-dependent, it changes with temporally and spatially changing temperature distribution, described for two dimensions by (Turcotte and Schubert, 1982):

$$\rho C_p(\partial T/\partial t) = \nabla\cdot(\kappa\nabla T)+H, \quad (2\text{-}7)$$

where ρ is the density, C_p is the specific heat, T is the temperature, t is the time, κ is the thermal conductivity, and H is the heat production per unit volume. Furthermore, the changing temperature field controls buoyancy forces through the temperature dependence of density, expressed as (Huismans, 1999):

$$\rho(T) = (1 - \alpha_v T)\,\rho_0, \quad (2\text{-}8)$$

where ρ_0 is the density of the layer material at 0°C and α_v is the volumetric expansion coefficient.

Factors controlling mechanics of rifting and breakup

The most commonly cited factors controlling the mechanics of rifting and breakup include:

1) composition of individual layers of the lithospheric multilayer (e.g., Buck, 1991; Hopper and Buck, 1996; Brun, 1999; Huismans and Beaumont, 2008, 2011);
2) layer thickness in this multilayer (e.g., Burov and Diament, 1995);
3) thermal regime (e.g., Goetze and Evans, 1979; Kirby, 1983, 1985; Carter and Tsenn, 1987; Ranalli and Murphy, 1987; Ord and Hobbs, 1989; Davis and Kusznir, 2004);
4) effective elastic thickness of the lithosphere, representing the long-term integrated strength of the multilayer, the layers of which vary from coupled to decoupled (e.g., Watts, 2001; Burov and Watts, 2006; Huismans and Beaumont, 2008);
5) temporal and lateral changes in rheology of lithospheric layers (e.g., Ziegler et al., 1995; Geoffroy, 2005; Nemčok et al., 2005b);
6) strain rate (Allemand and Brun, 1991; Bassi et al., 2003; Ranalli, 1995; Nemčok et al., 2012a);
7) length of the extension time period (e.g., Huismans, 1999; Davis and Kusznir, 2004);
8) presence of hot spots (e.g., Brozena and White, 1990; Feighner and Richards, 1995; Feighner et al., 1995; Ribe et al., 1995; Ribe, 1996; Ito, 1996, 1997; Ribe and Delattre, 1998; Smallwood and White, 2002; Mittelstaedt et al., 2008);
9) occurrence of depleted versus enriched mantle (e.g., Pollack, 1986; Hirth and Kohlstedt, 1996, 2004; Asimow, Dixon, and Langmuir 2004; Lee et al., 2005);
10) presence or absence of magma (Ebinger and Casey, 2001; Davison, 2005; Geoffroy, 2005);
11) preexisting anisotropy in individual layers (Rosendahl, 1987; Molina, 1988; Versfelt and Rosendahl, 1989; Younes and McClay, 2002; Nemčok et al., 2005b); and
12) loading stress regime (e.g., Mandl, 1988; Huismans and Beaumont, 2005).

Because the role of factors 1–3, 6, and 12 will be discussed in Chapter 5, and factor 11 will be covered by Chapter 6, this chapter focuses on the remaining factors: 4, 5, and 7–10.

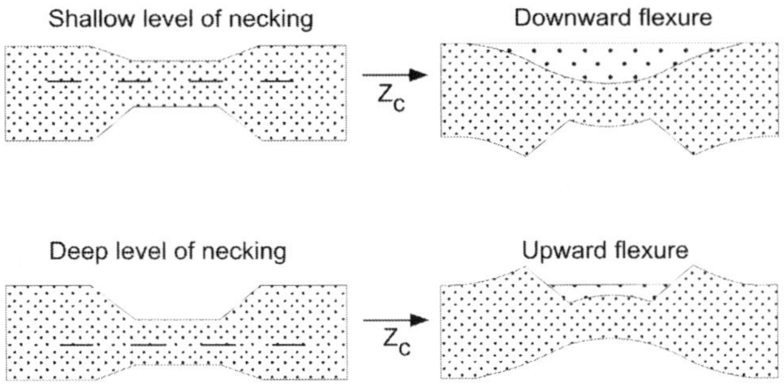

Figure 2.1. The expression of different necking depths on basin geometry (Kooi et al., 1992; Cloetingh et al., 1995). Z_c – level of necking, that is, the level of zero vertical movements in the absence of isostatic forces. A shallow level of necking creates a shallow basin and a high asthenospheric dome, which results in a downward state of necking. A deep level of necking creates a deep basin in combination with a shallow asthenospheric dome, resulting in an upward state of necking.

Effective elastic thickness

Lithospheric loads are compensated flexurally instead of in purely isostatic fashion, which can be understood with the help of effective elastic strength, EET (Kooi et al., 1992; Cloetingh et al., 1995). It reflects the integrated strength of the lithosphere that responds to long-term (> 10^5 yr) loading by flexure and controls the depth of necking in the rift process (Figure 2.1). EET has been defined as the combined effect of thicknesses of detached strong layers in the lithospheric multilayer (Burov and Diament, 1995):

$$\mathrm{EET} = \left(\Sigma^{n}_{i=1}\Delta h_i^{3}\right)^{1/3}, \quad (2\text{-}9)$$

where n is the number of layers with thicknesses h_1, h_2, and so on.

The depth of necking is defined as the depth without vertical movements that regional extension induces in the absence of isostatic restoring forces. It is correlated either with a depth level where the lithospheric multilayer achieves maximum strength (Braun and Beaumont, 1989; Kooi et al., 1992) or with a decoupling layer such as the brittle–ductile transition zone, where normal faults sole out (van der Beek et al., 1995).

The deep necking depth correlates to the strong, old, stable lithosphere described in Chapter 5. This scenario is dominated by stretching and is associated with:

1) the combination of the strongest crust with the strongest lithospheric mantle, most of the time without a ductile lower crust that would decouple the upper crust from the upper mantle;

2) a narrow asymmetric rift system with deep basins in syn-rift time;
3) the largest flexural stresses controlling pronounced flexural uplifts on one or both flanks; and
4) stretching-related normal faults dipping toward the axis of stretching.

The transition from regional crustal stretching to crustal necking in this scenario is typically characterized by an activity-timing shift toward the stretching axis. An example comes from the Moroccan passive margin (Le Roy and Piqué, 2001; Figures 2.2a–c). Figures 2.2a–c document how the syn-rift units active until Carnian were located in the easternmost onshore portions of the Tarfaya, Souss, and Essaouira Basins and the latter two underwent fluvial deposition. The syn-rift units active until Norian-Rhaetian can be found in the westernmost onshore portions of these three basins and the first and third underwent shallow marine deposition. The syn-rift units active until Rhaetian(?)-Hettangian are located in the offshore portions of the aforementioned basins and the former two underwent deeper marine deposition. Apart from the inboard progression of the controlling fault activity, the syn-rift fill recorded progressive deepening of the rift system. The inboard progression resembles the rolling hinge mechanism described in Chapter 1 (Figure 1.8).

The shallow depth of necking correlates to the mechanical stratigraphy of the weak, young, thickened lithosphere described in Chapter 5. This scenario is dominated by buoyancy forces and associated with:

1) strong upper crust and upper mantle layers decoupled along the lower crust;
2) a wide symmetric rift system with shallow basins in syn-rift time;
3) the smallest flexural stresses controlling most reduced flank uplift; and
4) buoyancy-related normal faults dipping away from the axis of stretching.

The regional crustal stretching and subsequent crustal necking in this scenario is characterized by the last activity-timing shift toward the stretching axis. An example comes from the Vøring passive margin (Geoffroy, 1994, 2005; Schlindwein and Jokat, 1999). Unlike the Moroccan example, which took about 32 Ma for crustal stretching and stretching–necking transition, the Vøring case took about 204 Ma, although characterized by discontinuous extension. The extension initiated during the late Permian in outboard locations and continued until the Paleocene–Eocene boundary in inboard locations (Ziegler, 1989; Brekke, 2000; Skogseid et al., 2000). The Paleocene was characterized by the emplacement of traps that buried the preexisting Late Cretaceous normal fault patterns (Geoffroy, 1994; Gernigon, 2002). The transition from crustal stretching to necking then took place rather quickly, during the Paleocene–Eocene transition, culminating with the breakup (Gernigon, 2002; Van Wijk and Cloetingh, 2002). While the Mesozoic stretching rate was as low as $7*10^{-16}s^{-1}$, taking place during 75 Ma (Gernigon, 2002), the Paleocene–Eocene stretching–necking transition was exceptionally fast (Hinz and Weber, 1976; Roberts et al., 1979).

Temporal and lateral changes in lithospheric layer rheologies

Temporal changes in the rheology of the lithospheric multilayer are affected by the rifting and subsequent cooling. They control significant changes in flexural behavior of the lithosphere under the rift system with time. For example, the post-rift cooling of the lithosphere controls the increase in its effective elastic thickness. This leads to the progressive widening of the subsiding sediment catchment region. This is indicated by an onlap of post-rift strata on the fault-constrained syn-rift strata and adjacent rift system flanks (Watts et al., 1982). The deformation of the multilayer changes the lateral and vertical distribution of crust and mantle materials and the temperature structure of the lithosphere, which both control lateral and vertical density variations. Lateral density variations then produce pressure differences between stretched and unstretched parts of the lithosphere (Le Pichon and Alvarez, 1984).

Even at a regional scale, the breaking up lithospheric multilayers can be characterized by significant lateral differences in their strength due to (Geoffroy, 2005):

1) small-scale lithospheric convection in the basal portion of the lower mantle;
2) magmatic reservoir distribution in the upper continental crust and distribution of mantle melting zones;
3) cratonic mantle lithosphere underplating; and
4) convection patterns in asthenosphere underneath lithosphere.

A small-scale convective instability in the lower mantle is shown by numerical models to potentially cause enhanced removal of the lower mantle lithosphere, which reduces the integrated lithospheric strength in each respective region, and to lead to the upwelling of hot asthenosphere, that is, causing further reduction of the integrated lithospheric strength (see Huismans and Beaumont, 2008). The lower mantle can become unstable upon the creation of significant geometrical and

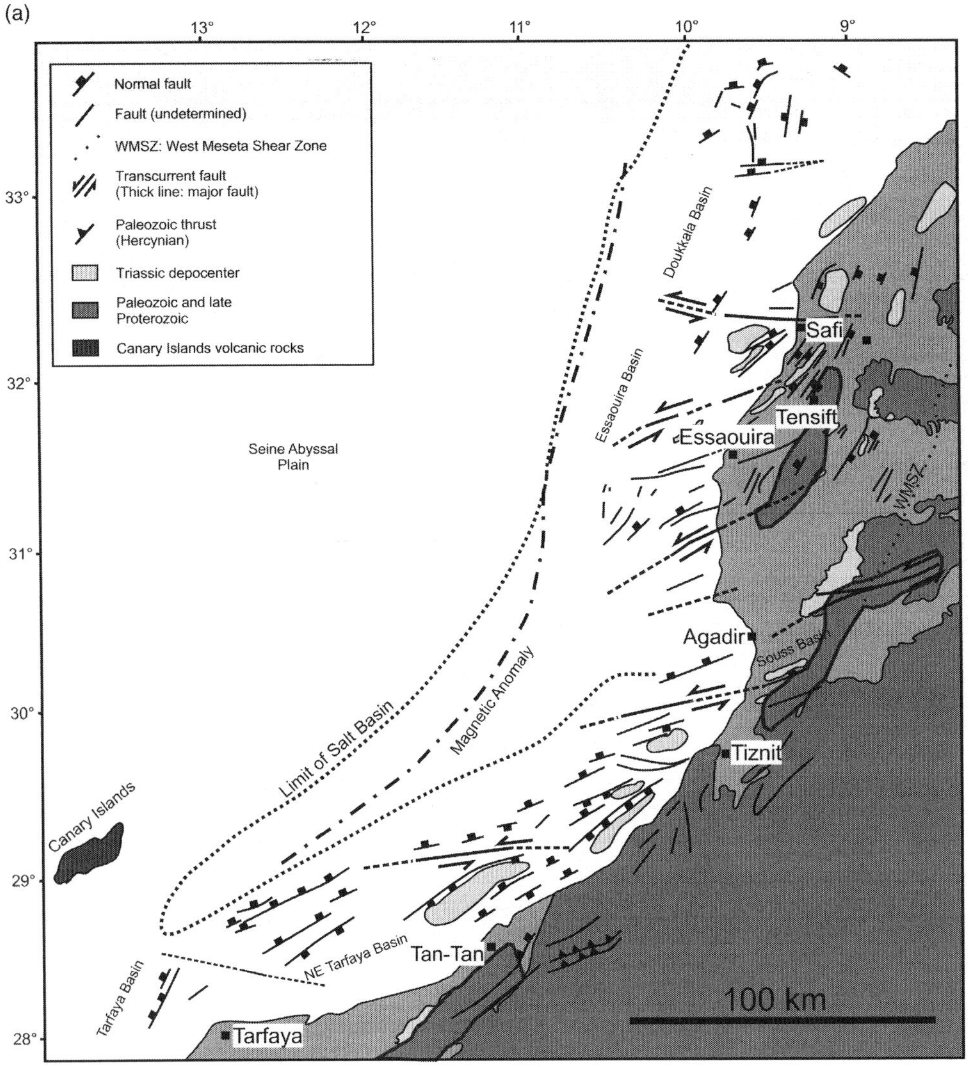

Figure 2.2. (a) Map of initial rifting locations in Morocco (dark gray) during Carnian (outlined as based on data in Le Roy and Piqué, 2001). The ancient High Atlas Massif supplied sediment for fluvial deposition in the Essaouira and Souss Basins. Tizi n'Test Fault in the Atlas underwent sinistral strike-slip reactivation. Zemmour thrusts in the Tarfaya Basin were reactivated as normal faults. (b) Map of main rifting locations in Morocco (dark gray) during Norian-Rhaetian (outlined as based on data in Le Roy and Piqué, 2001). The main rifting was coeval with CAMP basalt extrusion in the area. The Essaouira Basin area underwent silt and salt deposition in the shallow sea environments. The areas around the Souss Basin underwent salt deposition. The western Tarfaya area underwent silt and salt deposition in shallow sea environments. The eastern Tarfaya area remained unaffected by marine incursions. (c) Map of late rifting locations in Morocco (dark gray) during Rhaetian (?)–Hettangian (outlined as based on data in Le Roy and Piqué, 2001). The brittle extension in the Essaouira and Tarfaya Basin areas has ceased. However, both were undergoing significant subsidence. Their sedimentary section recorded Liassic marine transgression, recorded by carbonates unconformably lying over older continental and shallow-marine sequences.

thermal perturbations (Houseman et al., 1982; Fleitout and Yuen, 1984; Yuen and Fleitout, 1985). Examples are provided by a mantle delamination scenario taking place after significant lithosphere thickening (Bird, 1978a, 1978b; Houseman et al., 1982) and small-scale destabilization of the lithosphere–asthenosphere boundary (Buck, 1985; Yuen and Fleitout, 1985). An example of the geometrical perturbation causing convective instability comes from the places with strong thickness variations of lithosphere, resulting from preexisting rift events or an occurrence at craton boundaries (Buck, 1985; Mutter et al., 1988; Anderson et al., 1992; Anderson, 1994; King and Anderson, 1995, 1998; Keen and Boutilier, 1995; Huismans, 1999).

(b)

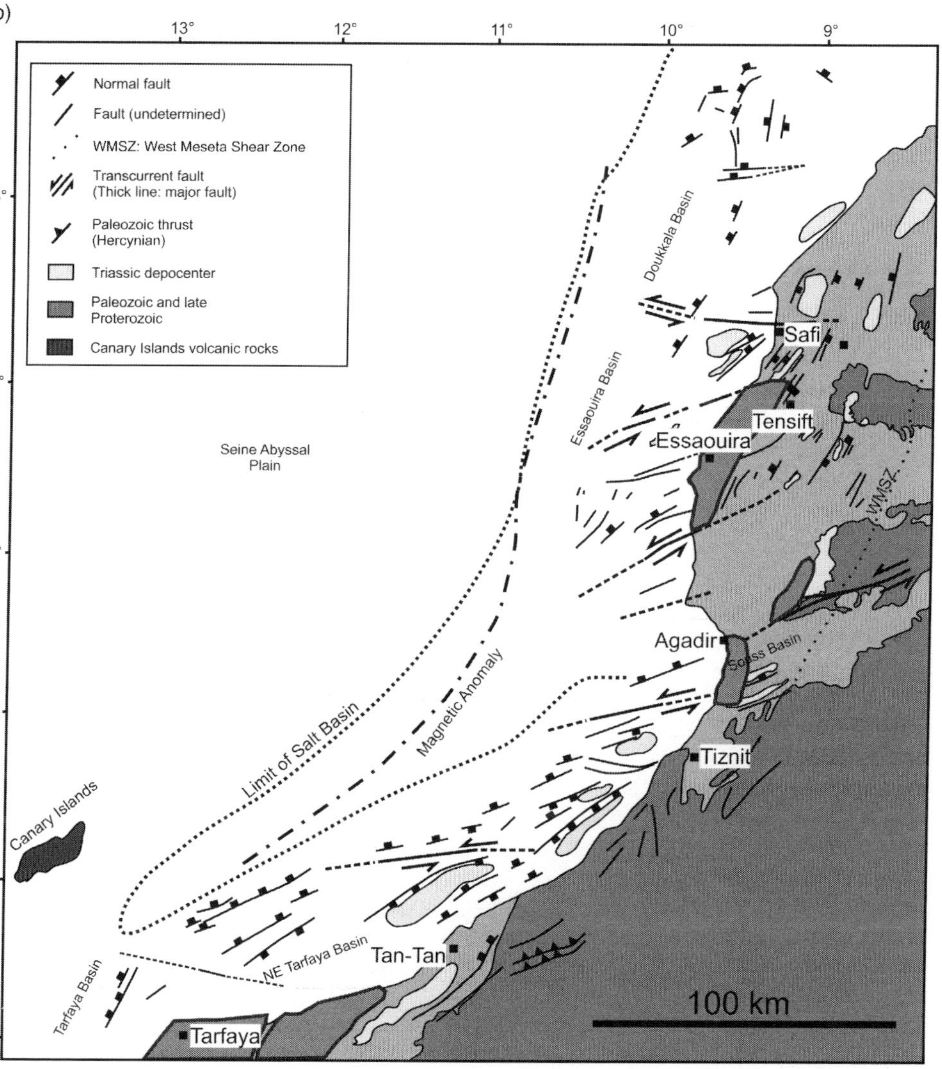

Figure 2.2. (*cont.*)

The characteristic time for mantle lithosphere upwelling can be estimated from Stoke's flow solution adapted for power law rheologies (Weinberg and Podladchikov, 1994, 1995):

$$V = (\Delta\rho g r^2)/3\mu_{eff}, \qquad (2\text{-}10)$$

where V is the velocity of an inviscid sphere rising through the infinite Newtonian fluid, $\Delta\rho$ is the density difference between the sphere and the Newtonian fluid, g is the acceleration of gravity, r is the sphere radius and μ_{eff} is the Newtonian fluid viscosity, and sphere and Newtonian fluids represent asthenosphere and mantle lithosphere materials, respectively.

The adaptation of lithospheric material to power law rheology is given by Weinberg and Podladchikov (1994), being somewhat similar to Equation (2-6). The substitution of this equation to Equation (2-10) yields (Weinberg and Podladchikov, 1994):

$$V = [3^{(n-1)/2}\cdot r\cdot A(\Delta\rho g r)^n]/[3\cdot 6^{n-1}\cdot \exp(Q/RT)], \qquad (2\text{-}11)$$

where velocity, V, is proportional to buoyancy force, $\Delta\rho g r$, and temperature, T. Finally, the characteristic time is then calculated as the time, t, during which the asthenospheric diapir undergoes rising one time its own radius (Huismans, 1999):

$$t = r/v_{dome} = [3\cdot 6^{n-1}\cdot \exp(Q/RT)]/\,[3^{(n-1)/2}\cdot A(\Delta\rho g r)^n]. \qquad (2\text{-}12)$$

The interpretation of whether upwelling takes place then comes from the comparison of this time with the

(c)

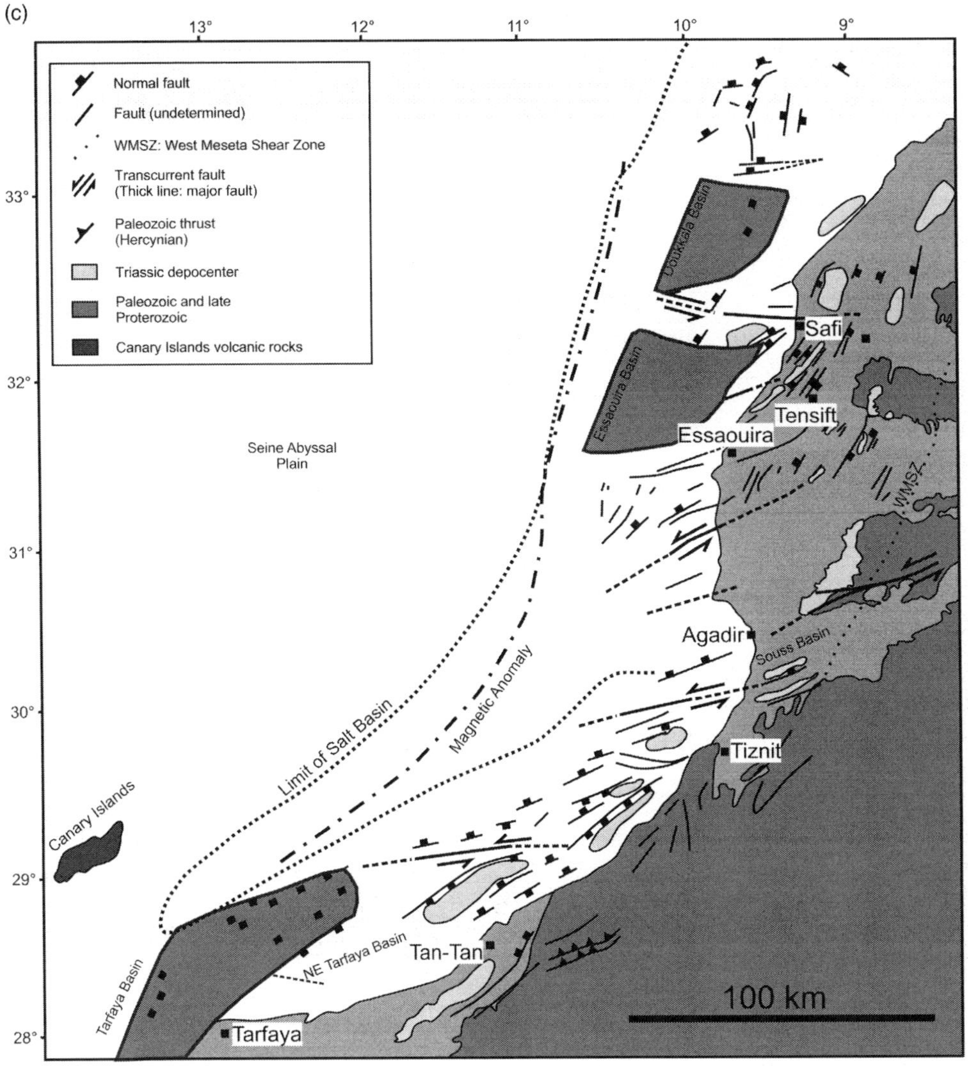

Figure 2.2 (*cont.*)

time needed for the post-rift thermal relaxation because the buoyancy force that drives the asthenospheric diapiric flow is a thermal feature.

Distribution of mantle-melting zones remains controversial in the literature (compare Anderson, 1994; Arndt and Christensen, 1992; Mahoney and Coffin, 1997; Saunders et al., 1997; Courtillot et al., 2003). However, trace elements of basalts in large igneous provinces together with seismic p wave modeling done for high-velocity zones indicate that adiabatically melting asthenosphere rises through the mantle lithosphere during rift–drift transition at future magma-rich margins (see Thompson and Morrison, 1988; Kerr, 1994; Saunders et al., 1997; Holbrook et al., 2001). In fact, the modeling shows that the inflow rate can exceed the half-rate of lithospheric stretching by ten times. The base of the lithosphere above the inflow can react by small-scale convection (see Fleitout et al., 1986; Griffiths and Campbell, 1991; King and Anderson, 1998; Geoffroy, 2001; Huismans et al., 2001; Morency et al., 2002; Huismans and Beaumont, 2008). Furthermore, geological and geophysical research in large igneous provinces and SDR wedges indicates that (Geoffroy, 2001, 2005; Figure 2.3):

1) they are both fed by a limited number of feeder dyke systems parallel to rift zones (Geoffroy, 1994; Callot, 2002; Callot et al., 2002);
2) their dykes are fed laterally from the upper crustal magma reservoirs, in which magma differentiates (Lister and Kerr, 1991; Callot et al., 2001;); and
3) flood basalts can extrude into distances of the first 100 km from their feeder systems (Self et al., 1997).

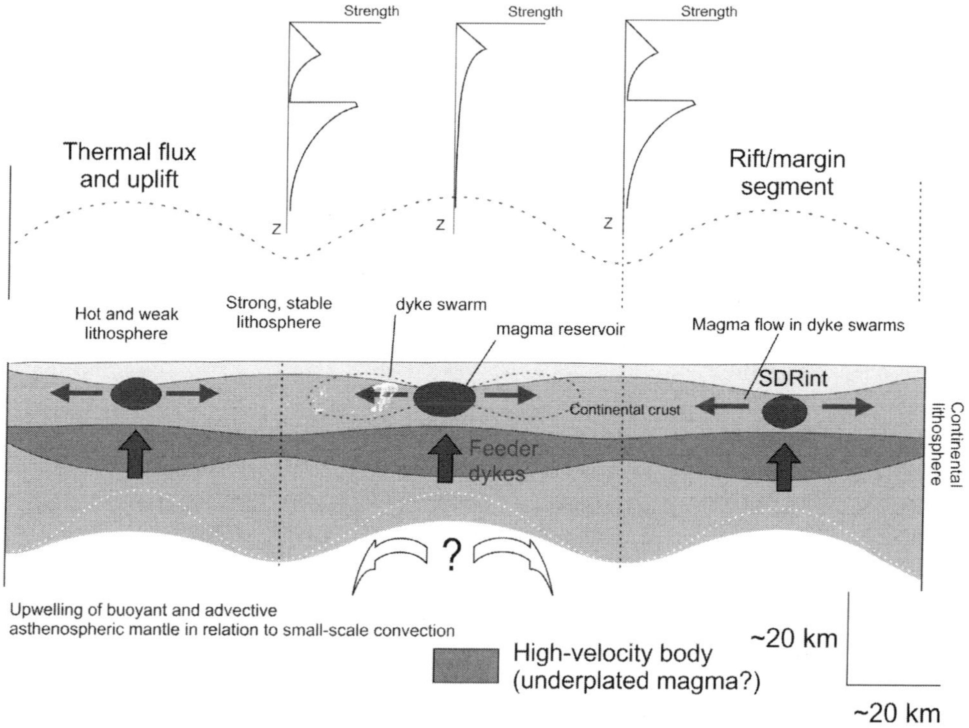

Figure 2.3. Strike-oriented schematic section through the magma-rich passive margin (Geoffroy, 2005).

The summarizing Figure 2.3 indicates that crustal magma reservoirs are situated above the mantle-melting zones corresponding to the tops of small-scale convection cells (Geoffroy, 2005). The undifferentiated mantle magma produced by small-scale convection cells in the mantle ascends into crustal reservoirs where it undergoes differentiation and laterally migrates into dyke systems of the transitional crust (Roberts et al., 1979; Callot, 2002). The example of the seaward-dipping reflector wedge penetrated by wells comes from the Vøring margin. Pb-isotope analyses made on DSDP Leg 81 samples show that lavas record crustal contamination (Morton and Taylor, 1987).

In the case of rifting taking place next to a craton, for example, Congo and Pilbara (Exmouth Plateau) cratons, the numerical models have shown that buoyant lower cratonic lithosphere preferentially flows into the subcrustal rift zones as the plates diverge from each other (Huismans and Beaumont, 2011). Its density and rheology are analogous to a partly depleted cratonic mantle, being about 15 kgm^{-3} less dense than the asthenosphere and three times more viscous (King, 2005; Sleep, 2005; Griffin et al., 2009).

Asthenospheric convection patterns have been shown in numerical simulations to progressively reduce the lithospheric mantle thickness or even eventually remove the mantle lithosphere completely, introducing significant lateral strength variations to the lithospheric multilayer (Huismans and Beaumont, 2008). Such convective lithospheric mantle removal introduces a misbalance of crustal and lithospheric mantle extensions. Furthermore, the lithospheric mantle thinning can cause decompression melting in upwelled asthenospheric material, introducing underplated melt. The asthenospheric upwelling also drives other secondary processes such as crustal/lithospheric magma underplating, eruption of continental flood basalts, and the development of large thermal gradients between extended and unextended regions (e.g., McKenzie and Bickle, 1988; Ruppel, 1995; Huismans et al., 2001), all contributing to lateral changes in lithospheric layer rheologies.

Asthenospheric convection tendency can be estimated by linear stability analysis (Turcotte and Schubert, 1982), yielding the Rayleigh number, Ra, indicating whether thermal and velocity perturbations in a viscous layer have a tendency to grow with time (Huismans, 1999):

$$Ra = (\rho_0 g \alpha_v \Delta T b^3)/\mu\kappa, \tag{2-13}$$

where ρ_0 is the layer density at 0°C, g is the acceleration of gravity, α_v is the coefficient of thermal expansion, ΔT is the vertical temperature variation in the layer, b is the layer thickness, μ is the viscosity, and κ is the thermal diffusivity.

The critical Rayleigh number for unstable convection is given by (Huismans, 1999):

$$Ra_{crit} = [\pi^2 + (2\pi b/\lambda)^2]^3/(2\pi b/\lambda)^2, \tag{2-14}$$

where $2\pi b/\lambda$ is the dimensionless wave number and λ is the wavelength of the perturbation. Values larger than Ra_{crit} indicate a tendency for convectional instability while the smaller ones indicate a stable system.

To conclude this subchapter on lateral changes in the lithospheric layer rheologies evolving with time, the numerical simulations indicate that lithospheric heterogeneities during rift–drift transition enhance a probability of asymmetric margin development (see Huismans and Beaumont, 2008).

Length of the extension period

The cumulative length of extension events eventually followed up by the continental breakup varies from margin to margin.

The rifting at the future magma-poor Goban Spur margin started during Permian-Triassic and continued during Jurassic, as documented by stratigraphic records of failed rifts (Cook, 1987), which indicate rift younging from the east to the west. The continental breakup, constrained by DSDP well data, took place during Aptian-Albian (Joppen and White, 1990). The youngest magnetic stripe anomaly of the oceanic crust has Albian-Santonian age (Scrutton, 1984). Although rather long, the extension did not develop a magma-rich margin at this location. One more example of the magma-poor margin, Galicia Bank, developed following the Berriasian–latest Aptian rifting (Boillot et al., 1989), having a rather short time span in comparison with that of the Goban Spur.

The cumulative length of rifting events at a magma-rich Vøring margin is long even in comparison with the Goban Spur case. As described earlier, the extension initiated here during the late Permian and ended by the Paleocene–Eocene breakup, comprising late Permian–Triassic, Late Jurassic–Early Cretaceous, and Late Cretaceous-Paleocene extensional events (Ziegler, 1989b; Brekke, 2000; Skogseid et al., 2000; Gernigon, 2002; Van Wijk and Cloetingh, 2002). On the contrary, the cumulative time interval for rifting in the magma-rich South China Sea includes Tertiary rifting of an unknown age interval (see Davis and Kusznir, 2004), which, based on ages of magnetic stripe anomalies, developed into seafloor spreading at about 24 Ma. The spreading was terminated at about 10 Ma, which was coeval with the emplacement of basaltic seamounts and other igneous bodies (Briais et al., 1993).

The comparison of short and long cumulative rifting activity periods in both magma-poor and magma-rich scenarios indicates that the cumulative length is just one of the factors controlling the breakup mechanism, which is combined with the other eleven factors listed earlier to control the breakup mechanism. Nevertheless, the role of several rifting events, one after another (see Table 9.1 in Chapter 9), in heat flow distribution prior to the rifting event that develops into the continental breakup is in their potential to cause an already perturbed thermal regime by previous rift events if there was no time long enough for the thermal equilibration. Associated thermal anomalies can be characterized as lithospheric-scale heat anomalies, having the order of magnitude of 100 km.

Presence or absence of hot spots

Although they do not provide exact analogs, studies of plume interactions with mid-oceanic ridges help to explain the plume role in continental breakup location and mechanism.

The analog and numeric models of ridge-centered plumes (Feighner and Richards, 1995; Feighner et al., 1995; Ribe et al., 1995; Ito et al., 1996) indicate that the plume–ridge interactions are independent of whether the plume buoyancy has thermal or chemical control. For the off-ridge plumes, the aligned patterns of constructional volcanism near the Galápagos and Reunion hot spots (Morgan, 1978) and systematic bathymetry and basalt geochemistry variations along ridge segments near hot spots (Talwani et al., 1971; Schilling, 1973, 1985) indicate that plumes can interact with a ridge even if they are as far as 1,400 km from each other (Ribe, 1996).

The following experiments (Kincaid et al., 1995a) and numerical modeling (Kincaid et al. (1995b) demonstrate that the plume material distribution underneath the ridge is controlled by:

1) gravitational spreading;
2) advection by corner flow;
3) upslope flow toward the ridge along the sloping lithospheric base;
4) incorporation into the lithosphere; and
5) plume-induced lithospheric thinning,

and the first two are dominant.

A migrating ridge is the most realistic assumption because the migration changes the distance over which the plume and ridge interact (Ito et al., 1997; Ribe and

Delattre, 1998). Given a uniformly thin lithosphere, the scaling law for the waist width, W, is written as (Ribe and Delattre, 1998):

$$W = W_0 A F_1(\Pi_b) F_3[\{(x'_p+B)/C\}/\{F_2(\Pi_b)\}], \quad (2\text{-}15)$$

where Π_b is the buoyancy number given by Ribe and colleagues (1995); A, B, and C are the scaling factors for the height, lateral shift, and width, which are controlled by the azimuth of the migration vector relative to the spreading direction; and the migration number $\Pi_b = U_m/U$, with U_m being the ridge migration rate. x'_p is given as x_p/W_0 and F_3 is defined as (Ribe, 1996):

$$F_3(s) = (1 - 1.25s^2)^{1/2}. \quad (2\text{-}16)$$

Given the migrating ridge scenario, a single value of the plume–ridge interaction distance, x_{max}, which is given for stationary ridge scenario in function (Ribe, 1996):

$$x_{max} = 0.89 W_0 F_2(\Pi_b)(1 - 0.34\,\Pi_b^{0.30}), \quad (2\text{-}17)$$

where F_2 is the function defined as (Ribe, 1996):

$$\log_{10} F_2 = 0.043p + 0.060p^2 - 0.0062p^2, \quad (2\text{-}18)$$

changes into smaller and larger values x- and x+, depending on the ridge advancing toward or retreating from the plume, which allows us to describe the ridge–plume interaction in three stages:

1) no interaction, W = 0;
2) start of the interaction at a plume-to-ridge distance of $|x_p| = x-$, nonzero W; and
3) end of the interaction at a plume-to-ridge distance of $|x_p| = x+$, W=0.

The interaction process includes a ridge jump (e.g., Mammerickx and Sandwell, 1986; Müller et al., 1993; Hardarson et al., 1997; Briais and Rabinowicz, 2002), controlled by processes such as:

1) lithospheric tension induced by a buoyant convecting asthenosphere (Mittelstaedt and Ito, 2005);
2) thermo-mechanical lithospheric thinning by the ascending plume (Jurine et al., 2005); and
3) lithosphere reheating due to magma introduction (Kendall et al., 2005).

Numeric modeling indicates that the rate of hot spot–related magmatic heating required for ridge jump initiation increases nonlinearly with an increasing spreading rate and age of the reheated lithosphere (Mittelstaedt et al., 2008). Observations from Iceland (Smallwood and White, 2002) and Ascension (Brozena and White, 1990) hot spots indicate that the ridge can undergo repeated jumps to stay approximately centered on the plume head. The jumps take several million years, during which the spreading takes place at both the old and new centers (Smallwood and White, 2002; Mittelstaedt et al., 2008). They result from a combination of (Brozena and White, 1990):

1) the disruption in the melt-focusing mechanism below the ridge;
2) the enhanced melt availability near the plume; and
3) the thinner and weaker lithosphere near the plume.

The plume presence near the potential breakup location of the continental lithosphere adds deviatoric stresses developed over upwelling asthenosphere convection systems to the deviatoric stresses developed in stress cycles by block/plate movements and modifies the frictional forces along boundaries of individual lithospheric layers between themselves and with rising asthenosphere. This stress field perturbation should affect a large area, if we look closely at the size of the regions, which are affected by transient thermal and dynamic uplifts and permanent isostatically compensated uplift (see Bertram and Milton, 1989; Nadin and Kusznir, 1995; Jones et al., 2002 for the Paleocene Iceland Plume and Sengör, 2001 for the Afar Plume).

There is a debate about the role of plume impingement on the continental breakup. For example, numerical simulations of Braun and Beaumont (1987) indicate that buoyancy forces resulting from domal uplift in response to plume impingement on the lithospheric base may not be sufficient to drive the whole lithospheric failure. This provides an interesting correlation with an interpretation that most rifts were initiated by regional extension (Sengör and Burke, 1978). This is true even for regions previously understood as classic examples of plume-driven breakup, such as the Central Atlantic and West India, which show indications of regional extension playing a significant role (McHone, 2000; Sheth, 2005; Nemčok et al., 2010). However, the buoyancy forces and thermal stresses in the Central Atlantic cannot be ruled out because the occurrence of surface compressional stresses at the sides of the assumed upwelled region was proven by contractional deformation observed in numerous areas (Withjack et al., 1998 and references therein).

This debate extends to the need of plume impingement for the development of magma-rich margins, in order to explain increased mantle temperatures (White and McKenzie, 1989; Skogseid et al., 1992; White, 1992). However, some magma-rich margins such as the northwestern Australian and eastern U.S. margins cannot be directly linked to mantle plumes (Hopper et al., 1992; Holbrook and Kelemen, 1993), and plume head dimensions are frequently too small to affect entire magmatic provinces (Griffiths and Campbell, 1991; Bijwaard and Spakman, 1999). The fact that increased mantle temperatures do not have to be associated with plume impingement (Anderson, 2000) has been indicated by numerical modeling (see van Wijk et al., 2001). It indicates that the melting potential is increased with small-scale mantle convection, where the late syn-rift melting can be controlled by lithospheric deformation evolution. For example, the increase in melt volume is proportional to the increase in extension rate. A similar tendency was concluded from a comparison of end-members in an extensional rate spectrum (Nemčok et al., 2012a). However, it is also the deformation history that has a controlling role, as concluded from a comparison of Møre and Vøring neighbor margins in Norway (Fernàndez et al., 2005). The differences of the magma-rich Vøring margin from the magma-poor Møre margin are:

1) the occurrence of the extra rifting event at the beginning of the rifting history;
2) two times thicker underplated body underneath the distal margin;
3) 30 km-thicker original Caledonide lithosphere;
4) a slightly smaller stretching factor;
5) larger thickness of adjacent oceanic crust; and
6) a 10 km-thinner lithosphere underneath the distal margin.

These differences were attributed to different rifting histories. This includes the enhanced heat transfer from the oceanic crust adjacent to the Møre margin to the continental crust of the Vøring margin. The transfer took place through the contact provided by transform. This includes the occurrence of the ridge jump responsible for the separation of the Jan Mayen micro-continent initially adjacent to the Møre margin (Fernàndez et al., 2005).

There seems to be a gradation from magma-rich to magma-poor margins, which can be implied from a comparison of passive margin segments (Figure 2.4) and strike-parallel differences along the southern Red Sea (Menzies et al., 2002). A complexity of magma-rich margin development controls is indicated by the variable relationships among (Table 2.1):

1) present-day thickness of large igneous province (LIP) products;
2) the length of the basalt volcanism activity period;
3) timing relationship of the basal volcanism to silicic volcanism;
4) timing relationship of LIP activity and rift stages;
5) scale of associated pre-magmatic uplift;
6) presence versus absence of seaward-dipping volcano-clastic wedges;
7) presence or absence of underplated bodies; and
8) presence or absence of magmatized crust at distal margin

for different margins.

Depleted versus enriched mantle

The density of the asthenosphere and mantle lithosphere is controlled by the thermal expansion of the rocks. The rifting brings the gravitational instability of both systems proportional to the importance of the thermal perturbation. Lateral density variations control buoyancy forces. Regardless of the extension-driving mechanism, the thinning of the lithosphere controls local gravity forces. These forces further enhance the lithospheric thinning by the active doming, creating the feedback mechanism progressively gaining importance with time during the rifting event.

The negative buoyancy of the lithospheric mantle studied at subduction zones is understood as one of the plate tectonic forces (e.g., Waschbusch and Beaumont, 1996; Beaumont et al., 1999; Ellis and Beaumont, 1999). The continental mantle is about 20 kgm^{-3} denser than the asthenosphere (Stacey, 1992). Although this average estimate is frequently accepted for Phanerozoic lithosphere (i.e., intermediately strong, relatively young, stable lithosphere described in Chapter 5), its application to magnesium-rich depleted cratonic lithospheres (i.e., strong, old, stable lithosphere discussed in Chapter 5) is a matter of ongoing debate.

Nevertheless, volumetric seismic velocities studied worldwide in the mantle, which are proportional to densities (e.g., Gardner et al., 1974; Christensen and Mooney, 1995; Brocher, 2005), are systematically higher than those of the asthenosphere (Burov and Watts, 2006). Furthermore, the average density of the continental lithosphere, thanks to the crust being 530 kgm^3 less dense than the mantle (values taken from Meyers et al., 1996), is also in the order of 150 kgm^3 less dense than that of the continental mantle forming the lithosphere of the proto-oceanic crust corridor adjacent to continental margins, which were developed by the crustal breakup followed by the mantle lithosphere breakup.

These density relationships for the mantle indicate that there will be more to the role of the mantle rheology in

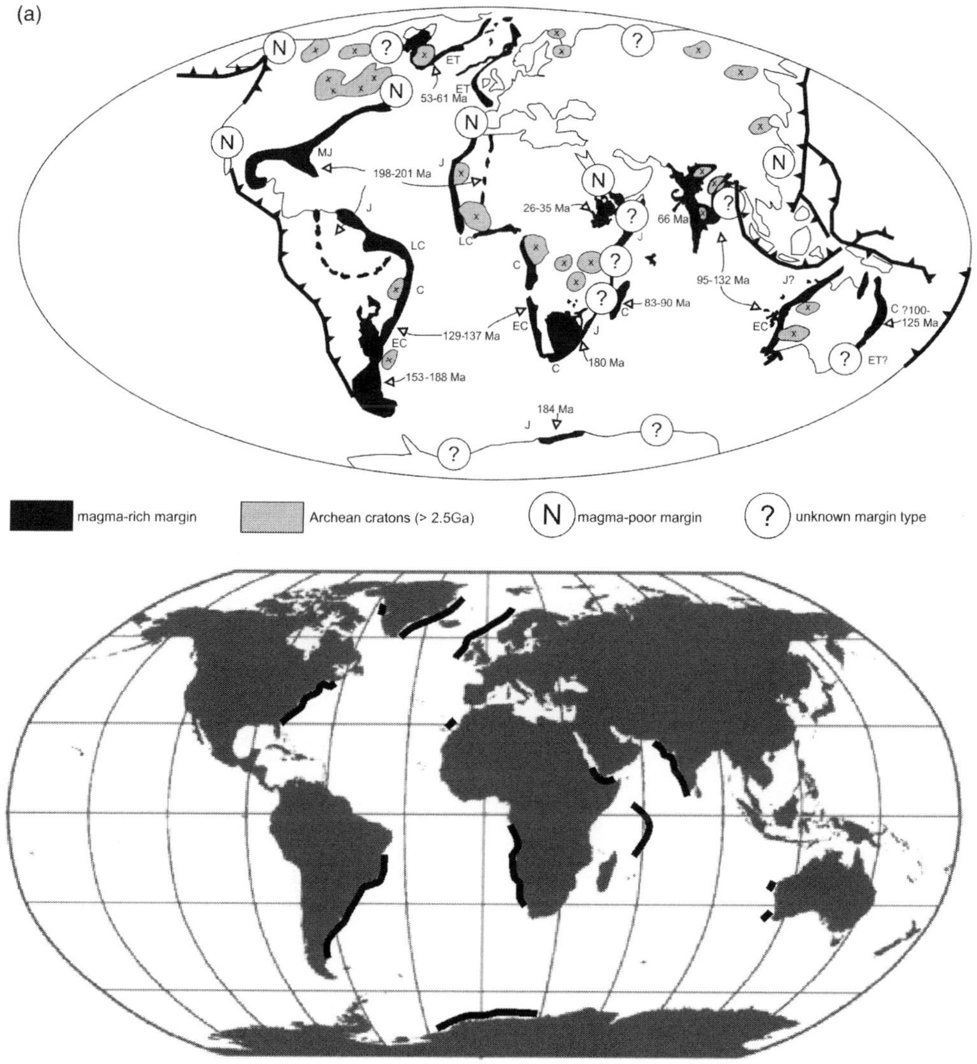

Figure 2.4. (a) Distribution of magma-rich margins younger than 200 Ma (Menzies et al., 2002). Approximate age of the oldest oceanic crust adjacent to the margin is given together with age of large igneous province (LIP) on the continental margin. Ages include J – Jurassic, C – Cretaceous, T – Tertiary, E – early, M – middle, L – late. Archaean cratons older than 2.5 Ga are shown to illustrate the existence versus lack of relationship between LIPs and cratonic edges. (b) Distribution of magma-rich margins (Geoffroy, 2005).

control of the continental breakup than variations in its original rheology and thickness, which vary with continental lithosphere thickness that ranges from more than 200 km in Archaean cratons, to 200±50 km in early Proterozoic lithospheres, and to about 140±40 km in the middle to late Proterozoic lithospheres (Artemieva and Mooney, 2001; Poudjom-Djomani et al., 2001). This will include the development history of the respective mantle, represented by mantle enrichment and depletion events, as they will modify the mantle density on top of the density controlled by thermal perturbation related to rifting.

The mantle enrichment event is, for example, represented by the subduction of the oceanic mantle together with the oceanic crust (e.g., Walter, 1999). The oceanic crust has been significantly enriched in large ion lithophile elements in relation to the mantle and has a lower melting point than the depleted mantle. Therefore, the subduction would re-fertilize the depleted mantle to some degree. The only depletion events in the subduction system are melt extraction events at island arc and that related to the emplacement of the continental andesite. Although the exact composition and amount of residue from these two partial melting events is not known in detail, it is reasonable to expect the residue to be more enriched than the original pre-subduction mantle.

Table 2.1. Characteristics of magma-rich margins (Menzies et al., 2002).

	LIP: Present-day thickness of subaerial volcanic rocks	LIP: Period of eruption of 70%–80% of the basaltic rocks	LIP: Age of silicic volcanic rocks: pre-basaltic, syn-basaltic, or post-basaltic eruptions	LIP and tectonics: pre-rift, syn-rift, or post rift?	LIP: Estimated scale of pre-magmatic uplift	Presence and/ or absence of a seaward-dipping reflector series	Presence or absence of high-velocity (~ 7,4 km/s) lower crust	HVLC: Presence of >10 km of new mafic igneous crust
1a Ethiopia	>2 km	Basalt-rhyolite (29–31 Ma] (base not dated)	Synchronous with basaltic eruptions 26–31 Ma	Pre-rift magmatism	Not known -buried by rift activity	Subaerial "inner" SDRS	Yes	Not known
1b Yemen	>2 km; originally ca. 4 km with ca. 2 km lost to erosion	Basaltic eruptions 29–31 Ma	Post-basaltic eruptions 26–29 Ma	Pre-rift magmatism	10–100 m (marine to continental transition in sediments)	denuded remnant of "inner" SDRS and buried "outer" SDRS	Yes	Not known
2a Greenland	5–7 km	53–56 Ma	Intrusions, no volcanics reported	Pre-rift, syn-rift, and post-rift magmatism	Hundreds of meters	SDRS – no sediments reported	Yes	Yes
2b United Kingdom Tertiary Volcanic Province	ca. 1 km: heavily denuded margin	58–61 Ma and 53–56 Ma	Syn-basaltic eruptions from 58–61 Ma and absent 53–56 Ma	Pre-rift magmatism then syn-rift and post-rift (i.e., SDRS)	Unknown but volcanics erupted onto subaerially weathered marine sediments	SDRS – mostly volcanic rocks	Yes – Rockall (5 km thick)	Yes – under continent-ocean transition
3a India	1 km, limited data; ca 1 km (SDRS)	Basalt/rhyolite ca. 95–118 Ma	Syn-basalts	Not known	Not known	SDRS-Sylhet province?	Not known	Not known

3b Australia	1–2 km: >1 km (SDRS) and 1 km (Wallaby Plateau)	Bunbury 123–132 Ma	Syn-basalts	Syn-rift and post-rift magmatism	Not known	SDRS- Wallaby Plateau	7.2–7.3 km/s Exmouth Plateau	Not known
4a Brazil Parana	1.8 km: originally ca. 4.8 km with 3 km lost to erosion	Basalt eruptions 129–133 Ma	Syn-basalts (and post-basalts)	? Pre-rift and syn-rift magmatism	? Possible pre-rift elevation ca. 500 m. (timing unclear)	Yes	Not known	Not known
4b Namibia Etendeka	0.9 km: originally ca. 3.9 km with 3 km lost to erosion	Basalt eruptions 131–133 Ma	Syn-basalts (and post-basalts)	? Pre-rift and syn-rift magmatism	? Possible pre-rift elevation ca. 500 m. (timing unclear)	SDRS mixture of volcanics and sedimentary rocks	Yes	Yes
5 Central Atlantic Magmatic Province	1 km	198–201 Ma	No silicic volcanic rocks	Post-rift magmatism (southeastern United States) and syn-rift magmatism (northeastern United States and Africa)	? 0.9–2.6 km syn-rifting or pre-volcanic	SDRS volcanics with some inter-bedded sediments	Yes, significant igneous intrusions	No, normal oceanic crust (7–8 km)

Abbreviations: LIP – large igneous province; SDRS – seaward-dipping reflector series, HVLC – high-velocity lower crust.

References: Ethiopia and Yemen: Berckhemer et al. (1975); Davison et al. (1994); Baker et al. (1996a, 1996b); Hoffman et al. (1997); Menzies et al. (1997a, 1997b); Al'Subbary et al. (1998); George et al. (1998); Baker et al. (2000); Ebinger and Casey (2001); Ukstins et al. (2002); Baker et al. (2002). Greenland and United Kingdom: Roberts et al. (1979); Mutter et al. (1982); White et al. (1987); White and MacKenzie (1989); Brodie and White (1994); Jolley (1997); Saunders et al. (1997); Larsen and Saunders (1998); Korenaga et al. (2000): Planke et al. (2000); Klausen and Larsen (2002). India and Australia: Storey et al. (1992); Von Rad and Thurow (1992); Colwell et al. (1994); Exon and Colwell (1994); Milner et al. (1995); Frey et al. (1996); Kent et al. (1997). Parana and Etendeka: Hawkesworth et al. (1992); Renne et al. (1992, 1996a, 1996b); Gallagher et al. (1994); Turner et al. (1994); Stewart et al. (1996): Gladczenko et al. (1997); Peate (1997); Clemson et al. (1999); Davison (1999); Jerram et al. (1999); Hinz et al. (1999) and references therein; Bauer et al. (2000); Corner et al. (2002); Mohriak et al. (2002); Trumbull et al. (2002); Watkeys et al. (2002). Central Atlantic Magmatic Province: McBride (1991); Holbrook and Kelemen (1993); McHone (1996); Lizarralde and Holbrook (1997); Withjack et al. (1998); Hames et al. (2000); Benson (2002); McHone and Puffer (2003); Schlische (2003).

Another example of the enrichment event is the subduction evolving to collision, depending on the extent of the overriding, controlled by the character of the subduction zone varying from the retreating to the neutral and the advancing types (see Doglioni, 1993; Waschbusch and Beaumont, 1996; Beaumont et al., 1999; Ellis and Beaumont, 1999). The advancing type controls the largest amount of overriding.

Depending on the scenario, the enrichment can involve (Workman and Hart, 2005):

1) depleted mantle plus continental crust;
2) depleted mantle plus continental crust and recycled oceanic crust; and
3) depleted mantle plus continental crust, recycled oceanic crust, and oceanic island source.

The enrichment is related to the water content in the mantle, which represents a factor controlling mantle melting and its viscosity (Hirth and Kohlstedt, 1996, 2004; Grove and Parman, 2004). As opposed to the dry mantle, the wet one is less buoyant and less viscous (Pollack, 1986; Hirth and Kohlstedt, 1996; Lee et al., 2005). It is the depleted end-member, in which dehydrated and melt-depleted layers are strong in a viscous sense and chemically buoyant to suppress the development of convective instabilities in the mantle during rifting. As a consequence, the heat transfer is assumed to function dominantly by conduction (Lee et al., 2005).

When we compare subduction and collision cases of enrichment with the types of lithosphere described in Chapter 5, it is the young, post-orogenic lithosphere that is supposed to have the most enriched mantle. On the contrary, it is the old, cratonic lithosphere that is typically underlain by thick mantle keels. They are significantly depleted in meltable components as documented by element composition studies of mantle xenoliths (Jordan, 1978; Boyd, 1989; Griffin et al., 1999a, 1999b). The cratonic mantle keels are significantly depleted of meltable content and have a larger molar ratio Mg/(Mg+Fe). The thermobarometric data from the continental mantle xenoliths indicate that a thick (up to 175 km underneath Siberia and South Africa; Lee et al., 2005) layer of chemically depleted mantle is underlain by mantle material with increased fertility (Rudnick et al., 1998; Griffin et al., 1999a, 1999b; Kopylova et al., 1999; Lee and Rudnick, 1999), potentially capable of small-scale convection. The convecting mantle is more enriched in terms of meltable content and has a smaller molar ratio Mg/(Mg+Fe).

Another case of mantle enrichment is the impingement of plume, which supplies fluids and heat controlling the enrichment in incompatible elements (e.g., Aulbach et al., 2008). The case example comes from the Tanzanian segment of the East African Rift system. Here the magma intrusion and heating events related to rifting are indicated by (Chesley et al., 1999; Lee and Rudnick, 1999; Rudnick et al., 1999; Burton et al., 2000; Watson et al., 2006):

1) metasomatic veins filled with iron-rich olivine, orthopyroxene, chromite, zircon, and rutile;
2) rutile with zirconium concentrations at its surface;
3) garnet breakdown into coronas; and
4) recent rhenium addition.

The occurrence of the plume is indicated by (Chesley et al., 1999; Vauchez et al., 2005; Becker et al., 2006; Aulbach et al., 2008):

1) seismic velocity and anisotropy data;
2) iron, lithium enrichment of olivines and their high δ^7Li character in deep-seated and more fertile iron-rich peridotites;
3) iron, lithium enrichment of olivines and their high δ^7Li character in fertile garnet lherzolites from the basal portion of the lithosphere;
4) radiogenic ^{187}Os/^{188}Os ratio in garnet lherzolites that is similar to that of volcanic rocks derived from the plume-like asthenospheric mantle;
5) non-mantle-like concentrations of platinum group elements; and
6) multiple osmium isotope components.

The example of a depletion event is a magmatic event. In the case of oceanic slabs, the most compelling evidence for the upper mantle depletion comes from systematic studies of heavy-element isotopic compositions (e.g., Workman and Hart, 2005 and references therein). The observed depletion in abyssal peridotites is related to melt extraction due to the latest spreading events. Most of the depletion is understood as a consequence of the extraction of the enriched continental crust. This process is responsible for the removal of up to 90 percent of the most incompatible elements and 80–85 percent of the elements undergoing radioactive decay.

A nice documentation of mantle depletion due to a succession of rift-related magmatic events comes from the Moroccan segment of the Central Atlantic margin. A complete record of the geochemical evolution of Moroccan basalts in time is preserved thanks to the four basalt units of the Central High Atlas, containing lower, intermediate, upper, and recurrent units (Marzoli et al., 2004).

We can observe a progressive decrease of an eruption rate from the lower unit to the recurrent unit. This decrease correlates with:

1) a change from intersertal to porphyric petrographic texture;
2) progressive decrease in the light-to-heavy rare earth element ratio (LREE/HREE); and
3) progressive decrease in incompatible element contents in the basalts;

which are all possibly linked to a progressive depletion of their mantle source (Marzoli et al., 2004; Deenen et al., 2010).

An important consequence of significant melt extraction during one or multiple rift events is a mantle density reduction (Lee et al., 2005).

It has to be said that chemical effects interact with thermal effects during rifting and early drifting. This is better illustrated on the oceanic mantle, which has no chemical discontinuity between the depleted and enriched layers as is the case in the continental mantle. After 20 Ma of spreading, the thermal effects exceed chemical effects and the entire oceanic mantle becomes negatively buoyant. Potential scenarios with different types of continental mantle can be described only qualitatively. We can expect an enriched mantle to be prone to small-scale convection and significant melt extraction, which will have a tendency to be a system prone to magma-rich margin development. Furthermore, the more advanced rifting stages and breakup stage could be potentially characterized as a buoyancy-dominated system. On the contrary, a depleted mantle would have a tendency to suppress small-scale convection and melt extraction. Furthermore, the more advanced rifting stages and the breakup stage could be potentially characterized as a stretching-dominated system.

Presence or absence of magma

Related to the mantle fertility described earlier comes the role of magma. The absence can be correlated to stretching-dominated breakup mechanisms, described in the following text on magma-poor margin development. The presence of magma plays an important role in the buoyancy-dominated breakup mechanism, especially during the necking stage. Its role is described in the following text on magma-rich margin development.

Crustal breakup followed by mantle lithosphere breakup

This breakup scenario (Figure 2.5) can be characterized by:

1) basin-controlling faults penetrating deep into the crust;
2) narrow regions affected by extension (less than 100 km wide), that is, abrupt crustal thinning;

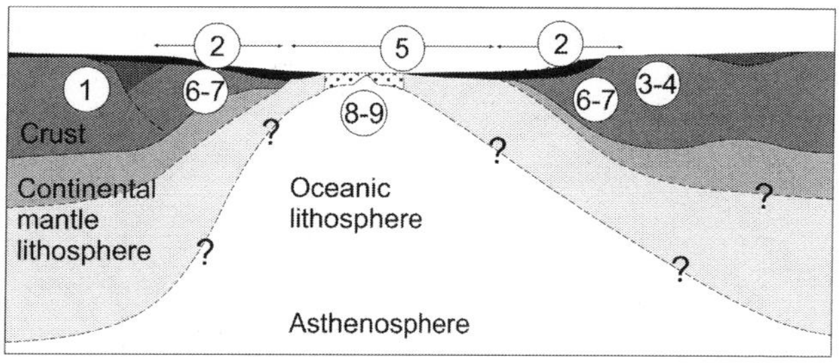

Figure 2.5. Characteristic properties of the margin originated by the crust-first–mantle-second type of breakup (Huismans and Beaumont, 2011). The example is based on the Iberia–Newfoundland conjugate margins. This break-up scenario can be characterized by: 1 – basin-controlling faults penetrating deep into the crust, 2 – narrow regions affected by extension (less than 100 km wide), that is, abrupt crustal thinning, 3 – most commonly asymmetric cross-sectional geometry of the rift system, 4 – rift flank uplift, 5 – exhumation of the continental mantle in the so-called proto-oceanic corridor, 6 – limited magmatism during rifting, 7 – magma-poor margin development, 8 – delayed seafloor spreading center initiation, and 9 – delayed start of organized seafloor spreading. See text for further explanation.

3) the most commonly asymmetric cross-sectional geometry of the rift system;
4) rift flank uplift;
5) exhumation of the continental mantle in the so-called proto-oceanic corridor;
6) limited magmatism during rifting;
7) magma-poor margin development;
8) delayed seafloor spreading center initiation; and
9) delayed start of organized seafloor spreading.

(1) Graben-bounding faults of the Newfoundland and Iberia margins frequently sole out at middle to lower crustal depths, while the base of the crust remains unfaulted, sometimes undergoing upwarping (e.g., Keen et al., 1987). Similar detachment depths and upwarped Moho structures can be seen in reflection seismic images in the proximal margin of northern offshore Gabon and eastern offshore India (own observations).

(2) As Alpine data (Manatschal, 2004) documented, the deformation of the future distal margin is controlled by the necking of the lithospheric mantle. The necking is associated with major detachment faults, which dip toward the necking point. This leads to the localized thinning of the crust. Data from the Iberia-Newfoundland conjugate margins document that they started to develop with distributed extension, which became subsequently narrowly focused (Huismans and Beaumont, 2007). Extreme crustal

thinning was achieved by the superimposed conjugate and detachment shearing (Manatschal, 2004 and references therein).

(3) The asymmetric cross-sectional geometry of the rift system, having lithospheric mantle necking located away from the extension maximum in the upper crust, has been indicated by Alpine data (Manatschal, 2004), analog material models (Brun and Beslier, 1996; Michon and Merle, 2003), and numerical models (Harry and Sawyer, 1992). In fact, this shift is typical for the lithospheric scenarios characterized by stronger mantle lithospheres and lower crusts (see Huismans and Beaumont, 2005). Further numerical modeling focused on this problem reveals that the rifting initiates with a subsidence of crustal segment controlled by frictional-plastic faulting underlain by the ductile necking of the lower lithosphere (Huismans and Beaumont, 2011). This phase is followed by asymmetric extension resembling a simple shear of Wernicke (1985), which finishes with crustal rupture. Then follow the continued extension, necking, exhumation, and exposure of the lower lithosphere.

(4) Uplifted flanks are associated with numerous passive margin segments, including:

a) South African Atlantic (Brown et al., 1990, 2000; Rust and Summerfeld, 1990; Ten Brink and Stern, 1992; Gallagher and Brown, 1999; Raab et al., 2001) and its conjugate South American Atlantic (Gallagher et al., 1994; Mohriak et al., 2008);
b) North American Atlantic (Royden and Keen, 1980; Watts and Thorne, 1984; Miller and Duddy, 1989; Steckler et al., 1994; Nemčok et al., 2005b);
c) Galicia Bank, North Atlantic (Davis and Kusznir, 2004 and references therein);
d) Red Sea (Morgan et al., 1985);
e) Southeastern Australian (Moore et al., 1986; Dumitru et al., 1991);
f) East Indian (Nemčok et al., 2007, 2008); and
g) West Indian (Gunnell and Radhakrishna, 2001; Gunnell et al., 2003) passive margin segments.

A comparison of these segments with the worldwide distribution of magma-rich margins (Figure 2.6) yields an interesting result. Out of seven listed margin segment groups, four are magma-rich margin-dominated groups and three are magma-poor margin-dominated groups, indicating that while this breakup scenario is characterized by the rift flank uplift, the rift flank uplift can occur in other breakup scenarios. This suggests that the rift flank uplift must have several different drivers.

Indeed, the development of rift shoulders has been explained by numerous models, which include transient and permanent uplift mechanisms. Transient uplift models are associated with the thermal effects of rifting (Eastern Canadian margins – Royden and Keen, 1980; topical study – Hellinger and Sclater, 1983; U.S. Atlantic margin – Watts and Thorne, 1984; Red

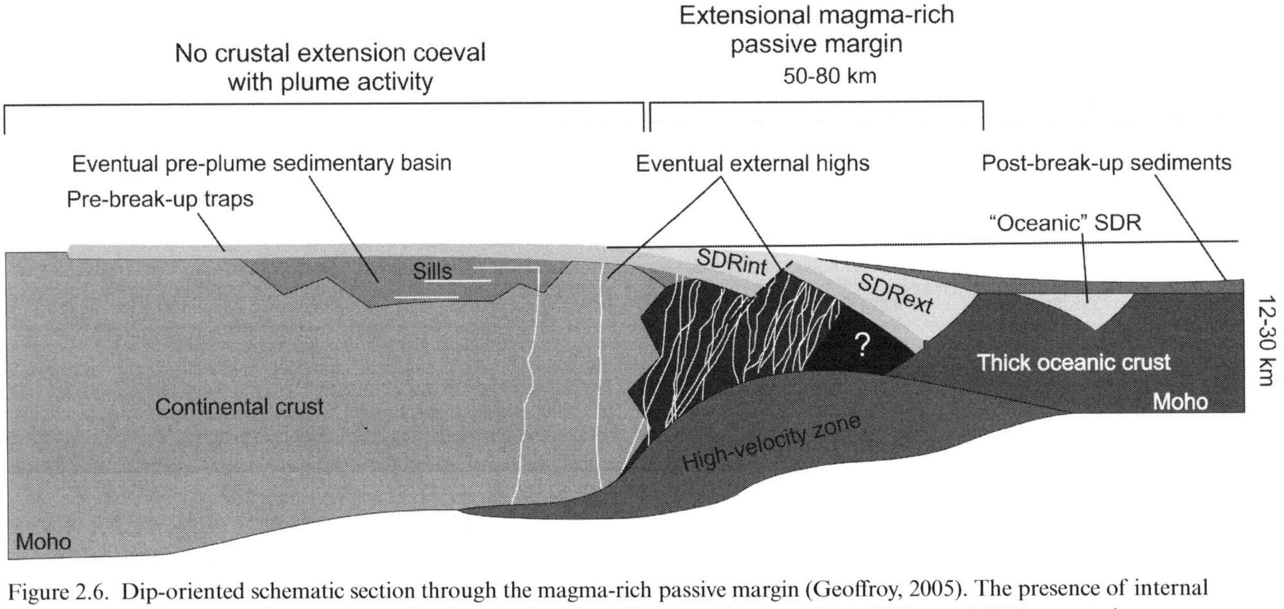

Figure 2.6. Dip-oriented schematic section through the magma-rich passive margin (Geoffroy, 2005). The presence of internal sedimentary basins is not the rule. Internal and external seaward-dipping reflector wedges, SDR_{int} and SDR_{ext}, contain seaward-dipping lava flow and other volcanic products and some sedimentary horizons.

Sea margins – Morgan et al., 1985), lateral heat flow (Atlantic margins – Steckler, 1981a; topical studies – Cochran, 1983; Alvarez et al., 1984; Buck et al., 1988) and secondary convection under rift shoulders (topical study – Keen, 1985; Gulf of Suez flanks – Steckler, 1985; topical study – Buck, 1986). These mechanisms, which can create up to 1,500 m-high shoulder uplift, operate only during the time period of elevated thermal regime. Mentioned positive topography subsequently decays over the time period equivalent to the thermal time constant of the lithosphere, which is roughly 60 Ma. Permanent uplift models are related to the magmatic underplating (topical studies – Cox, 1980; McKenzie, 1984; White and McKenzie, 1988; south and central Queensland, Australia – Ewart et al., 1980) and lithospheric unloading and/or plastic necking (topical studies – Zuber and Parmentier, 1986; Parmentier, 1987; Braun and Beaumont, 1989; Issler et al., 1989; Weissel and Karner, 1989; Chery et al., 1992). Subsequently, the processes of erosion, deposition, and gravity gliding started to be incorporated into existing models of the rift shoulder uplift (e.g., Van Balen et al., 1995; Van der Beek et al., 1995; Burov and Cloetingh, 1997; Mohriak et al., 2008; Figure 2.7).

(5) Supporting evidences for lithospheric mantle exhumation come from reflection seismic sections from the Iberian margin (Reston et al., 1995, 1996; Manatschal et al., 2001; Wilson et al., 2001), offshore Gabon (Rosendahl et al., 2005; own observations), offshore East India (own observations), offshore Brazil (Zalán et al., 2011), wells (Boillot et al., 1980, 1987; Beslier et al., 1996; Whitmarsh et al., 2001), geochemical studies (Abe, 2001; Hébert et al., 2001), field analogs (Manatschal and Nievergelt, 1997; Manatschal and Bernoulli, 1999; Manatschal, 2004 and references therein), analog material modeling (Brun and Beslier, 1996; Brun, 1999), and numeric modeling (Huismans and Beaumont, 2005).

Proto-oceanic crust made of exhumed continental lithospheric mantle in regions adjacent to continents has frequently poorly imaged base in refraction and reflection seismic data sets, indicating mantle exhumation and subsequent serpentinization (see Pickup et al., 1996). As a result, seismic velocities and densities of the proto-oceanic crust are not characteristic of oceanic crust (see, e.g., Meyers et al., 1996; Wilson et al., 2003; Rosendahl et al., 2005). For example, the serpentinization progressively decreasing downward has been interpreted from velocities increasing from about 7.2 to about 7.9 kms^{-1} in the Iberian Abyssal Plain and from additional data provided by along-strike correlation to the ODP drill holes (Whitmarsh et al., 1990; Discovery 215 Working Group, 1998; Chian et al., 1999).

Alpine transects also offer a chance to study proto-oceanic crustal assemblages adjacent to the continental margins. Neither Grisons ophiolites nor seafloor sequences drilled by an ODP well in offshore Iberia indicate normal oceanic crust (see Manatschal, 2004) and have to be defined as a proto-oceanic crust (see Rosendahl et al., 2005 and references therein) or as the zone of exhumed continental mantle (Whitmarsh et al., 2001) (Figure 2.8; Manatschal, 2004). Alpine ophiolites lack sheeted-dyke complexes, have relatively small volumes of gabbros and basalts, and contain predominately serpentinites. They are overlain by extensional allochthons and tectono-sedimentary breccias with continent-derived material (e.g., Manatschal and Niegervelt, 1997; Desmurs et al., 2001). The similarity of Alpine ophiolites to drilled mantle rock in front of the Iberian margin is supported by geochemical data (e.g., Rampone and Piccardo, 2000; Muentener et al., 2002). Grisons outcrops contain serpentinized peridotites derived from spinel lherzolites and harzburgites, intruded by gabbros and basalt dykes. Platta nappe outcrops document that gabbro bodies represent less than 5 percent of the mantle rock volume (Manatschal, 2004). U-Pb zircon dating indicates their emplacement in a relatively short time interval (Schaltegger et al., 2002).

The amount of massive basalts, pillow lavas, pillow breccias, and hyaloclastics increases in Platta nappe outcrops toward the ocean. Away from the edge of the continental crust, pillow lavas form isolated bodies less than 100 m in diameter and a few tens of meters thick. Further oceanward, these bodies become aligned and appear to be controlled by late, syn-magmatic, high-angle faults. Similar faults can be seen in ultra-deep reflection seismic images in offshore Gabon (own observations). The basalts stratigraphically overlie serpentinites, ophicalcites, gabbros, and associated breccias, indicating their emplacement following the mantle unroofing. Similar stratigraphic relationships can be seen in reflection seismic images in offshore Gabon and East India (own observations). The late, syn-magmatic, high-angle faults, formed by a serpentine gouge, truncate basaltic dykes documenting their age as younger than the age of basalt emplacement.

The proto-oceanic corridor adjacent to the Iberian margin is 40–130 km wide. Its seismic velocity structure is characterized by:

1) 2–4 km-thick upper layer with v_p velocities of 4.5–7.0 kms^{-1} with a high-velocity gradient of ca 1 s^{-1}; and
2) less than 4 km-thick lower layer with v_p velocity of 7.6 kms^{-1} with a low-velocity gradient of less than 0.2 s^{-1}.

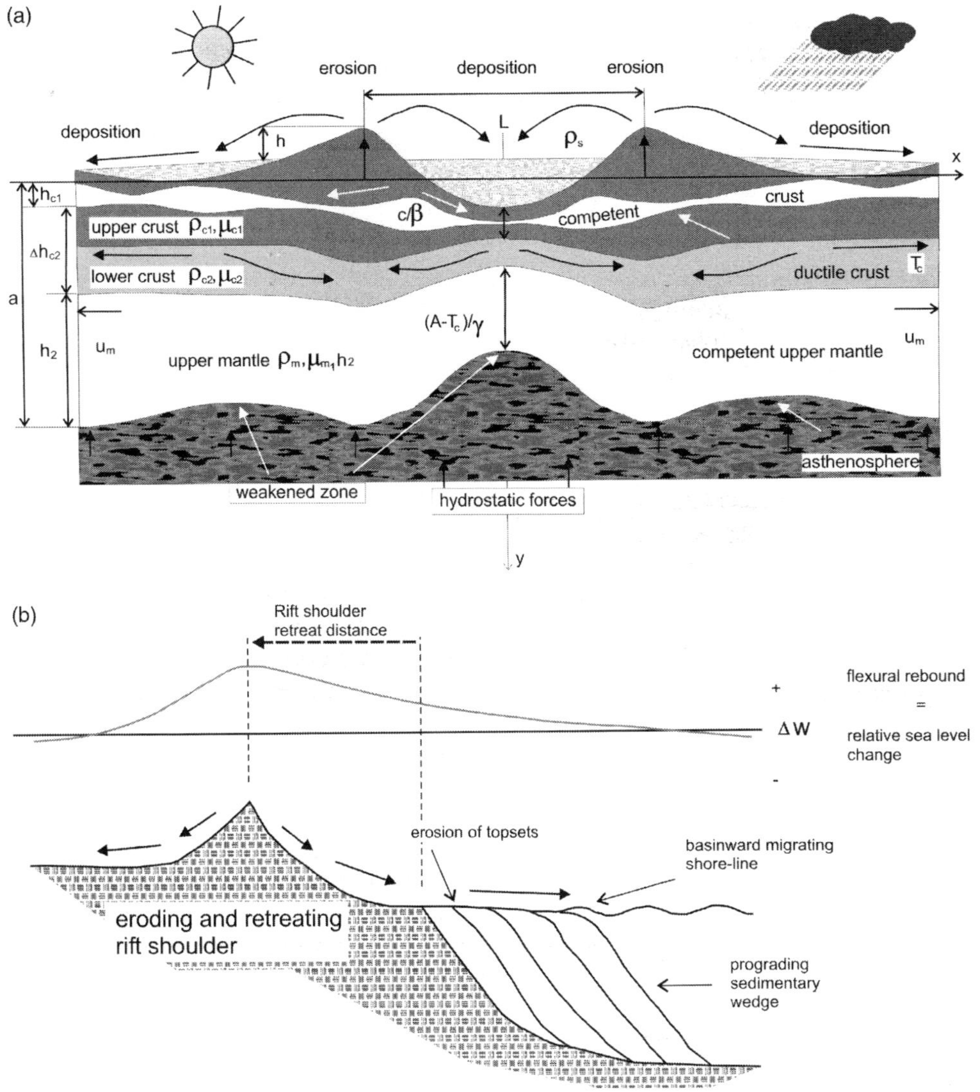

Figure 2.7. Rifting-related basin subsidence and rift shoulder uplift (after Van Balen et al., 1995; Burov and Cloetingh, 1997). (a) Figure schematically shows that initial lithospheric stretching brings hot mantle material closer to the surface, resulting in the increase of the thermal regime in the crust beneath the basin. Black arrows and ρ_s show sediments eroded from the rift shoulders, which exert sedimentary load on the basin. White arrows indicate strong parts of the crust and mantle lithosphere, which flex and weaken. This results in the decrease of the effective elastic thickness values beneath both basin and the rift shoulders further from initial extensional value. In response, ductile material of the lower crust flows from the basin center toward the rift shoulders, facilitating their uplift. Erosion of the rift shoulders enhances their uplift. (b) Figure shows the initial post-rifting coastal offlap and the rift shoulder retreat at the passive margin. The flexural rebound related to erosion of the rift shoulder causes an uplift, which affects the adjacent portion of the basin. The adjacent portion of the basin records this uplift by erosional truncation of topsets of the sedimentary wedge. Coastal onlap can occur when the uplift rate is smaller than basin subsidence, which is controlled by thermal contraction and the flexural response to the increase of the lithospheric rigidity controlled by cooling and sedimentary loading. $+\Delta w$ indicates the instantaneous flexural uplift and $-\Delta w$ the subsidence.

The velocity of the upper layer is less than the velocity of the continental crust, while the velocity of the lower layer has a velocity faster than the continental crust, heavily intruded continental crust, underplated continental crust, and normal oceanic crust. The best explanation of the described velocity structure is mantle rocks affected by serpentinization, which decreases with depth. The image of the proto-oceanic crust also either lacks a Moho reflector or contains it in a weak form.

The magnetic character of the Iberian proto-oceanic crust involves low-amplitude magnetic anomalies, which are considerably lower than the normal oceanic crust further oceanward. This is caused by a relatively small amount of magmatic material in the upper seismic layer, which progressively increases oceanward.

ODP cores of the Iberian proto-oceanic crust proved the upper continental mantle, represented by serpentinized peridotite. Primary-phase chemistry and clinopyroxene trace-element compositions indicate a heterogeneous mantle depleted by less than 10 percent partial melting and percolated by mafic melts (Abe, 2001; Hébert et al., 2001). Trace compositions are compatible with compositions from a continental lithospheric mantle. The exhumation of the mantle rocks is indicated by deformation under progressively decreasing pressure-temperature conditions, including olivine shear zones followed by serpentinization, then cataclasis of serpentinite mylonites, and then low-temperature replacement by calcite (Beslier et al., 1996). The serpentinized mantle is overlain by tectono-sedimentary breccias, reworking the adjacent exhumed basement (Manatschal et al., 2001). ODP Site 1070 penetrated sediments overlying the exhumed mantle and intruded by dykes. ODP Site 1069 penetrated the extensional allochthon, which is more than 10 km long and several hundred meters thick.

(6) The limited magmatism during rifting characterizing this scenario is only a qualitative description, as resultant magma-poor margins can have variable amounts of associated magmatism. For example, in the case of the Santos, Campos, and Espírito Santo margin segments in the South Atlantic, coming from the south to the north, the width of the proto-oceanic crust corridor in front of them is roughly the same but only the most distal southern Santos margin contains volcanic rocks (Zalán et al., 2011). A comparison of four transects through the southern Santos, northern Santos, Campos, and Espírito Santo segments indicates that the southern Santos segment has the widest continental margin that can be also characterized by a system of alternating thin and thick bands of stretched crust, indicating a system of several competing necking zones (Zalán et al., 2011). Further north, on the conjugate African side, the northern Gabon margin segment underwent continental mantle exhumation but the exhumation was coeval with the activity of the Loiret, Binhai, and Komo volcanoes, responsible for the development of volcano-sedimentary wedges located on both continental and proto-oceanic crusts (see Nemčok et al., 2012a). The northern portion of Nova Scotia, characterized by minimal magmatic products and having a proto-oceanic corridor (see Dehler and Welford, 2012 and references therein), represents the analog to the Campos and Espírito Santo segments lacking volcanic rocks. The Eastern Indian margin, with some areas of the distal margin depressions dominantly filled with volcano-sedimentary sequences, forms the margin with exhumed continental mantle and significant magmatic activity.

To conclude, the Campos–Espírito Santo and Eastern India segments are the opposite end-members of the volcanic occurrence at margins with exhumed continental mantle. For this type of margin, the amount of magmatism seems to increase with the extension rate (Nemčok et al., 2012a). Furthermore, the amount of magmatism appears to be inversionally proportional to the width of the exhumed mantle corridor.

(7) The classic example of the magma-poor margin segment with exhumed mantle corridor is the Iberian margin, containing the largest database on such margin types. It contains a relatively small amount of basaltic material (Whitmarsh et al., 2001), which can be consistent with the mantle exhumation during the early seafloor spreading that would focus melt into the ridge by a matrix pressure field associated with the divergent flow in the lithosphere and asthenosphere at the young seafloor spreading center (see Spiegelman and McKenzie, 1987; Spiegelman and Reynolds, 1999). However, other margins provide exceptions to this interpretation. For example, the only basaltic material in the Santos, Campos, and Espírito Santo margin segments, the southern Santos segment, occurs in small amounts in the most distal portion of the margin or in the volcano-sedimentary wedges that Nemčok and colleagues (2012a) mapped in offshore northern Gabon, which reside on both a distal margin and an adjacent proto-oceanic crust, as if the crust–mantle interface was another melt-focusing location. In fact, the Loiret volcano residing inside of the distal margin, about 15 km landward from the oceanic–continental crust boundary, which provided large volumes of volcanic material during the late syn-rift period combined with the period of the crustal breakup, continued its activity well into the early drift period, as mapped from reflection seismic sections (see Nemčok et al., 2012a). Although the termination of its activity requires further research, we can speculate that it is coeval with the termination of mantle exhumation, which stopped bringing new hot mantle, probably carrying some lower crustal material as well, on its way up to the exhumed corridor.

(8–9) The delayed seafloor spreading center initiation and the delayed start of organized seafloor spreading are two characteristic features of the breakup scenario with crustal breakup followed by mantle breakup. The crustal breakup is followed by the development of the proto-oceanic corridor until the initial seafloor spreading centers develop. Then they try to link to each other by initial transforms to eventually reach the stage of full linkage and the start of organized seafloor spreading (see Rosendahl et al., 2005). The width of the exhumed mantle corridors developed prior to the organized spreading varies among margins with proto-oceanic crust corridors. The values of about 106 and 95 km characterize the corridors in front of the Iberian and East Indian margins, respectively, which represent slower and faster ends of the extension rate interval. When these widths are compared with numerical simulation of continental breakup, done for colder and warmer scenarios and varying lithospheric mantle types (see Huismans and Beaumont, 2005), we can see that the weaker lithospheric mantle makes it to continental breakup and subsequent seafloor spreading faster than a strong mantle. This indicates that weaker scenarios have less time for mantle unroofing, as discussed in Chapter 5. The models document that the warmer lithospheric mantle speeds up the lithospheric mantle and lower crustal necking, making it to seafloor spreading faster. In this case the faster thinning of the lithosphere is enhanced by advanced buoyancy-driven flow. Also, the weaker the crust, the faster the necking instability of the lithospheric mantle in warmer scenarios is.

Details of the breakup development in this scenario

Details of the crustal breakup and continental mantle unroofing can be described using a comparison of outcrop analogs from the western Alps compared with well and seismic data from the conjugate Newfoundland-Iberian margins (see Whitmarsh et al., 2001; Manatschal, 2004; Figure 2.8).

Lithospheric stretching

Alpine transects document that rifting in the Alpine realm started with differential subsidence and the formation of the rift system. Rifting eventually localized in more distal portions of the future margins. As a result, proximal and distal margins have different crustal architectures (see Lemoine et al., 1986; Eberli, 1988; Manatschal and Bernoulli, 1998).

The onset of rifting introduces abrupt changes in facies, abrupt changes in thickness of sediments, syn-rift sediment thickening into footwalls of fault blocks, and syn-rift sediment onlaps onto rotating hanging walls. The rifting is typically diffuse and distributed before extension becomes localized along a few major faults. Major faults operate independently of one another and most of the movement along them occurs over a short time interval.

The proximal basins are larger and older than distal ones. They are bounded by listric faults soling out at mid-crustal levels. Faults are relatively steep in their uppermost portions. Proximal margins from Alpine transects contain the record of the lithospheric stretching stage without any overprint by subsequent stages of the breakup development. They are characterized by grabens and half-grabens, which are 10 to 30 km wide and filled by syn-rift fill ranging in thickness from several hundred to several thousand meters (Lemoine et al., 1986; Eberli, 1988; Manatschal and Bernoulli, 1998). They are bounded by relatively high-angle faults of both vergencies. A similar fault pattern is known from the Nova Scotia margin (Deptuck, 2011). Sedimentary fill, although variable, is dominated by turbidites and other mass deposits intercalated with hemipelagic sediments commonly fining and thinning upward. A nice example of the boundary fault is the Lugano–Val Grande Fault, bounding the Monte Generoso Basin (Bertotti, 1991). Its deformation changes from brittle, represented by cataclasites, to ductile, represented by quartz mylonites. The fault soles out at a depth of about 15 km, in greenschist metamorphic facies conditions.

Proximal margins of the Newfoundland and Iberia regions are formed by systems of half-grabens and grabens bounded by relatively high-angle faults (e.g., Montenat et al., 1988; Driscoll et al., 1995; Pérez-Gussinyé et al., 2003). They are further characterized by prominent structural highs separating different extensional provinces.

Lithospheric necking

Deformation of the future distal margin is controlled by the necking of the lithospheric mantle. Necking is associated with major detachment faults that dip toward the necking point. This leads to the localized thinning of the crust. The location of the lithospheric mantle necking does not have to coincide with the extension maximum in the upper crust, as documented by Alpine data (Manatschal, 2004), analog material models (Brun and Beslier, 1996; Michon and Merle, 2003), and numerical models (Harry and Sawyer, 1992). This shift is typical for the lithospheric scenarios characterized by stronger mantle lithospheres and lower crusts (see Huismans and Beaumont, 2005).

Distal margins from Alpine transects carry the record of the lithospheric necking. They also contain the structural grain of the proximal margins as the older generation of structures. However, the initial subsidence patterns are later replaced by uplift, during advanced rifting. The most prominent younger-generation rifting structure is a pattern of low-angle detachment

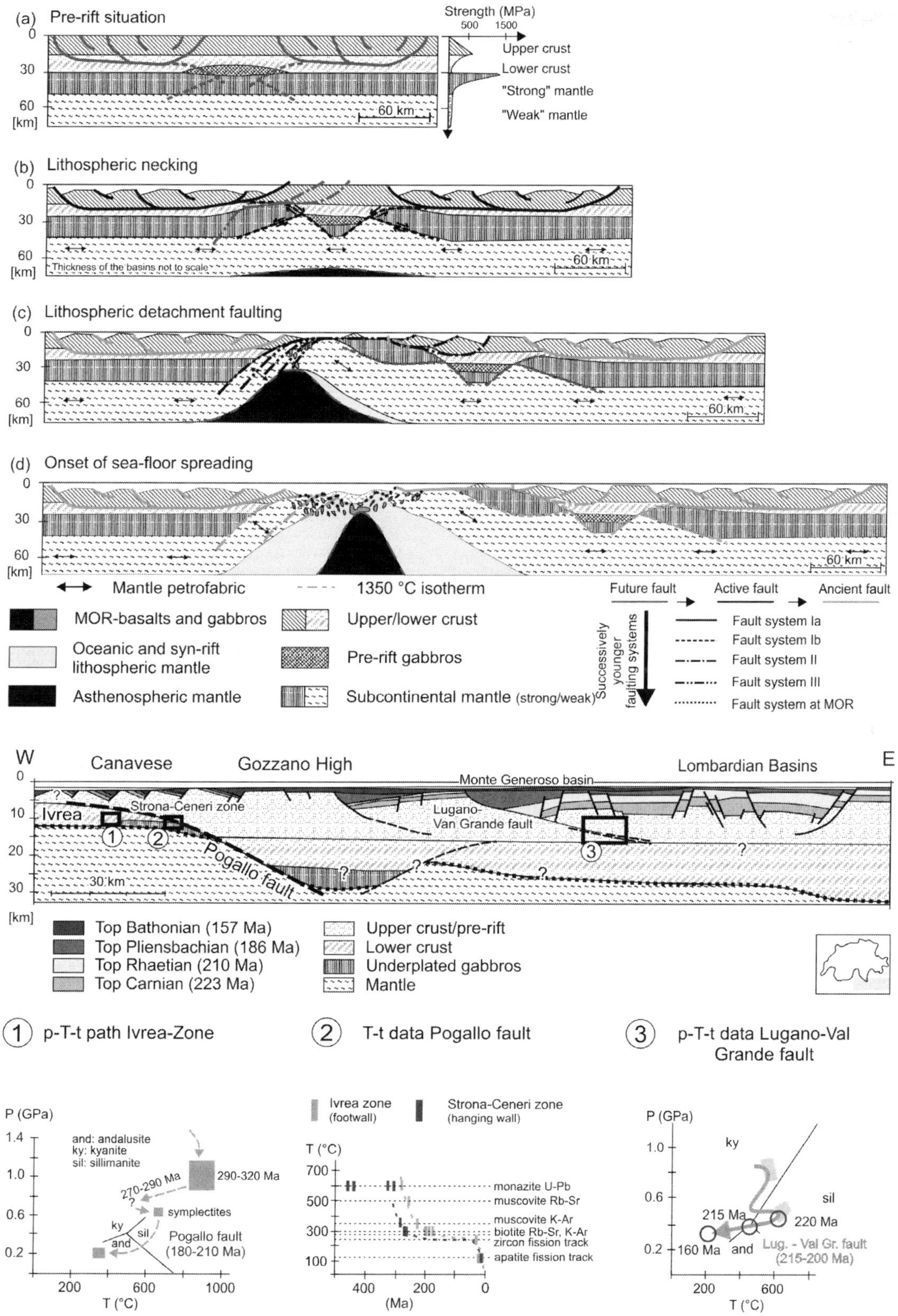

Figure 2.8. Conceptual model of the rift-drift development (Manatschal, 2004). Stages include: a) pre-rift stage, b) lithospheric necking, c) lithospheric detachment faulting, and d) onset of seafloor spreading.

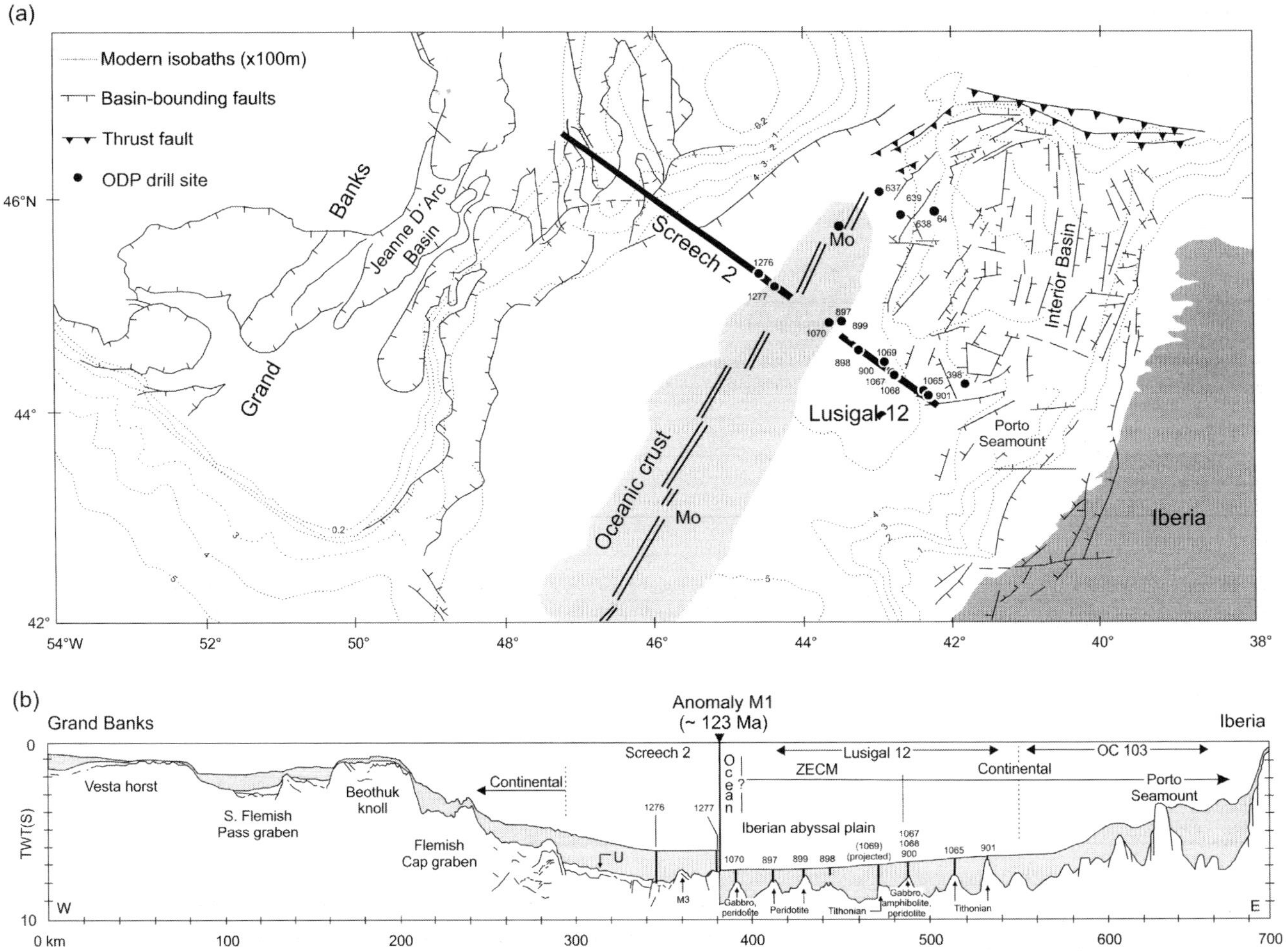

Figure 2.9. The Newfoundland–Iberia conjugate margins (Manatschal, 2004). a) Map of the Cretaceous Atlantic between Iberia and Newfoundland reconstructed to anomaly M0 time (ca. 121 Ma) based on the reconstruction pole of Srivastava and colleagues (2000) with the Newfoundland plate fixed relative to present geographic coordinates. b) Conjugate seismic reflection section from the Newfoundland and Iberia margins along profiles SCREECH 2 and Lusigal 12, juxtaposed at anomaly ca. M1. Sediments are shaded gray above the basement in the Iberian case or above either the basement or the U reflection in the Newfoundland case.

faults. They form breakaway faults in the continental crust, cut oceanward into the mantle, and are overlain by extensional allochthons. The lower crustal rocks are separated from the brittle upper crust by crustal-scale, continent-ward dipping faults (see Handy and Zingg, 1991; Muentener and Hermann, 2001).

The distal basins are smaller and younger than proximal ones. They are, in some cases, truncated by low-angle detachment faults.

The distal margin of the Iberia region is characterized by the tilted faults blocks of continental basement and pre-rift sediments overlain by relatively thin syn-rift and post-rift sedimentary sequences (Figure 2.9). Tilted blocks are underlain by a prominent reflector (de Charpal et al., 1978; Reston et al., 1995) imaging the top of the upper continental mantle, which reaches the surface just a bit closer to the margin from the unroofed continental mantle drilled by OPD wells (see Manatschal, 2004). Several other reflectors further landward image shallow-dipping faults under tilted blocks (e.g., Krawczyk et al., 1996). These reflectors form distinct breakaways further landward, also cutting an earlier detachment fault with continent-ward dip (Whitmarsh et al., 2000).

Analogs for the major detachment faults in the distal margin are the Alpine Pogallo and Margna Faults (Handy and Zingg, 1991; Muentener and Hermann, 2001), which separate lower crustal footwall rocks from upper crustal hanging wall rocks, and both accommodated the crustal thinning down to about 10 km. Both faults did not produce any distinct depocenters despite the significant crustal thinning, just like those from the

Porcupine Basin, Ireland, and the Galicia Interior Basin (see Reston et al., 2001; Pérez-Gussinyé et al., 2003).

Described observations indicate that future distal and proximal margins behave differently during the initial rifting. Therefore, the final place of the breakup seems to be determined relatively early in the rifting evolution, probably controlled by the lithospheric heterogeneities, described in Chapter 6. The places where the crust was thinned down to at least 10 km apparently made it later to breakup.

The lithospheric necking stage is characterized by two types of faults (Figure 2.8b). The first type cuts through the brittle upper crust and soles out at the middle to lower crustal levels. These faults bound rift basins, which are up to 4 km deep and 30 km wide. Their offsets are less than 10 km and total extension is limited. Both footwalls and hanging walls of these faults subside, which prevents deeper crustal rocks from exhumation along them. A nice example of such faults is the Lugano–Val Grande Fault from the Alps (Bertotti, 1991).

The second type cuts through the deeper crust and accommodates large amounts of extension. It is kinematically linked with the first fault type. This fault type is probably spatially related to the necking location in the lithospheric mantle. The second-type faults can accommodate for the crustal thinning below 10 km of thickness. Nice examples of these faults are the Pogallo and Margna Faults in the Alps (Manatschal, 2004 and references therein) and faults interpreted in reflection seismic images through the Iberia Abyssal Plain (Whitmarsh et al., 2000).

Propagation of both types of faults is guided by preexisting anisotropy, described in Chapter 6. Despite their kinematic linkage, they are controlled by different extension modes, operating in future proximal and distal margins differently. In accordance with depth-dependent stretching data published by Davis and Kusznir (2004), the former is characterized by upper crust-dominant extension, while the latter is characterized by lower crust-dominant extension. This coexistence can be visualized by a simple experiment where we pull apart two ends of a Teflon stick above a burning candle. At some point in time, the same amount of pull would control accelerated extension accommodated by a rapid localized thinning in the warmest spot.

Lithospheric detachment faulting

The lithospheric detachment faulting stage is characterized by two types of faults (Figure 2.8c). The first type of fault cuts through the entire crust, which is now extended, cooled, and brittle, and soles out at the crust–mantle boundary. It controls basins up to 4 km deep and 20 km wide. Both footwalls and hanging walls of these faults subside. Examples of these faults come from the reflection seismic profiles through the Iberia Abyssal Plain (Manatschal, 2004 and references therein). This type of fault controls well-known lower-extension basins (e.g., Bosence, 1998), characterized by listric faults, hanging wall subsidence, and hanging wall rotation leading to half-grabens. In these basins, the sediment catchment space increases by vertical or steeply inclined shear. As a consequence, footwall exposes only shallow crustal levels; sediments onlap the faults at high angle, thickening toward the fault.

The second type of fault has downward concave geometry. Such faults facilitate the exhumation of the continental lithospheric mantle at the sea floor. They can have offsets larger than 10 km without producing distinct fault-related topography, exhuming deeper crustal and mantle rocks from underneath a relatively stable hanging wall. An example of this is the Err detachment fault from the Alps (Manatschal, 2004 and references therein). Further examples come from the reflection seismic images through the Iberia Abyssal Plain and ODP Leg 900, 1067, 1068 penetrations (Manatschal, 2004 and references therein). Studied examples of downward concave faults indicate that they are rooted in the mantle and overlain by continent-derived extensional allochthons and tectono-sedimentary breccias. Numerical modeling by Lavier and colleagues (1999) indicates that the downward concave faults tend to develop in the scenario characterized by a thin, upper, brittle layer underlain by a weak layer. Although it may be tempting to compare the downward concave faults from the lithospheric detachment faulting stage to the extensional settings prone to core complex development, the amount of crustal extension here is much larger (see Manatschal, 2004). It is neither a steady-state nor at least a cyclic process like the core complex formation at slow and ultra-slow spreading ridges (see, e.g., Buck et al., 2005 and references therein).

The rooting-into-the-mantle aspect of the downward concave faults (Figure 2.8; Manatschal, 2004) has been supported by the fact that Alpine mantle rocks do not represent only the uppermost mantle rocks, which would be the case if the detachment was rooted along the crust–mantle boundary. Next to the continent, the mantle rocks are formed by spinel peridotite mixed with garnet-pyroxenite rocks, which equilibrated at lower temperatures than pyroxenite-poor peridotites located further oceanward (see Desmurs et al., 2001; Muentener and Piccardo, 2003).

A further aspect of the downward concave faults is their effect on the distribution of magmatism. While the hanging wall of the distal margin should contain volcanic complexes, the footwall does not. The specific feature of downward concave faults is their ability to accommodate an extreme amount of extension. The extension in the Iberian Abyssal Plain increases oceanward from a few kilometers along the first-type fault of the lithospheric detachment faulting stage to a few tens of kilometers along the downward concave faults (Manatschal, 2004). Downward concave faults control higher-extension basins, which totally differ from those controlled by the first-type faults (see Wilson et al., 2001). These basins develop by pulling the footwall from underneath a stable hanging wall along downward concave faults with large offsets. Their sediment catchment space increases by subhorizontal shear. They are floored by detachment faults, which expose deep crustal and mantle rocks. These basins have a fill subparallel to the detachment fault.

In accordance with the depth-dependent stretching data published by Davis and Kusznir (2004), both described faults propagated in the region that was extended during the lithospheric necking the most, the future most distal margin. The second type of faulting follows the propagation of the first-type faulting. While the first type soles out along the crust–mantle boundary, the second type cuts through the mantle. The propagation of both faults, just like the two fault types in the lithospheric necking stage, is controlled by a complex interaction of mechanical, chemical, and thermal processes involved in rifting and rifting–drifting transition.

Unlike the warming of the crust that contributes to the lithospheric necking development, the transition from lithospheric faulting stage to the lithospheric detachment faulting stage probably reflects the extent of embrittlement of the extending system (Whitmarsh et al., 2000) and the onset of serpentinization. The serpentinization initiates when the crustal thickness decreases below about 10 km and the lithospheric mantle arrives in the serpentine stability field (Pérez-Gussinyé and Reston, 2001).

The second type of faulting of the lithospheric detachment faulting stage apparently follows the first type with the onset of magmatic activity, which is indicated by the cross-cutting relationships between magmatic rocks and detachment faults in the Platta nappe (Figures 2.10g–h; Manatschal, 2004). The magmatic weakening is coupled with rising asthenosphere. Therefore, it is associated with a dynamic change in the thermal structure, not yet well understood (Minshull et al., 2001). As the numerical modeling of Huismans and Beaumont (2005) indicated, the thermo-magmatic weakening does not necessarily have a purely vertical structure and can have a small lateral component in a direction parallel to the unroofing direction along the downward concave fault.

Melts derived from the rising asthenosphere may also have had a triggering effect on the changing deformation mode from the pure-shear-dominant one during the lithospheric necking stage to a transient phase of the simple-shear-dominant one during the later lithospheric detachment faulting stage.

The changes of the rifting-faulting controlling mechanisms (deformation localization) can be characterized by a change from mechanically dominated rifting during the initial lithospheric necking stage to hydration/serpentinization-dominated control during the early lithospheric detachment faulting stage to thermally/magmatically dominated control during the later lithospheric detachment faulting stage.

Onset of seafloor spreading

The time needed for a transition from the lithospheric detachment faulting stage to an onset of organized seafloor spreading can be estimated from the Alpine and Iberian margins data.

The age of the crustal breakup in the Alps is constrained by the oldest sediments resting on the unroofed continental mantle and adjacent continental mantle and by the radiometric ages of emplacement and exhumation of magmatic bodies in the unroofed mantle (e.g., Peters and Stettler, 1987; Bill et al., 2001; Lombardo et al., 2002; Schaltegger et al., 2002). All published Alpine data indicate a rather short period of mantle exhumation and magma emplacement, implying exhumation-controlling detachment faulting linked to gabbro emplacement and exhumation. Stratigraphic relationships in ultra-deep reflection seismic images through the proto-oceanic-continental crust boundary near the Loiret volcano in offshore Gabon indicate much longer magmatic activity and exhumation time span (own observations).

The age of the crustal breakup in the Iberian region has not been dated directly, but is based on $^{40}Ar/^{39}Ar$ plagioclase ages (Feraud et al., 1996; Manatschal et al., 2001), which date the cooling of gabbros and amphibolites across the isotherm of 150°C, constraining the age of exhumation at around 137 Ma, about 9 Ma before the age of the first magnetic anomaly associated with seafloor spreading (see Whitmarsh and Miles, 1995; Russell and Whitmarsh, 2003).

A comparison of continental breakups controlled by the fast, intermediate, and slow extension rates, made by Nemčok and colleagues (2012a), indicates that the

Figure 2.10. Outcrops of the various rocks and structures involved in the rift–drift transition from the Alpine outcrops (Manatschal, 2004).
(a) Fault surface of the Err detachment ("d") at Piz Laviner.
(b) Err detachment ("d") separating cataclastically deformed granitic detachment in footwall ("f") from an extensional allochthon composed of Triassic dolomites and schists in the hanging wall ("h") (photo courtesy of Sudipta Tapan Sinha).
(c) Exhumed continental lithospheric mantle ("em") at Parsettens overlain by continent-derived breccias ("b") and dolomites ("d") (photo courtesy of Mainak Choudhuri).
(d) Tectonized and exhumed mantle at Falotta overlain by pillow basalts ("pb").
(e) Syn-rift sediments ("srs") onlapping onto the exhumed detachment formed by a brittle fault zone ("bfz") at Fuorcla Cotschna.
(f) Post-rift sediments ("prs") overlying tectonized exhumed mantle (em) in Tasna (photo courtesy of Atul Patidar).
(g) Relationships between high- and low-temperature deformation in the mantle rocks and its relationship to magmatic processes. A strongly localized serpentinite gouge zone ("sgz") with a top-to-the-ocean shear sense cut by high-temperature peridotite mylonites showing a top-to-the-continent shear sense. A dolerite dyke ("dd") cuts the high-temperature mylonite foliation ("mf"), cut by a serpentinite gouge zone and occurring in clasts of this zone.
h) Gabbro intrusions in serpentinized mantle (photo courtesy of Achyuta Ayan Misra).

Figure 2.10 (*cont.*)

exhumed continental mantle corridor is narrowest in fast-extension settings and widest in slow extension settings (Figure 2.11a). These observations, together with published numerical models (e.g., Huismans and Beaumont, 2005 and references therein), indicate that the timing of the organized seafloor spreading taking over the mantle unroofing is controlled by the development of the asthenospheric anomaly and associated thermal perturbation. The faster the extensional setting the quicker the onset of organized spreading is.

The magmatic weakening coupled with the rising asthenosphere, described earlier, should, following the simple numerical models made by Coblentz and colleagues (1994), Richardson and Coblentz (1994), Jones and colleagues (1996), and Huismans (1999), cause the compressional component of the buoyancy force to vanish and the pressure difference due to asthenospheric upwelling to control a significant ridge push toward the neighboring areas. With the continuing thinning of the mantle lithosphere the extensional buoyancy forces increase dramatically and the thinned area is uplifted to lower depths. This uplift is in accordance with observations of volcanic-dominated formations developed on top of the thinnest continental crust and unroofed continental mantle in subaerial conditions (e.g., South Atlantic, Cameroon, Equatorial Guinea, and Gabon – Rosendahl et al., 2005; Central Atlantic, northeastern United States – Keen and Potter, 1995; Jackson et al., 2000; Central Atlantic, southeastern portion of Nova Scotia – Dehler, 2010, OETR Association, 2011; Dehler and Welford, 2012).

These simple models allowed Huismans (1999) to estimate that initial rift push force in his model having been in a range of $1\cdot10^{-11}–1.45\cdot10^{-11}$ Nm^{-1} could increase to $1.85\cdot10^{12}–2.2\cdot10^{12}$ Nm^{-1} upon subsequent lithosphere thinning. This indicates an important role of convective lithospheric thinning in the significant increase of the rift push force with stress levels in the order of plate boundary forces. For comparison, the known ridge push forces fall into a range of $2\cdot10^{12}–3\cdot10^{12}$ Nm^{-1} (Kusznir, 1991).

At this point it is interesting to compare the widths of continental margin and proto-oceanic crust corridor that Nemčok and colleagues (2012a) discussed with the results of the numerical simulation of continental breakup for various lithospheric scenarios focused on colder and warmer rifting scenarios that Huismans and Beaumont (2005) undertook. Simulations indicate

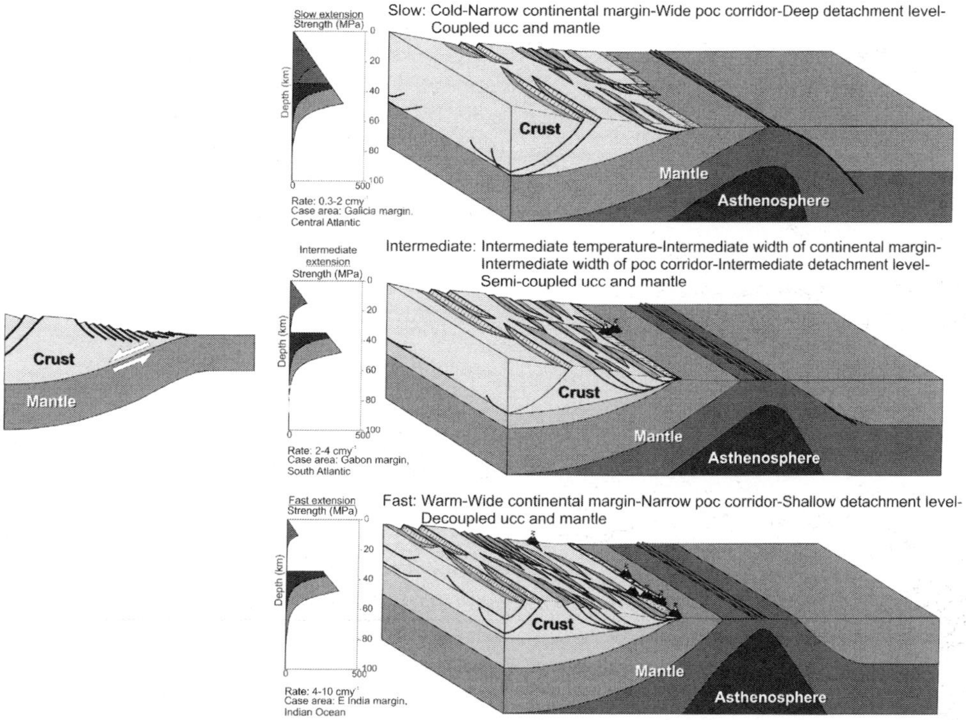

Figure 2.11. Sketch illustrating the general lithospheric break-up mechanism. It also shows the estimated strength profiles of continental margins and structural architectures of syn-break-up lithospheres for slow-, intermediate-, and fast-extension settings. Figures are composed using own data and data from Macdonald (1982), Bown and White (1994), Davis and Kusznir (2004), and Huismans and Beaumont (2005).

that the asymmetry of the lower mantle and asthenospheric bulge advancing in a direction away from the overall footwall of the rift zone is a common feature, developed best in a strong crust–weak mantle and least in a weak crust–weak mantle scenario. Hand in hand with this relationship goes the development of the "mantle-breaking," upward-convex, shallow-dip fault, which becomes fully developed at the end of necking of the lower crustal and upper mantle layers. The weak crust–weak mantle scenario develops this fault as the latest from all lithospheric scenarios, indicating a longer time of crustal thinning preceding its development. The role of the upward convex fault stops at the moment when the lower mantle breaks up and the asthenospheric anomaly starts to develop the seafloor spreading center.

The simulations document that a weaker lithospheric mantle makes it to continental breakup faster than a strong mantle, indicating that weaker scenarios have less time for mantle unroofing. The models document that the warmer lithospheric mantle speeds up the lithospheric mantle and lower crustal necking. In this case the faster thinning of the lithosphere is enhanced by enhanced buoyancy-driven flow. Also, the weaker the crust the faster the lithospheric mantle necking instability in warmer scenarios is.

The comparison of all modeled scenarios (see Huismans and Beaumont, 2005) documents that the fastest potential for the continental breakup occurs in the case of combined warmest thermal regime with the best crustal strain localization. Such localization would be provided by preexisting shear zones and preexisting orogens (e.g., Rosendahl, 1987; Nemčok et al., 2005b), such as those present in both the east coast of India and Gabon areas.

Mantle lithosphere breakup followed by crustal breakup

This breakup scenario (Figure 2.12) can be characterized by:

1) basin-controlling faults penetrated shallowly into the crust;
2) wide regions affected by extension (hundreds of km wide), that is, less abrupt crustal thinning;
3) more common relatively symmetric cross-sectional geometry of the rift system;
4) faulted early syn-rift basins;

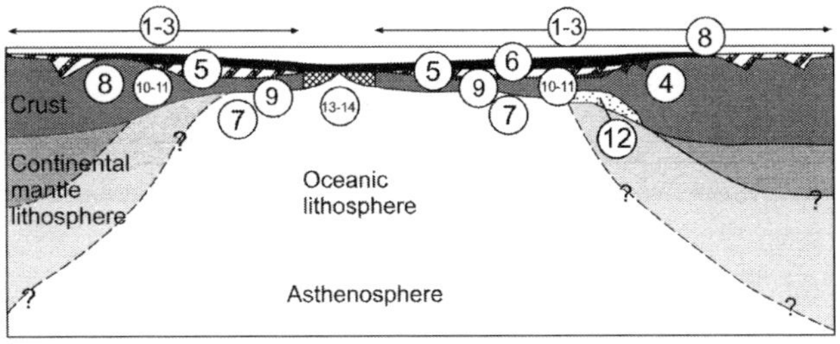

Figure 2.12. Characteristic properties of the margin originated by the mantle-first–crust-second type of breakup (Huismans and Beaumont, 2011). The example is based on the South Atlantic conjugate margins. See text for further explanation.

5) undeformed late syn-rift basins;
6) late syn-rift basin fill terminating with shallow-water sag basins with evaporites and other sediments;
7) limited syn-rift subsidence due to replacement of underlying continental mantle lithosphere by hot asthenosphere;
8) lack of distinct rift flank uplift;
9) no exhumation of the continental mantle in so-called proto-oceanic corridor;
10) some magmatism during rifting;
11) magma-rich margin or margin with some magma;
12) lower crustal regions with seismic velocities indicating magmatic underplating;
13) seafloor spreading center initiation relatively soon after crustal breakup; and
14) start of organized seafloor spreading relatively soon after crustal breakup.

1) In this scenario, represented by case examples such as the Congo and Angolan margin segments on the African side of the South Atlantic (Contrucci et al., 2004; Henry et al., 2005; Moulin et al., 2005), southern Santos, Pelotas segments on the South American side of the South Atlantic (Zalán et al., 2011), and the Exmouth Plateau (Driscoll and Karner, 1998; Karner and Driscoll, 1999), the controlling faults detach at the brittle–ductile transition zone. This zone is located higher than that in the scenario undergoing crust breakup followed by mantle breakup. Here, a viscous lower crust is sufficiently thick and weak to allow for the decoupling of the upper crust from the mantle lithosphere.

A comparison of passive margins, which underwent slow, intermediate, and fast extension leading to continent breakup (see Nemčok et al., 2012a), indicates that an increase in extension rate tends to increase the thickness of the ductile lower crust. This then tends to increase its capability to decouple the upper crust from upper lithospheric mantle and decrease the detachment depth of the basin-controlling faults. It needs to be said, however, that the extension rate is not the only critical factor controlling the development of varying crustal configurations. The other two critical factors are lithospheric composition, discussed in Chapter 5, and thermal regime, discussed in Chapters 9–13.

2–3) Wide and symmetric rift systems are discussed in Chapter 5 in detail. However, it is interesting to see how the stretched continental margin increases in width as we pass from the northern South Atlantic characterized by crust-first–mantle-second breakup scenarios to the central South Atlantic characterized by mantle-first–crust-second breakup scenarios. Serial transects published by Zalán and colleagues (2011) show this width as starting with 74 km in the Espírito Santo segment, and changing to 78 km in the Campos segment, to 276 km in the northern Santos segment, and to 510 km in the southern Santos segment (Figures 2.13a–d). Furthermore, the widening of a continental margin with increasing controlling rate of extension, that is, with a warmer breakup scenario, has been also concluded from a systematic study by Nemčok and colleagues (2012a), where the values of 135 and 202 km represent slow-extension scenarios of the Iberian margin and the Adria margin of the reconstructed Liguria-Piemonte Ocean (measurements taken from Manatschal, 2004), the value of 250 km represents intermediate-extension scenario of the Gabon segment, and values of 414 and 471 km represent fast-extension scenarios of the Mahanadi and Krishna-Godavari segments of the East Indian margin.

A close look at serial sections of Zalán and colleagues (2011) coming from the northern region characterized by crust-first–mantle-second breakup scenarios to the southern region characterized by mantle-first–crust-second breakup scenarios offers an interesting indication of why the mantle-first–crust-second breakup scenarios are characterized by wide regions affected by extension and less abrupt crustal thinning (Figures 2.13a–d). It is the thicker ductile lower crust that allows this scenario to avoid reaching the crustal necking and mantle exhumation stages too early and in several locations, where they would compete for the continental breakup localization. On the contrary, it is the northern cross-sections that have a lot of thinner lower ductile crust where this takes place. The northern cross-sections recorded at least one extra location with a complete lower crust withdrawal, that is, an attempt to initiate mantle exhumation, apart

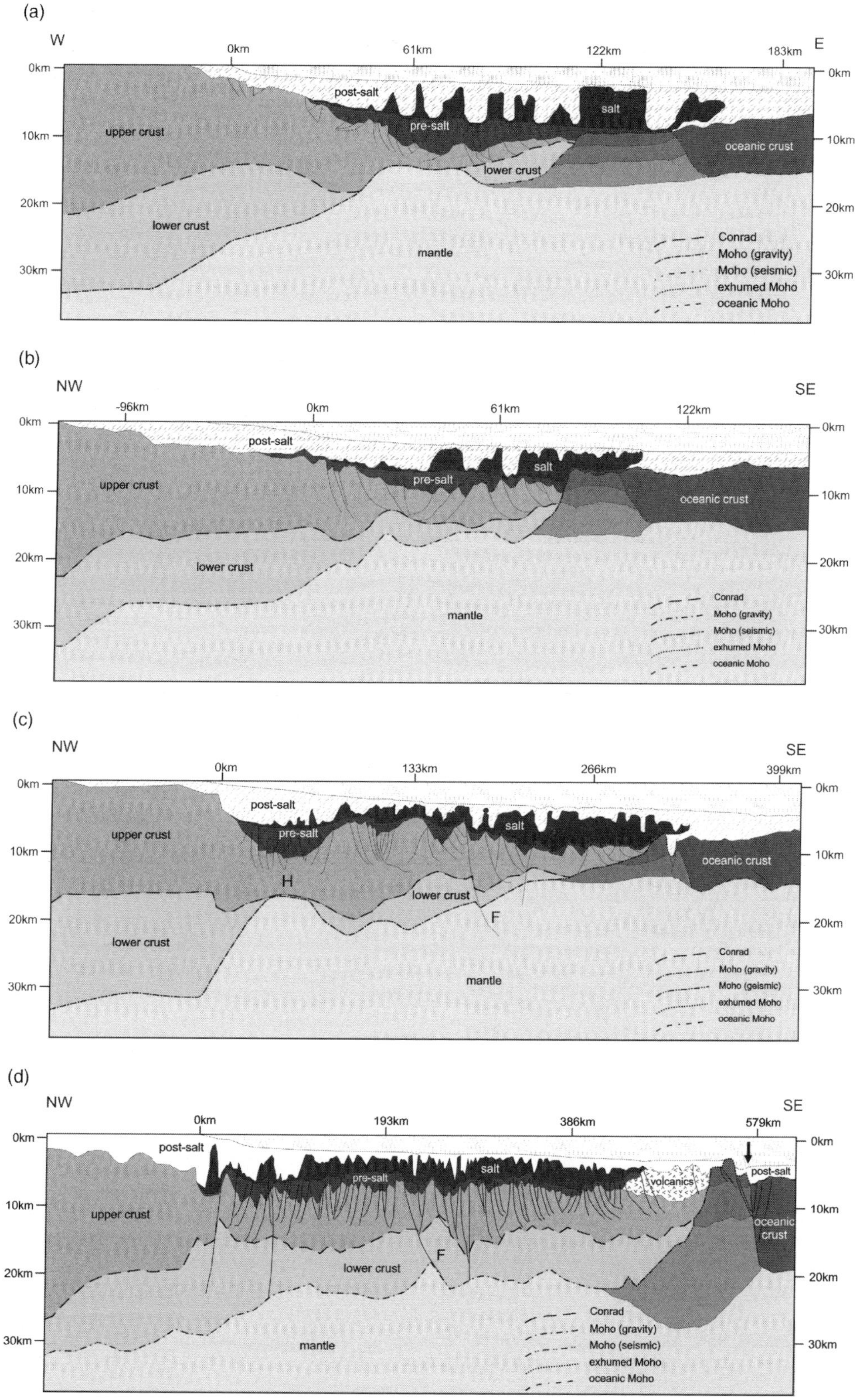

Figure 2.13. Geological cross-sections interpreted from reflection seismic and gravity data through the Espírito Santo (a) Campos, (b) northern portion of Santos, (c) and southern portion of Santos Basins (Zalán et al., 2011).

from the location that eventually reached the continental breakup. This behavior is similar to that in the southern North Atlantic, where the stretched continental crust developed several locations competing for the location of the continental breakup, which were divided by future micro-continents and continental ribbons (see Perón-Pinvidic and Manatschal, 2010; Figure 2.14).

The requirement of a ductile crust that would successfully decouple the upper crust from the lithospheric mantle is also indicated by Huismans and Beaumont's (2008) thermo-mechanical finite-element models. Their models reach the continental breakup after about 250 km of stretching, which leads to the development of an approximately 600 km-wide region affected by the stretching. The modeling indicates that it is the contact of the lower-strength crust with the mantle that controls the greater penetration of the initial decoupling and extension into the continental crust, that is, wider rift zone. This mechanism results in a combination of the wide-rifting style in the crust and narrow-rifting style in the lithospheric mantle. Furthermore, this rifting style controls diachronous crustal extension characterized by the extension termination event becoming younger toward the future continental breakup location and removal of the lithospheric mantle from the area underneath the rift axis.

4–5) Faulted early syn-rift basins and undeformed late syn-rift basins have been well demonstrated for this breakup scenario by an interpretation of both ultra-deep and standard industrial reflection seismic profiles through the Kwanza Basin, Angola (see Henry et al., 2005; Karner, 2008) (see also Figure 2.15). They show that the lithostratigraphy in this region contains

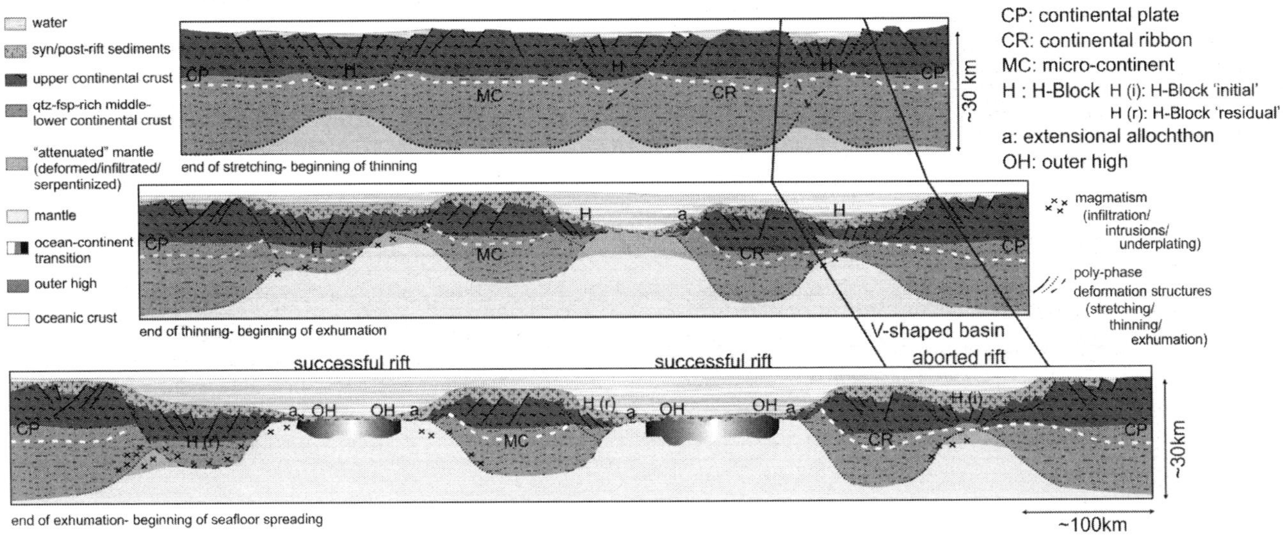

Figure 2.14. Rift evolution model illustrating the development of different types of crustal blocks (Perón-Pinvidic and Manatschal, 2010).

Figure 2.13 (*cont.*)

a) The Espírito Santo profile has a narrow character with a large crustal taper. It contains a distinct necking zone in its central portion, resulting in lithospheric mantle directly underlying the upper crust. Following forward gravity modeling, the exhumed mantle in the proto-oceanic corridor is divided into three layers with upward-decreasing densities.

(b) The Campos Basin profile is moderately wide with moderate crustal taper. It does not contain any necking zones. Following forward gravity modeling, the exhumed mantle in the proto-oceanic corridor is divided into three layers with upward-decreasing densities.

(c) The northern Santos Basin profile is relatively wide with moderate crustal taper. The crust is necked in the western portion of the margin. The necking resulted in mantle underlying the upper crust, represented by a hanging-wall block (sensu Perón-Pinvidic and Manatschal, 2010). A fault F that cuts through the entire crust separates the hyper-extended distal margin from the stretched crust of the proximal margin. Following forward gravity modeling, the exhumed mantle in the proto-oceanic corridor is divided into two layers with upward-decreasing densities.

(d) The southern Santos Basin profile shows a wide margin with low crustal taper and one necking zone with fault F cutting through the entire crust. The continental mantle is exhumed in the region next to the Florianopolis oceanic fracture zone (arrow). Gravity modeling requires three-layer division of the exhumed mantle with upward-decreasing densities.

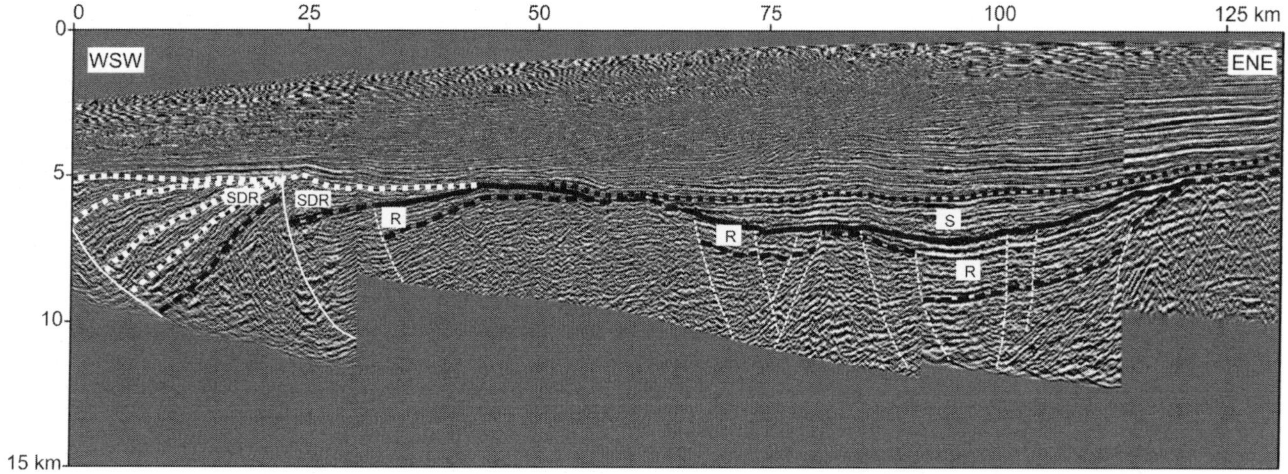

Figure 2.15. Uninterpreted and interpreted reflection seismic profiles through the Orange Basin, South Africa (modified from Thompson, 2009). Note the sag basin fill (S) extending beyond the fault-controlled early and main syn-rift (R) depositional areas. Most distal margin contains volcanoclastic wedge (SDR), interpreted by Thompson (2009) as predating the sag basin and by us as coeval with the sag basin. Rift-controlling faults are shown as dashed lines; faults of the SDR are represented by solid lines.

(Karner et al., 2003; McAfee, 2005; Moulin et al., 2005; Vagnes et al., 2005):

a) early and main syn-rift Neocomian–lower Barremian strata deposited in half-grabens indicating syn-depositional activity of graben-controlling faults by their thickening toward the faults.
b) thick late syn-rift middle Barremian–middle Aptian strata, which were deposited in a shallow-water broader basin reaching far beyond the pre-existing graben-controlling syn-rift faults, and which seal the graben-controlling faults;
c) late syn-rift middle-upper Aptian shallow-water evaporite strata; and
d) post-rift Albian-recent strata.

The early syn-rift sediments penetrated by industrial wells are represented by shallow-lake strata deposited along the rift axis, and alluvial fans prograding from basin margins to the lake. All of these facies are included in Jurassic–Neocomian Vandji formation. Individual half-grabens subsided, forming deep troughs with more than 2 km-thick lacustrine shale, such as the shale of the Neocomian Kissenda formation in the Gamba Horst region.

The main syn-rift sediments from wells recorded fault-controlled topography including structural highs and lows. The former are represented by the Gamba horst; the latter are indicated by localized distribution and variable thickness of organic shales of the Barremian Melania shale and the Marnes Noires and Bucomazi formations. The late main syn-rift sediments of the central Gabon-Angola rift region contain more than 1 km-thick sand-rich fluvial-lacustrine sediments, such as the Barremian Dentale formation in Gabon and the Red Cuvo formation in the Lower Congo Basin. The Dentale fluvial system flowed along the axis of the South Gabon Rift Basin. Regions distal to fluvial axes, such as southernmost Gabon and the Congo, were occupied by persistent shallow lakes with silt, shale, and locally carbonate deposition recorded by the Argil's Vertes, Viodo, Toca, and Cuvo carbonate formations.

The late-rift sediments undeformed by syn-rift faults record the regional denudation after the main syn-rift phase, which can be seen in well data from this region. Although there was a widespread regional denudation, a subtle topographic control over subsequent deposition continued, resulting in sedimentary fill division into the Aptian Gamba, Chela, and Grey Cuvo formations. The basal lowstand fluvial sediments, which were best developed in inner basins and near intra-basinal highs, and transgressive shallow lake facies, which include winnowed high-energy lake shoreline sandstones deposited along the axis of the outer hinge zone, represent proven reservoir facies.

The timing of the continental breakup is dated to be younger than the deposition of late syn-rift strata (see Karner, 2008; Karner et al., 2003; Moulin et al., 2005), indicating a lack of fault control during their deposition and ongoing extension eventually leading to the continental breakup.

6) The observations of late syn-rift basin fill terminating with shallow-water sag basins with evaporites and other sediments on the African side of the central South

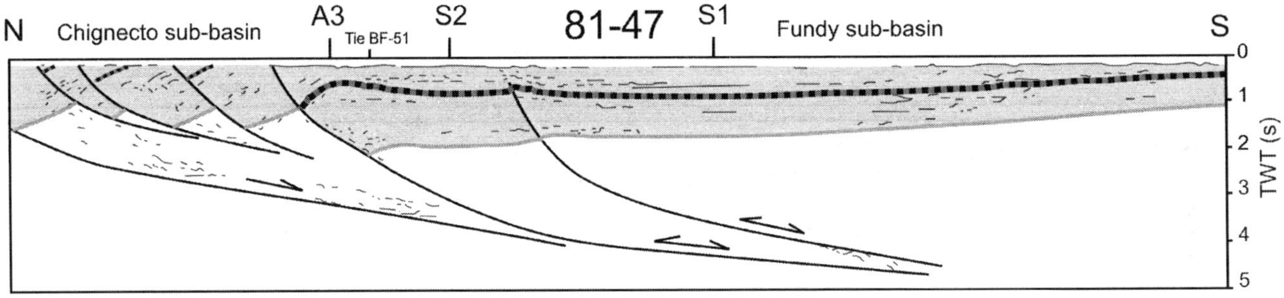

Figure 2.16. Late Jurassic–Early Cretaceous inversion anticlines of the preexisting syn-rift fill in the Fundy Basin, Canada (Withjack et al., 2009). Note Late Jurassic–Early Cretaceous folds associated with half-graben inversion. The shortening direction was north-northeast–south-southwest.

Atlantic (see Henry et al., 2005; Dupre et al., 2007; Karner and Gamboa, 2007; Karner, 2008) represent one of the most identifying features of this breakup scenario. The persistence of relatively shallow depositional conditions during rifting requires isostatic balance with sub-crustal material of relatively low density, and which is buoyant more than the underlying asthenosphere undergoing upwelling, and most likely represented by hot and depleted lower cratonic lithosphere (Huismans and Beaumont, 2011).

7) Limited syn-rift subsidence due to the replacement of underlying continental mantle lithosphere by hot asthenosphere characterizes this breakup scenario. Being characterized by faster extension rates and/ or increased thermal anomalies, associated systems behave as buoyancy-dominated, somewhat analogically to fast-spreading mid-oceanic ridges (see Buck et al., 2005). The limited subsidence is controlled by ongoing thinning of the mantle lithosphere prior to its breakup. The thinning enhances the extensional buoyancy forces and the region undergoing thinning undergoes uplift to shallower depths. This uplift simulated by modeling (Coblentz et al., 1994; Richardson and Coblentz, 1994; Jones et al., 1996; Huismans, 1999) provides an interesting correlation with:

a) subaerial to shallow-water conditions of the magma extrusion described later in point (11);
b) synchronous evaporite and volcanic formation in regions such as the Santos and Nova Scotia margins (Demercian, 1996; Meisling et al., 2001; Deptuck, 2011; OETR Association, 2011); and
c) diachronous syn- and early post-breakup inversion event on conjugate margins of the Central Atlantic (Withjack et al., 1998 and references therein).

Some bore hole-constrained reflection seismic sections through the Santos Basin, that is, surrounding the Avedis High, allow us to see volcanic rocks unconformably overlain, or progressively onlapped, by thin late syn-rift evaporites (Meisling et al., 2001), indicating a shallow depositional environment. Along some of the volcanic ridges in the Santos Basin, the volcanic activity continued until breakup and early post-breakup.

The diachronous contractional event at the Central Atlantic margin (Shaler and Woodworth, 1899; Behrendt et al., 1981; Hamilton et al., 1983; Lomando and Engelder, 1984; de Boer and Clifton, 1988; Lucas et al., 1988; Wise, 1992; Withjack et al., 1995, 1998; Roden-Tice and Wintsch, 2002; Baum et al., 2003, 2008; Bertotti et al., 2010; Figure 2.16) provides an interesting correlation with the concept of the underlying asthenosphere undergoing upwelling (Figure 2.17). This inversion event, which is better documented on the North American margin, was traveling in time from the south to the north (see Withjack et al., 1998). It was roughly coeval with late rifting and early drifting. While early and main syn-rift was characterized by distant plate-tectonic forces controlling divergent lithospheric displacements and widespread extension, that is, active asthenospheric upwelling had a little influence on the lithospheric displacements, the late syn-rift was characterized by substantially thinned lithosphere and gravitational body forces and traction forces associated with the hot, low-density asthenospheric upwelling increased substantially. In response, the lithospheric displacements near the upwelling exceeded those far from the upwelling, causing shortening in the intervening zone. This shortening may have been further enhanced by the fact that the ductile lower crust was initially thickest above the crest of the thermal anomaly, progressively thinning and/or pinching out in a direction away from it. The gravitational gliding of the upper crustal blocks down the slopes of the up-domed region, detached along the ductile lower crust, may have contributed to the local contraction at

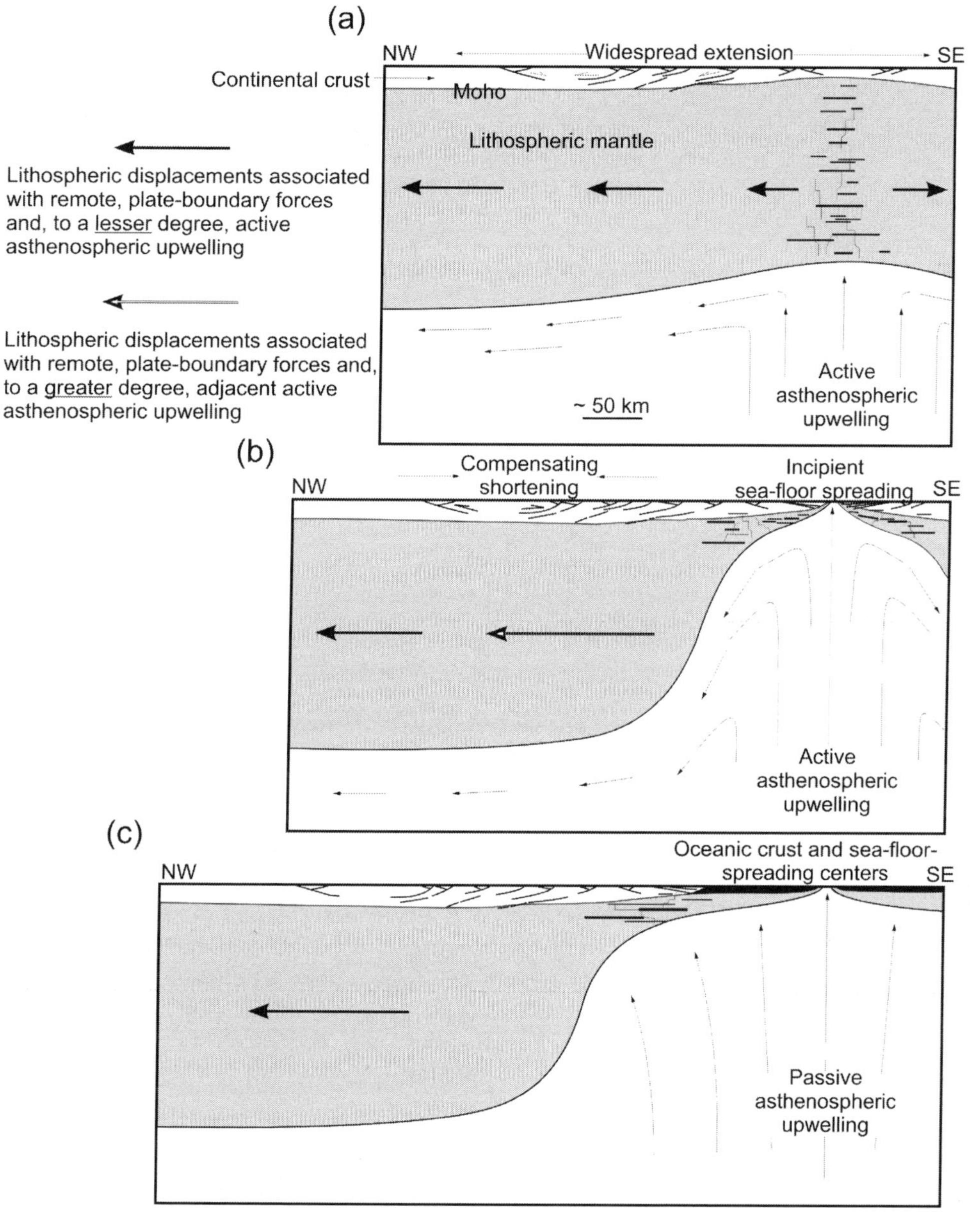

Figure 2.17. One of the models explaining the contractional events traveling in time along the Central Atlantic margins from south to north (Withjack et al., 1998). a) Early rifting characterized by distant plate-tectonic forces controlling divergent lithospheric displacements and widespread extension. Active asthenospheric upwelling has little influence on lithospheric displacements. b) Late rifting characterized by substantially thinned lithosphere. Gravitational body forces and traction forces associated with the hot, low-density asthenospheric upwelling increase substantially. In response, lithospheric displacements near the upwelling exceed those far from the upwelling, causing shortening in the intervening zone. c) Early drifting characterized by lithospheric displacements far from the upwelling increased, eventually equaling those near the upwelling. Most of the earlier contraction ceases, and the asthenospheric upwelling becomes passive.

the transition between regions with ductile and brittle lower crust (or, at least, thick and thin ductile lower crust), located somewhere at the boundary of the up-domed region. Because the Central Atlantic propagated from the south to the north, the timing of the contractional event would be diachronous from the south to the north along the former rift system.

The end of the contractional event was most likely caused by the post-breakup changes in asthenospheric upwelling and lithospheric displacements. The

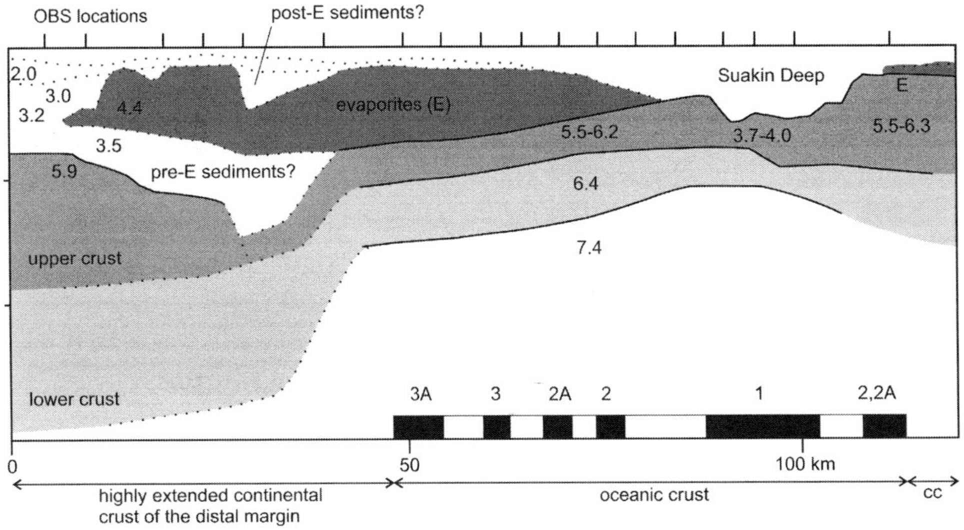

Figure 2.18. Interpreted refraction/reflection profile through the Suakin Deep and adjacent continental margins with seismic velocity (Vp) distribution (Egloff et al., 1991). The solid lines are boundaries indicated by refracted arrivals; reflections are indicated by dotted lines only. Small numbers are velocities in km/s. The solid bars at the base of the diagram are normal-polarity seafloor magnetic anomalies.

lithospheric displacements in the region far from the former upwelling progressively increased and eventually became as large as those in the region near the former upwelling, which became progressively passive. Furthermore, the extra gravity forces enhancing the upper crustal gliding progressively vanished with a cooling-controlled subsidence of the uplifted region and progressively increasing shear drag along the former detachment horizon.

At this point, it is interesting to list a few existing estimates of the height of the initial oceanic crust with respect to the surrounding continental margins to appreciate variations in the buoyancy-driven uplift of the breaking-up rift system. For example, the volcanic seabeds in the Tethys and Suakin Deeps of the central Red Sea occur at elevations of about 2 km and 2–3 km higher than those of the upper basement surface mapped on the continental margins at a distance of 20–40 km and more than 10 km from the rift axis, respectively (Egloff et al., 1991; Ligi et al., 2011; Figure 2.18). The depicted situation contrasts the situation from the magma-poor example of the conjugate Iberia-Newfoundland margins, where the initial oceanic crust was elevated 1–2 km above the surrounding exhumed continental mantle on the Newfoundland side and had about the same elevation as the exhumed mantle on the Iberian side (Perón-Pinvidic et al., 2007; Figure 2.19; Van Avendonk et al., 2009). The Newfoundland estimate is contradicted by Hopper and colleagues (2004), who estimated a lower elevation of the initial oceanic crust.

8) The lack of distinct rift flank uplift in this breakup scenario has been systematically studied by finite-element modeling (Huismans and Beaumont, 2005, 2008, 2011), which indicates the decrease in flexurally controlled flank uplifts with an increase in the thickness of the ductile lower continental crust. The weak system with thick decoupling lower crust is characterized by buoyancy-dominated behavior, while the strong system without effective decoupling along the lower crust is characterized by flexure-dominated behavior. The key factors controlling the occurrence versus lack of the flexure-driven uplift are discussed in Chapter 5 in detail.

9) This breakup scenario is characterized by no exhumation of the continental mantle in the so-called proto-oceanic corridor (Figure 2.12) because the breakup happens first in the lithospheric mantle and only subsequently in the crust.

10) This breakup scenario is characterized by some magmatism during rifting. For example, during the main syn-rift, the area now present on the Angolan margin was characterized by an extrusion of large volumes of lavas and tuffs during the main syn-rift phase. Volumes of volcanic products were large enough to fill local sediment catchment areas, sometimes forming shallow-water to emergent platforms. The overall shallow depth of the background topography was indicated by the deposition in the area to the east of the external hinge zone, which was characterized by a shallow lake to sabkha deposition, as recorded, for example, by

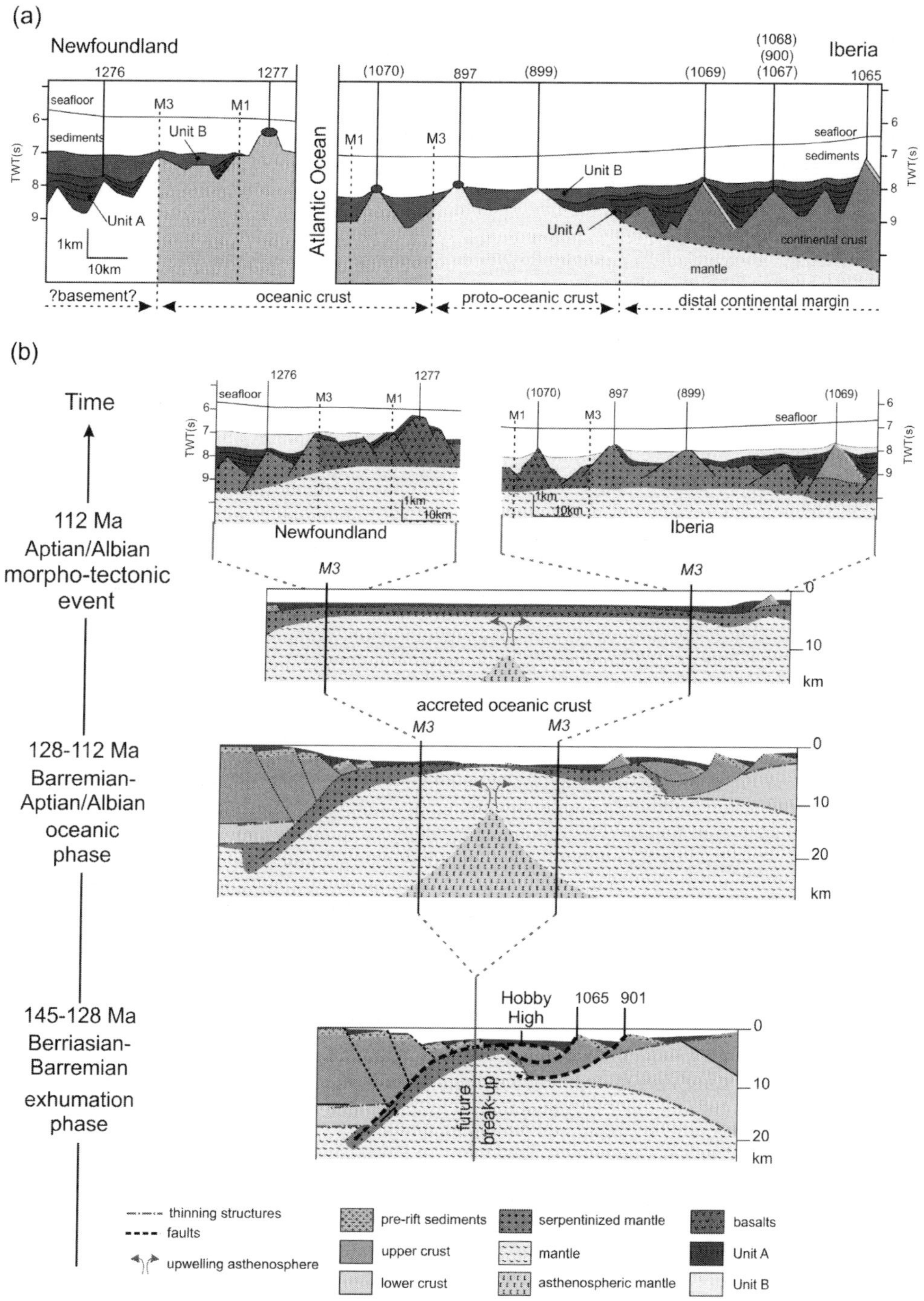

Figure 2.19. Crustal architecture of the Newfoundland–Iberia conjugate margins (Perón-Pinvidic et al., 2007). (a) Fit of present-day conjugate cross-sections. (b) Model of the continental breakup development.

the Barremian Maculungo salt member in the South Kwanza Basin (McAfee, 2005).

Another case area with syn-rift magmatism comes from the Santos margin of Brazil. Here, the main syn-rift strata represented by the Barremian lacustrine Lagoa Feia formation overlie the Neocomian–lower Barremian early syn-rift strata dominated by basalt rock suite (Guardado et al., 1990; Chang et al., 1992).

11) This scenario leads to the development of magma-rich margins or margins with some magma. Magmatic activity at these margins can:

1) predate (plateau basalt emplacement);
2) be synchronous with (seaward-dipping, volcanic-dominated wedges; underplating; intrusions in the transitional crust); and
3) postdate (anomalously thick initial oceanic crust; oceanic, seaward-dipping, volcanic-dominated wedges; underplating)

continental breakup.

In fact, the post-breakup activity is rather characteristic of most magma-rich margins (Geoffroy, 2005).

One of the key reasons for the magmatism, as indicated by modeling studies (e.g., McKenzie and Bickle, 1988; Pedersen and Ro, 1992; Bown and White, 1995), is an increased mantle thermal regime. A positive anomaly of an extra 50–200°C was suggested as sufficient (see Pedersen and Skogseid, 1989; White and McKenzie, 1989; Bown and White, 1995). An increased mantle thermal regime can also be from the anomalously thick initial oceanic crust (see McKenzie and Bickle, 1988; White and McKenzie, 1989), as its thickness is directly linked to potential mantle temperature (Parsons and Sclater, 1977; Watts, 1978, 2001; Watts and Zhong, 2000).

While such a temperature increase was originally explained by plume effect (e.g., White and McKenzie, 1989; Skogseid et al., 1992), a lack of plume effect in instances such as the northwestern Australia, portion of the Central Atlantic and West Indian–Seychelles system (Hopper et al., 1992; Holbrook and Kelemen, 1993; Sheth, 2005; Collier et al., 2008; Nemčok et al., 2010), indicates the existence of multiple controlling factors.

For example, the important control of the extension rate has been documented by 2-D finite-element modeling performed by Van Wijk and colleagues (2001). The decompression melting in their models took place in the head of the upwelling asthenospheric diapir in a roughly 175 km-wide zone, located between the depths of 20 and 50 km. It started about 5 Ma before the continental breakup, being analogical to magmatic activity timing from the Vøring magma-rich margin (see Skogseid et al., 1992). The sensitivity testing indicates that an increase in extension rate results in an increase of the total melt volume and a decrease in the rifting time span (Table 2.2). Furthermore, the importance of syn-rift cooling is indicated by the reduction in the total melt volume at slower extension rates. Calculated melt volumes are roughly comparable to those estimated from transects through the known magma-rich margins (Table 2.3).

Table 2.2. Total melt volumes calculated from 2-D finite-element models testing the effect of the extension rate (Van Wijk et al., 2001).

Extension rate (mmy^{-1})	Mantle temperature (°C)	Total melt volume (km^3)	Time span of rifting (Ma)
6	1333	0	>70
11	1333	±100	40
16	1333	±490	28
20	1333	±640	20
28	1333	±975	16
32	1333	±1400	15
16	1283	±25	30
16	1383	±1550	27

Table 2.3. Total melt volumes estimated from transects through the selected magma-rich margins (calculation made by Van Wijk et al., 2001 from transects of Eldholm and Grue, 1994 on various regions of the North Atlantic province).

Margin	Total melt in km^3 per margin length
Lofoten	900
Møre	1150
Vøring	1550
Hatton Bank	1700

Apart from the role of the extension rate, the modeling reinforces the role of the mantle temperature. The main difference between the models with and without prescribed thermal anomaly is given by the time scales and velocities of diapiric upwelling beneath the rift system (see Huismans, 1999). This is mainly due to the larger buoyancy force and the focusing effect resulting from the thermal anomaly at the base in the central part of the model. Theoretically, it should not matter whether the thermal anomaly is due to earlier uncompensated rift event or plume impingement. The effectiveness of the thermal anomaly during the stage before the lithospheric mantle breakup should be controlled by the presence of mantle lithosphere being able to extract magma, which can enhance the thermal anomaly by heat convection tied to magma ascent on top of the background heat conduction.

The most common magmatic products at magma-rich margins, such as those in the Barents Sea, Norway, southeastern Nova Scotia, Rockall, Orange Basin, and northeastern United States, are wedges composed of lava flows, tuffs, and some sediments (e.g., Hinz and Weber, 1976; Mutter et al., 1982; Austin et al., 1990; Sheridan et al., 1993; Oh et al., 1995; Barton and White, 1997; Skogseid, 2001; Thompson, 2009; Figures 2.15, 2.20). Well data, reflection seismic interpretation, and paleoenvironmental interpretation indicates subaerial to shallow-water conditions for their extrusion (e.g., Eldholm et al., 1989; Keen and Potter, 1995; Jackson et al., 2000; Rosendahl et al., 2005; Dehler, 2010; OETR Association, 2011; Dehler and Welford, 2012).

On top of mentioned wedges, magma-rich margins typically contain coeval dykes, sills, and larger intrusions within the subsided continental crust (e.g., White and McKenzie, 1989; Skogseid, 2001; Figure 2.21). These bodies together with wedges are formed during the late syn-rift, syn-breakup, and, commonly, post-breakup stages of margin development. The earliest oceanic crust is commonly anomalously thick, in comparison to a worldwide average of 7.08 km reported by White (1992), and contains a high-velocity zone (Geoffroy, 2005).

12) The interpretations of a lithospheric zone with high seismic shear velocity in association with this breakup scenario have been made on the African side of the central South Atlantic (see Ritsema and Van Heist, 2000; Sebai et al., 2006; Begg et al., 2009). It was interpreted as connected to the Congo craton (McKenzie and Priestley, 2008; Begg et al., 2009).

However, before we discuss the evidences for magmatism, it is useful to divide the margin into three zones (Huismans and Beaumont, 2008; Figure 2.22):

1) proximal;
2) sag; and
3) distal.

A substantial thermal subsidence during the early post-rift characterizes the sag zone. As mentioned earlier, this subsidence starts at the beginning of the late syn-rift and continues into the post-rift period. The finite-element modeling (Huismans and Beaumont, 2008) indicates that subsidence results from the lithospheric mantle thinning in the sag and distal zones. This thinning is significant enough to cause the decompression melting in the upwelling asthenosphere.

This interpretation provides an interesting comparison with bodies characterized by seismic velocities that suggest magmatic underplating, which were found in magma-rich systems such as the conjugate Camanu-Angolan margins of the South Atlantic, the U.S. margin of the Central Atlantic, and the West Indian and North Atlantic margins (e.g., Kelemen and Holbrook, 1995; Pandey et al., 1995; Miles et al., 1998; Singh, 1999, 2002; Holbrook et al., 2001; Radhakrishna et al., 2002; Krishna et al., 2006; Collier et al., 2008; Huismans and Beaumont, 2008).

The West Indian examples of the potential underplating represent cases with a density of 3,000 kgm^{-3} occurring underneath the Laxmi Ridge and the northwestern portion of the Deccan volcanic province (Miles et al., 1998; Radhakrishna et al., 2002). A basal portion of the continental crust of the Laxmi Basin contains an anomalous velocity layer, where seismic velocities reach values of 7–7.4 kms^{-1}, which can be seen in neither normal continental nor normal oceanic crust (Krishna et al., 2006).

The West Indian example indicates a varying distribution of the potential underplating. While it is present underneath:

1) the Laxmi Ridge (Miles et al., 1998);
2) some portion of the Laxmi Basin (Krishna et al., 2006);
3) the northwestern portion of the Deccan volcanic province (Radhakrishna et al., 2002); and
4) the Cambay-Koyana region of the Indian margin (Kaila et al., 1981, 1990; Kaila and Sain, 1997; Singh and Mall, 1998; Reddy et al., 1999),

it is absent in the highly stretched continental crust underlying the Seychelles Bank, forming a conjugate margin to the Laxmi Ridge and continental crust underlying a larger portion of the Deccan volcanic province (Francis and Shor, 1966; Francis et al., 1966; Krishna et al., 2006).

13) A good natural laboratory for a seafloor spreading center initiation relatively soon after the crustal breakup is the central Red Sea system, which apparently contains oceanic and continental crusts juxtaposed over relatively narrow transition zones (see Egloff et al., 1991; Ligi et al., 2011). Before we discuss these transition zones, it is useful to introduce the Red Sea system. It represents an intermediate stage of the transition from rifting to seafloor spreading (Bonatti, 1985), in the overall setting characterized by extension rates increasing from the north to the south in the Red Sea. The northern region undergoes initial continental breakup (Purser and Bosence, 1998; Cochran, 2005). Here, the positive heat flow anomalies (e.g., Lowell and Genik, 1972; Cochran, 1983, 2005; Bonatti, 1985; Purser and Bosence, 1998; Cochran and Karner, 2007) are currently centered on the depocenters (Buck et al., 1988). Seismic velocities characterizing regions outside of depocenters, representing neither characteristic oceanic nor continental crust values, most likely indicate significantly intruded continental crust (see Cochran, 1983, 2005; Cochran and Martinez, 1988; Gaulier et al., 1988). The southern region undergoes organized

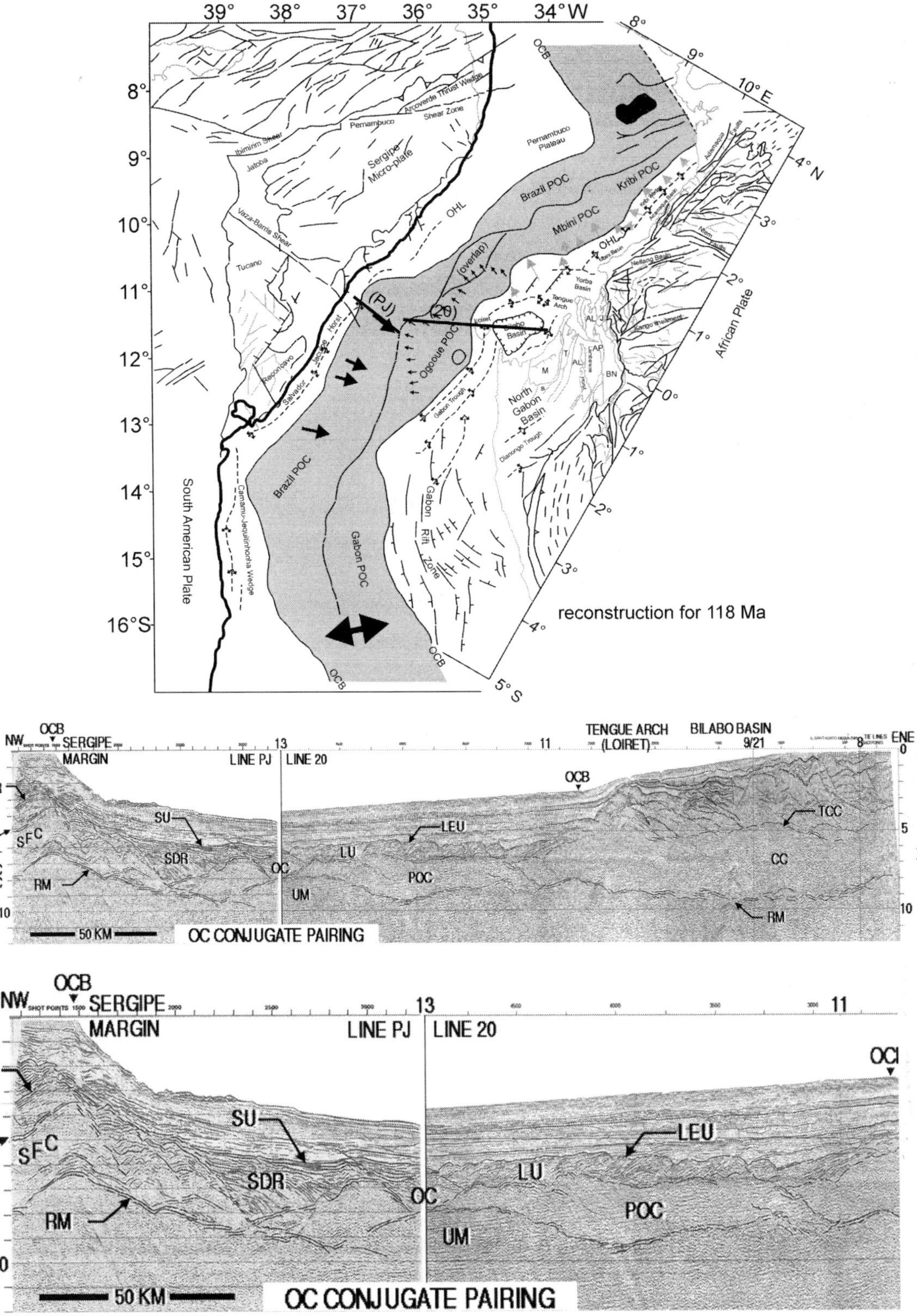

Figure 2.20. Conjugate pairing of the Sergipe and Gabon continental margins, illustrating the occurrence of volcanic rock-dominated wedge (SDR) on the Brazilian side and volcanic rock-dominated Libreville unit (LU) on the African side (Rosendahl et al., 2005). The Libreville unit is imaged as covering the distal continental margin (CC) and the proto-oceanic (POC) and oceanic (OC) crusts. The Brazilian SDR seems to reside on the continental and proto-oceanic crusts. The Brazilian continental margin is represented by São Francisco Craton (SFC). RM – Moho, LEU – top Libreville erosional unconformity, SU – Senonian unconformity (angular, erosional), OCB – ocean–continent boundary, UM – upper mantle, TCC – top continental crust, R – syn-rift sediments.

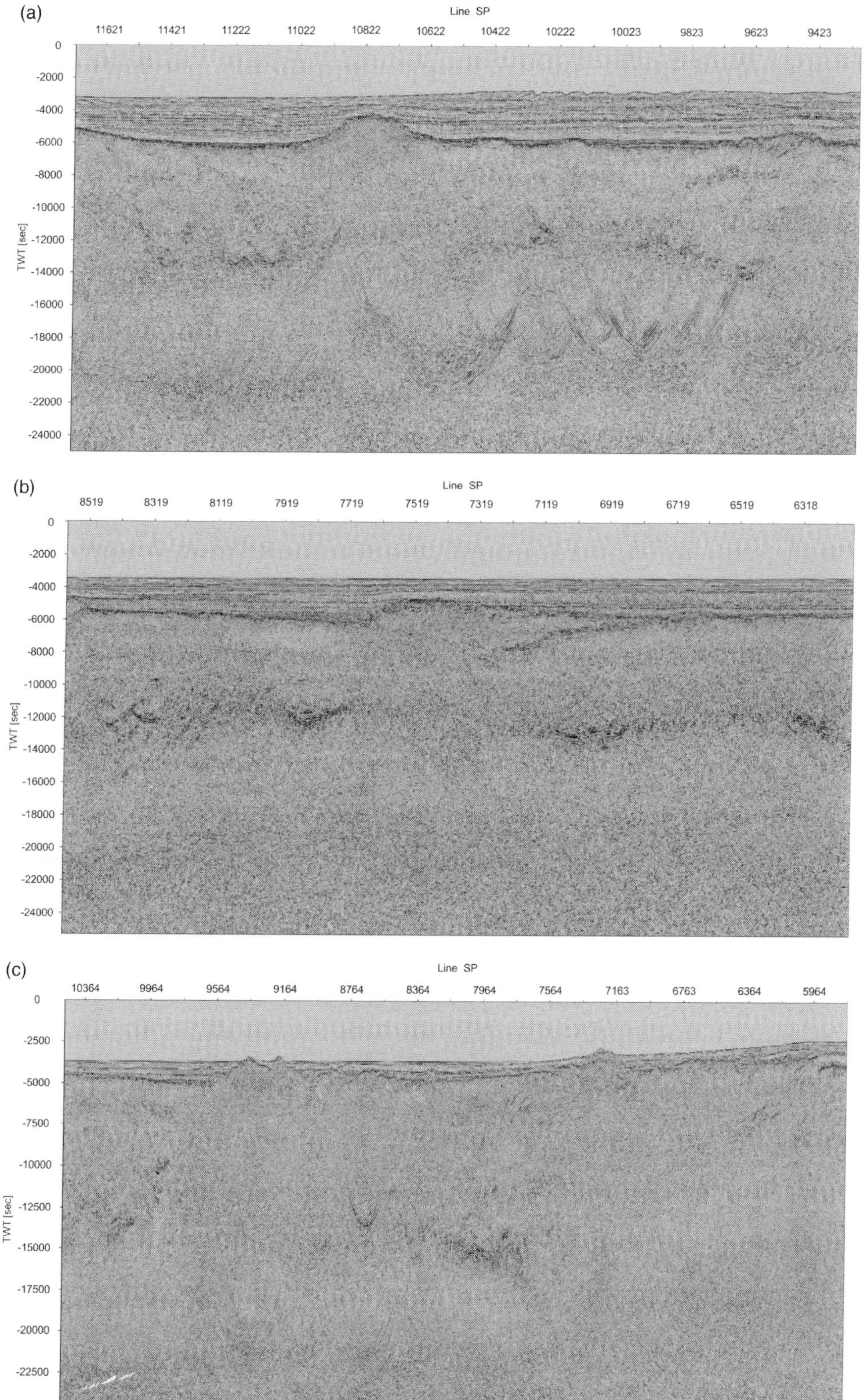

Figure 2.21. (a) Portion of the GXT profile 3000 through the Laxmi Basin. (b) Portion of the GXT profile 4000 through the Laxmi Basin. (c) Portion of the GXT profile 5000 through the Laxmi Basin.

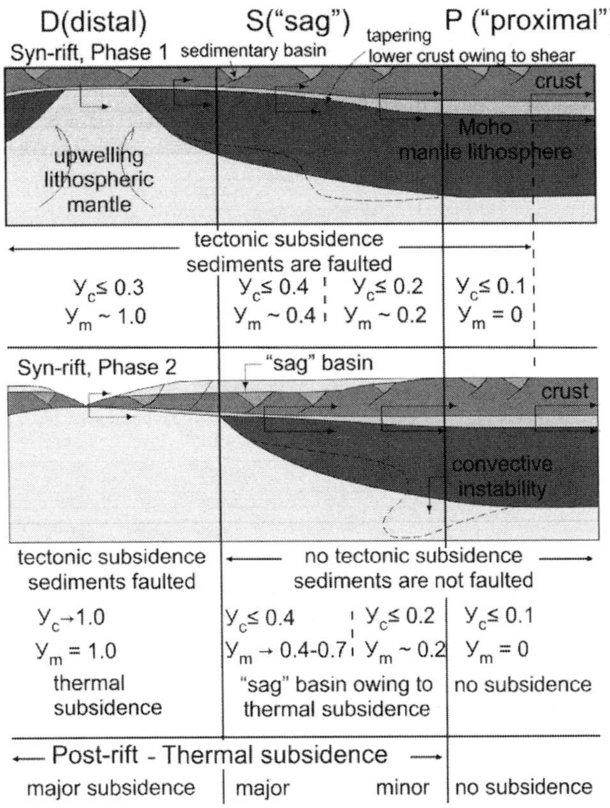

Figure 2.22. Figure summarizing main characteristics of dynamic models classified according to zones proximal (P), sag (S), and distal (D) (Huismans and Beaumont, 2008). γc and γm are crustal and lithospheric mantle attenuation factors. Phase 1 crustal extension is distributed, producing limited attenuation and subsidence; basins are faulted. Mantle lithosphere extends by focused necking and ruptures under D with some attenuation under S. Phase 2 crustal extension migrates to the rift axis, D. Additional faulting is confined to basins in D. Mantle lithosphere is advected laterally. Unfaulted "sag" basins develop where there is cooling and thermal subsidence in zone S; transient uplift in S may occur if mantle lithosphere is further attenuated. Post-rift thermal subsidence correlates with $\gamma m(x)$ and is confined to D and S. Incremental strain at the surface accumulated during each of the stages shown in A, B, and C indicates early deformed syn-rift sediments and late syn-rift sediments are undeformed.

seafloor spreading represented by the development of magnetic stripe anomalies (Cochran, 1983; Bonatti, 1985; Purser and Bosence, 1998). Some of the depocenters in the northernmost portion of the southern region, such as the Tethys Deep, contain magnetic data characterized by distinct axial dipole anomalies indicating that the organized seafloor spreading started as recently as around 0.78 Ma (e.g., Ligi et al., 2008). Individual depocenters along the axis of the divergent system are disconnected. They are separated by shallower areas lacking magnetic stripe anomalies and containing sedimentary cover. Reflection seismic profiles indicate that the cover is deformed (e.g., Bonatti et al., 1984).

A nice example of the narrow transition zone in the central Red Sea is a combined refraction/reflection seismic profile through the oceanic crust of the Suakin Deep and its adjacent continental margins (see Egloff et al., 1991) showing a less than 20 km-wide transition zone. Multibeam sonar bathymetry data of the oceanic crust of the Tethys Deep, which is located about 3° further north, and its adjacent continental margins indicate that this transitional zone contains several volcanic centers (Ligi et al., 2011). The seismic refraction experiments indicate a presence of syn-rift basins on distal margins not very far from the oceanic–continental crustal boundary.

14) The start of organized seafloor spreading relatively soon after crustal breakup in this scenario is indicated by numerical modeling (Huismans and Beaumont, 2011). This also can be implied from a comparison of breakup scenarios with different extension rates (Van Wijk et al., 2001; Nemčok et al., 2012a), which indicate a decrease in the width of the exhumed mantle corridor with the increasing thermal regime of a respective scenario.

Details of the breakup development in this scenario

In this scenario, including two sub-scenarios; the first one with asthenospheric inflow and the second one with cratonic inflow, the ductile lower crust of increased thickness serves as a decoupling horizon between the brittle upper crust and the sub-crustal lithospheric layers (Huismans and Beaumont, 2011; Figures 2.23a–b).

Stretching

The crustal stretching in this scenario affects a rather wide corridor of the brittle upper crust. The examples from the southern Santos and Kwanza margin segments indicate that this corridor is at least 276 km (northern Santos), ca. 300 km (Congo-Angola segments), 510 km (southern Santos) wide (see Contrucci et al., 2004; Moulin et al., 2005; Zalán et al., 2011).

The crustal stretching is coeval with localized necking of the mantle lithosphere (Huismans and Beaumont, 2008, 2011). A sum of significant stretching in the mantle necking zone and a small stretching in the surrounding mantle is comparable with a more homogeneously distributed small amount of stretching in the upper crust affecting a wide corridor. The overlap of these two stretching distributions in layers separated by viscous

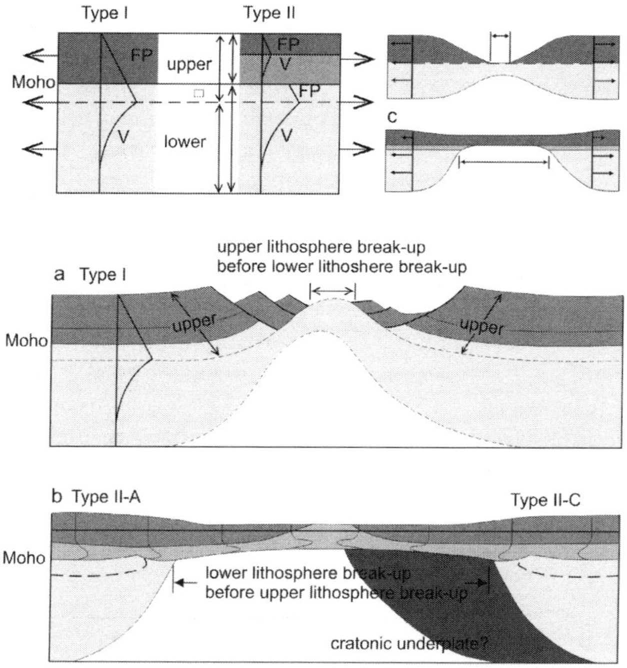

Figure 2.23. Characteristic properties of crust first–mantle second (a) and mantle first–crust second (b) break-up scenarios (Huismans and Beaumont, 2011). Scenarios (a) and (b) are based on observations from the Iberia–Newfoundland conjugate margins and central South Atlantic margins, respectively. See text for further description.

lower crust results in the rupture of the mantle lithosphere, which is overlain by the crust that underwent a distinctively smaller amount of stretching.

The stretching phase continues until the lithospheric mantle ruptures beneath a still moderately extended crust. This results in the juxtaposition of the middle crust and upwelled asthenosphere.

Necking

The crustal necking period starts with mantle lithosphere layers advected away from the zone of the future lithospheric breakup by divergent plate movements. Huismans and Beaumont's (2005, 2008, 2011) models indicate that the crustal stretching transitioning into the necking takes place either above the upwelled hot asthenosphere, which starts to form the oceanic lithosphere upon cooling, or above the buoyant lower cratonic lithosphere, which invades the region between diverging lithospheric mantle layers. The latter may occur in situations where the rifting takes place along the craton boundary. The examples are provided by rifting along the Congo Craton in the central South Atlantic and the Pilbara Craton at the Exmouth Plateau. The chemically controlled buoyancy of such a relatively depleted cratonic lithosphere was discussed earlier.

The interpretations of a lithospheric zone with a high seismic shear velocity have been made on the African side of the central South Atlantic (see Ritsema and van Heist, 2000; Sebai et al., 2006; Begg et al., 2009). It was interpreted as connected to the Congo Craton (McKenzie and Priestley, 2008; Begg et al., 2009). Similar evidence comes from the Exmouth Plateau (Simons et al., 1999; Fishwick et al., 2008; McKenzie and Priestley, 2008). Furthermore, the asthenosphere-sourced magmas from this region indicate their contamination by both continental crust and lithospheric mantle (Halliday et al., 1988; Rankenburg et al., 2005; Regelous et al., 2009). The African continental margins of the central South Atlantic are relatively magma-poor, which Huismans and Beaumont (2011) interpreted as indicating cratonic inflow suppressing asthenospheric decompression melting.

The characteristic feature of numeric models of this breakup scenario (Huismans and Beaumont, 2005, 2008, 2011) is the pressure-driven extrusion of some portion of the viscous lower crust toward the rift axis. However, it needs to be emphasized at this point that there is no natural example of the margin segment that provides direct evidence for the lower crustal flow toward the rift axis. There are indirect evidences, such as it is a logical interpretation that the lower crust on its way to the axis is used for underplating regions of localized upper crustal extension, represented by core complexes observed in several margin segments of the central Southern Atlantic (see Aslanian et al., 2009; Zalán et al., 2011). Models indicate that the crustal necking in this scenario is eventually reached after prolonged crustal stretching.

Both sub-scenarios, the first one with asthenospheric inflow and the second one with cratonic inflow, control shallow depositional environments for the syn-rift event at the rift axis in numerical models (see Huismans and Beaumont, 2011). The main factors causing this situation include the delayed crustal thinning and buoyant asthenospheric or cratonic inflowing material. While the numerical modeling rules out the lithospheric mantle exhumation in the first sub-scenario, the second one may be accompanied by the continental mantle exhumation if the cratonic inflow continues on its way to the rift axis and becomes exposed as the crust ruptures. This can be the case of the unusually narrow corridors of the continental mantle exhumed at the margin contact with oceanic crust in front of the Santos margin segment (see Zalán et al., 2011), having a width varying from 18 to 42 km, which is too narrow in comparison to the narrowest proto-oceanic corridors having a width in the order of 90–100 km for the fastest extension sub-scenarios of the crust-first–mantle-second breakup scenario (see Nemčok et al., 2012a).

Both sub-scenarios of the mantle-first–crust-second scenario can be characterized by the crustal extension diachronously migrating across the margin toward the future crustal breakup location. This can be seen in numerical models (Huismans and Beaumont, 2011) and in several magma-rich segments of the Central Atlantic Rift system (Nemčok et al., 2005b and references therein).

The necking of the crust in magma-rich scenarios is accompanied by the formation of seaward-dipping, volcanic-dominated (SDR) wedges. The wedges are deposited in the catchment areas controlled by syn-magmatic roll-over tectonic flexures controlled by landward-dipping normal faults (e.g., Brooks and Nielsen, 1982; Karson and Brooks, 1999; Geoffroy, 2001; Geoffroy et al., 2001 – onshore studies, Gibbson and Love, 1989; Tard et al., 1991; Eldholm et al., 1995; Planke et al., 2000; Gernigon, 2002; Nemčok et al., 2010 – offshore studies). Based on results of these studies, Geoffroy (2005) pointed out that the development of the seaward-dipping volcanic-dominated wedges is a sudden phenomenon. Furthermore, the oceanward-increasing thickness of these wedges in a system of half-grabens, oceanward-increasing displacement of their controlling normal faults, oceanward younging of the controlling fault initiation, and oceanward younging of the controlling fault activity termination in Baffin Bay, Iceland, and the Gop Rift (see Geoffroy, 2001; Doubre, 2004; Nemčok et al., 2010) indicates a progressive acceleration of the crustal necking process during this relatively short time interval. Apart from controlling normal faults, the SDR wedge development is accommodated by diking, which can locally accommodate as much as 80 percent of the horizontal stretching (Geoffroy, 2005).

As the crustal necking evolves and SDR wedges form, the magma-feeding dyke swarms form separate clusters parallel to the future breakup trajectory, each cluster apparently centered on the local magma reservoir (see Callot et al., 2001; Bromann-Klausen and Larsen, 2002; Doubre, 2004). This centering provides an interesting correlation with the locations of maximum crustal flexure (Geoffroy, 2001; Geoffroy et al., 2001). Data from Isle of Sky (Doubre and Geoffroy, 2003) indicate that some magma reservoirs remain active from the flood basalt stage to crustal breakup stage. Their length is apparently controlled by the thermal regime of the stretched lithosphere (Bosworth, 1985; Ebinger et al., 1999).

As can be seen on reflection seismic images across different parts of the Laxmi Basin evolving from a rift system into the seafloor spreading system (Nemčok et al., 2010), the amount of magmatic reworking of the stretching crust prior to breakup comes to the point that we cannot see the SDR wedge geometries with their controlling faults anymore and the seismic image shows just a heavily magmatized transitional crust inside narrow zones sandwiching the breakup trajectory. This is the evolution stage when all individual magmatic reservoirs try to link together and control the future breakup trajectory (Callot et al., 2002). The trajectory can have any geometry ranging from relatively straight to zigzag ones as it is in Greenland (e.g., Geoffroy et al., 2001; Callot, 2002).

Interpreted as associated with remnant magnetization of SDR wedges and intrusions, the heavily magmatized, thinned continental crust frequently records magnetic anomalies relatively similar to those at oceanic ridges (Schlindwein and Jokat, 1999; Perrot et al., 2003 – eastern Greenland, Behn and Lin, 2000 – U.S. Atlantic margin). It is the synthetic interpretation of gravity, magnetic, and reflection seismic data, such as that done in the northeastern Atlantic (Geoffroy, 2005) and offshore South Africa (Thompson, 2009), that allows us to determine the continental character of this transitional crust that thermodynamically behaves more like an oceanic one.

Onset of seafloor spreading

Simulated by analog material experiments (e.g., Callot et al., 2002), the described necking synchronous with increasing magmatization of the thinning continental crust eventually ends up with linking the system of local magmatic reservoirs and reaching the continental breakup. Along the breakup zone, the active and melting asthenosphere, which is flowing upward, reaches the surface (see also Geoffroy, 2005). This event is understood as associated with magma invasion into the breakup zone. This can be implied from the existence of high-velocity (underplated) bodies, which have been identified as underlying both distal margin and adjacent initial oceanic crust in regions such as offshore West India (Miles et al., 1998; Radhakrishna et al., 2002; Krishna et al., 2006; Nemčok et al., 2010) and most likely magma-related outer highs such as the ones along the western margin of the Laxmi Ridge, offshore West India (Nemčok et al., 2010).

It has to be noted that no well penetrations of the outer highs from magma-rich margins are available at the moment. In fact, ODP Sites 1070 and 1277, from the magma-poor margin related to the Iberia-Newfoundland breakup, provide the only well data on any outer high. Being characteristic for magma-poor scenarios, even they document an important tectono-magmatic event

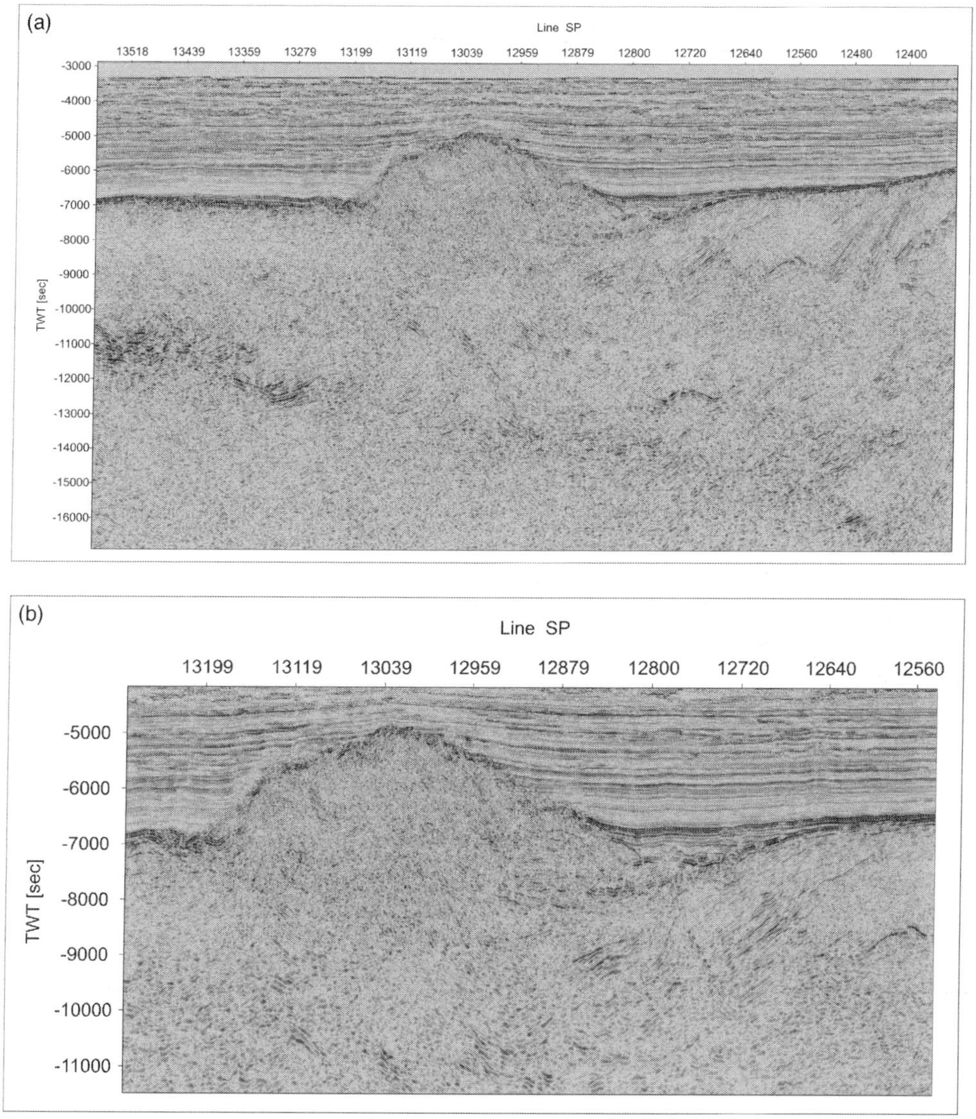

Figure 2.24. (a) Portion of the GXT profile 3000 through the Arabian Abyssal Plain (Nemčok et al., 2010). (b) Detail of the outer high is shown in (a). Note the large volcano located on the continental crust of the Laxmi Ridge right next to the boundary between the oceanic and continental crusts. Its load flexed the continental margin crust, developing a flexural basin filled with volcanic products and sediments. See text for further explanation.

(Perón-Pinvidic et al., 2007), recorded by the turbidites and debris flow interlayered with basalt flows resting on the uplifted exhumed continental mantle (see Shillington et al., 2004; Van Avendonk et al., 2006). Perón-Pinvidic and Manatschal (2010) interpret this event as a major magmatic event resulting in the underplating of a previously exhumed continental mantle.

Lacking the well control and assuming that the thermal regime is warmer than that interpreted for the Iberia-Newfoundland breakup, we interpret the seismically imaged outer highs of the western margin of the Laxmi Ridge as volcanic complexes, representing a significant magmatic event during the mantle-first–crust-second breakup scenario. This is based on the seismic images documenting that the complexes are younger than the distal margin. They flex it down by a material, which is, as indicated by reflector tracing, represented by a mixture of the volcanic products with the results of background deposition (Figure 2.24). Both Free-Air and Bouguer gravity anomaly maps associate respective volcanic complexes with significant, although small, positive anomalies on a moderately positive image of the oceanic crust and a negative image of the continental ribbon formed by the Laxmi Ridge (Nemčok et al., 2010).

3 Determination of unstretched continental, thinned continental, proto-oceanic, and oceanic crustal boundaries

Introduction

Crustal type determination is the approach used at the beginning of the exploration for hydrocarbons. It results in a distribution map of continental, oceanic, and proto-oceanic crust types, each of which hosts different petroleum systems due to different source rocks located on them. During this process, it is useful to know the strengths and weaknesses of the constraints we use for determination and how detailed results can be expected from different approaches.

Methods of crustal boundary determination

Methods using shelf break, magnetic stripe, heat flow, and pre-drift lithology distribution data

The simplest plate reconstructions use shelf break as a proxy for the continent–oceanic crust boundary. While they are useful as a first quick approach, they bring inherited errors (Figure 3.1) even at a regional scale.

Quick determination of the crustal type distributions can be done using digital maps of the oceanic crust age, based on magnetic stripe anomaly data (e.g., Müller et al., 1997; Figure 3.2). While these maps are accurate for oceanic crust formed by organized seafloor spreading, their regions near continental crust may include errors due to extrapolations made for the location of the boundary between continental and oceanic crusts in magnetically quiet zones (Figure 3.2). Magnetically quiet zones along the continent–ocean boundaries can be potentially explained as caused by (1) very slow extension rate; (2) the existence of superchrons instead of zones with magnetic reversals; or (3) the occurrence of the proto-oceanic crust zones formed by unroofed continental mantle (e.g., Cochran, 1982; Ebinger and Casey, 2001; Manatschal, 2004 and references therein; Rosendahl et al., 2005 and references therein). Further problems with continent–ocean boundary determination are caused by the continental crust heavily intruded by mafic bodies or the proto-oceanic crust covered by large panels of continental upper crust (e.g., Watts, 1978; Meyers et al., 1996; Manatschal, 2004 and references therein; Rosendahl et al., 2005 and references therein).

In settings with older oceanic crust, the heat flow data provide a rough way of distinguishing between oceanic and continental crusts. The reason is that, apart from old continental shields, the oceanic crust is colder than the continental crust (see Allen and Allen, 1990). It lacks the heat production characterizing the continental crust, which is an extra heat source on top of the heat coming from the mantle (e.g., Čermák and Rybach, 1982; Čermák and Haenel, 1988). For illustration, the heat production of typical upper continental crust is about 2.5 μWm^{-3} (Čermák and Rybach, 1982; Cloetingh et al., 1995). The typical lower continental crust has a heat production of 0.25 μWm^{-3} (Cloetingh et al., 1995). The continental mantle, representing the proto-oceanic crust (see Manatschal, 2004), can be characterized by the heat production of a depleted ultrabasic zone of 0.01 μWm^{-3} or by the heat production of a pyrolite zone of 0.084 μWm^{-3} (Čermák and Rybach, 1982), which are taken as 0 in large-scale thermal calculations (e.g., Cloetingh et al., 1995).

The known pre-drift lithology can be used for the determination of the oceanic and continental crust distributions. For example, the outer evaporite distribution margin of the pre-drift evaporites in the Central Atlantic lies either exactly at the ocean–continent boundary or it is slid down in a local gravity glide and lies on the oceanic crust.

Methods using total magnetic field data

Total magnetic field data are also useful for the crustal boundary determination. This is due to the fact that the overall magnetic susceptibility of rocks is roughly equivalent to the susceptibility of their magnetic minerals, times the percentage of these minerals, divided by

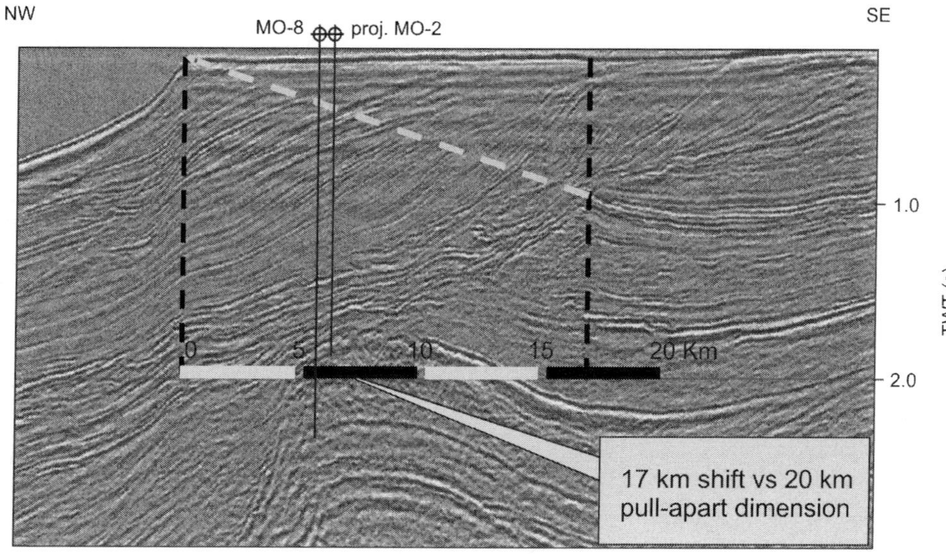

Figure 3.1. Reasons for a misfit of some reconstructions – problem of shelf break shifted basinward for younger stratigraphies (seismic profile from Fusion Oil & Gas, 2002). Note a shelf break shift of about 17 km because of the progradation. Compared to an average width of a pull-apart basin from the offshore Morocco or Equatorial Atlantic margins, which is about 20 km, the mistake approaching 20 km can result in ignoring a whole petroleum system based on the source rock associated with such a pull-apart basin.

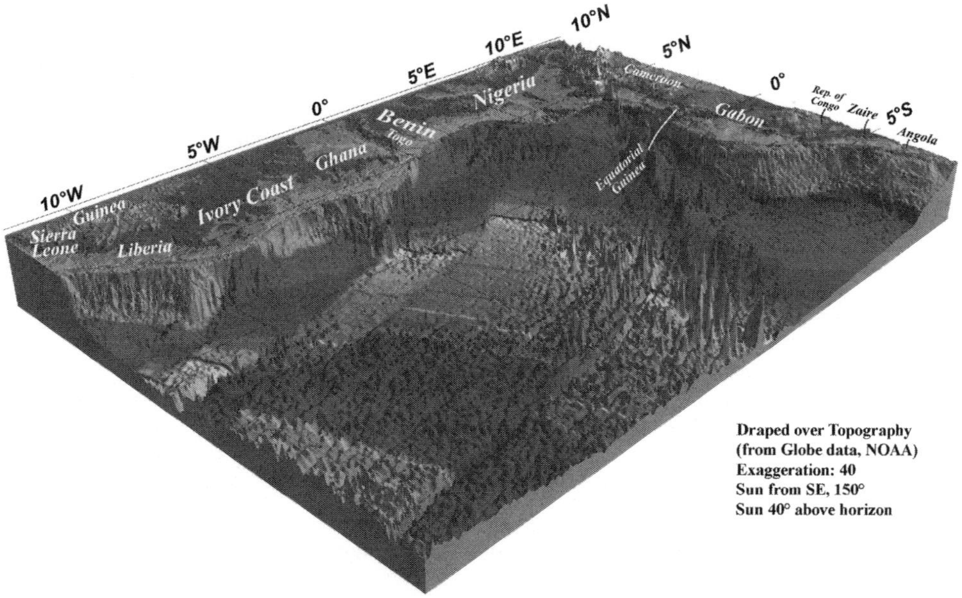

Figure 3.2. Age of the oceanic crust in 5 Ma resolution (Müller et al., 1997) draped over ocean floor and continent topography maps. Note the oceanic crust draped over clear continental crust in small areas in offshore Côte d'Ivoire and Benin.

100, and it is ultramafic and mafic rocks that are rich in magnetic minerals. Peridotites, basalts, and gabbros, most typical for oceanic and proto-oceanic crusts, have magnetic susceptibilities that are at least one magnitude larger than the susceptibilities of diorite and granite, which are typical for continental crust (Table 3.1). Furthermore, the proto-oceanic crust formed by rocks such as peridotite should be more magnetized than oceanic crust containing gabbro intrusions and basaltic layer (Table 3.1).

Because the total magnetic field anomaly is calculated by subtracting the magnitude of the ambient

Table 3.1. Typical magnetic susceptibilities of common Earth minerals and rocks (Lillie, 1999).

Material	Magnetic susceptibility
Magnetite	$10{,}000 \times 10^{-5}$
Peridotite	500×10^{-5}
Basalt/gabbro	200×10^{-5}
Diorite	20×10^{-5}
Sandstone	10×10^{-5}
Granite	1×10^{-5}
Salt	-1×10^{-5}

magnetic field from the magnitude of the total magnetic field (e.g., Lillie, 1999):

$$\Delta F = F - F_{amb}, \quad (3\text{-}1)$$

the same magnetic body would have different anomaly at different locations across the globe (Figure 3.3; Lillie, 1999).

The most prominent magnetic anomalies inside the oceanic crust are located along oceanic fracture zones (Cochran, 1973; Fainstein et al., 1975; Cande and Rabinowicz, 1979). It is due to the location of large quantities of serpentinized peridotites from the upper mantle along them (Thompson and Melson, 1972; Bonatti, 1976, 1978). Landward ends of these anomalies are typically used for the determination of the oceanic–continental crust boundary (e.g., Karner, 2000). Extension of oceanic fracture zones toward the continental margin is usually characterized by an isometric magnetic pattern, consisting of a mosaic of magnetic anomalies with complex arrangement, as can be seen in the Equatorial Atlantic (De Azevedo, 1991). Furthermore, the magnetic signal of the continental margin is usually modified by thick prograding sedimentary wedges (e.g., Chang et al., 1992; Karner and Driscoll, 1999). Another problem can be added to the determination of the fracture zone ends by the presence of a magnetically quiet zone around the ends of oceanic fracture zones (e.g., Karner, 2000). The ends of magnetic anomalies associated with oceanic fracture zones are apparently not a clear criterion for the continental–oceanic crust boundary because of the potential presence of basic intrusions inside the adjacent continental crust, controlled by either breakup processes or hot-spot effects (see Nemčok et al., 2005b; Figure 3.4).

Methods using gravity data

Free-air gravity data

Free-air gravity anomaly data are computed by correcting the measured gravity for the latitude and elevation of the measurement location relative to mean sea level. Because the free-air field is strongly influenced by changes in relief, it is most useful where the topography is flat.

The free-air gravity anomaly at the continent–ocean boundary can be described using a semi-infinite slab model at regions in isostatic equilibrium (see Lillie, 1999; Figure 3.5). Figure 3.5 illustrates a thin oceanic crust at a passive margin, which is underlain by the mantle at the same depth as approximately the middle crust of the adjacent continent. The mass excess of the mantle exerts a force that pulls the oceanic crust downward. By the Airy model, the resulting ocean basin subsides until it contains enough water so that the region is in isostatic equilibrium. The water deepening oceanward represents a mass deficit, which is a function of the water column multiplied by the density difference at the upper crustal level (Figure 3.6; Lillie, 1999). The mantle below the oceanic crust represents a mass excess, which is the function of the depth difference between oceanic and continental crusts multiplied by the density difference between the mantle and continental lower crust. The resulting free-air gravity anomaly for this model is the sum of described mass deficit and excess curves (Figure 3.6; Lillie, 1999). It is characterized by almost zero values for continent and ocean interiors, a maximum for the continent near the crustal boundary, and a minimum for the ocean near the boundary.

It needs to be emphasized that the described anomaly characterizes a very simple model where the densities of oceanic and continental crusts were represented by single values, which were made similar, and the role of sediment loading was ignored.

The role of sediment loading can be understood from an offshore Gabon study (Watts and Stewart, 1998), where the elastic behavior of the oceanic crust is constrained by the estimated total elastic thickness of 20 km. The distribution of sediments above the ocean–continent transition is determined from the reflection seismic profile. Each rock unit interpreted in the profile is flexurally back-stripped using a density inferred from the seismically derived stacking and interval velocities. This procedure allows for the separating of the sediment cover from the basement, which was given the depth it would have in the absence of the cover, in order to yield the total subsidence and uplift of the margin.

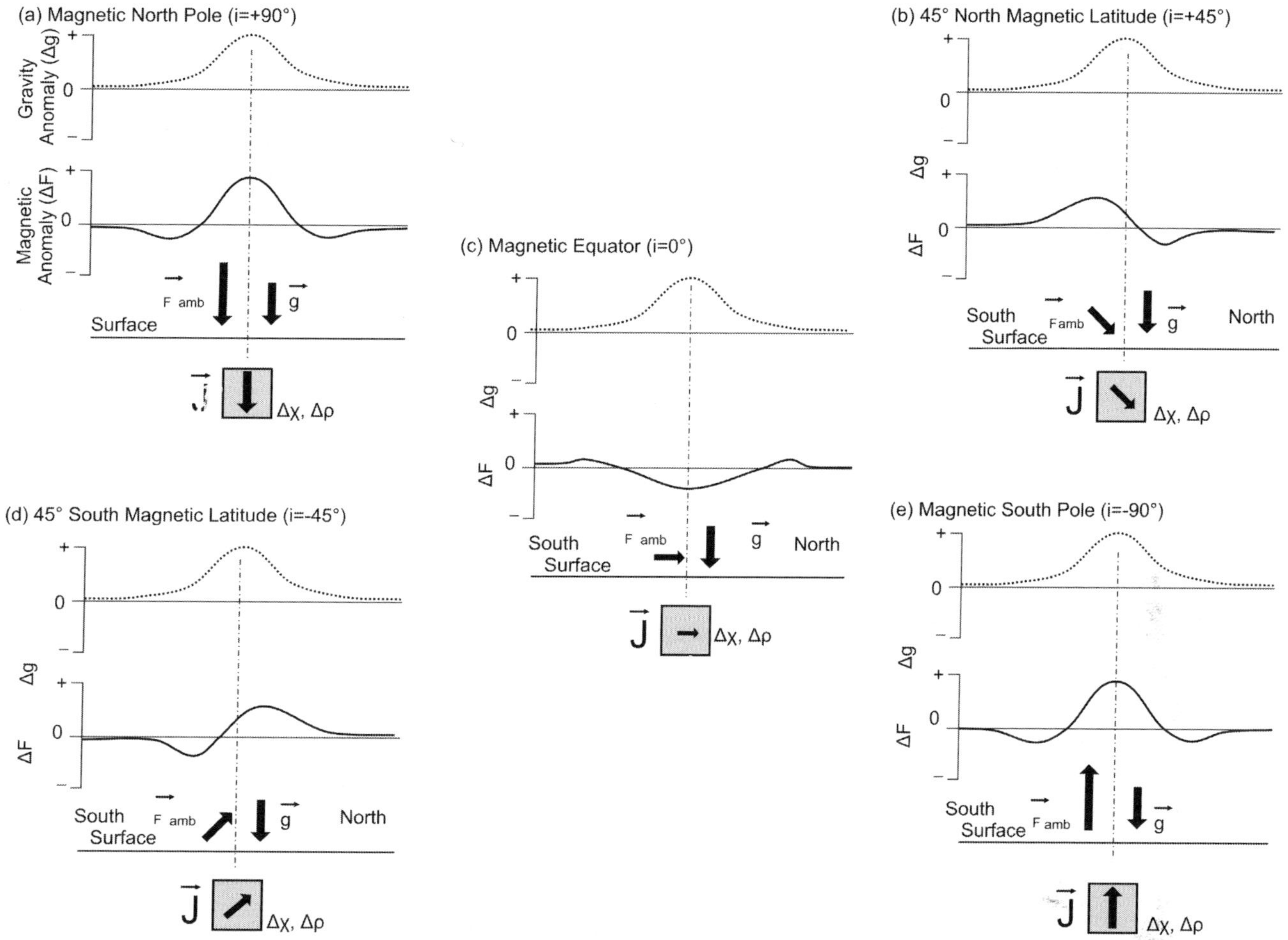

Figure 3.3. Gravity and magnetic anomalies from the same body at different magnetic latitudes (Lillie, 1999). At each latitude, the gravity anomaly, Δg, resulting from the body of density contrast, Δρ, is the same. For the same body, with magnetic susceptibility contrast, Δχ, the magnetic anomaly, ΔF, varies. At the magnetic north and south poles (a) and (e), the anomaly is a central high with flanking lows. A high-magnitude ambient field (F_{amb} of about 60,000 nT) induces high-magnitude magnetization, J, resulting in high-amplitude magnetic anomalies, ΔF. At the magnetic equator (c), the induced field opposes the ambient field at the surface directly over the body, leading to negative magnetic anomalies. F_{amb} of about 30,000 nT leads to low-amplitude ΔF. At intermediate latitudes (b) and (d), the anomaly is asymmetric, with intermediate amplitude ΔF.

The amount of crustal thinning required to explain the total subsidence and uplift is computed assuming Airy isostasy from (Watts and Stewart, 1998):

$$\beta = [t_c(\rho_m - \rho_c)]/[t_c(\rho_m - \rho_c) - Y(\rho_m - \rho_w)], \qquad (3\text{-}2)$$

where β is the amount of crustal thinning, t_c is the initial crustal thickness, Y is the total subsidence and uplift, and ρ_m, ρ_c, and ρ_w are densities of the mantle, crust, and water.

β is used to calculate the gravity anomaly associated with rifting (Figure 3.7; Watts and Stewart, 1998). The difference between Y and the present-day water depth helps to estimate the sediment load. The difference between Y and the present-day basement provides the amount of the sediment-controlled flexure. The sum of the excess mass due to the sediment loading with deficit mass due to flexure of the basement and underlying Moho provides the gravity anomaly associated with sediment cover (Figure 3.7; Watts and Stewart, 1998). The calculation is completed by the sensitivity analysis testing the behavior of the weak and strong oceanic crusts represented by the total elastic thicknesses of 5 and 35 km, respectively. The comparison of rifting, sediment-related, and total gravity curves with the observed one illustrates an important role of the

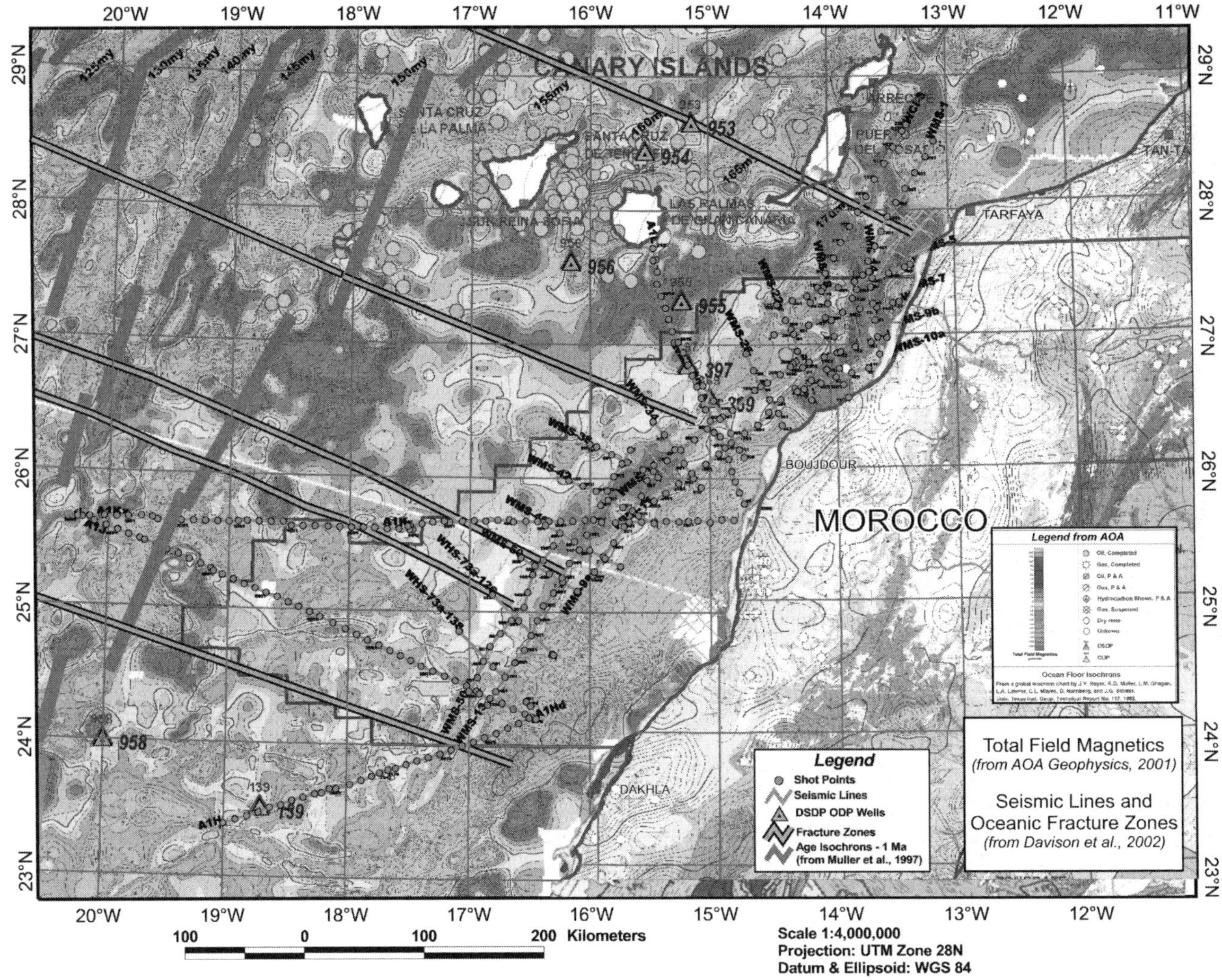

Figure 3.4. Total magnetic field map of Morocco (Nemčok et al., 2005b). The seismic grid used in this study and oceanic fractures are shown. Raw offshore magnetic data are from Getech, Canadian GSC, and Kerr-McGee Oil and Gas Corporation, calculated offshore magnetic data are from AOA Geophysics, Inc. (2001), and the onshore geological map is from Choubert et al. (1988). Note that the crustal boundary can be interpreted in the Dakhla-Boujdour segment of the margin. The presence of magnetic intrusive and effusive bodies of the Canary Island magmatic system makes this interpretation impossible in the Boujdour-Tarfaya segment. In this case, magnetic data have to be correlated with gravity data to make the interpretation.

syn-rift and passive margin sediments in modifying the gravity anomaly illustrated earlier in Figures 3.5 and 3.6. It documents the fact that the anomaly along the passive margin changes its character as it passes from sediment-starved segments through normal-sediment segments to segments affected by the sediment input from large rivers.

The free-air gravity map from the northern Brazilian continental margin (De Azevedo, 1991) illustrates how, from the ocean toward the shelf, anomalies increase up to +85 mGal, and then decrease toward the shore. This decrease of gravity values is interrupted along extensional margin segments by a broad positive anomaly (+30 mGal) over the shelf, which can be sometimes particularly wide. This anomaly cannot be assigned to the gravity effect of any bathymetric relief and is, therefore, related to mass distribution within the crust (Mello and Bender, 1988). Free anomalies can also form a characteristic belt of conjugate positive (+75 mGal) and negative (−30 mGal) anomalies caused by the combined influence of crustal boundary and thick sediment effects. Similarly to Bouguer anomalies, free-air anomalies related to the continental–oceanic crust boundary along extensional and transform margin segments look different. Northwest–southeast striking extensional and east–west striking transform margin segments have smaller and higher horizontal gradients, respectively. Each fan in front of the large river is indicated by a large positive anomaly of up to +45 mGal. Fans of the Amazon, Niger, Congo, and Ogooue Rivers (e.g., De Azevedo, 1991; Dailly, 2000) serve as typical examples but smaller sediment bodies are also imaged. In the oceanic domain, the free-air gravity field is characterized by:

1) gravity high with amplitude from +25 to more than +75 mGal near the shelf break;
2) anomalies directly related to bathymetric relief of the ocean floor (positive and negative over major ridges and trenches inside oceanic fracture zones); and
3) steep gradients at boundaries of ridges.

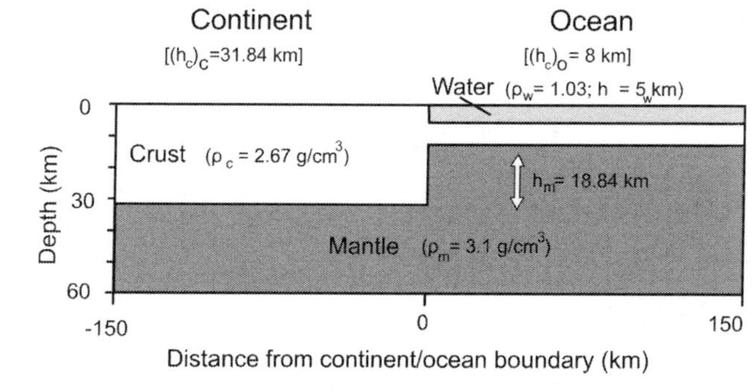

Figure 3.5. Airy isostatic model of the transition from the thick continental to thin oceanic crust at a passive continental margin (Lillie, 1999). Densities of crust and mantle are simplified so that reasonable contrasts result for the water versus upper continental crust (-1.64 gcm^{-3}) and the mantle versus lower continental crust ($+0.43$ gcm^{-3}). See text for definition of variables.

Because of being dominated by near-field density contrasts, a free-air gravity map is useful in defining the general bathymetry and fabric of both thinned continental and oceanic crusts (e.g., Cande et al., 1988). Fracture zone trends, the "edge-effect" anomaly of the shelf break, and hot-spot traces can be usually unambiguously identified (Karner, 2000).

Rosendahl and colleagues (2005) discuss how the maps of free-air, Bouguer, and isostatic gravity data for the oceanic crust indicate flow lines of oceanic crust generation. The flow lines are evident in the northern South Atlantic despite the overprinting by the younger Cameroon Volcanic Zone (Meyers and Rosendahl, 1991; Meyers et al., 1998). The lines can be traced from the external boundary of the oceanic crust to the Mid-Atlantic Ridge (see Sandwell and Smith, 1997; Smith and Sandwell, 1997). As seismic imaging indicates, most of the gravimetric flow lines correspond to crustal offsets or sediment-filled troughs associated

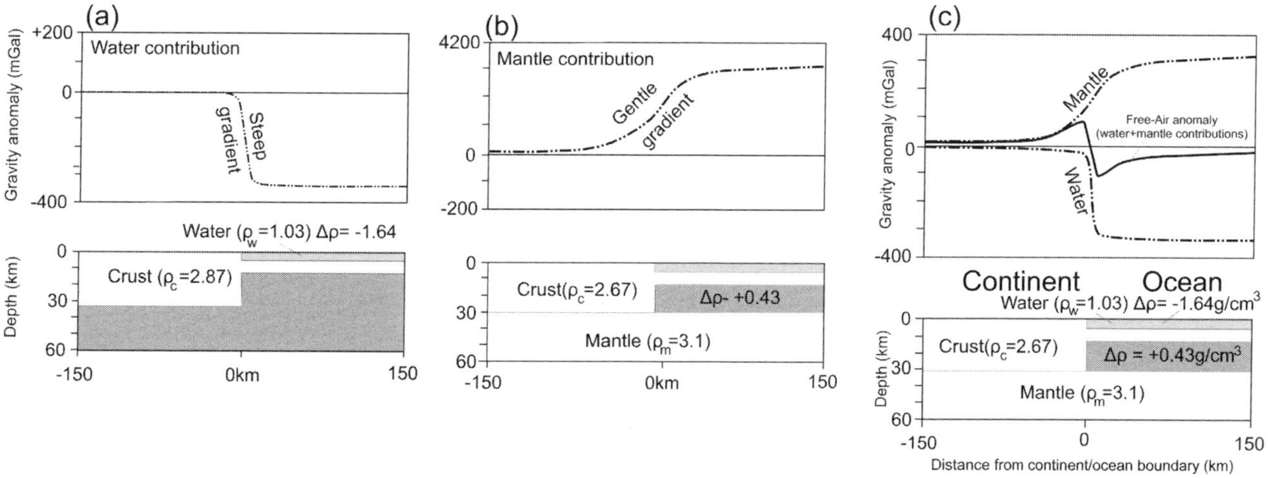

Figure 3.6. The main gravity contributions at a passive continental margin have equal amplitude but different gradient (Lillie, 1999). (a) The water effect is shallow, causing an abrupt change (steep gradient). (b) The extra mantle beneath the oceanic crust is a deeper effect, giving a less abrupt change in gravity (gentle gradient). (c) The free-air gravity anomaly at a passive continental margin is a positive/negative "edge effect," due to the summing of contributions that have equal amplitudes but different gradients.

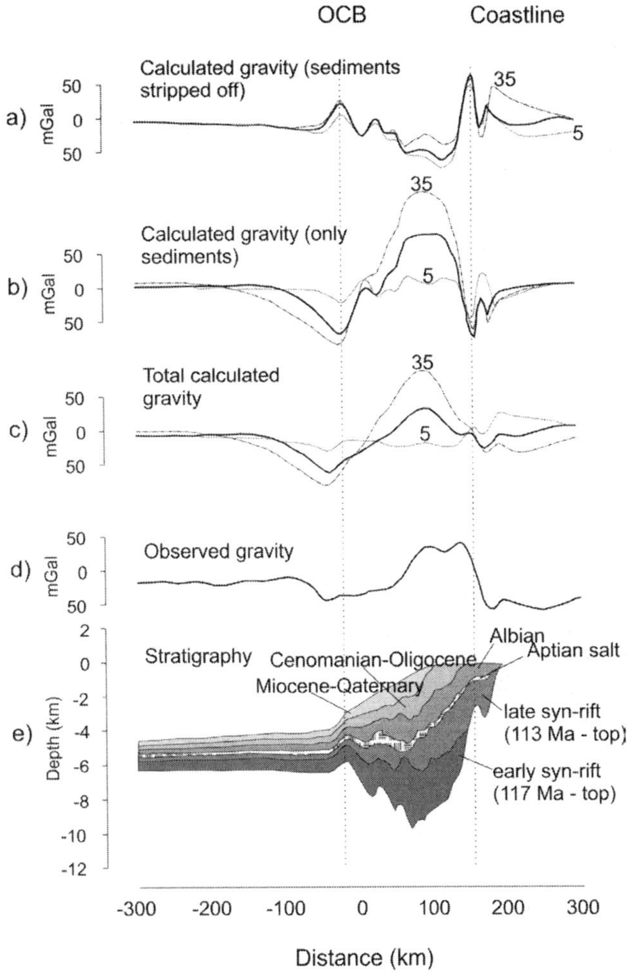

Figure 3.7. Gravity modeling along a section in offshore Gabon made for total elastic thickness of oceanic crust; Te = 20 km, Te = 5 and 35 km for comparison (Watts and Stewart, 1998). See text for explanation.

with oceanic fracture zones. Larger crustal offsets are associated with minor magnetic anomalies.

The use of free-air gravity data from the northern Brazilian margin (Nemčok et al., 2004b) illustrates a probabilistic approach in determining the continent–oceanic crustal boundary, using the Arc GIS project (Figure 3.8). The interpreted free-air data are stored in a separate Arc GIS layer, which allows their comparison with other interpreted data layers.

Normal or hill-shaded free-air anomaly data are used for the determination of oceanic fracture zones (Figure 3.8), interpreted as 3-D structures containing ridges and troughs. Coming from ocean to continent, the oceanic crust is indicated by the presence of oceanic fracture zones. Tracking the São Paulo oceanic fracture zone from east to west allows us to compare its free-air anomaly character in the flat region underlain by oceanic crust with the region characterized by complex topography, which collects a large amount of sediment transported by the Amazon River (Figure 3.8). As the São Paulo fracture zone extends westward underneath the Amazon fan, its positive anomaly progressively disappears. The last sign of it provides the westernmost evidence for the oceanic crust extent.

Coming from continent to ocean, the continental crust in front of the shoreline is characterized by a system of north-northeast–south-southwest trending positive and negative anomalies, imaging canyons, ridges, and elongate sediment bodies. Tracing these anomalies from south-southwest to north-northeast comes to the point where they disappear, providing the north-northeast–most evidence for the continental crust extent. Thus, instead of the exact oceanic–continental crust boundary, we have a probability zone. Finishing the synthetic Arc GIS approach, various data sets are interpreted and compared, helping to narrow down this probability zone.

As mentioned earlier, crustal boundaries at extensional and strike-slip margin segments have different gravity anomalies. The free-air gravity map of extensional example, the Nova Scotia–Georges Bank margin, contains a continuous positive anomaly behind the oceanic–continental boundary that is about 2,200 km long and 150 km wide (Figure 3.9). The strike-slip example, the southern Morocco margin, shown a bit later, contains a discontinuous system of positive anomalies; the average lengths and widths are about 100 and 50 km.

Bouguer anomaly gravity data

The simple Bouguer anomaly is calculated by first removing the effect of a rock slab between the measurement point and mean sea level. The slab for onshore and offshore settings has usually assumed densities of 2.67 and 2.2 gcm^{-3}. For the simple continental margin model (Figure 3.10; Lillie, 1999), the Bouguer anomaly results from correcting the water mass deficit to approximate that of the upper crust. The anomaly has nearly zero values at the continent interior. It then mimics the geometry of the Moho surface by having progressively larger values as the top of the mantle shallows oceanward. It then somehow mimics the bathymetry by increasing the values with the increase of the water column.

The broad regions of positive gravity anomalies in Bouguer maps associated with the thickened oceanic crust tend to obscure the location of the oceanic fracture zones that are obvious on the free-air gravity maps (Karner, 2000). Because it highlights density contrasts within the crust, the Bouguer anomaly map tends

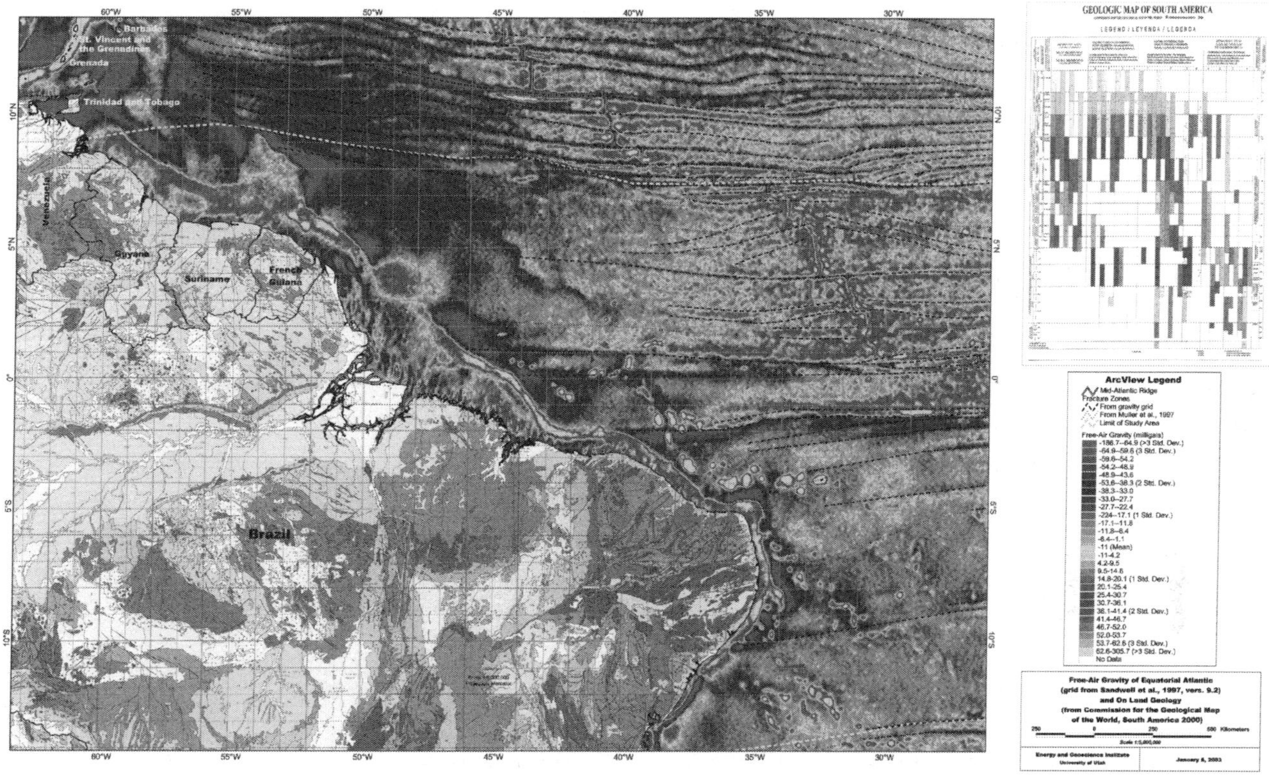

Figure 3.8. Interpreted free-air gravity data on the oceanic crust to the north of Brazil (data from Commission for the Geological Map of the World, South America, 2000; Sandwell, grid version 9.1). Medium gray and dark gray – ridge and trough associated with oceanic fracture zone, light gray – volcanic mount not associated with oceanic fracture zone.

to accentuate the location of sedimentary basins and flexurally compensated features such as rift shoulders (Karner, 2000). Thus, the most prominent features of Bouguer data should be:

1) a negative-positive gravity gradient defining the ocean–continent boundary;
2) positive anomalies corresponding to thickened oceanic crust and offshore rift shoulder zones; and
3) negative anomalies that correlate with the location of rift depocenters or sub-basins and oceanic fracture zones.

Bouguer gravity data calculated from the satellite-derived free-air gravity data help to highlight geologic features such as the oceanic fracture zones and major trend changes (Karner, 2000). Its derivative, crustal Bouguer gravity data (see Karner, 2000 for the method description), have proven useful in highlighting the rift structure of the oceanic margin and the location of the continent–ocean boundary in West Africa and Northwest Australia (Karner et al., 1997; Driscoll and Karner, 1998; Karner, 2000).

Because the Bouguer anomaly method subtracts mass from areas above sea level, large compensated topographic highs may appear as gravity lows, similar to basins. Such cases have been documented from the Cameroon-Gabon margin segments (Rosendahl et al., 2005). On the contrary, the Bouguer anomaly gravity map of the northern Brazilian margin (De Azevedo, 1991) contains negative anomalies, which correlate with major sedimentary depocenters. Its gravity highs correspond to 1) shallow basement areas over the shelf and onshore and 2) thinned continental crust oceanward from the shelf. Bouguer anomalies passing from continent to ocean increase from values of −85 to 0 mGal along the coast to values of +100–140 mGal on the continental rise, forming a several hundred km-wide belt of large horizontal gradients. Because the Equatorial Atlantic Ocean was initiated as a system of major strike-slip fault zones and a series of pull-aparts, where spreading centers developed first, the mentioned belt of horizontal gradients looks different at different margin segments. Extensional margin segments are characterized by a wide zone with a relatively small horizontal gradient. Strike-slip margin segments are characterized

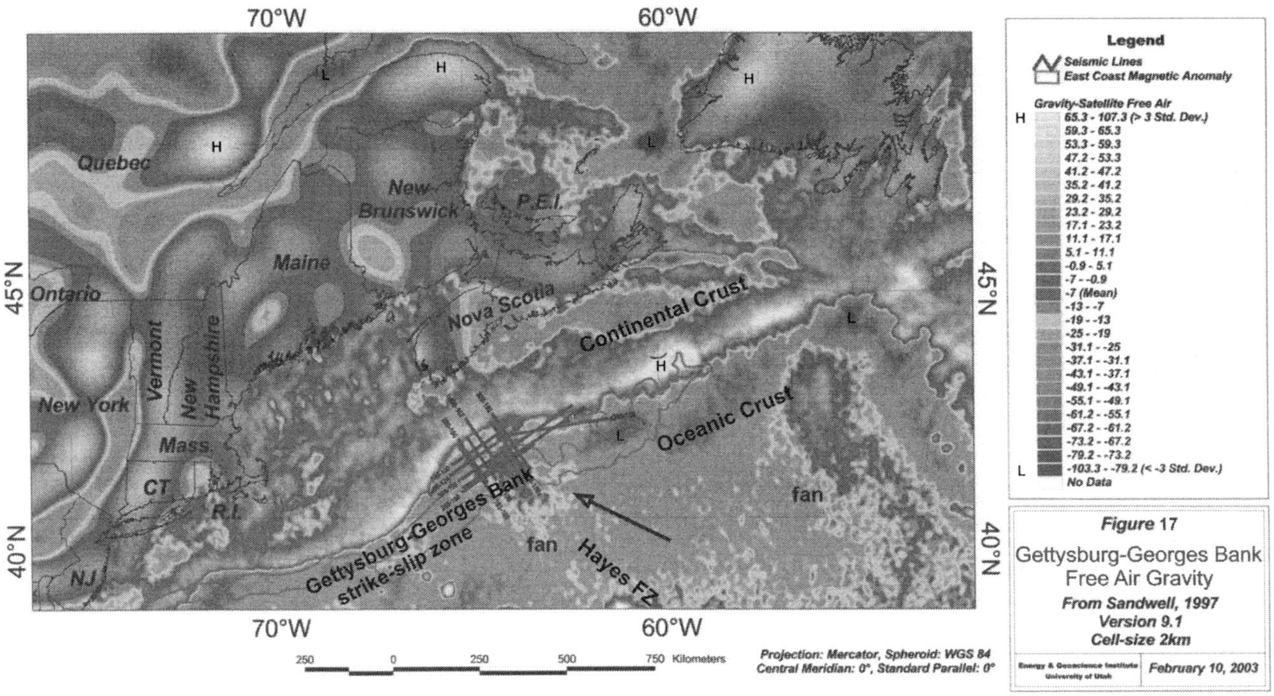

Figure 3.9. Free-air gravity map of the Nova Scotia–Georges Bank area (Nemčok et al., 2005b). The seismic grid used in this study is shown. A Quaternary submarine fan is imaged close to the seismic grid and is partly crossed by profiles of the grid. A second Quaternary fan is imaged southeast of the Laurentian Channel on the eastern side of the map. Data are from Sandwell and Smith (1997). H – gravity high, L – gravity low.

by a very narrow zone with an extremely large horizontal gradient.

Bouguer gravity anomaly map from the African side of the Equatorial Atlantic (AOA Geophysics, 2001 in Nemčok et al., 2004b) allows us to correlate Bouguer gravity imaging with crustal architecture, which is different at margin segments that originated by normal, oblique-slip and strike-slip faulting–controlled upper crustal breakup. The correlation shows that (Figure 3.11):

1) strike-slip continental margin segments with narrow or missing crustal stretching zone, characterized by steep underwater cliffs and rather narrow width of the oceanic–continental crust transition, coincide with steep Bouguer anomaly gradients;
2) oblique-slip continental margin segments with about a 30–100 km-wide crustal stretching zone, characterized by intermediately steep slopes, coincide with intermediate Bouguer anomaly gradients; and
3) normal fault continental margin segments with about a 50–300 km-wide crustal stretching zone, characterized by the least steep slopes, coincide with the least steep Bouguer anomaly gradients.

Isostatic residual anomaly gravity data

Theoretically, the isostatic residual gravity anomaly handles the onshore Bouguer anomaly problem by assuming that the excess mass of large topographic highs is Airy-compensated by a reduction in mass below the high. The isostatic residual gravity anomaly calculation elevates the Moho beneath highs to remove the compensating crustal roots, which in effect compensates for the elevated masses removed above mean sea level. Hypothetically, the isostatic residual gravity anomaly field onshore should illuminate large basins as gravity lows, unless they are compensated at depth, and large elevated highs as relative gravity highs, especially if they are incompletely compensated.

An example of the isostatic residual gravity anomaly map of the continental–oceanic crustal boundary comes from offshore southern Morocco (Nemčok et al., 2005b; Figure 3.12). As mentioned earlier, it contains a discontinuous system of positive anomalies. This system of anomalies does not exactly locate the uplifted continental blocks adjacent to the oceanic crust. Sediments in the offshore are thicker than those of Nova Scotia (Figure 3.9), and significantly modify the gravity signal. As a result, the center of the positive anomaly outlining the continental margin and the maximum of the negative

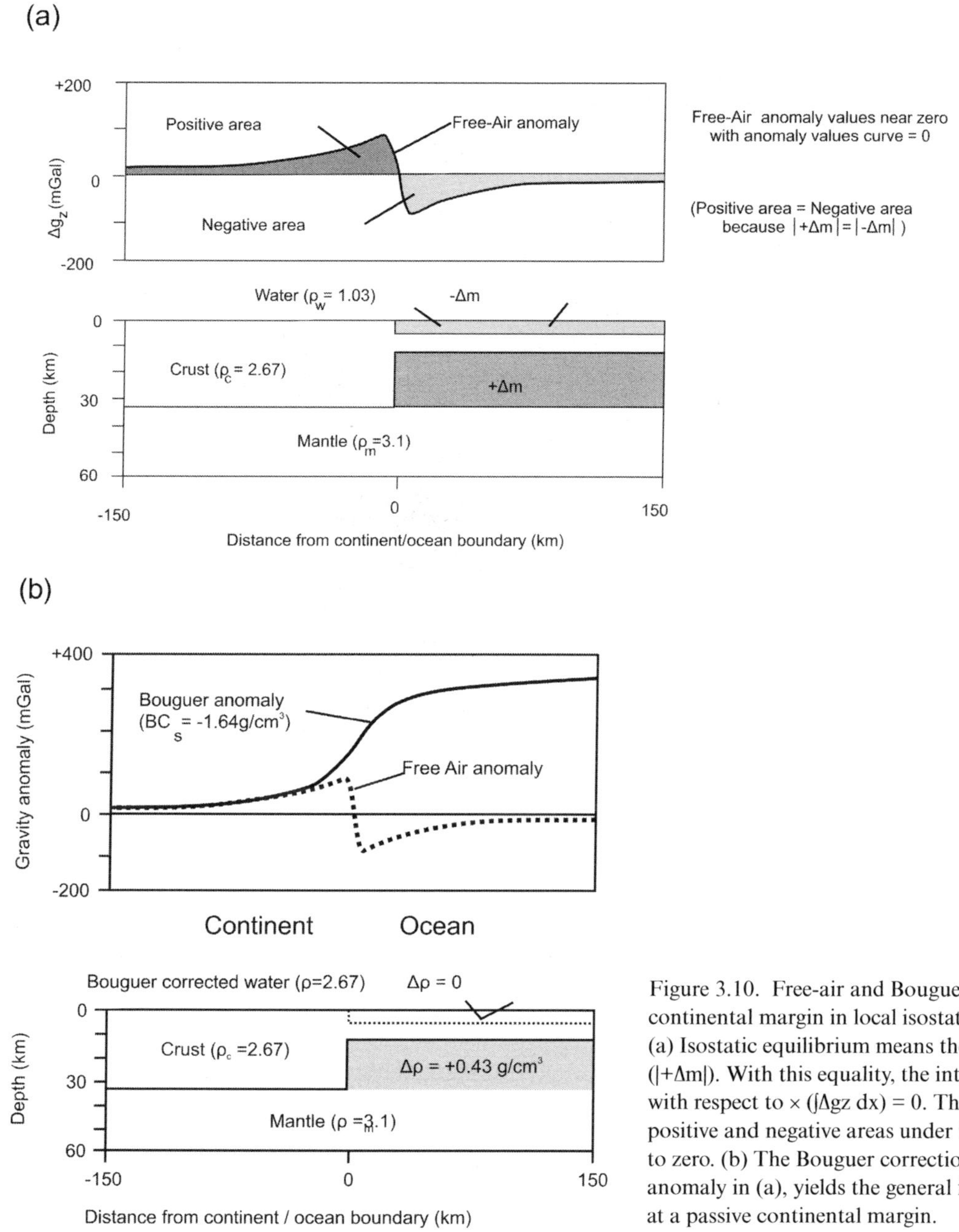

Figure 3.10. Free-air and Bouguer gravity anomalies for passive continental margin in local isostatic equilibrium (Lillie, 1999). (a) Isostatic equilibrium means the absolute value of excess mass (|+Δm|). With this equality, the integral of the change in gravity with respect to x (∫Δgz dx) = 0. The zero integral means that the positive and negative areas under the free-air anomaly curve sum to zero. (b) The Bouguer correction at sea, applied to the free air anomaly in (a), yields the general form of the Bouguer anomaly at a passive continental margin.

anomaly outlining the crustal boundary are shifted oceanward (see Watts and Stewart, 1998; Figure 3.7). This shift can be observed in seismic images, indicating that the crustal boundary occurs landward of the positive anomaly maximum. Furthermore, this positive gravity anomaly is segmented, suggesting that continental crust is broken into fault blocks, ranging in length from 23 to 227 km (Figure 3.13). In fact, the entire margin, landward of the first continental positive anomalies, is imaged as a mosaic of positive and negative anomalies (Figure 3.12). The anomalies have average widths and lengths of 20–30 km and 30–100 km, respectively. They are separated by very steep gravity gradients. Interpreted fault blocks have abrupt boundaries and are elongated in a northeast–southwest direction. Most are rather angular, typically bounded by northwest-southeast, northeast-southwest to east-northeast–west-southwest, and north-northeast–south-southwest-oriented trends. Northeast-southwest to east-northeast–west-southwest striking segments are parallel to syn-rift, sinistral, strike-slip faults mapped in onshore outcrops (Laville and Petit, 1984; Tayebi, 1989), and our offshore seismic grid.

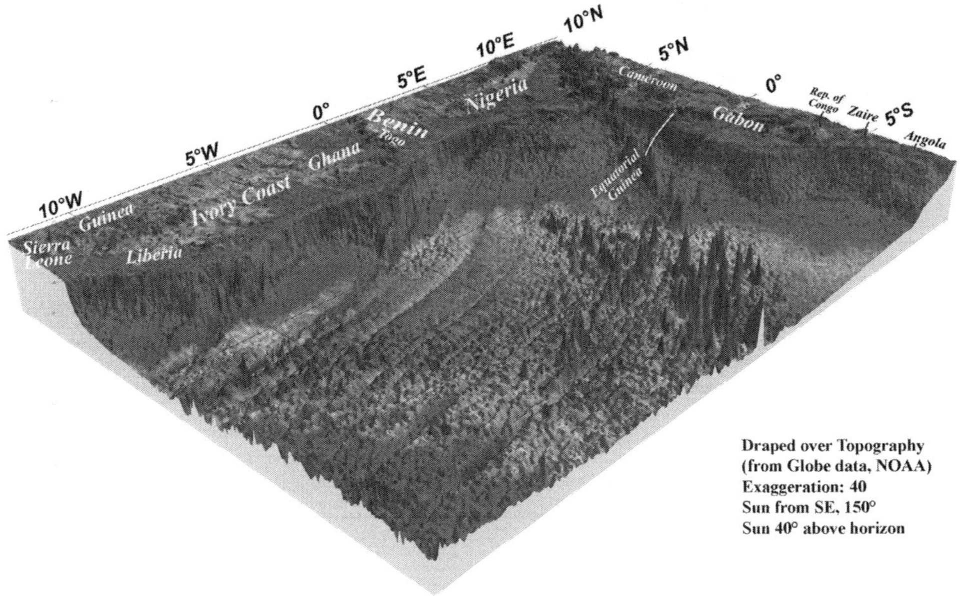

Figure 3.11. Bouguer gravity anomaly map of the Gulf of Guinea (AOA Geophysics, 2001).

Figure 3.12 also shows how the isostatic residual anomaly data can be used for fault pattern interpretation at the continental margin, once key faults correlated with those mapped in seismic images. The plate reconstruction (Nemčok et al., 2005b) allows us to see that the continental margin segment here developed along the strike of the transfer zone between two different rift zones. The transfer zone is characterized by an approximately 200 km-wide zone of east-northeast–west-southwest to northeast-southwest striking, sinistral, strike-slip faults (Figure 3.12). Fault blocks between them are bounded by north-northeast–south-southwest striking normal faults. They branch off the constraining, sinistral, strike-slip faults, progressively changing their displacement with changing geometry, from oblique-slip faults close to strike-slip faults, to normal faults further away. Northwest-southeast striking, dextral, strike-slip faults are also present; however, these are subordinate to the sinistral, strike-slip faults. The average fault block size is about 10 km wide by 20–50 km long.

Spatial derivatives of gravity data

In theory, spatial derivatives are designed to enhance the edges of bodies producing anomalies. The isostatic residual gravity anomaly first vertical derivative display is meant to enhance fault features and dipping boundaries, indicated by changes from negative to positive anomalies, or vice versa. The down-thrown or deeper side should be indicated by negative values represented by cool colors.

The total horizontal derivatives of Bouguer and isostatic residual gravity anomalies are calculated from a combination of simultaneous derivatives in orthogonal directions. Maxima in the total horizontal derivative field of the isostatic residual gravity anomaly, represented by hot colors, should indicate fault locations more exactly than the first vertical derivative display, although the displays are complementary to each other.

The dip-azimuth of the isostatic residual gravity anomaly field is computed by calculating the direction of the maximum gradient (or slope) of the anomaly field at each node on a grid. This is mapped using a 360° color spectrum. In theory, dip-azimuth display of the isostatic residual gravity anomaly field should be a powerful method for showing the locations and orientations of complex geological structures.

In practice, the geological utility of gravity derivatives is often unclear where the gravity fields are unconstrained by other data. Therefore, a synthetic approach is always recommended.

The example of the synthetic approach comes from the northern South Atlantic (Rosendahl et al., 2005). Figure 3.14 shows a set of conjugate margin closures made for the age of about 118 Ma. This is the time after which the production of the proto-oceanic crust by mantle exhumation was replaced by the production of the oceanic crust by organized seafloor spreading; that is, this is the time of the maximum width of the proto-oceanic crust corridor developed between two drifting continents (Figure 2.20a; Rosendahl et al., 2005).

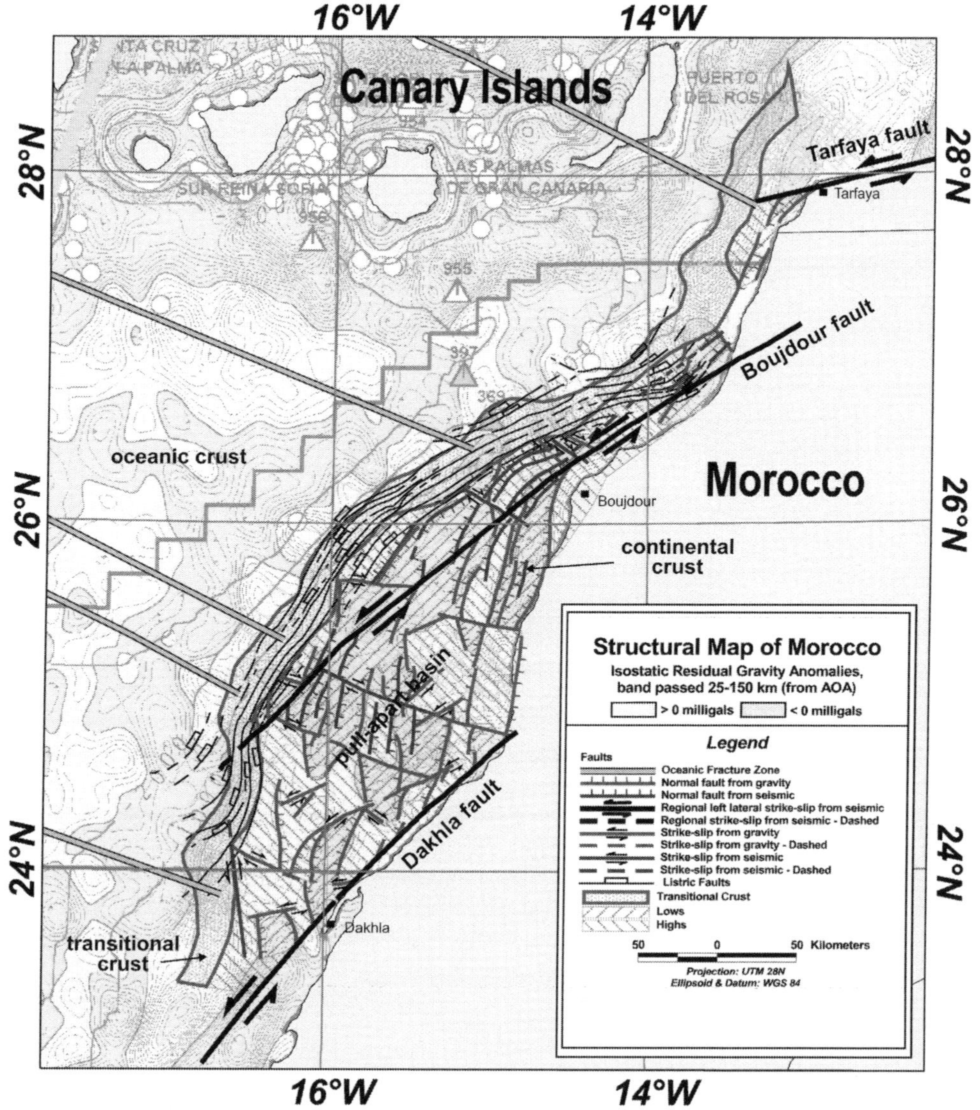

Figure 3.12. Tectonic elements and interpreted faults, offshore southern Morocco (Nemčok et al., 2005b). Elements include basement crustal types, syn-rift pull-apart basins, and syn-drift listric faults in Jurassic and Cretaceous strata. Raw offshore gravity data are from Getech and Kerr-McGee Oil and Gas Corporation; calculated offshore gravity data are from AOA Geophysics, Inc. (2001).

The isostatic residual gravity image of this model (Figure 3.14) shows that the proto-oceanic corridor can be divided into the north-northwest–south-southeast trending South Gabon segment and the northeast-southwest trending Sergipe-Alagoas segment. The former underwent continental breakup controlled by the rift branch, the latter by the transform. The isostatic residual gravity image of the South Gabon segment contains gravity highs on both continental sides, while the Sergipe-Alagoas segment has the high on the African side. Free-air and Bouguer data for offshore and onshore regions, respectively, indicate that the Sergipe-Alagoas segment can be divided into three major compartments.

The total horizontal derivative of Bouguer anomaly data contains maxima on both sides at both segments (Figure 3.14). Interpretation of the Sergipe-Alagoas segment indicates the major northeast-southwest striking bounding-fault system on the Brazilian side dominating over the one on the African side. The image also indicates a large number of north-northwest–south-southwest striking normal fault-controlled contacts on the African side of the South Gabon segment.

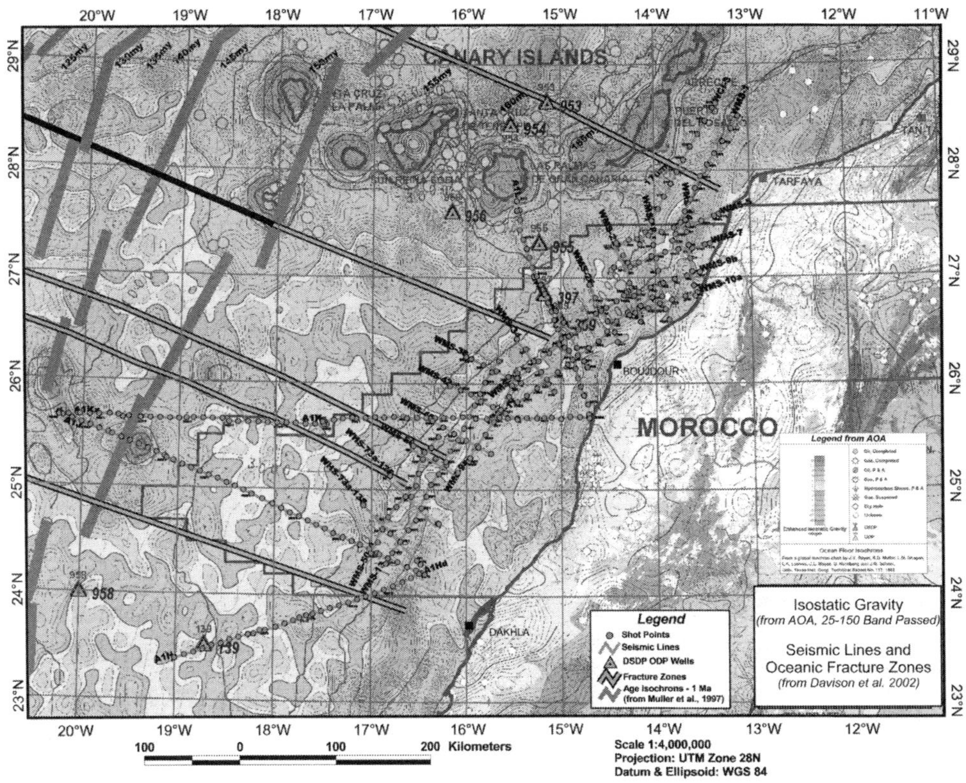

Figure 3.13. Isostatic residual gravity anomaly map of Morocco with location of seismic profiles used in the study and oceanic fracture zones (Nemčok et al., 2005b). Raw offshore gravity data are from Getech and Kerr-McGee Oil and Gas Corporation, calculated offshore gravity data are from AOA Geophysics, Inc. (2001), the onshore geological map is from Choubert et al. (1988), and the ocean bathymetry is from Smith and Sandwell (1997).

The total horizontal derivative of isostatic residual anomaly data highlights the strike-slip fault-controlled contacts on the Brazilian side of the Sergipe-Alagoas segment even better than total horizontal derivative of Bouguer anomaly data (Figure 3.14). On the other hand it does not highlight the stretched African margin of the South Gabon segment as well as the total horizontal derivative of Bouguer anomaly data. Mentioned observations highlight the need for a synthetic approach when interpreting various gravity derivatives.

Methods using reflection seismic data

A commonly used reflection seismic technique for interpreting different types of crust is to use "rift indicators." This indicator would be a wedge of sedimentary or volcanic rocks thickening toward the boundary fault. It would be assumed that it represents a fill of the failed rift unit in the continental crust. While this method works in certain cases, the ultra-deep seismic image from offshore Equatorial Guinea (Figure 3.15; Rosendahl et al., 2005) documents the risk involved. It does contain several "rift indicators." Thanks to its large imaging depth, however, the crust type can be determined from the seismic image of the crust itself, and from the crustal thickness between the top of the basement and Moho surface, apart from gravity and magnetic data modeling (see Rosendahl et al., 2005). It turns out that observed "rift indicators" occur in the proto-oceanic crust, instead of the continental crust. This observation once again emphasizes the need for using multidisciplinary data in the interpretation of crustal type distribution.

The ideal reflection seismic data for the interpretation of the crustal type distribution is the data deep enough to provide us with the image of the entire crust, such as those from offshore West Africa and East India (see Rosendahl et al., 1991, 2005; GX Technology, 2005a, 2005b, 2005c; ION/GX Technology, 2007; Nemčok et al., 2012a; Figures 3.15, 3.16).

Using a regional seismic grid, Rosendahl and colleagues (2005) conclude that the thickness of the oceanic crust generated by the paleo-Mid-Atlantic Ridge is rather uniform, having a thickness of about 4.2–6.5 km. While this thickness is less than the global average

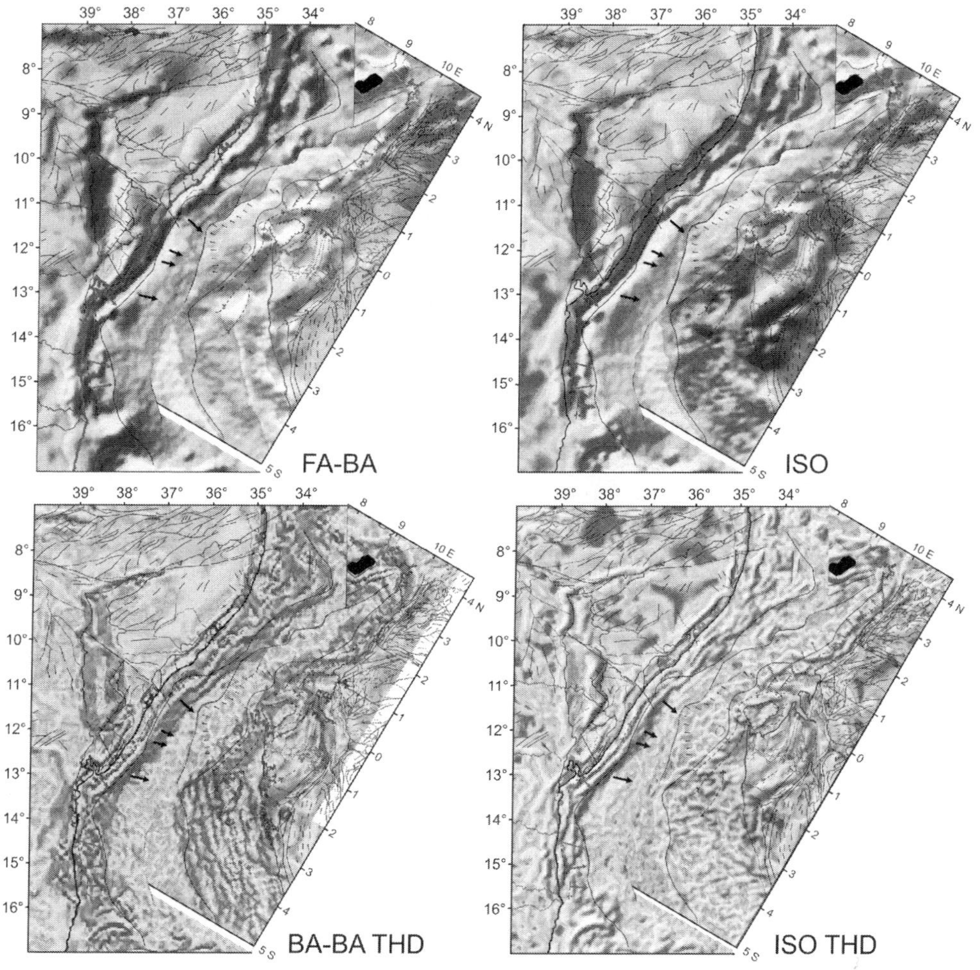

Figure 3.14. Composite gravimetric image panels corresponding to plate configuration at 118 Ma for the northern South Atlantic with onshore fault patterns (Rosendahl et al., 2005). FA-BA – combination of onshore Bouguer and offshore free-air gravity imaging, ISO – isostatic residual anomaly data, THD – total horizontal derivative.

of 7.08 km (White et al., 1992), it lies within published values for the normal oceanic crust of the Atlantic Ocean (Rosendahl and Groschel-Becker, 1999). Rarely does the thickness between oceanic fracture zones vary by more than 10 percent. The uniformity of the thickness in seismic profiles parallel to the flow lines of the crust is rather interesting. The acoustic character of the oceanic crust varies only minimally from area to area and not in any systematic way. The image of the oceanic crust can be characterized as seismically un-layered and structureless in the regions among oceanic fracture zones (Figure 3.15). Its top is indicated by a strong continuous reflector. Images of the oceanic fracture zones represent the only complexity in the oceanic crust image. Some of them are characterized by the offset of the Moho surface, which is interpreted as being represented by a fairly continuous reflector. It is absent under the hot-spot-controlled Cameroon Volcanic Zone.

Reflection seismic data can be used for the interpretation of the oceanic crust based on the presence of oceanic fracture zones, just like in the case of gravity and magnetic data interpretation described earlier (see Figure 3.8). Figure 3.17 shows the serial images of the Saint Paul Fracture Zone in offshore Brazil. They reiterate that the structure of the oceanic fracture zone is formed by a three-dimensional distribution of ridges and troughs, which can be interpreted as flow lines indicating the presence of oceanic crust.

The proto-oceanic crust has a different character than the oceanic crust (Rosendahl et al., 2005). Its thickness is larger, usually below 10 km. The seismic image documents its faulted and deformed character (Figures 3.15, 3.16). The base of the crust is usually defined by a continuous reflector with the exception of areas dissected by fracture zones. Fracture zones are defined by large faults that dissect the entire crust and

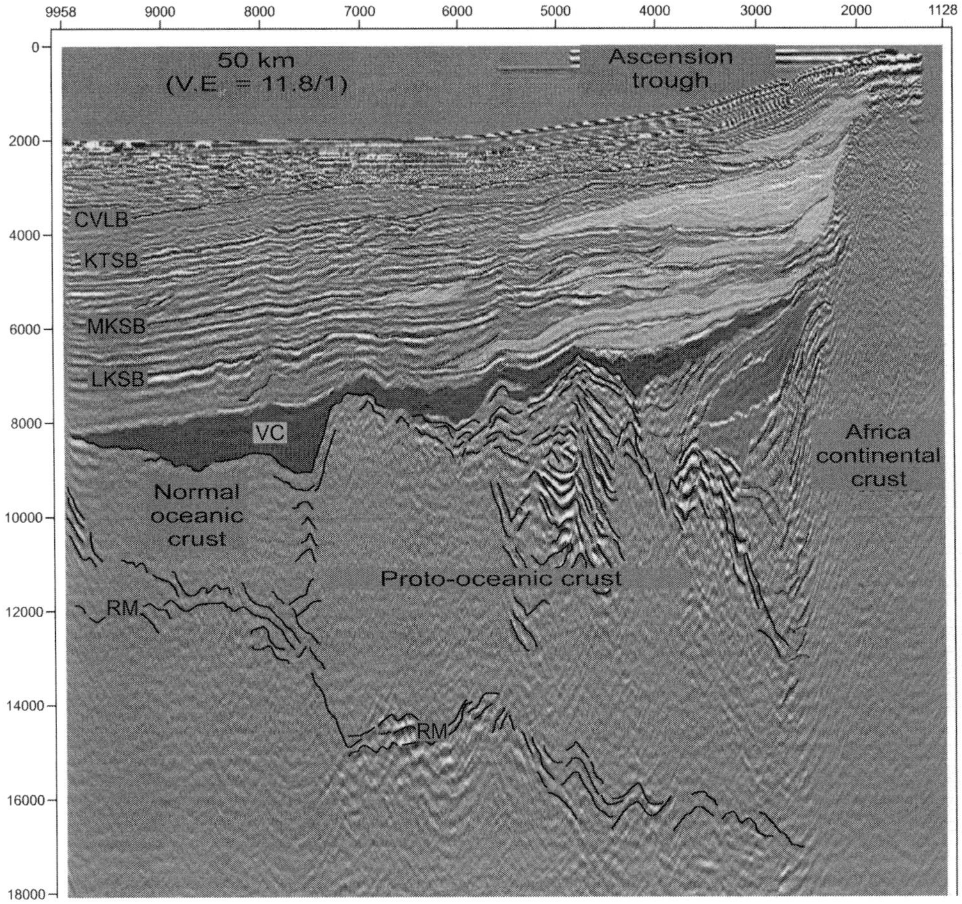

Figure 3.15. Segment of the PROBE profile 5 showing the occurrence of the "rift indicators" on the proto-oceanic crust (data from Rosendahl et al., 2005; depth migration by GX Technology, 2005, pers. comm.). Note that the imaging, available only down to 9–10 seconds TWT, would not provide the image of the Moho surface, and "rift indicators" would create a false illusion of stretched continental crust.

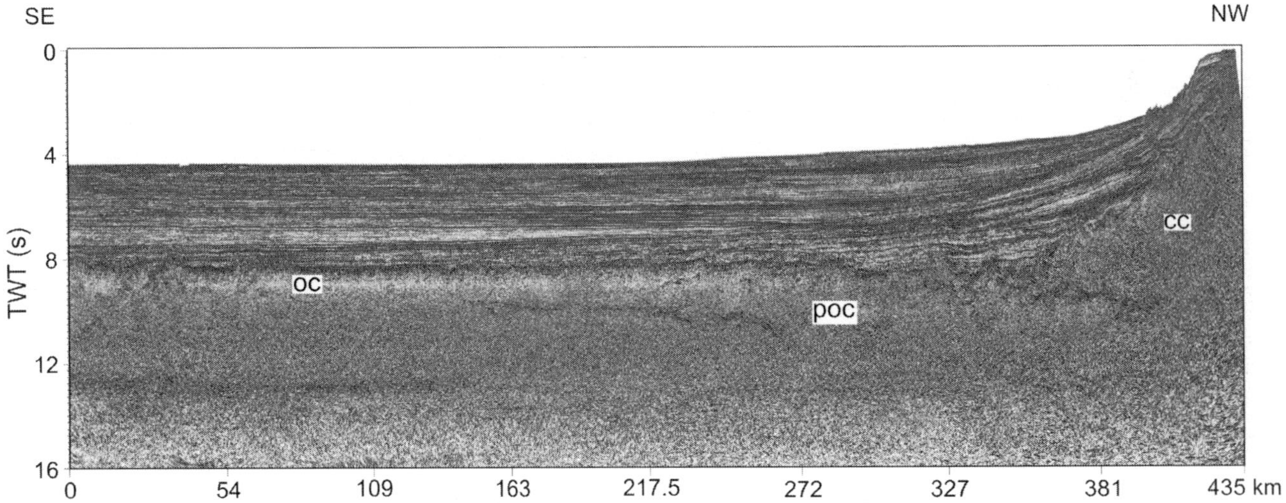

Figure 3.16. Reflection seismic section through oceanic (oc), proto-oceanic (poc), and continental crusts (cc), offshore East India (data from ION/GX Technology, 2007).

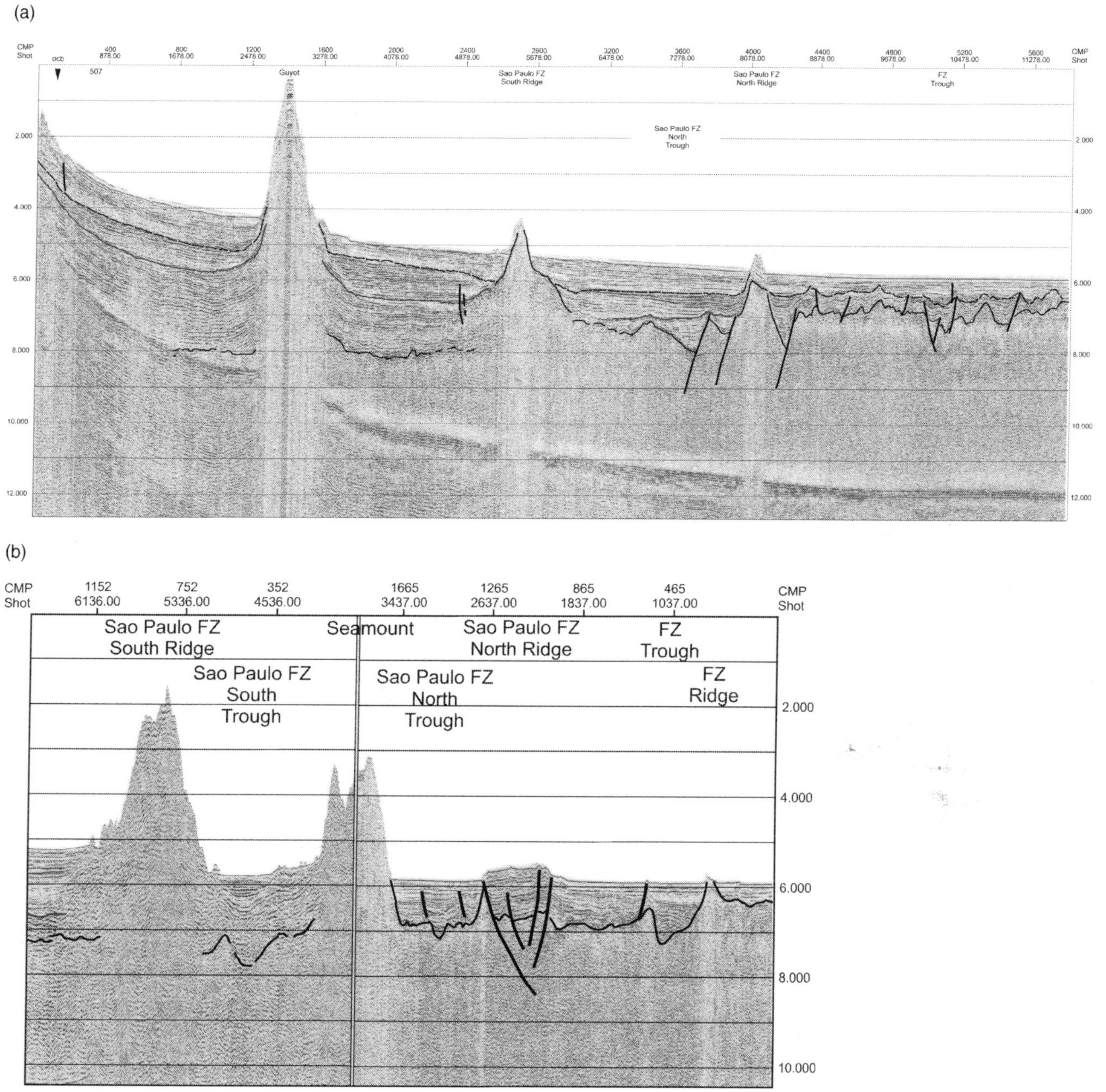

Figure 3.17. LEPLAC reflection seismic profiles through the Saint Paul oceanic fracture zone to the north of the Barreirinhas Basin, Brazil (Nemčok et al., 2004b).

usually coincide with large offsets of the Moho surface. The proto-oceanic crust is seismically layered.

Methods using the synthetic approach

One of the synthetic methods, called three-dimensional gravity inversion, combines the use of bathymetry, geomagnetic isochron, gravity, and sediment thickness data to calculate the residual gravity anomaly of the mantle (Kusznir, 2009). The sediment thickness, being the only interpretational input data set, is estimated from the available reflection seismic imagery in the region. The mantle anomaly is inverted in the three-dimensional spectral domain in order to determine the depth of the Moho surface, crustal thickness, thinning factor of the continental lithosphere, and residual thickness of the

(c)

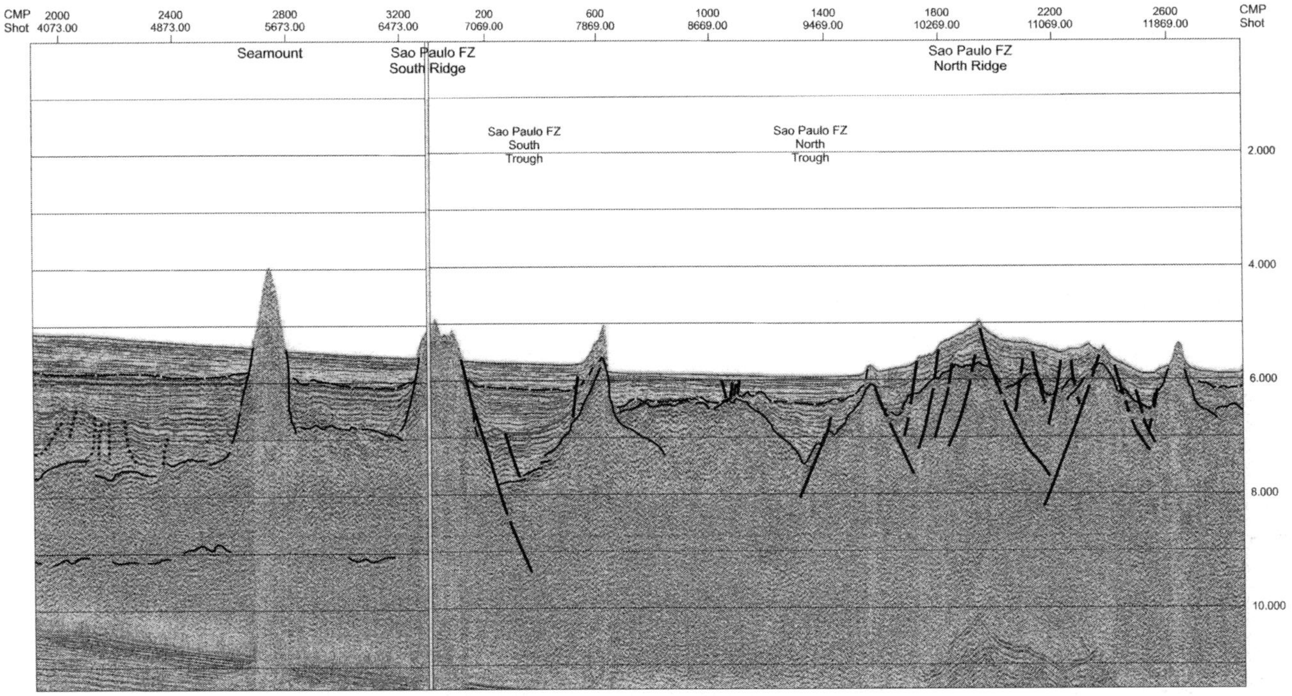

Figure 3.17 (*cont.*)

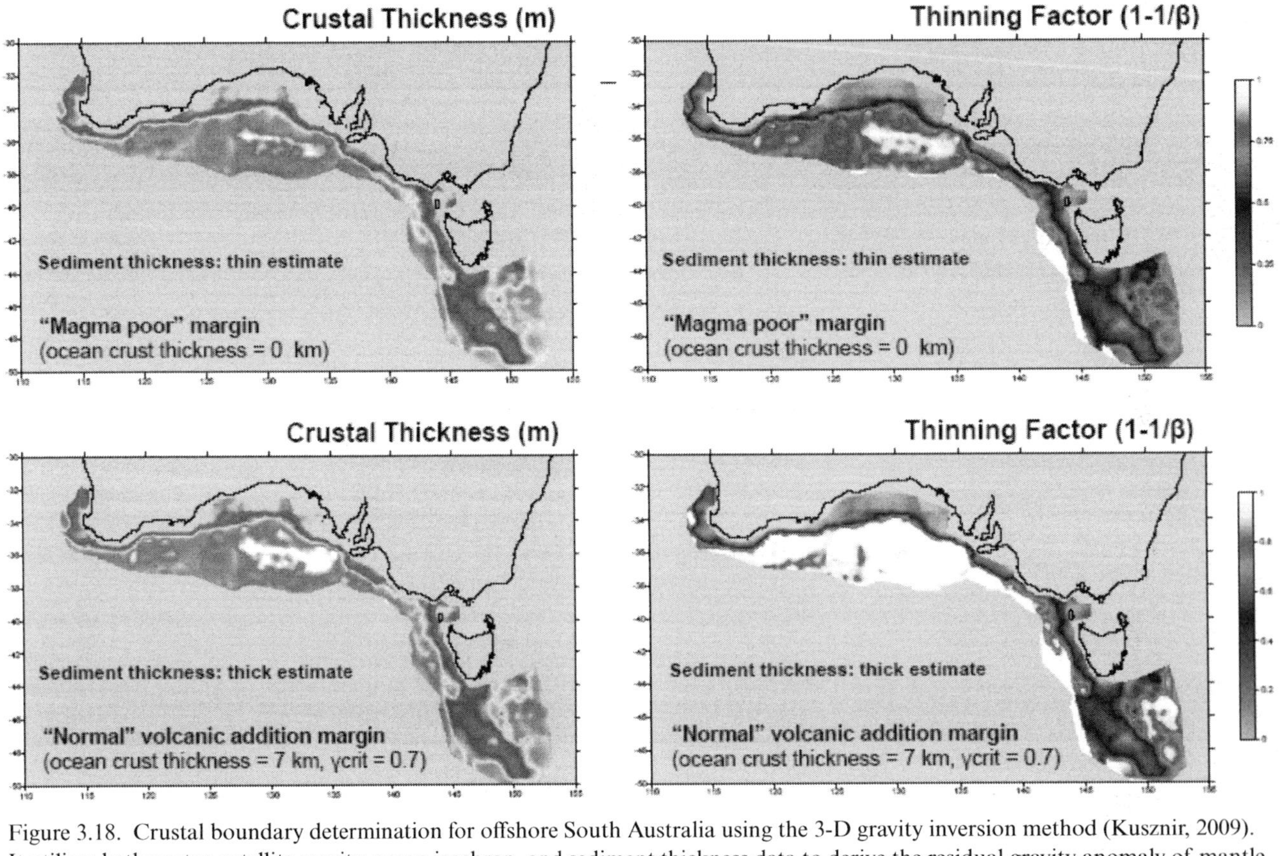

Figure 3.18. Crustal boundary determination for offshore South Australia using the 3-D gravity inversion method (Kusznir, 2009). It utilizes bathymetry, satellite gravity, ocean isochron, and sediment thickness data to derive the residual gravity anomaly of mantle, which is then inverted in the 3-D spectral domain to determine Moho depth, crustal thickness, continental lithosphere thinning factor, and residual continental crustal thickness. The method also incorporates a lithospheric thermal gravity anomaly correction. The upper figures show the example of the magma-poor margin. The zero oceanic crust limit represents the boundary between continental crust and oceanic crust, which is frequently represented by proto-oceanic crust. The lower figures show the example of the magma-rich margin, where a volcanic correction has been applied with ocean crust thickness of 7 km.

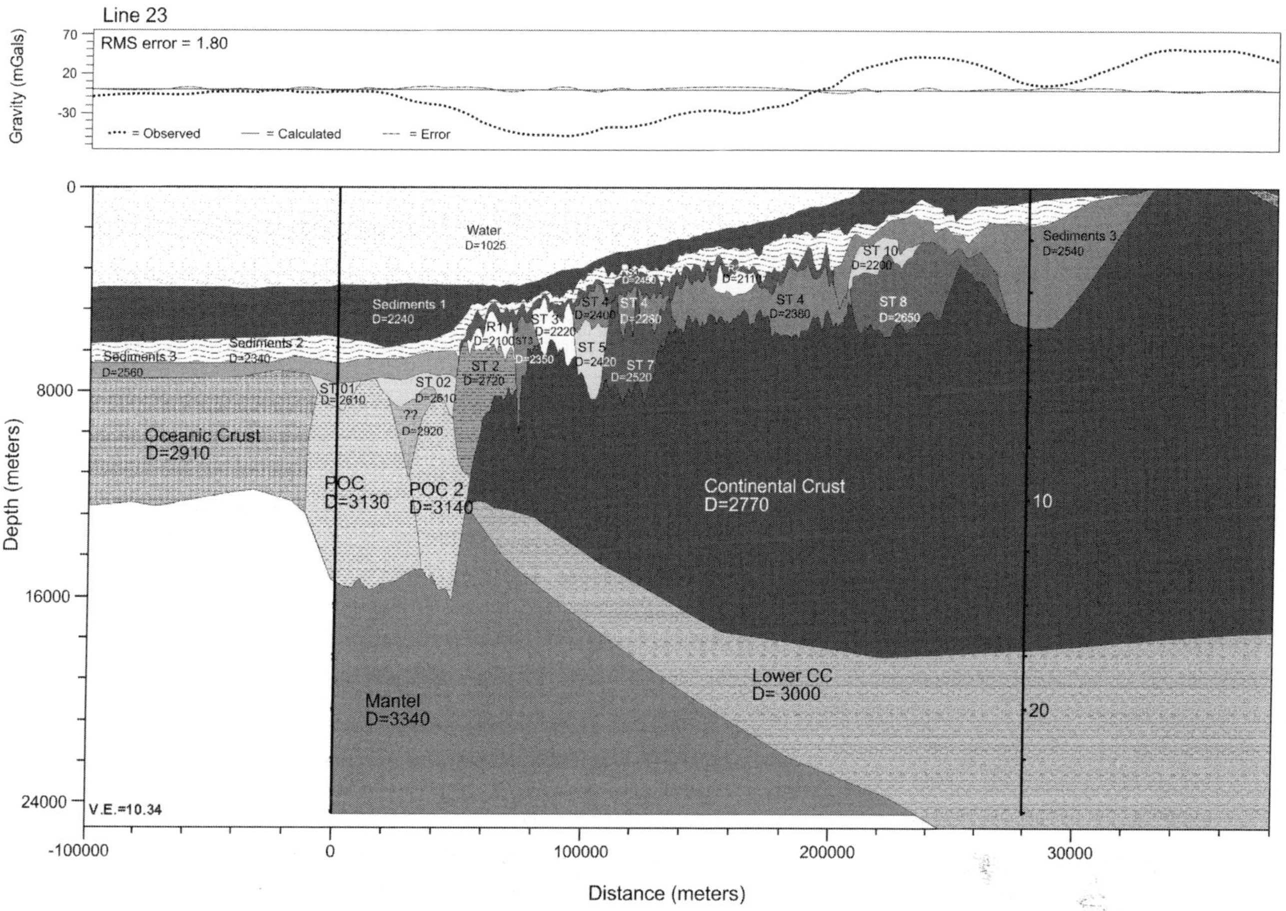

Figure 3.19. Interpreted PROBE seismic profile 23 in offshore Gabon with interpreted density distribution matching the observed gravity data by calculated ones (Meyers et al., 1996). Note that proto-oceanic crust in the South Gabon segment may contain slivers of lower crust or upper mantle, emplaced along detachments and unroofed.

continental crust (Figure 3.18). Each individual sediment layer is flexurally back-stripped using either a known or estimated layer density to calculate the lithospheric thermal gravity anomaly produced by basin margin uplift and subsidence, considering isostatic equilibrium. Stretching and thinning factors β and (1–1/ β) are computed to be used in the determination of the gravity anomaly due to rifting. Once thinning factor and crustal thickness distributions are determined, they are used in combination with oceanic crust isochrons for interpretation of the boundary between continental and oceanic, respectively proto-oceanic, crusts (Kusznir and Karner, 2007).

The offshore Gabon case (Figure 3.19; Meyers et al., 1998) documents the synthetic use of reflection seismic and gravity data for crustal type definition. The comparison of the reflection seismic section in Figure 3.19 with a free-air gravity map indicates that the flow lines of the gravity map are roughly perpendicular to the profile, which justifies the two-dimensional gravity modeling approach to match the interpreted density distributions with an observed gravity curve, measured along the profile by Rosendahl and colleagues (1991). The interpretation of the seismic profile allows placing the Moho surface on the lowest strong reflector, allowing differentiating the oceanic, proto-oceanic, and continental crusts according to their thickness. It also allows differentiating crust types according to their different seismic character. The images of the oceanic and proto-oceanic crusts are the same as those described earlier (Figures 3.15, 3.16). The continental crust is different, characterized by a relatively transparent brittle upper crust and a highly reflective ductile lower crust. The lower crust contains numerous undulated reflector packages, interpreted as potential detachment fault zones. The upper crust contains numerous faults, which can be interpreted as detached at the middle crustal level.

Table 3.2. Summary of densities interpreted at various study areas by Avdulov (1970), Ibrmajer et al. (1989), Peltier et al. (1989), Cloetingh et al. (1995), Meyers et al. (1996), Shaw and Lin (1996), Mohriak et al. (1998), Watts and Stewart (1998), Huismans (1999), Scheck and Bayer (1999), Smallwood et al. (1999), Canales et al. (2000), Burov and Poliakov (2001), Frederikse et al. (2001), Hansen and Nielsen (2002), Meredith and Egan (2002), Wilson et al. (2003), Davis and Kusznir (2004), Martínez-Martínez et al. (2004), Fernàndez et al. (2005), Huismans and Beaumont (2005), Rosendahl et al. (2005).

Continent	Density (kgm^{-3})
Seawater	1,030
Sedimentary cover	(1,750-) 2,110–2,650
Upper continental crust	2,500–3,000 (average 2,700–2,800)
Lower continental crust	2,800–3,000
Upper lithospheric mantle	3,000–3,340
Lower lithospheric mantle	3,260–4,500(?)
Asthenosphere	
Proto-oceanic crust	
Whole proto-oceanic crust	3,080–3,130
Oceanic crust	
Upper oceanic crust	2,500–2,990
Lower oceanic crust	2,920–3,200

The distribution of densities characterizing different crusts, which matches the observed gravity data, indicates different densities for all three types of crusts (Figure 3.19; Meyers et al., 1998). The densities of the upper and lower continental crusts, being 2,770 and 3,000 kgm^{-3}, overlap with the worldwide average (Table 3.2) determined from the density data taken from the current geophysical literature on passive margins. The density of the oceanic crust, being 2,910 kgm^{-3}, also overlaps with the worldwide average. The density of the proto-oceanic crust is, however, lighter than that of the typical upper continental mantle rocks (see Table 3.2), and the matching average density requires interpreting a combination of upper continental mantle rocks with lower continental crustal ones (Meyers et al., 1998).

Synthetic studies done in West Africa and East India (e.g., Meyers et al., 1998; Wilson et al., 2003; Nemčok et al., 2012a,c) indicate that we frequently need to honor lateral changes of lithologies. For example, in offshore Gabon, on top of combining seismic and gravity data, Wilson and colleagues (2003) used the seismic velocity distribution. In order to match the seismic and gravity interpretations with observed data, they had to assume various degrees of serpentinization of different proto-oceanic crustal blocks, using seismic velocity–density relationship for serpentinized peridotite based on specifying the serpentinized fraction (see Miller and Christensen, 1997). Modeled distributions also allow us to explain a transition from the serpentinized peridotite of the proto-oceanic crust to the upper mantle across the Moho being represented by a modest reflection coefficient compared to the situation for the oceanic crust on top of the mantle, characterized by larger acoustic impedance contrast (observe Figure 3.16). A mechanism for the serpentinization of the unroofed upper continental mantle composed of peridotite is discussed in Chapter 2, based on reflection seismic and well studies of the Iberian continental margin (e.g., Beslier et al., 1996; Pickup et al., 1996; Whitmarsh et al., 2000) combined with outcrop studies of the European and Adriatic passive margins exposed in the Alps (Manatschal, 2004 and references therein), as caused by the hydrothermal circulation of seawater.

Another way of interpreting the distribution of the oceanic and proto-oceanic crusts and their boundary with continental crust is possible in the case of robust regional seismic data coverage allowing detailed sediment thickness contouring (Rosendahl in Nemčok et al., 2004b). When done for early syn-drift stratigraphies, the thickness distribution (Figure 3.20) is controlled by the pattern of oceanic fracture zone–related troughs and ridges, providing the flow lines of the oceanic crust generation, discussed earlier. The landward end of the "flow line" distribution serves as a proxy for the boundary between continental and proto-oceanic crusts.

Seismic line marriage verification

Regardless of how sophisticated the method was that allowed us to define the distributions of oceanic, proto-oceanic, and continental crusts, and in spite of how rigorous the plate reconstruction method using determined crustal boundaries was, it is always useful to test the precision of our reconstruction. A very precise way of testing this is the seismic profile marriage technique (Rosendahl, 2001, pers. comm.), which takes interpreted reflection seismic profiles from conjugate continental margins and merges their external proto-oceanic ends or external continental ends together (Figures 2.20, 3.21; Rosendahl et al., 2005). The technique requires us to have seismic profiles from

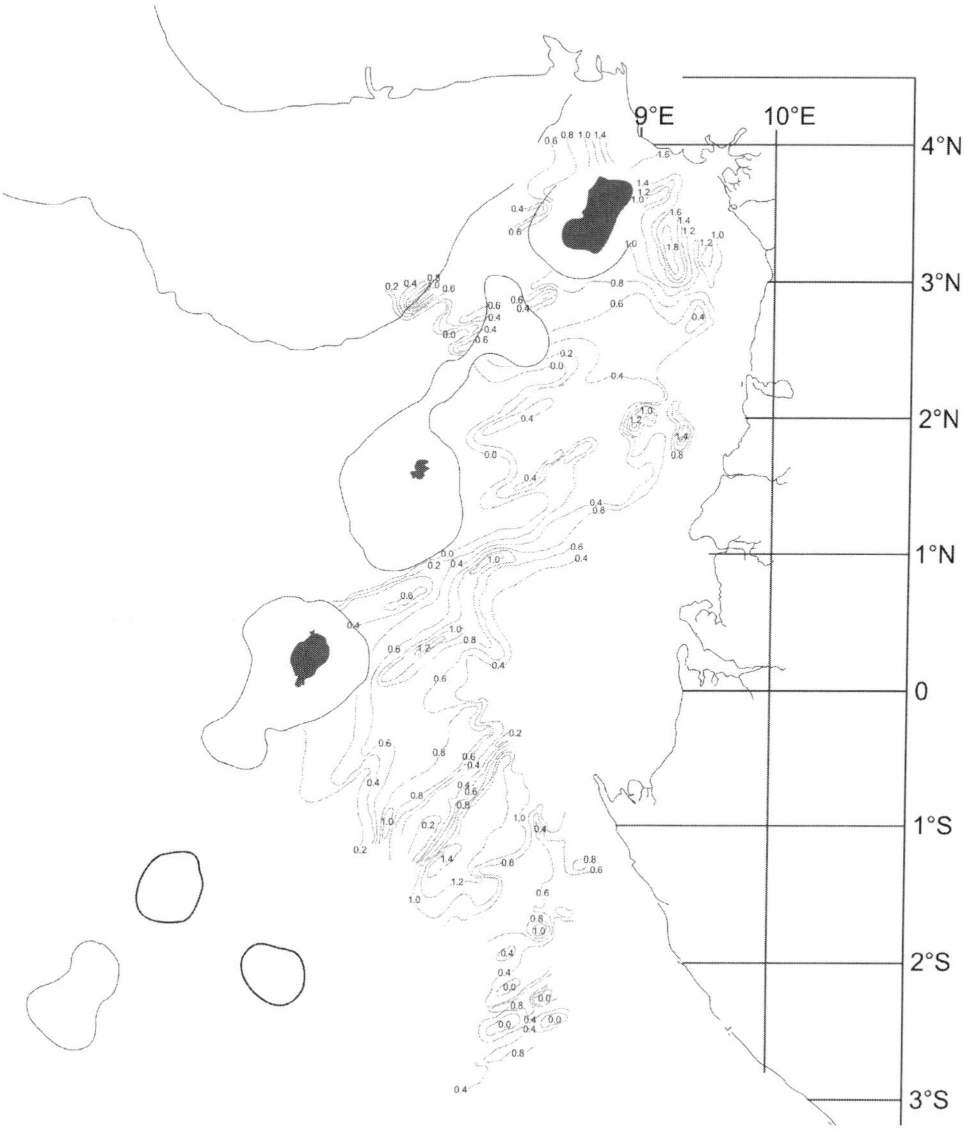

Figure 3.20. Isopach map of thickness of the Lower Cretaceous sediments overlying oceanic and proto-oceanic crust in offshore Cameroon-Gabon (Nemčok et al., 2004b). Units are in seconds of the two-way reflection time.

the paired opposite margins, which intersect at conjugate positions near the merged oceanic–proto-oceanic or proto-oceanic–continental crustal boundaries. While this is easier to achieve at ocean segments developed by roughly orthogonal drifting, the method is more complicated for segments with complex drift histories and it is very difficult for the combination of complex drift history with strike-slip or oblique-slip tectonics preceding the continental breakup.

From this perspective, the conjugate pairing of the PROBE profile 23 with the respective profile PR from the Brazilian margin is relatively ideal (Figure 3.21; Rosendahl et al., 2005). Paired sections are located in the South Gabon segment of the South Atlantic (Figure 2.20), which was initiated by slightly oblique rifting and then controlled by slightly oblique drifting. The conjugate pair provides a match for almost the whole extent of the proto-oceanic corridor, just prior to the onset of the organized seafloor spreading that formed the oceanic crust. Because the match is made for a snapshot of 118 Ma ago, the younger sediments do not match. However, the horizons that matter, the Moho surface, top of the basement, and earliest syn-drift sediments, are paired well.

One more seismic marriage example comes from the Sergipe-Alagoas segment (Figure 2.20) and represents

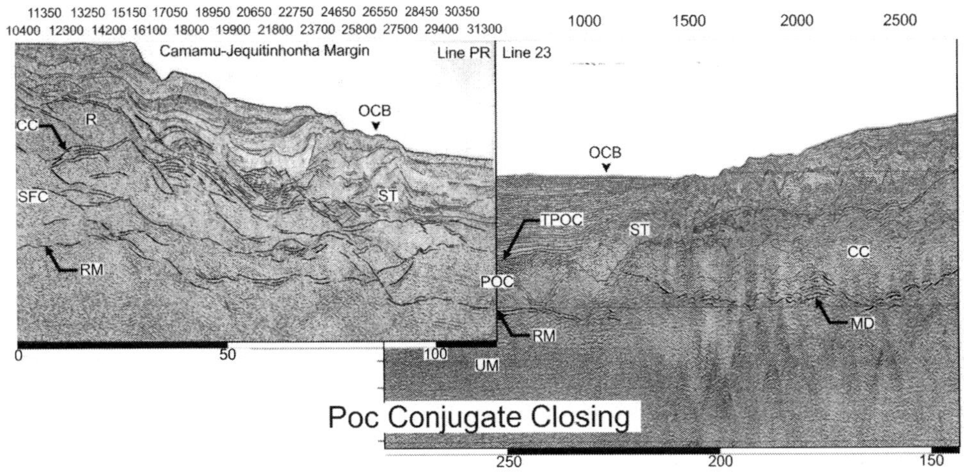

Figure 3.21. Conjugate pair of PROBE section 23 with section PR from the Brazilian and African sides of the South Gabon segment of the northern South Atlantic (Rosendahl et al., 2005). TCC – top of the continental crust, R – syn-rift sediments, ST – evaporites, MD – main detachment (close proxy of the gravity-defined Moho), DCF – decompressed Congo fold belt ductile lenses, DT – Dentale Trough, SB – base of evaporites. See Figure 2.20a for location.

the difficult end-member from the described pairing spectra. Figure 2.20 show the pairing of the PROBE section 20 with Brazilian section PJ, which are located in the segment, which was initiated by sinistral transform and then controlled by highly oblique drifting. Because of a lack of suitably located sections, the conjugate pair does not exactly provide a match for the whole extent of the proto-oceanic corridor, like the previous example (Figure 3.21), but matches for a bit younger time snapshot, as the presence of a very narrow oceanic crust corridor in the middle of the conjugated pair allows us to imply. However, the horizons that matter, the Moho surface, top of the basement, and early syn-drift sediments, are paired well.

4 Determination of timing of rift and continental breakup events

Introduction

As described in Chapters 1 and 2, the rift flank undergoes an uplift during rifting and the continental margin experiences an uplift following continental breakup (e.g., Moore et al., 1986; Rosendahl, 1987; Miller and Duddy, 1989; Brown et al., 1990; Rust and Summerfield, 1990; Dumitru et al., 1991; Ten Brink and Stern, 1992; Steckler et al., 1994; Van der Beek et al., 1994; Van Balen et al., 1995; Burov and Cloetingh, 1997).

Uplifted rift flanks, also called rift shoulders, are common features of rifts. Such a rift shoulder usually has an asymmetric geometry. It has a steep scarp facing the basin and a gentle slope facing the land. Rift shoulders have been explained in a variety of models based on transient and permanent uplift mechanisms. The transient uplift models are related to the thermal effects of rifting and include:

1) depth-dependent extension (Royden and Keen, 1980; Hellinger and Sclater, 1983; Watts and Thorne, 1984; Morgan et al., 1985);
2) lateral heat flow (Steckler, 1981a, 1981b; Cochran, 1983; Alvarez et al., 1984; Buck et al., 1988); and
3) secondary convection under the rift shoulders (Keen, 1985; Steckler, 1985; Buck, 1986).

These mechanisms, which can create a shoulder elevation of 500–1,500 m, operate only during the elevated thermal regime of the lithosphere. Formed positive topography subsequently progressively decays over a time period equivalent to the time of 50–100 Ma required for the thermal disturbance to propagate through the lithosphere (see Lachenbruch and Sass, 1977; Vasseur and Burrus, 1990; Deming, 1994). The permanent uplift models are related to (1) the magmatic underplating (Cox, 1980; Ewart et al., 1980; McKenzie, 1984; White and McKenzie, 1988) and (2) lithospheric unloading and/or plastic necking (Zuber and Parmentier, 1986; Parmentier, 1987; Braun and Beaumont, 1989; Issler et al., 1989; Weissel and Karner, 1989; Chery et al., 1992). Only relatively recently, the processes of erosion and deposition started to be incorporated into the existing models of the rift shoulder uplift (e.g., Van Balen et al., 1995; Van der Beek et al., 1995; Burov and Cloetingh, 1997; Figure 2.7a).

Dating the rift shoulder uplift, the post-breakup passive margin uplift, and their erosional decay allows us to determine the timing of the rift and breakup events. It can be done either indirectly, by dating the sediment eroded from the uplifted terrains and deposited in the surrounding depocenters, or directly, by dating the exhumation of the uplifted terrains. The depositional events of the syn-rift and syn-breakup sediments are most commonly dated by a chronostratigraphic approach. The exhumation events of uplifted terrains are usually dated by low-temperature thermochronology methods applied to apatite and zircon minerals and less direct approaches such as studies of vitrinite reflectance, T_{max}, or production index profiles.

The exact determination of rift and breakup events is critical for a correct thermal history model, basin model, and source rock maturation history. The aim of this chapter is to outline the most typical dating methods, using both method descriptions and case studies demonstrating their successful application.

Chronostratigraphy

The timing of rifting and continental breakup is best determined when pre-tectonic, syn-tectonic, and post-tectonic sedimentary sequences are all preserved. Examples of clear distinction of pre-tectonic, syn-tectonic, and post-tectonic sediments in cross-sections from the rift terrains come from the Pannonian Basin system (Nemčok and Lexa, 1990; Tari et al., 1992), northern Utah (Constenius, 1996) and the Transantarctic Mountains–Ross Sea Shelf area (Van Balen et al., 1995 and references therein), and analog material models (e.g., McClay, 1995). Examples from the pull-apart settings have been published from the Dead Sea region (Ben-Avraham and Zoback, 1992) and analog material models (e.g., Sims et al., 1999). Examples from passive margins come, for example, from eastern North America (Van Balen et al., 1995 and references therein; Withjack et al., 1998), offshore Santos and Campos Basins, Brazil (Karner, 2000), offshore Norway (Doré et al., 1997), Galicia Bank (Montadert et al., 1979a; Manatschal, 2004 and references therein), and the Goban Spur–Bay of Biscay region (Montadert et al., 1979a; de Graciansky et al., 1985; Peddy et al., 1989; Klemperer and Hobbs, 1991; Louvel et al., 1997).

Syn-tectonic sediments also provide information concerning the sequence of normal faulting, as is shown for the northern Kenya portion of the East African Rift system (Hendrie et al., 1994), the Gulf of Patras (Sorel, 2000; Figure 1.8), and western Tasmania (Noll and Hall, 2006). Provided that a complete syn-tectonic sedimentary record is available, the relative timing of the controlling faults can be obtained using the paleontological or other chronostratigraphic record. In the case of partially eroded syn-tectonic strata, we can try to project missing sedimentary layers onto cross-sections and three-dimensional models. In the case of strongly eroded or missing syn-tectonic sediments, we need to apply the uplift-dating techniques discussed later in this chapter.

Paleontological record in syn-tectonic sediments is the most robust data set that can be used for age determination, which can be accomplished by two methods including the traditional zonal approach and the composite standard database (Figure 4.1a). The former traditionally uses several fossil events, represented by first and last occurrences, to define a fossil zone, with separate zonations developed for different fossil types. Unfortunately, due to problems such as reworking, preservational bias, and facies change, different age interpretations by each discipline, such as palynomorphs, nannofossils, and foraminifers, may result for the same section. This variance occurs not only because each discipline analysis is conducted separately from the others, but also because the biostratigraphic zones are defined by a small number of microfossil events rather than by the total data set (Figure 4.1a).

The composite standard approach minimizes these problems by analyzing the entire biostratigraphic data set from all disciplines simultaneously, thereby giving a truly integrated and multidisciplinary chronostratigraphic interpretation (Figure 4.1a). Also, the greater number of biostratigraphic datums potentially available for age interpretation from any one level greatly increases the probability of successful correlation relative to the handful of datums that would define a biostratigraphic zone(s) for the same interval (Figure 4.1b). For example, the standard nannofossil zonation for a 3 million-year period in the Late Eocene of southwest Africa is defined by four nannofossil datums, whereas the same interval of time in EGI's Angolan Composite Standard database has potentially 260 datums calibrated to absolute age and available for correlation (Figure 4.1b).

The calibration of biostratigraphic ranges within a composite standard database has been accomplished via the method of graphic correlation (Figure 4.2). A section is calibrated to the composite standard via extinction (top) and origination (base) events in common between the subject section (on the y-axis) and the composite standard database (x-axis). Based on the graphic representation of these datums, the paleontologist constructs a time-depth curve (i.e., the line of correlation – LOC) that gives a one-to-one correlation between subject well depth and the composite standard time scale. A typical LOC records period of deposition (off-vertical segments), as well as intervals of time for which no surviving rock record is available in the subject well; that is, a hiatus, marked by horizontal offsets in the LOC (Figure 4.2). Such a graphic correlation chart can easily provide important parameters for basin analysis. The slope of the LOC is the average rock accumulation rate, which, when compaction is accounted, provides the average sediment accumulation rate. Hiatus durations are simply the length in millions of years of the horizontal offsets. Also, because well sections are calibrated in absolute age, rather than via a relative biostratigraphic zonation, intra- and inter-basin correlations can be done using absolute age and therefore independent of facies changes or biostratigraphic events.

The hiatuses portrayed in Figure 4.2 can have various causes:

1) erosional unconformity;
2) surface of non-deposition;
3) condensed zone; and
4) normal fault.

If an erosional surface and if multiple well analyses indicate that it is a regionally extensive surface, it would be a candidate for a depositional sequence boundary (see Vail et al., 1977; Posamentier et al., 1988). If it is a regionally extensive surface of non-deposition and/or condensed zone, it would likely mark a maximum flooding surface and, therefore, a genetic sequence boundary (Galloway, 1989; Posamentier and Allen, 1999; Catuneanu, 2002). However, it could also denote a purely local erosional event such as a channel downcut or slump. The high-resolution age control and hiatus definition that the graphic correlation method provides allows reliable determination of the regional extent and distributions of hiatuses, which, together with integration with log and seismic data, allows for accurate recognition and definition of sequence boundaries and their differentiation from local events.

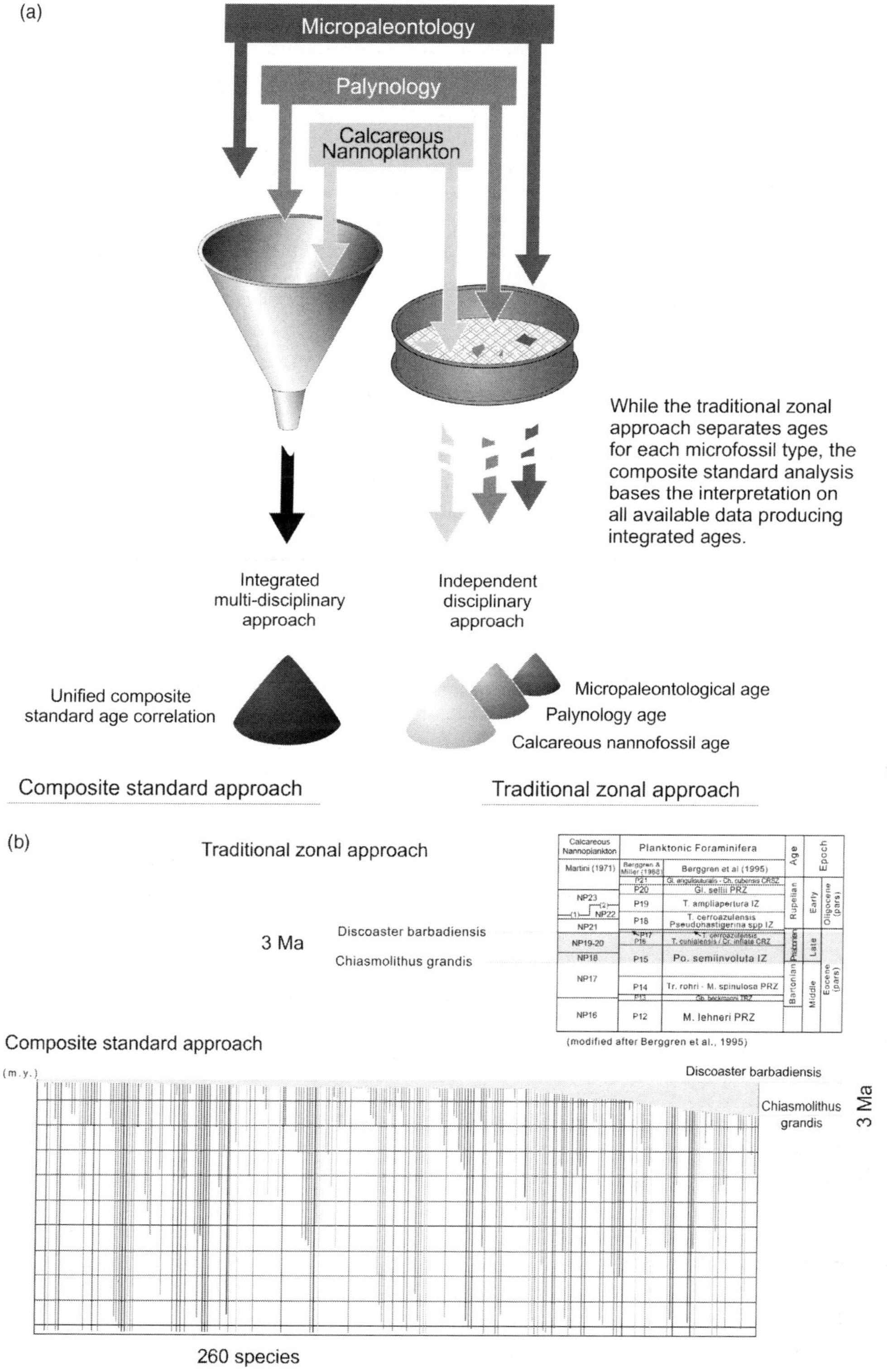

Figure 4.1. (a) Composite standard and traditional zonal biostratigraphic approaches (Sikora, 2003, pers. comm.). (b) Comparison of composite standard and zonal approaches on example of 3 Ma-long Oligocene time period from the Angolan passive margin (Sikora, 2003, pers. comm.).

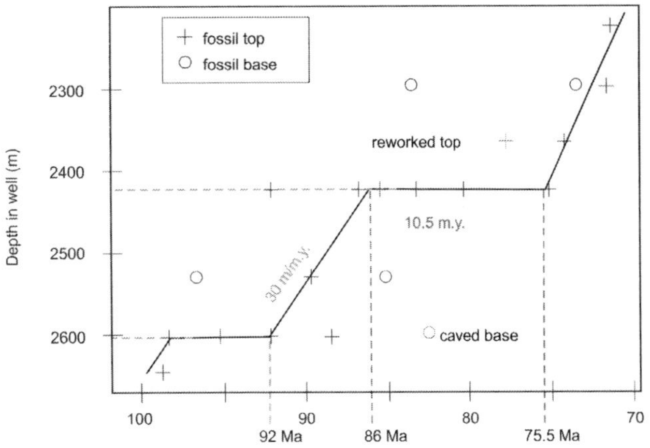

Figure 4.2. Time correlation of well record with composite standard (Sikora, 2003, pers. comm.).

Thus a relatively conformable succession of genetically related strata bounded at their upper surface and base by unconformities and their correlative conformities, which represent Vail's depositional sequence boundaries (Vail et al., 1977), can be defined via this chronostratigraphic method. Such a sequence is composed of a succession of genetically linked deposition systems (systems tracts) and is interpreted to be deposited between eustatic-fall inflection points (Posamentier et al., 1988). The sequences and the system tracts that they enclose are subdivided and/or bounded by a variety of "key" surfaces that bound or envelope these discrete geometric bodies of sediment. They mark changes in depositional regime "thresholds" across that boundary.

The correlation could also provide a genetic stratigraphic sequence of Galloway (1989b), which uses the maximum flooding surfaces as sequence boundaries. Such a sequence lies between the transgressive system tract and that of the highstand. The lowstand system tract in this model is associated with a relative fall and an early rise. Catuneanu (2002) points out that the importance of the maximum flooding surfaces is tied to the ease with which they can be mapped across a basin. On the negative side, the maximum flooding surface may be diachronous (Posamentier and Allen, 1999). On the positive side, genetic sequences have subaerial unconformities within the sequence, which can potentially cross from one genetic sediment package to a completely unrelated package (Catuneanu, 2002).

Drawing from experience with time correlations of well records with worldwide composite standards done at Amoco for years (Sikora, 2003, pers. comm.), we can conclude that chronostratigraphic hiatuses and depositional intervals in the correlation curves, such as the one shown in Figure 4.2, are usually low-order events and, therefore, are more likely to be regional events.

While traditional seismic interpretation usually bases its interpretation of rift and breakup events on lithostratigraphic tops, the integrated chronostratigraphic sequence stratigraphy allows interpreted seismic data to be compared with absolute geologic time (Figure 4.3a; Howe, 2003, pers. comm.; Sikora, 2003, pers. comm.). Figure 4.3b shows such a comparison of the well chronostratigraphic curve with the well log suite. The chronostratigraphic curve providing the information about absolute geological time, depositional rates, duration of hiatuses, paleobathymetry, and paleoenvironments thus becomes tied with additional information on facies distribution and distribution of erosional unconformities. Such integration helps to explain complicated problems, such as the vertical expansion of the section by salt diapir penetrated by well (Krebbs, 2003, pers. comm.).

Given a suite of wells, the described approach is suitable for determining complex problems such as the age of an unconformity, which was cut into different stratigraphic levels in different areas of the same rift system or passive margin and was subsequently covered by a sedimentary wedge with the diachronous age of its base (Figure 4.4; Sikora, 2003, pers. comm.). Figure 4.4a shows a suite of five hypothetic wells through the sedimentary wedge divided by seven sequence boundaries. The largest volume of sequences is characterized by progradation. Well 1 penetrates a sequence boundary, which was formed by subaerial erosion in this area, while its rough equivalents in deeper-seated wells 2 and 5 were formed by sediment starvation only. Figure 4.4b indicates that each of the five wells has a unique record of deposition intervals and hiatuses. The only similarity is the geologic time D recorded in each of them by a sequence boundary. Figure 4.4b also provides an interesting look at the sequence boundary formed at around time E in wells 2, 3, and 4. The hiatus associated with this boundary tends to increase in a basinward direction, documenting the progressively delayed arrival of the prograding wedge basinward. Diachronous boundaries like this can pose a major problem in determining timing of tectonic events associated with erosional events, such as the continental breakup event.

The example of breakup timing from the passive margin, where the upper crustal breakup was controlled by normal faulting, comes from offshore Nova Scotia (Nemčok et al., 2005b). Here, the earliest syn-rift basins, such as the Fundy and Orpheus Basins, opened along the transtensional, sinistral Cobequid–Chedabucto transfer fault zone during

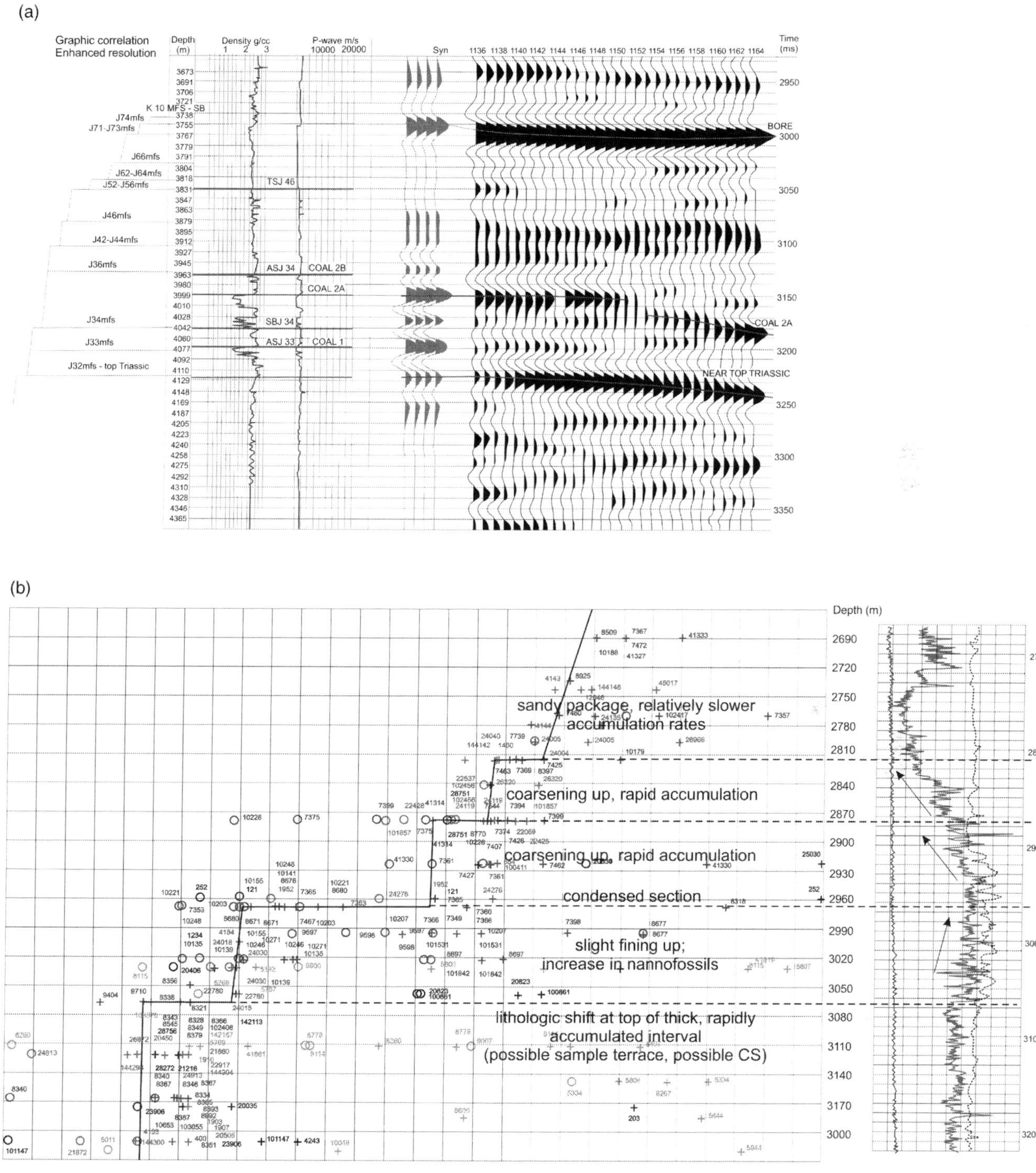

Figure 4.3. (a) Paleoenvironmental integration, allowing the characterization of depositional environments, facies trends, and downslope sediment transport (Sikora, 2003, pers. comm.). Note that the graphic correlation in this North Sea well identifies numerous depositional hiatuses, which are not associated with gamma kick or strong seismic expression. This indicates how quantitative biostratigraphy can improve the resolution of other geological and geophysical data. (b) Comparison of the well chronostratigraphic curve with the well log suite (Howe, 2003, pers. comm.).

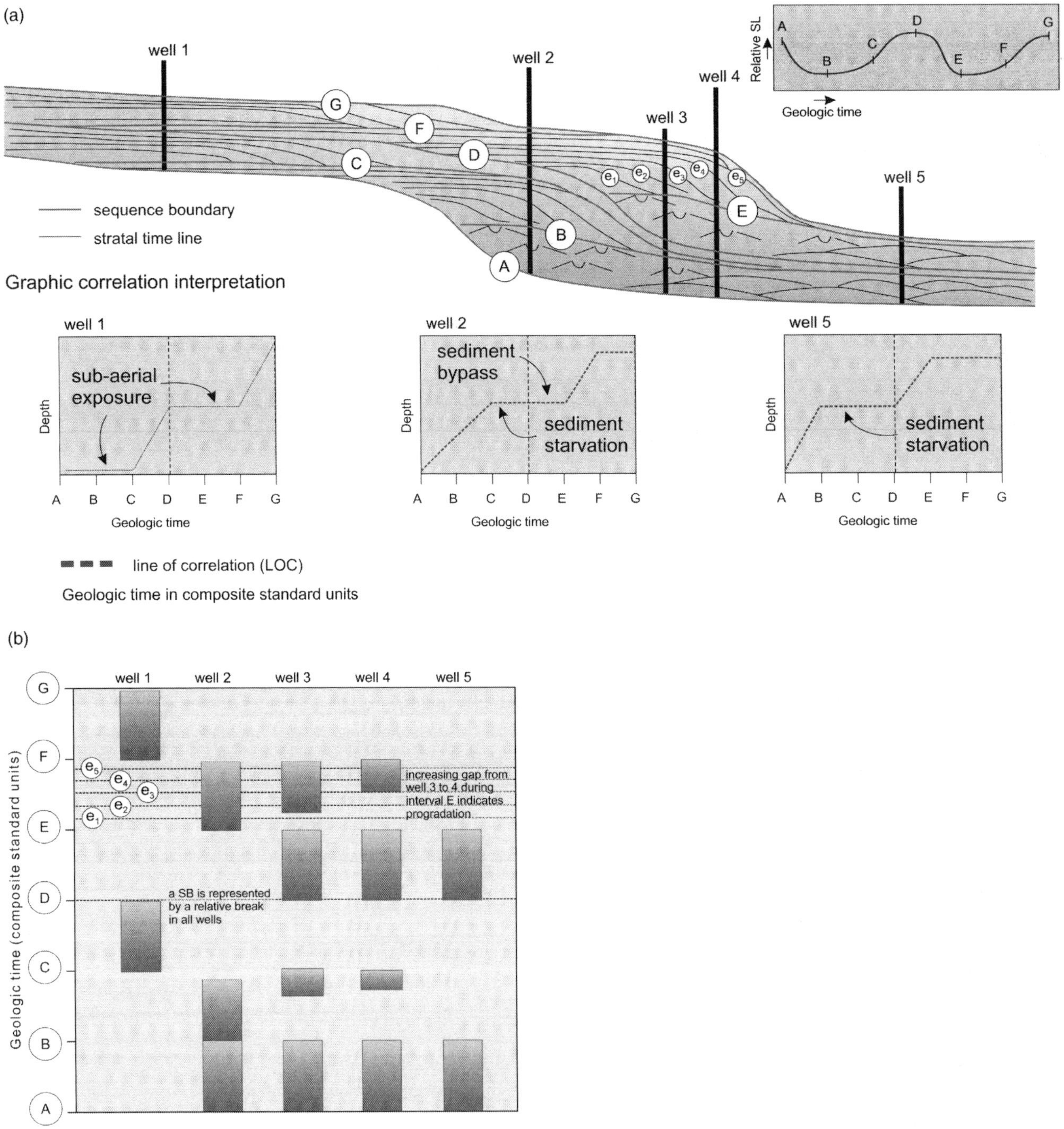

Figure 4.4. (a) Chronostratigraphic integration to sequence stratigraphy (Sikora, 2003, pers. comm.). (b) Wheeler diagram of the stratigraphic integration to sequence stratigraphy made for five wells shown in (a) (Sikora, 2003, pers. comm.).

the Triassic (Keppie, 1982). Most of the orthogonal rift basins south of this strike-slip fault zone contain Upper Norian–Sinemurian red beds, evaporites, and carbonates (Wade et al., 1995). K-Ar ages determined from basalt flows intercalated with red beds indicate a Hettangian–Early Sinemurian age (Wade et al., 1995). All these data provide constraints on the lower bracket for the probability interval of the breakup timing, constraining the age of the sediment located underneath the breakup unconformity.

The constraints on the upper bracket would be provided by the data on sediments located above the

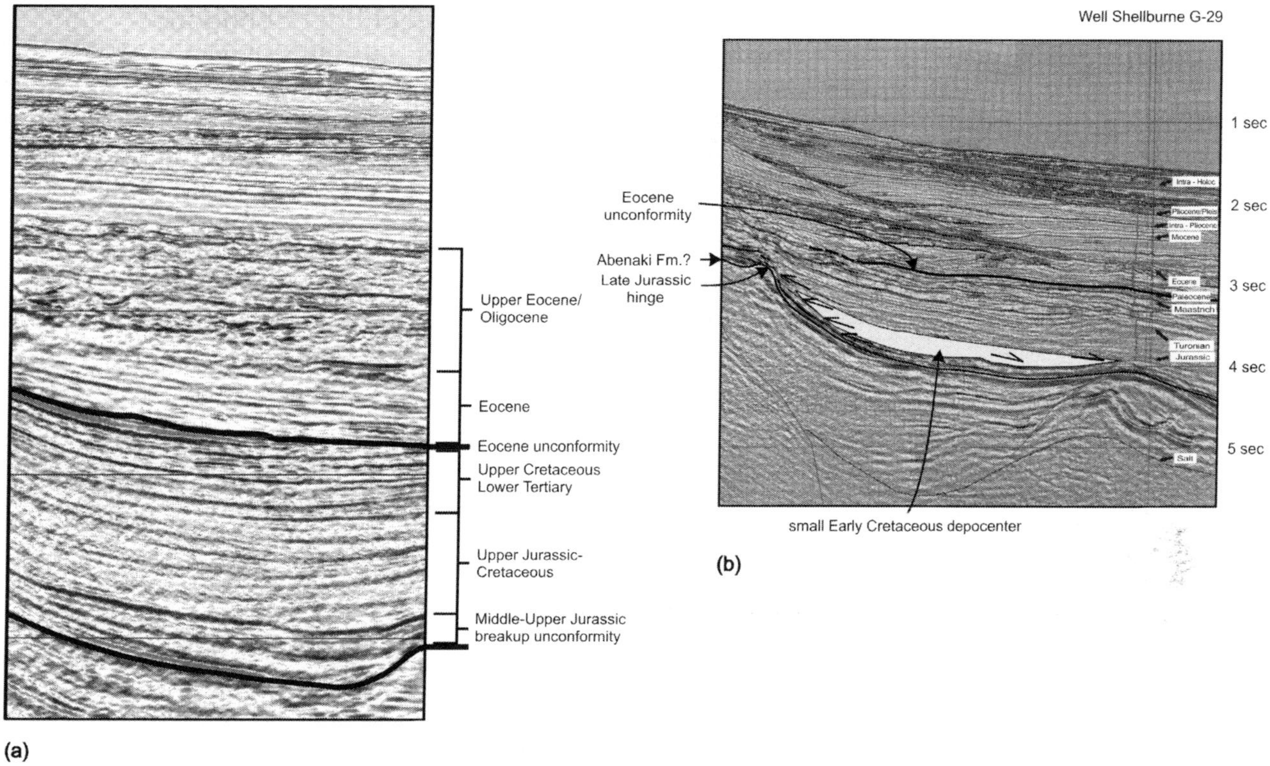

Figure 4.5. Seismostratigraphy of the continental slope, offshore Nova Scotia (Nemčok et al., 2005b). Original seismic data and reprocessed seismic data were provided by the Canada-Nova Scotia Offshore Petroleum Board and the Kerr-McGee Oil and Gas Corporation. a) Middle Jurassic to Oligocene units in the study area. b) Post-salt units around the Shelburne G-29 well.

breakup unconformity. The problem arises from the distribution of wells in the region (Nemčok et al., 2005b), which causes the constraint on the breakup unconformity timing determination to come only from a single deeper-offshore well, Shelburne G-29, while the rest of the wells are located more or less behind the present-day shelf edge. Although Shelburne G-29 provides us with a reasonable tie to surrounding seismic profiles where the breakup unconformity can be recognized (Figure 4.5; Nemčok et al., 2005b), a lack of a well location spread in deeper offshore prevents us from understanding the distribution of hiatuses between the various depths of erosional cuts combined with the various ages of the basal syn-drift sediments at different parts of the passive margin.

It is expected that deeper erosional cuts occur in more landward portions of the extensional passive margin because of a more pronounced isostatic rebound of the progressively thicker continental crust at the passive margin, which underwent the normal faulting-controlled upper crustal breakup. In the case of the first syn-drift sediments arriving to deeper basin in prograding wedges, we expect to find a progressively younger base of the syn-drift sediments basinward. Therefore, the ideal well data set for determining the breakup timing for a certain passive margin segment would be a well coverage in different parts of the passive margin. Only a narrowed gap between various stratigraphies forming the base of the breakup unconformity and various stratigraphies forming the base of the overlying syn-drift sediments would lead to relatively accurate breakup timing determination.

Thus, based on the ages of sediments below and above the breakup unconformity available to different scholars, the age of the breakup unconformity in the Scotian Basin is probably Aalenian (Cloetingh et al., 1989; Tankard et al., 1989; Welsink et al., 1989) or Sinemurian-Toarcian (Wade and McLean, 1990). The age of this unconformity similarly determined on the Moroccan counterpart, the d'Abda Basin, is Toarcian-Aalenian, and becomes as young as Callovian northward (Aajdour et al., 1999; Echarfaoui et al., 2002). Magnetic data for the northern Central Atlantic segment, between the Nova Scotia and central-northwestern Morocco conjugates, indicate that continental drifting here started either before 191 Ma (Toarcian) or 175 Ma (Bajocian-Bathonian) (Klitgord and Schouten, 1986; Srivastava and Tapscott,

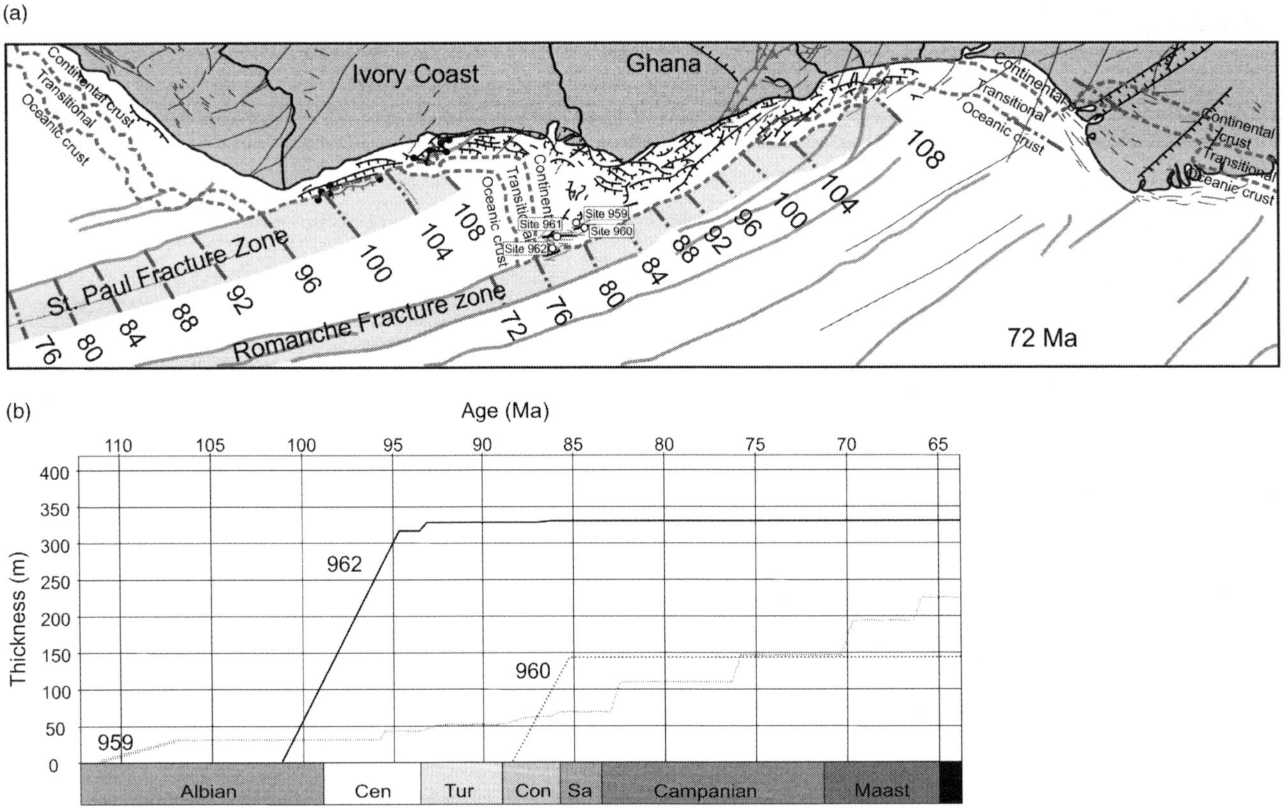

Figure 4.6. (a) Ghana Ridge area with locations of seafloor-spreading ridges in time (Ma) on their relative way out of contact with the broken up continent (Nemčok et al., 2012b). The youngest age shown is 72 Ma. Locations (thick dark gray lines) were derived from a map of the digital age of oceanic crust by Müller and colleagues (1997). Thick medium gray lines – oceanic fracture zones, thin dark gray lines – continental faults. Spreading centers traveling oceanward along ridges of the continental margin caused local heating and uplift of these ridges (e.g., Pierce et al., 1996), indicated by oceanward younging of the associated unconformity. (b) Comparison of biostratigraphic graphic correlation charts for ODP Sites 959, 960, and 962 from the Ghana Ridge done by a composite standard approach (Sikora in Nemčok et al., 2004b). Ramps represent periods of deposition; flats represent either hiatuses or erosional unconformities. Note a resolution of hiatuses/unconformities in Site 959, which is finer than that in Mascle et al. (1996) done by zonal approach.

1986; Verhoef and Srivastava, 1989). The discrepancy in discussed age determinations indicates the difficulty with which an exact timing of the continental breakup can be determined if the amount of well penetration in the deeper offshore continental margin is limited.

We would think that the determination of breakup timing should be easier at passive margin segments initiated by strike-slip tectonics, which tend to undergo much smaller isostatic rebound than segments initiated by normal faulting (see Karner and Watts, 1982; Holt and Stern, 1991; Kooi et al., 1992; Nemčok et al., 2012b). The reason would be the warmer thermal regime, which strongly weakens the continental lithosphere. This should bring the lithosphere close to local Airy isostasy, which prevents flexural and rebound effects from playing an important role. Therefore, while the extensional margins would be significantly eroded, the strike-slip margins would have a better chance of having a thicker sedimentary cover preserved for dating. In order to see whether such a lack of isostatic rebound is true for the strike-slip margin, we can discuss a following case example from the Equatorial Atlantic.

This example of the chronostratigraphic analysis of existing biostratigraphic data comes from DSDP Sites 959 and 962 penetrating the strike-slip margin, offshore Côte d'Ivoire and Ghana (Sikora in Nemčok et al., 2004b; Figure 4.6a). Both wells were studied with focus on determining the rift–drift transition timing. Site 959 is located in the southeastern corner of the Deep Ivorian Basin, to the north of the Côte d'Ivoire–Ghana Marginal Ridge (Ghana Ridge), which is parallel to the Romanche Fracture Zone. Site 962 is located on the Ghana Ridge, near the eastern edge of the Romanche Fracture Zone.

Site 959 contains the most stratigraphically complete section of all DSDP Leg 159 sites drilled in the area (Figures 4.6a–b). We will focus only on its Cretaceous portion.

The Upper Cretaceous section at a depth interval of 870–986 m, which was deposited at depths below the CCD, can be characterized by the occurrence of the agglutinated foraminifer *Rzehakina* in a diverse abyssal assemblage, which documents mildly anoxic conditions. It can be divided into three parts – Upper Maastrichtian, Lower Maastrichtian, and upper Middle Campanian – using dinoflagellate biostratigraphic indexes.

The underlying Lower Campanian carbonaceous sedimentary interval, penetrated at an interval of 986–1,027 m, is represented by anoxic facies. Its fossil record contains abundant radiolarians and rather rare, siliceous, agglutinated foraminifera, including the Early Campanian index *Uvigerinammina jankoi*. The interval represents the most significant Late Cretaceous oxygen depletion event in the well. The same event was indicated in an industrial well in the area (Nemčok et al., 2004b), as well as in ODP sites south of Côte d'Ivoire. This event has been described along both the eastern and western Atlantic margins and has been called Oceanic Anoxic Event 3.

The pre-Campanian sedimentary section penetrated by well is rather thin, but it is stratigraphically and depositionally complex. Between depths of 1,027 and 1,034 m, it contains the uppermost Coniacian–Lower Santonian section, composed of nannofossil-rich claystone with sporadic phosphate hardgrounds. This carbonaceous section is characterized by abundant buliminid benthic foraminifera with opportunistic behavior in low-oxygen environments. The section indicates a sediment starvation period on the middle-upper slope. The buliminid species, widespread in the whole Atlantic, indicate the development of a deepwater connection between the South Atlantic and Central Atlantic via the Equatorial Atlantic during this time period.

The uppermost Coniacian–Lower Santonian section is unconformably underlain by another thin sedimentary interval, a regressive sandy package, penetrated between depths of 1,034 and 1,046 m. The package contains rare inner-to-middle-shelf benthic taxa, but the entire interval is most likely allochthonous, being deposited in environments deeper than the middle shelf. The occurrence of foraminifers is very limited. Nannofossils indicate that the package was deposited during a large sea-level fall at the Turonian–Coniacian boundary.

Underneath lies the Lower–lower Middle Turonian carbonaceous interval, penetrated in an interval of 1,046–1,053 m. It is characterized by few species of low-oxygen, opportunist benthic foraminifera, representing endemic species known so far from the Benue Trough located at the eastern end of the Equatorial Atlantic (Kuhnt and Wiedmann, 1995). This foraminiferal assemblage indicates a locally restricted marine basin without a deepwater connection to the Central Atlantic.

A depth interval of 1,053–1,064 m provided clastic carbonates intercalated with sand. Carbonates can be characterized as skeletal grainstones. Poorly preserved nannofossils indicate a Middle Cenomanian age. Foraminifera are represented by rare, hedbergellid, planktonic species and shallow shelf nodosariid benthic taxa generally derived from sand beds. Carbonates contain abundant echinoid, molluscan, and algal debris, as well as carbonate platform foraminiferal taxa such as *Trocholina*. This sedimentary section must have been deposited in a narrow basin of unknown depth that was receiving allochthonous sediments from different provenances. The terrigenous clastic sediments were probably derived from the Ghana shelf located to the east. The debris flows were most likely channeled along the base to the northern slope of the Ghana Ridge. The crest of the Ghana Ridge, isolated from terrigenous deposition, may have supported a shallow-water carbonate platform that provided the carbonate debris to the Middle Cenomanian section in well 959. The base of this section is formed by a major stratigraphic discontinuity.

The section between 1,064 and 1,159 m depths, below the discontinuity, is composed of pyritic sandy siltstone with scattered phosphate claystone in its upper part. It also contains sideritic siltstone and argillaceous sandstone with common terrestrial plant debris in the lower part. Only very rare planktonic and benthic foraminifera occur in the upper part, including the Lower Albian *Blefuscuiana infracretacea*. The whole section likely represents a transitional marine facies with brief periods of more shallow normal marine deposition of Early Albian age, which is not very firmly documented but can be correlated with industrial wells in the area (Nemčok et al., 2004b). The lower part likely indicates fluvial to brackish paleoenvironments. Terrestrial pollen and spores indicate an Upper Barremian–Middle Albian age range. They are similar to pollen and spores determined from the uppermost Aptian–Lower Albian Bima sandstone in the Benue Trough (Allix et al., 1981).

Site 962 contains the least stratigraphically complete section of all DSDP Leg 159 sites (Figures 4.6a–b). Again, we will focus only on its Cretaceous portion.

A depth of 69 m in the well is characterized by the occurrence of an exceptionally large unconformity, represented by a missing Santonian-Oligocene sedimentary section. This unconformity caps a palygorskite-rich claystone present at 69–71 m depths. It contains radiolarians of the Coniacian age. This claystone is unconformably underlain by carbonaceous claystone with phosphate hardgrounds with radiolaria and rare nannofossils of the Late Cenomanian age, penetrated in a depth interval of 71–83 m.

This condensed section overlies a thick interval of interbedded sand and clastic carbonate, penetrated between depths of 83 and 393 m. The interval contains nannofossils, benthic foraminifera, and marine palynomorphs, allowing a good correlation with uppermost Albian–lower Upper Cenomanian. The interval is similar to the partially equivalent Cenomanian–Turonian interval from well 959, indicating rapid, allochthonous sedimentation from sediment flows derived from both terrigenous and carbonate provenances. The uppermost Albian to the Lower Cenomanian part of this section coarsens upward. It contains progressively less carbonate and more sand in the same direction and its facies trend is similar to that in the analogous section in wells 959 and 960. The exception here is that this regressive interval is overlain by the Middle-to-Upper Cenomanian section, characterized by claystone dominating over sandstone and by an increased amount of allochthonous carbonate beds, indicating a flooding of the margin. This transgression here likely culminated with low-oxygen facies, deposited at the time of maximum flooding. The facies are represented by the radiolarian-rich, non-calcareous claystone.

The chronostratigraphic integration of the determinations made in the described two wells with three more wells in the area indicates that it is the middle Cretaceous–middle Late Cretaceous depositional history that records a transition from syn-rift to post-rift conditions of this part of the Equatorial Atlantic.

The middle Cretaceous–middle Upper Cretaceous section exhibits the very heterogeneous chronostratigraphy that would be expected of such a transition. However, apparently the least variable stratigraphy represents the oldest section, which recorded a change from continental to shallow marine deposition. Few findings of spores and pollen from the continental facies in well 959 and the lacustrine section in well 960 indicate that the age of the section is younger than Late Aptian.

The overlying Lower Albian section in well 959 represents the earliest marine facies recognized in the area, being age equivalent to poorly dated near-shore brackish and fluvial/deltaic facies in well 960 and the industrial well in the area, which may, however, contain a minor marine influence. It needs to be emphasized that the poor age control represents a serious problem, probably masking greater chronostratigraphic variability in this oldest section in the area.

The overlying Upper Albian–Santonian sedimentary section is characterized by better chronostratigraphic control. A highly variable chronostratigraphy for this section is rather typical. When we take a look at the three-dimensional distribution of post-pull-apart lows and highs in the region (Figure 4.7; Nemčok et al., 2004b) and place all recognized hiatuses and erosional unconformities from the area on hypothetical places of interpreted seismic images (Nemčok et al., 2004b), it becomes apparent that a variable chronostratigraphy is most likely controlled by local structures.

None of the well data provide evidence about the Middle Albian section in the area. The Late Albian seems to represent the first development of the regionally extensive shallow-water marine facies. The industrial well in the deep basin, offshore Côte d'Ivoire, indicates a transgression event, recorded by brackish, coarse-grained facies giving way to shallow-water marine claystone and muddy limestone. Well 962, which unfortunately did not penetrate the sedimentary section older than Upper Albian, is characterized by rapid allochthonous sedimentation during the Late Albian. It deposited clastic carbonate debris, which was most likely sourced by carbonate platforms at structural highs isolated from terrigenous sedimentation such as the crest of the Ghana Ridge. Erosion of such structural highs is documented by Late Albian stratigraphic hiatuses in wells 959 and 960 located close to the Ghana Ridge crest. Terrigenous, sandy sediment flows at Site 962 may have derived from near-shore depositional environments such as the one represented by the lower, brackish portion of the transgressive section in the industrial well in the area (Nemčok et al., 2004b).

The Early-Middle Cenomanian time period in well 962 is characterized by a rapid regressive deposition, involving progressively larger amounts of the coarse-grained terrigenous sediment. This regression interpretation is in accordance with the observation of the extensive regional erosion in well 960 and one industrial well in the area (Nemčok et al., 2004b). No Lower Cenomanian section is penetrated by well 959, which, however, atypically contains a thin, allochthonous, sand/limestone Middle Cenomanian section.

The Late Cenomanian can be characterized by initiated transgression in the area controlled by well 962, followed by initiated deposition in the area around well 959 in the latest Cenomanian, and followed by the Early Turonian deposition initiation in one industrial well in the region (Nemčok et al., 2004b).

The character of the basin during this time was still narrow, leading to a stratified water column and to low-oxygen facies deposition recorded by wells 959 and 962. The industrial well in the area (Nemčok et al.,

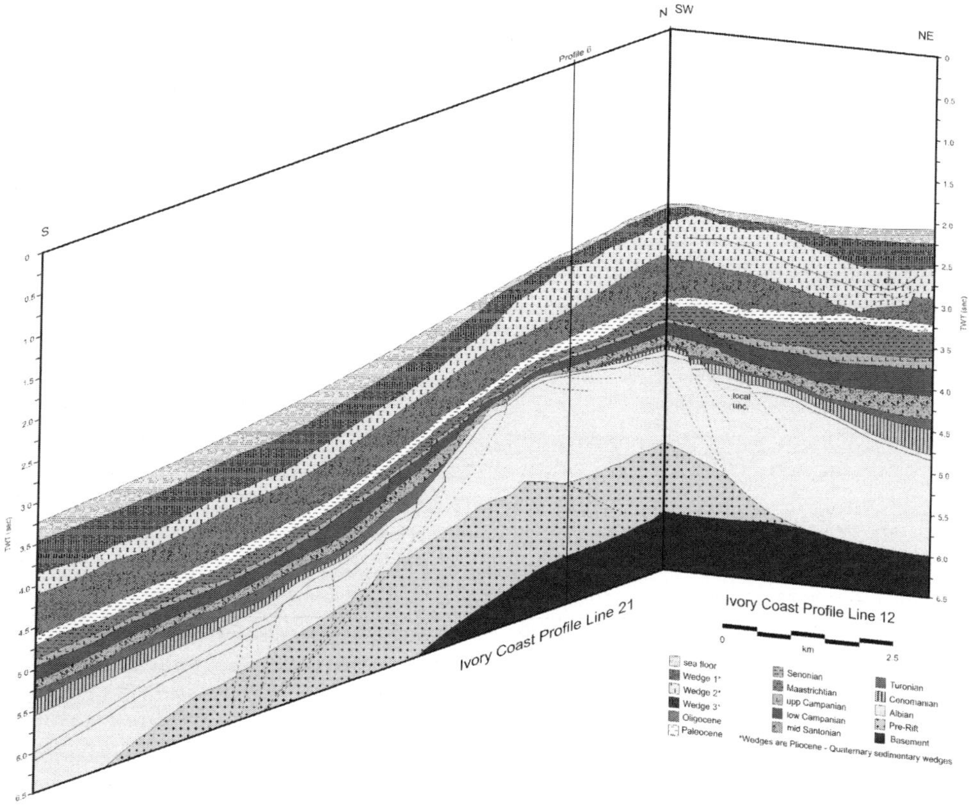

Figure 4.7. A block diagram constructed from portions of interpreted seismic profiles in offshore Côte d'Ivoire (Nemčok et al., 2004b). Note that it took the whole Late Cretaceous deposition to change the high and low topography the failed pull-apart basin system left after the continental breakup.

2004b) indicates that this transgression was quickly followed by progradation of sand wedges and renewed erosion before the end of the Early Turonian. This regression is responsible for the occurrence of continued coarse-grained sediment flows in some anoxic depocenters such as the one documented in well 962. The sources for this deposition were most likely eroding crestal regions of the isolated structural highs. This situation is indicated by a continuing stratigraphic hiatus in well 960 at the Ghana Ridge crest. The sediment flows must have been bypassing the area with well 959, which underwent sediment starvation and development of phosphate hardgrounds. The well 959 area was located at shallower water depths than the well 962 area. As a result, it was not affected by a complete anoxia. Its depositional area is characterized by the occurrence of a community of low-oxygen opportunistic benthic foraminifera known to be endemic to the Gulf of Guinea margin.

The terminal Turonian eustatic sea-level low is most likely responsible for erosional removal of the Middle-Upper Turonian sedimentary section. Following the event, deposition resumed during the latest Turonian in the well 959 area and continued into the Early Coniacian. It resulted in a small amount of sandy, lowstand deposits with allochthonous marine sediment with inner-to-middle shelf provenance. Well 962 and one industrial well in the area (Nemčok et al., 2004b) are indicative of the non-deposition in the same time.

The Early Coniacian–Early Santonian is also the time when the last well in the area records a deposition of thick, sandy, and clastic carbonate sediments, marking a first marine incursion in the region, which is recorded during the Late Albian in well 962, the Middle Cenomanian in well 959, and the Coniacian–Early Santonian in well 960. The marine incursion was rather close to a system of eroding structural highs located in a relief characterized by a relatively steep gradient. The combination of data from wells 959, 960, and 962 points to the extensive erosion of the Ghana Ridge and sediment transport occurring preferentially down the southern slope of the ridge and/or to the west toward the Romanche Fracture Zone and well 962 area, while the northern slope and well 959 area was mostly bypassed.

The Coniacian section in well 960 indicates the beginning of a flood in the Late Coniacian. This corresponds to the deposition initiation in areas with wells 959 and 962. Well 962 records a deepwater radiolarian mud with authigenic palygorskite. Well 959 penetrated dysoxic mud with abundant opportunistic benthic foraminifera. The Late Coniacian is a strong candidate for a fully developed syn-drift stage of the Equatorial Atlantic. Unlike the earlier Cenomanian assemblage, the benthic foraminifera of the Upper Coniacian section at well 959 are cosmopolitan, observed in the North Atlantic, the South Atlantic and the Caribbean, indicating a deepwater connection between the Central and South Atlantic Basins with water depths of 200 to 1,000 m. The connection could be initially narrow and located near the Romanche Fracture Zone.

Warm, dense, hypersaline water forming on the African shelf during the Late Coniacian to Early Santonian flood may have sunk and formed a vigorous bottom-water current through the well 962 area, similar to the movement of intermediate Mediterranean water through the Strait of Gibraltar today. This could also explain the palygorskite development in the well 962 area.

Low-temperature thermochronology

Apatite fission track thermochronology

The method yields information on the time-temperature history of the apatite-bearing rock (Laslett et al., 1982, 1987; Vrolijk et al., 1992). It is based on tracks formed by fission decay occurring in nuclei of U^{238} atoms, which are thermally instable. At temperatures above 60°C, the developed tracks are progressively shortened, that is, annealed. At temperatures of ~120°C, the tracks are fully annealed. This temperature is called total annealing temperature (e.g., Donelick et al., 2005). The t-T conditions for track annealing vary with apatite composition (Green et al., 1986a, 1986b) and apatite solubility (Donelick et al., 2005). The apatite represents a group name for a series of related minerals, which can be roughly described by the chemical formula $Ca_5(PO_4)_3(OH,F,Cl)$.

Fission tracks in the apatite grain (Figure 4.8; Geotrack International, 1999a) are continuously formed by the spontaneous fission decay of the radioactive isotopes U^{238}, which represent impurities within the apatite crustal lattice. The apatite usually contains about 10–100 ppm of uranium, although higher concentrations also have been reported (see Duddy and Kelly, 1999). The ratio of the concentration of the daughter products in the apatite grain, that is, the number of fission tracks per unit volume, with the concentration of the original isotope, that is, the number of U^{238} nuclei per unit volume, depends on the time during which the fission tracks have been forming. This time represents the time period when the apatite grain has resided at a temperature cooler than the total annealing temperature. This ratio is used for the calculation of the fission track age from the measured concentrations of the original and daughter products, where the fission track age represents the age of the oldest fission track in the apatite grain.

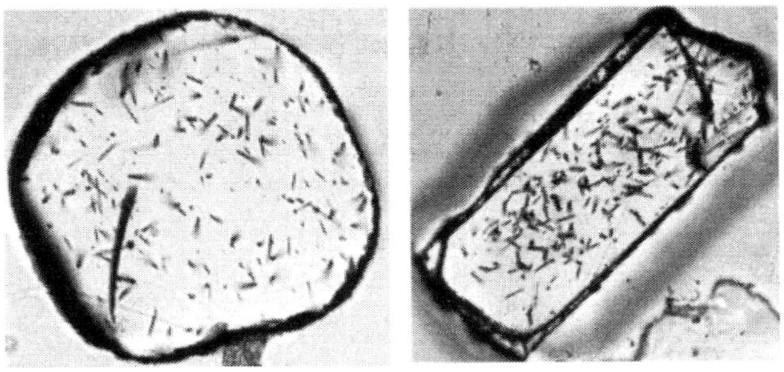

Figure 4.8. Apatite grain with fission tracks (Geotrack International, 1999a).

The development of a fission track can be explained using the process of induced fission in the case of the U^{235} isotope (Duddy and Kelly, 1999). A collision of a thermal neutron with U^{238} causes the atom to split into two subequal fragments, which have high positive charges. The positively charged atoms of the lattice repel each other. As they move apart, they damage the crystal lattice of the apatite by the ionization process, representing a removal of electrons from the lattice atoms. This produces a deformation zone in the crystal, that is, the fission track. The fission track can be characterized as a glass-like discontinuity zone.

Because of the low annealing temperature, that is, the closure temperature, of the system, the fission track age of a rock can be significantly younger than its other radiometric ages, for example, those determined by K-Ar or U-Pb methods. A typical example comes from slowly cooling magmatic rocks. Numerous laboratory and field studies (see Donelick, 1994 and references therein) indicate that apatite fission track ages should be understood as ages of cooling, and not as ages of crystallization. Furthermore, this cooling takes place at different temperature thresholds for different apatite chemistries. For example, fission track ages of the apatite rich in fluorine, in the cases of typical geological cooling rates of 1–10°C/Ma, indicate when the apatite cooled below the temperature of about 110–120°C during geologic history. Fission track ages of the apatite poor in fluorine indicate when the apatite cooled below the temperature of about 140–150°C.

The distribution of the etched fission track lengths in the apatite grain reflects the temperature history the apatite grain underwent during the fission track formation. The initial fission track length is about 16.3 μm (e.g., Donelick, 1994). At sufficiently high temperatures for times of 10^6–10^7 years (temperatures exceeding 60°C or in the 120–140°C interval), the initial fission track length undergoes progressive shortening because the disarrangement of the crystal lattice is instable and has a tendency to anneal. The degree to which the initial fission track length shortens depends on the thermal history the fission track underwent. In the case of simple thermal history, the older fission tracks are recognizable from the younger ones according to their shorter length because they underwent more from the total thermal history experienced by the apatite grain.

Apatite fission tracks are typically used for the uplift/exhumation studies of rocks buried above the 5 km depth, due to the fission track threshold preservation temperatures of 110–120°C and 140–150°C for the fluorine-rich and fluorine-poor apatites, respectively (e.g., Green et al., 1986a, 1986b; Donelick et al., 2005 and references therein). The timing of the uplift is then determined by the following approach. One of the possible situations is that the studied rocks were buried prior to their uplift at depths with temperatures above 110–120°C in the case of fluorine apatites and 140–150°C in the case of F-CL-OH apatites. These temperatures are high enough for forming fission tracks to anneal instantly after their development. As the sampled rock ascends above the depth with respective total annealing temperature conditions, developed fission tracks do not disappear but start to accumulate, undergoing the aforementioned time-dependent shortening. If the process of exhumation occurs rapidly, the fission track age indicates the cooling age. However, slow exhumation causes prolonged residence at partial annealing temperatures (i.e., partial annealing zone between a temperature of about 60°C and, depending on apatite kinetics, a total annealing temperature of 110–150°C), with attendant track shortening and hence, age reduction. In this case, the fission track age and the length distribution contain the information on the thermal history and inverse modeling of these data is needed for its extraction.

Another possible situation is when the studied rocks, which were once at surface, never got buried below the depth with temperatures causing total annealing. In this situation, the studied apatite grains will preserve the information about previous thermal histories of the grains. In the case of sedimentary rocks, the apatite grains would hold the record of the thermal history of the source areas. In the case of crystalline rocks, apatite fission track ages would provide information on previous exhumation events (i.e., cooling event) that brought the rock to the surface prior to burial.

Apart from the total annealing threshold temperatures, a different chemical composition controls a different annealing kinetics (Green et al., 1986a, 1986b; Crowley et al., 1991; Carlson et al., 1999; Barbarand et al., 2003). For example, data published by Geotrack International (1999b; Figure 4.9) document systematic control of the chlorine content in the apatite grain of the annealing kinetics, indicating a faster annealing with a decrease in chlorine content.

When the studied rocks are sedimentary rocks, detrital apatite grain populations can show variations in:

1) sediment provenance area;
2) thermal history before the deposition;
3) chemical composition; and
4) abundance of occurrence.

Routinely, either composition or solubility is determined to characterize annealing kinetics of each individual grain and used for the study of the thermal history. The composition represented by % Cl is analyzed by using Electron Microprobe Analysis. The size of the c-axis parallel diameter of the polygon marking the intersection of an etched fission track and the grain surface, Dpar, is used as a proxy for solubility.

The apatite grains for the method are gathered by gravity and magnetic separation from the sandstone, coarse siltstone, silty shale, and many igneous and metamorphic rock types, where they form one of the most resistant accessory minerals.

The grain preparation starts with revealing the fission tracks in apatite by etching with acid under controlled conditions. The acid preferentially etches deformation zones in the crystal lattice; that is, fission tracks, while the undeformed parts of the lattice undergo much less etching. Each etched track can be characterized by the intersection of the track with the grain surface (i.e., etch pit).

Subsequently, uranium must be quantified for each analyzed grain or crystal. Two methods are currently employed. Individual grains can be analyzed by laser ablation inductively coupled with plasma mass spectrometry (see Hasebe et al., 2004; Donelick et al., 2005). Alternatively, the external detector method (see Tagami and O'Sullivan, 2005 and references therein) entails the use of a low-uranium detector at each apatite mount for monitoring the uranium content through inducing fission of U^{235} with thermal neutrons in a reactor. Each grain is covered by a thin layer of muscovite with low uranium content. Extra muscovite layers are also placed separately on small glass plates, located on

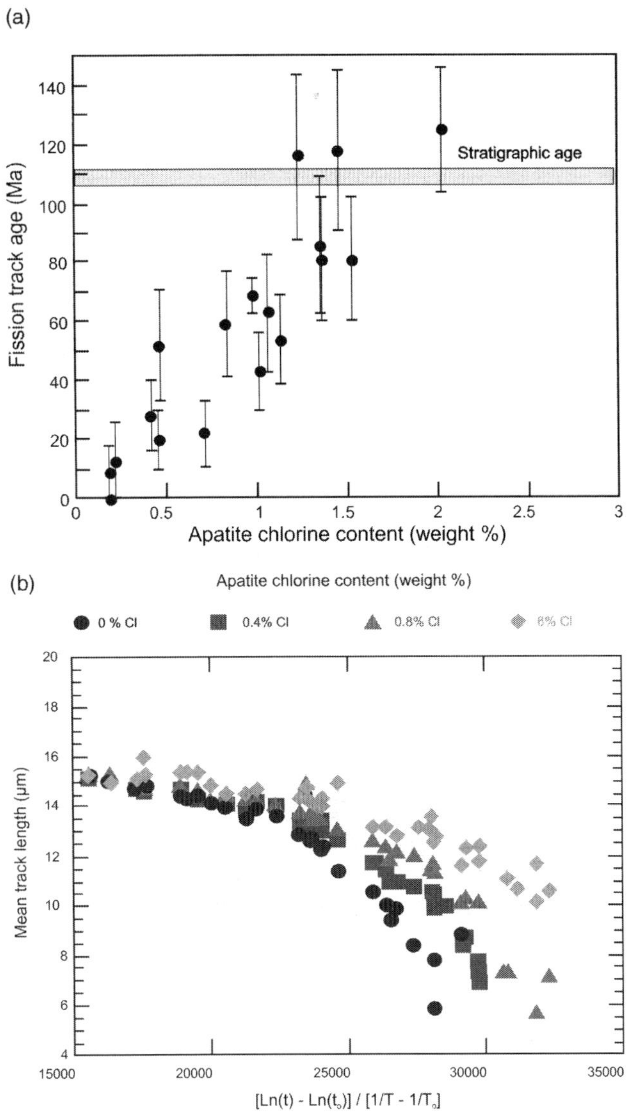

Figure 4.9. (a) Fission track age versus Cl content for individual apatite grains from a single sample of volcanogenic sandstone from a present-day temperature of about 95°C in the Flaxmans Well, Otway Basin (Geotrack International, 1999b). Fission tracks in fluorine-rich (Cl-poor) grains are more easily annealed than in grains with higher Cl contents. As a result, zero fission track ages (totally reset) are measured in Cl-poor grains while Cl-rich grains in the same sample have undergone little or no age reduction. (b) Laboratory annealing studies indicating a systematic chlorine content influence on annealing kinetics Geotrack International, 1999b. In this plot, measured mean lengths are plotted against a function of annealing temperature and time, designed to bring all data into common scale. At any given temperature-time combination, low-Cl apatites show a degree of annealing greater than Cl-rich apatites.

the sides of samples and used as dosimeters. Samples are then placed into small plastic containers and irradiated in a nuclear reactor. The irradiation is followed by etching of muscovite sample covers and dosimeters for induced fission tracks (see Donelick, 1994).

The fission track ages can be calculated by zeta calibration methods on apatite samples, providing mean, pooled, and central ages (see Hurford and Green, 1983; Galbraith, 1992; Gallagher et al., 1994). The pooled age can be calculated only in cases when all apatite grains form a monomictic population (Donelick, 1994), meaning that all fission track density counts from all the studied grains can be pooled. In more typical rift and passive margin situations, the apatite grains are polymictic, belonging to various sediment provenances with different thermal histories and have different chemical compositions causing the different annealing of individual grains (Green et al., 1989a). In this case, the standard χ^2 test allows us to determine whether the studied data set can be characterized by variations beyond the conventional Poissonian statistics used for monomictic cases. In polymictic cases (Figure 4.10; Gallagher et al., 1994), the central age handles the problem of non-Poissonian distributions and provides a more robust determination of the central tendency of the single grain age data. It is calculated as a mean of log distribution of single-grain ages, which is weighted by individual measurement precision. The precision of the central age can be estimated by the 1σ error (Green, 1981), which indicates the analytical precision, and by the age dispersion, which is represented by the relative standard deviation of the central age. The monomictic grain population is indicated by a deviation below 10 percent. In this case, the central and pooled ages converge and the data set passes the standard χ^2 test.

The spread in the individual grain age measurement can be evaluated in Galbraith's (1988) radial diagram (Figure 4.11; Gallagher et al., 1994). The diagram normalizes the grain ages by their specific precision estimates, plotting them against the reciprocal of their precision. The most precise determinations are plotted at the largest distances from the origin.

Typically, published results of fission track analyses on the onshore portions of passive margins indicate a younging trend of ages from the continent interior toward the shoreline, that is, with increasing elevation (e.g., Moore et al., 1986; Gleadow and Fitzgerald, 1987; Bohannon et al., 1989; Brown et al., 1990; Dumitru et al., 1991; Fitzgerald, 1992; Foster and Gleadow, 1994; Gallagher et al. 1994; Gallagher and Brown, 1999; Brown et al., 2000; Raab et al., 2001; Figure 4.12). This fission track age trend can be explained by a thermal influence of the rift process, characterized by a transient perturbation to the steady-state thermal regime, which causes the fission track annealing to some extent. It can also be explained by a post-rift erosional exhumation

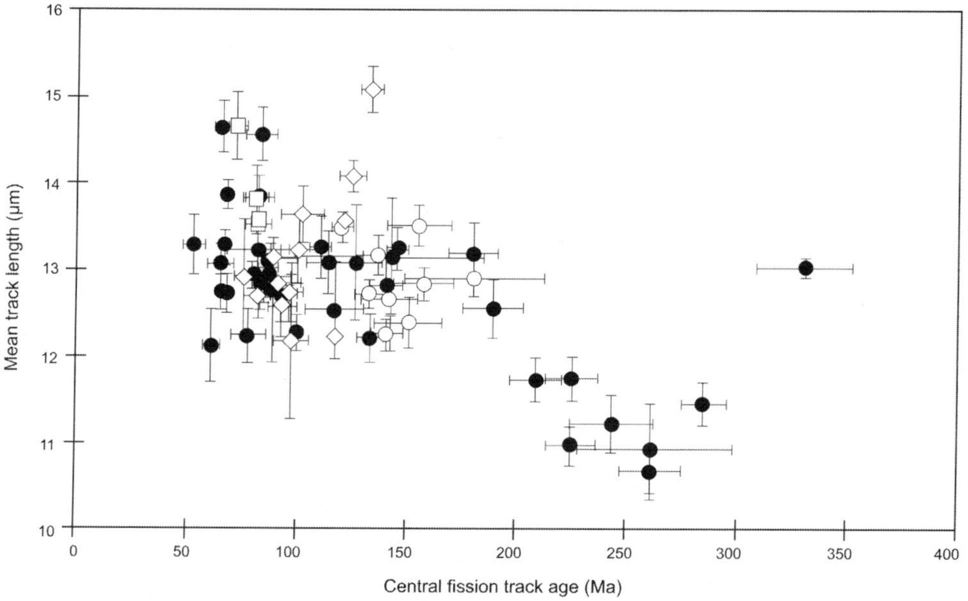

Figure 4.10. Measured fission track age and mean track length for the four subgroups studied in southeastern Brazil (Gallagher and Brown, 1999). The data are divided into four groups. Square – sample from recent intrusive rock, open circle – sample from sediment, diamond – sample from sediment-basalt contact, solid circle – sample from basement.

(Figure 4.13), characterized by the sample residing at a certain steady-state thermal regime, which causes a similar level of annealing prior to the sample exhumation.

Because the fission track annealing is temperature-controlled and not pressure-controlled, the influences of the rift-perturbed thermal regime and exhumation cannot be separated from the fission track data alone. However, Gallagher and colleagues (1994) used a 2-D finite-difference thermal model of the thermal history at the rift margin to evaluate the role of the rift-perturbed thermal regime on the fission track thermochronology. The model indicates that the relevant thermal history recorded by fission track data is controlled dominantly by exhumation. Furthermore, there will be a component of the denudational rebound in the present-day topography elevation (see Stephenson and Lambeck, 1985; Gilchrist and Summerfield, 1990; Brown, 1991; Ten Brink and Stern, 1992). Therefore, the fission track data from the rift margins tend to provide constraints for long-term topography development and for determining rebound contribution, rather than directly recording the surface uplift. In other words, the fission track data cannot directly constrain the vertical motion of the Earth's surface, but they can provide information on exhumation (see England and Molnar, 1990). Therefore, estimations of denudation and initial elevation can be related to the present-day topography by using a suitable isostatic model, provided the calculation is made for a large enough length scale.

An example of the use of the apatite fission track method in rift terrain comes from the Basin and Range province in the United States (Stockli et al., 2003). The study is focused on the White Mountains crustal block, representing a footwall to the steep-dipping White Mountains Fault Zone with total normal fault displacement of about 8 km. The footwall tilt is recorded by a 25° dip of the pre-extensional Miocene volcanic rocks resting on the basement. Here, the combination of apatite fission track and (U-Th)/He thermochronological data indicates footwall exhumation as rapid, and starting at about 12 Ma.

Examples of the use of the apatite fission track method in the passive margin terrains come from the Atlantic and Indian Ocean margins (Moore et al., 1986; Miller and Duddy, 1989; Rust and Summerfield, 1990; Dumitru et al., 1991; Gallagher et al., 1994; Steckler et al., 1994; Gallagher and Brown, 1999; Brown et al., 2000; Lisker and Fachmann, 2001; Raab et al., 2001; Cogne et al., 2011).

Using an apatite fission track method for dating the passive margin uplift after the continental breakup, it is advantageous to combine it with an independent data set, biostratigraphically defined prograding wedges in the adjacent deep basin, which are sourced by accelerated erosion of the uplifting margin (Figure 2.7b). Examples of such an integrated approach come from the Atlantic margins (e.g., Brown et al., 1990; Van Balen et al., 1995 and references therein; Henk, 2004,

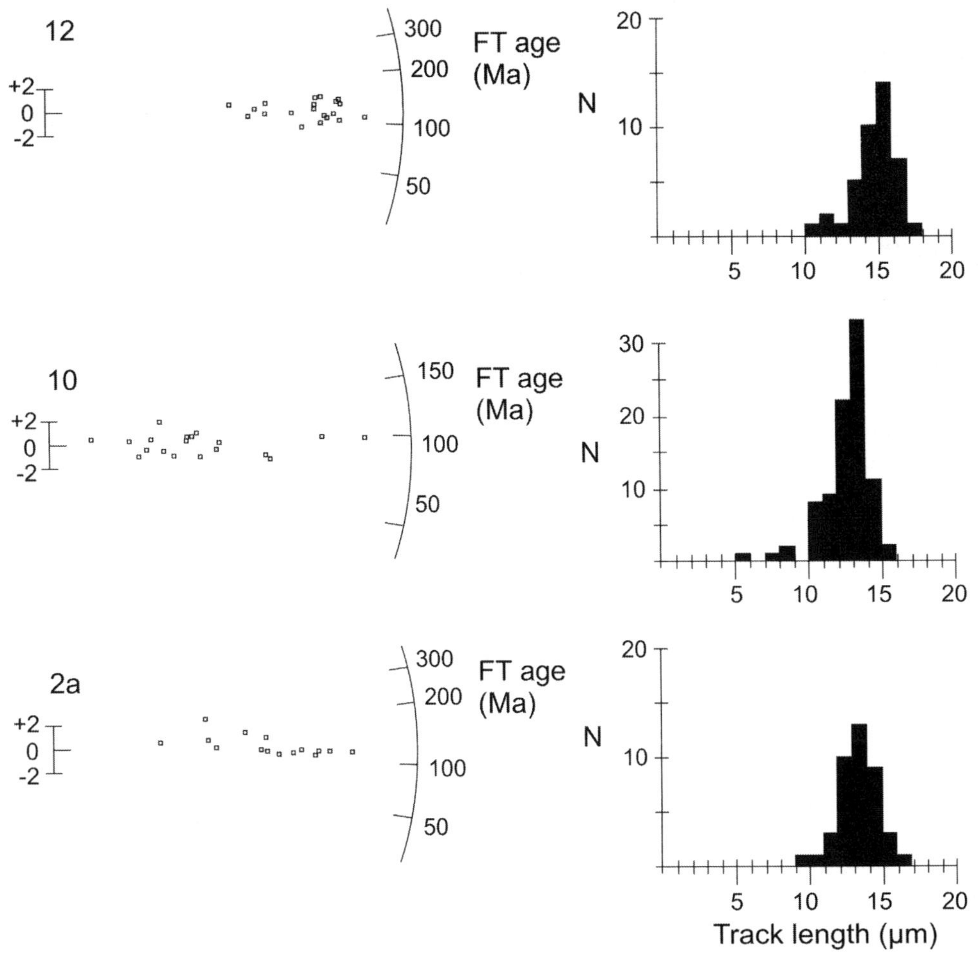

Figure 4.11. Radial and track-length distribution diagrams for one basalt sample and two samples from the sediment directly underlying the basalt, southeastern Brazil (Gallagher et al., 1994).

pers. comm.). We will focus only on the Santos Basin, Brazil, and the Walvis Basin, Namibia, examples, discussed in Chapters 1 and 7.

Zircon fission track thermochronology

The zircon fission track methodology works with fission track threshold preservation temperatures larger than those of apatite fission tracks. As such, the method can provide information on the timing of temperature events due to igneous activity, elevated heat flow and deep burial, due to its total annealing temperature higher than that of apatite.

Like apatites, zircons represent a common grain component of medium- to coarse-grained clastic sediments and they are also present as accessory minerals in igneous and metamorphic rocks. They are also rather erosion-resistant, capable of remaining in the transported sediment load after long transport.

The zircon fission tracks are approximately 50 angstroms wide and 12 μm long (Duddy and Kelly, 1999). The tracks shorten at high temperatures, although the annealing kinetics in zircons is less known than that in apatites. The data from super-deep borehole SG-3, Kola Peninsula, Russia, indicate that it takes temperatures higher than 200°C to cause a detectable annealing, while temperatures of about 300°C have been interpreted as controlling the total annealing in South Island, New Zealand (Duddy and Kelly, 1999).

Because of the high thermal stability of tracks in the zircon grain, the zircon fission track method can be used even in scenarios proven not to work in the case of standard radiometric techniques. A typical example would be trying to date magmatic rocks from highly weathered outcrops, which would prevent the usage of standard radiometric techniques. Nevertheless, we still need to count with minimum age determination because the cooling of pluton postdates its emplacement, while

(a)

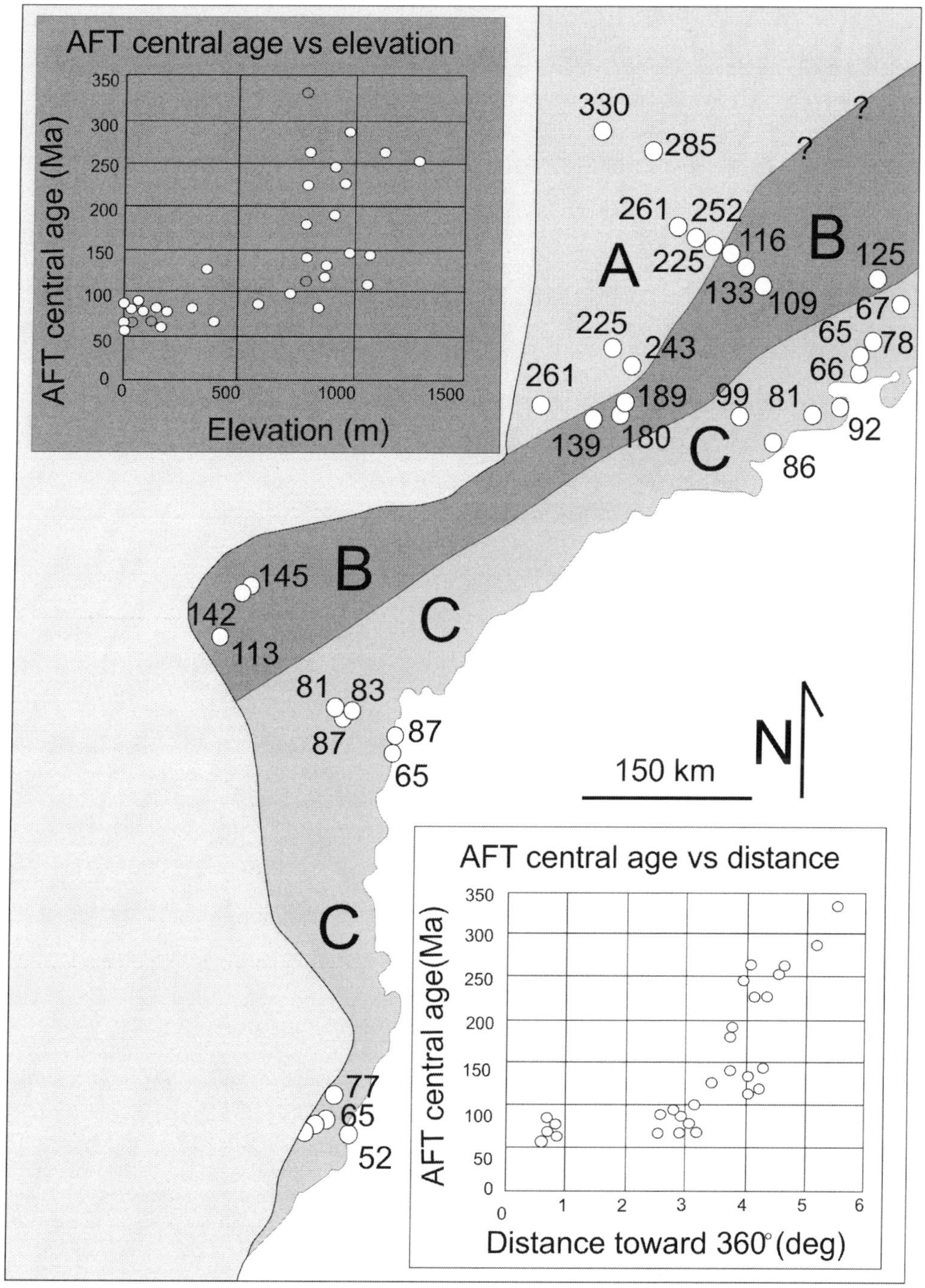

Figure 4.12. (a) Apatite fission track ages (Ma) for the onshore Santos Basin (Gallagher et al., 1994). (b) Map of the Santos Basin (combined from de Almeida, 1976; Ojeda, 1982; Pereira et al., 1986; Pereira and Macedo, 1990; Demercian et al., 1993; Gallagher et al., 1994; and Karner, 2000). The seafloor bottom contours indicate a southeast-facing double horseshoe shape of the basin. The basin lies predominantly on the thinned continental crust, as indicated by the position of the continent–ocean boundary determined from the gravity data and interpreted ends of the oceanic fracture zones. The distance between the Cretaceous hinge zone and the ridge of the Serra do Mar Mountains documents the rift shoulder retreat distance from the Cretaceous until the present. The lateral extend of the rift shoulder roughly corresponds to the width of the basin with salt tectonic. The distance between the Cretaceous hinge zone and the line indicating the upper extent of the salt shows the distance traveled by the upper salt portion during the southeast-directed gravity gliding. The uppermost portion of the salt horizon is very thin and the lowermost portion is significantly thickened. The distance between the continent–ocean boundary and the salt extent line documents the distance traveled by salt during gravity gliding. It ranges from 56 to 162 km in different locations. (c) Fission track ages as a function of the distance from the continent boundary along transect through the Santos Basin passive margin, southeastern Brazil (Gallagher et al., 1994).

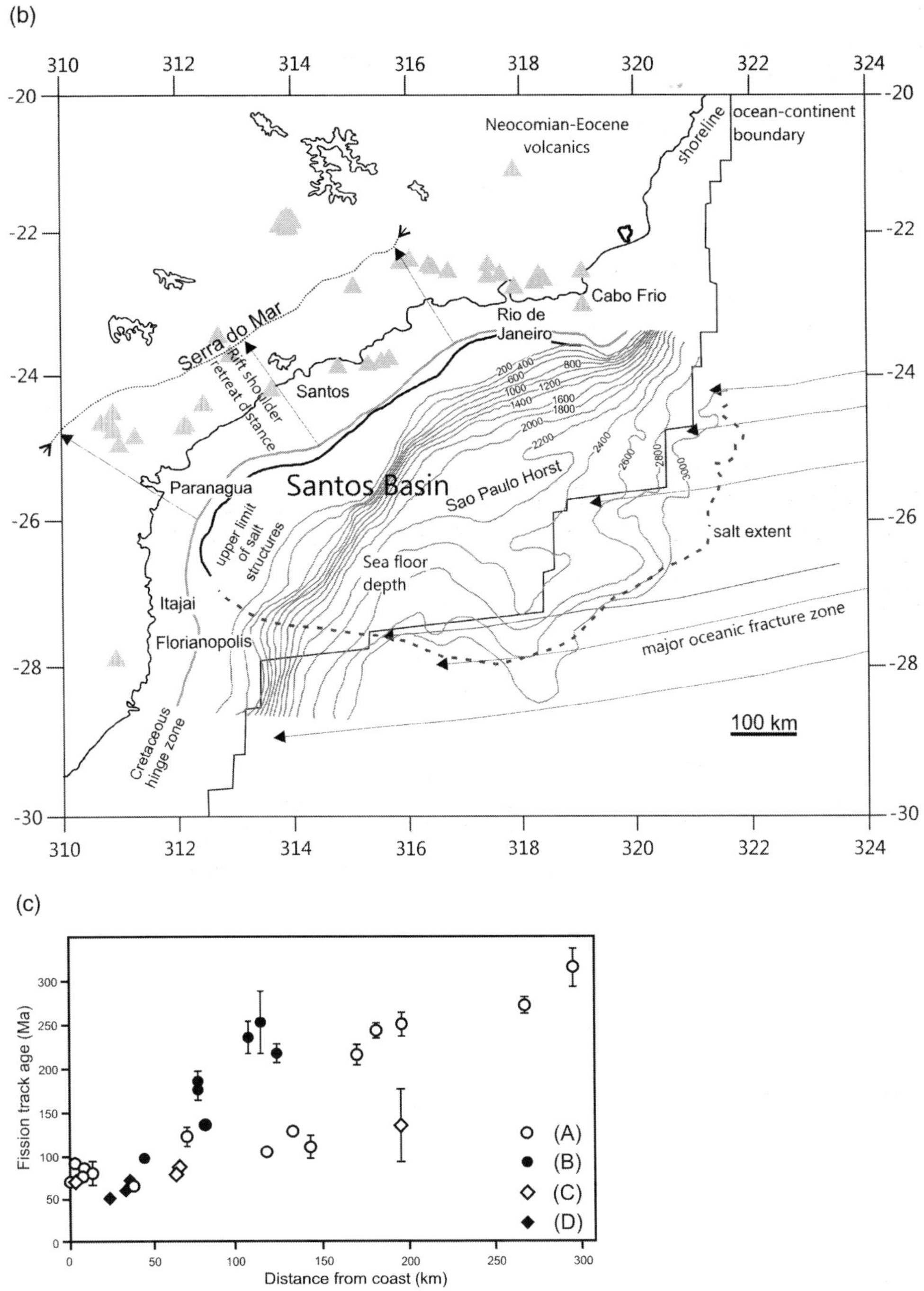

Figure 4.12 (*cont.*)

the determination for weathered volcanic rock, which cooled rapidly, is more or less accurate. It is also interesting that dating syn-rift and syn-breakup basalts not containing zircons can be made possible by working with zircons from surrounding metamorphic zones developed in rocks that contain zircons, such as sandstones and granitoids.

An example of the use of the zircon fission track technique in the rift settings comes from the Mahanadi Basin in onshore eastern India (Lisker and Fachmann, 2001). In order to (1) study variations in depth and the rates of denudation, and (2) determine the timing of rift events, the zircon fission track analyze was combined here with:

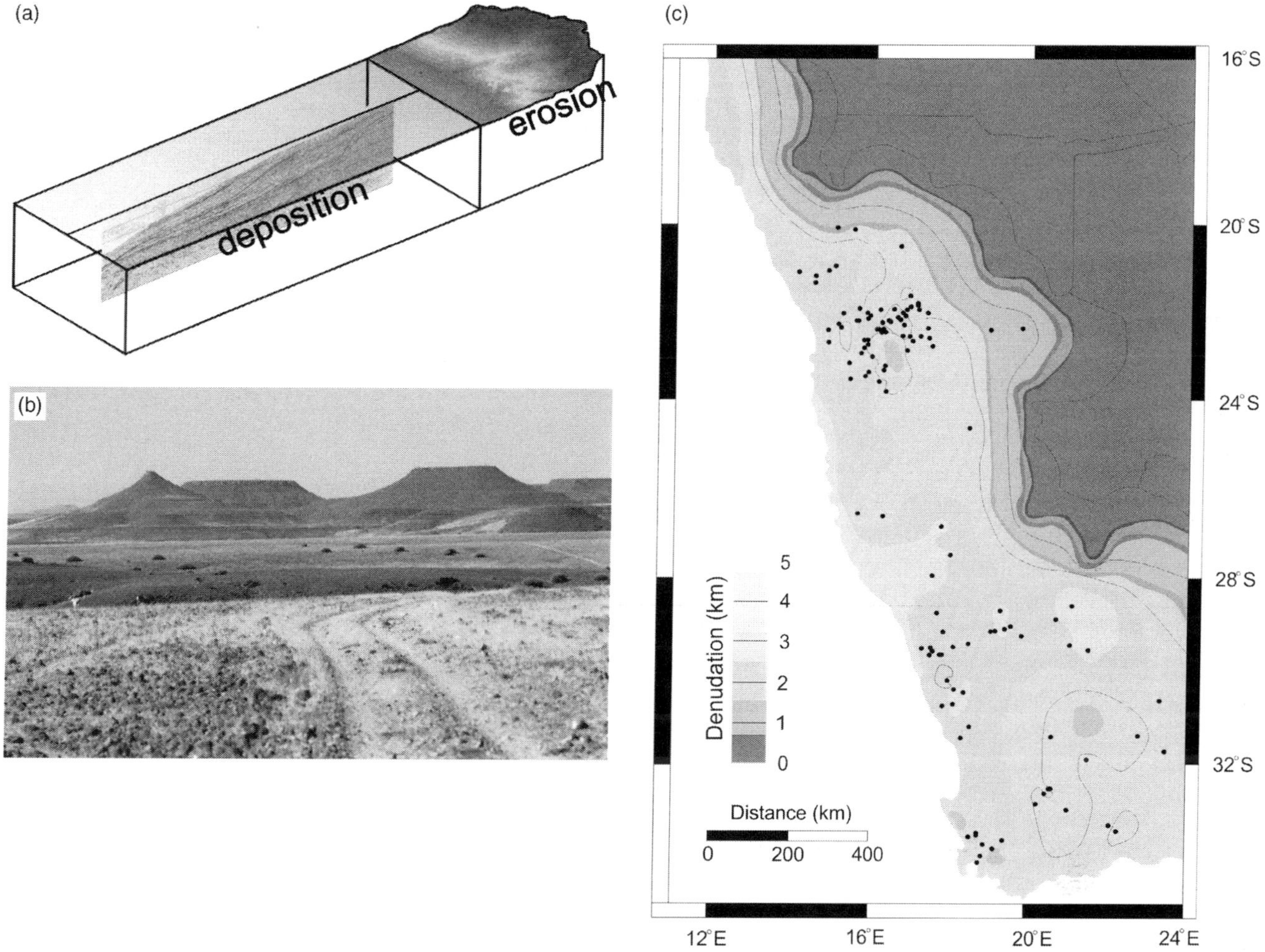

Figure 4.13. Mass balance study done for eroded sediment in onshore Namibia and deposited sediment in the offshore Walvis Basin (Henk, 2004, pers. comm.). (a) Studied profile with adjacent sediment provenance. (b) Table mountains in the sediment provenance area. (c) Missing burial calculated from apatite fission track data (Gallagher and Brown, 1999; Brown et al., 2000; Raab et al., 2001).

1) apatite fission track analysis;
2) an $^{40}Ar/^{39}Ar$ radiometric technique determining cooling ages of biotites and amphiboles from the metamorphic basement; and
3) the intrusion ages of pseudotachylites and mafic dykes.

To date the pseudotachylite and biotite from locations near large shear zones by laser fusion, the samples were prepared as thin sections. They were mapped by scanning electron microscope using a backscattered mode. In order to date the mafic dykes and metamorphic rocks, the samples were prepared as mineral concentrates of plagioclase, potassium feldspar, biotite, and hornblende. They were gathered from crushed and sieved rock powder using a magnetic separator, heavy fluid, and ultrasonic cleaning. The final mineral concentrate was then handpicked under the microscope. Covered in cadmium foil and placed in quartz glass containers, the mineral samples were then irradiated in the nuclear reactor.

In order to do the correction for interfering irradiated Ar isotopes, the $^{37}Ar/^{39}Ar$ analyses were done using a noble gas mass spectrometer. Argon release from the biotite thin sections was performed by laser ablation. The argon was released by the total fusion in the small grain case and by step heating using a power increase of the defocused laser during a time step of 30 s in the large grain case.

The results were corrected for:

1) mass discrimination;
2) radioactive decay of ^{37}Ar and ^{39}Ar;
3) isotopic interferences of calcium and potassium;

4) atmospheric contamination; and
5) systems blanks because of the measurements at each second step.

The isotopic ages for every step were calculated by using McDougall and Harrison's (1988) general equation. The error was determined by progradation of analytical mass discrimination, and standard measurement.

The calculated $^{40}Ar/^{39}Ar$ ages of the plagioclase from onshore tholeiitic basalt dyke and feldspar from offshore basaltic trachyandesite flow are about 116 Ma and 107 Ma, respectively, indicating Aptian and Albian crystallization ages, respectively (Lisker and Fachmann, 2001). Thermal demagnetization of the dyke (Fachmann, 2001) indicates a paleoposition identical to basalts of the Rajamahal Traps in northeast India (Sakai et al., 1997). These ages are younger than the breakup unconformity and are related to the intrusive and effusive activity in the Mahanadi and Bengal Basins, Rajmahal and Sylhet Traps. The offshore trace of this activity is represented by the 85°E Ridge. The gravity models of the ridge indicate a hot spot associated with magmatism, which includes a crustal underplating underneath the ridge (see Subrahmanyam et al., 1999).

The calculated $^{40}Ar/^{39}Ar$ ages from hornblende and biotite from the high-grade metamorphic rock of the basement are about 845 Ma, and 662 and 593 Ma, indicating Late Proterozoic metamorphism. Few samples from pseudotachylites with ages of around 515 Ma indicate a Cambrian, pan-African orogenic signature (Lisker and Fachmann, 2001).

When all applied methods are compared, it is apatite and fission track ages that allow us to recognize a development of the Mahanadi Basin containing both Carboniferous-Permian and Middle Triassic extensional events and the continental breakup, which occurred during the Early Cretaceous (Lisker and Fachmann, 2001).

Apatite (U-Th)/He thermochronology

(U-Th)/He dating (see Ehlers and Farley, 2003) offers a dating alternative associated with threshold temperatures lower than those of the fission track techniques. It is based on the accumulation and diffusive loss of helium produced by the alpha decay of uranium and thorium trace elements in the apatite crystal lattice. The amount of helium produced in a mineral is expressed as (Ehlers and Farley, 2003):

$$^{4}He = 8\,^{238}U(e^{\lambda 238t} - 1) + 7/137.88\,^{238}U(e^{\lambda 235t} - 1) + 6\,^{232}Th(e^{\lambda 232t} - 1), \quad (4\text{-}1)$$

where ^{4}He, ^{238}U and ^{232}Th are the present-day amounts, t is the accumulation time or He age, 1/137.88 is the present-day $^{235}U/^{238}U$ ratio, and λ is the radioactive decay constant, where:

$$\lambda_{238} = 1.511 \times 10^{-10}\ yr^{-1},\quad \lambda_{235} = 9.849 \times 10^{-10}\ yr^{-1},\quad \lambda_{232} = 4.948 \times 10^{-11}\ yr^{-1}. \quad (4\text{-}2)$$

The comparison of measured parent and daughter isotopes defines the time since the closure under the assumption of no additional helium sources.

Over the past decade, the method became an important method focused on rocks cooling as they pass through the upper 1–3 km of the crust, bridging the gap beyond the resolution of zircon and apatite fission track techniques. By measuring the amounts of radiogenic ^{4}He, uranium, and thorium in the apatite grain, we can determine the (U-Th)/He age.

This age is progressively reset by heating due to diffusive loss of ^{4}He. The total loss occurs at a temperature of about 60–75°C for timescales involving millions of years (Farley et al., 1996; Wolf et al., 1996). Both these characteristics make this method analogous to apatite fission track progressive annealing and total annealing for low temperatures.

Although the helium method was used for ages as young as a few hundred kyr of volcanic apatites (Farley et al., 2002) and as old as a few Ga of meteorites (Min et al., 2003), a secular disequilibrium in the ^{238}U series modifies Equation 4-1 for crystallization ages younger than 1 Ma.

Workflow of this method starts with the vacuum extraction of helium by furnace heating (Zeitler et al., 1987; Lippolt et al., 1994; Wolf et al., 1996) or heating with a laser (House et al., 2000; Reiners et al., 2002) and continues with purification and mass spectrometric analysis. Removed from the vacuum system, apatite grains are dissolved, allowing us to analyze uranium and thorium by inductively coupled plasma mass spectrometry. For apatite grains with common uranium, thorium, and helium contents, the helium age dating precision falls below 1.5 percent (Farley, 2002).

It needs to be said that the method is affected by the content of mineral inclusions in the apatite grain. Inclusions of minerals with rather high uranium and thorium concentrations, such as zircon and monazite, tend to yield artificially high helium ages (see House et al., 1997; McInnes et al., 1999). The problem can be treated by the microscopic determination of inclusions, unless the inclusions are below the detection threshold (Farley, 2002). Nevertheless, the indication of a problem can be routinely done by a test on the reproducibility of helium age from a number of apatite grains.

The method accuracy is further affected by the distance traveled by emitted α particles. Because they travel for about 20 µm through the apatite, each group of particles could either lose its own particles due to their ejection from the crystal edges or gain others injected from neighbor grains. This causes potential errors in the helium age determination of up to tens of % (Farley et al. 1996). A typical remedy for the problem is measuring dimensions of the crystal and calculating the α-ejection correction (see Farley et al., 1996). Such ejection relies on a homogeneous uranium and thorium spatial distribution assuming spherical geometry, which is not always the case.

The method's assessment of the cooling history relies on accurate quantitative knowledge of helium diffusivity, which comes from laboratory step-heating experiments. These experiments demonstrate that helium loss is consistent with thermally activated volume diffusion (Zeitler et al., 1987; Lippolt et al., 1994; Wolf et al., 1996; Warnock et al., 1997). After the necessary simplifications, the experimental results can be first converted into diffusion coefficients and then converted into kinetic parameters helping to describe the diffusivity by a temperature-dependent function (Fechtig and Kalbitzer, 1966; Ehlers and Farley, 2003). The simplest description of kinetic parameters uses the closure temperature, which ranges between 60 and 75°C, under cooling rates of about 10°C per million years (see Dodson, 1973; Ehlers and Farley, 2003). Although there seems to be no evidence for closure temperature variations with apatite chemistry, the closure temperature varies with grain size. For example, the grain size change in a minimum dimension from 50 to 150 µm results in a closure temperature change of about 10°C (Farley, 2000).

Yet another factor affecting the method is the helium concentration gradient, which affects the rate of the helium diffusive loss. Because the α ejection reduces the concentration gradient at grain edges, the helium loss rate will be slower, increasing the closure temperature determination in a few degrees Celsius (Farley, 2000).

The zonation of uranium and thorium also affects the measurements. Grains with high uranium and thorium concentrations near the rim lose more helium by diffusion than a homogeneous grain. This results in an apparently younger helium age (Meesters and Dunai, 2002a, 2002b).

The cooling rate of the sampled area is the last factor worth mentioning as affecting the method. A quickly cooled grain undergoes a small diffusive loss, while a slowly cooled one resides for a long time at temperatures under which small differences in loss rate transfer into substantial age differences.

It can be concluded from all of this information that the slower the geologic cooling rate of the sampled rock, the greater the sensitivity of the helium age to less significant factors controlling the helium loss. They include (Ehlers and Farley, 2003):

1) uranium and thorium zonation;
2) apatite grain size; and
3) kinetic parameter variability.

It can also be concluded that a correction, F_T, for the distance traveled by emitted α particles (sensu Farley et al., 1996) works for apatite grains larger than 75 µm, placing a lower limit on the size of sampled apatite grains. This represents a major handicap on sampled grain in comparison with the apatite fission track method, which can be successfully applied to small grains or even subhedral, fragmentary, needle-shaped, or rounded types. As shown by numerous applications of the method to case studies, the attempt to avoid apatites with typical zircon or monazite inclusions typically leads to favoring sampled rocks such as medium- to coarse-grained granodiorites and avoiding alkali granites, fine-grained granitoids, gneisses, volcanic rocks, and, especially, sedimentary rocks.

Vitrinite reflectance and rock-eval pyrolysis

It is advantageous to combine an apatite fission track timing technique with profiles of maturity data, such as vitrinite reflectance, R_o, and pyrolysis-derived temperature, T_{max}, to gain independent constraints on a depth of exhumation of sedimentary complexes. This helps to calibrate the exhumation history derived from the fission track analysis.

A nice example of the use of the R_o data set comes from Duddy and colleagues (1994) (Figure 4.14). Figure 4.14 shows a simple linear maturity–depth relationship in the region with a conductive thermal regime, which is not perturbed by any advective heat transfer. In this situation, the missing portion of the maturity-depth curve converging to the paleo-surface temperature of 15°C allows us to estimate the amount of original burial removed by erosion. As Duddy and colleagues (1994) discussed, the maturity-depth functions can be nonlinear, making the determination of the missing section complicated. For details of the data distributions, the reader is referred to Duddy and colleagues (1994). However, it is useful to discuss the determination of vitrinite reflectance data and T_{max} data, which can be used in the same manner as R_o data, in the following text.

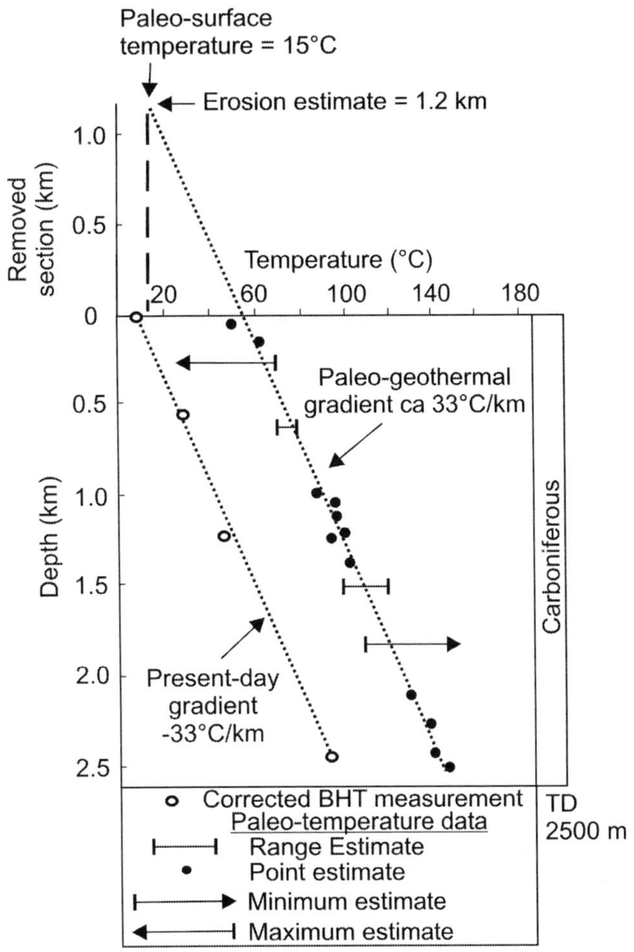

Figure 4.14. Paleotemperature profiles constructed using vitrinite reflectance and apatite fission track data (Duddy et al., 1994). The amount of uplift and erosion is estimated by construction, which is appropriate only for a pure conductive system.

Vitrinite reflectance

The thermal maturity of the hydrocarbon source rock is a useful paleotemperature indicator for sedimentary rocks, which underwent low-temperature metamorphosis in the 50–300°C temperature range (Middleton, 1982; Underwood and Howell, 1987).

The standard methods try to estimate two types of thermal maturity parameters:

1) hydrocarbon generation parameters, which are used to identify the stage of the hydrocarbon generation, and which are independent of the magnitude of the thermal stress;
2) parameters of the thermal stress, which are used for the description of the relative effects of temperature and time.

When working with the vitrinite reflectance method, the situation may arise when two source rocks containing different kerogen types produce the same amount of oil of the same type, but the determined R_o from respective samples can be different (Peters and Moldovan, 1993). In this hypothetical case, the oil type is related to the hydrocarbon generation while the measured R_o is related only to thermal stress. Therefore, R_o at the beginning of the oil generation window can vary among different hydrocarbon source rocks. This is why a R_o of about 0.6 percent is generally considered as indicating the onset of the oil generation window (Dow, 1977; Peters, 1986), while exceptions are possible. For example, the Monterrey shale of the San Joaquin Basin, California, is known to reach the onset at 0.3 percent.

Petroleum geologists generally accept the vitrinite reflectance method as the method providing estimates of the kerogen maturity in sedimentary rocks (Bostick, 1979). Despite the fact that R_o is associated with the thermal stress more than with the hydrocarbon generation, R_o values are generally used as indicators of the onset and end of the oil generation window (Table 4.1). An increase of R_o continues during the whole oil generation window, caused by complex, irreversible, aromatic chemical reactions.

The vitrinite occurs in a variety of sedimentary rocks and is, to a large extent, independent of the lithology. It originates from material derived from lingo-cellulose parts of higher plants. The vitrinite reflectance method can be used only for sediments younger than Silurian because higher plants did not exist in the pre-Devonian.

The samples for the vitrinite reflectance analysis are prepared as follows. Kerogens separated from the source rock samples are coated by epoxy and polished (Bostick and Alpern, 1977; Baskin, 1979). In this stage, they are ready for the measurement of the percent of the incident light of the certain wavelength (usually 546 nm) reflected from phytoclasts from the vitrinite submerged in oil (Stach et al., 1982). The measured vitrinite reflectances in percent represent the average of measurements made on 50–100 phytoclasts in each polished sample. Each R_o value is derived from a set of 50–100 individual measurements.

Measurements of R_o in fewer than fifty phytoclasts do not necessarily provide statistically representative values. Furthermore, complications may occur when:

1) oil-generating macerals retard a normal R_o dependence on maturity (Hutton and Cook, 1980; Price and Barker, 1985);
2) bitumen retards a normal R_o dependence on maturity (Hutton et al., 1980);

Table 4.1. Rock-Eval and other geochemical parameters describing the thermal maturity of the source rock (Peters and Moldowan, 1993).

Maturity stage	PI $[S_1/(S_1+S_2)]$	T_{max} (° Celsius)	Vitrinite reflectance (%)
Onset of the oil generation window	~0.1	~435–445	~0.6
Maximum maturity in oil generation window	~0.25	~445–450	~0.9
End of the oil generation window	~0.4	~470	~1.4

Table 4.2. Some factors influencing the indication value of the vitrinite reflectance (Peters and Moldowan, 1993).

Factor	Effect on vitrinite reflectance
cuttings fallen down below during drilling	decreased value
weakly polished vitrinite	decreased value
contamination by drilling mud (e.g., lignite filling)	decreased value (usually)
oxidized or recycled vitrinite	increased value (usually)
variations in natural reflectance in vitrinite subgroups	increased or decreased value
statistical errors (small measurement number)	increased or decreased value
incorrect maceral identification (e.g., liptinite, solid bitumen	increased or decreased value

3) the fluid overpressure in the sediment retards the maturity (Price and Wenger, 1992);
4) the reflectance becomes anomalously increased by shear deformation of the source rock (Levine and Davis, 1983; O'Hara et al., 1990).

Because of these reasons, it is important to measure R_o values rather carefully. Vitrinite reflectance can be affected by other problems, some of them listed in Table 4.2. Therefore, it is useful to combine several different types of maturity indication data (Table 4.1).

Unlike thermal maturity parameters derived from biomarkers, the vitrinite reflectance is less sensitive and precise at source rock maturities not reaching the oil generation window, that is, less than 0.6 percent. For this range of low maturity values, the biomarker parameters are more precise than vitrinite reflectance for interpretations (Mackenzie et al., 1988). Apparently, this relationship reverses for higher maturity values.

Rock-eval pyrolysis

T_{max}, and also the production index, PI, which represent direct and derived results of the Rock-Eval analysis (Figure 4.15), serve as other maturity data useful for estimations of removed overburden.

The Rock-Eval pyrolysis of the hydrocarbon source rock (Espitalie et al., 1977) is the programmed oven heating of organic matter without oxygen. The Rock-Eval method, in comparison with other extraction methods, is rather fast and requires only small sample volumes.

A detailed description of the pyrolitic conditions is provided by Clemetz and colleagues (1979). Source rock samples less than 100 mg large are pyrolyzed at 300°C for a period of three to four minutes. Then the programmed pyrolysis follows, at which the temperature increases at a rate of 25°C per minute up to a temperature of 550°C. Both parts of pyrolysis are accomplished in helium atmosphere.

The pyrolysis provides a range of measured/calculated parameters (e.g., Peters, 1986). The flame ionization detector detects all organic compounds generated during pyrolysis. The first generation peak, S_1 (Figure 4.15), represents a mass of hydrocarbons in milligrams, which are thermally distilled from one gram of source rock. The second generation peak, S_2, represents a mass of hydrocarbons in milligrams, which are generated by pyrolitic degradation of kerogen in one gram of the source rock. The third generation peak, S_3, represents a mass of CO_2 in milligrams, which is generated from one gram of the source rock during the whole programmed heating up to a temperature of 390°C, analyzed by the detection of thermal conductivity.

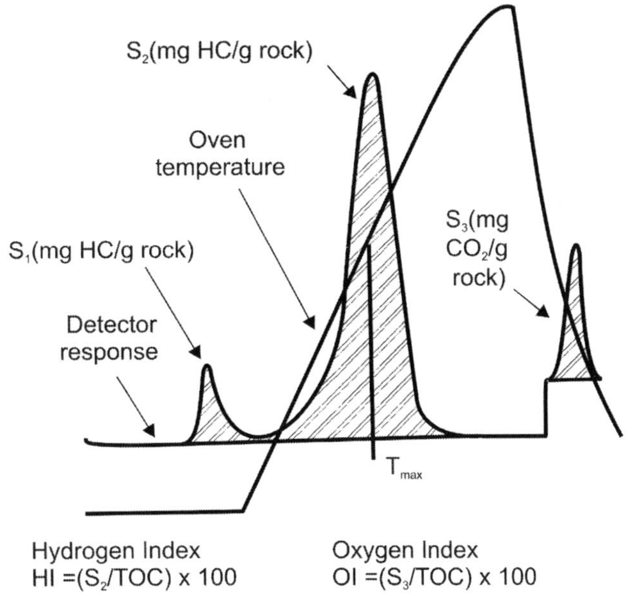

Figure 4.15. Scheme of the Rock-Eval pyrogram showing the evolution of organic compounds from a rock sample during progressive heating (Peters, 1986).

The temperature is constantly monitored during the pyrolysis. The temperature at the second generation peak, S_2, is recorded as T_{max}. This temperature generally increases with the depth of the source rock's burial, and, therefore, can be used as a source rock maturity indicator. Similar to the vitrinite reflectance profile in Figure 4.14, it represents an independent data set that can be used for the estimate of the amount of exhumation in combination with fission track data for the timing of tectonic events.

Although the pyrolysis provides other measured values such as hydrogen and oxygen indexes, described in detail in Peters (1986), Peters and Moldovan (1993), and Welte and colleagues (1997), another maturity indicator, which can be used just like vitrinite reflectance or T_{max}, is the production index, PI. It can be calculated as the ratio of generated hydrocarbons to the total amount of hydrocarbons:

$$PI = S_1/(S_1+S_2). \qquad (4\text{-}3)$$

Toward the end of this chapter, it is useful to mention several remaining dating methods, such as (Geyh and Schleicher, 1990):

1) correlations of studied sections with sections of known age on the basis of ^{18}O and paleomagnetic records;
2) K-Ar parent-daughter isotope ratio method; and
3) ^{14}C, ^{10}Be, ^{26}Al cosmogenic isotope dating methods.

The ^{14}C method works traditionally well with very young lacustrine sediments, being precisely calibrated by dendrochronology to the start of the Holocene (see Stuiver et al., 1998a). The use of methods other than those based on the parent-daughter isotopic ratio, which seems relatively straightforward, draws from their possible cross-calibration, due to their overlapping applicability ranges (Ivanov et al., 2002). For method details, the reader is referred to Geyh and Schleicher (1990) and Faure and Mensing (2005) and references therein.

While the $^{40}Ar/^{39}Ar$ method, which is a technical variant of the K-Ar method, was briefly mentioned earlier, we need to discuss the K-Ar method. It is frequently used for the direct dating of fault activity by being applied to, for example, syn-kinematic sericite, like in the West Carpathian case study (see Nemčok and Kantor, 1990) or to magmatic rocks coeval with various rifting events, like in the Pannonian Basin system case study (see Pécskay et al., 1995).

Both K-Ar and $^{40}Ar/^{39}Ar$ methods are based on ^{40}K isotope decay to ^{40}Ca and ^{40}Ar. Controlled by branching decay, the equation for K-Ar clock is given by:

$$t = 1/\lambda_{total} \times \ln(\lambda_{total} / \lambda_{EC} \times {}^{40}Ar_{rad}/{}^{40}K + 1), \qquad (4\text{-}4)$$

where t is the age, λ_{total} is the total decay constant, λ_{EC} is the decay constant of potassium for radiogenic ^{40}Ar by electron capture, and $^{40}Ar_{rad}/^{40}K$ is the ratio of radiogenic ^{40}Ar to present-day ^{40}K. Three sets of potassium decay constants are used in physical, chemical, and geological studies. They are statistically undistinguished from each other (Min et al., 2000). Furthermore, in the case of 0–25 Ma-old rock and mineral samples, there is no difference among age calculations, which use those three different potassium decay sets. Therefore, the error in the age calculation using K-Ar clock depends entirely on the accuracy of the $^{40}Ar_{rad}/^{40}K$ measurements. The $^{40}Ar/^{39}Ar$ method treats the problem by replacing the determination of potassium from the K-Ar method by the determination of the artificially created ^{39}Ar due to ^{39}K (n,p) ^{39}Ar reaction with fast neutrons (Merrihue and Turner, 1966) done using irradiation in a nuclear reactor. In the K-Ar method, the problem of the determination of $^{40}Ar_{rad}$ is most problematic in young volcanic rocks and minerals, where the low quantities of radiogenic argon compare to an overwhelming amount of the atmospheric argon. This problem is treated by using analytical techniques such as the isotope dilution technique by the atmospheric argon (Brandt, 1965; Rasskazov et al., 2000) and unspiked K-Ar method (Cassignol et al., 1978; Gillet and Cornette, 1986). Unlike the K-Ar method, problematic for young volcanic rocks, the inverse isochron approach of the step-wise heating technique in the

$^{40}Ar/^{39}Ar$ method provides a possibility to determine the isotopic composition of the initially trapped argon and yield an accurate age. On the contrary, there are no strict criteria to reveal the presence of extraneous argon by the K-Ar method. To demonstrate the problem with the K-Ar method, we can look at the precisions of $^{40}Ar_{rad}$ determinations from the Pleistocene and Holocene volcanic rocks from the Canary and Hawaii archipelagos published by Guillou and colleagues (1996a, 1996b, 1997), who used a specially constructed 180°, 6 cm radius, 620 V accelerating potential mass-spectrometer. The determination precision for the Pleistocene rocks is just 0.7–2.7 percent while the precision for the Holocene rocks is higher than 50 percent.

5 The role of lithospheric composition and compositional variations in evolving rift margin architectural development and in breakup locations

Introduction

Lithospheric type distribution apparently controls the location of rifting and subsequent continental breakup in the case of successful rifting. This can be implied from maps showing the location of cratonic blocks, mobile belts, rifts, and boundaries of continental and oceanic crust, such as those made for North America and its African counterpart (Davison et al., 2002 redrawn in simplified form in Nemčok et al., 2005b; Figure 6.5), central and southern Africa (Figure 6.1; Molina, 1988), and eastern India (Nemčok et al., 2012c).

As documented in synthetic publications, the lithospheric thickness varies from province to province (e.g., Allen and Allen, 1990; Christensen and Mooney, 1995 and references therein). Apart from the thickness, the rheology of the lithosphere separates various lithospheric types. This can be demonstrated using the regional transect made through the Carpathian orogen and its foreland (Figure 5.1; Lankreijer, 1998). Figure 5.1 shows how different lithospheric sections, characterized by different lithological sections and different thermal regimes, contain different lithospheric strength profiles.

As dictated by the analog material and numerical modeling need for simplified multilayers representing the various types of continental lithosphere (e.g., Faugère and Brun, 1984; Davy, 1986; Allemand et al., 1989; Allemand and Brun, 1991; Davy and Cobbold, 1991; Tron and Brun, 1991; Brun et al., 1994; Govers and Wortel, 1995; Benes and Davy, 1996; Brun and Beslier, 1996; Faccenna et al., 1996; Burov and Cloetingh, 1997; Brun, 1999; Huismans, 1999; Maggi et al., 2000a, 2000b; Mart and Dateuil, 2000; Davis and Kusznir, 2002, 2004; Corti et al., 2003), lithospheric variations can be grouped into three basic categories. The following text discusses how these different lithospheric types can control the type of rifting that deforms them; that is, which of them are prone to narrow rifting, wide rifting, and core-complex mode of rifting and how the 4-D mechanical stratigraphy of the lithosphere controls the rift style.

Lithospheric multilayer

Multilayer is the most typical representation of the rheologic distributions in the lithosphere. Various data sets help with multilayer definition. Starting from the top, apart from the earthquake data described in this chapter, resistivity distribution helps to separate the upper brittle crust from the ductile lower crust. Because the rheology is significantly affected by thermal regime and also by the presence of fluids, the zones of low electrical resistivity can serve as indicators of rheologically weak zones in the crust (e.g., Wannamaker and Hohmann, 1991; Jones, 1992; Jiracek, 1995). Evidence is accumulating worldwide that the top of a deep crustal conductor itself commonly follows a crustal isotherm of about 450°C, perhaps related to the brittle–ductile transition (Bailey, 1990; Jiracek, 1995; Wannamaker et al., 1997).

Well-defined crustal layering is the outcome of the 560 determinations of the seismic velocity-depth structure of crust worldwide compiled by Christensen and Mooney (1995) (Table 5.1). Using the seismic refraction method, the apparent seismic velocities were directly measured, having an accuracy of about 3 percent or about ± 0.2 kms^{-1}. This is based on the accuracy of seismic arrival time determination, chronometer correction, field surveying uncertainties, typical seismograph and shot point spacing, and the effects of lateral variations in near-surface low-velocity anomalies. The depths of refracting horizons, however, were successively calculated from the uppermost layer downward, having an accuracy of 10 percent. The average crustal velocity is 6.45 kms^{-1} ±0.21 kms^{-1} and the range is from 5.5 to 7.5 kms^{-1}. The top of the lithospheric mantle is defined as a depth with velocity exceeding 7.6 kms^{-1}. Thickness of the crust underlying active and inactive

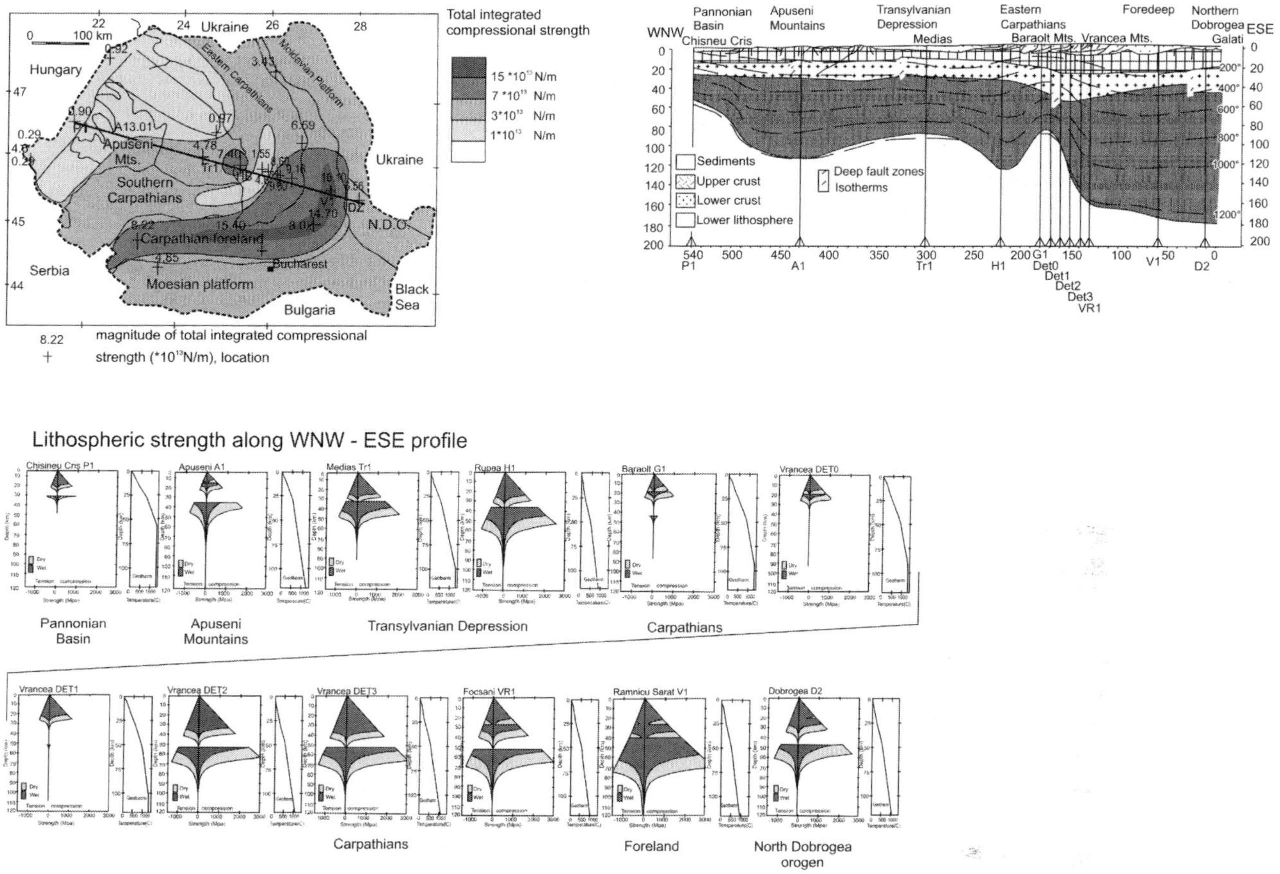

Figure 5.1. Integrated lithospheric yield strength profiles of various portions of the East Carpathians and their foreland (Lankreijer, 1998).

Table 5.1. Velocities and crustal thickness for tectonic provinces and average continental crust (Christensen and Mooney, 1995).

Crustal property	Orogens	Shields and platforms	Continental arcs	Rifts	Extended crust (Basin and Range)
v_p at 5 km	5.69 ± 0.67	5.68 ± 0.81	5.80 ± 0.34	5.64 ± 0.64	5.59 ± 0.88
v_p at 10 km	6.06 ± 0.39	6.10 ± 0.40	6.17 ± 0.34	6.05 ± 0.18	6.02 ± 0.45
v_p at 15 km	6.22 ± 0.32	6.32 ± 0.26	6.38 ± 0.33	6.29 ± 0.19	6.31 ± 0.32
v_p at 20 km	6.38 ± 0.34	6.48 ± 0.26	6.55 ± 0.28	6.51 ± 0.23	6.53 ± 0.34
v_p at 25 km	6.53 ± 0.39	6.65 ± 0.27	6.69 ± 0.28	6.72 ± 0.35	6.69 ± 0.30
v_p at 30 km	6.68 ± 0.43	6.80 ± 0.27	6.84 ± 0.30	6.94 ± 0.37	6.89 ± 0.40
v_p at 35 km	6.81 ± 0.40	6.96 ± 0.30	6.99 ± 0.29	7.12 ± 0.33	6.93 ± 0.46
v_p at 40 km	6.92 ± 0.44	7.11 ± 0.33	7.14 ± 0.25	7.12 ± 0.30	–
v_p at 45 km	6.96 ± 0.43	7.22 ± 0.39	–	–	–
v_p at 50 km	6.99 ± 0.52	–	–	–	–
Crustal thickness	46.3 ± 9.5	41.5 ± 5.8	38.7 ± 9.6	36.2 ± 7.9	30.5 ± 5.3
Average crustal velocity	6.39 ± 0.25	6.42 ± 0.20	6.44 ± 0.25	6.36 ± 0.23	6.21 ± 0.22
Upper mantle velocity	8.01 ± 0.22	8.13 ± 0.19	7.95 ± 0.23	7.93 ± 0.15	8.02 ± 0.19

rifts in the database ranges from 18 km to 46 km. For example, the crustal thickness underneath the Kenyan portion of the East African Rift system varies along the strike of the system between 20 and 36 km.

The internal layering of the crust determined from seismic velocities commonly uses a Conrad discontinuity, a gradational boundary, which separates typical upper crustal velocities of 6.0–6.3 kms^{-1} from the middle crust with velocities of 6.6–6.8 kms^{-1} (Christensen and Mooney, 1995).

Deep reflection seismic data allow differentiating relatively transparent brittle upper crust from the relatively reflective middle and lower crusts. These data are characterized by the initiation of major reflections at the middle crustal levels. They are controlled by large contrasts in the acoustic impedance between amphibolite and granitic/tonalitic gneisses. This interpretation draws from seismic modeling tied to the corehole penetrating the Phanerozoic crust of the Appalachian Piedmont (Christensen, 1989). Furthermore, the garnet-rich rocks with high densities and seismic velocities are expected to cause significant reflectivity (Christensen and Mooney, 1995). Yet another reason for prominent lower crustal reflectors can be a postulated presence of mafic horizontal sill-like intrusions (Meissner, 2006, pers. comm.), which are known to cause strong reflectors in the case of 1–20 m-thick doleritic sills (Juhlin, 1990). Although the evidence for these sills is abundant for the upper crust (Ferré et al., 2002), they are already very rare in the middle crust (Simancas et al., 2003). Examples of the laminated anisotropic character of the lower crust comes from the German portion of the West European Platform as based on the dispersion analysis of Love and Rayleigh waves (Bischoff et al., 2006).

The worldwide database of seismic velocities and densities of Christensen and Mooney (1995) documents that younger, Phanerozoic crusts of Western Europe and the North American coastal regions have a character different from the older, pre-Cambrian crust. It also shows the pre-Cambrian crust having an age between 0.6 and 3.8 Ga and varying in character depending on its age. For example, the Archaean crust is about 27–40 km thick with the exception of its collisional regions. It does not have a high-velocity layer of 7 kms^{-1} at its base. On the contrary, the Proterozoic crust has a thickness of 40–55 km and a high-velocity layer at its base (Drummond, 1988; Durrheim and Mooney, 1991, 1992). Such difference in crustal thickness and trace element content between the Archean and Proterozoic crusts is attributed to differences in mantle temperatures playing the major role in the magmatic and rheologic processes of crustal evolution. In comparison to pre-Cambrian crusts, the Phanerozoic crust of Western Europe has an average thickness of 35 km (Woollard, 1959).

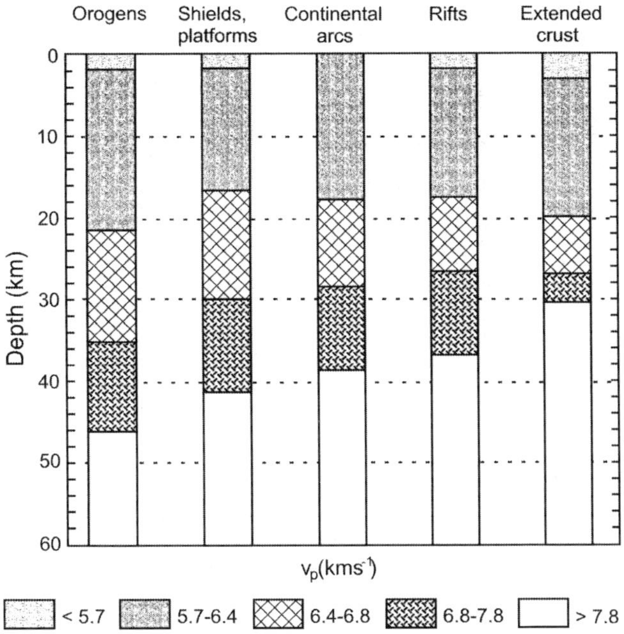

Figure 5.2. Average crustal structure divided into five basic tectonic settings (Christensen and Mooney, 1995).

Seismic velocity layering of the continental crust varies depending on the types of crust. The worldwide database allows us to divide the continental crust into five groups including (Christensen and Mooney, 1995; Figure 5.2):

1) orogens,
2) shields and platforms,
3) continental arcs,
4) rifts, and
5) extended crust.

The crustal thickness of these five globally averaged provinces decreases in listed order from about 46 km to about 30 km. Four defined velocity layers, with the exception of the uppermost low-velocity layer missing in the continental arc province, are present in every single province while their depth intervals change. Some crusts can contain melts in their deeper parts. This can be indicated by the significant departures of their seismic velocity-depth curves from those typical for "dry" rheology (Christensen and Mooney, 1995).

A comparison of the velocity structures of continental arcs with those of shields and platforms documents rather similar lower-crustal profiles (Christensen and Mooney, 1995). The arcs, however, differ by significantly higher upper-crustal velocities, indicating the modification of their velocity structure either by the

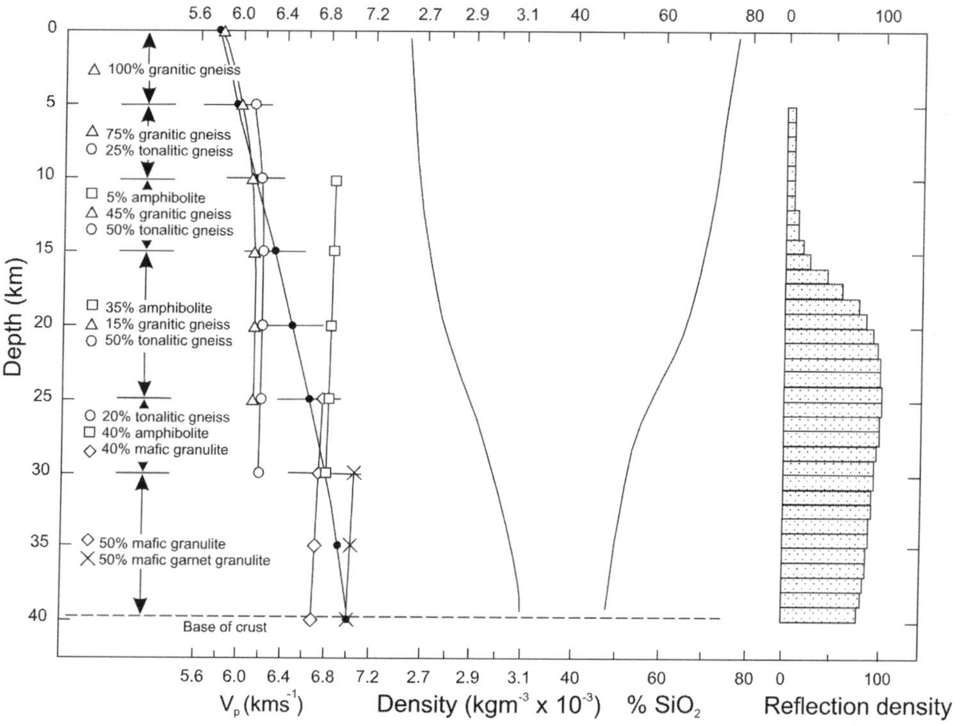

Figure 5.3. Comprehensive model of the crustal multilayer based on various geophysical data distributions (Christensen and Mooney, 1995).

emplacement of faster-velocity intrusive rocks or by the incorporation of metasedimentary rocks into the upper crust.

The orogens differ from shields and platforms by low velocities. This is because the cratonization introduces either emplacement of larger mafic magma volumes or erosion associated with isostatic uplift.

The worldwide database of seismic velocities and densities allows us to characterize the upper mantle layer by seismic velocities spanning from 7.6 to 8.9 kms^{-1} (Christensen and Mooney, 1995). The average value is 8.07 $kms^{-1} \pm 0.21$ kms^{-1}.

An example of the asthenosphere defined underneath the lithospheric multilayer comes from the German portion of the West European Platform, where it is indicated by a low-velocity zone of Love and Rayleigh waves (Bischoff et al., 2006).

Provided that we have the seismic velocity and density data on the lithospheric multilayer, the lithology of individual lithospheric layers can be implied from rock tables containing seismic velocity and density measurements from rock specimens (Christensen and Mooney, 1995). Some of the specimen properties, such as seismic velocity anisotropy, provide further insight on the structural characteristic for different layers. Data from Christensen and Mooney (1995) indicate that anisotropy controlled by preferred mineral orientation decreases for metamorphic rocks as we proceed from low- to high-grade rocks. Thus, seismic anisotropy is likely to be the significant property of the upper and middle crust, while its presence in deeper levels progressively decreases.

In order to compare several data constraints on the interpretation of lithospheric layering, it is useful to employ various functions combining these data such as the functions recalculating seismic velocity and density (e.g., Ludwig et al., 1970; Gardner et al., 1974; Godfrey et al., 1997; Brocher, 2005). Various data constraints can be recalculated and compared to imply a comprehensive model of the lithospheric multilayer, as demonstrated for the crustal multilayer by Christensen and Mooney (1995), who use seismic velocity, density, and reflection density data (Figure 5.3).

Lithospheric types

For the purpose of the following text, the continental lithosphere can be divided into three different types, including the strong, old, stable lithosphere, the intermediately strong, young, stable lithosphere, and the weak, thickened lithosphere.

The strong, old, stable lithosphere can be characterized as a lithosphere that was stabilized in pre-Cambrian

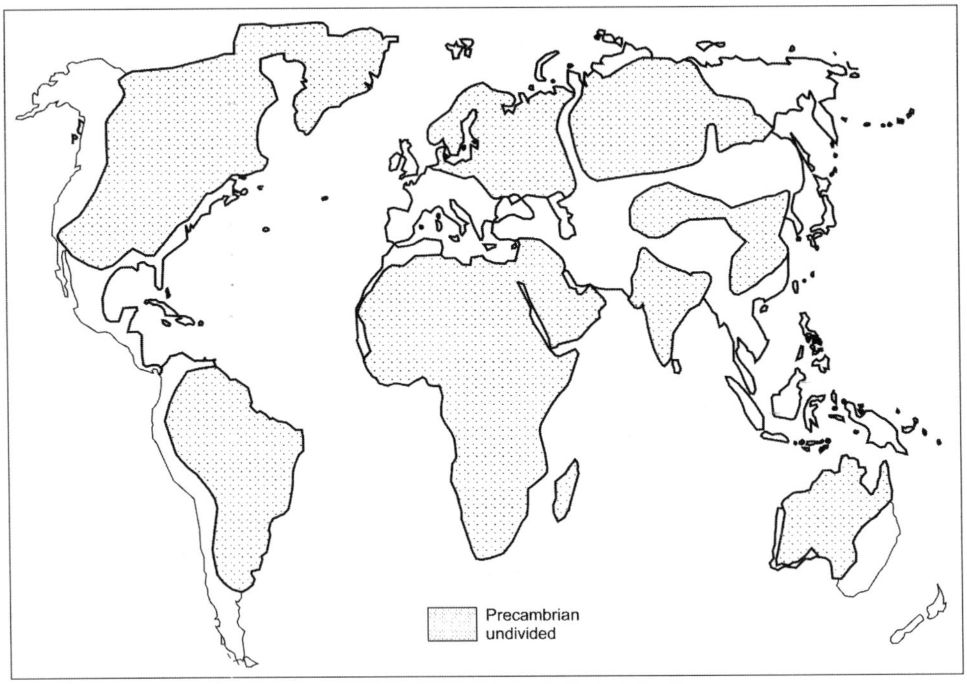

Figure 5.4. Worldwide distribution of Precambrian lithosphere (Christensen and Mooney, 1995).

times and that represents an old, cratonic lithosphere (Figure 5.4; Christensen and Mooney, 1995). Data from the worldwide lithospheric database on the platform and shield crusts of Russia (Christensen and Mooney, 1995) document that the average thickness of the crust of the old, stable, lithosphere is about 40–50 km. Drumond (1988) and Durrheim and Mooney (1991, 1992), however, present worldwide evidence about the differences between Proterozoic and Archaean crusts. Most Proterozoic crusts have a thickness ranging from 40 to 55 km and contain a high-seismic velocity layer, with a velocity of about 7 kms^{-1}, at their base. On the contrary, Archaean crusts are only 27–40 km thick, with the exception of collisional boundaries, and usually lack the basal high-velocity layer. Furthermore, these two crusts differ in the content of major and minor trace elements (Durrheim and Mooney, 1994). The differences also are in their lithospheric thickness. While it ranges from 140 km to about 350 km for Archaean lithosphere, it is less than about 200 km for Proterozoic lithosphere (Artemieva and Mooney, 2002).

It is entirely possible that the strong, old, stable lithosphere may not contain a weak lower crustal layer, which results in its integrated yield strength profile indicating a brittle strength from its upper surface to the base of the upper mantle and ductile strength controlling only the deformation of the lower mantle (e.g., Brun, 1999; Huismans and Beaumont, 2005; Figures 5.5, 5.6). This is the case when the numerical and analog material models choose a two-layer approximation of this lithospheric multilayer (e.g., Brun, 1999; Mart and Dateuil, 2000; Corti et al., 2003). Lithology of this lithosphere can be visualized as a quartz-controlled crustal rheology resting on top of the olivine-controlled mantle rheology under cold thermal regime, approximately characterized by a surface heat flow of 30 mWm^2 or less (Huismans et al., 2005). Another, more likely, alternative would be a higher heat flow regime controlling a similar lithosphere, the lower crust of which has a more mafic composition (see Christensen and Mooney, 1995). In fact, a slightly higher heat flow regime can be quite common, as documented by the values published, for example, for the Ukrainian, Baltic, and Canadian shields and West Australia with the exception of a narrow coastal region (Platte River Associates, 1995). This case would also represent the Archaean and Proterozoic lithospheres of India, the central African, south African, and northeastern Brazilian shields, cratons, and mobile belts shown, for example, in Figure 6.1 (own results; Ray, 1963; Molina, 1988; CGMW, 2000). Their crust most likely became dry and refractory because of preexisting orogenic and dehydration-melting events.

Rift examples developed in this type of lithosphere come from regions such as the Baikal Rift, the U.S. Mid-Continent Rift, and the Recôncavo-Tucano Rift. Examples of rifts developing into passive margins in this type of lithosphere come from segments of the

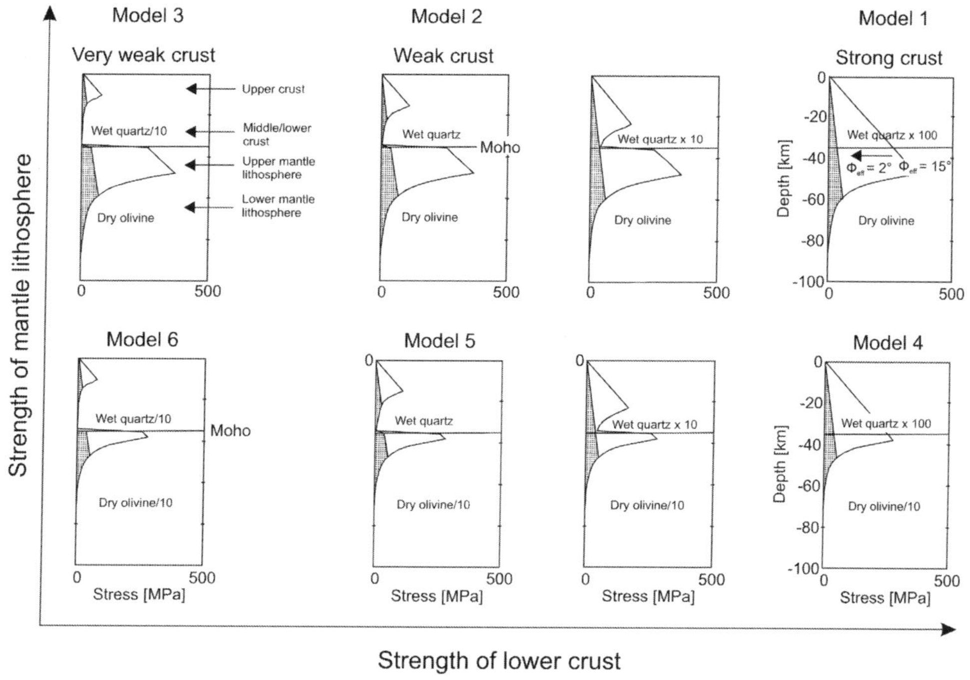

Figure 5.5. Integrated lithospheric yield strength profiles for various types of lithosphere (Huismans and Beaumont, 2005).

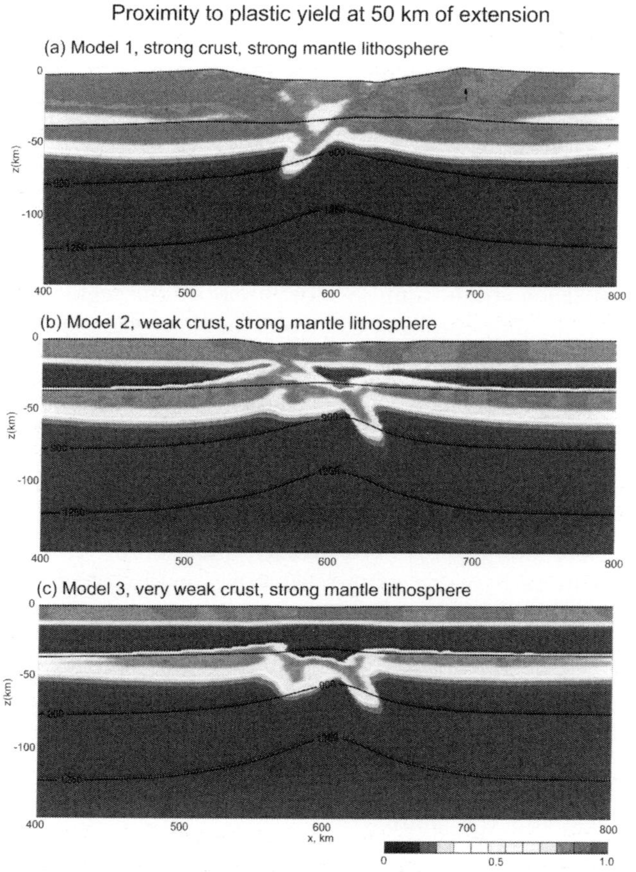

Figure 5.6. 2-D lithospheric strength profiles for various types of lithosphere (Huismans and Beaumont, 2005).

Labrador-Greenland passive margin, the East Indian margin, and the northern South Atlantic margins.

The intermediately strong, relatively young, stable lithosphere would represent a lithosphere that became stable during the Phanerozoic eon. Typical examples are lithospheres of the eastern coastal region of the U.S.-Nova Scotia region, the adjacent northwest African coastal region, and the West European Platform. The worldwide lithospheric database (Christensen and Mooney, 1995) documents that the average crustal thickness of all shield and platform crusts is 41.5 km. This puts an upper bracket on the interval on crusts of young, stable lithospheres.

The East European Platform near the Carpathians has, for example, a 45 km-thick crust. This is interpreted from the seismic refraction profiles (Guterch et al., 1976, 1994), while the Moesian Platform has a crustal thickness of 28–30 km, determined from deep seismic sounding (Rădulescu, 1988; Enescu et al., 1992; Mocanu and Rădulescu, 1994).

The lithospheric thickness ranges from 120 km to 200 km in regions such as the Bohemian Massif, the West European Platform, and the East European Platform in the vicinity of the Carpathians (Lankreijer, 1998 and references therein).

A German example of the West European Platform indicates a 27–30 km-thick crust and about a 50 km-thick mantle lithosphere, determined from the analysis of Love and Rayleigh waves (Bischoff et al., 2006).

Another examples come from the North Alpine foreland (Strehlau and Stange, 2006; Zappone et al., 2006). Here, the earthquake data indicate a 16 km-thick seismogenic upper crust, underlain by middle crust with minimum to zero seismicity in a depth interval of 16–20 km, which is underlain by the seismogenic lower crust. Lower crustal xenoliths in outcrops indicate that the lower crustal rheology is controlled by metapelites, granitic gneisses, and spinel-pyroxene granulites. Determined crustal velocities are within the 6.0–6.7 kms^{-1} interval. The crust–mantle transition zone is characterized by a velocity of about 7.2 kms^{-1}.

The intermediately strong, relatively young, stable lithosphere contains a weak lower crustal layer, which results in its integrated yield strength profile indicating four-layer strength distribution characterized by brittle upper crustal and upper mantle layers separated by ductile lower crust and lower mantle layers (e.g., Brun, 1999; Huismans and Beaumont, 2005; Figures 5.5, 5.6). This is the case when the numerical and analog material models choose four-layer approximation of this lithospheric multilayer (e.g., Faugère and Brun, 1984; Davy, 1986; Allemand and Brun, 1991; Davy and Cobbold, 1991; Tron and Brun, 1991; Brun et al., 1994; Brun and Beslier, 1996; Brun, 1999). The two weak layers control the level of decoupling between the upper crust and upper mantle.

The intermediately strong, relatively young, stable lithosphere would represent a fully thermally equilibrated lithosphere of a normal thickness, which would exist under a thermal regime characterized by a surface heat flow of 45 mWm^2 or higher. This lithosphere would be through at least 60 Ma of the thermal equilibration after the last orogenic event. Case examples include lithospheres lacking post-Variscan orogenic events such as those on the Appalachian side of the North American continent, the African side conjugate to the Appalachians, and the East European Platform of southern Poland. Such a lithosphere can be represented by a quartz-controlled crust. Another, more likely, alternative would be a higher heat flow regime controlling a similar lithosphere but with more mafic, feldspar-controlled, composition of the lower crust. Imagining a little higher heat flow regime would be encouraged by data from the eastern United States and Wales (Ungerer et al., 1990) and the East European Platform (Nemčok et al., 1998).

Rift examples developed in this type of lithosphere come from regions such as the Bristol Channel Rift and the rift system of Bohemia and southern Poland. Examples of the rifts developing into passive margins in this type of lithosphere come from the U.S.-Nova Scotia and opposite African margins of the Central Atlantic.

The weak, young, thickened lithosphere can be characterized as a lithosphere that did not have a sufficiently long time for its thermal equilibration after the last orogenic event and finds itself in the stage of thermal thinning (see Zoetemeijer, 1993). The example of progressive thermal thinning is illustrated by a comparison of twelve integrated yield strength profiles located along the transect, which runs from the most recent frontal portion of the East Carpathian thrustbelt to its oldest hinterland portion in Figure 5.1 (Lankreijer, 1998).

The integrated yield strength profile of this lithospheric type indicates four-layer strength distribution characterized by brittle upper crustal and upper mantle layers separated by ductile lower crust and lower mantle layers (see Brun et al. 1994; Brun, 1999; Corti et al., 2003), which, contrary to the intermediately strong, young, stable lithosphere, is characterized by strong decoupling along the ductile lower crust (Figures 5.5, 5.6; Huismans and Beaumont, 2005).

The northern sector of the Tyrrhenian-Apennine system offers an example of the weak, young lithosphere (Chiozzi et al., 2006). It is characterized by large lateral variations of its rheologic multilayer occurring as the result of different tectonic processes leaving their record in terms of thermal regime and crustal thickness. The lithosphere of this region has been shaped by events starting with rifting followed by drifting in the Ligurian-Provençal Basin. The controlling extension took part prior to or coevally with the subduction regime in the Apennines. It later extended eastward into the Tyrrhenian Sea, replacing an earlier contraction. As a result of these temporally and spatially different processes, the lithospheric thermal regime did not reach equilibrium in most areas of the region. In fact, the only equilibrated area is the lithosphere of the Variscan Corsica area. The calculation of the strength profiles indicates that the thickness of the brittle crust varies among areas, being 20 km in the geologically younger portion of Corsica, 10–15 km in the northern Tyrrhenian area, and 30 km in the external Apennines. These estimated thicknesses generally correlate with the distribution of upper crustal earthquakes.

Rift and passive margin examples developed in this type of lithosphere come from areas such as the Basin and Range province, the Aegean region, the Tyrrhenian region, the South China Sea, and the Pannonian Basin system.

Deformation controls in different layers of the lithospheric multilayer

Deformational behavior of different lithospheric rheologies in various subsurface conditions can be estimated

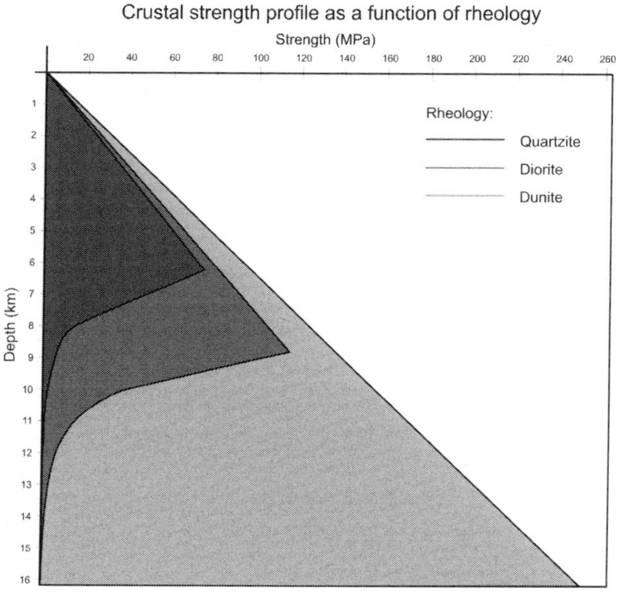

Figure 5.7. Lithologic control of integrated crustal yield strength profile.

using the extrapolation of laboratory failure criteria, describing where rock failure occurs either by cataclastic or ductile deformation (Figure 5.7; Goetze and Evans, 1979; Brace and Kohlstedt, 1980; Kirby, 1985; Carter and Tsenn, 1987; Ranalli and Murphy, 1987). The brittle failure is controlled by the Coulomb-Mohr criterion (Byerlee, 1978; Carter and Tsenn, 1987) and the ductile failure by the power-law creep law (Kirby, 1983, 1985). Rock failure occurs at a given strain rate when stresses reach the lower of the brittle and ductile yield strengths.

The yield strength integration over the one-dimensional vertical profile through the lithosphere produces a continuous strength profile with depth, such as those made along the transect through the Carpathians and their foreland (Figure 5.1; Lankreijer, 1998). A lithospheric multilayer represented by such a profile is defined with the use of various data constraints. In the Carpathian example, the lithology and thickness of the crust is derived from deep seismic sounding data (Cornea et al., 1981; Rădulescu et al., 1983; Tomek et al., 1987, 1989; Posgay et al., 1991, 1996) and gravity modeling (Bielik et al., 1994, 1995; Szafián, 1999; Pospíšil et al., 2006), providing a thickness range of 25–52.5 km. The thickness and rheology of the mantle is interpreted from the deep seismic sounding data (Chekunov, 1989; Mocanu and Rădulescu, 1994; Răileanu et al., 1994; Posgay et al., 1996), xenolith studies (Downes and Vaselli, 1995), thermal modeling (Demetrescu and Veliciu, 1991; Čermák, 1994; Lenkey, 1999), p seismic wave travel-time residuals (Babuška and Plomerová, 1988; Pajdušák et al., 1988), and magnetotelluric data (Ádám, 1976, 1997; Pečová et al., 1979), providing a thickness range of 60–120 km.

The depth of the brittle crust in the Carpathian example is defined by the distribution of the earthquake focal depths, indicating bottoming below 10 km in the warm Pannonian Basin system interior, below 20 km in the peripheral Pannonian Basin system areas including the surrounding orogenic belts, and below 40 km in the Bohemian Massif in the Carpathian foreland (Lankreijer, 1998 and references therein).

The critical data used for the extrapolation of the laboratory failure criteria along the discussed 1-D profiles are:

1) interpreted rock composition;
2) calculated temperature-depth function; and
3) estimated tectonic strain rate.

Although the calculation of strength profiles involves a number of intrinsic uncertainties, numerous studies indicate the usefulness of extrapolating microphysical models from the laboratory to the macro scale (e.g., Ranalli and Murphy, 1987; Cloetingh and Banda, 1992; Beekman, 1994; Burov and Diament, 1995; Lankreijer et al., 1997). Estimation of the strain rate represents such an uncertainty. In Lankreijer's (1998) study, a strain rate of 10^{-15} s^{-1} was chosen because it represents a commonly observed rate for both compressional and extensional settings (see Carter and Tsenn, 1987; Van den Beukel, 1990).

The brittle strength along the profile is calculated using Ranalli and Murphy's (1987) relationship modified from Sibson's (1974) relationship:

$$(\sigma_1 - \sigma_3) = \beta \rho g z (1 - \lambda), \qquad (5\text{-}1)$$

where σ_1 and σ_3 are the maximum and minimum principal compressional stresses, β is a constant (3 for thrusting, 1.2 for strike-slip faulting, and 0.75 for normal faulting, assuming a friction of 0.75), ρ is the rock density, g is the acceleration due to gravity, z is the depth, and λ is the assumed hydrostatic/lithostatic pressure ratio. If densities cannot be obtained on the subsurface rocks of the studied lithospheric multilayer, representative rock densities are taken from the values determined for the respective region or taken from analogs tabulated in rock mechanics tables such as those in Daly and colleagues (1966), Vutukuri and colleagues (1974), Toulokian and colleagues (1981), Ibrmajer and colleagues (1989), and Holbrook and colleagues (1992).

The ductile strength is calculated from the power-law creep expression (Kirby, 1983):

$$(\sigma_1 - \sigma_3) = K^{1/n} * \varepsilon^{1/n} \exp(E/nRT) \quad \text{and} \quad K = (1/A)^{1/n}, \qquad (5\text{-}2)$$

or

$$(\sigma_1 - \sigma_3) = (\varepsilon/A)^{1/n} \exp(E/nRT), \qquad (5\text{-}3)$$

where K and A are scaling factors, n is the power law exponent, E is the activation energy, ε is the strain rate, R is the universal gas constant, and T is the temperature, taken from the temperature-depth profile. Flow parameters for various lithologies can be obtained from tabulations by Carter (1976), Kirby (1983), Kirby and McCormick (1984), Suppe (1985), Carter and Tsenn (1987), and Ranalli and Murphy (1987).

The faulting instability is described by the Drucker-Prager failure criteria (Drucker and Prager, 1952):

$$\sigma_f = \beta J_1 + \sqrt{J_2}, \qquad (5\text{-}4)$$

where σ_f is the failure stress, J_1 and J_2 are the first stress invariant and the second invariant of deviatoric stress, respectively, and β is a coefficient (≥0 for compression) that may vary as a function of J_1. The axis of the conical failure envelope is defined by the hydrostatic stress relationship, $\sigma_V = \sigma_{H1} = \sigma_{H2}$, where both horizontal stresses and vertical stress are equal. For cohesive materials the apex of the conical failure envelope lies in the tension octant of the working stress space at a distance of τ_o * cot φ from the origin, where τ_o and φ are the cohesion and angle of internal friction, respectively. The half apex angle γ is related to the angle of internal friction:

$$\tan\gamma = \sqrt{3}\sin\varphi(3 + \sin^2\varphi)^{-1/2}. \qquad (5\text{-}5)$$

When the stress state is represented by a point inside the conical failure envelope, the material is stable. When the rock follows a burial path, points are inside the cone. Any addition of a tectonic load will result in a stress path moving toward the stability envelope.

Note that the described cone does not have an ideal shape. It has a rounded apex, located in the tension octant. The other deviation from an ideal cone occurs in the compression octant, in the area where normal stresses are high. The deviation results from the fact that inter-grain friction is a pressure-sensitive shear resistance of materials. It depends on the normal stresses that act across the particle boundaries, and these tend to increase either with progressive burial or with tectonic loading.

The effects of the cohesion, angle of internal friction, and fluid pressure on the critical strength of the rocks for faulting are controlled by the Coulomb-Mohr equation:

$$\tau = c + \tan\varphi(\sigma_n - p_f). \qquad (5\text{-}6)$$

The cohesion is typical for specific lithology (Figure 5.8). It is a pressure-insensitive rock property, understood as resulting from the electrostatic grain bonding. A good example of a rock type the strength of which is mostly controlled by cohesion is a weak limestone, whose strength can show only slight pressure dependence. Its cohesion is controlled mainly by cementation. Its strength is, therefore, close to the strength type of the ideal plastic material. Figure 5.8 shows a range of upper crustal rocks with their typical values of cohesion. Igneous rocks and sediments cemented by calcite or silica possess the largest cohesion. The cohesion can be reduced by the micro-cracking (e.g., Brace, 1961).

The brittle strength of the crust is dominated by the pressure-dependent term from Equation (5-6). Therefore, the loss of cohesion can result only in a limited reduction of the critical strength of brittle lithospheric layers for faulting. It is assumed to range from 2 percent to 10 percent for crust, depending on the acting dynamical pressure and the depth of the brittle–ductile transition zone (Huismans and Beaumont, 2005).

The angle of internal friction is also a material property (Figure 5.9). It is understood as controlled by the physical friction among grains, although it is strictly a mathematical function. Figure 5.9 documents several features of the angle of internal friction, such as its increase with grain size decrease. This function has been documented by numerous rock mechanic experiments (e.g., Karig, 1986; Maltman, 1994). The angle of internal friction also varies with the lithologies of the grains composing the rock aggregate as demonstrated by rock mechanics experiments of Byerlee (1968), Singh (1976), Briggs (1980), Radney and Byerlee (1988), and Theil (1995). The angle of internal friction also varies with the amount of deformation, which tends to reduce it to residual value. Rock mechanic experiments comparing peak and residual values of the angle of internal friction for dry and wet Triassic marl and Liassic shale in the Bristol Channel, United Kingdom, document the changes in angle from less than one degree to more than six degrees (Davies et al., 1991).

The relationship between the ratio of peak and residual shear strengths and normal stress made for several materials by Vutukuri and colleagues (1974)

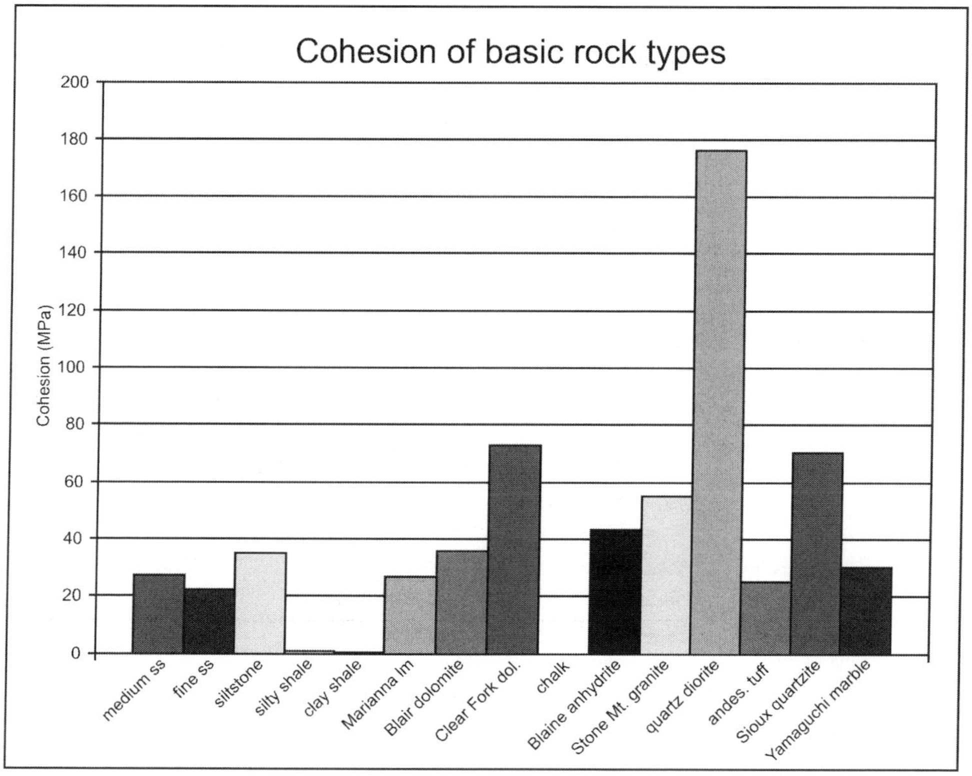

Figure 5.8. Cohesion of rock types (Nemčok et al., 2005a).

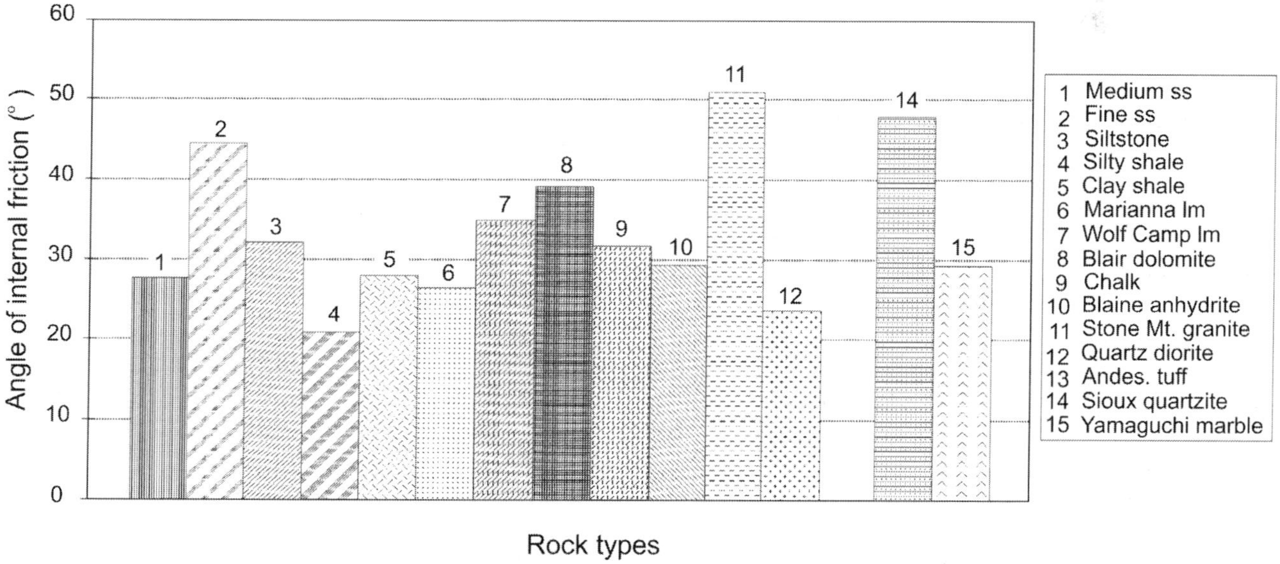

Figure 5.9. Angle of internal friction of basic rock types (Nemčok et al., 2005a). Values are taken from Lama and Vutukuri (1978).

demonstrates the same point. It illustrates differences for elastic/frictional plastic materials such as limestone, conglomerate, and sandstone, while the almost ideal plastic material represented by shale keeps the constant ratio of peak and residual shear strengths under increasing normal stress.

The results of the rock mechanic tests show that the dependence of the shear strength on the normal stress is not linear (Mandl, 1988 and references therein). Depending on a rock type, the departure from a linear relationship is progressively more pronounced with the increasing confining pressure (e.g., Nemčok et al. 2000). Tri-axial testing of various rocks indicates that it is a decrease in the angle of internal friction for intact rock with increasing effective normal stress, σ_n, that causes the decrease of the apex of the Drucker-Prager cone. The major control of this behavior is the suppression of dilatancy under higher confining pressure. The described reduction of the frictional apex angle in Equation (5-5) is significant for different materials under different confining pressures. Particularly significant friction angle reduction takes part when the confining pressure controlling the dilatancy suppression causes brittle micro-deformation to change to an intra-crystalline glide for a specific material.

It is assumed that the changes in the angle of internal friction can reduce the critical strength of brittle lithospheric layers for faulting within 50–90 percent (Huismans and Beaumont, 2005), depending on the formation of the gouge along faults within the respective lithospheric layer and mineral transformations. For example, the experimental work on deformation-assisted mineral transformation mechanisms in granitic upper crustal and mafic lower crustal and upper mantle rocks can produce shear zones characterized by a very low angle of internal friction (Bos, 2000; Bos and Spiers, 2002; Handy and Studenitz, 2002).

Fluid content exerts important control over the rock strength. As shown by the rock mechanic tests, the increase in fluid content decreases the shear strength of the rock (e.g., Arch et al., 1988; Theil, 1995). Transient and permanent pore fluid pressure variations can significantly weaken brittle rocks (Sibson, 1990b; Rice, 1992; Ridley, 1993; Streit, 1997; Ingebritsen and Manning, 1999; Connolly and Podladchikov, 2000). The decrease is not only achieved by the subtraction of the fluid pressure from the total normal stress but it also can be caused by the rock reaction to fluid saturation, which does not have to be a linear function. This can be demonstrated by comparing peak and residual values of the angle of internal friction for dry and wet Triassic marl and Liassic shale in the Bristol Channel. A decrease in the angle of internal friction due to wetting can result in a difference of 2–16° (Davies et al., 1991).

When the state of stress is below the frictional-plastic yield, the flow becomes viscous. The viscous behavior of crust and mantle is controlled by quartz- and olivine-rich lithologies and their deformational response to stresses at high temperatures. Their rheologies at these conditions can be characterized as temperature-dependent and nonlinear, as documented by laboratory experiments with wet quartzite and dry olivine (Karato and Wu, 1993; Gleason and Tullis, 1995). Their effective viscosity, η, can be described as (Huismans and Beaumont, 2005):

$$\eta = A^{-1/n} (I_2)^{(1-n)2n} \exp [(E+Vp)/nRT], \quad (5\text{-}7)$$

where I_2 is the second invariant of the deviatoric strain-rate tensor and V is the activation volume that makes the viscosity dependent on the pressure, p. The role of the activation volume, V, can be quickly demonstrated in the example of the lower crust with an activation volume of 0 m^3mol^{-1} used in Huismans and Beaumont's (2005) lithospheric models. This model is characterized by a weak, viscous, lower crust. It requires a multiplication of the resultant viscosity value by a factor of 100 to obtain a strong-crust model.

Furthermore, it would be interesting to run a sensitivity analysis on the role of several factors, such as the different rheologies, geothermal regime, stress regime, strain rate, and pore fluid pressure, on the integrated yield strength profile.

Figure 5.7 shows the effect of specific lithology on the integrated yield strength of the upper crust. Note that certain rheologies are not realistic for the upper crust but we used them anyway to demonstrate the strength contrasts among varying lithologies. Figure 5.7 demonstrates that the peak brittle strength is largest for dunite composition and smallest for quartzite composition. Because the conditions were equal for all three modeled materials, being 70 mWm^2 for the surface heat flow, 0.36 for the pore fluid pressure/lithostatic pressure ratio, 10^{-15} for the strain rate, and normal faulting type for the stress regime, we can compare how lithologies differ from each other in rheological control of the integrated strength profile. Figure 5.7 shows that while the brittle–ductile transition is at about 6 km and the maximum brittle strength is about 72 MPa for the quartzite crustal layer, the transition depth and maximum strength are about 9 km and 105 MPa for the diorite and the dunite remains brittle in the whole tested thickness of 16 km, reaching the maximum brittle strength of about 250 MPa at the base.

Layering can affect the deformation process by potential detachment localization inside the weakest layer (Kirby and Kronenberg, 1987; Davy and Cobbold, 1991). Using smaller-scale analogs for the role of layer rheology, detachments are affected by boundaries such as the original lithological boundaries, the smectite–illite transition zone, boundaries between various cementation zones, boundaries between various hydrothermal alteration zones, and boundaries between melt and host rock. The rheologies of layers are controlled not only by their "dry" petrological composition, but also by their fluid content and deformation patterns.

Apart from the control associated with the multilayer composition of the lithosphere, it is the thermal regime that controls the integrated yield strength profile of the lithosphere (Figures 5.10, 5.11). The thermal conditions control the transition between the brittle and ductile deformation regimes, which take place at different temperatures for different materials. They control the strength of involved rocks as shown by laboratory experiments (e.g., Goetze and Evans, 1979). The crust and the very upper lithospheric mantle may behave like elastic entities at a low geothermal gradient of about 10°Ckm^{-1} (Ord and Hobbs, 1989). However, as the geothermal gradient increases, the likelihood of deviation from elastic behavior for different materials increases. Creep processes become more dominant at temperatures exceeding roughly half of the melting temperature of rock (Carter and Tsenn, 1987). Therefore, the strength in the lower parts of both crust and lithosphere is strongly influenced by the thermal regime. Dislocation climb and dislocation glide are the main flow mechanisms occurring in the lower part of the lithosphere (Goetze and Evans, 1979).

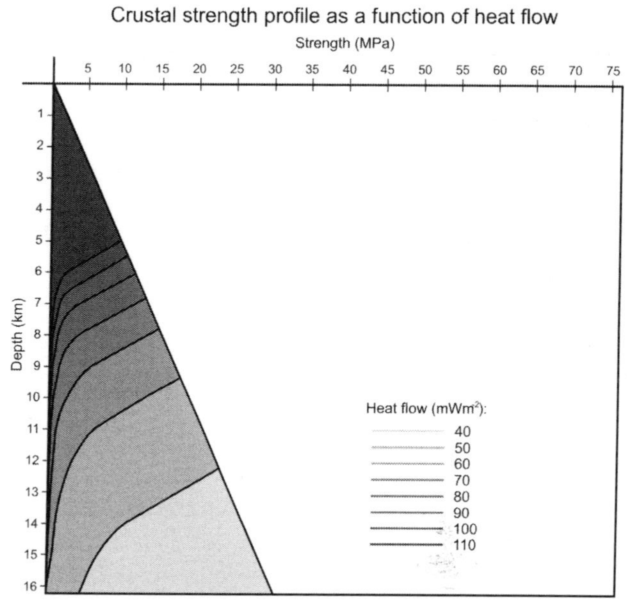

Figure 5.10. Heat flow control of integrated crustal yield strength profile.

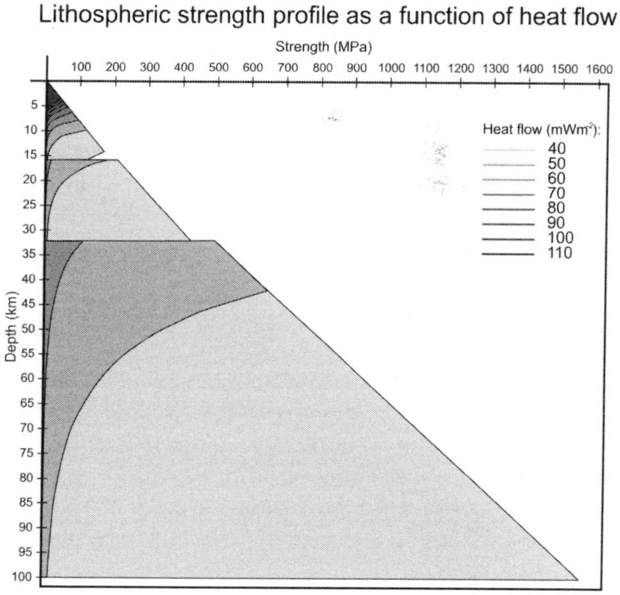

Figure 5.11. Heat flow control of integrated lithospheric yield strength profile.

Figure 5.10 shows how a hypothetical quartzite layer, which is under a pore fluid pressure/lithostatic pressure ratio of 0.36, normal faulting stress regime, and strain rate of 10^{-15}, behaves under various thermal regimes represented by surface heat flow values increasing from 40 to 110 mWm^2. It documents that the brittle–ductile transition ascends with warmer thermal regimes. Coincidentally, the value of the maximum brittle strength decreases. Figure 5.11 documents what the same variations in thermal regime do to the integrated yield strength profile of the lithosphere. The lithosphere is composed of upper crustal, lower crustal, and mantle layers represented by quartzite, diorite, and dunite, respectively, and it is affected by the same factors as those in Figure 5.10. Described sensitivity analysis provides us with all three main types of lithosphere described earlier. The coldest thermal regimes provide us with an integrated yield strength profile somewhat similar to the profiles of strong, old, stable lithospheres. Slightly warmer scenarios result in profiles somewhat similar to the profiles of the intermediately strong, relatively young, stable lithospheres. Distinctively warmer scenarios provide us with profiles somewhat similar to the profiles of weak, young, thickened lithospheres.

The Styrian Basin development history, discussed in Chapter 1, serves as a good example of how thermal regime variations control lithospheric strength. Calculating integrated yield strength profiles of the lithosphere, Lankreijer (1998) documents a relationship between the elevated Karpatian-Badenian heat flow regime associated with a significant volcanic event and weak lithospheric rheology during that time. Later, the lithospheric strength has increased, which correlates with volcanic quiescence.

Using a finite-element approach, Henk (2006) conducted a sensitivity analysis of the crustal rheology impact on the rift unit geometry and its subsidence

Table 5.2. Seafloor spreading rate categories (Macdonald, 1982; Bown and White, 1994). The spreading rate affects the melt generation in the seafloor-spreading process and hence the relief and geochemistry of the axial crust.

Spreading rate ($cmyr^{-1}$)	Typical topography of axial region*	Case example	Reference
0.3–1 – ultra-slow	deep axial valley	Gakkel Ridge	Edwards et al. (2001)
	high relief flanks	Southwest Indian Ridge	Cannat et al. (1999)
1–2 – slow	axial low	Mid-Atlantic Ridge	Macdonald (1982)
	rugged flanks		
2–4 – intermediate	variable	Pacific-Antarctic Ridge	Marks and Stock (1994)
4–10 – fast	axial high	East Pacific Ridge	Macdonald (1982)
	smooth flanks		

*The transition from axial valley to axial high typically indicates a difference between slow and fast spreading. However, variations do exist, e.g., the slow-spreading Reykjanes Ridge with axial high.

history. The modeled half-graben has geometry similar to that of the Jeanne d'Arc Basin from reflection seismic sections of Tankard and Welsink (1989). It is characterized by a graben-controlling normal fault detached along the upper crust–lower crust contact. The fault has listric geometry and is prescribed to the model as fully developed. The footwall of the model is fixed and the hanging wall is moving at a rate of 2 mmy^{-1}, which represents a rate similar to the present-day extension rate in the Gulf of Evia, Greece (Roberts, 2007 and references therein) and fits into the category of the fastest slow-spreading rates when compared to seafloor spreading rates (Table 5.2; MacDonald, 1982; Bown and White, 1994).

The same 58 km-thick lithospheric rheological multilayer was subjected to two thermal scenarios (see Henk, 2006). The first scenario was characterized by high crustal strength and a surface heat flow of 45 mWm^2. The second scenario was characterized by low crustal strength and a surface heat flow of 85 mWm^2. A comparison of their horizontal displacements, vertical displacements, and strain patterns shows significant differences (Figure 5.12). While the horizontal displacement of a strong scenario shows that the dominant boundary in the displacement distribution is the normal fault with its detachment fault, the weak scenario has a complex displacement distribution also inside the footwall. Furthermore, the vertical displacement distribution is laterally almost homogeneous in the strong scenario but characterized by a prominent uplift driven by ongoing footwall unroofing in the weak scenario. Strain patterns of the two scenarios also differ. The strong scenario contains increased strain values clustered inside the hanging wall near the curvature of the boundary fault and along the axial plane of the roll-over anticline. It also contains very small clusters in the footwall, in the broader area underneath the main curvature of the bounding fault. The weak scenario has a cluster of increased strain values located along the main curvature of the bounding fault. This cluster extends a bit along the subhorizontal detachment fault. However, this scenario also contains a similar, curved cluster, which indicates the potential existence of a new fault that wants to develop at a deeper level, honoring the deeper brittle–ductile boundary location calculated by the model and trying to ignore the shallower detachment initially prescribed to the model. While the Moho boundary in the strong scenario is flat, it undergoes upwarping underneath the unroofing footwall in the weak scenario.

The distributions of principal stress directions in scenarios differ by indicating significant perturbations in the lower crust of the weak scenario and somewhat stronger perturbations in the brittle footwall and hanging wall above the largest bounding fault curvature.

Calculated curves of vertical movements affecting the surface of models are characterized by minor flank uplift on the footwall side and major flank uplift on the flexural hinge side for the strong scenario, and exactly the opposite for the weak scenario (Figure 5.13). The basin is wider and subsiding more in the case of the strong scenario.

The strength of the lithosphere is further influenced by lithospheric stress, expressed as a differential stress (Figure 5.14) or an effective differential stress (Figure 5.15). Values of lithospheric stresses can be determined by several methods (McNutt, 1987 and references therein). Methods that use seismic data such as earthquake focal depths (e.g., Chung and Kanamori,

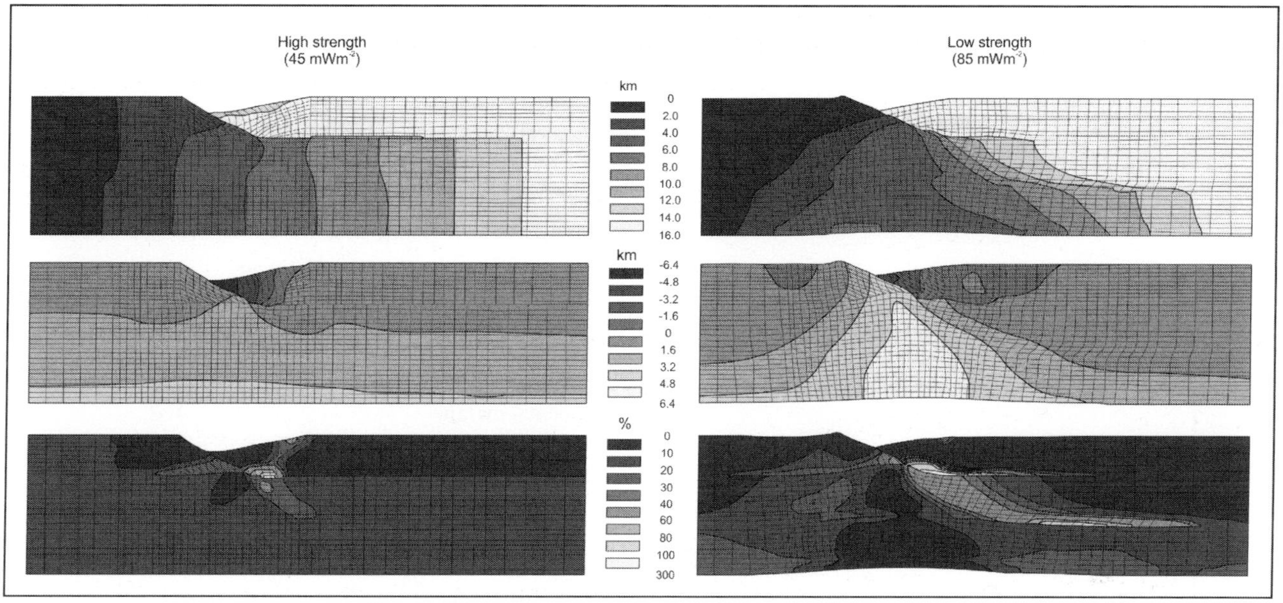

Figure 5.12. Horizontal displacements (upper row), vertical displacements (middle row), and strain patterns (lower row) of both strong and weak rifting end-members (Henk, 2006).

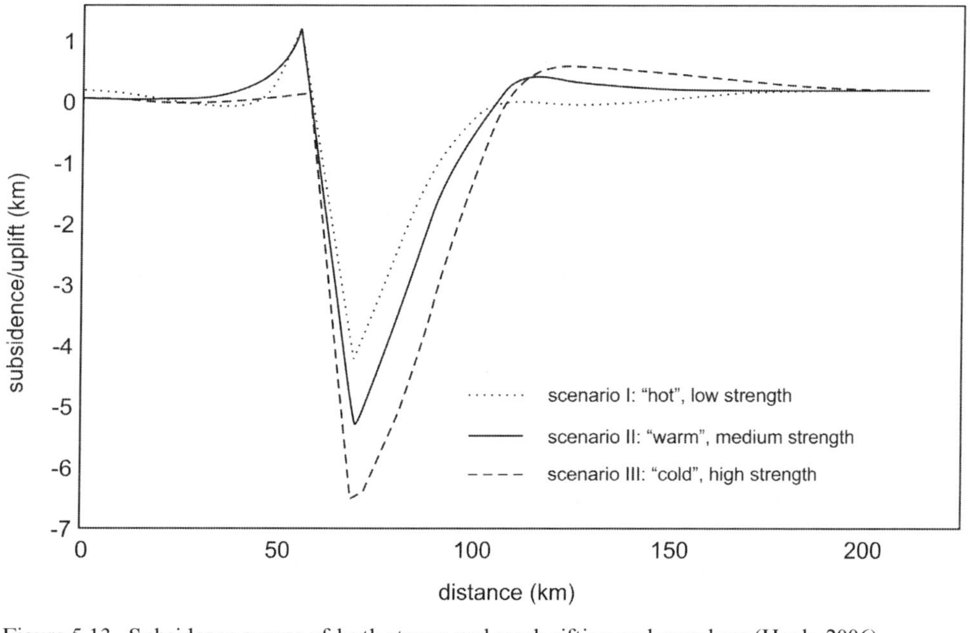

Figure 5.13. Subsidence curves of both strong and weak rifting end-members (Henk, 2006).

1980; Meissner and Strehlau, 1982; Chen and Molnar, 1983) provide estimates ranging from several tens to several hundreds of MPa. A similar range of possible stresses comes from laboratory experiments (e.g., Goetze and Evans, 1979; Brace and Kohlstedt, 1980; Kirby, 1985; Rutter and Brodie, 1988). Similar values come from the numerical simulations (e.g., Govers and Wortel, 1995; Burov and Cloetingh, 1997; Chéry et al., 2001). In fact, stresses generated by plate boundary processes are typically in the order of several tens to several hundreds of MPa (Bott and Kusznir, 1984). For comparison, Table 5.3 illustrates the magnitudes of major tectonic forces within lithospheric plates (Kusznir, 1991; Davis and Kusznir, 2002).

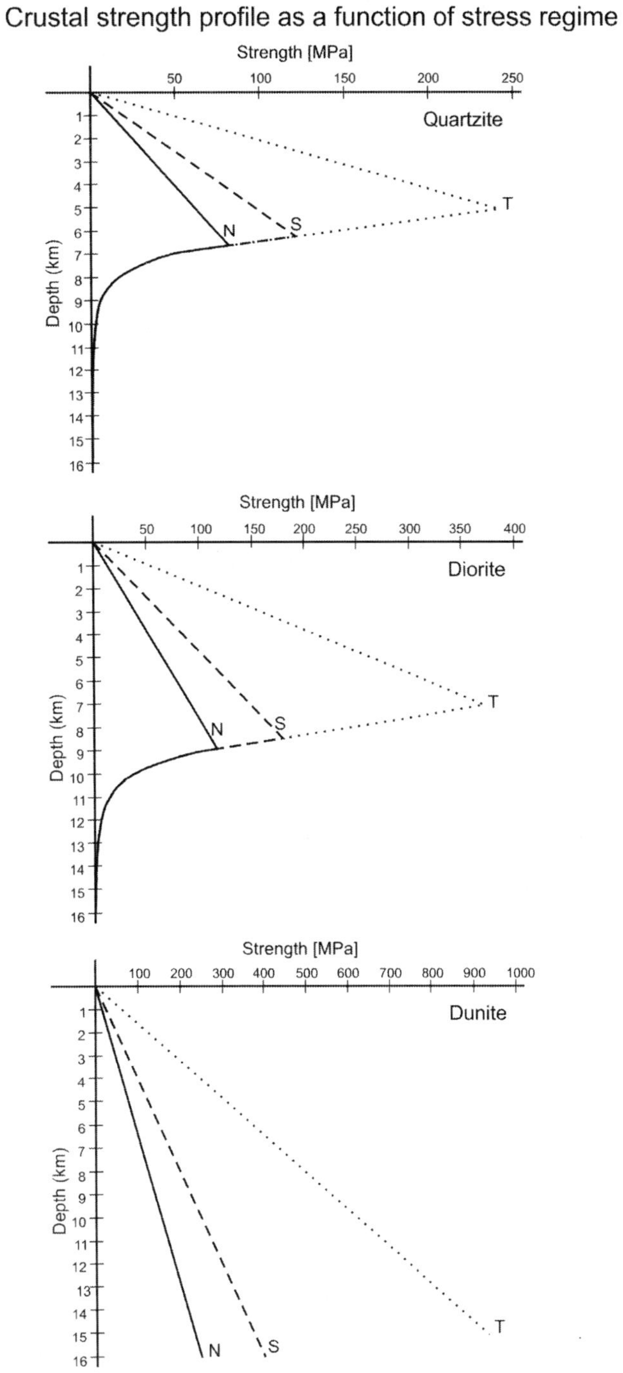

Figure 5.14. Role of the stress regimes on integrated crustal yield strength profile. T – thrust regime, S – strike-slip regime, N – normal faulting regime.

The differential stress controlling large-scale normal faulting is smaller than that controlling the strike-slip faulting, which is smaller than the stress controlling thrusting. The difference between the differential stresses controlling normal faulting and thrusting is about fourfold (see Mandl, 1988). About fourfold also is the difference between the maximum brittle strengths under normal and thrust stress regimes controlling 16 km-thick slabs of quartzite, diorite, and dunite, keeping the other factors, such as pore fluid pressure/ lithostatic pressure ratio of 0.36, strain rate of 10^{-15}, and surface heat flow of 70 mWm^2, fixed (Figure 5.14).

Table 5.3. Magnitudes of major tectonic forces within lithospheric plates (Kusznir, 1991; Davis and Kusznir, 2002).

Force	Magnitude (TNm^{-1})
Ridge push (Kusznir, 1991)	2–3
Ridge push (Davis and Kusznir, 2002)	0–3.5
Subduction suction	0–3
Subduction slab pull	0–5
Plateau uplift	0–4
Continental margins	1–2

The strain rate also affects the integrated yield strength of the lithosphere (Figure 5.16). Figure 5.16 documents that faster strain rates make the tested 16 km-thick slab of quartzite or diorite more brittle. The brittle–ductile transition depth descends with increasing strain rate and the maximum value of the brittle strength becomes larger. Because function 5-1 is strain-rate insensitive, function 5-2 controls the lower of the two tested strengths. To visualize how the ductile strength becomes larger with increasing strain rate, it is useful to simulate its behavior with a viscous material. Silicone putty used in analog material models provides a nice example. The putty has Newtonian viscosities of 10^3–10^4 Pa s (e.g., Brun, 1999 and references therein). If we put a Newtonian viscous liquid into a liquid-filled cylinder with a loosely fitting piston, any piston movement would require a fluid flow between the piston and the cylinder walls (see Mandl, 1988). The system behaves in such a way that the faster piston movements require larger stress to move the piston. As a result of the varying strain rates applied to the used Newtonian fluid, the material can have a broad range of strengths controlled by the function (e.g., Brun, 1999):

$$(\sigma_1 - \sigma_3) = \eta\varepsilon, \quad (5\text{-}8)$$

where η is the fluid viscosity and ε is the strain rate.

Higher strain rates can result in the enhanced mechanical coupling of the strong layers in the lithospheric multilayer. Strain rates are typically estimated with the

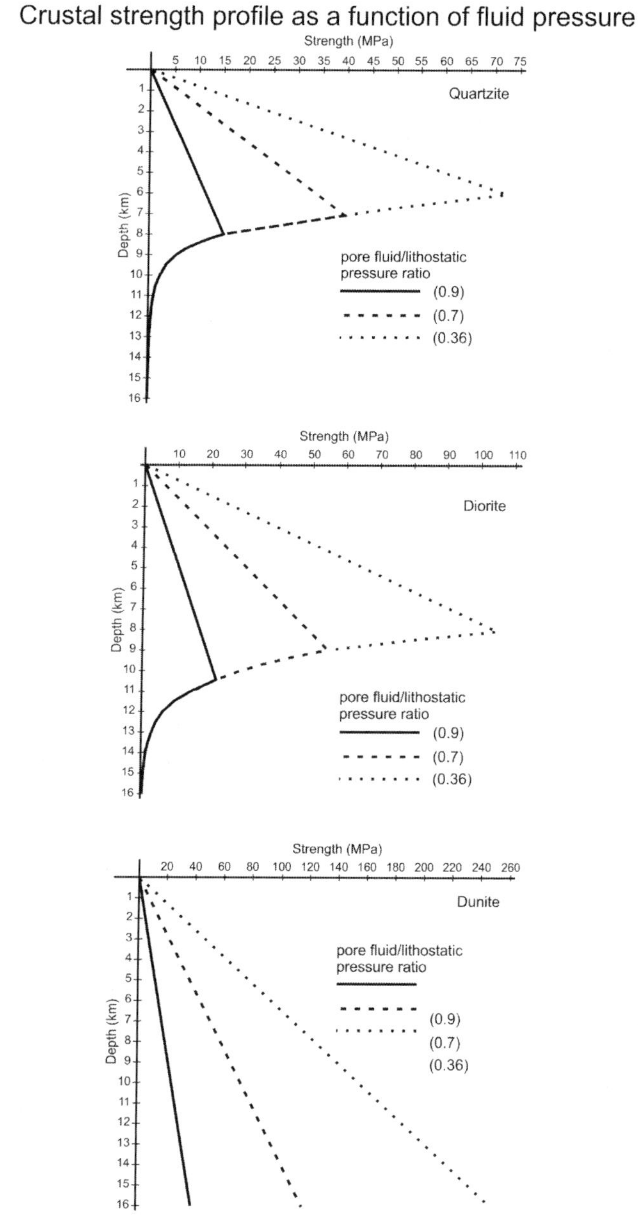

Figure 5.15. Role of the pore fluid pressure on integrated crustal yield strength profile.

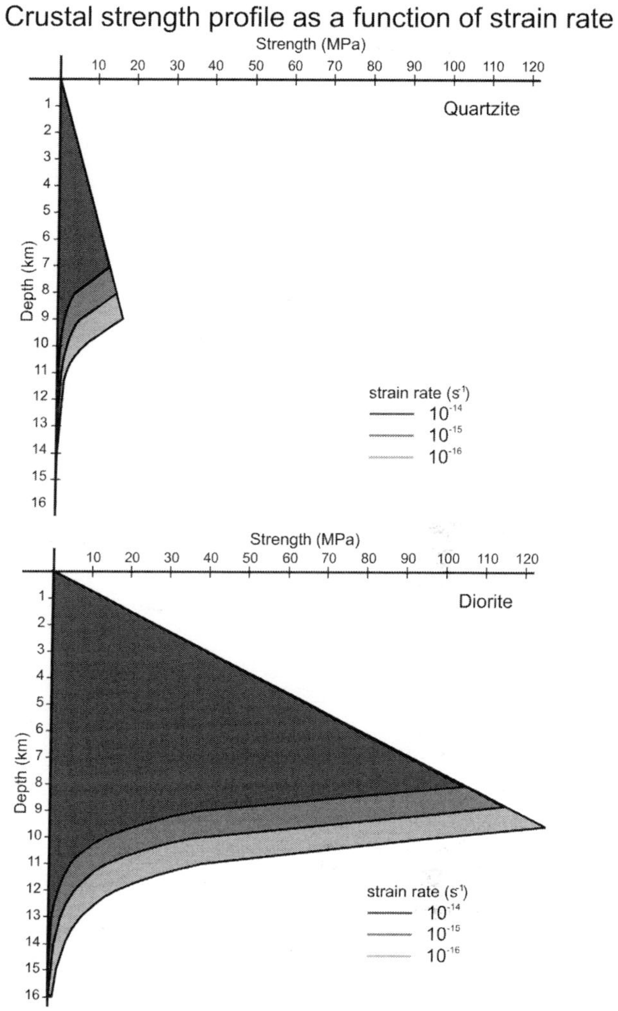

Figure 5.16. Role of the strain rate on integrated crustal yield strength profile.

accuracy of one order of magnitude. Such variation can cause strength changes of less than 10 percent.

The ratio of pore fluid pressure to lithostatic pressure is the remaining factor controlling the integrated yield strength of the lithosphere in this discussion. Figure 5.15 shows that higher pore fluid pressures cause the brittle–ductile transition to descend, making the upper brittle layer thicker. The tests on quartzite, diorite, and dunite, however, indicate that the thickening of the upper brittle layer is associated with its weaker brittle strength. While the increase of hydrostatic pore fluid pressure to almost lithostatic pressure causes a descent of the brittle–ductile transition zone of about 1–2 km for tested materials, the same change in the pore fluid pressure results in a four-to-six times smaller maximum strength of the upper brittle zone.

The most important implication of the lithospheric strength is that the lithospheric loads are compensated flexurally rather than in an isostatic manner. As alluded to in Chapter 2, the temporal changes in rheology of the lithospheric multilayer, which are induced by extension and subsequent cooling, control the variations in lithospheric flexural response in the region underneath the basin. This controls a complex stratigraphic record. Warming weakens the lithosphere, which results in the narrowing of the zone where the basin is compensated. Cooling strengthens the lithosphere, which results in the widening of the zone where the basin is compensated.

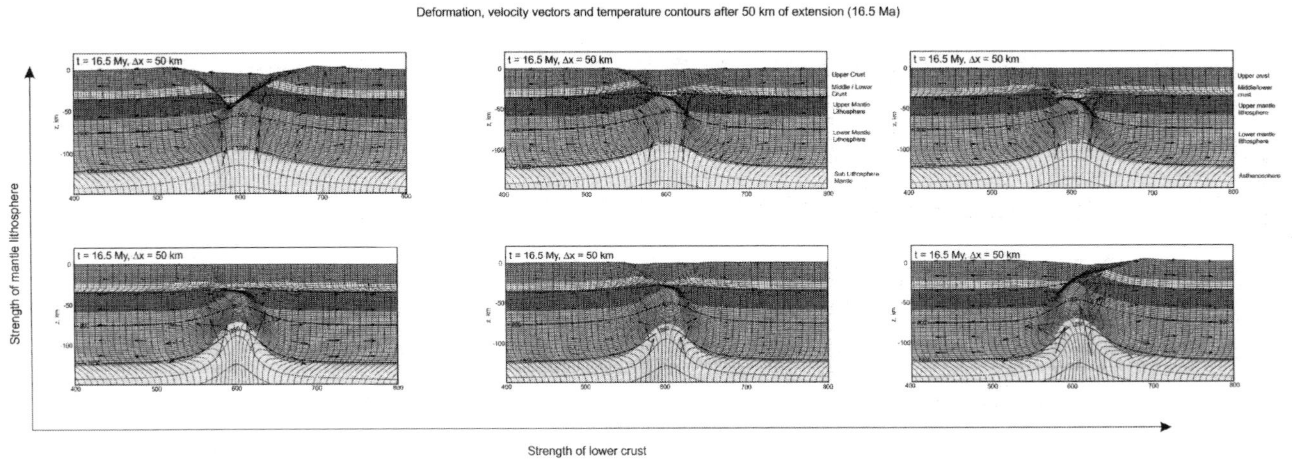

Figure 5.17. Rifting-controlled lithospheric structure simulated numerically for different lithospheric scenarios after first 50 km of extension (Huismans and Beaumont, 2005).

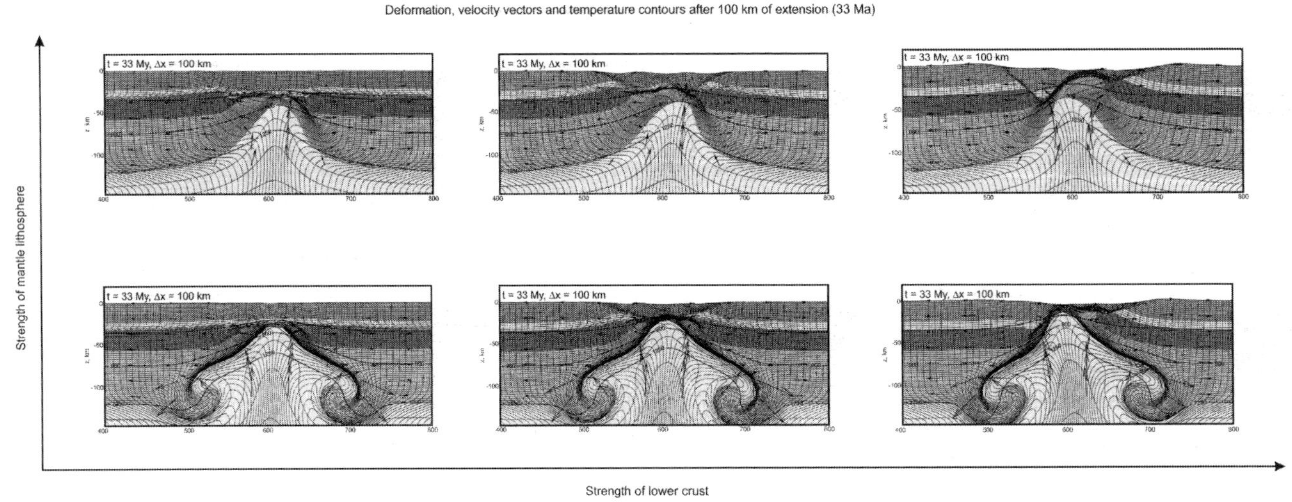

Figure 5.18. Rifting-controlled lithospheric structure simulated numerically for different lithospheric scenarios after 100 km of extension (Huismans and Beaumont, 2005).

The second most important implication of the lithospheric strength is that it controls the shape of extensional basins through the necking depth (Zuber and Parmentier, 1986; Figure 2.1), described in detail in Chapter 2. A shallow level of necking controls a shallow basin and an elevated asthenospheric dome. This results in downward flexed necking. A deep level of necking causes a deep basin and a less pronounced asthenospheric dome. This controls an upward flexed necking.

Rifting of the strong, old, cratonic lithosphere

Finite-element simulations of the rifting in the strong cratonic lithosphere performed by Huismans and Beaumont (2005) indicate that this is the lithospheric type that is prone to narrow rift development (upper right panel in Figures 5.17, 5.18, 5.19). The associated deformation is characterized by strain softening in the frictional-plastic regions of the lithosphere. It causes pronounced localization of a single system of shear zones controlling an asymmetric extension mode (Huismans and Beaumont, 2005). As the extension progresses, the lower crust gets cut out and the upper lithospheric mantle gets in contact with the upper crust (see Figures 2.8, 2.13a,c, 2.19b). Necking of the brittle upper lithospheric mantle accompanied by the development of the upward-convex shallow-dipping fault zone soon leads to lower mantle contact with the upper crust (Figure 2.8), its unroofing in the proto-oceanic crust corridor (Figures 2.20a,b, 3.16, 3.21), and the

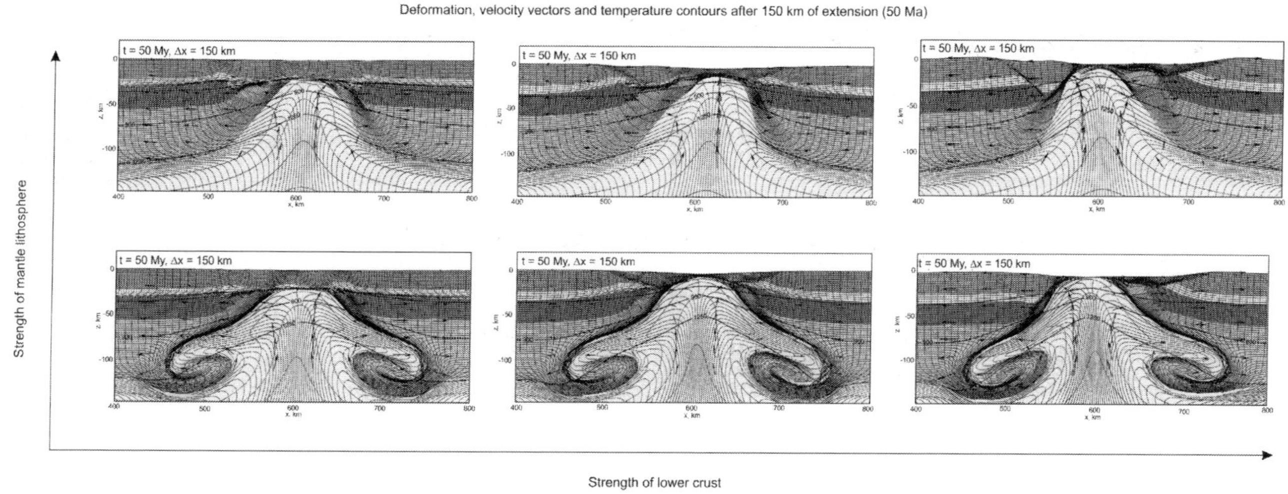

Figure 5.19. Rifting-controlled lithospheric structure simulated numerically for different lithospheric scenarios after 150 km of extension (Huismans and Beaumont, 2005).

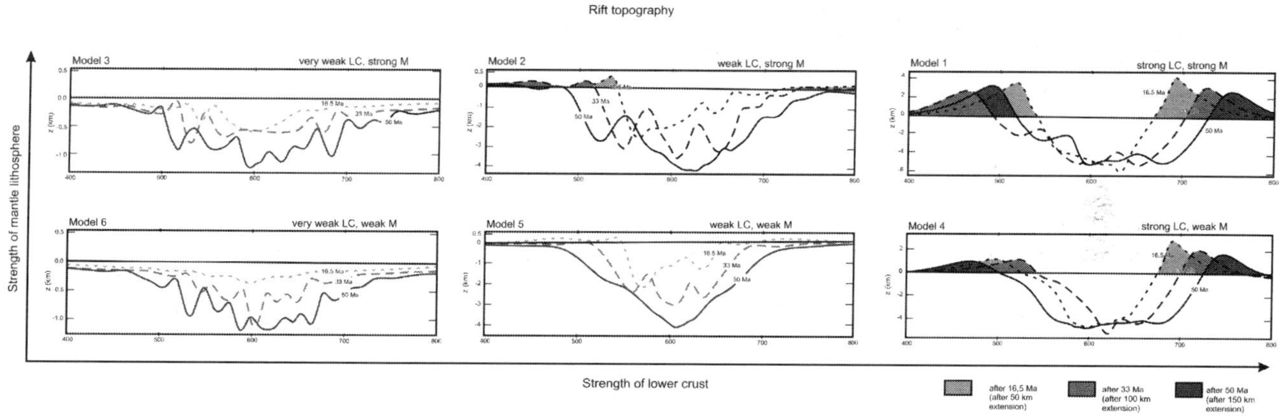

Figure 5.20. Role of lithospheric strength in the rift-related topography (Huismans and Beaumont, 2005).

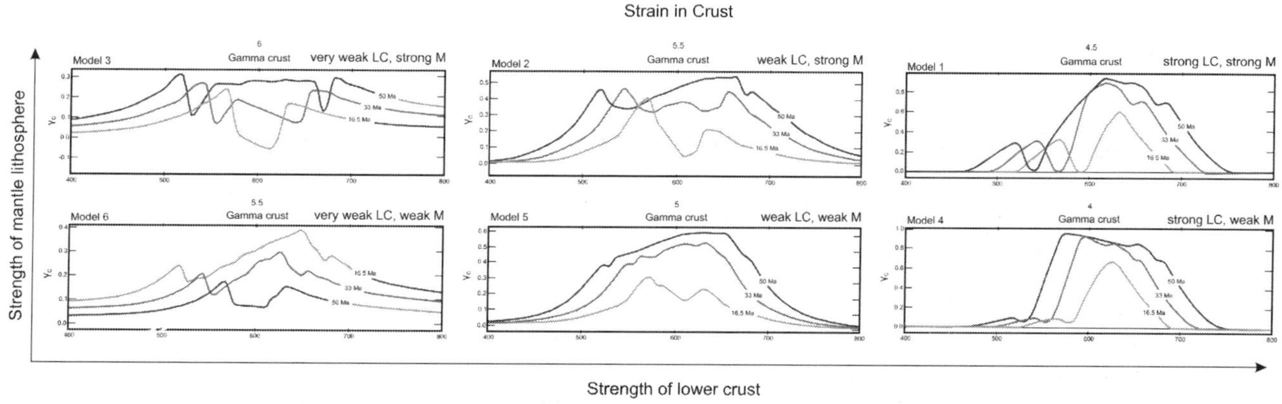

Figure 5.21. Role of lithospheric strength in the rift-related crustal strain distribution (Huismans and Beaumont, 2005).

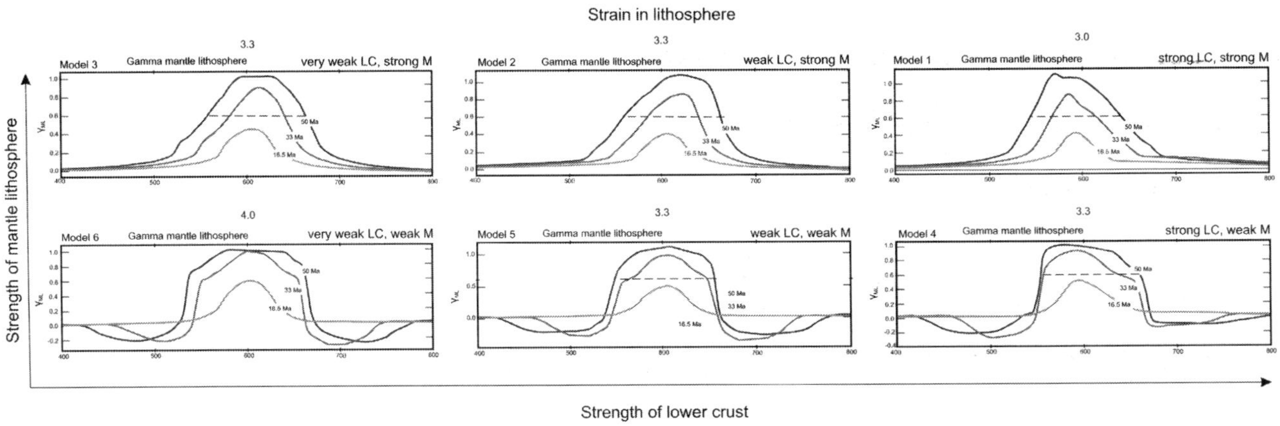

Figure 5.22. Role of lithospheric strength in the rift-related lithospheric strain distribution (Huismans and Beaumont, 2005).

subsequent establishment of the seafloor spreading center (Figures 2.8, 2.19b) as the asthenosphere makes it to the surface and controls the termination of the mantle exhumation and causes the onset of organized seafloor spreading.

The cause behind overall rift asymmetry is the strain softening, which preferentially weakens one of the conjugate shear zones. This weak shear zone remains the weaker one until the continental breakup, which controls the overall lithospheric-scale asymmetry (upper right panel in Figures 5.17, 5.18, 5.19). The asymmetry is expressed by the lateral offset of crust and mantle lithosphere thinning distributions. As the upper lithospheric mantle breaks up, the upwarping lower mantle and underlying asthenosphere develop an asymmetry characterized by a steeper "hanging-wall side" and a less steep "footwall side" of the asthenospheric dome. The asthenospheric dome advances toward the hanging wall with time. Its advance is accompanied by a master shear zone flattening above its axis.

Rifting of the intermediately strong, relatively young, stable lithosphere

Unlike the rifting of the strong, cratonic lithosphere, the rifting of the intermediately strong, relatively young, stable lithosphere is characterized by viscous flow in the lower crust separating brittle upper crust and upper mantle layers (e.g., Allemand and Brun, 1991; Brun, 1999; Huismans and Beaumont, 2005). This flow, which controls a certain amount of decoupling between the two (Allemand and Brun, 1991), acts against the asymmetry of the rift system and promotes a wider zone of rifting (central panels in Figures 5.17, 5.18, 5.19). The crustal asymmetry is suppressed because the conjugate shear zones detach along the ductile lower crust (e.g., Allemand and Brun, 1991; Huismans and Beaumont, 2005). As the extension progresses, semi-coupling of the upper crust with the upper lithospheric mantle controls the development of a system of synthetic crustal shear zones on either side of the rift zone. The result is lower crustal necking, which is a compromise between the one controlled by blocks bounded by frictional shear zones and the one controlled by pinching-out ductile material. The upper mantle necking, at least in a lower thermal scenario, keeps some asymmetry, highlighted by the development of the upward-convex shallow-dip fault, which is less well localized than that in the old, cratonic lithosphere case.

The cause behind the progressive loss of the rift asymmetry is the decrease in coupling between the upper crust and the upper mantle along the lower crust, which contributes to a more distributed style of upper crustal deformation. However, some asymmetry of the lower mantle and asthenosphere remains, as can be seen at the stage when the upper lithospheric mantle breaks up and the upwarping lower mantle and underlying asthenosphere can be, at least for the lower thermal scenario, characterized by a steeper "hanging wall side" and a less steep "footwall side" of the asthenospheric dome (central panels in Figures 5.17, 5.18, 5.19). The asthenospheric dome advances toward the hanging wall with time. Its advance is accompanied by the flattening of the overlying shear zone above its axis.

Rifting of the weak, young, thickened lithosphere

The rifting of the weak, young, thickened lithosphere further highlights the loss of rift asymmetry and its wider crustal character with weaker lithospheric strength. The presence of highly viscous lower crust essentially decouples the upper crust from the lithospheric mantle. While the mantle deformation keeps some asymmetry, especially during earlier rift stages and in colder scenarios, the upper crustal deformation is practically symmetric (lower left panel in Figures 5.17, 5.18, 5.19; Huismans and Beaumont, 2005). It is because flowing lower crust absorbs the imprint of the asymmetric mantle deformation. The decoupled upper crust undergoes extension in a distributed wide-rift mode. The wide-rift character of the upper crustal deformation results in wide and shallow rift architecture.

The colder scenarios of this rift mode provide very slight indications of the lower mantle and asthenospheric asymmetry, expressed by a very subtle difference between the dip of the "hanging wall side" and the "footwall side" of the asthenospheric dome. Similarly, relatively unimportant is the progression of the asthenospheric dome toward the hanging wall in time.

This lithosphere controls rifting with very pronounced depth-dependent stretching. Unlike the other two lithospheric scenarios, a weak lithosphere seems to develop the upward convex shallow-dip faults later, after the necking of the lower crust and upper mantle.

The role of various parameters on the rifting style

In order to understand the role of various rheological and rheology-controlling parameters in the rifting mode, it is interesting to leave the discussion of end-member lithospheric types and organize various numerical and analog material models into trend diagrams.

Role of crustal and mantle strength

An important message comes from a series of three trend diagrams comparing colder and warmer rifting scenarios, tracking the role of the increasing strength of either lower crust or mantle lithosphere (Figures 5.17, 5.18, 5.19; Huismans and Beaumont, 2005).

Figure 5.17 shows the trend diagram for early rifting. Close observation reveals that the asymmetry of the lower mantle and asthenospheric bulge advancing in the direction away from the overall footwall of the rift zone is a common feature. This feature is best developed in a strong crust–weak mantle scenario and least developed in a weak crust–weak mantle scenario. Hand in hand with this relationship goes the development of the upward-convex shallow-dipping fault, which becomes fully developed at the end of the necking of the lower crustal and upper mantle layers (Figures 5.18, 5.19). The weak crust–weak mantle scenario has it developed after the longest time period in comparison to the extension histories of the remaining rift scenarios. The role of the upward-convex shallow-dipping fault is finished at the moment when the lower mantle "breaks up" and the asthenospheric anomaly starts to develop the seafloor spreading center (Figure 5.19). Thanks to the enormous thinning at continental crust–proto-oceanic crust transition and pronounced post-breakup uplift along certain passive margin segments, portions of the upwarped detachment faults can be observed in deep reflection seismic images. The examples come from the cases of either more-or-less orthogonal rifting such as that in the Iberia margin (Manatschal, 2004 and references therein) or pull-apart rifting such as that in offshore Côte d'Ivoire (Nemčok et al., 2012a).

Figures 5.17, 5.18, 5.19 also indicate that the rifting of warmer lithospheric scenarios characterized by a weaker mantle has a tendency to develop unstable systems. The unstable systems are characterized by the occurrence of small-scale convection cells in the lower lithospheric mantle. The cells are driven by thermal buoyancy forces (see Huismans and Beaumont, 2005). On the contrary, the strong mantle suppresses a tendency for mantle instability. Furthermore, the weaker lithospheric mantle scenario makes it to continental breakup faster than the strong lithospheric mantle scenario. A combination of a strong crust with a weak mantle lithosphere undergoes mantle lithosphere thickening underneath rift shoulders, which results from the mentioned small-scale convection. Although the small-scale convection does not significantly alter the surface heat flow regime in the syn-rift phase, it speeds up the development rate of the deep-seated heat anomaly (Huismans and Beaumont, 2005).

It is also the warmer lithospheric mantle that speeds up the lithospheric mantle and lower crustal necking. The faster thinning of the lithosphere is enhanced by buoyancy-driven flow. Furthermore, the warmer lithospheric mantle scenario undergoes the change from the asymmetric crustal extension to the more symmetric one earlier than the colder lithospheric mantle. The weaker the crust the faster the lithospheric mantle necking instability for warmer scenarios.

A comparison of all the scenarios (Figures 5.17, 5.18, 5.19) documents that the fastest potential continental breakup occurs in the case of the warmest thermal regime combined with the best crustal strain location. The best crustal strain location can be provided by the preexisting anisotropy such as preexisting shear zones and mobile belts controlling the location of the East African Rift system (Rosendahl, 1987) and such as the fault architecture of the preexisting orogens controlling the location of rifting and subsequent continental breakup along the South Atlantic, and the East Indian and eastern North American coasts (Molina, 1988; Nemčok et al., 2005b, 2012c; Figures 6.1, 6.5).

It also seems logical to claim, because of the difference in fault propagation trajectory and the difference in inherent localization for normal and strike-slip fault zones (Figure 3.11), that the breakup in the case of the warm regime with a strong crust is faster for strike-slip faults than that for normal faults. Indirect support comes from the calculations of the rate of lateral breakup propagation for several ocean segments with either normal fault or strike-slip control, discussed in Chapter 1. While the examples of the former had a rate of 0.08 and 0.24 my^{-1}, the example of the latter had a rate of 0.62 my^{-1}.

A comparison of Figures 5.17, 5.18, 5.19 perhaps allows us to conclude that the asthenospheric diapirism develops toward the surface in the region, which is underneath the largest cumulative displacement along the upward-convex shallow-dipping fault; that is, probably its weakest part.

The aforementioned trends in crustal and mantle strength have an important impact on:

1) the rift topography;
2) thinning distribution in extending crust and mantle; and
3) the thermal regime of the rift.

Figure 5.20 documents the role of the lithospheric multilayer type on the rift-related topography along the profile through the developed rift zone. It indicates that a combination of the strongest crust with the strongest lithospheric mantle controls the deepest and narrowest rift system (upper right panel in Figure 5.20). Furthermore, the weakest crust–weakest mantle scenario is associated with the shallowest and widest rift system (lower left panel in Figure 5.20). The same trend is valid for the flank uplift. A combination of the strongest crust with the strongest lithospheric mantle supports the largest flexural stresses controlling the largest flank uplift. At this point, it is interesting to note that it is not just erosion that lowers the altitude of the rift-flank mountains in time and causes a later retreat of their crests away from the rift axis. The evolution of the rift system itself also lowers this altitude. Furthermore, it is our belief that the isostatic rebound at flanks should be increasingly more suppressed with an increasing obliquity of the extension to the rift axis. This implication comes from our experience about uplift of various normal fault- and strike-slip controlled passive margin segments along the entire Atlantic.

It is interesting to note, when observing the rift flank uplift in modeled pure rifts (Figure 5.20; Huismans and Beaumont, 2005), that the footwall of the upward-convex shallow-dipping fault is always characterized by an isostatic rebound larger than the rebound at the opposite side of the rift zone. This asymmetry is most obvious in the strong-crust case and it progressively pales toward depressed flanks in the weak crust case.

Figure 5.21 documents the role of the lithospheric multilayer type on the rift-related crustal strain distribution along the profile through the developed rift zone. It indicates that a combination of the strong crust with a weak lithospheric mantle controls the largest strain in the crust. A close inspection of the weak mantle cases for various crustal strengths indicates that the hanging wall of the upward-convex shallow-dipping fault contains the largest crustal strain. This feature is the most obvious in the weak mantle–strong crust scenario. It pales out progressively from this scenario toward a scenario with a strong mantle and a weak crust. The reason for this is that the strong mantle–weak crust case is characterized by pure shear in the crust. Furthermore, it is our belief that the strain should be more localized for progressively more oblique extension, as can be seen, for example, from the structural interpretation of the Bouguer gravity anomaly gradients along passive margin segments in offshore Côte d'Ivoire where the upper crustal breakup was controlled by strike-slip, oblique-slip, and normal faulting (Figure 3.11).

Figure 5.21 also indicates that the largest crustal strain is developed in the strongest mantle–weakest crust scenario, while the smallest is associated with the weakest mantle–strongest crust scenario.

Figure 5.22 indicates the role of the lithospheric multilayer type on the rift-related lithospheric strain distribution along the profile through the developed rift zone. A close inspection of the strong mantle cases for various crustal strengths indicates that the hanging wall of the upward-convex shallow-dipping fault contains the largest crustal strain. This strain distribution asymmetry is very weak in the cases with weak lithospheric mantle and progressively disappears along the trend from weak mantle–strong crust scenarios to weak mantle–weak crust scenarios, where the strain distribution is practically symmetric.

Figure 5.22 also indicates that the largest lithospheric strain, that is, largest lithospheric thinning, is developed in the weak mantle–weak crust scenario, while it is smallest in the strong mantle–strong crust scenario.

Figure 5.23 documents the role of the lithospheric multilayer type in the heat flow distribution at a 5 km-depth level along the profile through the developed rift zone. It indicates that the weak crust scenarios seem to have thermal regimes that are more symmetric than the regimes of the strong crust scenarios. Furthermore, the combination of the strong crust

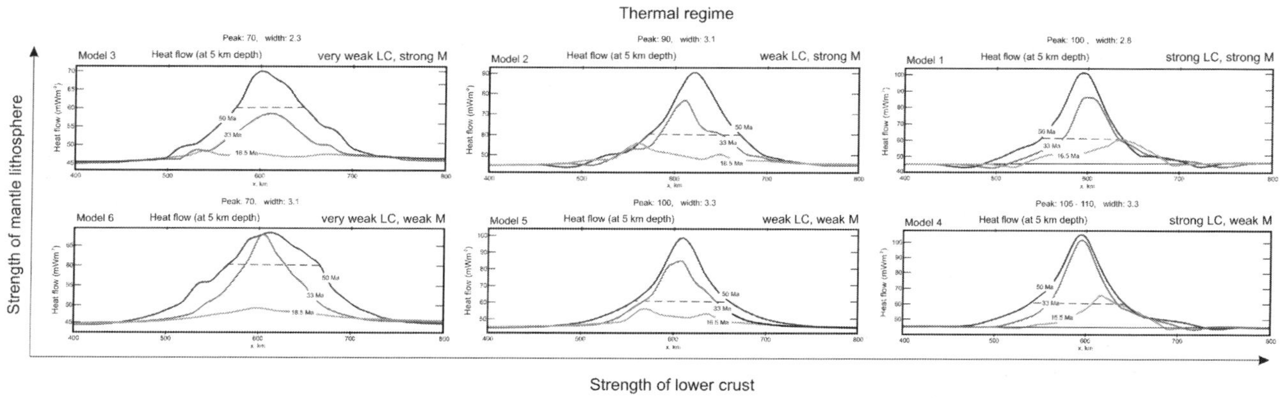

Figure 5.23. Role of lithospheric strength on the rift-related heat flow distribution (Huismans and Beaumont, 2005).

and weak lithospheric mantle is characterized by the warmest heat flow anomaly in the rift zone. The heat transfer through the footwall of the upward-convex shallow-dipping fault is faster than the transfer through the hanging wall of this fault. This phenomenon controls heat flow asymmetry in practically all scenarios.

Comparison of models from Figures 5.17, 5.18, 5.19 indicates that the lithospheric multilayers evolving in time control the rift style in particular by:

1) the level of anisotropy of the lithospheric multilayer;
2) the level of coupling among layers of the multilayer; and
3) the influence of the lithospheric mantle convection as it develops.

The results of different combinations of these factors can be grouped into three basic rift architecture groups:

1) narrow asymmetric rifting of the whole lithosphere;
2) narrow asymmetric rifting of the upper lithosphere combined with narrow symmetric rifting of the lower lithosphere; and
3) wide symmetric rifting of the crust combined with narrow symmetric rifting of the mantle lithosphere.

The initial frictional-plastic failure of the lithosphere is controlled by the textural flaw/rheological weakness. The detailed discussion of crustal and lithospheric mantle weaknesses is in Chapters 1 and 6. Thus, the competition between the theoretical conjugate systems of shear zones, modeled in Huismans and Beaumont's (2005), homogeneous multilayers, can be constrained by the presence and distribution character of the lithospheric weaknesses. If we ignore the role of anisotropy, we can correlate lithospheric types with characteristic rifting modes.

The strong, old, cratonic lithosphere would start to rift using conjugate shear zones until one of them is preferred by shear strain softening. The shear along it progressively develops into the upward-convex shallow-dipping fault zone, which is one of the main accommodators for the crustal and upper mantle necking and the ascent of the lower mantle and asthenosphere. The developed rift systems are narrow and characterized by deep syn-rift basins and pronounced flexural uplifts on one or both flanks. The rifting type would be the narrow asymmetric rifting of the whole lithosphere.

The intermediately strong, relatively young, stable lithosphere would start to rift using conjugate shear zones until one of them is preferred by shear strain softening. In this case the extension in the upper mantle is transferred into the upper crust via viscous drag in the ductile lower crust. The coupling is sufficient to transfer the localized strain into the overlying upper crust. Less coupling between the upper mantle and upper crust, however, allows for a wider rift mode and the reduction of the rift flank uplift due to the reduction of the flexural stresses. The rifting style would be the wider asymmetric rifting of the upper lithosphere combined with the narrow symmetric rifting of the lower lithosphere.

The weak, young, thickened lithosphere would start to rift using conjugate shear zones until one of them is preferred by shear strain softening inside the upper lithospheric mantle. In this lithospheric scenario the decoupling between the upper crust and upper mantle along the ductile lower crust is very advanced. As a result, the extension transfer to the upper crust results in a wide region as the upper mantle drives almost pure shear deformation of the upper crust via simple shear drag in the lower crust. Minimum coupling between the upper mantle and upper crust allows for a wide rift mode and the suppression of the rift flank uplift due to the suppression of the flexural stresses. The rift width is determined by the integrated strength of the decoupling lower crustal layer. The rifting style would be a wide symmetric rifting of the crust combined with narrow symmetric rifting of the mantle lithosphere. This rift mode can be associated with two important exceptions from the homogeneous stretching model discussed in Chapter 2.

The first exception is a rift basin that lacks a post-rift thermally controlled subsidence, such as the south-central portion of the East Slovakian Basin and the east-central portion of the Styrian Basin described in Chapter 1. Other examples come from the Danubian Basin and other basins from the interior of the Pannonian Basin system (Lankreijer, 1998). Their basinal behavior can be explained by a basin formation in the area where the basins are underlain by the intact lithospheric mantle.

The second exception is a rift basin that has recorded a subordinate syn-rift subsidence followed by prominent post-rift subsidence, such as the Eastern Black Sea Basin (Egan and Meredith, 2005). Such basinal behavior can be explained by basin formation in the area where the lithospheric mantle has been totally removed during rifting and replaced by asthenosphere.

The role of the crustal thickness

It is not only the degree of anisotropy of the lithospheric multilayer, its thermal regime, the degree of coupling between individual layers, and individual layer rheologies, but also an individual layer thickness that controls the rift architecture. Such control was documented by a set of analog material models focused on testing the brittle crust thickness control on the rift zone width (Vendeville et al., 1987; Allemand and Brun, 1991). Allemand and Brun (1991) have documented in the case of the brittle layer decoupled from the substratum along the ductile layer that the propagating rift-controlling faults eventually meet at the brittle–ductile transition in the crust. They do it irrespectively of the modeled rift asymmetry, which seems an inherent feature for a modeled two-layer system regardless of the fact that some tests were generated by one-sided extension and others by double-sided extension (see Allemand, 1988; Allemand and Brun, 1991). Allemand and Brun (1991) further gathered a set of natural examples, including five different narrow rift cases, which did not provide conclusive proof.

While it is difficult to compare five rift systems developed in five different rheological settings, and, especially, controlled by different thermal histories, which become more important as the rifts evolve into more mature stages, it is much easier to prove the point in the meso-scale scenario. A suitable example is provided by the spacing of the tensile fracture pattern in bioclastic calcilutite layers of various thicknesses in the outcrop in the Bristol Channel (Figure 5.24; Nemčok et al., 1995). Figure 5.24 indicates that the thicker layer affected by extension is stronger, which results in the larger spacing of the tensile fractures. A small-scale example, but for the normal faulting, is described in Chapter 1, where finite-element simulation (Mandl, 1988) shows that the fault spacing is controlled by shear softening, provided a certain host rheology, thickness, and pore fluid pressure.

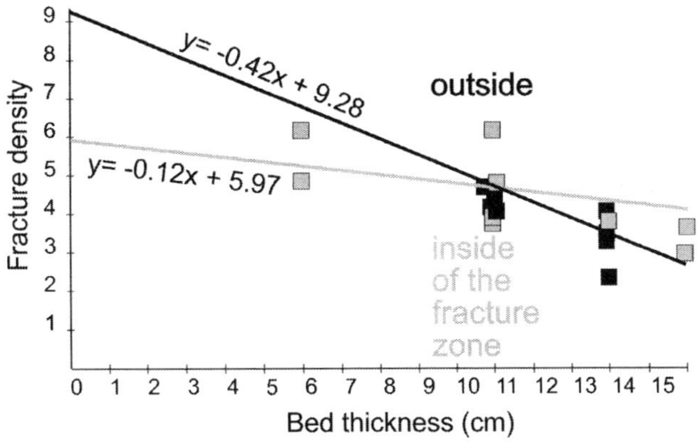

Figure 5.24. Graph demonstrating an inverse relationship between fracture density and bed thickness for extensional veins related to the extensional event at the end of the inversion-related strike-slip faulting (Nemčok et al., 1995). Data are from Trwyn-y-Witch. Separate relationships are shown for fractures developed inside and outside fracture zones. Equations represent linear regressions for the two data sets.

The best way of studying the brittle crust thickness–rift width relationship would be to find a wide rift region with a flat Moho surface, regionally consistent thermal regime, and rheologic distribution and do the measurements there. A suitable example of such a terrain comes from the Basin and Range province. The Nevada portion of the province has an exceptionally good database of seismoactive normal faults (USGS, 2006; Figure 1.23) that represent a new generation of faults detached at the present-day middle crustal level. They have formed in a region that:

1) was already significantly stretched;
2) has a flat Moho; and
3) is not very morphologically diversified.

Together with a publicly available crustal thickness database (Shevenell, 2005a; Figure 1.23), the province

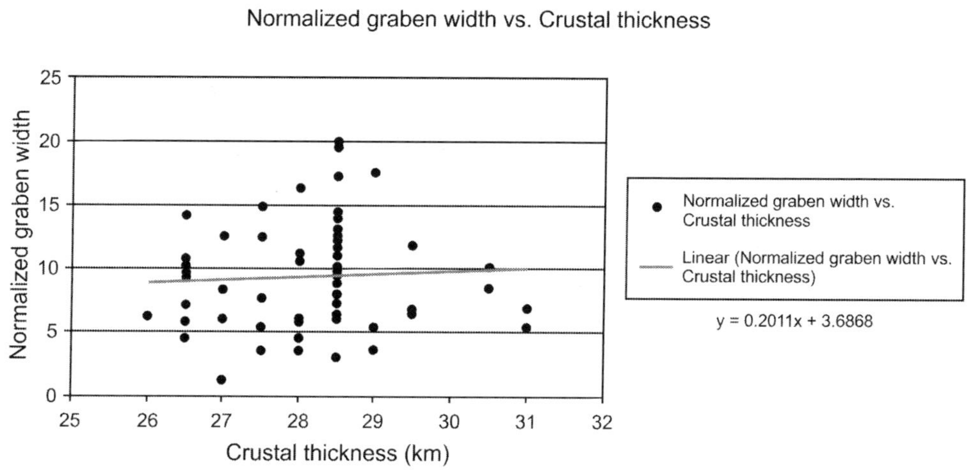

Figure 5.25. Relationship between crustal thickness and rift width determined from the Basin and Range province using the data from Figure 1.23.

should serve as a very good natural laboratory for testing the crustal thickness–rift width relationship (Figure 5.25). Figure 5.25, however, indicates that there is no well-defined thickness–rift width relationship such as exists in analog models, which use homogeneous crustal layers (see Allemand and Brun, 1991). This indicates that it is practically impossible to find a totally isotropic crust in nature.

6 The role of preexisting anisotropy in structural styles of rifts and passive margins

Introduction

The sinuosity of the Western rift branch of the East African Rift system is now generally interpreted as a deflected rifting propagated around the Tanganyika shield (Rosendahl, 1987). The complex pattern of the rift systems of various ages in the broader northern South Atlantic and Equatorial Atlantic regions seems to have followed, or at least been affected by, the distribution of the Late Proterozoic mobile belts among various cratons (Figure 6.1; Molina, 1988). The sinuous East Indian passive margin seems to have followed to a certain extent the areas of Archaean and Proterozoic folding (Nemčok et al., 2012a). The sinuosity of the rift systems controlling the location of the Central Atlantic is interpreted as controlled by the sinuosity of the doubly vergent Hercynian orogenic belt whose western side is known as the Appalachians and whose eastern side is known from remnants such as the Rokelides and the Mauritanides, to name a few (Nemčok et al., 2005b). These examples document the rift-system scale effects of the preexisting crustal anisotropy.

Examples of the effect of the preexisting anisotropy at the rift-unit scale come from:

1) the Bristol Channel, United Kingdom, where graben-bounding Jurassic–Lower Cretaceous faults reactivated preexisting Variscan thrusts (Mechie and Brooks, 1984; Brooks et al., 1988, 1993);
2) the Pannonian Basin system, where Badenian-Pannonian normal faults reactivated preexisting thrusts developed by the Upper Cretaceous and Eocene–Lower Miocene orogenic systems (e.g., Royden, 1985; Tari et al., 1992; Horváth, 1993); and
3) the Melut and Anza grabens of the East African Rift system, where controlling Paleogene faults reactivated the preexisting Mesozoic structures (Ibrahim et al., 1991; Bosworth, 1992; Hendrie et al., 1994).

Additional examples come from the rift basins of the Baikal Rift (Florensov, 1966), the Yadung-Gulu Rift (Armijo et al., 1986), the Rhine Graben (Illies, 1981), the Oslo Rift (Bederke, 1966), and the Rio Grande Rift (Faulds et al., 1990; Beck and Chapin, 1994).

As a result of the interaction of propagating rifts with preexisting anisotropies, the subsequent passive margins contain very different types of failed basins and continental edges in their various segments. The failed basin types vary between orthogonal rift units and pull-apart basins for failed basins. The margin types vary between normal fault-controlled continental margins and strike-slip-controlled ones (Figure 6.2).

Different failed basin types have different development histories. This results in different sedimentary records, which control all play concept elements of the syn-rift petroleum systems. This conclusion can be illustrated with the example of the Nova Scotia–Georges Bank–Morocco conjugated margins of the Central Atlantic (Figure 6.2; Nemčok et al., 2005b). A comparison of the three Central Atlantic segments, which include two segments initiated by pseudo-orthogonal rifting and one segment initiated by a sinistral strike-slip fault zone, shows that the sizes of potential petroleum systems and distributions of seal rock and traps are different for each segment type. While the best seal, represented by the late syn-rift evaporitic formation, is present in normal faulting-initiated segments, the evaporites are missing in the strike-slip-initiated segment. While the structural traps are controlled by major northeast-southwest striking normal faults in extensional segments (Figure 6.2), the strike-slip segment has its traps associated with east-northeast–west-southwest striking bounding strike-slip faults and north-south striking normal faults developed in pull-apart basins among

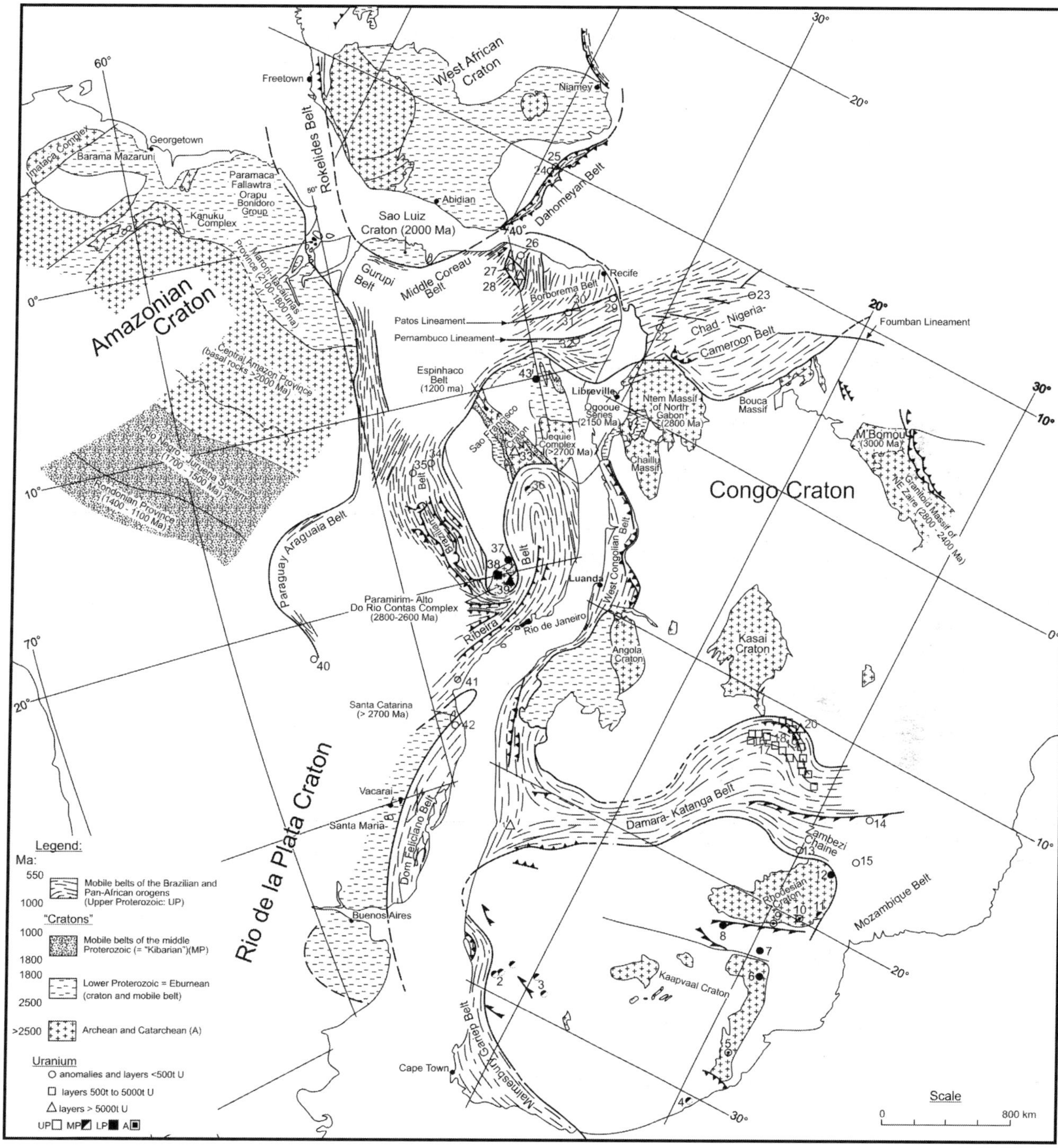

Figure 6.1. Map of West Africa and adjacent Brazil showing the distribution of mobile belts and cratons prior to rifting that resulted in continental breakup (Molina, 1988).

bounding strike-slip faults (Figure 3.12). The extensional segments are characterized by major block sizes, serving as a proxy for the size of the petroleum system, of 30 × 100 km, which are larger than the block sizes of 10 × 50 km typical for the strike-slip segment.

Effect of preexisting anisotropies on fault propagation mechanism

According to Coulomb-Navier's theory for isotropic rock, the fault geometry is controlled by the stress field, which is characterized by the orientation and

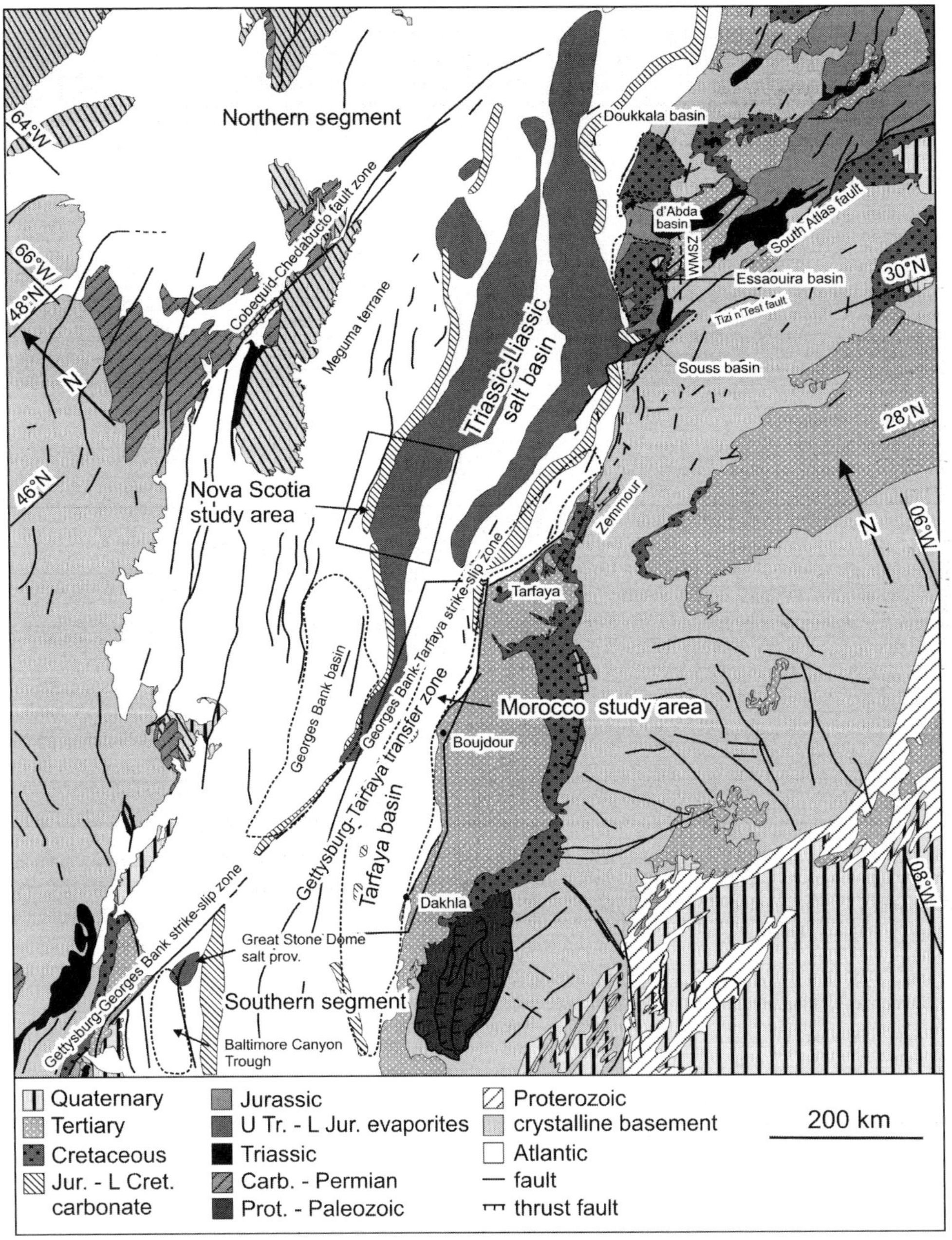

Figure 6.2. Palinspastic match of the Moroccan and American continental margins in the Central Atlantic Basin (modified from Davison et al., 2002) showing the location of the main sinistral Gettysburg-Tarfaya transfer zone, dividing the Central Atlantic into northern and southern segments, and a central segment formed by the transfer zone (Nemčok et al., 2005b). Note the location of the late syn-rift Upper Triassic–Liassic evaporites. The evaporites are present along the margins initiated by more-or-less orthogonal rifting while they are missing in the central segment characterized by margins initiated by strike-slip faulting. The northern and southern Central Atlantic segments have margins with first-order northeast–southeast striking normal faults. The central segment has margins with east-northeast–west-southwest striking sinistral strike-slip faults and north–south striking normal faults.

magnitudes of the principal stress axes, and rheology of the affected material (Mandl, 1988):

$$\alpha = \pm (45° - \varphi/2), \qquad (6\text{-}1)$$

where α is the angle between the fault plane and the maximum principal compressive stress axis, σ_1, and φ is the angle of internal friction characterizing the host rock.

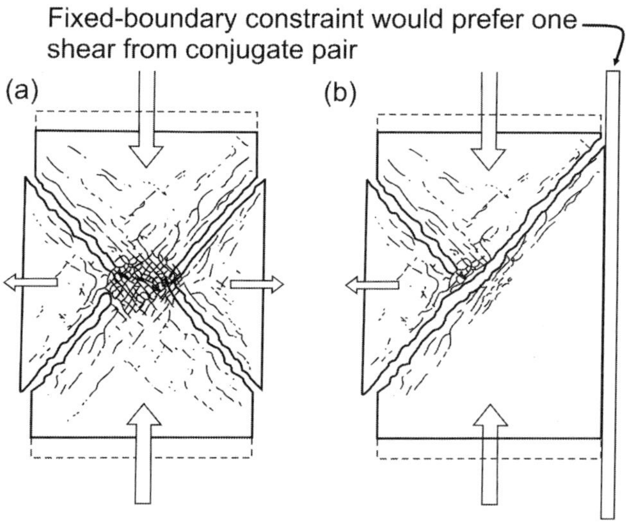

Figure 6.3. Conjugate shear fractures in a brittle rock cylinder loaded by three-axial stress apparatus (Mandl, 1988). Two conjugate ruptures develop and the angle between their surfaces is preserved during shortening in the case of the free interfaces around the cylinder. If we placed a fixed boundary constraint on the right side of the cylinder, we would prevent a lower conjugate rupture surface on the right side of the specimen from forming.

If a rock multilayer is affected by a stress field with vertical σ_1 and horizontal σ_3 stresses, a development of two conjugate systems of normal faults is expected. However, the strength anisotropy of the rock volume can cause preferred development in only one of them. The preexisting thrustbelt detachments inclined in one direction and underlying salt bodies are examples of such anisotropy. Their physical effect can be demonstrated by thinking about a rigid boundary that prevents the multilayer from stretching to both sides, and that just results in one-sided stretching exploiting the free interface (Figure 6.3; Mandl, 1988). The second conjugate system of normal faults may start to propagate but never makes it beyond an incipient state because of the energy balance reasons constrained by a rigid buttress.

A natural example of the geologic setting that favors one-sided shear system development comes from the Reitbrook Salt Dome, Germany (Horsfield, 1980). The ease of sediments, which rest on the salt body, to slip down the salt flanks controls the normal fault system over the salt crest. The stress axes in this scenario are asymmetrically oriented with respect to strength anisotropy planes and slip along only one potential conjugate; slip direction takes place because of the lower shear resistance, despite the stresses acting on potential conjugate planes being equal.

The preferred shear development can be caused by a difference in shear softening along conjugate slip planes. If the direction of the σ_1 stress from Equation (6-1) does not coincide with the direction of the maximum incremental shortening, e_1, but deviates within the interval $(-\varphi/2, +\varphi/2)$, one potential slip plane can be closer to the direction of the maximum shearing than the other one.

The influence of the preexisting anisotropies on the rift propagation can be approximated by the tensile fracture propagation in the highly anisotropic Triassic Mercia Mudstone Group in the Bristol Channel. This group crops out in the Watchet area, where it shows numerous phases of transient hydraulic fracturing. The rock consists of red marl with several evaporite-rich horizons (Warrington et al., 1994). These horizons can be interpreted as a record of the intra-cratonic rift unit, which was drying out, being close to complete infill (see Cosgrove, 2001). Several mudstone-evaporite intercalations in the Mercia Mudstone section represent renewed basin-deepening events and basin drying-up events. One more lithology present in outcrops is windblown sand, occurring in small bodies.

A system of hydraulic fractures, which we will use as an analog for the rift propagation affected by preexisting anisotropy, relates to the post–middle Cretaceous inversion of the Bristol Channel Basin. The orientation of syn-inversion stresses in the area included north-south oriented σ_1, east-west oriented σ_2, and vertically oriented σ_3. Under the stress conditions:

$$(\sigma_1 - \sigma_3) < 4T, \quad (6\text{-}2)$$

and criterion:

$$p_f = \sigma_3 + T, \quad (6\text{-}3)$$

where T is the tensile strength of the host rock and p_f is the pore fluid pressure, a system of hydraulic fractures formed, oriented parallel to bedding.

These fractures are preserved in the mudstone thanks to being filled by satin spar, a fibrous form of gypsum, indicating the direction of fracture opening by an orientation of fibers (see Durney and Ramsay, 1973; Figure 6.4). Not all the fractures are subhorizontal, having various geometries of dilatant shear fractures. However, outcrops show that the fibers in fractures of various geometries remain vertical regardless of their orientation, indicating sufficiently high fluid pressures during their activity (Figure 6.4).

The network of fractures in the Mercia Mudstone Group is preserved around the initial evaporite horizons. During the inversion-related uplift of the currently

Figure 6.4. Gypsum veins of various orientations, Watchet, Bristol Channel. All of them have an antitaxial character. The median line, formed by minor host rock fragments, records the plane of the initial opening. The gypsum fibers grew away from the median line. All veins have vertical fibers regardless of their orientation. Note that vein segments with fibers subparallel to walls developed by reactivation of preexisting normal faults, associated with initial rifting.

Figure 6.5. Layer of the Hettangian-Sinemurian calcilutite exposed on the wave cut platform at Quantoxhead, Bristol Channel. The figure shows a relatively straight preexisting joint, located roughly in the middle of the photo, and its effect on the younger joints.

outcropping section, they underwent volume changes during their hydration to gypsum when the low pressure and temperature conditions for this change occurred (see Jowett et al., 1993). The fluid pressure controlling the stress criterion for the hydraulic fracturing developed thanks to this hydration, which should result in a volume increase of 63 percent if the entire calcium sulfate remains in the rock volume (e.g., Shearman et al., 1972).

If we assume that the aforementioned hydraulic fractures are in a map view, they provide an analog for the effect of the preexisting crustal architecture on the propagating rift-bounding faults. In this case Figure 6.4 would illustrate what happens when the propagating rift unit-bounding faults meet the preexisting zone of weakness, which is located at an acute angle to the direction of the rift propagation. The stress buildup cycles building the stress conditions toward those necessary for further rift-bounding fault propagation will keep reactivating the zone of weakness, which has practically zero cohesion and friction along the zone reduced to residual value, at stress conditions lower than those needed for the rift-bounding fault propagation. The zone of weakness would get periodically reactivated as transtensional dextral strike-slip. Figure 6.4 thus serves as an analog for a rift zone composed of four rift units linked by three dextral transtensional faults. It provides an analog to rift zones such as the Malawi and Tanganyika Rift Zones in East Africa (see Versfelt and Rosendahl, 1989), zones of the Gulf of Suez–Red Sea Rift (see Younes and McClay, 2002), and zones of the Rhine Graben (see Brun et al., 1992).

One more analog of the preexisting weak zone effect on the rift-zone-bounding fault propagation comes from the Hettangian-Sinemurian sedimentary section of the Bristol Channel at Quantoxhead, composed of intercalated marine calcareous mudstones and bioclastic calcilutites. Here, the outcrop shows a joint formed by a pure opening reacting to the preexisting joint as it propagated into its close vicinity (Figure 6.5).

The joint propagated toward the preexisting weakness in a direction parallel to the maximum horizontal stress and perpendicular to the direction of the minimum horizontal stress. Because the controlling stresses underwent perturbation in the close vicinity of the weakness, the propagation trajectory of the joint has changed accordingly. If we compare the described joint trajectory (Figure 6.5) with stress perturbation around the weakness reactivated as a strike-slip fault (see Homberg et al., 1997), we can observe that opposite sides of the reactivated weakness have distinctively different stress trajectories and that the younger joint geometry in Figure 6.5 resembles them rather closely.

Examples of the effect of rift system-scale anisotropy on rift location

An example of the large-scale anisotropy effect on rift location comes from the East African Rift system,

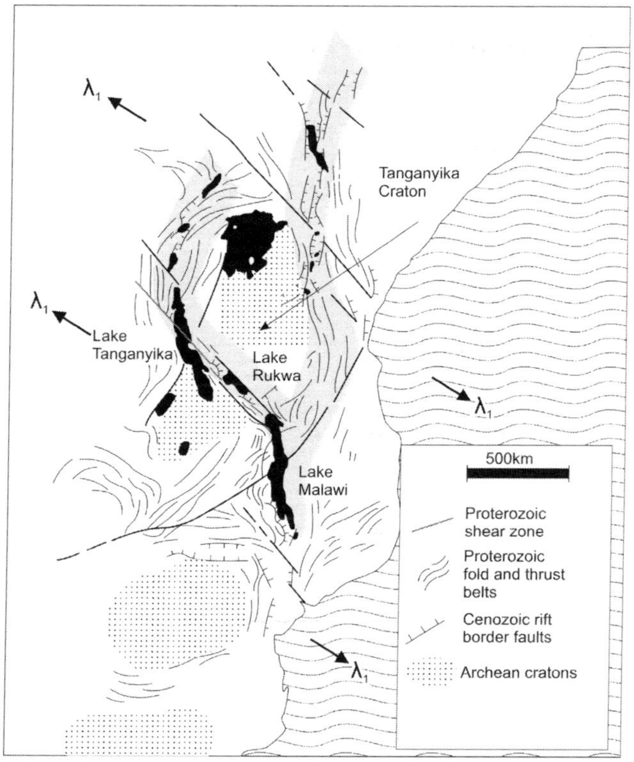

Figure 6.6. Tectonic framework of East Africa showing a rift bifurcation around a craton (Versfelt and Rosendahl, 1989).

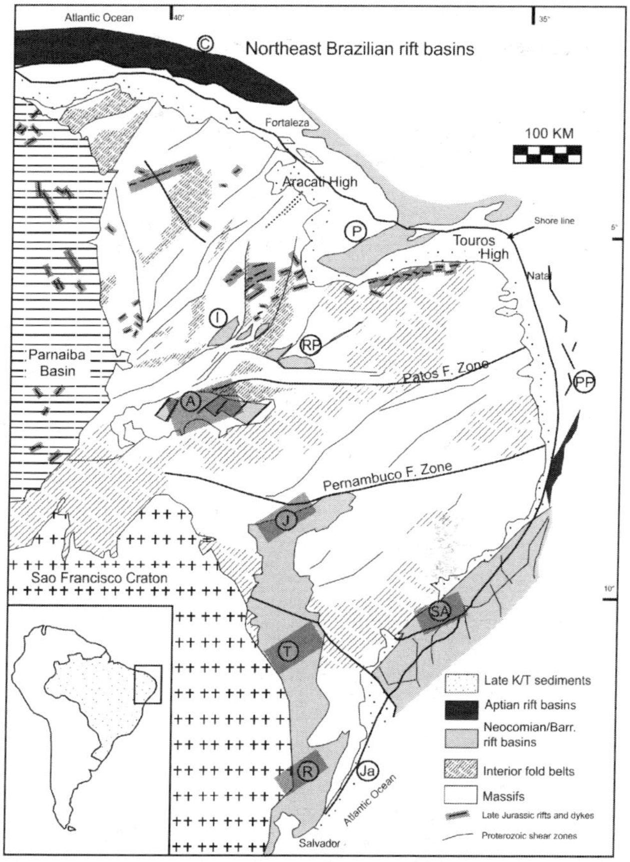

Figure 6.7. Map of northeastern Brazilian basins showing the timing of various rift phases (de Matos, 1992). R – Recôncavo Basin, T – Tucano Basin, J – Jatobá Basin, Ja – Jacuipe Basin, Sa – Sergipe-Alagoas Basin, A – Araripe Basin, I – grouped Iguatu, Malhada Vermelha, Lima Campos, and Icó Basins, RP – Rio de Peixe Basin, P – Potiguar Basin, PP – Pernambuco-Paraiba Basin, C – Ceará Basin. Note that numerous small basins such as the Araripe, Iguatu, Malhada Vermelha, Lima Campos, Icó, and Rio de Peixe Basins, located along the southeastern continuation of the onshore Potiguar Basin trend, form the Cariri Valley portion of the Cariri Valley–Potiguar Rift Branch.

which bifurcates into the east and west rift branches around the Tanganyika Craton (Figure 6.6; Versfelt and Rosendahl, 1989). The craton acted as a resistant core, forcing rift propagation to follow surrounding Proterozoic mobile belts. The craton is formed by a stable Archaean nucleus, which is composed of greenstone belts sandwiched between ensialic plutons. The surrounding mobile belts are gently arcuate, extensively folded, Proterozoic rocks wrapped around cratons (e.g., Anhaeusser et al., 1969). They contain a variety of structures controlling their anisotropy, including the overall fabric of the rocks, faults, fractures, axial planar fabrics, trends of igneous bodies, contrasts in metamorphic grades, schistosity, and compositional layer boundaries.

Other anisotropies, which were present in the region prior to the development of the East African Rift system, were the earlier rift systems of the Permo-Triassic and Cretaceous ages and the transcontinental dislocation zones. The zones were formed by an activation of major sutures in the pre-rift crust by various African orogenies and rift events. They are usually located within mobile belts.

A nice large-scale example of the anisotropy affecting the rifting and subsequent continental breakup comes from the Brazilian side of the northern South Atlantic, comprising the Northeast Brazilian Basin system (Figure 6.7), which can be divided into three rift groups: the Jatobá-Tucano-Recôncavo Rift Zone, the Sergipe-Alagoas Rift Zone, and the Cariri Valley–Potiguar Rift branch. The development history of these rift groups documents the opening of the northernmost South Atlantic and its late Berriasian-Aptian interaction with the eastward-propagating dextral transtensional rift system of the future Equatorial Atlantic.

The existence of preexisting fault zones made the earlier rifting in this region quite complex. The

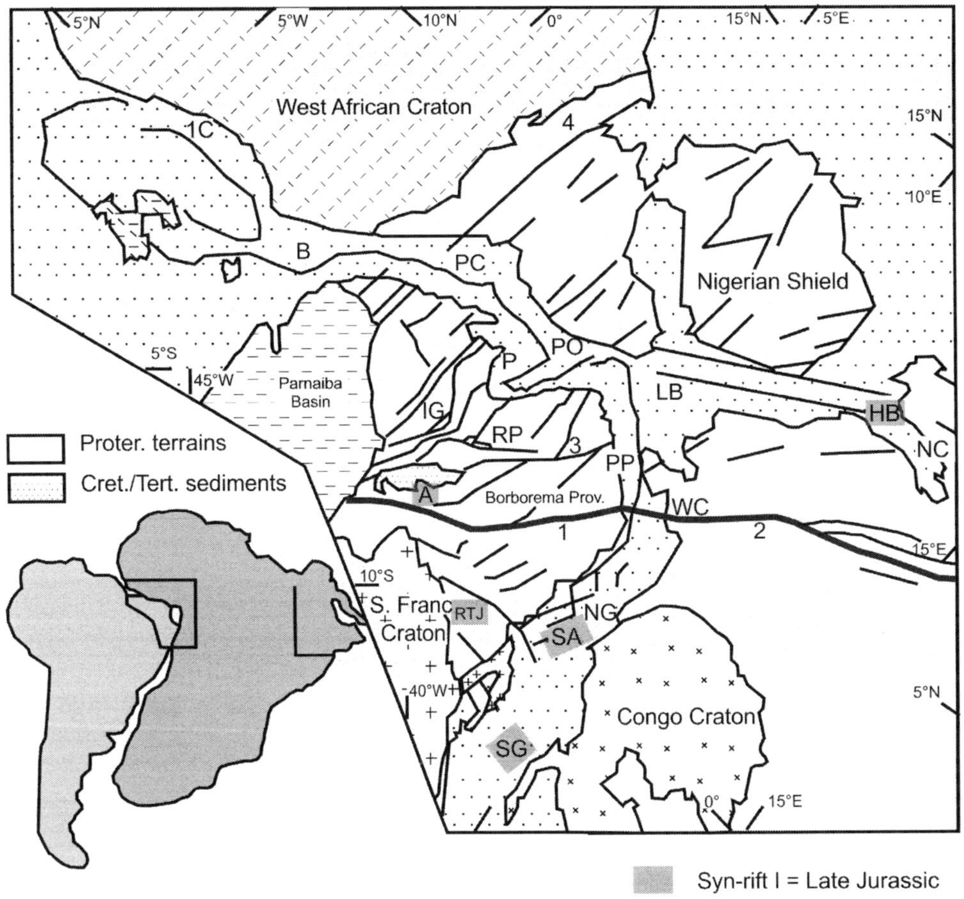

Figure 6.8. Pre-drift reconstruction of the northeast Brazil and adjacent West African continents (de Matos, 1992). Thick lines – main faults and hinge zones, thin lines – local faults and other geological boundaries. Main Proterozoic faults contain: 1 – Pernambuco strike-slip fault, 2 – Ngaoundere strike-slip fault, 3 – Patos strike-slip fault, 4 – Dahomeyan thrust fault. Basin codes are explained in Figure 6.7.

preexisting fault zones were part of the structural grain of the Braziliano–Pan-African terrains reworked by Proterozoic orogeny, represented by the Borborema province in Brazil and the Nigerian Shield–Nigerian Pan-African belt in Africa (Figure 6.8). The structural grain was characterized by northeast-southwest trending fold belts and east-west striking strike-slip fault zones (e.g., Jardim de Sá, 1984). Most of the outcropping structures in the Borborema province were formed in the ductile deformational environment and document the considerable erosion that took place after the time of their origin.

East-west striking strike-slip fault zones are usually 1–5 km wide, mostly having a mylonitic and occasionally a cataclastic character. The most prominent ones are the Patos and Pernambuco Fault Zones (Figure 6.7). Their extension to the African continent has been documented (see de Matos, 1987; Figure 6.8).

Initial rifting started during the Late Jurassic (Chang et al., 1988). This rifting related to the northward propagation of the South Atlantic Rift systems never made it through the Patos Fault Zone as indicated by the presence of Upper Jurassic sediments in the Jatobá-Tucano-Recôncavo, Sergipe-Alagoas, and Araripe Basins and their lack in wells located in basins north of the Patos Fault Zone such as the Rio de Peixe and onshore Potiguar Basins (Figure 6.9). Furthermore, Figure 6.9 indicates that the Pernambuco Fault Zone, located 150 km south of the Patos Fault Zone, acted as a boundary important enough to control the northern distribution of Late Jurassic rifts.

All Late Jurassic rifts formed a wide and shallow depression terminated to the north of the Araripe Basin in the northwest and along the Pernambuco Fault Zone in the northeast, marking the Late Jurassic propagation front of the South Atlantic Rift system.

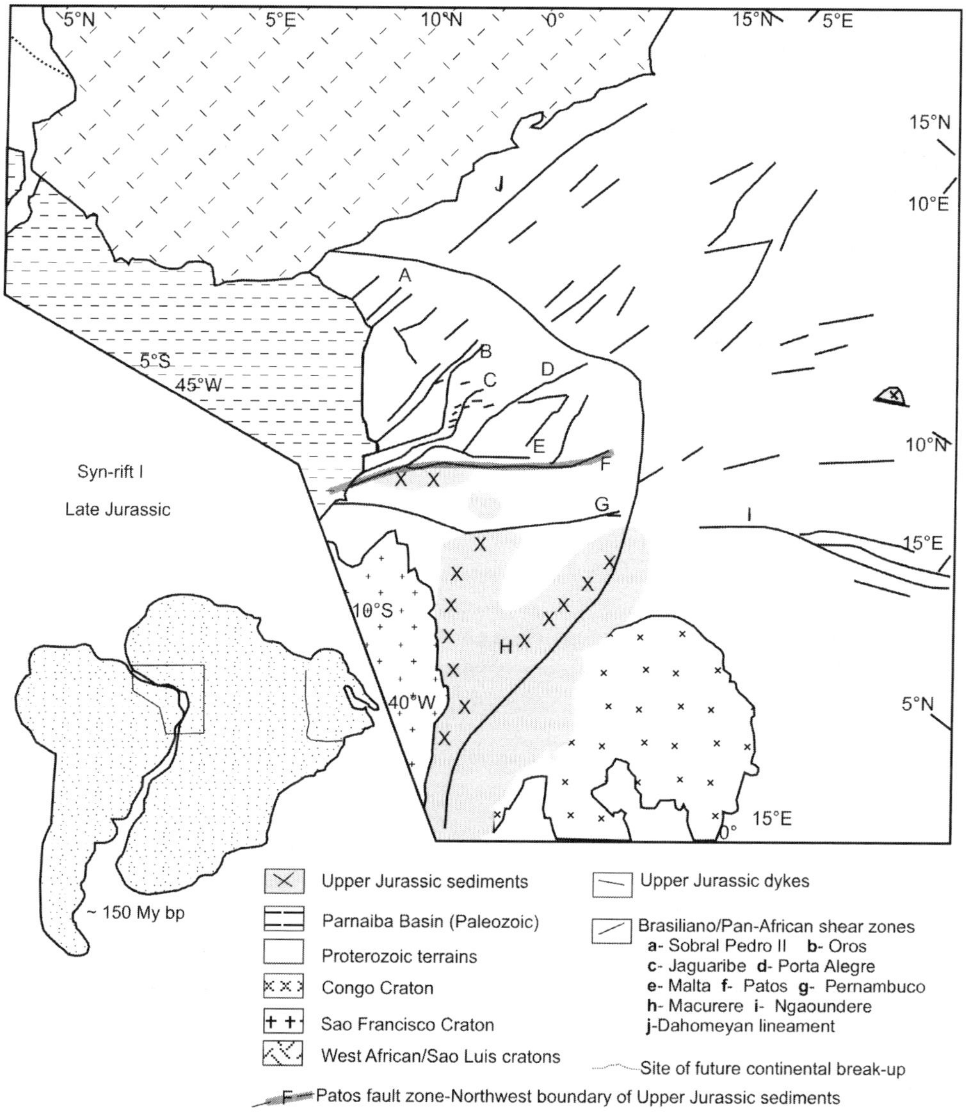

Figure 6.9. Reconstructed map showing the location of the Late Jurassic syn-rift basins of northeastern Brazil and adjacent Africa (de Matos, 1992).

Although the local controlling mechanisms for the Late Jurassic rifting do not seem completely clear, this rifting was relatively widespread and reached a sufficiently mature stage to be accompanied by magmatism (e.g., Conceicao et al., 1988; Figure 6.7).

While the Late Jurassic magmatic activity did not develop to the south of the Patos Fault Zone, it was quite widespread to the north of it, represented by mafic dykes of northwest-southeast and almost east-west strikes. The tholeiitic dyke swarms to the southeast of the Potiguar Basin are typical representatives. They have K/Ar and fission track ages of 124–167 Ma (Sial, 1976; Gomes et al., 1981; Sial et al., 1981; Horn et al., 1988), which indicates their activity during the Late Jurassic–Neocomian time period.

A large width of the shallow Late Jurassic basin to the south of the Patos and Pernambuco Fault Zones indicates an extension over a large area. To the north of these zones, the extension resulted only in dyke development in the Borborema province and the Paleozoic Parnaiba Basin (Almeida et al., 1988; de Matos, 1992). This dramatic north-south change indicates that the Patos and Pernambuco Fault Zones had to be active as accommodation zones and weak enough to stop the propagation of the South Atlantic Rift system into the region north of them.

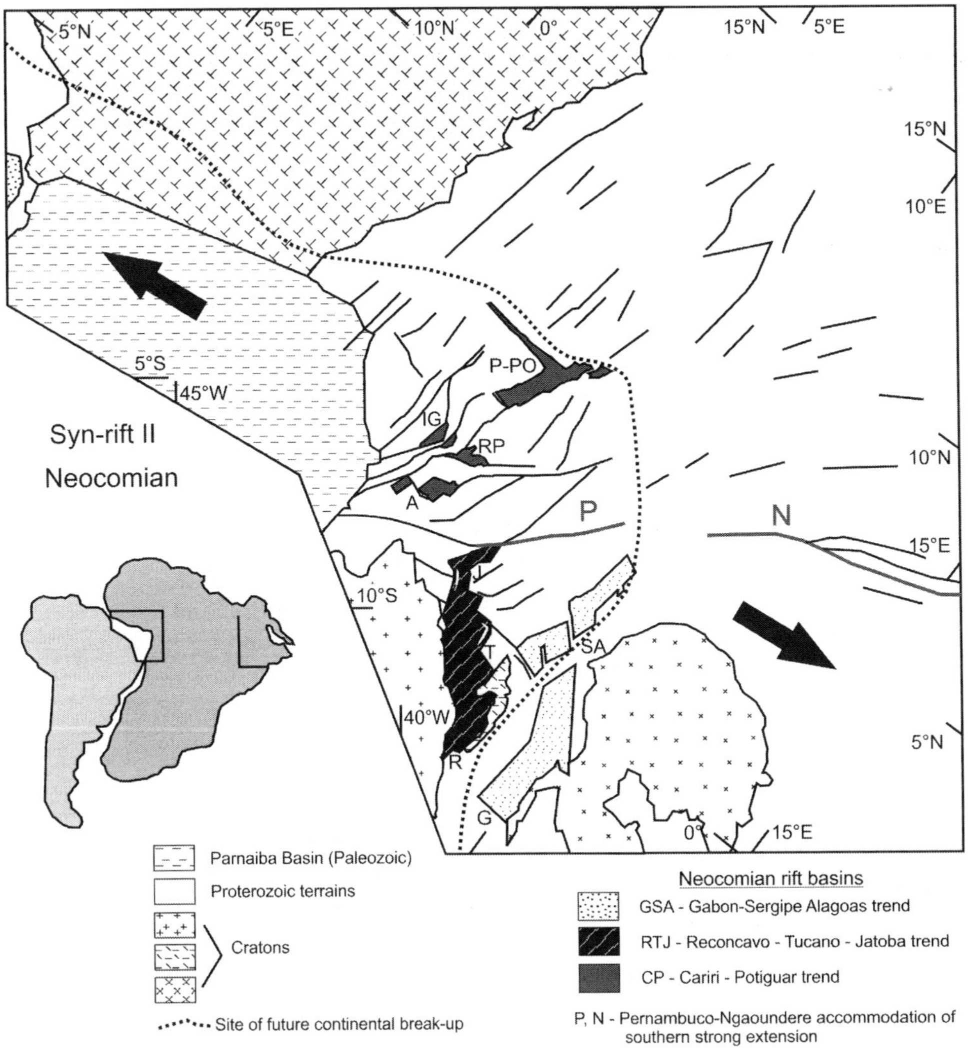

Figure 6.10. Reconstructed map showing the location of the Neocomian–early Barremian syn-rift basins of northeastern Brazil and adjacent Africa (de Matos, 1992). Large arrows show the direction of the regional extension.

The main rift phase, which resulted in a system of major rift valleys associated with areally-extensive fracturing, took part during the Neocomian–early Barremian (Chang et al., 1988; Figure 6.10). This phase did not succeed in propagating the rift system along the future South Atlantic beyond the Pernambuco-Ngaoundere Fault Zone either. However, it developed a few more northeast-southwest trending rifts in the Cariri Valley to the northeast of the Araripe Basin, such as the Iguatu, Malhada Vermelha, Lima Campos, Icó, and Rio de Peixe Basins, and developed the onshore Potiguar Basin (Figures 6.7, 6.10). This indicates that the extensional deformation jumped into a northwestern region, forming a number of intra-cratonic basins along the Cariri–onshore Potiguar trend to bypass the problem with propagating through the preexisting Pernambuco, Ngaoundere, and Patos Fault Zones. This also documents that the main east-west trending strike-slip fault zones such as the Pernambuco, Ngaoundere, and Patos continued behaving as a large complex accommodation zone, balancing extensional deformation in the Jatobá-Tucano-Recôncavo and Sergipe-Alagoas Rift Zones with coeval extension in the intra-cratonic rift branch of the Cariri-Potiguar trend. This trend is exactly parallel to the Pan-African foldbelt and its thrust faults, which are occurring in the Borborema province.

The pre-rift faults of the Cariri Valley illustrate a kinematic connection of the northeast-southwest striking thrust faults with the east-west striking Patos strike-slip fault zone. At its western boundary, the splay of faults branches from the Patos zone. At some distance from the zone, the faults eventually reach northeast-southwest strikes. The opening of all basins of the Cariri-Potiguar trend was coeval and all of them indicate that their controlling faults reactivated preexisting structural grain (de Matos, 1992).

The southernmost basin of this trend, the Araripe Basin, is the only one that has the main syn-rift, Neocomian–lower Barremian, sedimentary package deposited on top of the initial syn-rift, Upper Jurassic package. The Upper Jurassic strata rest either directly on the basement or on Devonian sediments of the Paleozoic Parnaiba Basin (Ghignone, 1986). Neocomian–lower Barremian strata are up to 1.8 km thick in depocenters, represented mainly by fluvial/deltaic facies (Ghignone, 1986; Campanha, 1987). This deposition was controlled by northeast-southwest striking normal faults and northwest-southeast striking transfer faults.

Central in this trend, the Iguatu, Malhada Vermelha, Icó, and Lima Campos Basins start their sediment fills with Neocomian–lower Barremian strata. They are a system of relatively shallow and asymmetric half-grabens controlled by northeast-southwest striking normal faults. They were controlled by northwest-southeast trending extension. Their basin floors are generally tilted to the southeast. Around the western terminus of the Patos strike-slip fault zone, the Neocomian–early Barremian extension of the northwest-southeast strike reactivated Proterozoic northeast-southwest striking thrusts as normal faults and a portion of the Patos strike-slip fault zone as a sinistral strike-slip fault zone. It resulted in a pull-apart extension in the Araripe Basin system (de Matos, 1987). The Rio de Peixe Basin system also forms a system of pull-apart basins controlled by northeast-southwest striking normal faults branching off the east-west striking fault that also belongs to the Patos strike-slip fault zone, located just 50 km to the north of the strike-slip fault controlling the Araripe Basin system.

The northernmost basin of the Cariri Valley–Potiguar rift branch is the onshore Potiguar Basin. Its main syn-rift, Neocomian–lower Barremian strata are covered by the Aptian–Campanian fluvial/marine transgressive sequence (Souza, 1982). Neocomian–lower Barremian sediments are represented by fluvial, deltaic, lacustrine, and deltaic fan sediments. The syn-rift deposition in the offshore Potiguar Basin lasted until the late Barremian, differing from the onshore rifting a bit.

The deposition in the onshore Potiguar Basin took part in half-grabens, controlled by northeast-southwest striking normal faults. Half-graben systems are linked by northwest-southeast striking strike-slip faults that serve as transfer zones, balancing differential extension between individual graben systems. Reflection seismic data indicate that normal faults reactivated preexisting ductile shear zones of the Braziliano orogen (de Matos, 1992). Residual gravity anomaly data contain linear positive anomalies adjacent to controlling normal faults, potentially indicating the presence of dykes intruding during rifting (de Matos,1992).

The offshore Potiguar Basin was developed by a slightly oblique transtension controlling the opening of narrow northwest-southeast trending basins.

The Neocomian–early Barremian rifting was followed by the late Barremian rifting, which was characterized by extension culmination in the northern rift zones of the South Atlantic Rift system (Chang et al., 1988). It represented the final rifting stages in the Jatobá-Tucano-Recôncavo and Sergipe-Alagoas Rift Zones, and intra-cratonic rift basins located to the south of the Pernambuco-Ngaoundere strike-slip system (Figure 6.11). It also represented a major change in the controlling dynamics and the rift initiation in the eastern Equatorial Atlantic domain. Figures 6.10 and 6.11 document a change from the northwest-southeast striking regional extension to the west-northwest–east-southeast extension during this phase. During this time, the rifting in the Cariri Valley and onshore Potiguar rift basins was aborted.

The rifting initiation in the eastern Equatorial Atlantic domain was coeval with continuous deposition in the offshore Potiguar Basin. Evidences for this initiation come from the Benue Trough in Nigeria. The west-northwest–east-southeast extension, characteristic for the northeast Brazilian corner during this time, reactivated earlier, northeast-southwest striking, normal faults of the Potiguar Basin as sinistral transtensional strike-slip faults. New grabens opened during the late Barremian on the Aracati Platform to the west of the Potiguar Basin, being controlled by normal faults perpendicular to the new direction of extension.

The rifting initiation in the Benue Trough is recorded by basal upper Barremian conglomerates lying below the Aptian Lower Bima Formation (Benkhelil, 1988, Popoff, 1988). These conglomerates are present also in the North Cameroon Basin (Popoff, 1988).

Late Barremian rifting was followed by Aptian rifting that reached the region of all future Equatorial Atlantic basins (Souza, 1982; de Matos, 1992). The most interesting Aptian situation was present in the Pernambuco–Paraiba Basin region. This region was affected by a large number of roughly east-west striking dextral strike-slip faults, which formed a 340 km-wide deformation zone to the south of the west-east

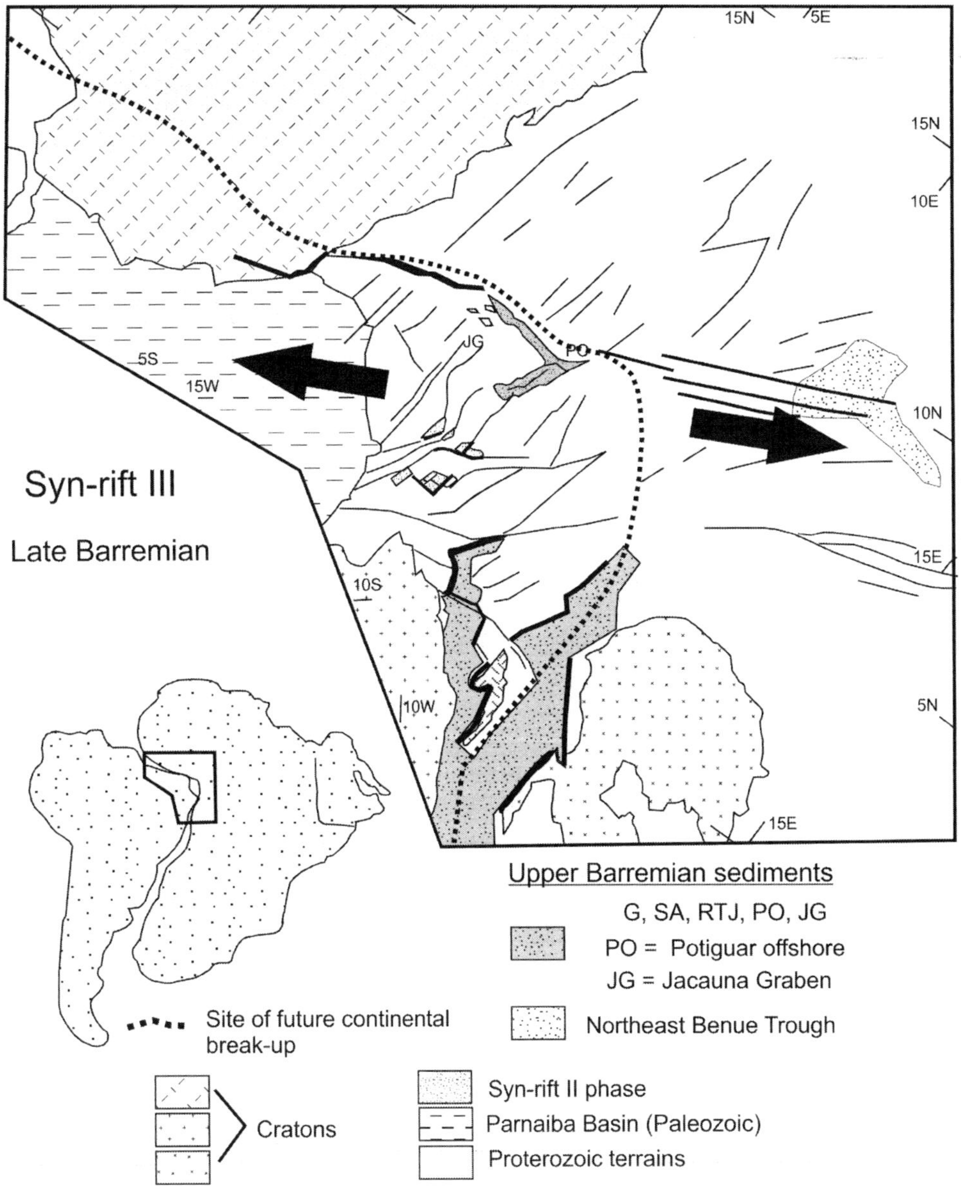

Figure 6.11. Reconstructed map showing the location of the late Barremian syn-rift basins of northeastern Brazil and adjacent Africa (de Matos, 1992). Large arrows show the direction of the regional extension.

trending future breakup in the offshore Potiguar Basin (Figure 6.12). At least some of them were reactivated preexisting transfer faults of the Braziliano orogen. This region was also deformed by numerous coeval northeast-southwest striking sinistral strike-slip faults. They were formed along preexisting thrusts of the Braziliano orogen. Both sets of faults acted as a conjugate set. They were linked by smaller normal faults and oblique-slip faults.

It was during the Aptian when the northern tip of the South Atlantic Rift system, which was locked during the Late Jurassic–late Berriasian roughly to the south of the Pernambuco-Ngaoundere strike-slip fault zone, resumed its northward propagation (Figure 6.7). This is documented by syn-rift seismostratigraphy of small half-grabens in seismic cross-sections cutting through the Pernambuco Plateau (Gomes et al., 2000; Nemčok et al., 2004b) and proven Aptian sediments in a small north-northeast–south-southwest trending half-graben in the Cabo sub-basin of the Pernambuco–Paraiba Basin (de Matos, 1987). The entire Touros Platform between the onshore Potiguar Basin and the Patos Fault Zone apparently behaved as the accommodation zone between the rifting in the northernmost South Atlantic

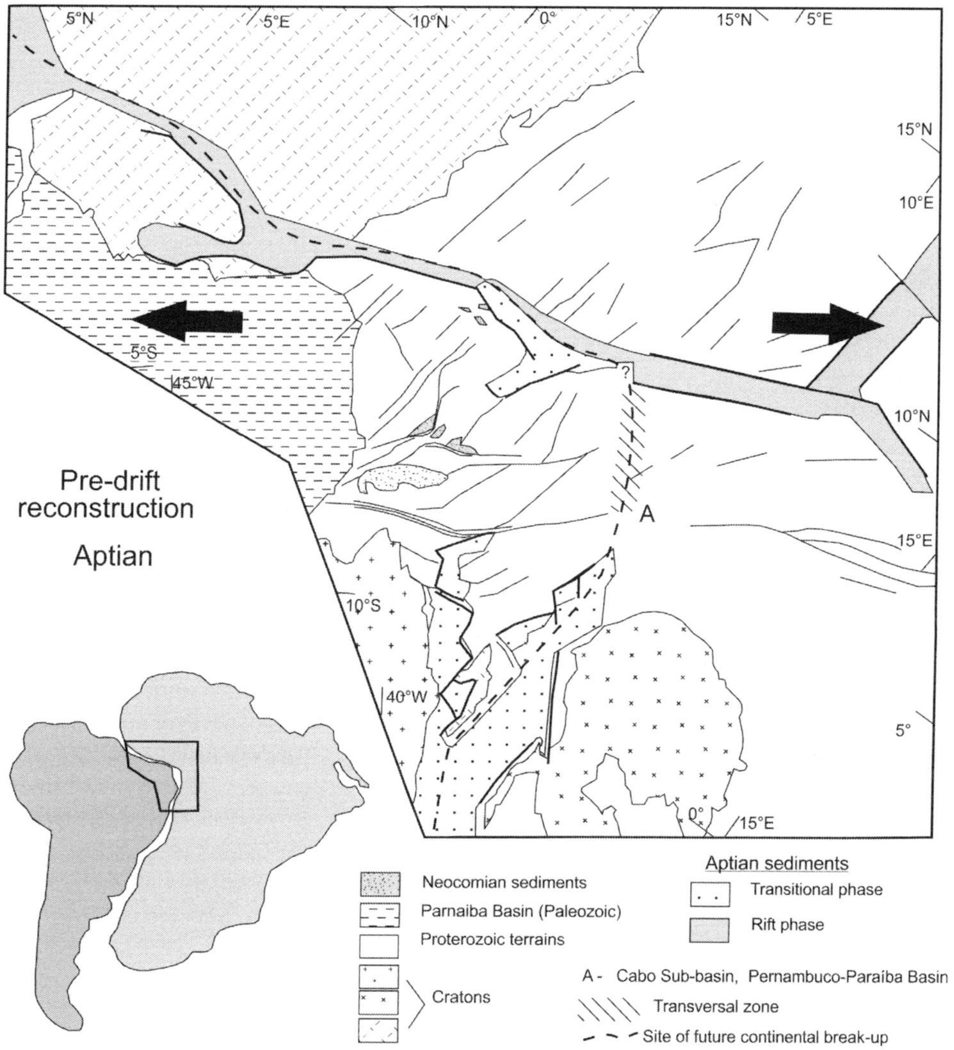

Figure 6.12. Reconstructed map showing the location of the Aptian basins of northeastern Brazil and adjacent Africa (de Matos, 1992). Large arrows show the direction of the regional extension.

system and the Equatorial Atlantic rift system for a little longer (de Matos et al., 1987).

We believe that the aforementioned delay in rift propagation in this region was caused by a dextral strike-slip linkage of the Potiguar Basin with the Benue Trough (Figure 6.12), which prevented the remaining continental bridge from undergoing rifting. It was the northward propagation of both the Sergipe-Alagoas sinistral strike-slip system and the system of roughly north-south trending normal faults that made it through this last continental bridge located between the offshore Potiguar-Benue trend and the Patos dextral strike-slip fault zone, along the future Pernambuco-Paraiba segment of the South Atlantic.

Note that the Aptian rift was very deep and narrow and it is located oceanward from the modern continental slope. These facts, along with a lack of exploratory wells in the basin, make the exact timing of the rift propagation through the Pernambuco-Paraiba segment difficult to determine. The rifting was not accompanied by any volcanism (Gomes et al., 2000) that could be dated.

Several tens of kilometers of the sinistral displacement have been determined along northeast-southwest striking faults right at the northeast Brazilian corner (Popoff, 1988). Various branches of this sinistral northeast-southwest trending rift system were located in a vast region between the northeast Brazilian corner and the Trans-Saharan domain (Fairhead, 1988a, 1988b; Unternehr et al., 1988; Guiraud and Maurin, 1991, 1992).

The rift–drift transition in the Pernambuco-Paraiba Basin did not happen until the Aptian–Albian transition. The basin formed the northern boundary for the South Atlantic salt province. However, the data on the continental breakup timing in this area is not direct. The timing is implied, using a marine transgression, which initiated at about the Aptian–Albian transition, according to available biostratigraphic data (Noguti and Santos, 1972; Dias-Brito, 1987), as a constraint. This rift–drift transition completed the described rifting of the Southern Atlantic affected by preexisting anisotropy.

The Georges Bank-Nova Scotia-Morocco segment of the Central Atlantic provides another example of the effect of the preexisting anisotropy on the rift system and subsequent breakup location and character. Rift-fault systems in this region (Figure 6.2; Nemčok et al., 2005b) were controlled by the amount of extension, pre-extensional basement rheology, and pre-extensional faults such as the Late Paleozoic Hercynian fault system.

The Georges Bank region underwent relatively strong extension over a broad zone, while the Nova Scotia–Morocco region experienced mild extension over a relatively narrow zone (Royden and Keen, 1980; Piqué et al., 1998; Le Roy and Piqué, 2001). Published structural analyses suggest that rifting-related extension was oriented northwest-southeast (Ratcliffe and Burton, 1985; Venkatakrishnan and Lutz, 1988; Olsen et al., 1990; Olsen and Schlische, 1990; de Boer, 1992; Schlische, 1993).

Appalachian tectonic maps (Cohee et al., 1962; PennWell, 1982; Hatcher et al., 1986) contain fault patterns demonstrating that the Late Paleozoic orogenic belt situated between the future North America and Africa had numerous curved segments, such as the Tennessee salient, the Roanoke recess, the Pennsylvania salient, the New York recess, and the Meguma terrane salient (e.g., Macedo and Marshak, 1999). Their sides were formed by faults with significant strike-slip components (see Hatcher et al., 1986). All major orogenic salients and recesses are outlined by juxtaposed positive and negative anomalies on the Appalachian Bouguer gravity anomaly map (see Haworth et al., 1980). The most distinctive gradients outline the Meguma terrane and basement-involved thrust sheets in the Pennsylvania salient.

The west-northwestern and northern sides of the Meguma terrane (Figure 6.2) are formed by syn-orogenic, sinistral and dextral, transpressional, strike-slip faults, respectively. The African affinity Meguma terrane collided with the North American Plate along these transpressional sutures (Le Roy and Piqué, 2001). The northern side of the suture is formed by the east-west striking dextral transpressional fault system that includes the Cobequid and Chedabucto Faults. The Meguma terrane is characterized by west-verging thrusts (e.g., Brown, 1986; Keen et al., 1991), in contrast to the east-verging Zemmour thrusts in the northeastern Tarfaya Basin in Morocco (Le Roy and Piqué, 2001).

Mapped fault patterns along the west-northwestern and north-northwestern sides of the Pennsylvania salient have Late Paleozoic sinistral and dextral displacement components, respectively (Hatcher et al., 1986). Both of these fault systems were transpressional. The north-northwestern side of the salient is lined by the Gettysburg and Newark Basins of the Triassic–Jurassic Central Atlantic rift system. They lie along the Late Paleozoic Martic dextral, transpressional, strike-slip fault, and several other faults, which are subparallel to it. The faults are characterized by an overall en echelon pattern (Cohee et al., 1962). The northern margin of both basins coincides with the east-northeast–west-southwest striking boundary between the thin-skin thrustbelt in the north, which deforms only Paleozoic sediments, and the orogenic belt in the south, which also deforms Precambrian crystalline rocks (Cohee et al., 1962; PennWell, 1982).

Different rift fault systems were developed within different parts of the preexisting, doubly vergent Paleozoic orogenic belt and within the adjacent portion of its eastern foreland. It was the preexisting structures and related basement rheology that affected the orientation and size of syn-rift fault blocks. Block size in the southeastern part of the studied rift system, which developed partly on the West African shield, was barely affected by syn-orogenic deformation, and is characterized by a width of 20–30 km and a length of 50–100 km. The northern part of the studied rift system, which developed above contractional portions of the orogen, contains 20 km-wide and 20–50 km-long blocks. Southwestern and central parts of the studied rift system, which developed by sinistral transtensional reactivation of preexisting dextral transpressional faults, are characterized by the smallest block sizes with lengths and widths of 20–50 km and 10 km, respectively.

The roughly east-west striking, sinistral, transtensional, Cobequid–Chedabucto strike-slip fault (Figure 6.2) was developed by the reactivation of the dextral, transpressional suture between the American Plate and the Meguma terrane. Thrusts within the Meguma terrane were reactivated as normal faults in the Nova Scotia rifts (Brown, 1986; Keen et al., 1991; Withjack et al., 1995).

Similar evidence for fault reactivation comes from the Moroccan side of the Central Atlantic (Figures 2.2, 3.12, 6.2). The preexisting syn-orogenic thrusts and strike-slip faults have different orientations in different areas and syn-rift fault patterns change from basin to basin. East-vergent Zemmour thrusts in the eastern onshore Tarfaya Basin control the 030° strike of rift-related normal faults. Normal faults in the Souss Basin, next to the High Atlas, have strikes of 050°. Further north, in the Essaouira Basin, they have 020° strikes. Even further north, in the Doukkala Basin, the strike of normal faults remains the same. The presence of syn-rift, sinistral, strike-slip faults on the African side is demonstrated in the western High Atlas, where they reactivated Hercynian dextral, transpressional faults of 070° strike (Tayebi, 1989). The Doukkala and Tarfaya Basins contain 070° striking, sinistral, transfer faults (Le Roy and Piqué, 2001), which could be the result of similar reactivation. A typical example of Hercynian dextral, transpressional faults on the Moroccan side is the northeast-southwest striking West Meseta shear zone (Piqué et al., 1980).

Faults in the Pennsylvania salient were reactivated as the sinistral Gettysburg–Georges Bank strike-slip zone, which forms a portion of the Georges Bank–Tarfaya strike-slip zone (Figure 6.2). The African part of this zone includes the east-northeast–west-southwest to northeast-southwest trending Dakhla, Boujdour, and Tarfaya sinistral, transtensional faults in Morocco (Figure 3.12), and possibly even the sinistral South Atlas Fault (Figure 6.2) (Tayebi, 1989). Two distinctive curvatures of Middle–Upper Jurassic shelf break carbonates suggest the presence of two more strike-slip faults between the Tarfaya and South Atlas Faults (Figure 6.2): an unnamed fault and the Souss Basin Fault. An exploration well and seismic data document that additional northeast-southwest to east-northeast–west-southwest trending, sinistral, strike-slip faults are present in the area north of the Souss Basin Fault, including the Amsittene, Essaouira, and Tensift Faults (Le Roy and Piqué, 2001; Figure 2.2).

Other sinistral, strike-slip faults that strike east-northeast–west-southwest have been discovered further southeastward in onshore Morocco, such as the Tize n'Test Fault, which has its syn-rift activity dated by deformed syn-tectonic sediments (e.g., Laville and Petit, 1984). The strike of the Gettysburg–Georges Bank strike-slip zone is similar to the Cobequid-Chedabucto Fault Zone, mentioned earlier, which was dextral during the Late Paleozoic Hercynian orogeny and sinistral during Late Triassic to Early Jurassic rifting.

Sinistral transfer faults formed a system of "rails" that prevented the Georges Bank–Tarfaya Central Atlantic segment from rifting orthogonally, resulting in the development of several pull-apart basins in western Morocco between the "rails" (Figures 2.2, 3.12). The Georges Bank Basin shown on the PennWell (1982) map is a pull-apart counterpart to these on the African side. Basins north and south of the Georges Bank–Tarfaya strike-slip fault underwent almost orthogonal rifting controlled by reactivation of thrusts having almost ideal orientations for normal fault reactivation.

As alluded to in the introduction, the distribution of the main rift zones and transfer zones, honored by the continental breakup, controls the hydrocarbon habitat of the subsequent passive margin segments, simply by being characterized by basins with different tectonic and deposition histories (see Figure 6.2). Apart from the example shown in Figure 6.2, further examples can be discussed using Figure 6.13.

Transfer zones from Figure 6.13 divide passive margins of the northern South Atlantic and Equatorial Atlantic to a system of segments where the continental breakup was controlled by normal, oblique, and strike-slip faulting (see Nemčok et al., 2012b). As discussed in Chapter 3 (see Figure 3.11), while the strike-slip fault end-member of the passive margin segments started the drifting period characterized by large underwater cliffs, the normal fault end-member of the passive margin segments started the drifting period characterized by a slope relatively close to the gravitational instability angle for its sedimentary cover. As a result, the gravity gliding at normal fault-controlled margin segments starts relatively early after the breakup while some of the strike-slip fault-controlled margin segments need a lot of time until the deposition covers the initial large underwater cliffs and develops a slope available for the gravity gliding (Nemčok et al., 2012b). Therefore, for the same source rock stratigraphy in adjacent segments, the match between the migration timing and trap timing can be dramatically different.

Examples of rift zone-scale anisotropy effects on rift and controlling fault locations

Examples of the rift zone-scale anisotropies come from the Tanganyika and Malawi Rift Zones of the East African Rift system. The most typical reason for the complex rift zone architecture, including rift units changing their polarity, is transcontinental dislocation zones intersecting the trend of the rift zone. Here, the Henga Valley–North Chamaliro, West Chamaliro–East Chamaliro, and Rukwa-Ufipa-Luama transcontinental dislocation zones are all also associated with the occurrence of major kinks in the rift zone trend (Figure 6.14; Versfelt and Rosendahl, 1989). While not all rift unit polarity reversals along the rift zone trends in the area

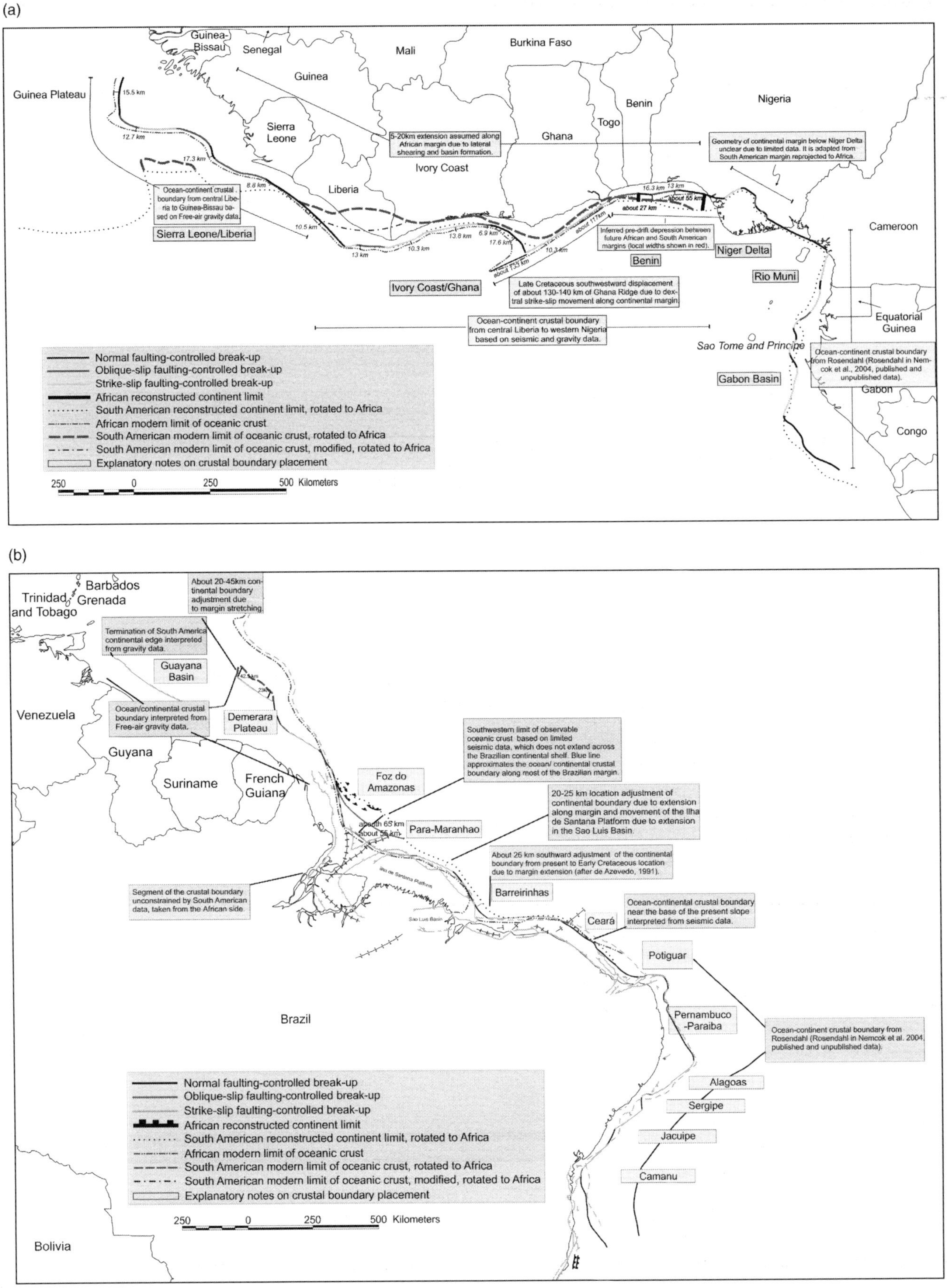

Figure 6.13. (a) Map of the African passive margins of the Equatorial Atlantic and northern South Atlantic divided into segments based on their initiation (Nemčok et al., 2004b). Light gray – strike-slip initiation, medium gray – oblique-slip initiation, dark gray – normal fault initiation. (b) Map of the South American passive margins of the Equatorial Atlantic and northern South Atlantic divided into segments based on their initiation (Nemčok et al., 2004b). See Figure (a) for explanation.

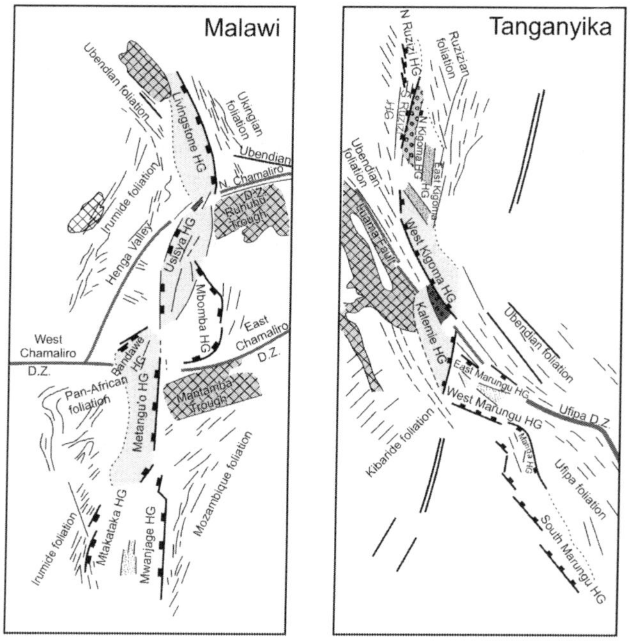

Figure 6.14. Generalized structure maps of the Tanganyika and Malawi Rift Zones superimposed on the pre-rift structure (Versfelt and Rosendahl, 1989).

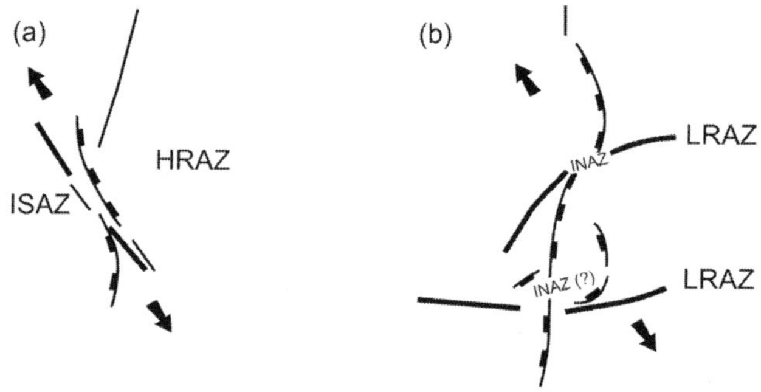

Figure 6.15. Selected linking architectures in the Tanganyika (a) and Malawi (b) Rift Zones superimposed on the pre-rift structure (Versfelt and Rosendahl, 1989). Arrows indicate inferred direction of extension for the East African Rift system. a) Kavala Island model of the isolation accommodation zone (high-relief accommodation zone) linking half-grabens characterized by polarity alternation. The pre-rift fabric and/or preexisting weak zones are subparallel to the direction of extension. b) Malawi model of the interference accommodation zone (low-relief accommodation zone) linking half-grabens characterized by polarity alternation. The pre-rift fabric and/or preexisting weak zones are sub-perpendicular to the direction of extension.

are associated with linkages made by reactivated transcontinental dislocation zones, each reactivated continental dislocation zone divides rift units with different polarities.

The orientation of the transcontinental dislocation zone with respect to the trend of the propagating rift zone and its controlling extension direction seems to affect the type of linkage it would be used for between the two adjacent rift units (Versfelt and Rosendahl, 1989).

The Henga Valley–North Chamaliro transcontinental dislocation zone in the Malawi Rift Zone is the example of the weak zone oriented almost perpendicular to the propagation trend of the Malawi Rift Zone. In this case, the rift unit-controlling faults splay off the reactivated transcontinental dislocation zone and progressively change their geometry to become closer to being parallel to the controlling extension (Figure 6.14; Versfelt and Rosendahl, 1989). This type of rift unit linkage forms a very small and narrow low-relief accommodation zone (Rosendahl, 1987; Figure 6.15).

The West Chamaliro–East Chamaliro transcontinental dislocation zone in the Malawi Rift Zone is yet another example of the weak zone oriented almost perpendicular to the propagation trend of the Malawi Rift Zone but controlling extension is oblique to both rift unit-bounding faults and the transcontinental dislocation zone (Figure 6.14; Versfelt and Rosendahl, 1989). Here, rift unit-bounding faults practically die out in the vicinity of the transcontinental dislocation zone. This type of rift unit linkage forms a broad and diffuse low-relief accommodation zone (Rosendahl, 1987; Figure 6.15).

The Rukwa-Ufipa-Luama transcontinental dislocation zone in the Tanganyika Rift Zone is an example of the weak zone oriented almost parallel to the propagation trend of the Tanganyika Rift Zone. In this case, the rift unit-controlling faults splay off the reactivated transcontinental dislocation zone and the strike-slip-dominant displacement component gets progressively replaced by a normal displacement component with a distance from the transcontinental dislocation zone. At some distance from it, the geometry is perpendicular to the controlling extension and faults are normal faults (Figure 6.14; Versfelt and Rosendahl, 1989). This type of rift unit linkage forms a high-relief accommodation zone (Rosendahl, 1987; Figure 6.15).

A study by Versfelt and Rosendahl (1989) concludes that there is a general parallelism between the trends of rift-unit border faults and the pre-rift structural grain with few exceptions. All of them seem to occur when the pre-rift fabric is at a higher angle to the rift propagation trend, which is more or less in accordance with observations Donath (1961) made from rock mechanics tests of rock specimens with planar anisotropies at a different angle to the direction of maximum shortening, described later in this chapter. However, despite the aforementioned general parallelism, rift unit-controlling faults do not seem to follow major lithological boundaries for larger distances, which is the case of low-relief accommodation zones (Versfelt and Rosendahl, 1989).

A study of the Jeremboabo transfer fault from the Late Jurassic–Early Cretaceous Jatobá-Tucano-Recôncavo Rift Zone, northeastern Brazil (Destro et al., 2003a), documents that not all anisotropies reactivated by regional extension as transfer faults must have pure strike-slip or transtensional character. Their character is determined by the relationship between the preexisting fault geometry and the reactivating stress field.

Here, the outcrop studies of striated faults determined the controlling stresses, which were characterized by east-west oriented minimum principal compressional stress σ_3 during the early Neocomian and northwest-southeast oriented σ_3 during the late Barremian–early Aptian (Magnavita, 1992). Based on the deformation of the syn-rift sediments around one of the studied faults, its activity was determined as Berriasian-Valanginian. It is located on the flexural side of the rift zone where it links smaller rift-bounding faults. Its displacement sense is consistent with the displacement sense of linked bounding faults on both of its sides. Its displacement was sinistral oblique with a strong reverse component. Its core and damage zones are about 5 km wide, characterized by folded and overturned layers in the footwall next to the fault. The transfer character of the Jeremboabo Fault is further documented as based on its linkage with a major dextral Caritá transfer fault located between rift units of the Jatobá-Tucano-Recôncavo Rift Zone with opposite polarity.

At this point in this chapter it would be interesting to look at the strike-slip reactivation of the preexisting weaknesses in a more systematic way, using analog material experiments. Richard and Krantz (1991) did it by first creating a reverse fault, a normal fault or a vertical fault related to uplift in the modeling material and then loading the system by stresses capable of fault reactivation as a strike-slip fault. The initial reverse, normal, and vertical uplift faults had dips of 54°, 45°, and 90°. Each of the studied tectonic scenarios was modeled using three rheologic scenarios:

1) scenario characterized by a 6 cm-thick sandstone layer, simulating brittle layer;
2) scenario characterized by 5 and 1 cm-thick sand and silicone layers, simulating a brittle layer with less coupling to its substratum; and
3) scenario characterized by 4 and 1 cm-thick sand and silicone layers, simulating brittle layer with rather weak coupling to its substratum.

Strike-slip reactivation of the reverse fault in a pure brittle system resulted in an intermediately broad fault system, located roughly parallel to the preexisting weakness and rooting in it (Figure 6.16; Richard and

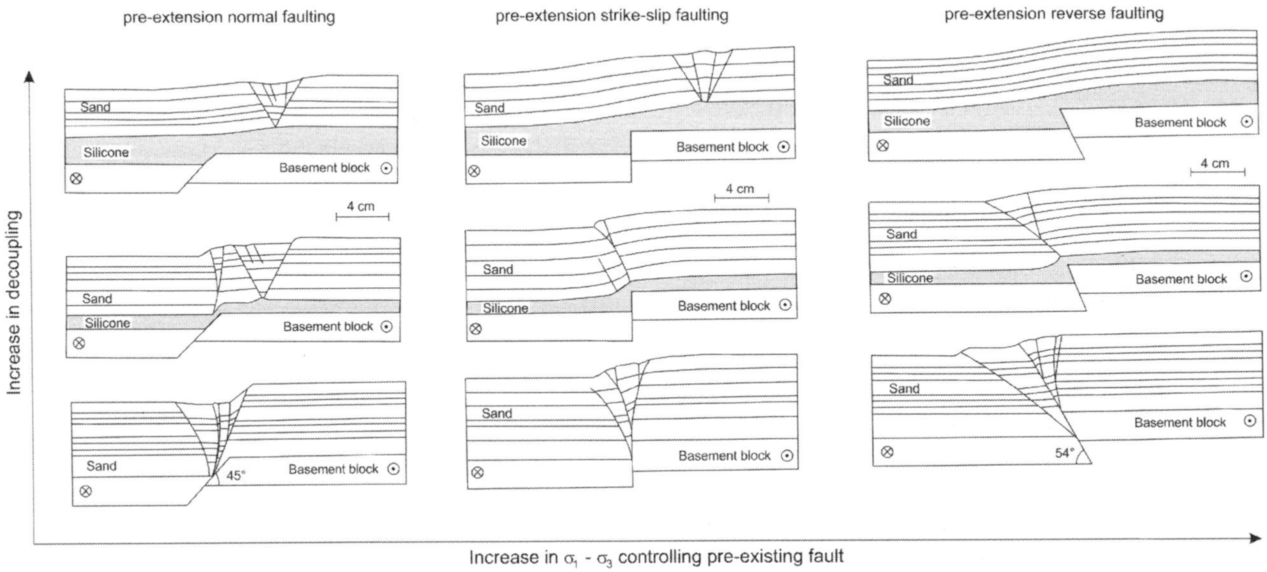

Figure 6.16. Reactivation in pure strike-slip mode of a 45° normal (left), 90° uplift (middle), and 54° reverse basement fault (Richard and Krantz, 1991). See text for explanation.

Krantz, 1991). The bulk shearing did not occupy a very wide zone. Reactivation in a weakly decoupled system resulted in a narrower fault system, located more in the footwall of the system than parallel to the preexisting fault and rooted in the preexisting fault. The bulk shearing zone in this case was wider than the zone in the previous one. Reactivation of the strongly decoupled system showed that most of the block movements were accommodated by bulk shearing, which occupies a wide zone. No preexisting reverse faults were created in the cover. The only failure, created only during the strike-slip reactivation, is a single fault not rooted into the preexisting fault.

Strike-slip reactivation of the vertical uplift fault in a pure brittle system resulted in an intermediately broad fault system, which is narrower than the one made by reserve fault reactivation (Figure 6.16; Richard and Krantz, 1991). The system is roughly parallel to the preexisting fault and roots in it. The bulk shearing did not occupy a very broad zone. The preexisting fault pattern was composed of a reverse fault in a down-dropped block and a normal fault in the uplifted block. Reactivation in a weakly decoupled system resulted in the narrower fault system, located more in the footwall of the system than parallel to the preexisting fault and rooted in the preexisting fault. The bulk shearing zone in this case was wider than the zone in the previous one. The pre-reactivation fault pattern consisted of the only reverse fault formed in the down-dropped block. Reactivation of the strongly decoupled system showed that most of the block movements were accommodated by bulk shearing, which occupies a wide zone. The failure is represented by a set of faults developed in the footwall rather far from the preexisting fault and without rooting in it. This failure was formed already during pre-reactivation deformation and it was subsequently reactivated.

Strike-slip reactivation of the normal fault in a pure brittle system resulted in an intermediately broad fault system, which is roughly parallel to the preexisting fault and roots in it (Figure 6.16; Richard and Krantz, 1991). The bulk shearing did not occupy a very broad zone. Reactivation in a weakly decoupled system resulted in a broader fault system, which was formed, in fact, already during the pre-reactivation deformation. It is characterized by a graben in the footwall a bit further from the basement fault and a fault right above the basement fault. The bulk shearing zone in this case was wider than the zone in the previous one. Most of the younger strike-slip movement was located along the fault right above the basement fault. Reactivation in the strongly decoupled system showed that most of the block movements were accommodated by bulk shearing, which occupies a wide zone. The pre-reactivation deformation developed a graben located in the footwall rather far from the basement fault. Later strike-slip reactivation used the newly formed faults inside the earlier graben.

These experiments show that coupled systems have a tendency to link newly developed faults with reactivated preexisting faults. On the contrary, decoupled systems have a tendency to develop new faults and reactivate deep preexisting faults without necessarily linking them.

The sensitivity analysis Richard and Krantz (1991) conducted involved not developing pre-reactivation faults in the sedimentary cover and keeping only a preexisting fault in the basement. It showed fault patterns different from those described earlier but still different from typical Andersonian faults, which would develop in the homogeneous system unaffected by preexisting faults.

The next example of the rift zone-scale anisotropy effects comes from the Gulf of Suez–Red Sea Rift system, from between 30°N and 26°N latitudes (Figure 6.17a; Younes and McClay, 2002). The late Oligocene–Miocene Rift system here is represented by four rift units, which include the Darag, October, Zeit and Duwi half-graben basins displaying floor tilt polarities to the southwest, northeast, southwest, and northeast, respectively. The basins are separated by Zaafarana, Morgan, and Duwi accommodation zones (Moustafa, 1997), which can be found in extensive outcrops of basement and pre-rift rocks.

The basement belongs to the strong, old, stable lithosphere of the Arabian-Nubian shield, which was stabilized during Precambrian times. Its upper crust is composed of granitic intrusions surrounded by zones of metamorphic rocks with intense foliations. The upper crust is dissected by a large-scale northwest-southeast striking pre-rift faults and shear zones.

During late Oligocene–Miocene rifting, faults propagated from the basement upward through the pre-rift and syn-rift sediments (Younes and McClay, 2002). North-south to northwest-southeast striking faults developed perpendicular to the direction of regional extension (Angelier, 1985; Steckler et al., 1988; Khalil, 1998). They have normal fault displacements. Faults with other strikes propagated being affected by the Precambrian structural architecture (Lyberis, 1988; Jarrige et al., 1990; Younes, 1996). Northeast-southwest and west-northwest–east-southeast ones are characterized by sinistral and dextral oblique displacements (Prat et al., 1986; Chorowicz, 1989; Abu El Karamat and Fouda, 1990). As documented at the Zaafarana accommodation zone region, some of the normal faults have also been affected by preexisting weakness. In this region, the border faults are parallel to the gneissic foliation (Younes and McClay, 2002).

A comparison of the two northern Zaafarana and Duwi accommodation zones shows the following similarities (Younes and McClay, 2002):

1) neighboring graben-bounding faults undergo arrest and die out in splays where they intersect the preexisting Precambrian shear zone;
2) faults in accommodation zones have a strike-slip component; they are relatively straight and coincide with the preexisting Precambrian shear zone; and
3) faults in accommodation zones were kinematically linked with normal faults in neighboring basins.

Described data allow us to surmise that the development of accommodation zones in the study area started with the reactivation of the northwest-southeast striking zones of weakness prior or coeval with the propagation of the rift-bounding faults and finished with the arrest of the border faults and transfer of displacement, in a fashion somewhat similar to the small-scale example from the Bristol Channel (Figure 6.5). The arrest of the propagating rift-bounding faults is documented by a profile with fault throws in relation to the location of the Hamrawin shear zone in the Duwi accommodation zone area (Figure 6.17b; Younes and McClay, 2002). Figure 6.17b shows how they die out in the vicinity of the reactivated shear zone.

Interaction of the propagating rift-bounding fault with a preexisting zone of weakness depended on the strike-slip sense of the reactivated shear zone and dip-slip direction of the propagating fault (Figure 6.18; Younes and McClay, 2002). If the slip components were divergent, that is, plunging away from each other, the local tensile stresses at the intersection became enhanced, allowing the rift-bounding fault to kink or jog. The kinks had the potential to evolve into transfer faults linking the main intersecting faults. If the slip components were convergent, that is, plunging toward each other, the propagation of the rift-bounding fault either terminated abruptly at the reactivated shear zone or cut right through it.

While the discussed Gulf of Suez–Red Sea examples of the rift zone-scale anisotropy effect indicate that the accommodation zone development was preceding and coeval with the propagation of the rift-bounding faults, the following example of the rift system-scale anisotropy effect from East India documents the development of an accommodation zone between the two rift zones as coeval with and postdating their propagation. The zone is a dextral Coromondal strike-slip fault zone that is parallel to the boundary between the Archaean cratons and Proterozoic orogenic belt in the onshore and mezo-scale structural grain of the orogen (Nemčok et al., 2012a).

(a)

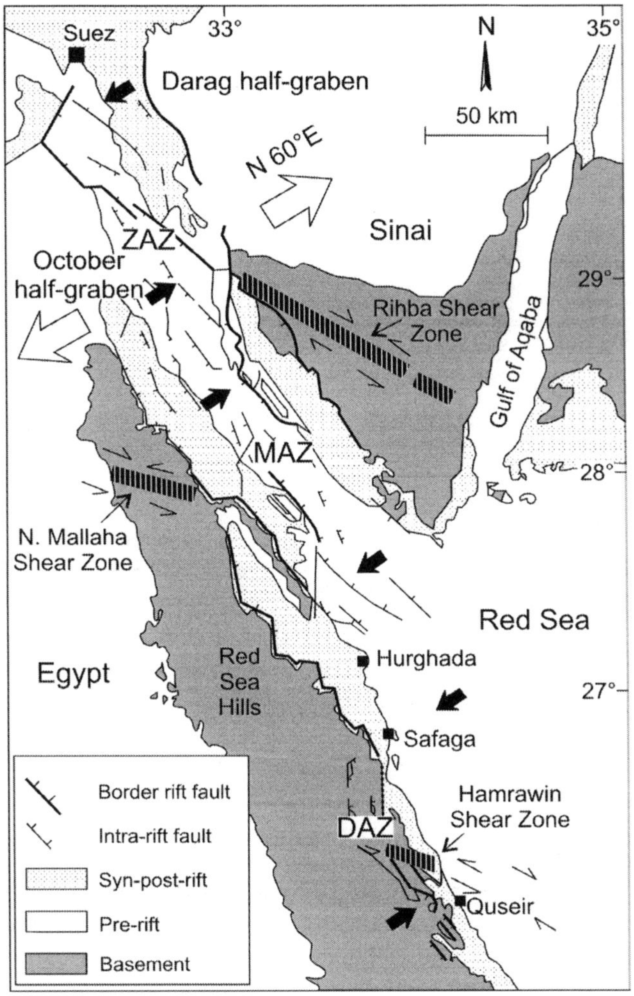

Figure 6.17. (a) Simplified tectonic map of the study area (Younes and McClay, 2002). Black bold arrows indicate dip direction of half-grabens; large gray arrows indicate direction of late Oligocene–early Miocene extension. Solid lines are normal faults. Major Precambrian shear zones are shown in stippled lines. ZAZ is the Zaafarana accommodation zone, MAZ is the Morgan accommodation zone, and DAZ is the Duwi accommodation zone. (b) Fault displacement profile along the main fault system of the Duwi accommodation zone (Younes and McClay, 2002). Shaded area indicates overlap between the Duwi transfer zone and the northernmost tip of the Hamrawin shear zone. Displacement is measured from cross-sections, taking into account thickness variation or pre-rift strata and slope of the top of the basement. Note that these are minimum values, as in most places the pre-rift sedimentary section has been eroded from the footwall.

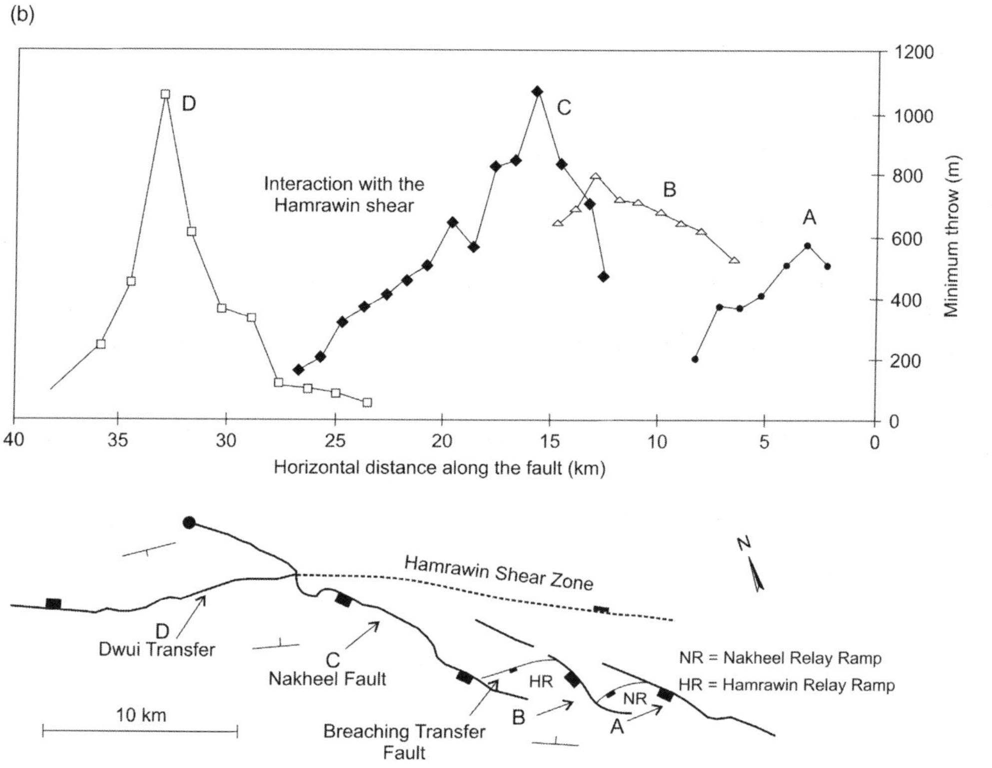

Figure 6.17 (*cont.*)

Having north-northwest–south-southwest strike, it connects the east-northeast–west-southwest striking Krishna-Godavari Rift Zone in the northeast with the east-northeast–west-southwest striking Cauvery Rift Zone in the southwest. Faults of its northern and southern horsetail structures deform the faults of both rift zones (Nemčok et al., 2012c; Figure 6.19). Detailed mapping of reflection seismic images tied to wells (Sinha, 2013) reveals that the rift zones were competing to capture the continental breakup location. While the breakup along the Cauvery Rift Zone propagated during Valanginian-Hauterivian, the breakup along the Krishna-Godavari Rift Zone propagated from the east to the west during the Barremian-Aptian to eventually reach the arrest leading to the breakup propagation along the Coromondal accommodation zone during the Aptian (Sinha, 2013).

Examples of small-scale anisotropy effects on fault location

Small-scale anisotropies affect the evolving structural style of a rift terrain by influencing local fault and fracture development in brittle and transient environments. They include inhomogeneities in the original rock structure such as bedding, foliation planes, and preexisting fractures. Their role in the propagation of new faults was determined by calculations and rock mechanic testing of relatively small cylindrical specimens.

The calculations focusing on the role of a single plane of weakness, such as schistosity or a similar group of S-planes (Jaeger, 1960), indicate that the shear strength in the studied rock system varies continuously from a maximum on planes perpendicular to S-planes to a minimum on planes parallel with S-planes. They also suggest that the rock mass would not be disrupted by movement along a single group of fault planes F, obliquely cutting S-planes, and having an orientation similar to that in the isotropic rock. Furthermore, the S-planes themselves show no movement under any of the tested scenarios. The newly propagated faults F are always formed between S-planes and the nearest potential fault planes predicted by Coulomb-Navier's theory for the isotropic rock (Figure 6.20). If α and β are the angles between the maximum principal compressive stress σ_1 and the S and F planes, respectively, then low β angle values will result in an angle α being relatively smaller than α in a corresponding isotropic system (Figure 6.20a). The high values of the angle β will be accompanied by an angle α somewhat greater than α in a corresponding isotropic system (Figure 6.20b).

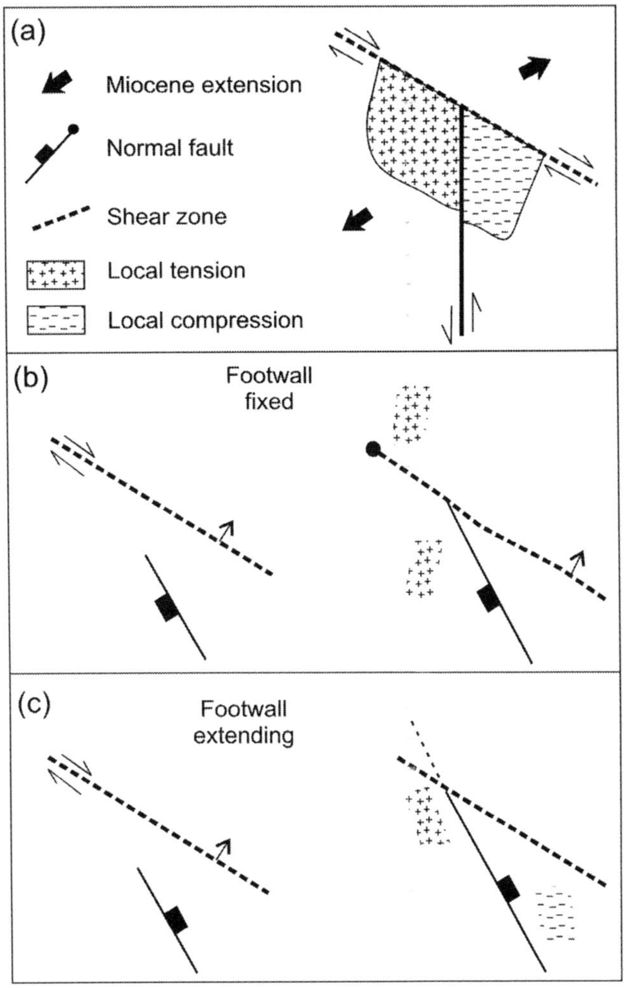

Figure 6.18. Kinematics of the reactivation of preexisting Precambrian shear zones during late Oligocene–early Miocene rifting in the Gulf of Suez–Red Sea region (Younes and McClay, 2002). Obtuse angles between rift-bounding faults and reactivated shear zones promote extension, acute ones contraction. See text for explanation.

These calculated results were later confirmed and refined by rock mechanic testing (Donath, 1961). The testing focused on several kinds of anisotropy, such as bedding, cleavage, and schistosity. It was performed at room temperature and under gradually increasing confining pressure. Anisotropy effects were simulated in brittle, brittle–ductile, and ductile deformation environments. The compressive tests were carried out on jacketed cylinder-shaped specimens submerged in kerosene to simulate the confining pressure.

The testing results indicate that the anisotropy orientation distinctively affects the differential stress at failure (Figure 6.21a). Preexisting planes located close to potential fault planes predicted by Coulomb-Navier's theory for the isotropic rock reduce the differential stress to a value about five or six times smaller than its value in the case of unfavorably oriented anisotropies, parallel or perpendicular to the maximum principal compressive stress, σ_1. Furthermore, Figure 6.21a shows that increased confining pressure increases differential stress.

In an anisotropic rock, as shown by these tests, the angle α between the fracture and the maximum principal compressive stress axis σ_1 will depend on the attitude of the anisotropy relative to σ_1. The rock mechanics tests (Donath, 1961), however, show that, in certain specific cases, newly formed faults can be parallel to the preexisting anisotropy and can have the orientation as predicted by Coulomb-Navier's theory. This phenomenon happens when the angle β between anisotropy and σ_1 is above 0° and below 30°, and, in some cases, when it equals 45° (Figure 6.21b). Some influence of the anisotropy orientation on the failure plane can be noted even at angles of β = 60° and 75°. This latter phenomenon occurs when the deviation angle β attains 90° and shears are formed according to Coulomb-Navier's theory.

The increasing confining pressure tends to reduce the effects of anisotropy on shear fractures. Ductility, being a function of confining pressure and rock type, also reduces the effect of anisotropy. The intensity of the influence depends on the type of anisotropy. Cleavage and schistosity were shown to be more favorable than irregular bedding.

Normal and shear stresses acting on a newly formed shear plane are also influenced by anisotropy (Donath, 1961). Experimental results made at a confining pressure of 350 bar on Martinsburg slate indicate a linear relationship between the normal stress σ_n and shear stress τ on the planes of anisotropy, governed by the empirical formula $\tau = 89.8 + 0.329\sigma_n$. The formula yields an angle of internal friction, φ, of 18° and an average cohesive strength, c, of 89.8 bars. In cases of cleavage inclinations β being 30°, 45°, 15°, and 60°, points indicating τ and σ_n on the Mohr diagram lie very close to the failure envelope along cleavage. If the cleavage inclination β amounts to 60°, a shear plane is formed at an angle α of 49° in relation to the maximum principal compressive stress σ_1. The shear stress τ is zero on cleavage planes parallel and perpendicular to σ_1 and, therefore, a shear failure along cleavage cannot take place.

The maximum principal compressive stress σ_1 at failure and the shear resistance τ_o also vary with the orientation of the planar anisotropy in the rock (Donath, 1961). All discussed results indicate that anisotropy strongly affects the strain geometry of rocks displaying at least moderately brittle behavior. Rocks with a clear

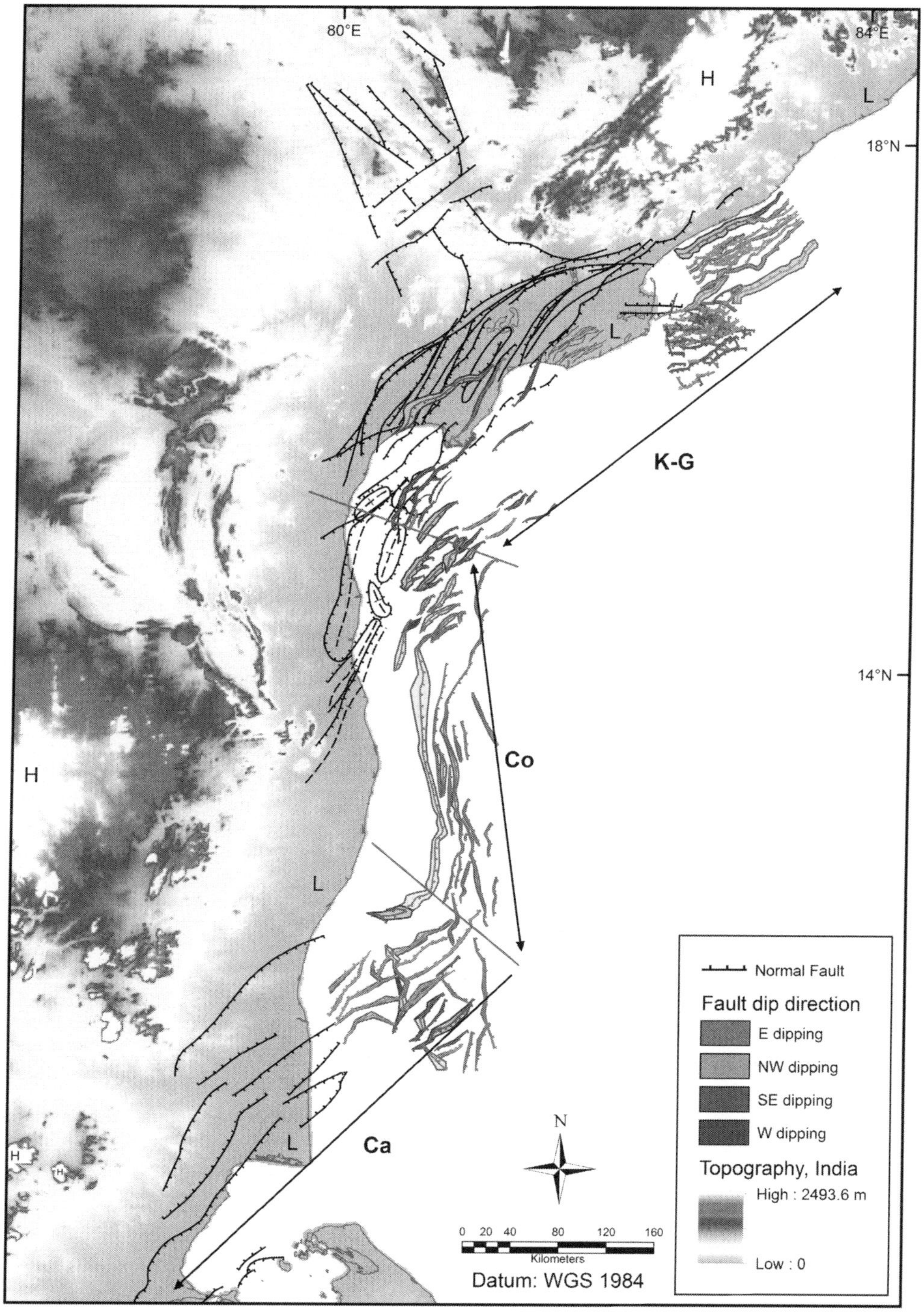

Figure 6.19. Fault map of the East Indian continental margin (Nemčok et al., 2012a). Ca – fault pattern of the Cauvery Rift Zone, Co – fault pattern of the dextral Coromondal accommodation zone, K-G – fault pattern of the Krishna-Godavari Rift Zone. Note that faults of the Coromondal zone deform faults of the Cauvery Rift Zone.

planar anisotropy and having at least moderately brittle behavior can be cut by shear fractures at any angle α of up to 60° relative to the maximum principal compressive stress σ_1. The described evidence, however, does not contradict Coulomb-Navier's theory. Although a planar anisotropy can cause a large deviation of a single shear fracture from the σ_1 direction, it does not affect newly developed conjugate shears equally.

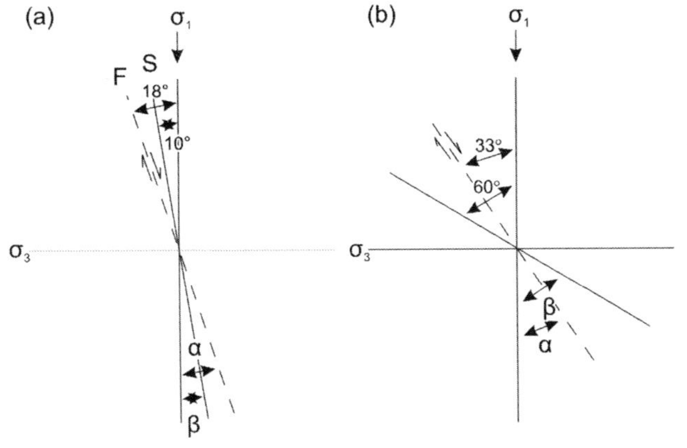

Figure 6.20. Schematic diagram of the failure F in an anisotropic rock system, in which the shear stress τ deviates from its minimum value along preexisting S planes (Turner and Weiss, 1963). Further explanation in text.

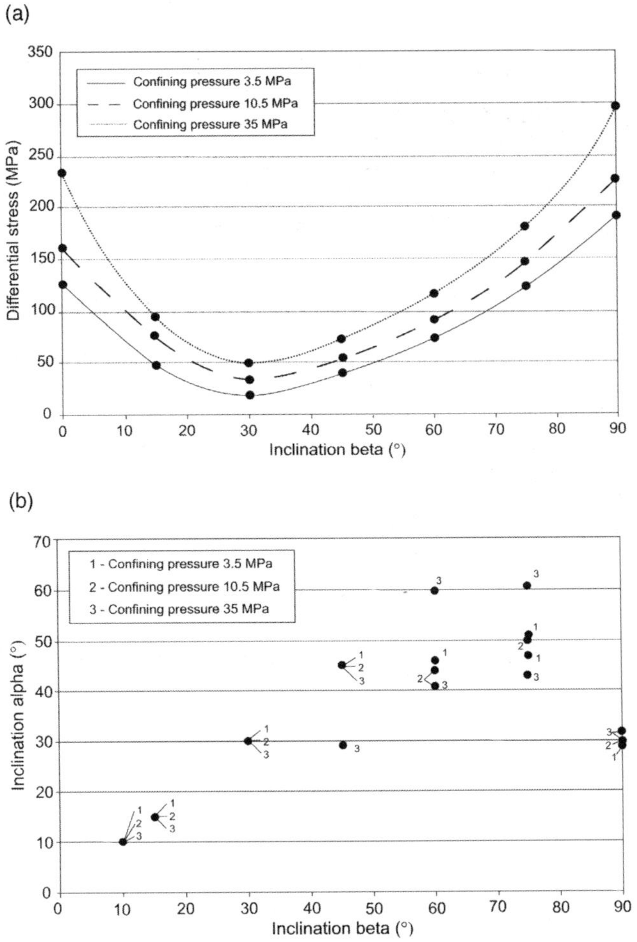

Figure 6.21. (a) Differential stress necessary for failure plotted as function of the inclination β of the cleavage from the maximum principal compressive stress σ_1 and confining pressure in the Martinsburg slate (Donath, 1961). (b) Influence of the cleavage on the angle α of the shear plane from the maximum principal compressive stress σ_1 in the Martinsburg slate (Donath, 1961). Note that the inclination of the failure α only approaches zero because the shear stress τ does not exist on cleavage planes parallel to σ_1 and therefore a shear failure along cleavage with an inclination β of 0° cannot take place.

These conclusions have been documented by numerous natural examples of shear deformation structures in slates indicating that such rocks are dissected by faults oblique with respect to the cleavage (e.g., Turner and Weiss, 1963). Shear along S-planes has a tendency to take place only in a fractured rock system that was pre-weakened in the direction parallel to the S-planes, but it is more or less equally strong in all other directions (Jaeger, 1960). Even here, S-planes will be preferentially used rather than other anisotropies only in a particular range of orientations; that is, when their deviation β from the σ_1 direction varies from 10° to 40°.

Natural examples of the preexisting anisotropy effect on rift fault patterns come from the North Sea and the Bristol Channel. The North Sea provides the example of the interpreted reflection seismic section, where the Early Cretaceous rift unit-bounding fault undergoes a modification of its geometry due to the influence of the planar anisotropy in the Rattray volcanic formation, which was located in the uppermost portion of the pre-rift stratigraphy (Stewart and Reeds, 2003; Figure 6.22). The Bristol Channel offers the outcrop example, where the Jurassic–Lower Cretaceous local-scale normal fault patterns measured at outcrops 1, 2, 4, 5, and 8 along the northern flank are similar but the remaining pattern at location 7 is different from them (Nemčok et al., 1995; Figure 6.23). The dominant fault pattern, characterized by west-northwest–east-southeast striking faults, has a dip-slip normal fault character thanks to being perpendicular to the direction of regional extension. The unusual fault pattern, represented by east-northeast–west-southwest to east-west striking faults, has a dextral transtensional character thanks to being oblique to the direction of regional extension. While the dominant normal faults were propagated through the rock section lacking any prominent preexisting anisotropy as true Andersonian faults, the propagation of oblique faults was affected by the preexisting Variscan thrust faults.

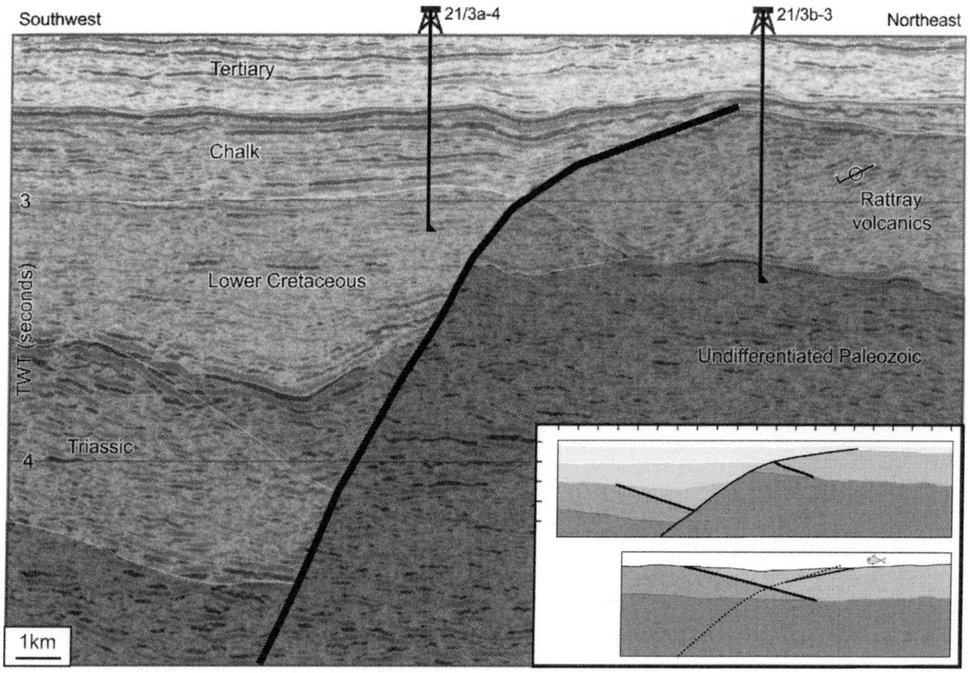

Figure 6.22. Example of the normal fault with geometry modified because of the effect of the preexisting anisotropy, North Sea (Stewart and Reeds, 2003).

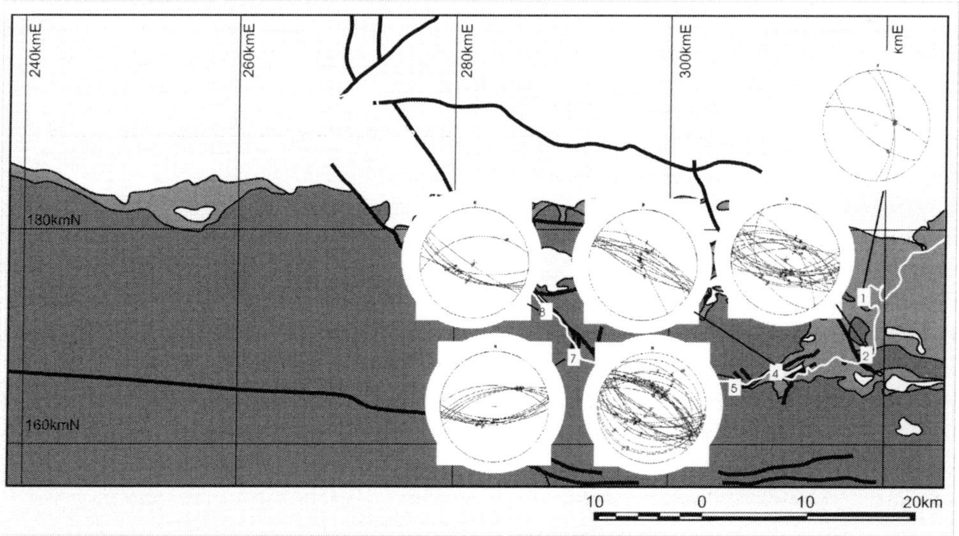

Figure 6.23. Map of the studied northern margin of the Bristol Channel with rifting-related normal faults shown as great circles on lower hemisphere equal area stereographic projections at the numbered localities (Nemčok et al., 1995).

7 The role of syn-extensional deposition and erosion in evolving structural styles of rifts and passive margins and the effects of tectonics on deposition and erosional patterns

Introduction

The sequence stratigraphic tools and related concepts were initially developed for passive continental margins where subsidence and deposition rates generally pale in comparison with those from active rift and pull-apart basins (see, e.g., Vail et al., 1977; Posamentier et al., 1988; Van Wagoner et al., 1988). Subsidence curves, discussed in Chapter 1, indicate that this rate difference should increase from rift to pull-apart basins (Figures 1.14–1.17), which commonly reach subsidence values above 1 mmy^{-1} (Johnson et al., 1983; Christie-Blick and Biddle, 1985; Pitman and Andrews, 1985).

Initially developed sequence stratigraphic models were subsequently modified, honoring the concept that the regional basin subsidence can dominate the accommodation space production attributed to eustatic sea-level fluctuations and interact with variations in sediment supply (Galloway, 1989a, 1989b; Embry, 1990; Coakley and Watts, 1991; Jordan and Flemings, 1991; Sinclair et al., 1991; Posamentier and Allen, 1993). Scholars have recognized that tectonically active settings can contain basins that undergo a continual sea-level rise and that have stratigraphic sequences without erosive boundaries represented by composite surfaces of marine flooding, maximum transgression, and downlap (Dart et al., 1994; Gawthorpe et al., 1994, 1997; Burns et al., 1997; Hardy and Gawthorpe, 1998).

A large number of studies have been conducted relatively recently on the tectonics–deposition interactions in rift and passive margin settings, describing the structural styles and facies distribution. These studies were done on:

1) rift basins with gravity gliding-controlled stretching (e.g., Trudgill and Cartwright, 1994; Cartwright et al., 1995; Trudgill, 2002);
2) rift basins with tectonically controlled stretching in both continental (e.g., Sharp et al., 2000; Ehlers and Farley, 2003; Harris et al., 2004) and marine settings (e.g., Dawers et al., 1999; Davies et al., 2000; Sharp et al., 2000);
3) pull-apart basins in both continental (e.g., Decker et al., 2005) and marine settings (e.g., Dorsey et al., 1995; Dorsey and Umhoefer, 2000);
4) post-rift basins developed on failed rifts (e.g., Argent et al., 2000);
5) passive margins (e.g., Gallagher et al., 1994; Van Balen et al., 1995; Manatschal, 2004);
6) passive margin basins with gravity gliding-controlled stretching (e.g., Van Heijst et al., 2002); and
7) mantle plume-influenced basins (e.g., Faleide et al., 2002; Jones et al., 2002).

Drawing from literature, the goal of this chapter is to discuss the interaction of deposition and erosion with tectonics in evolving rift and passive margin settings.

Rift basins with gravity gliding-controlled stretching

Canyonlands, Utah's natural laboratory for normal fault propagation development in extensional terrains allows us to discuss the role of normal fault growth and linkage in the evolving facies distribution. The region consists of numerous grabens formed in a 460 m-thick competent layer of Pennsylvanian–Lower Permian sandstones and limestones gliding on the inclined 300 m-thick Pennsylvanian evaporitic layer toward the Colorado River in the northwest (Trudgill and Cartwright, 1994). This started after the Pliocene-Pleistocene incision of the river (Baker, 1933; Lewis and Campbell, 1965; McGill and Stromquist, 1975, 1979; Huntoon, 1982) into the stratigraphic section dipping at 2–4° to the northwest (Trudgill, 2002). The grabens have developed during the

Figure 7.1. Displacement/fault-length diagram for the southern relay ramp of the Devil's Lane graben, Canyonlands, Utah (Trudgill and Cartwright, 1994). Explanation in text.

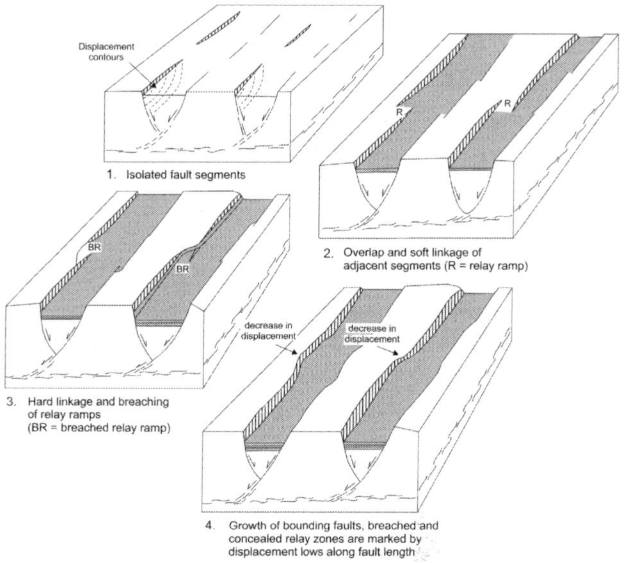

Figure 7.2. Fault-growth model by segment linkage in the Canyonlands graben system (Trudgill and Cartwright, 1994). 1) Isolated fault growth. 2) Fault propagation reaching overlaps leads to soft linkages and relay ramp development. 3) Continuing fault growth reached the relay ramp breaching stage, which could be achieved by footwall (FBR) or hanging wall breaching (HBR) – modify the rift case to hanging wall breaching. 4) Local lows in the total displacement profiles along the fault strikes indicate earlier locations of relay ramps in the stage of fully developed fault zones.

past 500,000 years, determined from the ages of their fill (Biggar and Adams, 1987; Schultz and Moore, 1996). Most of the grabens are bounded by overlapping or linked fault segments. All faults in the region have normal displacements less than 150 m large and strikes of 100–6,500 m (e.g., McGill and Stromquist, 1975, 1979). They control grabens 200–300 m wide and maximally 6,500 m long. The graben spacing is about 750–1,000 m.

Overlapping fault segments are connected via relay ramps (sensu Larsen, 1988; see Figure 1.7). Relay ramps are defined as soft linkages, linking the neighbor normal faults by a zone of ductile deformation, without any structural discontinuity at the observation scale (Walsh and Waterson, 1988). Their function in linking is illustrated in Figure 7.1. Although the relay zone in the profile made along the fault strikes is a zone where the displacement along each fault progressively dies out (see Figure 1.7), the sum of their displacement in the relay zone area keeps the accommodated displacement at levels characteristic of the neighbor faults in areas outside of the relay ramp (Figure 7.1).

As the neighbor faults propagate toward each other, the deformation regime in the relay ramp progressively changes, eventually reaching the breached relay ramp stage (Figure 7.2). At the breaching stage, neighbor faults become directly linked via either footwall breaching of the relay ramp or hanging wall breaching.

As a result of the fault development history, described in Chapter 1, fault displacement amounts and relay ramps among them vary in time and space. Soft-linked and hard-linked neighbor faults (sensu Walsh and Waterson, 1988) are just snapshots of the fault development in time in a gradational series, which initiates with two independent faults propagating into a future overlap position, follows with gradational development of the soft-linking relay ramp, and finishes with their hard linkage via the breached relay ramp.

The role of the relay ramp is to accommodate new increments of strain in the area of interacting propagation tip regions until rotational deformation and fracturing in the relay ramp transitions into full breaching and a direct connection of the interfering faults (Figure 7.2). This fault development is scale-independent, being first observed on a small scale in the Bristol Channel, United Kingdom (Peacock and Sanderson, 1991) and working in a variety of extensional settings (Morley et al., 1990; Machette et al., 1991a, 1991b; Anders and Schlische, 1994; Childs et al., 1995; Dawers and Anders, 1995; Cartwright et al., 1996; Crider and Pollard, 1998; Rowan et al., 1998; Dawers and Underhill, 2000; McLeod et al., 2000).

Because normal faults control the stratigraphic fill of the fault-controlled space (e.g., Leeder and Gawthorpe,

1987; Prosser, 1993; Gawthorpe et al., 1994; Schlische and Anders, 1996; Ravnås and Steel, 1998; Gawthorpe and Leeder, 2000), the fault development history from Figure 7.2 controls complex spatial and temporal variations in associated syn-rift facies distributions.

The value of the Canyonlands study area located in an arid climate is represented by rather detailed control on fault growth interaction with sediment migration pathways. Although similar studies have been conducted in central Greece (Gawthorpe and Hurst, 1993; Leeder and Jackson, 1993; Eliet and Gawthorpe, 1995), East Africa (Crossley, 1984; Frostick and Reid, 1987; Morley and Wescott, 1999; Lezzar et al., 2002) and the Basin and Range province, United States (Leeder and Jackson, 1993; Jackson and Leeder, 1994; Peakall, 1998), they did not manage to reach the same level of data detail.

A combination of surface imagery interpretation with fieldwork in Canyonlands allows us to define the pre-extensional and present-day drainage patterns (Figure 7.3), and, in numerous cases, subsequent generations of temporary drainage patterns reacting to the evolving fault block system (Trudgill, 2002). One of the reasons for the good preservation of drainage patterns is the annual rainfall in the area averaging only at 15–20 cmy^{-1}, exceeded by evaporation (Nuckolls and McCulley, 1987). The drainage pattern is mostly dendritic, with localized trellis networks represented by both dip and strike streams, actively flowing about twice a year in the form of intense flashfloods.

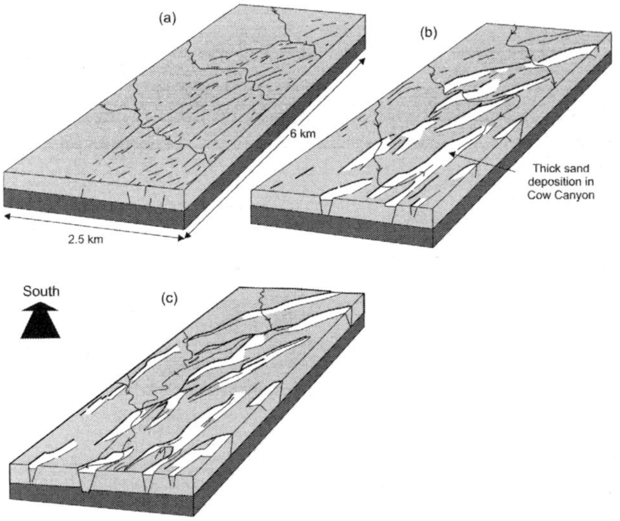

Figure 7.3. Model of the Canyonlands fault growth and deposition evolution (Trudgill, 2002). (a) The pre-extensional southeast-to-northwest drainage system starts to interact with the initial segmented fault system. (b) Faults grow by segment linkage, starting with soft linkage, and control several shallow grabens, which capture and redirect some of the initial streams. Captured streams start to develop first thick alluvial sedimentary bodies in some grabens. (c) Faults reaching the hard linkage stages of development further modify stream pattern. Many of the grabens own their own localized drainage systems. The main redirection direction of the drainage pattern is to the northeast, bypassing most of the extensional region.

The pre-extensional drainage pattern was relatively simple, connecting the Abajo Mountains to the southeast with the Colorado River in the northwest (Figure 7.3). The syn-extensional drainage pattern documents various stream interactions with a growing normal fault pattern and the development of the horst-graben topography. Various interaction examples indicate complete stream capture or various kinds of stream diversion.

A complete capture can be observed in Red Lake Canyon, which captured the Butler Wash stream, which, having no exit, kept ponding on its bottom until the graben became filled after it became inactive. Localized minor drainage systems with no outlet were, in fact, established in numerous grabens in the area. This pattern develops a system of localized sediment bodies in each of the respective grabens, followed by a gradual bypass of the grabens as drainages modify their courses trying to find friendlier topography in areas away from major grabens (Figure 7.3; Trudgill, 2002).

Y and the Cross Canyons provide data illustrating both stream capture and diversion. None of the drainage paths in this area keeps its pre-extensional course. Drainage segments, which became parts of the footwall blocks, got disconnected from their preexisting upper and lower segments. They remain as a system of abandoned meanders on footwall highs, similar to systems that can be observed in the wide-rift Basin and Range province. A significant portion of the present-day drainage pattern is now diverted from the original southeastern-to-northwestern course, along graben northeast-southwest strikes (Figure 7.3). Also erosional curves of streams are strongly modified, characterized by low incision rates in their lower segments caused by fault growth. Much of the alluvial deposition is now localized within closed graben drainages and only a minor amount of sediment from the Abajo Mountains reaches the Colorado River (Trudgill, 2002).

Other examples of the stream diversion come from southeast of the area. Here, several streams started to use relay ramps in the developing fault system as entries into developing grabens (Figure 7.4; Trudgill, 2002). As some of the relay ramps became breached and fully linked normal fault systems continued their growth, the associated streams underwent two alternative modifications. If the stream incision kept up with the uplift associated with footwall rotation next to the

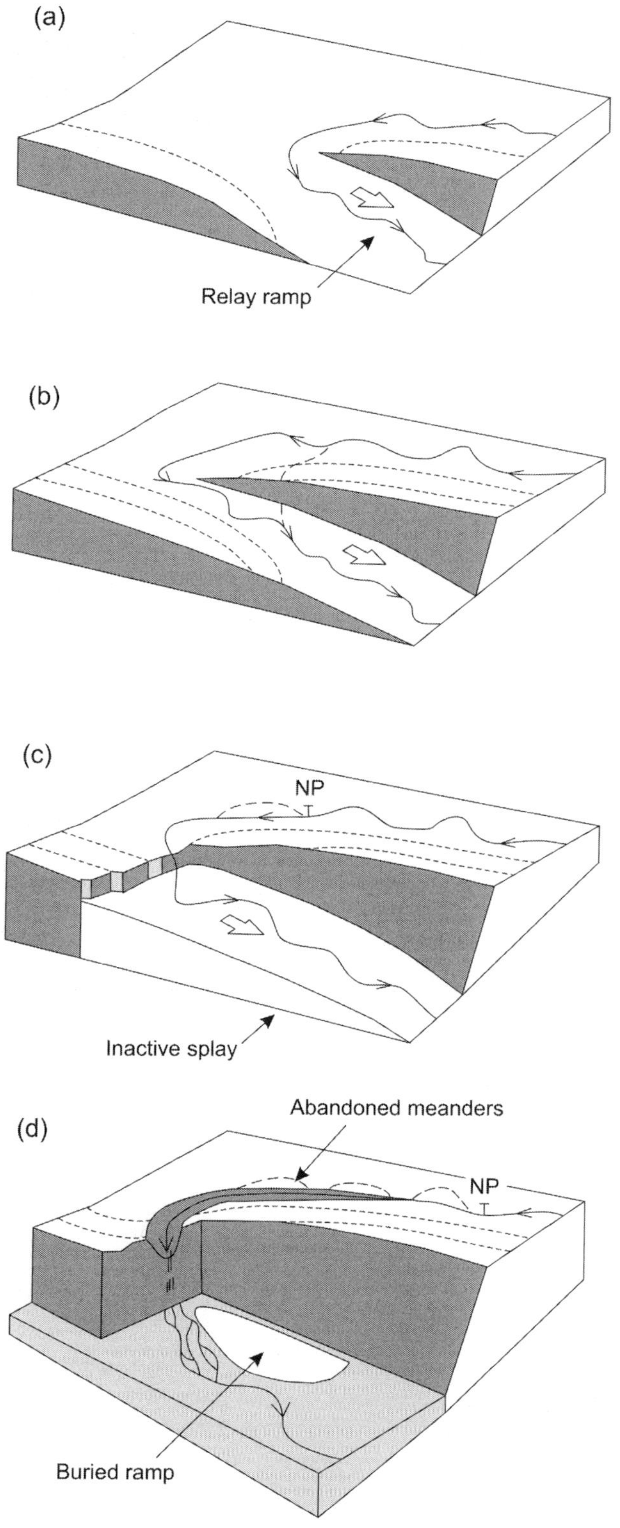

Figure 7.4. Relay ramp and associated stream evolution (Trudgill, 2002). (a–b) Soft-linkage stages. (c–d) Hard-linkage stages. NP – upstream knickpoint developed as the stream incises in an attempt to reach the new base level in the graben. While waterfall persists in the breached relay ramp, the knickpoint migrates upstream and the incised channel cuts through earlier meanders.

Figure 7.5. River flowing through the small step near the end of the normal fault, progressively loosing its energy on a flat graben floor; and developing alluvial fan, Basin and Range province, United States. The normal fault has its trace going diagonally down from the upper left corner and having a propagation tip just a bit to the right of the alluvial fan. Then there is a gap and a second normal fault continuing toward the lower right corner.

graben-bounding fault, the stream entered the graben as a waterfall. Continuing across the flat graben floor, it rapidly lost its energy and lost its stream load, which resulted in the development of an alluvial fan. Similar examples can be found in the present-day Basin and Range province (Figure 7.5). If the stream incision could not keep up with the uplift associated with footwall rotation next to the graben-bounding fault, the stream diverted, trying to bypass the problem by flowing parallel to the growing obstacle. Yet another case of stream diversion is the stream flow against the growing normal fault, which also results in graben parallel flow into the region of possible graben exit.

As Figure 7.4 documents, it is not exactly the fault propagation that pushes the bypassing stream further in front of the propagating fault tip. Some of the faults in Canyonlands indicate that their propagation to the surface was preceded by a development of the monoclinal fold (Cartwright and Mansfeld, 1998), which was on its own an important stream diverter. Similar fault-propagation folds have been described from the narrow rift domain of the Miocene Gulf of the Suez Rift (Gawthorpe et al., 1997; Sharp et al., 2000).

Rift basins with tectonically controlled stretching

As mentioned earlier, the syn-rift sedimentary successions are a result of the interaction among fault-controlled subsidence, sediment supply, and sea-level fluctuations (e.g., Leeder and Gawthorpe, 1987;

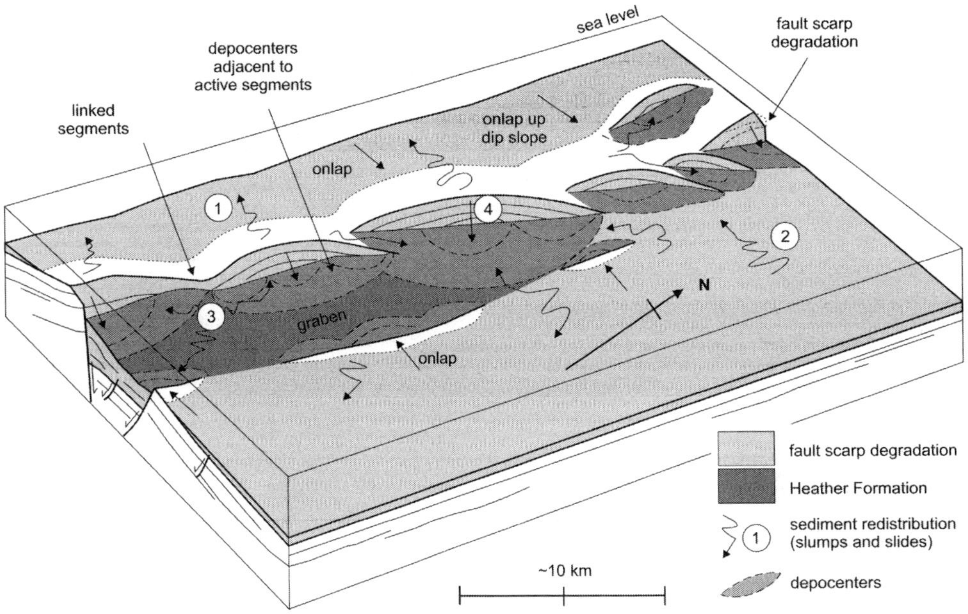

Figure 7.6. Facies distribution of the syn-rift Heather Formation in the Strathspey-Brent-Statfjord half-graben, North Sea (McLeod and Underhill, 1999).

Schlische, 1991; Prosser, 1993; Gawthorpe et al., 1994). As a consequence, in active rift regions, fault segment growth controls the subsidence rate, depocenter migration patterns, and facies distribution in the basin.

A nice example of the initial fault pattern control comes from the early-rift Middle Jurassic Tarbert Formation from the East Shetland Basin, North Sea (Davies et al., 2000). Here, the sand-rich Tarbert Formation represents the transgression of the delta complex. Its base can be recognized as the occurrence of the marine shoreline sediments above or laterally equivalent to the nonmarine Ness Formation (Vollset and Doré, 1984). The whole Tarbert deposition, that is, the isolated fault growth, took 3–4 Ma.

Reflection seismic profiles from the area show that not all of the isolated faults consequently grew up by linkage. While the faults of the future Statfjord East Fault did, the other faults, located to the east-southeast of it, did not. Their propagation terminated either at the end of the deposition of the Tarbert Formation or during the deposition of the subsequent Heather Formation. Similar cases of ceased propagation of the initial fault pattern are known from the Gulf of Suez (Sharp et al., 2000).

Three-dimensional seismic mapping in the East Shetland Basin constrained by well data indicates that the Tarbert Formation can be characterized by significant spatial variability in thickness, being deposited in numerous isolated depocenters. In fact, a somewhat similar pattern of neighboring depocenters is also typical for the immediately overlying Heather Formation (McLeod et al., 2002) with an important difference being that the facies distribution indicates that previously isolated faults propagated far enough to develop a system of relay ramps in the Strathspey-Brent-Statfjord half-graben, used as the sediment entry points into the graben (Figure 7.6). It is also apparent that the earlier generation of isolated faults, which deforms only the Tarbert Formation or the lower portion of the Heather Formation, contains short faults with small displacements. On the contrary, the lengths and displacements of faults, which made it to the development of the relay ramps among propagating neighbor faults, are larger (Davies et al., 2000).

The spatial variation in the Tarbert Formation facies observed in the study area records both the inherent variability of paralic to shallow marine depositional environments and the distribution of depositional space controlled by the early syn-rift fault pattern (Figure 7.7). The normal fault initiation enabled the initial marine transgression and flooding of coastal swamps, which terminated significant peat accumulation and established marine bays, later followed with landward migration of a barrier island system (Figure 7.7). Facies variations are reflecting proximity to sediment entry points and 3-D geometry of the individual fault-controlled depocenters, both competing with the overall basin subsidence, also expressed by the deposition on footwalls. The initial faults were apparently promoting the development of aggradational

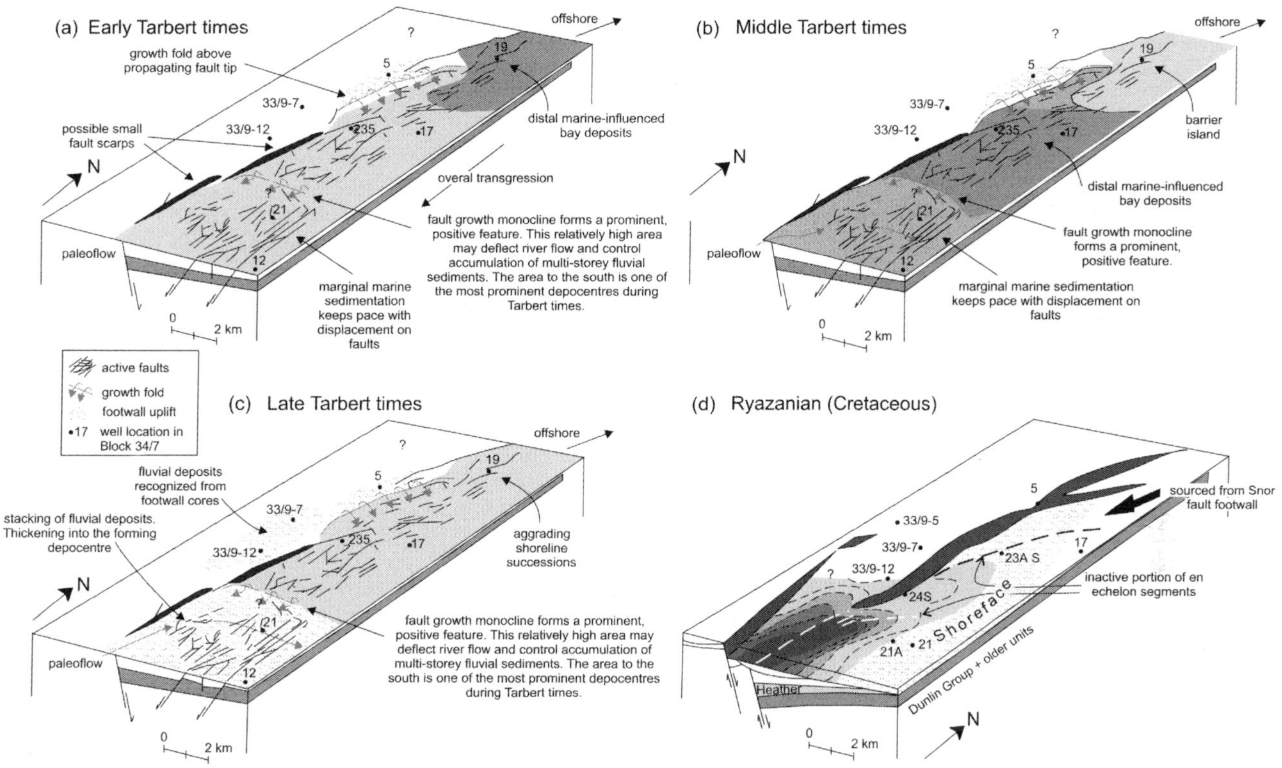

Figure 7.7. Four paleogeographic snapshots showing various development stages of the isolated fault pattern and associated facies distribution in the East Shetland Basin, North Sea, during the middle Late Jurassic–Early Cretaceous (Davies et al., 2000). Note that the Ryazanian snapshot is shown to highlight a contrast between the early-rift isolated fault development and rift-climax fault interaction and linkage development.

shoreface successions and confining the river systems in the southeast of the area. The observed facies distribution implies that deposition rates were comparable to the fault-controlled subsidence.

Comparison of the initial fault study in the Tarbert Formation (Davies et al., 2000) with soft-linked fault study in the Heather Formation (McLeod et al., 2000) indicates that the fault development was characterized by increasing displacement rates in time. As a result, the smaller Tarbert Formation thickening next to bounding faults compares to the much larger Heather Formation thickening.

This phenomenon has been observed in numerous extensional basins (e.g., Prosser, 1993) and explained by fault localization, fault interaction, and linkage (Cowie, 1998; Gupta et al., 1998, Cowie et al., 2000; Noll and Hall, 2006; see Figure 1.7). The ability to predict fluvial reservoirs associated with early depocenter history in small areas controlled by initial isolated faults in the southeastern portion of the area, which ceased their propagation and never made it to linkage, can make an interesting addition to the exploration target in other rift analogs.

A nice example of the interacting and linking fault pattern control of deposition comes from the Upper Jurassic–lowermost Cretaceous Draupne Formation of the East Shetland Basin (Dawers et al., 1999). Here, a number of spatially discontinuous stratigraphic units is spatially related to depocenters, which were controlled by interacting and linking en echelon faults of the Statfjord East Fault system. The main control of the facies distribution is the along-strike variation in fault displacements. The sediments record fault activity not only by thickness variations related to displacement variations but also by facies shifts along fault strikes.

The rifting in the East Shetland Basin reached its climax during the Kimmeridgian-early Volgian. The pattern of overall northward fault system propagation in the basin is documented by the northward younging of associated late Volgian and Ryazanian shoreface sediments. Most of the relay ramps between overlapping fault segments served as entry points into basins for sediments, which were supplied from eroding footwalls, in a fashion similar to other rift regions (e.g., Gawthorpe and Hurst, 1993; Underhill, 1994; Gawthorpe et al.,

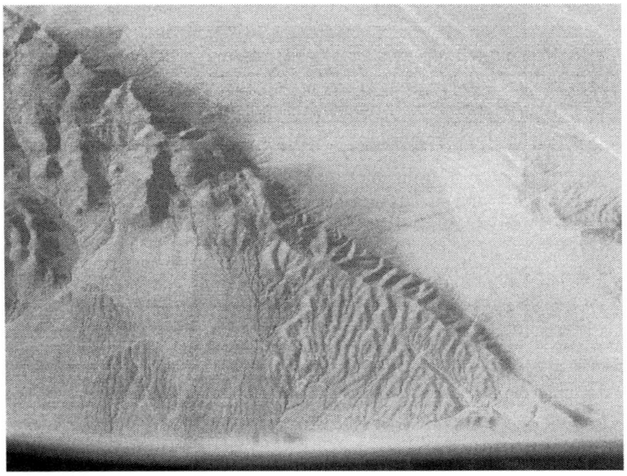

Figure 7.8. Long erosional portions of streams on the gently tilted footwall side versus very short portions on the normal fault scarp. Former ones are dominant graben-filling agents; latter ones pale as the same length ratio applies for associated alluvial fans.

1997). The most significant source of sediment was the large Snorre footwall, where the extensive area of the dip-slope made it a major sediment source for the adjacent basin.

A comparison of fault-slope and dip-slope sides of footwall blocks in their importance for deposition in adjacent grabens can be easily assessed from those aerial photos of the Basin and Range province:

1) where some of the footwalls are large enough to develop doubly sided climate and associated differing erosional patterns or
2) where the large size of the dip-slope with a less incised but more extensive drainage pattern makes it a more significant sediment source than the fault-slope with a more incised but less extensive drainage pattern (Figure 7.8).

Processes and products of the footwall degradation were also studied in detail in the North Sea region (e.g., McLeod and Underhill, 1999; Stewart and Reeds, 2003). These studies highlight the gravitational instability of steep normal fault scarps, recognized also in other extensional settings (e.g., Witkind et al., 1962; Hamblin, 1976; Keefer, 1984; Ferentinos et al., 1988; Stewart and Hancock, 1988; Koukovelas et al., 1996). However, it is only the northern North Sea basin where the degraded fault scarps together with the debris of the ancient slope failure were fully preserved (Livera and Gdula, 1990; Yielding et al., 1992; Lee and Hwang, 1993; Schulte et al., 1994; Coutts et al., 1996; Sawyer and Keegan, 1996; Faerseth et al., 1997; Underhill et al., 1997).

Such preservation allowed for the description of a staged model of the development of the fault scarp and its degradation in the northern portion of the Brent block (Figure 7.9; McLeod and Underhill, 1999). The mechanism of the scarp failure was controlled by the environmental controls and by varying rheologies.

The mass wasting was later studied in a more systematic way thanks to readily available 3-D reflection data sets from the Moray Firth Basin, North Sea (Stewart and Reeds, 2003). It is dependent of factors such as:

1) the original fault geometry;
2) rheology, preexisting anisotropy, and fluid regime; and
3) environment.

Reflection seismic profiles constrained by wells from the Moray Firth Basin indicate several nice examples of fault geometries and their subsequent degradation in both marine and subaerial environments. One of the marine examples shows a case of the Jurassic normal fault controlling a half-graben, which had its upper part exposed on the basin floor for about 50 Ma until the Late Cretaceous (Figure 7.10a). Preservation of the condensed Upper Jurassic sediment on the footwall indicates that the footwall remained below the wave base. The upper portion of the main fault scarp shows a preserved scarp degradation complex. The slope failure mode must have been represented by small-scale flow, listric failures, and raveling.

One of the subaerial examples shows a difference in the erosional modification of opposing fault-slopes of the same horst, which was subaerially exposed for about 70 Ma (Stewart and Reeds, 2003; Figure 7.10b). This, together with the aforementioned observations from the Basin and Range province, indicates that if the horst-graben province with subaerially exposed horsts experiences a climate characterized by prevalent winds, a large enough horst can develop doubly sided erosional regimes, which modify fault-controlled slopes differently.

The last example taken from the Moray Firth Basin study is a marine example, showing a footwall that underwent continuous, although reduced deposition during rift- and early post-rift time (Figure 7.10c; Stewart and Reeds, 2003). Following the faulting, the fault slope underwent a gravity gliding on top of a listric fault, developing an extensional domain in the rear portion of the glide and a contractional domain in the frontal portion of the glide.

As a relatively modern counterpart to the discussed Jurassic tectonic–deposition interactions in the North Sea (e.g., Dawers et al., 1999; Davies et al., 2000), the

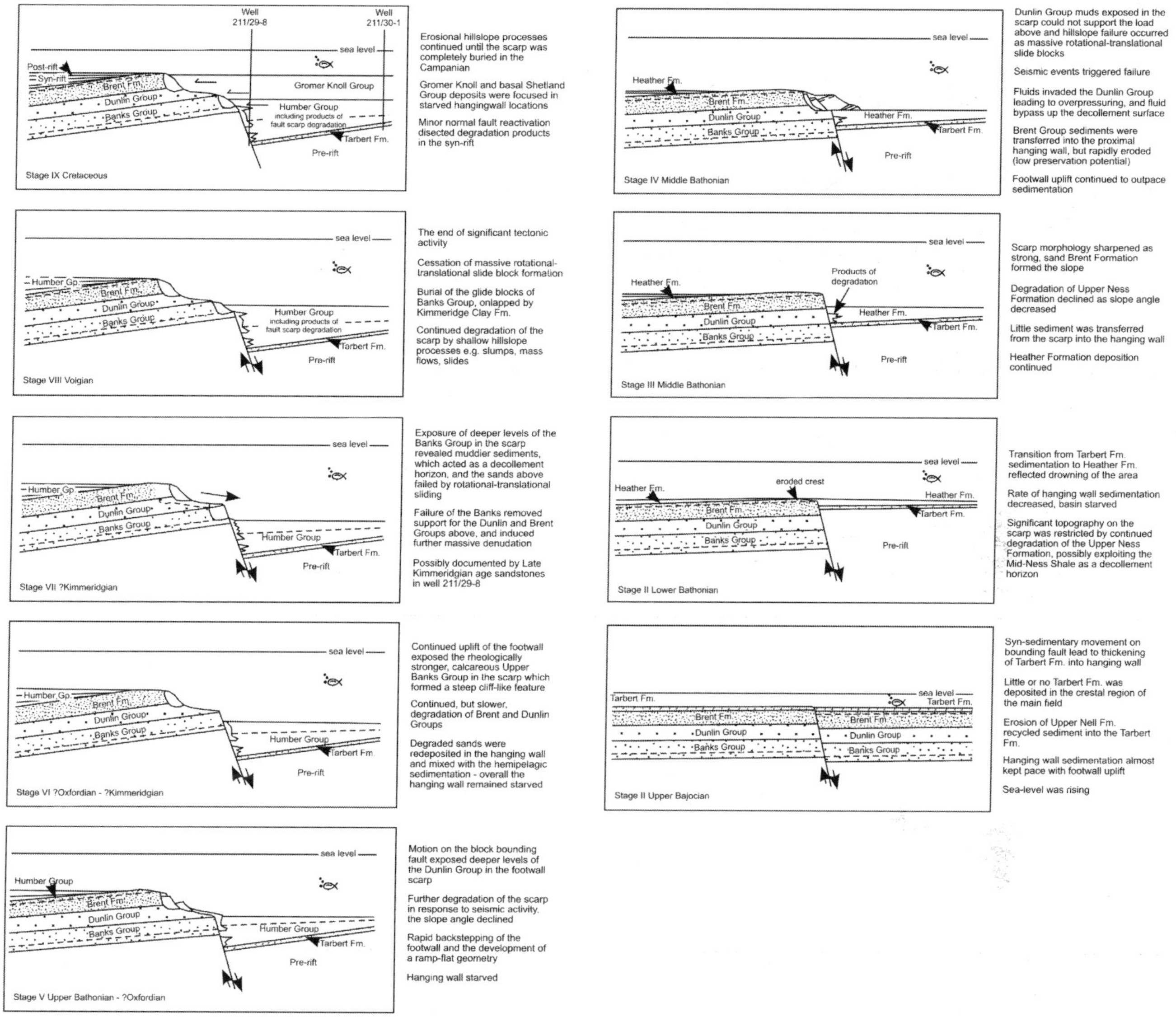

Figure 7.9. Staged model of the fault development and its degradation, northern North Sea (McLeod and Underhill, 1999). 1) Syn-sedimentary movement of the fault controlling thickening of the Tarbert Formation in the hanging wall. Little or no Tarbert Formation was deposited on the footwall crest. Hanging wall deposition almost kept up with footwall uplift. The sea level was rising. 2) Transition from Tarbert to Heather deposition indicates drowning of the area. Hanging wall deposition rate decreased to control basin starvation. Significant topography of the footwall scarp did not develop because of the continued degradation. 3) Fault scarp morphology steepened as strong sandy sediments formed the slope. Scarp degradation declined as slope angle decreased. Little sediment was transferred into the hanging wall. Heather Formation deposition continued. 4) Dunlin Group muds formed fault scarp, unable to support the load above, and experienced hill-slope failure in the form of rotational-translational slide blocks. Failure probably triggered by seismic events. Brent Group sediments have been transferred into the proximal hanging wall, but rapidly eroded. Footwall uplift continued to outpace deposition. 5) Continued slip of the block-bounding fault exposed deeper levels of the Dunlin Group in the fault scarp. Further slope degradation occurred in response to seismic events and the slope declined. Rapid back-stepping of the footwall occurred, developing a ramp-flat fault geometry. The hanging wall underwent starvation. 6) Continuous footwall uplift resulted in exposure of the competent calcareous upper Banks Group in the fault scarp, causing a steep cliff. Continuous but slower degradation of the Brent and Dunlin Groups took place. Degraded sands were redeposited in the hanging wall and mixed with the hemipelagic deposition. However, the overall character of the hanging wall is starvation. 7) Exposure of the deeper section of the Banks Group brought muddier sediments to the scarp. They controlled the development of the detachment horizon for the rotational-translational sliding of the sandy overburden. Failure of the upper Banks Group cancelled the support for the Dunlin and Brent Groups above it, triggering a massive slope degradation. 8) This time brings an end to significant fault activity. The period is characterized by cessation of the massive rotational-translational sliding and burial of the earlier glide blocks of the Banks Group by the onlapping Kimmeridge Clay Formation. Slope degradation continued by shallow hillslope processes such as slumps, mass flows, and slides. 9) Erosional hillslope processes continued until the scarp became buried during Campanian. Hanging wall deposition can be characterized as starved.

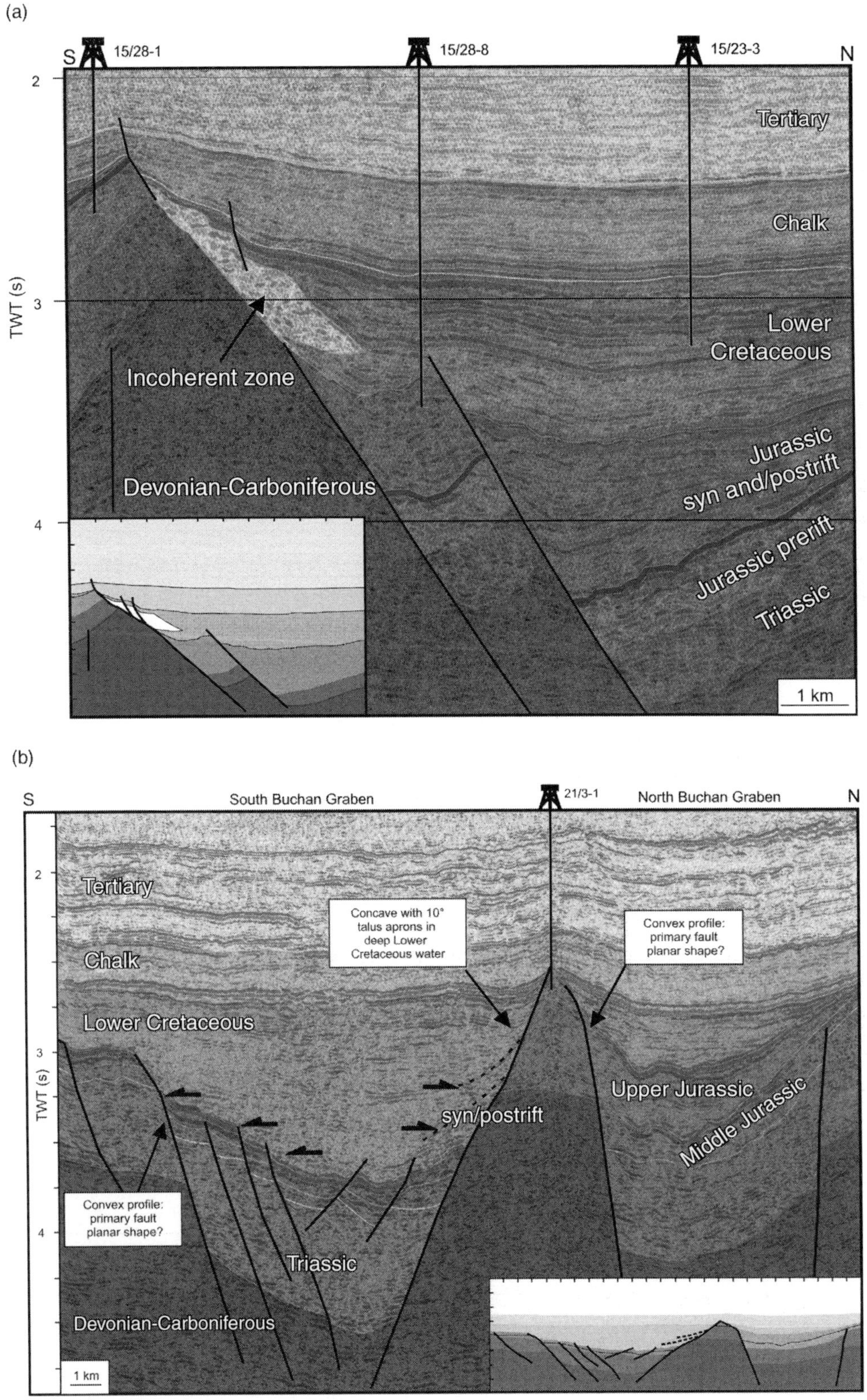

Figure 7.10. (a) Long-exposed submarine fault scarp undergoing mass wasting, Moray Firth Basin, North Sea (Stewart and Reeds, 2003). Note that the initial fault geometry was controlled by the rock section characterized by weak rock and strong rock. (b) Long-exposed subaerial horst, bounded by two fault-controlled slopes, each undergoing different erosional modification, Moray Firth Basin (Stewart and Reeds, 2003). (c) Never-exposed submarine scenario with normal fault geometry cut by subsequent gravity glide on top of listric fault, Moray Firth Basin (Stewart and Reeds, 2003).

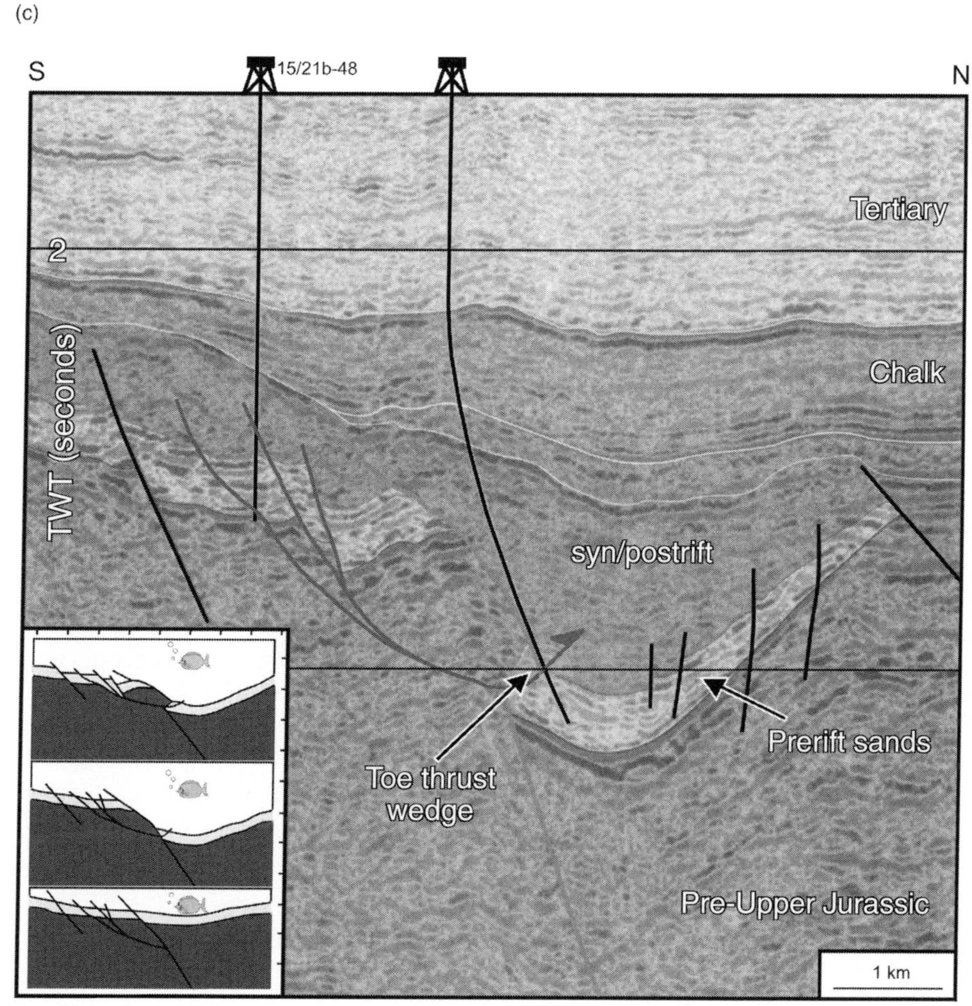

Figure 7.10 (*cont.*)

Miocene syn-rift Alaqa delta complex from the Gulf of Suez Rift provides a chance to add:

1) structural measurements of fault kinematics on top of geometries, which could only be determined from seismic imaging in the North Sea case; and
2) measurements of paleotransport vectors on top of facies distribution, which could only be determined from seismic imaging in the North Sea case.

The Alaqa delta complex represents a coarse-grained syn-rift clastic wedge deposited along the footwall-facing margin of the El Qaa half-graben (Garfunkel and Bartov, 1977; Gawthorpe et al., 1990; McClay et al., 1998). The graben is about 40 km long and 15–20 km wide, formed by eastward tilting during the early Miocene (Gupta et al., 1999). The delta was developed during the Burdigalian episode of basin deepening following the Aquitanian initial rifting (Garfunkel and Bartov, 1977; Krebs et al., 1997). The basinal equivalents of the deltaic complex were marine basinal marls and clastic sediments of the Rudeis-Mheiherrat Formation.

Measurements of the fault plane striations document a dominant normal faulting component of slip along graben-bounding faults, controlled by northeast-southwest extension and characterized by along-strike variations of the total displacement (Moustafa, 1992; Sharp et al., 2000). The study area allows one to observe the development of both the 3–5 km-wide and 12 km-long relay ramp between the Sidri and Gebah Faults and the fault-propagation fold in front of the propagation tip of the Sidri Fault. The fault-propagation fold, represented by monocline, developed as a consequence of folding, which preceded the upward propagation of a blind normal fault, in accordance with the observations from other rift settings (e.g., Patton, 1984; Withjack et al.,

1990; Schlische, 1995; Gawthorpe et al., 1997; Sharp et al., 2000).

The relay ramp between the Sidri and Gebah Faults is a location for the transverse footwall-sourced delta complex (Gawthorpe et al., 1990). It comprises an aggradational coarsening-up succession about 600 m thick, which is located in the immediate hanging wall of both overlapping faults (Gupta et al., 1999). The delta complex thins and becomes finer basinward, transitioning into marine marls westward over a distance of about 2.5 km. The large-scale coarsening-up succession forms an overall shallowing-up package as a result of the overall delta progradation over the basinal marls. Distal prodelta sandy turbidites and marls coarsen upward into proximal prodelta turbidites and debris flow units, passing eventually into coarse-grained proximal delta deposits.

Individual 15–30 m-thick delta wedges form a distinctive foreset unit, which shows rapid down-transport facies transition into proximal prodelta bottomset sandstone turbidite and conglomerate debris flow deposits, indicating its Gilbert delta character (see Postma, 1984, 1995; Postma and Roep, 1985; Colella, 1988a, 1998b). Individual Gilbert deltas are separated by transgressive intervals recording times of sediment starvation during which the underlying delta was drowned and submerged. These vertically stacked deltas record cycles of delta progradation terminated by an abrupt transgression of the delta top and drowning of the prograded delta (Gupta et al., 1999). To produce the described facies pattern, the shoreline during this time must have been fairly mobile, moving repeatedly seaward and landward within a several kilometers-wide zone.

The studied outcrops indicate that, apart from the east-west facies changes, the delta complex is characterized by graben strike-parallel facies changes, such as the fining and thinning toward both south and north. On top of it, the syn-rift stratal relationship in the zone adjacent to the Sidri Fault records the interaction of growing structures, such as the fault and its initial fault-propagation fold, and delta deposition. Syn-rift strata in the hanging wall syncline and adjacent to the fault-propagation monocline form a wedge characterized by (Gupta et al., 1999; Figure 7.11):

1) West-to-east and along-strike south-to-north onlap onto the monocline forelimb formed by pre-rift strata;
2) stratal convergence and thinning to the north on the crest and flank of the monocline;

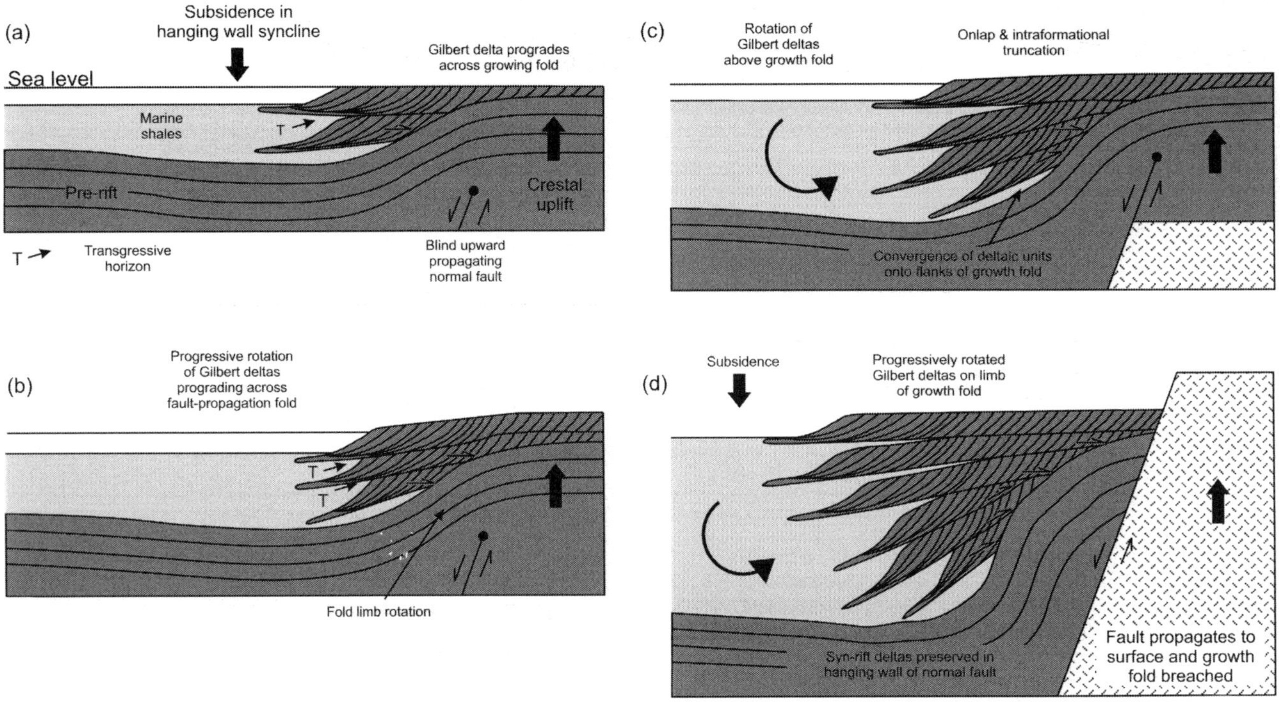

Figure 7.11. Simplified cross-sections illustrating the deposition interaction with the tectonic during the development of the graben-bounding fault, El Qaa graben, Gulf of Suez (Gupta et al., 1999). Note that the hanging wall syncline continuously deforms as the fault propagation monocline and its subsequent truncated form achieve progressively larger amplitude. Syn-tectonic units are separated by local intra-formational unconformities in the fold crestal region.

3) progressive shallowing of the syn-rift dips with younger stratigraphy; and
4) Southward stratal expansion and dip decrease away from monocline toward hanging wall syncline.

Most reliable shallowing syn-rift bedding readings come from transgressive horizons separating individual Gilbert deltas, which undergo dip changes from 40° to 15°. This indicates that deltas were progressively rotated by a fault-controlled tilt of the hanging wall. Furthermore, paleotransport vectors indicate transverse delta progradation in the relay ramp area between the Sidri and Alaga Faults and longitudinal progradation toward the most mature fault portions. Basinal offshore marine marls deposited on a hanging wall represent interfan sediments between this point-source delta and the next one to the north, entering the basin at another relay ramp location.

Observed stratal relationships (Figure 7.11) clearly indicate that the fault was blind in the early stages of its growth and the deltaic deposition reacted to the growth of its propagation fold. The hanging wall tilt kept deltaic deposition constrained within several kilometers from the evolving graben-bounding fault. Along-strike onlap and convergence of syn-rift strata away from the relay ramp area suggest that there was an along-strike variation of displacement along the deposition-controlling fault.

Initially, the deposition rate could not keep up with the forelimb uplift. Later, the deposition rate balanced and exceeded the uplift, which is indicated by onlap onto the folded pre-rift strata during the monocline growth. As deformation continued, the graben-bounding fault eventually propagated through the fault-propagation monocline, allowing the preservation of the immediate hanging wall.

Overall decreases in fault throws and associated sedimentary relationships merit the conclusion that the propagation of neighbor faults was converging, starting with underlaps progressively evolving to overlap zones of relay ramps, which became eventually breached and the fault reached the stage of hard linkage, analogically to some already described examples (see Figures 1.7, 7.3, 7.4). With fault evolution, the erosional and depositional scenarios evolved (Figures 7.12a, b). The fault scarp aprons were sourced by erosion in short steep drainage basins of the footwall edge (see also Figure 7.8); a delta complex was fed by erosion in a larger drainage basin developed in the evolving relay zone (see also Figure 7.12c).

As Figures 7.12b and 7.12c (see also Leeder and Jackson, 1993; Eliet and Gawthorpe, 1995) indicate, major sediment entry points from the footwall develop either at relay ramps or via antecedent drainage cuts in regions where erosion keeps up with footwall uplift. Because the Alaqa delta progradation occurred in cycles, separated by abrupt transgressions, the detailed pattern of the shoreline migration was most likely controlled by (e.g., Dorsey et al., 1997; Gupta et al., 1999):

1) high-frequency sea-level fluctuation; or
2) episodic fault-controlled subsidence; or
3) climatically and lithologically controlled variations in sediment supply.

Two end-members of the drainage evolution scenarios after the fault linkage development depend on how river incision compares to the uplift of the relay zone. If the footwall drainage incising keeps up with relay zone uplift, sediment dispersal pattern and deposition in hanging wall depocenters will be maintained despite the breaching of the relay ramp (Figure 7.12b – **stage C**). If the incision cannot keep up with footwall uplift associated with relay ramp breaching, the drainage may become reversed and sediment dispersal switched off (Figure 7.12b – **stage D**). Among other factors, such as structurally controlled uplift rates, bedrock erodibility, rates of fluvial incision, and the efficiency of bedrock land-sliding (e.g., Tucker and Slingerland, 1996), the aforementioned balance depends on a type of relay ramp breaching (Figure 7.2), whether it is a footwall or hanging wall type.

Because fault growth events and linkage events along the whole evolving fault array do not occur coevally, the deposition should be characterized by spatial and temporal variations in sediment entry point development and sediment flux. Activation and deactivation of entry points during the development of a linked fault pattern explains the onsets and termination of coarse clastic progradation observed in many rift basins and passive margins (see Schlische, 1992; Gupta et al., 1999).

Deposition in rift basins frequently has to react to incremental rotation of normal faults and fault blocks, which controls fault angles progressively dipping less with increasing extension (e.g., Wernicke and Burchfiel, 1982; Jackson and McKenzie, 1983; McKenzie and Jackson, 1986; Barr, 1987; Mandl, 1987; Jackson et al., 1988; Davison, 1989). The scale can be any, from meso-scale to regionally significant, as observed in the North Sea and Gulf of Suez Rift systems (Colletta et al., 1988; Roberts et al., 1990, 1993; Patton et al., 1994; Bosworth, 1995; McClay et al., 1998).

The following example of block rotation comes from the Hammam Faraun and El Qaa fault blocks of the Gulf of Suez (Sharp et al., 2000). Here, both

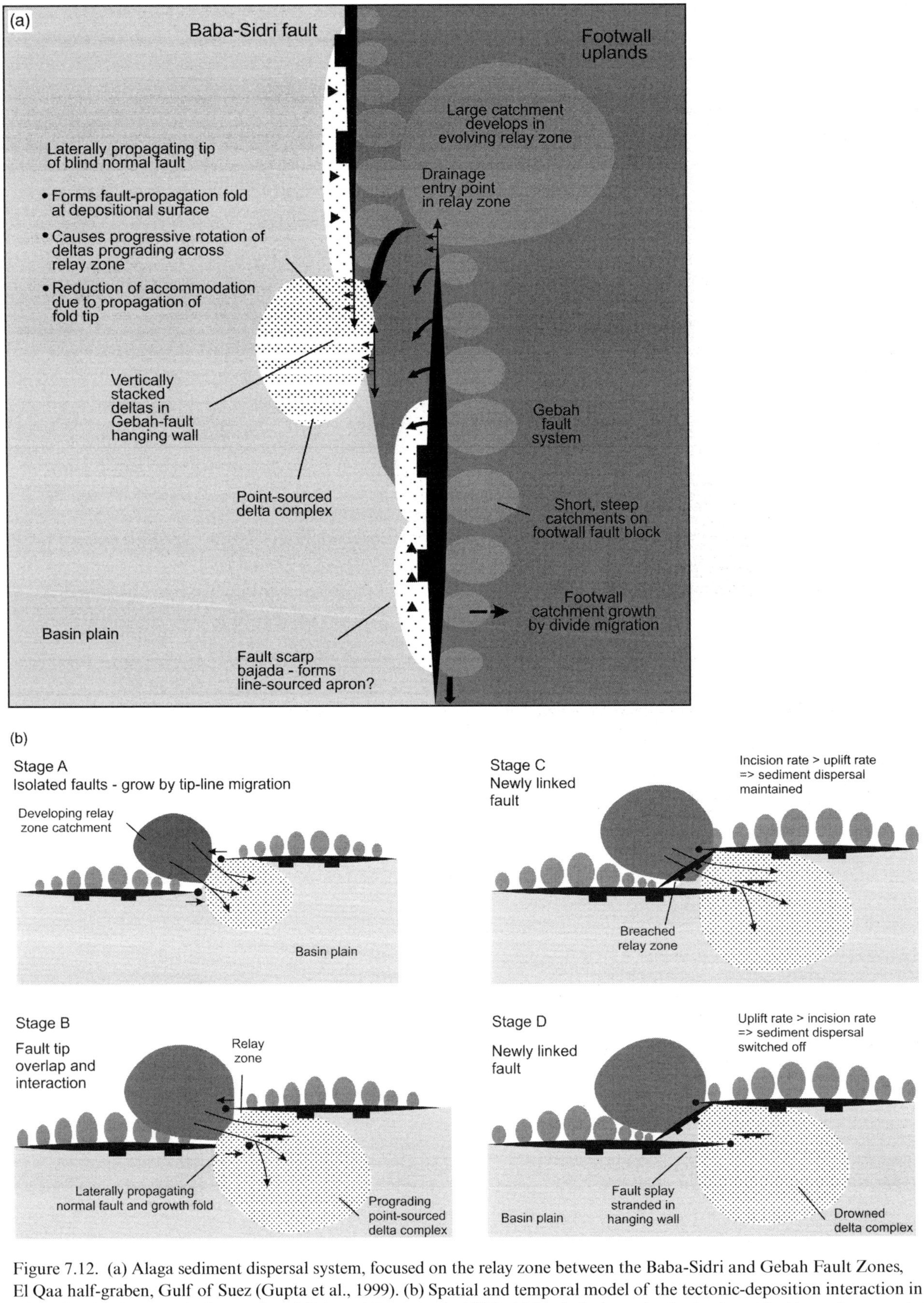

Figure 7.12. (a) Alaga sediment dispersal system, focused on the relay zone between the Baba-Sidri and Gebah Fault Zones, El Qaa half-graben, Gulf of Suez (Gupta et al., 1999). (b) Spatial and temporal model of the tectonic-deposition interaction in the rift basin with focus on fault growth and linkage (Gupta et al., 1999). (c) Little drainage basin developed inside a relay ramp between two normal faults in the Basin and Range province. Its occasional river antecedently cuts through the normal fault closer to us.

blocks dip at about 20°–30° toward the east, bounded by west-dipping faults with displacements of 2–5 km. Fault blocks did not behave as totally rigid as indicated by folding and flexing in immediate footwalls and hanging walls. Apart from bounding faults, both blocks contain a system of 50–350 m displacement faults. These faults are west-dipping, parallel to the main bounding faults, and east-dipping.

Because of block rotation, west-dipping faults in certain regions became subvertical or even overturned, now having an apparent reverse displacement (Figure 7.13a). The older syn-rift fill in their hanging walls is represented by lacustrine, fluvial, and initial shallow-marine sediments. They form wedges, which thin away from the faults and thicken and amalgamate toward the faults in cross-sections. The immediate hanging walls are characterized by channelized bodies, similar to the example from the Basin and Range province, which allows a full 3-D view (Figure 7.13b). Measured sediment transport vectors are parallel to the bounding faults.

Because of rotation, east-dipping faults in certain areas of the Gulf of Suez example became rather shallow-dipping. Their activity is, like in the case of west-dipping faults, recorded by syn-rift sediments (Figure 7.14a), which also show strike-parallel changes in a relationship with decreasing fault displacement (Figure 7.14b) and also contain fault-propagation folds.

The amount of eastward rotation about the horizontal axes determined from various paleo-horizontal sedimentary structures after the deposition time of the Nukhul Formation reaches about 25° (Sharp et al., 2000). The rotation includes two mechanisms:

1) long-lived activity and eastward rotation about horizontal axes on the down-to-the-west, main block-bounding faults; and

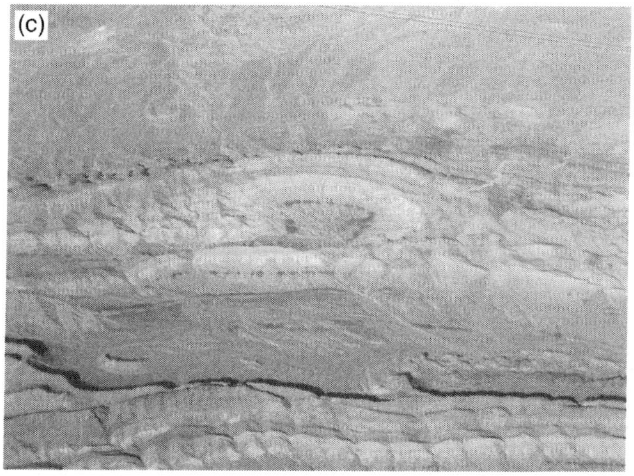

Figure 7.12 (*cont.*)

2) footwall flexing and flexural domino faulting local to bounding fault (sensu Anders et al., 1993).

Figure 7.13a shows that rotated smaller-scale faults cease their activity much earlier than main bounding faults for various reasons, such as loosing in competition with weaker faults (e.g., Mandl, 1988) and rotation into frictionally unfavorable states for further development. Similar normal fault development abandonment can be frequently seen in the Jurassic–Early Cretaceous strata outcropping along the Bristol Channel coast. Here, long, continuous cliff sections through normal faults allow one to observe that some of them never developed any further from their initial stages, being represented only by en echelon array of Riedel shears.

That the activity termination of internal block faults of the Gulf of Suez example is earlier than the activity of the block-bounding faults is clearly indicated by their propagation tips "sealed" by younger syn-rift sediments of the Rudeis Formation in outcrops, documenting that faults were early-formed features, subsequently abandoned and rotated (Sharp et al., 2000).

There are important spatial variations in facies and thickness distributions between the early-rift Abu Zemina and Nukhul Formations and the rift-climax Rudeis Formation (Figure 7.14c). These variations were controlled by a slow subsidence during the rift initiation, associated with non- and shallow-marine deposition and rapid subsidence during the rift climax, associated with deep marine deposition.

The early-rift formations are often thickest in the middle of the studied block, locally thickening toward small-scale faults. Hanging wall basins associated with meso-scale faults initiated with fluvio-lacustrine deposition, passing upward into tidal and later fully marine deposition. Accordingly, the Thal Ridge fault area shows lateral facies transitions from fluvial to tidal to open-marine facies parallel to the fault, in accordance with measured sediment transport vectors. The increasing section thickness correlates with an increase in the cumulative fault displacement.

The rift climax formation is thickest in the immediate hanging wall of the block-bounding faults. It thins and onlaps toward the footwall highs (Garfunkel and Bartov, 1977; Patton et al., 1994). The depositional system is characterized by stacked aggrading footwall-derived fan deltas and pro-delta muds and turbidites (Garfunkel and Bartov, 1977; Gawthorpe et al., 1990; Patton et al., 1994; Moustafa, 1996; Gupta et al., 1999; Sharp et al., 2000). They transition into a deeper-water central basin fill, which thins, onlaps, and amalgamates up the hanging wall dip-slope toward the

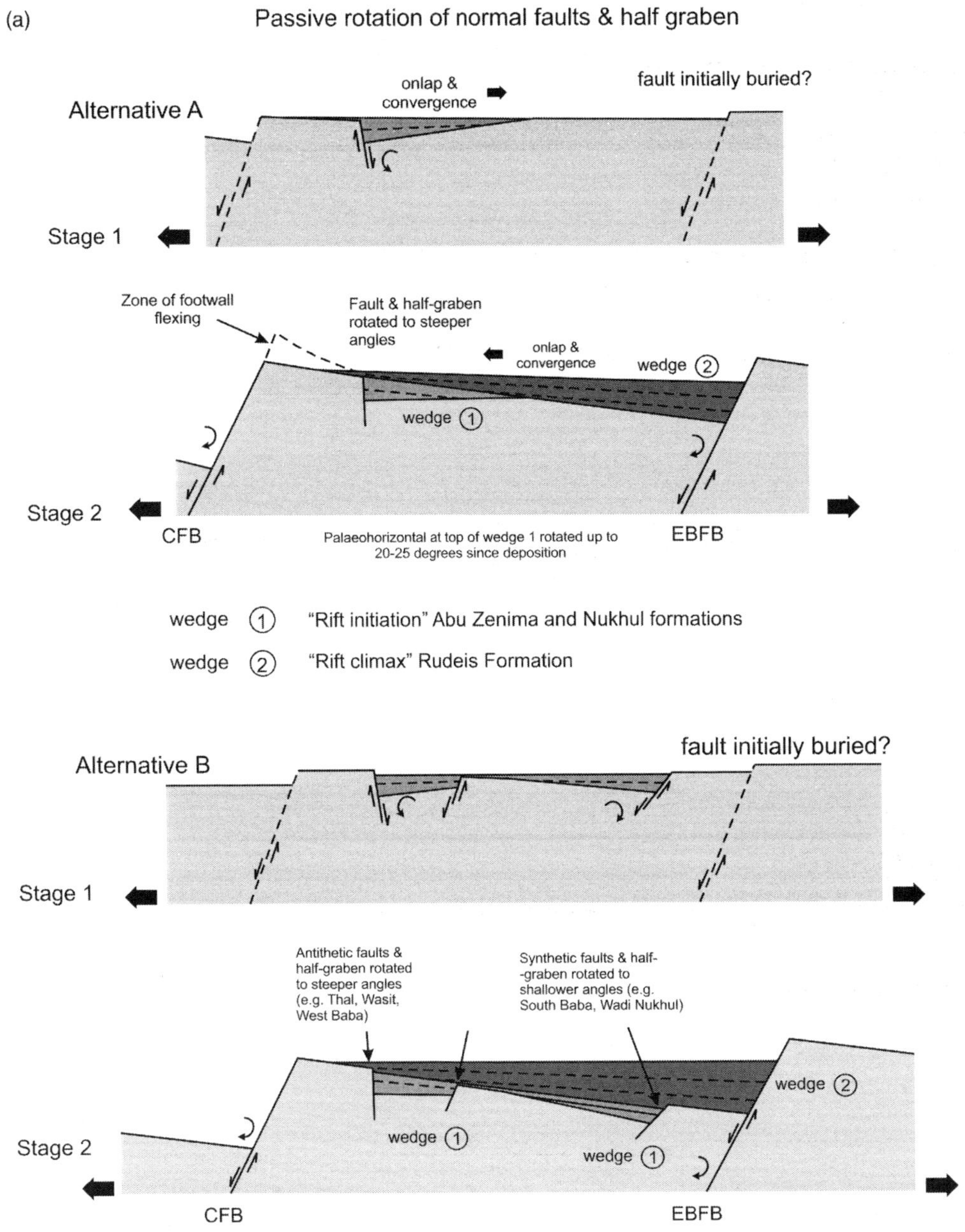

Figure 7.13. (a) Passive tilting of the early-formed faults in the Hammam Faraun and El Qaa blocks, Gulf of Suez (Sharp et al., 2000). Rotation model alternative A based on east-dipping faults only. Rotation model alternative B made for both fault types. (b) Miniature erosional streams in the upper portion of the normal fault scarp above almost nonexistent alluvial fans. Footwall uplift is faster than erosion and sediment transport. It results in the starvation of the adjacent hanging wall, which undergoes development of the longitudinal braided occasional river located right next to the normal fault scarp.

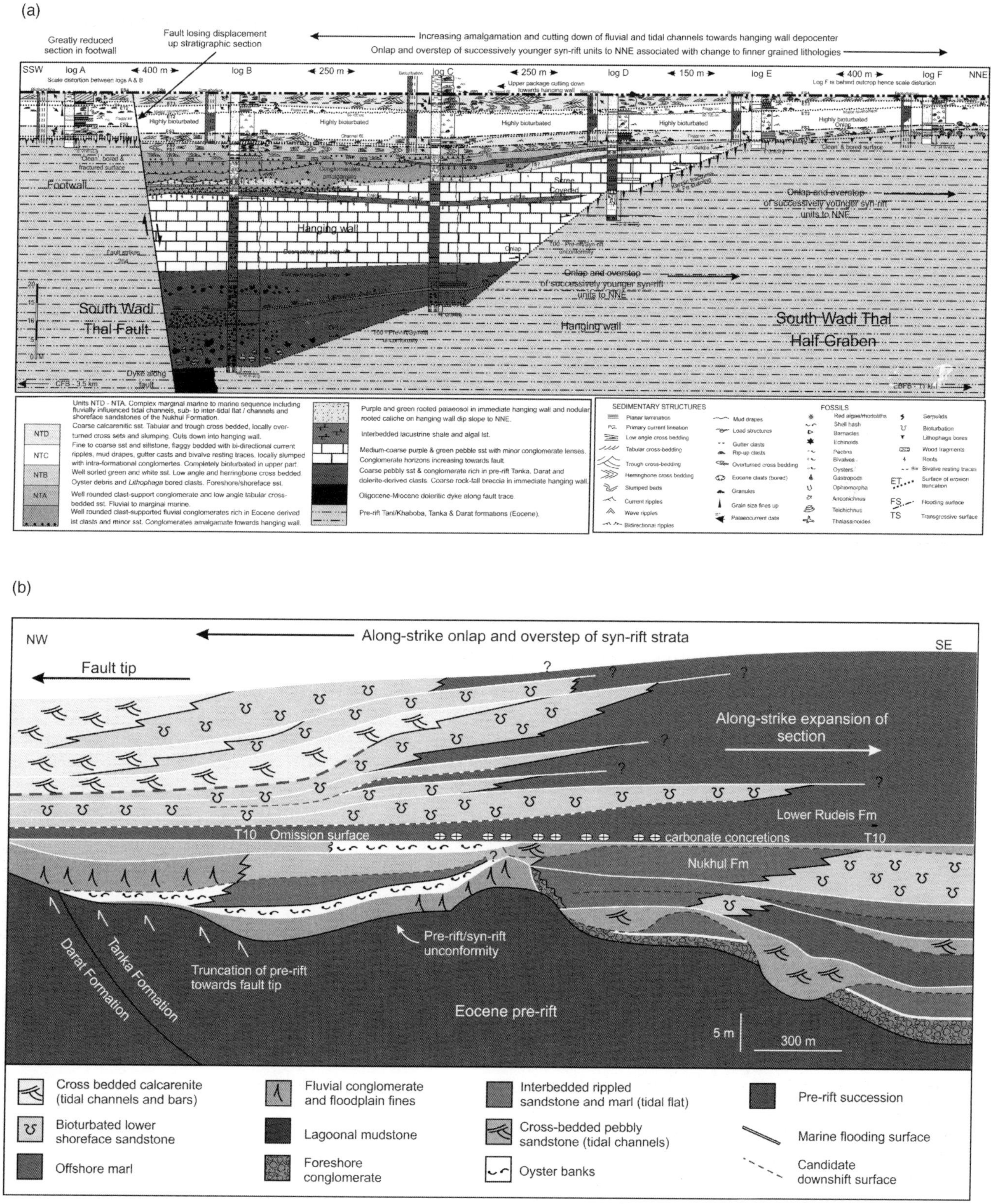

Figure 7.14. (a) Geological cross-section through the South Wadi Thal half-graben (Sharp et al., 2000). Note that the Abu Zemina Formation is present only in the hanging wall and thins dramatically eastward. The Nukhul Formation is present in both the hanging wall and footwall. Thinning to the northeast is associated with onlap and overstep of progressively younger syn-rift units. Divergence and section expansion toward the southwest are associated with preferential stacking of fluvial and tidal channels. These channels are parallel to the fault trace. (b) Fault strike-parallel facies changes, showing the rift-initiation Abu Zemina and Nukhul Formations and the rift-climax Rudeis Formation, Thal Ridge Fault, Gulf of Suez (Sharp et al., 2000). Decreasing amount of the total fault displacement

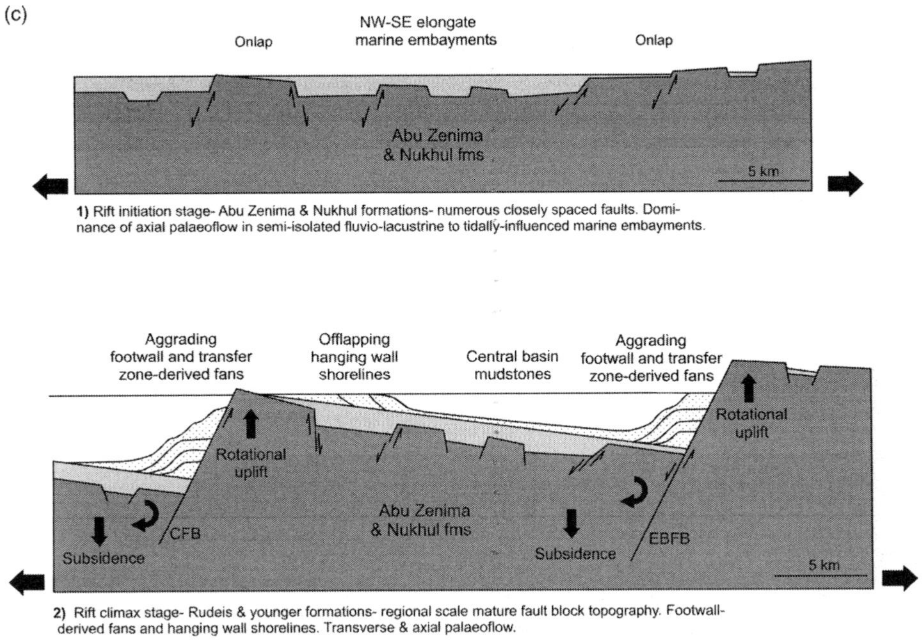

Figure 7.14 (*cont.*)
is indicated by the arrow pointing toward the fault propagation tip. The vertical facies transitions indicate a back-stepping transgressive sequence from non-marine to open-marine environments. The lateral facies transitions also indicate a southeast-northwest onlapping and overstepping transgressive relationship toward the fault propagation tip. Note the erosion-based, attached shoreface sandstones at the northern end of the section. (c) Structural and depositional styles typical for rift-initiation and climax, drawing from fieldwork in the Gulf of Suez (Sharp et al., 2000). The initiation is characterized by numerous closely spaced meso-scale faults and dominant axial sediment transport in semi-isolated lacustrine, tidal, and marine embayments. The climax is characterized by regional-scale fault blocks and facies distribution with thick aggrading hanging-wall fans, a central mudstone-dominated basin, and fringing hanging-wall shorelines.

footwall high. Overall, the rift-climax deposition results in block-wide facies relationships typical of mature rift basins (see Leeder and Gawthorpe, 1987; Gawthorpe and Leeder, 2000). This indicates that it is the main block-bounding faults that define the regional paleogeography on a 25 km-wide and 40 km-long scale.

Source rock studies on failed rift basins of passive margins such as the Congo Basin with the lacustrine Lower Cretaceous syn-rift section (Harris et al., 2004) also indicate an important interaction of deposition and tectonics. Here, the syn-rift section was systematically sampled and analyzed for content of the total organic carbon and source rock maturity. The results indicate that both overall organic carbon content and bacterial fraction of organic matter increase upward, reaching a maximum in the lower portion of the late syn-rift section, which represents a period when normal faulting more or less terminated and the rift basin was filling up (Figure 7.15). This was also the time when water depths became shallower than the depths during the earlier rift phases characterized by the deposition of sediments leaner in organic content. This was also when the deposition rates became slower.

The slower deposition rates partially controlled the higher organic carbon content. The deposition of the richest organic sediments was linked to a very high flux of dissolved carbonate into the rift lake. This is indicated by a systematic decrease in the ^{13}C isotope in carbonate minerals. The enhanced flux can be explained by the argument that a reduced relief of drainage basins feeding the rift lake caused more efficient cycling of carbon derived from plants into soil carbonate.

Good control over the footwall uplift/exhumation patterns next to main graben-bounding faults can be achieved by the systematic sampling along profiles perpendicular to graben and apatite (U-Th)/He or fission track thermochronometry performed on samples (see Ehlers and Farley, 2003). The exhumation determination is, however, affected by complex thermal processes, which deviate from a one-dimensional thermal field in footwall ranges, and include (Figure 7.16):

1) lateral heat flow across large, range-bounding fault due to the juxtaposition of a cool hanging wall and a relatively warmer footwall;

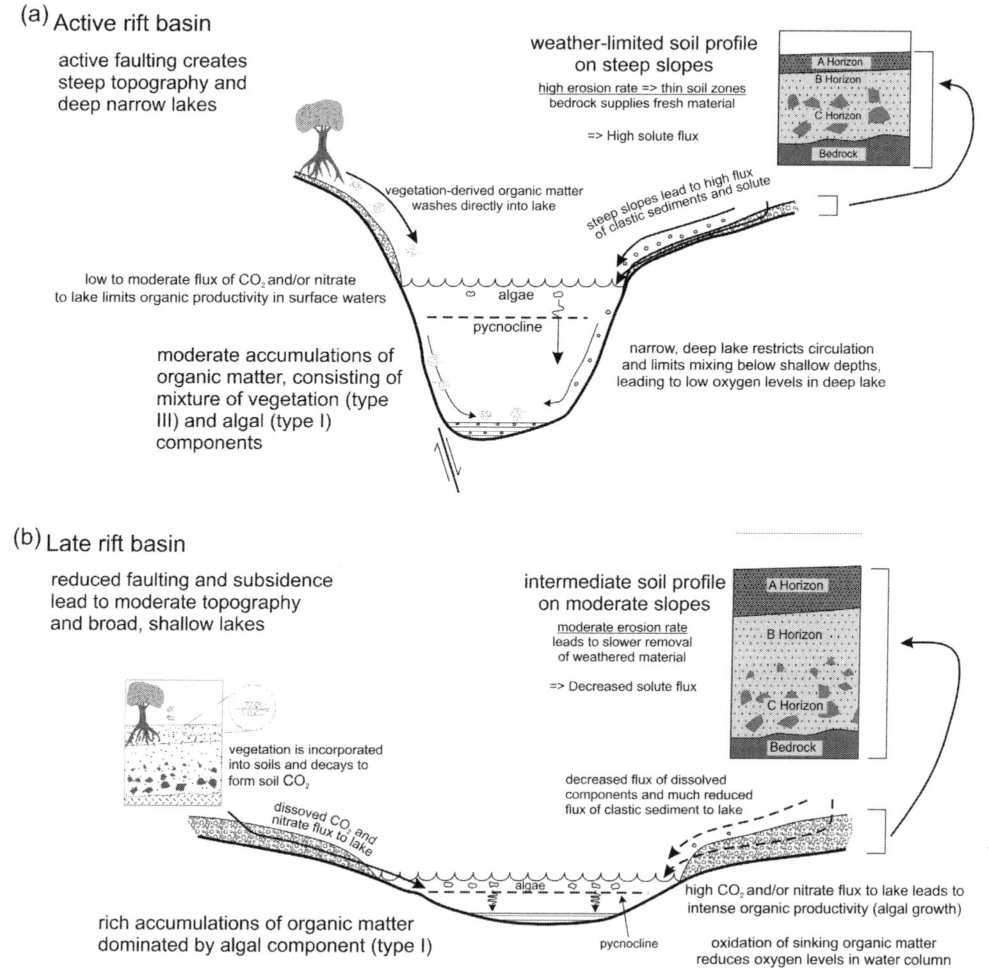

Figure 7.15. Factors controlling the organic matter deposition in the early- and late-rift stages of the lacustrine Congo Basin (Harris et al., 2004).

2) uplift and erosion of the footwall combined with deposition and burial in the hanging wall;
3) footwall tilt; and
4) topography.

An example of the exhumation pattern in the footwall range of the wide-rift province comes from the Wasatch Mountains, Utah (Ehlers and Farley, 2003). Results of the 2-D numerical model of the thermal field and predicted apatite helium ages are shown in Figure 7.17. The calculated thermal model assumes maximum footwall exhumation and maximum hanging wall burial right next to the Wasatch Fault due to footwall rotation and uplift, and hanging wall subsidence and rotation. The fault has a dip of 45°. The hinge is located at a distance of 30 km from the fault. The exhumation rate is assumed to decrease linearly across the footwall and hanging wall, being 0 mmy^{-1} behind the hinge.

Isotherms in the footwall are swept upward because of the exhumation and topography, being warmer than isotherms of the equilibrium thermal regime. They are swept downward in the hanging wall because of deposition, being cooler than the isotherms of the equilibrium thermal regime. For a better appreciation of thermal regimes, which are out of equilibrium, one can have a look at geothermal gradients along the shore of northeastern Brazil, which vary from west to east as a function of a decreasing deposition rate, that is, decreasing deposition role in thermal perturbation (see Gallagher et al., 1994).

Juxtaposition of increased and depressed thermal gradients in the Wasatch Fault footwall and hanging wall, respectively, results in curved isotherms across the bounding fault (Figure 7.17a). The lateral variation in exhumation rates and thermal gradients with a distance from the fault controls the cooling paths of

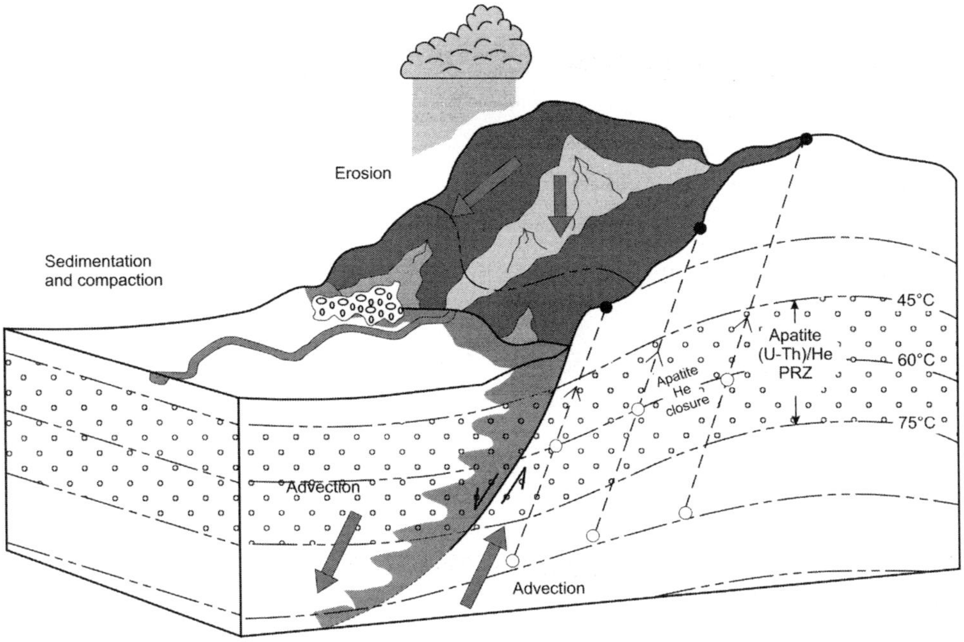

Figure 7.16. Thermal processes in a normal-fault-bounded footwall range, which control the interpretation of apatite-based thermochronometry (Ehlers and Farley, 2003). Isotherms are curved from heat and mass advection (arrows) in the footwall and the hanging wall and by topographic relief. Rocks in the subsurface (open circles) are exhumed in the outcrop (filled circles). The stippled zone between 45° and 75°C indicates the helium partial retention zone where helium is neither fast enough to maintain a zero concentration, nor slow enough for complete helium retention.

footwall rocks. Apatite (U-Th)/He ages calculated on these paths start at 35 Ma, which is the assumed age of the granitoid intrusion forming the footwall in the study area. The onset of exhumation was assumed to initiate at 10 Ma.

Rocks exhumed from depths shallower than 1.5 km show ages rapidly decreasing with paleodepth (Figure 7.17b). These rocks are associated with a helium partial retention zone developed between 35 and 10 Ma and are now displaced to the surface. The steep part of the curve at depths below 1.5 km represents rocks originated below the helium partial retention zone and rapidly exhumed at the surface. The break in slope of the curve in Figure 7.17b, thus, provides the age of onset of rapid exhumation controlled by faulting. Below the break in slope, the age-depth profile only roughly approximates the exhumation rate because rocks exhume at different rates and through different temperature gradients across the footwall. Numerical modeling (see Ehlers et al., 2001, 2003) can provide the translation of this apparent rate into the true exhumation rate on the fault surface.

The young ages between 0 and 25 km from the fault represent rock samples exhumed from depths below the helium partial retention zone (Figure 7.17c). The rapid increase in ages, at a distance beyond the 25 km well into the footwall range, represents the exhumed helium partial retention zone. The preservation of the helium partial retention zone at a greater distance from the fault reflects the smaller degrees of exhumation in areas toward the hinge.

Most of the studies of interaction between deposition, erosion, and tectonics in the rift setting so far focused mainly on high-frequency sea-level fluctuation and episodic fault-controlled subsidence aspects of control over deposition and erosion evolution in rift systems. Note, however, that climatic and lithologic controls of the sediment distribution are also important.

A set of deltas entering one of the half-grabens of the Pleistocene Lake Bonneville in Utah helps to illustrate the point. Here, the coarse-grained deltas vary along the graben-controlling fault depending on factors such as (Currey et al., 1984):

1) the size of the drainage basin, partially controlled by the erodibility of local rocks; and
2) the presence versus absence of glaciation in the drainage basin.

Features such as a large Stockton bar (Gilbert, 1890, cited in Burr and Currey, 1992) indicate an important impact of the prevalent wind, that is, wave direction,

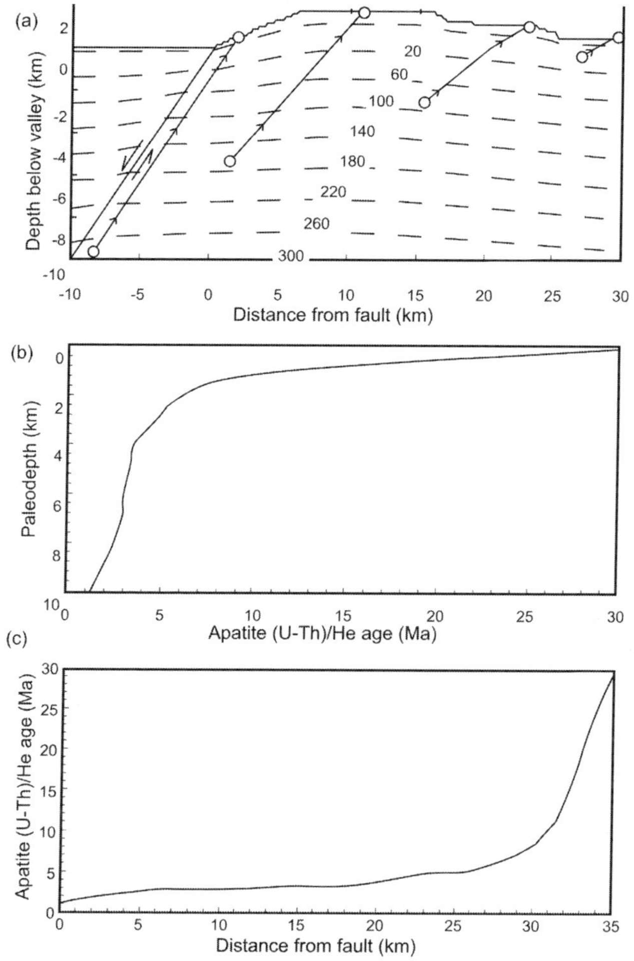

Figure 7.17. Thermal field and predicted helium ages in the footwall of the Wasatch normal fault, Utah, after 10 Ma of exhumation (Ehlers and Farley, 2003). (a) Isotherms are in ° C. Open circles with solid lines represent a simplified exhumation trajectory of the rock sampled at the surface. (b) Predicted helium age versus paleodepth after applying a tilt correction to restore samples to their initial depth. (c) Predicted footwall helium ages versus distance from the Wasatch Fault.

on the redistribution of the coarse delta products in the graben.

Climatic and lithologic aspects can be further highlighted by looking at erosion and deposition rates at different climatic and lithologic conditions. Change in climate can directly affect nonmarine deposition by altering:

1) stream discharge;
2) the ratio of chemical to physical detritus eroded from the source area; and
3) lake levels (Beer et al., 1990).

In arid regions, the lack of water hinders faster erosion. On the other hand, the vegetation cover in humid zones prevents faster erosion. A typical example of vegetation control comes from the Bengal Fan. Here, ODP sites 717C and 718C indicate a decreasing deposition rate since 7 Ma (Burbank et al., 1993). The decrease took place in response to a decline in mechanical denudation, which was a result of the increasing slope stabilization by the dense plant cover. The decrease happened despite the intensifying monsoonal climate (Harrison et al., 1993 and references therein).

Seasonality plays an important role in climatic control of erosion (e.g., Sinclair, 1997). The larger the seasonal changes, the faster the average erosional rate (e.g., Ahnert, 1970; Kukal, 1990). The long, dry season of the monsoon climate, the Mediterranean climate, and the tropical savanna and steppe climates leave the ground bare and vulnerable to erosion by the often heavy rains of the wet season.

A comparison of climatic effects on erosion in the same tectonic setting was made in the South American Andes (Scholl et al., 1970). They are characterized by:

1) erosion rate of 0.05 mmy^{-1} in the tropical arid climate;
2) erosion rate of 0.5 mmy^{-1} in the temperate humid climate; and
3) erosion rate of 1.1–1.65 mmy^{-1} in the polar climate;

highlighting the accelerated erosion by glacier and glacial melt waters. Despite the fact that comparisons of denudation rates with either mean annual temperature or mean annual precipitation made for the world's major drainage systems (Pinet and Soriau, 1988) show certain relationships, it seems that basic climatic distinction is not sufficient for a division of erosional rate categories. It turns out that climatic distinction contains about forty factors controlling the erosion, such as:

1) distribution of precipitation throughout the year;
2) the number of exceptional events such as storms;
3) torrential rains;
4) floods;
5) the duration of snow cover; and
6) the number of days with temperatures below freezing point.

Lithology is another rather important control over erosion and deposition, controlling the grain size together with climate. Certain rocks such as mudstone, for example, do not yield coarse detritus when weathered (e.g., DeCelles et al., 1991; Keith et al., 1991). Other rock types yield coarse detritus in direct proportion to their relative durabilities during erosion and transport (e.g., DeCelles et al., 1991; Kováč et al.,

1992). Furthermore, unconsolidated rocks erode faster than solid rocks.

The effect of lithology on erosional control is coupled with climatic control. For example, solution rates of carbonate and siliceous rocks vary significantly as a function of climate (Saunders and Young, 1983). Other controlling factors of erosion and deposition include differences due to structure, texture, and fracturing.

Pull-apart basins

In comparison to rift basins, pull-apart basins are characterized by faster tectonic subsidence rates (see Figures 1.14–1.17), and, as a result, by even more dramatic tectonic effects on deposition rates and facies distribution. For example, deposition in the Loreto Basin, Baja California, reacts to pull-apart-bounding fault activity by shelf- and Gilbert-type deltas in the marine setting (Dorsey et al., 1995; Dorsey and Umhoefer, 2000).

The southern portion of this basin fill, which is outcropping at Baja California, shows:

1) the southernmost part of the normal fault system branching off the northern controlling dextral strike-slip fault (southern Sierra Microondas fault system);
2) the northern portion of the southern controlling dextral strike-slip fault (Pliocene Loreto Fault);
3) the southern portion of the normal fault branching off the southern controlling dextral strike-slip fault (Quaternary Loreto Fault); and
4) the southernmost portion of the through-going strike–oblique-slip fault connecting opposite corners of the pull-apart basin (unnamed fault along the west side of the Sierra Microondas).

The outcrop study of the pull-apart basin was done in two areas, which cover a pull-apart center and southeastern corner. The stratigraphic correlations between these two areas were made possible because of the existence of reliable marine sections with traceable shell beds and four marker tuff horizons with plagioclase and biotite dated by the $^{40}Ar/^{39}Ar$ method. This allowed dividing the fill into four sequences.

Sediment transport data measured using cross-bedding, clast imbrications, and flute casts indicate that (Dorsey and Umhoefer, 2000):

1) the sequence 1 is characterized by westward transport in both the central area and the southeastern corner;
2) the sequence 2 is characterized by north-northeast-ward transport in the central area, while the southeastern corner is characterized by north-, northeast-, east-, and southeast-ward transport; and
3) the sequence 3 is characterized by north-northwest-ward transport.

This sediment transport history, after comparing it with the pull-apart evolution models in Figures 1.9–1.11, indicates:

1) sediment transport in the initial pull-apart controlled by disconnected normal and strike-slip faults characterizing the sequence 1 (see also Figure 7.18); and

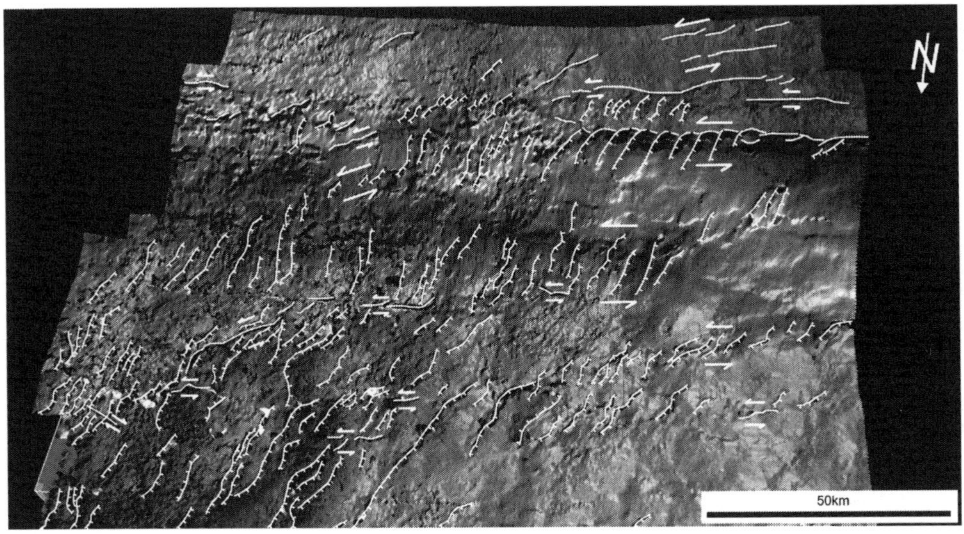

Figure 7.18. Surface in 3-D seismic cube, unspecified passive margin showing a fault pattern associated with initial pull-apart development in its northern half. Note disconnected normal and strike-slip faults.

2) transport in the fully connected pull-apart basin, which undergoes the evolution of progressively large overlap characterizing sequences 2 and 3 associated with the merging of the asymmetric depocenters moving toward each other into one symmetric depocenter.

This interpretation of the data that Dorsey and Umhoefer (2000) described is also in accordance with their subsidence curve (Figure 7.19a), which shows:

1) a relatively slow subsidence of 0.4 mmy^{-1} controlling the deposition of sequence 1 and some portion of sequence 2;
2) followed by a fast subsidence of about 8 mmy^{-1} controlling the deposition of the remaining portion of sequence 2 and all of sequence 3; and
3) followed by a relatively slow subsidence of 0.4 mmy^{-1} typical for sequence 4.

The accelerated subsidence is coeval with the merger of the pull-apart-controlling faults and the merger of the two independent asymmetric depocenters from both propagating pull-apart sides into one depocenter bounded by a fully linked fault system. The slow-down of the subsidence during the deposition of sequence 4 indicates when the pull-apart basin underwent development of the through-going strike-/oblique- slip fault, which links both opposite corners, in accordance with analog material models (Sims et al., 1999; see Figure 1.10). The previously described subsidence pattern is similar to subsidence histories in other pull-apart basins (e.g., Johnson et al., 1983; Christie-Blick and Biddle, 1985; Pitman and Andrews, 1985) and is expressed in the Loreto Basin by a tripartite of facial distributions (Figure 7.19b).

Structural fieldwork reveals different timing of various faults in the Loreto Basin. Outcrop structural data indicate that:

1) the southern Sierra Microondas fault system began its activity early during the deposition of sequence 3 and ceased at about 2 Ma, during the deposition of sequence 4;
2) the southern portion of the normal fault branching off the southern controlling dextral strike-slip fault has Quaternary activity; and
3) the northern portion of the southern controlling dextral strike-slip fault has Pliocene activity,

allowing us to conclude that two dextral northern and southern strike-slip faults were propagating southeastward and northwest-ward till they developed an overlap zone, where they started to curve toward each other, changing their geometries and respective kinematics toward oblique-slip faults and normal faults at regions close to the other strike-slip fault.

Sequence 1 deposited in a slowly subsiding system (Figure 7.19a) is represented by alluvial fans (Figure 7.19b – **stage A**). This nonmarine deposition was initiated during the latest Miocene–earliest Pliocene, the time when evidences show the first significant strike-slip faulting in the Gulf of California (see Lonsdale, 1989).

Sequence 2 deposition, which started with increased subsidence and experienced an onset of the accelerated subsidence (Figure 7.19a), indicates by its basal marine incursion an increase of the tectonic subsidence that was probably amplified by an eustatic sea-level rise at 2.62 Ma. The onset of the accelerated subsidence in the central area controlled deposition of fourteen Gilbert delta parasequences occupying a section about 650 m thick, which had accumulated during a short time interval of less than 100 k.y. (Dorsey et al., 1995; Dorsey and Umhoefer, 2000; Figure 7.19c). Deposition of sequence 2 also witnessed rapid westward tilting producing significant westward thickening and lithology coarsening toward the boundary fault (Figure 7.19b – **stage B**). The adjacent footwall produced a correspondingly rapid coarse sediment influx, which kept the basin in a constantly filled state despite the rapid subsidence.

Initiation of the sequence 3 deposition recorded propagation and activity of the Sierra Microondas fault system, documenting the end of the propagation of the northern bounding fault to the southeast and its linkage with the southern bounding fault. It also recorded a slowing slip on the southern portion of the southern bounding fault and overall change in subsidence geometry, highlighted by a change in the central area from asymmetric to symmetric geometry (Figure 7.19b – **stage C**).

To highlight the short length of the pull-apart basin "life," which makes it different from the "longer-living" rift basin, we can conclude that the Loreto pull-apart basin was filled up in about 2 Ma, and then started to undergo uplift and erosion.

The deposition of its sequence 2 allows us to look at the interaction of deposition and tectonics in the fully developed pull-apart basin (Figure 7.19c). The activity of a main bounding fault can be studied in detail using fourteen Gilbert deltas occurring between the $^{40}Ar/^{39}Ar$ – dated tuff horizon 2 and the base of sequence 3, which both represent the basin-wide marker horizons. A sedimentary section between them has recorded fourteen episodes of deltaic progradation and marine flooding (Dorsey et al., 1995; Dorsey and

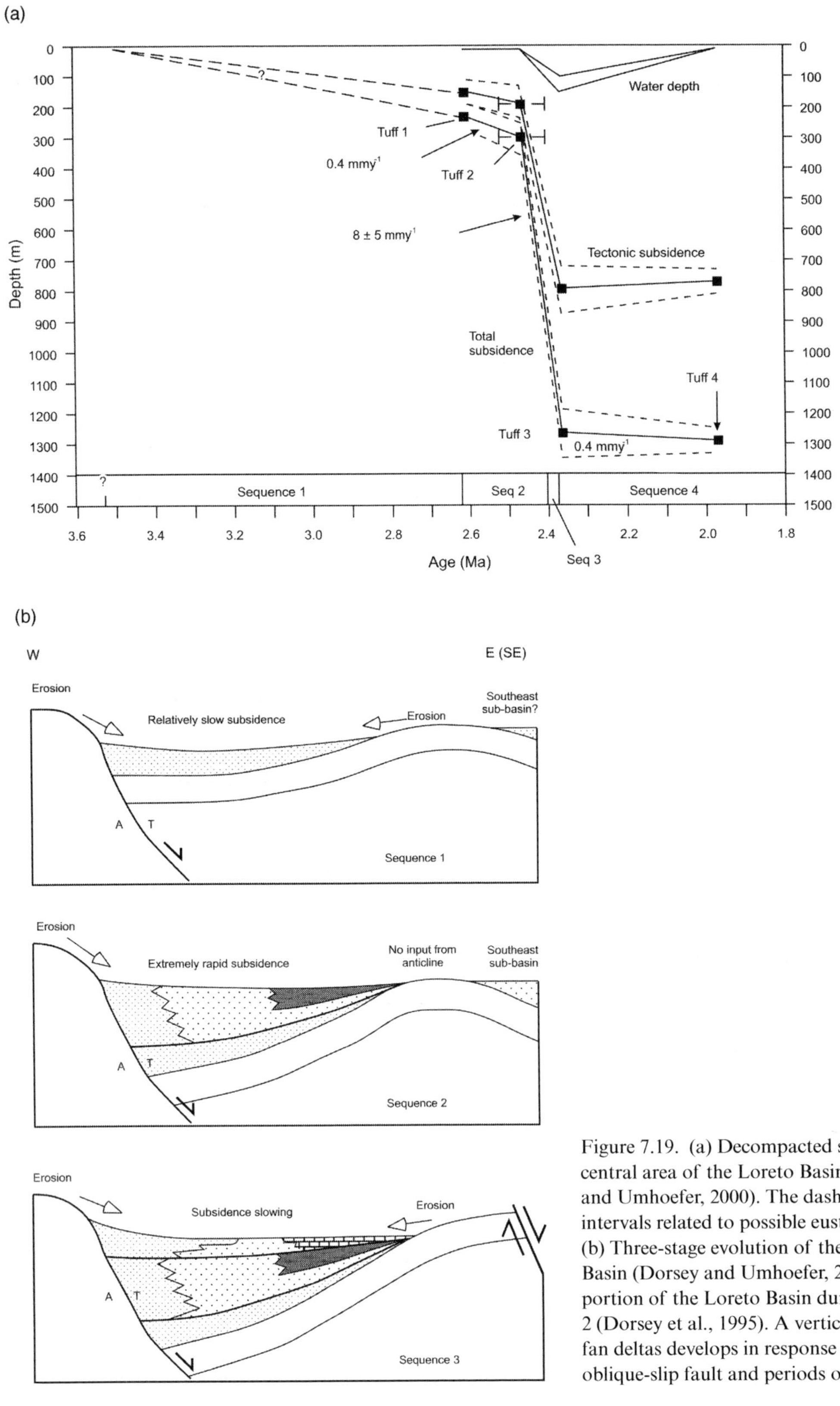

Figure 7.19. (a) Decompacted subsidence curve for the central area of the Loreto Basin, Baja California (Dorsey and Umhoefer, 2000). The dashed lines represent uncertainty intervals related to possible eustatic sea-level fluctuations. (b) Three-stage evolution of the central area of the Loreto Basin (Dorsey and Umhoefer, 2000). (c) Sketch of the central portion of the Loreto Basin during deposition of the sequence 2 (Dorsey et al., 1995). A vertically stacked system of Gilbert fan deltas develops in response to episodic activity of the dextral oblique-slip fault and periods of quiescence.

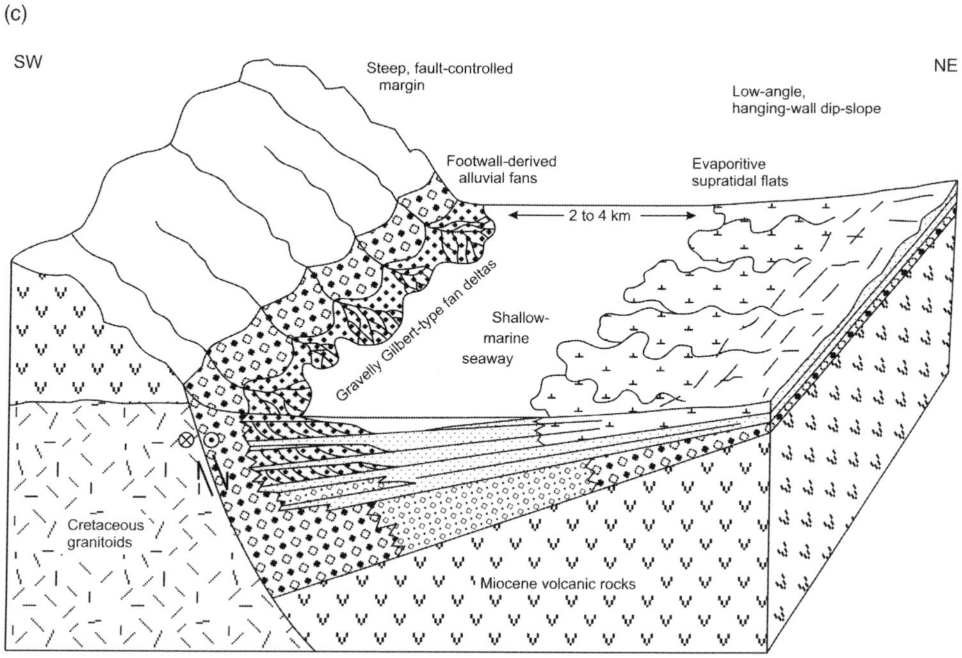

Figure 7.19 (*cont.*)

Umhoefer, 2000). Although this sedimentary sequence could be explained by:

1) eustatic sea-level fluctuation accompanied by uniform basin subsidence to produce accommodation space for each successive delta; or
2) episodic fault-controlled subsidence where the drowning of each Gilbert-type delta represents an abrupt fault displacement followed by tectonic quiescence and next delta progradation; or
3) uniform rapid subsidence accompanied by avulsion of deltaic distributary channels, lateral switching of delta lobes, and autocyclic delta-lobe abandonment during background subsidence,

the sequence 2 delta system was controlled by a combination of factors (2) and (3) for the following reasons. Facies mapping has shown that each separate Gilbert delta cycle is just 2–3 km wide in a direction perpendicular to transport. The size analog is a small steep alluvial fan that fed the delta system. This indicates that the entire delta-plain system of each individual delta was drowned during the delta abandonment phases, rather than just a small fraction of a larger delta as would be expected from autocyclic lobe switching. The section of the central basin portion with fourteen progradational parasequences compares to just four parasequences in the southeastern corner of the basin. When compared to the temporal variations in oxygen isotopes from the North Atlantic (Raymo et al., 1992), which can be used as a proxy for eustatic sea-level changes, it becomes obvious that while four laterally continuous shell beds in the southeastern basin corner appear to match four highstands, fourteen paracycles in the central basin portion with significant fault activity have accumulated during one and a half cycles of the eustatic sea-level change (Dorsey and Umhoefer, 2000).

Similar coarse-grained basin margin facies passing over short distances to basin-axial lacustrine and marine deposits are common in a variety of pull-apart basins (e.g., Crowell, 1974a, 1974b, 1982; Johnson et al., 1983; Christie-Blick and Biddle, 1985; Pitman and Andrews, 1985; Karner and Dewey, 1986). A combination of very rapid subsidence and very rapid deposition apparently helps to keep a "positive imbalance" between the sediment supply and the effects of water depth in the basin, that is, conditions typical of Gilbert deltas (Nemec, 1990). Slower overall rates of basin subsidence and deposition are expected to cause more uniform sediment distribution and low depositional gradients in the basin, being responsible for shelf-type fan deltas (Colella, 1988a; Leeder et al., 1988; Massari and Colella, 1988). Knowing this, it is not surprising to see the fast subsidence period from Figure 7.19a being associated with Gilbert-type deltas and the preceding slow subsidence period with shelf-type deltas.

Post-rift basins developed on failed rifts

Previously active rift basins can exert a control on the early post-rift deposition. An example comes from the distribution of the Lower Cretaceous Punt Sandstone in the Inner Moray Firth Basin, North Sea (Argent et al., 2000). The prerequisite for its architecture and facies distribution is a presence of the preexisting basin topography that provided control of the sand deposition within the available depositional accommodation. This model applies to similar basin settings characterized by mass-flow deposition with a sand-rich load accumulating in a deepwater marine basin with preexisting basin-floor topography.

The Inner Morey Firth Basin was characterized by (Argent et al., 2000):

1) unfilled Jurassic fault-controlled depressions;
2) differential compaction in Jurassic grabens; and
3) local Lower Cretaceous faulting.

Sands entering the basin progressively filled depressions in the basin floor starting with those closest to the sediment entry points (Figure 7.20). Sand-rich sediments filled the first depression via stacking as they onlapped the inter-basin highs in the form of laterally extensive sheets. After the first depression became filled, the bypass situation developed on the basin floor where the feeding channel propagated by incision across the filled depression and inter-depression high, allowing the sediment transport further basinward into the next depression to be filled. This depositional mechanism created (Argent et al., 2000):

1) some depressions on the fill-and-spill trend filled with stacked sheets or lobes, incised by channels;
2) some depressions on the fill-and-spill trend, which are underfilled or unfilled with sands; and
3) some inter-basin highs incised by channels that can be backfilled with sand.

The previously described depositional model is somewhat analogous to the fill-and-spill process for deepwater massive sand deposition in the Gulf of Mexico (Rowan and Weimer, 1998; Weimer et al., 1998), where the depressions, however, are associated with salt withdrawal. The model results in the upward transition from relatively unconfined or ponded sand sheet deposits to incised, erosional channel elements that have subsequently been back-filled. Back-filled erosional channel cuts connect individual depressions of ponded sheet sands where there is an upward transition from sheet-type depositional elements to leveed channels as the depression fills to spill-point.

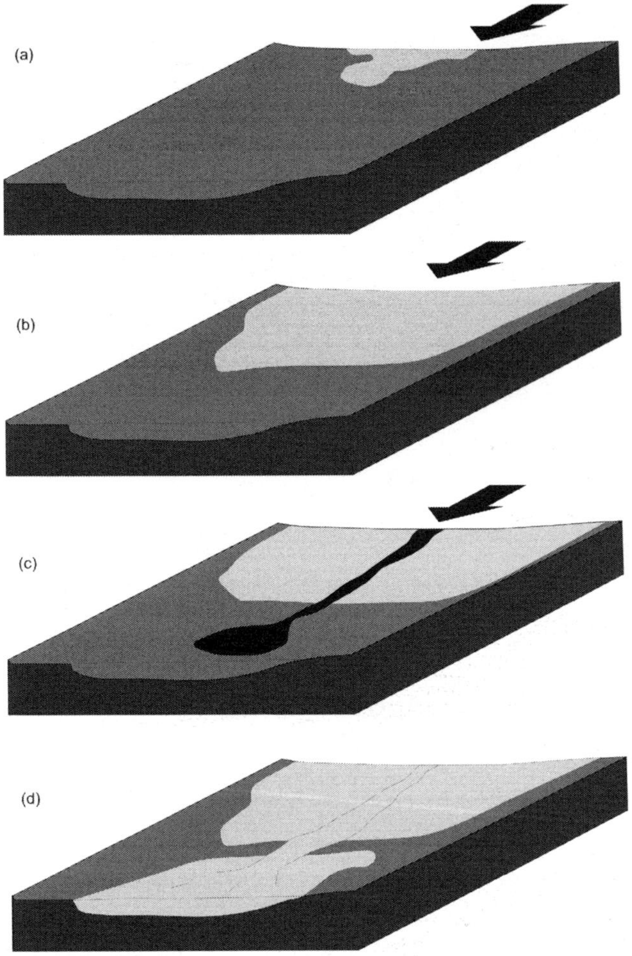

Figure 7.20. Depression fill-and-spill model for the progressive fill of the deepwater Inner Moray Firth Basin floor topography in Lower Cretaceous (Argent et al., 2000). a) Fill by channel sands feeding into depressions where terminal mass flow lobes stack and pond. b) Stacking and ponding resulting in depression fill. c) The floor shape changes characterized by feeder channel prograding further basinward until it encounters the next depression. d) Stage when the depositional system gets abandoned and the channel is backfilled.

Passive margins

One of the most prominent erosional/depositional responses of the passive continental margins to the early syn-drift tectonics is margin erosion coupled with accelerated basin deposition. Apatite fission track studies indicate erosional removal of volumes of burial as large as 3–4 km from the uplifting and exhuming passive margins on the African and Brazilian sides of the South Atlantic (Rust and Summerfield, 1990; Gallagher et al., 1994; Gallagher and Brown, 1999; Brown et al., 2000; Raab et al., 2001), the American side of the Central Atlantic (Miller and Duddy, 1989; Steckler et al., 1994), and southeastern Australia (Moore et al., 1986; Dumitru et al., 1991).

Such uplifted flanks, or shoulders, are common features of rifts. The rift shoulder is typically asymmetric. It has a steep scarp facing the sea and a gentle slope facing the land. Flank uplifts have been explained by models based on transient and permanent uplift mechanisms. Transient uplift models are related to the thermal effects of rifting and include depth-dependent extension (Royden and Keen, 1980; Hellinger and Sclater, 1983; Watts and Thorne, 1984; Morgan et al., 1985), lateral heat flow (Steckler, 1981a, 1981b; Cochran, 1983; Alvarez et al., 1984; Buck et al., 1988) and secondary convection under rift shoulders (Keen, 1985a; Steckler, 1985; W. R. Buck, 1986). These mechanisms, which can create a shoulder elevation of 500–1,500 m, operate only during the elevated thermal regime of the lithosphere. Created positive topography will decay over the time period equivalent to the thermal time constant of the lithosphere, which is roughly 60 Ma. Permanent uplift models are related to the magmatic underplating (Cox, 1980; Ewart et al., 1980; McKenzie, 1984; White and McKenzie, 1988) and lithospheric unloading and/or plastic necking (Zuber and Parmentier, 1986; Parmentier, 1987; Issler et al., 1989; Braun and Beaumont, 1989; Weissel and Karner, 1989; Chery et al., 1992). Only relatively recently, processes of erosion and deposition started to be incorporated into existing models of the rift shoulder uplift (e.g., Van Balen et al., 1995; Van der Beek et al., 1995; Burov and Cloetingh, 1997; Figure 2.7).

These uplifts have different character at different margin segments. Therefore, before we discuss erosion, uplift, and deposition characteristics of passive margins, we need to divide them into extensional and transform segments. Furthermore, we need to divide extensional segments into magma-rich and magma-poor ones, and magma-poor ones into colder and warmer ones.

When we compare extensional and transform margins, it becomes apparent that removal of the "hanging wall" of the former unloads its "footwall" in a gross sense, making it prone to post-breakup uplift, the amount of which depends on its rheology. On the contrary, a lack of any such distinct unloading at transform margin prevents it from any important uplift after the continental breakup. In fact, numerical modeling and case studies (Karner and Watts, 1982; Holt and Stern, 1991; Kooi et al., 1992; Pierce et al., 1996; Nemčok et al., 2012b) indicate that it is not just the geometry of unloading at transform margins that suppresses their post-breakup uplift, but it is also their increased thermal regime due to their juxtaposition with hot new oceanic crust that prevents them from any distinct flexural behavior. The high temperatures weaken the continental lithosphere, which results in very low flexural rigidity. Thus, the transform margin is always close to local Airy isostasy, such that flexural and rebound effects are unlikely to play a major role.

Examples of two contrasting transform margin groups come from the southern terminus of the Central Atlantic in offshore Guyana, the Gettysburg-Tarfaya Transform in offshore Morocco, the St Paul Transform in offshore Côte d'Ivoire, and the Romanche Transform in offshore Côte d'Ivoire, Ghana, Togo, and Benin. The former three are examples of transforms that more or less did not undergo any tectonic events after the continental breakup. The fourth example represents the case when it did.

The Guyana margin represents an example of a transform margin that did not have to accommodate any distinct flexural uplift and transpression-driven uplift due to post-breakup spreading vector changes, and that remained quiescent for a very long time after continental breakup. Comparison of its shelf break lines mapped for different time periods from the breakup time in the Jurassic until the Oligocene period always shows no oceanward shift, indicating its long-term stability. Being located far from entries of any major rivers into deep basin, this margin undergoes practically no modification by prograding sedimentary wedges in regions among local rivers during the Late Jurassic–Paleocene and only a minimal one during Paleocene-Oligocene.

The second example of such relatively quiescent transform margin comes from the African side of the Gettysburg-Tarfaya Transform (Nemčok et al., 2005b; Figure 6.2). A lack of flexurally driven uplift along this margin segment and the occurrence of flexurally driven uplift at neighbor extensional segments at both sides during the post-breakup time period resulted in very different facies distributions between the two types of segments (Figure 7.21a). The transform margin, represented by a system of failed pull-apart basins (see Figures 3.12, 2.2a–c), is the only margin segment where siliciclastic sediments sourced from the West African block can enter the deep basin during the Middle–Late Jurassic (Figure 7.21a). They advanced by the fill-and-spill mechanism described earlier in the subchapter on post-rift basins developed on failed rifts. Neighboring extensional margins affected by flexural post-breakup uplift were cut off from any distinct terrestrial sediment input by a system of uplifted zones and only a limited amount of gaps among them allowed for some siliciclastic sediment entry points into the deep basin. As a consequence, carbonate platform development dominated deposition at extensional margin segments.

Eventually, after the erosion removed most of the positive topography of elevated flank uplifts, the siliciclastic sediment was readily accessing the deep oceanic basin. Most of the carbonate production was terminated, apart from local occurrences of small area extent. This

(a)

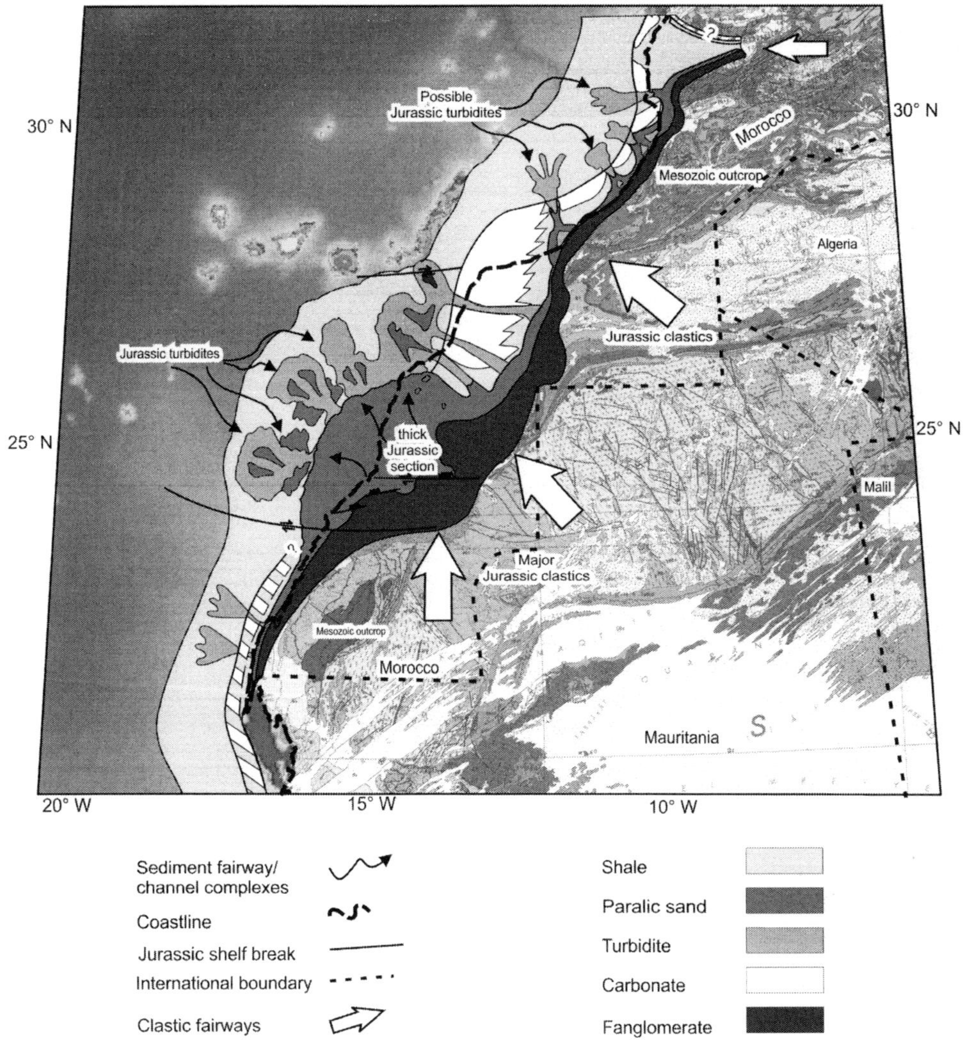

Figure 7.21. (a) Middle to Late Jurassic paleogeography of Morocco. The map is sourced primarily from Lynx Western Sahara GIS project (Nemčok et al., 2005b). (b) Early Cretaceous paleogeography of Morocco. The map is sourced primarily from Lynx Western Sahara GIS project (Nemčok et al., 2005b).

situation can be implied from the interpreted Early Cretaceous paleogeography (Figure 7.21b).

A somewhat similar, fill-and-spill depositional history picture comes from the St Paul Transform, offshore Côte d'Ivoire (Figures 3.2, 3.11, 6.13a). The Middle Albian breakup here (De Azevedo, 1991 and references therein) brought marine environments onto oceanic crust and adjacent failed pull-apart basins. It also generated an unconformity due to an isostatic rebound of the extensional and oblique-slip continental margin segment located between the St Paul and Romanche Transform segments (see Figure 6.13a for location of different segments). Seismic profiles document onlapping Upper Albian reflectors onto a Middle Albian unconformity (Figure 4.7). The Cenomanian-Turonian period was characterized by the existence of eroding structural highs among failed pull-apart basins on the margin (Figure 4.7). They formed a system of barriers and constraints for the clastic sediments entering the deep basin. The main clastic input into this region was from the corner between the northeast-southwest striking main dextral strike-slip faults and the main northwest-southeast striking normal faults (Nemčok et al., 2004b). Eroding preexisting highs formed a system of local sediment provenances together with the continental crustal Ghana Ridge located between the Côte d'Ivoire and Ghana and associated still further south with the Romanche Fracture Zone.

(b)

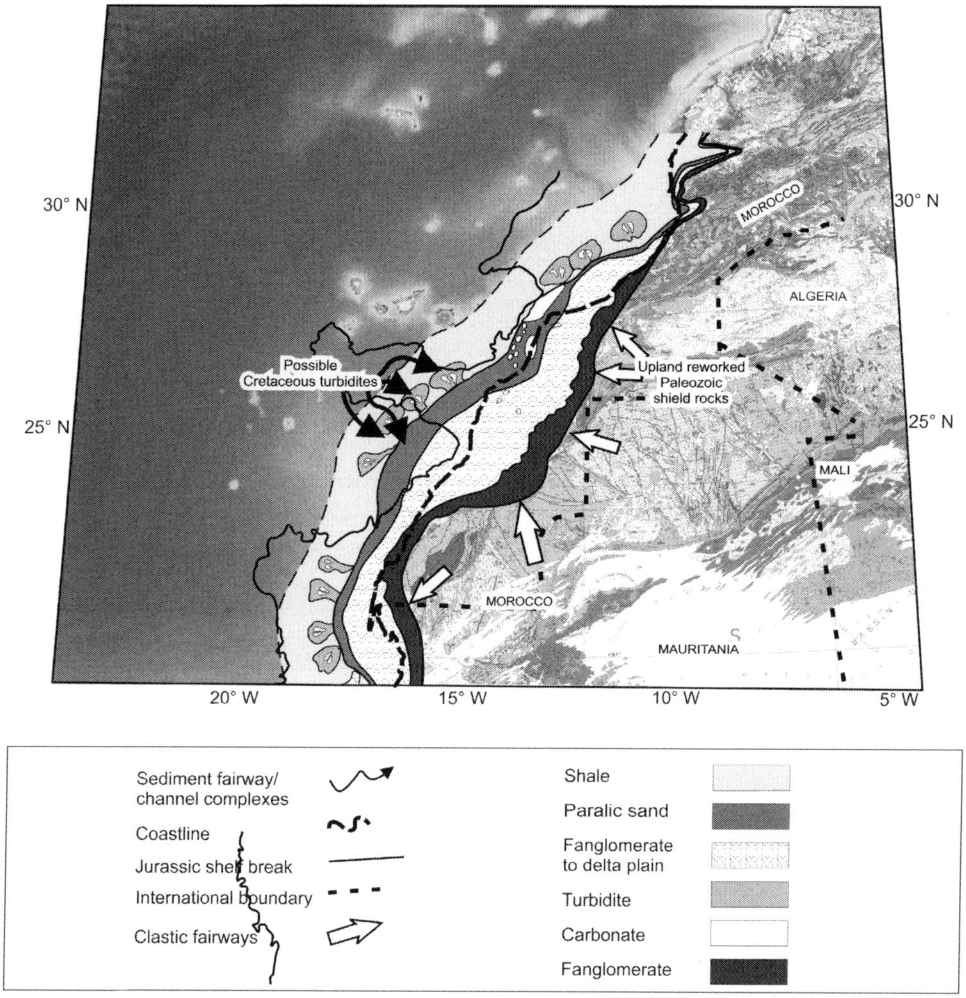

Figure 7.21 (*cont.*)

A quick look at the Romanche Transform margin (Figures 1.13, 4.6a, 6.13a) reveals that the adjacent oceanic crust indicates changes in spreading vector during Albian-Thanetian (Nemčok et al., 2012b; Figure 7.22). Spreading directions during the Albian-Cenomanian, affected by the preexisting block structure of the African and Brazilian continental crusts, were striking from east-northeast to west-southwest, as interpreted from the magnetic anomalies related to seafloor spreading (Müller et al., 1997). Later, this early northeast-southwest divergence progressively changed into an almost east-west divergence. It can be recognized through a change in flowline directions reflected in fracture zone geometry in the data from Müller and colleagues (1997). The changes took place in two stages, during the Coniacian-Maastrichtian (89–65 Ma) and Danian-Thanetian (65–59 Ma), the former being more important.

Seismic imaging constrained by well data (Nemčok et al., 2012b) indicates that the Romanche Transform itself experienced two phases of transpressionally driven deformation: 108–92 Ma and 65–52 Ma. As shown by numerical modeling done by the same authors, the former event could have been controlled by a 9° change in divergence direction, which would generate a convergence component of 2,970 m per 1 Ma (m Ma^{-1}) across the Romanche Transform. Such convergence is capable of driving an uplift of about 2,500–4,500 m.

Numerical models indicate that, apart from the transpression-controlled uplift of the transform margin, there was thermally driven uplift related to the passing-by spreading ridge, as it moved from east to west along the Ghana Ridge (Figures 4.6a, b). As the thermal anomaly decayed in time, subsidence progressively replaced the uplift. Shifting location in a westerly direction, both earlier uplift and subsequent

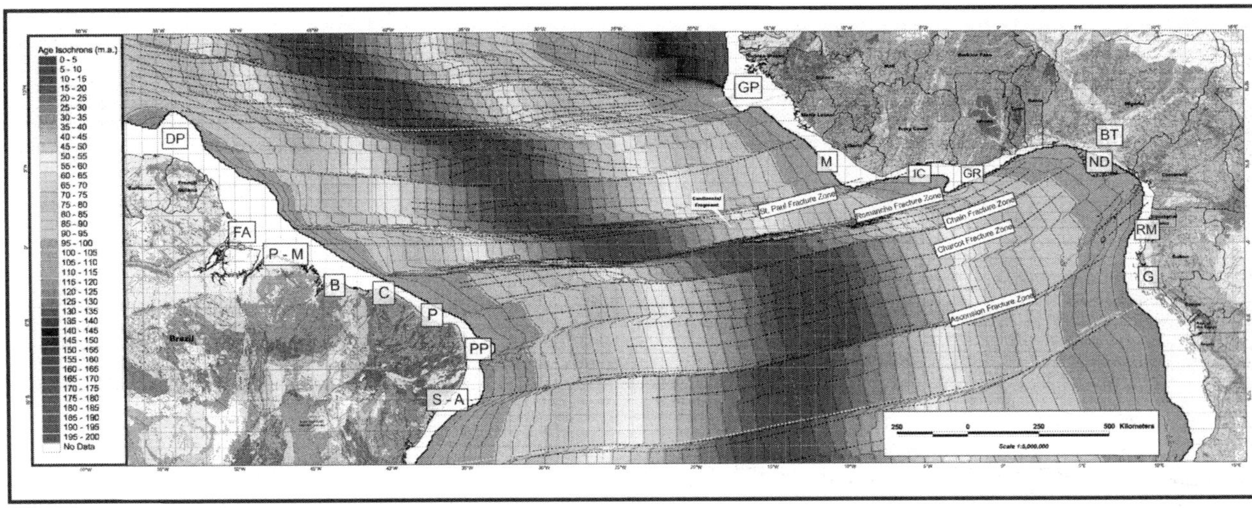

Figure 7.22. Map of the Equatorial Atlantic oceanic crust age based on magnetic anomaly data and the geology of adjacent continents (maps are from CGMW, Africa, 1990, South America, 2000; Müller et al., 1997). Abbreviations: GP – Guinea Plateau, M – Monrovia Basin, IC – offshore Côte d'Ivoire, ND – Niger Delta, BT – Benue Trough, RM – Rio Muni Basin, G – offshore Gabon, DP – Demerara Plateau, FA – Foz do Amazonas Basin, P-M – Para-Maranhão Basin, B – Barreirinhas Basin, C – Ceará Basin, P – Ptiguar Basin, PP – Pernambuco Paraiba Basin, S-A – Sergipe-Alagoas Basin. Note that initial spreading in the region between the St. Paul Transform and the boundary between the North and South Gabon Basins had a northeast–southwest trend that progressively changed to east-northeast–west-southwest by Eocene.

subsidence traveled as westward-migrating waves. Such migrating subsidence would explain the abrupt end of Turonian-Coniacian carbonate deposition sometime in the Coniacian, and the more rapid subsidence of the ridge that Basile and colleagues (1998) noted. Note, however, that thermal expansion-driven uplift of maximally 340 m is an order of magnitude less important to the control of the Romanche Transform margin uplift (Nemčok et al., 2012b).

The development of the prominent zone of uplift along the Romanche Transform margin overlapped in time with a very important event in sediment transport in the broader northern Gulf of Guinea region. Prior to this uplift, during Albian-Turonian, this region contained a system of north-south trending rivers. Their drainage basins were located on the southern slopes of the West African Shield. Deep canyons were cut into the shelf during periods of sea-level fall. Canyons funneled sediment into deepwater, concentrating it in large turbidite fans. One large, Abeokuta delta was located at the Benin–Nigeria border. It deposited the reservoir sands present in the Seme Field, Benin, and the Aje Field, Nigeria. Coeval and postdating this uplift, during the mid-Turonian, the drainage basins on the southern slopes of the West African Shield started to feed the Niger River that flowed first to the east-southeast, behind the uplifted region, and then turned more than 90° clockwise, utilizing the Bida and Benue failed rift basins to feed its delta (Doust and Omatsola, 1990). The sediment input was more or less continuous since the mid-Turonian, although episodic transgressions have occasionally interrupted regressive sequences.

Cutting local north-south trending rivers from a majority of their drainage basins resulted in a decreased coarse sediment supply into offshore Benin. For example, the sedimentary record of the Abeokuta Delta indicates that it became inactive during the Santonian. It seems that the Niger River captured most of the drainage systems feeding the Abeokuta Delta after the Santonian. Abandoned canyons in some regions of the offshore Benin were later filled with shale. Therefore, the petroleum systems of this transform margin, which was affected by post-breakup transpression, relies heavily on early syn-drift reservoir rocks.

Similar early drift vector changes, potentially responsible for transpression, described as occurring as late as the Turonian–early Campanian, were observed by numerous workers in the region between the Romanche Transform and the boundary between the North and South Gabon Basins, and its South American conjugate margin (Rodrigues et al., 1984; Seiglie and Baker, 1984; Figueiredo, 1985; Haxby, 1985; Zalán et al., 1985; Aguiar et al., 1986; Klitgord and Schouten, 1986; Benkhelil, 1988; Fairhead, 1988a, 1988b; Teisserenc and Villemin, 1990; De Azevedo, 1991; Fairhead and Binks, 1991; Binks and Fairhead, 1992; Mascle et al., 1997; Dailly, 2000; Nemčok and Rosendahl, 2006; Nemčok et al., 2012b).

With respect to the flank uplift at extensional margins, numerical modeling, described in Chapter 5 (Huismans and Beaumont, 2005; Figure 5.18), indicates that a lithospheric multilayer characterized by the strongest possible crust and strongest lithospheric mantle supports the largest flexural stresses controlling the largest flank uplift. At this point, it is interesting to note that it is not just erosion that lowers the altitude of the rift-flank mountains in time and causes a later retreat of their crests away from the rift axis (e.g., Mohriak et al., 2008 and references therein). The evolution of the rift system itself also lowers this altitude.

The fact that post-breakup flexural behavior of extensional continental margins is affected to a large extent by the geometry of unloading is highlighted by the following. When one observes the rift flank uplift in modeled pure rift (Figure 5.18; Huismans and Beaumont, 2005), it is always the footwall of the upward-convex shallow-dipping fault, which is characterized by the isostatic rebound larger than the rebound at the opposite side of the rift zone. This asymmetry is most obvious in the strong crust case, which has distinct flexural behavior, and it progressively pales toward depressed flanks in the weak crust case, which is less capable of any distinct flexural behavior.

In this respect, it would be interesting to find a long stretch of extensional margins, which are characterized by a transition from the magma-poor to the magma-rich type, such as the Espírito Santo–Campos–Santos margins of Brazil (see Zalán et al., 2011), to see a systematic variation in their flexural behavior. Looking at their differences in direction from a magma-poor breakup at the Espírito Santo margin toward a magma-rich breakup to the south of the Santos margin allows us to make the following observations. While the width of the proto-oceanic corridor remains roughly the same, the width of the stretched continental margin becomes wider toward the south. This width starts with 74 km in the Espírito Santo segment, follows with 78 km in the Campos segment, then 276 km in the northern Santos segment, and ends with 510 km in the southern Santos segment (Figures 2.13a–d). Furthermore, among those four margins, only the most distal southern Santos margin contains volcanic rocks. The southern Santos margin also contains a system of alternating thin and thick bands of stretched crust, indicating a system of several competing necking zones, which should translate into distributed unloading driven by hanging wall removal as opposed to more localized unloading characterizing the remaining three margins.

When we now look at lithostratigraphic columns of respective margins (see Pereira et al., 1986; Pereira, 1990; Pereira and Macedo, 1990; Cainelli and Mohriak, 1999; ANP Web site, 2013) in search for the timing of the flank uplift and its erosion indicated by the occurrence of major prograding wedge, we can make the following observations. A continental breakup and the onset of prominent proximal sandstone deposition took place at 117 Ma (mid-Aptian) and 115.5 Ma (early Albian) at the Espírito Santo margin. A continental breakup and proximal sandstone deposition took place at 122.5 (early Aptian) and 105 Ma (mid-Albian) at the Campos margin. These two events happened at 119.2 Ma (early-middle Aptian) and 86 Ma (mid-Coniacian) at the Santos margin. This indicates that the coldest margin, the Espírito Santo margin, experienced flank uplift after 1.5 Ma from the continental breakup. It took 17.5 Ma for the intermediate Campos margin to undergo flank uplift after continental breakup. Finally, the warmest margin, the Santos margin, waited 33.2 Ma to undergo flank uplift after the continental breakup.

This, perhaps, indicates that it takes warmer continental margins a certain time to cool down to gain stronger rheology capable of flexural response to their unloading. Regardless of whether this is true, this certainly indicates rather different timing of the flank uplift of these three margin segments, which translates into different timing of margin flank exhumation, which feeds the earliest reservoir rock deposition after the continental breakup. This flank uplift timing difference between colder, intermediate, and warmer margin segments is further highlighted by the different onset of the major basinal erosion event that follows the deposition of such sandstone-rich formation at the proximal margin. The event took place during the Cenomanian, Campanian, and Paleocene at the Espírito Santo, Campos, and Santos margins, respectively.

The interesting spoon-shaped geometry of the Santos margin could have further affected the timing of the flank uplift at this segment, as discussed by Mohriak and colleagues (2008), who argue that the effect of erosion, deposition, and even gravity gliding can be implied from the data on the Santos Basin, southeastern Brazil. Here the apatite fission track data from the exhumed rocks of the flank uplift indicate the ages of 60–90 Ma on the coastal plain, which increase landward to ages of above 300 Ma in the continental hinterland (Gallagher et al., 1994; Figure 4.12a, c). The coastal region contains the Serra do Mar Mountains, located about 56–162 km landward from the Cretaceous hinge zone (Figure 4.12b). Their length is approximately the same as the width of the offshore Santos Basin, located in front of it, characterized by spoon-shaped geometry converging in the southeastern direction.

The post-breakup lithostratigraphy of the basin fill starts with the Aptian Ariri Formation represented by shallow marine evaporitic sediments (Figure 7.23). The section follows with the carbonate Albian Guarujá Formation, the Albian Florianopolis Formation, and the Albian regressive wedge of the Itanhaém Formation buried underneath the Upper Cretaceous shaly Itajaí Formation and the sandstone Ilhabela Member. Senonian-Paleocene is characterized by the large regressive wedge of the Santos Formation, which represents a prograding sedimentary wedge in the basin adjacent to the uplifting margin, which records this uplift by erosional truncation of topsets of the sedimentary wedge (see Figure 4.1b), while the Serra do Mar Mountains can be interpreted as an erosionally retreated uplifted shoulder.

The basin dip southeast-ward converging direction is also a direction of the areally-extensive gravity gliding that affects most of the Santos Basin, taking place on top of the Aptian Ariri evaporitic formation. The architecture of the gravity glides is affected by the overall spoon-shaped geometry of the basin, as can be seen from the thickest gravity glides running down to the southeast along the axis of the basin and the thinnest ones located along its northeastern and southwestern sides (Mohriak et al., 2008).

A separation of the evaporite removal events from the continental slope combined with the evaporite accumulation in the deep basin for time periods coeval with the deposition of main seismically mappable sequences allows for the following observations (Figure 7.24). The gravity gliding and removal of evaporites from the margin and their accumulation in the deep basin started relatively early, in the Albian, after a sufficiently thick carbonate layer was deposited on top of the Aptian evaporitic Ariri Formation to trigger gravitational instability. However, it was the Cenomanian when salt removal from the margin and addition to the basin started to become significant. The culmination of this mechanism took part during the Campanian-Thanetian, followed by considerable slowdown starting in the Eocene. The acceleration timing was coeval with the progradation of the massive Santos prograding wedge, which controlled the development of significant counterregional faults in front of its thickest portion (Mohriak et al., 2008).

Numerical modeling of the isostatic rebound responsible for the marginal uplift (Gallagher et al., 1994) indicates that the continental margin next to the Santos Basin underwent a protracted denudation since the continental breakup. The results indicate more than 3 km of exhumation on the coastal plain and as little as 1 km farther in the hinterland (Figure 7.25). When we plot the apatite fission track ages with the hinterland-ward distance from the coast, a diagram similar to that of Figure 7.17c can be obtained. Analogically to the interpretation of Figure 7.17, the break in the slope of the curve in Figure 7.25, thus, provides the age of onset of rapid exhumation controlled by faulting, yielding about the start of the Senonian, which is also the onset time of both the Santos wedge deposition and accelerated salt removal/accumulation.

It could be useful to now compare the magnitude of flank uplift and its exhumation at magma-poor extensional margins described so far with that of magma-rich extensional margins. The example comes from the Walvis Margin, Namibia (Henk, 2004, pers. comm.). It is a magma-rich margin located further south from the aforementioned transitional Espírito Santo–Campos–Santos margins and their African conjugates. After its addition to the south, the Espírito Santo-Walvis margins represent a complete transition from magma-poor margins in the north to magma-rich margins in the south.

The onset of continental rifting along the Walvis Margin is not well defined, as to date no wells have penetrated the syn-rift deposits. Rifting is assumed to start in the Late Jurassic, sometime between 160 and 140 Ma. An intense magmatic activity related to the Tristan da Cunha hot spot caused the magma-rich character of this margin. Etendeka and Paraná volcanic rocks in Namibia and Brazil, respectively, are associated with this hot spot. They were emplaced at the rift–drift transition in a short time span, between 132 and 130 Ma.

Nearshore outcrops located in the Albin Ridge show a landward-tilted fault block, developed by rifting. It stands up to 200 m above the coastal plain. The ridge top is formed by 132–130 Ma-old Etendeka effusive rocks. The coastal plain is formed by the late Proterozoic-Cambrian metasediments and plutonic rocks, which in large areas form the pre-rift basement.

Seismic and well data from offshore image the two wedges of seaward-dipping reflectors, an inner wedge close to the coast and an outer wedge at the transition to the normal oceanic crust (Henk, 2004, pers. comm.; Thompson, 2009). The inner wedge is composed of subaerial lava flows interbedded with syn-rift sediments. Following the breakup unconformity, two megasequences of Cretaceous and Tertiary to recent sediments form the post-rift sediments. Maximum thickness of the spindle-shaped deposit is 5 km.

Seismic images of the post-rift succession document a truncation of older strata during the Late Cretaceous–Early Paleocene at distal margin close to the transition to the oceanic crust and the onlap pattern

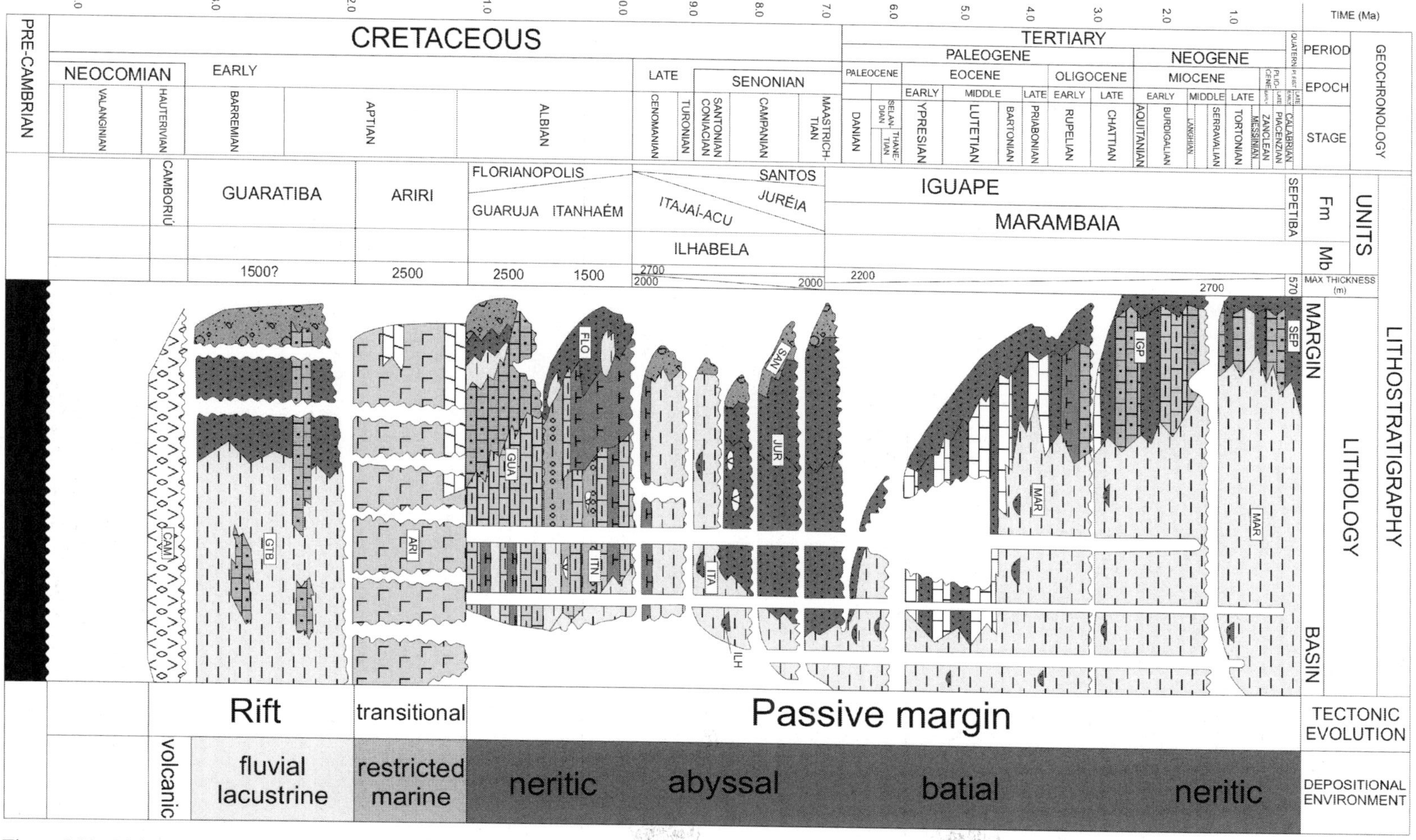

Figure 7.23. Lithostratigraphic column of the Santos Basin (modified from Pereira et al., 1986; Pereira, 1990; Pereira and Macedo, 1990; Cainelli and Mohriak, 1999). CAM – Camboriú Formation with basalts, GTB – alluvial and lacustrine Guaratiba Formation, ARI – shallow marine evaporitic Ariri Formation, GUA – carbonate Guarujá Formation, FLO – Florianopolis Formation, ITN – regressive wedge of the Itanhaém Formation, ITA – shaly Itajaí Formation, ILH – sandstone Ilhabela Member, SAN – regressive wedge of the Santos Formation, JUR – basinal shale of the Juréia Formation, MAR – regressive wedges of the Marambaia Formation, IGP – Iguape Formation, SEP – Sepetiba Formation. Note that significant Santos prograding wedge, reacting to the accelerated sediment transport from uplifting margin, developed during Senonian-Paleocene.

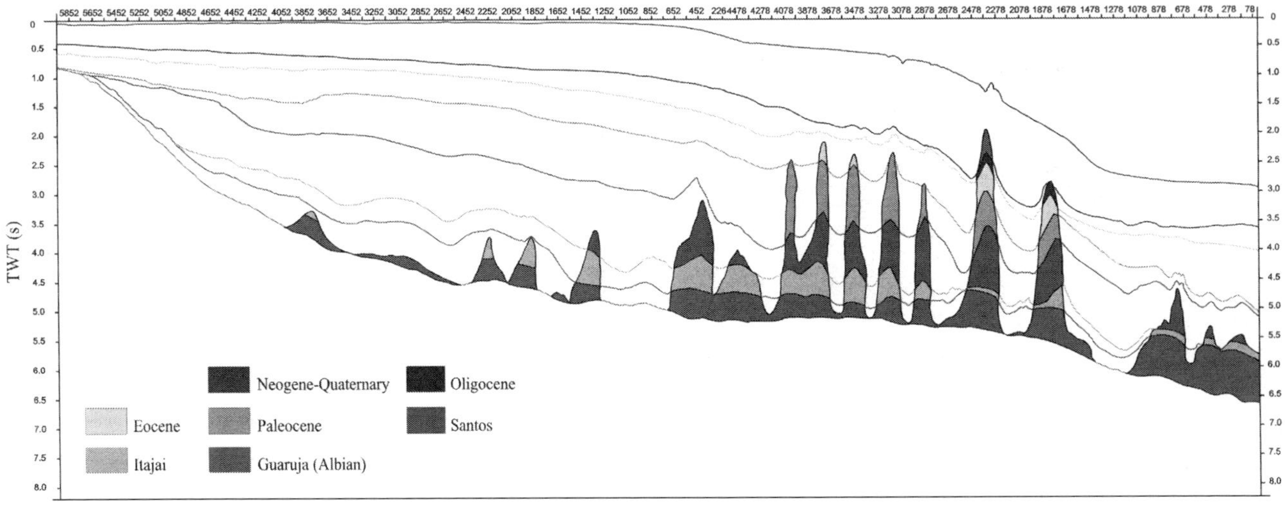

Figure 7.24. Stepwise history of the evaporite layer removal from the slope and accumulation in the deep basin. Note that the removal/accumulation went on rather early during the Albian, but at a rather slow pace, and that the significant removal/accumulation started in Cenomanian and culminated during Campanian-Thanetian.

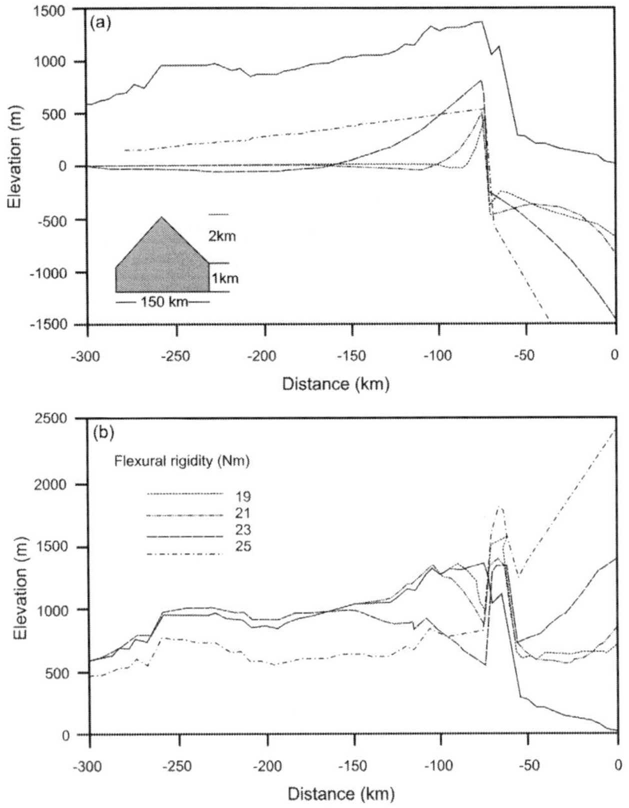

Figure 7.25. Calculated rebound (a) and initial elevation (b) for the polygonal load 150 km wide with a maximum of 3 km denudation, along a profile through the Santos Basin passive margin, southeastern Brazil (Gallagher et al., 1994). The left boundary of the load is approximately coincident with the escarpment at -75 km. Different model curves are for different flexural rigidities (in Nm). The solid line is the averaged topographic profile from the previous figure.

onto the pre-rift basement. The latter indicates substantial erosion of the seaward-dipping reflector sequence during the early post-rift phases, but from Cenomanian times onward this hinge zone was covered by sediments and thus protected from erosion.

The youngest preserved reference level for the reconstruction of the eroded strata onshore provides a minimum estimate of erosion of about 1,600 m. It can be increased by 600 m, representing the thickness of preserved volcanic rocks in the Etendeka and Paraná Plateaus. Higher amounts of overburden have been suggested from an interpretation of apatite fission track data (Gallagher and Brown, 1999; Brown et al., 2000; Raab et al., 2001). The erosion close to the shore was on the order of 3.6 km, decreasing toward the higher elevations further inland. Such erosional cut makes this margin comparable to the aforementioned Santos margin. Lacking a distinct flexural component of the flank uplift, it must have been controlled by transient components related to thermal effects associated with an elevated thermal regime of the lithosphere due to hot spot activity. A permanent component due to magmatic underplating is also possible.

Passive margin basins with gravity gliding-controlled stretching

An important difference between extensional and transform margins is their affinity for gravity gliding. A comparison of transform and extensional margin segments of the Equatorial Atlantic (see Figure 6.13a–b), made by Nemčok and colleagues (2012b; Figure 7.26) indicates that extensional margins frequently undergo gravity gliding and it takes place relatively soon after

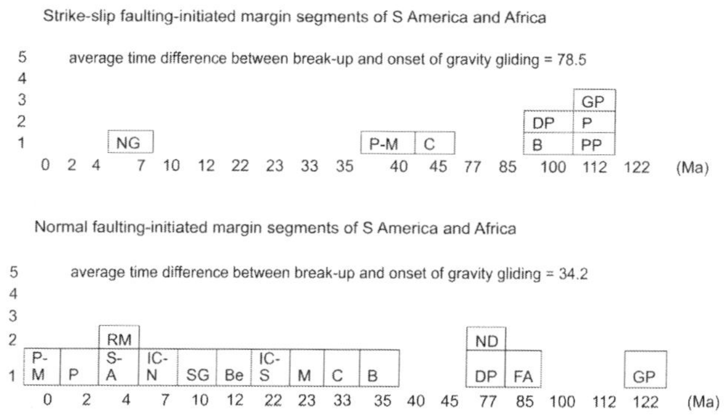

Figure 7.26. Time difference between continental break-up timing and onset of gravity gliding in the respective margin segment in the study area split to strike-slip faulting- and normal faulting-initiated segments (Nemčok et al., 2012b). Basins: GP – Guinea Plateau, M – Monrovia Basin, IC-N and IC-S – Ivory Coast northern and southern portions, ND – Niger Delta, BT – Benue Trough, RM – Rio Muni Basin, DP – Demerara Plateau, FA – Foz do Amazonas Basin, P-M – Para-Maranhão Basin, B – Barreirinhas Basin, C – Ceará Basin, P – Ptiguar Basin, PP – Pernambuco Paraiba Basin, S-A – Sergipe-Alagoas Basin, NG – North Gabon, SG – South Gabon.

continental breakup. It also indicates that transform margins do not necessarily experience gravity gliding. If they do, it takes place rather late after breakup and it rather occurs in areas where strike-slip fault zones transition into oblique-slip or normal faults.

These two margin types have in common the capability of major rivers at both of them to develop continental slope that would undergo cycles of gravitational instability and cycles of its consolidation by gravity-gliding events.

Because of the extensive literature on the subject, it is not our intention to focus in detail on the effect of gravity gliding on passive margins and mechanisms of deposition and sediment distribution at gravity glide systems. We would just briefly discuss the effect of growth faulting on deposition. The previously discussed example comes from the Miocene Imo River field of the Niger Delta (Van Heijst et al., 2002).

The analog material experiment in Van Heist and colleagues' (2002) study simulated four sea-level change cycles, each of them showing similar timing of drainage evolution on the modeled shelf-margin delta. All modeled highstand episodes produced delta progradation. During the subsequent sea-level fall, several canyon incisions developed in the exposed delta front, eroding toward the highstand delta apex.

The deposition of the transgressive systems tract followed the retreating shoreline. At late rise to highstand, the retrogradation changed into aggradation, resulting in the restoration of the incised highstand delta on the footwall.

Despite the numerous geological differences, the model and seismic/well data from the Imo River field share common features, especially in the delta and fluvial architecture in both footwall and hanging wall blocks. For example, the model shows incised valleys on the footwall block of the growth fault, which transported sediment to the hanging wall lowstand fan at times of the sea level fall and lowstands. The seismic imaging in the field indicates the presence of similar valleys in the footwall in the middle-upper Miocene section (Van Heijst et al., 2002). The model also manages to reproduce the hanging wall preservation of all tracts, with coarse-grained deposits formed during the eustatic fall, lowstand, and early rise of the sea level. It also manages to reproduce the footwall succession, which consists of stacked late transgressive and highstand sediments, which were repeatedly subject to erosion and sediment bypass as the shoreline moved seaward from the growth fault during lowstands.

An important outcome of the modeling is the recognition of diachronous erosion and deposition along the stream profile, as a consequence of the sea level change (see also Van Heijst and Postma, 2001). The stream profile at a maximum sea-level lowstand is an analog to the type 1 unconformity (sensu Vail and Todd, 1981), characterized by subaerial exposure and associated erosion from a down-cutting stream. During the subsequent rise, a deposition develops strongly diachronously, commencing on the shelf and delta plain, while erosion continues in a landward direction. This diachroneity is further enhanced in growth fault scenarios (Figure 7.27; Table 7.1). For example, the upper delta and fluvial domain incisions continue during early sea-level rise and coincide with the deposition of the transgressive systems track in a hanging wall. Thus, the upstream part of the unconformity on the footwall does not record the lowstand hiatus but originates from early rise and becomes progressively younger up the dip.

According to the interplay model of eustatic sea-level change and local subsidence (Posamentier and Allen, 1993), the interplay determines the type of unconformity and lowstand basin physiography, which would, in turn, control the morphology of the following transgressive systems tract. The interplay in the Imo River field resulted in three type 1 unconformities and one type 2 unconformity in the hanging wall and four type 1 unconformities in the footwall.

The type 1 unconformity is characterized by few cross-shelf bypass valleys that act as point sources for lowstand fans (Figure 7.28a). These valleys transfer significant volumes of freshly eroded footwall sediment to the lowstand fans on the hanging wall, in addition to the sediment the fluvial valley supplies. The type unconformity 2 is characterized by little or no incision of river valleys into

Table 7.1. Systems tract development variation in the analog material model for growth fault/eustatic sea level/deposition interaction (Van Heijst et al., 2002).

Sequence stratigraphic unit	Footwall*	Hanging wall**
Highstand systems tract	highstand progradation, highstand delta, fluvial, and deltaic deposits	condensed section (incidental: toe of highstand delta)
Falling stage systems tract	type 1: erosion/bypass by incised valleys on delta plain	type 1 sequence: lowstand delta progradational wedge
	type 2: bypass by braided fluvial system on whole delta plain	type 2 sequence: shelf margin delta systems tract
Lowstand systems tract	bypass	thin progradational to aggradational wedge
Early transgressive systems tract	bypass	backstepping lobes that cover lowstand wedge and valleys on lower delta plain
Middle-late transgressive systems tract	incised valley fill with backstepping geometries	condensed section

*accommodation creation by regional sea level rise only
**accommodationcreation by regional sea level rise and local subsidence

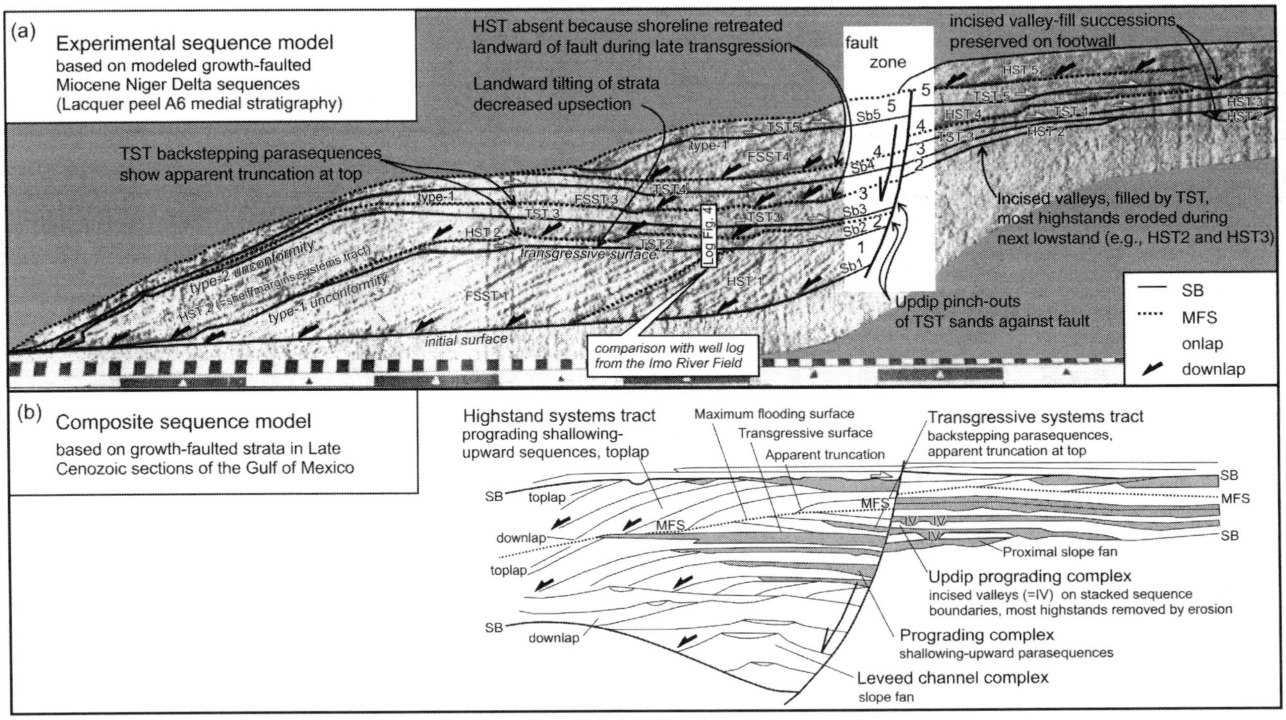

Figure 7.27. (a) Sequence model based on the across-fault correlation of the medial stratigraphy of the completed model (Van Heijst et al., 2002). The strata close to the fault are not representative of natural settings and need to be excluded. (b) Conceptual systems tract model for Cenozoic fourth-order sequences along the growth of the Gulf of Mexico (Van Heijst et al., 2002).

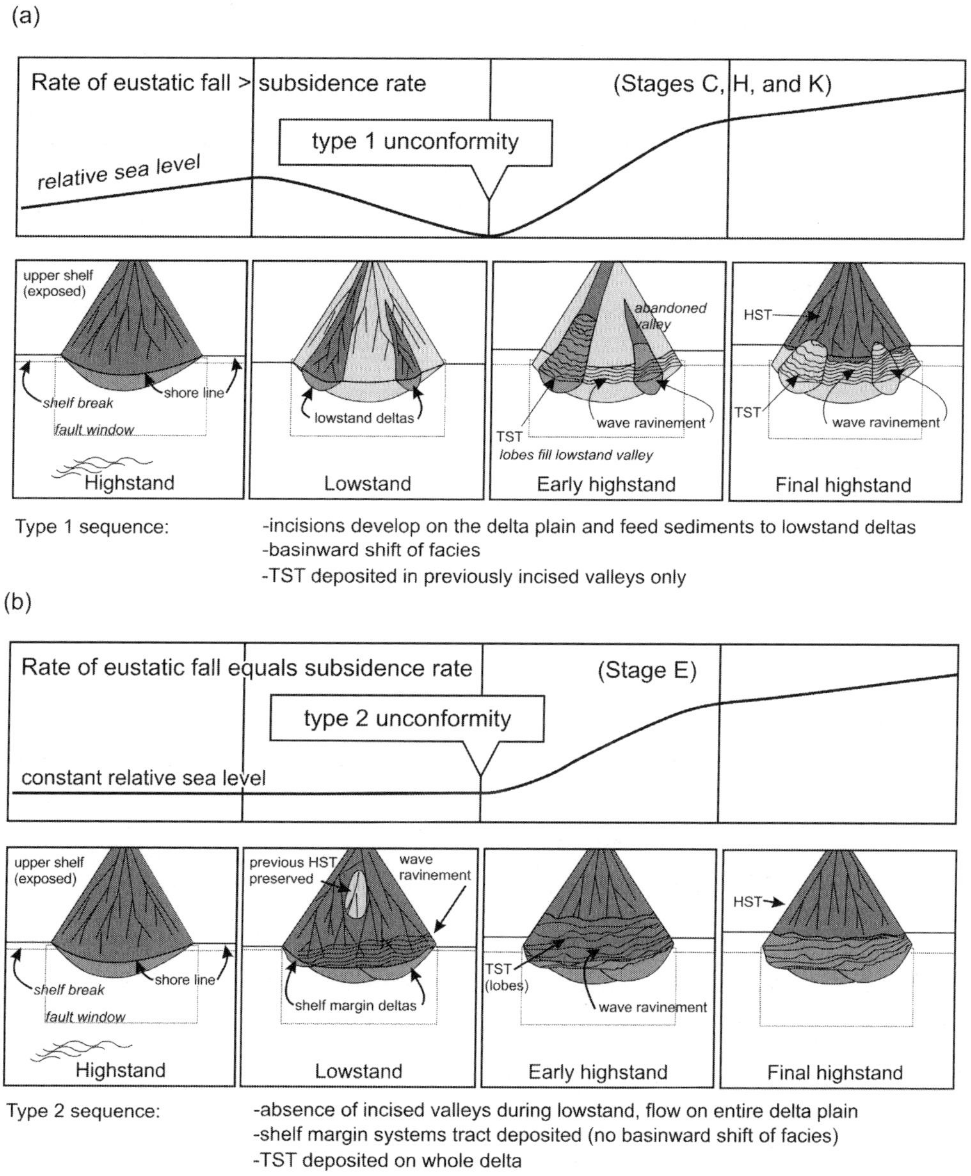

Figure 7.28. Main differences in delta evolution from analog material experiments on the interaction of tectonics, eustatic sea-level changes, and deposition around a growth fault on a passive margin (Van Heijst et al., 2002). (a) Type 1 sequence results in valley incision migrating progressively upslope until one of the valleys reaches the fluvial valley and captures all the discharge. The subsequent transgressive systems tract (TST) is only present in this main valley. (b) Type 2 sequence develops a braided system on the entire delta plain and, consequently, the subsequent TST covers the entire delta-shelf system.

the delta plain and shelf, so that the feeder system of the delta can potentially behave as a line source (Figure 7.28b).

Mantle plume-influenced basins

The last category of tectonic/deposition/erosion interactions to discuss in this chapter is basins influenced by plumes because of their specific deposition/erosion characteristics.

A hot, ascending mantle plume below the lithosphere controls the formation of a broad swell above the flattened plume head (e.g., Griffiths and Campbell, 1991; He et al., 2003). It could reach up to 2,000–2,500 km in diameter (see also Sengör, 2001; Nemčok et al., 2007). The mechanisms involved in uplift include dynamic, permanent, and isostatic uplifts.

The dynamic uplift takes place when an abnormally hot mantle is emplaced beneath the lithosphere. It

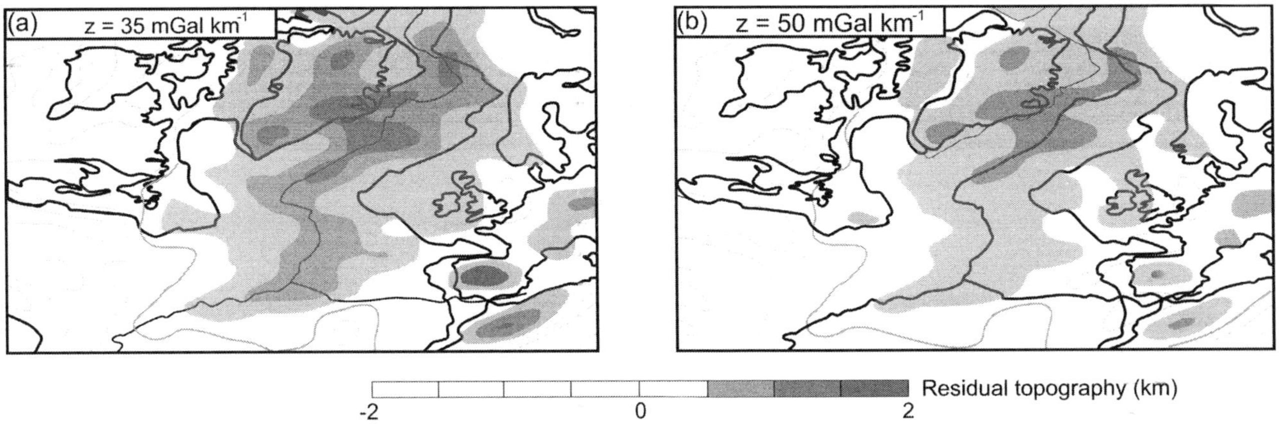

Figure 7.29. Present-day dynamic support in the North Atlantic region, calculated by dividing the long wavelength free-air gravity anomaly by a constant admittance of 35 (a) and 50 (b) mGal km^{-1} (Jones et al., 2002). K – Kangerlussuaq, S – Scoresby Sund, bold lines – continent–ocean boundary.

includes effects such as the plume impact, buoyancy caused by plume material ponding underneath lithosphere (Guillou-Frottier et al., 2007), and thermo-mechanical erosion of the lithospheric base (Davies, 1994; d'Acremont et al., 2003). This uplift has a transient character. It disappears when the thermal anomaly dissipates (Nadin et al., 1995, 1997; Jones et al., 2002). The Iceland plume in the North Atlantic region serves as a typical example (Figure 7.29).

The permanent uplift is controlled by the crustal thickening in places where plume-related thermal anomalies result in melting the products, which are injected either into or underneath the crust. The magmatic underplating results in isostatic uplift of the crust (Brodie and White, 1994, 1995; Clift, 1997; White and Lovell, 1997; Clift and Turner, 1998). The example of permanent uplift, discussed in detail a bit later, comes from the Paleocene evolution history of Britain and Ireland and from the regions flanking the continental breakup trajectory at the Paleocene–Eocene boundary (Jones et al., 2002).

The isostatic thermal uplift takes place as controlled by heating and associated decrease in lithospheric density due to ascending hot plume material. It has a transient character and dissipates with the plume-head disappearance (Sleep, 1990; Clift and Turner, 1998).

The literature provides several direct estimates of dynamic support, done for the Afar Dome (Sengör, 2001), the Iceland plume (Jones et al., 2002), the Emeishan flood basalts in southwest China (He at al., 2003), and basalts associated with 85°E Ridge, offshore East India (Nemčok et al., 2007, 2012c).

The estimate can be made from the erosional thickness reduction of a regionally extensive sedimentary unit. For example, the Maokou Formation underlying the Emeishan flood basalts shows a progressive thinning toward the center of the flood basalt province (He et al., 2003; Figure 7.30). Only the lowermost of the three biostratigraphic zones of this formation is present over the dome crest. Biostratigraphic and erosional analysis indicates that the rapid crustal uplift here took place in less than 3 Ma. The uplift was active for 1.0–2.5 Ma. The erosion associated with uplift removed 300 m of Maokou limestone from the dome crest. Based on maximum thickness of conglomerate layers and underlying basalts, and possible maximum thickness of the eroded strata, which is about 500 m, the total uplift was estimated at more than 1,000 m (He et al., 2003).

The accelerated Paleocene tectonic subsidence in the North Sea accompanied by uplift of surrounding areas has been explained by vertical motions associated with crustal breakup in the future North Atlantic and activity of the Iceland plume (Faleide et al., 2002). During the same time period, two major magmatic events occurred at 59 and 55 Ma (White and Lovell, 1997; Ritchie et al., 1999).

The uplift in North Sea surroundings and magmatic events were synchronous with deposition of Paleocene sandstone reservoirs along the axis of the Faeroe-Shetland Basin (White and Lovell, 1997; Naylor et al., 1999). Similar synchronicity has been observed between uplifts and sandy sediments of the southern North Sea (Knox, 1996).

The data show that the North Atlantic is unusually shallow around Iceland, where anomalous topography culminates, rising up to about 2 km above the sea, representing a 4.5 km difference from the worldwide average ridge elevation (Jones et al., 2002). The plume-related uplift, estimated by:

1) subtracting the anomalous topography from the depth of the oceanic crust, which varies with age

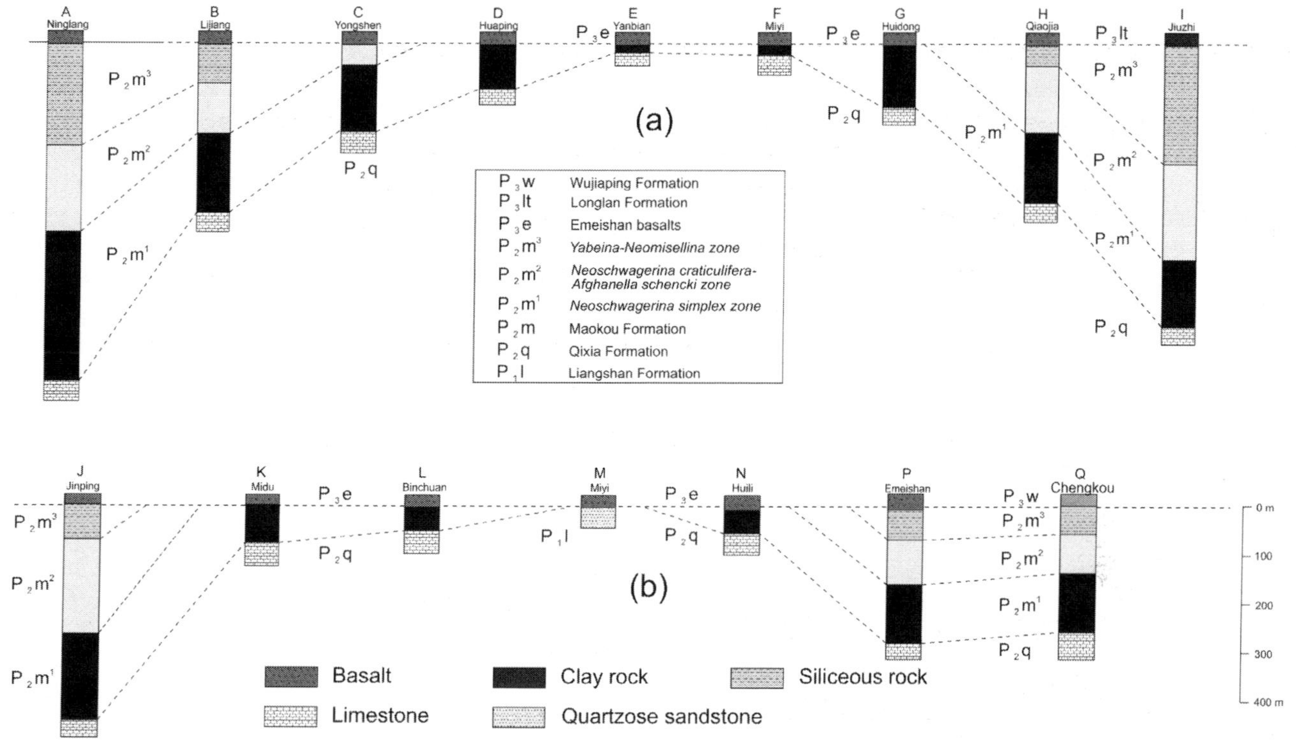

Figure 7.30. Thickness distribution of the Maokou Formation of the Emeishan igneous province along east–west (a) and north–south (b) transects (He et al., 2003). Numbers at Maokou sections indicate thickness in meters.

in a systematic manner in a direction away from the plume; and

2) linking the dynamic support with a long-wavelength, free-air gravity anomaly,

provides the following estimates.

Because the circum-Iceland region has a 7–10 km-thick crust generated by an anomalously hot mantle (White, 1997), the thickness difference of about 3 km from the standard oceanic crust controls a permanent uplift of 0–0.5 km, which is much smaller than the uplift necessary for the development of the anomalous topography. Therefore, a portion of the difference has to be attributed to a dynamic uplift, estimated from topography calculations to represent about 1.5–2 km (Jones et al., 2002). Yet another share in the difference is controlled by the active seafloor-spreading ridge itself. This can be seen when one compares the highly elevated active Kolbeinsey Ridge north of Iceland with the less elevated extinct Aegir Ridge south of Iceland.

Gravity data-based estimations provide a value of the peak dynamic support of 1.8 km for Iceland (Jones et al., 2002). Here, the active spreading controls the long-wavelength gravity field. The dynamic support is centered on the ridge, having values at the active ridge segment larger than those at the extinct one. The surrounding continental margins undergo dynamic support of about 0.5–1 km. Although the results derived by topography and gravity data-based calculations are not exactly the same, they are in reasonable agreement.

Determining the Paleocene shape of the Iceland plume is much more difficult. Independent data sets used for the estimate include (Jones et al., 2002):

1) geochemical evidence from the British magmatic province about Paleozoic magmatic underplating underneath the crust;
2) vitrinite reflectance, apatite fission tracks, and other data on Paleocene uplift in Britain and Ireland; and
3) seismically and well-derived sediment thickness data from surrounding basins.

The geochemical data from flood basalts document the fractional crystallization of their liquid mass, requiring the residuum left somewhere at depth (Thompson, 1974). The depth of the underplated body, estimated using geobarometry on basalts and their phenocrysts, indicates a depth of about 30 km (Thompson, 1974; Brodie and White, 1995). This can be explained by basaltic melt less dense than the surrounding mantle rocks ascending to basal crustal levels where the densities are similar. When such an underplated body is added, isostatic balance results in permanent uplift.

Mentioned geochemical and geobarometric indications are in agreement with seismic experiments in the area, which image high-velocity bodies located along the base of the continental crust (Barton and White, 1997a).

The second data set, missing portions of the vitrinite reflectance profiles combined with apatite fission track data from uplifted provinces, similar to the one shown in Figure 4.14 from an unrelated area, indicates a Paleocene–earliest Eocene denudation of 1–2.5 km in the center of the uplifted region in northwest England.

The third data set, data on coeval deposition rates and sediment volumes in the uplift-surrounding regions, indicates that the sediment flux into surrounding basins progressively increased during the Paleocene, then reached a maximum at about 59–58 Ma, and then decreased during the Eocene (Reynolds, 1994; Clarke, 2002). Deposition data were directly compared to denudation estimates because the time lag between the acceleration time of denudation and time of the increased sediment flux in the offshore is less than 100 ka (Reading, 1991; Burgess and Hovius, 1998).

The temporal and spatial coincidence works for deposition, denudation, and magmatic activity, all peaking at 59–58 Ma, suggesting that they were all associated with the injection of the hot vertical asthenosphere sheet beneath the Faeroes-Irish Sea-Lundy zone, controlling the permanent uplift (Jones et al., 2002).

Apart from the permanent uplift, the dynamic uplift is indicated by the occurrence of a major regression-transgression cycle recorded in all basins of the region (Nadin et al., 1997; Jones et al., 2001). The maximum regression corresponds to the lowermost Eocene Flett Formation (Milton et al., 1990; Ebdon et al., 1995).

8 Fluid flow systems associated with oceanic hot spots, oceanic transforms, continental transforms, and rifts

Introduction

The understanding of how fluid flow associated with hot-spot- or hot-line-controlled volcanic chains, oceanic fracture zones, and large continental faults affects local thermal regimes and diagenetic histories requires an understanding of fluid flow mechanisms operating in these settings. After identification of flow mechanisms, we need to understand:

1) their importance in control of the local fluid regime;
2) fluid migration pathways;
3) typical fluid reservoirs; and
4) controlling factors of various flow mechanisms, which determine where and when some mechanisms dominate over others.

After finishing this homework on fluid flow mechanisms and their controlling factors, we can build local thermal models and distinguish which of them are purely conductive, and which of them are partially controlled or dominated by the heat flow transported by fluids.

The goal of this chapter is to discuss the fluid flow mechanisms, fluid sources, fluid migration pathways, and flow-controlling factors in geological settings characterized by the occurrence of:

1) hot-spot and hot-line volcanic chains;
2) oceanic fracture zones; and
3) large continental faults.

The discussion is done using current literature drawing from field studies, and numerical and analog material models. It tries to develop a foundation for the assessment of the effect of different types of fluid flow mechanisms on local thermal regimes and diagenetic histories.

Fluid flow associated with oceanic hot spots

As an oceanic crust moves over a hot spot, ascending heat and fluids from the mantle bulge the oceanic lithosphere, forming a mid-plate swell. Basaltic eruptions build marine volcanoes. The load of each volcano warps down the elastic oceanic lithosphere, forming a moat-like depression with an outer bulge seaward of the moat forming in response to plate loading, called the arch (Turcotte and Schubert, 1982). Examples come from the Hawaiian, Canary, and Marquesas Islands (Figure 8.1). Despite the similar processes of formation of these three hot-spot volcanic systems, there are many geometric differences between them.

During their growth, volcanoes experience continuous degradation by large slumps, debris flows, and relaxation and lateral creep of their flanks. Plate motions eventually carry each such volcano across the hot spot and a new volcano forms in its wake.

In Hawaii, the axes of the flexural moat and arch are about 140 km and 200–250 km from the center of the island chain (Figure 8.2). The depth to the seafloor is roughly 5 km at the moat axis, and about 4.5 km at the arch axis (Moore et al., 1994). The Marquesas Islands of the South Pacific have no bathymetric expression of a moat or an arch. The surrounding "archipelagic apron" (sensu Menard, 1956) slopes gently away from the island, being the result of overfilling of the moat by extensive mass wasting (Filmer et al., 1994). Some of the youngest Canary Islands exhibit no moat, arch, or apron, perhaps because of their small size and young age (Gee et al., 2001).

The sedimentary fill of the moat contains three units (Rees et al., 1993; Leslie et al., 2002). The basal pelagic unit, which accumulated on the seafloor prior to eruption, overlies the oceanic crust both at the bottom of the moat and below the volcanic island (Rees et al., 1993). The pelagic unit can have variable thickness depending on the age of the oceanic crust and the deposition rates. It is less than 200 m thick in the Hawaiian moat (Winterer, 1989) and about 3 km thick surrounding Gran Canaria of the Canary Islands (Ye

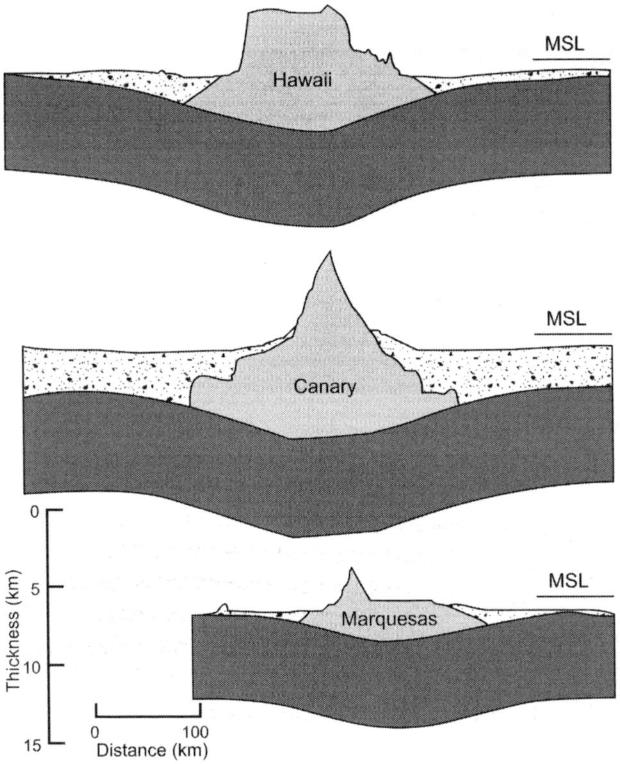

Figure 8.1. Cross-sections of the Hawaiian, Canary, and Marquesas Islands (Watts et al., 1997). Volcanic islands have different sizes, amounts of sediment, and depth below sea level. MSL is the mean sea level. The dark gray indicates flexed oceanic crust, the light gray represents volcanic load, and the dotted area is sediment infill.

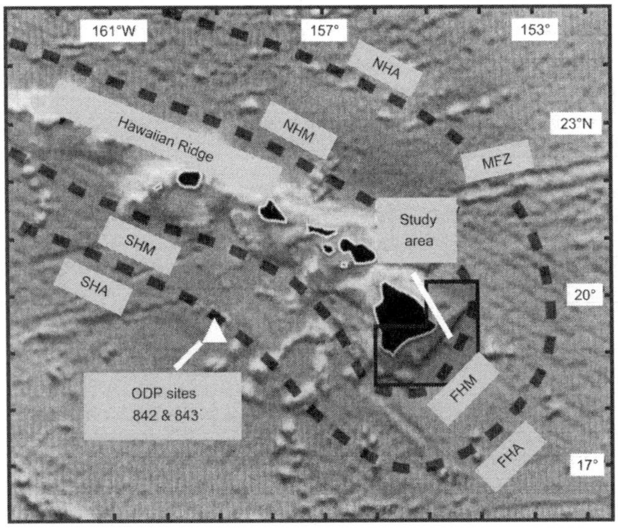

Figure 8.2. Shaded bathymetric map of the Hawaiian Islands (Leslie et al., 2002). NHA – North Hawaiian Arch, FHA – Front Hawaiian Arch, SHA – South Hawaiian Arch, NHM – North Hawaiian Moat, FHM – Front Hawaiian Moat, SHM – South Hawaiian Moat.

et al., 1999). The second unit consists of a series of volcanic slope failures, the most voluminous of which are slumps and debris flows. These mass wasting events can be several thousand cubic kilometers in volume and travel distances larger than 200 km (Moore et al., 1994). The majority of these events likely occur during the late stages of shield building when pore pressures are highest (Moore et al., 1994). If multiple volcanic episodes occur, such as at the Gran Canaria and the Marquesas Islands, there will be multiple accompanying periods of mass wasting (Wolfe et al., 1994; Funck and Schmincke, 1998). The third unit is a blanketing turbidite and pelagic cover that fills depressions in rough topography after volcanism has ceased (Rees et al., 1993; Leslie et al., 2002).

Two-dimensional finite-element models were constructed to evaluate how the volcanic geometry and geologic history of the Hawaiian, Canary, and Marquesas Islands affect the interplay of buoyancy and compaction-driven flow patterns (Figure 8.3) in the volcanic edifice and sedimentary apron (Christiansen and Garven, 2004a, 2004b).

Seafloor bathymetry at a constant temperature produces nonhorizontal isotherms in the subsurface, resulting in buoyancy-driven convection, with recharge at topographic lows and discharge at highs (Phillips, 1991). Compaction within the sedimentary apron, surrounding and beneath the volcanic edifice, results in fluid expulsion. This type of focused flow can alter buoyancy-driven patterns (Christiansen and Garven, 2003).

Hawaii, being the largest of the three sets of islands, has the largest amount of lithospheric flexure, the largest volume of mass wasting, and the greatest bathymetry (Figure 8.1). The fluid modeling shows that the Darcy velocities and excess pressure here are the greatest of the three localities modeled (Christiansen and Garven, 2004b).

After 0.48 Ma of volcano building (Figure 8.4a), compaction of the pelagic sediments expels fluid through the permeable upper oceanic crust toward the flexural arch, and to a lesser extent through the thin, newly deposited mass wasting unit (Christiansen and Garven, 2004b). After 1.20 Ma of volcano construction (Figure 8.4b), similar compaction-driven fluid flow is still dominant. However, fluid velocities have decreased beneath the edifice as a result of compaction. The increase in lithospheric flexure has caused the first buoyancy-driven convection cells to develop beneath the volcanic edifice (Christiansen and Garven, 2004b). After volcanism termination and before deposition termination at 1.92 Ma (Figure 8.4c), the buoyancy-driven flow has a strong influence on the fluid flow regime

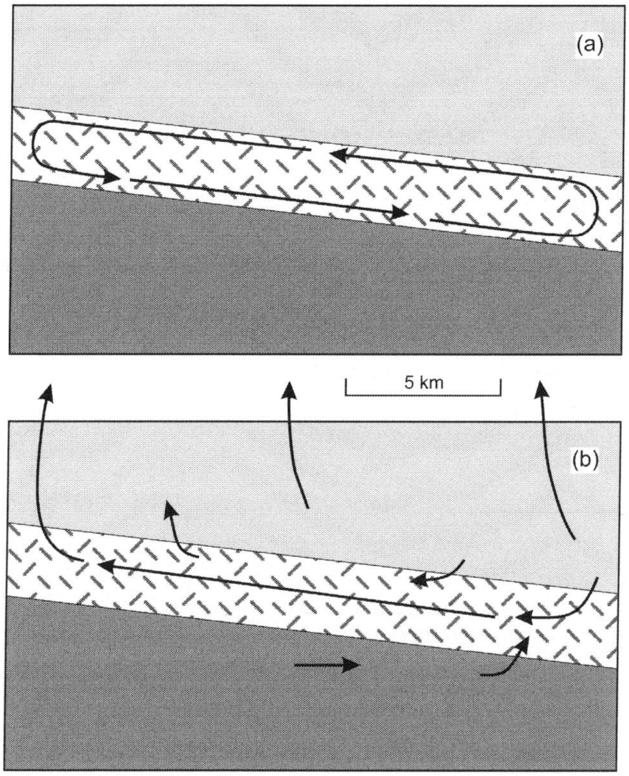

Figure 8.3. Conceptual model for coupled processes of convection and compaction when the system is joint-dominated (Christiansen and Garven, 2003). (a) With only buoyancy-driven flow, a regional convection cell forms in the permeable layer of the crust, and little fluid movement occurs in the sediment layer. (b) When there is a combination of convection-driven and compaction-driven flow, the sediment layer contributes to the fluid flow patterns. Fluid drained from the consolidation sediment is directed to the surface as well as to the permeable ocean crust. This added source of fluid to the crust layer changes the flow regime from a convection cell to a flow-through pattern. The fluids flow laterally across the basin and out to the surface at the topographic basement high.

beneath the edifice, but compaction-driven flow is still dominant in the sedimentary units beyond the weight of the volcano (Christiansen and Garven, 2004b). At 5.8 Ma after volcanism initiation and 3 Ma after volcanism termination, the fluid flow patterns have stabilized (Figure 8.4d). The majority of fluid flow occurs as buoyant convection cells beneath the edifice. Flow through the sediments at bathymetric lows has changed direction now that compaction-driven flow has halted. Now recharge occurs in the depth of the moat and discharges on the flanks of the volcanic edifice and toward the arch in buoyancy-driven flow patterns (Christiansen and Garven, 2004b).

The large amount of pre-volcanic deposition, which is about 3 km thick and which underlies the Canary Islands, contrasts with the other islands modeled (Figure 8.1). The combination of continental and pelagic sediments results in lower porosities and higher permeabilities (Giambalvo et al., 2002). When the edifice begins to build at 2.5 Ma (Figure 8.5a), fluid is focused through the upper oceanic crust toward the arch as well as up and out of the fringing sediments, as it was in Hawaii. Flow beneath the edifice is directed downward toward the oceanic crust and outward to the sedimentary apron (Christiansen and Garven, 2004b). After 7.0 Ma (Figure 8.5b), fluid flow continues in a fashion from 2.5 Ma ago, but at reduced velocities beneath the edifice because of compaction and buoyancy-driven flows having no effect (Christiansen and Garven, 2004b). By 14.0 Ma (Figure 8.5c), conditions are suitable for buoyancy-driven flow to begin beneath the volcano. Because of deposition rates by mass wasting, which are lower than those in Hawaii, the compaction-driven flow velocities in the Canary Islands are reduced to 60 percent, although they have similar orientation (Christiansen and Garven, 2004b). At 18.0 Ma after volcanism initiation, the flow patterns are in a steady state (Figure 8.5d). The flow pattern is similar to that in Hawaii. It has convection cells beneath the volcano. Its recharge occurs in the moat depocenter and discharge takes place at the flank and arch (Christiansen and Garven, 2004b).

The Marquesas Islands are much smaller than the previous two archipelagos, reducing the effects of compaction, and are formed on younger crust, reducing the amount of pre-volcanic pelagic sediment (Christiansen and Garven, 2004b; Figure 8.1). Once volcanism initiates, fluid is pushed from under the volcano toward the flanking sediments and arch as it is in the previous examples (Figure 8.6a). At 0.80 Ma (Figure 8.6b), compaction-driven flow continues to push fluids toward the arch, while the fast-building, overfilled apron directs fluids up through the sediments (Christiansen and Garven, 2004b). At 1.60 Ma, a convection cell forms beneath the edifice (Figure 8.6c). After 2.5 Ma, flow vectors are no longer transient (Figure 8.6d), and buoyancy-driven flow dominates. Recharge occurs at a distance greater than that in the previous simulations, due to the apron sloping away from the island, characterized by no moat. Discharge occurs at the flank and arch (Christiansen and Garven, 2004b).

The previously discussed hydrologic evolution was found to depend primarily on (Christiansen and Garven, 2004b):

1) edifice height;
2) deposition represented by mass wasting;
3) edifice growth rates; and
4) the amount of pre-volcanic sediment.

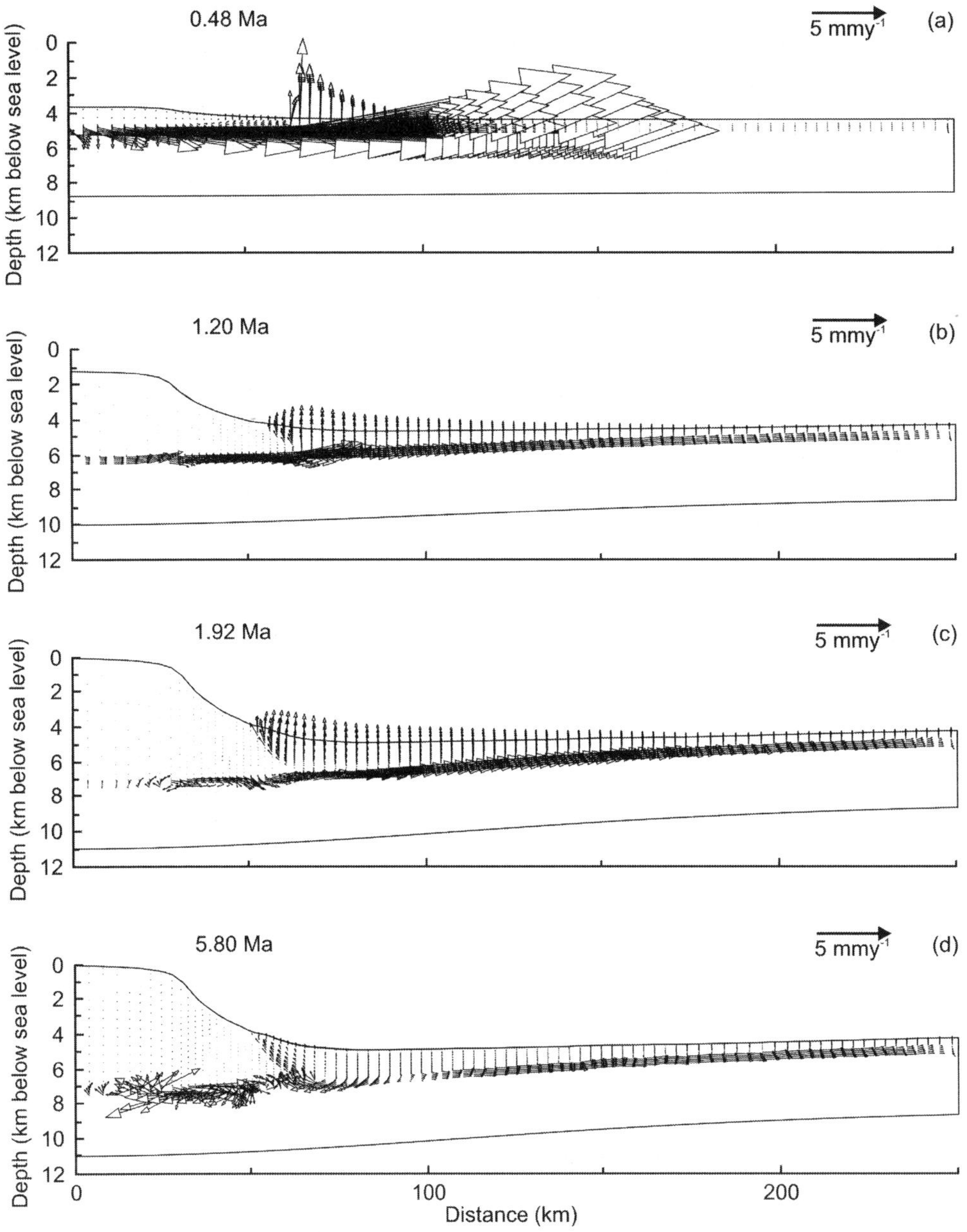

Figure 8.4. Fluid flow vectors from the Hawaiian simulation (Christiansen and Garven, 2004b). Time steps shown are at 0.48, 1.20, and 1.92 Ma after volcanism initiation and when the system reaches a steady state. The arrowhead size is proportional to the length of the arrow.

Pore fluid pressure measurements near La Palma of the Canary Islands (Figure 8.7) indicate that pore pressures were primarily negative, suggesting downward fluid flow in the order of a few $mmyr^{-1}$. Only one distal measurement showed positive values, indicating upward flow. This pressure regime is not typical for a passive margin, where excess pore fluid pressures expel fluids (Urgeles et al., 2000). On the contrary, the data here are interpreted to represent a wide hydrothermal convection cell, with recharge of about 100 km from the discharge area on the island flanks (Urgeles et al., 2000; Figure 8.8).

Heat flow measurements in the Hawaiian moat near Oahu and at Maro Reef show variability around an average heat flow of 10–20 mWm^2 (Harris et al., 2000). This variability in heat flow measurements could be the result of differences in porosity, permeability, and conductivity within different mass wasting units/blocks in

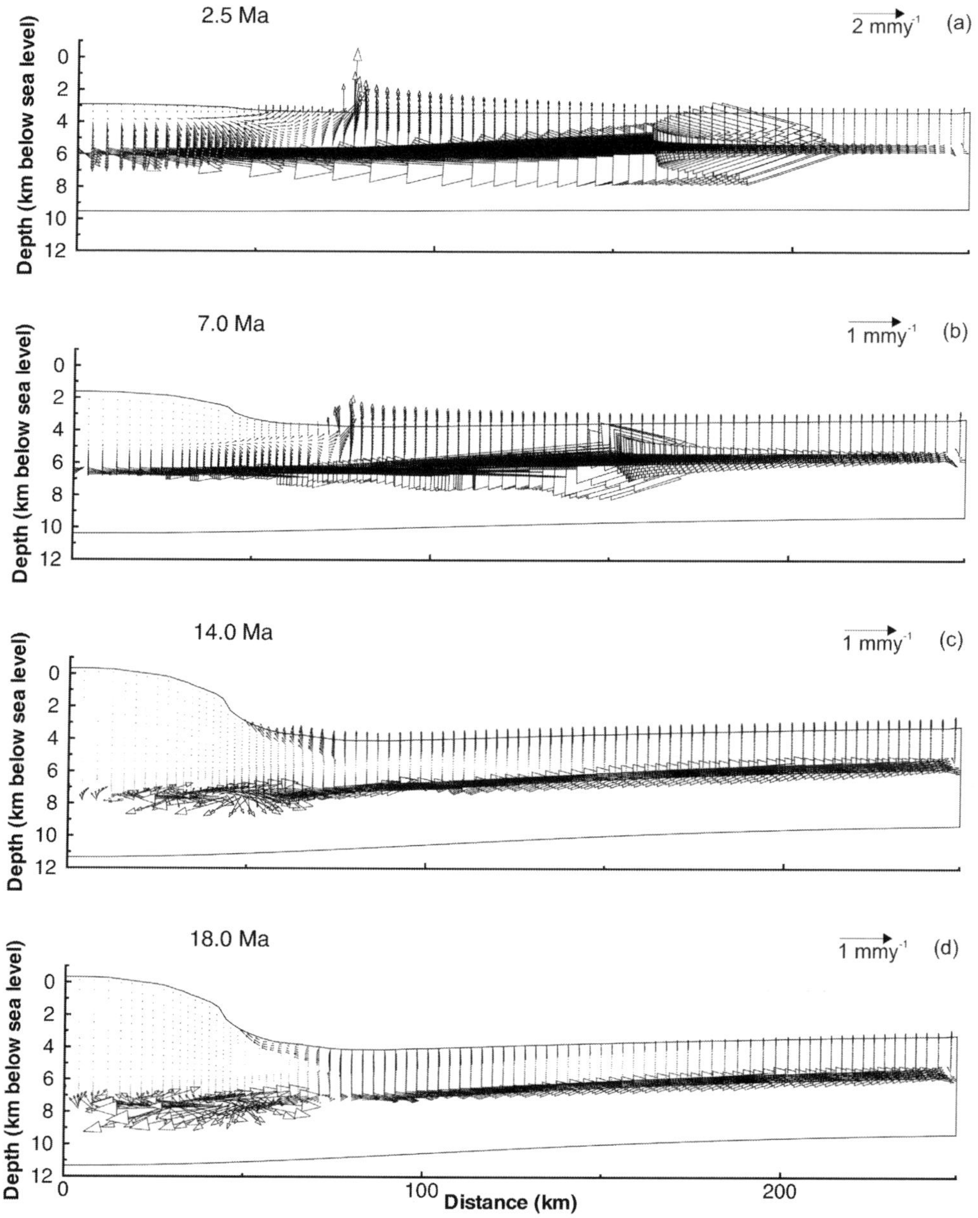

Figure 8.5. Fluid flow vectors from the Canary simulation (Christiansen and Garven, 2004b). Time steps shown are at 2.5, 7.0, and 14.0 Ma after volcanism initiation and when the system reaches a steady state. The arrowhead size is proportional to the length of the arrow.

the moat (Christiansen and Garven, 2004b). There is no sign of moat-wide heat flow variations. This indicates that large-scale fluid migration could be occurring at velocities too small to affect heat flow values (Christiansen and Garven, 2004b).

Vesicles and pores lined with hydrothermal alteration products, such as phyllosilicates, zeolites, and chlorite and olivine replacements, occur in samples taken from the toe of a slump near Hawaii by submersible. They indicate that vigorous hydrothermal circulation has occurred, most of which took place in the lower thrust faults (Morgan and Clague, 2003), closest to the oceanic crust. Fluid temperatures of about 150–200°C are required for observed chloritization to occur (Morgan and Clague, 2003). A fluid flow model requires fluids expelled from beneath the volcano to have permeated the toe of the slump before intermixing with ocean waters (Morgan and Clague, 2003; Figure 8.9).

A second hydrothermal system may exist on the summits and flanks of oceanic hotspots. It is associated

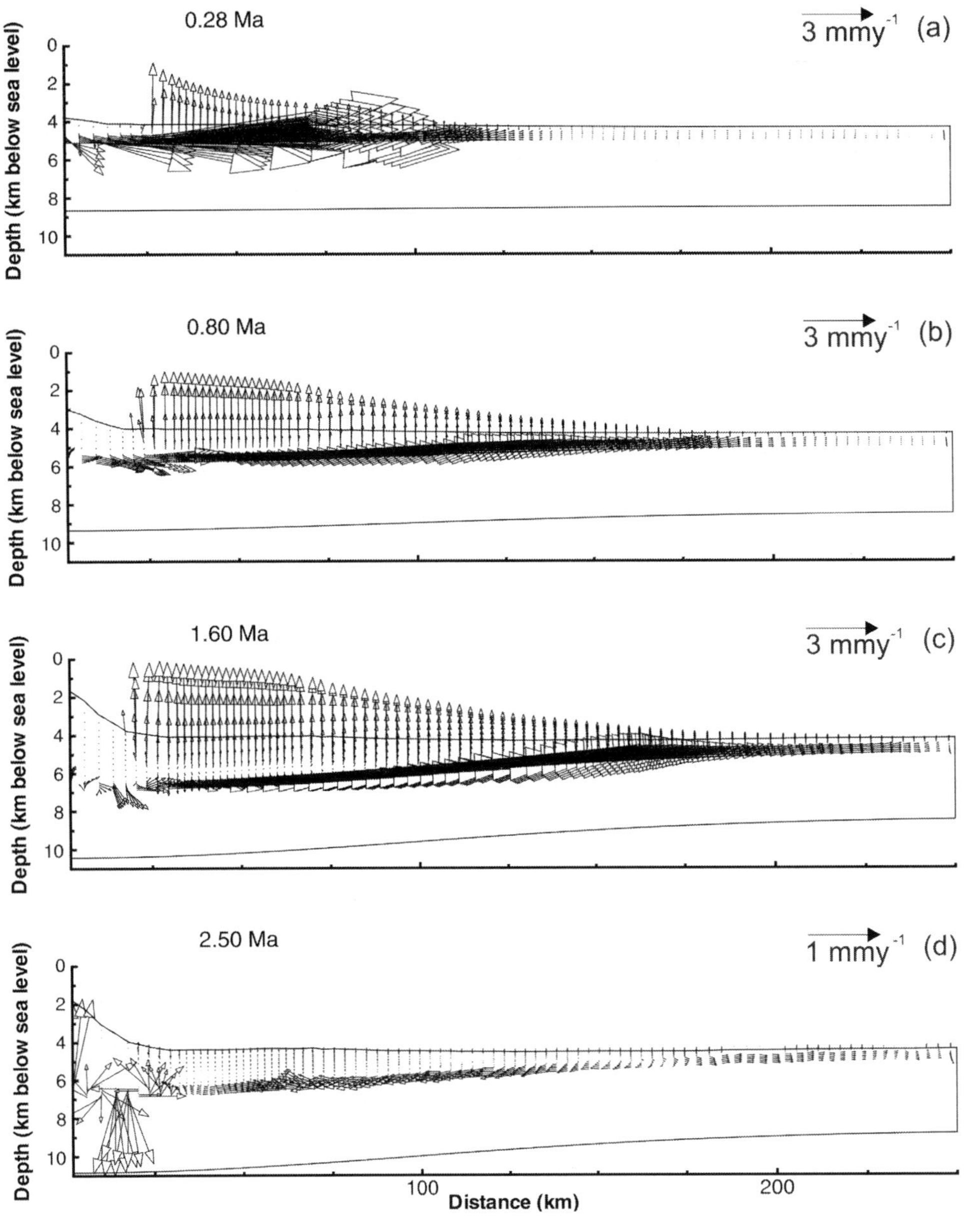

Figure 8.6. Fluid flow vectors from the Marquesas simulation (Christiansen and Garven, 2004b). Time steps shown are at 0.28, 0.80, and 1.60 Ma after volcanism initiation and when the system reaches a steady state. The arrowhead size is proportional to the length of the arrow.

with hydrothermally deposited iron and manganese crusts, which are found at mid-ocean ridges, back-arc basins, and crests of still submerged hot spots such as Hawaii (Malahoff et al., 1982; DeCarlo et al., 1983; Karl et al., 1988), Tahiti, the Macdonald Seamount, and the Pitcairn hot spots (Puteanus et al., 1991; Hekinian et al., 1993; Stoffers et al., 1993; Hodkinson et al., 1994; Glasby et al., 1997). This system starts with seawater recharge. As the cold seawater moves through the volcanic rock, its temperature increases, while oxygen content and pH drop. It is now able to leach such elements as copper, zinc, iron, and manganese from the volcanic rocks (Alt, 1995). If there is no mixing of this hydrothermal fluid with seawater as it rises, it will precipitate copper, zinc, and iron sulfides on the seafloor. If mixing occurs, iron-manganese oxides will precipitate (Scholten et al., 2004). Chemical analysis of gas-rich fluids from Loihi, Hawaii, suggests that hydrothermal fluids at temperatures in excess of 200°C have mixed with cold unaltered seawater prior to venting (Sedwick

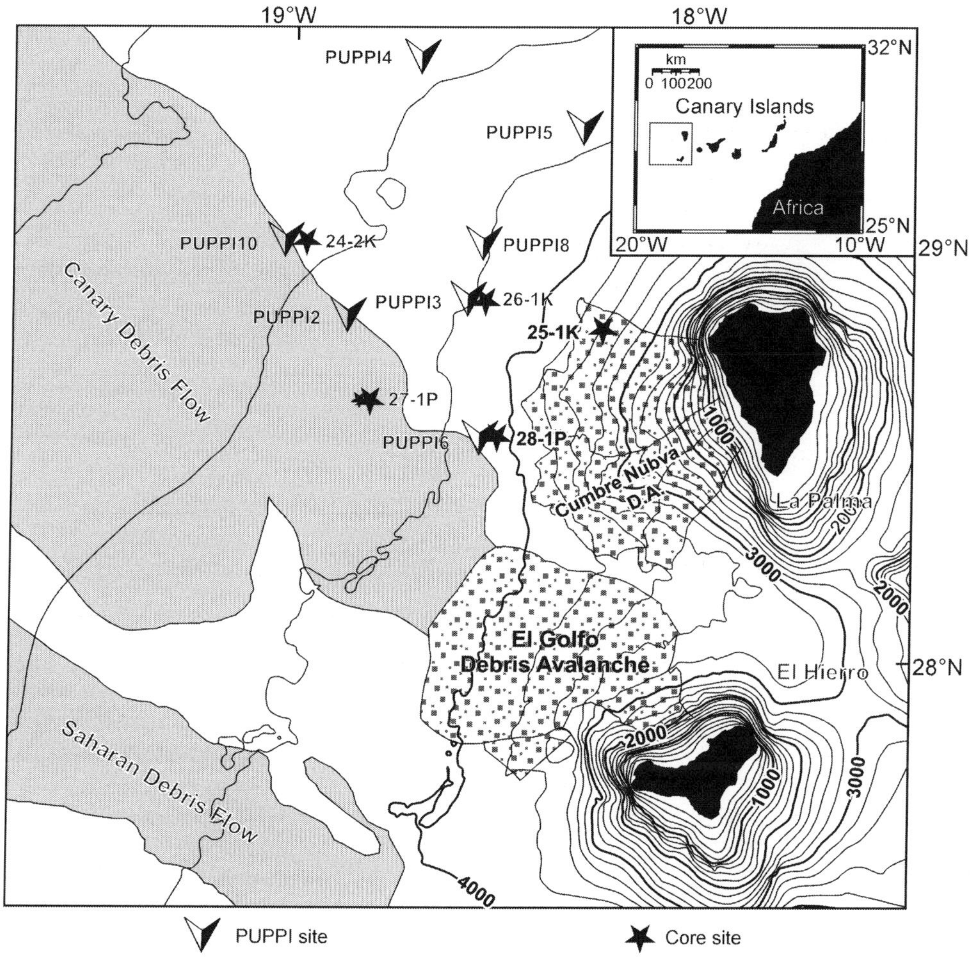

Figure 8.7. Bathymetry of the La Palma study area with main facies distribution, and location of pore fluid pressure measurements and cores (modified from Masson, 1996; Roberts and Cramp, 1996; Urgeles et al., 1997, 1999, 2000). Contour interval is 200 m.

et al., 1992). It is because the volcanoes are dormant for about 99 percent of their lifetimes (Latter, 1987), during which the circulation of seawater at their summits represents a significant heat loss process (Puteanus et al., 1991).

Fluid flow associated with oceanic transform faults

Observations of methane plumes in the water column over a 1,200 km segment of the Mid-Atlantic Ridge demonstrate that hydrothermal activity occurs over its entire length. The activity is most vigorous near the ridge segment ends where oceanic transform faults offset the ridge (Charlou and Donval, 1993; Figure 8.10).

The methane is produced by inorganic reactions. These reactions involve mantle- and serpentinization-derived carbon and hydrogen gas, respectively. Serpentinization is the reaction of crustal water with ultramafic rocks, during which hydration and oxidation reactions cause low oxygen fugacity, which allows reduced species, such as H_2, Fe^{2+}, or Fe^0, to be produced (Janecky and Seyfried, 1986; Charlou et al., 1991).

No methane was measured directly above the transform fault zones at the base of the Mid-Atlantic Ridge (Charlou and Donval, 1993). Measured methane concentrations were highest in the vicinity of serpentinized ultramafic rocks (Charlou and Donval, 1993) exposed in the footwalls of detachment faults within the axial valley (Rona et al., 1992). Ridge-parallel extensional faults, together with the shear faults of the fracture zones perpendicular to them, create a permeable pathway for fluid flow (Rona et al., 1992). Seawater is recharged at the base of the ridge. It undergoes reactions with ultramafic rocks (Charlou and Donval, 1993) and gets expelled in the axial valley as a diffuse low-temperature discharge (Rona et al., 1992).

Observations of active hydrothermal vent sites have been reported within extensional transform zones in

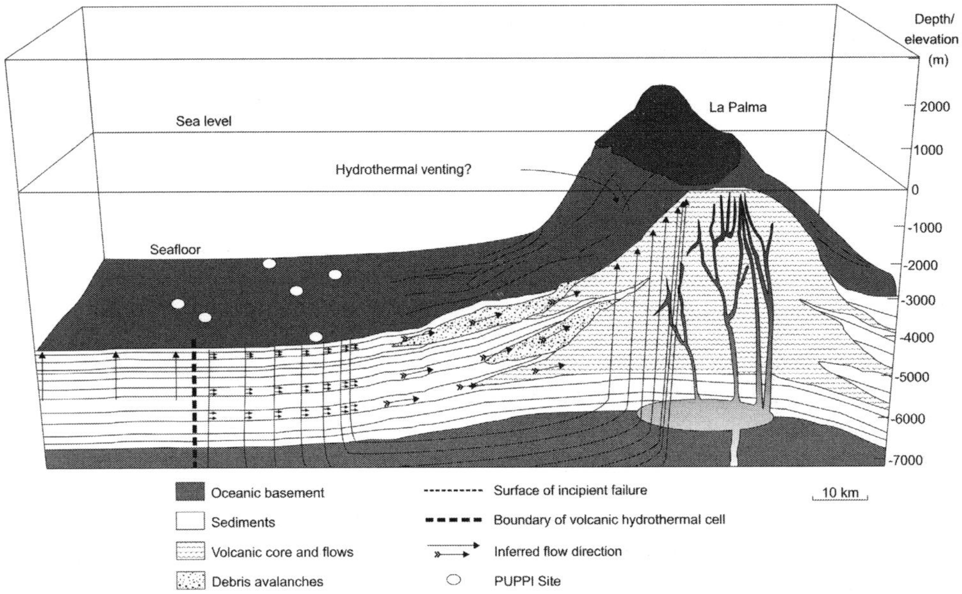

Figure 8.8. Fluid flow model for the flanks of an oceanic island based on a distribution of measured ambient pore pressures (Urgeles et al., 2000). Depths of shallow magma chambers and uplifted basement are taken from Kluegel and colleagues (1997) and Urgeles and colleagues (1998).

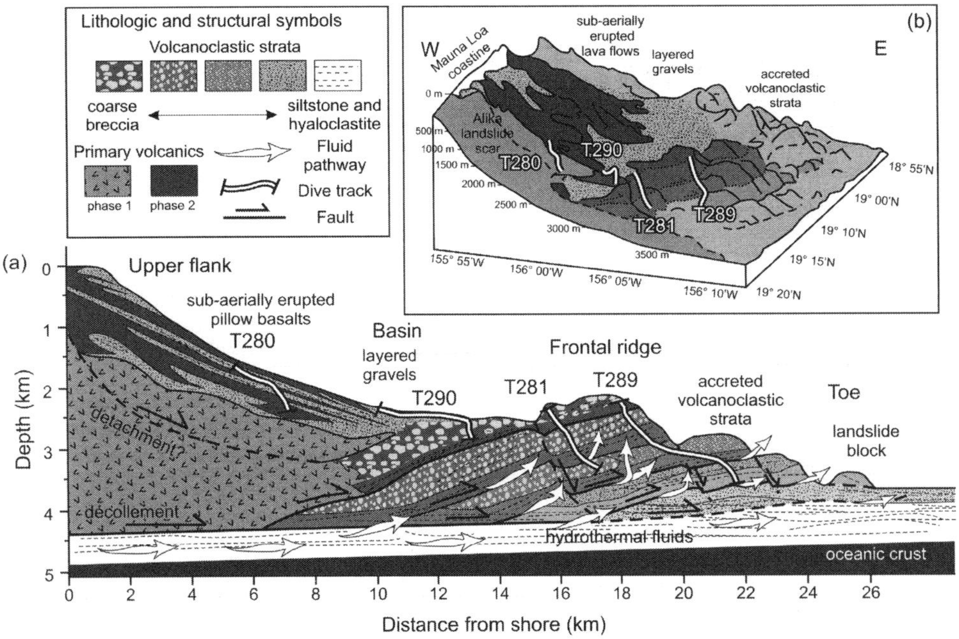

Figure 8.9. (a) Interpreted cross-section through the west flank of Mauna Loa, with submersible traverses projected into a plane of a section (Morgan and Clague, 2003). Note an imbricate stack of landward-coarsening volcanoclastic thrust sheets piled up in front of the sliding flank, which subsequently collapsed along high-angle fractures (short dashes). Hydrothermal fluids expelled from buried sediments migrated along décollement and thrust faults before venting to the seafloor. Young, subaerially erupted lava flows and gravels lap down onto the bench. Fault-zone locations are schematic. (b) Perspective view showing lithologic distribution of the South Kona slump (Morgan and Clague, 2003). Laterally continuous volcanoclastic thrust sheets form the frontal ridge. Upper flank is draped by pillow flows interbedded with shoreline-derived gravels, which fill the mid-slope basin.

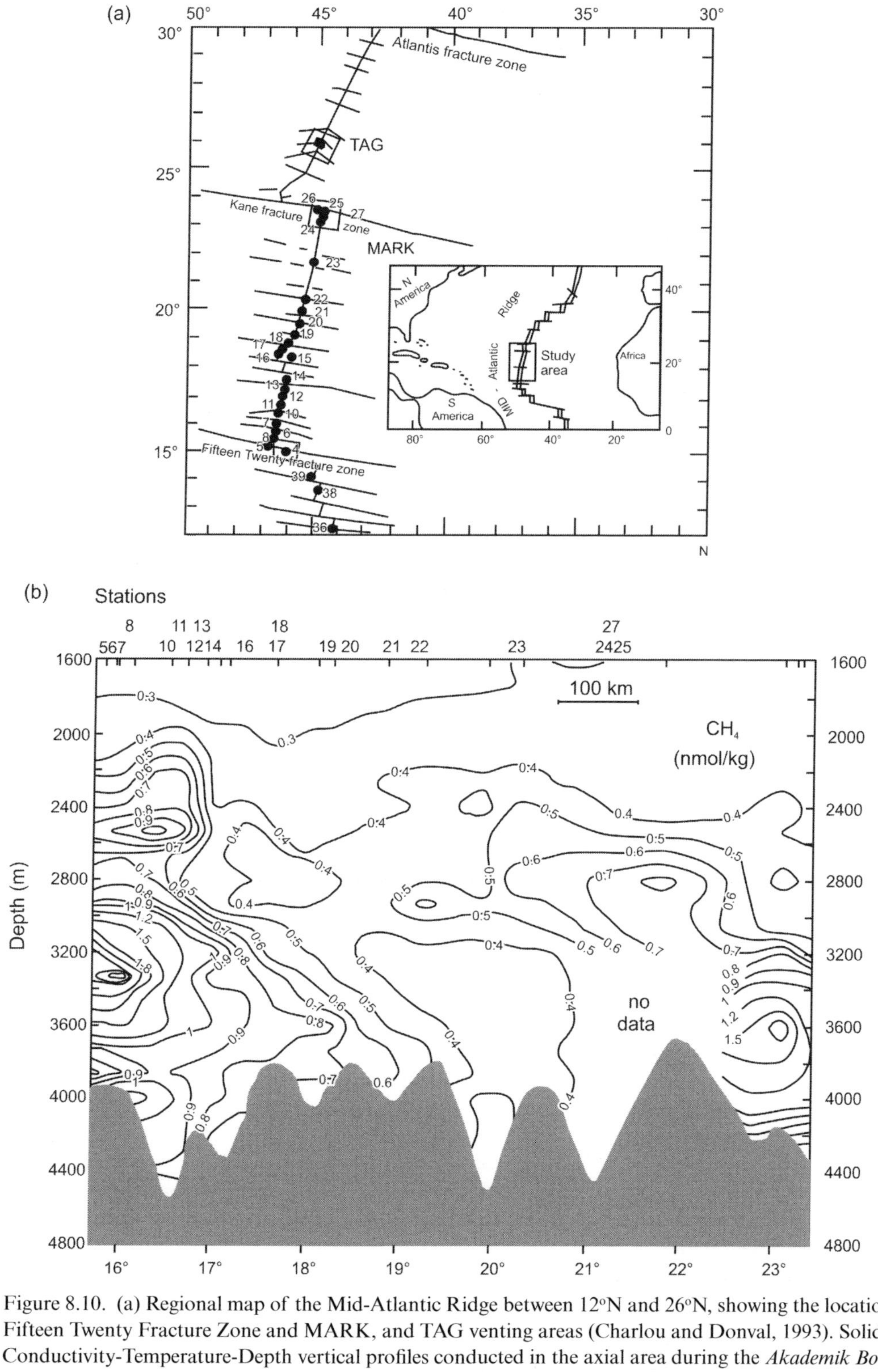

Figure 8.10. (a) Regional map of the Mid-Atlantic Ridge between 12°N and 26°N, showing the locations of the Fifteen Twenty Fracture Zone and MARK, and TAG venting areas (Charlou and Donval, 1993). Solid circles mark the Conductivity-Temperature-Depth vertical profiles conducted in the axial area during the *Akademik Boris Petrov* (1985) (stations 36, 38, 39) and *Ridelente* (1988) (stations 4 to 27) cruises. Squares mark known TAG (26°N) and MARK (23°N) venting sites studied during the *Atlantis III Alvin* cruise in 1990. The eastern intersection of the rift axis with the Fifteen Twenty Fracture Zone where hydrographic surveys were conducted during the *Ridelente* (1988) and MAR/88 (1988) cruises is also marked. (b) Methane fields between 1,600 m and the seafloor along the section 15° 36'N / 23°N of the Mid-Atlantic Ridge (Charlou and Donval, 1993). CH_4 plumes in the inner floor are clearly identified near the Fifteen Twenty and Kane Fracture Zones. A CH_4 plume emitted at the top of rift mountains (2,500 m) is defined around 16° 30'N. Sample depths are indicated by dots (8-L Niskin-type bottle data). The bathymetry of the ridge axis is schematic and very expanded.

two case areas. The extensional transform zones here are areas of overlapping volcanic systems and/or faults arranged in an en echelon pattern at oblique angles to the spreading direction (Taylor et al., 1994). Both sites occur within pull-apart basins, characterized by crustal thinning and moderate volcanic activity.

In the first case area, in the northeast Pacific, the Blanco Fracture Zone separates the Juan De Fuca Ridge from the Gorda Ridge. The Blanco Fracture Zone is formed by a northwest-southeast trending group of en echelon transform faults, which are separated by pull-apart basins (Hein et al., 1999). One of them is the East Blanco pull-apart basin. It was the site of a T-wave swarm in January 1994, which was interpreted to have volcanogenic origin (Dziak et al., 1996).

No evidence of new volcanic material was found in this basin (Dziak et al., 1996). Measurements in the water column at 800 m above the valley floor showed positive anomalies of ^{3}He and temperature (Dziak et al., 1996). Submersibles and towed cameras discovered an active hydrothermal field at the base of the central pillow-lava volcano, the last eruption of which was estimated to be less than 100 years old (Dziak et al., 1996). Samples taken from the hydrothermal deposits consisted primarily of barite mounds. They contain some silicified bacterial mats and mineralized sediments (Hein et al., 1999).

Hydrothermal fluids at this field are discharging through the pillow basalts and sediments. Their maximum temperature must reach the precipitation temperature of barite, estimated at 200°C–250°C (Hein et al., 1999). The hydrothermal deposits contain minor amounts of sulfides. They are greatly enriched in metals (Hein et al., 1999).

Hein and colleagues (1999) did not formulate a model for this hydrothermal system because of too many unanswered questions. Although the driving mechanism for circulation is unknown, it is likely to be the heat from a plutonic body. The fluid migration pathway is unclear, but a partial mantle fluid source is likely, considering the high concentration of ^{3}He in the basin.

In the second case area, the Tjornes Fracture Zone links the subaerially exposed neovolcanic zone of northeast Iceland with the Kolbeinsey Ridge, which is located to the north in the Arctic Ocean. The extensional transform zone is 75 km wide. It is formed by transform faults and grabens. A hydrothermal field has been discovered at a water depth of 400 m, in a heavily sedimented basin floor (Hannington et al., 2001). The field consists of around twenty mounds. Having diameters of about 10 m, they are composed of anhydrite and talc (Hannington et al., 2001). These mounds occur in a 1 km-long north-south trending zone at the basin floor. Being parallel to the orientation of the basin-bounding normal faults, this trend indicates that the field is fault-controlled (Hannington et al., 2001). Anhydrite, which is unstable at temperatures below ~150°C (Shikazono and Holland, 1983), allows us to imply that the site has undergone continuous hydrothermal activity for a relatively long time (Hannington et al., 2001).

The ^{87}Sr/^{86}Sr isotope ratios of anhydrite and its parent hydrothermal fluid are very similar (Kuhn et al., 2003). The ^{87}Sr/^{86}Sr of seawater is higher than that (Kuhn et al., 2003). On the contrary, the ^{87}Sr/^{86}Sr of local basalts, which hydrothermal waters are assumed to be in equilibrium with, is lower than that (Schilling et al., 1999). This indicates that anhydrite is precipitated from a mixture of hydrothermal fluid with seawater. Such mixed fluid does not undergo the fractionation of ^{87}Sr/^{86}Sr when anhydrite is precipitated (Mills and Tivey, 1999).

Calculations from the aforementioned ratios allow us to define that ~60 percent and ~40 percent of the mixture are formed by seawater and hydrothermal fluid (see Mills et al., 1998; Kuhn et al., 2003). A strontium/calcium partitioning coefficient between hydrothermal fluid and anhydrite implies that precipitation is from a non-evolved fluid (Mills et al., 1998). Furthermore, sulfur isotopes from the anhydrite indicate a seawater source of SO_4^{2-} as well (Kuhn et al., 2003).

Temperature measurements of the venting fluid and temperature determinations from fluid inclusion analyses averaged at 250°C, which represents a maximum possible temperature at this depth before the phase separation occurs by boiling (Hannington et al., 2001). Because these waters are at maximum temperature, no mixing with cold seawater can occur near the vent site (Kuhn et al., 2003). The lack of metals in the hydrothermal emissions is likely due to boiling in the subsurface (Hannington et al., 2001), which is consistent with the low measured salinity of the venting fluids (Kuhn et al., 2003).

A model for fluid flow in this case area (Kuhn et al., 2003) requires hydrothermal fluid rising from far below the seafloor (Figure 8.11). This fluid reaches the boiling point as it ascends, sending up the vapor phase to mix with cooler percolating seawater causing it to condense, controlling the temperature of the ascending fluid. There is not enough heat available for transfer from the rising fluid to the percolating fluid to keep the fluids near the boiling point. Therefore, 60 percent of shallow circulating seawater must be conductively heated to an estimated 200°C. This requires a large heat flow, which is confirmed by high temperature gradients of 52.3–61.27°Ckm^{-1} recorded near the vent sites (Devey and shipboard scientific party, 2002).

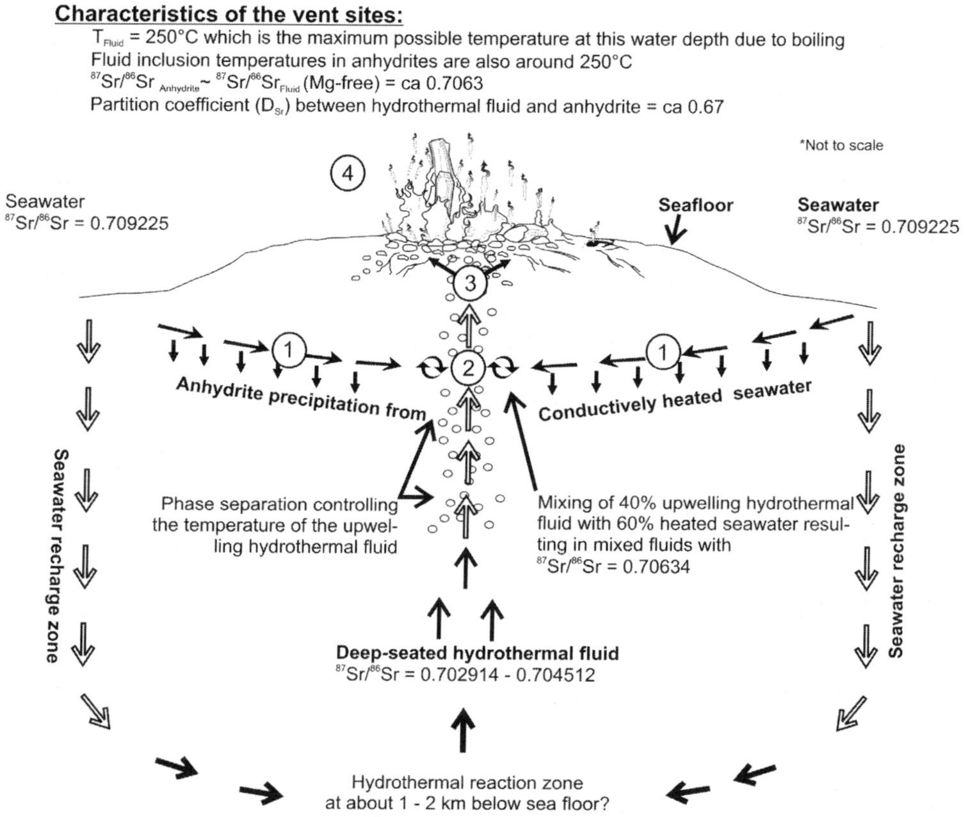

Figure 8.11. Schematic drawing showing the main processes related to strontium-calcium geochemistry at the Gimsey hydrothermal field (Kuhn et al., 2003). Seawater ($^{87}Sr/^{86}Sr$ ratio = 0.709225) entraining the seafloor to a shallower depth is heated to >150°C, which leads to anhydrite precipitation (1). The heating of seawater is either done by heat conduction from the host rock or caused by mixing with upwelling hot hydrothermal fluid. It is assumed that the fluid has equilibrated with sediment and underlying basalt resulting in a $^{87}Sr/^{86}Sr$ ratio of 0.702914–0.704512 (Schilling et al., 1999). The temperature of the ascending hydrothermal solution is controlled by phase separation and cannot exceed 250°C at a shallow depth. The mixed fluid being further conductively heated to 250°C has a $^{87}Sr/^{86}Sr$ ratio of 0.70634 (2). This fluid rapidly ascends to the seafloor (3) and precipitates anhydrite at and beneath the seafloor (4).

Fluid flow associated with continental transform faults

Prominent fluid flow systems associated with continental transform faults are rather a characteristic, instead of exceptional, feature of this tectonic setting, as testified by studies of transforms such as the Gettysburg-Tarfaya Transform (Steckler et al., 1993) (see Nemčok et al., 2005b for location), the Dead Sea Transform (Ritter et al., 2003), the San Andreas Transform (Unsworth et al., 1999; Bedrosian et al., 2002; Unsworth and Bedrosian, 2004; Li et al., 2006), and the Romanche Transform (Bouillin et al., 1998; Clift and Lorenzo, 1999). Data from the Ghana segment of the Romanche continental transform that evolved into the stage when it bounded continental and oceanic crusts from each other, indicate that fluid flow systems can occur even after continental breakup. It is documented here, for example, by the existence of calcite veins in the post-breakup Turonian strata. The veins have precipitated from the hydrothermal fluid about 20–30°C warmer than the temperature of normal seawater (see Marcano et al., 1998; Clift and Lorenzo, 1999). Further documentation comes from quartz veins occurring in the lower Coniacian strata of the ODP 960 site. Their fluid inclusion data indicate fluid temperatures of 160–170°C (Lespinasse et al., 1998). Such temperatures match the expected precipitation of penetrated hydrothermal kaolinite (Holmes et al., 1998). Hydrothermal fluids of even higher temperatures have been indicated along the southern portion of the Ghana transform margin, based on correlation with the occurring chlorite and mica stacks, indicating temperatures of 200–300°C (Benkhelil et al., 1996).

The first example of the fluid flow regime around the continental transform fault comes from the Dead Sea Transform in the Arava Valley, Jordan. Although the understanding of fluid flow associated with this transform does not reach the detail of the San Andreas

Transform studies discussed later, which study different types of fault segments separately, it allows us to understand the role of total slip on the fluid flow system.

Here, a joint magnetotelluric and seismic tomography study shows strong conductivity and velocity contrasts across the transform (Ritter et al., 2003). The transform fault here appears to act as a fluid flow barrier between the two juxtaposed lithologies represented by the fluid-saturated sedimentary fill of the western fault block and Precambrian rocks of the eastern fault block.

There was no fault zone conductor imaged in this region, although the resolution of the magnetotelluric and seismic surveys may not have been fine enough to image a fault zone several tens of meters wide (Ritter et al., 2003).

Modeled recordings from in-fault explosions recorded with seismometer arrays aligned with the surface trace of the transform fault have confirmed the presence of a very narrow low-velocity wave guide, 3–12 m in width and with a velocity reduction of 10–60 percent in comparison to the velocities of the adjacent rocks (Haberland et al., 2003). The narrow nature of the damage zone here is attributed to:

1) lower slip rates;
2) less recent seismicity; and
3) a smaller amount of total slip in comparison to the San Andreas Fault (Ritter et al., 2003).

The second example of the fluid flow regime around a continental transform fault comes from the San Andreas Transform, California. Here, individual segments of the San Andreas Fault with diverse seismicity patterns have been found to have differing fluid volumes and distributions within the damage zone. These segments include:

1) locked segment;
2) aseismically creeping segment; and
3) transitional segment.

Fluid system associated with a locked fault segment

The Carrizo Plain segment in central California represents a locked segment of the San Andreas Fault. It is controlled by transpression and, therefore, its deformation record contains associated thrust faults. The last rupture of this segment was the M 7.9, 1857 Fort Tejon earthquake. There has been little micro-seismicity recorded since the advent of modern instrumentation (Hill et al., 1990).

A magnetotelluric survey allows us to image the distribution of fluids in fractured rocks of this fault segment. Crystalline basement rocks are imaged as generally highly resistive, while the interconnected poor spaces filled with saline water dominate the conductance of natural low-frequency electromagnetic waves. Accordingly, conductive bodies imaged in the Carrizo Plain segment of the San Andreas Fault (Figure 8.12) allow us to infer two ways faults influence fluid flow.

First, the faults act as barriers to flow perpendicular to the fault strike a few kilometers down from the surface (Unsworth et al., 1999). This can be inferred from the existence of springs, which were observed to the northeast of the San Andreas Fault, where shallow topography-driven flow is impeded by the impermeable thrust faults (Arrowsmith, 1995). This phenomenon is illustrated by Figure 8.12, where isolated conductive bodies are located within the sedimentary section, up-slope of the thrust faults.

Second, down to the depths of at least 4 km, the San Andreas Fault may act as a conduit for fluid flow parallel to the fault trend, having fluids migrating through the fault gouge and breccia (Unsworth et al., 1999). This conductive area centered below the subsurface trace of the San Andreas Fault imaged in magnetotelluric surveys (Figure 8.12) is called the fault zone conductor. Compared to fault zone conductors imaged in other San Andreas Fault segments, discussed later, the one at the Carrizo Plain is rather a subtle feature. Multiplying conductivity by fault zone width, it has a calculated fault zone conductance of 20 S, which is an order of magnitude smaller than those of the creeping and transitional fault segments. Below the conductive body resistive, crystalline rocks are present on both sides (Unsworth et al., 1999).

Fluid system associated with an aseismically creeping fault segment

The fault segments of the San Andreas and Calaveras Faults near Hollister represent aseismically creeping fault segments. Here, the aseismic creep reaches a rate of about 10–15 mmyr^{-1} (Burford and Harsh, 1980; Evans et al., 1981). Basement rocks of the southwestern fault block are represented by the Gabilian granites. The northeastern block is formed by the Franciscan Complex, which is represented by a tectonic mélange and overlain by the sedimentary Cretaceous/Tertiary Great Valley Sequence (Dibblee, 1980).

Magnetotelluric surveys here have imaged a 7 km-wide zone of low resistivity, which extends from the San Andreas Fault to the Calaveras Fault (Bedrosian et al., 2002, 2004; Figure 8.13). The fault zone conductor below the San Andreas Fault extends to a depth of 10 km, which is much deeper than that in the locked and transitional fault segments, and which is abruptly

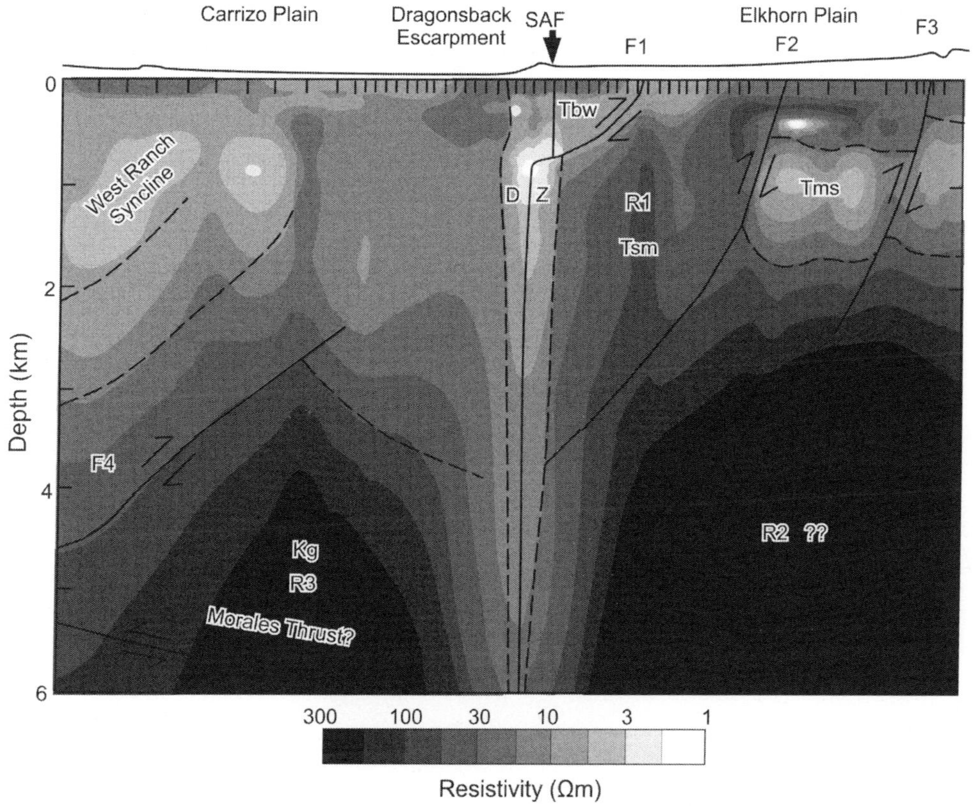

Figure 8.12. Resistivity model of the locked San Andreas Fault segment at the Carrizo Plain (Unsworth et al., 1999). R – resistive body, F – thrust fault, DZ – damage zone. Maximum resistivity is indicated by wavy hachure, maximum conductivity by black.

terminated to the southwest against the granite. To the northeast, between the San Andreas and the Calaveras Faults, conductors take the form of discrete pockets divided by impermeable northeast-dipping faults within the Great Valley Sequence. These isolated conductors reach depths of 4–6 km. Fault zone conductance beneath the San Andreas Fault has been calculated to be 400–600 S, while the total conductance of the shallow discontinuous fault zone conductors is 1,000–1,500 S (Bedrosian et al., 2002).

The porosity of the San Andreas Fault zone conductor has been calculated using:

1) the salinity of a 600 m-deep well located 1 km southwest of the San Andreas Fault (Stierman and Williams, 1984), which is 17 mgl^{-1}; and
2) the average resistivity of the Great Valley Sequence between the faults at a depth of 1.5 km, which is 5 Ωm.

Using Archie's law for electrical resistivity to porosity conversion, porosities of 15–35 percent were calculated. These values represent porosities of rocks where electrical conduction is dominated by an interconnected pore fluid, which utilizes fractures rather than spherical pore spaces in this case (Bedrosian et al., 2002). Fifteen to thirty-five percent represents the upper limit of possible porosity range because the calculations do not take into account the conducting effect of clay minerals, which occur in the fault zone. The salinity values of the sedimentary units in areas to the northwest of the faults are likely higher than that from the aforementioned well. However, measurements were available there. The reason for this comment is that potential increase in salinity values would translate into reduction of calculated porosities (Bedrosian et al., 2004).

At a depth of 10 km, the resistivity of the fault zone conductor increases to 10 Ω m. This value corresponds to porosities of 2–10 percent (Bedrosian et al., 2004), where the lower limit is probably more realistic. This image could represent a fluid pathway from an overpressured zone beyond the brittle ductile boundary, constantly supplying upward-migrating fluids (Rice, 1992). Alternatively, overpressured fluids may be located below the seismogenic depth. They would be sucked upward when dilatational fractures open during a rupture (Sibson et al., 1988).

To compare locked and aseismically creeping fault segments, we can discuss the two-dimensional primary

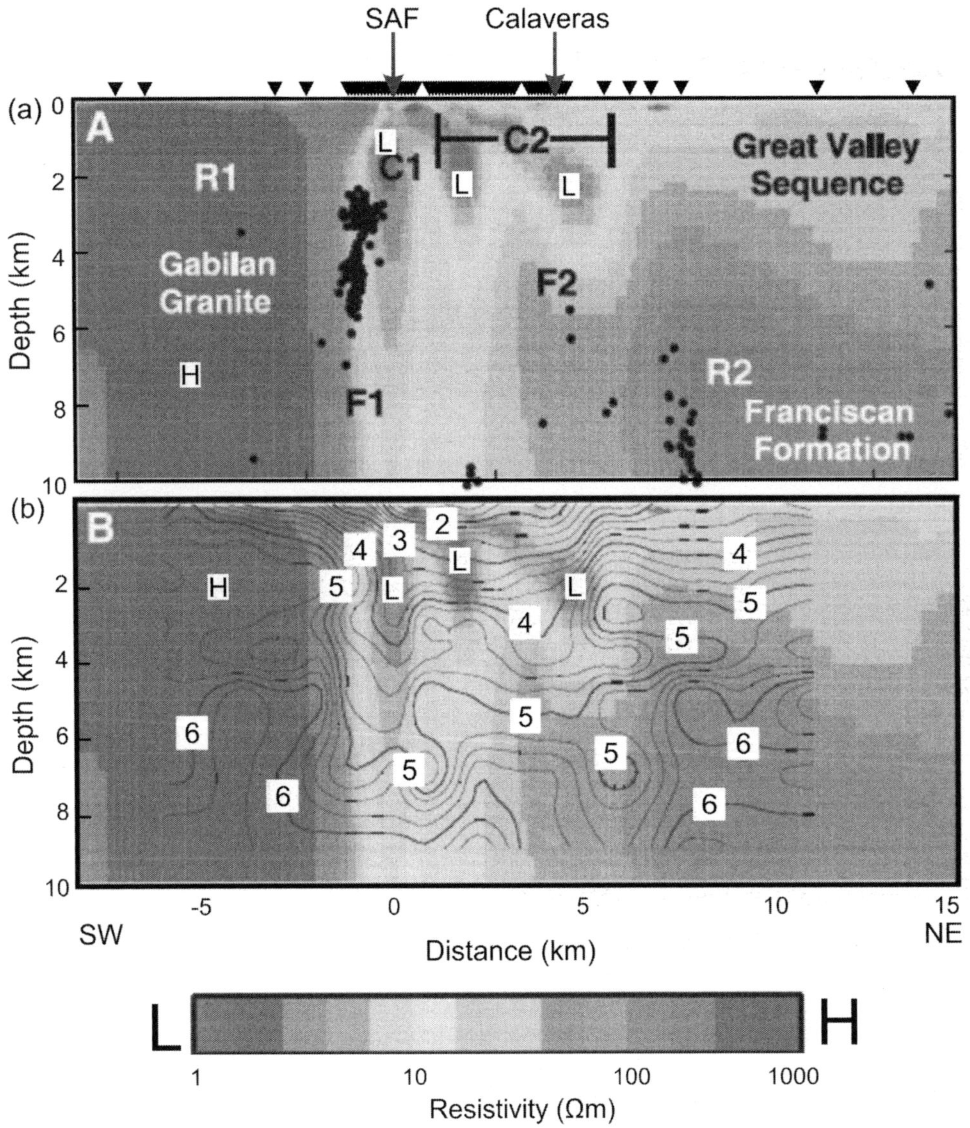

Figure 8.13. (a) Resistivity model of the aseismically creeping segments of the San Andreas and Calaveras Faults near Hollister, California (Bedrosian et al., 2002). Earthquake hypocenters within 2 km of the profile are shown as black circles. Inverted triangles indicate magnetotelluric site locations. Black arrows indicate the surface traces of the San Andreas and Calaveras Faults. Maximum resistivity is indicated by letter H, maximum conductivity by letter L. (b) The same model overlain by P-wave velocity contours from Thurber and colleagues (1997). Velocities are labeled in kms^{-1}.

wave velocity (Vp), and primary wave velocity secondary wave velocity ratio (Vp/Vs) models, which have been constructed for the locked segment of the San Andreas Fault near Hollister (Figure 8.14). This would be useful because porosity is the parameter most correlated with seismic velocity, with cracks having a larger effect than more spherical pore spaces (Toksoz et al., 1976). An increase in crack density would result in a decrease of both Vp and Vs, and in an increase of Vp/Vs ratio (O'Connell and Budiansky, 1974). Increased pore pressure would further decrease seismic velocity by reducing the effective confining pressure on cracks, with the strongest effect taking place at lithostatic pressure (O'Connell and Budiansky, 1974; Toksoz et al., 1976).

Figure 8.14 indicates a zone of low Vp between the two creeping faults (see Figure 8.13 for fault location). Furthermore, there is a zone of high Vp/Vs beneath the surface trace of the San Andreas Fault, and in isolated areas to the northeast of the Calaveras Fault (Thurber et al., 1997). Velocity models have geometry of the low Vp and high Vp/Vs patterns similar to that of the low-resistivity zone seen in the magnetotelluric surveys

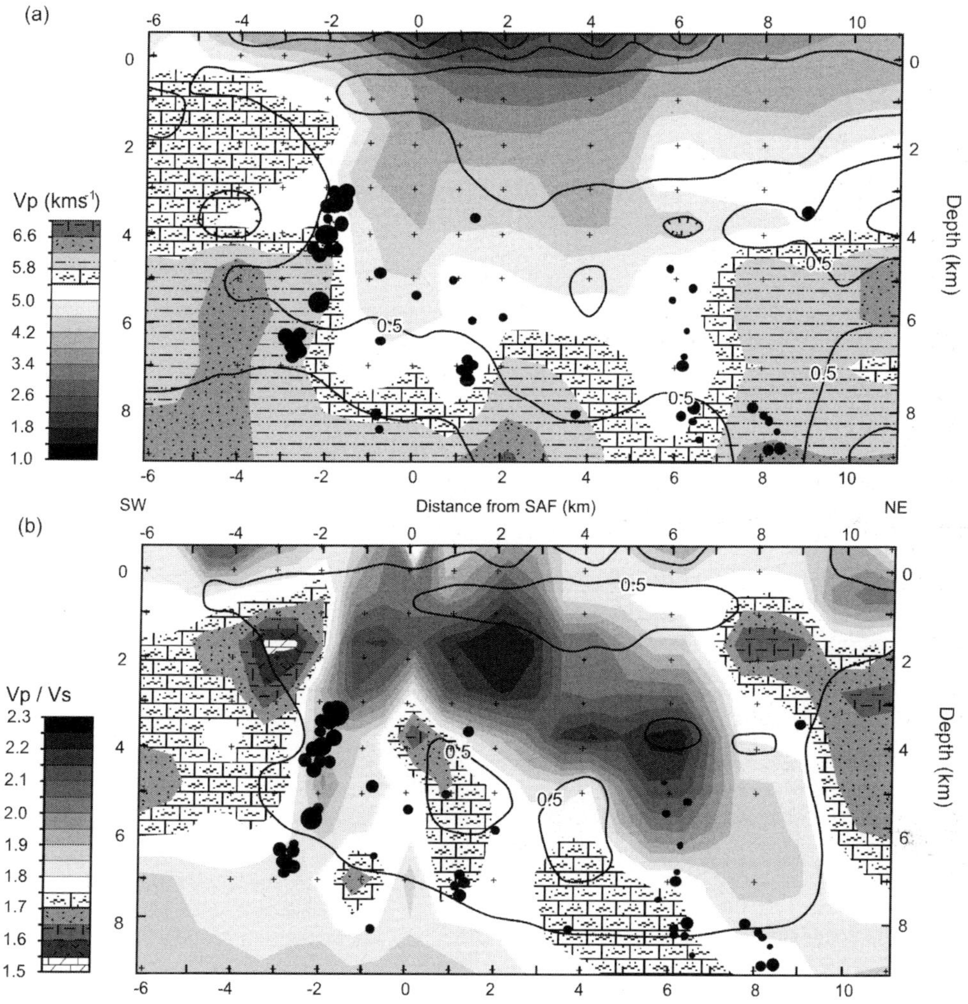

Figure 8.14. 2-D models of (a) Vp and (b) Vp/Vs in a plane normal to the San Andreas Fault in the case of an aseismically creeping fault segment (Thurber et al., 1997). Figure shows earthquakes with magnitudes of 0.5–3.0, which were included in the inversion, indicated by the circles. Note that the earthquakes are concentrated in areas adjacent to or beneath high Vp/Vs zones. Grid nodes are indicated by "+" and resolution contours are shown in 0.25-contour intervals.

(Bedrosian et al., 2002, 2004; Figure 8.13). The high Vp/Vs ratios cut areas of differing Vp values, suggesting that the high ratios are indicative of fluids, not lithology (Thurber et al., 1997).

To continue comparing locked and aseismically creeping fault segments, we can discuss the synthetic fault zone trapped waves, which have been recorded in the creeping section of the San Andreas Fault near Hollister, analysis of which provides information on the structure of the San Andreas Fault. Within the damage zone, the seismic velocity is much lower than that of the surrounding country rocks, trapping and transmitting seismic wave trains (Li et al., 1997). A width of the shallow fault zone at the Carrizo Plain is 120 m, with a Vs of 0.7–0.8 kms^{-1} and seismic attenuation of 30–40 (Li et al., 1997). The seismic attenuation here increases with:

1) permeability (Biot, 1956);
2) increased crack density (O'Connell and Budiansky, 1974); and
3) lower-aspect ratio cracks (Mavko and Nur, 1979).

Fluid system associated with transitional fault segment

The segment of the San Andreas Fault near Parkfield represents a transitional fault segment connecting the locked and creeping fault segments discussed earlier. This fault segment experiences moderate earthquakes of about M6 about every twenty-two years. It is characterized by extensive micro-earthquake activity at a depth of about 3 km (Bakun and Lindh, 1985). A magnetotelluric image of this segment shows that there is a resistive block of Salinian granite overlain by

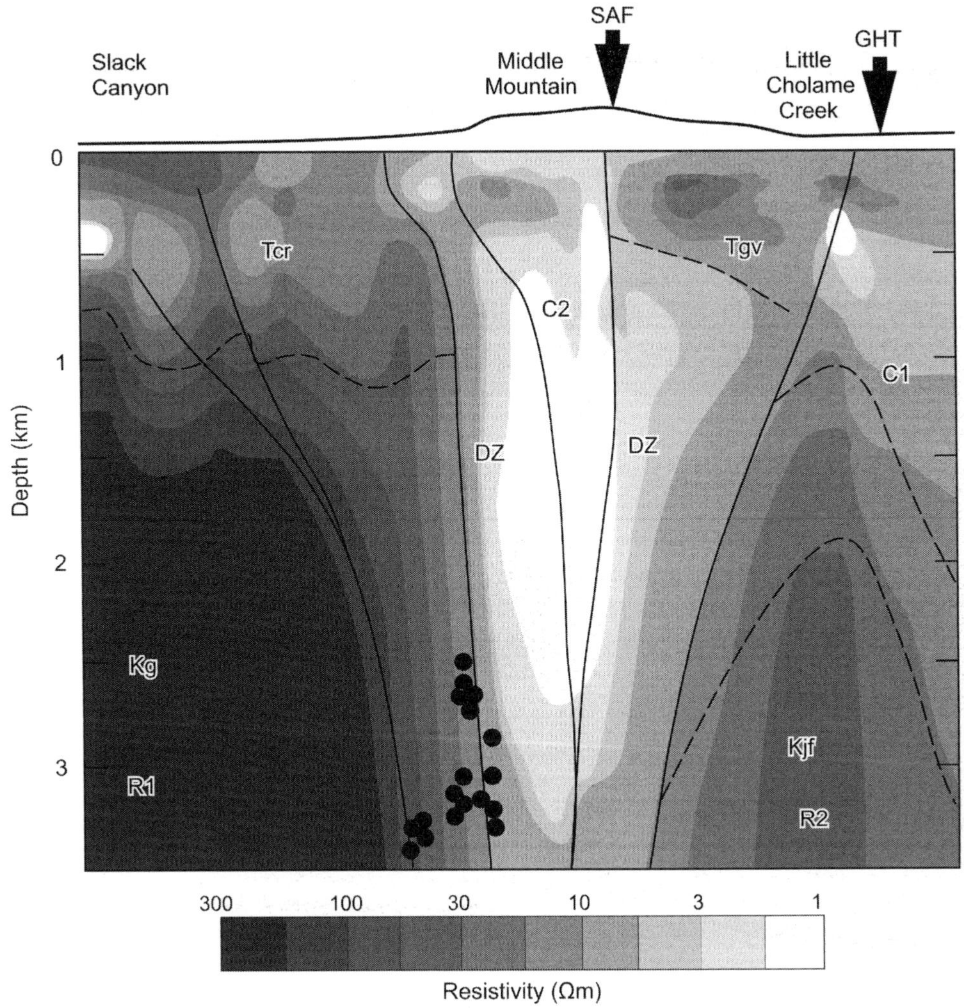

Figure 8.15. Resistivity model of the transitional segment of the San Andreas Fault at Parkfield (Unsworth and Bedrosian, 2004). R – resistive body, C – conductive body, T_{cr} – Tertiary Coast Range units, T_{gv} – Great Valley sequence, K_g – Salinian granite, K_{jf} – Franciscan Formation, and DZ – damage zone. GHT and SAF indicate surface traces of the Gold Hill thrust and San Andreas faults. Maximum resistivity is indicated by dark gray, maximum conductivity by white.

Tertiary-Quaternary sediments to the southwest of the fault. Furthermore, there is the conductive, fluid-rich Franciscan Complex covered by Tertiary and Holocene sediments to the northeast of the fault (Dibblee, 1980).

Magnetotelluric imaging shows that the fault zone conductor at Parkfield is around 600 m wide at its maximum near the surface (Unsworth et al., 2000), and extends to a depth of 2–3 km (Unsworth and Bedrosian, 2004; Figure 8.15). Seismic refraction data show that transpressional deformation here has produced a flower structure that is 5 km wide and extends to a depth of 10 km (McBride and Brown, 1986). Furthermore, there are lateral resistivity contrasts associated with thrust faults located within the faulted Quaternary sediments to the southwest of the fault (Figure 8.15). This situation is most likely a result of groundwater impedance perpendicular to the fault, creating compartmentalized reservoirs (Unsworth et al., 1997) similar to those seen at the locked Carrizo Plain fault segment.

At 1.4 km northeast of the fault, the Phillips-Varian well penetrated 1,700 m of briny sediment above the faulted contact with the Franciscan basement. Overpressured brine with a salinity of 30,000 ppm was encountered at this fault zone (Jongmans and Malin, 1995). This salinity value and aforementioned Archie's law allowed for the calculation of a porosity of 9–30 percent for the fault zone conductor located at a depth of 3 km (Unsworth et al., 1997). When the expected geometry of pore spaces is that of a crack, it is the minimum value of the interval that represents this porosity. However, the estimate has an optimistic value because the conducting effect of clay occurring in the fault zone was not considered. The conductance of the fault zone at a depth of 3 km is calculated to be 200 S. This value, characterizing the transitional fault segment, is an order of magnitude higher than that of the locked fault segment at the Carrizo Plain, and roughly half of that calculated for the aseismically creeping fault segment near Hollister (Unsworth et al., 1999).

Three magnetotelluric surveys perpendicular to the strike of the transitional fault segment near Parkfield, and within 8 km of one another, have been conducted in order to see along-strike variations in the fault zone conductor (Unsworth et al., 2000; Unsworth and Bedrosian, 2004). They show that a fault zone conductor is continuous, with similar widths and variable, but well constrained, depths of 2–3 km (Figure 8.16), becoming shallower to the northwest. Variations in the depth of the fault zone conductor here are controlled by changes in (Unsworth and Bedrosian, 2004):

1) fracture density;
2) fluid composition; and
3) fluid concentration.

Figure 8.16 depicts two hydrologic interpretations of the magnetotelluric models for the transitional fault segment at Parkfield. The first one suggests that meteoric fluids accumulate in the damage zone (Unsworth et al., 1999). The second one suggests that the fault zone conductor receives metamorphic fluids from the Franciscan Complex (Irwin and Barnes, 1975).

Seismicity decreases along the fault strike to the southeast as the connection between the fault zone conductor and the reservoir of metamorphic fluids increases (Figure 8.16). At a depth of 5 km, resistivity to the east of the San Andreas Fault reaches 30 Ω m. Using Archie's law, calculated porosities range from 1 to 10 percent, demonstrating a potential fluid pathway between the fault and the fluid reservoir of the Franciscan Complex (Unsworth and Bedrosian, 2004).

The Vp contrast across the transitional fault segment at Parkfield is the dominant velocity feature, including higher velocities to the southeast of the fault (Figure 8.17). However, there is relatively little change in Vp along the fault strike (Thurber et al., 2003). A low Vp structure, located just northeast of the San Andreas Fault, somewhat correlates with the low-resistivity structure discussed earlier (see Figure 8.16; Unsworth and Bedrosian, 2004). The low Vp volume extends to as deep as 8 km and shallows to the northwest. The low-resistivity zone also shallows to the northwest, but is not as deep (Thurber et al., 2003).

The Vp/Vs structure varies significantly both across and along the strike of the fault at this segment. These high Vp/Vs zones are found adjacent to the San Andreas Fault and are virtually aseismic (Thurber et al., 2003). Movement in these zones may be accommodated either by fault creep or by dynamic rupture during large earthquakes (Thurber et al., 2003), or perhaps it is only below this fluid-saturated zone that fault zone rocks can accumulate enough stress for brittle failure to occur (Unsworth and Bedrosian, 2004).

Finite-difference 3-D simulations of 2–5 Hz fault zone–trapped waves recorded at the transitional fault segment at Parkfield show evidence of a damaged core zone. It is about 150 m wide and extends to a depth of 10 km (Li et al., 1997). A seismic attenuation of 10–50 and Vs reductions of 30–40 percent are seen compared to the country rock (Li et al., 2004). Both are indicative of increased crack density (O'Connell and Budiansky, 1974). A width of the low-velocity wave zone likely represents both the macroscopic damage zone with dilatational fractures and the zone of microscopic fault processes, represented by mechanical, chemical, thermal, and other kinematical processes (Li et al., 2004).

One week after the M 6.0 Parkfield earthquake on September 28, 2004, the aforementioned seismic networks were redeployed in the same configuration, and explosives detonated at the same locations as those used in 2002. This was done to examine the effect of coseismic damage, post-main shock healing, and shear-wave splitting on the San Andreas Fault.

Data from a repeated detonation done on December 30, 2004, roughly three months after the rupture, show a decrease in shear wave velocity of about 1.25 percent in a roughly 150–200 m-wide rupture zone centered asymmetrically on the San Andreas Fault (Li et al., 2006). Because data acquisition began one week after the rupture, an extrapolated coseismic velocity decrease of 2.5 percent was calculated, which corresponds to a 0.035 percent increase in crack density (Li et al., 2006).

Recordings of repeated aftershocks during the three-month-long post-rupture study show that the seismic velocity was increasing with differing magnitudes in three dimensions. Perpendicular to the fault, seismic velocity increased by 1.1 percent within the rupture zone, with more modest velocity recoveries of less than 0.5 percent outside of the rupture zone (Li et al., 2006). This seismic velocity increase suggests that the crack density decreased by 0.016 percent (Li et al., 2006). Seismic velocity recovery along the strike was found to be greater in areas that underwent larger coseismic displacement (Li et al., 2006). Comparisons of velocity increases determined for clusters of similar-magnitude aftershocks occurring at different depths allow us to imply that there is more velocity recovery occurring at shallower depths (Li et al., 2006).

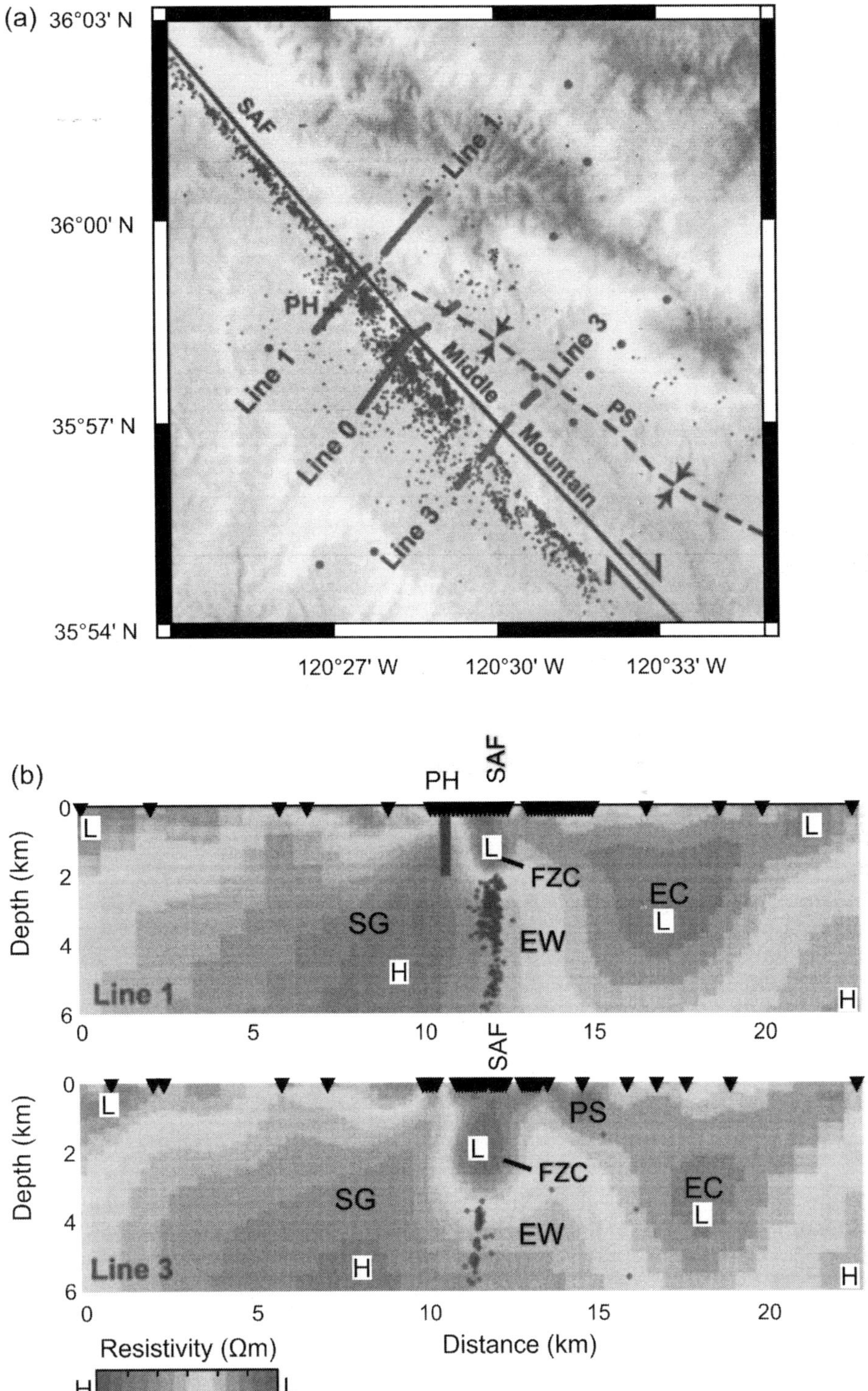

Figure 8.16. Magnetotelluric surveys perpendicular to the strike of the transitional segment of the San Andreas Fault near Parkfield (Unsworth and Bedrosian, 2004). (a) Location map of the magnetotelluric stations (large gray circles). The small gray dots denote earthquakes taken from Thuber and colleagues (2003, 2004) and the High Resolution Seismic Network catalog (Nadeau and McEvilly, 1997). See (b) for abbreviation captions. (b) Resistivity models for profiles 1 and 3 from inversion of TE, TM, and vertical magnetic field data. In each inversion error floors were applied (ρTM 10%, ρTE 20%, phase 5%, Tzy 0.02). The larger error floors on apparent resistivity allow the estimation of static shift coefficients. These were generally found to be in the range of 1.2 to 0.8. H and L indicate resistivity highs and lows. SAF – San Andreas Fault surface trace, PH – SAFOD Pilot Hole, PS – Parkfield syncline, SG – Salinian Granite, EC – Eastern conductor, EW – East Wall of the San Andreas Fault, FZC – Fault zone conductor. VE = 1:1 in both models.

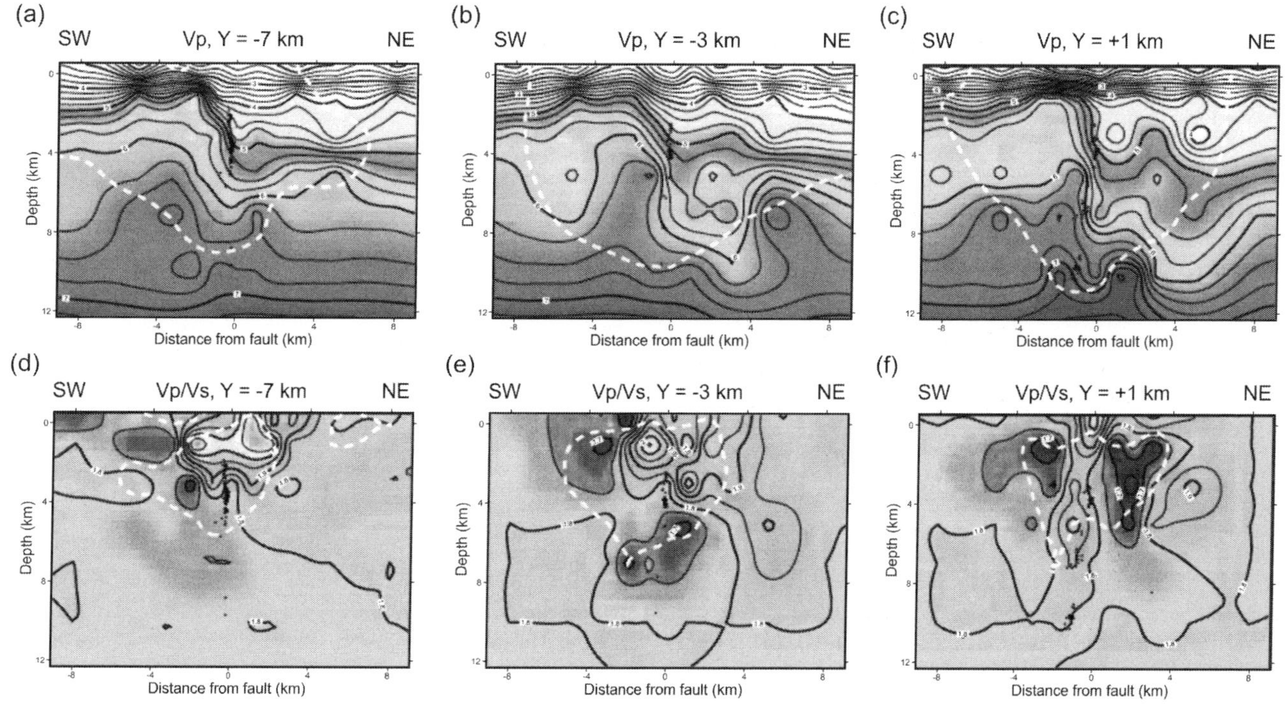

Figure 8.17. Southwest–northeast cross-sections through the 3-D models of Vp (a–c) and Vp/Vs (d–f) structure through the transitional segment of the San Andreas Fault at Parkfield (Thurber et al., 2003). The San Andreas Fault trace is at 0 km. Depths are relative to the sea level. Vp values are in kms^{-1}, and Vp/Vs values are dimensionless. Earthquakes within 1 km of the section are shown as filled circles. The white dashed line indicates the region considered to be resolved adequately (resolution matrix diagonal element >0.25).

The 2004 Parkfield earthquake nucleated at a depth of 7.9 km, indicating that cracks in greater depths are not closing as quickly as those at shallower levels. Cumulative velocity increases with aftershock determined for focal depth between 3 and 7 km demonstrate that healing is occurring throughout this interval. Less well-constrained data suggest that velocity recovery may persist to depth levels as deep as 9.2 km (Li et al., 2006).

A change in the ratio of P-wave to S-wave travel times ($\Delta t_p/\Delta s_p$) within the rupture zone was observed in data from a fault-perpendicular seismic array (Li et al., 2006). This phenomenon has been interpreted to indicate that there has been a post-seismic increase in the fluid content of the rupture zone (Li et al., 2006; Figure 8.18). P-wave velocity is controlled by fluids in pore spaces more strongly than S-wave velocity. A $\Delta t_p/\Delta s_p$ of 1.64 is expected for completely dry cracks and 0.17 for totally saturated cracks (Li et al., 2006). However, a $\Delta t_p/\Delta s_p$ value of 0.57 was found within the rupture zone and 0.65 in the surrounding rocks. This suggests that shallow cracks within the fault zone are wetter than those outside, which is possibly due to high-pressure water coming up through the fault zone (Li et al., 2006).

Another seismic study using data from both before and after the 2004 Parkfield earthquake shows that shear-wave polarization is fault-parallel for stations on or ~ 100 m southwest of the surface trace (Cochran et al., 2006). This polarized zone has similar dimensions to the fault zone wave guide (Li et al., 2004, 2006). Shear-wave splitting aligns the fast orientation with the long dimension of cracks (Crampin, 1990). The apparent crack density was calculated to be about 3 percent for both before and after the 2004 earthquake (Cochran et al., 2006). Anisotropy measurements are relatively insensitive to temporal changes because both fast and slow shear waves are similarly affected (Peng and Ben-Zion, 2005). Microcracks contribute to anisotropy most in the shallow crust where confining pressures are the lowest, but travel time delays for events at depths of 7–8 km were observed, suggesting that anisotropy persists to these depths (Cochran et al., 2006). Anisotropy is interpreted to be due to aligned microcracks outside the polarized zone and highly fractured shear fabric within (Boness and Zoback, 2004).

To study transitional fault segments, researchers have drilled scientific boreholes in close proximity to the San Andreas Fault at the Cajon Pass and Parkfield. The first

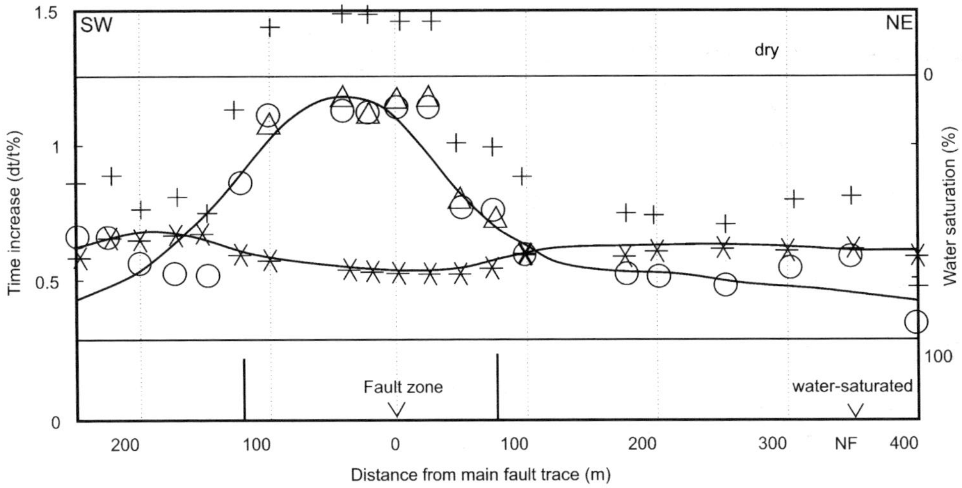

Figure 8.18. Travel time increases in percent for P (crosses), S (circles), and trapped (triangles) waves measured from cross-correlations of seismograms at array across the transitional segment of the San Andreas Fault for repeated shot PMM in 2002 and 2004 (Li et al., 2006). They show that seismic velocities decreased after the 2004 M 6.0 earthquake because of rock damage within the fault zone. Some stations did not work in both experiments because of weather and battery problems. The black line is a polynomial fit to travel-time increases of S waves in 2004. A pair of vertical gray bars denotes a zone with greater travel-time increases. Ratio of travel-time changes for P waves to S waves (gray stars with the curve) indicates a degree of water saturation in cracks. Two horizontal light lines indicate the ratios predicted for a range of water percentage for a Poisson solid. NF – north fault strand.

of them, the Cajon Pass research borehole, is a vertical well that reached a depth of 3.5 km, at a distance of approximately 3 km from the surface trace of the San Andreas Fault near San Bernardino, California. Geochemical and geophysical methods identified multiple fracture systems between depths of 1,829 and 2,115 m (Barton and Moss, 1988; Hammond et al., 1988; Kharaka et al., 1988a, 1988b; Lachenbruch and Sass, 1988; Silver et al., 1988). Their fracture apertures reached dimensions of 10–30 cm (Barton and Moos, 1988), and fractures with apertures exceeding 1 cm typically occurred every 1–50 cm (Barton and Zoback, 1992). In situ bulk permeability tests found that the permeability was 1–3 orders of magnitude greater than that determined from cored host rocks because of fractures acting as conduits (Morrow and Byerlee, 1988). Permeabilities were generally low, having values of 0.5×10^{-18}–1.67×10^{-18} m^2 at pressures of 5 percent above lithostatic. However, there was a single fracture with localized relative high permeability of 1.8×10^{-16} m^2 (Coyle and Zoback, 1988).

Encountered fracture zones were found to differ in:

1) fluid conductivity (Kharaka et al., 1988a);
2) temperature (Lachenbruch and Sass, 1988); and
3) isotope chemistry (Hammond et al., 1988; Torgensen and Clarke, 1992; Kharaka et al., 1988b).

These differences indicate poor fracture connectivity. The distinct fluid geochemistry in these fracture systems suggests a local origin for each fluid volume (Kharaka et al., 1988a), while δD and $\delta^{18}O$ values indicate that fluids are of meteoric origin, having undergone little water–rock interaction (Kharaka et al., 1988b).

Furthermore, a past fluid circulation at the Cajon Pass borehole site caused low-grade zeolite metamorphism in:

1) the fracture systems;
2) along subsidiary fault zones; and
3) in basement rock (James and Silver, 1988).

Outcrop studies in the vicinity of the borehole determined that this mineralization is pervasive within 1 km of the surface trace of the San Andreas Fault and diminishes to practically no mineralization at about 4 km from the fault (Vincent and Ehlig, 1988). The outcrops are a result of distinct exhumation, as the rocks now exposed at the surface near the borehole have been uplifted by 2.5–3.5 km since early Quaternary.

Zeolite minerals such as laumontite and stilbite occur here as plagioclase mineral replacements and as fracture fillings (James and Silver, 1988). Some fractures contain single-colored zeolites, while others contain multiply-colored and textured bands parallel to the fracture walls, which were caused by multiple stages of dilatation and mineralization (James and Silver, 1988).

This kind of hydrothermal alteration is associated with the circulation of geothermal fluids at or below 300°C (Irwin and Barnes, 1975; Wang, 1984; Wang et al., 1986).

Isotope concentrations from fluids sampled during drill-stem tests of the Cajon Pass borehole found helium to be of crustal origin, and ^{4}He model ages of 33,000 – 5x10^{-6} years were calculated (Torgensen and Clarke, 1992). Taking into account the depth interval in which fluids were sampled, a vertical transport velocity of 0.04–6 cmyr^{-1} is evoked (Torgensen and Clarke, 1992). This contrasts with vertical velocities obtained from pressure head and in situ permeability measurements, amounting to about 10^{-5} cmyr^{-1} (Coyle and Zoback, 1988). To satisfy both of these conditions, fluid exchange must have occurred at least once in 33,000 – 5x10^{-6} years. This suggests that fluid flow had a transient character and it was generally sealed within the fracture system with punctuated intervals of rapid flow (Torgensen and Clarke, 1992).

The second scientific borehole, SAFOD, included an initial vertical well providing the following petrologic data from two cores (Schleicher et al., 2006):

1) a clay-rich shear zone at 3,067 m; and
2) a mud rock at 3,436 m.

Both have striated slickensides and mineralized veins. They contain thin films of smectite, which has been formed by dissolution–precipitation reactions (Schleicher et al., 2006). Fluids flowing through open fractures must have been at temperatures of 130°C, as indicated by smectite precipitation (Moore and Reynolds, 1997; Robinson and Santana de Zamora, 1999). The creeping movement of the San Andreas Fault at this location likely keeps fractures open longer, so that slow processes such as nucleation and crystal growth are possible (Schleicher et al., 2006).

Leaving the exact division of San Andreas Fault segments into locked, aseismically creeping and transitional, we can discuss the associated fluid flow using the geochemical data in the remaining text of this subchapter. We will be looking at geochemical studies of regional scope, which have sampled springs and wells, especially those that show signs of deep water circulation, in the vicinity of faults of the entire San Andreas Fault system.

Samples from forty-one thermal and saline springs, several wells, and twenty cold springs were collected between San Francisco and Los Angeles (Kharaka et al., 1999). A quick look at the entire discharge system indicates a conspicuous lack of high-discharge thermal springs associated with the main surface trace of the San Andreas Fault, despite the distinct topography. Some of the factors controlling this behavior include low precipitation, which is below 50 cmyr^{-1} (Planert and Williams, 1995), and impermeable lithologies that bound the fault in many locations. However the main factor seems to be the regional permeability structure. This is suggested by the fact that the majority of high-discharge thermal springs is associated with ancillary transform, normal, and reverse faults on the flanks of the Traverse and Coast Ranges and only minor fluid circulation occurs in the shallow crust along the San Andreas Fault itself (Kharaka et al., 1999).

δD and δ^{18}O values show contribution neither from the mantle nor from the magmatic sources. The majority of analyzed samples plots on the Global Meteoric Water Line, indicating a meteoric origin and low-residence times (Kharaka et al., 1999). Low tritium isotope concentrations of less than 0.1 tritium units show that there is no modern water present in spring discharge (Kharaka et al., 1999). Samples from sedimentary sequences, however, show shifts from the Global Meteoric Water Line, having increased δ^{18}O values accompanied by increased salinity. While small shifts are most likely the result of evaporation, larger shifts are the result of either fluid exchanges with minerals or the mixing of fluids with waters with a different origin and evolutionary history (Kharaka et al., 1999).

Chemical geothermometry analyses focused on the concentrations of SiO_2 and certain cations were used to estimate both the equilibrium subsurface temperature and maximum circulation depth (Fournier et al., 1974). About half of the samples give equilibrium temperature values above 80°C, with a maximum of 150°C (Kharaka et al., 1999). Circulation depths calculated using geothermal gradients and mean annual surface temperatures reveal circulation depths of 2.5–6 km. This indicates the presence of the topography-driven fluid flow system. It is characterized by meteoric waters recharged in high topography and percolated down until they encountered a permeable pathway, such as a fault zone, where the fluid flowed relatively rapidly to the surface (Kharaka et al., 1999).

Furthermore, springs, seeps, and wells of the San Andreas Fault system have been analyzed for helium isotopes. ^{3}He/^{4}He ratios of 0.12–4.0 Ra have been documented, where Ra is the ^{3}He/^{4}He value of air and the ^{3}He/^{4}He value of mantle helium is 8 Ra. This indicates a 1–50 percent mantle contribution in the sampled waters (Kennedy et al., 1997). These helium fluxes are observed in waters near the subsidiary faults of the system as well as along the main San Andreas Fault (Kennedy et al., 1997).

A transport model has been proposed from the mantle to the San Andreas Fault, which requires that (Kennedy et al., 1997):

1) the brittle ductile transition is a permeable boundary; and
2) fluids passing through this boundary are at or near lithostatic pressure.

From ^3He measurements obtained in a borehole near the transitional San Andreas Fault segment at Parkfield, a CO_2 flux was extrapolated for the entire San Andreas Fault system. This was accomplished by using the proportions of volatiles measured for Mid-Ocean Ridge basalts to ^3He, resulting in an estimated CO_2 flux of 10^{11} molyr^{-1} (Kennedy et al., 1997). Because CO_2 has the same density, specific heat, and viscosity as water at seismogenic depths (e.g., Duan et al., 1996), CO_2 could lower the effective stress at seismogenic depths (Irwin and Barnes, 1980; Bredehoeft and Ingebritsen, 1990).

The CO_2 flux in the fault zone probably plays an important role in mineral dissolution and precipitation and also affects the fault dynamics (Kharaka et al., 1999). For example, the thermochemical data indicate that any water coupled with mantle helium and carbon would likely be incorporated into hydrous minerals during its ascent in the lower crust (Yardley, 1986; Frost and Bucher, 1994). However, while the presence of mantle helium is well resolved, note that the physical-chemical transport processes focusing mantle fluids into the fault zone are speculative (Kharaka et al., 1999).

Fluid flow associated with rifts

Fluid systems in hot-spot, transform, and rift settings contain fluid flow mechanisms controlled by the tectonic and sedimentologic character of the respective setting. This character controls the permeability structure and water-table configuration, which determine whether the fluid system undergoes forced or free convection. Forced convection is due to differences in hydraulic head; free convection is due to buoyancy. Hydrologically "open" settings develop forced convection by means of topography-driven or pressure-driven flow, while "closed" settings develop free convection by means of buoyancy-driven flow.

Forced convection takes place in settings characterized by:

1) development of significant topographic gradients in subaerially exposed areas (e.g., Garven and Freeze, 1984a, 1984b); and
2) development of the compaction-produced hydraulic heads increasing with depth in areas with fast deposition rates and fine-grained sediments (e.g., Bethke, 1985; Neuzil, 1986; Bethke and Corbet, 1988).

Free convection takes place in settings characterized by:

1) thermal perturbations within the upper crust induced by emplacement of intrusive bodies (Norton and Knight, 1977);
2) temperature variations at seafloor-spreading ridges inducing seawater circulation (Anderson et al., 1979; Friedman, 1999);
3) salinity variations in the vicinity of salt bodies (Hanor, 1987; Evans and Nunn, 1989);
4) uneven production of radiogenic heat within the lithosphere (Person and Garven, 1992);
5) thermal conductivity contrast among different lithologies (Person and Garven, 1992); and
6) localized crustal thinning (Person and Garven, 1992).

Fluid flow mechanisms

The fluid flow in a porous matrix under the effects of external loading, aquathermal pressuring, time, dehydration, and precipitation can be formulated as (Biot, 1941; Cooper, 1966; Bear, 1972; Sharp, 1983; Shi and Wang, 1986):

$$[\alpha_n/(1-n) + n\alpha_s + n\beta]dp/dt = k/\eta\nabla^2(p - \rho_f\phi) + [\alpha_n/(1-n) + n\alpha_s]d\sigma_m/dt + [n(n\gamma_f-\gamma_s) + (\gamma_s-\gamma_b)]T/dt + (1/1-n)(-\partial n/\partial t) + h/\rho_f, \quad (8\text{-}1)$$

where p is the pore fluid pressure, ρ_f is the fluid density, ρ_s is the density of solid grains, $\alpha_n = -\partial n/\sigma_e$ is the porosity compressibility, $\alpha_s = 1/\rho_s(\partial\rho_s/\sigma_e)$ is the compressibility of solid grains, β is the fluid compressibility, n is the porosity, γ_f is the thermal expansion coefficient of the fluid, γ_s is the thermal expansion coefficient of the solid, γ_b is the bulk thermal expansion coefficient, σ_m is the total mean stress due to overloading and tectonic compression, $\sigma_e = \sigma_m - p$ is the effective mean stress, $\phi = gd$ is the gravitational potential, g is the acceleration due to gravity, d is the depth below the surface, T is the temperature, k is the absolute permeability, η is the fluid viscosity, and h is the fluid source or sink due to dehydration of clays and other causes. d()/dt denotes a quantity-time derivative.

The term on the left side of the equation is a product of the rate of change in pore pressure with specific fluid storage. From left to right, the first term on the right side of the equation describes the pore fluid diffusion while the others describe various fluid sources. The first source term describes fluid sourcing by the porosity reduction driven by rate of change of mean stress, that is, sediment compaction. The second source term describes control by the thermal expansion coefficients of fluid and rock, the bulk thermal expansion coefficient, and the rate of change in temperature; that is, the contribution of aquapressuring to pore fluid generation. The third source term defines the control by porosity changes with time, however, in this case associated with creep of the rock skeleton and chemical processes. The fourth term describes the fluid source or sink associated with clay dehydration, hydrocarbon generation, gypsum-to-anhydrite transformation, mineral diagenesis, and oil-to-gas cracking (Osborne and Swarbrick, 1997).

The sum of fluid diffusion and fluid sourcing determines how the pore fluid pressure of the specific fluid flow system is generated, maintained, and dissipated. If the fluid sourcing dominates over fluid dissipation, the system develops extra fluid pressure. If the fluid sourcing and dissipation are equal, the pore fluid pressure regime will be in a steady state. If the diffusion is dominant over sourcing, the system reduces its pore fluid pressure.

Depending on which flow-controlling factors dominate, the extensional basins host various fluid flow mechanism and their combinations.

Topography-driven fluid flow

If the recharge of the meteoric water in elevated permeable highs next to extensional basins manages to support the groundwater discharge, the flow is maintained. The time t_d determining the diffusion of transients in such flow system is (Lachenbruch and Sass, 1977):

$$t_d = L^2/4D, \tag{8-2}$$

where L is the diffusion distance and D is the hydraulic diffusivity, related to:

$$D = K/S_s, \tag{8-3}$$

where K is the hydraulic conductivity and S_s is the sediment specific storage. The hydraulic conductivity can be defined either for fault/fracture (Kohl et al., 2000):

$$K = 4\cdot\log\,[1.9/(\varepsilon/d_h)]\cdot\sqrt{(a/\rho_f)}, \tag{8-4}$$

where ε/d_h is the relative fracture roughness defined as the ratio of the mean height of asperities and the hydraulic diameter, a is the fracture aperture, and ρ_f is the fluid density, or for homogeneous and isotropic porous matrix (Deming, 1994b):

$$K = (S_s\nabla^2 h)\,\partial h/\partial t, \tag{8-5}$$

where h is the hydraulic head, t is the time, and $\nabla^2 h$ is the curvature of the head field.

The specifics of extensional basins is that they do not only have anisotropic hydraulic conductivity, which contains the maximum conductivity parallel to bedding and minimum conductivity perpendicular to bedding, but these principal directions of the conductivity ellipsoid also rotate and not necessarily only about horizontal axes. This three-dimensional problem can be reduced to the two-dimensional problem when we look at it using the cross-section through an orthogonal rift, which is parallel to the tectonic transport. Here the calculations of horizontal and vertical hydraulic conductivities become (Mailloux et al., 1999):

$$K_{xx} = K_{max}\sin^2\theta + K_{min}\cos^2\theta$$
$$K_{zz} = K_{max}\cos^2\theta + K_{min}\sin^2\theta$$
$$K_{xz} = K_{zx} = (K_{max}-K_{min})\sin\theta\cos\theta, \tag{8-6}$$

where θ is the amount of block rotation.

The fluid diffusion time, t_d, is normally several magnitudes smaller than the rift development time, that is, the time of active maintenance of the topographic gradients. An illustrative example for the former comes from the Rhine Graben, where ^{14}C dating of the fluids in the discharge area revealed a residence time of 31,000 years (Plothner, 1988), while the typical time span for rifting discussed in Chapter 1 ranges:

1) from 10 to 120 Ma for narrow rifts, being sometimes continuous and sometimes separated by periods of tectonic quiescence;
2) from 0.1 to 9 Ma for wide rifts, being sometimes continuous and sometimes separated by periods of tectonic quiescence; and
3) from 0.25 to 10 Ma for pull-apart basins, being sometimes continuous and sometimes separated by periods of tectonic quiescence or periods of contraction.

Because of the large difference between diffusion and rift development times, the flow system can be characterized as a roughly steady-state one (Deming, 1994); that is, occurring everywhere topographic relief exists. The fluid recharging the system at topographic high moves down seeking the path of least resistance or a combination of the highest hydraulic conductivity K with the highest hydraulic head gradient ∇h. The hydraulic head h is defined as the mechanical energy per unit mass of fluid, equivalent to the height above a reference height to which fluid will rise in a tube (Hubbert, 1940):

$$h = z + p/\rho_f g, \tag{8-7}$$

Table 8.1. Fluid transport lengths from selected basins affected by topography-driven flow.

Area, basin	Max fluid transport length (km)	Max downward transport under the recharge area (km)	Max subhorizontal transport within the aquifer (km)	Max upward transport toward the discharge area (km)	Source
Pechelbronn area, central Rhine Graben	35	5.0	28	2.0	Person and Garven (1992)
Initial rifting stage, Socorro, La Jencia sub-basins of the Rio Grande Rift	21.0	3.0	15.0	3.0	Mailloux et al. (1999)
Sea of Galilee pull-apart basin, Dead Sea Valley	17.5	1.0	15.0	1.5	Gvirtzman et al. (1997)
Worms area, Upper Rhine Graben	16	3.0	10.0	3.0	Lampe and Person (2002)
Mature rifting stage, Socorro, La Jencia sub-basins of the Rio Grande Rift	15.0	7.0	1.0	7.0	Mailloux et al. (1999)
Larderello area, internal Northern Apennines	11.0	5.0	1.0	5.0	Bellani et al. (2004), Ranalli and Rybach (2005)

where z is the elevation term and remaining pressure term contains p being the pore fluid pressure, ρ_f being the fluid density, and g being the acceleration due to gravity.

Aquifer satisfying the least resistance or highest combined K and ∇h conditions can be anyone from the potential choices in the entire rift fill. Therefore, the fluid transport length can include almost the entire rift fill thickness plus the entire rift unit width, which is estimated to range from 20 to 70 km, based on data from the South Atlantic region (own data), the East African Rift system (Cornford et al., 2005), and worldwide comparison of continental rifts (Lippman et al., 1989). Table 8.1, which lists selected transport lengths from rift and pull-apart case studies ranging from 11 to 35 km, indicates that transport lengths tend to be affected by the internal fault block architecture.

Fluid flow velocity, q, in the topography-driven system is proportional to the hydraulic head gradient and host rock permeability according to Darcy's law (Darcy, 1856; Hubbert, 1969):

$$q = -[(k\rho_{f0}g/\eta)\nabla h], \tag{8-8}$$

or

$$q = -K\nabla h, \tag{8-9}$$

where k is the permeability tensor, ρ_{f0} is the reference fluid density, for example, freshwater, g is the acceleration due to gravity, η is the fluid dynamic viscosity, ∇h is the hydraulic head gradient, and K is hydraulic conductivity tensor.

Topography-driven fluid flow velocities are relatively high, ranging from millimeters to meters per year (see Table 8.2a.), indicating that this flow is important for both mass and heat transport.

Natural examples of topography-driven systems come from the Rhine Graben (Person and Garven, 1992; Lampe and Person, 2002), the Rio Grande Rift (Mailloux et al., 1999), and the Sea of Galilee pull-apart basin (Gvirtzman et al., 1997).

Both numerical simulation studies of the northern and central Rhine Graben (Lampe and Person, 2002; Person and Garven, 1992, respectively) indicate a major change of the fluid flow system during the transition of the initial submarine rift stage to the subsequent subaerial rift stage regardless of the timing of the rift development being a little different. To focus on a topography-driven flow in this text, we focus only on the subaerial stage.

The northern Rhine Graben simulation (Lampe and Person, 2002) indicates a phenomenon also described later in the case of the Rio Grande study. The best-fitting model to natural data requires modeled

Table 8.2. a) Calculated flow rates for topography-driven fluid flow in selected basins.

Area	Rate (my^{-1})	Source
Rhine Graben	0.0084	Plother (1988), Person and Garven (1992)
Medium to tight faults, Upper Rhine Graben	0.001–100000	Lampe and Person (2002)
Rio Grande Rift zone	0.01	Mailloux et al. (1999)
Rhine Graben	0.12–0.13	Person and Garven (1992)
Southern Po Basin	1.0	Wygrala (1989)
Middle Miocene-Quaternary upper aquifer, Upper Rhine Graben	1.0–10.0	Lampe and Person (2002)
Reasonable representative rate interval	***0.0084–10.0***	

b) Calculated flow rates for compaction-driven fluid flow in selected basins.

Area	Rate (my^{-1})	Source
North Sea	0.00003	Hermanrud (1986)
Sea of Galilee, Dead Sea Rift Zone	0.01–0.1	Gvirtzman et al. (1997)
Reasonable representative rate interval	***0.00003–0.1***	

c) Calculated flow rates for buoyancy-driven fluid flow in selected basins.

Area	Rate (my^{-1})	Source
Rhine Graben	0.0032–0.0045	Person and Garven (1992)
Open fault, Rhine Graben	0.0032–0.032	Bachler et al. (2003)
Kurnub Aquifer, Sea of Galilee, Dead Sea Rift Zone	0.1	Gvirtzman et al. (1997)
Lau Basin, Tonga	0.32	Schardt et al. (2006)
Arad Aquifer, Sea of Galilee, Dead Sea Rift Zone	1.0	Gvirtzman et al. (1997)
HYC ore deposit, Mt. Isa Basin, North Australia	1.0	Yang et al. (2006)
Century ore deposit, Mt. Isa Basin, North Australia	1.5–8.0	Yang et al. (2006)
Kidd Creek ore deposit, Mt. Isa Basin, North Australia	1.75	Yang et al. (2006)
Lau Basin, Tonga	31.5	Schardt et al. (2006)
Reasonable representative rate interval	***0.0032–8.0***	

faults to have moderate- to low-permeability character. This long-time permeability average for faults indicates that their opening time periods are not significant in geologic time. This simulation documents a recharge in the uplifted footwall edge located at the eastern main graben boundary. Lateral flow into the basin utilizes shallow and deep aquifers, although the flow rate in the deep one pales in comparison with a fast rate of 1–10 my^{-1} in the shallow Tertiary-Quaternary one. Discharging faults are located in the western to central portion of the basin, having flow rates in millimeters per year.

The central Rhine Graben fluid flow system is recharged also in the east. A discharge is located in the western to central portion of the basin, at the set of faults bounding a local horst structure, known for the location of geothermal areas such as Soultz-sous-Forets (Person and Garven, 1992). The numeric simulation shows that the fastest flow rates, which reach 0.13 my^{-1} in this area, are located around the main graben-bounding fault because of the largest hydraulic head gradient occurring here.

The Rio Grande Rift study demonstrates how the tectonic development of several fault blocks in La

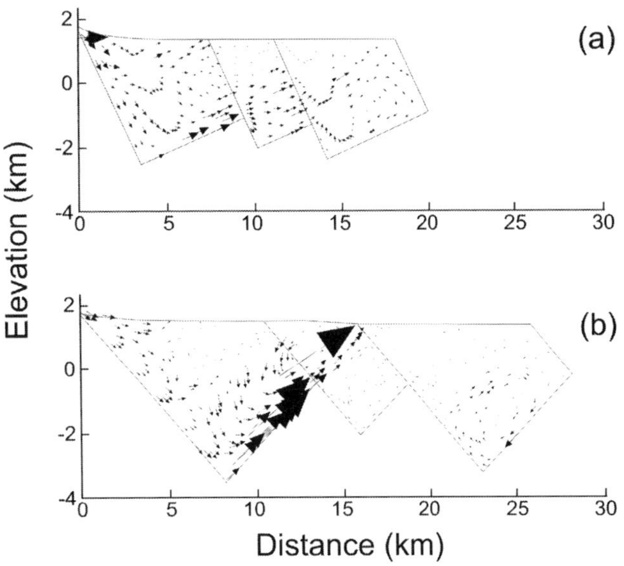

Figure 8.19. Velocity vector evolution within the Rio Grande Rift at (a) 14 Ma and (b) 0 Ma (Mailloux et al. 1999). Shaft lengths and vector heads are proportional to the Darcy velocity with vector heads placed on the element. Recharge occurs adjacent to the Magdalena Mountains in the western La Jencia sub-basin throughout rift evolution. Discharge occurs in the eastern Popotosa Basin from the beginning of the simulated time period to 5 Ma with recharge from the Magdalena Mountains flowing beneath the playa deposits. The area of discharge shifts to the present-day location of Socorro Springs in the western Popotosa sub-basin at approximately 3 Ma.

Jencia and Socorro sub-basins kept changing the fluid flow system in the region (Mailloux et al., 1999). The flow system was controlled by the interaction of changing geometry of several aquifers and faults. While the early rift stage between 14 and 3.5 Ma experienced a basin-wide, topography-driven flow, the advanced rift stage between 3.5 and 0 Ma underwent discharge shift to the center of the basin (Figure 8.19a, b). Both rift stages had a recharge area in the adjacent Magdalena Mountains in the west, having a rate of 0.2 mmy^{-1} and an infiltration effectiveness of 5 percent.

Another interesting conclusion is that the fluid flow model for long-term development stages requires moderately permeable faults to match the natural data. In reality, the model represents the entire stage history characterized by stress cycles controlling episodes of fault activity characterized by high permeability and episodes of tectonic quiescence characterized by low permeability faults, which apparently even up in the model in geological time.

The early rift-related topography-driven flow system utilized only a basal aquifer made of Oligocene tuffs and discharged at the very eastern flank of the basin. The advanced rift underwent block faulting, basal aquifer rotation in fault blocks, uplift and erosion of footwall edges of rotated blocks, and deposition of the upper aquifer. A juxtaposition of the basal aquifer with the upper aquifer through the hydrologic window in the footwall edge in the Socorro Springs area resulted in a shortcut and reduction of the flow system length to half of its original size.

Numerical simulation of the Sea of Galilee pull-apart basin (Gvirtzman et al., 1997) documents a complex interplay of topography- and buoyancy-driven fluid flows depending on the interaction of two aquifer systems with a complex fault pattern, which varies from area to area. Topography-driven flow dominates the study area underneath the basin itself. It has recharge in the Galilee Mountains adjacent to the pull-apart from the west and discharge along eastern boundary faults as a system of hot springs (Figure 8.20). Lateral flow utilizes shallower and deeper systems of aquifers. The former is characterized by the fastest rate of 40 my^{-1} at its top. The latter is characterized by slower rates, that is, 5 my^{-1} and 0.8 my^{-1} in the Lower Cretaceous Kurnub and Jurassic Arad aquifers, respectively. Discharging springs are formed by a mixture of freshwater from the shallower Cretaceous Judea aquifer with hotter and saline fluids from the Kurnub and Arad aquifers. The Kurnub and Arad aquifers, located further away from the pull-apart on both sides where they rest undeformed by faults, developed free convection cells described a bit later in the subchapter on buoyancy-driven flow.

Person and Garven (1994) conducted an interesting parametric study of the topography-driven flow and its competition with buoyancy- and compaction-driven flows. It numerically tested a set of three continental rift basins (Figure 8.21):

1) first characterized by highly permeable basin fill;
2) second with basin fill having a moderate permeability of 10^{-15} m^2; and
3) third characterized by basin fill with low permeability of 10^{-16} m^2.

Furthermore, each of these basins is characterized by a zone of alluvial sediments with a high permeability of 10^{-14} m^2 located along the main boundary fault and missing along the opposite basin boundary.

The modeling indicates that the first rift basin scenario underwent topography-driven fluid flow. Meteoric waters recharged through the uplifted footwall edge next to the main boundary fault, descended through the fault core and the damage zone of the boundary fault, and migrated laterally into the basin to discharge somewhere in the basin center (Figure 8.22a). Highly permeable sediments between the basin center and the

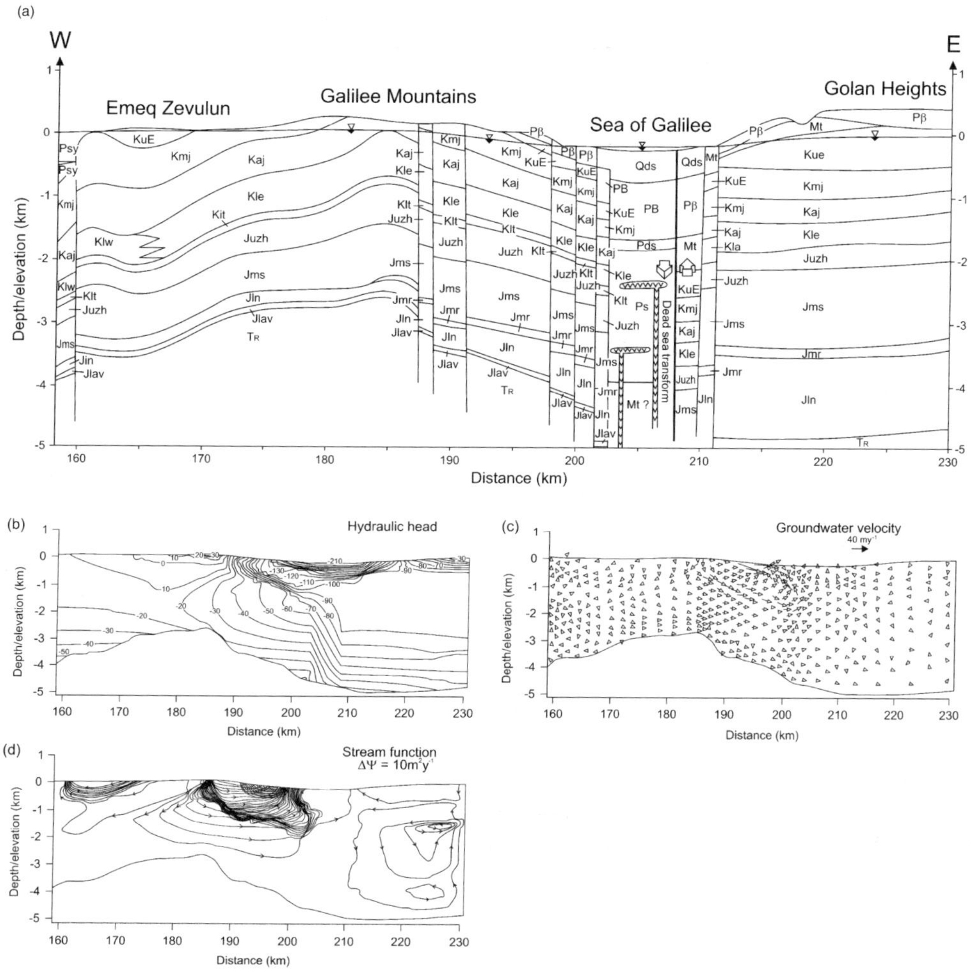

Figure 8.20. Fluid flow simulation for the Sea of Galilee pull-apart basin (Gvirtzman et al., 1997). (a) Geologic cross-section through the Galilee Mountains, the Sea of Galilee, and the Golan Heights. Formations were grouped into hydrological units on the basis of their lithology and relative permeability. The thick line beneath the Sea of Galilee is the transform and the nearby arrows give a sense of motion. Note gabbro intrusions within the basin. Quaternary: Qds – marly Lisan Formation, Pliocene: Pb – Upper basalt formation, Pds – unnamed marly formation, Psy – Yafo marly formation, Ps – Sedom Formation containing salt and gabbro, Miocene: Mt – Lower basalt formation, Upper Cretaceous–Eocene: KuE – Mount Scopus, Adulam, and Maresha Formations with chalk and marl, Albian-Turonian Judea Group: Kmj-Kamon, Dir Hana, Sakhnin, and Bina Formations with carbonates, Kaj-Zalmon, Yahini, Hidra, Asfuri, and Rama Formations with marl and limestone, Lower Cretaceous Kurnub Group: Kle-Hatira, Nebi Said, En el Assad Formations with sand, limestone, and marl, Klw – Helez, Telamin, and Yavne Formations with limestone and sand, Klt – Tayassir volcanic rocks, Jurassic Arad Group: Juzh – Zohar and Halutza limestone formations, Jms – Sederot carbonate formation, Jmr – Rosh Pina marly formation, Jln – Nirim carbonate formation and Jlav – Asher volcanic rocks. (b–d) Results of the coupled variable-density groundwater flow and heat transfer model, carried out along the geologic cross-section from (a). Results include (b) equivalent freshwater hydraulic head distribution; and (c) computed groundwater velocities (m/yr), where the vector length is linearly proportional to flow rate. Note that high velocities are found in the two upper aquifers (Judea group) emerging as springs at the Sea of Galilee western coast (Fuyla Springs). (d) Stream function exhibiting the deep, buoyancy-driven, free convection cell beneath the Golan Heights, and the shallow, gravity-driven convection beneath the Galilee Mountains.

subordinate boundary underwent free convection by several convection cells.

The second rift basin scenario underwent topography-driven flow with similar characteristics, only the decreased permeability of the basin fill prevented buoyancy-driven convection cells from developing (Figure 8.22b).

The third rift basin scenario underwent topography-driven flow restricted to a narrow zone including the fault zone and adjacent coarse alluvial and sand flat

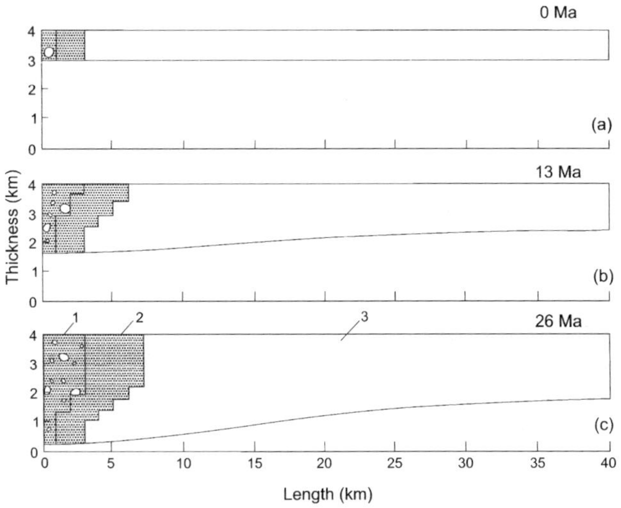

Figure 8.21. Hydro-stratigraphic evolution during the initial stage of continental-rift basin formation from one of the numerical models used in the two-dimensional sensitivity study (Person and Garven, 1994). The numbers identify the different hydro-stratigraphic units used in the simulations (1 – alluvial fan; 2 – sandflat; 3 – lacustrine deposits). The rock properties associated with each of the hydro-stratigraphic units are the following: unit 1, horizontal and vertical permeabilities are 10^{-14} and 10^{-15} m^2, porosity 50 percent and porosity fit coefficient -5×10^{-3} 1/m; unit 2, horizontal and vertical permeabilities are 10^{-15} and 10^{-16} m^2, porosity 55 percent and porosity fit coefficient -5.5×10^{-3} 1/m; unit 3, horizontal and vertical permeabilities are 10^{-16} and 10^{-18} m^2, porosity 60 percent and porosity fit coefficient -6×10^{-3} 1/m.

sediments (Figure 8.22c). The entire remaining portion of the basin experienced compaction-driven fluid flow. A comparison of modeled scenarios indicates a critical role of the permeability structure of the rift basin in controlling the combination of occurring fluid flow mechanisms.

Compaction-driven flow

The sediment compaction is described by the first fluid source term from Equation (8-1), $([\alpha_n/(1-n) + n\alpha_s]d\sigma_m/dt)$. The subtraction of the pore fluid pressure, p, from the total mean stress, σ_m, results in the effective stress, σ_e, which controls the porosity reduction by its increase. One of the empirical equations describing this relationship is given by Athy (1930):

$$n = n_o \exp(-b\sigma_e), \qquad (8\text{-}10)$$

where n_o is the porosity at the surface and b is expressed as:

$$b = 1/(\rho - \rho_f)gH, \qquad (8\text{-}11)$$

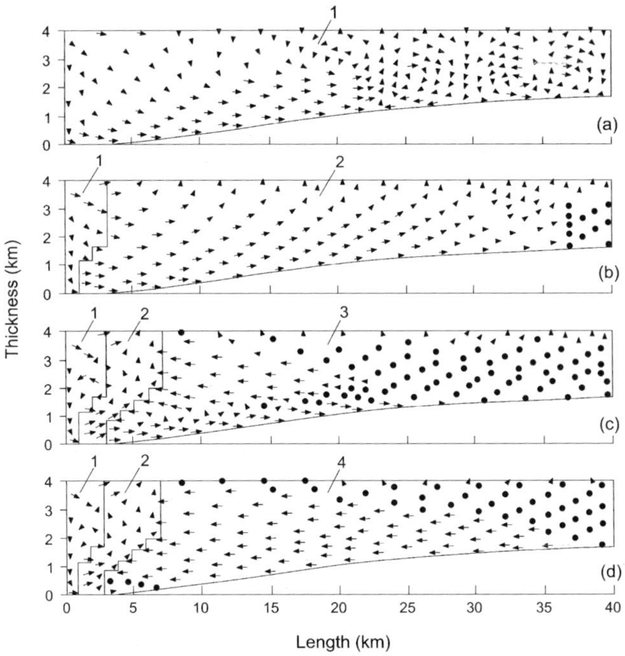

Figure 8.22. Effect of lateral variations in permeability on computed groundwater velocity vectors during the initial phase of rift basin formation represented by three possible facies distributions shown in Figure 8.21 (Person and Garven, 1994). The numbers identify the different hydro-stratigraphic units used in Figure 8.21. Number 4 indicates modified properties of lacustrine sediments having horizontal and vertical permeabilities of 10^{-17} and 10^{-19} m^2, a porosity of 60 percent, and a porosity fit coefficient of -6×10^{-3} 1/m. The model results present hydrologic conditions 26 Ma after the onset of basin formation. The length of the shafts of the velocity vectors are scaled to the log of the magnitude of the computed Darcy velocity for each plot. Direct comparisons in shaft lengths between velocity vector plots, however, should not be made. The closed circles indicate that the computed velocity was >1,000 times lower than maximum velocity.

where ρ is the bulk density of a porous medium, ρ_f is the fluid density, g is the acceleration of gravity, and H is a material parameter. The bulk density can be determined from (Shi and Wang, 1986):

$$\rho = \rho_f n + \rho_s(1-n), \qquad (8\text{-}12)$$

where ρ_f and ρ_s are fluid and sediment densities, respectively.

Parameter b has values of 10^{-8}-10^{-7} Pa^{-1} for shale and $5*10^{-9}$–$5*10^{-8}$ Pa^{-1} for sandstone (Athy, 1930; Hoholick et al., 1984). The porosity changes with compaction that Smith and Wiltschko (1996) compiled for various sandstones, shales, and carbonate indicate that relationship 8-10 is apparently controlled by additional factors, such as rock age, degree of lithification, and other lithology-related parameters.

The porosity–effective stress relationship also depends on stress path (e.g., Lambe and Whitman, 1969). If the loading stress is smaller than stresses that loaded the sediment in the past, the porosity change with stress will be much less significant than in the normal case. Therefore, in the absence of tectonic stress, in the case when the only loading is imposed by burial, Athy's relationship (8-10) can be modified to:

$$n = n_{max}\exp[-b'(\sigma_e - \sigma_{max})]. \quad (8\text{-}13)$$

The compaction contributes to pore fluid pressure generation by the amount resolvable from Equation (8-1):

$$dp = [\alpha_n/(\alpha_n + n\beta(1-n))]\, d\sigma_m. \quad (8\text{-}14)$$

Upon decreasing porosity due to compaction, pore fluids are expelled from the shallow-seated sediments. For example, freshly deposited sandstones have initial porosities in the 39–49 percent range (Lundegard, 1992), which change to porosities of 15–25 percent after a 2–3 km burial because of the grain rearrangement and certain contributions of chemical dissolution at grain contacts (Sclater and Christie, 1980). Below a depth limit of 2–3 km, the mechanical compaction pales down and the remainder of compaction is taken by diagenetic cementation. This can be illustrated by clay compaction history, which starts with porosities of 65–80 percent at the surface and finishes at depths of 4–6 km with porosities of 5–10 percent due to grain rearrangement and subsequent ductile deformation.

Because of the layered character of the extensional basin fill, the compaction-driven fluid flow is controlled by the permeability structure of the basin fill, which is usually anisotropic. The low-permeability layers of the permeability structure tend to be dominated by fluid perpendicular to bedding planes, while the high-permeability layers tend to undergo a fluid flow parallel to their bedding planes. As a result, pore fluids expelled from a shale horizon tend to flow vertically until they encounter a reservoir layer where they follow its geometry (Bethke, 1985). The sedimentary sequence with few reservoirs isolated inside the shale-dominated section is characterized by a vertical flow-dominant fluid system (Mann and Mackenzie, 1990).

A natural example of the compaction-driven flow comes from the Sea of Galilee pull-apart basin (Gvirtzman et al., 1997) where the 500 m-thick Miocene-Quaternary fill right under the lake is characterized by steep hydraulic head gradient toward the lake, driving the compaction-driven fluid flow inside of the aquitard perpendicularly to bedding planes (Figure 8.20).

The syn-rift situation in the La Jencia and Socorro sub-basins of the Rio Grande Rift resulted in a similar effect during the deposition of Playa sediments forming the uppermost 1 km of the 26-5 Ma-old Popotosa Formation (Mailloux et al., 1999). Compaction of Playa sediments created about 20 m of excess hydraulic head at 1 km depth in the rift unit center, resulting in flow upward and across bedding planes. Because Playa sediments represented only less than 20 percent of the syn-rift fill at the time of their deposition and compaction, their compaction-driven flow was insignificant in comparison with topography-driven flow active at the same time.

Another case of insignificant compaction-driven flow comes from the spreading ridge of the Lau back-arc basin, Tonga, where the low-permeability sedimentary cover undergoes upward-directed flow (see Schardt et al., 2006). This flow has no impact on dominating buoyancy-driven flow in the area.

It can be concluded in general that the basin, which contains the sedimentary sequence with areally-extensive reservoirs, is characterized by a lateral flow-dominant fluid system (e.g., Magara, 1976; Mann and Mackenzie, 1990). A natural example comes from the Pechelbronn area of the central Rhine Graben, where the first, submarine, stage of the Oligocene-present rift development was characterized by subordinate compaction-driven flow combined with dominant buoyancy-driven flow. The compaction contributed by a hydraulic gradient of only 0.13 m per km to the control of the flow system, assisting buoyancy-driven flow in focusing flow into the basal aquifer and then into faults in the discharge area. The subhorizontal flow in the basal aquifer reached a rate of about 0.0045 my^{-1}, as the numerical model indicated. During the subsequent subaerial rift development stage, both compaction and buoyancy-driven fluid flows became overwhelmed by topography-driven flow.

Another example of the subhorizontal flow comes from the set of numerical models loosely related to extensional basins such as the Angolan Cretaceous Rift Basin (see Person and Garven, 1994). In these models, a rift fill is schematically represented by high-permeability alluvial fans and sand flats located along the boundary faults on a deeper side of the half-graben and by low-permeability lacustrine sediments occupying the remainder of the rift unit. A prescribed subsidence of 0.1 mmy^{-1} represents a worldwide average for rift basins (see Tiercelin and Faure, 1978). The numeric simulation indicates that as the permeability contrast between high- and low-permeability rift facies reaches certain critical value, topography-driven flow becomes

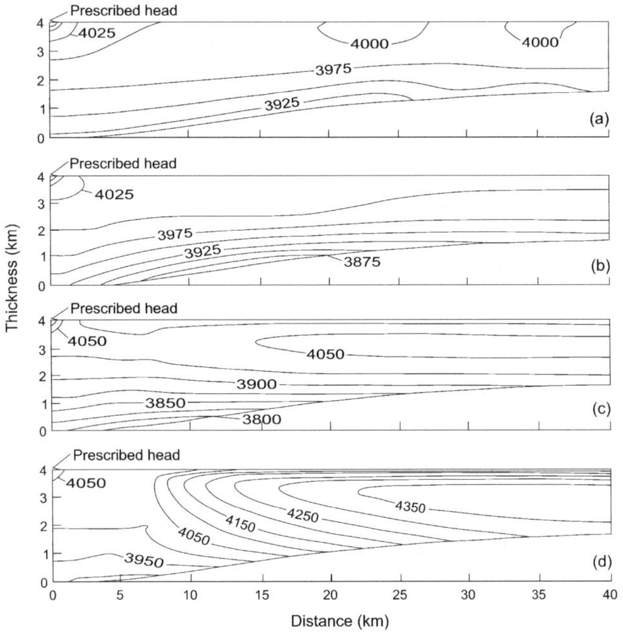

Figure 8.23. Effect of lateral variations in permeability on computed hydraulic heads (in meters) during the initial phase of continental-rift basin formation (Person and Garven, 1994). The contour interval is 25 m. The imposed hydraulic-head boundary condition is plotted above the land surface. The position of the water table was gradually increased from 0 to 100 m over the 26 Ma-long simulation period along the left side of the basin. Facies distribution is the same as that in Figure 8.22.

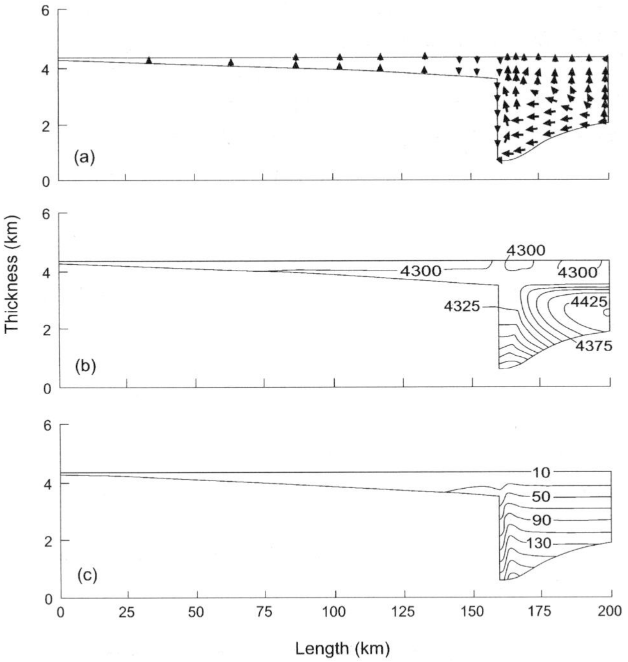

Figure 8.24. Effect of permeability variations on computed groundwater velocity vectors, hydraulic heads (in meters), and temperatures (in degrees Celsius) 5 Ma after the start of the flexural stage of rift basin formation (Person and Garven, 1994). Onlap sedimentary layers above the syn-rift fill include a basal aquifer overlain by an aquitard. The rock properties associated with onlapping hydro-stratigraphic units are: aquifer, horizontal and vertical permeabilities are 10–14 and 10–15 m^2, porosity 50 percent and porosity coefficient –5×10-3 1/m; aquitard, horizontal and vertical permeabilities are 10–17 and 10–19 m^2, porosity 60 percent and porosity coefficient –6 × 10–3 1/m.

isolated in the rift margin, utilizing fault zones, alluvial fans, and sand flats, while the rest of the rift fill undergoes compaction-driven fluid flow. Assuming homogeneous sediments in the rest of the basin, flow takes place across layers in both the central and shallower end of the half-graben, while it is subhorizontal in the region where lacustrine sediments are in close proximity to alluvial fans and sand flats (Figure 8.22c, d). The former is driven by excess pressures increasing toward the surface, that is, in the region where the compaction relationship with depth has the largest gradient (Figure 8.23d).

A numerical simulation, which was done for the thermal subsidence stage of the same rift development, indicates that the compaction-driven fluid flow continues in the syn-rift section buried underneath the thermal subsidence-related section (Person and Garven, 1994; Figures 8.24a, b). As the thermal subsidence stage proceeds, the importance of compaction vanes as the thermal anomaly within the lithosphere relaxes.

Disequilibrium compaction is a typical mechanism generating overpressure in numerous extensional basins and passive margins. It tends to occur in basins characterized by combinations of facies such as thick clay, mud, marl, and shale in the sedimentary sections undergoing rapid burial. Examples come from the Gulf of Mexico (Dickinson, 1953), the Mahakam Delta (Burrus et al., 1994), the North Sea (Mann and Mackenzie, 1990; Audet and McConell, 1992), and the Williston Basin (Burrus et al., 1994). Overpressuring acts against the sediment compaction and can preserve higher-than-normal porosities in situations where cementation is insignificant (Osborne and Swarbrick, 1997).

Buoyancy-driven flow

Equation (8-7) can be rewritten as:

$$h = z + p/\rho_{f0}g, \tag{8-15}$$

to express an equivalent hydraulic head using reference density ρ_{f0}. One of the examples can be equivalent freshwater hydraulic head. z is the elevation term for the hydraulic head h. The second term on the right side is the pressure term for the hydraulic head, h, p is the pore fluid pressure, and g is the acceleration due to gravity.

Velocity of the laminar flow of an inhomogeneous fluid inside anisotropic porous material under pore pressure balance and ignoring inertial forces can be calculated from (Bear, 1972):

$$q = -(k/\eta)[\nabla p + \rho_f \nabla z], \quad (8\text{-}16)$$

where q is the Darcy velocity or specific discharge, k is the permeability tensor, η is the fluid dynamic viscosity, ∇p is the pore fluid pressure gradient, ρ_f is the fluid density and ∇z is the elevation gradient. The specific discharge represents the volumetric flow rate per unit time divided by the cross-sectional area, which is perpendicular to the flow direction. The Darcy velocity, q, is related to the linear velocity, v, by:

$$v = q/n, \quad (8\text{-}17)$$

where n is the porosity.

Equation (8-16) can be rewritten as (Deming, 1994):

$$q = -(k\rho_{f0}\, g/\eta)[\nabla h + (\rho_f - \rho_{f0}/\rho_{f0})\nabla z], \quad (8\text{-}18)$$

where k is the permeability tensor, ρ_{f0} is the reference fluid density, g is the acceleration due to gravity, η is the fluid dynamic viscosity, ρ_f is the fluid density, and ∇h and ∇z are hydraulic head and elevation gradients. This equation demonstrates that fluid flow can result from a hydraulic head gradient ∇h and fluid density gradients, giving rise to buoyancy forces. The density gradients develop from thermal expansion or salinity gradients (Hanor, 1979; Frape and Fritz, 1987; Ranganathan and Hanor, 1987).

If the fluid system occurs in the homogeneous porous layer defined by two impermeable boundaries, one-dimensional mathematical analysis allows us to indicate free convection due to thermal expansion when the Rayleigh number R is higher than 27.1 (Turcotte and Schubert, 1982):

$$R = (\alpha g \rho_f^2 C k \Delta z^2 \gamma)/\eta\lambda, \quad (8\text{-}19)$$

where α is the fluid coefficient of thermal expansion, g is the acceleration due to gravity, ρ_f is the fluid density, C is the heat capacity of the fluid, k is the permeability of the porous medium, Δz is the height or thickness of the convection cell, γ is the thermal gradient, η is the dynamic viscosity of the fluid, and λ is the thermal conductivity of the porous medium. Thermal conductivities of sedimentary and other rocks are described in detail in Chapter 10.

It is reasonable to expect fluid density increase with depth, which occurs in response to the increase in salinity, observed in various basins (e.g., Hanor, 1979; Frape and Fritz, 1987; Ranganathan and Hanor, 1987). If the salinity gradient is flat, the geothermal gradient itself can be high enough to cause density inversion and convective overturn, provided high enough permeability of host rocks (Deming, 1994).

The analysis using the Rayleigh number criterion is designed for scenarios where the top and basal surfaces of the medium potentially undergoing convection are parallel to each other and perpendicular to the gravity vector. A brief look at the structural architecture of extensional basins described in Chapter 1 (Figures 1.4, 1.7–1.9, 1.11–1.12, 1.16, 1.18–1.20) reveals that, because of block tilting associated with tectonic development of extensional basins, the reservoir layers in question can depart from this assumption by several tens of degrees. In fact, the convection is rather likely in horizons tilted more than 5° and its presence is not necessarily indicated by Rayleigh-type analysis (see Criss and Hofmeister, 1991).

Equation (8-18) indicates that vertical and lateral fluid flows are likely to occur when any lateral fluid density gradient exists. Therefore, it is reasonable to assume that buoyancy-driven fluid flow is rather common in the extending upper crust in scenarios that do not allow topography-driven flow to prevail. These include:

1) extensional scenarios without subaerial exposure such as seafloor-spreading ridges and their surroundings (Lowell, 1975; Lowell and Rona, 1985; Fisher and Becker, 1995; Yang et al., 1998; Schardt et al., 2006);
2) submarine environments in extensional basins (Person and Garven, 1992; Lampe and Person, 2002; Yang et al., 2006); and
3) special subaerial extensional scenarios lacking distinct topographic gradients such as deep isolated permeable aquifers (Gvirtzman et al., 1997).

A natural example of the spreading ridge comes from the Lau Ridge of the Lau back-arc basin, Tonga (Schardt et al., 2006). Numerical simulation of the flow regime at Lau Ridge studied the effect of a system of faults by simulating their different permeabilities, numbers, and depths of soling out (Schardt et al., 2006). The rest of the hydrological architecture contained sediment cover acting as an aquitard while the only semi-aquifers in the deeper section of the oceanic crust were represented by a basalt breccia zone and an interface between sheeted dyke and gabbro layers.

An interesting outcome is the fact that changing the permeabilities of seven faults of the model results in complex response because faults apparently compete among themselves for position of recharge and discharge areas and also for the rate of their fluid flow. However, one thing was common for all sensitivity tests. The increased fault permeability seemed to shift discharge activity along faults further away from the heat source, represented by a magma chamber under the ridge.

Changing the amount of faults also triggered a complex response due to their competition (Figure 8.25). A three-fault system developed a simple flow system involving recharging central fault and discharging external faults. A five-fault system used three internal faults for recharge and shifted discharge to the most external ones. A seven-fault system continued in the trend of shifting discharge to the external faults. It also changed the central fault from recharger to discharger, presumably when sufficient heat from the heat source became extracted to produce buoyant hydrothermal fluids.

Generally, this modeling indicates that significant discharge can be expected at central valley flanks and structural highs. Furthermore, an important observation is the fact that numerically simulated flow rates are significantly lower than modern measured ones because they are long-term averages while the modern ones are transient.

The calculated duration of the water discharge with temperatures above 200°C ranges from 100,000 to 800,000 years, depending on the long-term average fault permeabilities and existence of the cap rock for the system.

Note that the described example comes from the fast-spreading ridge. As Buck and colleagues (2005) documented, spreading ridges have different topographies depending on the spreading rate. Therefore, the described geothermal system is assumed to represent the fast ridges, while geothermal systems of ridges from other spreading-rate categories could be different.

Natural examples of rift basins in marine conditions come from the Rhine Graben and the Mount Isa Basin in northern Australia (Yang et al., 2006). The 40–15 Ma period in the development of the central Rhine Graben was characterized by submarine conditions and the development of a weak, density-driven flow system. The system was replicated by numerical simulation to form a single broad convection cell (Person and Garven, 1992). The flow had clockwise sense with direction from the basin center toward the southeast graben flank, and a rate that was slower than 0.0045 my^{-1}. The free convection was apparently influenced by the presence of a southeast-dipping basal aquifer and the location of a permeable zone along the southeastern main boundary fault (Wood and Hewett, 1982). It was also influenced by a contribution from compaction-driven flow. Because of its small flow rate, the described free convection had too little of an effect to disturb the thermal conduction-dominant regime.

Numerical simulation of the fluid flow in the Mount Isa Basin was done in an attempt to understand the role of faults interacting with a system of aquifers in controlling the occurrence of all Australian stratiform Zn-Pb-Ag deposits next to major faults active at the time of their deposition (Yang et al., 2006). Sensitivity testing focused on controls of the fluid system such as:

1) impact of a scenario with an upper aquifer only versus a two-aquifer system;
2) impact of fault depth;
3) impact of fault permeability;
4) impact of permeability of the uppermost layer in the section; and
5) impact of the basal heat flow.

Results indicate that the number of aquifers does not dramatically change the fluid flow system. Two-aquifer scenarios recharge cold seawater and discharge mineralized and heated fluids along the same faults. The single-aquifer scenario only increases the temperature of discharging fluids.

If a particular discharging fault is deeper than the other faults, it captures the dominant amount of the discharge from the system while discharge along the remaining discharging faults pales out.

The fault permeability reduction, simulating their period of inactivity, results in the reduction of the convection in the basin.

The impact of the permeability of the upper portion of the section is quite significant. Buoyancy-driven convection in the basin is at its best when there is no cap preventing it from a connection to the surface (Yang et al., 2006).

Increased basal heat flow at one end of the study area enhances the fluid discharge at that end, suppressing the discharge along faults located at the other end.

It is generally valid that (Yang et al., 2006):

1) discharging faults have isothermal regimes like those common in commercially used geothermal systems such as Karaha–Telaga Bodas (see also Nemčok et al., 2006a);
2) interplay among active faults and aquifers controls the basin fluid flow and hydrothermal discharge;
3) seawater recharges through faults, moves along clastic aquifers where it gets heated and enriched in metals, and discharges through faults;

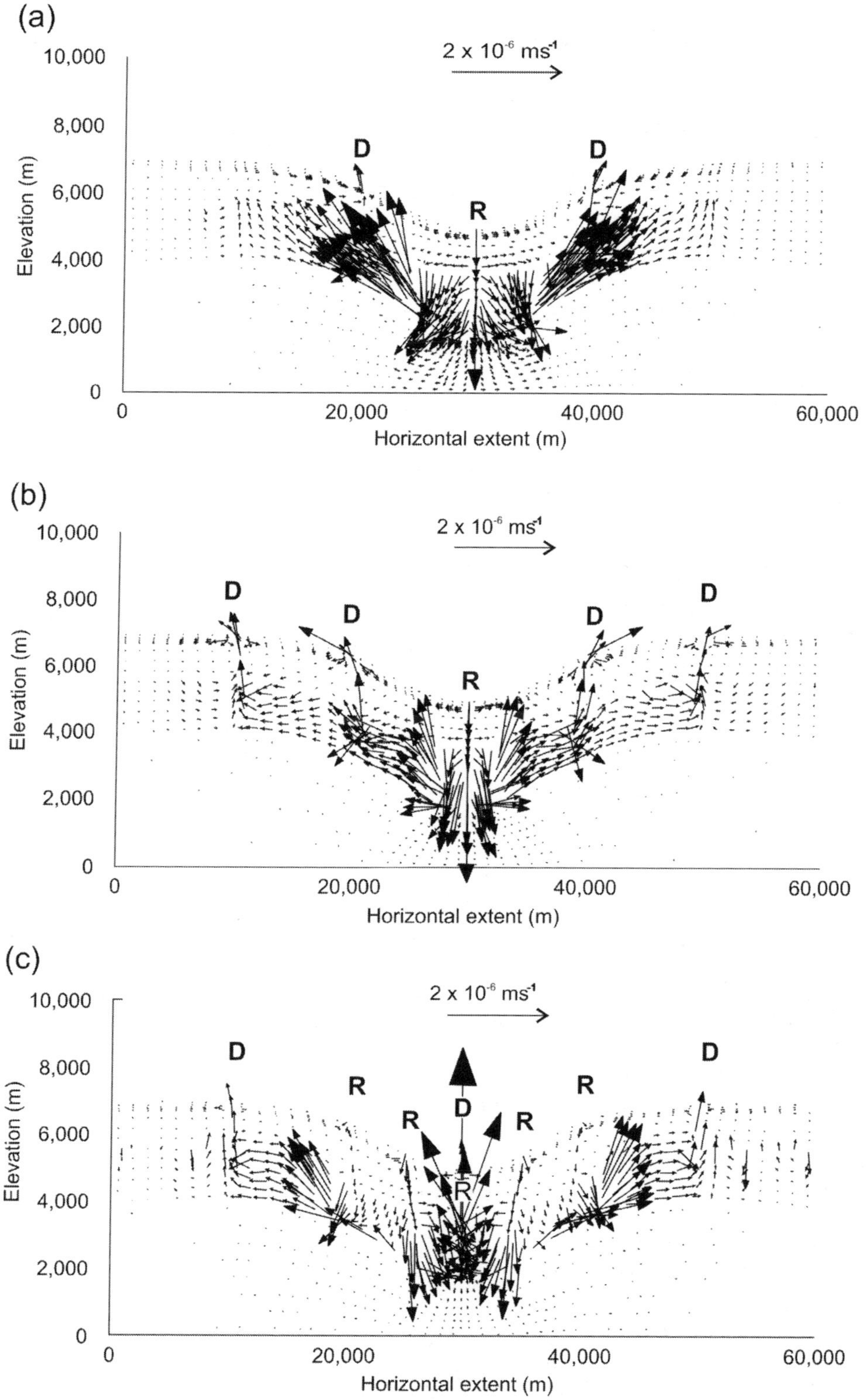

Figure 8.25. Fluid flow velocity and distribution patterns for different fault setups in the Mount Isa Basin model after 10,000 years from fluid flow initiation (Schardt et al., 2006). (a) three-fault pattern including faults 2, 4, and 6. (b) five-fault pattern including faults 1, 2, 4, 6, and 7. (c) seven-fault pattern including all faults (1–7). Major fluid flow occurs within the breccia zone as well as the sheeted dike-gabbro boundary because of the large permeability difference between the two units (one order of magnitude). D – fluid discharge, R – fluid recharge.

Table 8.3. Typical behavior of permeability on production and geologic time scales (Ungerer et al., 1990).

Permeability range	Production time scale	Geologic time scale
>1 d (coarse sand, fractured limestone)	highly permeable	highly permeable, hydrostatic pressure gradients
10^{-3} to 1 d (sandstone)	moderately permeable	highly permeable, hydrostatic pressure gradients
10^{-3} to 10^{-6} d (cemented sandstone, silt)	poorly permeable	moderately permeable, hydrostatic pressure gradients unless rapid burial
10^{-6} to 10^{-9} d (marl, shale)	practically impermeable	poorly permeable, often over pressured, sometimes hydraulic fracturing
$<10^{-9}$ d (compacted shale, tight limestone)	impermeable (measurements difficult)	practically impervious, frequent hydraulic fracturing, frequently over pressured

4) the size of convection cells is controlled by fault spacing; and
5) tectonic quiescence shuts down convection cells.

While the discussed simulation was made in two dimensions, note that the fluid system is truly three-dimensional. The fluid discharge is focused on fault portions with maximum permeability, which usually occurs in:

1) most dilatant fault segments thanks to the relationship of their geometries to controlling stress field;
2) fault–fault intersections; and
3) fault-permeable horizon intersections.

A natural example of rifting-related basins in subaerial conditions without a distinct topographic gradient comes from the surroundings of the Sea of Galilee pull-apart basin (Gvirtzman et al., 1997). This basin serves as an example where the topographic fluid flow dominates but two free convection cells developed underneath the western and eastern sides of the Galilee Mountains at depths of 3–5 km in the deeper Lower Cretaceous Kurnub and Jurassic Arad aquifers (Figure 8.20). The cells have a size of 6–10 km and counterclockwise flow. Their mechanism is driven by unstable conditions created by an increasing temperature with a depth where cold high-density fluid lies above hot low-density fluid. Their occurrence is possible thanks to the existence of the thick and permeable aquifer, and the respective combination of two aquifers. Flow rates reach about 1 and 0.1 my^{-1} in the Kurnub and Arad aquifers, respectively.

The reason for the counterclockwise flowing cell is the following. The subsiding pull-apart basin was filled with evaporites having a high value of thermal conductivity. Their juxtaposition with carbonates with lower conductivity results in heat refraction, that is, heat moving up through the higher-conductivity material. Therefore, at a given depth, a lateral temperature gradient is created in the flank carbonates in a direction toward the evaporites. It results in flank carbonate groundwater sinking with respect to groundwater in the pull-apart fill.

Sensitivity tests indicate that the previously discussed free convection cells become shut off when the permeability of the aquifers decreases below a certain threshold.

Gvirtzman and colleagues (1997) observed that hydraulic conductivities of sandstone and carbonate forming Kurnub and Arad aquifers at the basin scale are almost two orders of magnitude larger than conductivities measured from core plug samples and one order of magnitude larger than conductivities determined by a pumping test in the field. Furthermore, conductivities also change depending on the time scale (see Ungerer et al., 1990; Table 8.3).

There are also special scenarios highly prone to large density gradients such as areas around salt bodies. In this scenario the density gradient evolves from:

1) the thermal conductivity of the salt, which is much higher than the conductivity of the surrounding sediments; and
2) the salinity changes controlled by dissolution of the salt body (Evans et al., 1991).

Examples of the latter come from the Gulf of Mexico where the dissolution of the salt body crests is caused by meteoric groundwater flow combined with effective salt body flank sealing by shale sheets (Atwood and Forman, 1959; Bodenlos, 1970). This mechanism results in the development of a salt plume above the less dense deep groundwater.

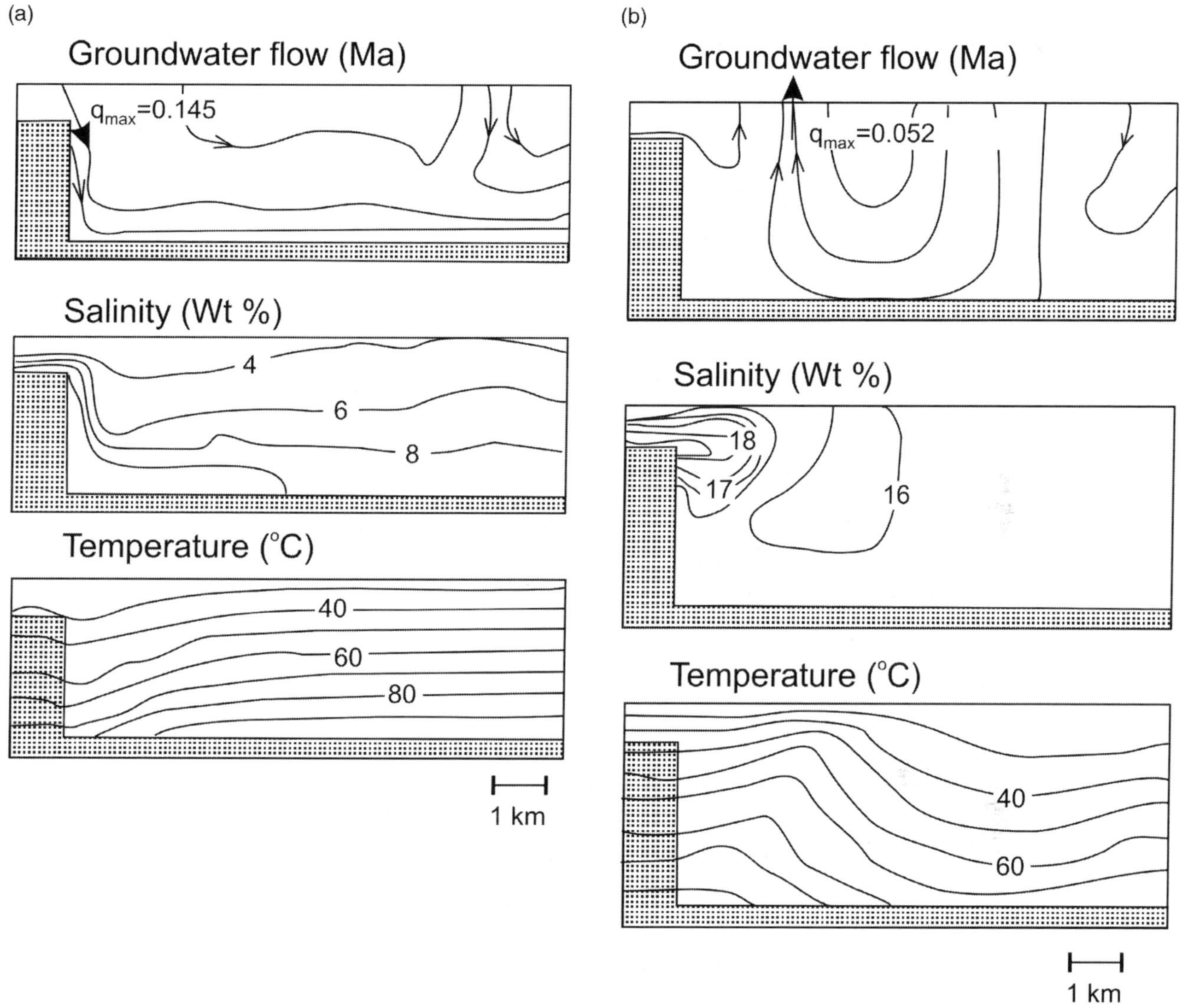

Figure 8.26. Models of fluid-driving mechanisms around salt bodies (Evans et al., 1991). (a) Streamlines, salinity contours, and isotherms after 10 Ma for a geological scenario with a salt dome, in which the salt (shaded) builds into a homogeneous medium and ambient groundwater in surrounding sediments has low salinity. Downward buoyancy forces resulting from high salinities in the crestal area control the groundwater flow. Isotherms are depressed near the salt dome, leading to low surface heat flow above the dome. The upper boundary condition is the specified head, which drops 4 m off the flank of the dome. Modifying head or recharge on the upper boundary did not affect the general flow patterns. (b) Similar simulation of results after 14 Ma, but the diapir builds into sediments containing groundwater with high background salinity. Model parameters and hydraulic boundary conditions are the same as in (a). When the background salinity is high, thermal effects control groundwater density, leading to free convection near the salt dome.

Fluid-driving mechanisms around salt bodies were studied in detail by numerical modeling (Evans et al., 1991), which focused on the influence of the low (Figure 8.26a) versus high salinity (Figure 8.26b) of the background fluid. The simulation was capable of coupling groundwater flow with heat transport and the transport of dissolved salt. Its results indicate that a large salinity gradient develops by salt dome dissolution in the case of the low-salinity background fluid (Figure 8.26a). The density gradient drives the fluid into the basin, which carries the extra salt further, increasing the salinity of the background fluid in direction away from the salt body. The groundwater transports heat downward, cooling the thermal regime above and adjacent to the body.

Results also indicate that the thermal effect around the salt body in the case of the high background salinity (Figure 8.26b) is more significant than the effect in the low-salinity case (Evans and Nunn, 1989). The sensitivity analysis indicates that when the salinity exceeds 15 weight %, the thermal effects dominate over salinity effects, resulting in the upward groundwater flow near the salt body (Evans et al., 1991; Figure 8.26b). As a result, the thermal regime above and adjacent to the body undergoes warming. As a further result, the convection cell in this scenario is clockwise in contrast with the counterclockwise cell from the lower background salinity case (compare Figures 8.26a, b). Similar fluid circulation with upwelling along the salt flank was reported from the area near salt diapirs in the Louisiana portion of the Gulf Coast (Hanor, 1987) and the Danish Central Trough in the North Sea (Jensenius and Munksgaard, 1989).

9 The role of pre-rift heat flow in thermal regimes of rifts and passive margins

Introduction

Typical features of passive continental margins, which host sedimentary basins accounting for 45 percent of recoverable hydrocarbon reserves, include (Makhous and Galushkin, 2005):

1) a complex thermodynamic history of the continent–ocean transition zone;
2) fast deposition rates and potentially thick sedimentary cover, which includes pre-breakup rift fill;
3) potential gravity tectonics on top of evaporitic or shale horizons;
4) potential occurrence of extensive carbonate platforms; and
5) the existence of large deltaic cones.

Hydrocarbon exploration in this setting occurs in depths of 100–4,000 m and offers the promise of discovery of new oil fields in the near future. However, the complex thermodynamic history of this setting represents enough of a challenge even without starting at different pre-rift thermal conditions.

The pre-rift heat flow, present in an area prior to rifting, plays an important role in establishing the subsequent thermal regime in individual rift blocks. It sets the stage for the syn-rift thermal regime by determining its initial character, although its influence progressively diminishes during the sequence of rift development stages potentially followed by passive margin development. The discussion of possible heat flow scenarios existing at the initiation of rifting and their role in subsequent thermal regimes is the goal of this chapter.

Plate settings of the developing rift systems

Rifting can develop in all three types of lithosphere described in Chapter 5. They include:

1) a strong, old, stable lithosphere that can be characterized as a lithosphere, which was stabilized in Precambrian times and represents an old cratonic lithosphere (Figures 5.4, 5.5, 5.6, 6.1);
2) intermediately strong, relatively young, stable lithosphere that represents a lithosphere that became stable during the Phanerozoic (Figures 5.5, 5.6); and
3) a weak, young, thickened lithosphere that can be characterized as a lithosphere that did not have a sufficiently long time for its thermal equilibration after the last orogenic event and finds itself in the stage of thermal thinning (Figures 5.1, 5.5, 5.6).

Rifting can occur here as controlled by (e.g., Sengör, 1976, 2001; Turcotte and Schubert, 1982; Cloetingh et al., 1995; Spadini et al., 1996):

1) regional extension;
2) hot spot activity;
3) local extension in the rear of the orogen; and
4) local extension associated with transform fault activity.

The thermal regime in all these settings is dominated by the heat flow coming into the crust from the mantle (see Chapman and Rybach, 1985). Other important energy sources include:

1) heat loss from extruded effusive rocks;
2) energy from volcanic explosions;
3) heat associated with metamorphic reactions;
4) strain energy released in earthquakes;
5) potential energy in uplifted mountains; and
6) changes in the kinetic energy associated with rotation of the Earth.

The distribution of the Earth's heat flow field allows one to infer the heat transfer processes responsible for its individual features. One of the most reliable quick diagnostic features is the size of heat flow anomaly. The heat flow distribution feature is generally called an anomaly if it departs from the heat flow field predictable by the simple Earth model (Chapman and Rybach, 1985).

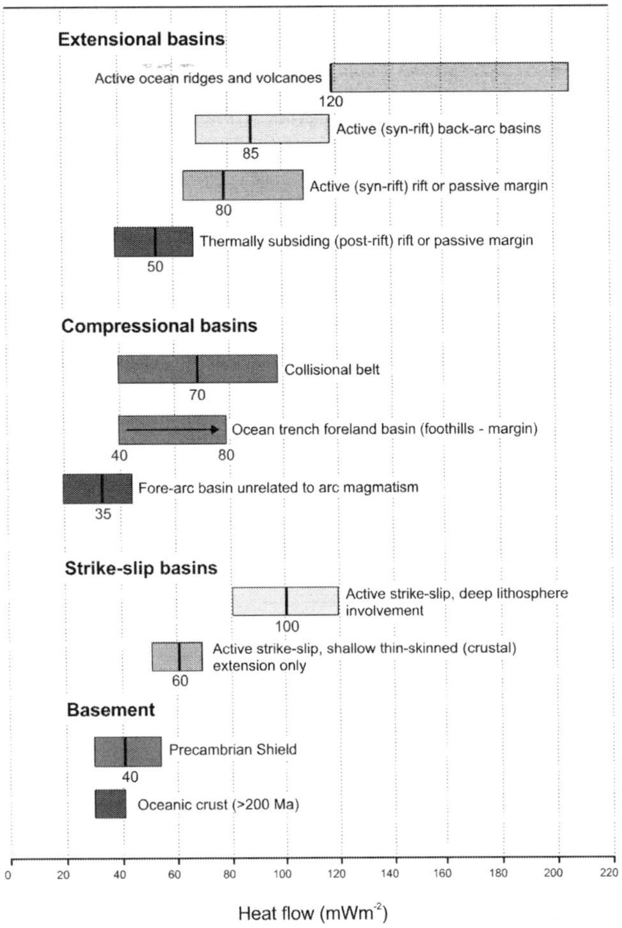

Figure 9.1. Summary of typical heat flow values associated with sedimentary basins of various types (Allen and Allen, 1990).

Thermal regimes of pre-rift plate settings

The coolest continental thermal regimes are present in regions formed by old continental shield (Figures 9.1, 9.2; Table 9.1). This can be demonstrated by the surface heat flow contours of the broader Equatorial Atlantic region (Figure 9.2). Here, one can compare the thermal regimes of several tectonic settings. Figure 9.2 shows that the coolest regions are represented by the West African Shield, the Guyana Shield, and the São Francisco Craton. The oldest portions of the South Atlantic, the Equatorial Atlantic, and the Central Atlantic oceanic crust include areas that form the second coolest region.

Old passive margins, such as the Atlantic margins (Figure 9.2), can be described as having near-normal geothermal gradients (sensu Robert, 1988). They have present-day geothermal gradients of about 25–30°Ckm^{-1}. Typical values are, for example, 27°Ckm^{-1} in Congo, 25°Ckm^{-1} in Gabon, and 25°Ckm^{-1} in the Terrebone Parish borehole in the Gulf Coast (Figure 9.3). Typical vitrinite reflectance profiles of these geothermal gradients show that a vitrinite reflectance, R_o, is about 0.5 percent at depths of about 3 km (Figure 9.3). These profiles have sub-linear geometry.

Continental cratons contain 1,000 km-scale heat flow anomalies, representing the signature of mantle dynamics- and plate-evolution-controlled heat flux (Chapman and Rybach, 1985). Examples come from the continental cratons of North America,

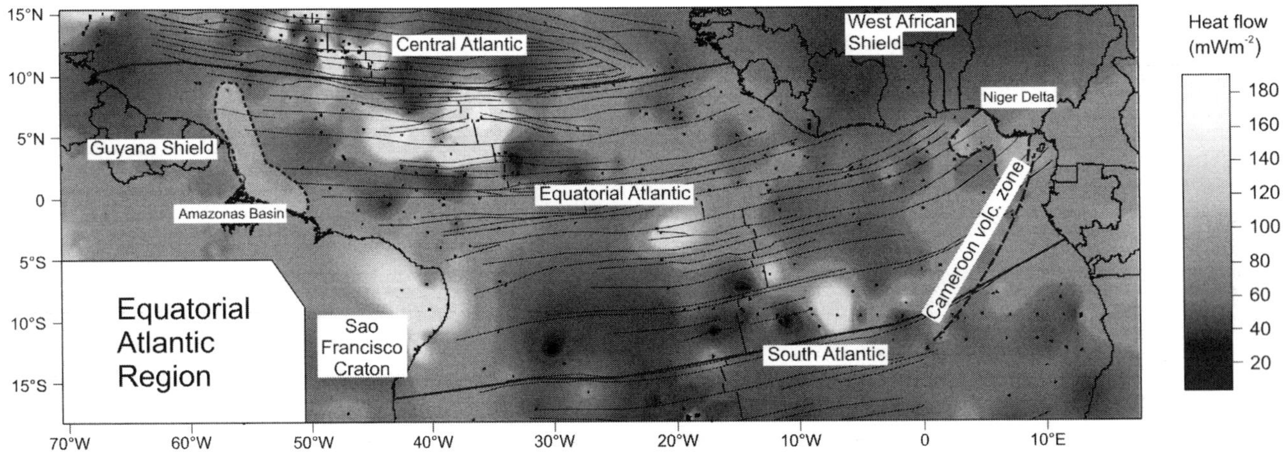

Figure 9.2. Surface heat flow contour map of the broader Equatorial Atlantic region. Data are taken from the database maintained by the International Heat Flow Commission. The coolest regions are represented by the West African Shield, the Guyana Shield, and the São Francisco Craton. The oldest portions of the South Atlantic, the Equatorial Atlantic, and the Central Atlantic oceanic crust include areas that form the second coolest region.

Table 9.1. Typical values of surface heat flow for geologic provinces (Platte River Associates, 1995). Some generalizations can be made within a given province, but there are numerous local anomalies. Therefore, listed values serve only as guidelines in determining appropriate values for further studies.

Geologic province / specific example	Heat flow (Heat Flow Units)	Heat flow (mWm^{-2})
Average oceanic	1.6	67.0
Average continental	1.4	58.0
Eastern United States	1.36	57.0
England/Wales	1.41	59.0
Great Plains, western United States	1.5–2.5	62.0–104.0
Young or active extensional basins	>1.9	> 80.0
Pannonian Basin, central Europe	2.0–2.5	85.0–105.0
Pattani Trough, Gulf of Thailand	1.9–2.4	80.0–100.0
Gulf of Leon margin, western Mediterranean	1.6–2.1	65.0–90.0
Gulf of Suez, Egypt	1.5–2.0	60.0–84.0
Viking Graben, North Sea	1.3–2.0	56.0–82.0
Uinta Basin, Utah	1.0–1.6	40.0–65.0
North Biscay margin, East Atlantic	0.8–1.0	36.0–43.0
Old shield areas	< 1.0	< 40.0
Balkan-Ukrainian Shield	0.7–1.0	30.0–40.0
Canadian Shield	0.8	34.0
West Australia	0.9	39.0

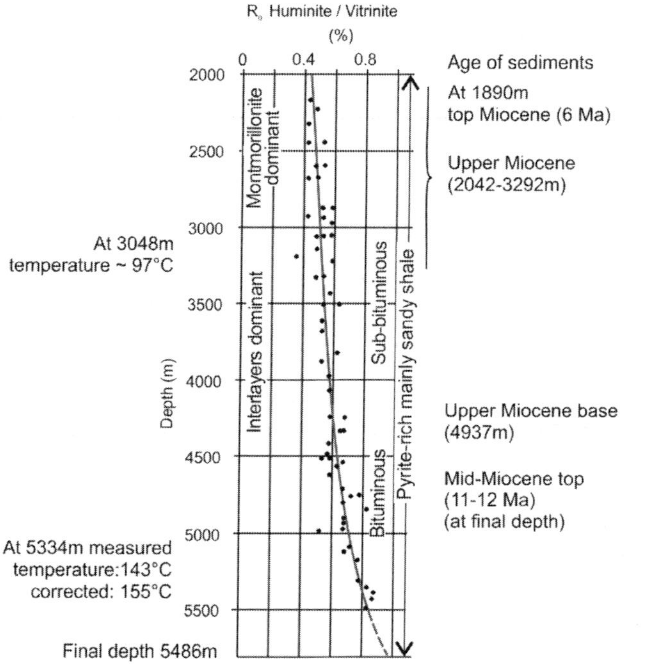

Figure 9.3. Vitrinite reflectance profile for the Terrebone Parish, Point au Fer bore hole in Louisiana (Heling and Teichmüller, 1974). The profile is sublinear and continuous, indicating a near constant geothermal gradient through time.

northeastern South America, West Africa, West Australia, and northern Eurasia, all characterized by low heat flow anomalies with a characteristic value of about 40 mWm^{-2}, which represents only half of the global average (see Chapman and Rybach, 1985; Blackwell and Richards, 2004; Figure 9.1). Apart from recharge areas of geothermal systems (Coolbaugh et al., 2005; USGS, 2006; Figure 11.17), cratonic heat flow values of 35–40 mWm^{-2} represent the terrestrial heat flow minima. Strong, old, stable lithospheres of the Archaean age (see Figure 5.4 for location) have managed to lose all transient orogenic heat associated with heat sources in orogen-formed crust because of deep erosional removal of a significant portion of the upper crustal radioactive heat sources, which usually produce 40–50 percent of the surface heat in the young orogenic crust.

As can be implied from the existence of plate settings characterized by repetitive rifting cycles (see Table 9.2), the heat flow distribution prior to subsequent rifting can be already perturbed by previous rifting if there was no time long enough for the thermal equilibration. Associated thermal anomalies can be characterized as lithospheric-scale heat anomalies, having an order of

Table 9.2. Characteristics of selected rift basins (Huismans et al., 2001).

Basin	Crustal thinning	Sub-crustal thinning	Timing of the first rift event (Ma)	Timing of the second rift event (Ma)	Timing of calc-alkaline volcanism (Ma)	Timing of alkaline volcanism (Ma)	Timing of post-rift doming (Ma)
Pannonian Basin	1.6–1.8	8.0–10.0	18.0–14.0	12.0–11.0	18.0–6.0	12.0–0.0	12.0–11.0
Baikal Rift	1.4–1.6	3.0–4.0(?)	24.0–6.0	4.0–0.0	?	20.0–14.0	0.0
Oslo Graben	1.3–1.4	4.0–5.0	300.0–270.0	270.0–240.0	?	280.0–270.0	240.0–220.0
North Sea	1.3–1.6	5.0–5.5	248.0–219.0	166.0–118.0	?	176.0–163.0	183.0–156.0
Upper Rhine Graben	1.1–1.2	3.0–5.0(?)	40.0–23.0	18.0–0.0	?	12.0–0.0	10.0–0.0
Gulf of Lion margin	1.2–2.0	3(?)	23.0–16.0	?	36.0–30.0	12.0–0.0	no
Valencia Trough	1.5–2.5	4.0–8.0	23.0–16.0	?	23.0–18.0	10.0–0.0	15.0–0.0(?)
Alboran Sea	1.5–2.5	4.0–8.0	23.0–16.0	9.0–5.0	?	10.0–0.0	11.0
Dnieper-Donets Basin	1.1–1.5	1.1–10	379.0–362.0	345.0–340.0	?	363.0	330.0

magnitude of 100 km. Discarding heat convection, typically this scale of anomaly is controlled by:

1) the thermal conductivity structure of the lithosphere;
2) the amount and distribution of heat-producing radioactive elements; and
3) the heat into the base of the lithosphere.

Scientists have described numerous examples of such anomalies from the western United States (Figure 9.4).

An example of the positive heat flow anomaly of the Rio Grande Rift Zone has a length of 50 to 100 km. Its surface heat flow values are larger than 100 mWm^{-2} (Blackwell and Richards, 2004; Figure 9.4a). It represents a relatively typical example of the rift valley anomaly. The simplest interpretation of the heat source here is the isotherm ascent in relation to lithospheric thinning combined with heat flow from intrusion emplaced into the crust (Chapman and Rybach, 1985).

The Battle Mountain High and Snake River Plain anomalies in the western United States, including a hot Yellowstone area (Blackwell and Richards, 2004; Figure 9.4a), are another example of the positive heat flow anomaly. It has distinct elongated geometry. This stripe anomaly is driven by extensional tectonics and Tertiary magmatism associated with a combination of (Blackwell, 1978; Brott et al., 1981; Chapman and Rybach, 1985):

1) the current Yellowstone hot-spot effects;
2) not equilibrated hot-spot effects in the Snake River Plain; and
3) extension in the Basin and Range province.

An example of the negative anomaly comes from the Colorado Plateau, which is characterized by a heat flow of about 60 mWm^{-2} (Keller et al., 1979; Reiter et al., 1979; Bodell and Chapman, 1982; Blackwell and Richards, 2004; Figure 9.4a). It represents a remnant thermal state of the whole region, which existed prior to the Cenozoic tectonic development (Chapman and Rybach, 1985).

Good comparisons of the discussed heat flow anomalies can be done along the east-west striking transect through the western United States, made along the 39° N latitude (Chapman and Rybach, 1985; Figure 9.4b). Particularly distinct are the aforementioned Battle Mountain High positive and Colorado Plateau negative anomalies, labeled as C and D, respectively. The former, formed by several local lows and highs, is characterized by heat flow values ranging from 75 mWm^{-2} to more than 100 mWm^{-2}. The latter, relatively homogeneous, is characterized by values around 50–60 mWm^{-2}. The easternmost anomaly in Figure 9.4b, labeled as E, is located in the southern Rocky Mountains. It is associated with a thickened low-density crust, which typically has relatively high heat production controlled by radioactive decay.

Figure 9.4b demonstrates that the boundaries among anomalies C, D, and E, and remaining ones, are not sharp but transitional. Such transitions are typical for areas located between regions with different lithospheric histories, as heat flow studies done along the European Geotraverse point out (see Mueller, 1983; Balling, 1985; Mongelli, 1985). However, it needs to be said that the relationship between various anomalies is

(a)

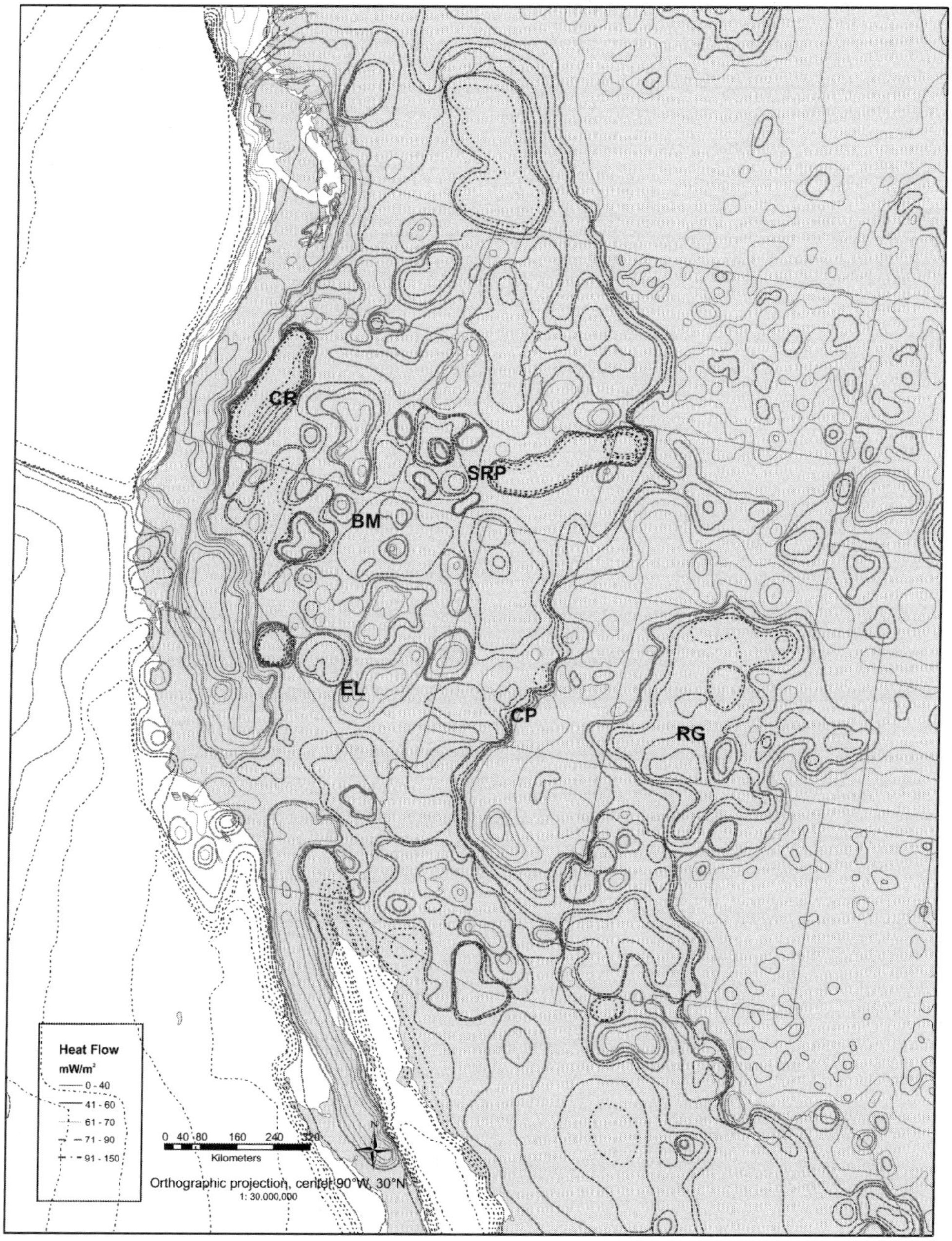

Figure 9.4. (a) Positive heat flow anomalies of the Rio Grande Rift Zone (RG), the Cascade Range (CR), Battle Mountain High (BM), and the Snake River plain (SRP) of the western United States, including the Yellowstone area, negative anomalies of the Colorado Plateau (CP), and Eureka Low (EL) (Blackwell and Richards, 2004). (b) Heat flow, Bouguer gravity, and topographic profiles across the western United States along the 39° N latitude (Chapman and Rybach, 1985).

not just spatial but also temporal because of low conductivities of lithospheric rocks. Thermal anomalies in the lithospheric mantle, which developed just 10 Ma ago or later, most likely did not have sufficient time to perturb the thermal regime at the surface, if one assumes the conductive heat transport being dominant (Rybach, 1985).

Furthermore, note that the rift-, intrusion-, hot-spot-related heat flow anomalies of the earlier events have some time to become somewhat reduced

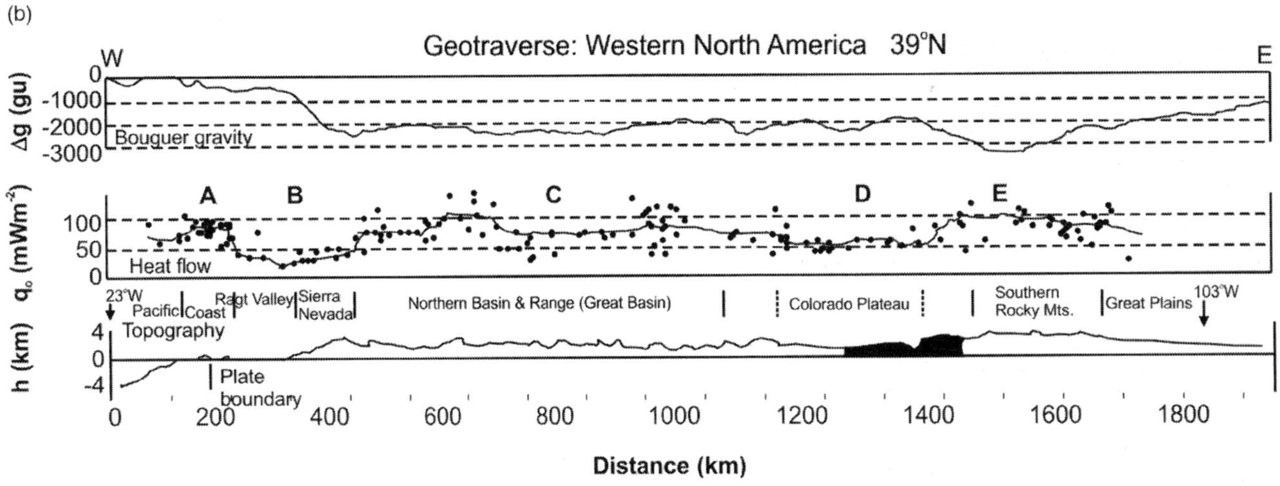

Figure 9.4 (*cont.*)

from their peak values between their initiation and the start of the next rift event. Therefore, this discussion describes heat flow anomalies that have surface heat flow values higher than those occurring prior to the next rift event.

Preexisting thermal perturbations can be associated with regions affected by lithospheric extension such as back-arc basins (Figure 9.5), continental rift systems (Figure 9.6), some strike-slip-related basins (Figure 9.7), and the internal arcs of zones of B-type subduction (Figure 9.8).

Some Californian strike-slip basins have very high geothermal gradients, for example, 200° Ckm^{-1} that characterizes the Imperial Valley. This means that even very young sediments in these basins can be highly mature. Figure 9.7 shows a prominent system of high surface heat flow anomalies associated with a system of pull-apart basins occurring between the Salton Sea and Baja California. These pull-apart basins are developed inside releasing oversteps between the San Andreas, Brawley, Imperial, and San Jacinto dextral strike-slip faults (Fuis and Kohler, 1984; Moore and Adams, 1988; Larsen and Reilinger, 1991). The pull-apart basins are progressively closer to developing oceanic crust by seafloor spreading in the direction from the Salton Sea to Baja California and all are associated with tholeiitic basalt and rhyolite volcanism.

Continental rifts have high present-day geothermal gradients, for example, more than 50° Ckm^{-1} in the Red Sea or up to 100° Ckm^{-1} in the Upper Rhine Graben (Allen and Allen, 1990). Typical heat flow values for young rifts cluster around 60.5–90.5 mWm^{-2} (Morgan, 1984).

Oceanic measurements and deep boreholes in the Red Sea (Girdler, 1970) indicate that high surface heat flows, generally more than 126 mWm^{-2}, occur in a broad band at least 300 km wide (Figure 9.6a, b). This band is centered along the rift axis. The organic maturation shown by vitrinite reflectance profiles and by oil, gas, and condensate occurrences indicates that the highest maturity is present in the southern portion of the Red Sea. Intermediate values are located in the center of the Red Sea and the lowest values occur in the Gulf of Suez (Figure 9.6c). This cooling trend can be correlated with different amounts of extension, which is largest in the south of the Suez–Red Sea Rift system (Allen and Allen, 1990). Thus, the Oligocene–Miocene elevated heat flows of the Gulf of Suez have now decreased to near normal values, while high heat flow values are present in the still actively rifting southern Red Sea.

Heat flows of internal arcs are elevated because of magmatic activity (Allen and Allen, 1990; Figure 9.8). The Tertiary anthracites of Honshu, Japan, with a vitrinite reflectance of 2–3 percent, mentioned earlier, serve as an example. Similar patterns are found in ocean–continental collision zones such as the Andean Cordillera, and are also found, but less commonly, in parts of continent–continent collision zones such as the Alps.

A case area for lithospheric extension controlling a development of back-arc basins is the orogenic hinterland of the Carpathians, the Pannonian Basin system, where the surface heat flow values exceed 100 mWm^{-2} in several basins (Figure 9.5). The extension started here during the Eggenburgian (20 Ma). The region experienced several local basin inversion events that took place during the Late Miocene and Quaternary. The subduction that drove the extension continues to the present day only along a small segment located at the

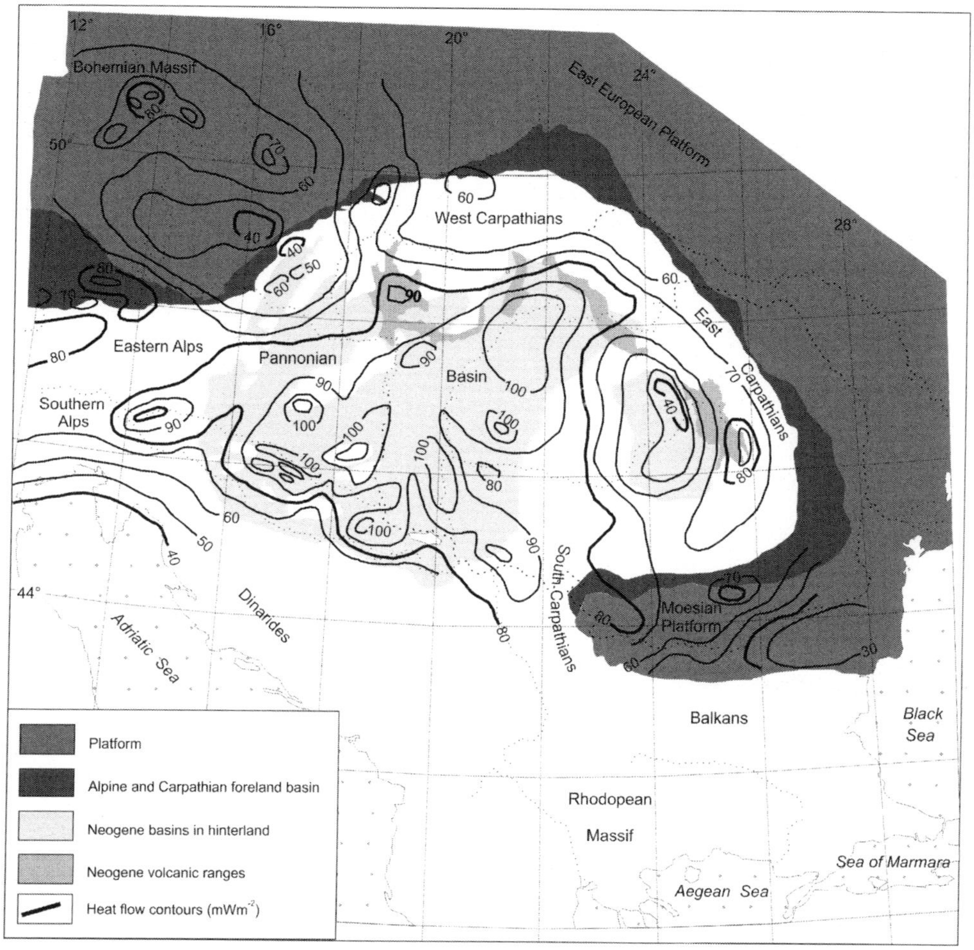

Figure 9.5. Heat flow map of the Carpathian-Pannonian region (Nemčok et al., 1998). The number of boreholes used for the heat flow calculations by Čermák and Hurtig (1977) and Krus and Šutora (1986) varies among different countries. The largest numbers come from Hungary (100), the Czech Republic (89), and Slovakia (59). Boreholes are concentrated mainly in basins with industrial activity.

junction of the Eastern and Southern Carpathians (e.g., Nemčok et al., 1998).

Weak, young, thickened lithospheres have their largest heat flow anomalies having a size comparable to lithospheric thickness. The sizes of these 100 km anomalies are comparable to the size of orogenic belts, geologic terrains, and large physiographic provinces.

An example of such an anomaly comes from the Cascade Range of the northwestern United States, associated with Pliocene and younger magmatic activity (Blackwell et al., 1981) (Figure 9.4a). The magmatism and heat flow anomaly is driven by the subduction of the Juan de Fuca Plate underneath the North American Plate. Our study of the Medicine Lake area located near the boundary of the Cascades with the Basin and Range province indicates a present-day formation of the pull-apart basin along the northwest-southeast striking dextral strike-slip zone. Its fill is dominated by magmatic rocks and characterized by a local positive thermal anomaly driven by a local geothermal system.

The weak, young, thickened lithosphere is also characterized by having crustal-scale heat flow anomalies that are practically related to crustal heterogeneity caused by past orogenic events, and that did not have sufficient time to thermally equilibrate. The factor controlling the thermal perturbation most is the lateral and vertical heterogeneity of the crust. Vertical layering controls the distribution of felsic and mafic rocks. The more felsic upper crustal rocks are characterized by heat production controlled by radioactive decay, which is ten times larger than heat production of more mafic lower crustal rocks (see Buntebarth and Rybach, 1981; Buntebarth, 1984; Meissner, 1986; Rybach, 1986; Ibrmajer et al., 1989). Lateral heterogeneity is usually

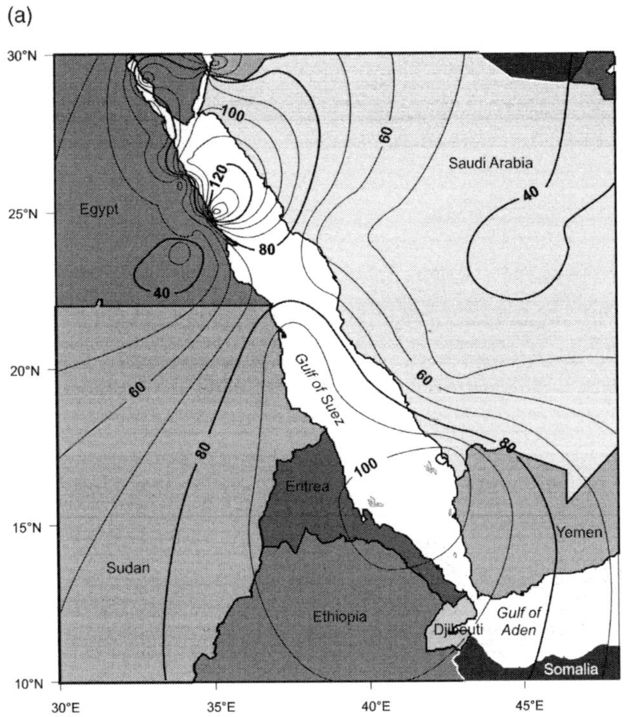

Figure 9.6. (a) Surface heat flow contour map of the broader Red Sea region. Data are taken from the database maintained by the International Heat Flow Commission. (b) The present-day heat flow in the Red Sea Rift, with the location of boreholes illustrated in the reflectance profiles in (c) (Girdler, 1970). (c) Subsidence and vitrinite reflectance curves along several boreholes from (b). Black dots indicate subsidence history of prominent evaporitic layers of the late Miocene age; open dots indicate vitrinite reflectance. Reflectance values in borehole A, located in the south of the Red Sea where present-day surface heat flow values are high, are elevated. The Red Sea rift becomes colder in the direction from south to north.

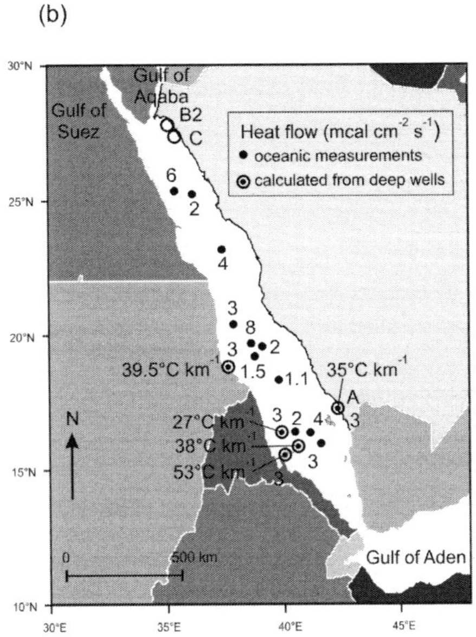

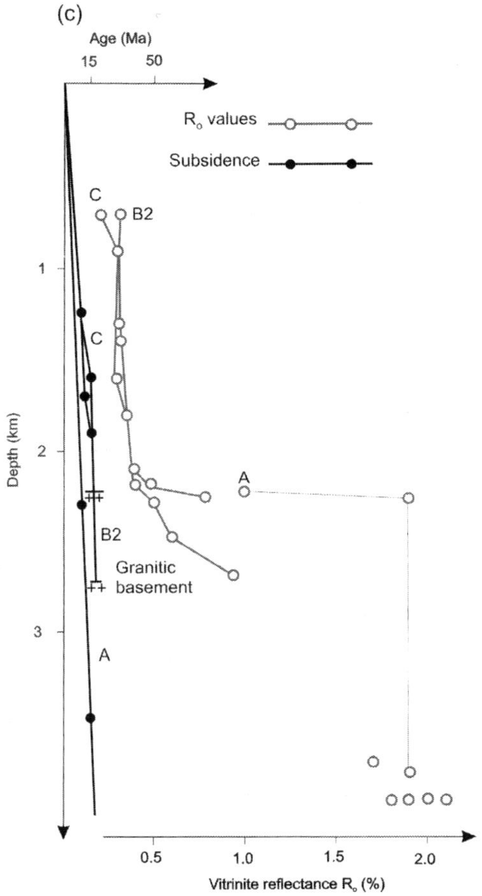

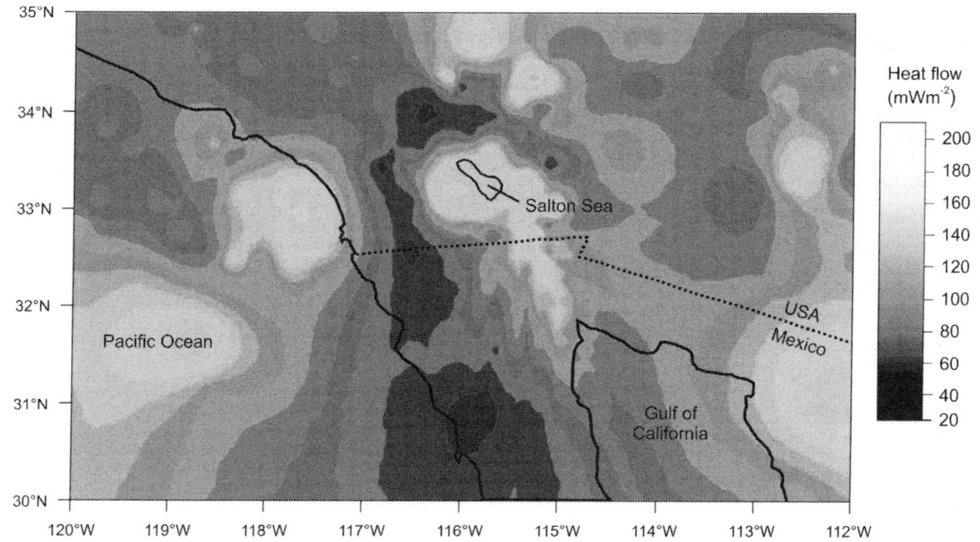

Figure 9.7. Surface heat flow contour map of the broader Salton Sea and the northern Baja California regions. Data are taken from the database maintained by the International Heat Flow Commission.

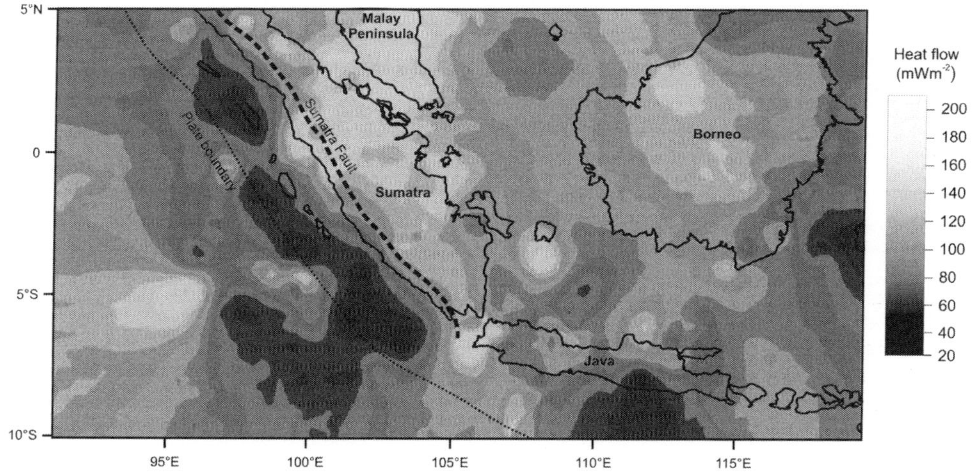

Figure 9.8. Surface heat flow contour map of the Sumatra-Java region. Data are taken from the database maintained by the International Heat Flow Commission.

associated with an introduction of deep-seated normal faulting, emplacement of intrusions, deep sedimentary basins, and orogenic belts.

A nice example of the heat flow perturbation controlled by lateral heat production heterogeneity comes from Great Britain, characterized by two zones of increased heat flow. The first one, trending northwest-southeast, is located in central Great Britain and the second one, trending east-west, is located in southern Great Britain. They were both assigned to the presence of rock belts with increased radioactive decay, the first one associated with Caledonian and the second with Variscan orogeny (see Richardson and Oxburg, 1979).

This example illustrates the point that it is not just relatively short-lived orogen-related thermal regime perturbation that can control the pre-rift thermal regime but the distribution of orogenic zones with different heat productions that can cause long-lived thermal perturbations, existing long after the dynamic effect by the orogen had sufficient time to thermally equilibrate.

An example of the negative anomaly caused by the same effect comes from the Ivrea zone of northern Italy, which is composed of mafic metamorphic rocks and ultrabasic rocks, which are characterized by extremely low heat production (Hohndorf et al., 1975).

The effects of orogens on the heat flow distribution due to the thickening of the heat-producing section can be relatively significant, especially as time progresses and the initially colder crust starts to warm up because of increased heat production driven by enhanced radioactive production.

Immediately after thrusting, the thickened lithosphere becomes stronger than the non-thickened lithosphere and colder because of the thickening of the colder crust (Figure 9.9). Subsequently, the developing thrustbelt elevates geothermal conditions within the thickened lithosphere by a burial of radiogenic-enriched crustal rocks. This results in the weakening of the thickened lithosphere after 10–20 million years following the initiation of thrusting (see Figure 9.9; Zoetemeijer, 1993 and Figure 5.1; Lankreijer, 1998). Typical examples

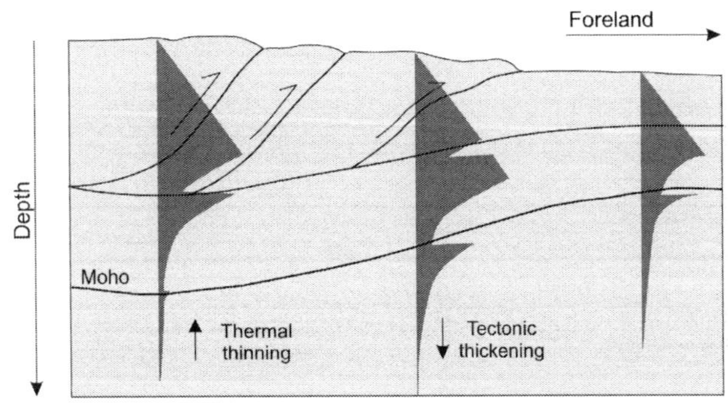

Figure 9.9. A cartoon illustrating different parameters controlling the lithospheric strength in a compressional region (after Zoetemeijer, 1993). The lithosphere is simulated as a three-layer system.

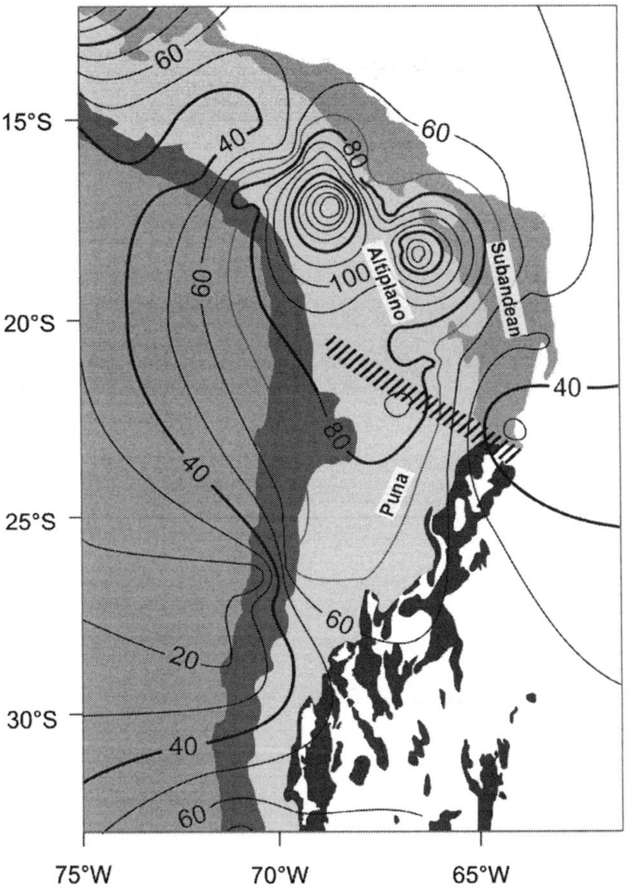

Figure 9.10. Surface heat flow contour map of the Andes between 10° S and 35° S longitudes. Data are taken from the database maintained by the International Heat Flow Commission.

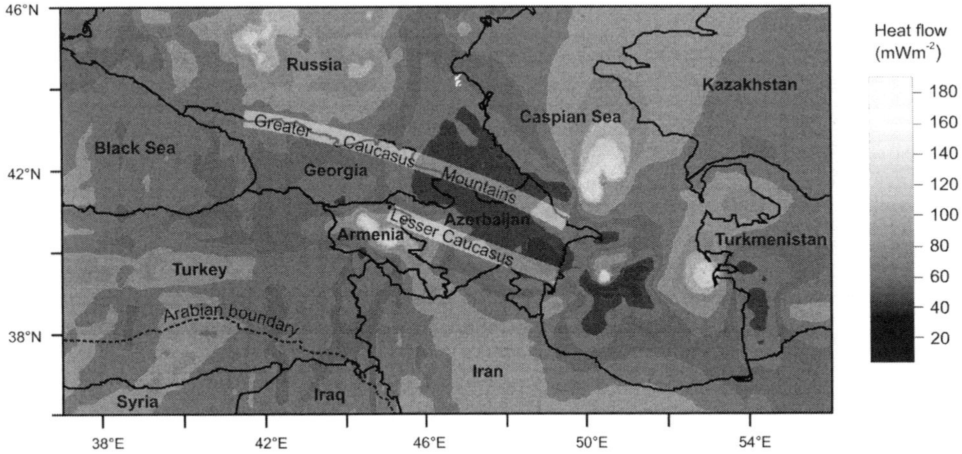

Figure 9.11. Surface heat flow contour map of the broader region around the colliding Eurasian and Arabian Plates. Data are taken from the database maintained by the International Heat Flow Commission.

of thermal weakening come from the Aegean and Carpathian-Pannonian regions (Figures 9.5, 5.1).

Examples of the subduction- and collision-related orogenic belts come from the Andes and Greater Caucasus, respectively. These examples document that if the belts are rather long-lasting like the Andes or the Greater Caucasus, their syn-tectonic sediments are deposited under their own thermal regimes and subsequently incorporated into the orogenic belt. The thermal regime of the Andes is shown in Figure 9.10. Although the Andes are rather complex, the surface heat flow contours indicate that some orogenic zones, such as the Altiplano Plateau, have significantly hot thermal regimes.

The thermal regime of the collisional system in front of the Arabian Plate, which includes the Greater Caucasus, shows a moderate warming in the contractional zone directly in front of the indenting Arabia, reaching surface heat flow values of 80–90 mWm^{-2}, and a moderate cooling along the orogenic zone toward the east, characterized by a decrease from values of 80–90 mWm^{-2} to values of 50 mWm^{-2} (Figure 9.11). This pattern is in accordance with expectations that the geothermal conditions should be most elevated directly in front of the indenter because here the crust undergoes the greatest thickening by the burial of radiogenic-enriched crustal rocks.

10 The role of structural and stratigraphic architecture in thermal regimes of rifts and passive margins

Introduction

The stratigraphic architecture controls the thermal regime of rifts and passive margins through lithology-dependent thermal conductivity (Figures 11.9, 11.21; Ehlers and Chapman, 1999), specific heat, radioactive heat production, and thermal diffusivity, the description of which is the goal of this chapter. Because the extension modifies the stratigraphic architecture of rifts by developing fault-controlled juxtapositions of blocks with different lithostratigraphies, the structural architecture also represents an important control over the thermal regime. Lateral changes of lithofacies cause thermal perturbations, as shown in the Salt Lake Valley in Utah, on the opposite side from its boundary fault (Figures 11.19, 11.21; Ehlers and Chapman, 1999). Development of the sedimentary fill of the rift unit also modifies the temperature, pressure, porosity, density, and composition of the saturant of the original lithostratigraphy, thereby further affecting the thermal regime.

Before discussing the specific roles of thermal conductivity, specific heat, radioactive heat production, and thermal diffusivity on the thermal regime, it is useful to discuss the principles of heat conduction.

Heat conduction

A rock volume that undergoes temperature change, ΔT, also experiences change in its heat content, ΔQ, which is expressed by:

$$\Delta Q = cm\Delta T, \qquad (10\text{-}1)$$

where c is the heat capacity and m is the mass of the rock volume.

Given a rock layer, the heat flowing from the basal surface characterized by the temperature T_2 to the upper surface characterized by temperature T_1 can be calculated by Fourier's first equation of heat conduction:

$$Q = -\lambda[(T_2 - T_1)/h]At = -\lambda(\Delta T/h)At, \qquad (10\text{-}2)$$

where λ is the thermal conductivity of the layer, h is the thickness of the layer, A is the area of the layer, and t is the time span of the heat flow.

Using the unit area and unit time, the heat flow density, q, can be either expressed as (e.g., Buntebarth, 1984; Haenel et al., 1988):

$$q = -\lambda(dT/dh), \qquad (10\text{-}3)$$

or formulated in three dimensions as:

$$q = -\lambda \mathrm{grad}T = -\lambda\nabla T, \qquad (10\text{-}4)$$

where $q = (q_x, q_y, q_z)$ and $\nabla T = (\partial T/\partial x, \partial T/\partial y, \partial T/\partial z)$ are vectors. ∇T is the temperature gradient and λ is the thermal conductivity tensor written as:

$$\lambda = \begin{pmatrix} \lambda_{xx} & \lambda_{xy} & \lambda_{xx} \\ \lambda_{yx} & \lambda_{yy} & \lambda_{yz} \\ \lambda_{zx} & \lambda_{zy} & \lambda_{zz} \end{pmatrix}, \qquad (10\text{-}5)$$

Neglecting lateral heat flow, the thermal conductivity tensor reduces to λ_{zz} and Equation (10-4) turns into:

$$q = \lambda_{zz}(dT/dz). \qquad (10\text{-}6)$$

Thermal conductivity, λ_{zz}, in typically anisotropic rift and passive margin rock sections has to be determined in a vertical dimension for one-dimensional calculation 10-6. It is quite common that anisotropy represented by either local heat sources or complex patterns of rocks with various thermal conductivities (Figure 10.1) is significant enough to break down simple vertical heat flow

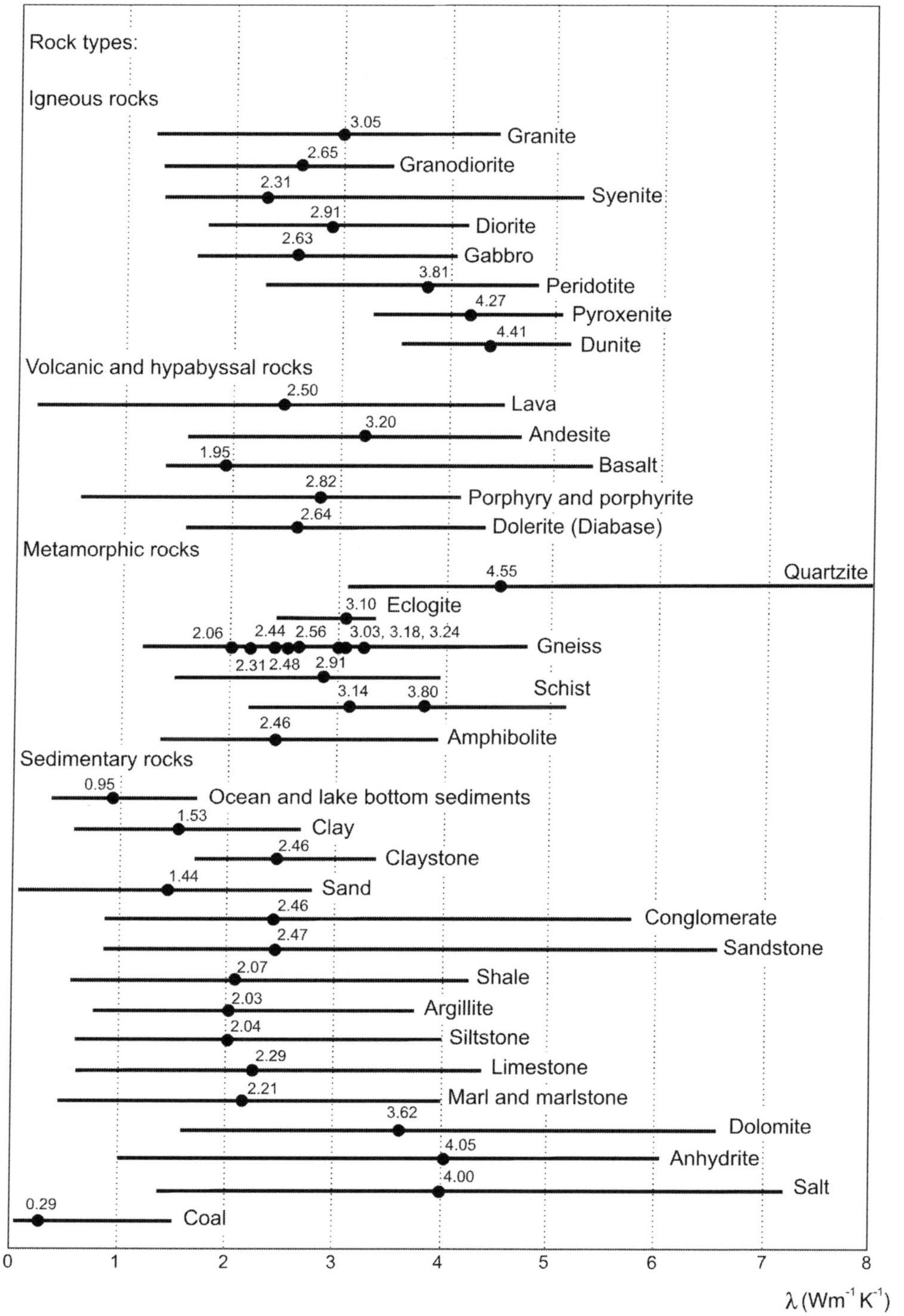

Figure 10.1. Thermal conductivity of rocks at room temperature (Kappelmeyer and Haenel, 1974; Čermák and Rybach, 1982; Zoth et al., 1988). Listed average values in $Wm^{-1}K^{-1}$ are: 2.06 for gneiss in a direction perpendicular to its schistosity, 2.31 for plagioclase gneiss, 2.44 for all samples of gneiss (measured in randomly oriented samples), 2.48 for gneiss in direction parallel to its schistosity, 2.56 for potassium feldspar gneiss, 3.03 for granite gneiss, 3.18 for amphibolite gneiss, 3.24 for quartz diorite gneiss, 2.91 for schist in direction perpendicular to its foliation, 3.14 for all samples of schist (measured in randomly oriented samples), and 3.80 for schist in direction parallel to its foliation.

density. Such rock patterns include the sequences of different lithostratigraphic units (see Figures 7.9, 7.27a), facies inter-fingering (see Figures 7.10a, 7.11, 7.12a, b, 7.19b) and quick lateral pinch-outs (see Figures 7.11, 7.19c) discussed in Chapter 7.

The thermal conductivities of these highly variable sections further depend on temperature (Čermák and Haenel, 1988):

$$\lambda(T) = \lambda_o/(1 + CT), \quad (10\text{-}7)$$

where C is the constant, having a value around 10^{-3} K, which differs from material to material and has to be experimentally determined.

Each material that undergoes heat flow has a different behavior, resulting in a difference between inflowing and outflowing heat flows, which represents a thermal energy that remains in the rock volume dV. The difference calculated by the second Fourier's equation along the z axis is expressed as:

$$dq_z = (\partial q_z/\partial z)dz. \quad (10\text{-}8)$$

The thermal energy stored in the rock volume dV during unit time is then:

$$\rho c dV(\partial T/\partial t), \quad (10\text{-}9)$$

where ρ is the rock density, c is the specific heat capacity of the rock, and ∂T/∂t is the temperature change during unit time.

A full Fourier's second equation for heat conduction is expressed as:

$$-(\partial q_x/\partial x + \partial q_y/\partial y + \partial q_z/\partial z) = \rho c(\partial T/\partial t). \quad (10\text{-}10)$$

Substituting for q with corresponding equations from (10-4) is written as:

$$\{[\partial(\lambda_x(\partial T/\partial x))/\partial x] + [\partial(\lambda_y(\partial T/\partial y))/\partial y] + [\partial(\lambda_z(\partial T/\partial z))/\partial z]\} = \rho c(\partial T/\partial t). \quad (10\text{-}11)$$

Under conditions of isotropic homogeneous rock, which is characterized by its thermal conductivity equal in all directions, Equation (10-11) reduces to:

$$\lambda\,(\partial^2T/\partial x^2 + \partial^2T/\partial y^2 + \partial^2T/\partial z^2) = \rho c(\partial T/\partial t)$$

or

$$\partial T/\partial t = \lambda/\rho c\Delta T = \kappa\Delta T = \kappa\,\mathrm{div}(\mathrm{grad}T) = \kappa{*}\nabla^2T, \quad (10\text{-}12)$$

where $\kappa = \lambda/\rho c$ is the thermal diffusivity, $\Delta = \nabla{*}\nabla = (\partial^2/\partial x^2 + \partial^2/\partial y^2 + \partial^2/\partial z^2)$ is the Laplace operator, and $\nabla = (\partial/\partial x + \partial/\partial y + \partial/\partial z)$ is the Nabla or Del operator.

Fourier's second equation can be extended into a form that describes thermal processes by the advection-diffusion equation in the homogeneous isotropic solid characterized by temperature-independent thermal diffusivity (e.g., Carslaw and Jaeger, 1959):

$$\partial T/\partial t = \kappa\nabla^2T - v\nabla T + H/\rho c, \quad (10\text{-}13)$$

where T is the temperature, t is the time, κ is the thermal diffusivity, *v* is the extension rate, ρ is the rock density, c is the specific heat of the rock, and H is its radioactive heat production rate, expressed as the heat produced by volume element of the unit mass during unit time. The first term on the right side expresses the heat conduction, the second term represents the heat advection, and the third term is the radioactive heat production.

Under steady-state conditions, characterized by zero temperature change and assuming no heat advection, Equation (10-13) reduces to:

$$\Delta T = -H/\lambda. \quad (10\text{-}14)$$

In order to appreciate a temperature-dependent thermal conductivity and a depth-dependent radioactive heat production rate, Equation (10-14) needs to be expressed as (Čermák and Haenel, 1988):

$$H(z) + d/dz[\lambda(T)dT/dz] = 0. \quad (10\text{-}15)$$

When the boundary conditions are defined by T_o being the $T_{surface}$ and q_o being equal to $\lambda(dT/dz)_{z=0}$, analytical solutions for this equation vary with the thermal properties of the lithological section, which can be represented in different ways (Čermák and Haenel, 1988).

The first example of the lithological section contains n sediment layers, each characterized by an insignificant heat production rate, H(z) = 0, and thermal conductivity, λ_i, which is constant along the entire thickness, Δz_i, of each layer. In this scenario, the temperature at depth z is:

$$T(z) = T_o + q_o\Sigma^n_{i=1}\,\Delta z_i/\lambda_i. \quad (10\text{-}16)$$

If the entire lithological section is formed by a single layer, characterized by a constant heat production rate, $H(z) = H_o$, and constant thermal conductivity, $\lambda(T) = \lambda_o$, the temperature at depth z becomes:

$$T(z) = T_o + (q_o/\lambda_o)z - (H_o/2\lambda_o)z^2. \quad (10\text{-}17)$$

The third example contains a lithological section similar to the section of the second example with a single difference, temperature-dependent thermal conductivity obeying the function 10-7. In this example, the temperature at depth z is:

$$T(z) = (1/C)\{(1 + CT_o)\exp[(C/\lambda_o)(q_o z - H_o z^2/2)] - 1\}, \quad (10\text{-}18)$$

where C is the material parameter introduced in Equation (10-7).

The fourth example is also represented by a single layer. However, its heat production rate is exponentially dependent on depth, $H(z) = H_o\exp(-z/D)$, and thermal conductivity remains constant, $\lambda(T) = \lambda_o$. The temperature at depth z is then calculated from the equation:

$$T(z) = T_o + (q_o - H_o D)z/\lambda_o + H_o D^2 [1 - \exp(-z/D)]/\lambda_o. \quad (10\text{-}19)$$

If the lithological section is also represented by a single layer, but its heat production is exponentially dependent on depth, $H(z) = H_o\exp(-z/D)$, and thermal conductivity is a function of temperature obeying Equation (10-7), the temperature at depth z becomes:

$$T(z) = (1/C)[(1 + CT_o)\exp\{(C/\lambda_o)[H_o D^2 (1 - \exp(-z/D)) - H_o Dz + q_o z]\} - 1]. \quad (10\text{-}20)$$

The sixth example of the lithological section is formed by a group of n layers with heat production H_i and a thermal conductivity constant across each layer of thickness Δz_i. In this scenario the temperature at depth z is:

$$T(z) = T_o + (q_o - \Sigma^{n-1}_{i=1} H_i\Delta z_i)(z - \Sigma^{n-1}_{i=1} \Delta z_i)/\lambda_1 - H_n(z - \Sigma^{n-1}_{i=1} \Delta z_i)^2/2\lambda_1. \quad (10\text{-}21)$$

Thermal conductivity

Apart from several exceptions, the thermal conductivity of rocks varies between 1 and 6 $Wm^{-1}K^{-1}$ as can be seen in numerous rock tables (e.g., Clark, 1966; Kappelmeyer and Haenel, 1974; Čermák and Rybach, 1982; Buntebarth, 1984; Poelchau et al., 1997; Yalcin et al., 1997; Figure 10.1). The thermal conductivity is a scalar quantity. All listed data document that it is a material property controlled by:

1) the rock type;
2) the rock-forming minerals;
3) the preferred mineral orientation; and
4) the crystal structure of individual minerals.

Both preferred mineral orientation and crystal structure can result in the thermal conductivity anisotropy due to macroscopic orientation of mineral grains and ion arrangement in the crystal lattice, respectively (e.g., Schon, 1983; Meissner, 1986). This anisotropy then controls different rates of heat diffusion in different directions (see Equations (10-11) and (10-12)). As a consequence, the heat flow direction does not have to coincide locally with the direction of the largest temperature gradient. The thermal conductivity anisotropy is more rule than exception because sedimentary fill and post-rift cover of rifts and passive margins are typically anisotropic, as expressed in Equations (10-4) and (10-5).

Three components of thermal conductivity are measured along three mutually perpendicular coordinate axes on either crystal or rock samples. Two coordinate axes are chosen to be parallel to the layering of the sedimentary rock. The remaining axis is perpendicular to layering. The layering and compositional changes are frequently responsible for significant anisotropy of thermal conductivity of sedimentary rocks (see Schon, 1983).

On the contrary, magmatic rocks, which form:

1) the crystalline basement in deeply eroded rift flanks;
2) bypassed structural highs on passive margins; and
3) rift-related magmatic bodies,

are frequently characterized by relatively insignificant anisotropy. As a result, their thermal conductivity can be taken as isotropic for thermal regime computations.

As discussed earlier, the rock-forming minerals define the thermal conductivity of a rock depending on their content. So-called maximum and minimum thermal conductivities are calculated as weighted arithmetic and harmonic mean values, respectively:

$$\lambda_{max} = \Sigma^n_{i=1} p_i\lambda_i,$$
$$1/\lambda_{min} = \Sigma^n_{i=1} p_i/\lambda_i, \quad (10\text{-}22)$$

where p_i is the fraction of the i-th mineral having a conductivity λ_i and $\Sigma^n_{i=1} p_i = 1$.

Different minerals have different thermal conductivities. Quartz, for example, is a good conductor while plagioclase is not. When the rock contains quartz, it tends to have its thermal conductivity controlled by variations in the quartz fraction (e.g., Roy et al., 1981). For example, depending on the quartz content varying between 20 and 35 percent of the rock composition, the thermal conductivity of granite can vary from 2.5 to 4 $Wm^{-1}K^{-1}$ (Buntebarth, 1984). On the contrary, the increase in the anorthite-rich plagioclase lowers the thermal conductivity of the rock (e.g., Schon, 1983). Various ratios of quartz and plagioclase, further complicated by the presence of other minerals such as

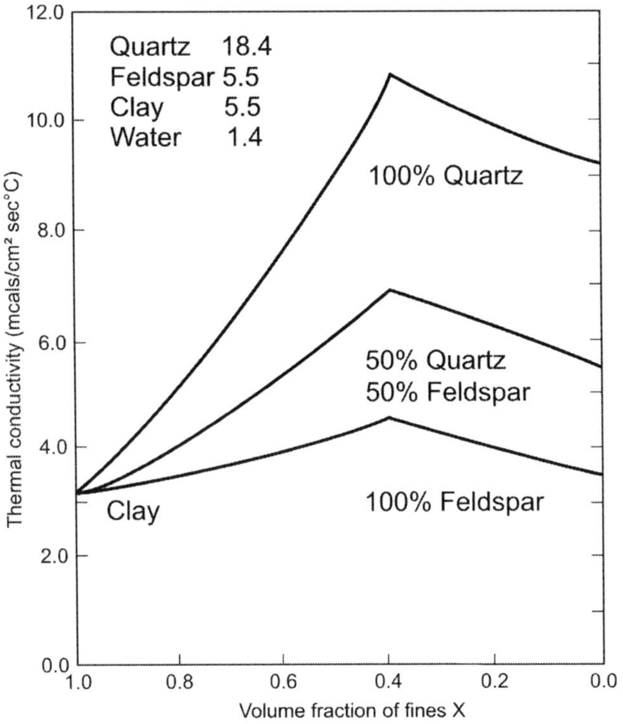

Figure 10.2. The effective thermal conductivity of clastic rock-water mixtures as a function of the volume X occupied by fine grains (Palciauskas, 1986). The figure also contains thermal conductivities of individual components.

clays, control the thermal conductivity of sandstone (Palciauskas, 1986).

Further important for the thermal conductivity of sediments is the grain size (Palciauskas, 1986; Figure 10.2). Various grain-water mixtures shown in Figure 10.2 represent analogs for various rift and passive margin sediments, which have coarser grain fraction formed by quartz and feldspar, while the finer fraction is formed by clay minerals. The volume fraction of coarser grains equals to one minus the inherent porosity of the coarse-grained sediment, ϕ_c. The volume fraction of finer grains equals to one minus the inherent porosity of the fine-grained sediment, ϕ_f, which is multiplied by the volume occupied by the fines, X. Thus, the sediment can be divided into two end-members, a coarse-grained end-member with zero fraction of fine grains and a fine-grained end-member with zero fraction of coarse grains. The saturating fluid volume can be then calculated from (Palciauskas, 1986):

$$\phi = \phi_c - X(1 - \phi_f). \qquad (10\text{-}23)$$

If the coarse-grained end-member, which is characterized by a volume of fines equal to zero, has a porosity of 40 percent, the fine-grained member, characterized by a volume of fines equal to one, has a porosity of 60 percent. The first end-member is coarse-grained sandstone. The second end-member is a clay-water system. Figure 10.2 indicates that if the coarse-grained sandstone volume is below 0.4, the important thermal conductivity changes are controlled by both lithology and grain size. A similar effect can be implied from a comparison of sandstone and quartz sandstone data.

Grain-water mixture calculations of the effective bulk thermal conductivity have been used in numerous variations in various studies (e.g., Palciauskas, 1986; Brigaud et al., 1990; Bachu, 1991; Poelchau et al., 1997). Because these calculations assume circular grains and infinitely large sediment volumes, their design serves its purpose for simplifying studies. More heterogeneous systems would require a different approach.

In search for controls of thermal conductivity, an indirect calculation method, designed for bounded rock volumes with random-shape particles or layers, indicates that the conductivity is controlled by (Bachu, 1991):

1) conductivity contrasts;
2) fraction of components;
3) interconnectivity of components; and
4) distribution and shape of heterogeneities.

The important outcome of this approach is the effect of complex facies distributions, which are, for example, typical of alluvial and deltaic rift settings, on thermal conductivity anisotropy. If a sandstone section contains horizontally oriented lenses of shale, the sedimentary section has its horizontal thermal conductivity larger than vertical conductivity. As a result, sedimentary sections having less than 45–65 percent of shale have their thermal conductivity controlled predominately by sand fraction. However, if the shale fraction is higher than 65 percent, it mostly controls the thermal conductivity of the system.

The thermal conductivity of rift and passive margin rocks is controlled by two mechanisms (e.g., Meissner, 1986): lattice and phonon conductivity by radiation. They are both temperature-dependent. Numerous experiments with evaporites, sandstones, and limestones (e.g., Birch and Clark, 1940; Kappelmeyer and Haenel, 1974; Kopietz and Jung, 1978; Zoth, 1979; Roy et al., 1981; Čermák and Rybach, 1982; Mongelli et al., 1982; Morin and Silva, 1984; Haenel et al., 1988) have documented their thermal conductivity dependence on temperature, characterized by a decrease with temperature. As shown in Figure 10.3, this function has different slope and limiting value for different rocks. The thermal conductivities of

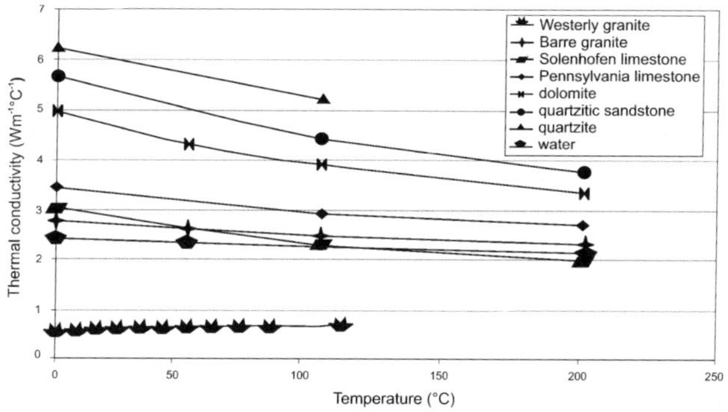

Figure 10.3. Thermal conductivity temperature dependence of several rock types and water (data from Clark, 1966) (Nemčok et al., 2005a).

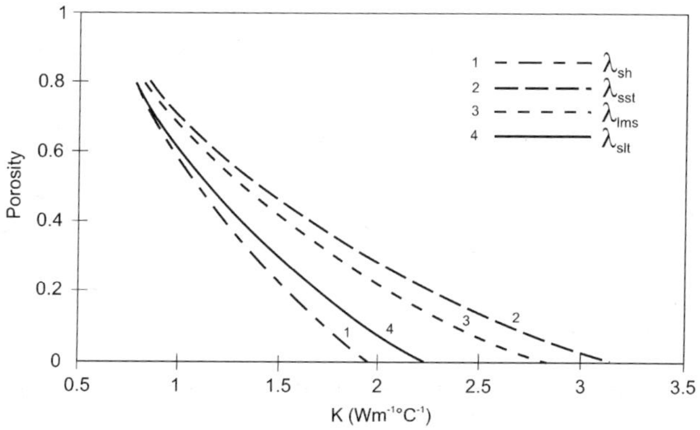

Figure 10.4. Changes in thermal conductivity of sediments (grains + pores) represented by shale (λ_{sh}), sandstone (λ_{sst}), limestone (λ_{lms}), and siltstone (λ_{slt}) due to varying porosities (Yalcin et al., 1997). Thermal conductivities of the pure lithotypes and water (λ_w) used for the calculation are: λ_{sh} = 1.98 Wm^{-1}deg C^{-1}, λ_{sst} = 3.16 Wm^{-1}deg C^{-1}, λ_{lms} = 2.83 Wm^{-1}deg C^{-1}, λ_{slt} = 2.22 Wm^{-1}deg C^{-1}, λ_w = 0.60 Wm^{-1}deg C^{-1}.

well-conducting sediments such as sandstone, quartzite, or salt show a distinct decrease with temperature. The thermal conductivities of poor conductors such as shale with porosity smaller than 25 percent are almost temperature-independent (see Kappelmeyer and Haenel, 1974; Poelchau et al., 1997). These trends result in lower thermal conductivity contrasts of multilayers at higher temperatures.

The other factor controlling the thermal conductivity is pressure. Unlike temperature, its increase leads to thermal conductivity increase although its effect is by far not as important as the effect of temperature (Morin and Silva, 1984). The thermal conductivity increase is caused by both grain deformation and pore-volume compaction during the increasing pressure.

Under larger pressure, the crystal lattice becomes deformed, and elastic properties of individual crystals influence the thermal conductivity.

Under pressure the sediment characterized by a relatively high initial porosity consisting of pore spaces located among and within the grains undergoes compaction. With increasing pressure this porosity progressively decreases, resulting in increasing thermal conductivity. As a result one can see thermal conductivity increasing trends with both decreasing porosity (Figure 10.4; Yalcin et al., 1997) and increasing density (Bachu et al., 1995) (Figure 10.4). This function is linear until reaching the elastic limit and can be expressed by the equation (Buntebarth, 1984):

$$\lambda = \lambda_0(1 + ap), \qquad (10\text{-}24)$$

where p is the pressure and a is a constant that needs to be determined experimentally. It has a value different for different rocks and varying between 1* 10^{-5} and 5 * 10^{-5} MPa^{-1}.

The effect of porosity on thermal conductivity can be demonstrated on igneous and metamorphic rocks, the porosity of which is only about 1 percent, significantly lower than the porosity of sedimentary rocks (see, e.g., Toulokian et al., 1981). This porosity is associated with fractures. Despite this low porosity, the total closure of fracture apertures leading to practically nonexistent porosity controls about 10 percent change in thermal conductivity.

In the case of sediments, the porosity changes can control the thermal conductivity changes by a factor of two to four, depending on the lithology (Yalcin et al., 1997). For example, sandstones buried at depths of 1–2 km can retain a porosity of up to 15 percent or more (Wygrala, 1989). Because of their saturation by fluids such as oil, water, and gas, which all have thermal conductivities lower than the conductivities of the rock matrix (see Figures 10.3 and 10.5a), the effective thermal conductivity of the saturated sandstone can be reduced significantly (Figures 10.5b–c). A typical example comes from sediments saturated by gas hydrate, which reduces their thermal conductivity (Stoll and Bryan, 1979).

The effect of porosity on thermal conductivity depends on the character of pores; whether they are characterized by isolated volumes or interconnected network. Models treating porosity as isolated pore spaces and interconnected pore volumes have a tendency to yield maximum and minimum values of

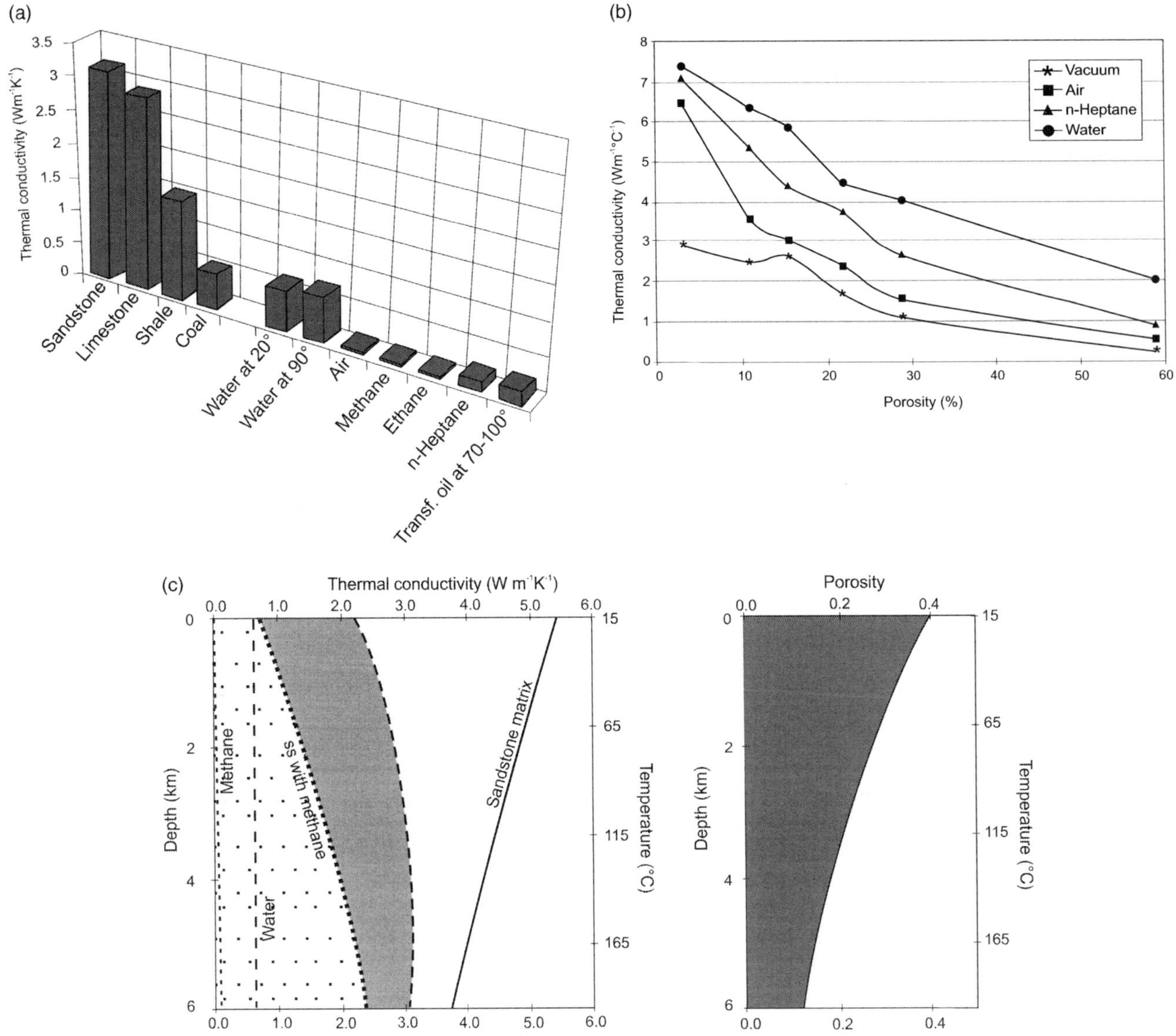

Figure 10.5. (a) Thermal conductivities of selected rocks and fluids (Poelchau et al., 1997). The values for rocks are matrix conductivities for 0 percent porosity at 20°C. Data are from Clark (1966), Weast (1974), Blackwell and Steele (1989), and IES (1993). (b) Bulk thermal conductivity of quartz-rich sandstone saturated by various fluids as a function of its porosity (data from Clark, 1966). (c) Influence of porosity, temperature, and pore-filling fluids on the effective thermal conductivity of sandstone (Zwach et al., 1994). Note the large difference between sandstone filled with methane and sandstone filled with water. Note also that the thermal conductivity of gas-filled sandstones increases to a depth much greater than the depth of water-filled sandstones.

thermal conductivity, respectively. The function for the calculation of thermal conductivity, λ, assuming isolated pore volumes, is (Buntebarth, 1984):

$$\lambda = \lambda_m\{1-[(3\phi(1 - \lambda_f/\lambda_m))/(2 + \phi + \lambda_f/\lambda_m)]\}, \quad (10\text{-}25)$$

where λ_f is the fluid conductivity, λ_m is the rock matrix conductivity, and ϕ is the porosity. Assuming interconnected pore volumes, the function for thermal conductivity calculation is (Buntebarth, 1984):

$$\lambda = \lambda_m\{1-[(\phi(1 + 2\lambda_f/\lambda_m)(1 - \lambda_f/\lambda_m))/(\phi(1 - \lambda_f/\lambda_m) + 3\lambda_f/\lambda_m)]\}. \quad (10\text{-}26)$$

Various factors control thermal conductivity through their interactive effect on porosity. Some of the most

Table 10.1. Common diagenetic reactions in sandstones and their temperature ranges (Bjorlykke et al., 1989).

Source mineral(s)	Pore water	Reaction temperature (°C)	Product mineral(s)
Aragonite and high-Mg calcite		20–50	Low-Mg calcite (sparry)
Feldspar	$-Na^+$, $-K^+$	20–100	Kaolinite
Fine-grained low-Mg calcite		60–70	Low-Mg calcite (poikilotopic)
Amorphous silica and opal CT		60–80	Quartz
Smectite	$+K^+$	50–100	Illite + quartz
K-feldspar	$+Na^+$, $-K^+$	>65	Albite
Quartz (pressure solution)		100–150	Quartz
K-feldspar + kaolinite		120–130	Illite + quartz

Table 10.2. Thermal conductivities and specific heat capacities of cements (Mann et al., 1997).

Cement type	Density (kgm^{-3})	Thermal conductivity ($Wm^{-1}K^{-1}$)		Specific heat capacity ($kJkg^{-1}K^{-1}$)	
		(at 20°C)	(at 100°C)	(at 20°C)	(at 100°C)
Silica cement	2,650	7.70	6.00	0.177	0.212
Calcite cement	2,721	3.30	2.70	0.199	0.232
Dolomite cement	2,857	5.30	4.05	0.204	0.238
Anhydrite cement	2,978	6.30	4.90	0.175	0.193
Halite cement	2,150	5.70	4.85	0.206	0.214
Clay cement	2,810	1.80	1.60	0.200	0.220

important factors are (Chilingarian, 1983; Poelchau et al., 1997 and references therein):

1) fluid pressure;
2) temperature; and
3) diagenetic processes.

For example, the overpressure retards the porosity reduction. The temperature controls the porosity reduction through its effect on fluid viscosity, solution, and cementation (e.g., Gregory, 1977; Stephenson, 1977). The diagenetic control of the thermal conductivity via effects on porosity is rather complex because diagenetic processes can either reduce or enhance porosity. Minerals in contact with pore fluids can undergo chemical alteration or dissolution, depending on the ion balance in the rock matrix, fluid, and temperature (Table 10.1). Certain minerals can precipitate from the pore fluid, depending on the relative saturations of ions needed for the development of respective cements. Such critical saturation is temperature-dependent (Table 10.1; Bjorlykke et al., 1989). In the case of cementation, the solid cement properties affect the bulk material thermal conductivity (Mann et al., 1997). For illustration, thermal conductivities of several cements are listed in Table 10.2.

The previous discussion of interactive control of thermal conductivity allows us to conclude that the conductivity is a function of:

1) porosity
2) state of diagenesis;
3) mineralogy; and
4) pore fluid,

which can be characterized as $\lambda_{\text{rock matrix}} > \lambda_{\text{saturated rock}} > \lambda_{\text{dry rock}}$.

As illustrated earlier, all pore fluids have thermal conductivities lower than rock conductivities (see Figures 10.3, 10.5). Therefore, porosity, lithology, and pore fluid type are the most important controlling factors of thermal conductivity because the conductivity of the saturated sediment can increase by a factor of two to five during its compaction history (Woodside and Messmer, 1961; Palciauskas, 1986). Figure 10.5c illustrates the effect of the interacting factors on the sandstone as it undergoes progressively greater burial and temperature. While thermal conductivities of minerals and rock matrix decrease with depth as a function of increasing temperature, the conductivity of the pore fluid increases in the same time with increasing temperature. The proportion of the pore fluid relative to the rock matrix decreases with advancing compaction because of decreasing porosity. The resultant effective thermal conductivity at each point of the compaction history depends on interaction among:

1) temperature;
2) pore fluid; and
3) porosity.

The initial trend starts with thermal conductivity increase due to rapidly declining porosity. As burial proceeds and temperature increases, the effect of decreasing mineral conductivity starts to prevail and the effective thermal conductivity of the rock system undergoes reversal toward reduced values. If the rock system includes saturation by gas, a trend of the thermal conductivity increase continues to a greater depth than the depth in the gas-unsaturated case; that is, reversal point occurs at greater depth. This effect is controlled by the rock plus gas conductivity increasing linearly with temperature, while the rock plus water conductivity levels out. This porosity-temperature dependence of the thermal conductivity is known for most common rocks from rift and passive margin settings such as shale, limestone, and sandstone (Yalcin et al., 1997). The effect of the porosity-dependent reduction of the thermal conductivity is, however, suppressed in well-conductive rocks such as salt and quartzite, thermal conductivities of which are dominated by temperature-dependence (Ungerer et al., 1990). The effect of interacting factors can be well demonstrated on different thermal conductivity-depth profiles of different types of sandstone (Palciauskas, 1986).

It is interesting to observe how the effective thermal conductivity of pure quartz sandstone remains almost unaffected by an increasing burial because of the fact that increasing temperature effect compensates for the effect of compaction. On the contrary, feldspathic sandstone and clay have different behavior. Feldspar and certain clay minerals do not undergo a pronounced temperature effect on their thermal conductivities. Therefore, their thermal conductivity-depth profiles can be characterized by the increase of thermal conductivity with depth, which is more significant than in the case of quartz sandstone. The clay-water mixture compacts as shale while the feldspar-water mixture compacts as sandstone. Because of the higher initial porosity, the clay-water mixture will have a somewhat smaller thermal conductivity than the feldspar-water mixture. However, because of its compressibility being larger than the compressibility of the feldspar-water mixture, its thermal conductivity will increase with depth faster.

Table 10.3. Thermal diffusivity of selected rocks (Roy et al., 1981; Bunterbarth, 1984).

Reference	Buntebarth (1984)	Roy et al. (1981)
Material	κ (10^{-6}m^{2}s^{-1})	κ (10^{-7}m^{2}s^{-1})
bituminous coal	0.15	
dolomite		8.5 ± 2.0
gneiss	1.2	
granite	1.4	
gypsum		3.0 ± 0.3
limestone	1.1	
rock salt	3.1	
slate	1.2	
sandstone	1.6	
tuff		6.5 ± 2.5

Thermal diffusivity

Thermal diffusivity, κ (Table 10.3) is related to thermal conductivity, λ, as follows:

$$\kappa = \lambda/\rho c, \qquad (10\text{-}27)$$

where ρ is the rock density and c is the specific heat capacity. Its change with pressure in the range 0–300 MPa can be expressed by the equation (Buntebarth, 1984):

$$\kappa = \kappa_o(1 + a'p), \qquad (10\text{-}28)$$

where κ_o is the initial thermal diffusivity, a' is a material constant in the range of 1 to 5 * 10^{-4} MPa^{-1}, and p is the pressure. The material constant varies with lithology and needs to be determined experimentally for each rock type.

Like thermal conductivity, the thermal diffusivity decreases with the temperature (Figure 10.6; Roy et al., 1981).

Specific heat capacity

Equation (10-1) expresses that the increase of the internal energy, ΔQ, of a volume element is proportional to its mass, m, its temperature, and its capability to store heat, which is characterized by specific heat capacity. The specific heat capacity is a material property (Table 10.4). Rocks, which have minimal effective porosity, have their specific heat capacity-depth profile dominated by

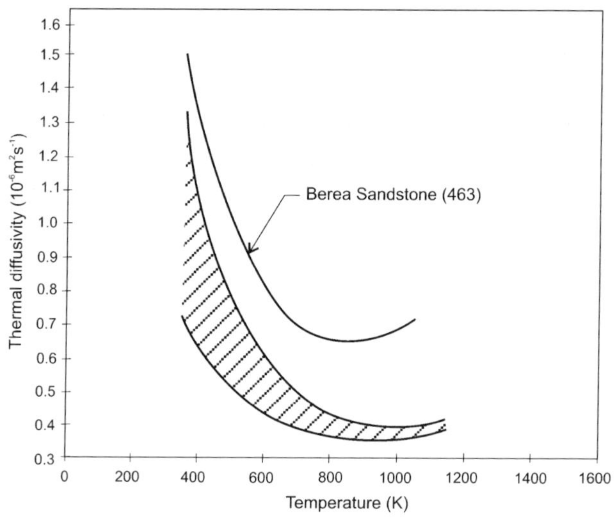

Figure 10.6. Thermal diffusivity dependence on the temperature of sandstone (Roy et al., 1981).

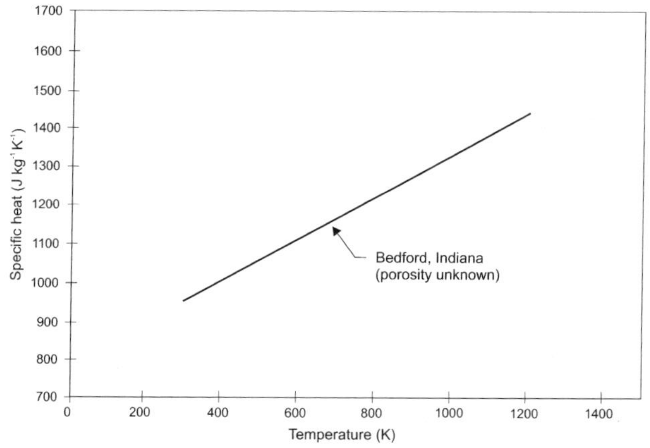

Figure 10.7. Specific heat dependence on the temperature of limestone (Roy et al., 1981).

temperature-dependence (Roy et al., 1981; Mongelli et al., 1982; Figure 10.7). This temperature-dependence for crystalline rocks at constant pressure is expressed by the equation (Buntebarth, 1984):

$$c = 0.75(1 + 6.14 * 10^{-4}T - 1.928 * 10^4/T^2). \qquad (10\text{-}29)$$

Sediments, often characterized by distinct porosity, behave as matrix-fluid mixtures. Their specific heat changes with depth as a sum of changes in the specific heat of both the rock matrix and pore fluid. When the fluid is water, it tends to increase the effective specific heat because of its high specific heat (Shi and Wang, 1986; Yalcin et al., 1997). In the case of matrix-fluid mixtures, the specific heat can be calculated by the formula using the volume average from the values of the rock matrix and pore fluid (e.g., Bear, 1972; Van der Kamp and Bachu, 1989; Ferguson et al., 1993):

$$\rho c(z) = \rho_f c_f \phi(z) + \rho_s c_s(1 - \phi(z)), \qquad (10\text{-}30)$$

where ρ is the bulk density, c is the bulk specific heat, ρ_f is the fluid density, c_f is the specific heat capacity of the fluid, $\phi(z)$ is the depth-dependent porosity, ρ_s is the solid grain density, and c_s is the specific heat of the solid grain.

The bulk-specific heat capacity is further affected by the solid cement properties (e.g., Mann et al., 1997). Some of them are listed in Table 10.2. Therefore, the specific heat capacity of the sediment becomes a function of:

1) temperature;
2) porosity;
3) state of diagenesis;
4) mineralogy; and
5) pore fluid.

Radiogenic Heat Production Rate

Most of the rocks involved in rift and passive margin settings contain minerals with radioactive elements. Although all naturally occurring radioactive isotopes produce heat, the only important heat generation comes from the series of uranium, thorium, and the potassium isotope ^{40}K (Kappelmeyer and Haenel, 1974; Schon, 1983; Rybach, 1986a, 1986b; Table 10.5). A portion of the mass of each decaying nuclide in the radioactive decay process is converted into energy. Most of this energy carries the form of kinetic energy or emitted particles or the form of electromagnetic radiation (γ-rays; Van Schmus, 1984). In the case of β^- decay, a portion of the energy is carried away by neutrinos and the remaining energy is absorbed by the Earth and converted to heat.

Basic data on the four most important heat-producing isotopes are listed in Table 10.6. The radioactive heat production, H, is calculated with the help of heat generation constants as the amount of heat released per unit mass of U, Th, and K per unit time, if the concentrations of uranium, C_U, thorium, C_{Th}, and potassium, C_K, are known (Schmucker, 1969):

$$H = \rho(3.35\ C_U + 9.79\ C_{Th} + 2.64\ C_K)10^{-5}, \qquad (10\text{-}31)$$

where ρ is the density. Different calculations have been described by Rybach (1976, 1986a, 1986b) and Buntebarth (1984). Rybach (1986b), for example, shows that uranium generates an amount of heat four times larger than the amount produced by the same amount of thorium.

Table 10.4. Specific heat capacity of selected rocks (Roy et al., 1981; Schon, 1983; Buntebarth, 1984; Hutchinson, 1985; Yalcin et al., 1997).

Reference	Buntebarth (1984)	Yalcin et al. (1997)	Roy et al. (1981)	Hutchinson (1985)	Schon (1983)
Material	c ($Wsg^{-1}K^{-1}$)	c ($calg^{-1}°C^{-1}$)	c ($Jkg^{-1}C^{-1}$)	c ($Jkg^{-1}K^{-1}$)	c ($Jkg^{-1}K^{-1}$)
basalt					762...2,135(1,231)
bituminous coal	1.26				
calcareous sandstone	0.84				
clay	0.86				
clay/siltstone					753...3,546(1,240)
diabase					791...829(860)
diorite					1,118...1,168(1,136)
dolomite					648...1,465(1,088), 756
gabbro			775 ± 60		879...1,130(1,005)
gneiss					754...1,176(979)
granite					741...1,548(946), 672
gypsum			1,010 ± 150		
ice	2.1				
gravel					756
limestone			860 ± 50	1,004	753...1,712(887), 714
marble			883 ± 20		753...879(857)
obsidian			920 ± 100		
oil	2.1				
peridotite					961...1,088(1,005)
pyroxenite					879...1,214(1,005)
quartz diorite					1,214
quartzite					718...1,331(991)
salt		0.206 (20°C) 0.212 (100°C)		854	1,474...4,651(2557)
sandstone	0.71			1,088	670...3,345(972)
sandy shale		0.205 (20°C) 0.248 (100°C)			
shale				837	
serpentinite					963...1,130(1,005)
peat					1758
shale		0.213 (20°C) 0.258 (100°C)			754...1,729(1,096)
water	4.2			4,180	

Depending on the radioactive isotope concentration, the heat production is lithology-dependent, varying over several orders of magnitude (Tables 10.5, 10.7). Evaporites can be characterized by generally low heat production with the exception of potassium salt. The same is true for carbonates. Sandstones are generally low to medium heat generators. Siltstones and shale are characterized by high heat generation. Among the rift

Table 10.5. Heat production H in μWm^{-3} of sediments and heat proportions generated by U, Th, and K decay (Rybach, 1986b).

Rock	U	Th	K	Density	Heat generation
	(ppm)	(ppm)	(%)	(kgm^{-3} *10^3)	(μWm^{-3})
BLACK SHALES	20.2	10.9	2.6	2.4	5.5
SHALE/ SILTSTONE	3.7	12.0	2.7	2.4	1.8
SANDSTONES					
graywacke	2.0	7.0	1.3		0.99
arkose	1.5	5.0	2.3		0.84
quartzite	0.6	1.8	0.9		0.32
DEEP SEA SEDIMENTS	2.1	11.0	2.5	1.3	0.74
CARBONATES				2.6	
limestone	2.0	1.5	0.3		0.62
dolomite	1.0	0.8	0.7		0.36
EVAPORITES					
anhydrite	0.1	0.3	0.4	2.9	0.090
salt	0.02	0.01	0.1	2.2	0.012

Table 10.6. Energy data on major heat-generating isotopes (Van Schmus, 1984).

Element isotope	Potassium	Thorium	Uranium	Uranium
	^{40}K	^{232}Th	^{235}U	^{238}U
Isotopic abundance (Wt %)	0.0119	100.0	0.71	99.28
Decay constant (year^{-1})	5.54 * 10^{-10}	4.95 * 10^{-11}	9.85 * 10^{-10}	1.551 * 10^{-10}
Total decay energy (MeV/decay)	1.34	42.66	46.40	51.70
Beta decay energy (MeV/decay)	1.19	3.5	3.0	6.3
Beta energy lost as neutrinos (MeV/decay)	0.65	2.3	2.0	4.2
Total energy retained in earth (MeV/decay)	0.69	40.4	44.4	47.5
Specific isotopic heat production (cal/g-year)	0.220	0.199	4.29	0.714
Present elemental heat production (cal/g-year)	26*10^{-6}	0.199	-	0.740

and passive margin sediments, the black shale is associated with highest heat generation (Rybach, 1986b). Studies of fine sandy and silty sediments indicate the effect of the gain-size on the specific heat production (Pereira et al., 1986). γ-activity has been documented to increase with increasing content (Schon, 1983). The concentration of U, Th, and K apparently correlates with the amount of quartz in the rock (see Meissner, 1986). The rock density also affects the heat production of each rock type. The greater the rock density, the smaller the space is for large ions of the heat-producing elements. This controls an inverse relationship between heat production and density (Buntebarth and Rybach, 1981; Čermák et al., 1991). Because both quartz amount and rock density control the radiogenic heat production rate, and the rocks located progressively deeper in the continental crust and mantle are progressively more mafic and denser (see Figure 5.3; Christensen and Mooney, 1995) as discussed in Chapter 5, the heat production decreasing with the depth is to be expected. In fact, the typical heat production profile through the thinned continental crust and uppermost mantle in the Basin and Range province documents exactly this trend (Figure 10.8). For modeling purposes, this decrease

Table 10.7. Heat production rate of specific rocks (Pollack and Chapman, 1977; Buntebarth, 1984; Hutchinson, 1985; Meissner, 1986; Rybach, 1986b; Ibrmajer et al., 1989).

Reference	Rybach (1986b)	Buntebarth (1984)	Meissner (1986)	Pollack and Chapman (1977)	Hutchinson (1985)	Ibrmajer et al. (1989)
Material	H (μWm^{-3})	H (μWm^{-3})	H (μWm^{-3})	H (μWm^{-3})	H (μWm^{-3})	H (μWm^{-3})
amphibolite		0.3				
anhydrite	0.090					
arkose	0.84					
black shale	5.5					
carbonate			0.7			
chondrite		0.026				
deep sea sediments	0.74			0.5		
diorite		1.1				
dolomite	0.36					
dunite		0.0042	0.004			
eclogite-low U		0.034	0.04			
eclogite-high U		0.15				
gabbro		0.46				
gneiss		2.4				
granite		3.0				4.5
granodiorite		1.5				
graywacke	0.99					
igneous rocks – silicic			2.5			
igneous rocks – mafic			0.3			
limestone	0.62				0.84	
mica shist		1.5				
oceanic lherzolite			0.01			
olivinfels		0.015				
peridotite		0.0105				
quartzite	0.32					
salt	0.012				0.0	
sand (beach)			1.2			
sandstone		0.34–1.0			0.84	
shale			2.1		1.05	
shale/siltstone	1.8					
slate		1.8				
pore water					0.0	

with depth can be simplified as the exponential decay of the heat production with depth (Lachenbruch, 1968),

$$A = A_s \exp(-z/D), \qquad (10\text{-}32)$$

where A_s is the volumetric heat production at the Earth's surface, z is the depth, and D is the depth at which the heat production drops down to the value of 1/e of the surface heat production.

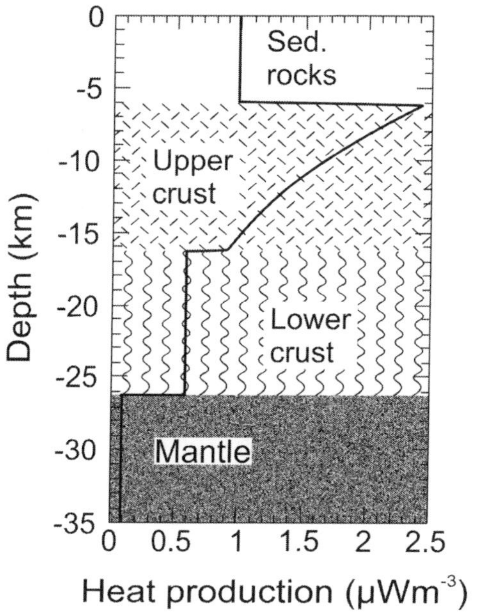

Figure 10.8. Typical Basin and Range province heat production profile through the crust and portion of the mantle (Ehlers et al., 2001).

The effect of heat production on the thermal regime can be illustrated by 1-D calculation (Rybach, 1986b):

$$T(h_z) = T_o + h_z[(q + Hh)/\lambda] - h_z^2(H/2\lambda), \qquad (10\text{-}33)$$

where h_z is the depth, T_o is the surface temperature, q is the basal heat flow, H is the average heat production, h is the thickness, and λ is the thermal conductivity of sediments. The results of this calculation indicate that the impact of different values of heat production is relatively significant, although not as strong as the impact of different values of thermal conductivity.

Effect of thermal properties on heat distribution

The thermal regime in rift and passive margin settings is controlled to a large extent by the thermal conductivity structure, heat production structure, and heat flow from the mantle. It is the distribution of the thermal conductivities and heat capacities that to a large degree controls the heat distribution in the rift and passive margin settings and associated maturation history of source rocks. As indicated earlier, distinct differences appear in thermal conductivities even among saturating fluids. For example, water, characterized by a conductivity of 0.6 $Wm^{-1}K^{-1}$, has its conductivity significantly higher than the conductivity of gas, air, or oil (see Figure 10.5a). As a consequence, differences in temperature distribution would depend on the level of hydrocarbon saturation in the case of both source and reservoir rocks (Poelchau et al., 1997; Figures 10.5b, c).

As a rule, most sedimentary rocks are anisotropic. They have horizontal thermal conductivity greater than vertical conductivity (Gretener, 1981). However, because the heat flow follows the temperature gradients that have the greatest vertical component, the increased horizontal thermal conductivities do not necessarily have a significant impact inside the sedimentary cover (Poelchau et al., 1997; Yalcin et al., 1997). However, at least one special case may exist. Being rather typical for an emergent stage of the rift settings, it involves an extra fluid flow mechanism-related thermal system added to the background thermal regime. In the presence of the fluid flow in the layered section composed of highly conductive sandstone and low-conductivity shale, the lateral heat flow can become enhanced.

The thermal regime perturbation, also common in both rift and passive margin settings, is associated with abrupt lateral conductivity contrasts across normal faults. The juxtaposition of an enhanced thermal regime of the footwall, which is characterized by an increased thermal gradient and surface heat flow values, with a depressed thermal regime of the hanging wall, which is characterized by decreased thermal gradient and surface heat flow values, can cause lateral heat flow. Such a situation is documented in the case of the Wasatch normal fault of the Basin and Range province (Ehlers et al., 2001). Another abrupt conductivity contrast has been documented in the Sea of Galilee pull-apart basin in the Dead Sea region (Gvirtzman et al., 1997). Here, the basin is filled by highly conductive evaporites, having a conductivity of 4 $W°C^{-1}m^{-1}$, while the basin flank is formed by carbonates, having a conductivity of 3 $W°C^{-1}m^{-1}$. The lateral thermal conductivity difference results in the heat refraction causing a larger amount of heat reaching the surface of the basin fill, which is documented by a robust surface heat flow database in the area (see Ben-Avraham et al., 1978; Eckstein and Simmons, 1978; Levitte et al., 1984; Galanis et al., 1986).

The role of lithology in the thermal regime can be further documented by a simple 1-D model developed from data from Alberta, Canada (Bachu, 1985; Figure 10.9). This model compares the regional function of temperature with depth with the local function of temperature with depth through individual layers of the multilayer, which is rather different. Further examples come from the North Sea (Bachu, 1985). They show the difference between calculated sandstone and shale thermal gradients, where the less conductive shales have a higher temperature gradient.

The effect of the specific heat is important in the transient thermal regime. Because it gives the energy used or released in a temperature change, it has an obvious effect on heat conduction under transient conditions while it does not play a role under steady-state conditions characterized by constant temperatures.

The effect of the heat production is directly proportional to the depth and to the amount of heat generated. As heat generated by radioactive decay is dissipated by heat conduction, the temperature increase with time and depth is also controlled by thermal conductivity (Yalcin et al., 1997). It remains to be demonstrated how important a role in heat sourcing is provided by crustal heat sources. It has been demonstrated that 20–60 percent of the surface heat flow originates from crustal radioactive heat production (Rybach, 1986b) and even more in areas of England, Wales, the Central Australian Shield, and the French Hercynian Shield (Ungerer et al., 1990). Equation (10-33) illustrates how an uneven distribution of radioactive heat sourcing in the lithospheric profile controls the nonlinear temperature-depth profile. This profile is characterized by a geothermal gradient decrease with depth. Note that time is also an important factor in heat production as the internal radioactive heat sources increase the thermal regime of the sediment section.

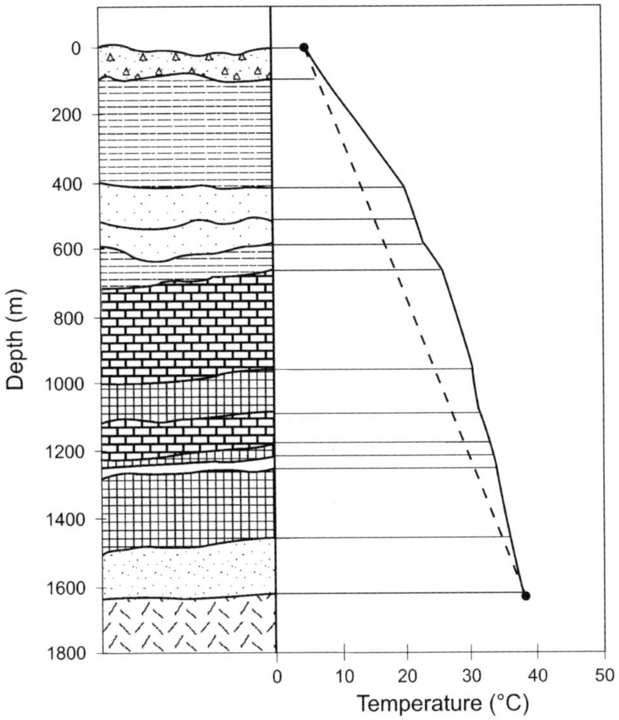

Figure 10.9. Comparison of the average temperature/depth profile in a sedimentary section (dashed curve) with temperature/depth profiles through individual layers (solid curve) (Bachu, 1985).

The heat production becomes rather unevenly distributed because the resultant crustal architecture in rift terrains and passive margins is frequently characterized by juxtapositions of segments with high and low crustal attenuation (Figures 2.13, 2.14, 2.21, 10.10). Furthermore, it is useful to discuss Equation (10-32) one more time because this simplified model seems to satisfy the observed heat flow generation distribution even in extensional terrains (Lachenbruch, 1970; Čermák and Bodri, 1992). It is further consistent with heat generation data from batholiths (Swanberg, 1972). In this model, the D parameter serves as a measure of the rate of vertical concentration of heat production. However, note that depth distribution of the radiogenic heat sources in the crust can be evaluated only indirectly, for example, using the empiric function converting seismic velocities, v_p, into heat production, A (Rybach and Buntebarth, 1984):

$$\ln A = 12.6 - 2.1v_p \text{ for Precambrian lithospheres and}$$
$$\ln A = 13.7 - 2.17v_p \text{ for Phanerozoic lithospheres.} \qquad (10\text{-}34)$$

Apart from the localized thinning affecting the upper portion of the crust in the brittle and lower portion of the crust in a ductile regime, which controls the heat

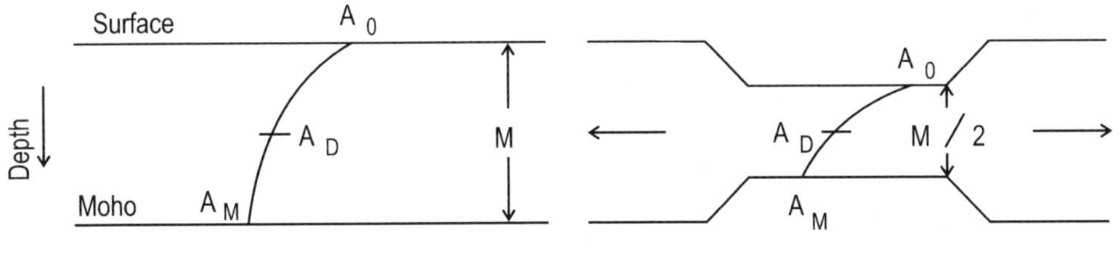

Figure 10.10. Variation of parameter D produced by the mechanical thinning of the lithosphere during rifting (Čermák and Bodri, 1992). A_o, A_D, and A_M are the heat production at the surface, depth of D, and Moho depth, respectively.

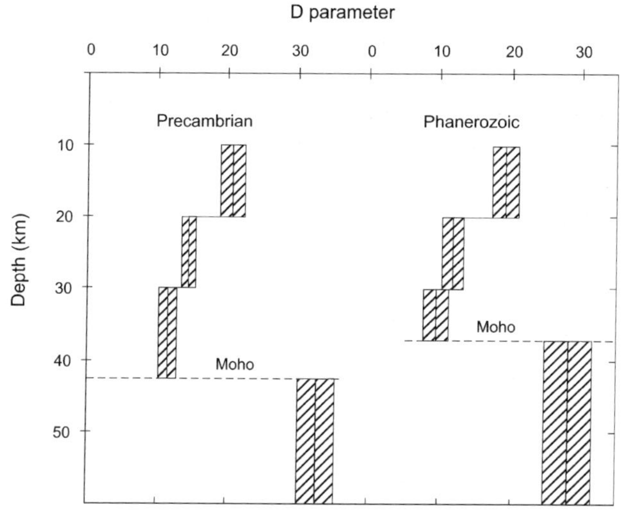

Figure 10.11. Standard values of the D parameter for selected crustal and upper mantle depth sections in Precambrian and Phanerozoic terrains (Čermák and Bodri, 1992). Heavy lines indicate mean values of D; shaded areas correspond to one standard deviation.

production distribution, the age of the lithosphere also is in control (Figure 10.11). This has been determined from a study of 200 selected seismic velocity profiles worldwide, cutting through the Archaean to Alpine lithospheres (Čermák and Bodri, 1992). The difference in calculated mean value of the heat production at a 10 km depth between the Precambrian and Phanerozoic terranes is almost fourfold. The Phanerozoic value is about 2.23×10^{-6} Wm^{-3}, while the Precambrian value is about 0.6×10^{-6} Wm^{-3}. A decrease of parameter D with lithospheric age seems to indicate more differentiated radioactive elements in the younger crust in comparison to the more homogeneous older crust (Čermák and Bodri, 1992).

When we compare the values of the D parameter, which were calculated as based on seismic velocities from different rift zones (see Čermák and Bodri, 1992), with average values for the lithospheric type, we can see that the D parameter can perhaps provide some correlation with the amount of stretching for the respective lithospheric layer (Čermák and Bodri, 1992). For example, in the Upper Rhine Graben, a depth range of 10–20 km is associated with the D value, which is two times smaller than the worldwide average, while D is already 4.7 times smaller than the worldwide average for Moho at a depth of 20–25 km. A similar, depth-dependent trend can also be observed for the upper mantle underneath the Upper Rhine Graben. Furthermore, an interesting D-parameter trend comes from the Basin and Range province, where the upper crustal values do not differ

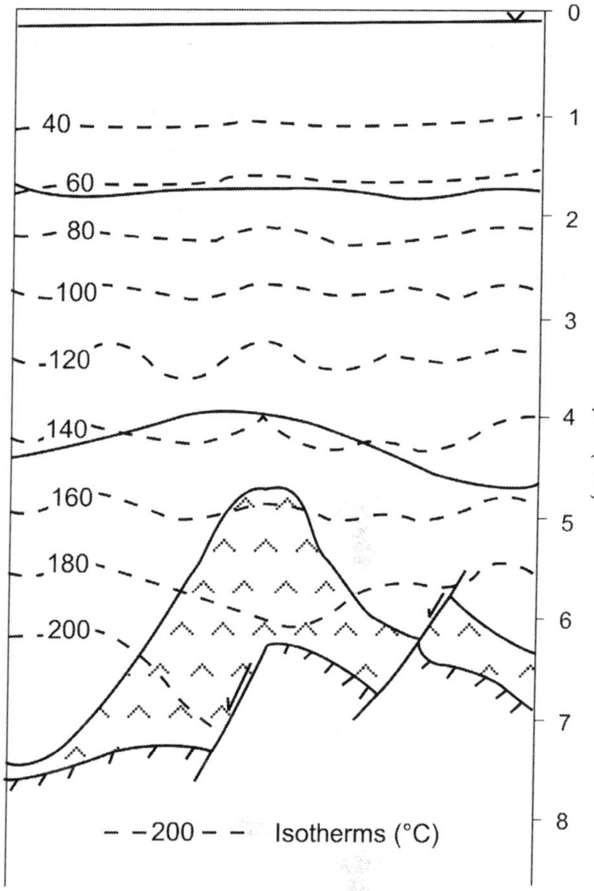

Figure 10.12. Effect of a salt diapir on temperature distribution (Yalcin et al., 1997). Note the higher temperature gradient above the salt diapir.

dramatically from the average value. The lower crust is associated with only one-third of the average value. This may, perhaps, indicate different extension mechanisms and rates for upper and lower crustal layers.

This interesting study of heat production distribution in extensional terrains practically documents a relationship between extensional processes and vertical distribution of radioactive heat production. The rate of heat production decrease underneath rift zones is larger than the rate in terrains remaining unextended. A differentiation of the heat production associated with rifting is more pronounced in the mantle than in the crust and, sometimes, there is a dramatic difference between upper and lower crust heat productions. Note that a certain redistribution of radioactive elements due to rifting remains observable for a long time after the end of rifting.

As mentioned earlier for a graben-horst scale, the lateral conductivity contrasts can cause a lateral component of heat conduction (Yalcin et al., 1997). On a smaller scale, most prominent lateral contrasts in thermal conductivity occur around basement highs and evaporite bodies, which are both quite common in rift and passive margin settings. The resulting lateral component of heat conduction causes an isotherm deflection. The magnitude of this perturbation depends on the thermal conductivity of the disturbing body.

It is known that salt diapirs behave as heat pipes controlling positive temperature anomalies over their roofs (Figure 10.12). As a result of anomalous salt conductivity, the isotherms in the underlying sediments become depressed, while the overlying sediments undergo the opposite effect, an isotherm ascent.

At the end of this chapter, we need to highlight that practically any lateral facies changes, any lateral inhomogeneity in the sedimentary formation made by thinning or thickening, porosity changes, and saturation by hydrocarbons are frequently indicated by thermal perturbations caused by respective lateral differences in thermal conductivity and concentrations of heat-producing elements (e.g., Bachu, 1985). Further effects can be imposed by chemical changes caused by diagenesis and dehydration of clay minerals, which can develop a system of heat sources and sinks in young basins, which will further perturb the background thermal regime.

11 The role of syn-rift deposition and erosion in thermal regimes of rifts and passive margins

Introduction

Apart from their impacts on structural architecture and tectonics, deposition and erosion affect the thermal regimes of rifts and passive margins to a considerable extent when their rates are significant (Table 11.1). A depositional rate faster than 0.1 mm yr^{-1} exerts a noticeable cooling effect on the surface heat flow while a significant erosional rate has the opposite effect, resulting in advection of hotter material toward the surface.

A comparison of deposition rates from the East African Rift system (Tiercelin and Faure, 1978) indicates that rates of 0.07–3.5 mmy^{-1} are typical, that is, frequently reaching values above the 0.1 mmy^{-1} threshold. Erosion studies of rift settings such as the footwall of the Wasatch normal fault (Ehlers et al., 2003) yield erosion rates of about 0.8–1.2 mmy^{-1}, proving that erosional rates in rift settings can reach significant values capable of thermal regime perturbations.

All these evidences point toward erosion and deposition having an important impact on the thermal regime and maturation history of hydrocarbon source rocks. The discussion of this impact will start with basic physical principles and finish with natural examples.

The role of deposition on thermal regimes

Deposition tends to cool down the thermal regime. This effect lasts for a long time after the termination of deposition (De Bremaeker, 1983; Ungerer et al., 1990). The main factors associated with cooling are the time span of the depositional event and thermal conductivities of deposited sediments (e.g., Jessop and Majorowicz, 1994; Yalcin et al., 1997).

The basic physical principles of the cooling effect can be explained using a simple description of deposition, sediment compaction, and pore fluid flow in an oceanic basin (Hutchinson, 1985). The description uses a two-layer model of the rock section, which does not have lateral boundaries defined (Figure 11.1). This allows us to use a one-dimensional calculation of the vertical heat flow (Hutchinson, 1985):

$$\partial_t Q = (q - \partial_z q^*(dz/2)) - (q + \partial_z q^*(dz/2)) + H, \quad (11\text{-}1)$$

where the rate of loss or gain of heat, Q(z, t), from an element, dz, is given by the difference between heat flows, q(z, t), at the top and basal surfaces of the element summed together with internal heat production, H(z), of the element but depth and time arguments of the heat, heat flow, and heat production are ignored for simplicity.

Both depth and time arguments can be omitted in the equation (Hutchinson, 1985):

$$\partial_z(\lambda\partial_z T) - \partial_z\{[\rho_f c_f v_f \phi + \rho_s c_s v_s(1 - \phi)]^*T\} + H = [\rho_f c_f \phi + \rho_s c_s(1 - \phi)]^*\partial_t T, \quad (11\text{-}2)$$

where λ(z, t) is the composite thermal conductivity at depth z, T(z, t) is the temperature at depth z and time t, $\rho_f(z, t)$ is the pore fluid density, $\rho_s(z, t)$ is the sediment particle density, $c_f(z, t)$ is the pore fluid heat capacity, $c_s(z, t)$ is the sediment particle heat capacity, $v_f(z, t)$ is the pore fluid flow velocity, $v_s(z, t)$ is the sediment particle transport velocity, and ϕ is the porosity.

The active deposition in the discussed example controls:

1) the sediment porosity;
2) the composite thermal conductivity;
3) the velocity of sediment particles; and
4) the flow rate of fluid escaping from compacting pores.

The sediment porosity change is described as (Hutchinson, 1985):

$$\phi(z) = \phi_o \exp(-z/C_{const}), \quad (11\text{-}3)$$

Table 11.1. Factors determining the thermal regime in a sedimentary basin (Deming, 1994b). Limiting deposition rates can be compared with data in Chapter 7 and worldwide examples discussed by Tiercelin and Faure (1978).

Factor	Importance (order)	Qualifications
overburden thickness	first	always important
heat flow	first	always important
thermal conductivity	first	always important
surface temperature	second	always important
deposition	first	>0.1 mmy^{-1}
	second	0.1–0.01 mmy^{-1}
	third	<0.01 mmy^{-1}
gravity-driven fluid flow	first–second	foreland basins
compaction-driven fluid flow	third	unless focused
free convection	unknown	
initial thermal event	first (0–20 Ma)	rift basins only
	second (20–60 Ma)	
	third (>60 Ma)	

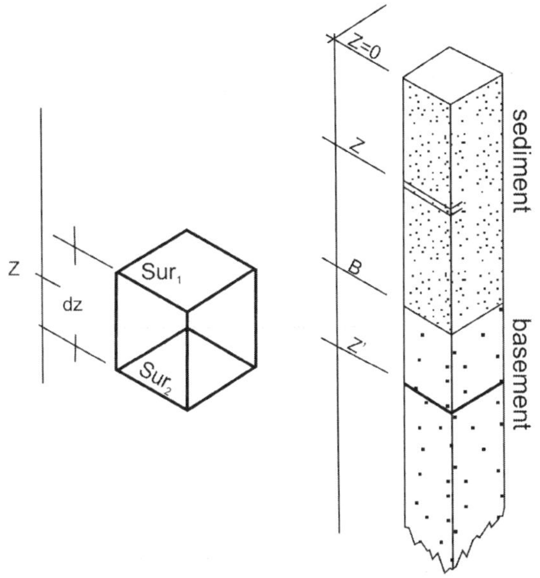

Figure 11.1. Simplified two-layer model represented by a sediment/basement column (Hutchinson, 1985). Sediment is added at the surface ($z = 0$) with a deposition rate, v_o. The porosity is described as a function of depth z ($\phi(z)$). It has a surface value of ϕ_o. The basement interface occurs at depth B. Elements of depth dz at Z and Z' within the sediment and basement layers are discussed in the text; the difference in heat flow across Sur_1 (in) and across Sur_2 (out) plus the radioactive heat production H is related to the rate at which heat is gained or lost from dz.

where C_{const} is the compaction constant defined by Rubey and Hubert (1960) and ϕ_o is the sediment porosity at the basin floor.

The change in composite thermal conductivity is defined as (Hutchinson, 1985):

$$\lambda(\phi) = [-\alpha + \sqrt{(\alpha^2 + 8\lambda_s\lambda_f)}]/4, \qquad (11\text{-}4)$$

where λ_s is the thermal conductivity of the sediment, λ_f is the thermal conductivity of the pore fluid, and α is the substitute for:

$$\alpha = 3\phi(\lambda_s - \lambda_f) + \lambda_f - 2\lambda_s. \qquad (11\text{-}5)$$

The change in velocity of sediment particles is described as (Hutchinson, 1985):

$$v_s = [v_o(1 - \phi_o)]/(1 - \phi), \qquad (11\text{-}6)$$

where v_o is the rate of deposition.

The flow rate change of fluid escaping from compacting pores is defined as (Hutchinson, 1985):

$$v_f = \{[v_o(1 - \phi_o)]/[1 - \phi_o \exp(-B/C_{const})]\} * \exp[(z - B)/C_{const}], \qquad (11\text{-}7)$$

where B is the basement depth.

Figure 11.1 shows that a depth, z, ranges from 0 to B. Equation (11-7) indicates that the flow rate loses importance as the value of depth B approaches the values of the compaction constant C_{const}.

To compare the importance of the fluid flow and sediment particle velocities to heat flow, both their contributions can be expressed in a ratio (Hutchinson, 1985):

$$\text{fluid contribution / sediment contribution} = [\rho_f c_f \phi_o] / [\rho_s c_s(\exp(B/C_{const}) - \phi_o)]. \qquad (11\text{-}8)$$

It indicates that the pore fluid flow contribution is larger than the contribution of the sediment particle velocity in a depth interval ranging from the basin floor to a depth at which the top basement reaches a depth equal to the compaction constant of the sediment cover.

An interesting insight into the contribution of pore fluid flow and sediment particle velocity into the heat flow regime is provided by a sensitivity study (Hutchinson, 1985) that varies the value of the initial sediment porosity. Its results are displayed as the ratio of the heat flow, Q, to its initial equilibrium value, Q_o, being calculated for the chosen thermal conductivity and density of the sediment (Figure 11.2). The initial sediment porosity value and compaction constant vary from 50 to 80 percent and 1 to 4 km, respectively. The study indicates that

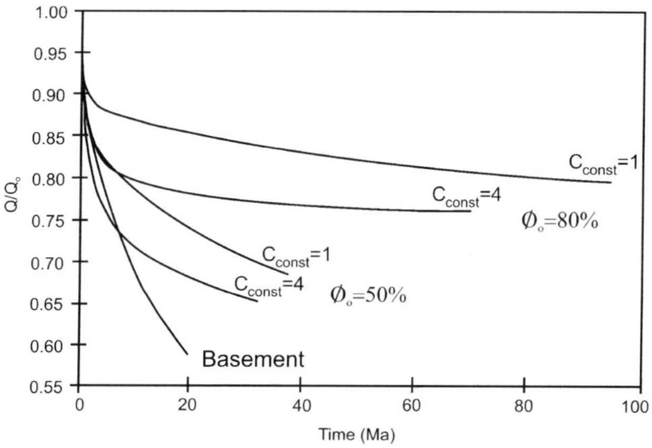

Figure 11.2. Dependence of the fractional heat flow alteration Q/Q_o on the compaction constant, C_{const}, and initial sediment porosity, ϕ_o (Hutchinson, 1985). The deposition rate is 0.5 mmy^{-1}, sediment conductivity is 2.5 $Wm^{-1}K^{-1}$, heat capacity of sediment particles is 900 $Jkg^{-1}K^{-1}$, and density of sediment particles is 2,700 kgm^{-3}. Basement density is 3,300 kgm^{-3}, thermal conductivity is 3.1 $Wm^{-1}K^{-1}$, specific heat capacity is 1,160 $Jkg^{-1}K^{-1}$, and heat production is 0 μWm^{-3}. Pore fluid density is 1,030 kgm^{-3}, thermal conductivity is 0.67 $Wm^{-1}K^{-1}$, specific heat capacity is 4,180 $Jkg^{-1}K^{-1}$, and heat production is 0 μWm^{-3}. The analytical solution for a basement half-space with the same sedimentation rate is also shown. All curves terminate after a deposition of 10 km of sediment.

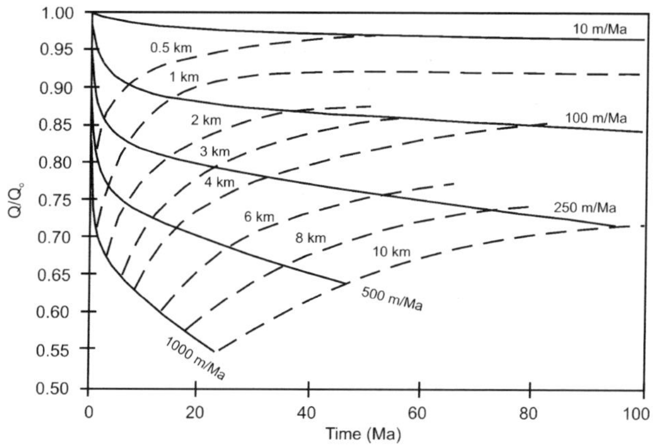

Figure 11.3. The variation of the fractional heat flow alteration Q/Q_o through time for deposition of shale with a deposition rate of 0.1, 0.25, 0.5, and 1 mmy^{-1} (Hutchinson, 1985). The dashed lines link points of equal basement depth.

the bulk of the heat flow value is controlled by the initial sediment porosity. The most porous sediments affect the heat flow the least. The compaction constant role is subordinate to the role of porosity, and only large values control a slight heat flow increase.

The same sensitivity study done by Hutchinson (1995) further focused on the role of the specific sediment in heat flow, apart from the role of the initial porosity and the compaction constant. As to be expected from the discussion in Chapter 10, it indicated that the thermal conductivity and heat capacity of various sediment grains vary significantly, documenting distinct sediment lithology control over the heat flow. The smallest heat flow variations are associated with the highest conductivity sediments such as sandstone and salt. The largest heat flow variations are associated with low-porosity, low-conductivity sediments such as limestone.

The magnitude of the thermal perturbation driven by a depositional event is controlled by its rate and history (Hutchinson, 1985; Figure 11.3). An example from deep oceanic depositional settings using several different deposition rates for shale illustrates the effect of the rate on the heat flow regime. A sensitivity study documented in Figure 11.3 shows the impact of the deposition rate on heat flow alteration, expressed as the ratio of the heat flow, Q, to its initial equilibrium value, Q_o. For depositional rates between 0 and 0.5 mmy^{-1} and time periods apart from the first 10 Ma of the depositional event, the deposition rate impact on heat flow dominates over the impact of the depositional event duration. For these depositional rates, the accelerated heat flow change from the first 10 Ma years slows down for the remaining portion of the observed time period, although the amount of slowing down is inversionally proportional to the deposition rate. However, this characteristic breaks down for fast deposition rates, as shown by the example of the depositional rate of 1 mmy^{-1}. This case indicates that a significant heat flow alteration can continue for a long time in the case of fast deposition.

Note that it already takes as little as 1 km of slowly deposited sediment to cause an important cooling effect on the surface heat flow. Depositional rates over 0.1 mmy^{-1} are known to cause significant perturbations of the heat flow regime (e.g., Hutchinson, 1985; Doligez et al., 1986; Deming, 1994b). A nice example of significant perturbation comes from the north Brazilian passive margin, where both depositional rate and the sediment thickness reach their maxima in front of the Amazon River mouth and both progressively decrease in an easterly direction. A present-day geothermal gradient map (Coolbaugh et al., 2005) shows a nice eastward cooling trend in a direction away from the Amazon fan, starting with a value of 22°Ckm^{-1} for the offshore Amazon Basin, going through 28° Ckm^{-1} in the Para Maranhao Basin, 31° Ckm^{-1} in the Barreirinhas Basin, and finishing with 33° Ckm^{-1} in the Potiguar Basin.

It is valid that the lower the thermal conductivity of the sediment deposited by the depositional event, the greater is the heat flow reduction that happens under the same depositional rate. Therefore, if the sediment provenance starts to feed the deposition in the adjacent basin by sediment with increased thermal conductivity while the depositional rate remains the same, the surface heat flow in the basin increases. The heat flow is still reduced in comparison with a no deposition scenario, but the magnitude of reduction is smaller than in the case of previous low-conductivity sediment. That is why the sediment provenance feeding the deposition by high-conductivity siliciclastic sediments following the earlier low-conductivity carbonates can affect the heat flow regime even if the deposition rate remains the same. Such lithological switches are known from passive margins, as, for example, from the Bulgarian segment of the Western Black Sea between the Middle and Late Paleocene when siliciclastic sediments replaced carbonate sediments (e.g., Nemčok et al., 2004c; Stuart et al., 2011) or, for example, from numerous margin segments of the Central Atlantic between the Late Jurassic and Early Cretaceous when siliciclastic sediments replaced carbonate sediments (e.g., Poag, 1979; Sawyer et al., 1982; Brownfield and Charpentier, 2003; Davison, 2005; Nemčok et al., 2005b; Eliuk, 2010; Eliuk and Wach, 2010; Deptuck, 2011).

High deposition rates are known to result in the development of pore fluid pressures higher than normal in compacting sediments, which can retain porosity higher than the porosity under normal compaction without the effect of overpressure (see Osborne and Swarbrick, 1997). Examples come from the Caspian Sea (e.g., Bredehoeft et al., 1988), the Gulf Coast (Dickinson, 1953), the Mahakam Delta (Burrus et al., 1994), the North Sea (Mann and Mackenzie, 1990; Audet and McConell, 1992), and the Williston Basin (Burrus et al., 1994). A large database of present-day depositional rates that Kukal (1990) completed for various depositional settings indicates that most of the present-day analogs to flood plain, delta, bay, gulf, inland sea, and marginal sea settings can reach important depositional rates, which gives them potential for overpressure development. The presence of overpressured zones can retard the heat flow through the sediment section by decreasing the thermal conductivity of sediment in these zones (e.g., Yalcin et al., 1997). An example of this surface heat flow–decreasing mechanism has been published from the Kura Basin in the Republic of Georgia (Yukler and Erdogan, 1996).

As Figure 11.3 documents, the transient effect of the depositional event can last tens of millions of years and only then does the surface heat flow return to its background value. For example, the basin scenario characterized by the 20 Ma-long depositional event with a deposition rate of 1 mmy^{-1} undergoes very long thermal perturbation characterized by cooling in comparison to the background thermal regime. It takes 40 Ma years after the end of this event just to reduce the heat flow perturbation induced by the event to 50 percent of its value reached at the end of the event (Deming, 1994b). Note, based on the analogy with numerical simulations described in Chapter 13, that if the thermal effect of the depositional event lasts long after the event, the induced maturation history perturbation lasts even longer because of the temperature- and time-dependence of the maturation processes (see Person and Garven, 1992).

Because rift and passive margin basins are typically characterized by histories composed of various deposition, nondeposition, and erosion events, it is useful to think in terms of complex basin histories controlling complex unequilibrated heat flow regimes. An example of composite depositional history composed of two rates can help to illustrate this point. We can discuss a history that contains the first rate of 0.1 mmy^{-1}. It is followed by the second rate of 0.5 mmy^{-1}, which is followed by nondeposition. The calculated heat flow change, expressed as the ratio of the heat flow, Q, to its initial equilibrium value, Q_o, would then indicate that while the surface heat flow responds to the effect of deposition, it takes a long time for the perturbation to vanish. The time constant for the disappearance of the perturbation reaches the values typical of the thermal response of the entire lithosphere (see Hutchinson, 1985).

This indicates that a heat flow perturbation driven by a depositional event takes a certain period of time to vanish. The approximate life spans of heat flow perturbations caused by depositional events are estimated, for example, by Lachenbruch and Sass (1977), Hutchinson (1985), Vasseur and Burrus (1990), and Deming (1994b), who calculate the time periods necessary for thermal perturbations to affect the entire lithosphere. Their results calculated for a lithosphere with thickness and thermal diffusivity of 100 km and 1.04×10^{-6} m^2s^{-1}, respectively, range from 50 to 100 million years. The results indicate that the lithosphere is characterized by a relatively large thermal inertia. This means that the background thermal regime has the ability to persist for a period of time as long as the life span of the petroleum system. Taking properties characterizing the oceanic lithosphere, a depositional event shorter than 50 Ma does not cause a thermal perturbation reaching a depth of 100 km. Such an event would reach a depth of only 10 km.

Note, for the sake of simplicity, that the previous discussions (Figures 11.2, 11.3) did not include the radiogenic heat production that exists in sediments. However, as discussed in Chapter 10, radiogenic heat production in numerous types of sediments cannot be neglected.

Hutchinson (1985) studied the effect of radiogenic heat production involved in thermal perturbation by deposition. He describes that when thermal perturbations shown in Figure 11.3 are compared with those assuming heat production of 10^{-6} Wm^{-3}, their histories look distinctly different. The effect of the radiogenic heat production on the transient thermal regime is negligible in basin scenarios with very high background heat flow. On the contrary, the effect is important in basin scenarios characterized by low heat flow.

The effects of sediment blanketing on the thermal regime of the rift-related basin in conjunction with various extension rates have been modeled by finite difference simulation by Ter Vorde and Bertotti (1994) (Figures 11.4a, b). The models illustrate two mechanisms involved in thermal blanketing:

1) the effect of the low-temperature sediments, which produce the shorter-term effect of increasing the rate of cooling of the crust below them; and
2) the effect of thermal conductivity lower than the conductivity of the underlying crust, which in time controls a development of the high-temperature gradient in sedimentary cover causing a decreasing rate of the crustal cooling.

The second effect is simulated in thermal diffusivity being progressively lower from Figure 11.4a-1 to Figure 11.4a-3.

The modeling shows that the deposition rate controls which of the two mechanisms dominates because the deposition has to compete with the rate of heat conduction from the underlying crust into the sedimentary cover (Ter Vorde and Bertotti, 1994; Figure 11.4b). Of course the threshold deposition rate varies with the thermal conductivity of the sedimentary fill of the graben. For example, if the thermal conductivity of the sedimentary fill reaches 50 percent of the thermal conductivity of its basement, the threshold deposition rate is 1.8 mmy^{-1}. While the faster deposition rates for this particular conductivity result in the cooling of the crust, the slower rates blanket the underlying crust, keeping it warmer. For illustration, if one prescribes thermal conductivities of 1.05–1.45 $Wm^{-1}K^{-1}$, 0.8–1.25 $Wm^{-1}K^{-1}$, and 1.7–2.5 $Wm^{-1}K^{-1}$ to shale, siltstone, and sand, their blanketing effect would start at deposition rates of 2, 2.2, and 1.4 mmy^{-1} (Ter Vorde and Bertotti, 1994). Looking at deposition rates in Kukal (1990), such rates are often known from flood plain, delta, bay, and gulf settings, and occasionally known from inland and marginal sea settings, which documents that thermal blanketing is relatively common.

While the upwelling of isotherms caused by the diffuse deformation of the lower lithosphere has no influence on the thermal regime at very shallow levels, the effect of the thermal blanketing perturbs the thermal regime at depths as deep as 30 km (Figure 11.4b; Ter Vorde and Bertotti, 1994). This suggests that shallow factors dominate control of the thermal regime at basin depth levels. This is an important point explaining the dramatic effect of syn-rift magmatism and fluid flow systems on shallow thermal regimes.

Because, as discussed in this chapter, the local thermal regime is typically not affected by a single factor, the paleotemperature-depth curves provide a first insight on likely mechanisms of cooling and heating (Duddy et al., 1994; Green et al., 2002). This insight is based on the assumption that the thermal conductivity anisotropy of the sedimentary section results in a roughly linear distribution of temperature with depth through the section (see Figure 10.9; Bachu, 1985). Assuming this, the increased burial should result in a roughly linear paleotemperature profile with depth, characterized by a temperature gradient similar to the present-day one (see Figures 1.16, 9.3; Heling and Teichmuller, 1974; Ebner and Sachsenhofer, 1995). However, the heating exerted by increased heat flow coming from the basement, which also results in a roughly linear profile, is characterized by a temperature gradient steeper than the present-day one (see Figures 11.5a–d; Green et al., 2004). On the contrary, a nonlinear temperature profile would be associated with the contribution of the geothermal waters or magmatism (see Figures 1.16, 13.6; Person and Garven, 1992; Ebner and Sachsenhofer; 1995).

The temperature changes in sediments are dynamically linked with temperature changes in their surroundings. Therefore, as finite-element simulation (Lin et al., 2000) shows, the basement is capable of buffering the temperature changes in overlying sediments. This buffering has the potential to reduce the development of negative and positive temperature anomalies in sediments affected by transient mechanisms. Some of the heat changes that would be spent on temperature changes in the sedimentary fill are used for cooling or warming the basement; that is, reducing the anomalies in sediments. For example, buffering tends to reduce both recharge and discharge thermal anomalies, as can be shown in numeric models of the Arkoma Basin thermal regime (Figure 11.6).

While most of the buffering effect in the Arkoma Basin model is caused by changes in vertical temperature gradient, in some areas, the heat actually flows laterally out of sediments and into the basement. Among various tectonic settings, this flow, caused by heat refraction, is important in extensional basins, characterized by an interface between sediments and basement

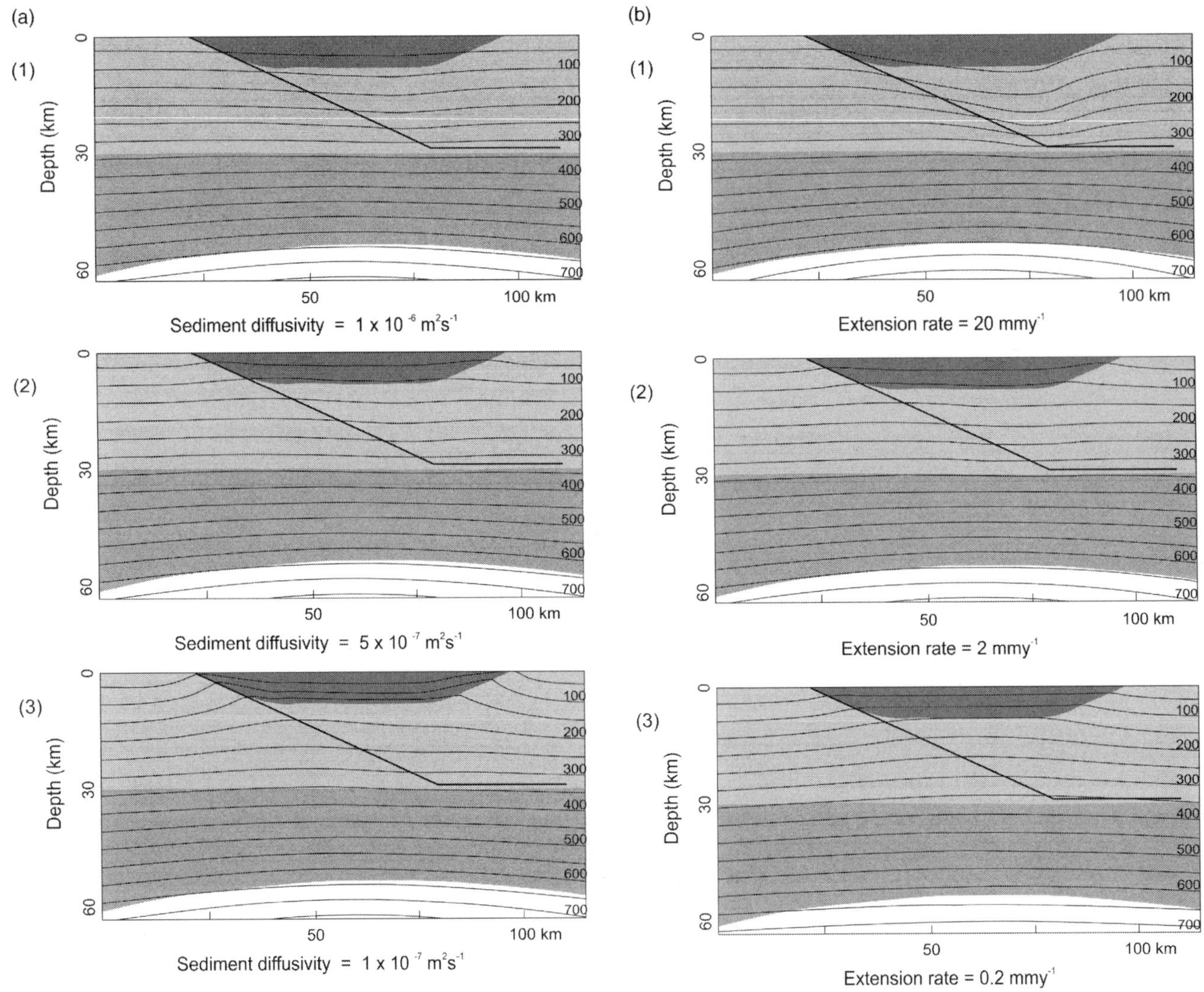

Figure 11.4. (a) Isotherms in the basin after 20 km of extension (Ter Voorde and Bertotti, 1994). Thermal diffusivity of sediments is 10^{-6} m^{2}s^{-1} (a), 5×10^{-7} m^{2}s^{-1} (b), and 10^{-7} m^{2}s^{-1} (c). (b) Isotherms in the basin after 20 km of extension (Ter Voorde and Bertotti, 1994). Extension rates are 20 mmy^{-1} (a), 2.0 mmy^{-1} (b), and 0.2 mmy^{-1} (c). Thermal diffusivity of the sediments is 5×10^{-7} m^{2}s^{-1}.

dipping at angles considerably higher than 10° (Lin et al., 2000).

Figure 11.6b also indicates that the difference between buffered and unbuffered sedimentary basin tends to disappear with time as the system transitions toward the steady-state regime. Therefore, buffering is presumably important for geological problems solved at a scale of millions of years and not important at a scale of tens of millions of years.

While the buffering simulations of Lin and colleagues (2000) focus on transients caused by fluid flow, the principle is applicable to other transients such as those caused by deposition and erosion. For example, a deposition rate as fast as 1–2 mmy^{-1} in the Gulf of Mexico and the South Caspian Basin (Woodbury et al., 1973; Nunn, 1985; Bredehoeft et al., 1988) or an erosion rate as fast as 5 mmy^{-1} in Taiwan (Li, 1976; Lan et al., 1990) would qualify for the transient causes. Such fast deposition or erosion events could control the development of thermal transients with amplitudes larger than 10°C at depths deeper than 6 km (Nunn and Sassen, 1986; Deming et al., 1990b).

The role of erosion in thermal regimes

Before discussing erosion's effects on thermal regimes, it is useful to distinguish uplift and exhumation as separate mechanisms (see England and Molnar, 1990),

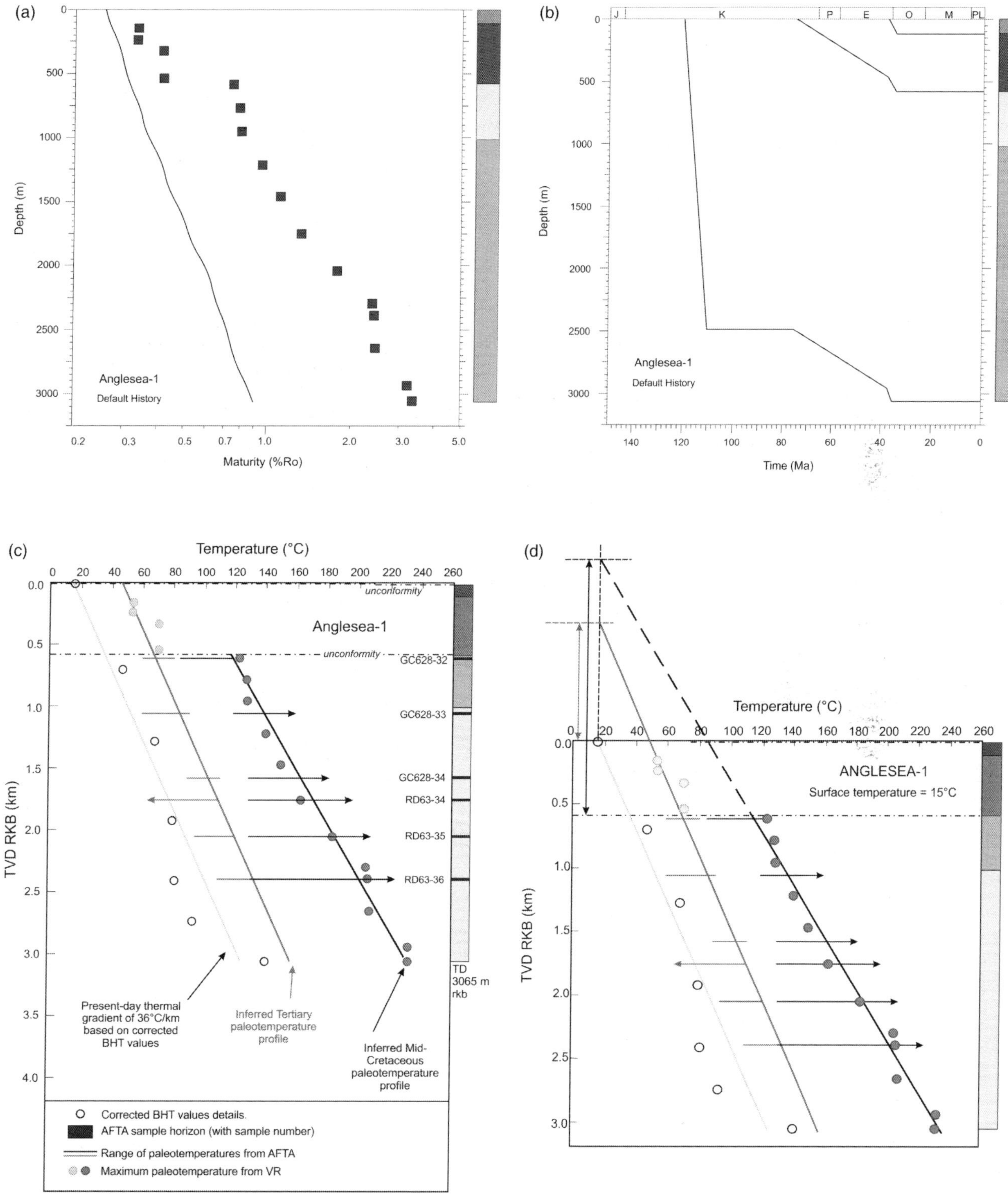

Figure 11.5. (a) Vitrinite reflectance values in samples from the Anglesea-1 well, Otway Basin, plotted against depth (Green et al., 2004). The solid line represents the profile predicted from the default thermal history, that is, the history calculated from the assumption that all units throughout the well are currently at their maximum temperatures since deposition. The default thermal history was constructed using the burial history derived from the section penetrated by the well (see Figure 11.5b), combined with a present-day thermal gradient of 36° Ckm^{-1}, derived from corrected bottom hole temperature data and a surface temperature

despite their frequent confusion in literature. The uplift causes the ascent of the reference surface with respect to geoid, causing an overall cooling of the uplifting area. On the contrary, the exhumation causes the ascent of material points with respect to reference surface, causing an upward heat advection. The sum of both mechanisms yields the uplift of material points with respect to the geoid.

Note that the exhumation history of a footwall in an extensional province may be complex, involving several tilt events, as indicated by a complex relationship between age determined by zircon and apatite fission track and (U-Th)/He analyses and distance from the fault in the Wasatch Mountains, Utah (Figure 11.7; Armstrong et al., 2003). Temperature histories modeled from the track length distribution for samples located at various distances from the fault indicate the occurrence of different exhumation histories in different parts of the tilted footwall. The simplest way of explaining this pattern of cooling histories dependent on distance from fault and three patterns of zircon fission track, apatite fission track, and (U-Th)/He ages, requires interpretation of two footwall tilt events (Figure 11.7; Armstrong et al., 2003).

During the initial extension and Wasatch Fault activity, the hinge position was located at least 33 km east of the fault, allowing for the accelerated exhumation of samples located 22–33 km away from the fault. Later, the hinge location shifted to a distance of about 25 km from the fault, which explains the coincidence of the accelerated exhumation between 0 and 25 km and the lack of accelerated exhumation beyond the hinge at 25 km.

The effect of erosion on the thermal regime in a one-dimensional solution for the continental crust can be expressed as (England and Richardson, 1977):

$$\partial T/\partial t = \partial/\partial x[\kappa(\partial T/\partial x)] + H(x,T,t)/\rho c + u_x(\partial T\partial x), \qquad (11\text{-}9)$$

where T is the temperature, t is the time, κ is the thermal diffusivity, H is the heat production, ρ is the density, c is the specific heat, and u_x is the erosion-controlled ascent of the studied rock column. This equation describes a thermal energy balance using the transient term on the left side, and diffusion, source, and mass advection terms on the right side.

Note that it is difficult to determine the erosion-controlled ascent of the studied rock column for Equation (11-9). The only natural scenario where calculating the exhumation rate from the cooling history is relatively simple is the scenario characterized by the following conditions (Parrish, 1983):

1) the critical isotherms must have been horizontal;
2) the critical isotherms must have remained at a constant depth with respect to the surface; and
3) the uplift rate equals the erosion rate.

However, tectonically active and rapidly eroding regions with pronounced topographic relief, such as those in rift or passive margin settings, are not characterized by these conditions.

A simple analytical calculation of the Peclet number, P_e, shows that the erosion-driven thermal perturbation depends on both the uplift rate and the length of the event (see Kohl et al., 2001):

$$P_e = q_{adv}/q_{diff} = [(\rho c_p)V_z L]/\lambda_{zz}, \qquad (11\text{-}10)$$

where q_{adv} is the heat flow due to mass advection, q_{diff} is the heat flow due to diffusion, ρc_p is the thermal capacity of the saturated rock, V_z is the uplift/exhumation rate, L is the characteristic uplift length of the system, and λ_{zz} is the vertical component of the thermal

Figure 11.5 (*cont.*)

of 15°C. Measured vitrinite reflectance values plot above the profile, showing that the sampled units have been hotter in the past. A distinct break is evident across the Late Cretaceous–Eocene unconformity, suggesting a major difference in the magnitude of paleothermal effects across this unconformity. (b) Burial history derived from the preserved section penetrated by the Anglesea-1 well, Otway Basin, used together with a present-day thermal gradient of 36° Ckm^{-1} to predict default thermal history for individual samples and profiles shown in Figure 11.5a (Green et al., 2004). (c) Paleotemperature constraints from apatite fission track and vitrinite reflectivity data in the Anglesea-1 well plotted against depth and attributed to one of the two paleothermal episodes (Green et al., 2004). Tertiary constraints define a linear depth profile, subparallel to the present-day temperature profile derived from corrected bottom hole temperature values and offset to higher temperatures by approximately 30 to 40°C. This suggests that heating was due primarily to deeper burial, with little or no difference in heat flow compared to the present day. In contrast, while mid-Cretaceous paleotemperature constraints also define a linear profile, the slope of the profile is distinctly higher than that of the Tertiary paleotemperature profile and the present-day temperature profile. This suggests that a major component of heating in this episode was due to elevated basal heat flow. (d) Fitting a linear paleotemperature profile to the constraints for each episode derived from apatite fission track and vitrinite reflectance data as a function of depth (see Figure 11.5c) (Green et al., 2004). This and the extrapolation to an assumed paleosurface temperature of 15°C allows for the estimation of the range of paleogeothermal gradients and removed section that are consistent with the data within 95 percent confidence limits.

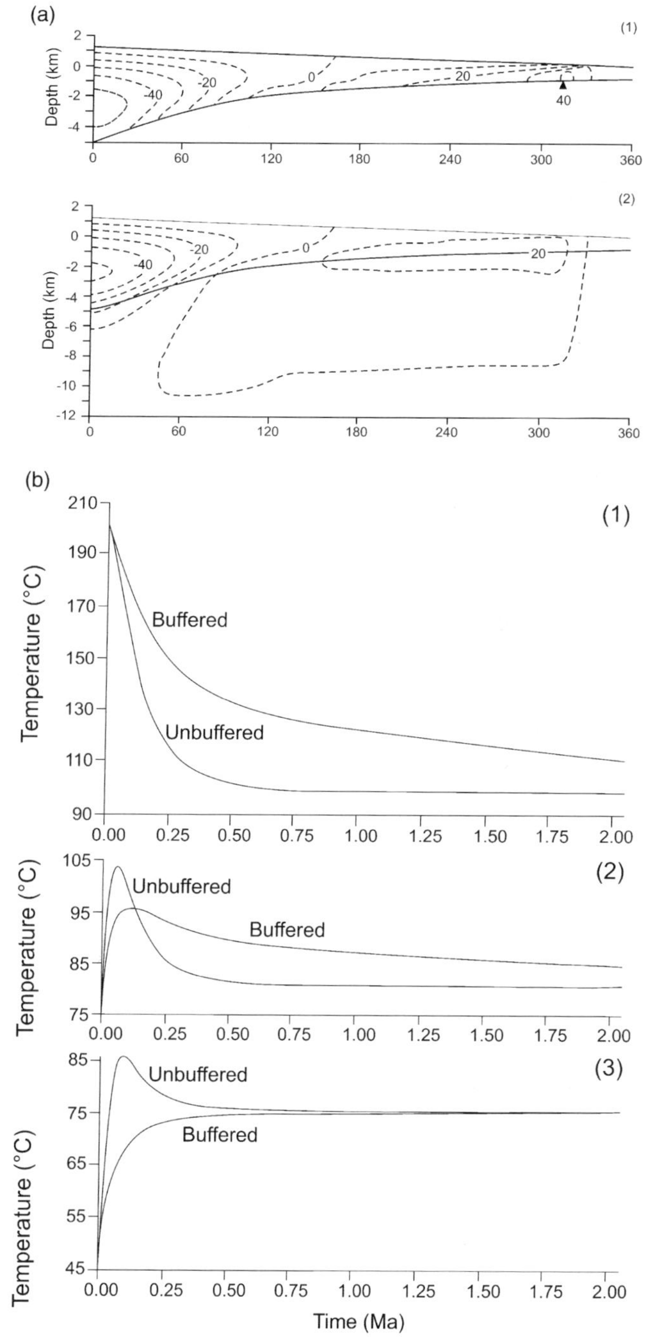

Figure 11.6. (a) Temperature and heat flow distributions at an elapsed model time of 0.1 Ma (Lin et al., 2000). 1) Temperature change for unbuffered model. 2) Temperature change for buffered model. (b) Temperature with time for unbuffered and buffered models at 1) recharge area, 2) basin center, and 3) discharge area (Lin et al., 2000).

conductivity tensor. Peclet numbers above one characterize advection-dominated thermal regimes. Peclet numbers below one represent diffusion-dominated thermal regimes. Equation (11-10) indicates that an uplift involving a short depth range must be associated with a fast uplift rate to control a thermal perturbation similar to a slow uplift involving a long depth range. For example, given the thermal diffusivity, κ, being equal to 10^{-6} m^2s^{-1}, and calculated from (Kohl et al., 2001):

$$\kappa = (\rho c_p)/\lambda, \tag{11-11}$$

where λ is the thermal conductivity tensor of the saturated rock, a Peclet number of >0.1 can be associated

1) with uplift, which has rate and depth values of 0.25 mmy^{-1} and 12 km, respectively;
2) or with uplift and depth values of 0.5 mmy^{-1} and 6 km, respectively.

Doing a sensitivity study for the mass advection effect on the thermal regime of the planned Gothard base tunnel in the Sedrun area, Switzerland (Figure 11.8), Kohl and colleagues (2001) calculated that, for the given subsurface heat flow regime and an assumed uplift equal to exhumation, the mass advection rate of 0.75 mmy^{-1} results in a $10 mWm^{-2}$ increase in the surface heat flow with respect to no mass advection, that is, a pure heat diffusion scenario.

It is important to discuss the time dependence of the erosion-controlled thermal regimes. For example, if one assumes the system undergoing erosion at a steady rate, u, and ignores radioactive heat production, the temperature, T, at the upper planar surface can be calculated from (Mancktelow and Grasemann, 1997):

$$T = c_1 \exp(-uy/\kappa) + c_2, \tag{11-12}$$

where c_1 and c_2 are constants determined by the boundary conditions, u is the vertical velocity that is positive for erosion, y is the vertical distance, and κ is the thermal diffusivity. Equation (11-12) indicates that as vertical distance, y, approaches infinity, the temperature approaches the temperature at infinity, which equals c_2. Therefore, the steady state can be reached with time only if the temperature approaches some constant value at depth.

When the radioactive heat production is not ignored, yet another mechanism acting against reaching the steady state with time is added to the discussed scenario. Erosion of radiogenic material leads to the progressive decrease of surface heat production with time as the erosion removes the radiogenic material, which exponentially decays with depth (see Equation (10-32) in Chapter 10). Right at the start of calculating erosion-affected thermal regimes, the inclusion of depth-dependent decay of heat production introduces a realistic nonlinear temperature gradient (Figure 11.9; Mancktelow and Grasemann, 1997).

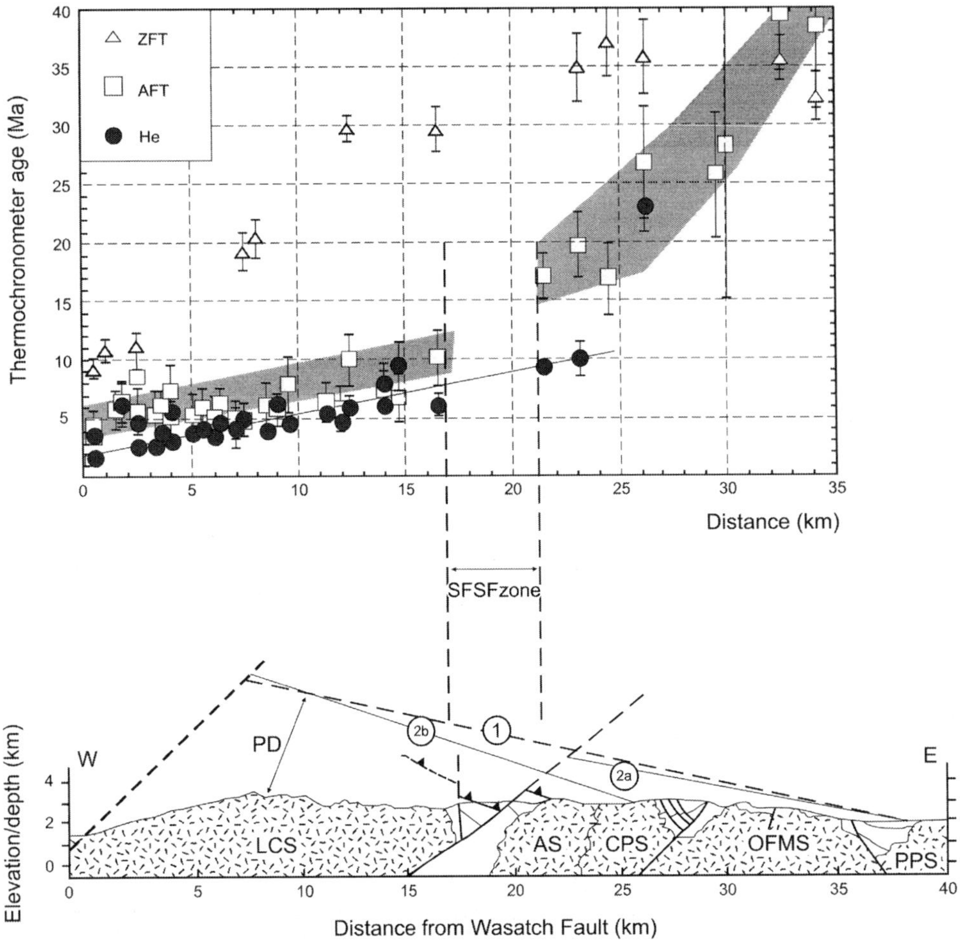

Figure 11.7. Combined apatite/zircon fission track and (U-Th)/He data sets with the distance from the Wasatch Fault (Armstrong et al., 2003). The solid line shows the best linear fit to the (U-Th)/He data. Shaded zones represent the age trends for the apatite fission track data. The cross-section shows the positions of thermochronologic data in fault blocks. Numbered lines above the section indicate simplified locations of paleohorizontal surfaces for single-tilt (dashed and labeled as 1) and for two-tilt events (solid hachured and labeled as 2). Tilt 2a and 2b are the earlier and later tilt events, respectively. PD – paleodepths of samples measured relative to the tilted surfaces. SFSF – Silver Fork-Superior Fault.

An interesting exercise is to compare the thermal perturbations caused by different erosion rates (Figure 11.10; Mancktelow and Grasemann, 1997). Figure 11.10 shows the three heat advection regimes associated with erosion rates of 0.1, 1, and 5 mmy^{-1}, typical for settings such as (see Kukal, 1990):

1) weathering/slope retreat/denudation/solution/chemical denudation;
2) soil creep/land sliding; and
3) dry-land cliff retreat, respectively,

and compares them to the case affected by 1 mmy^{-1} erosion but ignores the heat advection for the illustration of the effect. For simplicity, a series of models from Figure 11.10 ignores the radiogenic heat production. This comparison documents that even as slow an erosional rate as 0.1 mmy^{-1} causes an elevation of the temperature gradient (Figure 11.10a). Figures 11.10b–d illustrate that the heat advection effect in time causes slower cooling at the beginning of the studied time period and faster cooling at the end of the studied period because of the elevated temperature gradient near the surface. The figures further illustrate that in a tectonically active rift terrain, the isotherms will be constantly changing their depth because of heat advection and thermal relaxation. For example, a curve A in a scenario without heat advection passes through the 100°C isotherm at a depth of 4 km (Figure 11.10a) while it passes through the depths of 2.31 km and 1.24 km in scenarios with erosion of 1 mmy^{-1} and 5 mmy^{-1} (Figure 11.10c, d).

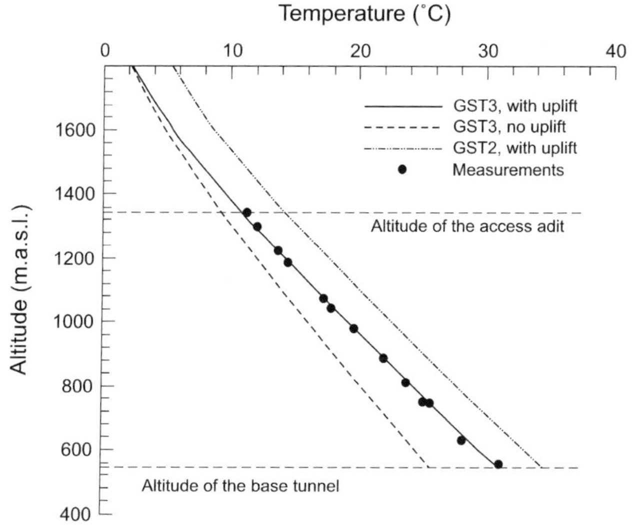

Figure 11.8. Temperature distribution along the Gothard base tunnel shaft in the Sedrun area, Switzerland, for model GST2 with uplift, model GST3 without and with uplift (Kohl et al., 2001). GST represents local ground surface temperature. The GST2 model is characterized by the ground surface temperature at sea level, T_0, equal to 13.5°C and the atmospheric lapse rate of 0.0048 Km^{-1}. The GST3 model is characterized by the ground surface temperature at sea level, T_0, equal to 10.2°C and the atmospheric lapse rate of 0.0048 Km^{-1}.

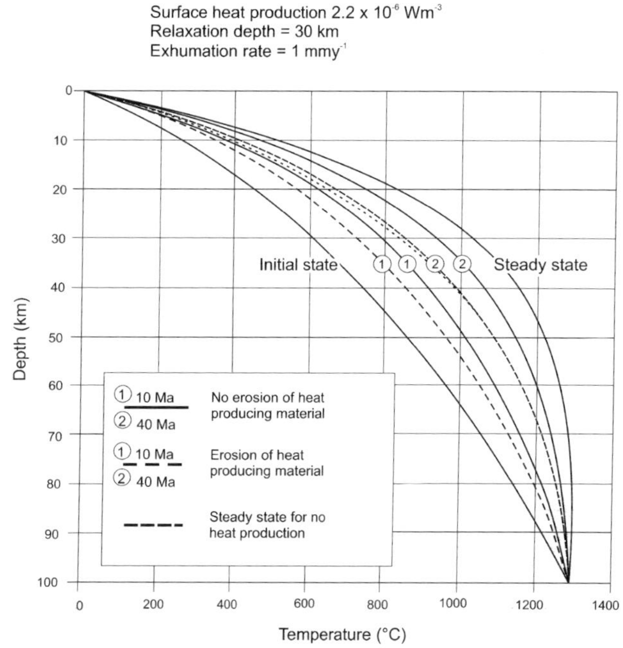

Figure 11.9. Variation in the lithospheric geotherm with time, with and without the effects of erosion of heat-producing material (Mancktelow and Grasemann, 1997). The base of the lithosphere, taken as fixed, is assigned a constant temperature of 1,300°C. Other assigned values include a constant temperature of 0°C for the Earth's surface, an initial volumetric heat production at the surface of 2.2×10^{-6} Wm^{-3}, and exponential decay of the heat production down to 1/e of its surface value at a depth of 30 km. Selected parameters establish an initial geotherm of 25° Ckm^{-1} in the upper 20 km of the lithosphere. Note that long times and deep exhumation would be required to approach a near steady state.

If the effect of radiogenic heat production is added to the discussed model, the initial static geotherm is characterized by a downward-decreasing temperature gradient. However, the heat advection effect in time also produces slower and faster cooling at the beginning and the end (Figure 11.11; Mancktelow and Grasemann, 1997).

To add an effect of the surface uplift that adds an extra cooling mechanism to the discussed scenario, one can look at the two following models; the first one with an instantaneous uplift of 3 km and the second one with an uplift of 3 km occurring over a 10 Ma period of time (Figures 11.12a, b; Mancktelow and Grasemann, 1997). A comparison of both figures documents that even a significant uplift accompanying the exhumation exerts only a minor imprint at the cooling history of exhumed rocks. An example of such regional uplift characterized by long-spatial wavelength superposed on the localized differential exhumation associated with local or short-wavelength extensional tectonics comes from the Inner Moray Firth in the British North Sea (Argent et al., 2002).

In the overall approach of adding progressively more complexities, note that the natural cases characterized by a constant exhumation rate are rare. It is more typical to see an initially faster rate, which dies off exponentially. This phenomenon adds the extra effect of relaxation of elevated isotherms as the advection effect decays (Grasemann and Mancktelow, 1993). Such an effect can be seen in Figure 11.13, where the relaxation of isotherms back to steady-state conditions can take a significant length of time, especially at greater depths.

Topographic influence on the subsurface temperature field

Erosion's effect on the thermal regime is further complicated by the resultant topographic relief, which, on its own, modifies the thermal regime (see Stüwe et al., 1994). The regime is characterized by isotherms elevated under the ridges and depressed under the valleys (Figure 11.14). This phenomenon results in a lower apparent thermal gradient under the ridges and a higher gradient under the valleys. The effect of topography on the subsurface temperatures decreases with

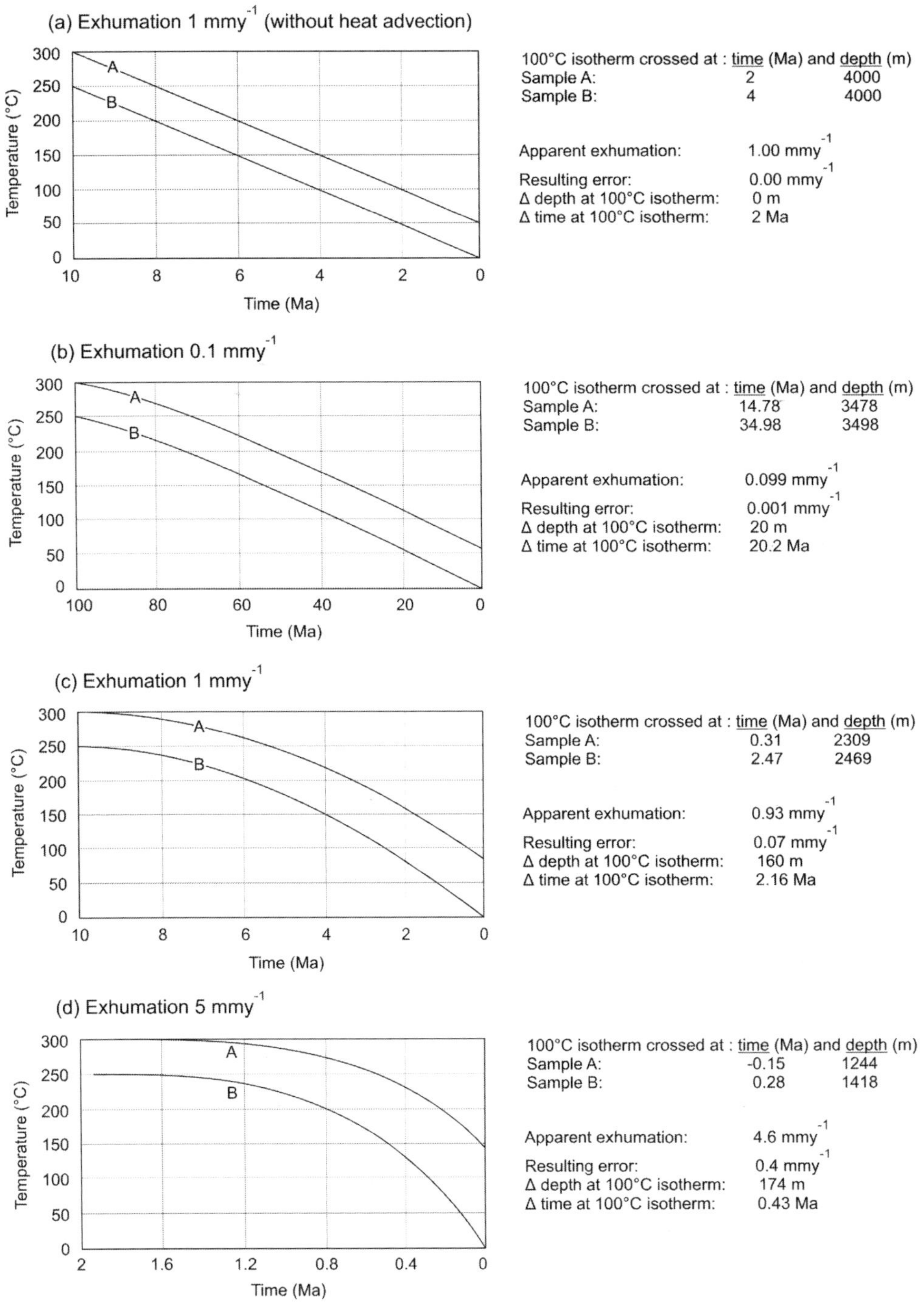

Figure 11.10. Calculated cooling curves for two rock samples, A and B, initially at 12 km and 10 km depth, respectively, exhumed at different rates (Mancktelow and Grasemann, 1997). An initial constant geothermal gradient of 25° Ckm^{-1} is assumed (assuming no internal heat production). The time and depth at which the two samples pass through the 100°C isotherm, as well as the difference in these depths and times, are given. The error in estimation of the exhumation rate that would arise using the altitude dependence method with apatite fission track ages, assuming blocking temperature of 100°C, is also calculated. Note that even at low exhumation rates the shape of the cooling curves is clearly influenced by the advection of heat, although the error in the apparent exhumation rate for real exhumation rates less than 1mmy^{-1} is too small to be geologically significant.

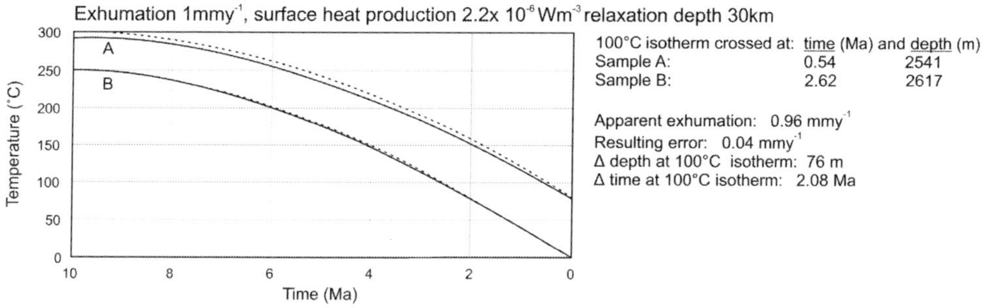

Figure 11.11. A comparison of the cooling history with (solid lines) and without (dashed lines) the effects of the radiogenic heat production (Mancktelow and Grasemann, 1997). Parameters for the solid lines are as in Figure 11.10c but with a value of 2.2×10^{-6} Wm^{-3} for the initial volumetric heat production at the surface. For these parameters, the near surface (≤12 km) geothermal gradient is a little different from Figure 11.10c, but the deeper temperature distribution is more realistic. However, for moderate exhumation (A and B from 10 and 12 km, respectively) in the range where the geothermal gradients are still similar, the apatite fission track altitude-dependence method yields results that differ little from Figure 11.10c, where the effects of heat production were ignored.

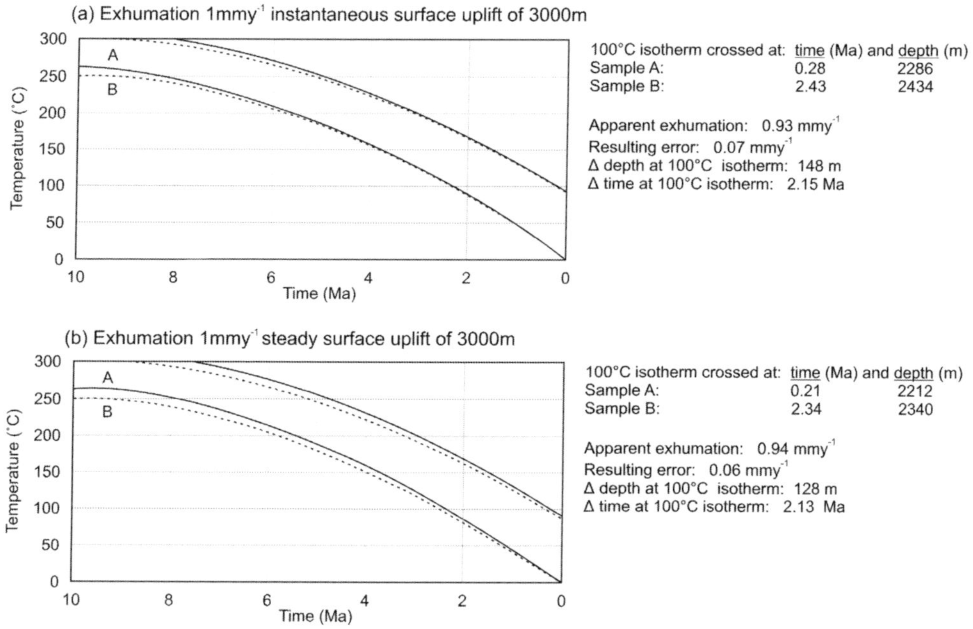

Figure 11.12. A comparison of the effects of surface uplift on cooling curves (solid lines) compared to a history without surface uplift (dashed lines) (Mancktelow and Grasemann, 1997). Parameters are as in Figure 11.10c but considering the effect of a 3 km surface uplift during exhumation, with a lapse rate of 4.5°C per 1 km of altitude. a) Here the surface is uplifted instantly. b) Here the surface uplift rate is constant at 0.3 mmy^{-1} over the entire period of 10 Ma. Although the effect is again negligible for the altitude-dependence method using apatite fission track ages, note that two distinct tectonic processes such as exhumation with and without surface uplift can result in very similar cooling histories.

depth such that, for example, at a depth of about 3.5 km the high-frequency lateral variations in isotherm depths due to canyon and ridge geometries are damped out and only broad isotherm perturbations are visible (Figure 11.14).

Further complexity is caused by non-steady-state isotherms that change their depths during exhumation. The topographic effects perturb the thermal regime only to a depth in the order of the width of isolated positive topographic features (e.g., Lees, 1910; Turcotte and Schubert, 1982).

The steady-state expression of the effect of the isolated topographic perturbation on the thermal regime in an eroding slab with a thickness, L, that is large in comparison with topographic relief has been formulated by Stüwe and colleagues (1994), providing a depth

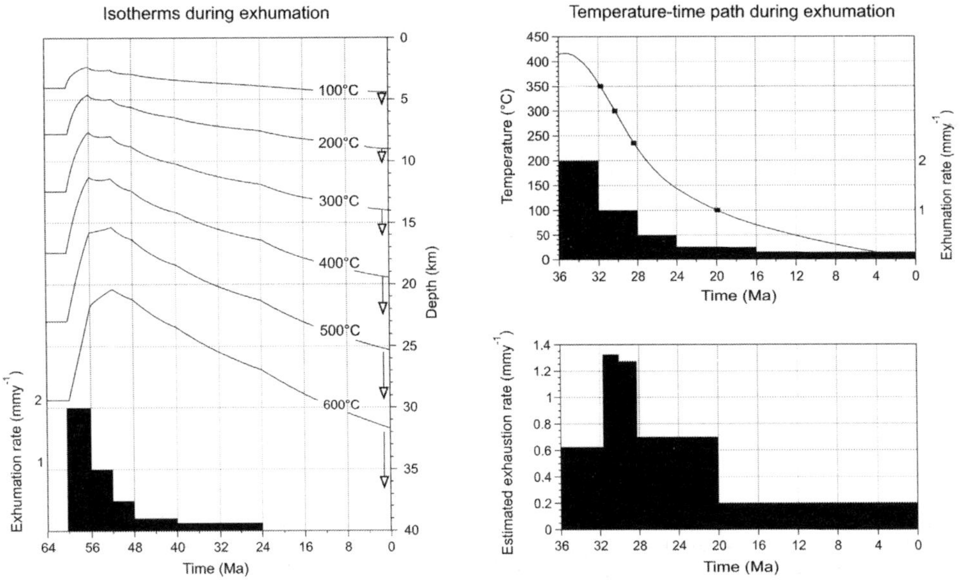

Figure 11.13. Isotherms during and following a period of rapid exhumation, initially at 2 mmy^{-1} and decreasing progressively to a slow rate of 0.125 mmy^{-1}, before stopping (Mancktelow and Grasemann, 1997). Heat production parameters and initial geotherm are the same as in Figure 11.11. Note that, because of the erosion of radiogenic material, the steady-state isotherm depth after the period of exhumation, represented by the arrow tip for each isotherm, is deeper than it was initially and that long times are required for the reestablishment of a steady state, particularly at a depth below 20 km.

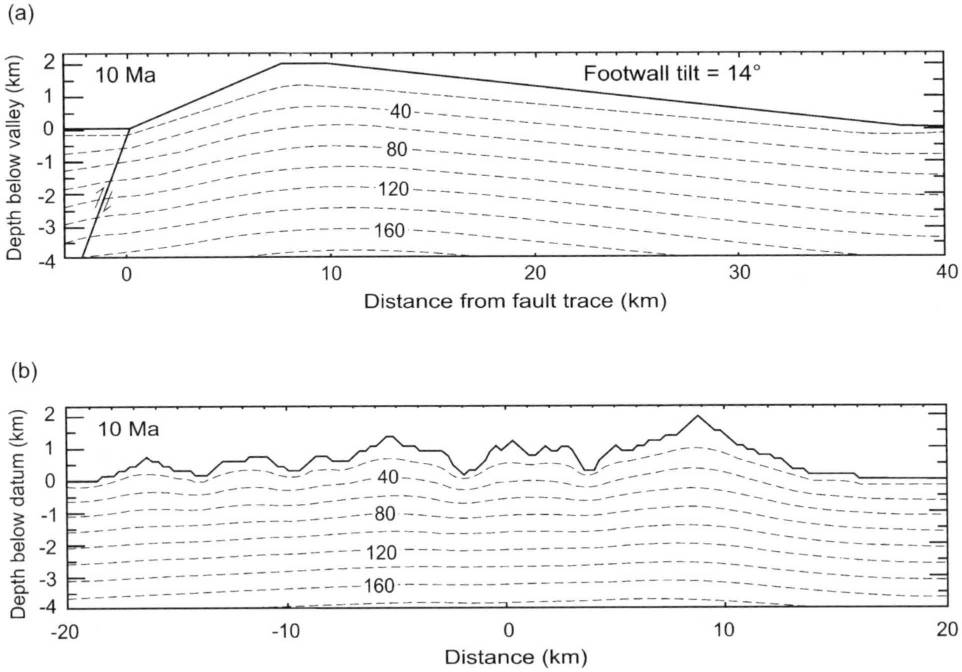

Figure 11.14. Isotherms in dip- and strike-oriented profiles through the Wasatch Mountains after 10 Ma of the boundary fault activity and associated exhumation (Ehlers et al., 2001). (a) Dip-oriented profile showing a 60° dipping fault with a maximum exhumation rate of 1.1 mmy^{-1} achieved in the tilted footwall next to the fault. The shaded region represents a sedimentary basin with low thermal conductivity. (b) Strike-oriented profile parallel to the boundary fault located in the footwall about 8 km behind the fault. A uniform exhumation rate of 0.7 mmy^{-1} was used.

of isotherms for any given location underneath any chosen topography:

$$Z = X + \varepsilon\varsigma_{(X,Y)}, \quad (11\text{-}13)$$

where Z is a dimensionless depth of a chosen isotherm expressed as Z = z/L, where z is a depth that is positive downward, from the lowest point of topographic relief, and L is some great depth such as a depth of base lithosphere. ε is the topography scale expressed as H/L where H is the maximum relief. X describes the depth–temperature relationship in the flat region undergoing erosion, which is given as:

$$X_{(\theta)} = -(1/\zeta)\ln(1 - \theta(1 - e^{-\lambda})), \quad (11\text{-}14)$$

where θ is the dimensionless temperature expressed as $\theta = T/T_L$, where T is the temperature of the chosen isotherm and T_L is the fixed temperature at a great depth. Z is the dimensionless denudation rate expressed as $\zeta = uL/\kappa$, where u is the denudation rate and κ is the thermal diffusivity.

A simplified steady-state function represents an upper limit on possible scenarios, achievable only after a long period of existence of the constant topography and a long activity period of a specific exhumation rate. Despite its limited use for natural cases, it illustrates the effects of the main factors controlling the thermal regime affected by topography:

1) amplitude of topographic perturbation;
2) wavelength of topographic perturbation; and
3) exhumation rate.

It expresses that the magnitude of thermal perturbation is proportional to the amplitude of topographic perturbation. The remaining factors, wavelength of topographic perturbation and exhumation rate, control the depth of thermal perturbations caused by topography effects as described by the following function, which expresses a depth, y, at which the surface disturbation decays to its 1/e subsurface value (Mancktelow and Grasemann, 1997):

$$y = 2/\{-u/\kappa - \sqrt{[(u/\kappa)^2 + (4\pi/\lambda)^2]}\}, \quad (11\text{-}15)$$

where u is the exhumation rate, κ is the thermal diffusivity, and λ is the wavelength of topographic perturbation.

While the increase in the wavelength of topographic perturbation increases the depth of thermal perturbation caused by topography, the increase in exhumation rate has the opposite effect (Figure 11.15; Mancktelow and Grasemann, 1997). In order to see the effect of both the wavelength of topographic perturbation and exhumation rate, the interacting effect of heat advection is ignored in Figures 11.15a, b. The comparison of these figures indicates that the wavelength of topographic perturbation exerts an effect stronger than the effect of the exhumation rate. Its effect causes isotherm expansion underneath ridges and condensation underneath valleys due to the cooling effect of the topography.

Because it is erosion that results in complex topography, the net effect of exhumation described in the previous subchapter finds itself in a complex interplay with the net result of topography in control of the thermal regime (Mancktelow and Grasemann, 1997). The opposing effects of associated heat advection and cooling, respectively, are especially obvious at topography with a short wavelength where steep slopes enhance the cooling effect of topography (Figure 11.16). The numeric model in Figure 11.16a indicates that given the short wavelength of the positive topographic perturbation, the 100°C isotherm cannot "penetrate" far into the ridge, even at a high exhumation rate of 5 mmy^{-1}, because of the strong cooling effect of the adjacent steep valleys. The increase in wavelength decreases this cooling effect, allowing the 100°C isotherm elevation inside the positive topographic feature (Figures 11.16b, c). Note that exhumation rates larger than 4–5 mmy^{-1} are relatively rare in nature. Therefore, they are used as extreme end-members in the models. The only known exhumation rates of this magnitude come from Taiwan, the Southern Alps of New Zealand, and parts of the Himalayas (Ehlers et al., 2001), none of which is located in extensional settings.

Now we can think about the rock element exhumed at a rate of 1 mmy^{-1} from the depth of 13 km underneath the valley (Mancktelow and Grasemann, 1997), which reaches the 100°C isotherm at a depth of 4.5 km after 1.5 Ma for various values of wavelength of topographic perturbation greater than 6 km. It is then interesting to note that the rock element exhumed in the same time from underneath the ridge reaches a progressively shallower depth at a shorter time with increasing wavelength. This indicates that a valley sample does not react to different morphologies while the ridge becomes warmer with size of the ridge. This documents that the geothermal gradient beneath valleys is relatively insensitive to a change in topographic wavelength and exhumation rate, in contrast to changes induced below ridges. In the case of narrow ridges, no matter how significant the heat advection is due to some fast exhumation rate, the cooling effect of steep narrow valleys adjacent to sharp ridges is too great to overcome.

The previously discussed numerical models of Mancktelow and Grasemann (1997) indicate that the effect of topography on the estimation of exhumation

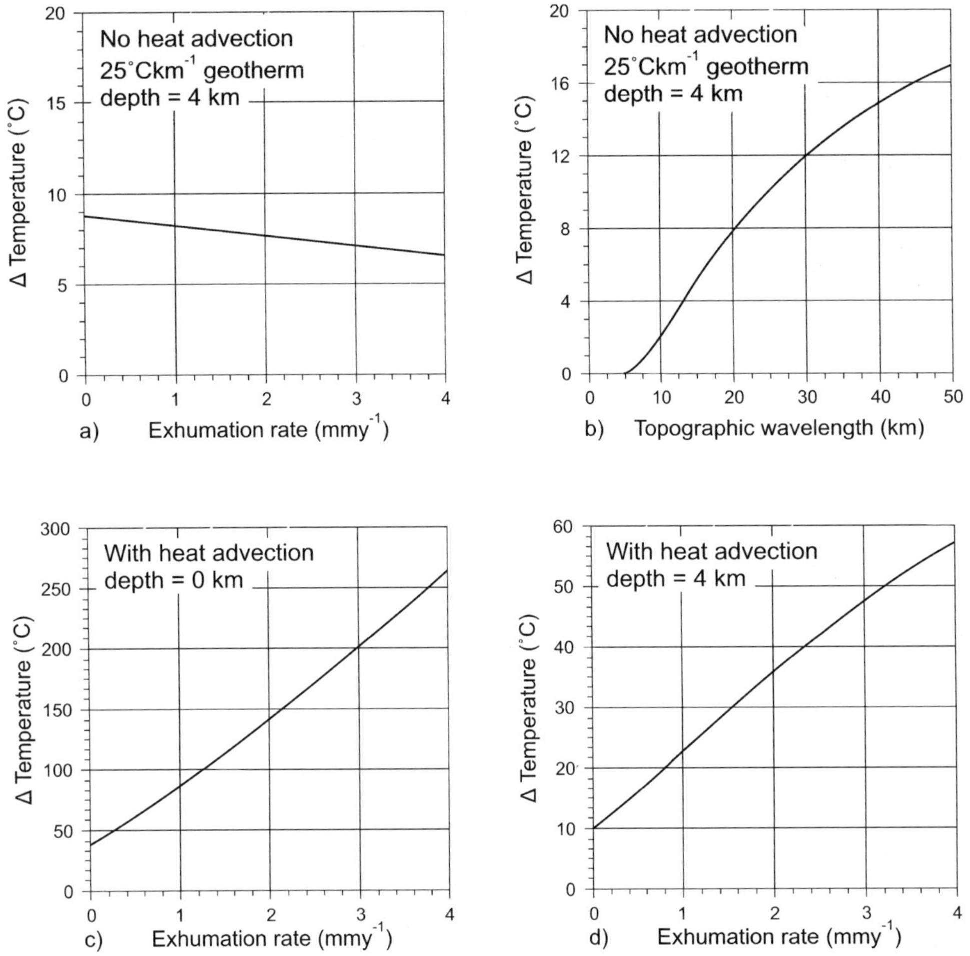

Figure 11.15. The influence of different parameters on the amplitude of the periodic disturbance in isotherms due to topography for the steady-state solution (Mancktelow and Grasemann, 1997). In all cases, the amplitude of the periodic topography is taken as 1,500 m, which is the maximum height difference between the ridge crest and valley floor of 3 km, a lapse rate of 4.5° Ckm^{-1} altitude, a reference temperature at y = 0 of 0°C, and, where required, a density of 2,800 kgm^{-3}, heat capacity of 1,100 $Jkg^{-1}K^{-1}$, surface heat production of 2.2×10^{-6} Wm^{-3}, and a relaxation depth of 30 km. (a) Effect of exhumation rate on the maximum temperature disturbance at a 4 km depth due to topography with 20 km wavelength, for an assumed constant geothermal gradient of 25° Ckm^{-1} and no heat advection. (b) Effect of wavelength of topography for the same conditions as in (a) using an exhumation rate of 1 mmy^{-1}. (c) Effect of exhumation rate on the magnitude of the periodic temperature disturbance at the reference plane y=0, including the effects of heat advection and with topographic wavelength of 20 km. d) The same as in (c) but considered at a depth of 4 km.

rates can be rather critical in the case of high exhumation rates and relatively broad topography, which are both characteristic for wide rifts and core complexes, provided this topography remained relatively stable during a prominent portion of the exhumation history.

The aforementioned modeling indicates that, for example, a convex-concave shape of the cooling curve can be achieved not only by exhumation rate change but also by a more complex geothermal gradient influenced by the topography effect. Furthermore, the local thermal regime can be a result of the interplay of numerous factors, including competing:

1) heat advection;
2) topography-controlled cooling;
3) inhomogeneous thermal conductivity structure of the area; and
4) heat convection,

which are all rather typical for extensional provinces such as the Basin and Range province in the western United States, as discussed, for example, in Chapter 9.

When we now think about areas subject to erosion on passive margins in the time period following the continental breakup, most of the exhumation in this setting, represented by examples such as the southeastern

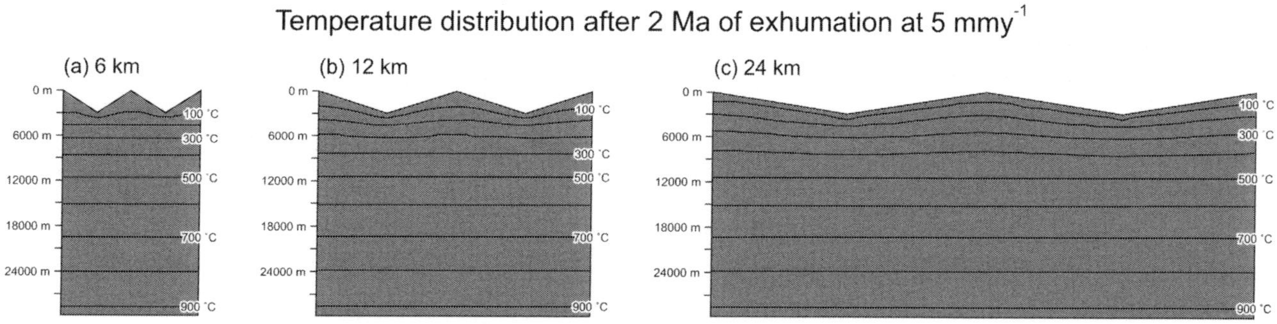

Figure 11.16. 2-D temperature distribution calculated for a snapshot of 2 Ma after the initiation of exhumation at a rate of 5 mmy^{-1} (Mancktelow and Grasemann, 1997). Note that the 100°C isotherm only penetrates into the ridges at larger wavelengths (see b and c), whereas at short wavelengths the lateral cooling effect of the steep V-shaped valleys precludes this isotherm geometry (a).

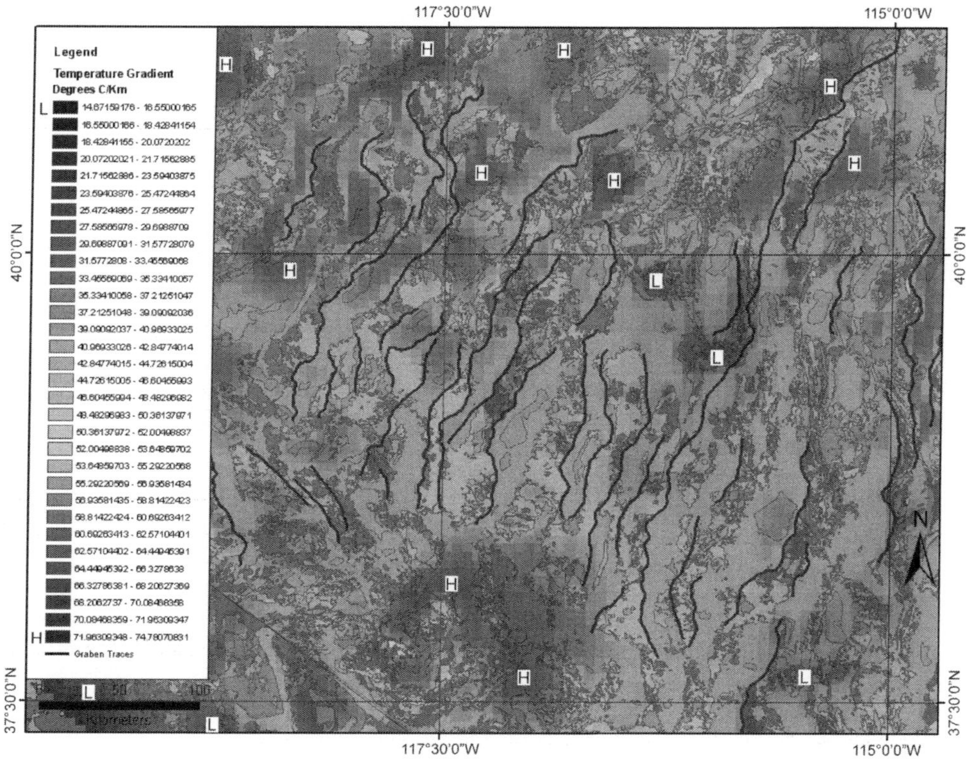

Figure 11.17. Map of the temperature gradient in ° Ckm^{-1} of the Basin and Range province, western United States (Coolbaugh et al., 2005; USGS, 2006). Note the cold footwalls and warm hanging walls of the main basin-controlling normal faults.

Brazilian margin (Gallagher et al., 1994; Mohriak et al., 2008), the Namibia margin (Gallagher and Brown, 1999; Brown et al., 2000; Raab et al., 2001), the eastern Canada margin (Royden and Keen, 1980), and the eastern United States margin (Watts and Thorne, 1984), occurs at uplifted continental flanks of the oceanic basin.

For comparison, as documented by cases in rift settings such as the Basin and Range province (Ehlers et al., 2003), the Gulf of Suez (Steckler, 1985; Sharp et al., 2000), the Moray Firth Basin (Argent et al., 2002; Stewart and Reeds, 2003), the northern North Sea (McLeod and Underhill, 1999), and the Loreto Basin, Baja California (Dorsey et al., 1995; Dorsey and Umhoefer, 2000), most exhumation in rift settings occurs in tilted footwalls next to controlling normal faults.

Thermal gradient data from the Basin and Range province (Figure 11.17; Coolbaugh et al., 2005; USGS, 2006) indicate that footwalls are relatively cold and basins are relatively hot. In order to understand the reasons, and to isolate the effects of erosion on the horst and deposition in the adjacent basin, it is useful to discuss case studies (Ehlers and Chapman, 1999; Ehlers et al., 2001, 2003) of the Wasatch Mountains in Utah, forming a footwall to the easternmost boundary fault of the Basin and Range province.

The finite-difference model of the thermal regime constrained by apatite and zircon fission track and (U-Th)/He data (Ehlers et al., 2003) indicates that the background thermal regime is controlled by heat flow from the mantle and by thermal properties of the crust such as thermal conductivity, heat production, and heat capacity. The important assumption for this model is a steady-state topography, which has been chosen based on the fact that landform modeling studies from the Basin and Range province indicate that a topographic steady state developed after the first 2 Ma of faulting (e.g., Densmore et al., 1998; Allen and Densmore, 2000), which is just a fraction of the life span of the Basin and Range rifting.

The factors expected to control the local thermal regime of the footwall include:

1) upward heat advection due to exhumation (Figure 7.16);
2) topography-controlled cooling;
3) thermal conductivity higher than the conductivity of the adjacent basin fill;
4) cooling due to meteoric water recharge;
5) distance from the mantle longer than the distance of the adjacent basin from the mantle; and
6) heat production higher than the production in the adjacent basin.

The factors expected to control the local thermal regime of the hanging wall include:

1) downward heat advection due to deposition (Figure 7.16);
2) topography-controlled cooling;
3) thermal conductivity lower than the conductivity of the adjacent footwall;
4) warming due to geothermal fluids discharge;
5) a distance from the mantle shorter than the distance of the adjacent footwall from the mantle;
6) warming due to potential rift-related magmatism; and
7) heat production lower than the production in the adjacent footwall.

Before returning to the Wasatch Fault, note, because there is no chapter devoted to thermal effects of magmatism, that the magmatism is capable of increasing the surface heat flow rather considerably, which takes part either on a more regional or a more local scale. The first example of a regional positive heat flow anomaly comes from the Rio Grande (Chapman and Rybach, 1985; Figure 9.4a), where the surface heat flow reaches values above 100 mWm^{-2}, in comparison to a background heat flow of about 90 mWm^{-2} (see Ehlers and Chapman, 1999). The anomaly is attributed to the combined effect of lithospheric thinning and intrusion (Chapman and Rybach, 1985). The second example comes from the northern part of the Taranaki Basin in New Zealand (Armstrong and Chapman, 1998), associated with a Quaternary volcanic province associated with crustal intrusions emplaced in the past 1 Ma (Allis et al., 1995; Armstrong et al., 1997). These intrusions increase the surface heat flow by 15–20 mWm^{-2}. The third example comes from the easternmost positive surface heat flow anomaly of the Carpathian-Pannonian region (Nemčok et al., 1998 and references therein; Figure 9.5). The increase in heat flow beyond the background value reaches about 20–30 mWm^{-2}. A further example of the local heat flow anomaly comes from the Salton Buttes of the Salton Sea geothermal system (Newmark et al., 1988), where a combination of cooling rhyolite domes and venting steam and gases controls a heat flow anomaly with values of 600–1000 mWm^{-2} located in a region with background values higher than 120 mWm^{-2}.

Returning to Utah, the Wasatch Fault, having a length of about 370 km, initiated about 11–18 Ma ago (e.g., Ehlers and Chapman, 1999). It is seismoactive and capable of 7.5–7.7 magnitude earthquakes (e.g., Arabasz et al. 1992). The fault undergoes a relatively vertical displacement rate of 0.7–1.5 mmy^{-1} and is associated with a total exhumation of about 11 km (e.g., Zoback, 1983; Parry and Bruhn, 1987, 1990; Parry et al., 1988; Bruhn et al., 1990; Machette et al., 1991a, 1991b).

The fault separates the footwall, the Wasatch Mountains, which can be characterized by an exhumation rate of 1 mmy^{-1} adjacent to the fault and decaying to zero at a hinge located at a distance of 40 km from the fault, from the Salt Lake Basin, which can be characterized by a deposition rate of 0.5 mmy^{-1} adjacent to the fault and decaying to zero at a hinge located at a distance of 20 km from the fault (Ehlers and Chapman, 1999).

The finite-difference modeling indicates that the impact of the fault dip on the transient local thermal regimes is rather minor. Far from the fault, beyond the hinge, the isotherms remain in their initial geometry. However, the thermal regime in the zone along and

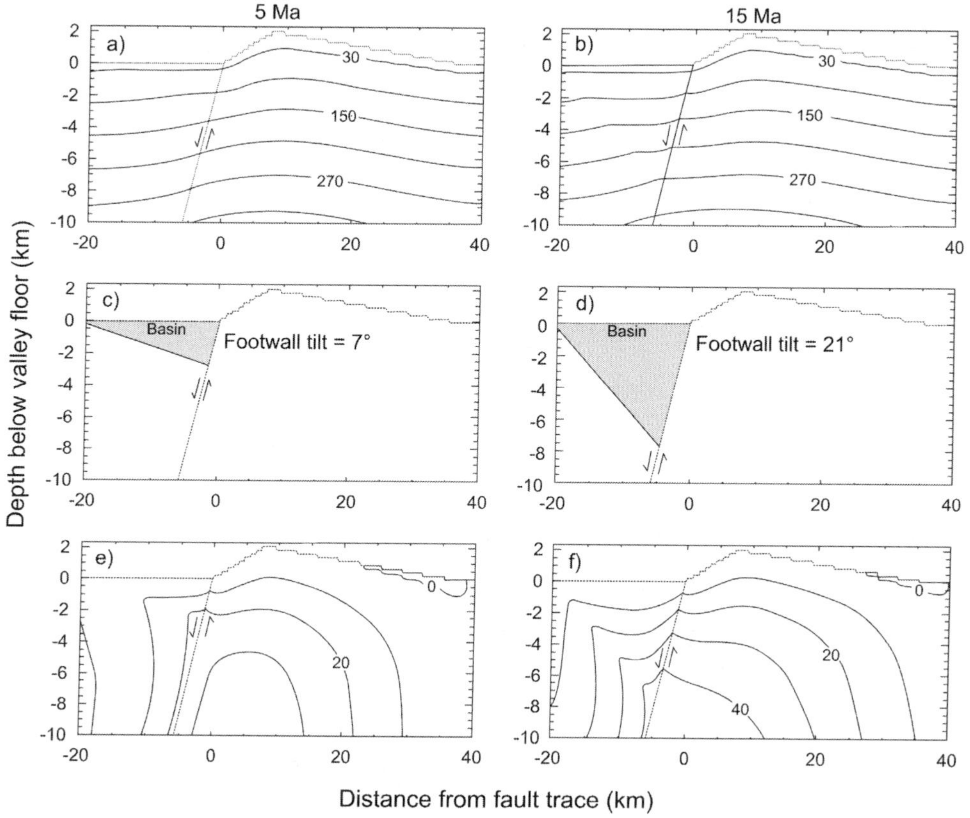

Figure 11.18. Transient isotherms, cross-section geometries, and residual temperatures after 5 Ma (figures a, c, e) and 15 Ma (figs. b, d, f) since the onset of activity on a normal fault with 60° dip (Ehlers and Chapman, 1999). Figures a and b illustrate isotherms of the hanging wall and the footwall perturbed by heat advection downward and upward, respectively. Cross-section geometries (c, d) show the footwall tilt geometry for the respective snapshot in its tilt and exhumation history. Residual temperatures in e and f are calculated by subtracting a 2-D static thermal model from the 2-D advective model. Isotherms are in °C.

immediately adjacent to the fault is relatively complex (Figure 11.18). This is due to:

1) the heat advection effect, which shifts isotherms in the hanging wall and the footwall downward and upward, respectively;
2) a low thermal conductivity of the new sedimentary fill, which controls an increased thermal regime of the deeper basin and its underlying crust;
3) a crust underneath the basin being heated by the heat advection in the footwall residing below the normal fault; and
4) the lateral flow from the footwall to the hanging wall.

The extra heat arriving into the hanging wall from footwall can be seen relatively well in a display of residual temperatures, derived from the subtraction of the static temperature field from the advective temperature field (Figures 11.18e, f). High residual temperatures in the footwall are controlled by the footwall tilt bringing the rock to the surface at the fastest pace in the area adjacent to the fault.

The effect of exhumation and fault angle on the surface heat flow can be seen in a model with isolated advective and topographic surface heat flows (Figure 11.19). Both of them in Figure 11.19 have heat flow components associated with basal heat flow and internal heat production removed to isolate the effect of heat advection and topography. Figure 11.19a illustrates the static topographic effect, characterized by a minimum-maximum couple. The maximum is located at the fault trace on the surface and formed by the isotherm pull-up toward the structural high formed by the footwall edge. The minimum is located along the footwall crest. The advective heat flow is characterized by a couple of small maximums and prominent maximums, divided by a broad and subtle minimum. A small maximum is controlled by the thermal conductivity contrast along the hinge of the basin where

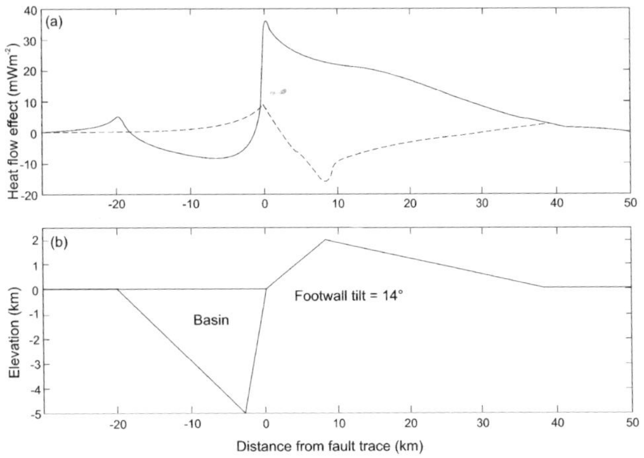

Figure 11.19. Calculated advective and topography-controlled components of the surface heat flow after 10 Ma-long displacement history of 60° dipping normal fault (Ehlers and Chapman, 1999). (a) Solid and dashed lines indicate advective and static topographic components of the heat flow. Advective curve includes the heat flow effect of a sedimentary basin and ignores the effect of topography. Topography does the opposite, ignoring the effect of material advection. (b) Cross-section showing the geometry associated with results shown in (a).

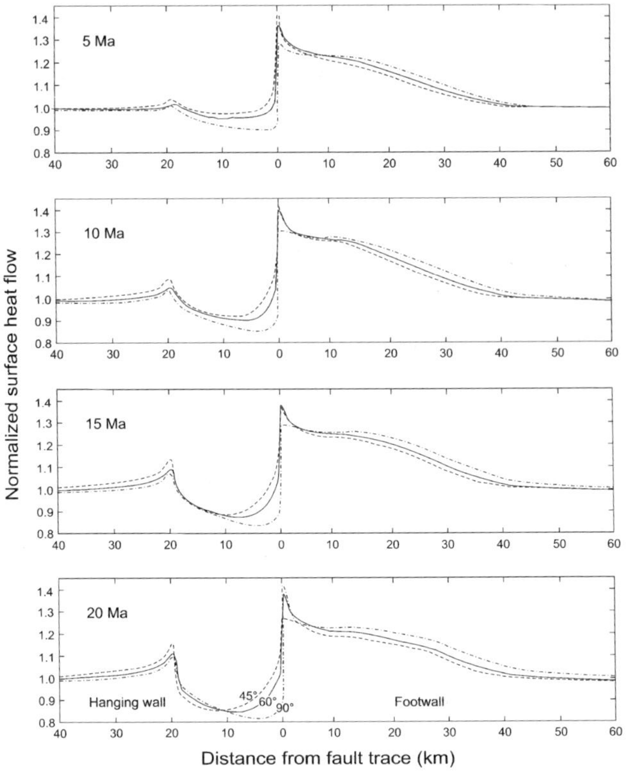

Figure 11.20. Transient surface heat flow from Figure 11.19 shown with distance from fault as a function of the fault dip in 5 Ma increments (Ehlers and Chapman, 1999). Heat flow was normalized using the 2-D static model heat flow. The curves in each time period represent models with fault dips of 45°, 60°, and 90°.

pinching-out sedimentary fill meets the basement on the surface. A subtle minimum in the basin is controlled by downward heat advection characterizing the basin undergoing deposition of low-conductivity sediment. A prominent maximum characterizes the footwall adjacent to the fault, representing the area with the fastest exhumation rate. The tapering-out advective heat flow in the footwall in a direction away from the fault correlates with fault tilt-controlled exhumation rate.

As alluded to earlier, the fault dip does not modify the thermal regime significantly (Figure 11.20). Figure 11.20 illustrates that decreasing fault dip controls a heat flow profile deviation from that of the vertical fault. The shallower the fault dip, the larger the lateral heat flow is from the footwall into the basin.

Both the magnitude of the heat flow perturbation and the offset across the fault increase with the fault activity time. If one takes a surface heat flow value of 90 mWm^{-2} as a background heat flow, the effect of the discussed transient along the Wasatch Fault shifts the heat flows of footwall and hanging wall to values of 110 mWm^{-2} and 80 mWm^{-2}, respectively.

Having a "warm" footwall and a "cold" hanging wall, however, represents an exact opposite to the heat flow regime observed in the Wasatch region (see Ehlers and Chapman, 1999). In searching for the explanation, we need to look at factors expected to control the local thermal regime of the footwall listed earlier and find out whether some of them could become dominant enough to cause the thermal regime of the footwall to become opposite to what is expected from the previously discussed numerical models.

The Wasatch Mountains rise 1–2 km above the Salt Lake Valley and their annual precipitation is about 14 cm greater than the precipitation in the valley. The presence of numerous hot springs along the Wasatch Fault proves the presence of the topography-driven fluid flow, which recharges via the Wasatch Mountains. The occurrence of a topography-driven fluid flow system indicates a heat loss in the recharge area. The geothermal power balance calculation indicates that discharge along the Wasatch Fault brings just a portion of the geothermal fluids sourced by recharge in the Wasatch Mountains to the surface, and that the remaining portion of the geothermal fluids has to flow further west (Ehlers and Chapman, 1999).

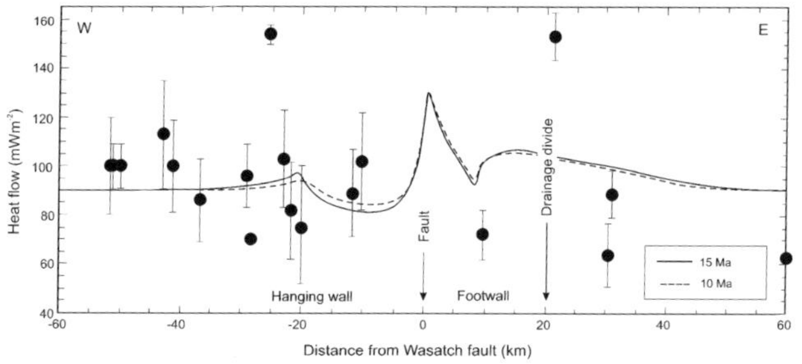

Figure 11.21. Modeled and observed heat flow values along the profile across the Wasatch Fault (Ehlers and Chapman, 1999).

When one looks at modeled and observed surface heat flows along the cross-section through the Salt Lake Valley and Wasatch Mountains (Figure 11.21), the following observations can be made. The conductivity contrast between:

1) the high-thermal-conductivity basement rocks and low-conductivity sediments (see Figure 10.1), and
2) the high-heat-production basement rocks and low-production sediments (see Figure 10.8)

results in the development of the subordinate heat flow maximum. This maximum can be further enhanced by the presence of a discharge area. The isotherm-depressing effect of the deposition controls the subtle minimum in the valley. The most interesting distribution of the surface heat flow occurs along the Wasatch Fault and the adjacent portion of the footwall. The former is characterized by a distinct maximum associated with a discharge of geothermal fluids. The latter shows a downward cusp in the otherwise broad maximum, indicating a recharge-related cooling in portion of the footwall, which is otherwise affected by the isotherm rise due to upward heat advection controlled by exhumation (Figure 7.16).

12 The role of deformation in thermal regimes of rifts and passive margins

Introduction

Factors controlling heat flow variations include:

1) radioactive decay;
2) deposition rate;
3) lithostratigraphy;
4) erosion rate;
5) preexisting heat flow;
6) thermal blanketing;
7) fluid flow;
8) intrusion; and
9) deformation rate.

While the first eight factors are discussed in other chapters written on thermal regimes, this chapter examines the role of the deformation rate on the thermal regimes of rift and passive margin settings. Furthermore, the following text also documents that the effects of factors such as heat production, deposition rate, and erosion rate are not independent of the deformation rate.

Deformation effects on thermal regimes

The kinematics of the rift area can be separated into horizontal extension and footwall and hanging wall tilts (e.g., Ehlers et al., 2003). The horizontal extension in the footwall can be assumed to reach a half of the horizontal displacement rate affecting a rift-boundary fault. The horizontal extension, $v_{extension}$, combined with fault dip defines the exhumation rate, v_{zmax}, of the footwall:

$$v_{zmax} = v_{extension} * \tan\alpha, \qquad (12\text{-}1)$$

where α is the fault dip. The footwall tilt is described using the vertical velocity component, v_z, which decays from the fault to the hinge (Figure 12.1) according to:

$$v_z = v_{zmax}\,(1 - x(1/x_{hinge})),\ (0 \le x \le x_{hinge}), \qquad (12\text{-}2)$$

where x_{hinge} is the distance from the fault to the hinge. V_z at the fault and the hinge equal v_{zmax} and 0, respectively.

The previously described kinematic model includes internal strain in both the footwall and the hanging wall. The internal strain prescription is based on the observation of meso-scale faults and joints in the first 10 km of the footwall in a direction from the boundary fault and the presence of a seismoactive antithetic normal fault in the hanging wall (Hecker, 1993; Keaton et al., 1993).

Friction along the fault zone is understood as a localized planar or narrow zonal heat source controlled by the shear stress magnitude and the slip rate. The temperature portion associated with this mechanism, T_{fr}, can be calculated using the formula (Carslaw and Jaeger, 1959):

$$\Delta T_{fr} = \tau v/\lambda\,(\surd(\kappa t/\pi)), \qquad (12\text{-}3)$$

where τ is the shear stress along the fault plane, v is the slip rate, λ is the thermal conductivity, κ is the thermal diffusivity, and t is the time.

While slip rates are known from a variety of normal faults (e.g., Jackson et al., 1982; Roberts, 2007), the opinion about shear stress magnitude along the faults varies from study to study. However, for illustration, if one takes a shear stress equal to 100 MPa, a surface temperature of 0°C, a thermal conductivity in the order of 1–5 $Wm^{-1}K^{-1}$, a slip rate of 2 mmy^{-1}, and a thermal diffusivity of 10^{-6} m^2s^{-1}, the resultant temperature portion due to fault friction is in the order of 2.5°C, which does not modify the background thermal regime significantly. However, faster slip rates can enhance the contribution of the heat generated by fault friction.

The role of the extension rate in the thermal regime of extensional terrain is often ignored and deformation is frequently set up as instantaneous, and followed by the subsequent conductive cooling of an involved rock section (e.g., McKenzie, 1978; Issler and Beaumont, 1989; Kusznir and Egan, 1989; Kusznir and Ziegler, 1992). Whether it is a generally appropriate assumption can be tested by a Peclet number, Pe, calculation (Ter Voorde and Bertotti, 1994):

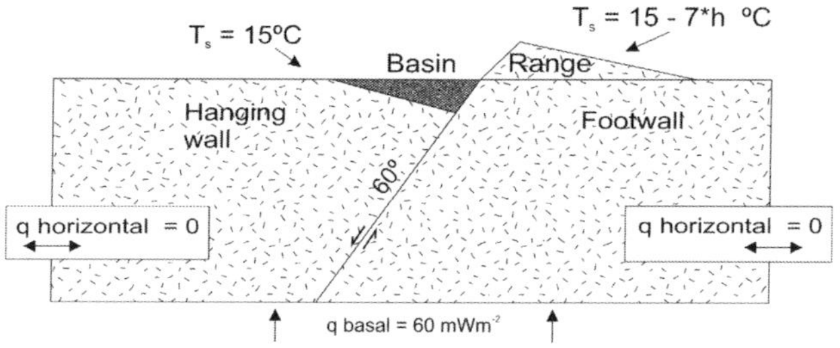

Figure 12.1. Model geometry and boundary conditions used for thermo-kinematic modeling (Ehlers et al., 2003). q indicates a flux boundary condition, T_s is the surface thermal boundary condition, and h is the footwall elevation from the reference level.

$$Pe = (v_{ext}h)/\kappa, \qquad (12\text{-}4)$$

where h is the typical length scale, v_{ext} is the extension, and κ is the thermal diffusivity. If h is taken as a crustal thickness of 30 km, κ is equal to 10^{-6} m^2s^{-1}, and the extension rate, v_{ext}, is slower than 1 mmy^{-1}, heat conduction can keep up with heat advection. This indicates that the complete conductive cooling takes place during the deformation period of time instead of during the post-rift phase if the extension rate is slower than 1 mmy^{-1}. On the contrary, if the extension rate is faster than 1 mmy^{-1}, a portion of cooling takes part during the post-rift stage. Although the latter case is more rule than exception, as indicated by rates derived from global positioning system (GPS) measurements in various rift regions such as $<6\pm2$ mmy^{-1} for the Apennines (D'Agostino et al., 2001; Hunstad et al., 2003), and 10 mmy^{-1} for the Gulf of Corinth and 2–3 mmy^{-1} for the Gulf of Evia (Billiris et al., 1991; Davies et al., 1997; Briole et al., 2000), the concept of instantaneous deformation and subsequent conductive cooling is invalid in slow rift cases and problematic in faster rift settings.

The extension rate's role in affecting the thermal regime is in the listric fault, which is detached at the top surface of the lower crustal levels, modifying the thickness of the upper crust, thus:

1) modifying the thickness of the layer with concentrated heat-producing elements; and
2) modifying the layers that reside between the top of the mantle, that is, the source of the mantle heat flow, and Earth's surface.

When one thinks about the heat advection-diffusion Equations (10-13) and (11-9), the radioactive heat production, H, can become a block function where:

$$H(z) = H_o \text{ if } h_1 \leq z \leq h_2, \text{ and} \qquad (12\text{-}5)$$

$$H(z) = 0 \text{ if } z < h_1, z > h_2,$$

where h_1 is the top of the heat-producing layer and h_2 is the base of the heat-producing layer. The values h_1 and h_2 depend on horizontal distance, x, and time, t, being controlled by the hanging wall movement.

The role of the extension rate is also in controlling the competition between heat conduction and heat advection (Figure 12.2). Finite difference models shown in Figure 12.2 compare the results of this competition taking place under the extension rates of 0.2, 2, and 20 mmy^{-1}. They indicate that if the extension rate is high, the heat advection dominates over the heat conduction, and isotherms move in the direction of material element trajectories. At slow and intermediate rates, however, the isotherms manage to equilibrate either during the movement of the material elements or the first portion during the element movement and second portion after it.

Figure 12.3 shows the cooling curves of four checkpoints in the footwall and four checkpoints in the hanging wall, which are located in Figure 12.2. It indicates that the hanging wall checkpoints have a tendency to become warmer at low extension rates. It further suggests that the downward movement of the hanging wall does not seem to cause any distinct cooling of the footwall, even at high movement rates. Although such footwall cooling has been observed in the Simplon Zone in the Central Alps of Switzerland (Grasemann and Mancktelow, 1993) and the Lugano Fault Zone of the southern Alps of Italy (Bertotti and ter Voorde, 1994), it had to be explained by the effect of additional cooling mechanisms such as regional uplift and exhumation in the first case, and the cooling of the earlier magmatic event in the second case.

Apart from the extension rate itself, the fault dip and hinge position affect the exhumation rate, that is, affecting the importance of heat advection (Ehlers et al., 2003). Figure 12.4 illustrates how particle paths are controlled by fault dip and footwall tilt, which together with exhumation rate affect the thermal regime. The steeper the fault dip for the same extension rate or the faster the extension rate for the same fault dip, the faster the exhumation rate is. The faster the exhumation rate, the stronger the effect of the heat advection in control of the local thermal regime is. As Figure 12.4 shows, the particle paths are upward-diverging, controlled by different location-dependent combinations of horizontal movement rate and vertical movement rate. As a result, the deepest rocks are exhumed in the

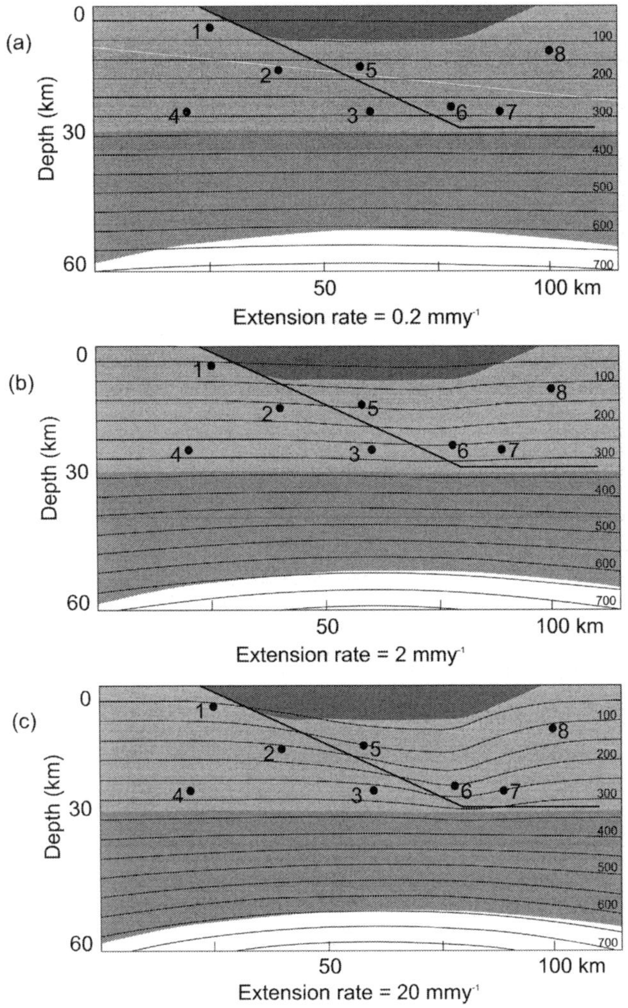

Figure 12.2. Temperature distribution in a basin after 20 km of extension (Ter Voorde and Bertotti, 1994). Extension rates are (a) 0.2 mmy^{-1}, (b) 2 mmy^{-1}, and (c) 20 mmy^{-1}.

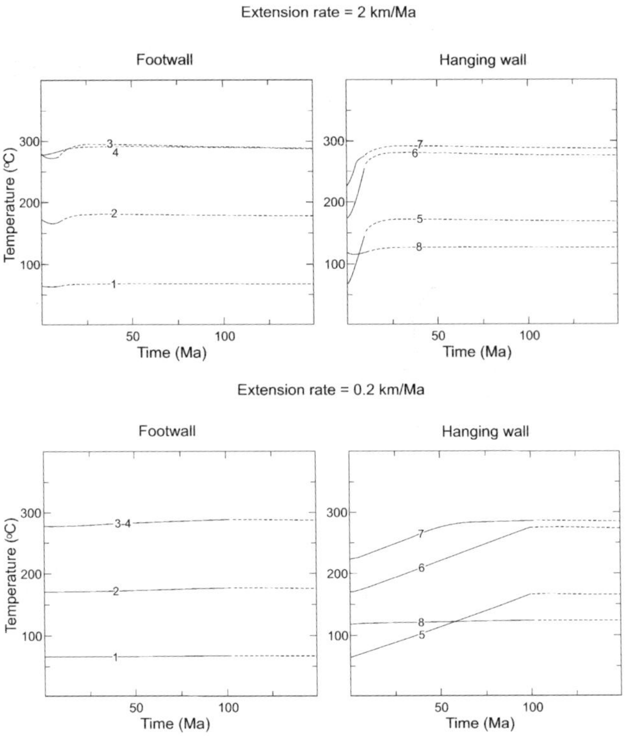

Figure 12.3. T-t curves measured at different points in the basin shown in Figure 12.2 after 20 km of extension (Ter Voorde and Bertotti, 1994). Extension rates are 0.2 and 2 mmy^{-1}. The solid part of a line gives the syn-rift phase; the dashed part gives the post-rift phase.

footwall right next to the boundary fault. The amount of total exhumation decays from fault toward hinge. Furthermore, for a 45°-dipping fault, a particle outcropping at a distance of 20 km from the fault has traveled 10 km laterally to get there. For a 60°-dipping fault, the same particle would have traveled only 5 km laterally. Note that as a result of a lateral component, particles travel obliquely with respect to isotherms.

Figure 12.4 further illustrates that heat advection-affected isotherms become curved and heat flow gains the two-dimensional character. For example, a checkpoint located 5 km away from the fault in the footwall at a depth with an isotherm of 60°C undergoes vertical and lateral heat flow components of 88 mWm^{-2} and 15 mWm^{-2}, respectively. Sensitivity

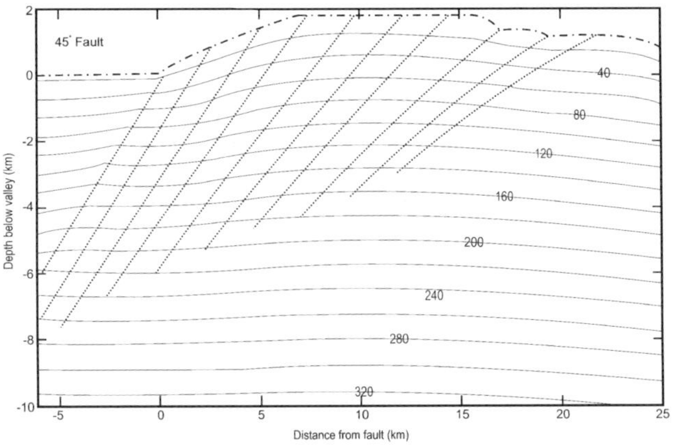

Figure 12.4. Example output from the coupled thermo-kinematic model for a 45° dipping fault (Ehlers et al., 2003). The bold dashed line represents the average topographic surface through the central Wasatch Mountains. Thin solid lines show the isotherms in ° Celsius after 16.5 Ma of the controlling normal fault activity, characterized by a maximum vertical velocity, V_{zmax}, of 0.6 mmy^{-1} and a footwall hinge location at 30 km from the fault. Pluses denote particle paths in 0.5 Ma increments.

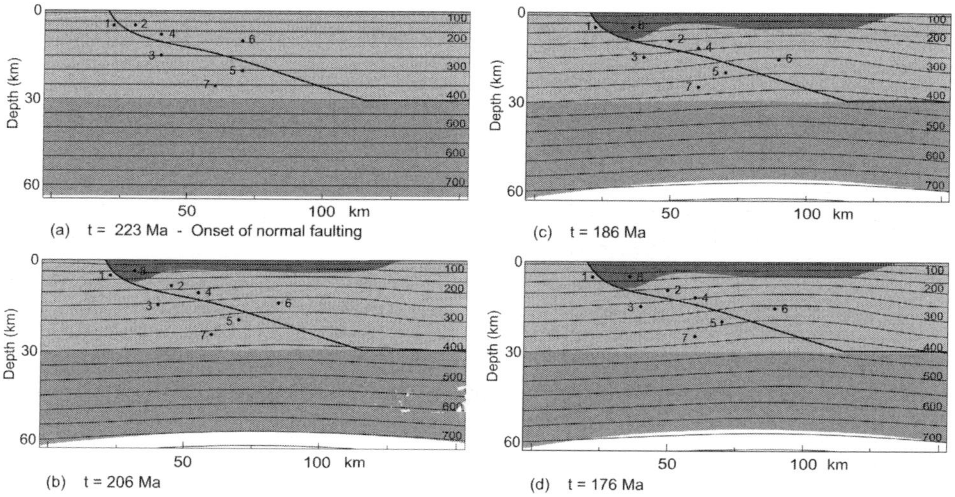

Figure 12.5. Modeled thermal evolution of a crustal segment affected by a crustal normal fault with geometry of the Lugano–Val Grande Fault (Bertotti and ter Voorde, 1994). The numbered points indicate positions for which the synthetic T-t curves shown in Figures 12.6 and 12.8 have been constructed.

analysis of controlling parameters indicates that increased exhumation rates enhance the role of lateral heat flow (Ehlers et al., 2003). This effect takes place because of isotherms being swept closer to the surface with an increase in exhumation rate. The faster the exhumation, the less time is available for traveling particles to cool by conduction. This situation results in a warmer footwall. The importance of the hinge location is in its defining the size of the area undergoing accelerated uplift.

The next good example of the deformation effect on thermal regimes comes from the Monte Generoso Basin in the southern Alps of Italy (Bertotti and ter Voorde, 1994). It is bounded by the Lugano–Val Grande normal fault, which is listric, having a dip of 50–70° in its uppermost 5–7 km and characterized by total vertical and horizontal displacements of about 8 and 20 km, respectively. Its reconstructed displacement rates characterizing three subsequent rift intervals were 1.08, 0.25, and 0.54 mmy^{-1}.

The thermal regime of the Monte Generoso Basin during various time intervals during rifting was simulated by finite-difference modeling (Bertotti and ter Voorde, 1994; Figure 12.5). The modeling indicates that the extension rate, which is as slow as those three rates determined for the Lugano–Val Grande normal fault, does not perturb the thermal regime significantly. Therefore, the main character of the thermal anomaly (Figure 12.5) is controlled by the lithospheric thinning. The control of the thermal regime in the case of pure shear scenario is apparently stronger than control in the simple shear scenario.

The thermal regime does not seem to be affected by the downward movement of the hanging wall either. It seems that extension rates, which are as slow as the ones along the Lugano–Val Grande normal fault, are associated with thermal equilibration, which is faster than mass movement, and, therefore, the hanging wall becomes progressively heated upon its arrival to greater depths.

At this point, it is worth looking at a function of temperature with time for four hanging wall and four footwall checkpoints (Figure 12.6) to understand the thermal behavior of both the hanging wall and the footwall in detail. The T-t curves shown in Figure 12.6 indicate that footwall points that did not undergo any subsidence felt the increased thermal regime due to the lithospheric thinning. Although each of the footwall points started at a specific pre-rift temperature controlled by its depth, the net increase in their temperatures during rifting is a function of their distance from the center of thinning. The hanging wall points, which are affected by both the subsidence and impact of deposition in the Monte Generoso Basin, indicate that their thermal regime is not affected by normal faulting very much. However, they are affected by a downward movement.

Note that, contrary to discussed modeling, the cooling data from both the fault rocks (Bertotti, 1991; Bertotti et al., 1993) and footwall rocks (Mottana et al., 1985) document a footwall cooling during rifting. The explanation for this phenomenon comes from the existence of the magmatism-related thermal anomaly developed shortly before rifting, the evidences for which include intrusion emplacement dating (e.g., Stahle et al., 1990;

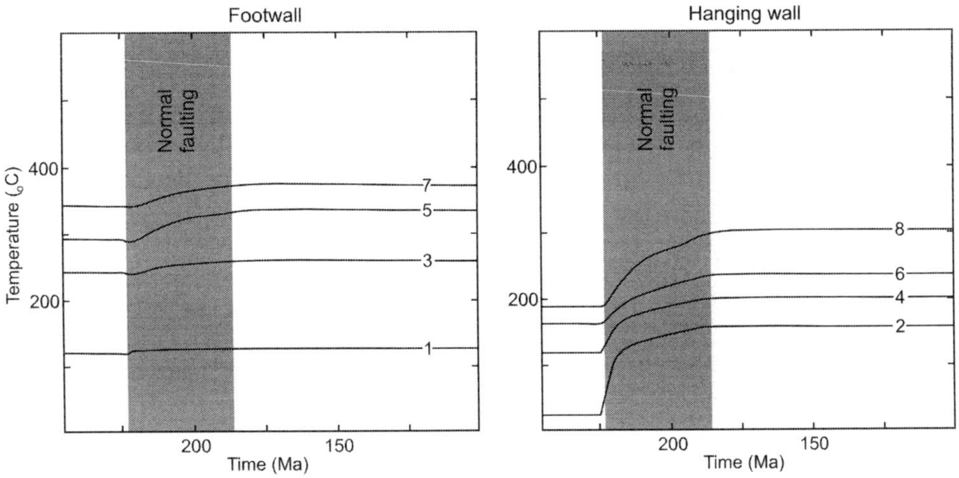

Figure 12.6. Synthetic T-t curves for representative points of the South Alpine thrust as predicted by a simple extension model (Bertotti and ter Voorde, 1994). Points are localized in Figure 12.5. The shaded zone indicates the fault activity time period.

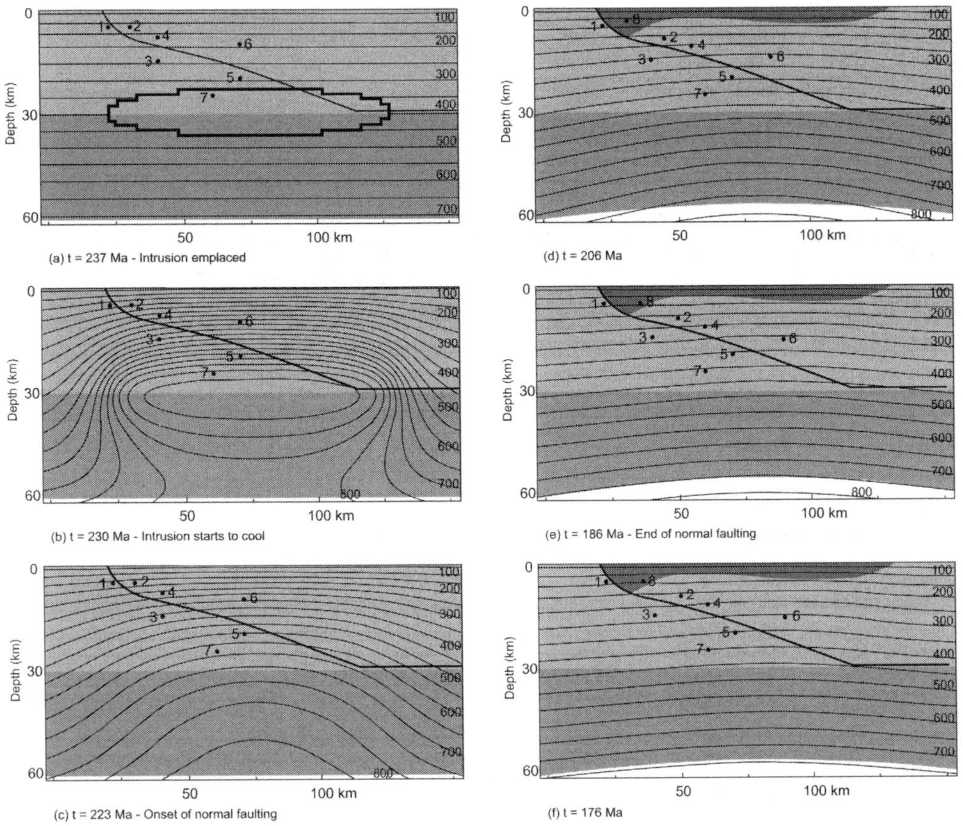

Figure 12.7. Thermal evolution of the Mount Generoso crustal segment as predicted by the model including the initial thermal anomaly (Bertotti and ter Voorde, 1994).

Gebauer, 1993) and reconstructions of volcanic material ages (Bertotti and ter Voorde, 1994 and references therein). Inclusion of this anomaly in the numeric simulation satisfactorily explains the mentioned cooling (Figure 12.7). Indeed, the temperature-time curves of checkpoints from the footwall indicate the overall history dominated by the post-intrusion cooling (Figure 12.8). Checkpoints from the hanging wall,

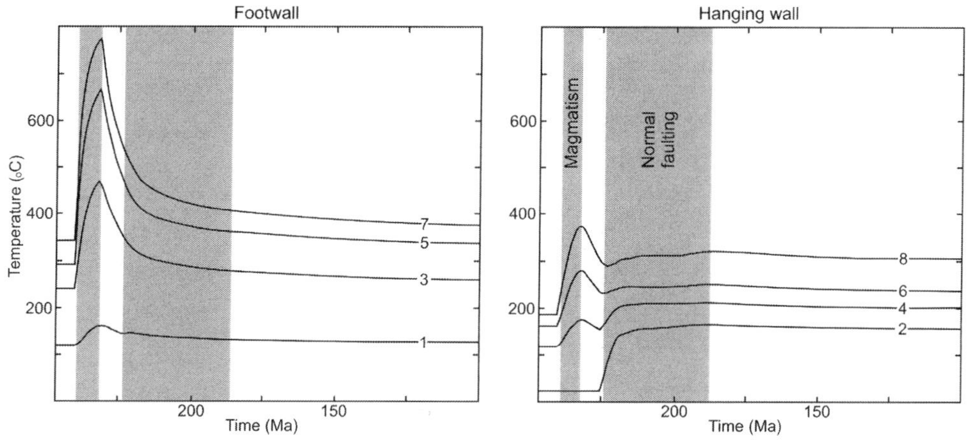

Figure 12.8. Synthetic T-t curves for the representative points located in Figure 12.5 (Bertotti and ter Voorde, 1994). The shaded zone on the left-hand side shows the emplacement and deactivation of the intrusion. The shaded zone on the right-hand side shows the activity interval of the normal faulting.

however, document a more complex temperature-time history as the post-intrusion cooling becomes overpowered by syn-rift warming due to lithospheric thinning.

Finite-element modeling of the thermal regimes of simple rift scenarios

The role of deformation in thermal regimes of simple rift scenarios can be now discussed in conjunction with the roles of varying crustal thicknesses and extension rates. In order to do this, we can discuss the numerical models of thermal perturbations in the rift setting, which were developed by Andreas Henk (2008, pers. comm.) using the finite-element program ANSYS® (Ansys Inc., Houston, USA) for a system of following $3 \times 3 \times 3 = 27$ scenarios:

1) three basic scenarios simulating different types of lithosphere (strong, intermediately strong, and weak), defining the initial crustal thicknesses as 30, 40, and 50 km (keeping the upper/lower crust ratio at 1:1);
2) each scenario from (1) undergoing an initial heat flow of 40, 60, and 80 mWm^{-2}; and
3) each individual scenario resulting from combinations of (1) and (2) undergoing extension at rates of 10, 30, and 50 mmy^{-1}.

The $3 \times 3 \times 3 = 27$ system of scenario design is identified in model codes that start with a specification of crustal thickness, followed by surface heat flow and extension rate specifications. For example, a model with crustal thickness, heat flow, and extension rate of 40 km, 60 mWm^{-2}, and 30 mmy^{-1}, respectively, has a code c40h60v30 (Tables 12.1, 12.2).

Because the modeling results are unpublished, we will devote part of our discussion to the description of the method and its constraints. All models have been done for the three-layer models including the upper crust, the lower crust, and the mechanically strong part of the lithospheric mantle. The thickness of the strong part of the lithospheric mantle varied depending on the thermal regime of the respective model. Its base in almost all models has been located within the ductile zone of the mantle lithosphere. The length of the model was 200 km. The respective mechanical behavior of model layers was governed by the Coulomb-Mohr law and temperature- and strain rate-dependent creep laws, described by Equations (5-2), (5-3), and (5-6) in Chapter 5. The definition of preexisting discontinuities, for example, faults, was done using contact elements. This approach handles large differential movements between parts of the model, but does not describe the fault propagation itself. Contact elements are defined at opposing sides of the preassigned fault and stiffness values similar to the Young's moduli of the rocks in contact are used to enforce compatibility between adjacent fault surfaces. Contact elements are also capable of describing frictional sliding, so that the influence of different friction coefficients can be studied.

The thermal calculations are based on heat transport by conduction and advection.

Boundary conditions of the thermal model include:

1) fixed temperature value at Earth's surface; and
2) fixed heat flow value at the base of the lithosphere.

Table 12.1. List of the rift model scenarios. Note that the warmest models mathematically collapsed because of excessive deformation (failed because of e.d.).

Crustal thickness (km)	Surface heat flow (mWm^{-2})	Extension rate (mmy^{-1})	Scenario	Comment
30	40	10	c30h40v10	
		30	c30h40v30	
		50	c30h40v50	
	60	10	c30h60v10	
		30	c30h60v30	
		50	c30h60v50	
	80	10	c30h80v10	
		30	c30h80v30	
		50	c30h80v50	
40	40	10	c40h40v10	
		30	c40h40v30	
		50	c40h40v50	
	60	10	c40h60v10	
		30	c40h60v30	reference model
		50	c40h60v50	
	80	10	c40h80v10	
		30	c40h80v30	
		50	c40h80v50	
50	40	10	c50h40v10	
		30	c50h40v30	
		50	c50h40v50	
	60	10	c50h60v10	
		30	c50h60v30	
		50	c50h60v50	
	80	10	c50h80v10	failed because of e. d.
		30	c50h80v30	failed because of e. d.
		50	c50h80v50	failed because of e. d.

Table 12.2. List of extra parameter variations applied to selected models.

Crustal thickness (km)	Surface heat flow (mWm^{-2})	Extension rate (mmy^{-1})	Scenario	Comment
40	60	3	c40h60v03	extension rate 3 mmy^{-1}
40	60	5	c40h60v05	extension rate 5 mmy^{-1}
40	60	30	c40h80v30mu02	coefficient of friction 0.2
40	60	30	c40h80v30mu04	coefficient of friction 0.4
40	60	30	c40h80v30mu06	coefficient of friction 0.6

Thermal material parameters include:

1) temperature-dependent thermal conductivity;
2) specific heat;
3) radiogenic heat production; and
4) temperature-dependent density.

As already alluded to, the mechanical calculations are based on the Coulomb-Mohr law in the brittle domain, whereas a power-law creep governs the ductile deformation.

Boundary conditions for the mechanical model include:

1) displacement-defined tectonic stresses;
2) isostasy;
3) gravity forces; and
4) a load of sediments in the developing half-graben.

The mechanical material parameters include:

1) Young's moduli;
2) Poisson's ratios;
3) densities;
4) stress–strain relationship governed by Byerlee's law (extensional regime and hydrostatic pore pressures);
5) power-law creep parameters (pre-exponential factor, creep activation energy, exponent); and
6) fault friction coefficients.

Thermal and mechanical calculations are coupled via temperature-dependent material behavior and thermal stresses as well as heat advection due to changes in model geometry.

Note that although the developing half-graben looks "empty" in the documenting figures, the load of its sediment fill is included in mechanical calculations. Similarly, the resulting modification of the temperature at the top of the basement underneath the basin fill is calculated in thermal models.

The structural architecture of each model includes the upper crust dissected by a single listric normal fault, which is prescribed to each scenario as detached at the top of the lower crust. The detachment is located along the top-lower crust in the entire model. Stresses in the models are built up by applying a constant extension rate, which is prescribed by the displacement boundary condition, to the right side of the model. Each model is run either to its failure or until the excessive deformation of its individual elements causes the numerical problems.

Modeling outcomes include:

1) total horizontal displacement distribution;
2) total vertical displacement distribution;
3) initial differential stress distribution;
4) total strain distribution;
5) principal stress vector pattern;
6) burial history plot for the deepest part of the basin, that is, the subsidence history of the nose of the hanging wall; and
7) temperature distribution.

Models of the structural behavior trends

Interpretation of the structural behavior trends have been made when scenarios have been grouped according to identical input extension rate, heat flow, and crustal thickness.

Grouping according to identical extension rates allows us to see the interaction of the initial heat flow with the crustal thickness. All three groups characterized by an extension rate of 10, 30, and 50 mmy^{-1} are characterized by the scenario with a heat flow of 80 mWm^{-2} combined with a crustal thickness of 50 km as being unstable, that is, having no mathematical solution, given a too large total deformation (see Table 12.1).

The numerical problems with models, which include a hot and 50 km-thick crust, result from the fact that particularly the lower crust is so weak that it behaves almost like a fluid. This results in very rapid deformation and excessive distortion of the corresponding model elements. Therefore, it is not possible to achieve a stable start situation for applying the extension rate. The obvious geological implication of this situation is that such a hot crust would have never gained a thickness of 50 km. Having a vertically integrated strength so low, the crust would have collapsed gravitationally long before reaching such a thickness.

All three extension rate groups indicate that:

1) increasing crustal thickness correlates with an increase of mantle uplift in the region underneath the developing extensional basin;
2) increasing crustal thickness correlates with an increase of the lower crustal uplift in the region underneath the developing extensional basin;
3) the thinnest crust scenario is characterized by the mantle uplift located underneath the roll-over anticline crest and lower crust uplift located under the first ramping-up portion of the normal fault in a direction away from its detachment, while the thicker crust scenarios are characterized by both uplifts being located above each other, and located underneath the center of the normal fault ramp;
4) a scenario having a 50 km-thick crust combined with a heat flow of 80 mWm^{-2} reaches an unstable flow; that is, rapid deformation and

excessive element distortion, which the numerical approach cannot handle;
5) an increase in both crustal thickness and initial heat flow tends toward controlling not just a progressively more pronounced core complex unroofing but also developing the train of ductile folds in the footwall underneath a laterally retreating hanging wall;
6) increasing crustal thickness correlates with an increase of the footwall tilt;
7) increasing crustal thickness correlates with an increase of the subsidence in the immediate hanging wall; and
8) increasing crustal thickness correlates with a widening of the footwall area at the surface behind the normal fault, which is affected by subsidence, associated with progressively less broad and more localized core complex

and that:

1) increasing initial heat flow correlates with a widening of the subsiding area located on the footwall surface behind the controlling normal fault;
2) increasing initial heat flow correlates with an increase in mantle uplift;
3) increasing initial heat flow correlates with an increase in lower crustal uplift;
4) increasing initial heat flow results in a shift of the mantle uplift maximum from the area underneath the crest of the roll-over anticline in the coolest scenario to the area underneath the center of the normal fault ramp in the warmest scenario;
5) increasing initial heat flow results in a shift of the lower crustal uplift maximum from the area underneath the lowermost part of the normal fault ramp to the area underneath the mid-ramp of the normal fault;
6) increasing initial heat flow results in changing the unaligned mantle and lower crustal uplifts of the coolest scenario to the aligned mantle and lower crustal uplifts of the warmest scenario;
7) increasing initial heat flow decreases the subsidence in the immediate hanging wall;
8) increasing initial heat flow correlates with increasing footwall tilt; and
9) increasing initial heat flow changes the broader uplift of the cool scenarios into the more localized uplift of the warm scenarios.

Although the two extension rate groups with 10 and 30 mmy^{-1} rates do not indicate the following phenomenon, a scenario with an extension rate of 50 mmy^{-1} apparently indicates that increasing crustal thickness correlates with more prominent upward-convex complex development on the normal fault ramp, that is, a more advanced core complex development.

Most distinct trends in the described set of models are captured by Figures 12.9 and 12.10, which show the stress vector and total horizontal displacement outputs. Figure 12.9, which focuses on the effect of increasing crustal thickness on the model, demonstrates a clear trend of progressively larger uplift in the lower crust with increasing crustal thickness. The maximum uplift is localized roughly underneath the tip of the laterally moving-out hanging wall. The increasing crustal thickness also correlates with broader uplift becoming more focused.

Figure 12.10, which focuses on the effect of increasing initial heat flow, shows a clear trend of steeply increasing lower crustal uplift with increasing initial heat flow value. The increase in initial heat flow also correlates with broader uplift becoming more focused.

Grouping according to identical initial heat flow allows us to see the interaction of crustal thickness with extension rate. Only the group characterized by the initial heat flow of 80 mWm^{-2} is characterized by three scenarios with a crustal thickness of 50 km combined with extension rates of 10, 30, and 50 mmy^{-1} as being unstable, that is, having no mathematical solution given a too large total deformation (see Table 12.1).

All three initial heat flow groups indicate that:

1) increasing crustal thickness correlates with an increase of mantle uplift in the region underneath the developing extensional basin;
2) increasing crustal thickness correlates with an increase of the lower crustal uplift in the region underneath the developing extensional basin;
3) the thinnest crust scenario is characterized by mantle uplift located underneath the roll-over anticline crest and the lower crustal uplift located underneath the first ramping-up portion of the normal fault in a direction away from its detachment while the thicker crust scenarios are characterized by both uplifts being located above each other, underneath the center of the normal fault ramp;
4) scenarios having a 50 km-thick crust combined with all three modeled extension rates reach unstable flow, that is, rapid deformation and excessive model element distortion, which cannot be handled by the numerical approach;
5) increase in crustal thickness tends toward controlling more pronounced core complex unroofing, which is localized underneath the center of

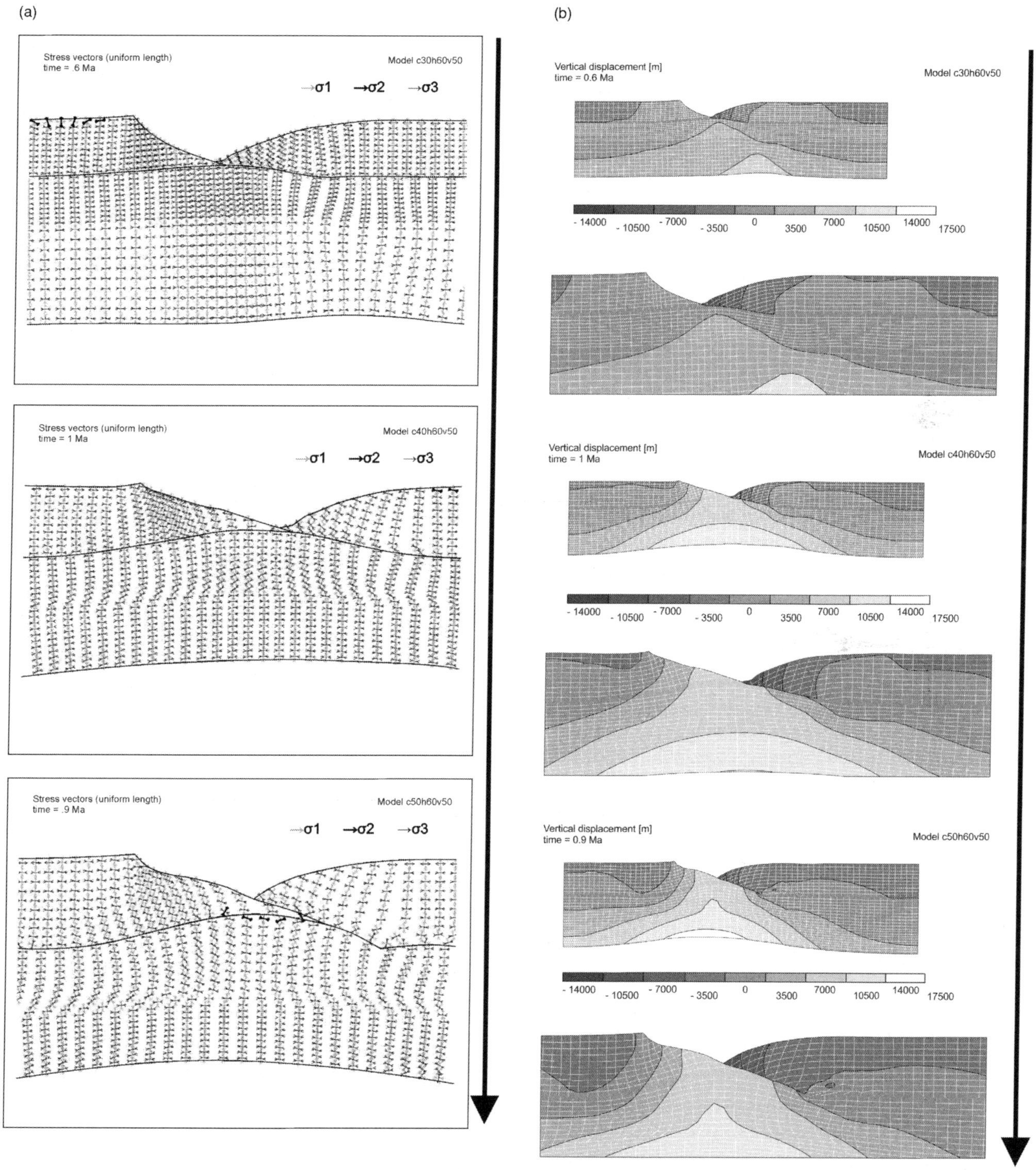

Figure 12.9. (a) A series of models c30h60v50 – c40h60v50 – c50h60v50 showing the behavior of scenarios characterized by an intermediate initial heat flow of 60 mWm^{-2} and a fastest extension rate of 50 mmy^{-1}, and an increase in crustal thickness from 30, through 40, to 50 km (modified from Henk, 2008, pers. comm.). Modeling results are shown after 0.6 Ma of deformation in a stress-vector format. Note a clear correlation of increasing crustal thickness with 1) an increase in lower crustal uplift and 2) the change of a broadly distributed uplift to a localized one. (b) A series of models c30h60v50 – c40h60v50 – c50h60v50 showing the behavior of scenarios characterized by an intermediate initial heat flow of 60 mWm^{-2} and a fastest extension rate of 50 mmy^{-1}, and an increase in crustal thickness from 30, through 40, to 50 km (modified from Henk, 2008, pers. comm.). Modeling results are shown after 0.6 Ma of deformation in total vertical displacement format. Note a clear correlation of increasing crustal thickness with 1) an increase in lower crustal uplift and 2) the change of a broadly distributed uplift to a localized one.

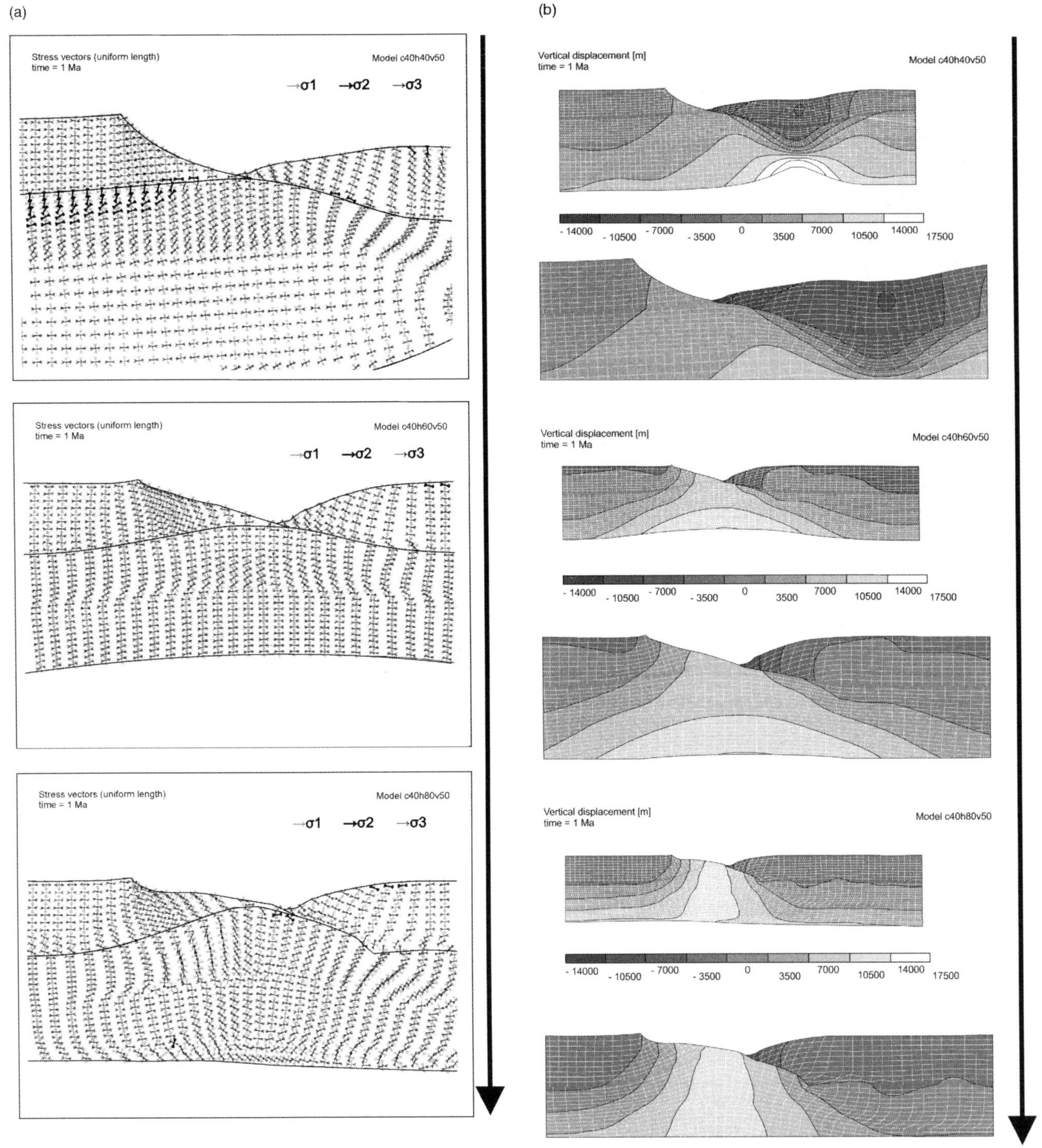

Figure 12.10. (a) A series of models c40h40v50 – c40h60v50 – c40h80v50 showing the behavior of scenarios characterized by an intermediate crustal thickness of 40 km and a fastest extension rate of 50 mmy^{-1}, and an increase in initial heat flow from 40, through 60, to 80 mWm^{-2} (modified from Henk, 2008, pers. comm.). Modeling results are shown after 0.6 Ma of deformation in a stress-vector format. Note a clear correlation of increasing initial heat flow with 1) an increase in lower crustal uplift and 2) the change of a broadly distributed uplift to a localized one. Both trends seem more expressive than in Figure 12.9 showing the influence of thickening crust. (b) A series of models c40h40v50 – c40h60v50 – c40h80v50 showing the behavior of scenarios characterized by an intermediate crustal thickness of 40 km and a fastest extension rate of 50 mmy^{-1}, and an increase in initial heat flow from 40, through 60, to 80 mWm^{-2}. (modified from Henk, 2008, pers. comm.). Modeling results are shown after 0.6 Ma of deformation in total vertical displacement format. Note a clear correlation of increasing initial heat flow with 1) an increase in lower crustal uplift and 2) the change of a broadly distributed uplift to a localized one. Both trends seem more expressive than in Figure 12.9 showing the influence of thickening crust.

the normal fault ramp in the warmest scenarios and being progressively more localized rather than broad from the coolest to the warmest scenarios;

6) increasing crustal thickness correlates with an increase of the footwall tilt;
7) increasing crustal thickness does not cause any change in subsidence in the immediate hanging wall;
8) increasing crustal thickness correlates with a widening of the footwall area at the surface behind the normal fault, which is affected by subsidence, associated with progressively less broad and more localized footwall tilt location; and
9) increasing crustal thickness correlates with a brittle–ductile transition zone occupying a progressively shallower depth (indicated by the top of preassigned lower crust undergoing larger particle flow)

and that:

1) increasing extension rate correlates with increasing subsidence in the immediate hanging wall;
2) increasing extension rate correlates with increasing footwall tilt;
3) increasing extension rate correlates with a widening of the footwall area at the surface behind the normal fault, which is affected by subsidence, associated with progressively less broad and more localized footwall tilt location;
4) increasing extension rate correlates with a core complex upward-convex geometry, which is located in the center of the normal fault ramp and becoming more expressive; and
5) increasing extension rate correlates with a brittle–ductile transition zone occupying a progressively shallower depth (indicated by the top of preassigned lower crust undergoing larger particle flow).

Note that, while both the modeled initial heat flow and crustal thickness represent the usual values in natural extension settings (see Chapters 5 and 9), the extension rates have been increased almost a magnitude to enhance our chance to see their effects on model behavior. This difference can be demonstrated, for example, by comparing the modeled extension rates with those of the present-day Basin and Range province in the western United States (Kreemer, 2007), which have a maximum rate of 5–6 mmy^{-1}.

The most distinct trends in the previously described set of models are captured by Figure 12.11, which shows the stress vector and total horizontal displacement outputs. Figure 12.11, which focuses on the effect of the extension rate on the model, demonstrates a clear growing tendency for core complex formation with increasing extension rate.

Grouping according to identical crustal thickness allows us to see the interaction of extension rate with initial heat flow. Only the group characterized by a crustal thickness of 50 km is characterized by three scenarios with an initial heat flow of 80 mWm^{-2} combined with extension rates of 10, 30, and 50 mmy^{-1} as being unstable, that is, having no mathematical solution given its total deformation is too large (see Table 12.1).

All three crustal thickness groups indicate that:

1) increasing extension rate correlates with increasing subsidence in the immediate hanging wall;
2) increasing extension rate correlates with increasing footwall tilt;
3) increasing extension rate correlates with a widening of the footwall area at the surface behind the normal fault, which is affected by subsidence, associated with progressively less broad and more localized footwall tilt location;
4) increasing extension rate correlates with increasing mantle uplift, which is located beneath the crest of the roll-over anticline in the thinnest crust scenario, below the flank of the anticline in the intermediately thick crust scenario and underneath the wedged-out end of the hanging wall in the thickest scenario;
5) increasing extension rate correlates with increasing lower crustal uplift, which is located underneath the wedging-out hanging wall (between a crest of the roll-over anticline and half-graben-bounding fault) in all crust thickness scenarios;
6) an increasing extension rate in the thickest crust scenario can shift initially unaligned mantle and lower crustal uplifts to uplifts located above each other; and
7) an increasing extension rate correlates with the area of zero vertical movements located in the hanging wall getting smaller and vanishing, from both the hinterland and foreland, in favor of the overall subsidence

and that:

1) increasing initial heat flow correlates with a widening of the subsiding area located on the footwall surface behind the controlling normal fault, that is, footwall tilt getting more localized behind the controlling normal fault (see Figure 11.7)

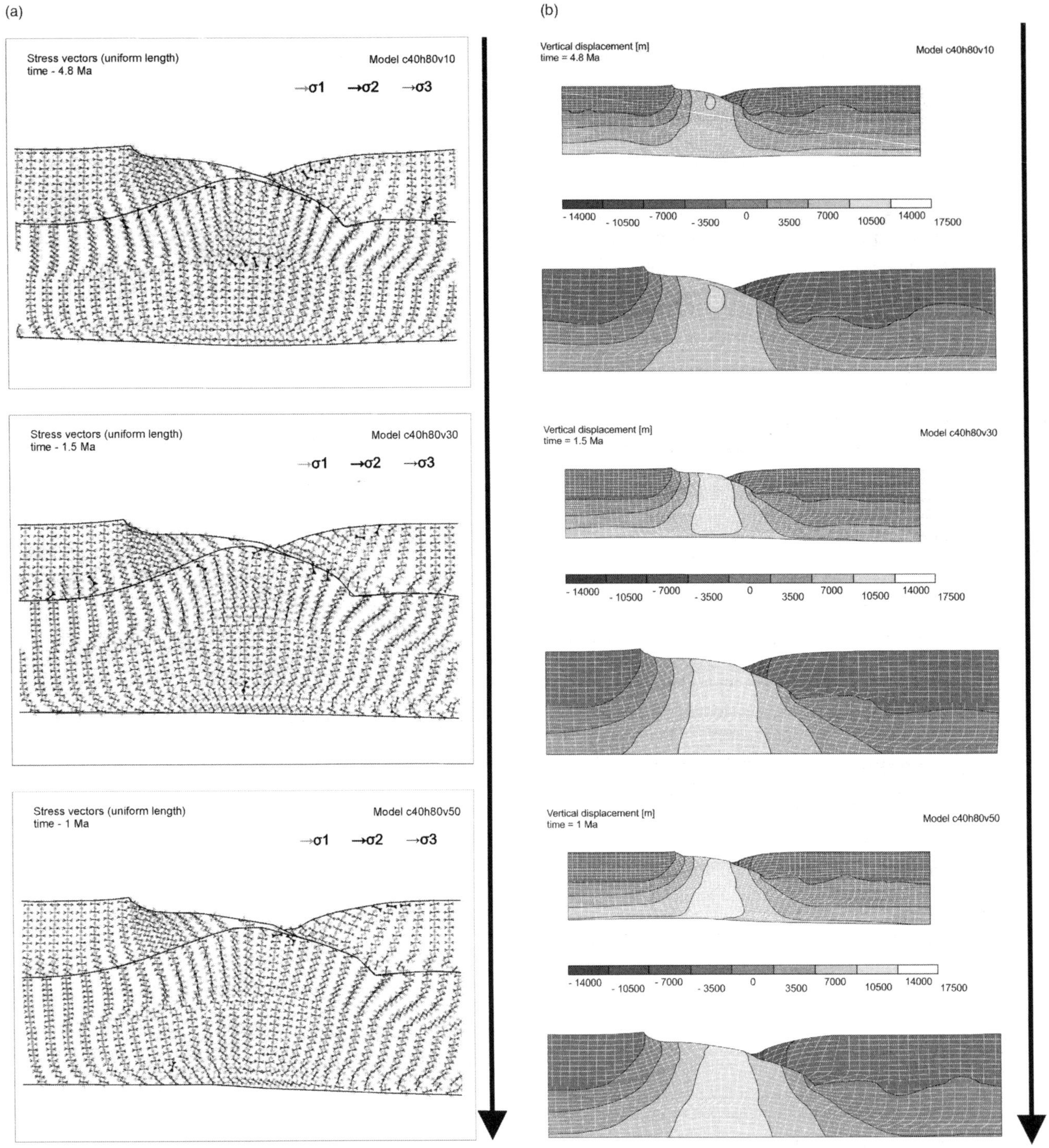

Figure 12.11. (a) A series of models c40h80v10 – c40h80v30 – c40h80v50 showing the behavior of scenarios characterized by an intermediate crustal thickness of 40 km and a high initial heat flow of 80 mWm^{-2}, and an increase in extension rate from 10, through 30, to 50 mmy^{-1} (modified from Henk, 2008, pers. comm.). Modeling results are shown after 0.6 Ma of deformation in a stress-vector format. Note a clear correlation of increasing extension rate with the development of a progressively more pronounced core complex. (b) A series of models c40h80v10 – c40h80v30 – c40h80v50 showing the behavior of scenarios characterized by an intermediate crustal thickness of 40 km and a high initial heat flow of 80 mWm^{-2}, and an increase in extension rate from 10, through 30, to 50 mmy^{-1} (modified from Henk, 2008, pers. comm.). Modeling results are shown after 0.6 Ma of deformation in total horizontal displacement format. Note a clear correlation of increasing extension rate with the development of a progressively more pronounced core complex.

(although the thickest crustal scenarios do not allow us to see it clearly, maybe because of the disadvantageous automatic choice of contoured colors);

2) increasing initial heat flow correlates with an increase in mantle uplift;
3) increasing initial heat flow correlates with an increase in lower crustal uplift;
4) increasing initial heat flow results in a shift in the mantle uplift maximum from the area underneath the crest of the roll-over anticline in the coolest scenario to the area underneath the center of the normal fault ramp in the warmest scenario;
5) increasing initial heat flow results in a shift in the lower crustal uplift maximum toward being closer to the normal fault;
6) increasing initial heat flow correlates with a narrowing of the subsiding area in the immediate hanging wall, that is, basin narrowing;
7) increasing initial heat flow correlates with a decreasing subsidence in the lower crust below the roll-over anticline crest;
8) increasing initial heat flow correlates with the increasing of the footwall tilt (see Figure 11.7);
9) increasing initial heat flow correlates with a decrease of the no-subsidence area, which is located to the right of the roll-over anticline (away from the controlling normal fault), from both sides, and which is replaced by subsidence also from both sides;
10) increasing initial heat flow correlates with an increase of the lower crustal area of subsidence, which is located between the crest of the roll-over anticline and the adjacent model boundary; and
11) scenarios having an initial heat flow of 80 mWm^{-2} combined with all three modeled extension rates reach unstable flow, that is, rapid deformation and excessive model element distortion, which cannot be handled by the numerical approach.

Only in two thinner crustal scenarios can we observe:

1) increasing extension rate correlating with the area of subsidence at lower crustal levels located underneath the roll-over anticline getting larger (and, perhaps, shifting a bit basinward from its location in the thinnest crust scenario); and
2) increasing extension rate correlating with increasing amplitudes of the lower crustal folds forming a train of folds in a footwall, which is located underneath a laterally retreating hanging wall,

while only in the thickest crust scenarios can we see that the increasing extension rate correlates with subsidence at lower crustal levels underneath the roll-over anticline that is getting progressively larger.

Only in the thicker crust scenarios (starting with a thickness of 40 km) can we observe an increasing initial heat flow correlating with the progressive development of the flowable lower crust that develops a train of ductile anticlines. This train of anticlines starts with a large anticline located underneath the steep portion of the boundary fault. It follows with a second anticline located to the right of the roll-over anticline, which has its amplitude smaller than that of the first anticline. Only in the thickest crust scenario can we clearly observe how a warm initial heat flow brings the brittle–ductile transition zone into the shallower depth levels. The brittle–ductile transition zone in the thickest scenario is interpreted as lying above the base of the upper crustal area of the prescribed model. This fact is indicated by a significant mesh distortion at depth levels, which were originally prescribed to be brittle, and which means that they are undergoing ductile flow as the core complex starts to develop in the area unroofed by the lateral movement of the hanging wall.

It is interesting to compare the models, which indicate that an increasing initial heat flow correlates with a widening of the subsiding area located on the footwall surface behind the controlling normal fault, despite the fact that one cannot see it in the model outputs made for the group of the thickest crust scenarios where the apparent lack is most likely caused by the disadvantageous automatic choice of contoured colors, with the case study of the Wasatch Fault footwall done by Armstrong and colleagues (2003) (see Figure 11.7). Based on their detailed timing and tilt determinations done for several fault blocks in the footwall of the Wasatch normal fault, we can see that the tilt value and its narrower localization adjacent to the main controlling fault develop with rift-activity time in the area, that is, with increasing thermal regime, which is in accordance with our models.

The most distinct trends in the described set of models are captured by Figures 12.10 and 12.11, which show the stress vector and total horizontal displacement outputs. Figure 12.11, which focuses on the extension rate's effect on the model, demonstrates a clear increase in both the core complex growth tendency and core complex-related uplift with extension rate. Figure 12.10, which focuses on the initial heat flow's effect on the model, shows a significant increase of both the core complex growth tendency and the uplift tendency with increasing initial heat flow. It also documents a change from a broadly distributed core complex uplift in scenarios with low initial heat flow to a localized core complex uplift in scenarios with high initial heat flow. This demonstrates that the extra heat localizes the core complex uplift.

To summarize the structural discussion, the most important observations from models focused on the structural development of simple half-grabens include:

1) The initial heat flow and crustal thickness are equally important factors in driving the model toward the unstable solution (i.e., excessive core complex deformation).
2) The initial heat flow is a more important factor than the extension rate in model control.
3) The crustal thickness is a more important factor than the extension rate in model control.
4) The unstable mathematical solution occurs when the core complex develops too quickly, which is characterized by a fast inflow of the lower crust from its flanks to its apex. This function agrees with the observed trend of uplift alignment between the lower crust and mantle depth levels with an increase in both crustal thickness and initial heat flow, also including an increase in the respective uplift values.

Models of thermal behavior trends

The strategy of interpreting the role of factors in thermal regimes was similar to the strategy applied to the interpretation of observations made on the structural development and architecture. Behavior trends have been possible to identify when scenarios have been grouped according to identical input extension rate, initial heat flow, and crustal thickness.

A grouping done according to identical extension rate allows us to see the interaction of initial heat flow with crustal thickness. All three groups characterized by an extension rate of 10, 30, and 50 mmy^{-1} are characterized by the scenario with a heat flow of 80 mWm^{-2} combined with a crustal thickness of 50 km as being unstable, that is, having no mathematical solution because of the too large total deformation (see Table 12.1).

All three extension rate groups indicate that (Figure 12.12):

1) crustal thickness increase correlates with a reduction of the thermal anomaly values in the mantle;
2) crustal thickness increase correlates with a reduction of the thermal anomaly values in the lower crust;
3) crustal thickness increase correlates with an increase in the thermal regime of the shallow footwall;

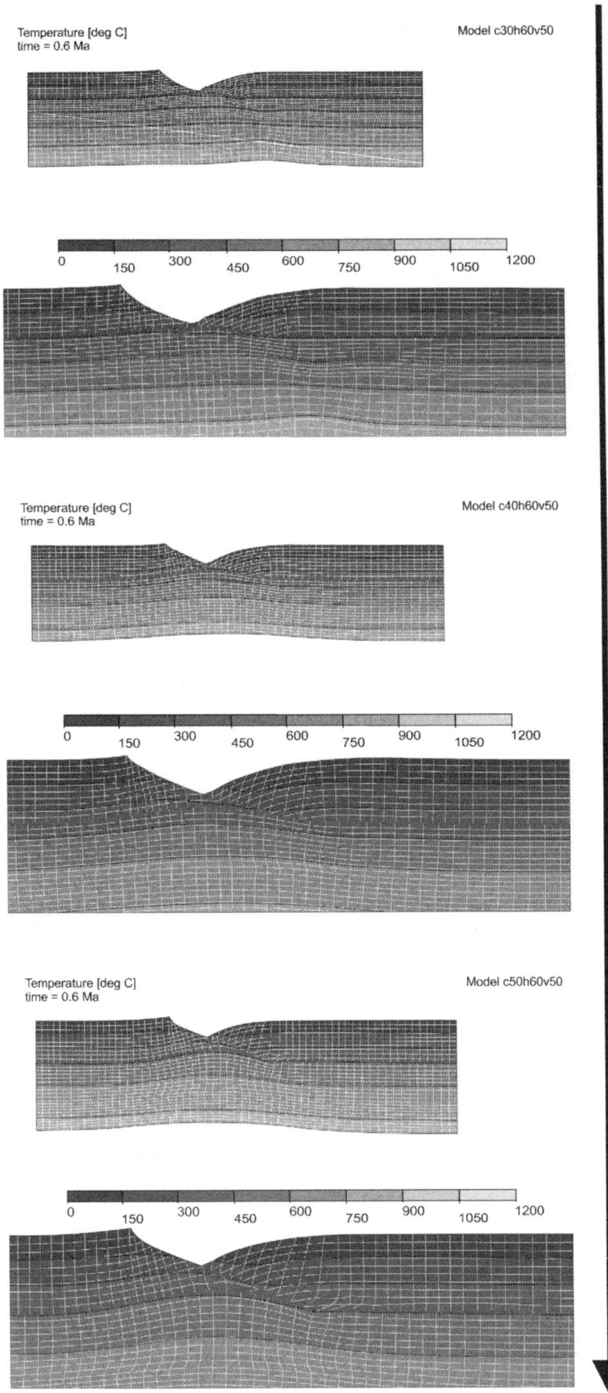

Figure 12.12. A series of models c30h60v50 – c40h60v50 – c50h60v50 showing the thermal behavior of scenarios characterized by an intermediate initial heat flow of 60 mWm^{-2} and fastest extension rate of 50 mmy^{-1}, and an increase in crustal thickness from 30, through 40, to 50 km (modified from Henk, 2008, pers. comm.). Modeling results are shown after 0.6 Ma of deformation in temperature isolines in ° Celsius. Note the progressive cooling of the extensional basin, mantle, and lower crust with an increase in crustal thickness.

4) crustal thickness increase correlates with an increase in the thermal regime of the shallow hanging wall;
5) crustal thickness increase correlates with a decrease in the thermal regime of the extensional basin;
6) crustal thickness increase correlates with a thermal anomaly in mantle traveling from underneath the roll-over anticline to underneath the very tip of the wedging-out hanging wall; and
7) crustal thickness increase correlates with a thermal anomaly in the lower crust traveling from underneath the wedging-out portion of the hanging wall to the area located a bit toward the normal fault ramp from there

and that (Figure 12.13):

1) increase in initial heat flow correlates with an increase of the thermal anomaly values in mantle;
2) increase in initial heat flow correlates with an increase of the thermal anomaly values in lower crust;
3) increase in initial heat flow correlates with an increase in the thermal regime of the shallow footwall;
4) increase in initial heat flow correlates with an increase in the thermal regime of the shallow hanging wall;
5) increase in initial heat flow correlates with an increase in the thermal regime of the extensional basin;
6) initial heat flow increase correlates with a thermal anomaly in the lower crust traveling from underneath the thinned wedge of hanging wall to underneath its very tip; and
7) initial heat flow increase correlates with a thermal anomaly in the mantle traveling from underneath the roll-over anticline toward underneath the thinner portion of the hanging wall.

A further observation we can make for a group of scenarios with a fastest extension rate of 50 mmy^{-1} is that an increase in both the crustal thickness and heat flow tends to change the system of unaligned mantle and lower crustal thermal anomalies to the system having them located above each other.

Grouping according to identical initial heat flow allows us to see the interaction of crustal thickness with extension rate. Only the group characterized by an initial heat flow of 80 mWm^{-2} is characterized by three scenarios with a crustal thickness of 50 km combined with extension rates of 10, 30 and 50 mmy^{-1} as being

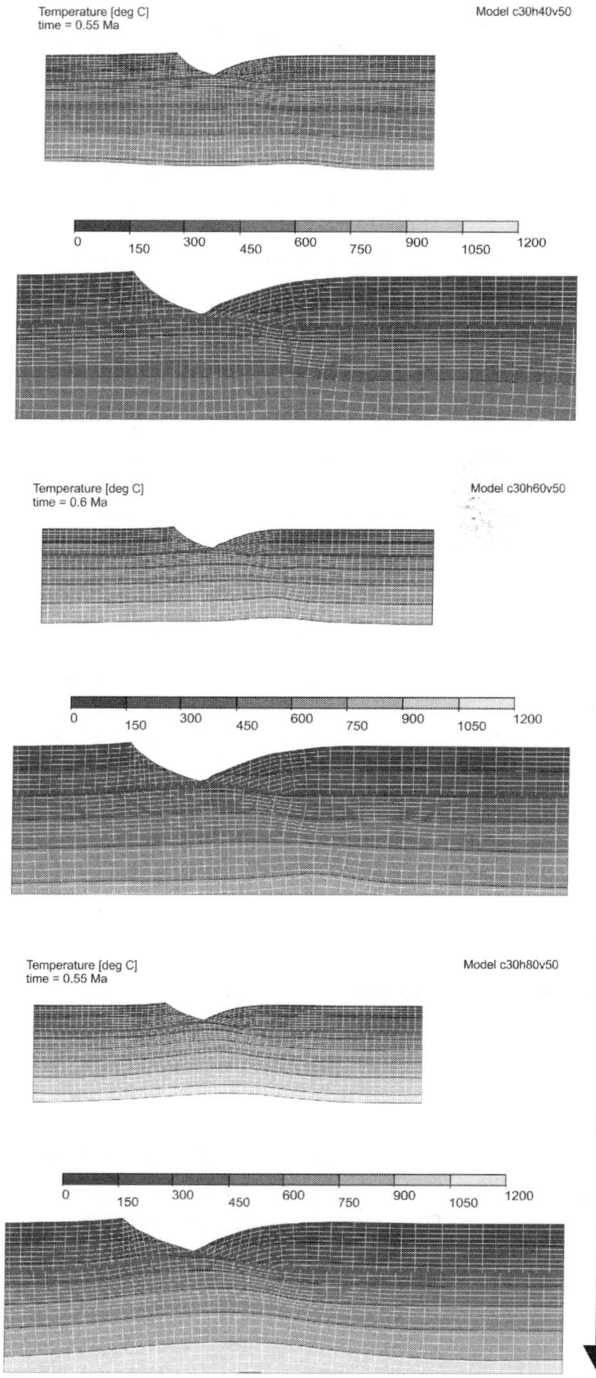

Figure 12.13. A series of models c30h40v50 – c30h60v50 – c30h80v50 showing the thermal behavior of scenarios characterized by a thinnest crust of 30 km and fastest extension rate of 50 mmy^{-1}, and an increase in initial heat flow from 40, through 60, to 80 mWm^{-2} (modified from Henk, 2008, pers. comm.). Modeling results are shown after 0.6 Ma (or after 0.55 for computational reasons) of deformation in temperature isolines in ° Celsius. Note the progressive warming of the extensional basin, mantle and lower crust with an increase in initial heat flow.

unstable, that is, having rapid deformation and excessive model element distortion, which cannot be handled by the chosen numerical approach (see Table 12.1).

All three initial heat flow groups indicate that:

1) crustal thickness increase correlates with a reduction of the thermal anomaly values in mantle;
2) crustal thickness increase correlates with a reduction of the thermal anomaly values in lower crust;
3) crustal thickness increase correlates with an increase in the thermal regime of the shallow footwall;
4) crustal thickness increase correlates with an increase in the thermal regime of the shallow hanging wall;
5) crustal thickness increase correlates with a decrease in the thermal regime of the extensional basin;
6) crustal thickness increase correlates with a thermal anomaly in the mantle traveling from underneath the roll-over anticline to underneath the very tip of the wedging-out hanging wall; and
7) crustal thickness increase correlates with a thermal anomaly in the lower crust traveling from underneath the wedging-out portion of the hanging wall to the area located a bit toward the normal fault ramp from there

and that:

1) increase in extension rate correlates with an increase in the thermal anomaly values in mantle;
2) increase in extension rate correlates with an increase in the thermal anomaly values in lower crust;
3) increase in extension rate correlates with an increase in the thermal regime of the extensional basin; and
4) increase in extension rate does not have any clearly observable effect on the thermal regimes of both the shallow footwall and the hanging wall.

Grouping according to identical crustal thickness allows us to see the interaction of extension rate with initial heat flow. Only the group characterized by a crustal thickness of 50 km is characterized by three scenarios with an initial heat flow of 80 mWm^{-2} combined with extension rates of 10, 30, and 50 mmy^{-1} as being unstable, that is, having rapid deformation and excessive model element distortion, which cannot be handled by the numerical approach (see Table 12.1).

All three crustal thickness groups indicate that (Figure 12.14):

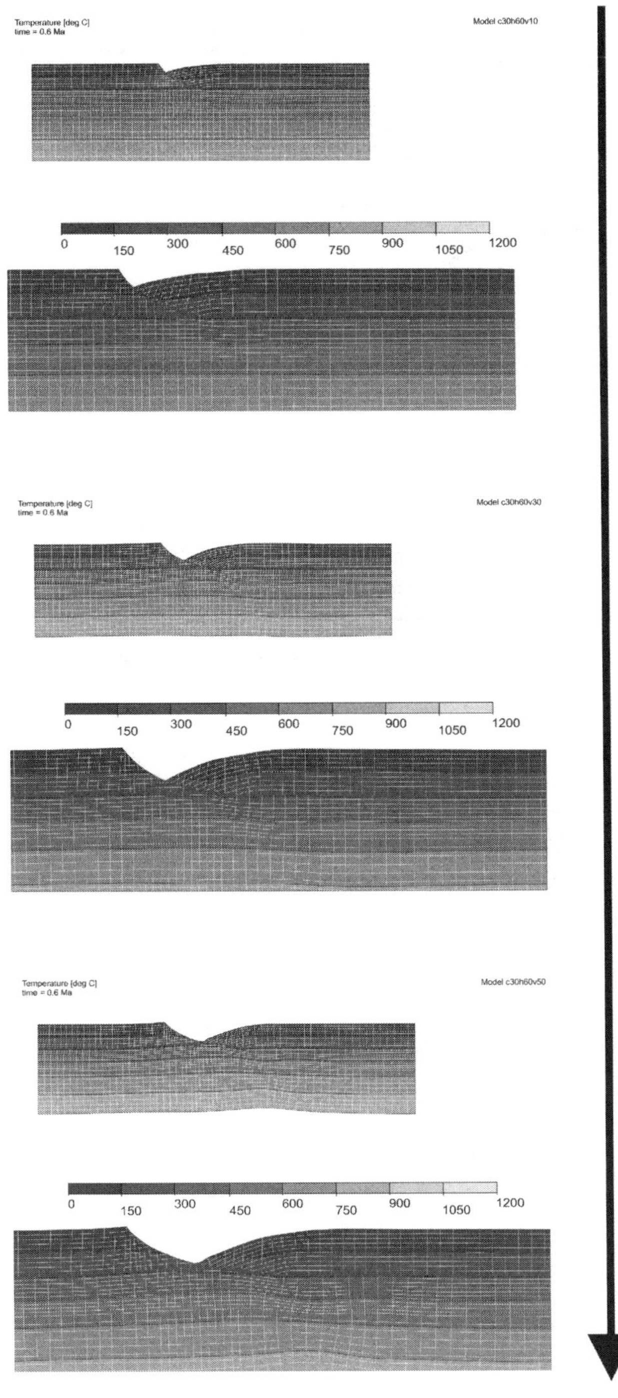

Figure 12.14. A series of models c30h60v10 – c30h60v30 – c30h60v50 showing the thermal behavior of scenarios characterized by a thinnest crust of 30 km and an intermediate initial heat flow of 60 mWm^{-2}, and an increase in extension rate from 10, through 30, to 50 mmy^{-1} (modified from Henk, 2008, pers. comm.). Modeling results are shown after 0.6 Ma of deformation in temperature isolines in ° Celsius. Note the progressive warming of the extensional basin, mantle, and lower crust with an increase in extension rate.

1) an increase in extension rate correlates with an increase in the thermal anomaly values in mantle;
2) an increase in extension rate correlates with an increase in the thermal anomaly values in lower crust; and
3) an increase in extension rate correlates with an increase in the thermal regime of the extensional basin

and that:

1) an increase in initial heat flow correlates with an increase of the thermal anomaly values in mantle;
2) an increase in initial heat flow correlates with an increase of the thermal anomaly values in lower crust;
3) an increase in initial heat flow correlates with an increase in the thermal regime of the shallow footwall;
4) an increase in initial heat flow correlates with an increase in the thermal regime of the shallow hanging wall;
5) an increase in initial heat flow correlates with an increase in the thermal regime of the extensional basin;
6) initial heat flow increase correlates with a thermal anomaly in the lower crust traveling from underneath the thinned wedge of the hanging wall to underneath its very tip; and
7) initial heat flow increase correlates with a thermal anomaly in the mantle traveling from underneath the roll-over anticline toward underneath the thinner portion of the hanging wall.

To summarize the thermal discussion, the most important observations from models focused on the thermal regime of simple half-grabens include:

1) Initial heat flow and crustal thickness are more important factors than extension rate in controlling the thermal regime of extensional setting.
2) They have the opposite effect in controlling associated mantle and lower crust thermal anomalies and thermal regime of the basin. An increase in heat flow makes them warmer while an increase in crustal thickness makes them cooler.
3) An increase in both crustal thickness and initial heat flow drives the unaligned lower crust and mantle thermal anomalies toward being located above each other.
4) The effect of the extension rate on the thermal regime is at a faster rate being able to "cut through" the initial isotherms, which do not have too much time to equilibrate, and at a slow rate being able to almost mimic the crustal architecture by the geometry of isotherms, which have almost enough time to equilibrate.

13 The role of fluid flow in thermal regimes of rifts and passive margins

Introduction

An important source of information about the role of fluid flow in thermal regimes of rifts and passive margins comes from hundreds of power plants worldwide that are developing electricity from present-day geothermal systems. Summarizing the information about their characteristics allows us to conclude that the geothermal fields comprise:

1) high-temperature fields, having temperatures higher than 225°C;
2) medium-temperature fields, with temperatures between 125°C and 225°C; and
3) low-temperature fields, characterized by temperatures below 125°C.

Researchers have conducted only a few studies on which exact factors control the existence of the geothermal system strongly enough to control an important positive thermal anomaly, based on the exploration and production data gathered by the geothermal industry. So far, geothermal systems have been categorized on the basis of geochemical criteria (Hochstein and Browne, 2000) or heat source criteria combined with somewhat limited reservoir criteria (Brophy, 2003). For our understanding of factors capable of controlling an important thermal anomaly we need a systematic way of describing the existing geothermal fields. Such a systematic description should help with understanding geothermal systems with respect to their size, evolution, characteristics, and life span.

This chapter tries to identify a systematic way based on a combination of elements necessary for the existence of the important geothermal field, in a way similar to the hydrocarbon system concept. Following this approach, our geothermal system elements include:

1) heat source;
2) fluid source;
3) reservoir;
4) cap rock; and
5) migration.

The following text discusses the interaction of system elements in control of the important geothermal system. Subsequently, the most important fluid flow mechanisms involved in important systems are discussed and documented by published case studies.

Geothermal system elements

Heat source

The heat sources of commercial geothermal systems in extensional settings include elevated temperature settings due to magmatism and/or shallow mantle–crust boundaries such as:

1) oceanic spreading centers (Icelandic geothermal systems such as Krafla, Reykjanes, and Svartsengi, and the southernmost Salton Sea–Baja California geothermal fields such as Cerro Prieto, Mesa de Andrade, and Mesa de San Luis);
2) rift systems (geothermal systems in Djibouti, Ethiopia, Kenya, Tanzania, and the Basin and Range province in the United States);
3) transtensional settings (Coso geothermal field in California, the northern Salton Sea–Baja California fields such as Brawley, Dunes, East Mesa, Glamis, Heber, Salton Sea, Tulecheck, and Westmorland); and
4) hot spots (geothermal systems in Hawaii and Yellowstone, United States).

Settings 1–4 can contain either the deep and large magmatic heat sources (e.g., The Geysers in California and Larderello in Italy) or the shallower and smaller volcanic system–associated heat sources (Dallol in Ethiopia, Eburru and Olkaria in Kenya, and Karaha–Telaga Bodas in Java), but settings 2 and 3 do not necessarily include a magmatic heat source. Settings 2 and 3 can reach increased thermal regimes due to the thinned crust, where a larger proportion of mantle heat makes it to the surface. The Basin and Range province provides examples of both magma-dependent and magma-independent cases. The eastern and western margins of the province contain geothermal systems heated by magmatic bodies, such as those at the Long Valley Caldera, Medicine Lake, Roosevelt Hot Springs, and Steamboat Hot Springs. The magma composition apparently does not play a role in controlling the differences among these geothermal systems. These differences are caused primarily by differences in their hydrologic and tectonic regimes. Several geothermal systems within the interior of the Basin and Range province appear to be heated by warmer rocks overlying a shallow crust–mantle boundary, such as Beowawe, Soda Lake, and Stillwater, representing magma-independent systems.

The presence of magmatism in rift settings depends on their development stage, as they evolve from continental rift basins such as Lake Tanganyika in the East African Rift system, Tanzania, where the continental crust begins to thin, to rifts with developed volcanic systems such as the Turkana Depression in the East African Rift system in Kenya.

A quick look at the characteristics of geothermal systems worldwide reveals that high- and medium-temperature systems form 49 percent and 21 percent, respectively. High-temperature areas are known from divergent plate settings that are active long enough for magmatism to develop, but they can also occur in settings that are not active long enough for magmatism to develop (see Hochstein and Browne, 2000). Medium-temperature systems seem to overlap only with rift settings that did not yet develop magmatism.

Fluid source

The fluid sources of explored and commercially used geothermal systems worldwide include:

1) magmatic fluids (geothermal systems such as Long Valley in California, Krafla in Iceland, the Phlegrean Fields in Italy, Karaha–Telaga Bodas, Kawah Ijen, and Tangkubanprahu in Java, Tulecheck and the Salton Sea in the United States, and Sibayak and Sorik Marapi in Sumatra);
2) meteoric fluids (roughly 88 percent of the geothermal systems with known fluid sources);
3) saltwater (geothermal systems such as Lake Assal in Djibouti or Reykjanes and Svartsengi in Iceland);
4) connate fluids from compacted sediments, old glacial waters, or metamorphic fluids (The Geysers and Clear Lake in California, and probably Roosevelt Hot Springs in the Basin and Range province);
5) highly saline fluids produced by chemical processes (geothermal systems such as Cesano in Italy, Lake Assal in Djibouti, Dallol in Ethiopia, and some of the Salton Sea–Baja California systems).

Note that the term *fluid* is used here in a loose sense because it covers both fluid and vapor. To our knowledge, fluid sources such as:

a) fluids from clay dehydration;
b) fluids provided by gypsum to anhydrite transformation; and
c) fluids provided by mineral diagenesis without dehydration

have never been recognized in explored and produced geothermal fields, which places them among only theoretically possible contributing fluid sources. They are either masked by dominant (1) and (2) type fluid sources or do not contribute to any existing geothermal field at all.

Among fluid sources of type (1)–(5), magmatic and meteoric ones are clearly prevalent (making roughly 4.5 percent and 62.5 percent of the geothermal systems with known fluid sources).

Magmatic fluids accumulate inside the magma column by a combination of diffusion-driven fractionation (McBirney, 1995 and references therein), compaction-driven fractionation (Sparks et al., 1985; Shirley, 1986, 1987; Ortoleva et al., 1987), and convection-driven fractionation (Tait et al., 1984; Morse, 1986; Tait and Kerr, 1987; Shinohara et al., 1995). Fluids released from magma can have a volume much larger than the volume of the parent intrusion, as indicated by mass balance calculations done for ore deposits (e.g., Keith and Shanks, 1988).

Magmatic sources can be further subdivided into magmatic sensu stricto, phreatic, and fumarolic, the latter two frequently found but rarely involved in producing geothermal fields. Worldwide data on fluid sources of commercial geothermal systems indicate that a magmatic fluid source is characteristic of geothermal systems located in tectonically active settings, which include a long-lasting magmatism associated with rift zones that are transitional to areas of seafloor spreading and developed seafloor-spreading zones.

Meteoric waters enter a fluid system at recharge areas. The recharge rate controls the water volume. The waters flow downward, acquiring heat and dissolved salts. The downward flow eventually encounters a permeable aquifer. Within the aquifer, the fluid moves laterally and up toward the discharge area, where it flows out. Typical distances between known recharge and discharge areas fall into an interval from a few to 70 km. The travel time for the fluid flow starting at the recharge area and finishing at the discharge area usually spans between 1,000 and 100,000 years (De Marsily, 1981; Smith and Chapman, 1983). Maximum flow rates of 1–10 meters per year can develop in aquifers, while much slower seepage occurs in aquitards (Garven and Freeze, 1984a, 1984b).

Rifting on the verge of seafloor spreading involves one more specialty: the possibility of saltwater recharge into the geothermal system, as documented by the Lake Assal system in Djibouti. A saltwater infiltration is quite common in the seafloor spreading regimes, as documented at the Reykjanes and Svartsengi geothermal fields in Iceland.

A type (4) fluid source was studied in the Salton Sea geothermal field (Muffler and White, 1968). Here, the prograde metamorphic changes with increasing temperature and the interaction between the hot brines and the Colorado Delta sediments result in a series of reactions, releasing abundant carbon dioxide. At temperatures of 150° to 200°C, dolomite reacts with kaolinite to produce chlorite, calcite, and carbon dioxide. In the zones of the most intense alteration at temperatures of 300° to 320°C, carbon dioxide is liberated by the breakdown of calcite and the complementary formation of epidote.

Numerical modeling has recognized connate waters released from the compacting sediments as an important fluid source in rifts with fine-grained sediments (e.g., Bethke, 1985). The example from the Pechelbronn area in the central Rhine Graben shows that compaction-driven flow acted on top of more dominant density-driven flow when the entire area was undergoing marine transgression and before the rift zone became subaerially exposed and gained a complex relief characterized by pronounced lows and highs (Person and Garven, 1992).

The example from the Rio Grande Rift indicates that compaction-driven fluid flow can combine with topography-driven fluid flow in rifts where alluvial fans deposited in a narrow zone along the graben-controlling faults change into thick playa sediments over a short distance toward the graben center (Mailloux et al., 1999).

Another scenario for a more significant compaction-driven flow is the thermal subsidence stage of the failed rift zone, dominated by fine-grained lacustrine onlap facies (Person and Garven, 1994). The role of compaction in sourcing and driving the fluids progressively wanes as the thermal anomaly within the lithosphere relaxes.

Metamorphic fluid contributions have been recognized at The Geysers, California (Goff, 1980). This indicates that a metamorphic source could add to fluid budgets in some geothermal fields but has not yet been widely recognized. The same reasoning can apply to fluid sources (a)–(c) mentioned earlier.

Highly saline brines seem to form a rather special case among fluid sources. The information on explored and produced commercial geothermal systems worldwide indicates that all such geothermal systems known to us are associated with meteoric water encountering evaporite deposits along its descent from the recharge area. Their resultant salinity is much greater than the salinity commonly found in meteoric fluid systems and, thus, more effective in transferring heat. In fact, all these areas, whether they are from the East African Rift system or from the transtensional Salton Sea–Baja California area, host high-temperature geothermal systems. Also most of their evaporite deposits have accumulated during quiescent periods in rift and pull-apart basins.

Reservoir

Reservoirs of explored and producing commercial geothermal fields include two main categories: matrix and fractured reservoirs. Matrix reservoirs include any porous rocks of sedimentary and volcanic origin that have a large enough matrix permeability to host geothermal fluids, which is 10^{-14} m^2 and higher. Permeability ranges for typical rocks present in geothermal areas are (Deming, 1994a, 1994b; Smith and Wiltschko, 1996):

1) 10^{-12} to 10^{-22} m^2 for carbonates;
2) 10^{-12} to 10^{-17} m^2 for sandstones;
3) 10^{-15} to 10^{-20} m^2 for siltstones;
4) 10^{-16} to 10^{-23} m^2 for shales and mudstones; and
5) 10^{-20} m^2 and lower for igneous and metamorphic rocks.

Note that anomalously high permeability, for example, the permeability of karst limestone, can reach a value of 10^{-9} m^2 (Deming, 1994a, 1994b). The list of permeability ranges of common rocks indicates that upper permeability categories of carbonate and sandstone reservoirs are large enough to host geothermal fluids. The same implication applies to tuffs and tuffites, known to serve as reservoirs in the Karaha–Telaga Bodas geothermal prospect (Christensen et al., 2002). Carbonates and sandstones do form reservoirs in known geothermal systems (carbonates and quartzite in Larderello and sandstones in the Salton Sea), although, because of mineral deposition and the development of fractures in them, it is difficult to separate their matrix and fracture permeabilities.

The fracture permeabilities cover a spectrum of 10^{-7}–10^{-19} m^2, representing fracture apertures on the order of 1–1,000 μm (Caine, 1999). Fractures with permeabilities at least equal to 10^{-13}–10^{-14} form the fluid-conductive fractures (e.g., Deming et al., 1990a; Mailloux et al., 1999; Lampe and Person, 2002; Schardt et al., 2006; Yang et al., 2006). Note, however, that matrix or fracture permeability is one of the most difficult parameters to determine for a geothermal reservoir because they are both volume-dependent quantities. Furthermore, Table 8.3 documents that permeability is also a time-sensitive quantity.

A matrix permeability usually pales in comparison with fracture permeability (e.g., Bachler et al., 2003; Belani et al., 2004). Whenever the fracture permeability exceeds the matrix permeability by three orders of magnitude, it captures the fluid flow almost entirely (e.g., Moore et al., 1990; Caine et al., 1996; Davidson et al., 1998; Caine, 1999). Therefore, fractures are much more effective for convective heat transfer than matrix permeability. Examples of tight elliptical surface projections of the heat flow anomalies located along fault–fracture zones come from:

1) Hvalfjordur in western Iceland (Flovenz and Saemundsson, 1993);
2) Solvholt in southern Iceland (Flovenz and Saemundsson, 1993);
3) Opal Mound and several smaller faults in the Roosevelt Hot Springs geothermal field, the eastern margin of the Basin and Range province (Wilson, 1980);
4) several faults bounding the central Rhine Graben (Illies, 1981); and
5) the Socorro Springs area in the Rio Grande Rift (Mailloux et al., 1999; Figure 13.1).

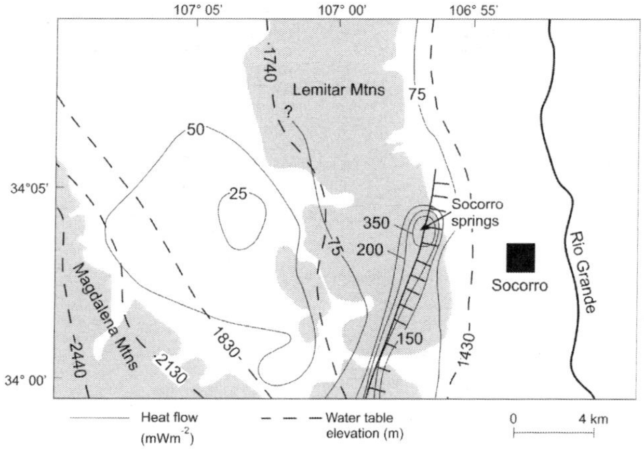

Figure 13.1. Hydraulic head in meters (dashed line) and surface conductive heat flow in mWm^{-2} (solid line) across the Socorro and La Jencia sub-basins (Anderholm, 1983, 1987; Gross and Wilcox, 1983; Barroll and Reiter, 1990; Mailloux et al., 1999). The shaded area represents the Lemitar and Magdalena Mountains. Conductive surface heat flow values decrease across the western La Jencia sub-basin, increase in the Lemitar Mountains, and dramatically increase near Socorro Springs before suddenly dropping off in the Socorro sub-basin. The hydraulic head exponentially decays in the Magdalena Mountains, is flat across the La Jencia sub-basin, and abruptly drops off in the western Socorro sub-basin.

In high-temperature systems, the matrix permeabilities will be reduced by mineral deposition more effectively than those in low-temperature systems. This phenomenon further enhances the contrast between the matrix and fracture permeability in high-temperature systems. Furthermore, the low-permeability lava flows are more prone to fracturing than highly permeable tuffs, which further enhances the difference between matrix and fracture permeabilities, as can be observed in the Medicine Lake geothermal prospect in California (Clausen et al., 2006).

Research at Karaha–Telaga Bodas (Christensen et al., 2002) indicates that combinations of both types of permeability are possible in high-temperature geothermal systems. Here, the matrix-controlled tuff and tuffite reservoir horizons coexist with fracture-controlled andesite flow and pyroclastic reservoirs.

Fracture systems in various geothermal systems are extensive enough to store large fluid volumes between the area of fluid recharge and the area of fluid discharge. Apart from exceptions such as those in the Brady, San Emidio Desert, Soda Lake, and Stillwater geothermal systems in the Basin and Range province, which appear to consist of a single fault zone, the fractured reservoirs are characteristic of geothermal fields occurring in tectonic settings, which can provide both the extensive fracture volumes and complexity such as:

1) relay ramps between extensional faults and linkage zones of extensional faults (Desert Peak East, Desert Peak, and Dixie Valley geothermal systems in the Basin and Range province);
2) extensive synthetic and antithetic fault systems at major extensional faults (San Emidio Desert and Stillwater geothermal systems in the Basin and Range province);
3) pull-aparts in strike-slip settings (geothermal systems such as Coso in California; Beowawe in the Basin and Range province; East Mesa, Heber, Mesa de San Luis, Tulecheck, and Westmorland in the Salton Sea–Baja California area);
4) contractional bridges among strike-slip faults (The Geysers geothermal field in California);

Table 13.1. Hydrogeological and thermal properties of units defined in cross-section in Figure 13.2 (Lampe and Person, 2002). Φ is the porosity, k is the permeability, λ_s is the thermal conductivity of the rock, λ_f is the thermal conductivity of the pore fluid, λ_{pm} is the thermal conductivity of the porous medium, and k_x/k_z is the anisotropy ration for each rock unit. Fault properties are listed under number 8.

Unit	Stratigraphy	Age	Lithology	Hydrostratigraphy	Φ (%)	k(m^2)	λ_s (W/m°C)	λ_f (W/m°C)	λ_{pm} (W/m°C)	k_x/k_z
1	Permian	Permian	clastics, pyroclastics, volcanics	Aquifer	15	10^{-13}	2.1	0.58	1.73	100:1
2	Pechelbronn beds	Priabonian	marl, calcareous shale, sand	Aquitard	20	10^{-18}	2.1	0.58	1.62	100:1
3	Septaria shale	Rupelian	shale	Aquitard	25	10^{-18}	2.1	0.58	1.52	100:1
4	Cyrenia marl, Meletta beds	Rupelian	calcareous shale	Aquitard	25	10^{-16}	2.1	0.58	1.52	100:1
4	Cerithia beds	Chattian	calcareous shale	Aquitard	25	10^{-16}	2.1	0.58	1.52	100:1
4	Upper Hydrobia beds	Aquitanian	calcareous shale	Aquitard	25	10^{-16}	2.1	0.58	1.52	100:1
5	Corbicula Beds	Chattian	evaporites	Aquitard	1	10^{-20}	2.1	0.58	2.07	100:1
6	Colored Niederoderm beds	Rupelian	marl	Aquitard	25	10^{-18}	2.1	0.58	1.52	100:1
6	Lower Hydrobia beds	Aquitanian	marl	Aquitard	25	10^{-18}	2.1	0.58	1.52	100:1
7	Middle Miocene–Quaternary	Middle Miocene–Quaternary	coarse sandstone	Aquifer	35	10^{-12}	2.1	0.58	1.34	10:1
8	fault		fault core, damage zone and debris appron	Aquifer	30		2.1	0.58	1.43	1:10

5) complex extensional or strike-slip systems (geothermal systems such as Karaha–Telaga Bodas; Bulalo, Tiwi, and Tongonan in the Philippines; and Brawley, Cerro Prieto, Dunes, Glamis, Mesa de Andrade, and the Salton Sea in the Salton Sea–Baja California area);
6) junctions of complex fault-fracture systems (Rye Patch and Steamboat in the Basin and Range province);
7) gravitational collapse–driven fault systems associated with calderas, gravity glides, orogenic-collapse areas (e.g., caldera-based geothermal systems such as Medicine Lake and the Long Valley Caldera in the Basin and Range province, although the former is not yet firmly proven, Valles Caldera in New Mexico, Los Humeros in Mexico, Berlín in El Salvador, Yellowstone in Wyoming, Newberry in Oregon, gravity glide-hosted geothermal systems such as Cove Fort, which is hosted by sediments below the glide body, and part of the Dixie Valley system in the Basin and Range province); and
8) areas affected by combinations of denudation-driven fractures and tensile fractures in submerged and emerged faulted blocks (the Roosevelt Hot Springs in the Basin and Range province).

Note that each such fault-fracture system has to have a large fracture volume and suitable geometry to allow for fluid inflow, upflow, and outflow. If it does not, its geothermal system is represented by a hot dry rock system.

Matrix reservoirs and fractured reservoirs can form various combinations, as indicated by the research at the Karaha–Telaga Bodas geothermal prospect, the Imperial Valley–Heber and East Mesa fields in the Salton Sea–Baja California area, and the Medicine Lake prospect, which is external to the western boundary of the Basin and Range province. Permeabilities of both types of reservoirs are totally stress-dependent (e.g., Carlsson and Olsson, 1979; Fisher et al., 1996) and fluid pressure-dependent (e.g., Knipe, 1993; Fisher et al., 1996), but the permeability relationships to effective stress for the host rock and fracture system are quite different. Unlike the host rock permeability, the fracture permeability can collapse abruptly when certain stress conditions are met (e.g., Knipe, 1993). The fracture permeability changes with effective stress within a relatively narrow range of permeability values (Sigal, 1998, pers. comm.), from fractures with fully opened apertures to collapsed fractures with residual apertures. This narrow permeability range can be used as a tool for distinguishing fracture and rock permeabilities, the latter changing with effective stress within a relatively broad interval of permeability values (Sigal, 1998, pers. comm.).

Research at the Karaha–Telaga Bodas indicates that fractured reservoirs are controlled by the relative timing of fracturing versus cementation and that the size of faults can make up for geometries preventing them from being dilational (Christensen et al., 2002). A combination of cementation and deformation processes related to fracturing controls the interaction of host rock and fault permeabilities and the resulting fluid flow pattern. The timing of different stages of the faulting cycle and the type of lithologies affected by these individual stages controls both the pore fluid flow and the pressure regime. Only when the right combination of fault event magnitude, fracture geometry, and fracture linking allows for the:

1) creation of a pathway between the fluid source and sink volumes; and
2) production and preservation of permeable fault zone fabrics,

does the extensive fluid flow between the host rocks and fault result. A fluid flow also depends on the location and size of fluid sources and sinks, where fluids enter and escape from the fault zone.

In places where matrix permeabilities are low, the fracture system may not be extensive enough to sustain a reservoir. The Long Valley Caldera at the western edge of the Basin and Range province may provide us with such an example. The Mammoth Mountain area of the Long Valley Caldera contains numerous tree-kill areas, steam vents, and springs with a significant amount of dissolved magmatic carbon (Sorey et al., 1999). Most of these features are located near normal faults. Here, maxima of helium isotope ratios in gas from steam vents relative to its ratio in air and carbon dioxide flux maxima can be correlated with seismic activity in the area, thus documenting a transient fluid flow along active normal faults (McGee and Gerlach, 1998; Priest et al., 1998; Sorey et al., 1999). A direct correlation of fluid flux and seismic activity, however, also implies the lack of a reservoir that would store carbon dioxide as it comes from its source. This implies that fault zones act as migration pathways and do not act as a reservoir for carbon dioxide.

A permeability of the reservoir controls the type of reservoir fluid in the case of a meteoric fluid source (Hochstein and Browne, 2000). The fluid type is controlled by the relationship between the fluid flow out of the reservoir and the fluid flow into the reservoir, that is, ratio k:

$$k = \text{permeability}_{\text{recharge}}/\text{permeability}_{\text{reservoir}} \quad (13\text{-}1)$$

If the reservoir permeability is great and permeability in the recharge area is intermediate, the reservoir hosts a liquid-dominated system. Geothermal systems such as Beowawe, Soda Lake, and Stillwater in the Basin and Range province and Cisolok and Cisukarame in Java provide examples of such liquid-dominated systems. The Manikaran geothermal system in India is an example of a high recharge rate controlling a liquid-dominated reservoir. This scenario is frequently controlled by topography-driven fluid flow, described in Chapter 8, which tends to have a sufficient meteoric water recharge.

If the permeabilities of both the reservoir and recharge area are intermediate, the reservoir contains a liquid-vapor system. Examples of liquid-vapor systems fed by meteoric fluid sources come from the Eburru, Namarumu, and Olkaria geothermal systems in Kenya, and the Aluto system in Ethiopia (Hochstein and Browne, 2000). This scenario is frequently controlled by topography-driven fluid flow, described in Chapter 8, which tends to have sufficient meteoric water recharge.

If the reservoir permeability is great and permeability in the recharge area is small, the reservoir hosts a vapor-dominated system. Geothermal areas such as The Geysers in California, Larderello, and Darajat and Kawah Kamojang in Java provide examples of such vapor-dominated systems fed by meteoric fluid sources (Hochstein and Browne, 2000). This scenario can be associated with the buoyancy-driven fluid flow described in Chapter 8, where free convection cell boils the remaining fluid trying to reach total dryness.

Apart from the primary rock permeability, the reservoir permeability is controlled by the reservoir rock rheology that influences how the hydrothermal alteration enhances/represses the permeability and how faults/fractures are formed and distributed. Most volcanic rocks and sediments can develop considerable secondary permeability, whereas dense intrusive rocks, metamorphic rocks, previously altered rocks containing high clay contents, and clay-bearing sediments are less prone to develop significant secondary permeability. Interbedding of these two rock types can prevent a significant development of secondary permeability from taking place even where one of the rocks has a high inherent permeability, as can be seen in the cap rock section of the Karaha–Telaga Bodas geothermal prospect.

Cap rock

A cap rock of the geothermal system capable of developing an important thermal anomaly can be provided by:

1) lithology characterized by low permeability prior to the development of a geothermal system;
2) lithology characterized by low permeability prior to the development of a geothermal system that underwent further improvement of its sealing character because of alteration/cementation that took place during the development of a geothermal system; and
3) initially permeable rock that gained its low-permeability character by alteration/cementation during the development of a geothermal system.

An example of a cap rock with initial low-permeability character is provided by the Salton Sea geothermal field. Here, about 400 m-thick Late Pleistocene–Holocene Lake Coahuilla sediments, represented dominantly by shale and subordinate sandstone lenses, overlie older and more permeable deltaic sediments deposited by the Colorado River (Herzig et al., 1988; Hulen and Pulka, 2001). Other examples of initially low-permeability sediments have been described in geothermal systems such as Brawley and Heber in the Salton Sea region, and Larderello, where the cap rock is represented by low-permeability argillites, evaporites, and carbonates.

Apart from relying on the initial permeability structure of the rocks, the geothermal systems have a tendency to make their own seals due to alteration to clay in regions where temperatures range from less than 250°C to 100°C. Well-developed "clay" zones have been described as sealing numerous geothermal systems, including Olkaria in Kenya, the Salton Sea, and Yellowstone, to name a few.

Northwest-southeast and northeast-southwest oriented profiles through the Salton Sea geothermal system (McKibben and Hardie, 1997; Hulen et al., 2003) indicate that the system is elongated in the northwest-southeast direction. Apart from the initially 400 m-thick low-permeability cap rock, the present-day cap rock of the reservoir is controlled by clay alteration. The thickness of the clay-alteration zone varies from about 2,600 m on the flank of the system to about 500 m over its crest (McKibben and Hardie, 1997). The cap rock geometry in the crestal area conforms to the 250°–260°C isothermal surface, while it progressively departs from this surface in areas located further down on flanks. The clay alteration zone is underlain by a 300–500 m-thick chlorite zone that overlies the biotite zone. The Salton Sea represents a case where an efficient initial cap rock was further enhanced by alteration associated with geothermal activity. Elsewhere, as at the Karaha–Telaga Bodas geothermal field, initially less effective sealing rocks have been changed to low-permeability caps by alteration. The initial cap rock of the Karaha–Telaga Bodas system consisted of permeable tuffs, pyroclastic, and epiclastic horizons intercalated with low-permeability lava flows (Nemčok et al., 2004a). The tuffs could have had initial porosities as high as 22 percent, as indicated by the analogy with rock mechanics data compiled by Lama and Vutukuri (1978). This would have made them good reservoir rocks prior to their alteration and compaction. No porosity data for the pyroclastic and epiclastic rocks were available in the aforementioned rock mechanics tables, but their densities of approximately 2,360 kgm^{-3} compared with the lava flow densities of about 2,430 kgm^{-3} indicate higher porosities than those of andesitic lava flows. As with the tuffs, the pyroclastic and epiclastic rocks were prone to penetrative shallow argillic alteration and later compaction, which progressively changed them into effective seals during the liquid-dominated stage of this geothermal system.

Unfractured brittle lava flows were sandwiched among the less-competent tuffs, pyroclastic flows, and epiclastic deposits at the Karaha–Telaga Bodas. The lava flows initially formed seals among porous horizons. At present, the lava flows show little alteration, indicating that their matrix porosity was too low to allow them to host the early penetrative fluids. Rock tables with data on andesitic lava flows (e.g., Lama and Vutukuri, 1978) would allow us to infer an initial porosity of 8 percent or less. The surrounding porous horizons must have absorbed the deformation, allowing the lava flows to remain relatively unfractured and retain their sealing properties.

An example of cap rock growth into the upper portion of the reservoir is also provided by the research of the Karaha–Telaga Bodas system (Nemčok et al., 2004a). Here, the formation and expansion of the vapor-dominated region led to the formation of condensate that percolated downward from the top of the steam zone. Anhydrite, calcite, and pyrite were deposited as the descending condensate was heated. This mechanism caused the cap rock–reservoir boundary to progressively shift downward as a result of mineral deposition in the fractures. Although voids are present in the present-day cap rock, the thermal profiles are indicative of low overall permeabilities and thermal gradients typical for areas with conductive heat transfer. The top of the present-day reservoir is indicated by the first occurrence of fractures with interconnected voids.

Potential mechanisms that could breach a typical "clay cap" include large gravity glides or faulting, as documented by historical and present-day events, for example, from Mount Rainier and Mount St. Helens in Washington (Brantley and Glicken, 1986; Reid et al., 2001) and from the Karaha–Telaga Bodas geothermal system (Katili and Sudradjat, 1984).

Migration

The migration of hot liquid or vapor into the reservoir rock is the last element necessary for the existence of an effective geothermal system. The reservoir can be hot, but without migrated warm fluid the area contains just a hot dry rock system. The fluid must have a temperature of 150°C or higher to make a distinct geothermal field.

Heat migration mechanisms include conduction and convection. Heat conduction dominates in low-permeability rock sections, whereas convection is typical for highly permeable rock sections or fracture systems saturated with fluids. The magnitude of thermal anomaly caused by the movement of fluids in a geothermal system depends on the velocity and depth of fluid circulation, and the thermal conductivity of the host rock section (e.g., Nunn and Deming, 1991). It requires Darcy flow velocities of 1 mmy^{-1} and greater to make a noticeable impact on the thermal regime of a geothermal system (e.g., Bredehoeft and Papadopulos, 1965). Numerical simulations of the competition between heat conduction and convection (e.g., Deming, 1994a, 1994b) indicate that a minimum rock permeability capable of hosting strong enough fluid flow to perturb the thermal regime is of about 10^{-15} m^2. However, to make a thermal perturbation important for the geothermal system, the fluid flow rate should reach a meter per year or more (see Garven and Freeze, 1984a, 1984b; Bethke and Marshak, 1990).

Conduction in rift-related geothermal systems is associated with cases where a certain part of the rift basin has its hydrodynamic conditions dominated by horizontal flow, which results in conduction-dominated upward heat transport (Person and Garven, 1992). Other conduction-dominated cases are represented by systems associated with (Walters and Combs, 1989; Flovenz and Saemundsson, 1993; Gvirtzman et al., 1997; Liotta and Ranalli, 1999; Ranalli and Rybach, 2005):

1) partially molten magma bodies;
2) cooling and solidifying intrusions;
3) mechanical energy released by crustal movements of seismic or aseismic character;
4) radioactive heat generation;
5) aquitard unit covering the aquifer undergoing convection;
6) aquitard separating two aquifers; and
7) thermal balance of alteration/metamorphic processes.

Most of the convection in rift-related geothermal systems is associated with fluid flow mechanisms such as:

1) fluid flow driven by topography or gravity (e.g., geothermal systems such as Newcastle, Gerlach, and Salt Wells in the Basin and Range province and Hippo Pool in Ethiopia);
2) fluid flow driven by sediment compaction (e.g., narrow rift settings such as the Pechelbronn area in the central Rhine Graben (Person and Garven, 1992), the Worms area in the northern Rhine Graben (Lampe and Person, 2002), the La Jencia and Socorro sub-basins of the Rio Grande Rift (Mailloux et al., 1999); post-rift thermally subsiding basin settings such as the Cretaceous Angolan post-rift basin (Person and Garven, 1994); and pull-apart basin settings such as the Sea of Galilee pull-apart basin (Gvirtzman et al., 1997)); and
3) fluid flow driven by buoyancy forces resulting from density gradients due to temperature or salinity gradients (temperature gradients in wide rift systems such as Beowave, Soda Lake, and Stillwater in the Basin and Range province, in narrow rift systems such as Aluto and Dallol in Ethiopia; Eburru, Lake Bogoria, Namarumu, and Olkaria in Kenya; Lake Natron and Songwe in Tanzania (Hochstein and Browne, 2000); in seafloor-spreading ridge settings such as the Lau ridge, the Lau back-arc basin, Tonga (Schardt et al., 2006), and in pull-apart basins such as the Sea of Galilee pull-apart basin (Gvirtzman et al., 1997); and salinity gradients in systems such as Lake Assal in Djibouti and Ethiopia, Lake Afrera in Ethiopia, and Cesano in Italy).

Some geothermal systems can contain other fluid flow mechanisms in subordinate form, such as the fluid flow driven by seismic pumping that was recognized as active in the earlier history of the Karaha–Telaga Bodas geothermal system (Nemčok et al., 2007).

The main fluid flow systems responsible for important thermal anomalies

Topography-driven fluid flow

A topography-driven fluid flow develops in areas affected by crustal, that is, thermal, doming and associated shoulder uplift around rift flanks under subaerial conditions. This scenario is characterized by a topographic gradient associated with a roughly parallel water table configuration that controls the flow from high-elevation areas to low-elevation areas (e.g., Freeze and Cherry, 1979). Physical details of the topography-driven fluid flow are discussed in Chapter 8.

Natural examples come from the northern and central Rhine Graben areas (Person and Garven, 1992; Lampe and Person, 2002), the Rio Grande Rift (Mailloux et al., 1999), and the Sea of Galilee pull-apart basin (Gvirtzman et al., 1997).

The northern Rhine Graben study and numerical simulation focus on an apparent paradox of a combination of fairly high basal heat flow in the area (Clauser and Villinger, 1990) with the low maturity of organic matter in the area (Lampe et al., 1999). Maturation modeling along the Nordheim wells (Figure 13.2) indicates that

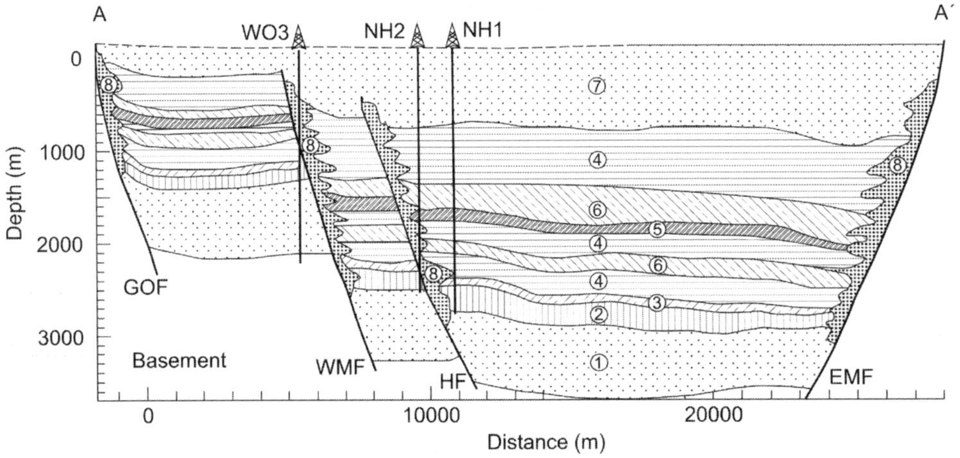

Figure 13.2. Cross-section showing the lithologic units and the major fault zones (GOF: Grustadt-Oppenheim Fault; WMF: Western Main Fault; HF: Hofheim Fault; EMF: Eastern Main Fault) in the Upper Rhine Graben (Lampe and Person, 2002). The cross-section was interpreted along the seismic profile tied to three exploration wells. The circled numbers depict the individual lithologic units. For a detailed description of the units, see Table 13.1.

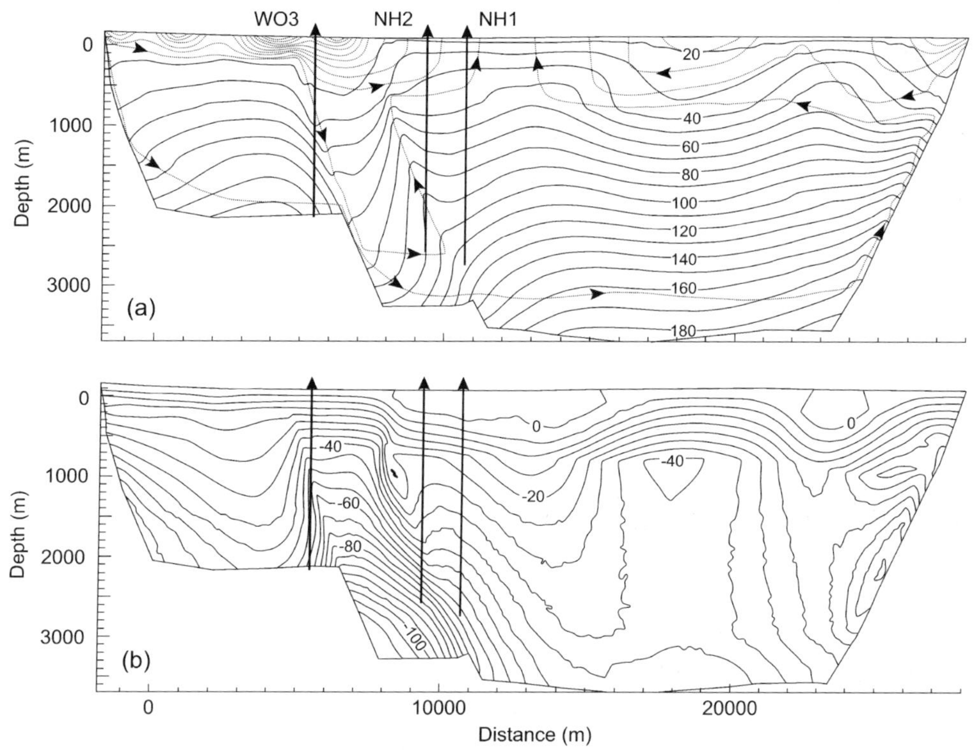

Figure 13.3. Temperature (solid lines) and stream functions (dotted lines) within the Upper Rhine Graben (Lampe and Person, 2002). Temperatures are given in ° C. (a) Open faults with permeabilities of $k_x = 10^{-14}m^2$ and $k_z = 10^{-13}m^2$ allow vigorous fluid flow in the order of cmy^{-1} to my^{-1} along the faults. (b) The temperature difference between the basal conductive temperature and the convective hydrothermal temperature (T_{conv}–T_{cond}). The model shows a net-cooling effect of the conductive temperature field by groundwater flow with values of up to –115 °C in the fault vicinity.

a present-day surface heat flow of 100 mWm^{-2} is too high to explain the vitrinite reflectance data from the core while a normal continental heat flow of 65 mWm^{-2} is too low. The numerical simulation tested the impact of different permeabilities of graben faults on the topography-driven flow and associated thermal regime. Tested permeabilities included:

1) large permeabilities, with horizontal and vertical permeabilities of 10^{-14} and 10^{-13} m^2, respectively, simulating highly transmissive faults (Figure 13.3);
2) intermediate permeabilities, with horizontal and vertical permeabilities of 10^{-16} and 10^{-15} m^2, respectively, simulating neither highly transmissive nor barrier faults; and
3) low permeabilities, with horizontal and vertical permeabilities of 10^{-19} and 10^{-18} m^2, respectively, simulating fluid flow barrier faults (Figure 13.4).

All three scenarios led to convective cooling along the fault-controlled rift margins while the basin center developed a positive heat flow anomaly. Particularly strong cooling of the basin develops if the shallowest sedimentary section is a good aquifer. When the conductive-only and conduction-convection-controlled thermal regimes are compared, it is apparent that fluid flow has a cooling effect on the basin. Even discharge areas are characterized by lowered heat flow values with respect to values characteristic for a conductive regime. The scenario that fits the present-day thermal regime the best is a topography-driven fluid flow utilizing medium- to low-permeability faults, indicating that opening episodes of faults on geologic time are not important for thermal effects. The medium- to low-permeability scenario explains a relatively high present-day conductive basal heat flow of 100 mWm^{-2} combined with a cooling effect of the convective groundwater flow. Similar advective cooling systems would be able to explain the thermal regime of the Medicine Lake geothermal prospect occurring in the pull-apart setting in northern California and the Upper Miocene–Pliocene East Slovakian Basin of the pull-apart-rift setting in the Western Carpathians. The numerical simulation further indicates that the advective cooling of the basin must have been operating for several million years, in fact up to 19 Ma. Note that in certain rift basins the fluid flow

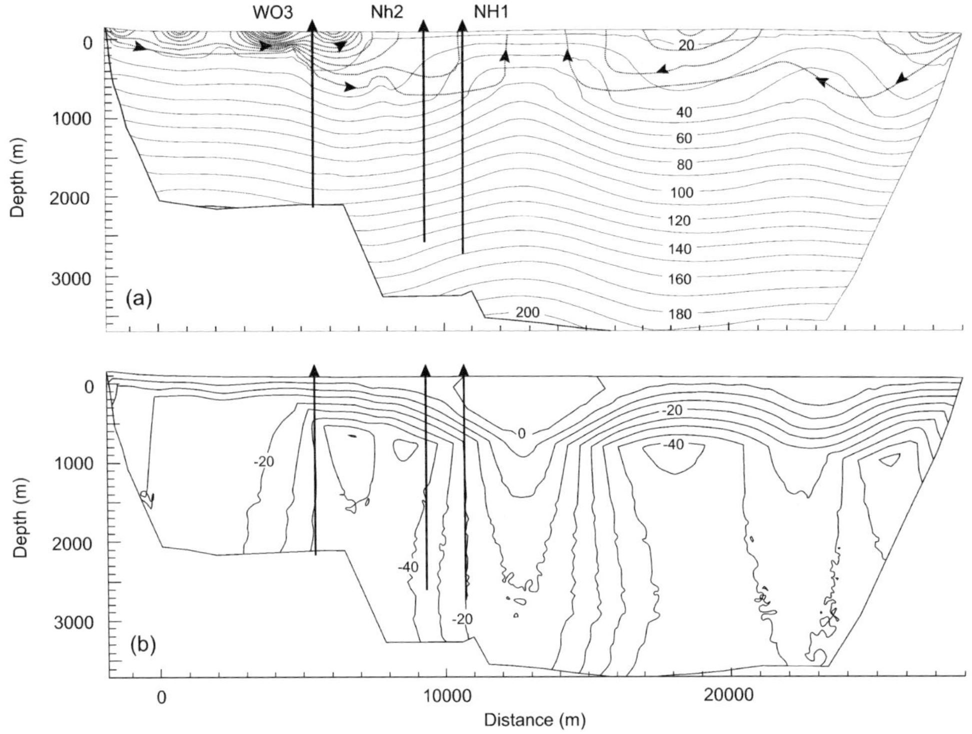

Figure 13.4. Temperature (solid lines) and stream functions (dotted lines) within the Rhine Graben (Lampe and Person, 2002). Temperatures are given in ° C. (a) Tight faults with permeabilities of $k_x = 10^{-19}m^2$ and $k_z = 10^{-18}m^2$ produce virtually the same temperature pattern as transmissive faults in Figure 13.3, yet flow rates are only in the order of 10^{-5} my^{-1}. As in Figure 13.3, flow is largely restricted to the shallow aquifer of younger Tertiary–Quaternary age. (b) Temperature differences between the basal conductive temperature and the convective hydrothermal temperatures ($T_{conv}-T_{cond}$). The pattern shows a net-cooling effect of the conductive temperature field by groundwater flow with values of up to −45 ° C in the fault vicinity.

mechanism can stop, source rock maturation is under a time lag of more than 8 Ma, and the basin has a high present-day heat flow, but it does not mean that the new extension or magmatic phase started.

The northern Rhine Graben simulation shows that a fluid flow along faults is capable of a significant, up to 115°C, reduction of the temperature of the adjacent wall rocks. A similar scenario, but on a much shorter time scale, has been documented for the Karaha–Telaga Bodas geothermal prospect, located in a strike-slip–extensional setting of the fore- to back-arc province (see Nemčok et al., 2007).

The central Rhine Graben simulation studies the transient thermal effects in fluid recharge and discharge areas on the background thermal regime and, by doing so, on hydrocarbon source rock maturation histories (Person and Garven, 1992).

Here, the studied topography-driven flow system had a significant effect on the thermal regime of the studied region after the establishment of the subaerial exposure of rift margins. Pliocene recharge areas around basin flanks were characterized by negative thermal anomalies. A present-day surface heat flow in these areas is still only as high as 30 mWm^{-2}. Pliocene discharge areas in the basin center had surface heat flow values as high as 300 mWm^{-2} along discharging fault zones while present-day values cluster around 110 mWm^{-2} (Person and Garven, 1992). A perturbation of the thermal regime had an important impact on the maturation history of the hydrocarbon source rocks (Figure 13.5). For example, the Landau 2 well in the discharge area had a geothermal gradient of 80° Ckm^{-1} and a top-of-the-oil window at a depth of 700 m. The Sandhausen 1 well in the recharge area had a geothermal gradient of 40°Ckm^{-1} and a top-of-the-oil generation window at 2,600 m. Well Landau 2, together with well Harthausen 1 from the discharge area, is characterized by elevated vitrinite reflectance values while well Sandhausen 1 has low values of vitrinite reflectance (Figure 13.6).

A numerical simulation shows the following trends. For the Peclet number (see Chapter 8) of 0.6, a difference between the tops of oil windows in recharge and discharge areas amounted to 2,300 m. The width of the respective oil windows also varied. The descending fluid

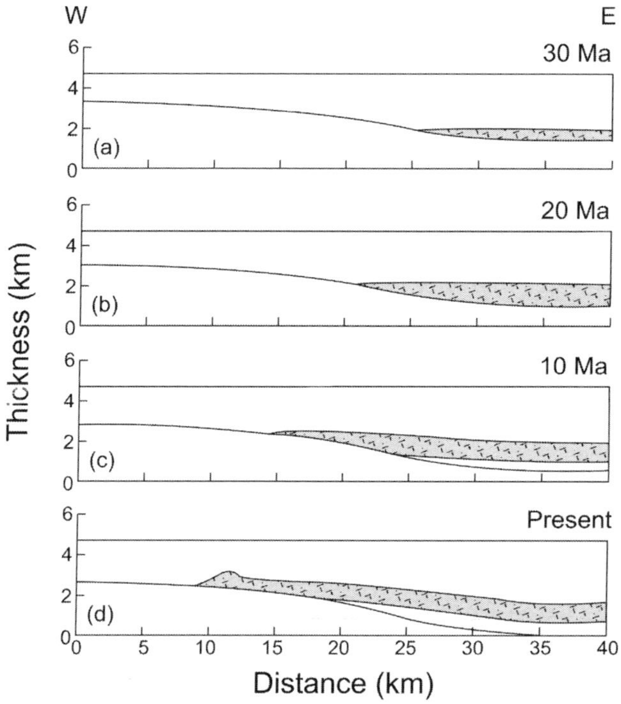

Figure 13.5. Location of oil window (15≤TTI≤160) along the cross-section through the central Rhine Graben near the Pechelbronn area, calculated for time intervals starting with 30 Ma ago (Person and Garven, 1992). Changes in oil window location are primarily due to convective heat transfer effects induced by regional topography-driven fluid flow.

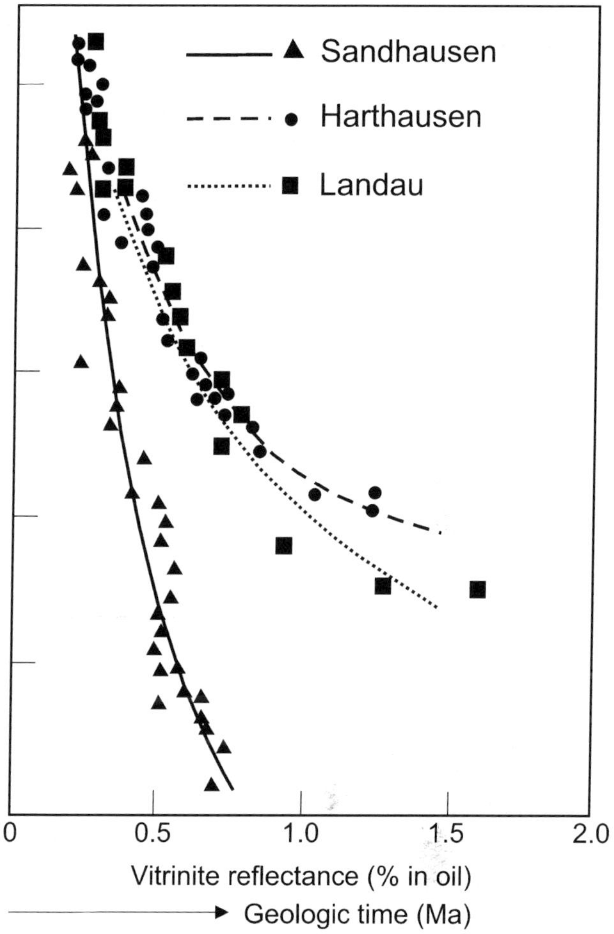

Figure 13.6. Observed vitrinite reflectance data from three wells within the central Rhine Graben in the broader Pechelbronn area (Person and Garven, 1992). Wells Harthausen 1 and Landau 2 with higher heat flow are located in the center of the rift zone in the fluid discharge area. Their vitrinite samples have consistently higher levels of reflectance than samples from the well Sandhausen 1 with lower heat flow, which is located within the fluid recharge area near the eastern fault escarpment.

flow in the recharge area resulted in a thermal gradient smaller than the conductive gradient, which controlled the existence of a broader window. The ascending fluid flow in the discharge area resulted in a thermal gradient larger than the conductive gradient, which controlled the existence of a narrower window. For a Peclet number of 0.6, the oil windows had thicknesses of 700 m and 1,300 m in discharge and recharge areas, respectively (Figure 13.7). A similar effect on thermal gradient can be seen in the Salton Sea and Medicine Lake geothermal areas when we look at both the present-day temperature contours and the upper surfaces of mineral distribution (Figures 13.8, 13.9). The inspection of both figures indicates the occurrence of thicker mineralization zones and less dense temperature contours in recharge areas, while the discharge areas are characterized by thinner mineralization zones and denser temperature contours.

The central Rhine Graben study also focused on simulating how long the transient effect of the fluid flow is felt by the region after the flow stops. The important outcome is the difference in thermal and source rock maturation histories. While the thermal regime equilibrates within about 2 Ma and changes back to a steady-state conductive regime in both former recharge and discharge areas, the perturbation in the maturation history has a longer life span than 2 Ma and it differs between the recharge and discharge areas (Figure 13.10). The time-temperature index (see Lopatin, 1971; Waples, 1980) profile in the recharge area returns to the profile solely controlled by the heat conduction after about 8 Ma. The time period necessary for such a return in the discharge area amounts to about 25 Ma. Both changes are proven to significantly modify the maturation histories of respective areas in comparison to the rest of the rift basin. The difference in transient effects on thermal versus maturation

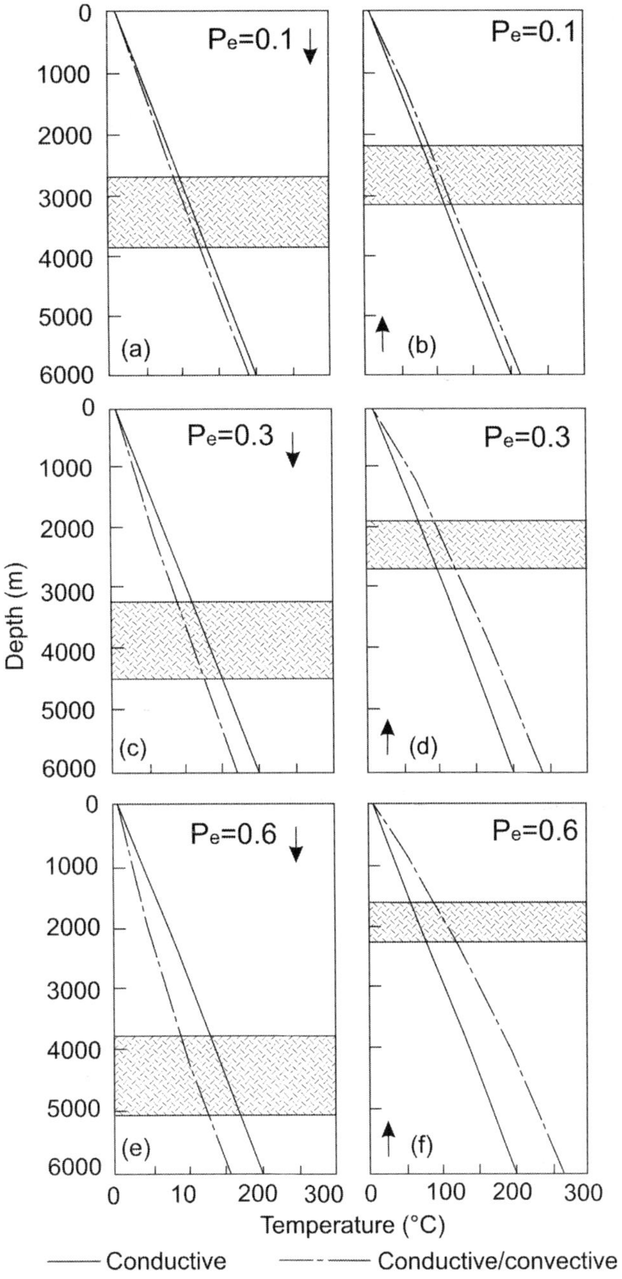

Figure 13.7. One-dimensional steady-state conductive (solid line) and quasi-steady-state conductive/convective (dashed line) temperature profiles and oil window (15≤TTI≤160; shaded region) computed for recharge and discharge areas of the topography-driven fluid flow using conductive/convective temperatures for thermal Peclet numbers ranging from 0.1 to 0.6 (Person and Garven, 1992). Arrows depict the direction of groundwater flow. The profiles were obtained by allowing the solution domain to grow from 1 to 6 km using a constant subsidence rate of 1×10^{-5} m/yr.

histories can be seen in the time-temperature index calculation, TTI (Waples, 1980):

$$TTI = \int_0^t 2^{n(T)} dt, \tag{13-2}$$

where the index increases exponentially with respect to the temperature-dependent fit coefficient n(T) and linearly with time, t. Although Equation (13-2) describes a model that is simpler than more recent petroleum generation models (e.g., Forbes et al., 1991), it successfully indicates a difference in the thermal and maturation history reaction to transient effects of the convective heat transfer.

This reaction is further controlled by the rate of tectonic subsidence (see Person, 1990). The slower the subsidence, the longer the time lag is between recharge and discharge areas. While the reinstated conductive regime in the recharge area erases the evidence of the former convection effects, such erasing in the discharge area requires subsidence of sediments down to depths with temperatures higher than those characterizing the former discharge area. Therefore, the most typical region where one can find the signal of the past convective effects is the preexisting discharge area of the slowly subsiding rift unit.

The Rio Grande Rift simulation studies the thermal regime dominated by topography-driven flow (Mailloux et al., 1999), described in detail in Chapter 8. The surface heat flow data in the study area have values of 77–135 mWm^{-2} (Reiter et al., 1975; Morgan et al., 1981; Witcher, 1988; Barroll and Reiter, 1990; Wade and Reiter, 1994). The present-day heat flow maxima occur in depressions adjacent to perennial drainage features with discharging geothermal fluids (Witcher, 1988), which are located in the basin center. Apart from the present-day discharge areas, partial resetting of apatite fission track ages, age of the barite-fluorite deposits and bleached fracture zones document the existence of an earlier discharge zone located at the eastern flank of the basin.

The present-day heat flow maxima in the discharge area are developed thanks to the hydrological window between the partially eroded footwall edge of the block containing the tilted lower aquifer and the overlying upper aquifer, the development of which caused the topography-driven flow discharge shift from the eastern flank of the basin to its center. The effect of the heat flow focusing through a narrow window results in a significant surface heat flow, characterized by values in excess of 200 mWm^{-2}. The area is slightly wider than 1 km and at least 8 km long (see Figure 13.1). The numerical model of the study area suggests that the

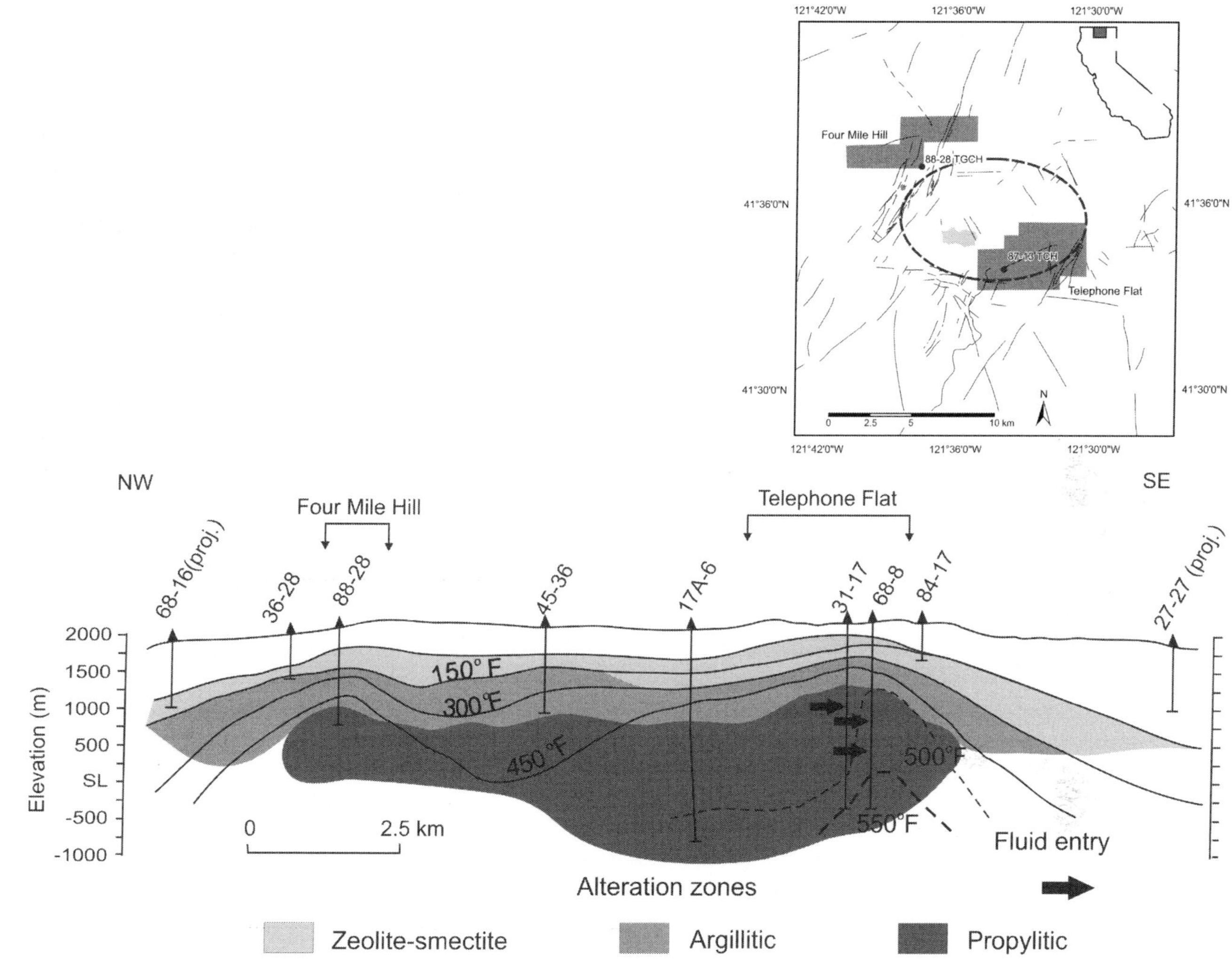

Figure 13.8. Relationship between alteration, temperature, and permeability from the Medicine Lake geothermal system (Clausen et al., 2006). See text for further explanation.

discharge area for the older flow system was about 40°C warmer than the case with only conduction involved.

While faults were interpreted as controlling the juxtaposition of aquifers, the numerical model does not seem to indicate that they formed either barriers or conduits for fluid flow. An interesting outcome of the study is the activity time period for the earlier topography-driven fluid flow system that discharged at the eastern margin of the basin. It acted between 15 and 5 Ma, representing a 10 Ma-lasting system capable of a major impact on the maturation history.

The Sea of Galilee pull-apart basin simulation tries to explain a combination of various surface heat flow values determined for various areas of the studied region (see Gvirtzman et al., 1997 and references therein). They include:

1) values of 70–80 mWm^{-2} in the central portion of the Sea of Galilee;
2) values of about 36 mWm^{-2} along the southern coast; and
3) values of about 135 mWm^{-2} in the southern Golan Heights, located 6–8 km east of the pull-apart margin.

The simulation indicates (Figure 13.11) that the lower heat flow values in the Galilee Mountains can be explained by cold water percolation downward in the recharge area, which results in the lowering of the local geothermal gradient. On the contrary, the upward flow of the heated water at the western margin of the pull-apart basin controls the increased temperatures of the rock section, increasing the thermal gradient. Discharge areas match with known hot springs

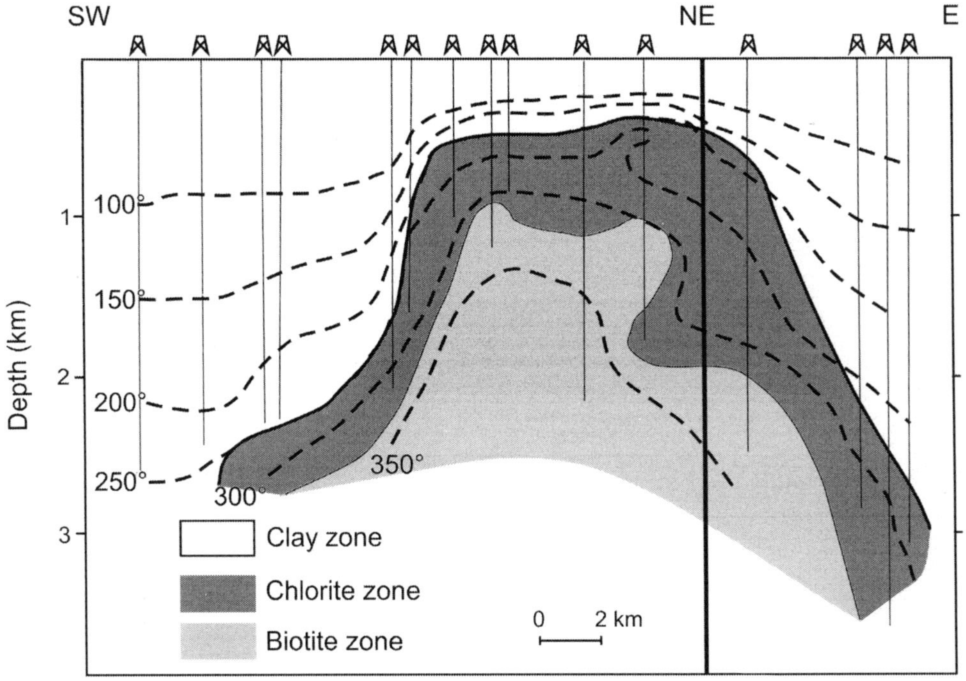

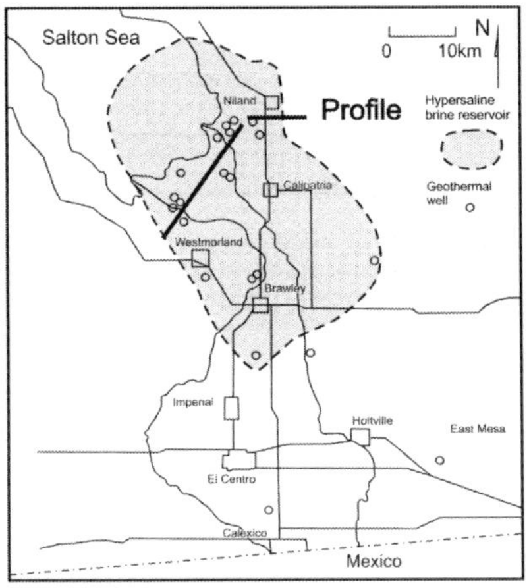

Figure 13.9. The top of the surface of the high-salinity brines of the Salton Sea system is elevated across the geothermal system, reaching a 500 m depth at the crest of the system and greater depths away from the geothermal axis (McKibben and Hardie, 1997). The shape is similar to the 250°–260°C isothermal surface.

in the area. Because the western boundary faults link the deeper and shallower aquifers with surface, different hot springs have different salinities. While the Hammat Gader springs and Mezar wells contain hot water with low salinity, springs and wells located along the western shoreline are highly saline (see Arad and Bein, 1986).

Compaction-driven flow

The compaction-driven fluid flow is a consequence of the mechanical loading of the sedimentary fill of the extensional basin during its ongoing subsidence and filling (e.g., Sharp, 1978; Palciauskas and Domenico, 1982; Neuzil, 1986). Typical sediments prone to significant compaction are syn-rift fine-grained lacustrine and playa sediments or fine-grained sediments associated with the thermal subsidence stage of the rift development. The coexistence of fast subsidence, low-permeability sediments and their pore-space collapse can result in overpressure generation in certain zones of compacting basins (see, e.g., Dickinson, 1953 on the Gulf of Mexico, Bredehoeft et al., 1988 on the South Caspian Basin, and Mann and MacKenzie, 1990 and Audet and McConell, 1992 on the North Sea). On the regional scale, a compaction typically produces hydraulic heads increasing with depth, which results in fluid driving upward and toward the higher-located

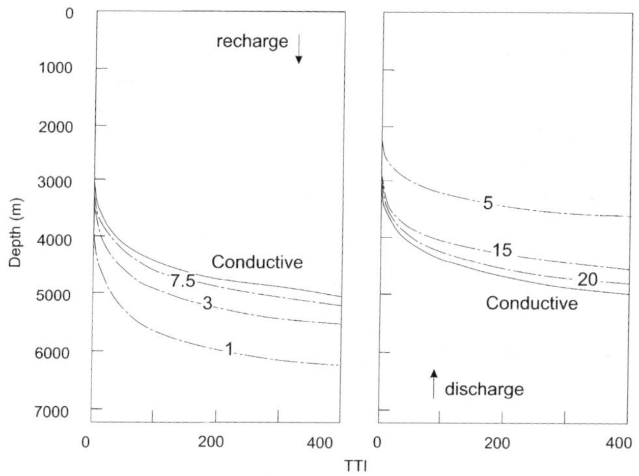

Figure 13.10. Transient conductive/convective TTI profiles (dashed lines) and steady-state conductive TTI profile (solid line) in recharge and discharge areas of the topography-driven fluid flow at different times after groundwater motion stopped using a tectonic subsidence rate of 1×10^{-4} my^{-1} (Person and Garven, 1992). Arrows depict the direction of groundwater flow. The numbers associated with the dashed lines indicate the time (in Ma) since groundwater flow stopped.

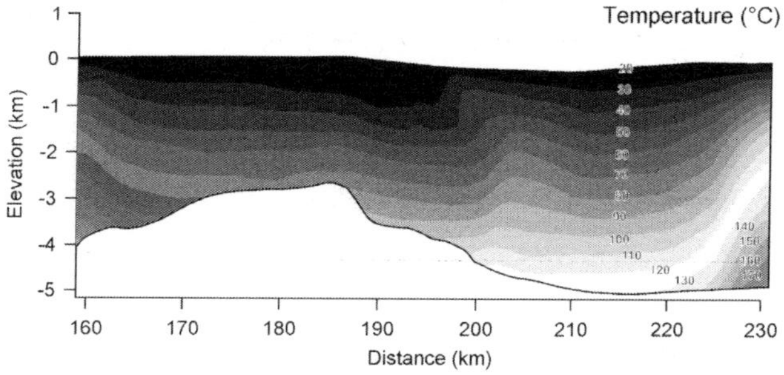

Figure 13.11. Temperature distribution (° Celsius) for the area and fluid flow models shown in Figure 8.20 (Gvirtzman et al., 1997). Note that the heat anomaly at the Fulya Springs results from the rising of deep hot groundwater, and that beneath the Golan Heights (Mezar wells) results from the free convection cell.

graben flank (e.g., Bethke, 1985). Physical details of the compaction-driven fluid flow are discussed in Chapter 8.

Natural examples of the compaction-driven fluid flow systems come from the central Rhine Graben (Person and Garven, 1992) and the Angolan Cretaceous rift basin (Person and Garven, 1994).

The central Rhine Graben simulation documents a marine transgression stage of the central Rhine Graben that took place from 40 Ma to 15 Ma, when the fluid flow was driven by a dominant density-driven mechanism and a subordinate compaction-driven mechanism

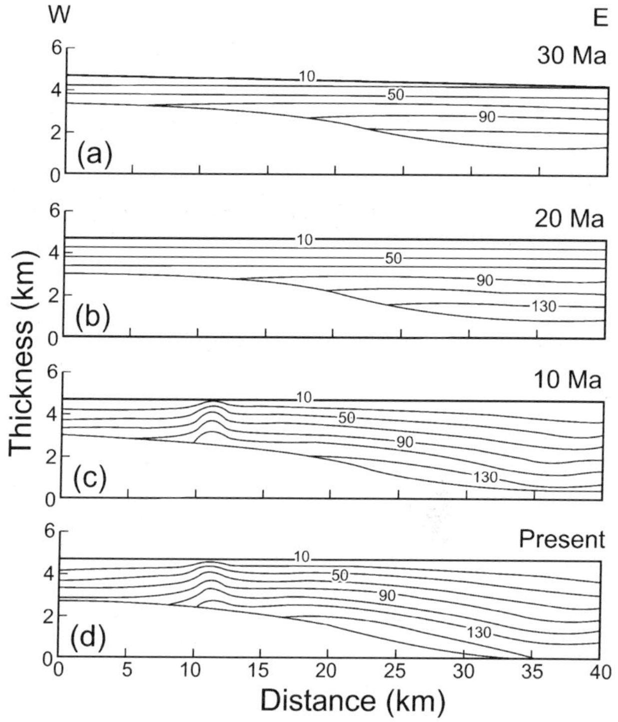

Figure 13.12. Computed isotherms within the central Rhine Graben along the section from Figure 13.5 since 30 Ma (Person and Garven, 1992). The contour interval is 20°C.

(Person and Garven, 1992). The simulation illustrates that the hydraulic gradients due to sediment compaction were too small to modify the density-driven fluid flow and that both flow mechanisms had too small velocities to disturb the overall conductive thermal regime (Figures 13.12, 13.13). Even the simulation using a highly transmissive boundary fault did not change this outcome of the thermal modeling.

The Angolan Cretaceous rift basin simulation (Person and Garven, 1994) takes the average syn-rift thickness and rift width values of 4 and 40 km, respectively, from the collection of rift data made by Lippman and colleagues (1989); represents various stages of the rift basin development by characteristic facies distributions; and describes the competition of all potential fluid flow driving forces for control of the flow during these development stages. The simulation documents that compaction-driven flow is important in the center of the rift unit during the syn-rift stage, if the deposition rate and permeability are sufficiently significant and low, respectively. The deposition rate is considered sufficiently significant if it is faster than 0.1 mmy^{-1} (see Deming, 1994b). The permeability is considered sufficiently low if its horizontal and vertical components are approaching 10^{-17} m^2 and 10^{-19} m^2,

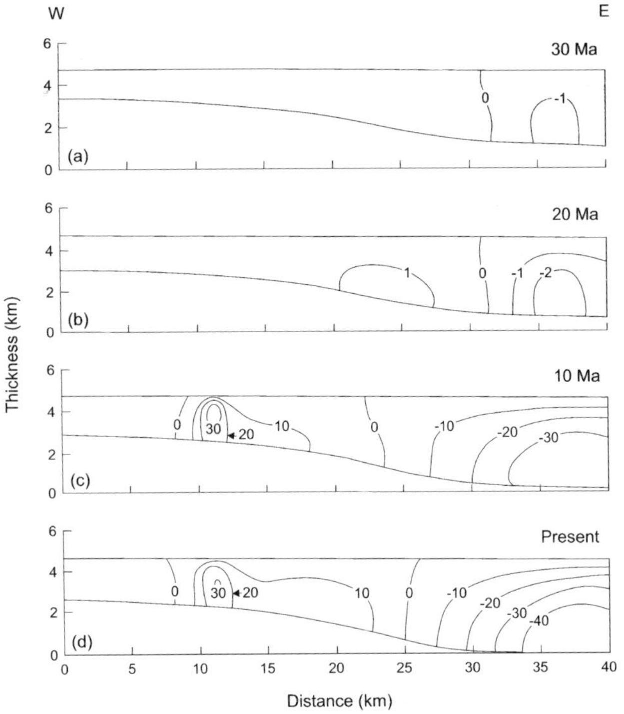

Figure 13.13. Computed deviations from a conductive thermal regime within the central Rhine Graben along the section from Figure 13.5 since 30 Ma (Person and Garven, 1992). The contour interval is 10°C. Negative contour lines are situated in groundwater recharge areas while positive contour lines represent regions of groundwater discharge.

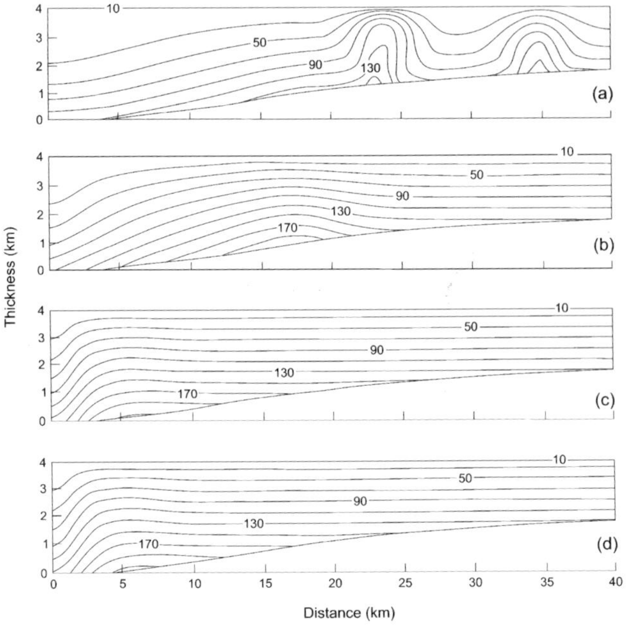

Figure 13.14. Effect of lateral variation in permeability on computed temperatures (in ° Celsius) during the initial phase of the continental-rift basin formation (Person and Garven, 1994). Thermal models are associated with hydraulic head distribution models from Figure 8.23. The contour interval is 20°C. Basement heat flow increased from 60 to 96 mWm^{-2} during the 26 Ma simulation period because of stretching and necking of the lithosphere.

respectively (Person and Garven, 1994; Figure 8.23). The compaction-driven fluid flow in the syn-rift stage flows laterally into more permeable units distributed along the graben margin, combining its effects on thermal regime with topography-driven fluid flow, which is restricted into more permeable units (Figure 13.14). As a result of the lateral flow, no modification of the conductive background heat flow regime is apparent in the basin center.

If the deposition is sufficiently significant, even the subsequent flexural development stage of the rift can be characterized by important compaction-driven flow. Being sealed from the surface by the flexural subsidence-related sedimentary package, the syn-rift sediments maintain the compaction-driven fluid flow in the central portion of the graben if they were prone to compaction-driven flow during the syn-rift stage (Figure 8.24). In this stage, the compaction-driven flow in the buried syn-rift basin center combines with buoyancy-driven flow in more permeable buried units of the basin margin. As a result of the lateral flow, no modification of the conductive background heat flow regime is apparent in the buried basin center.

Buoyancy-Driven Flow

Buoyancy-driven flow is controlled by density gradients developed because of the temperature or salinity gradients (e.g., Combarnous and Bories, 1975; Wood and Hewett, 1982, 1984; Rabinowicz et al., 1985; Person and Garven, 1992). The temperature gradients can develop by:

1) uneven production of radiogenic heat within the lithosphere;
2) occurrence of magmatic intrusions;
3) localized crustal thinning; and
4) thermal conductivity contrasts among different lithologies.

A thermally driven fluid flow in rift systems can play a more important role than in other basin types because of the rift characteristics of high basal heat flow, with a tendency for late syn-rift magmatism and accumulation of coarse sediments in zones adjacent to rift-boundary faults (see Andrews-Speed et al., 1984). Another typical region for this flow is represented by the seafloor-spreading ridges (e.g., Lowell, 1975; Lowell and Rona, 1985; Fisher and Becker, 1995; Yang et al., 1998).

Furthermore, the salinity gradients can develop from the effects of basinal brine salinity, especially important in rifts with thick evaporite horizons and those that are hydrologically closed (Duffy and Al-Hassan, 1988). Physical details of the buoyancy-driven fluid flow are discussed in Chapter 8.

Natural examples of the buoyancy-driven fluid flow systems come from:

1) a seafloor-spreading ridge represented by the Lau seafloor-spreading ridge, the Lau back-arc basin, Tonga (Schardt et al., 2006);
2) a submarine rift basin represented by the central Rhine Graben (Person and Garven, 1992) and the Mount Isa Basin, northern Australia (Yang et al., 2006); and
3) a subaerial pull-apart basin without a distinct topographic relief represented by the Sea of Galilee pull-apart basin (Gvirtzman et al., 1997).

The numerical simulation focused on the Lau seafloor-spreading ridge studies the buoyancy-driven fluid flow perturbing the conductive thermal regime at a fast-spreading ridge (Schardt et al., 2006). Hydrothermal fluids extract the heat from the central intrusion and transport it within available aquifers, represented by both the basalt breccia zone and the interface between the gabbro and sheeted dyke layers, and fault systems, toward the outer topographic highs on both sides of the central valley, and, in certain stages of the hydrothermal system development, also toward the high located right above the central magma chamber (Figure 13.15). The simulation shows that the discharge of hydrothermal fluids with temperatures of above 200°C can last about 100,000 to 800,000 years, depending on the fault permeabilities and on the existence and effectivity of the cap rock of the geothermal system.

The sensitivity analysis illustrates that the thermal regime perturbation by hydrothermal fluids mainly depends on:

1) host-rock permeability structure;
2) fault distribution; and
3) fault permeability.

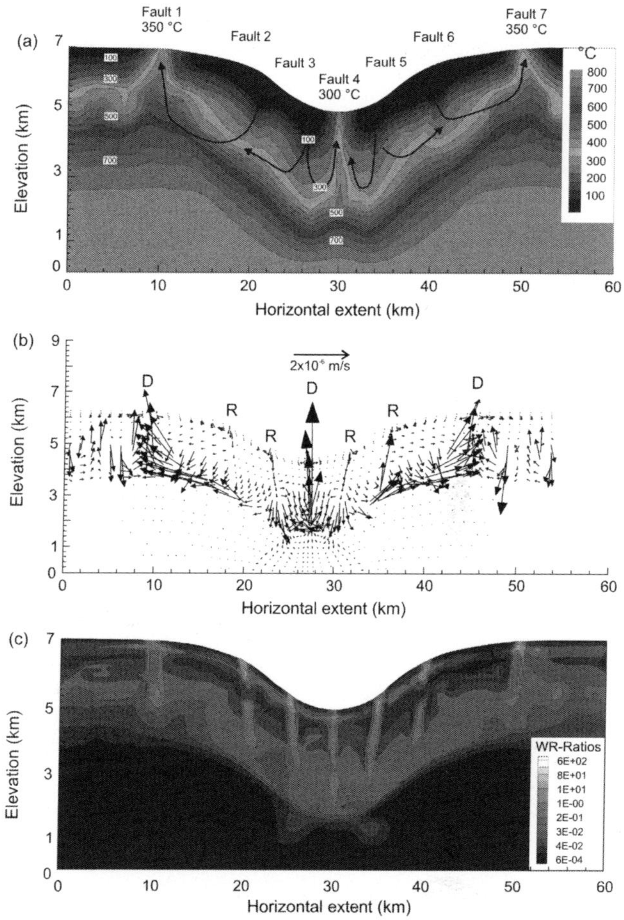

Figure 13.15. Snapshot of (a) heat distribution, (b) fluid flow, and (c) cumulative water-rock ratios after 10,000 years of average conditions (average rock and fault permeabilities, see Table 13.2) in the Lau Basin (Schardt et al., 2006). Main fluid recharge occurs through faults 2, 3, 5, and 6 in the topographic depression. The outermost faults 1 and 7 experience significant fluid discharge, while fault 4 is only active at lower rock permeabilities. Cumulative water-rock ratios suggest that the most intense fluid flow occurs within and around faults as well as within the breccia zone and sheeted dike layer. Note that the shown temperature distribution reflects conditions after 10,000 years, when the hydrothermal system has started to cool down and does not necessarily show maximum discharge temperatures. D is fluid discharge and R is fluid recharge.

The host rock permeability has an important impact on the distribution of recharge/discharge areas, rates of discharge, and discharging fluid temperatures. Low permeability correlates with high discharged fluid temperatures and discharge locations at distant topographic highs. An increase of host rock permeability results in lowering the discharge temperatures and the location of discharge at closer locations from the heat source. High rock permeability results in the central fault becoming a recharge area instead of discharge area because of the increase in fluid discharge rates allowing for a fast cooling of the heat source, thus reducing the life span of the geothermal system.

The fault distribution also has an important impact on the geometry of the geothermal system. A three-main-fault pattern controls a simple system of recharge above the heat source and discharge on adjacent topographic elevations, accordingly developing a system

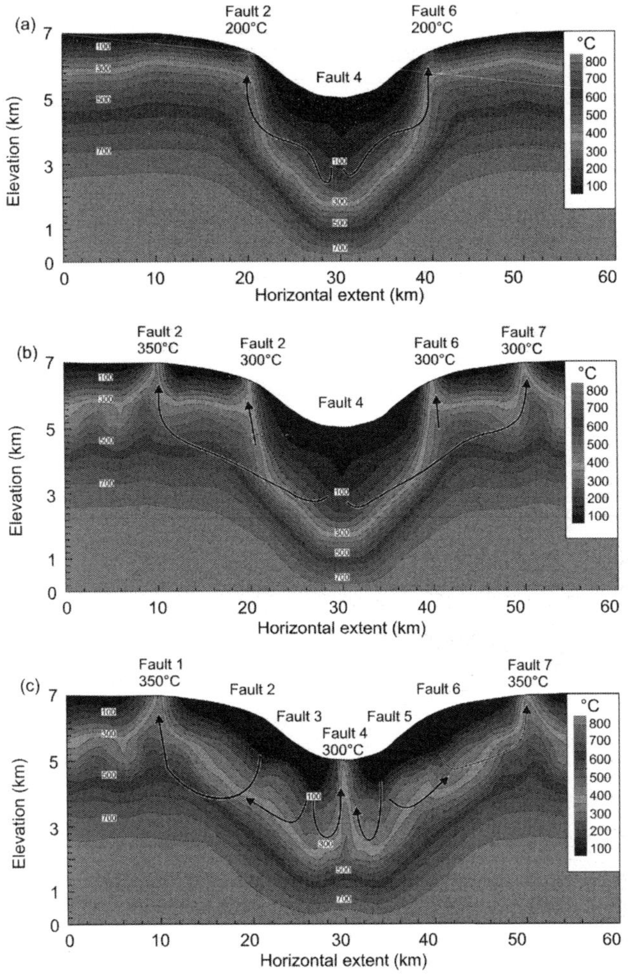

Figure 13.16. Modeled temperature distribution for simulations with different fault setups: (a) three faults, (b) five faults, and (c) seven faults for a time interval of 10,000 years in the Lau Basin (Schardt et al., 2006). In all cases, major fluid discharge occurs at elevated topographic positions as well as through fault 4 directly overlying the magma chamber, while fluid recharge is concentrated on the slopes of the depression. Note that discharge temperatures reflect conditions after 10,000 years while maximum discharge temperatures of individual faults may be higher during the lifespan of the hydrothermal system.

of thermal anomalies (Figure 13.16a). A five-main-fault pattern utilizes both external fault couples located at elevations as discharge areas, characterized by faster rates at external sites and slower rates at internal sites controlling associated thermal anomalies (Figure 13.16b). A seven-main-fault pattern controls an even more complex geothermal system. It is characterized by the central fault, located above the heat source, switching from the initial recharger to a later discharger, which, together with discharge zones at external locations, develops a characteristic system of thermal anomalies for both the initial and later development stages (Figure 13.16c). The switch from recharge to discharge develops after the central magma chamber produces enough heat to start the buoyant hydrothermal fluid flow.

The most complex modifications of the geothermal systems are controlled by different permeabilities of the main faults. The sensitivity testing indicates that a lack of clear correlation trends has to be due to the fault competition among individual faults, which controls selective fault responses. However, it can be concluded that the discharge-related positive thermal anomalies shift away from the heat source with increased fault permeability. The simulation further indicates that the fault permeability changes control the geothermal system geometry. The intermediate fault permeability develops a recharge-dominated geothermal system characterized by four recharge and three discharge faults. The high fault permeability develops a discharge-dominated system characterized by four discharge and three recharge faults.

The fluid velocity conversion Lowell and Germanovich (1994) defined allows us to calculate the heat flow, Q, from a single discharging vent of the described geothermal system:

$$Q = \rho_f c_f v (T - T_o) \pi d^2/4, \tag{13-3}$$

where ρ_f is the fluid density, c_f is the specific heat of the fluid, v is the discharge rate, T is the discharge temperature, T_o is the temperature of the seawater, and d is the vent diameter.

The life span of the discussed geothermal system with fluid temperatures of above 200°C depends on the rock and fault permeabilities and the existence of the cap rock, providing minimum and maximum of 100,000 and 800,000 years (see Cathles et al., 1997; Schardt et al., 2006). It is also logical that the geothermal system develops together with rift development. As the rift evolves, the geothermal system expands, reaching into the deeper zones, into the external portion of the central intrusion, and alternating initial central recharge area into discharge area, even though this discharge pales in comparison with the role of the external discharge areas on thermal regime.

The central Rhine Graben simulation for the submarine stage of the graben development that took place between 40 and 15 Ma shows a single broad convection cell with a clockwise fluid flow pattern directed from the graben center toward its western flank (Person and Garven, 1992). The buoyancy-driven fluid flow system is coupled with a weak compaction-driven flow in

Table 13.2. Ranges of physical rock properties used in models of fluid and heat migration at Lau Ridge shown in Figures 13.15 and 13.16 (Schardt et al., 2006).

Rock unit	Permeability range (m^2)	Porosity range (%)	Thermal conductivity range (W/m°C)
Sediments	5×10^{-18}	50	1.5
Breccia zone	2.5×10^{-15}–2.5×10^{-17}	1–20	1.2–3.7
Andesite	2.5×10^{-16}–2.5×10^{-18}	1–20	1.2–3.2
Sheeted dyke	1×10^{-17}	2	2.0
Gabbro	1×10^{-18}	1	2.0
Fault	1×10^{-12}–1×10^{-16}	30	2.0

the thickest part of the graben and controlled by the east-tilted basal aquifer (see Wood and Hewett, 1982). However, fluid rates are too small to have an important impact on the background conductive thermal regime (Figures 13.12a, b and Figures 13.13a, b).

The Mount Isa Basin simulation (Yang et al., 2006) studies the Mount Isa–McArthur Basin in north Australia, which underwent buoyancy-driven flow when it was under open marine conditions. The geothermal system was controlled by both the seawater recharge and the interplay between active syn-sedimentary faults and aquifers. Just like in the case of the Lau seafloor-spreading ridge, the numerical simulation indicates complex geometry changes of the geothermal system in response to variations in:

1) permeabilities of aquifers and cap rock of the system;
2) fault pattern; and
3) fault permeabilities.

The larger the aquifer depths and permeabilities, the larger the heat flow anomalies are at discharge areas. The effective cap rock tends to repress the development of significant thermal perturbations. The deep-seated faults tend to capture the significant thermal perturbations from shallower-seated faults. The convective cells are established and controlled by the fault spacing, the depth of fault penetration, and the fault-connecting aquifer depth.

Note that the discharge and recharge areas occupy segments along faults rather than entire faults, focusing on areas of the highest permeability. These areas coincide with (Betts et al., 2003):

1) dilatant fault segments where the fault geometry changes control an enhanced extensional component of the displacement under the controlling stress regime;
2) fault–fault intersections; and
3) fault intersections with especially permeable rock unit.

The calculated life spans of the geothermal systems that formed the Century and Kidd Creek ore deposits in the Mount Isa Basin are about 650,000 years long (Yang et al., 2006). These systems apparently changed their geometry during different development stages, as indicated by the change of discharge to recharge along the Termite Range Fault with the age of the geothermal system.

The Sea of Galilee pull-apart basin simulation (Gvirtzman et al., 1997) further emphasizes the occurrence of the buoyancy-driven flow in submerged areas lacking the influence of complex topography. In this case, it develops in the region dominated by the topography-driven fluid flow. However, it develops in the deeply located aquifers inside of the subhorizontally lying rock section of the Golan Heights, which is sealed by tight-permeability faults from the adjacent side of the Sea of Galilee pull-apart basin. The buoyancy-driven flow dominating in the section under the Golan Heights develops a positive thermal anomaly even warmer than the anomaly associated with the discharge area of the neighboring topography-driven flow (Figure 13.11).

Combination of different fluid flow mechanisms

This chapter documents that different depositional and tectonic stages of the extensional basins are prone to different fluid flow mechanisms. For example, subaerial stages of the rift system development are prone to topography-driven fluid flow while it vanishes in the submarine stages. Although the gravity-driven flow regimes in subaerially exposed rift systems, having typical flow rates from 0.0084 to 10 meters per year (Table 8.2), tend to overwhelm sediment-compaction- and buoyancy-driven flows, which achieve rates of 0.00003-0.1 my^{-1} and 0.0032-8 my^{-1}, respectively, different combinations of fluid flow mechanisms can coexist in different development stages of extensional settings. This is valid for:

1) a combination of forced convection driven by gravity-driven flow and free convection driven by buoyancy forces due to temperature and salinity gradients in rift systems (Person and Garven, 1994; Raffensperger and Garven, 1995);
2) a combination of topography- and compaction-driven fluid flows in the main rift stage in grabens characterized by a narrow zone of alluvial deposits along boundary faults and basinal low-permeability lacustrine sediments (Person and Garven, 1994); and
3) a combination of buoyancy-driven flow due to salinity gradients and compaction-driven flow in the late-rift flexural stage in the basin characterized by marginal coarse sediments and basinal fine-grained lacustrine and younger onlapping sediments (Person and Garven, 1994; Yang et al., 2006).

The fluid flow driven by seismic pumping generally pales in comparison with other flow mechanisms. It is important on the local scale inside the fault zones, but not at the reservoir scale.

The seismic pumping can develop when faulting occurs in a rock section with overpressured layers and a supra-hydrostatic vertical gradient in fluid pressure (Sibson, 1981b). Pumping depends on faults' ability to behave as seals when they are tectonically quiescent and as conduits during and immediately following rupturing. Fault sealing during inter-seismic periods can happen because of the formation of clays and the presence of cataclastic gouge or mineral precipitation within the fault core. Rupturing of these seals that form barriers to overpressured zones leads to fluid flow toward regions of lower pressures. Seismic pumping can occur in any tectonic setting where overpressuring has developed, but the extreme pumping action tends to be associated with steep reverse faults in areas undergoing collision and subduction and not with rift settings (Sibson, 1990a).

Interactions of geothermal system elements

The previous text describes a group of basic parameters that controls the existence of the geothermal system. Each of them can be evaluated and graded from excellent through fair, good, and poor, to unacceptable in different cases. It can be implied from this text that for the occurrence of the geothermal system it is important how all system elements interact. It is not imperative for a significant geothermal system to have the highest grading for all system elements because excellent system elements can make up for the poor grades of others. For example, a poor reservoir consisting of a single fault zone can be balanced by an effective fluid source. The examples come from the Brady, San Emidio Desert, Soda Lake, and Stillwater geothermal systems in the Basin and Range province (Hochstein and Browne, 2000 and references therein). The Soultz-sous-Forets case in the Rhine Graben indicates that an effective source does not always make up for a poor reservoir and the area becomes a hot-dry-rock system (see Kohl et al., 2000). A total lack of reservoir in combination with effective fluid source degrades a geothermal system from having fluids "stored" in the reservoir (e.g., Christensen et al., 2002; Yang et al., 2006) down to a system that just discharges fluids along migration pathways without storing them. An example comes from the Long Valley Caldera in the Basin and Range province (McGee and Gerlach, 1998; Priest et al., 1998; Sorey et al., 1999). This indicates that an absence of the system element cannot be saved by other system elements having excellent grades.

A poor fluid source can be balanced by a volumetrically extensive reservoir, that prevents fluid discharge from directly competing with the fluid source. A weak heat source can apparently be balanced by other system elements, too, otherwise no high-temperature systems would be listed as associated with developing rifts instead of advanced ones (see Hochstein and Browne, 2000). One of the most dependable system elements is the cap rock because geothermal systems can make up for an initial lack of an impermeable cap rock by pore-destructive alteration (see the Karaha–Telaga Bodas example discussed earlier). However, even if the cap rock is poor because of the breaching catastrophic events (e.g., Katili and Sudradjat, 1984; Brantley and Glicken, 1986; Reid et al., 2001), the ongoing fluid and heat production can usually make up for a "leaky" cap rock.

Because of the ongoing heat and fluid discharge in most geothermal systems, the geothermal system evaluation, unlike the hydrocarbon system evaluation, is trap-independent. Theoretically, geothermal systems would include any stratigraphically or structurally controlled region allowing entrapment of heated fluids that would otherwise migrate to the surface. Stratigraphic traps would be controlled by either the original stratigraphic relationship or by its post-alteration state. Primary stratigraphic traps could include:

1) porous volcanoclastic rocks with culminations at centers of volcanic cones;

2) caldera deposits limited by caldera-bounding faults;
3) sandstone bodies interbedded with deltaic and lacustrine shales; or
4) any permeable rocks sealed at the top by low-permeability rocks (see e.g., Rook and Williams, 1942; Herzig et al., 1988; Nemčok et al., 2004a, 2007).

Secondary stratigraphic traps could include bell-shaped alteration-capped reservoirs of the geothermal fields (e.g., McKibben and Hardie, 1997; Raharjo et al., 2002 and references therein). Structural traps could be controlled by reservoir culminations generated by tectonics such as roll-over anticlines in extensional systems or sealed-off fault blocks (see Mailloux et al., 1999).

14 Introduction to hydrocarbons in rift and passive margin settings

Oil and gas deposits, which are the subject of this book, are found in a wide range of sedimentary basin settings. These include:

1) rift setting, such as that of the Gulf of Suez;
2) rift-sag basin combination, represented by West Siberia and the North Sea;
3) passive margin settings, represented by basins in the Gulf of Mexico and West Africa regions;
4) passive margin–foreland basin couplets, such as those in East Venezuela and Alberta; and
5) pull-apart basin setting, such as that in Sumatra.

The extensional structuring in this range of settings may be:

1) preceding the basin development on the passive margin, passive margin–foreland basin couplet, and sag scenarios;
2) intimately associated with the continuing basin development in the rift, rift-sag, and pull-apart scenarios; or
3) even superposed at some later time, being genetically unrelated to previous tectonics, in the case of the orogenic collapse scenario.

Furthermore, rift basins can be subsequently inverted when subjected to far-field intra-plate stress (Ziegler, 1989; Lowell, 1995; Macgregor, 1995), deforming not just the rift sediment fill, but also any overlying sag and/or foreland basin strata (de Graciansky et al., 1989).

Hydrocarbon deposits in all aforementioned basin types are transient features. This is because the long-term preservation of hydrocarbons is difficult because of erosion, fault leakage, gas flushing, and biodegradation, which tend to destroy entrapped hydrocarbon accumulations. For example, oil and gas seep from traps at some finite rate, regardless of seal quality (Miller, 1992).

The generation, migration, and trapping of hydrocarbons in sedimentary basin involve a sequence of geological processes linking a source rock to one or more reservoir/seal couplets and result in a genetically related family of hydrocarbon accumulations (Biteau et al., 2003), which is called the petroleum system (Perrodon, 1980). The petroleum system integrates a broad range of geologic, geodynamic, chemical, thermal, and hydrodynamic phenomena that control these genetic relationships. Magoon and Dow (1994) emphasize the interdependence of elements related to rock strata, such as source, reservoir, seal, and overburden, and the processes of trap formation and generation-migration-accumulation. The petroleum system has a specific duration in geologic time. It reaches its peak when generation-migration-accumulation is most active. The geographic extent of the petroleum system encompasses the volume of active source rock and all discovered hydrocarbon accumulations, seeps, and shows that originate from that source rock volume.

Magoon and Dow (1994) formalized the criteria for identifying, naming, and mapping petroleum systems. Their system name identifies the source rock–reservoir pair with a symbolic qualifier to express the level of certainty in the oil/gas-source correlation supporting the association. Based on such an association, a basin with either more than one source rock or more than one reservoir contains multiple petroleum systems. Each of them requires individual characterization and mapping.

The petroleum systems are usually described by:

1) a map and cross-section (Figure 14.1) showing the geographic extent and relative position of the active source rock volume and traps at the critical moment;
2) a burial history diagram (Figure 14.2) representing the burial and modeled maturation history of the source rock; and
3) a petroleum systems chart (Figure 14.3) displaying the temporal relationships of the elements and processes over the duration of the petroleum system.

(a)

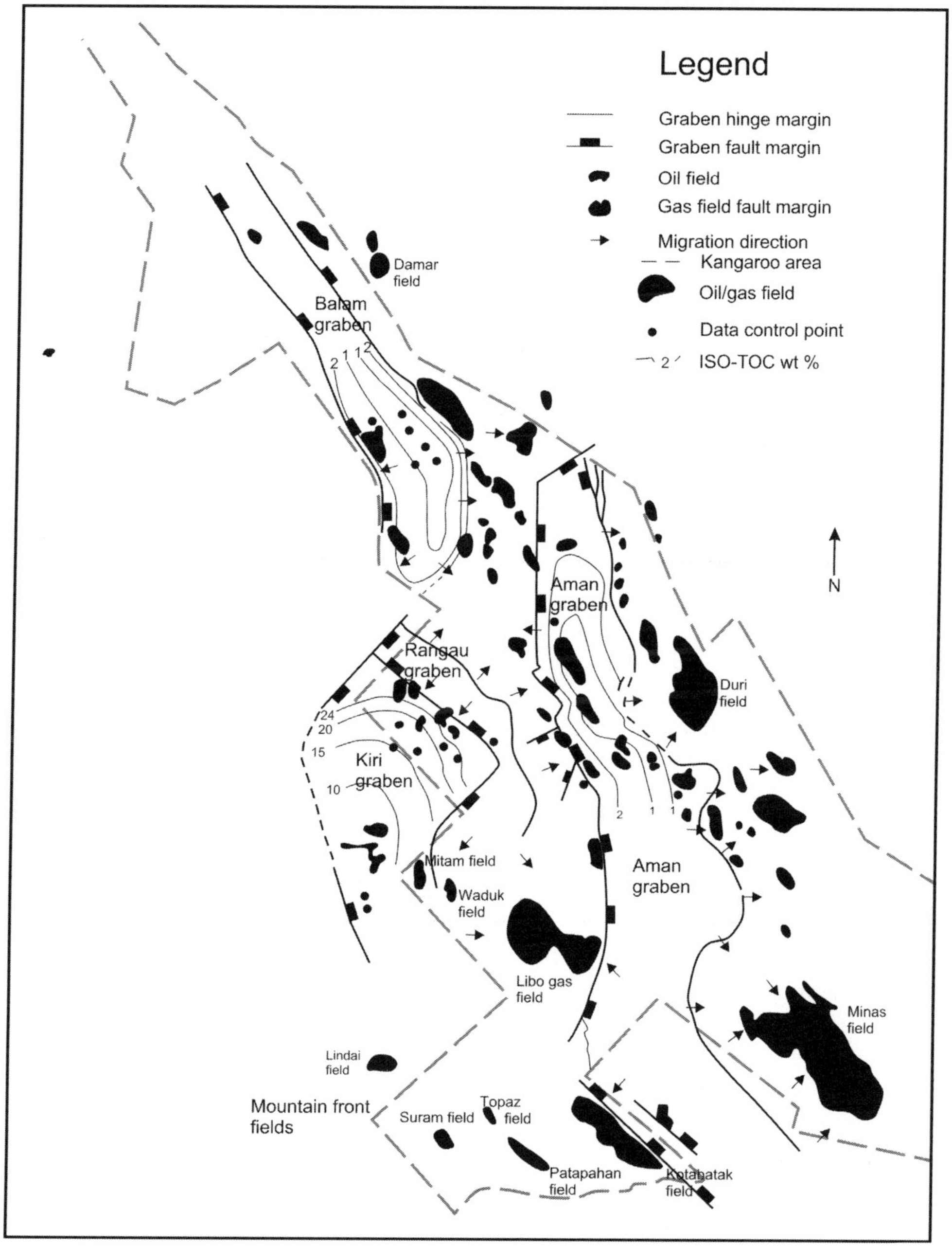

Figure 14.1. (a) Map illustrating petroleum system based on the Pematang Brown Shale Formation and Coal Zone Formation source rock units of the Central Sumatra Basin (modified from Williams and Eubank, 1995). The map shows a combination of data on the total organic carbon, migration directions, and major hydrocarbon occurrences. Migration was controlled by steeply dipping normal and strike-slip faults together with fluvial and deltaic coarse clastic sediments inter-fingering with and overlaying the Pematang Formation. Oil migrated laterally and vertically to both traps occurring at structural highs dividing individual pull-apart basins and traps formed in external zones of pull-aparts. (b) Schematic profile through the Guinea Bissau–southern Senegal salt province, showing potential reservoirs and source rocks with a zone of oil generation and possible migration pathways (Brownfield and Charpentier, 2003).

(b)

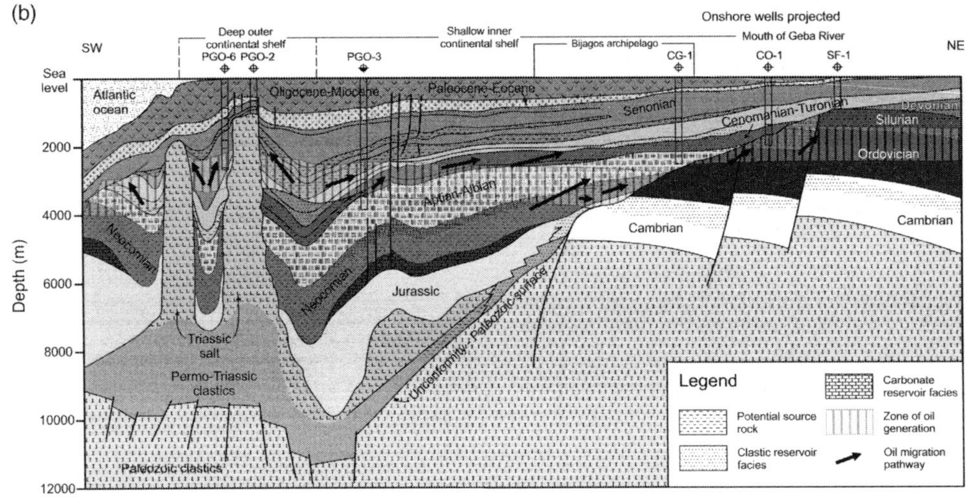

Figure 14.1 *(cont.)*

The actual operation of a petroleum system is usually complex. Its full characterization requires more than just these few mentioned descriptors. Numerous workstation routines are available to model and map the workings of petroleum systems with great specificity.

Demaison and Huizinga (1991) suggested an approach for characterizing elements of petroleum systems that help in hydrocarbon exploration. It is based on three criteria:

1) charge factor, which is determined from the richness and volume of mature source rocks in the basin;
2) migration drainage style, which is determined from the stratigraphic and structural framework of the basin; and
3) entrapment style, which integrates the structural framework of the basin with the distribution of quality seals.

The charge factor can be characterized as supercharged, normally charged, or undercharged. The migration drainage style includes vertically drained or laterally drained styles.

Vertical drainage dominates in basins with seals breached by faults. Basins undeformed by faults are prone to lateral drainage. Entrapment style comprises high-impedance or low-impedance styles. A good example of vertical and lateral migrations prevailing in different parts of the sedimentary fill comes from the Guyana continental margin. Here, normal faults of the extensional domain of the gravity glide system change the lateral-dominated migration occurring below their detachment level to vertical-dominated migration along them.

The known accumulated petroleum volumes of a basin can be compared to an estimate of the total volume of hydrocarbons generated, yielding a ratio called the petroleum system yield, which assists in evaluating the exploration potential of a basin (Biteau et al., 2003). This ratio has its interpretational limitations because (Biteau et al., 2003):

1) even the richest petroleum province does not necessarily have to be the most efficient in retaining hydrocarbons generated;
2) high retaining efficiency can enhance the prospectivity of a basin that contains only lean source rocks; and
3) the calculation is scale-dependent, as indicated by prospects/fields having higher ratios than ratios of entire sub-basins.

These limitations indicate that the oil richness of a province has more controls than just a petroleum system yield; for example, traps, reservoirs, and seals in particular (Perrodon, 1995; Figure 14.4). On the contrary, Figure 14.4 indicates that petroleum system size, basin type, and source potential index (sensu Demaison and Huizinga, 1991) do not represent important factors. A big role in hydrocarbon province richness is the addition of several types of systems. For example, an important enhancement of the hydrocarbon richness of the Sirte Basin, the Gulf of Suez, and the Viking and Central Grabens of the North Sea provinces is the superposition of the sag basin on the rift system (Perrodon, 1995). Similarly, the platform-type petroleum systems, like the Saharan Triassic and West Siberian Basins, which underwent younger subsidence, experienced their richness enhancement. In some cases, like in the Vienna

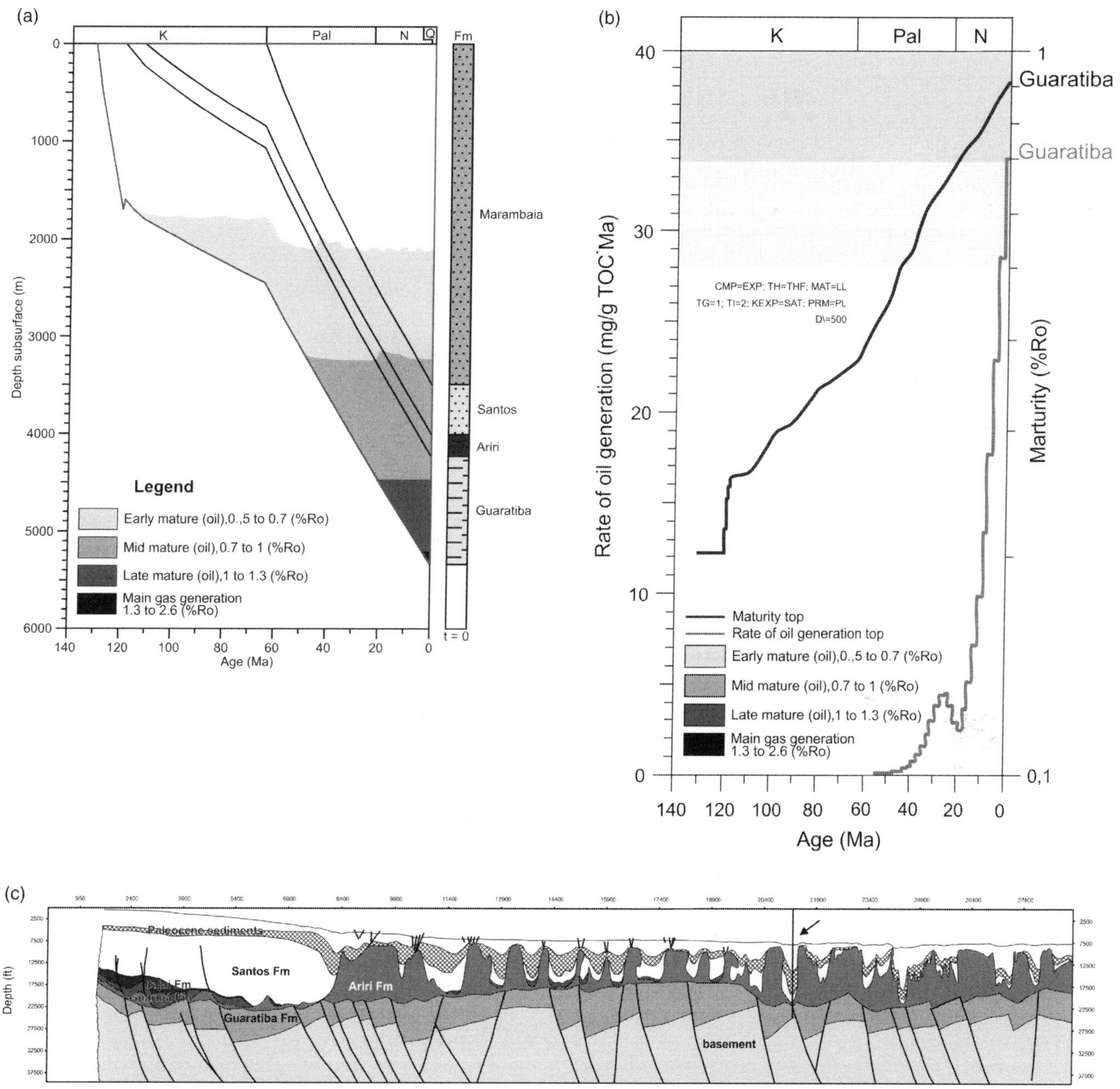

Figure 14.2. (a) Burial history diagram for the Guaratiba-Guaratiba(!) petroleum system in the Santos Basin, Brazil, done for its deep basin portion. The diagram represents the burial and maturation histories of a source rock unit, represented by the upper portion of the Guaratiba Formation, characterized by Type-I Kerogen. Parameters characterizing individual formations in the model are shown in Table 14.1. (b). Expulsion timing of the Guaratiba Formation in the deep portion of the Santos Basin for the pseudo-well shown in (a). Its location is indicated in (c). Note that this source rock reaches a peak expulsion of oil and gas at present day, which postdates both the reservoir deposition and trap formation. In comparison with other parts of the Santos Basin, this modeled pseudo-well has the overburden rather thin. The Guaratiba Formation reached a mid-mature (oil) window (0.7–1% Ro equivalent). Reservoir rocks in the basin include: Guaratiba Formation (alluvial fan/fan deltas), Guaruja Formation (carbonates – only underneath upper slope hinge), Santos Formation (coarser horizons of the slope), Marambaia Formation (coarser horizons of the slope, maybe turbidites in the Paleocene, Eocene, and Oligocene sections in the western part and Eocene in the eastern part of the deep basin). (c). Geological cross-section interpreted along the reflection seismic profile through the deep portion of the Santos Basin. The location of the modeled pseudo-well shown in (a) and (b) is indicated by an arrow.

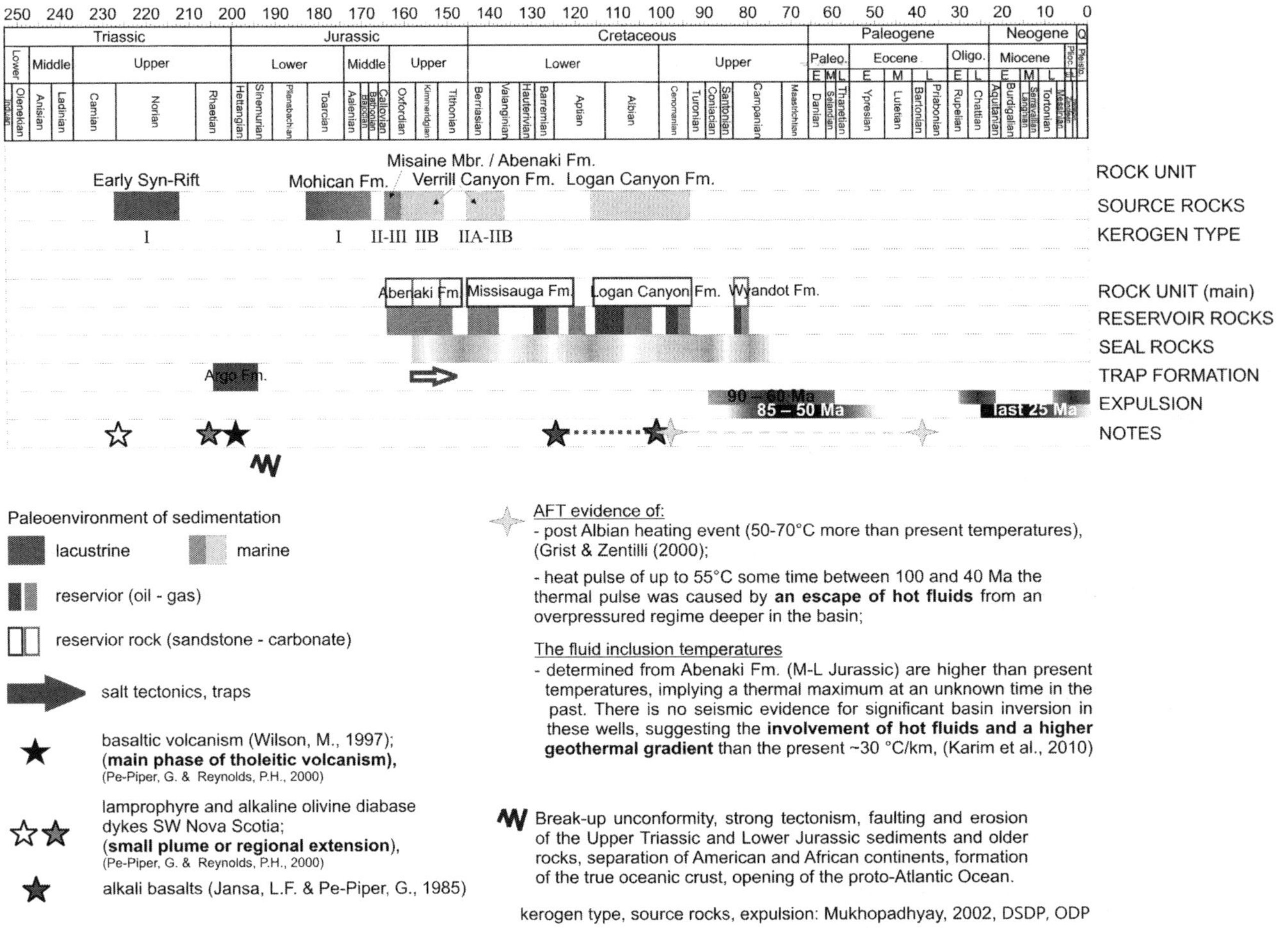

Figure 14.3. Petroleum systems chart for the Nova Scotia continental margin (Kotulová, 2012, pers. comm.). The stratigraphic chart is taken from Gradstein et al. (2004); data for the chart are taken from Jansa and Pe-Piper (1985), Mukhopadhyay (1994, 2002), Wilson (1997), Grist and Zentilli (2000), Pe-Piper and Reynolds (2000), Karim et al. (2010).

and Suez Basins, a younger extension-controlled subsidence can improve the source and reservoir rock quality of the preexisting platform sediments (Perrodon, 1995). Apparently, the richest petroleum provinces are those that comprise petroleum systems of rift type, continental platform type, and orogenic/deltaic type superimposed on top of each other. The examples come from the Maracaibo, East Venezuela, Putumayo, Alberta, and Arabian–Persian Gulf provinces (Perrodon, 1995).

The exploration objective is to identify opportunities for petroleum discoveries assembled in a portfolio of drillable prospects (White, 1993). The specific prospects, in turn, are derived from play concepts based on families of geologically related fields, prospects, and leads, all of a similar geological origin and charged from common petroleum source beds (Rose, 2001). Petroleum system analysis is a critical component in developing and risking hydrocarbon plays and prospects (Smith, 1994; Rose, 2001; Naylor and Spring, 2002). Together with commercial parameters, such as product prices and operating costs, and exposure to the exploration and production project to the fiscal, political, and business environments (Wood, 2003), assessment of the petroleum system is now just one of many components in the appraisal of prospects and ranking by investment efficiency or profitability (Rose, 2001; Wilson, 2002).

Current methods for risking plays and prospects more or less draw from a pioneering appraisal system developed by Shell (Sluijk and Nederlof, 1984) for evaluation of specific exploration prospects based on simple geochemical and geologic models. This system uses Monte Carlo simulations of three factors:

1) hydrocarbon generation, expulsion, and migration leading to the hydrocarbon charge;
2) trapping and retention of oil and gas; and
3) recovery.

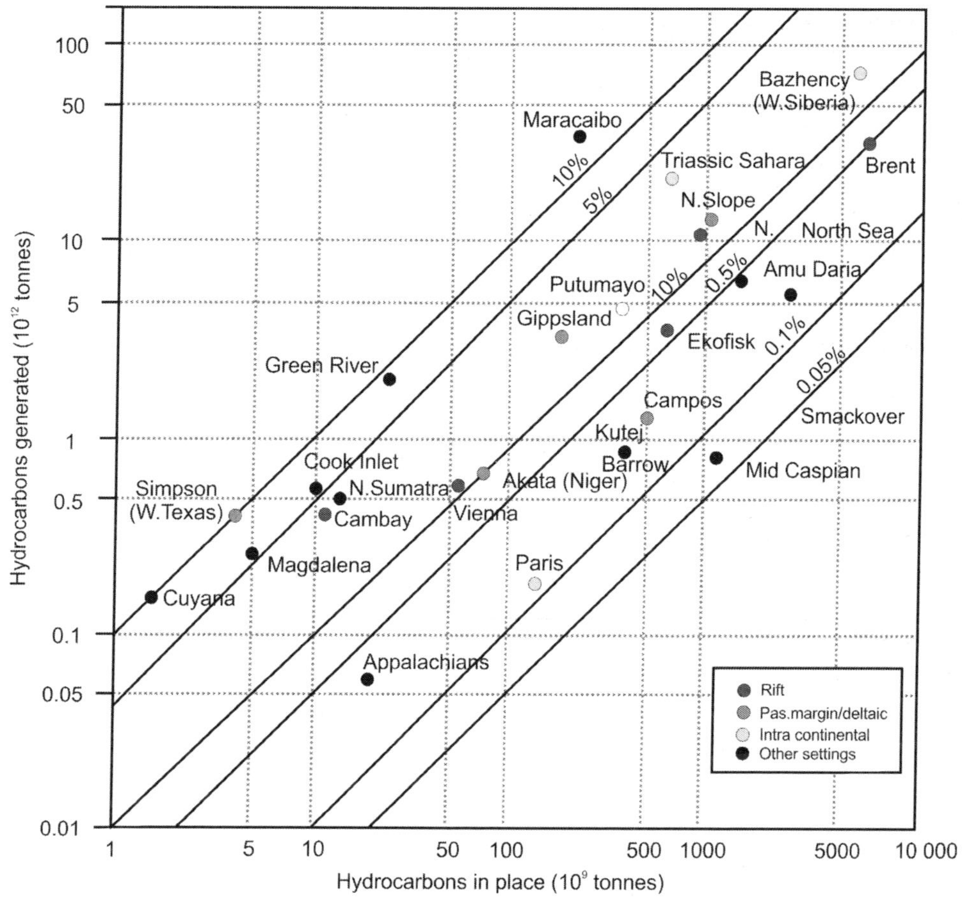

Figure 14.4. Diagram showing the petroleum system yield for a selection of petroleum provinces (Perrodon, 1995 and references therein).

The first two factors are derived directly from an understanding of the petroleum system. The third factor honors the reservoir properties and production technologies. Appraisal systems more advanced than that Naylor and Spring (2002) described retain these factors as prospect-specific and add other Monte Carlo simulations related to the state of technology that characterizes the uncertainties in the structural interpretation as a function of seismic data quality and coverage and play- and area-specific risk factors.

To see the importance of the rift and passive margin settings for oil and gas deposits in comparison with other settings, it is interesting to look at the known distribution of the world's 877 giant oil and gas fields (see Mann et al., 2004; Figure 14.5, Table 14.2), which are those with 500 million bbl of ultimately recoverable oil or gas equivalent, and which hold about 50 percent of the world's petroleum reserves. This distribution includes:

1) 304 fields at continental passive margins fronting major ocean basins;
2) 271 fields in continental rifts and overlying sag or "steer's head" basins;
3) 173 fields at collisional margins produced by terminal collision between two continents;
4) 71 fields at collisional margins produced by continental collision related to terrane accretion, arc collision, and/or shallow subduction;
5) 50 fields at strike-slip margins; and
6) 8 fields at subduction margins not affected by major arc or continental collisions.

Further updated by Horn (2003, 2004), the distribution of giant fields indicates that rift and passive margin settings must have the most favorable conditions for entrapment and preservation of oil and gas deposits; they contain 66 percent of the known giant hydrocarbon fields. Giant fields at continental passive margins represent 35 percent of the entire list, while giant fields at continental rifts and overlying sag basins, especially failed rifts at the edges or interiors of continents, form 31 percent. Explanations for the

Table 14.1. Description of modeled formations from Figure 14.2.

Formation	Time of deposition (Ma)	Type of sediment
Marambaia	0.0–65.0	50% sandstone/25% siltstone/25% shale
Santos	85.0–65.0	25% sandstone/25% siltstone/50% shale
Ariri	119.2–112.0	evaporite
Guaratiba	130.8–120.0	50% sandstone/25% siltstone/25% shale

Table 14.2. Giant oil and gas fields in rift and passive margin settings (data from Horn, 2003, 2004; Mann et al., 2004).

Settings	Province	Number of fields		
rift and sag setting	West Siberian Basin	66	1	66
rift and sag setting	North Sea	43	2	43
rift and sag setting	North Africa	33	3	33
rift and sag setting	North Caspian-North Caucasus-Black Sea	24	4	24
rift and sag setting	Southern part of Persian Gulf	ca. 15% of 151 fields	5	23
rift and sag setting	Siberian Platform	13	6	13
rift and sag setting	Canadian Arctic islands	5	7	5
rift and sag setting	Barents Sea	3	8	3
rift and sag setting	Irian Jaya	3	9	3
				213
passive margin setting	Southwestern part of Persian Gulf	ca. 53% of 151 fields	1	80
passive margin setting	Gulf of Mexico	60	2	60
passive margin setting	offshore West Africa	20	3	20
passive margin setting	offshore Northwestern Australia	17	4	17
passive margin setting	Sunda region	16	5	16
passive margin setting	offshore Brazil	9	6	9
passive margin setting	North Sea	9	7	9
passive margin setting	North Slope Alaska	7	8	7
passive margin setting	offshore West India	4	9	4
passive margin setting	Bass Strait	3	10	3
passive margin setting	North Africa	3	11	3
passive margin setting	offshore circum-East India	2	12	2
				230
		total number of fields		443

dominance of these two settings include (Horn, 2003, 2004; Mann et al., 2004):

1) the existence of high-quality source rocks in lacustrine and restricted marine facies during the early syn-rift stage (e.g., Central Sumatra source rocks – Williams and Eubank, 1995; source rocks of the Espírito Santo and the Campos Santos Basins, Brazil – Trinidade et al., 1995, Koike et al., 2007; source rocks in offshore Angola-Congo – Burwood et al., 1995, Katz and Mello, 2000; North Sea source rocks – Kubala et al., 2002);
2) a high potential for the post-rift sag and passive margin sedimentary packages to contain

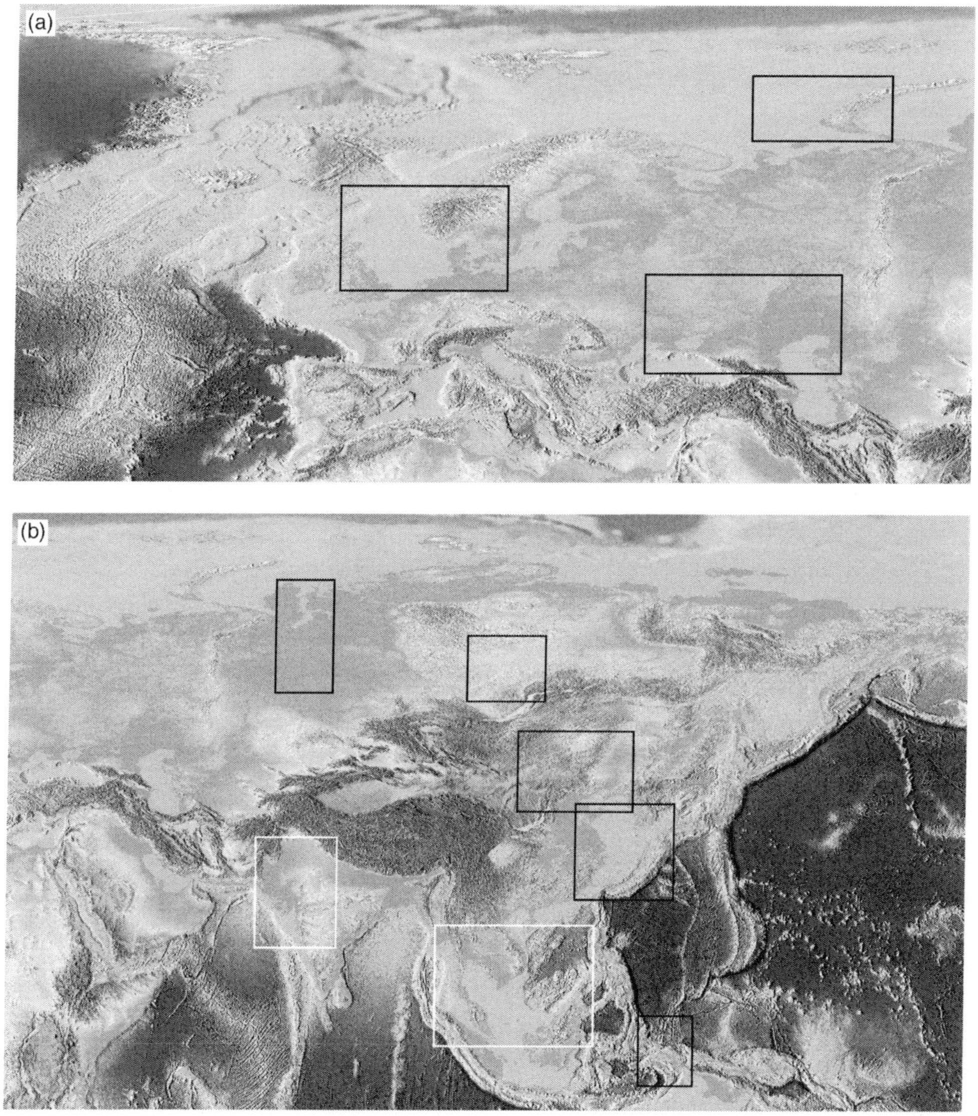

Figure 14.5. Main hydrocarbon provinces holding the world's giant oil and gas fields in rift and passive margin settings (modified from Mann et al., 2004). (a) Europe with the North Sea, North Caucasus, and Barents Sea provinces. Black and white boxes outline provinces in rift and passive margin settings. (b) Asia with the West Siberia, Siberian Platform, India-Pakistan, Sunda, Irian Jaya, Northeastern China, and East China Sea provinces. Black and white boxes outline provinces in rift and passive margin settings. (c) Africa and Arabian Peninsula with North Africa, West Africa, and the Persian Gulf provinces. Black and white boxes outline provinces in rift and passive margin settings. (d) North America with the Gulf of Mexico, the North Slope, and Canadian Arctic Islands provinces. Black and white boxes outline provinces in rift and passive margin settings. (e) South America with the Brazil province. Black and white boxes outline provinces in rift and passive margin settings. (f) Australia with Northwest Australia and the Bass Strait provinces. Black and white boxes outline provinces in rift and passive margin settings.

good reservoirs for hydrocarbons generated by syn-rift source rocks (e.g., Santos Basin – Koike et al., 2007; Campos, Lower Congo, and Gabon Basins – Katz and Mello, 2000);

3) a high potential for the post-rift sag and passive margin sedimentary packages to contain effective seals to help trap the hydrocarbons in underlying syn-rift sediments (e.g., Santos Campos and Espírito Santo Basins – Izundu, 2007; Watkins, 2008); and

4) a tectonic stability following the early rifting that allows the hydrocarbon source rocks and reservoirs to remain undisturbed by subsequent tectonic events acting on distant plate boundaries

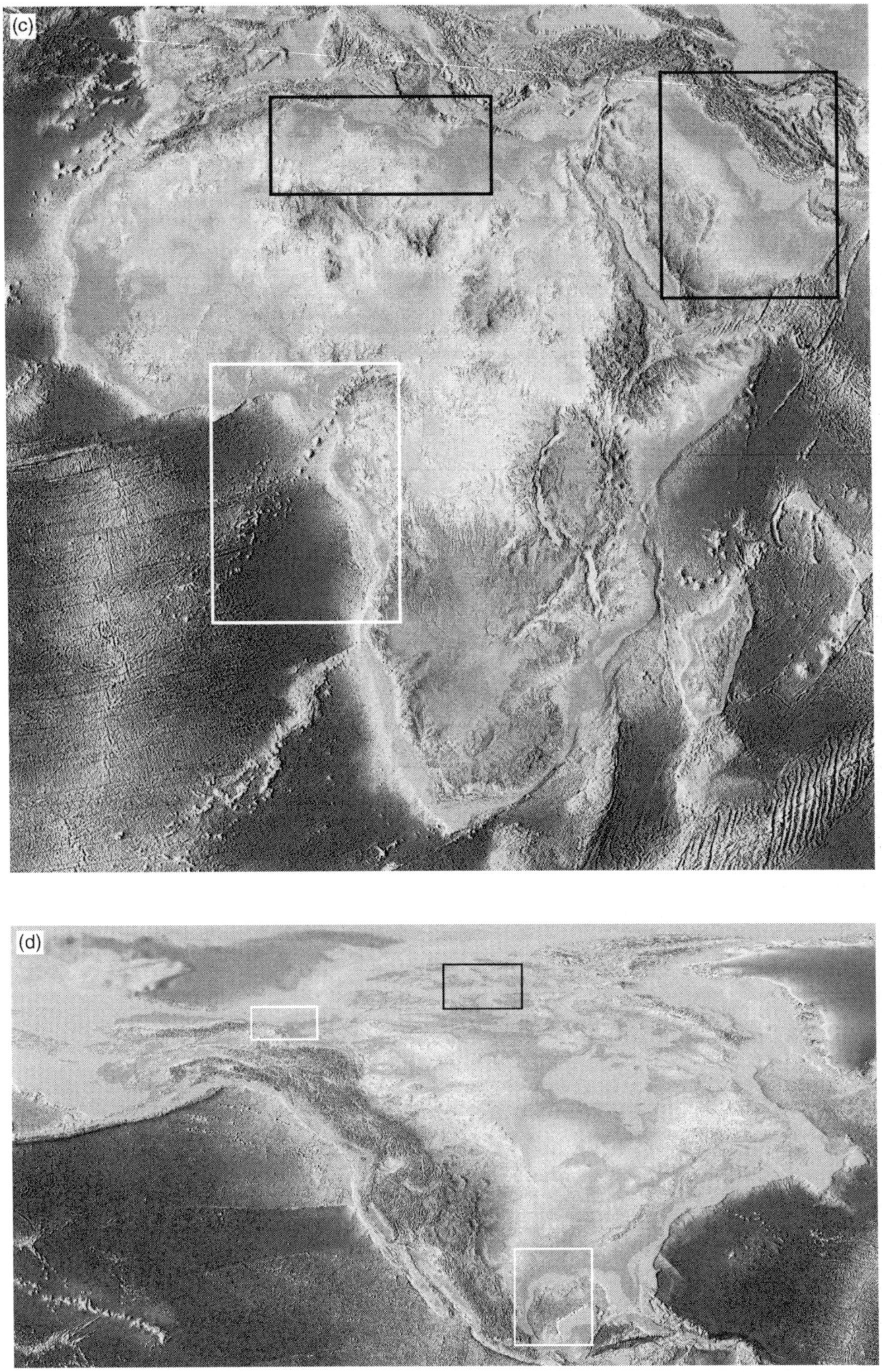

Figure 14.5 (*cont.*)

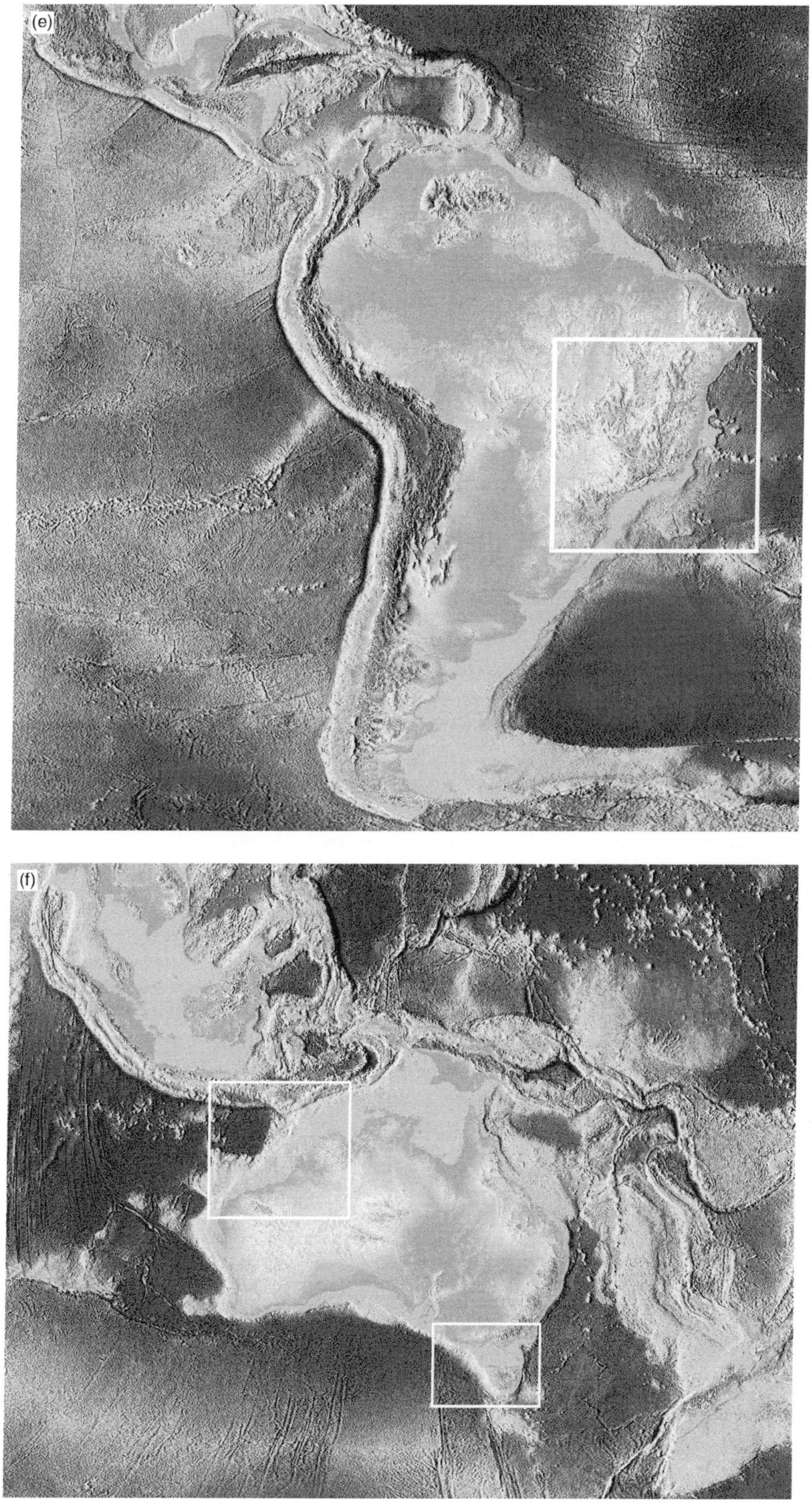

Figure 14.5 (*cont.*)

(e.g., interbedded source and reservoir rocks within the Kimmeridge Clay Formation of the North Sea province – Leythaeuser et al., 1988; Kubala et al., 2002).

A quick inspection of Table 14.2 allows one to see the key provinces that are the main contributors to the list of giant fields in extensional settings. As discussed earlier, the passive margin settings slightly dominate over the rift and sag basin settings.

The leading province in passive margin settings is the Persian Gulf, which is an interesting province because it comprises rift/sag, passive margin, and continental collision settings (Horn, 2003, 2004; Mann et al., 2004). It is practically a large Late Cenozoic foreland basin associated with Arabia–Eurasia collision. Its northeastward deepening is related to the southwestward movement of the overriding Zagros Mountains developed by shortening, which culminated during the Miocene-Pliocene. The continental collision-related traps controlled by shortening are distributed along the northeastern side of this province, while its southeastern side and southern end remain relatively undisturbed by shortening, preserving more-or-less passive margin and rift/sag character of traps, respectively. The main source rocks of the province have Cambrian-Permian and Middle Jurassic ages, while the best reservoirs occur in the Permian and Upper Jurassic sections (Pierce and Khalifa bin Mbarak, 1993). The main strength of this province is in its multiple source rocks and excellent evaporitic seals (Biteau et al., 2003). The petroleum system yields of this province range from 1 percent to 5 percent.

The second most important province in the passive margin settings is the Gulf of Mexico, which is a monomictic passive margin province. This province started to develop with Middle Jurassic rifting in the region between the future North America, the Yucatan Peninsula of Mexico, and South America (e.g., Diegel et al., 1995; Golonka, 2000 and references therein). The resultant crustal architecture contains a relatively small region of oceanic crust in the basin center surrounded by continental margins. The entire province is dominated by gravity glides detached mostly along evaporitic horizons, but sometimes also shale horizons (Diegel et al., 1995; Karlo and Shoup, 2000). Its traps are related to the gravity tectonics, the most important seals are associated with autochthonous and allochthonous salt bodies in various parts of the basin fill, marine source rocks have Upper Jurassic and Neogene ages, and reservoirs are in the Tertiary section.

The third most important passive margin province is the offshore West Africa province, which is, however, not by far larger than its neighbors seen in the fourth and fifth places in Table 14.2. Its development started with the Neocomian-Aptian rifting stage, which was followed by a passive margin development stage (e.g., Meyers et al., 1996 and references therein; Rosendahl et al., 2005). The transition between these two stages in most areas of this province is marked by an evaporitic section that supplies the most important seal. Source rocks include both syn-rift and post-rift lithostratigraphies (Doust and Omatsola, 1990; Katz and Mello, 2000). The majority of proven reservoirs occurs in the post-rift section (Mann et al., 2004). The main strength of some areas in this province is in their very high petroleum system yield. A combination of Cenomanian-Campanian turbiditic channels and lobes of the Cap Lopez, Azile, and Pointe Clairette Formations with good source rocks characterized by the marine organic matter in the same section in Gabon serves as an example, having a petroleum system yield of 32 percent (Biteau et al., 2003). The Angola Iabe petroleum system of this province reaches a petroleum system yield of 4 percent. An extra strength of this province is in the superposition of different basins on top of each other. This is documented by the existence of rich rift zones, such as those having the Lower Congo pre-salt section and the Angola Bucomazi section underneath the passive margin section (Biteau et al., 2003).

According to the number of giant fields, the leading province in rift and sag basin settings is the West Siberian province, followed by the North Sea and North Africa provinces in the second and third places, respectively.

The West Siberian province is controlled by the Jurassic-Quaternary sag basin overlying the Permian-Triassic failed-rift system. Its source rocks and reservoirs are in the Upper Jurassic-Cretaceous shale facies and the Cretaceous sag fill, respectively (Mann et al., 2004). The main strength of this province is in the superposition of different basins on top of each other in the same province, although, surprisingly, the petroleum system yields in the province do not exceed 5 percent (Biteau et al., 2003).

The North Sea province is a failed-rift province as well, initiated by the Middle Jurassic rifting (e.g., Helland-Hansen et al., 1992; Fjellanger et al., 1996; Johannessen et al., 1996). This rifting continued roughly until the middle Volgian in most parts of the North Sea, apart from the continuing rifting in the Moray Firth Basin, which did not stop during the deposition of the Kimmeridge Clay Formation and its equivalents (Fraser et al., 2002). The best source rocks are distributed in the Middle and Upper Jurassic stratigraphic intervals (Larsen and Jarvik, 1980; Kubala et al., 2002). The reservoirs are in the Jurassic and older strata, which are affected by widespread faulting, and the Cretaceous-Paleogene strata with minimum faulting (Moss et al., 2002). The main strength of this province is in the interbedded source rocks and reservoirs or in, at least, a short fault-controlled contact between them (Biteau et al., 2003). Note that some of the petroleum systems such as the Kimmeridge-Brent (!) and Mandal-Ekofisk (!) petroleum systems (Klemme, 1994; Biteau et al., 2003) in this province are rich but not especially efficient.

The North African province is a composite of the Algerian Illizi province and the Libyan Sirte rift branch. Its development started with the development of the Paleozoic orogen, which was followed by the Neocomian-Barremian rifting (Harding, 1984; Gras and Thusu, 1998; Ambrose, 2000). The subsequent time period was characterized by the subsidence of the entire province and later convergence. Source rocks occur in the pre-rift Ordovician and Silurian strata, and in the post-rift Upper Cretaceous shale facies. The best reservoirs are represented by the carbonate buildups on structural highs and traps are both structural and stratigraphic (Mann et al., 2004). The main strength of the Libyan Murzuk Basin from this province, for example, is in the superposition of source rock and reservoir, which is, thus, responsible for very short migration pathways. Here, the Silurian Graptolite shales either cover or laterally onlap the Ordovician Hawaz–Memunyat reservoirs.

The interpretation of the giant oil and gas field distribution and its controlling factors allows us to anticipate further discoveries (Horn, 2003, 2004; Mann et al., 2004; own data):

1) in deepwater portions of numerous passive margins, such as the margins of eastern and northern South America, West and East Africa, East and West India, and the Gulf of Mexico, which are associated with older petroleum systems also including syn-rift systems;
2) by expanding the existing Persian Gulf province further south into Yemen and the Arabian Peninsula, and further north into Iraq;
3) by expanding the West Siberian Basin into the Arctic offshore area;
4) by radially expanding the Illizi Basin of Algeria;
5) by continuing discoveries in Southeast Asia, where the Cenozoic rift, passive margin, and strike-slip environments coexist around the South China Sea or within the largely submerged Sunda continent; and
6) by expanding the recognized petroleum provinces of the Black Sea–Caspian region associated with the closure and burial of the northern Tethyan passive margin.

If one characterizes the rift and passive margin exploration settings, and differentiates them from the continental platform and orogenic settings, it turns out that these two settings are relatively unique. Rift petroleum systems and petroleum systems of passive margin regions, which are not affected by any important delta, have the highest values of the source potential index among all systems (sensu Demaison and Huizinga, 1991). This means that they are prone to hydrocarbon generation as opposed to reservoir-prone continental platform settings and trap-prone orogenic and deltaic settings. Only the passive margin regions with large deltas in the rift and passive margin settings are trap-prone.

Rift petroleum systems and petroleum systems of passive margin segments, which are not affected by any important delta, usually rely on the source rocks with type I kerogens, in comparison to the continental platform and orogenic/deltaic settings relying on type II and type III kerogens, respectively (Perrodon, 1995).

The extremely important character of rift petroleum systems and petroleum systems of passive margin segments, which are not affected by any important delta, is their dominating vertical migration, as opposed to lateral-dominant and combined lateral/vertical migrations of continental platform and orogenic/deltaic settings, respectively (Perrodon, 1995). Vertical migration seems to be one of the key factors in controlling the "oil richness" of petroleum provinces and systems (Perrodon, 1995; Figure 14.6). As can be seen in Figure 14.6, provinces with vertical migration are in the richer end of the spectrum of provinces while provinces with lateral migration occupy the poorer end of the spectrum. Although there are important exceptions to this trend, the general understanding is that lateral migration to intermediate-long distances may result in significant hydrocarbon losses along the migration path. This is illustrated by the occurrence of significant tar belts such as those located in the Alberta, East Venezuela, and Volga-Urals provinces (Perrodon, 1995).

Note that the richest provinces with vertical drainage separate themselves from the other provinces with vertical drainage by (Perrodon, 1995):

1) exceptionally efficient seal; or
2) recent hydrocarbon charge; or
3) source rocks acting also as seals.

Examples of the first case come from the West Texas, Saharan Triassic, and Arabian–Persian Gulf provinces (Weeks, 1952).

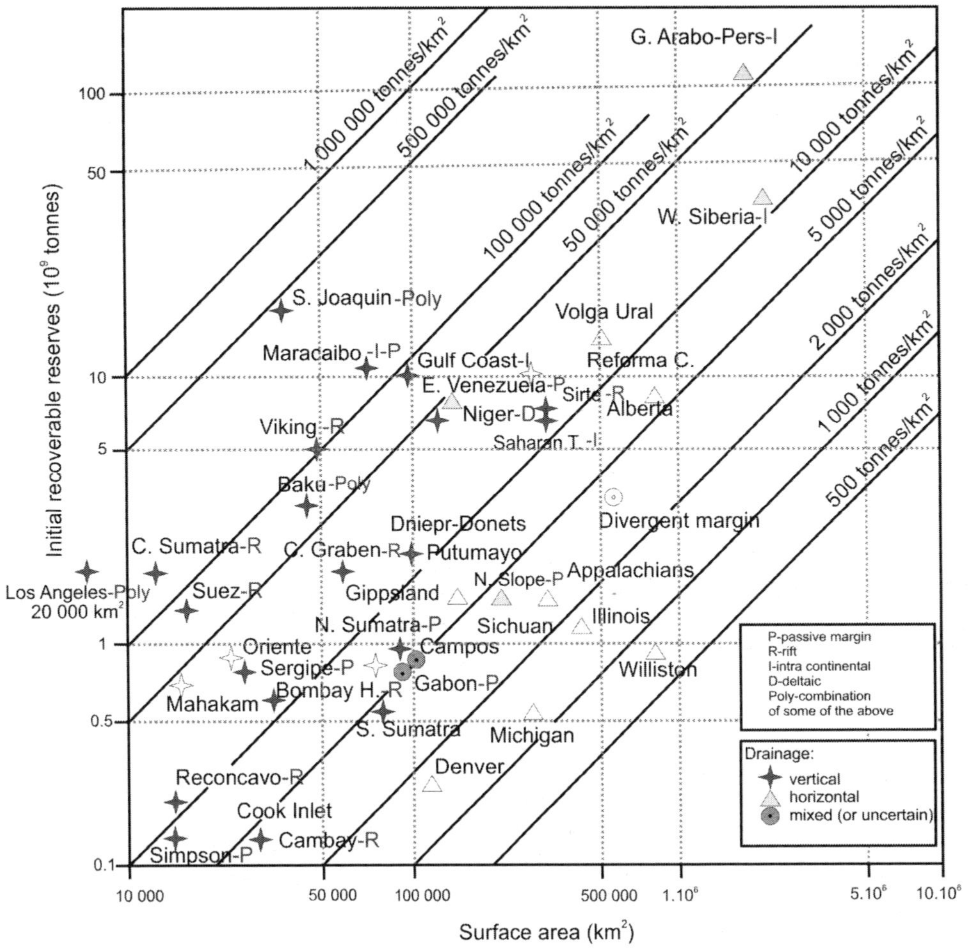

Figure 14.6. The amount of recoverable reserves per area of the petroleum province expressing the oil richness (Perrodon, 1995).

Portions of the Taranaki Basin, the South Caspian Basin, the Bohai Bay Basin, and the North Sea and northern Gulf of Mexico provinces represent the second case (e.g., Anderson et al., 1994a, 1994b; Roberts and Carney, 1997; Feyzullayev et al., 2001; Guliyev et al., 2001; Kubala et al., 2002; Losh et al., 2002; Hao et al., 2009).

Examples such as the upper Oxfordian–upper Ryazanian Kimmeridge Clay Formation in the North Sea (Kubala et al., 2002), the Senonian section of the Gabon Basin (Biteau et al., 2003), and the Upper Cretaceous shale of the Sirte Basin document the third case (Perrodon, 1995). Locally in the Gabon Basin, the petroleum system yield is as high as 60 percent. The global average Biteau and colleagues (2003) published for the third case reaches a petroleum system yield of 31 percent.

15 Models of source rock distribution, maturation, and expulsion in rift and passive margin settings

Source rock characteristics

Sediment with a content of organic matter higher than 0.5 percent is a candidate for being a petroleum source rock. The dead organic matter starts to undergo changes already during its transport to the depositional environment of the future source rock. The changes, which take place during the transport and early burial, are called diagenesis. They are characterized by organic matter transformation into large molecules; the largest of which are known as kerogen, representing precursors to oil and gas produced by subsequent catagenesis.

The earliest gas production, however, already occurs during diagenesis. It results in biogenic methane, converted from the organic debris by microorganisms called methanogens. Natural examples of biogenic methane come from various extensional basins of the Pannonian Basin system, developed behind the Carpathian Neogene accretionary wedge.

With advancing burial, the decreasing permeability and increasing temperature progressively stop the organic diagenesis and enhance thermally controlled reactions during catagenesis. Thus, given a sufficient amount of burial, the kerogens break down through catagenesis to form hydrocarbons: oil under temperatures between 110 and 150°C and gas under temperatures between 140 and 250°C. The chemical alteration of the organic matter, which started with the death of the respective organism and its burial, ends with thermal conversion forming a carbon residuum (Figure 15.1).

Note that only less than 1 percent of the annual photosynthetic production does not get decomposed and undergoes burial in sediments, and only a portion of it may feed the future petroleum system (Hunt, 1979; Waples, 1985). The productivity is controlled by many factors, among which the following ones are the main:

1) nutrient availability;
2) predators;
3) temperature;
4) light intensity;
5) carbonate supply; and
6) water geochemistry.

The organic richness in the future source rock is controlled by productivity, preservation, and dilution. Thus, the organic matter type in hydrocarbon source rock and type of derived hydrocarbon are controlled by the precursor organic matter type, depositional environment, and climatic conditions.

The terrestrial organic material is composed mainly of plants. They are rich in cellulose and lignin, which are typical for higher plants, and contain small amounts of other carbohydrates, proteins, and lipids. The marine organic material contains abundant proteins, carbohydrates, and lipids.

During the early diagenesis, microorganisms decompose organic matter through enzymatic processes. Proteins and carbohydrates undergo hydrolyze, yielding amino acids and sugars that microorganisms use as a source of energy. Lipids and lignins are less degradable. Diagenesis converts dissolved organic matter to a particular form and integrates it via polymerization and condensation into kerogen.

Because of a high degree of microbial nutritive selectivity, the lipids and proteins from the organic matter are consumed first. Then they are followed by carbohydrates, which are converted into individual amino acids and sugars. Lignins are consumed last.

The source material for hydrocarbons comes from four organic matter types (Tissot and Welte, 1984; Berkowitz, 1997):

1) lipids;
2) amino acids;
3) carbohydrates; and
4) lignins.

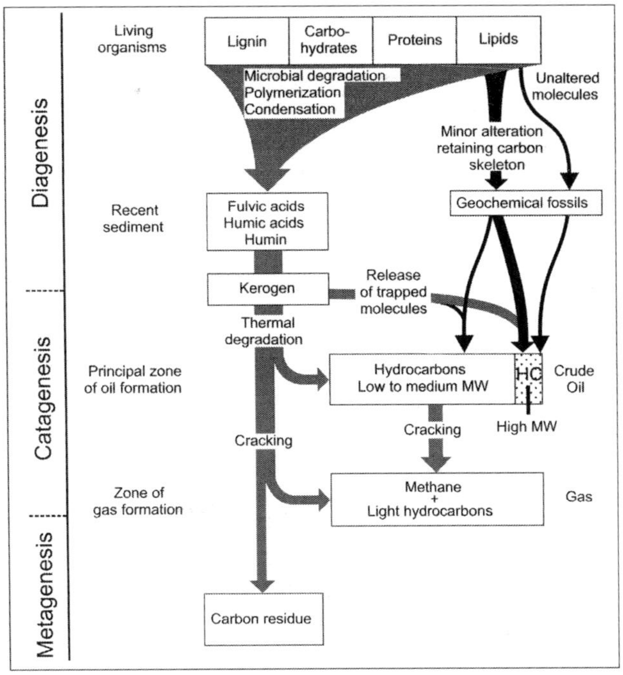

Figure 15.1. Stages of organic matter transformation from living organisms to hydrocarbons and inert carbon (Tissot and Welte, 1984).

Lipids are the most important type. They are a source of energy in living organisms, represented by aliphatic hydrocarbons, which can be further divided into fatty acids, waxes, terpenes, and steroids. Being buried, lipids become altered to carboxyl acids and later to alkanes through a process of putrefaction and low-temperature diagenesis.

Amino acids are present in sediments, either bound in peptides, proteins, humic acids, and protein fragments or existing in a free state. They are relatively abundant in seawater and pore water in the uppermost 10 cm of sediment section. Polar organic matter, such as amino acids and sugars, could be adsorbed to mineral particles, especially clay minerals.

Carbohydrates are hydrogen-poor, oxygen-rich compounds that have a characteristic O/C ratio of 2. They are a collective term for various sugars and their polymers, formed by polysaccharides such as cellulose and chitin. The higher plants synthesize the largest amount of cellulose. Polysaccharides are largely water-insoluble substances but they are converted by hydrolysis into water-soluble sugars.

The carbohydrates represent one of the most abundant components of plants and animals, serving as a source of energy and forming the supporting tissues of plants and certain animals. During extensive early diagenesis, carbohydrates can be strongly re-mineralized by microorganisms under aerobic and anaerobic conditions.

Residual carbohydrates not used by microbes, together with mostly plankton-derived other autochthonous and land-derived allochthonous material, undergo transformation by poly-condensation in the water column and subaquatic sediments into fulvic and humic acids and kerogen.

Lignins are complex polymers characterized by aromatic units and phenols. Aromatic compounds are usually not synthesized by animals but they are typical for plant tissues. From an evolutionary point of view, lignins appeared in a distinct quantity in association with the arrival of land plants during the late Silurian (e.g., Glasspool et al., 2004).

Lignin occurs as a three-dimensional network located between the cellulose micelles of supporting tissues of plants. Lignin together with cellulose forms the two organic matter components that are decomposed into humic substances during peatification and humification. Lignin is prone to degradation mainly in the presence of oxygen. It is predominately decomposed by wood-destroying fungi, which carry enzymes capable of attacking the lignin complex that would be otherwise hydrolyzed only with difficulty (Taylor et al., 1998).

Because of their relatively high chemical stability and resistance to microbial degradation, lignins can survive early diagenesis and have a good chance of remaining preserved. They are often considered the main input into vitrinite formation (Stach et al., 1982).

The content of organic matter in sediments is controlled by (Hunt, 1979):

1) organic deposition rate;
2) mineral deposition rate;
3) presence of oxygen; and
4) level of biological activity feeding on organic matter.

For example, the organic matter in the coastal region and upwelling areas is derived from competing marine and terrestrial environments. If the oxygen is lacking, the marine organic matter, represented by dead organisms such as phytoplankton, algae, diatoms, and dinoflagellates, is partly decomposed by fermentation. The organic matter is more or less intensively microbially decomposed into sapropel and diagenetically transformed into bituminite, which represents unstructured organic matter, and is characterized by a low-oxygen and high-hydrogen content.

The retexturing of the primary organic matter in upwelling areas can be accompanied by the incorporation of sulfur that leads into petroleum generation at rather low temperature conditions (Taylor et al., 1998).

On the contrary, the terrestrial organic matter, represented by cellulose and lignins, is degraded by anaerobic bacteria with different intensity during early diagenesis. It is subsequently diagenetically transformed into high-oxygen and low-hydrogen kerogen, which can be petrographically characterized by the presence of vitrinite and inertinite.

In an oxidizing environment, the organic matter destruction can go very far because aerobic bacteria use molecular oxygen for oxidizing organic compounds to carbon dioxide and water, apart from respiration. The oxidization can persist until the entire amount of organic matter is destroyed.

The marine organic matter particles falling through the well-oxidized water column are consumed by zooplankton and aerobic organisms. The remaining most resistant particles can be eventually consumed by benthic organisms. Burrowing organisms increase the oxygen diffusion below the seafloor by bioturbation. This leads to organic matter oxidization and subsequent destruction.

A highly oxygenated depositional area, characterized by favorable conditions for destruction of the autochthonous plankton, develops a "nonmarine" character of kerogen, resulting in its high-oxygen and low-hydrogen content. This kind of more oxic environment is characterized by various organisms feeding on lipids and other hydrogen-rich molecules.

In an oxidizing environment, the terrestrial organic matter, which is already characterized by high oxygen and low-hydrogen contents, undergoes extra oxygen enrichment and hydrogen decrease due to oxygenation. In some scenarios, type III organic matter can obtain an organic matter structure representing type IV organic matter.

The principal organic matter components in petroleum source rock are kerogen and bitumen. The former is represented by a solid organic matter, which does not dissolve in organic solvents. The latter is formed by soluble liquid hydrocarbons and gas, which are generated from the kerogens. Another component of the organic matter in source rock is the natural gas, typically adsorbed on molecular or micro-fracture surfaces within the kerogen or dissolved within the bitumen. The content of bitumen and natural gas is controlled by the progress of original sediment transformation. If the organic matter in sediment did not enter the catagenesis (see Figure 15.1), it is represented by kerogen.

Kerogen is formed by a combination of macerals and reconstituted degraded products of organic

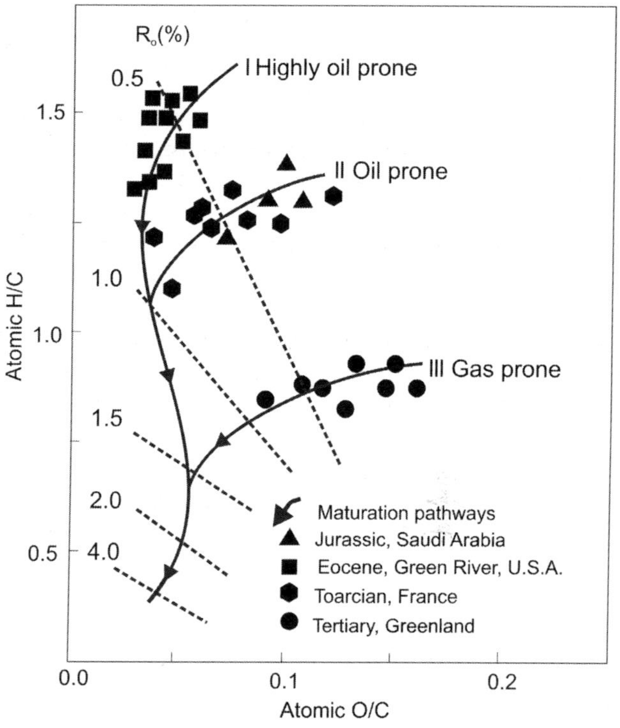

Figure 15.2. Van Krevelen plot of atomic H/C and O/C ratios for kerogens (Peters, 1986). It is used to distinguish the main hydrocarbon-generative types: types I, II, and III. The dashed lines with numbers indicate measured vitrinite reflectance, Ro, used as a proxy for thermal maturity of kerogen, described in the text.

matter, which maintain a portion of their original plant structure. The kerogen chemistry correlates to specific precursor organisms and the mechanism of their diagenesis and catagenesis. It can be characterized by the content of oxygen in relationship to hydrogen content (see Figure 15.2). The kerogen classification contains five primary types (Tissot et al., 1974; Orr, 1986; Peters, 1986; Baskin and Peters, 1992).

Type I kerogens are characterized by high hydrogen content, and contain Alginite maceral. The original organic matter of this maceral is freshwater algae. Typical H/C ratios representing paraffinic kerogens are higher than 1.25 and O/C ratios are less than 0.15. These kerogens are strongly oil-prone, having hydrocarbon yields as high as 80 weight percent and producing light oils. They are typically found in boghead coals and shales. Type I kerogens principally derive from lacustrine algae such as Botryococcus. It is associated with lacustrine depositional environments and few unusual marine environments such as that producing tasmanite. Examples of this kerogen come from the Middle Jurassic Parrot Boghead coal of the Inner Moray Firth Basin, the North Sea, which contains abundant freshwater Botryococcus algae and reaches TOC values as high as 6 percent (Kubala et al., 2002) and from the Middle Eocene Green River shale, Wyoming, Utah, and Colorado. Using the discussed characteristics as a template, some Persian Gulf Cretaceous limestones would also qualify for this category of kerogens.

Type II kerogens contain an amount of hydrogen smaller than the amount of hydrogen in type I kerogen, and contain macerals such as Exinite, Cutinite, Resinite, and Liptinite. The organic material of these macerals is pollen and spores, land-plant cuticles, land-plant resins, and land-plant lipids and marine algae, respectively. Associated H/C ratios are smaller than 1.3 and O/C ratios are between 0.03 and 0.18. These kerogens are both oil- and gas-prone, and derive from sources including marine algae, pollen, spores, leaf waxes, fossil resin, and bacterial-cell lipids. Despite their various origins, type II kerogens are placed into the same category based on their capability to produce oil. They are associated with marine depositional environments. Their hydrocarbon yields are up to 60 weight percent. Typical examples come from the Lower Toarcian shale of the Paris Basin, France, and the Aptian-Turonian sections of the Atlantic margins.

Type III kerogens are characterized by a small amount of hydrogen and contain Vitrinite maceral. It comes from woody and cellulosic material of land plants, that is, terrestrial material lacking fat and wax components. Associated H/C ratios are smaller than 1 and O/C ratios lie among 0.03 and 0.3. These kerogens are gas-prone. They associate with terrestrial depositional environments, having planktonic remains practically absent. Examples come from the Upper Cretaceous section of the Douala Basin, Cameroon, and the Pannonian section of the Danube Basin of the Pannonian Basin system.

Type IV kerogens are practically inert residual organic matter incapable of hydrocarbon generation. Their hydrogen contents in their immature state, expressed as atomic H/C values, are less than 0.65 and correlate with a kerogen type. They contain reworked organic debris and oxidized material of various origins. Known examples come from several marginal facies of the upper Oxfordian–upper Ryazanian system of the Kimmeridge Clay Formation and its lateral equivalents in the North Sea (Brown, 1990). Some marginal facies are dominated by terrigeneous Exinite macerals such as type IV Sporinite and Cutinite, and known to produce some oil, although not in commercial quantities. Furthermore, certain areas of the Norwegian-Danish Basin and Central Graben, North Sea, contain type IV kerogens dominating over local type II kerogens thanks to the transport of terrestrial material from land by turbidity currents far into the deep basin (Kubala et al., 2002).

Type V kerogens are represented by type II-S kerogens. They have an elemental composition similar to type II kerogens but contain at least 6 percent organic sulfur. Many of these kerogens also have high nitrogen contents. Type II-S kerogens start generating heavy oils of 10–20° API gravities at temperatures lower than generation temperatures of type II kerogens (Baskin and Peters, 1992). High-sulfur kerogens are mostly produced by marine depositional environments because freshwater environments have typically low sulfur content. Such sulfur-prone environments are typically characterized by extensive sulfate reduction and lack of Fe^{+2} ions, including organic-rich, anoxic, marine, non-clastic environments. The δ34 S of type II-S kerogens is controlled by sulfur incorporation during the early diagenesis at relatively low temperatures and reactions during catagenesis at higher temperatures (Amrani et al., 2005). At the beginning of the catagenesis, II-S kerogens are rich in thermally unstable, organically bound sulfur. This low thermal stability causes the early generation of bitumen and heavy oil and the release of an increased amount of H_2S.

Worldwide, the occurrence of type II-S kerogen should be considered wherever potential source rocks are in thick carbonate, marl, diatomaceous, and evaporitic sequences (Orr, 1993). Mentioned oils are generated at much lower maturities than those observed for other kerogen types. Although different from type II kerogens by increased sulfur content, they are not visually different. A typical example comes from the Miocene Monterrey Formation, California (e.g., Hold et al., 1998). Its kerogen is known to decompose at a faster rate than type II marine kerogen (see Tomic et al., 1995). Further differences from type II kerogen are:

1) earlier generation of hydrocarbons higher than C_{14};
2) increased H_2S content; and
3) dominant sulfur moieties in the pyrolysis extracts; i.e. alkylthiophenes.

Each mentioned kerogen-forming maceral group is characterized by a specific kerogen pathway on the Van Krevelen plot (Figure 15.3). Regardless of the pathway steepness, organic matters of all types advance toward lower values of H/C and O/C with an increasing level of organic matter transformation. In fact, the organic matter gains the characteristics of type III and IV kerogens at a high degree of its maturity regardless of its original composition.

Note that kerogens usually consist of different particles, that is, macerals, which coal petrologists have

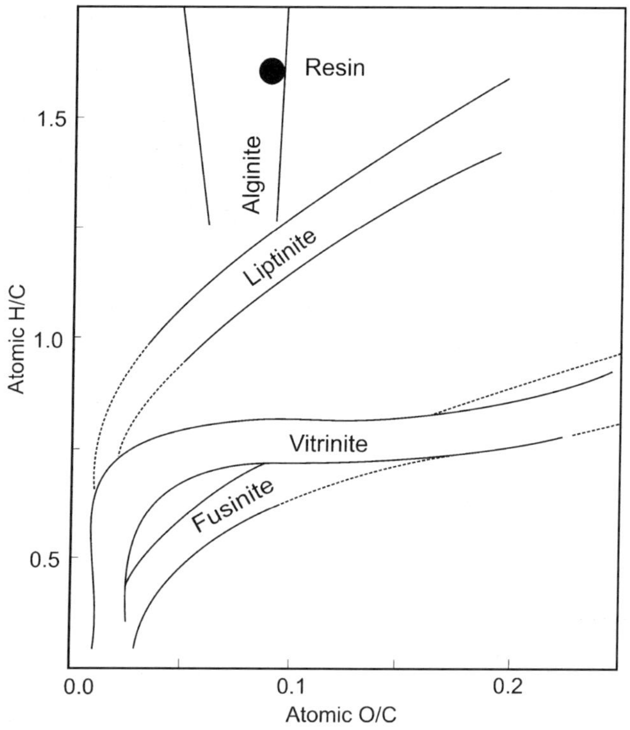

Figure 15.3. Kerogen-forming maceral groups shown in the Van Krevelen plot (after Powell and Boreham, 1994).

defined to allow correlating chemical characteristics of a certain kerogen type with its visual appearance. Such correlation is not perfect because there is no perfect biological separation of various types of living organic matter.

One more problem with identification is that kerogens are present in source rocks frequently in mixtures, as discussed later in the text on source rock deposition and preservation.

While macerals allow inferring the original depositional environment and provide some clue about organic matter maturity, the molecular indicators, known as biomarkers, help to determine the depositional environments, precursor organisms, and transformation pathways leading to the buried organic matter (Philp and Lewis, 1987; Waples and Machihara, 1991; Dahl et al., 1994). They are routinely used to identify oil-sourcing sediment and allow us to perform oil-oil correlations (Curiale, 1994).

For example, the depositional environments can be distinguished based on the chemical identification of their organisms. While the terrestrial organisms synthesize mainly C_{27}, C_{29}, and C_{31} normal paraffins, marine organisms synthesize mainly C_{15}, C_{17}, and C_{19} (Hunt, 1979). The chemistry can help with a more detailed distinction of terrestrial environments. The relative abundance of C_{27}, C_{28}, and C_{29} sterols (Volkman, 1988) can separate higher land plants, which are C_{29}-dominant, from marine phytoplankton, which is C_{27}-dominant. For example, the Middle Jurassic Brora Coal Formation of the Inner Moray Firth Basin is characterized by a very high pristane/phytane ratio (8.67), and a high proportion of C_{29} steranes, indicating the terrestrial organic material derived from the higher plants (Kubala et al., 2002). Oils of the Harald field in the Danish North Sea coming from the Middle Jurassic coal-bearing Bryne Formation are characterized by a high pristane/phytane ratio of 4.09–4.24 and sterane distributions dominated by C_{29} compounds (Petersen et al., 1996).

As shown by a systematic study of the Santos Basin oils, a C_{26}/C_{25} triciclic terpanic ratio is a good differentiator of marine and lacustrine depositional environments; the former indicated by values below one and the latter indicated by values higher than one (Koike et al., 2007). The relative abundance of C_{34} and C_{35} hopanes most probably indicates an anoxic depositional environment, which is further indicated by decreased pristane/phytane ratios (Kubala et al., 2002). Yet another indication of anoxic conditions is the presence of C_{27} trisnorhopane (Mello et al., 1988; Peters and Moldovan, 1993).

The controls of carbon isotope ratios are poorly understood but seem to reflect kerogen type variations and the influence of the maturation process within the specific formation as documented by source rock character synthesis done for the North Sea petroleum province (Kubala et al., 2002).

Biomarkers indicate different depositional conditions for the Kimmeridge and Heather Formations in the North Sea (Chung et al., 1992). For example, anoxic depositional conditions for deep graben facies of the late Oxfordian–late Ryazanian Kimmeridge Clay Formation are characterized by lighter carbon-isotope ratios, higher hydrogen index values, and generally lower pristane/phytane ratios, which are below 1.5. On the contrary, facies of the upper Callovian–lower Kimmeridgian Heather Formation deposited under dysaerobic or only intermittently anoxic conditions yield heavier carbon-isotope ratios, lower hydrogen index values, and generally higher pristane/phytane ratios, which are higher than 1.5.

Depletion in ^{13}C can also be used as an environment indicator but it is not straightforward because it is further influenced by the maturation process (Kubala et al., 2002). Hydrocarbons generated within and expelled from highly mature source rocks, exhibiting high API gravities, are known to associate with enrichment in ^{13}C (Schoell, 1984; Sofer, 1984). North Sea oils depleted in ^{13}C with carbon-isotope ratios around −30‰ were correlated with the upper Oxfordian–upper

Ryazanian Kimmeridge Clay source rocks deposited as condensed facies in the marine environment with highly anoxic conditions (Burwood et al., 1997). They were further characterized by increased C_{30} sterane and C_{28} bisnorhopane contents. On the contrary, oils enriched in ^{13}C came from the upper Callovian–lower Kimmeridgian Heather Formation characterized by terrestrially influenced organic matter and a more oxic depositional environment. Furthermore, oils coming from the coal-bearing Middle Jurassic Bryne Formation in the Danish sector have carbon-isotope ratios of about –26‰ heavier than the ratios characterizing the formation belonging to the marine source rock system of the Kimmeridge Clay Formation and its equivalents (Petersen et al., 1996).

When it comes to tracing precursor organisms, the triterpane, gammacerane, for example, is associated with protozoa, possibly bacteria, and indicative of a stratified water column, commonly in hypersaline environments (Schoell et al., 1994). Although the prominence of C_{30} gammacerane is commonly associated with hypersaline environments (Mello et al., 1988), the compound-specific isotopic analyses start to indicate that gammacerane may record an input from bacterial communities living at or below the chemocline in stratified lakes (Schoell et al., 1994; Sinninghe Damste et al., 1995).

The 4-methylsteranes are generally considered as representing biological markers for dinoflagellates (Habermehl and Hundrieser, 1983; Robinson et al., 1984; Wolff et al., 1986a, 1986b) and indicative of a significant contribution of the lipid constituents of dinoflagellate phytoplankton to kerogens of the Kimmeridge Clay Formation of the North Sea. A presence of β-carotane is attributed to the carotenoid pigment in algae (Peters et al., 1989; Bailey et al., 1990).

The third set of examples illustrates the role of biomarkers in unraveling transformation pathways. Chlorophyll, the molecular basis for photosynthesis in plants, undergoes complex microbial and geochemical transformations in aquatic environments and during diagenesis (Dydik et al., 1978). Generally, it transforms to the acyclic isoprenoid i-C_{19} (pristine, Pr) under oxidizing conditions, or i-C_{20} (phytane, Ph) under reducing conditions (Hunt, 1979; Collister et al., 1992). Their ratio, as derived from standard gas chromatography analysis of crude oil or bitumen extract, is used to distinguish between oxic (>1) and anoxic (<1) conditions in paleodepositional environments and also serves as an indicator of salinity in depositional environments. Hughes and colleagues (1995) have shown that Pr/Ph ratios below one represent marine conditions while ratios exceeding three characterize a terrestrial input. Ratios from 1 to 3 indicate lacustrine/marine environments.

The isoprenoid distribution is also influenced by maturity and biodegradation as indicated by the preservation of pristane over phytane during thermal diagenesis (Tissot and Welte, 1984; Peters and Moldowan, 1993). The ratios of isoprenoids to normal alkanes (Pr/n-C17 and Ph/n-C18) decrease with maturation because of the increasing prevalence of the n-paraffins. Ph/n-C18 values higher than one indicate biodegradation.

A cross plot of the ratios Pr/n-C17 and Ph/n-C18 reflecting the major factors (source, maturation, migration, and biodegradation) responsible for differences in crude oil composition is widely used for the classification of oils and rock extracts into different groups (Moldowan et al., 1993).

Furthermore, a hopane/sterane ratio also serves as an indicator of degradation. For example, a high hopane/sterane ratio of the Middle Jurassic Brora Coal Formation documents intense bacterial degradation in a peat-forming environment (Kubala et al., 2002).

Distributions of pentacyclic triterpanes such as hopanes show variation mainly related to the maturation level, although specific biomarkers occur, assisting in oil-source rock correlation and recognition of degradation (Kubala et al., 2002). Increased maturation level can be indicated by:

1) increased trisnorhopane ratio, Ts/Tm;
2) decreased relative abundance of C_{30} moretane and C_{29} normoretane; and
3) increased relative abundance of C_{30} diahopane and C_{29} rearranged norhopane,

although these parameters are not always well-correlated in the North Sea region. Furthermore, an increased maturation level tends to correlate with decreased relative abundances of C_{31}-C_{35} extended hopanes (Schou et al., 1985).

There are numerous controls of the source rock quality, including:

1) average organic richness;
2) organic matter type;
3) degree of organic matter maturity;
4) source rock thickness; and
5) lateral extent of source rock.

They control the source rock quality in an interactive way. They control whether source rock is capable of generating a sufficient quantity of bitumen, the quantity large enough to break out of the source rock pore space to migrate away as hydrocarbon phase. For example, the already mentioned sapropelic kerogens of marine and lacustrine environments can experience such breakout only at organic richness higher than 1.0

Table 15.1. Organic richness and maturity level evaluation guidelines for source rock (modified from Peters, 1986). Production Index (PI) is calculated as $S_1/(S_1 + S_2)$.

Organic richness			
TOC (wt. %)	S_1 (mg HC/g rock)	S_2 (mg HC/g rock)	Organic richness level
0–0.5	0–0.5	0–2.5	Poor
0.5–1	0.5–1	2.5–5	Fair
1–2	1–2	5–10	Good
>2	>2	>10	Very good
Maturity level			
T_{max} (°C)	Ro (%)	PI	Maturity level
435–445	0.6	0.1	Start of the oil generation (top oil window)
470	1.4	0.4	End of the oil generation (base of the oil window)
>470		0.4–1	Wet gas zone

wt.% of total organic carbon (TOC) (Bissada, 1982). Regardless of differences in source rocks, there are several general guidelines for organic richness categorization in the industry (see Table 15.1).

Organic matter type also plays an important role because different types of kerogens are prone to transformation into different hydrocarbons. For example, as natural gas requires permeability smaller than oil for its escape from the source rock, it can be produced from kerogens of very low concentrations.

The degree of organic maturity controls how far the organic matter is on its transformation path to oil and gas. Regardless of their complexities, there are some general industry guidelines for estimating maturity (Table 15.2).

One of the industry standards in determining the source rock maturity level is an open system programmed pyrolysis, such as Rock-Eval (Espitalié et al., 1977a; Peters, 1986; Peters and Cassa, 1994). The method uses programmed oven heating, generating products that are swept by a helium carrier gas and analyzed by a flame ionization detector. The heating is applied to pulverized source rock samples in an inert atmosphere. The bitumen is released from the rock by distillation and kerogen cracked into pyrolytic products. The entire procedure results in a variety of determined and calculated parameters (Figure 15.4), which are:

1) S1, representing the hydrocarbons that can be distilled from the rock, measured in mg HC/g rock;
2) S2, characterized by the hydrocarbons generated by pyrolytic degradation of kerogen, measured in mg HC/ g rock;
3) T_{max}, representing the temperature at which the maximum amount of pyrolysate, S2, is generated;
4) S3, characterized by the carbon dioxide generated during heating up to 390°C as measured by thermal conductivity detection, measured in mgCO_2/ g C_{org};
5) Production Index, PI, representing the ratio of bitumen to total hydrocarbon or S1/(S1+S2);
6) Hydrogen Index, HI, being the quantity of pyrolysable organic compounds from S2 relative to the quantity of organic carbon, measured in mg HC/ g C_{org}; and
7) Oxygen Index, OI, representing the quantity of carbon dioxide from S3 relative to the quantity of organic carbon, measured in mg CO_2/ g C_{org}.

HI and OI can be used as proxies for atomic ratios in the Van Krevelen plot described earlier (Figure 15.2), correlating kerogen compositions (Figure 15.5). Genetic potentials (S_1+S_2) indicate the source rock quality. Values greater than 6 mg HC/g rock and less than 2.5 mg HC/g rock indicate source rocks that are good-to-excellent and poor, respectively.

The degree of organic maturity, apart from Rock-Eval parameters, T_{max}, and production index, is commonly gauged by vitrinite reflectance, determined by optical microscopy. It is one of the best parameters for the determination of the coalification and maturation stage of the source rock, as well as the paleotemperature of its diagenesis. Vitrinite reflectance increases at varying rates with increasing source rock maturation. The reflectivity does not retrograde if the temperature is subsequently lowered. High temperatures resulting

Table 15.2. a) Geochemical character of algal-terrestrial source rocks from and oils derived from lacustrine fluvial-lacustrine facies association (Carroll and Bohacs, 2001).

Basin and formation	Ro (%)	TOC (%)	HI	Pristane/ Phytane**	β-carotane	Steranes	Tricyclic Index**	Gammacerane index	Hopane/ Steranes
rock, Luman Tongue of the Green River Formation, Wyoming	0.35–0.45	1.0–59.2	55–985	1.5–4.8	not detected	$C_{29}>C_{27}\geq C_{28}$ abundant 4-methyl	not reported	negligible	1.0–4.0
rocks, Hongyanchi Formation, Junggar Basin, China	0.86–1.09	0–8	58–365	1.5–4.1	trace	$C_{29}>C_{27}>C_{28}$	48–71	4–11	2.6–7.7
rocks and oils, lacustrine formations in failed rifts of the eastern Brazilian margin	0.4–0.7	≤6.5	<779	>1.3	not detected	$C_{29}>>C_{27}$	30–100	20–40	5–15
rock, Kissenda Formation, Gabon Basin	0.65–0.72	0.6–2.2	93–458	0.8–1.2	not detected	$C_{27}>C_{29}=C_{28}$ abundant 4-methyl	not reported	7–35	4.2–17.5
oils, Phisanulok Basin, Thailand	na	na	na	2.7–4.0	not detected	$C_{29}>>C_{28}>C_{27}$	negligible	negligible	28.1–47.5
rocks and oils, Central Sumatra Basin	0.4–0.8	0.7–9.2	110–946	2.2–3.2	not detected	$C_{29}>C_{28}\approx C_{27}$ abundant 4-methyl	12–27	4–6	3.8–7.7

b) Geochemical character of algal source rocks from and oils derived from lacustrine fluctuating profundal facies association (Carroll and Bohacs, 2001).

Basin and formation	Ro (%)	TOC (%)	HI	Pristane/ Phytane**	β-carotane	Steranes	Tricyclic Index**	Gammacerane index	Hopane/ Steranes
rocks, Laney member of the Green River Formation, Wyoming	0.4	1.5–17.1	52–1003	0.1–0.5	present	$C_{29}>C_{27}>C_{28}$ abundant 4-methyl	not reported	high	0.4–0.8
rocks, Lucaogou Formation, Junggar Basin, China	0.77–0.88	0–23	481–766	1.0–1.6	present	$C_{29}\geq C_{28}>>C_{27}$	79–85	19–36	2.5–5.2
rocks and oils, lacustrine formations in failed rifts of the eastern Brazilian margin	0.4–0.8	≤9	≤900	>1.1	present	$C_{29}>C_{27}$	100–200	20–70	5–15
rocks, Bucomazi Formation, Angola Basin	0.56–0.81	2–24	>700	not reported	not reported	$C_{27}>C_{29}\geq C_{28}$	not reported	13–150	1.5–7.7

c) Geochemical character of hypersaline-algal source rocks from and oils derived from lacustrine evaporative facies association (Carroll and Bohacs, 2001).

Basin and formation	Ro (%)	TOC (%)	HI	Pristane/ Phytane**	β-carotane	Steranes	Tricyclic Index**	Gammacerane index	Hopane/ Steranes
rocks, Wilkins Peak member of Green River Formation, Wyoming	0.2–0.45	4.1–19.0	32.0–1001	0.1–1.1	abundant	$C_{29}>C_{28}>>C_{27}$	6–21	8.0–82	0.2–2.0
rocks, Jingjingzigou Formation, Junggar Basin, China	0.88–0.91	0.8–2.0	129–477	0.8–1.1	abundant to dominant	$C_{29}>C_{28}>C_{27}$	79–245	15–34	1.7–7.1
oils, Karamay trend, Junggar Basin, China	na	na	na	0.9–2.3	abundant to dominant	$C_{29}\geq C_{28}>C_{27}$	260–580	31–56	0.7–2.2
rocks and oils, Jianghan Basin, China	0.45–0.55?	≤6.6	712–838	0.1–0.5	present	$C_{27}>C_{29}>C_{28}$	7–28	35–216	0.3–2.4

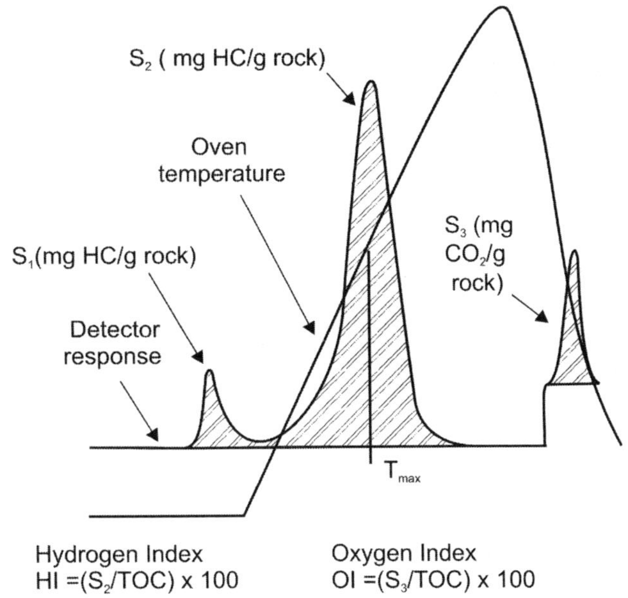

Figure 15.4. Simplified Rock-Eval pyrogram showing the evolution of products from a source rock during programmed heating (Peters, 1986).

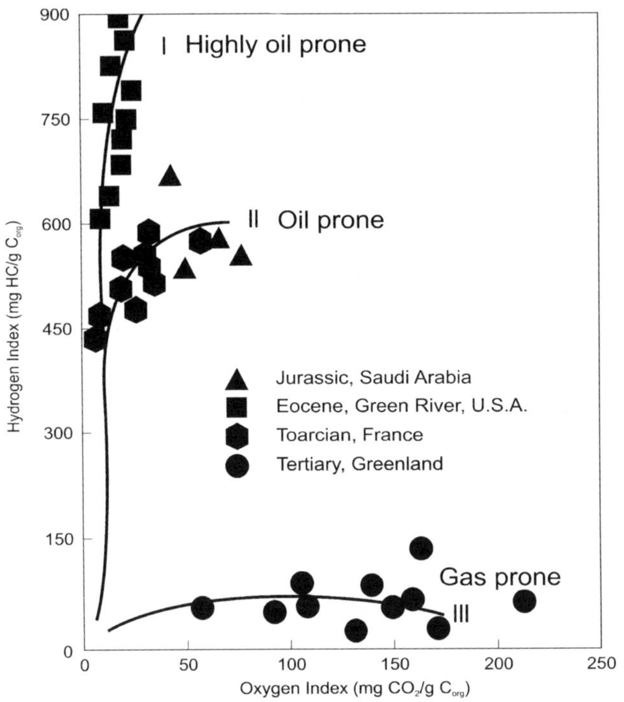

Figure 15.5. Main hydrocarbon-generative types of kerogens shown on a plot of hydrogen index (HI) with oxygen index (OI) for the source rock samples from Figure 15.2 (Peters, 1986).

from frictional heating during faulting can cause a local increase of coalification rank at thrust planes. There are a few other parameters such as time, heat flow, pressure, and thermal conductivity of rocks, which more or less affect the vitrinite reflectance. High rock conductivity and overpressure are known to retard the coalification trend.

Vitrinitic kerogen is most abundant in sedimentary rocks containing type III organic matter and may be absent in type I. It is absent in rocks older than Devonian. However, in pre-Devonian rocks, the reflectance of certain microfossils may also serve as a parameter of maturity (Bertrand and Achab, 1990). Such fossils include chitinozoans, graptolites, scolecodonts, and spores. A solid bitumen reflectance also can be used to determine paleotemperature and maturity.

Other kerogen macerals, such as Liptinite, inertinite, and solid bitumen, are not as reliable for temperature determinations as vitrinite. This is because the vitrinite undergoes a more regular increase in reflectivity with increasing temperature than other macerals. Liptinite, for example, usually gives reflectance values lower than those of vitrinite. The inertinite, on the other hand, has reflectivity values higher than those of vitrinite.

The most usual reasons for inaccuracy in the determination of vitrinite reflectance include:

1) low-quality polishing of the sections, which decreases values;
2) mistakes in maceral identification; and
3) very small size of particles,

which leads to measurement errors.

Generally, the vitrinite reflectance correlates with other parameters of source rock maturation as follows:

1. Ro < 0.5% (0.7%). These values indicate a diagenesis stage characterized by immature source rock.
2. Ro = 0.5% (0.7%) – ca. 1.3%. This interval indicates a catagenesis stage characterized by the main zone of oil generation. The onset of the oil window corresponds to a reflectance of about 0.5 percent in source rocks with kerogen type II and a reflectance of about 0.7 percent in source rocks with kerogen type I.
3. Ro = ca. 1.3% – 2%. This interval indicates a catagenesis stage characterized by wet gas and condensate generation.
4. Ro > 2%. These values indicate a metagenesis stage characterized by dry gas generation starting above the indicated value.

Although the vitrinite reflectance determination is the most common method for estimating the source rock thermal maturity, the production index and T_{max} are also frequently used (Table 15.2). Typically, a hydrocarbon generation is understood as associated with PI and T_{max} values of 0.2–0.4 and 436–460°C, respectively.

Note that Rock-Eval methods have inherent analytical and interpretation limitations (see Katz, 1983; Peters, 1986), but they require a minimum of time, money, and sample volume. This allows their use for systematic source rock screening in the explored area.

The source rock thickness's control over the source rock quality is addressed by the source potential index, SPI, designed as a measure of source rock volumetrics integrating the average source rock richness and stratigraphic thickness (see Demaison and Huizinga, 1991). This index provides the amount of hydrocarbons that can be generated in a source rock column with 1m^2 base. It is calculated in metric tons from a formula:

$$SPI = h(S_1 + S_2)d/1000, \quad (15\text{-}1)$$

where h is the net source rock thickness, $S_1 + S_2$ is the average genetic potential in kg HC/ton rock determined by Rock-Eval analysis, and d is the average source rock density, normally of about 2.5 t/m^3. Statistical research, which was done for basins in western North America and the Andes (see Demaison and Huizinga, 1991), documents a correlation between the large estimated petroleum reserves and rich source rocks with high source potential index.

Note that the source potential index method has certain limitations. For example, the index does not treat the variability in source rock quality, which is rather typical for extensional basins (see Katz, 1983) being controlled by the rapid facies changes described in Chapter 7 (see Gupta et al., 1999; Sharp et al., 2000; McLeod et al., 2002). A nice example of the facies variability from extensional setting comes from the Early Cretaceous Congo–Angola Rift Zones. It is based on a robust comparison of facies penetrated by wells analyzed by CoreLab (McAfee, 2005). The early-rift depositional record indicates that the inner-hinge zone had a low amplitude relief. The rift axis was occupied by a shallow lake. Alluvial fans prograded from the basin margins to the lake. All these facies are included in the Vandji Formation. Some inner rift hinge segments subsided, forming deep troughs with more than 2 km-thick lacustrine shale such as that in the Kissenda Formation in the Gamba Horst region. These shales abruptly overlie the early-rift fluvial sandstones and cause their rapid burial.

The main-rift sedimentary picture includes some segments of the inner hinge system becoming major topographic highs, such as the Gamba Horst, and some becoming depressions, such as those focusing the deposition of organic shale. As a result, the entire system is characterized by a highly variable thickness of organic shales such as the Melania Shale and the Marnes Noires and Bucomazi Formations in Congo. During the same time, the Angolan margin was characterized by extrusion of large volumes of lavas and tuffs. The volumes were large enough to fill local catchment areas, forming shallow to emergent platforms along the outer hinge zone. Shallow lake to sabkha deposition, such as the Barremian Maculungo salt member in the South Kwanza Basin, took place to the east of the external hinge zone.

The late main-rift depositional record shows the following scenario. The central rift region contains sand-rich fluvial-lacustrine sediments more than 1 km thick, such as those belonging to the Dentale Formation in Gabon and to the Red Cuvo Formation in the Lower Congo Basin. The Dentale fluvial system flowed axially from northwest to southeast along the axis of the Gabon Rift Basin. It was laterally confined by the inner hinge zone to the east and the outer hinge zone to the west. Regions distal to fluvial axes, for example, the southernmost Gabon and Congo, were occupied by persistent shallow lakes with silt, shale, and locally carbonate deposition recorded in the Argilles Vertes, Viodo, Toca, and Cuvo carbonate formations. Although there was widespread regional denudation after the main rift phase, a subtle topographic control over the subsequent deposition continued. One can see this in the Gamba, Chela, and Grey Cuvo Formations. The basal lowstand fluvial sediments were best developed in the inner basins and near intra-basinal highs. The transgressive shallow lake facies, including winnowed high-energy lake shoreline sandstones, were deposited along the axis of the outer hinge zone.

Source rock deposition and preservation

The source rock accumulation requires bio-productivity and anoxia, which would lead to the preservation of organic matter at the sediment-water interface (Katz, 1995a). Regardless of numerous scenarios providing the deposition and preservation of organic-rich sediments, the sediments rich enough in organic matter to become source rock are relatively rare, having limited temporal and spatial extent.

The organic matter in source rocks can be in situ or transported to the depositional environment. Examples of autochthonous organic matter come from peat bogs (mires), mangrove swamps, and microbial mats. The remaining depositional environments of future source rocks have their organic matter allochthonous, transported to its place typically with inorganic components (Huc, 1988). Such types of organic matter include:

1) phytoplankton growing in the water mass above or lateral to the deposition environment;

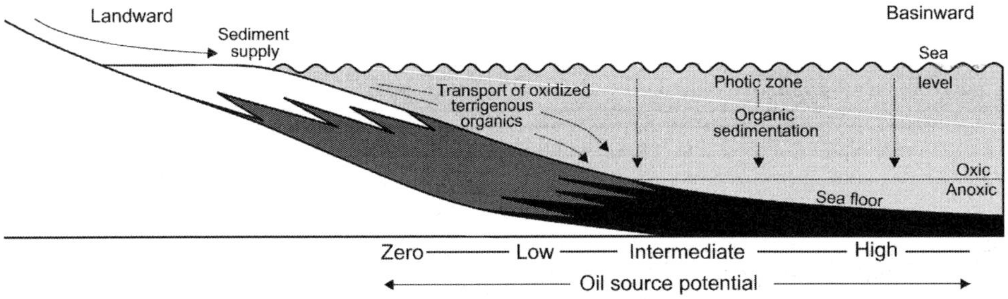

Figure 15.6. Model of marine oil-prone source rock deposition (Creaney and Allan, 1992).

2) terrigeneous plant matter carried to the sea environments by rivers; and
3) matter from coastal mires destructed by wave activity and transported.

Terrigeneous plant matter can undergo a considerable transport before deposition, as documented by organic matter of terrestrial origin determined in marine black shales deposited hundreds of kilometers from the shoreline (Habib, 1982; Simoneit and Stuermer, 1982). It is known that most coastal and shelf areas have a bio-productivity higher than average. This is due to nutrient loading from the land (Kruijs and Barron, 1990; Littke et al., 1997). In seas, except those with a very low energy, the sea currents transport terrestrial organic matter to quiet areas, where it gets deposited in silt and clay facies. If the basal portion of the water column in such area is anoxic, the organic matter gets preserved (Figure 15.6). This mechanism controls the correlation between shale and marl facies, and a high content of organic matter. In fact, Huc (1988) has reported a concentric geometry of organic richness maxima correlating with grain-size distribution.

Anoxia

The simplest way to imagine anoxia is to limit the supply of oxygen. It is because most of the oxidation in water, soils, and sediments is biological, requiring molecular oxygen. Numerous mechanisms restrict the occurrence of molecular oxygen to various extents. For example, the study of modern deposition in the Black Sea (Huc, 1988; Figure 15.7) documents a critical role of the water fill hydrodynamics in maintaining anoxic conditions during the process of source rock deposition. The modern Black Sea sediments have a concentric grain size distribution centered on depocenters in the Western and Eastern Black Sea. Almost the entire basin is characterized by anoxic condition on the bottom.

Note that the best way of determining anoxic conditions for a respective sedimentary formation is a synthetic approach, comparing the following indicators, because most of them are inconclusive on their own:

1) TOC content;
2) dark sediment color;
3) presence of pyrite;
4) presence of fine-laminated sediment; and
5) presence of undegraded marine organic matter.

While a TOC content of anoxic sediments is increased, the oxic sediments with numerous woody elements can have it increased as well. While the anoxic organic-rich sediments are generally dark, some rocks lean in organic content can be dark as well because of the higher pyrite or dark chert contents. While the pyrite precipitates under anoxic conditions, it can also develop way after the deposition, during the subsequent diagenesis.

However, fine laminae do document that burrowing organisms were absent at the bottom of the depositional environment, indicating a decreased dissolved oxygen content. Furthermore, the presence of undegraded marine organic matter, despite the general trend of its preferential consumption by predators, does serve as a relatively reliable identifier.

The organic matter preservation in sediment is controlled by two competing rates:

1) the rate of bio-productivity and organic matter accumulation at the sediment–water interface; and
2) the rate of organic matter destruction by oxidation and bottom-feeding organisms.

These two processes can be dependent on each other. This can be documented by cases where a very high bio-productivity associated with a high nutrient supply to the photic zone and coastal oceanic current upwelling (Parrish, 1982; Piper and Link, 2002) leads to the oxygen deficiency through a decay of the abundant organic matter (Pedersen and Calvert, 1990). This mechanism can develop anoxic bottom waters or an oxygen-minimum zone in the stratified water fill.

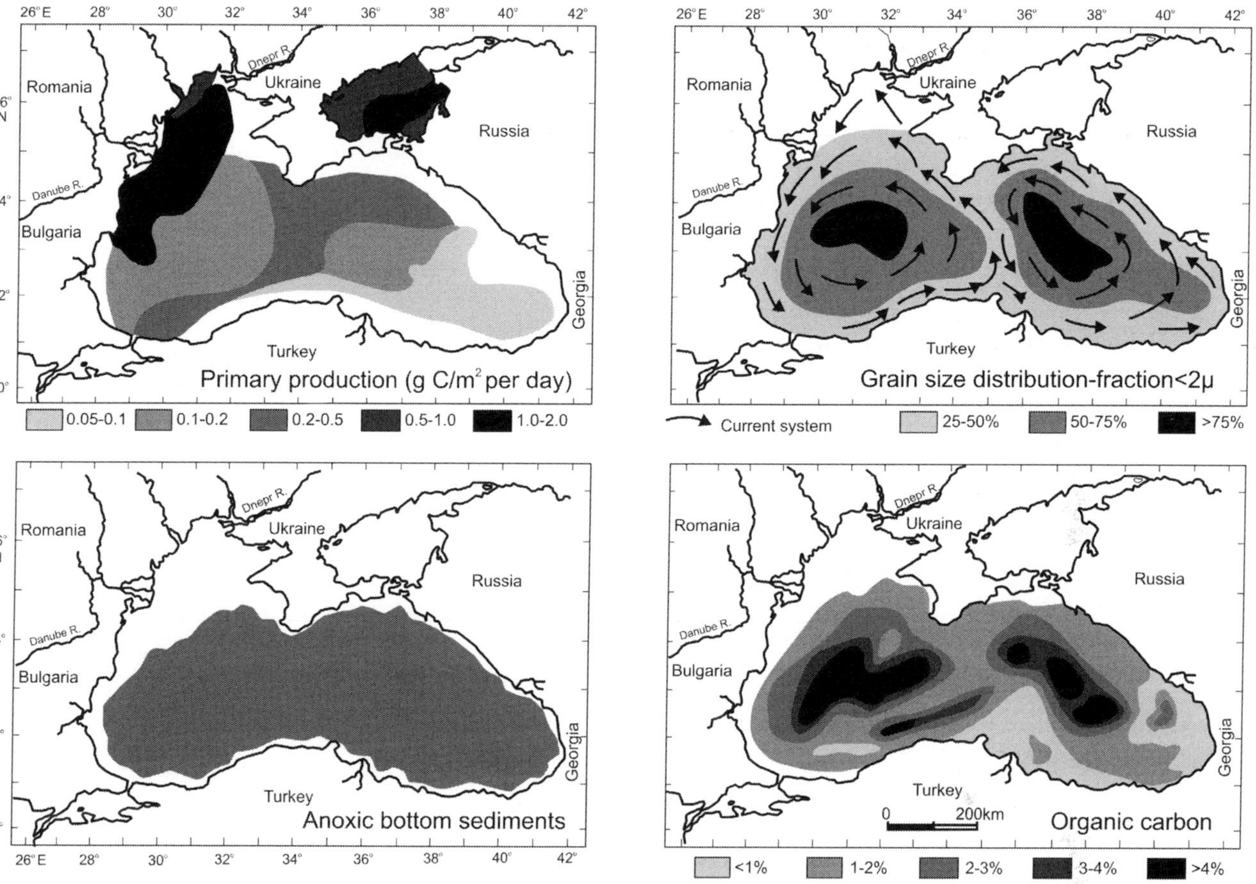

Figure 15.7. Organic carbon content of modern Black Sea sediments in relationship to primary organic productivity, area of anoxic bottom sediments, distribution of fine-grained sediments, and water fill current patterns (Huc, 1988).

Some source-rock deposition models describe the movement of the oxygen minimum zone onto a shelf during a relative sea-level rise to produce temporary basin anoxia (Katz, 1995a). A good organic matter preservation takes place in basins characterized by fast deposition rates where a rapid burial isolates the organic matter from processes of alteration or destruction at the sediment–water interface.

Stratigraphic association of known source rocks

It is interesting to look at the stratigraphic association of known source rocks. It shows that certain stratigraphies are more favorable for source rock deposition than others. A set of controls behind such distribution include:

1) secular changes in the world's flora through time; and
2) depositional and diagenetic conditions favoring preservation.

It is then interesting to find out that a small number of six stratigraphic intervals contains the source rocks responsible for more than 90 percent of the world's petroleum reserves, and that these intervals represent only one-third of the Phanerozoic time (see Ulmishek and Klemme, 1990; Klemme and Ulmishek, 1991).

These six intervals have the following share in their contribution to petroleum reserves:

1) Aptian-Turonian – 29.0 percent;
2) Upper Jurassic – 25.0 percent;
3) Oligocene-Miocene – 12.5 percent;
4) Silurian – 9.0 percent;
5) Upper Carboniferous-Lower Permian – 8.0 percent; and
6) Upper Devonian-Tournaisian – 8.0 percent.

Sixty-six and a half percent of the summed shares come from stratigraphies younger than the Middle Jurassic. Four of the six stratigraphies including Silurian, Upper Devonian–Tournaisian, Upper Jurassic and Aptian-Turonian, which provided 71 percent of the global reserves, coincide with global transgression events (see Sloss, 1979, 1988; Hallam, 1984; Haq et al., 1987).

The paleo-relief on continents flooded by such events is controlled by:

1) tectonic activity, which is often severe in rift and pull-apart basins, or subtle in broad intra-cratonic depressions; and
2) erosion immediately preceding the transgressive event.

The Silurian transgression event was caused by the melting of the late Ordovician ice cap. It was responsible for the deposition of the widespread Lower Silurian black shales, which fed almost all Paleozoic hydrocarbon systems in North Africa. Despite being widespread, the shale facies has gaps associated with highs in the post-glacial topography (Lüning et al., 2000).

The Late Devonian–Tournaisian transgression event coincided with active rifting in the foreland of the Urals in the Timan-Pechora Basin (Schamel, 1993). Contrary to the Silurian flooding controlling widespread black shale deposition in North Africa, the flooding in the Timan-Pechora region resulted in a system of localized source rocks associated with rifts and the foreland-basin of the Urals.

The Late Jurassic transgression event is well described in the Viking Graben and the East Shetland Basin of the North Sea (see Fraser et al., 2002). The event was characterized by progressive, highly asymmetric, and diachronous flooding of the intracontinental Late Jurassic North Sea – by the drowning of a failed rift system. The diachroneity and differential rates of marine flooding probably reflect inherent controls in crustal thickness and strength in different parts of the rift system. Linked to the history of basin drowning was the accumulation of organic-rich mudstones of the Kimmeridge Clay Formation and its lateral equivalents (Miller, 1990). The circulation of marine waters was restricted by the significant hanging-wall bathymetry and segmented nature of the basin floor. This, together with rising sea levels, diminishing rates of sediment supply, and stratified water fill, controlled anoxic conditions (Miller, 1990).

The Aptian-Turonian transgression event can be described reasonably well from the margins of the Equatorial Atlantic and the northern South Atlantic (see Katz and Mello, 2000; Nemčok et al., 2004b). Despite the spatial and temporal variability of the onset of seafloor spreading during this time interval in this region, this transgression event led to a large area expansion of an upward-widening oxygen-minimum layer in each of the local areas. This led to the future development of a regionally extensive system of source rock facies. A bit simplified, they can be divided into three groups (Katz and Mello, 2000):

1) source rocks of marine hypersaline environments;
2) source rocks of marine carbonate environments; and
3) source rocks of marine anoxic environments.

The first group is typed to oil pools located in basins such as Ceará, Potiguar, Sergipe-Alagoas, Bahia-Sul, and Gabon. Examples come from formations Paracuru in the Ceará Basin, Upanema in the Potiguar Basin, Muribeca in the Sergipe-Alagoas Basin, and Gamba in the Gabon Basin. Produced naphtenic oils have more than 60 percent of saturates, less than 40 percent of naphtenes, a medium sulfur content of 0.1–0.5 percent, and a low to medium oil/gas ratio of less than 200 m^3/m^3.

The second group has been deposited in the Sergipe-Alagoas, Bahia-Sul, Espírito-Santo, Lower Congo, Kwanza, and Gabon Basins in marl and calcareous black shale facies. Typical examples are the Regencia and Iabe Formations from the Espírito Santo and Lower Congo Basins. Produced naphtenic-aromatic oils have less than 50 percent of saturates, more than 25 percent of naphtenes, a medium sulfur content of 0.1–0.5 percent, and a low oil/gas ratio of less than 60 m^3/m^3.

The third group of Cenomanian-Turonian marine black shales, represented by examples such as the Azile and Anguille Formations of the Gabon Basin and the Itajaí Formation of the Santos Basin, produced oils in the Santos, Espírito Santo, Sergipe-Alagoas, Lower Congo, and Kwanza Basins. Mentioned oils have moderate naphthene content, a medium sulfur content of 0.1–0.5 percent, and high oil/gas ratios of more than 200 m^3/m^3.

Depositional settings typical for source rock development

Conditions suitable for deposition and preservation of organic-rich sediments occur in four main settings of relevance to petroleum resources in extensional domains:

1) lakes, especially large, long-lived, and stratified lakes;
2) passive continental margins and epi-continental seas with a developed oxygen-minimum layer;
3) tectonically active silled basins and continental depressions having restricted circulation of marine waters; and
4) coastal peat mires and swamps, the sites of coal formation.

Lakes are a good example for starting a discussion of a stagnant basin, although the occurrence of

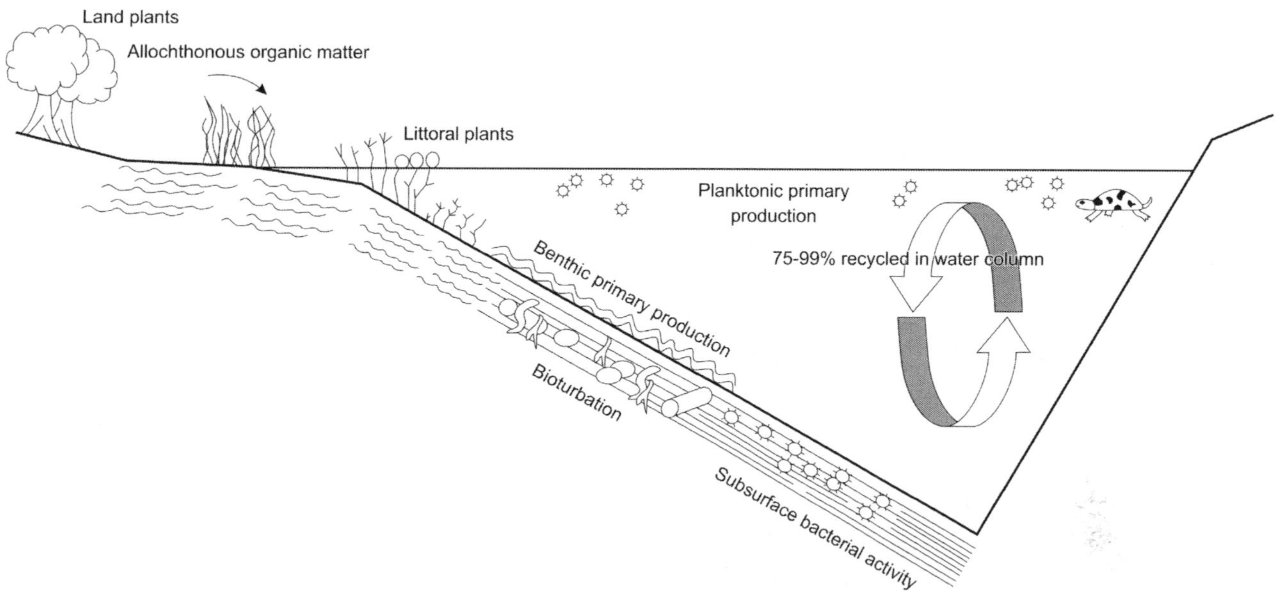

Figure 15.8. Model of distribution of organic material in a lacustrine basin (Bohacs et al., 2000).

a completely stagnant basin is rare. If an isolated lake is deeper than about 200 m, and if the climate is tropical or subtropical, the density stratification develops as a result of temperature differences in the water fill. Described depth and climate prevent the influence of storms and overturns, respectively. The dense water residing at the base of the water fill leads into a pycnocline development. It prevents the mixing of the divided layers, which would be the only mechanism bringing the oxygen to the lake bottom. This results in a progressive development of anoxic conditions.

Note that the described stratified lake is an end-member, while lakes cover a variety of depositional environments, ranging from overfilled through balanced-fill to underfilled lakes (Bohacs et al., 2000).

The lacustrine source rocks are known to be deposited in tectonically controlled lakes and flood plains in humid, semiarid, and arid environments. Preservation of organic matter in this setting depends on the previously discussed maintenance of anoxic or micro-oxic bottom-water layers taking part in thermally or density-stratified waters (Powell, 1986; Figure 15.8). Algae and higher plants form the organic matter, which is significantly modified by bacteria under slightly oxidizing conditions and supplemented by bacteria remains.

Various lake mechanisms can be grouped into the three categories, controlled by different geographic settings (Caroll and Bohacs, 2001):

1) fluctuating-profundal lakes;
2) fluvial lacustrine; and
3) evaporative.

The underfilled lakes, characterized by minimum terrestrial sediment input, develop a source rock with type I kerogen, prone to the generation of paraffinic oil with medium-high sulfur content. These occasionally deep lakes are characterized by relatively homogeneous source rock. Examples come from the Pematang Brown shale formation in the Central Sumatra Basin (see Williams and Eubank, 1995).

The overfilled lakes, characterized by maximum terrestrial sediment input, develop a source rock with mixed type I and III kerogens, prone to oil and gas generation. These lakes are characterized by highly heterogeneous facies distribution.

The lakes with balanced fill typical for varying terrestrial sediment input develop a dominantly type I kerogen with type I–III mixtures near flooding surfaces, prone to oil generation.

Evaporative lakes are characterized by relatively homogeneous source biota. Its preservation is enhanced by the distinct water stratification and repressed bacterial decay (Hite and Anders, 1991; Aizenshtat et al., 1999). Hypersaline, sulfate-rich lakes may preserve a distinctive type I-S kerogen that can generate oil at thermal maturities as low as that represented by 0.45 percent vitrinite reflectance (Caroll and Bohacs, 2001).

Relative to marine source rocks, lacustrine facies exhibit a high degree of spatial geochemical heterogeneity, as documented with the example of the Middle Eocene Green River Formation of the Uinta Basin in Utah (see Ruble and Philp, 1998; Ruble et al., 2001; Figure 15.9). Other examples of source rocks assumed

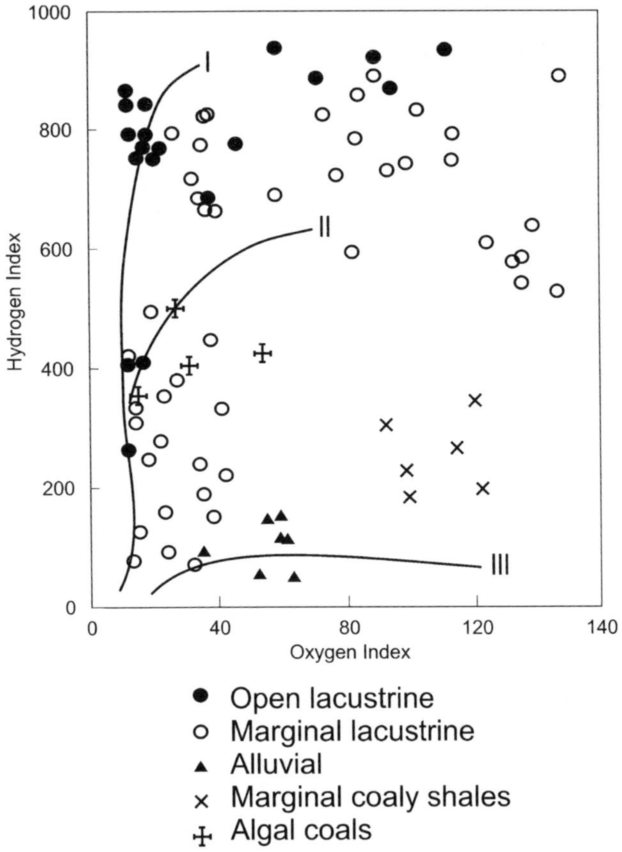

Figure 15.9. Highly variable lacustrine organic facies within the Eocene Green River Formation of the Uinta Basin, Utah. Rock-Eval pyrolysis data are from Anders and Gerrild (1984) and Hunt (1991).

to be deposited in stagnant basins come from the Eocene–Oligocene Elko Formation, Nevada, and several basins in China such as the Bohai Basin.

The temporal geochemical variability has been documented as controlled by the dying-out rift tectonics on the example of the failed Congo Rift Basin filled with lacustrine Lower Cretaceous syn-rift section (Harris et al., 2004). The basin is currently located on the Congo passive margin. Systematic sampling of the syn-rift section and analyses for total organic carbon content and maturity (Figure 15.10a) document that both overall organic carbon content and bacterial fraction of organic matter increase upward. They reach their maximum in the lower portion of the late syn-rift section, which coincides with practically terminating normal faulting and filling of the basin (Figure 15.10b). During this time the water depths became shallower than during the initial and main rift phases and the basin was characterized by relatively slow deposition rates.

The slower deposition rates correlate with higher organic carbon content. The deposition of the richest organic sediments was linked to a very high flux of dissolved carbonate into the rift lake, indicated by a systematic decrease in ^{13}C isotope in carbonate minerals in sediments. The enhanced flux can be explained by the reduced relief of drainage basins feeding the rift lake, causing a more efficient cycling of carbon derived from plants into the soil carbonate.

Passive continental margins and epi-continental seas are a good example for starting a discussion of oxygen-minimum layer, although this layer is not a must for these settings, as these settings can be reasonably oxygenated.

The oxygen-minimum layer located somewhere under the water surface is characterized by an amount of oxygen lower than its amount above and below. Its development is controlled by oxygen consumption faster than oxygen input. Such consumption can be caused by a decay of dead organisms that have sunk from the photic layer located above the oxygen-minimum layer.

The oxygen-minimum layer frequently develops right below the photic layer, at depths too deep for the influence of photosynthesis and turbulence, which would provide oxygen. The importance of the oxygen-minimum layer varies among seas and oceans. Important layers develop in regions with high organic productivity, occasionally in regions with restricted circulation. The deposition of sediments under low-oxygen conditions takes place in zones of the intersection of an oxygen-minimum layer with the sediment–water interface. Typical examples of such regions come from bottom set beds associated with prograding delta systems into oceanic basins. Those laid down inside a well-developed oxygen-minimum layer can be rich in organic matter. On the contrary, foresets in the same system would then be less rich if deposited above this layer.

It was discussed earlier that four global transgression events during the Silurian, Upper Devonian–Tournaisian, Upper Jurassic, and Aptian-Turonian provided 71 percent of the global reserves. It is known at least for the later three that these were times of severe depletion of dissolved oxygen in world oceans. This depletion was controlled by a combination of factors, paleoclimate and water circulation most likely being the dominant ones. Mentioned time periods were characterized by the downward and upward expansion of an oxygen-minimum layer, due to restricted oxygen input into a deeper portion of the water column. In combination with a major transgression event during the same time, the areal extent of the oxygen-minimum layer reached an important regional level in numerous basins worldwide (see, e.g., Creaney and Passey, 1993).

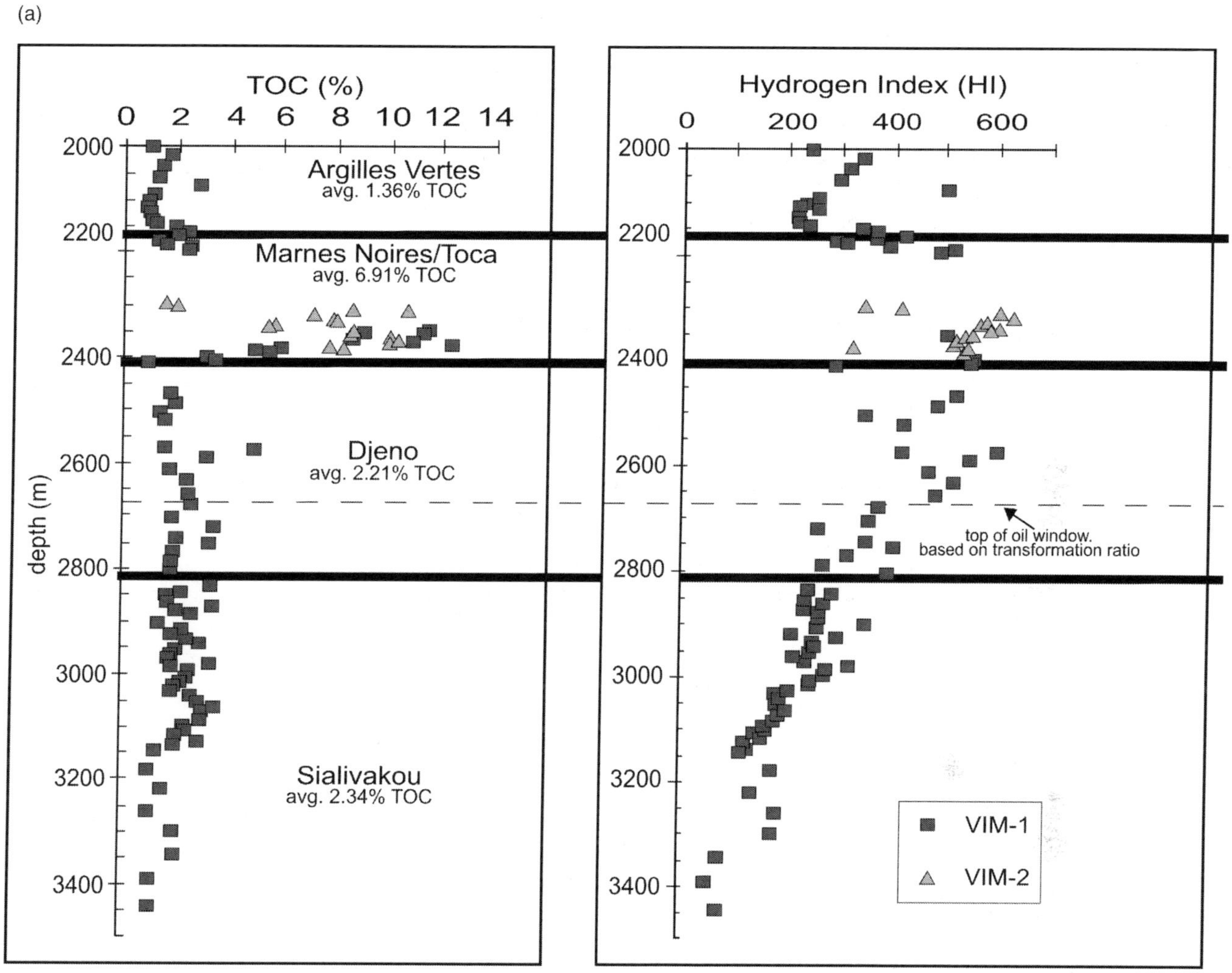

Figure 15.10. (a) Organic carbon content and hydrogen index data from the cuttings of VIM-1 and VIM-2 wells, Congo Basin (Harris et al., 2004). (b) Factors controlling the organic matter deposition in the early- and late-rift stages of the lacustrine Congo Basin (Harris et al., 2004).

Regarding the paleoclimate, the lower latitudes are favored in developing the oxygen-minimum layer because of the higher productivity of organic matter and the greater stability of water stratification. The differences among source rock provinces are largely related to the thickness and spatial distribution of the source rock facies. They are typically thicker in rift basins but laterally restricted. Rift basins are prone to concentration of high-quality oil-prone source rocks because of their low width/depth ratios and fast subsidence (Katz, 1995b).

Marine black shales with high organic content associated with transgression can have two possible sequence stratigraphic associations (Wignall and Maynard, 1993):

1) basal transgressive shale; and
2) maximum flooding shale.

The basal transgressive shale is associated with the initial flooding surface at a sequence boundary. It does not have prominent facies variations and it does not connect laterally to coeval shoreline facies. Instead, it passes proximally to sediment-starved facies such as phosphatic lags, as is known from the Duwi-Dakhla Formation in eastern Egypt (see Robinson and Engel, 1993).

The maximum flooding shale is usually a distal condensed facies located within a sedimentary sequence, which passes proximally to coarse-grained near-shore facies.

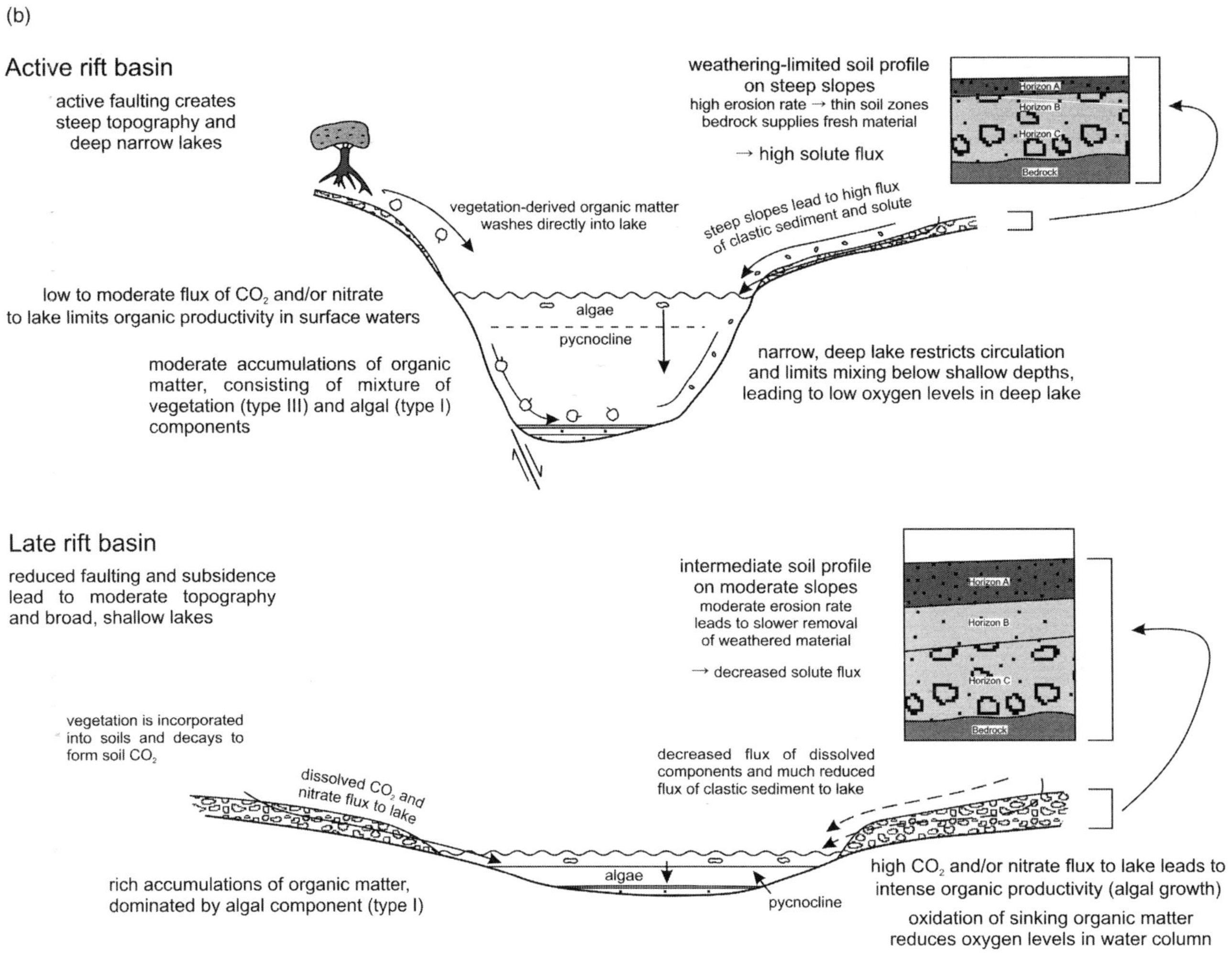

Figure 15.10 *(cont.)*

The third group of settings typical for the development of organic-rich sediments includes tectonically active silled basins and continental depressions having restricted circulation of marine waters. Note that a setting with restricted circulation is much more common than a stagnant basin. Furthermore, basins with restricted circulation are frequently connected to the open marine environment. Therefore, settings of the third group often include facies associated with oxygen minimum-controlled environments.

One of the reasons for the restricted circulation is the presence of a sill, which forms a dividing zone between the restricted area and the open sea. One of the most classical examples of a sill is the Apsheron sill dividing the South Caspian Basin from the rest of the Caspian Sea (e.g., Devlin et al., 1999; Guliyev et al., 2001a). If the sill has its crest at shallow water depth, waters entering or leaving the restricted basin are located near the surface. If the restricted basin is evaporitic like the Karabogaz Bay in the Caspian Sea, the water flows from the open sea into the restricted basin. If the restricted basin is fluvially dominated, as the remnant depressions in the Black Sea were, for example, the water flows from the restricted basin into the open sea. Regardless of the water flow direction over the sill, if the restricted basin is deep enough to avoid the storm effects, the water fill stratification develops, maintaining the basal water portion under anoxic conditions. Furthermore, the silled basin is prone to the loading of nutrients and organic detritus from the adjacent land mass.

Numerous silled basins may undergo an insignificant turnover of the basal water. However, it is too slow to disturb the anoxic conditions maintained at basal levels. Shallow silled basins can enter evaporitic conditions, being filled with evaporitic sequence, which includes excellent hydrocarbon source rocks.

Evaporitic environments combine the opportunity for distinct algae growth with excellent preservation

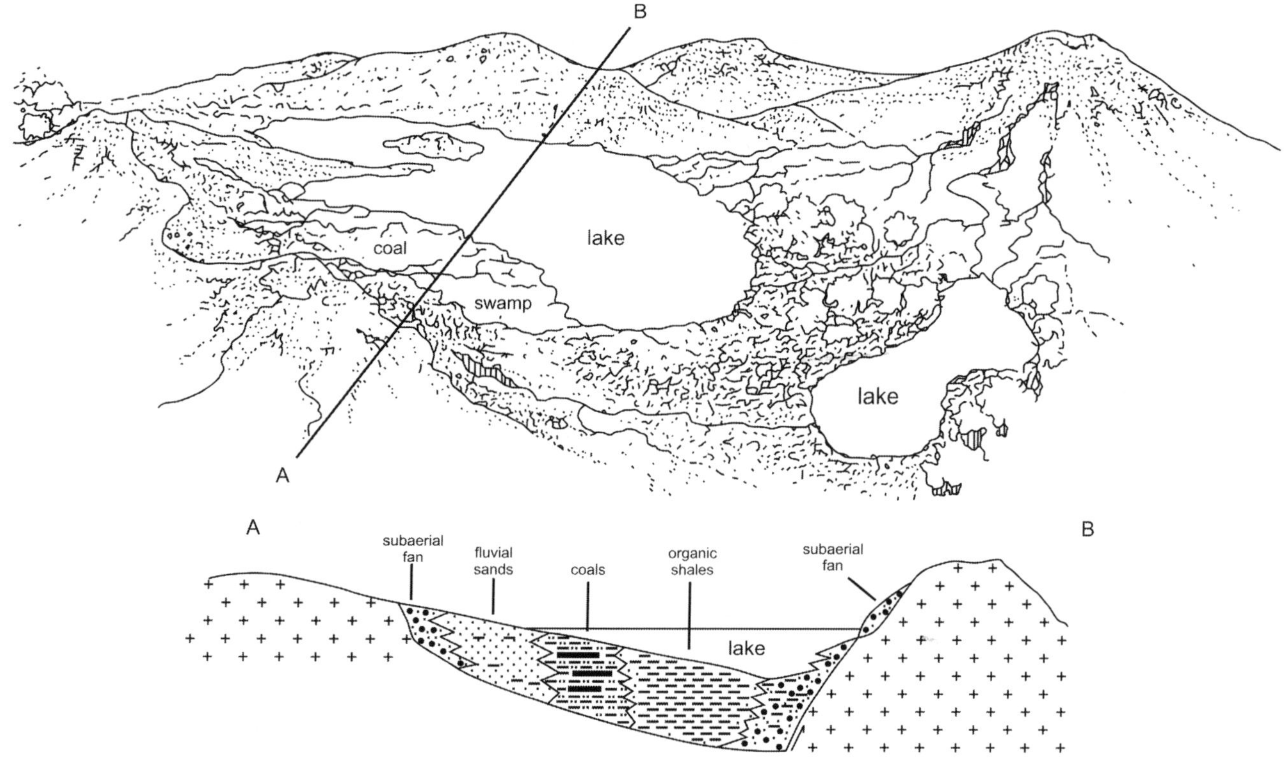

Figure 15.11. Shallow lacustrine depositional model for the Coal Zone formations in the Central Sumatra Basin (Williams and Eubank, 1995). From A to B: subaerial fan, fluvial sands, coals, organic shales, subaerial fan.

conditions. While evaporation concentrates nutrients, the high salinity eliminates predators. High organic matter production in turn reduces the oxygen content and the resultant distinct hydrogen-sulfide concentration controls an environment hostile to predators. The outcome is a sedimentary sequence characterized by multilayers of evaporite and organic-rich laminae, which can also laterally transition to each other.

The last group of settings typical for the development of organic-rich sediments is represented by coastal peat mires and swamps.

Peat preservation leading to coals is delicately controlled by the water table movements associated to the relative sea-level rise (McCabe and Parrish, 1992; Petersen et al., 1996). The requirement is to have the increase in accommodation space roughly equal to peat production rate (Bohacs and Suter, 1997). The modern data on peat production range from 1mmy^{-1}, which is characteristic for cool temperate regions, to about 6 mmy^{-1}, which is typical for a tropical climate. If the relative sea-level rise in the respective climate is higher than peat production, the coastal mires undergo inundation and burial by clastic sediments. If the rise is slower than production, the mires dry out and undergo oxidation and erosion.

A shallow water table is critical for controlling the anoxic environment required for organic matter preservation in coastal peat mires. A natural example of the hydrocarbon source rock deposited in a coal swamp is the Coal Zone formations of the Central Sumatra Basin, deposited in shallow pull-apart basin environment (Williams and Eubank, 1995; Figure 15.11). While the deepest portion of the tilted basin facing the controlling normal fault underwent the deposition of fans with coarse material and passing basin-ward into basinal organic shales, it is the hinge-ward side of the pull-apart that experienced coal swamp conditions.

Coals can be subdivided into the following types. More than 80 percent of them are humic and they are gas-prone (Hunt, 1991). Examples of these source rocks come from the Pannonian Basin system. The remaining ones are sapropelic and they are known to be oil-prone in numerous basins (Katz et al., 1991; Snowdon, 1991; Philp, 1994; Powell and Boreham, 1994). Important examples of these source rocks come from the exceptionally sapropel-rich Upper Permian source rocks of the Junggar Basin, which were deposited in an intra-mountain basin (Lawrence, 1990). These Upper Permian oil shales are among the richest and thickest source rocks in the world (Carroll et al., 1992), having

TOC and S_2 of up to 34 percent and 200 mg HC/kg rock, respectively.

Sapropelic coals are associated with two paleobotanical/paleoclimatic settings:

1) Tertiary angiosperm assemblages located within 20° zone around equator; and
2) Late Jurassic–Eocene gymnosperm assemblages developed on the Australian and associated plates.

Source rocks in rift and passive margin settings

Source rocks of rift and passive margin settings are represented by sediments deposited in a broad variety of environments; both terrestrial and marine in rift settings and marine in passive margin settings. The following text will discuss all environments, starting with rift settings, and their impact on organic facies and types of produced hydrocarbons.

Failed rifts

Lacustrine source rock facies of syn-rift settings can be divided into three groups:

1) algal-terrestrial;
2) algal; and
3) hypersaline algal.

The first type of the syn-rift lacustrine source rock facies is the algal-terrestrial organic facies, deposited in the fluvial-lacustrine environment association of overfilled lake basins (see Caroll and Bohacs, 1995, 1999, 2001). Facies association of these environments can be either predominantly silicic or carbonate-rich. Their characteristic features are:

1) progradational stratal geometries;
2) insignificant or absent depositional cyclicity associated with lake fluctuation; and
3) missing evidence of prolonged subaerial exposure.

Indicating hydrologically open lake systems, examples of facies of this type come from the Triassic–Jurassic Richmond Basin of the North American margin of the Central Atlantic (Olsen, 1990), the Upper Permian Hongyanchi Formation of the Junggar Basin, China (Carroll et al., 1992, 1995), the Lower Eocene Luman Tongue of the Green River Formation of the Uinta Basin, Wyoming (Roehler, 1992; Horsfield et al., 1994; Bohacs, 1998), the Paleocene portion of the Fort Union Formation, Wyoming (Liro and Pardus, 1990), some of the Cretaceous formations of the Songliao Basin, China (Xue and Galloway, 1993), some of the Paleocene intervals in the Central Sumatra Basin (Kelley et al., 1995), and some Cretaceous formations in the Doba and Doseo Basins of the Central African Shear Zone (Genik, 1993).

The algal-terrestrial organic facies of the Luman and Niland Tongues of the Green River Formation (see Table 15.2a) are characterized by the presence of two distinct organic matter populations; the first one with average TOC and hydrogen index values of 4.87 percent and more than 500, respectively, the second one with respective averages of 0.34 percent and less than 500. Kerogen macerals also contain two populations; the first one dominated by alginite, the second one represented by a mix of alginite and vitrinite (Horsfield et al., 1994). Biomarkers from these facies indicate relatively oxic freshwater environments having a mixed input of aquatic and terrestrial organic matter. Pyrolysis-gas chromatography analyses of kerogens of the representative Lower Eocene Luman Tongue mudstone indicate that these source rocks generate waxy oils at higher thermal maturities (Horsfield et al., 1994). Generally, algal terrestrial organic facies have intermediate TOC contents of 1–10 percent and mixed type I-III kerogens. High molecular weight n-alkanes, that is, waxes, dominate in the source rock extracts and produced oil, due to the input of protective tissues of higher land plants (Tissot and Welte, 1984) and membrane lipids of some freshwater algae (Goth et al., 1988; Tegellar et al., 1989). Enrichment in this type of organic matter can also be a result of selective bacterial degradation of other components in oxic to sub-oxic depositional environments (Powell, 1986).

Pyrolysis experiments with type III kerogens have shown that hydrocarbon generation occurs over a range of temperatures broader than that of type I kerogens (Tegelaar and Noble, 1994; Figure 15.12). This is consistent with data from the Lower Eocene source rocks of the Uinta Basin. The generation of oils of the Altamont-Bluebell fields coming from the deep section of the Uinta Basin occurred under thermal maturity equivalent to a vitrinite reflectance of 0.7 to 1.35 percent (Anders et al., 1992).

Oils from the Uinta Basin are characterized by a very high wax content (Caroll and Bohacs, 2001), which is also characteristic for oils of the Central Sumatra Basin (Kelley et al., 1995), the Songliao Basin, China (Yiang et al., 1985), basins of the Central African Shear Zone (Genik, 1993), and some Brazilian rift basins (Mello et al., 1988; Mello and Maxwell, 1990). Those organic facies that contain mixtures of type I–III kerogens generate subordinate amounts of gas (e.g., Clem, 1985; Rice et al., 1992; Genik, 1993; Kelley et al., 1995).

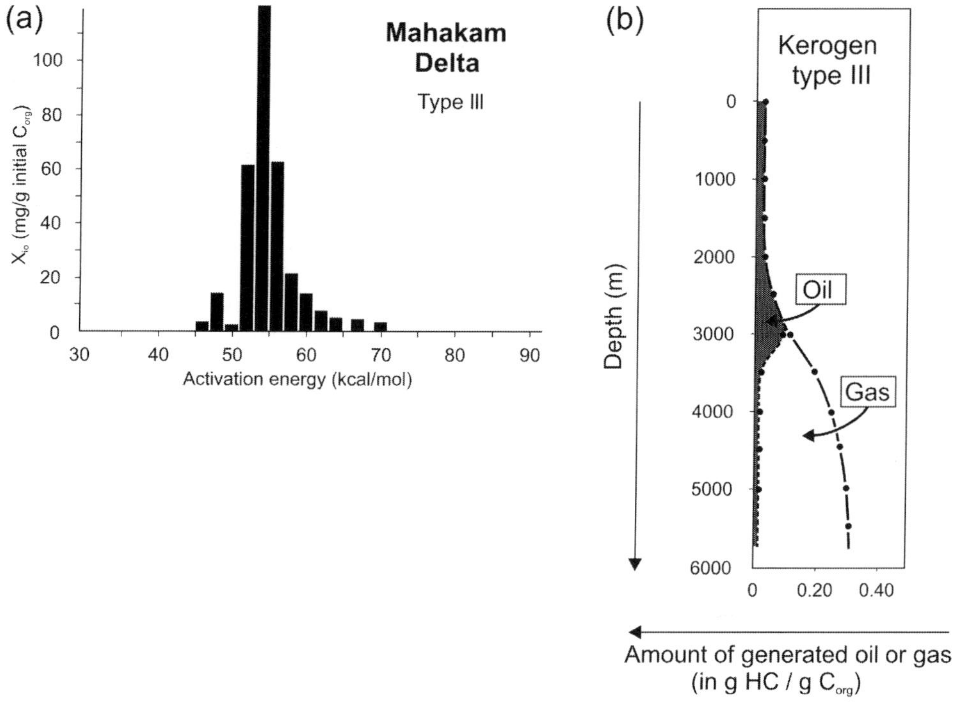

Figure 15.12. (a) Histogram of the activation energy of Mahakam Delta source rock with type III kerogens (Tissot and Espitalié, 1975; Ungerer and Pelet, 1987). (b) Hydrocarbon generation of type III kerogen for geologic subsidence rate and geothermal gradient comparable with those of Figures 15.13, 15.14 focused on type I and II kerogens (Tissot and Espitalié, 1975, reproduced in Robert, 1988).

The second type of the syn-rift lacustrine source rock facies is the algal organic facies, which is deposited in the fluctuating-profundal lacustrine environment association of lake basins with balanced fill (see Caroll and Bohacs, 1995, 1999, 2001). Facies association of these environments can be characterized by:

1) mixed aggradational/progradational stratal geometries;
2) significant depositional cyclicity associated with lake fluctuation; and
3) evidence of major lake desiccation and flooding.

Examples of fluctuating profundal facies come from the Triassic–Jurassic Newark Basin of the North American margin of the Central Atlantic (Olsen, 1990), a portion of the Middle Eocene Laney Member of the Green River Formation of the Uinta Basin, Wyoming (Roehler, 1992; Horsfield et al., 1994), the Upper Permian Lucaogou Formation of the Junggar Basin, China (Carroll et al., 1992, 1995), the Barremian Bucomazi Formation, the Lower Congo and Angola Basins (e.g., Burwood et al., 1995), and the Lower Cretaceous Lagoa Feia Formation of the Campos Basin, Brazil (Trinidade et al., 1995).

The algal organic facies is characterized by relatively homogeneous alginite-dominated type I kerogens, with subordinate vitrinite and inertinite (Table 15.2b). Biomarkers indicate predominantly aquatic organic matter input in saline lakes with anoxic bottom waters. TOC values are higher than those of algal terrestrial facies. Algal organic facies represent one of the richest lacustrine source rocks with a hydrogen index typically ranging from 600 to 900 (Espitalié et al., 1977a). Source rock extracts and produced oils are typically rich in n-alkanes derived from membrane lipids of aquatic organisms.

Pyrolysis experiments with type I kerogens have shown that hydrocarbon generation occurs over a very narrow range of temperatures, reflecting a little variation in chemical bond type within these homogeneous kerogens (Waples, 1980; Sweeney et al., 1987; Figure 15.13). The generation of oils from the Green River type I kerogens occurred under a thermal maturity equivalent to a vitrinite reflectance of 0.8 to 1.05 percent (Anders et al., 1992).

Oils from the Karamay and associated fields from the Junggar Basin in China are paraffinic (Clayton et al., 1997). They are low in sulfur and contain abundant n-alkanes with molecular weights below C_{30}. Similar n-alkane distribution is typical for the oils from type I kerogens in the Campos Basin, Brazil (Mello et al.,

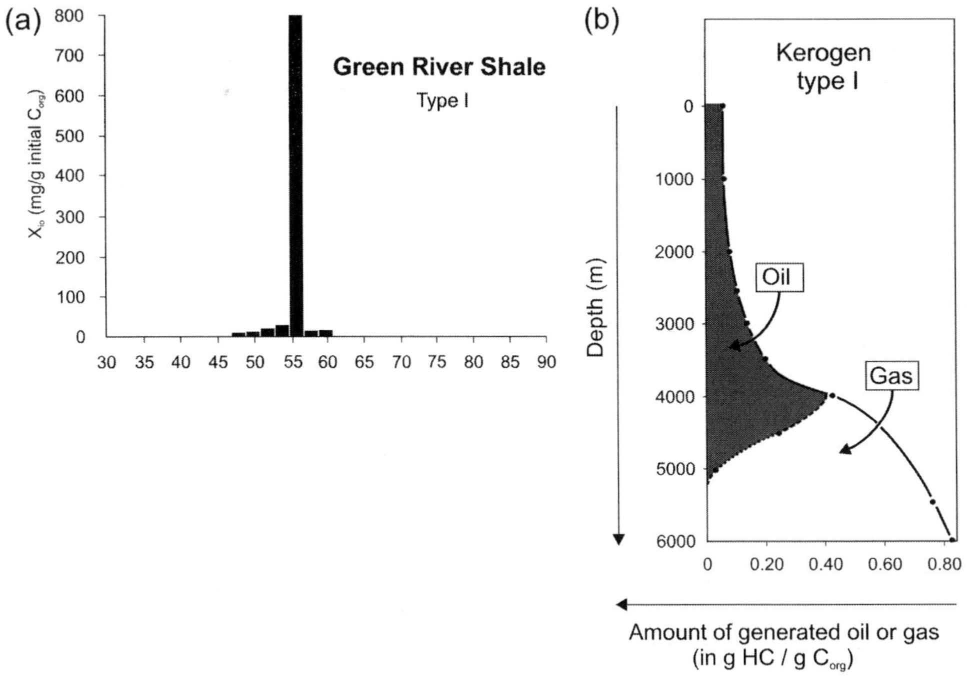

Figure 15.13. (a) Histogram of activation energy of Green River source rock with type I kerogens (Tissot and Espitalié, 1975; Ungerer and Pelet, 1987). (b) Hydrocarbon generation of type I kerogen for geologic subsidence rate and geothermal gradient comparable with those of Figures 15.12, 15.14 focused on type III and II kerogens (Tissot and Espitalié, 1975, reproduced in Robert, 1988).

1988; Mello and Maxwell, 1990). Note, however, that n-alkane content is known to vary significantly, being controlled by thermal maturity and biodegradation (Caroll and Bohacs, 2001).

The third type of the syn-rift lacustrine source rock facies is the hypersaline algal organic facies deposited in the evaporative environment association of underfilled lake basins (see Caroll and Bohacs, 1995, 1999, 2001). Evaporative facies association can be characterized by:

1) dominantly aggradational stratal geometries; and
2) numerous evidences of subaerial exposure and desiccation.

Examples of evaporative facies come from the Triassic–Jurassic Fundy Basin of the North American margin of the Central Atlantic (Olsen, 1990), the Lower–Middle Eocene Wilkins Peak Member of the Green River Formation of the Uinta Basin, Wyoming (Bradley and Eugster, 1969; Eugster and Hardie, 1975, Smoot, 1983), the Upper Permian Jingjingzigou Formation of the Junggar Basin, China (Carroll et al., 1992, 1995), the recent fill of the Great Salt Lake, Utah (Schamel and Ressetar, 1998), the Pleistocene Blanca Lila Formation, Argentina (Vandervoort, 1997), and some Tertiary formations of the lacustrine basins in China (Fu et al., 1986).

The hypersaline algal facies typically contains TOC values (Table 15.2c) lower than those of the brackish-saline facies, although intervals with enhanced TOC become deposited during periods of maximum lake extent. Maximum hydrogen index values are comparable to or higher than those of algal organic facies. Hypersaline algal organic facies has been reported as associated with I-S kerogen. Biomarkers indicate low-diversity assemblages of highly specialized organisms living in a difficult environment.

On the contrary from the previous two types of lacustrine organic facies associations, oils from several Tertiary lacustrine basins with hypersaline algal facies have been generated at low thermal maturity equivalent to a vitrinite reflectance of about 0.45 percent. Examples come from the Shengli oil field of the onshore Bohai Basin (Shi et al., 1982; Chen et al., 1994), the Dongpu Basin (Li et al., 1988, 1992), the Jianghan Basin (Fu et al., 1986; Sheng et al., 1987; Brassell et al., 1988; Peters et al., 1996), and the Qaidam Basin (Huang et al., 1991).

Paraffinic oils from the Jianghan Basin have an increased sulfur content of 3.59–12.91 percent (Sheng et al., 1987) and aromatic organosulfur compounds are abundant (Philp and Fan, 1987; Sheng et al., 1987). The sulfur compounds developed through diagenetic reactions between elemental sulfur and sulfides with phytol, fatty acids, and alcohols. Experiments with I-S kerogens indicate that this type reacts quickly to thermal stress, similarly to marine type II-S kerogens of the Monterey Formation, California (see Orr, 1986; Baskin and Peters, 1992).

Note that kerogens of the aforementioned lacustrine environments vary in space and time even in the same formation, as documented by the example of the lacustrine Lower Cretaceous syn-rift section from the Congo Basin discussed earlier (Harris et al., 2004). Yet another example of varying environments comes from the modern East African lakes, which have deposited organic matter, which ranges from type I to type III (Katz, 1988, 1990; Talbot, 1988). As generation analyses show, the mentioned organic matter variations are critical not only for calculating the quality and quantity of generated hydrocarbons, but also for calculating the timing of this generation (e.g., Anders et al., 1992; Tegelaar and Noble, 1994; Peters et al., 1996). Nevertheless, the syn-rift lacustrine source rocks combined with lacustrine, coastal plain, fluvial, and shallow marine syn-rift reservoirs, or younger syn-drift reservoirs, represent important petroleum systems on passive margins such as the Santos, Campos and Espírito Santo Basins, Brazil, where they produce oils of 29–30° API gravity, or basins at their Angolan counterpart (see Koike et al., 2007; Izundu, 2007; Berman, 2008; Watkins, 2008).

To discuss the source rock facies evolution in time and space in the rift setting, one can look at the following example of the evolving North Sea province. Its Middle Jurassic tectonic setting represented an early-rift setting (see Helland-Hansen et al., 1992; Johannessen et al., 1996; Fjellanger et al., 1996) characterized by uplift and deflation of the North Sea Dome. The Bathonian-Callovian rifting led to magmatism over the dome center (Husmo et al., 2002) and the Viking rift branch propagation first reaching the dome center, which was the place where all three propagating rift branches eventually met during the Late Jurassic (see discussion in Ziegler, 1990a; Underhill and Partington, 1993).

Associated early Middle Jurassic marine environments were passing into the coastal plain and alluvial environments. The coal-bearing formations are represented by the Aalenian–lowermost Bathonian alluvial plain Pentland Formation of the UK sector and the Aalenian-Bajocian coastal-plain and coal-swamp Sleipner and shallow-marine Hugin formations of the Norwegian sector (see Milner and Olsen, 1998). The Bathonian-Callovian rifting subsequently developed a set of tilted blocks in the North Viking Graben. The resultant complex relief with horst-and-graben-related highs and lows controlled significant facies variation over short distances. The middle Jurassic coal swamps in the future Viking Graben became inundated by sea from the north by the Oxfordian time period, characterized by widespread marine environments (Cockings et al., 1992; Husmo et al., 2002).

While the coal-bearing formations were typically gas-prone, an isolated hydrocarbon system of the Sogne Basin at the Norwegian–Danish boundary of the North Sea serves as an example where the Middle Jurassic coals of the Bryne Formation contain oil-prone vitrinitic components. The stable carbon-isotope data suggest that oils of the Harald and Tym fields come from this source (see Husmo et al., 2002).

The marine transgression in the North Sea province ended the deposition of coal-bearing formations and started the deposition of the marine organic facies association. As the rift system developed, marine facies have been deposited in spatially and temporally changing environments characterizing early syn-rift to late syn-rift settings.

First of the important marine source rocks is the upper Callovian–lower Kimmeridgian Heather Formation and its lateral equivalents such as the Haugesund, Egersund, and Lola Formations. It contains dark gray, frequently silty mudstones with numerous carbonate bands. TOC contents range from 2 to 2.5 percent (Goff, 1983; Field, 1985; Brosse and Huc, 1986) and pyrolysis-derived hydrogen indices are low. Both data types indicate an increased content of terrestrially derived humic kerogens of vitrinitic and inertinitic compositions. The Horda Platform contains the richest mudstones of the Heather Formation, which are thought to have contributed to pooled hydrocarbons in fields of block 35/8 in the Norwegian North Viking Graben and the Huldra field located further south (Ferriday and Hall, 1993).

The Heather Formation was deposited during the growth, linkage, and interactions of extensional faults (Underhill, 1991a, 1991b; Rattey and Hayward, 1993). The associated facies distribution was characterized by complex spatial and temporal variations, as discussed in Chapters 1 and 7 (see Davies et al., 2000; McLeod et al., 2000, 2002; Figures 7.6, 7.7). Undergoing a rapid rise in sea level, the region developed a plethora of coastal-shelf depositional systems on flanks of individual basins experiencing basinal deposition (Fraser et al., 2002). Detailed geometries of facies belts were controlled by rift branch-parallel graben and horst geometries (see Rattey and Hayward, 1993; Kadolsky et al., 1999).

Early to Late Oxfordian was the time of the continuing propagation of all three rift branches and advancing marine incursions along developed depressions (Fraser et al., 2002). The rifting advanced into the South Viking Graben, having an accelerated rate. The sedimentation responded by depositing the deep-water and organic-rich Kimmeridge Clay Formation in the North Viking Graben and organic-leaner offshore mudstones of the Heather Formation in the South Viking Graben. The late Oxfordian–early Kimmeridgian interval was the time of drowning of fully developed failed rift segments and accelerated rift subsidence, which resulted in the formerly important intra-basinal highs being draped by condensed mudstone facies of the Heather Formation.

As subsidence of the North Sea rift system continued, the depositional environments eventually deposited the Kimmeridge Clay Formation, an excellent source rock, and its lateral equivalents. The upper Oxfordian–upper Ryazanian Kimmeridge Clay Formation of the UK sector and its stratigraphic equivalents, such as the Draupthne, Tau, and Mandal Formations in the Norwegian sector and the Farsund Formation in the Danish sector, represent the main source rock in the North Sea. It contains dark non-calcareous mudstones. Its marginal facies contain a high portion of terrestrially derived organic matter that decreases toward the deeper graben axes (Brown, 1990). Terrigenous Exinite macerals, such as type IV sporinite and Cutinite, may produce some oil, although not in commercial quantities. One of the keys for its world-class source rock character is its inter-fingering with high-quality sandstone reservoirs. This makes it both source and seal, and controls the short distances of migration pathways (Cayley, 1987; Johnson and Fisher, 1998).

Marine-algal organic matter characterizes the facies of the deeper parts of the basin and restricted sub-basins away from the influence of terrestrially derived humic kerogens (Barnard and Cooper, 1981; Barnard et al., 1981; Demaison et al., 1983; Cooper and Barnard, 1984; Baird, 1986). Unstructured or amorphous liptinic type II kerogen, interpreted as mostly a product of degraded marine phytoplankton, is the primary oil-generative component (Fisher and Miles, 1983; Cornford et al., 1986). It contains only minor structured humic particles that are usually represented by small inertinite shards and vitrinite. High organic-carbon contents and preservation of Liptinite-rich, type II kerogens are likely to reflect both high phytoplanktonic productivity and efficient preservation under anoxic bottom-water conditions.

A TOC content of the Kimmeridge Clay Formation ranges from 2 percent to in excess of 10 percent. The hydrogen index varies according to kerogen type, which is controlled by specific sedimentary facies. Oil-prone, type II amorphous kerogens have a hydrogen index of up to 600 mg/g TOC. Mixed type II and III kerogens have an index of only about 200–400 mg/g TOC, which can be attributed either to an increased content of terrestrially derived humic material or to an increased bacterial degradation of the proto-kerogen in the depositional environment, possibly under dysaerobic rather than anaerobic bottom-water conditions. Gas-prone type III kerogens of vitrinitic and inertinitic composition have indexes of around 150 mg/g TOC.

Leaner source rocks of the Kimmeridge Clay Formation have been deposited over platform areas away from the grabens, such as those in the Stord and Norwegian-Danish Basins (Baird, 1986). One more instance of leaner source rocks is those in areas where turbidity currents brought terrestrial sediments far into the deep basin.

The paleogeography of the North Sea during the deposition of this formation was characterized by the regional development of restricted oceanic circulation, irregular and fragmented syn-rift basement floor bathymetry, and high deposition rates, all contributing to widespread anoxic conditions in different depositional areas up to those as shallow as at 200 m (Stow and Atkin, 1987; Miller, 1990; Fraser et al., 2002). Horst and graben-related facies distribution was further complicated by deep marine sandstone bodies, for example in the Viking Graben and the East Shetland Basin, reacting to major rift-related footwall uplifts and erosion in the area.

The rift-related bathymetric deepening advanced into all segments of the rift system during the late Kimmeridgian–early Volgian, coeval with the widespread occurrence of dysaerobic facies of the Kimmeridge Clay Formation and its lateral equivalents. An example of the North Viking Graben documents that pelagic mudstone facies dominated large basins, while shoreface sandstones became restricted to narrow zones rimming highs, sediment input from flanks was minimal, and depocenters underwent sediment starvation.

The rifting progressively terminated and became replaced by thermal subsidence in the entire North Sea by the middle Volgian, apart from continuing rifting in the Moray Firth Basins, which did not stop during the deposition of the Kimmeridge Clay Formation and its equivalents (Fraser et al., 2002).

Passive margins

The rift setting that makes it to the continental breakup has an important difference between depositional

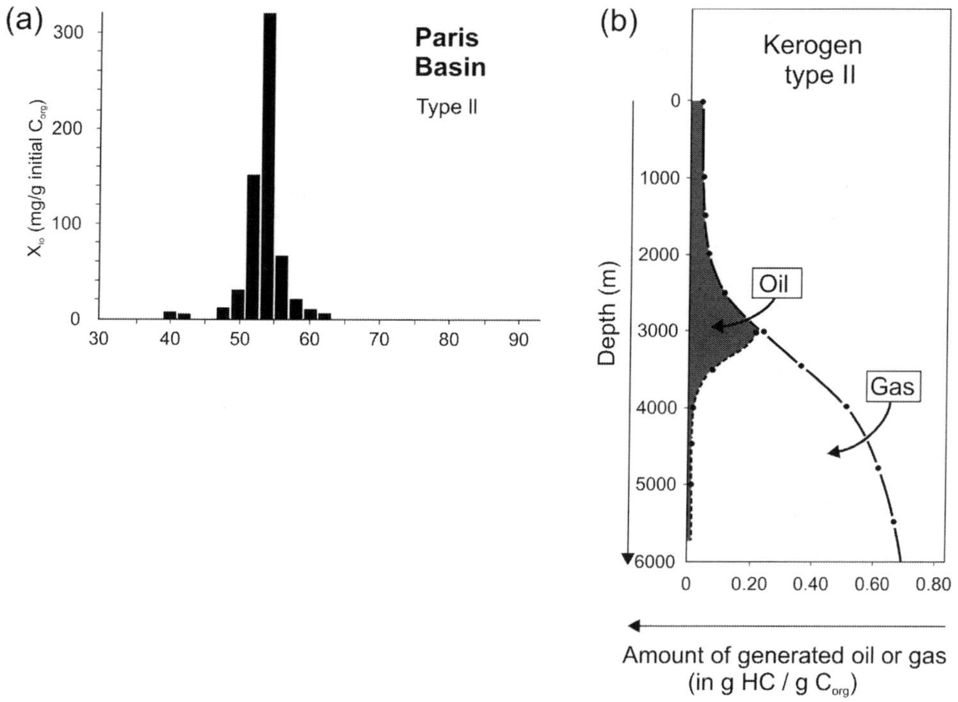

Figure 15.14. (a) Histogram of activation energy of Paris Basin source rock with type II kerogens (Tissot and Espitalié, 1975; Ungerer and Pelet, 1987). (b) Hydrocarbon generation of type II kerogen for geologic subsidence rate and geothermal gradient comparable with those of Figures 15.12, 15.13 focused on type III and I kerogens (Tissot and Espitalié, 1975, reproduced in Robert, 1988).

environments of the future proximal margin and environments of the future distal margin, described in Chapter 2. While the facies associations of the future proximal margin are comparable to continental rift associations, being characterized by grabens bounded by high-angle normal faults, those of the future distal margin are very different, being characterized by relatively thin syn-rift sedimentary sequences in shallow basins bounded by low-angle detachment faults. Furthermore, the future distal margin can undergo uplift instead of subsidence and the already limited deposition in individual basins can compete with volcanism trying to utilize the catchment areas for volcanic products, as one can see, for example, in offshore South Africa, Namibia, and East India. With respect to a certain predictability of the source rock type distribution, the passive margins need to be divided into segments affected by major rivers and segments undisturbed by deltas of major rivers entering the deep basin, which are characterized by oil-prone marine source rocks.

Type II kerogens of undisturbed segments, like type III kerogens, mature over a wide temperature/depth range (Figure 15.14) and are more likely to continue to generate during multiple episodes of maturation. An example, which we borrow from a non-passive margin setting, comes from the material and mass balance calculations of expulsion efficiencies done for the oil-prone marine source rock, the Toarcian Posidonia shale from the Hils syncline area in Germany (Leythaeuser et al., 1987). For samples taken from the source rocks with various maturity, the expulsion efficiencies vary from 63 percent at vitrinite reflectance of 0.68 percent, to 86 percent at 0.88 percent and then 96 percent at 1.45 percent. Their depth plots of generated hydrocarbons further indicate that the generation from type II kerogen started earlier than the generation from type III kerogen. This is in accordance with the knowledge about maturity thresholds indicated by the vitrinite reflectance of 0.5 and 0.7 percent for the onset of generation from type II and III kerogens (see Tissot and Welte, 1984).

Mentioned analytical data from the Posidonia shale allow us to conclude that the higher the original hydrocarbon generation potential, that is, better quality kerogen, the higher the degree of expulsion is (Leythaeuser et al., 1987). Because the texture of the shale and marlstone source rock was extremely fine-grained, very homogeneous, and lacking any laminae with increased permeability that would allow its draining, the expulsion was fracture-controlled. It is indicated by the fact that only a few vertical fractures are present in the immature source rock samples, while, in striking contrast, each mature sample is highly fractured by both meso- and micro-scale fractures parallel to bedding (Littke et al., 1988). Furthermore, these fractures do not occur in underlying Pliensbachian and overlying Aalenian mudstones in the areas where they are mature, because of their different permeability and kerogen structure.

Returning back to passive margins, the organic rich shales with type II kerogens are known to have been deposited either on the oceanic crust located far from the continental margin, as is known from several wells in the Central Atlantic or on the margin itself, as is known from the entire northern South Atlantic and the Equatorial Atlantic.

The Central Atlantic black shales on the oceanic crust have been documented by wells such as Light well No. 1 on the Cape Hatteras, DSDP well 367 in the Senegal Basin, and DSDP well 368 on the Cape Verde Rise (Segall et al., 2003). The stratigraphic age of the shales ranges from upper Aptian to lower Turonian. Well ties to reflection seismic profiles such as that with Valdivia 10–1975 profile indicate a large lateral extent of these black shales. Their extent is large enough to locate them on both oceanic and proto-oceanic crusts.

Despite the general view of an ocean basin being outboard of areas interesting for exploration because of its thermal regime cooler than that of continental margin and not having sufficiently thick overburden, the discovery of an increasing number of hydrocarbon fields on the oceanic crust in the Gulf of Mexico and offshore Nigeria tempts one to rethink this paradigm.

A potential improvement of the thermal regime in the proto-oceanic case would be its regime warmer than that of the oceanic crust. Furthermore, extra heat in the case of oceanic crust can be hypothetically locally added from:

1) the geothermal fluids migrating inside the oceanic fracture zone; or
2) the cooling intrusive body; or
3) the radioactive decay inside the thin continental block traveled far from the continental margin in a form of extensional or strike-slip allochthon.

Examples of geothermal fluid mechanisms are provided in Chapter 8. An example of the intrusive body comes from DSDP well 368 on Cape Verde Rise. It documents how a horizon of Albian black shale about 70 m thick reached a higher maturity thanks to the influence of the mafic sill intrusion. The shale, deposited below the carbonate compensation depth, is rich in organic matter, characterized by TOC values of up to 11.7 percent.

The presence of exotic blocks of continental crust is known from numerous oceans. It is usually associated with a block release from a contact zone between neighbor rift zones. An example is provided by the Elan Bank released from East India (see Gaina et al., 2003; Nemčok et al., 2008). An example of a block released from the contact zone between adjacent pull-apart basins systems comes from a block located almost at the contact of the Equatorial Atlantic spreading ridge with the Romanche Fracture Zone, which was released from the Ghana Ridge (Nemčok and Allen, 2001; Nemčok et al., 2004b).

Geochemical data from well penetrations of the upper Aptian–lower Turonian black shales in offshore Côte d'Ivoire and Ghana indicate good source rock properties. The TOC values range from 8 to 28 percent. These shales contain marine organic matter deposited in a reducing environment with very minor terrestrial input. Their oxygen indexes between 20 and 90 combined with hydrogen indexes of 180–650 plot respective kerogens as type II on OI/HI diagram (Figure 15.15).

An example of organic-rich shales deposited in continental margin environments away from the entry point of large deltas comes from the Ghana Ridge region (Nemčok et al., 2004b). Well penetrations document three stratigraphic intervals with source rock facies associations, the upper Albian, the upper Cenomanian–lower Turonian, and the lower Campanian intervals.

The late Albian time interval in the Ghana Ridge region marks the onset of extensive shallow marine conditions. The exploration well in the Deep Ivorian Basin, being the most landward of the studied wells that include ODP sites on the Ghana Ridge, documents this transgression by brackish, coarse-grained facies replaced by shallow marine claystone and muddy limestone. The earliest marine conditions are indicated to the south of the exploration well in the lower Albian, in ODP well 959, being the stratigraphic equivalent of poorly dated near-shore brackish facies in ODP well 960 and a fluvial-deltaic package in the exploration well.

The upper Albian section penetrated by the exploration well is indicated by planktonic foraminifera and dinoflagellates but almost no benthic species. The planktonic assemblage has very low species diversity, being composed dominantly of Hedbergella and Whiteinella, indicating a low productivity in abnormal marine conditions on the inner to middle shelf. This stressed assemblage was most likely controlled by slightly lowered surface water salinity and/or moderately high turbidity in a partially restricted basin. Although a low organic

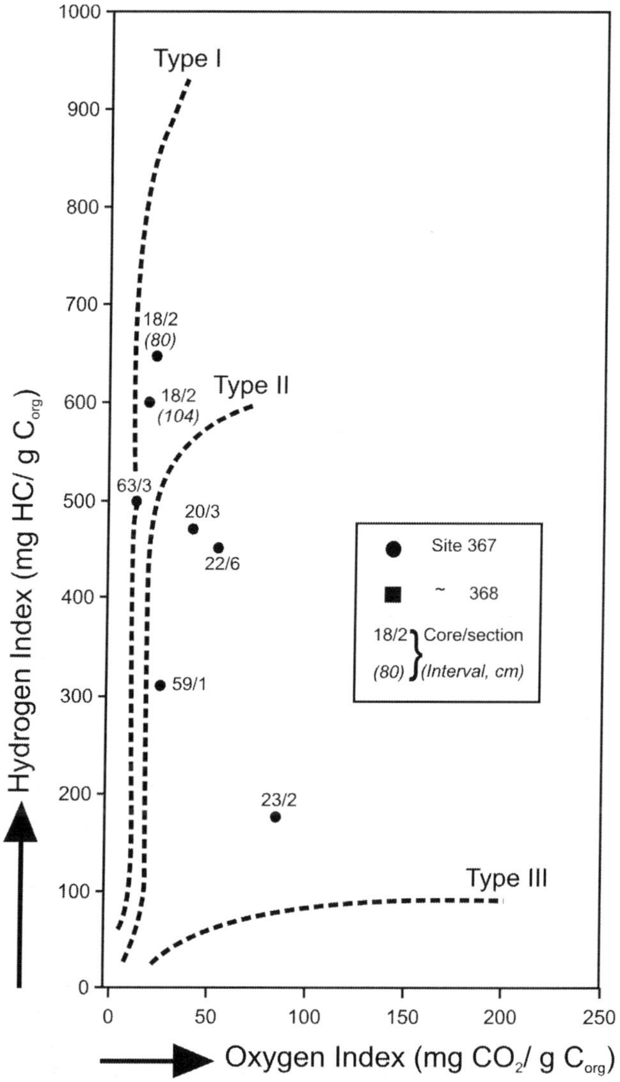

Figure 15.15. Pyrolysis results from the upper Aptian–lower Turonian black shale sample form the DSDP well 367 in Cape Verde basin (Segall et al., 2003).

influx is indicated, a very slow circulation in this basin likely resulted in low oxygen conditions. A combination of low organic input and low oxygen conditions explain the lack of benthic foraminifera in this region.

The late Cenomanian–early Turonian interval is marked by yet another transgression event, recorded by ODP well 962 on the ridge during the late Cenomanian, by ODP well 959 on the plateau north of the ridge in the latest Cenomanian, and by an exploration well in the Deep Ivorian Basin to the north of the plateau during the early Turonian, being the most landward of the studied wells. This more extensive flooding affected a still relatively narrow restricted basin developed by drifting following a continental breakup of the pull-apart basins system. This led to a stratified water column and the deposition of low-oxygen facies in the area penetrated by ODP wells 959 and 962.

The Cenomanian section in the exploration well in the Deep Ivorian Basin contains abundant planktonic foraminifera but almost no benthic ones. This section in ODP well 962 is characterized by claystone dominating over sandstone and by an increase in the number of allochthonous carbonate beds, indicating flooding of the margin. This transgression culminated with low-oxygen facies at maximum flooding represented by the radiolarian-rich non-calcareous claystone.

The lower Campanian sedimentary section records severe oxygen depletion located in the region between the oceanic basin floor and the upper slope of the Deep Ivorian Basin, as indicated by the exploration and ODP 959 wells. The exploration well penetrates facies with a very low foraminiferal diversity, with dominance by a few species of Recurvoides that are probably allochthonous, indicating stressful paleoecologic conditions under likely anoxia located on the upper to middle slope. The anoxic facies of ODP well 959 is characterized by a flood of radiolarians, but only very rare siliceous agglutinated foraminifera including the index fossil Uvigerinammina jankoi. This depositional interval marks the greatest oxygen depletion event in this well. Similar low-oxygen facies from this time interval are recognized across the entire Atlantic. Examples come from the Lower Congo Basin in Angola, the Santos Basin in Brazil, and offshore Morocco.

The Cenomanian-Turonian open marine source rocks represent the most significant source rocks for deep exploration in the Ghana Ridge region. Further south, in the Rio Muni Basin of the South Atlantic, the drift section starts with the Upper Aptian transgressive muds, limestones, and thin sands. They are progressively replaced by the deposition of organic-rich mudstones having excellent source rock potential and acting as a detachment horizon for gravity glides having Albian carbonate horizons at their base. The upper Aptian transgressive organic mudstones represent brackish to marginal marine facies (Bottero et al., 1989), finishing with maximum flooding surface at their top, forming the first marine flooding of the basin. This transgressive sequence is overlain by the aforementioned Albian carbonates, which are about 400 m thick in the Matondo well but vary considerably in thickness throughout the basin. The Albian carbonates are overlain by the Cenomanian marine sands and shales, penetrated by the Benito 1 well, and the younger Cenomanian (and probably also Turonian) section, which onlaps structural highs in the seismic sections and thickens away

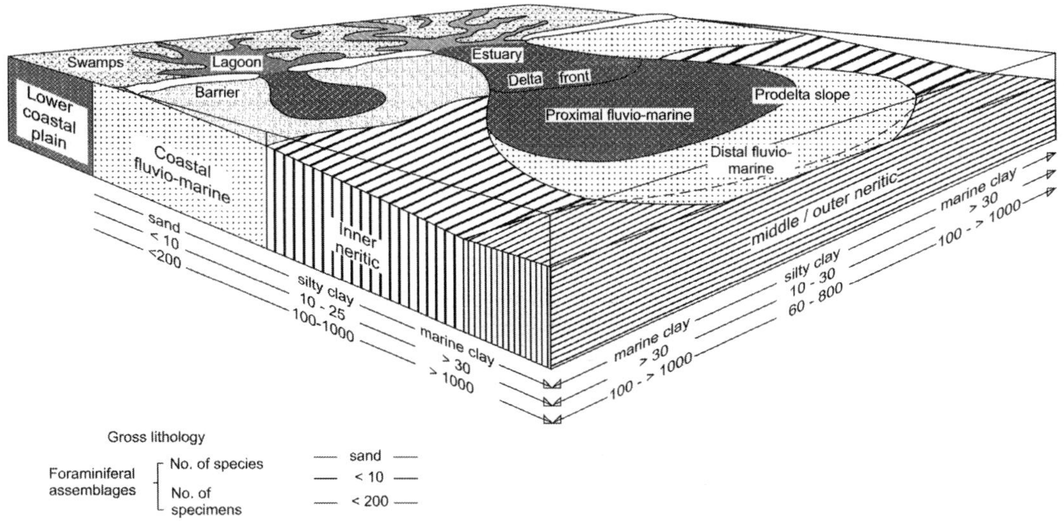

Figure 15.16. Depositional environment of the Niger Delta (Doust and Omatsola, 1990). The paralic sequence is primarily of lower coastal plain, coastal, and fluvio-marine origin, whereas the marine shales are thought to have accumulated in open-marine neritic environments. The continental or alluvial sands are believed to have been deposited in upper coastal-plain environments.

from them. Potential source rocks of these described sections include:

1) Aptian-Albian post-rift section; and
2) Cenomanian-Turonian drift section.

The Aptian-Albian post-rift source rock section of up to 200 m thick was penetrated by the Matondo-1, Benito-1, and East Eviondo-1 wells. The source facies is lacustrine and/or restricted marine type II/III kerogen. Determined TOC and hydrogen index values are as high as 2–4 percent and 200–500, respectively.

Even further south, the early syn-drift source rocks in Gabon, where evaporites mark the transition between the syn-rift and syn-drift section, are more difficult to assess. It is due to the fact that their organic facies vary over a short distance because of the distance from the clastic source and because of salt tectonics developing intra-domal depressions with restricted circulation and water with high salinity. Amongst the Albian Madiela, Cenomanian Cap Lopez, Turonian Azile, and Senonian Anguille Formations, the Azile Formation is most frequently typed to hydrocarbons in the offshore Port Gentil area. Its TOC ranges from 3 to 5 percent. Kerogens are comprised by mixtures of I and II or locally II/III mixes.

The Azile Formation in the whole area can be roughly divided into three facies. The southern one is richest in mixed I/II kerogen types and has the highest TOC. The northern one is characterized by mixed II/III or III type kerogens. Its TOC is lower than that in the south. The central area has a less favorable character, having low TOC content and a high content of the humic type III kerogen. The reason for such distribution of these three facies is the distribution of turbidites in the central area and marine shales in the north and south.

This brings us to the discussion of the influence of large deltas on organic facies of passive margins. The first example comes from the Niger Delta, formed by syn-drift regressive sedimentary sequence comprising a series of offlap cycles. The delta is one of the largest hydrocarbon provinces, with proven ultimate recoverable reserves of about 26 billion bbl, described by Haack and colleagues (2000). Stratigraphic columns of wells indicate a diachronous tripartite sequence starting with marine shales, going through paralic deltaic sands, and finishing with alluvial sands (Figure 15.16). These sequences replace each other at younger times in the direction to the south.

The Akata Formation represents marine shales composed of shale, clay, and silt. It includes turbiditic horizons deposited in the delta front to deeper marine environments. The stratigraphy ranges from the Paleocene to the present.

The Abgada Formation represents paralic deltaic sediments composed of intercalated sands, silts, and clays. Intercalations represent cyclic sequences of offlap units. Each unit has well-developed cyclicity on a thickness scale of typically 15–60 m or rarely 100 m. The thickness of these cycles is highly influenced by syn-depositional faults. Depositional environments of offlapping units include delta front, delta topset, and fluvio-deltaic.

The Benin Formation represents alluvial sediments composed almost entirely of sand. Apart from alluvial, its depositional environments also include the upper coastal plain.

The syn-drift source rocks here are represented by two stratigraphic intervals (Haack et al., 2000):

1) the Upper Cretaceous–lower Paleocene section; and
2) the middle Eocene–Pliocene section.

The Upper Cretaceous–lower Paleocene source rocks were deposited in a marine environment. They were found in the northwestern part of the delta, present in the Araromi, Awgu, and Imo Formations. However, they should be present under the whole delta and in the Anambra Basin, the ancestral Niger Delta. The Upper Cretaceous portion of these source rocks was also studied along the eastern margin of the delta in outcrops (Inyang et al., 1995) and wells (Haack et al., 2000). The associated oil was also discovered in the Anambra Basin, having a similar character to the oil from the Upper Cretaceous source rock in the Seme Field in offshore Benin (Haack et al., 2000). The Upper Cretaceous–lower Paleocene source rocks are characterized by type II and II-III oil-prone kerogens. Oils produced from the Shango-1 well are inferred to be produced by them. These source rocks are quite common in West African petroleum systems such as the Lower Congo Basin system (Schöellkopf and Patterson, 1997), although no large accumulations are typed to them in the Niger Delta.

The Middle Eocene–Pliocene source rocks form a mixture with different signatures. They are the principal source for hydrocarbons in the Niger Delta. Their richness varies. Some of them are dry-gas prone and some of them oil-prone. Laterally and temporally varying source rock facies are dominated by terrigeneous organic matter. They contain regional differences in thermal maturity, related to spatially and temporally complex deposition. All mentioned complexities play a role in the complex regional distribution of oil and gas in various depobelts of the Niger Delta.

The Aroh-2 well contains an example of thermally mature oil- and gas-prone middle Eocene source rocks. The richer facies, containing type II oil-prone kerogen, increases with depth. These source rocks are fingerprinted to oils in wells in the area. Gas chromatography indicates that these oil-prone deltaic facies generate primarily light to normal oils, a small portion of wet gas, and a minimum of dry gas. Foraminiferal assemblages indicate that the deeper portion of the source rocks from the Aroh-2 well was deposited in a neritic to bathyal environment. The concentration of C29 sterane derived from land plants is as high as 42–57 percent. It indicates an effective terrigeneous transport to these environments.

A conceptual model of Haack and colleagues (2000) explains that the more gas-prone end members of the middle Eocene–Pliocene source rock suite were deposited in oxic depositional environments with poorly preserved organic matter. The oil-prone organic facies were deposited in sub-oxic environments with enhanced preservation of the terrigeneous organic matter. As can be implied from the understanding of the primary structural and secondary depositional controls of depositional processes in the Niger Delta (see Hooper et al., 2002), mentioned source facies distribution is temporarily and spatially complex.

The second example of a large river influence comes from northern Brazil where the marine deltaic source rocks existing in the Para-Maranhao Basin in front of the Amazon River indicate that the terrestrial load from the river overwhelmed the marine environment to the northeast of the river mouth (Mello et al., 1991).

It can be concluded that the source rock successions are not merely passive elements within a sedimentary section of the passive margin in the regions around large deltas. As documented by the examples of the Upper Cretaceous shale in the Foz do Amazonas Basin, Brazil (e.g., Cobbold et al., 2004), the Upper Cretaceous–Paleocene shale in the Krishna-Godavari Basin (e.g., Choudhuri et al., 2006), Paleocene and younger shale in the Niger Delta region (e.g., Hooper et al., 2002), and Eocene shale in the southern Texas portion of the Gulf of Mexico province (e.g., Diegel et al., 1995; Karlo and Shoup, 2000), the organic-rich shales and marls, which are normally the weaker component of the mechanical stratigraphy and which are prone to overpressure development due to restricted compaction-driven dewatering, clay dewatering, and sometimes even hydrocarbon expulsion, control the location and deformational style of the gravity glide systems of passive margins.

16 Models of reservoir quality distribution

Introduction

Rocks that have permeabilities in the range of 10^{-13} to 10^{-14} m^2 are generally considered good matrix-permeability reservoirs (Deming et al., 1990a; Deming and Nunn, 1991; Thomas and Clouse, 1995). This range characterizes the higher portion of carbonate and sandstone matrix permeability spectra (e.g., Deming, 1994; Smith and Wiltschko, 1996).

The grain size and grain sorting have the largest effect on matrix permeability. The sediment deposition process controls both. The larger the grain size, the higher the permeability of the same lithology is (Prince, 1999). The better the sorting, the higher the permeability is. The permeability is controlled by the proportion of fine-grained particles decreasing the pore space among the coarse-grained particles.

Diagenetic and chemical processes further control matrix permeability. They include dissolution, recrystallization, cementation, and mineral transformations. All of these processes are controlled by the overall energy balance during rifting and subsequent events, mechanical stratigraphy of sediments, syn-tectonic deposition, erosion, and fluid flow.

This chapter discusses the reservoir rocks according to their depositional environments in order to develop an understanding of how reservoirs and their permeabilities vary in time and space in evolving rift systems.

Reservoir facies according to settings

Basement reservoirs

Basement reservoirs are relatively common in rift and passive margin settings as documented by about 110 producing fields and discoveries worldwide. The best scenario is the fractured basement covered by the source rock that also represents a seal. An example comes from the White Tiger oil field in the Cuu Long Basin, offshore South Vietnam (Areshev et al., 1992; Dmitriyevskiy et al., 1993). Its oil column is unusually high, exceeding 1,000 m. Here, the Oligocene argillite source rock covers the fault blocks of the uplifted footwall made of granite and granodiorite. Different mechanisms that have controlled development of the producing fractures formed in the hydrothermally altered igneous rock include:

1) fracturing inside the damage zone of the main fault, associated with various stages of fault propagation;
2) fracturing inside the fault block away from the main fault; and
3) fracturing controlled by weathering.

The Banat Depression of the Pannonian Basin system with its oil fields is a typical example where the main reservoirs are in the sedimentary cover while the fractured basement just adds to the total reserves (Filjak et al., 1969; Tari and Horvath, 2006). This region is one of many that document that a combination of weathering and tectonic fractures is fairly common in rift terrains characterized by subaerially exposed footwall edges and horsts (see P'An, 1982; Kiss and Toth, 1985; Matyas, 1996; Nelson, 2001). Furthermore, the tectonic fracture porosity can become enhanced by subsequent weathering and erosion, as it is in the Wilmington oil field in the Los Angeles Basin (Hubbert and Willis, 1955; Aguilera and Van Poollen, 1979; P'An, 1982).

Examples such as the Badejo, Linguado, Pampo, and Title oil fields in the offshore Campos Basin in Brazil (Tigre et al., 1983; Nelson, 2001) document that the "basement" reservoir can be represented, in fact, by the first syn-rift rock, the Neocomian fractured basalt. It forms an addition to the syn-rift Aptian Lagoa Feia formation coquinas, post-rift Albian Macal formation carbonates, and Eocene sandstone of the Carapebus member.

Good well core examples of the tectonic fracture porosity come from geothermal fields, such as the Geysers, northern California, and Karaha–Telaga Bodas, Java (Figures 16.1a–c). They document that a plethora of fractures is represented by open cavernous spaces that remained uncemented in most dilatant fracture segments. The open space can be supported by fracture breccia that never became completely cemented.

The fractures made solely by weathering are the producing fractures in the Hurghada heavy oil field

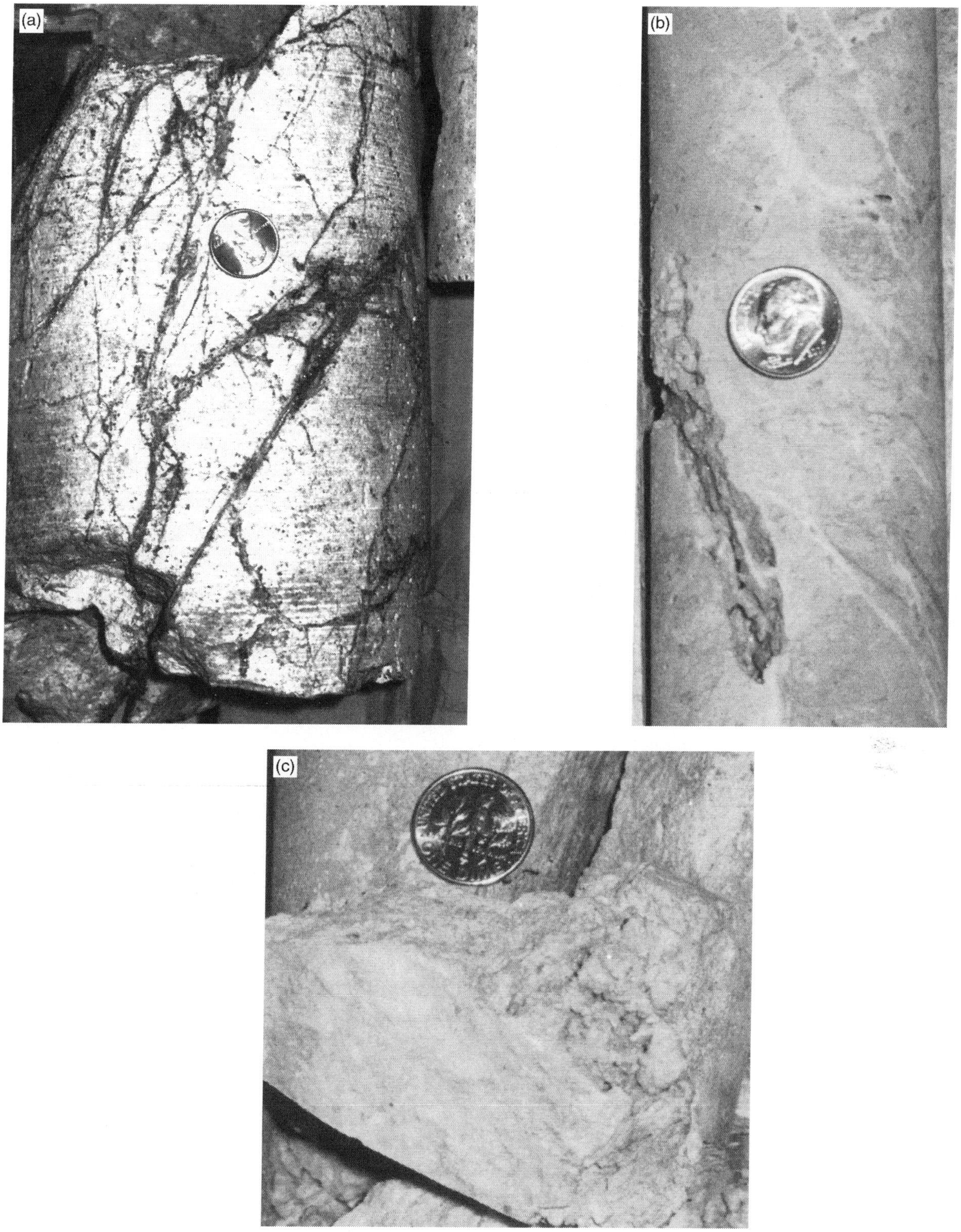

Figure 16.1. (a) Example of a producing geothermal fracture pattern from The Geysers geothermal field, California. (b) Example of a producing geothermal fracture from a depth of 2,001 m in well K33 in the Karaha–Telaga Bodas geothermal prospect, Java. Note an open/vuggy fracture with quartz lining. (c) Example of producing geothermal fracture from a depth of 1,344 m in well K33 in the Karaha–Telaga Bodas geothermal prospect, Java. Note a 10 cm-thick breccia inside the fracture. The breccia neither contains a wear matrix nor is fully cemented. It is barely "glued" together by calcite lining.

on the western shore of the Gulf of Suez, Egypt (P'An, 1982).

Existing fractured basement reservoirs allow us to conclude the basement reservoir can represent a locally important play, as proven by about 90 percent of Vietnam's oil production coming from fields in basement reservoirs. Furthermore, the rocks in paleo-highs and adjacent slopes have enhanced reservoir properties in comparison to those of depressions because of the more extensive weathering. The rocks in paleo-depressions have a high risk of reduced reservoir properties because of fractures clogged by allochthonous clay derived from weathered adjacent highs. Where the younger source rocks, or volcanic rocks and then younger source rocks, directly covered the weathered zone in igneous and metamorphic rocks, the fissures and fractures have remained open. If the igneous and metamorphic rocks are covered by the fine-grained sediments older than the source rock, most of the fissures can be filled with mud and iron oxides.

Pre-rift reservoirs

Pre-rift reservoirs contain various lithostratigraphies (Figures 7.10a–c), many of them discussed in the text on basement reservoirs. If we use some mature and large rift provinces, such as the North Sea, as a case area, we can imagine how many different lithologies the syn-rift sediments can onlap (Stewart and Reeds, 2003). Because the youngest pre-rift sediments can form important reservoirs in various types of footwall traps, it is important to understand their facies distribution regardless of how different their setting is from the subsequent rift setting.

The Lower–Middle Jurassic pre-rift facies distribution in the North Sea was controlled by the preexisting Triassic rift topography and Middle Jurassic doming and erosion (Husmo et al., 2002). The package is thin or absent in the southern areas of the North Sea Rift System, and thick in the north.

The Lower Jurassic is represented in the northern petroleum region by the Dunlin Group, which is a system of mudstone up to 600 m thick with some sandstone horizons (Husmo et al., 2002). The Middle Jurassic in this region is represented by the Brent Group, a system of facies up to 500 m thick ranging from shallow-marine to near-shore and nonmarine sandstones representing various delta-related facies and their surrounding environments.

The Lower Jurassic sedimentary section in the northern portion of the north central petroleum region, represented by fields such as Beryl, Bruce, and Sleipner Vest, is incomplete below the intra-Aalenian erosional unconformity (Husmo et al., 2002). The least complete strata are present in the southern portion of this region, which was located at the site of the future junction of all three rift branches of the North Sea Rift System, and which was the center of the pre-rift doming.

The Middle Jurassic strata in the north central region belong to the Fladen Group, consisting of Pentland and the overlying Hugin formations (Husmo et al., 2002), represented by the coastal plain sediments and marine sandstone, respectively. The Fladen Group in the southern portion of this region is markedly different from that of the northern portion. Its Pentland Formation contains the Rattray and Ron volcanic members reaching a maximum thickness of 1.2 km (see Figure 6.22).

The west and east petroleum regions are represented by isolated areas with only a few fields (Husmo et al., 2002). Their stratigraphic successions are incomplete. The Middle Jurassic Vestland Group, located in the eastern region, consists of the Sleipner and overlying Hugin formations. The former is coastal plain facies; the latter is marine sandstone.

The field reservoir core and outcrop analog studies, done to understand reservoir properties of the shoreface-shelf sandstone reservoirs such as those mentioned earlier, have documented that permeability structure depends on heterogeneities of various type and scale (e.g., Hampson et al., 2004; Sixsmith et al., 2008; Figure 16.2).

Figure 16.2 documents how a good understanding of the wave-dominated progradational shoreline systems, such as those exposed at Book Cliffs, Utah, can help to optimize the production in hydrocarbon fields such as Wytch Farm, southern United Kingdom, which produces from the upper Lower Jurassic Bridport sands (Hampson et al., 2004; Figures 16.3).

Although the porosity contrasts among various facies deposited at the transition from the terrestrial to the shallow marine realms may not be significant, the permeability contrasts are. The example comes from the Eocene clastic sediments of the Maracaibo Basin, Venezuela (Escalona, 2003), where porosities oscillate around 18–22 percent while the respective permeabilities vary from 161 to 700 mD (Table 16.1).

Syn-rift reservoirs

Early syn-rift reservoirs

Syn-rift reservoirs vary from rift settings to rift settings depending on the overall depositional character, which can be either terrestrial or marine.

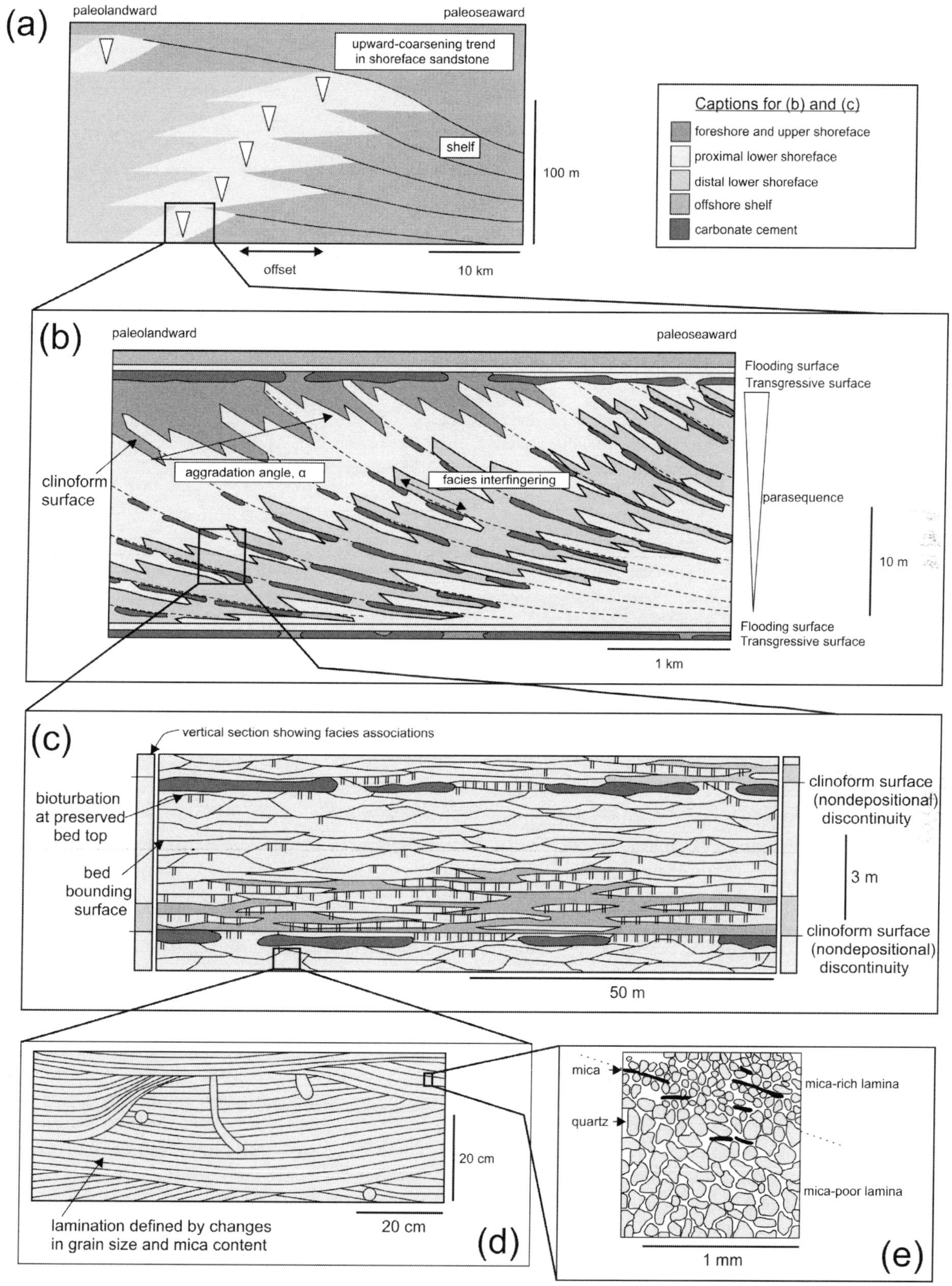

Figure 16.2. (a–e) Heterogeneities in shoreface-shelf reservoirs (Hampson et al., 2008), including (a) parasequence stacking patterns; (b) intra-parasequence facies architecture, carbonate cement distribution, and shale lengths and distributions; (c) bioturbation type and intensity, cross-stratification; (d) laminae and mica distribution; and (e) grain size and sorting (f–g). Impact of heterogeneities on reservoir permeability structure (Hampson et al., 2008). Note how the understanding of reservoir architecture varying between classical layer-cake model (f) and clinoform-bearing model (g) affects the ability to plan production properly.

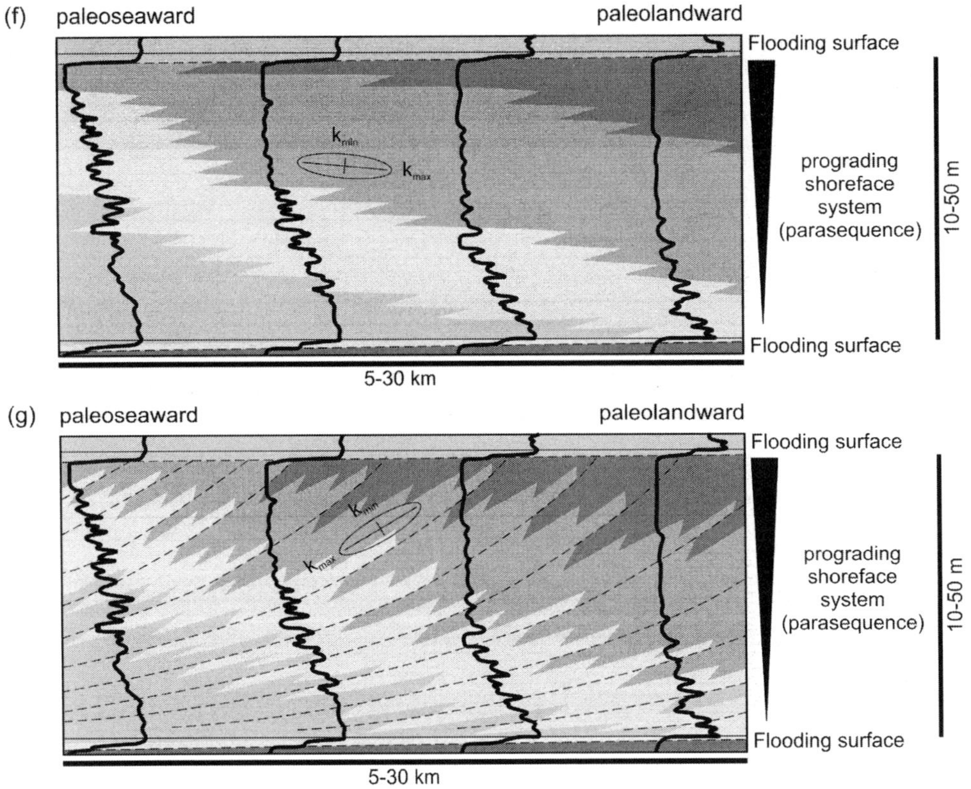

Figure 16.2 (*cont.*)

A good example of encroaching marine conditions, described in Chapter 15, comes from the North Sea province, where the Middle Jurassic early syn-rift stage starts with the uplift and deflation of the North Sea Dome (Helland-Hansen et al., 1992; Johannessen et al., 1995; Fjellanger et al., 1996) and Bathonian-Callovian rifting, characterized by marine environments passing into the coastal plain and alluvial environments in easterly and southerly directions.

The paleogeography then got progressively complicated by fault block tilting characteristic of Bathonian-Callovian rifting, which operated in the North Viking Graben region. Horst and graben topography controlled rapid facies changes over short distances. While the marine conditions in the central and eastern North Sea regions were coming from the north, they were coming from the west in the western region (Husmo et al., 2002).

The rifting continued during the Callovian-Kimmeridgian. This time period can be characterized by the most widespread normal faulting in the North Sea province (see Underhill, 1991a, 1991b; Rattey and Hayward, 1993). As discussed in Chapters 1, 7, and 15 (e.g., Davies et al., 2000; McLeod et al., 2000, 2002; Figures 7.6, 7.7), the growth, linkage, and interaction of normal faults controlled spatially and temporally variable facies distribution. Undergoing a rapidly rising sea level, the topography controlled basinal environments rimmed by a plethora of coastal-shelf environments located along margins of individual depressions (Rattey and Hayward, 1993; Kadolsky et al., 1999; Fraser et al., 2002). The Callovian–Late Jurassic marine transgressions were advancing into all three developing rift branches. The directions were from the north in the Viking rift branch and from the southwest in the Moray Firth branch (Fraser et al., 2002).

The Oxfordian was when the southward propagation of the Viking rift branch reached the South Viking Graben area. This was the time of deepwater marine conditions in the North Viking Graben and shallow marine conditions in the South Viking Graben.

A good example of the terrestrial conditions comes from the Gabon-Angola region (McAfee, 2005). Here a shallow lake occupied the axial zone of the rift branch. Its margins underwent deposition of alluvial fans, which prograded into the lake. The inner hinge zones were initially characterized by a low amplitude

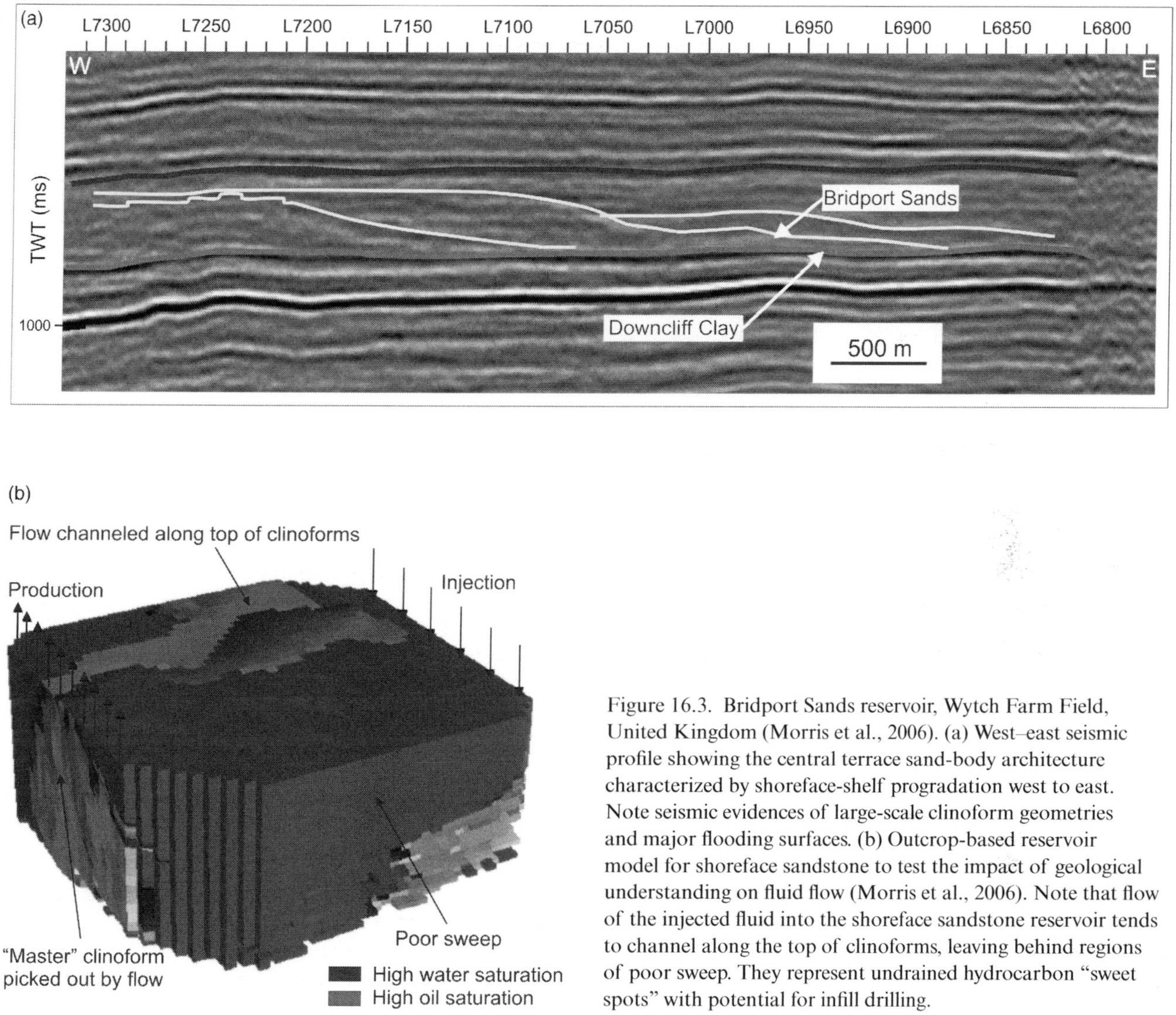

Figure 16.3. Bridport Sands reservoir, Wytch Farm Field, United Kingdom (Morris et al., 2006). (a) West–east seismic profile showing the central terrace sand-body architecture characterized by shoreface-shelf progradation west to east. Note seismic evidences of large-scale clinoform geometries and major flooding surfaces. (b) Outcrop-based reservoir model for shoreface sandstone to test the impact of geological understanding on fluid flow (Morris et al., 2006). Note that flow of the injected fluid into the shoreface sandstone reservoir tends to channel along the top of clinoforms, leaving behind regions of poor sweep. They represent undrained hydrocarbon "sweet spots" with potential for infill drilling.

positive topography hosting fluvial sandstone deposition on its slopes. Some of its segments subsequently subsided, developing into deep depressions with fast lacustrine shale deposition that replaced the earlier fluvial deposition rather abruptly, as the Neocomian Kissenda Formation in the Gamba Horst region documents. The rapid sandstone burial resulted in its silica and clay cementation, degrading its reservoir quality. On the contrary, the other segments of the initial inner hinge zone that escaped burial, footwall-related structural highs, and basin margins have this sandstone with good quality. This is documented by the Kissenda Formation hosting more than 500 MBO of recoverable oil in numerous fields, which do not exceed a size of 100 MBO (see Nemčok and Rosendahl, 2006 and references therein).

Main Syn-rift reservoirs

Late Oxfordian-Kimmeridgian was when the fully developed failed-rift segments were undergoing drowning and accelerated rift subsidence. The previously important intra-basinal highs were progressively losing their role in sediment supply.

The Late Oxfordian–late Ryazanian paleogeography of the evolving North Sea rift system was characterized by the vast regions of restricted oceanic circulation, irregular and fragmented bathymetry, and fast deposition rates (Stow and Atkin, 1987; Miller, 1990; Fraser

Table 16.1. Average reservoir properties of the Eocene clastic sediments deposited in a variety of environments located at the transition from the terrestrial to the marine region in the Maracaibo Basin, Venezuela (data taken from Escalona, 2003).

Depositional setting	Depositional environment	Porosity (%)	Permeability (mD)
middle-lower delta plain (affected by both fluvial and marine processes)	crevasse splay/tidal bar	20	260
middle-lower delta plain (affected by both fluvial and marine processes)	crevasse splay/tidal bar	20	390
middle-lower delta plain (affected by both fluvial and marine processes)	crevasse splay/tidal bar	22	418
fluvial/tide-influenced delta	fluvial channel, distributary channel	19	161
fluvial/tide-influenced delta	fluvial channel, distributary channel	18	515
fluvial/tide-influenced delta	fluvial channel, distributary channel	20	700

et al., 2002). While the rifting was already well advanced in the southern and northern rift branches during the Late Oxfordian, it only initiated in the Outer Moray Firth Basin of the western branch. The horst and graben-related reservoir facies distribution was complicated by the existence of deep marine sandstone bodies, for example in the Viking Graben and East Shetland Basin, which developed in association with major rift-related footwall uplifts and erosion.

The rift propagation along all three branches intensified during early-late Kimmeridgian. This was the time when all three propagating rift branches finally met (Fraser et al., 2002). Episodic graben-bounding fault activity, such as that in the North Viking Graben, controlled local sea-level falls taking part on the background of the general sea level rising, which resulted in basin-floor fan development. The other fan-controlling mechanism was the episodic activity of faults bounding the intra-basinal highs (e.g., Partington et al., 1993).

Based on all of these factors, the producing reservoirs deposited during the main syn-rift period can be divided into shallow and deep-marine reservoirs.

Typical examples of the shallow-marine reservoirs come from the Callovian–Volgian Fulmar Formation of the Central Graben (Johnson et al., 1986; Van der Helm, 1990; Turner, 1993; Currie et al., 1999; Fraser et al., 2002; Figure 16.4), which underwent retrogradational evolution in response to sea-level rise. Its producing sandstone reservoirs were deposited in the shoreface system (see Fraser and Tonkin, 1991; Figure 16.5). However, fields such as the Clyde (see Smith, 1987; Stevens and Wallis, 1991), Fulmar (Johnson et al., 1986; Glennie and Armstrong, 1991) and Puffin fields (Fraser et al., 2002) provide evidences for salt withdrawal and extensional faulting influence on its thickness. For example, the Fulmar field provides a number of lens-shaped sandstone bodies and an abundance of dewatering structures (Fraser et al., 2002. The Puffin field allows us to see a short distance accommodation space variation in space and time. The result is a system of elongated pods or primary rim synclines, partly infilled by the Triassic fluvial Skagerrak Formation sandstones, and interpods or secondary rim synclines containing broad lenses of the shallow-marine Upper Jurassic Fulmar Formation sandstones.

Stratigraphically equivalent shallow-marine sandstones from the northern North Sea region were part of a large westward prograding delta complex about 400 m thick. The sandstones are poorly consolidated with porosities of up to 34 percent and permeabilities of up to 10 Darcy (Gray, 1987). A producing example comes from the Troll field.

The shallow-marine Fulmar Formation of the Western Central Shelf contains sandstones 200–300 m thick deposited in a sand-dominated shoreface environment (Armstrong et al., 1987; Glennie and Armstrong, 1991; Stewart et al., 1999). The Kittiwake field documents that thick sandstone occurrences result from syn-depositional tectonics. Their high reservoir quality results from wave and storm reworking. The significant lateral changes in reservoir thickness reflect a combination of depositional processes, that is, erosional topography and topography controlled by salt/fault movements, and post-depositional, for example, truncation, processes (Glennie and Armstrong, 1991; see Figure 16.6).

While dip closure in the Kittiwake area appears to reflect later salt movements, the reservoir distribution reflects interpod geometry, characterized by sandstones

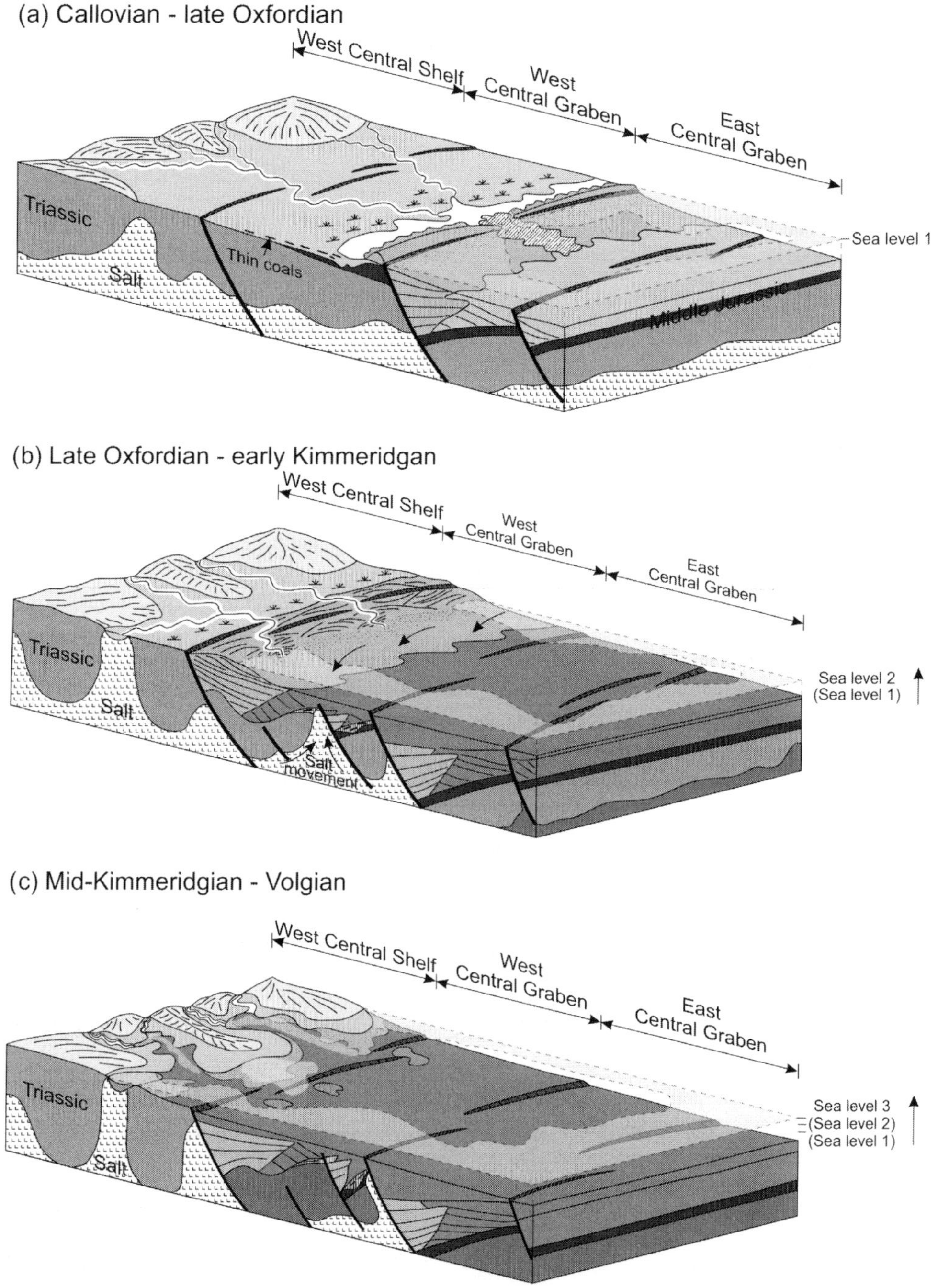

Figure 16.4. Schematic models of the Fulmar Formation sandstone deposition, Central Graben (Fraser et al., 2002). The models illustrate the retrogradational evolution of coastal facies belts in response to the rise in relative sea level from the graben center to the platform areas. Note the effect of fault development, syn-depositional Zechstein salt migration, and pre-transgression topography on reservoir thickness and preservation. (a) The Callovian–late Oxfordian time period showing the depositional scenario typical for the Fulmar Formation in the Kittiwake and Dunward/Dauntless fields. (b) The Late Oxfordian–early Kimmeridgian time period showing the low- to high-energy shoreface, representative of the Fulmar Formation in the Fulmar and Clyde fields. (c) The Middle Kimmeridgian–Volgian time period showing a low- to moderate-energy shoreface, representative of the Fulmar Formation in the Puffin, Elgin, and Franklin fields.

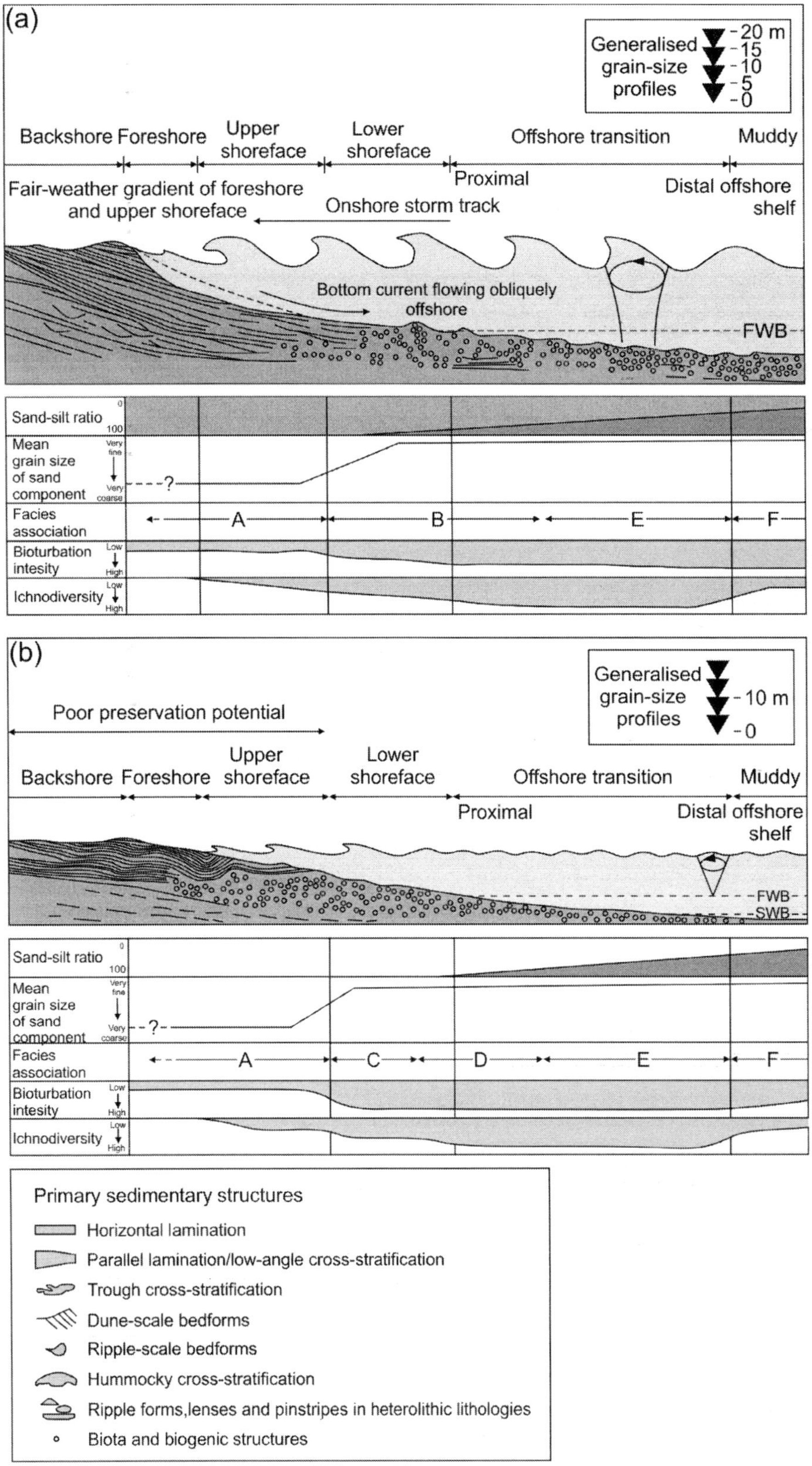

Figure 16.5. Shoreface depositional models represented by two end-member scenarios defining shelf and shoreface profiles in the Central Graben (Fraser et al., 2002). (a) High-energy, storm-influenced shoreface scenario illustrating widespread physical reworking during a period of intense storm activity. (b) Low-energy, bio-turbated shoreface scenario illustrating widespread biogenic reworking during fair-weather conditions.

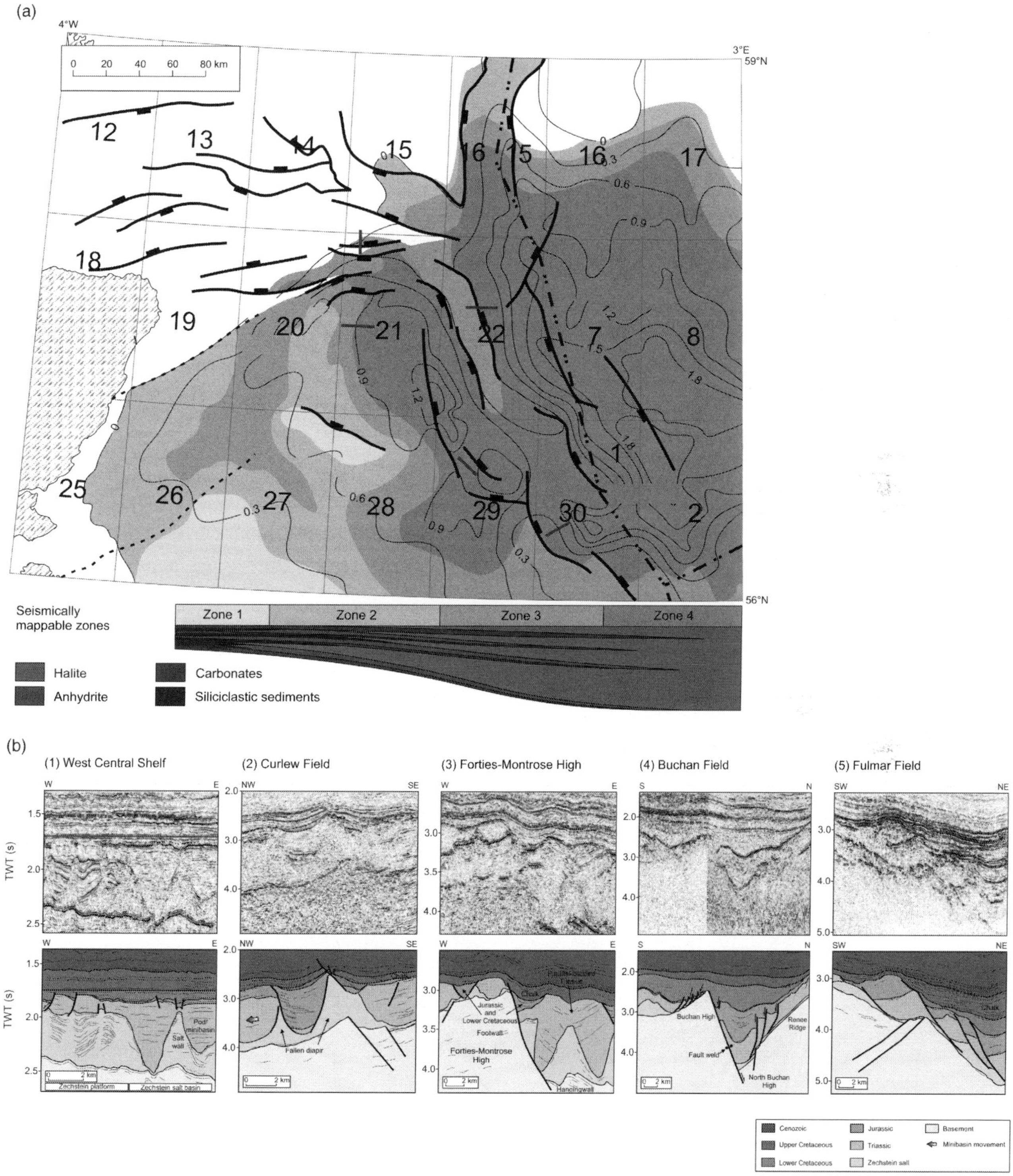

Figure 16.6. Tectonostratigraphic development of the West Central Graben and Central Shelf (Fraser et al., 2002). (a) Map and cross-section depicting the thickness of Zechstein sediments and the distribution of seismic facies and lithofacies determined from seismic data tied to wells (Clark et al., 1998). The thickest Zechstein formation is dominated by halite. The map shows locations of profiles shown in (b). Isopachs are after Smith et al. (1993). (b) Seismic profiles and their interpretation showing the role of salt in the Jurassic extensional architecture and the development of sediment catchment areas. 1. The contrast between Zechstein platform facies and the halite-dominated salt basin controls the mini-basin and salt-high architecture, which underlies the thin Jurassic section. The halite contains sulphate and carbonate interbeds that were not prone to salt movements during the Triassic (Clark et al., 1998). On the contrary, this sedimentary section is flattened along the base of the Upper Cretaceous section. 2. Deflated diapir at the top of a major basement relay ramp generating the space for thick Upper Jurassic sediments in UK 29/7. 3. Contrasting footwall and hanging wall

(c)

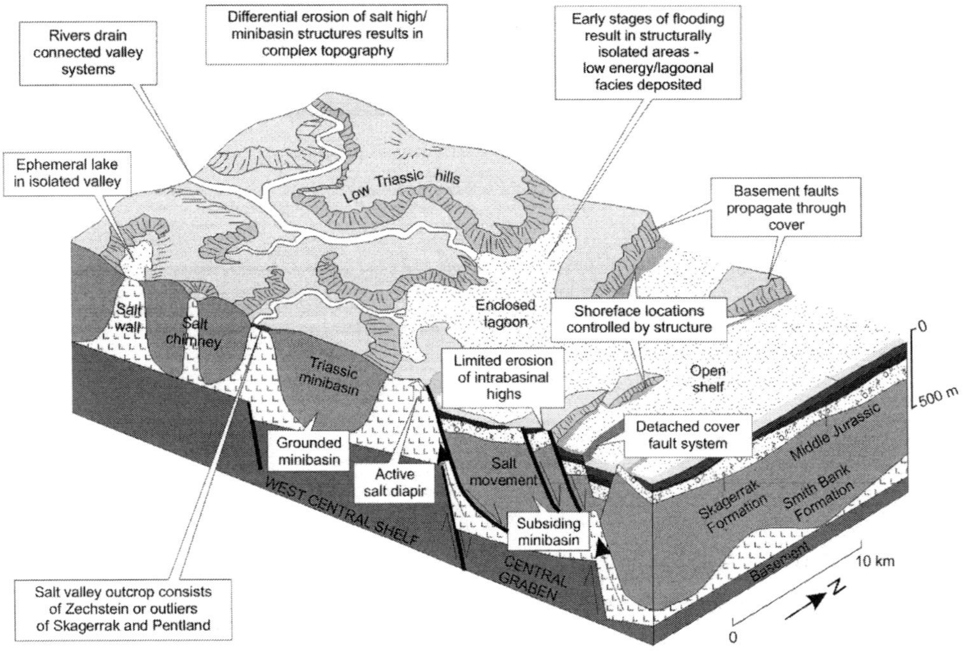

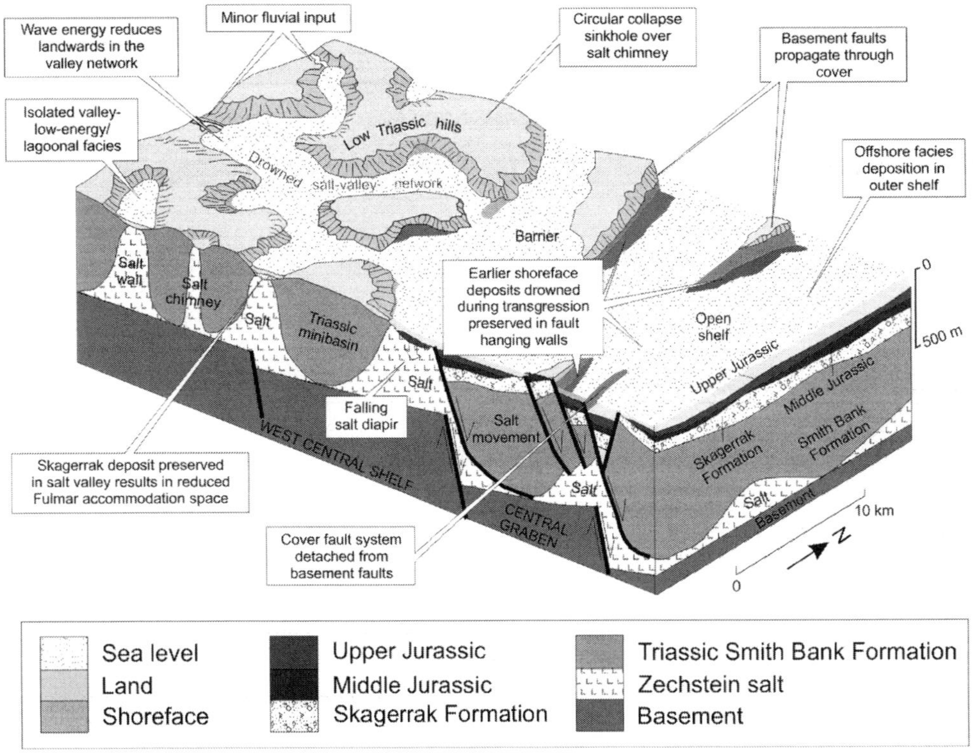

Figure 16.6 (*cont.*)

geometries of the Forties-Montrose High. 4. Fault weld close to the Zechstein pinch-out in the North Buchan Graben. 5. Tilted, perched fault blocks on steeply dipping basement in the Fulmar field area. (c) 1. Paleogeographic reconstruction of the West Central Shelf and Central Graben during the deposition of Oxfordian sedimentary sequences. It shows initial marine and shoreface retrogradation incursion along the axis of the graben. 2. Paleogeographic reconstruction of the West Central Shelf and Central Graben during deposition of Kimmeridgian sedimentary sequences. It shows continued marine transgression of the preexisting topography on the West Central Shelf, which resulted in the deposition of laterally restricted shoreface facies within the preexisting salt-valley mini-basin pattern.

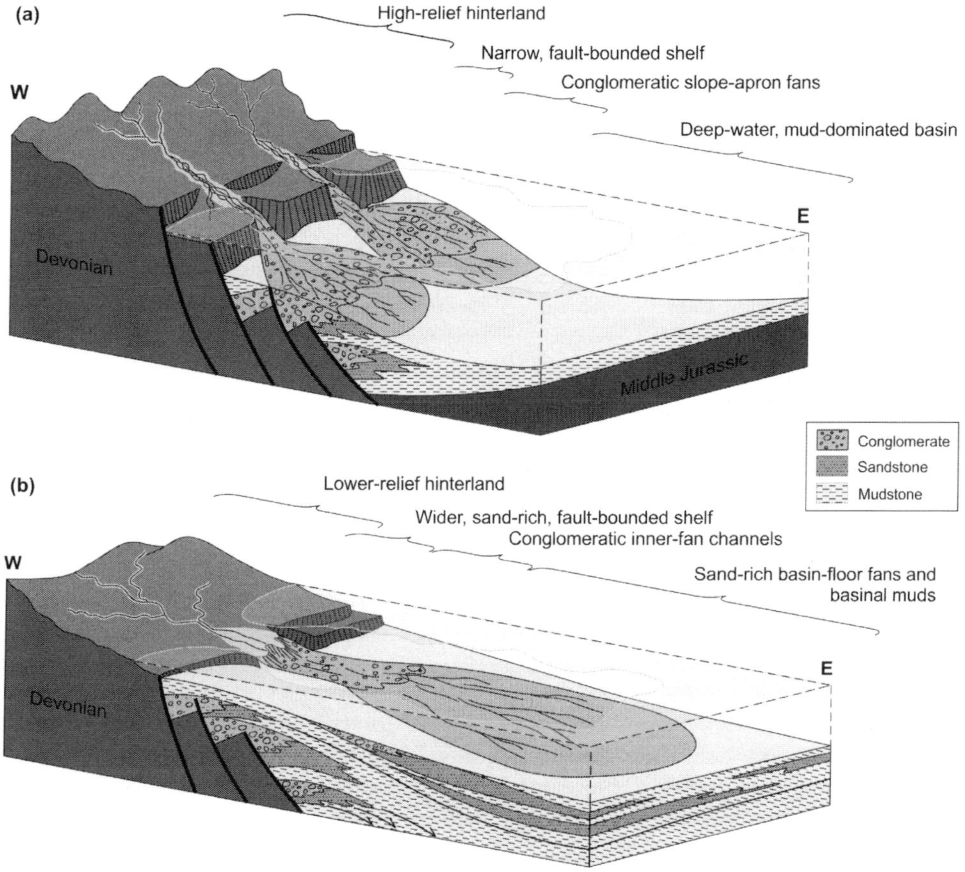

Figure 16.7. Scheme of the submarine fan development in relation to fault-scarp activity in the South Viking Graben (Fraser et al., 2002). (a) Initial conditions characterized by gravel-rich slope-apron fans, reflecting steep, active fault margins and high stream gradients. Examples come from the Brae, Tiffany, Thelma, and Toni field reservoirs, deposited during the early Volgian. (b) Subsequent conditions characterized by sand-rich basin-floor fans indicating reduced fault-scarp topography, lower stream gradients, and increased efficiency of the sandier gravity flows. Examples come from the Central Brae, Miller, and Kingfisher reservoirs, deposited during the middle Volgian.

wedging against pods comprised of Triassic Smith Bank Formation mudstones. The distribution of the Fulmar Formation here is influenced by its transgression over narrow belts that defined preexisting Early–Middle Jurassic valley topography located over dissolving salt bodies (Wakefield et al., 1993), or filling inherited Triassic topography dissected by an irregular system of fluvial valleys influenced by salt dissolution (Stewart et al., 1999).

Thickness and facies distributions in both the Kittiwake and Dunward–Dauntless fields illustrate that the Fulmar Formation of the Western Central Shelf is an extremely complex play characterized by high risk, which are both results of the irregular geometry of the pre-Fulmar fairway interpod/valley systems, their irregular lithological fill, and the ensuing irregularly connected hydrocarbon migration routes. Complicated sub-seismic facies distribution prevents even 3-D seismic imaging from being completely successful in reservoir prediction. In fact several interpods in the area have proven to be either filled by mud or not charged (Fraser et al., 2002).

Examples of the deep-marine reservoirs come from the northern North Sea region. They were deposited in the basin-floor fan environment. A typical example is the producing reservoir in the Statfjord field (Gradijan and Wiik, 1987). The sandstones here were point-sourced locally from intra-basinal highs. They were derived by erosion and slumping of weakly consolidated, pre-rift, mainly Middle Jurassic sediments, during the syn-rift fault-block rotation and footwall scarp degradation (Johnson and Eyssautier, 1987; McLeod and Underhill, 1999).

The reservoir sandstones of the Brae Formation in the Southern Viking Graben (Figure 16.7) were deposited in sand-rich basin-floor fans as documented in the

Brae, Tiffani, Toni, and Thelma fields (McClure and Brown, 1992) containing:

1) submarine-fan sandstones located in the hanging wall of the graben's western boundary fault;
2) slope-apron fans; and
3) proximal channel systems of basin floor fans.

They trap hydrocarbons through a combination of stratigraphic pinch-out with fault-seal and dip closure (Fraser et al., 2002). Stratigraphic trapping reflects the primary depositional geometry of the small-diameter (1–5 km) slope-apron fans, which is demonstrated by hydrocarbon column lengths considerably greater than structural closure. For example, the South Brae field has a maximum hydrocarbon column of 500 m, of which only 90 m is the result of structural closure (Turner et al., 1987; M. J. Roberts, 1991). Similar column conditions are controlling around a dozen slope-apron fan and proximal fan-channel accumulations along the Brae trend of fields (Stephenson, 1991; Turner and Allen, 1991; Cherry, 1993).

Another example of the producing basin-floor fan comes from the Miller field located basinward from the Brae, Tiffani, Toni, and Thelma fields. The field is a good example of the combination trap, in which closure is provided not only by compaction-related dip closure around the sand-rich fan complex, but also by stratigraphic pinch-out or reservoir-quality deterioration north-westward toward the lateral edge of the fan (Rooksby, 1991; Turner and Connell, 1991; Garland, 1993).

A good example of the terrestrial conditions comes from the Gabon-Angola region (McAfee, 2005). The main syn-rift sediments here were deposited in a highly fragmented depositional area characterized by some segments of the inner hinge system being major topographic highs, such as the Gamba Horst, and others being depressions. This setting controlled a highly variable facies distribution. The Angolan portion of the rift branch also underwent volcanism recorded by lava flows and tuffs, filling up some of the local depocenters. Such development resulted in the occurrence of shallow to emergent platforms located along the outer hinge zone and shallow lakes and sabkhas further east.

Fluvial-lacustrine environments of the late main syn-rift stage in both Gabon and Angola underwent a deposition of thick sand-rich sediments. The examples come from the Barremian Dentale formation in Gabon, which produces from the 5–6 MMBO Rabi field (Teisserenc and Villemin, 1990; Molnar et al., 2005; see also Nemčok and Rosendahl, 2006 and references therein), and the Red Cuvo Formation in the Lower Congo Basin. The Dentale fluvial system flowed from the northwest to the southeast along the axis of the South Gabon Rift Zone. It was laterally confined from the east and the west by the inner and outer hinge zones, respectively. Regions located far from the axis of fluvial deposition, such as those in the southernmost Gabon and Congo, were occupied by persistent shallow lakes undergoing silt, shale, and locally carbonate deposition recorded by the carbonate Argil's Vertes, Cuvo, Toca, and Viodo Formations. The Toca Formation is known to produce in both the Congo and Cabinda/Zaire Basins (Molnar et al., 2005). The porosities known for the Dentale Formation reservoirs reach values of 13–15 percent (see Nemčok and Rosendahl, 2006 and references therein).

Late syn-rift reservoirs

The rift-related bathymetric deepening eventually advanced into all three rift branches during the late Kimmeridgian–early Volgian. A typical example is provided by the North Viking Graben whose basinal region underwent pelagic mudstone deposition. It was rimmed by shoreface sandstones depositing in narrow zones around structural highs. The sediment input from graben flanks was subordinate and depocenters were characterized by sediment starvation.

The end of rifting in the North Sea was temporally variable. It took place during the Early–middle Volgian time interval with the exception of the Moray Firth Basins, where rifting continued until the end of Ryazanian (Fraser et al., 2002). The termination occurred first in the oldest rift units.

A good example of the terrestrial conditions comes from the Gabon-Angola region (McAfee, 2005). The late-rift sediments were associated with regional denudation that took place after the main rift phase. Despite the regional denudation, there was still a subordinate topographic control over facies distribution. It controlled their division into the Aptian Gamba, Chela, and Grey Cuvo Formations. The basal lowstand fluvial sediments were best developed in inner basins and areas around intra-basinal highs. The transgressive shallow lake facies, which include winnowed high-energy lake shoreline sandstones, were depositing along the axis of the outer hinge zone. Both of them host hydrocarbon accumulations. The Gamba Formation produces in the South Gabon Basin; the Chela Formation produces in the Congo Basin (Teisserenc and Villemin, 1990; Molnar et al., 2005). The Gamba Formation is characterized by an average reservoir porosity of 24–26 percent and a permeability of 700 mD at the Malongo West and Limba fields (see Nemčok and Rosendahl, 2006 and references therein).

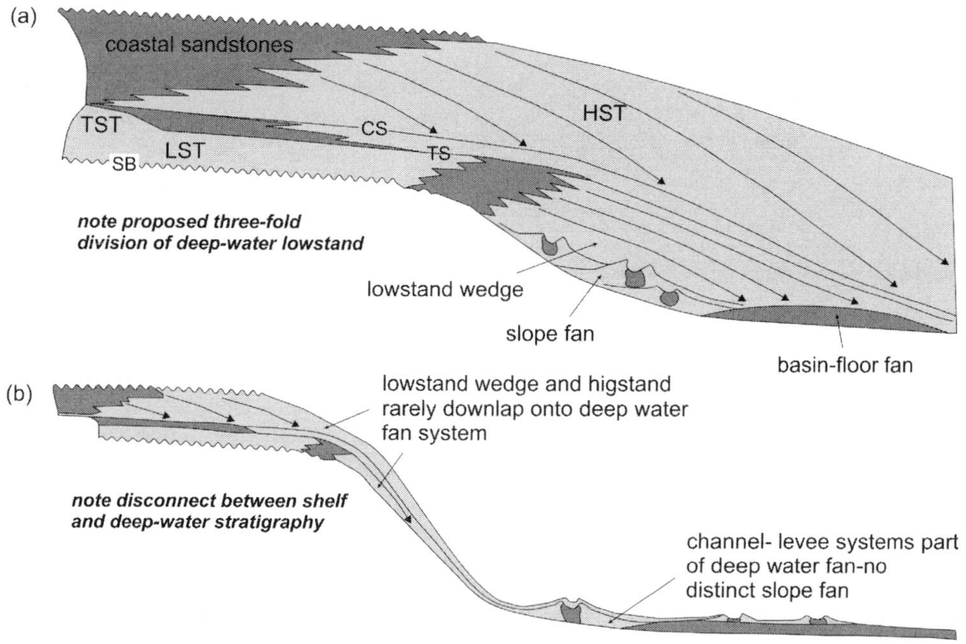

Figure 16.8. Comparison of deepwater sequence stratigraphic models representing high rates of sediment accumulation (Vail, 1987; Posamentier and Vail, 1988) (a) with those representing low rates of sediment accumulation (Haughton, 2008; Kendall and Haughton, 2008). Note the proposed threefold division of deepwater lowstand in (a) and the disconnect between shelf and deepwater stratigraphy in (b).

Syn-rift reservoir models

Using the North Sea examples, we can try to summarize the marine syn-rift reservoir discussion. We can conclude that the deepwater sandstones form volumetrically smaller and stratigraphically more complex accumulations than shallow-water sandstones (Fraser et al., 2002). They are characterized by rapid facies changes within individual submarine fan systems. Submarine fans have a diameter of 1–5 km and a maximum possible thickness of 1.5 km. Their sandstone bodies are partially encased in organic-rich muds and juxtaposed up-dip against boundary faults of the rift architecture, which are commonly sealing. These reservoirs are mostly located in the Viking Graben and Outer Moray Firth fields. Apart from the Statfjord field example, and other examples from the Brent-Statfjord-North Alwyn trend, which represent intra-basinal sediment sourcing, the submarine fans are sourced from external provenances. Relatively sporadic examples come from the extra-basinal sources along the main boundary faults, such as the Beryl Embayment (Knutson and Munro, 1991) and the region adjacent to the Crawford Spur (Yaliz, 1991).

Models and analog data on the deep-marine sandstone reservoir properties for our discussion come from the Ross formation, from the study area located to the west of Ireland (Kendall and Haughton, 2008), the Congo fan (UenzelmannNeben et al., 1997; Droz et al., 2003), and the Gulf of Mexico (Camp, 2000; Ostermeier, 2001; De-Hua and Batzle, 2006). They have shown that the deepwater sequence stratigraphic models of Vail (1987) and Posamentier and Vail (1988) can represent the end-member of the sediment accumulation spectra for this setting, representing high rates of sediment accumulation, while the opposite end-member is characterized by low rates of sediment accumulation (see Figure 16.8).

The stratigraphic complexity of the deepwater sandstone accumulations is primarily caused by (Posamentier and Kolla, 2003; Kendall and Haughton, 2008):

1) auto-cyclic processes related to local changes in depositional setting and source terrain; and
2) broader-scale changes including the eustatic movement of the sea level.

The result is a complex architecture where reservoir candidates can be found in (Figure 16.9; Posamentier and Kolla, 2003):

1) channel sands;
2) levee sands;
3) distal overbank supra-fan sand sheets; and
4) sands involved in down-slope slumps.

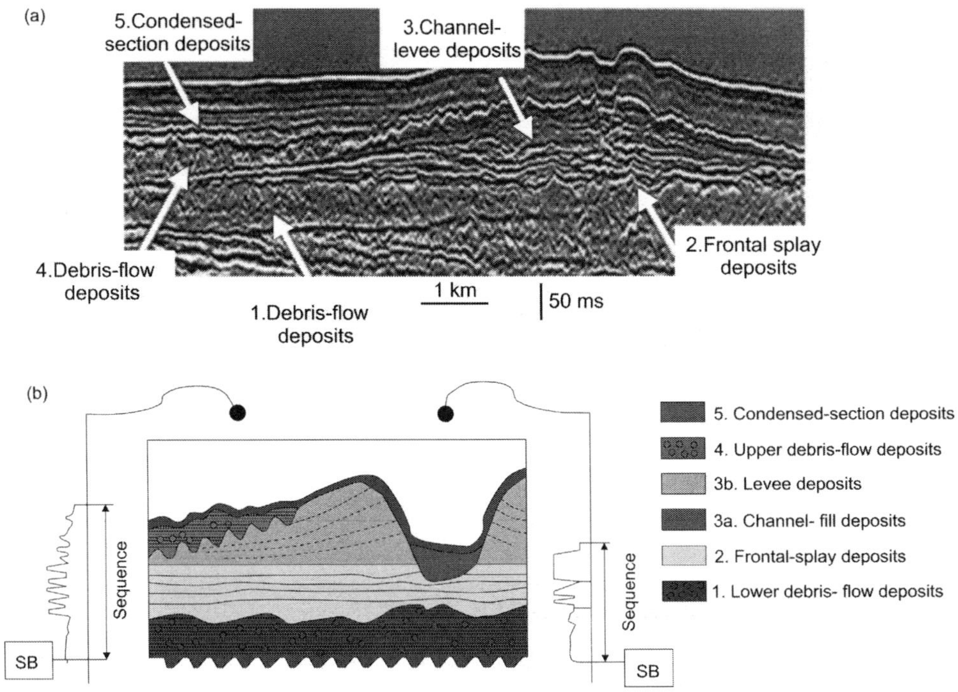

Figure 16.9. (a) Reflection seismic profile through offshore Indonesia imaging the stratigraphic succession of a deepwater sequence, including, from older to younger, sediments such as: 1) debris-flow sediments, 2) frontal-splay sediments, 3) channel-levee sediments, 4) debris-flow sediments. The entire mentioned succession was interpreted as covered by a thin layer of condensed-section sediments (Posamentier and Kolla, 2003). (b) Idealized deepwater depositional sequence, together with two hypothetical logs (Posamentier and Kolla, 2003).

Seismic, well log and core sample-based studies of the reservoir properties indicate that marine sandstones, such as those from the Miocene-Pleistocene section of the Gulf of Mexico, can have variable porosity and permeability (e.g., Ostermeier, 2001; De-Hua and Batzle, 2006; Figure 16.10a), which depend on numerous factors including:

1) geological age;
2) sand morphology including:
   a. grain shape;
   b. grain roundness;
   c. grain surface texture;
   d. grain orientation;
   e. grain packing;
3) sand composition;
4) sorting;
5) grain size distribution;
6) fluid saturation; and
7) stress conditions.

The first three factors are the primary ones. The geological age serves as a proxy for the degree of compaction, which is a very rough indicator because a variety of factors enhances or retards the normal compaction history. All three key factors controlling the matrix compressibility can be discussed using two end-member cases Ostermeier (2001) described (Figures 16.10b–c).

The first, shallow-water end-member is characterized by a somewhat higher initial compressibility (Figure 16.10d). Its material softening with increasing effective stress takes place at a maximum compressibility value several times larger than the initial compressibility value. Eventually, the strain softening is replaced by strain hardening as the stress keeps increasing after a certain value. This sand usually has a young geological age. Its thin section (see Figure 16.10b) shows that this sandstone can be characterized by a point-contact morphology and a significant amount of load-bearing ductile grains.

The second end-member is characterized by relatively low initial compressibility that decreases linearly with increasing effective stress (Figure 16.10e). This sand usually has an older geological age. Its thin section (see Figure 16.10c) shows that this sandstone can be characterized by long-contact morphology and the minimum amount of load-bearing ductile grains. The hard grains are, for example, formed by quartz, chert, and feldspar. The ductile material includes clays, micas, and rock fragments, which are about thirty to forty times softer than quartz (Hurlbut and Klein, 1977).

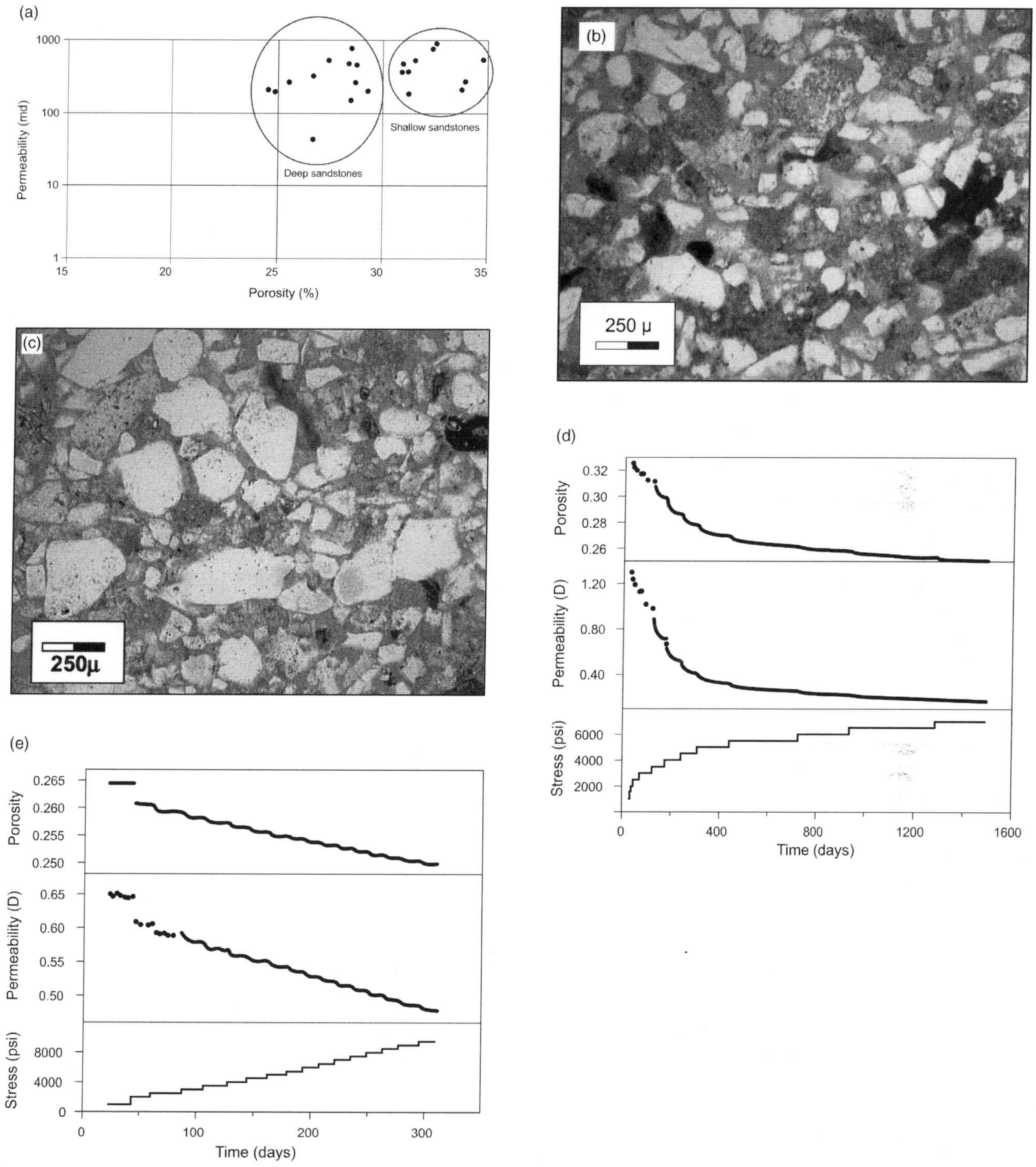

Figure 16.10. (a) Porosity–permeability relationship for shallow and deep sands, Gulf of Mexico (De-Hua and Batzle, 2006). (b) Thin-section photomicrograph of the shallow-depth reservoir sandstone from the Miocene-Pliocene deepwater sediment section of the Gulf of Mexico (Ostermeier, 2001). Its compressibility mechanisms include significant creep. It is characterized by rapid compressibility increase with stress and strain hardening at high stress levels. It is younger, softer, and less compacted than the sandstone in figure (c). It has a lower initial stress as well. It has a point-contact morphology and a significant amount of load-bearing ductile grains. (c) Thin-section photomicrograph of the deep-depth reservoir sandstone from the Miocene-Pliocene deepwater sediment section of the Gulf of Mexico (Ostermeier, 2001). Its compressibility mechanisms include essentially no creep. It is characterized by linear compressibility increase with stress. It is older, harder, and more compacted than the sandstone in figure (b). It has a long-contact morphology and a minimum amount of load-bearing ductile grains. (d) Porosity and oil permeability at initial water saturation, S_{wi}, as a function of isostatic stress increasing in steps, determined for the first end-member (Ostermeier, 2001). (e) Porosity and oil permeability at initial water saturation, S_{wi}, as a function of isostatic stress increasing in steps, determined for the second end-member (Ostermeier, 2001).

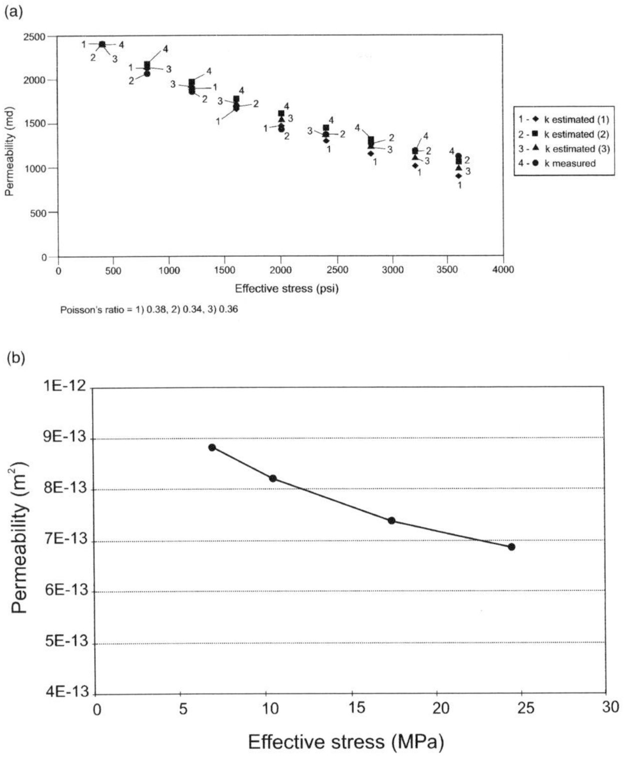

Figure 16.11. (a) Measured and calculated data of rock permeability dependence on effective stress for hard consolidated sandstone (Sigal, 1998, pers. comm.). (b) Measured data for fracture permeability dependence on effective stress for volcanic rocks (Sigal, 1998, pers. comm.).

A comparison of the two end-members indicates that the older sand has a higher chance of being more compacted because of its longer compaction and diagenetic history. Their respective composition and sorting affect the compaction function with stress because of mechanical stratigraphy controlling different proportions of various strain mechanisms. Physical considerations suggest that permeability should vary as the square of some characteristic grain size. For this grain size, the permeability decreases as the grain distribution broadens (see also Krumbein and Monk, 1943).

Factors 6 and 7 mentioned earlier describe permeability dependence on both the fluid pressure and the total stress (e.g., Wilhelm and Somerton, 1967; Mordecai and Morris, 1971; Daw et al., 1974). The rock permeability–effective stress (total stress minus pore fluid pressure) relationship (Figure 16.11a) is distinctively different from the fracture permeability–effective stress relationship (Figure 16.11b), which can be used as an indicative tool for the distinction of these two end-member permeabilities in a reservoir. In order to derive a rock permeability–effective stress relationship, we need to combine permeability–porosity and porosity–effective stress relationships.

The change in rock permeability with porosity can be expressed as (Thompson et al., 1987; Sigal, 1998, pers. comm.):

$$k = A r_{eff}^{2} n^{m}, \qquad (16\text{-}1)$$

where k is the permeability, A is the pore shape factor, r_{eff} is the characteristic pore throat size, which controls access to the primary flow network, n is the porosity, and m is the cementation factor. The pore shape factor, A, is calculated as the average pore area from the thin section. Its most common value for sandstone is 1.16, and can be as high as 1.39 for sandstone with calcite cement (Sigal, 1998, pers. comm.). A good estimate of r_{eff} is the pore throat size on a high-pressure mercury injection curve at which maximum injection rate occurs (Sigal, 1998, pers. comm.). The cementation factor, m, for well-connected pore spaces generally ranges from about 1.7 for poorly consolidated sediments to about 2.1 for consolidated sediments (Sigal, 1998, pers. comm.).

The change in rock porosity with effective stress, involving reasonable simplifications, can be obtained using a bulk modulus in the equation (Sigal, 1998, pers. comm.):

$$n(\sigma) = 1 - (1 - n_o)e^{(\sigma - \sigma o)/K}, \qquad (16\text{-}2)$$

where σ is the effective stress, K is the static bulk modulus, n_o is the initial porosity, and σ_o is the initial effective stress.

By combining Equations (16-1) and (16-2), the rock permeability–effective stress relationship can be determined. The pore shape factor, A, and cementation factor, m, can be treated as constants with effective stress changes (Sigal, 1998, pers. comm.). However, it is necessary to include the calculation of the pore throat size, r_{eff}, reduction with increasing effective stress (Zimmerman et al., 1986):

$$r_{eff}^{2}(\sigma) = r(\sigma_o)^2 e^{-cpc(\sigma - \sigma o)}, \qquad (16\text{-}3)$$

where c_{pc} is the pore compressibility and r is the initial characteristic pore throat size.

Mentioned deepwater reservoir properties are very important for the following three production aspects (Ostermeier, 2001):

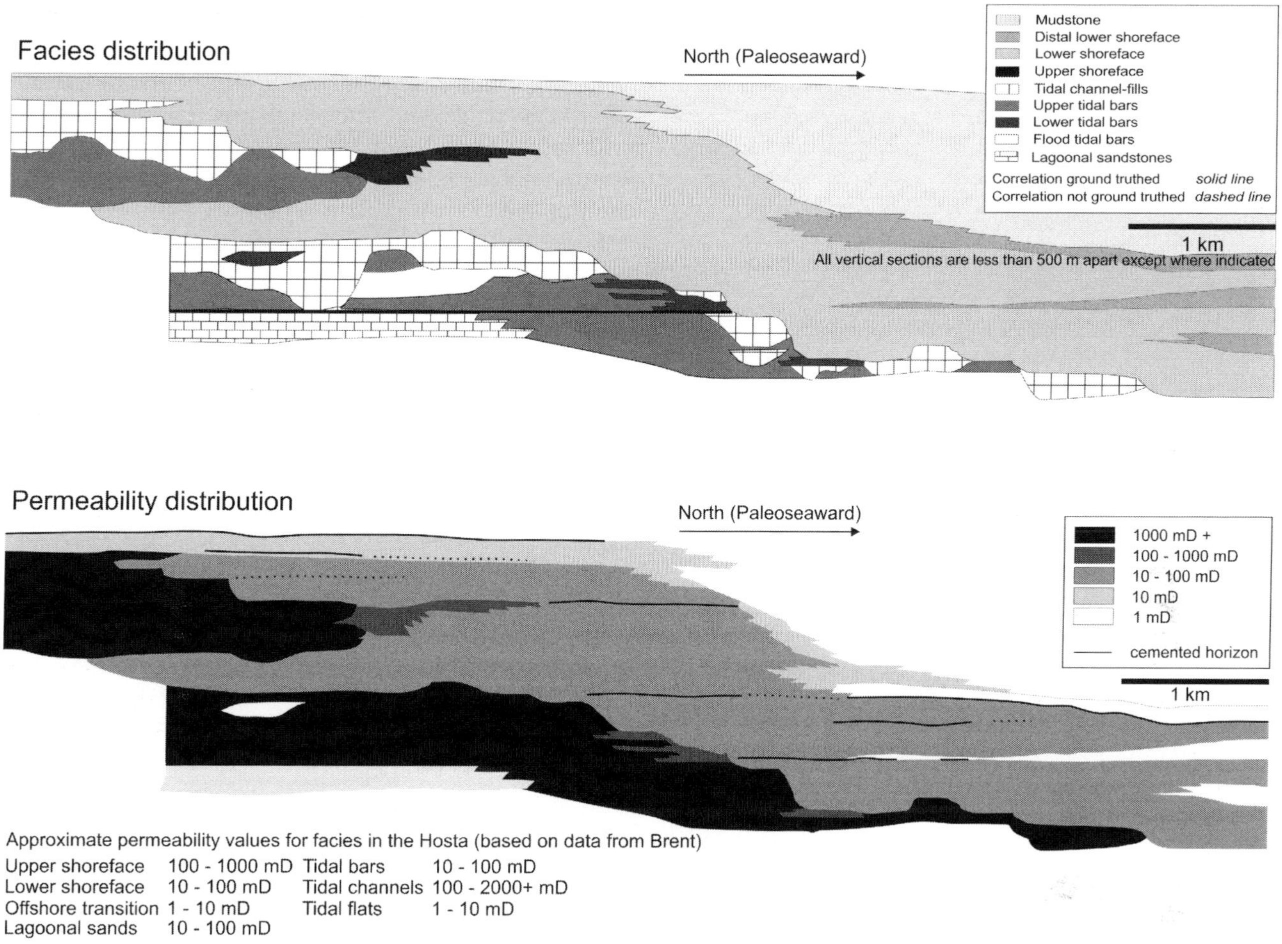

Figure 16.12. Permeability distribution controlled by facies distribution in the studied area of the Cretaceous Hosta Tongue, New Mexico (Hampson, 2004).

1) Syn-production compaction controls the seafloor subsidence, which impacts the structure design and performance.
2) Syn-production compaction controls the stresses loading the production casing.
3) Syn-production compaction affects the reservoir productivity.

The reservoir properties of the shallow-marine reservoir system are not solely controlled by their depositional environments (Fraser et al., 2002). Fluid influence on their properties has been documented from the high-pressure–high-temperature region of the Central Graben. Here, high pressures are controlled by processes such as depositional rates of overlying strata, undercompaction, kerogen transformation, oil cracking, smectite-illite transformation, and aquathermal processes (Cayley, 1987; Gaarenstrom et al., 1993). Bodies of shoreface sediments including reservoir sandstones have a width and length of 15–20 km and several tens of km, respectively (Fraser et al., 2002).

Case data on how shallow-marine depositional environments control reservoir properties come from outcrop analog studies such as those done on the shallow-marine Cretaceous Hosta Tongue sandstone in New Mexico (Hampson et al., 2004; Sixsmith et al., 2008). It represents a complex arrangement of wave- and tide-dominated facies associations deposited in a transgressive pattern; each of them controlling its own permeability range (Figure 16.12). The extensive channel and tidal bar systems represent the potentially best reservoir units. An important constraint on reservoir performance is the distribution of cementation because the cemented horizons break up shoreface units. As the Hosta study documented, the coarse lag sediments unaffected by cementation form high-permeability zones at the base of shorefaces but form low-permeability zones when cemented (Hampson et al., 2004).

We can also try to summarize the terrestrial syn-rift reservoir discussion. The case data on how some terrestrial depositional environments control reservoir properties come from:

1) outcrop analogs such as:
   a) the Atchafalaya Basin in Louisiana;
   b) the Po River in Italy;
   c) the Dauterive and Fausse Lakes in Louisiana (Figure 16.13);
   d) the Susitna Valley in Alaska;
   e) alluvial systems in offshore Malaysia;
   f) the Triassic Moenkopi Formation in southern Utah;
   g) the Brazos River in Texas;
   h) the lower Mississippi alluvial basin (Figure 16.14);
   i) the Congaree River in South Carolina (Miall, 1996; Bridge, 1999, 2003; Tye and Coleman, 1989b; Tye, 2004a); and
2) studies such as those done on the reservoirs such as:
   a) the fluvio-deltaic ones in the Prudhoe Bay field on the Alaskan Slope (Figure 16.15a);
   b) the fluvial Travis Peak Formation in North Appleby field, Texas (Figure 16.15b);
   c) the fluvial Officina Formation in Venezuela; and
   d) the fluvial Sterling Formation in the Beluga River field in Alaska (Figure 16.15c)

(Tye, 1991, 2004a, 2004b; Bridge and Tye, 2000).

All of the aforementioned studies indicate that all involved facies are primarily controlled by their depositional environments. The result is a complex arrangement of facies associations in the overall fluvio-deltaic system (see Figure 16.14). Each of the facies controls its own porosity and permeability range (Figure 16.15c). Their porosity and permeability distributions indicate that fluvial systems contain their best reservoirs in:

1) the channel fill; and
2) the channel bar sediments,

while deltaic systems contain their best reservoirs in:

1) the distributary channel; and
2) the distributary-mouth bar sediments.

Each of the reservoir bodies can be further graded as based on the model of the grain-size distribution implied from the deposition dynamics. Such a model for the distributary mouth bar, for example, indicates that the best reservoir portion lies in the apex and frontal margins of the bar (Figure 16.15d). Going further down in scale, one can add further details into the interpreted sedimentary architecture of such a bar, assigning specific lithologies and their associated permeabilities to individual strata of the mouth bar. The example of such an approach comes from the study of the Romeo delta unit of the Ivishak Sandstone in the Prudhoe Bay field, Alaska (Tye, 2004b).

Fracture permeability-controlled reservoirs

Because active rifting controls a development of faults and fractures, it is worth discussing fracture-permeability-controlled reservoirs, which occur in numerous rift settings containing more competent reservoir strata.

Fracture permeability includes: 1) permeability related to fault zones, which bound fault blocks, and 2) permeability related to fracture and fault patterns located inside fault blocks.

The fluid flow properties of bounding fault zones can be described with the help of architectural fault categories, using fault zone components such as the fault core, damage zone, and host rock (sensu Chester and Logan, 1986; Smith et al., 1990; Figure 16.16a). The fault cores may include:

1) single slip surfaces;
2) unconsolidated clay-rich gouge zones;
3) brecciated and geochemically altered zones; and
4) highly indurated, cataclasite zones

and damage zones may include:

1) small faults;
2) veins;
3) fractures;
4) cleavage; and
5) folds.

A fault core is a part of the fault zone where comminution, fluid flow, geochemical reactions, and other fault-related processes can alter the original lithology and its permeability. For example, progressive grain reduction, dissolution, reaction, and mineral precipitation during fault zone evolution typically cause the core to have reduced permeability relative to the adjacent damage zone and host rock. A damage zone is a part of the fault zone where the host rock permeability also can be changed. The fault core and damage zones are surrounded by relatively undeformed host rock (Figure 16.16a).

The four architectural styles of the fault zone include (Caine, 1999; Figure 1.25):

1) single fracture fault;
2) disturbed deformation zone;

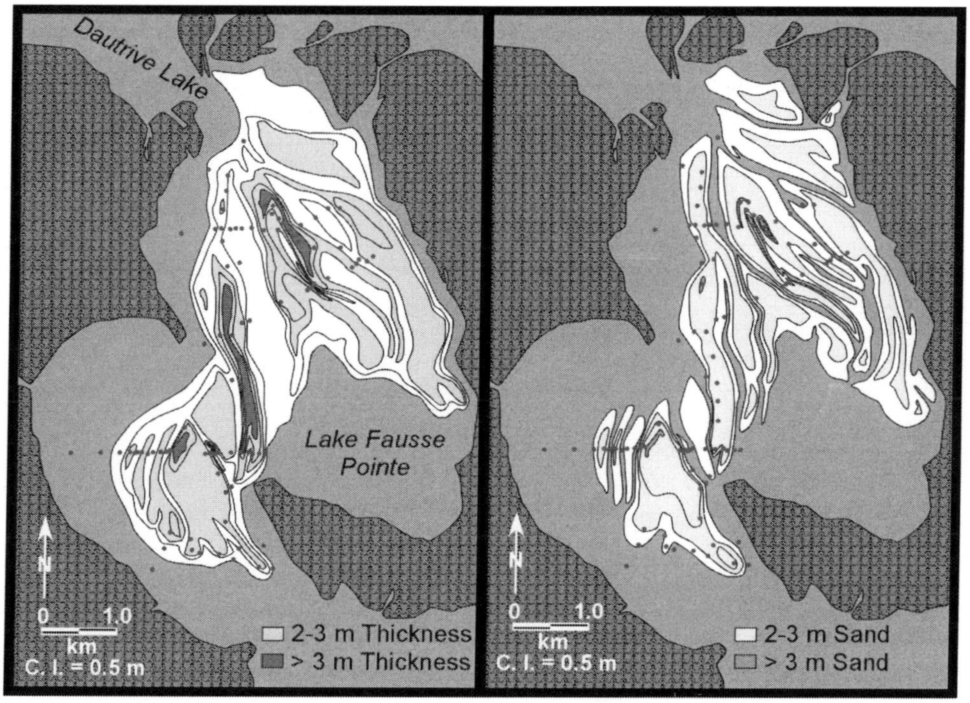

Figure 16.13. The lacustrine delta of the Dauterive–Fausse Pointe Lakes, covering approximately 100 km^2 (Tye and Coleman, 1989a). The left panel shows an isochore map of the total deltaic sediment thickness. Light gray and dark gray indicate a thickness of 2 to 3 m and >3.0 m, respectively. The right panel shows an isochore map of the net deltaic sand thickness. Light gray and dark gray indicate a thickness of 2 to 3 m and >3 m, respectively. The areas of little to no sand are the locations of the distributary channels. Note that sand deposits can approach 10 km in length in direction parallel to depositional dip, while they are only a few 100 m wide.

3) composite deformation zone; and
4) localized deformation zone.

Architectural styles determine the fluid properties of a fault zone, comprising localized conduit, distributed conduit, combined conduit-barrier, and localized barrier (Caine, 1999; Figure 16.16b).

Factors controlling the permeability of the fault zone include lithology, fault scale, fault type, deformation style, deformation history, fluid chemistry, pressure-temperature history, component percentage, and component anisotropy. For example, granitic rocks deformed by mature fault zones frequently have feldspars affected by fluid–mineral interactions. This results in fault core lithologies rich in clay gouge materials (e.g., Blanpied et al., 1995; Hickman et al., 1995; Goddard and Evans, 1995), which can significantly reduce a fault core permeability. Quartz-rich sandstones deformed by distributed deformation zones with small displacements have a low permeability fault core developed by the grain size reduction and the formation of deformation bands (e.g., Antonellini and Aydin, 1994). The faulting in sandstone/shale sequences or in granitic rocks with a sedimentary cover with combined mechanical and chemical processes leads to the formation of low-permeability fault cores (e.g., Chester and Logan, 1986; Caine, 1999).

As mentioned earlier, the permeability–effective stress relationships for the host rock and fracture system are quite different. The fracture permeability, which also depends on both the total stress (e.g., Carlsson and Olsson, 1979; Fisher et al., 1996) and fluid pressure (e.g., Knipe, 1993; Fisher et al., 1996), which are coupled (e.g., Oliver, 1986; Sibson, 1990b; Fisher et al., 1996; Connolly and Cosgrove, 1999), can collapse abruptly when certain stress conditions are met. The fracture permeability, which spans from fully opened apertures to collapsed fracture space, changes with effective stress within a relatively narrow interval of values (Figure 16.11b). This narrow range can be used as an indicative tool to distinguish fracture and rock permeabilities, the latter changing with effective stress within a relatively broad interval of values (Figure 16.11a).

The relationship between the fracture permeability and stress has been determined by (Walsh, 1981; Sigal, 1998, pers. comm.):

$$k^{1/3} = A + B \log_e(\sigma), \qquad (16\text{-}4)$$

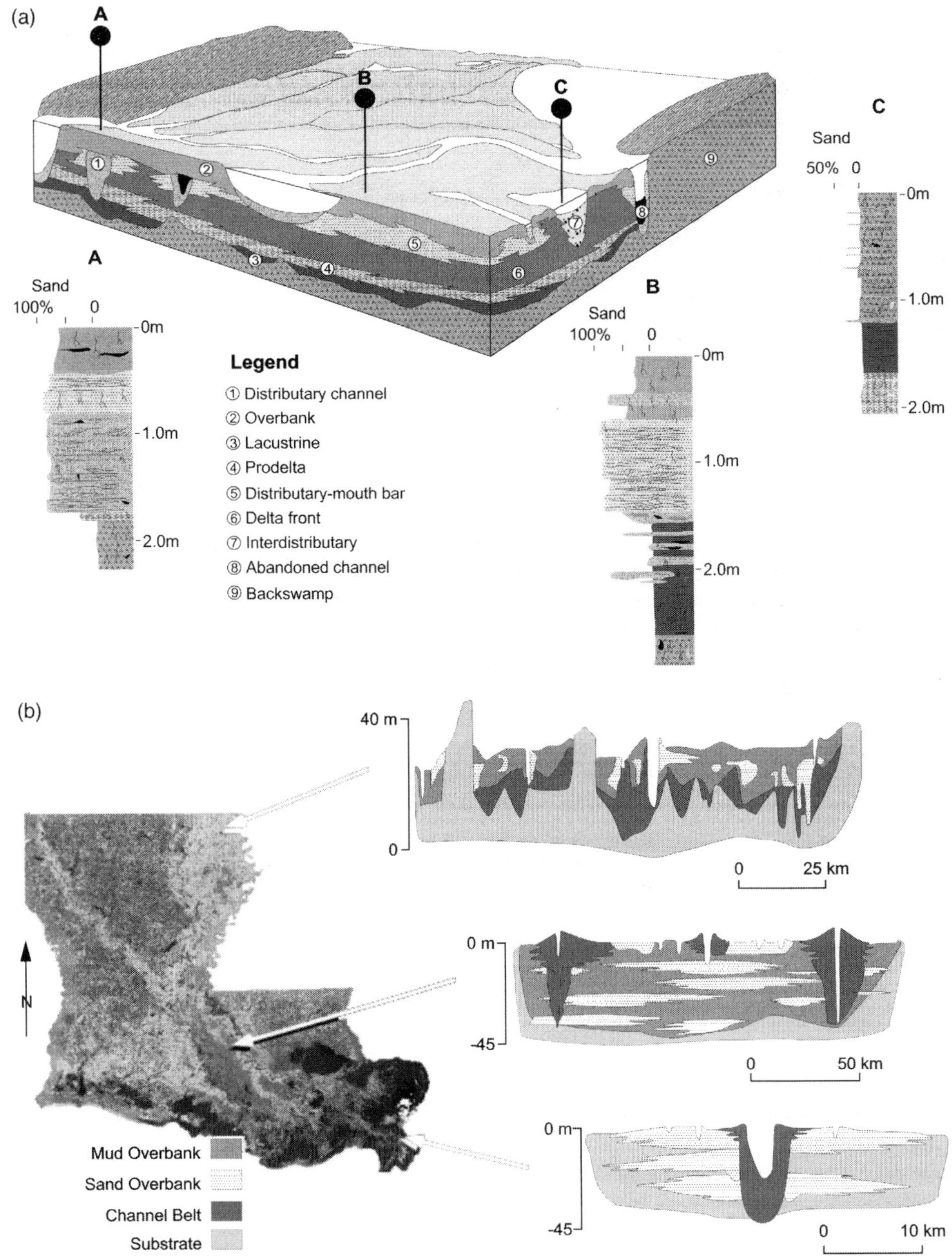

Figure 16.14. (a) Block diagram showing the lacustrine delta facies associations (Tye and Coleman, 1989a). The dip-elongate body of the lacustrine delta contains upward-coarsening lobes of prodelta, delta-front sediments, and distributary-mouth bar sediments that overlie lacustrine and backswamp mudstones. Individual deltaic lobes are separated by distributary channels. Note that the updip channel portions are sand-prone while their downdip portions are filled with fine-grained sediments. Note pseudo-wells through different regions of the lacustrine delta illustrating the quality of the expected reservoir. Pseudo-wells A, B, and C penetrate the distributary channel, distributary mouth bar, and interdistributary, respectively. (b) A series of cross-sections showing the along-valley changes in alluvial-basin stratigraphy (Tye and Coleman, 1989a). The proximal region is characterized by substrate/structural control of channel-belt positions. Note that both spacing between channel belts and the amount of fine-grain sediments increase with the width of the flood plain. Given a source of sand, the abundance and connectedness of the channel belts will be controlled by: 1) avulsion periodicity, 2) channel-belt width/floodplain width ratio, 3) channel-belt and floodplain aggradation rates and slopes, and 4) base-level changes.

(a)

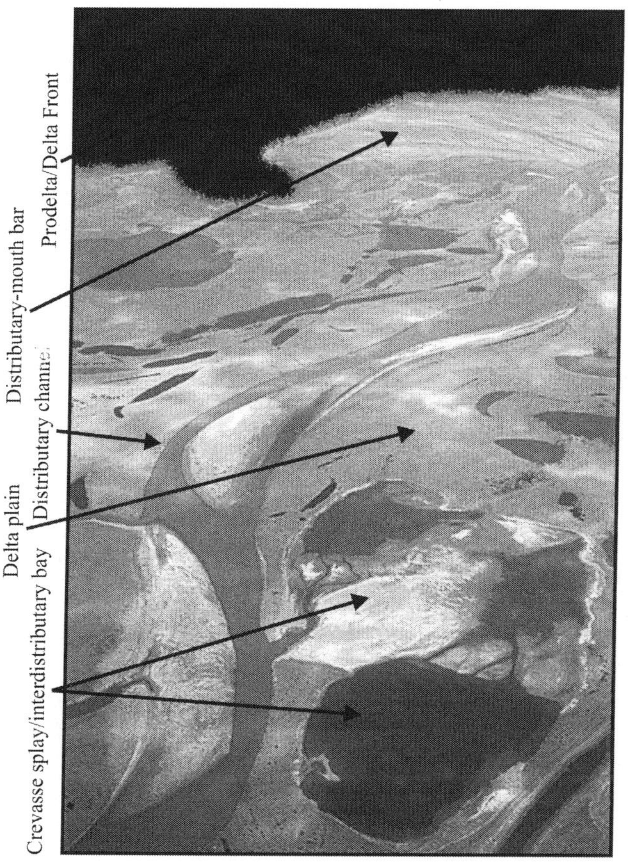

(b)

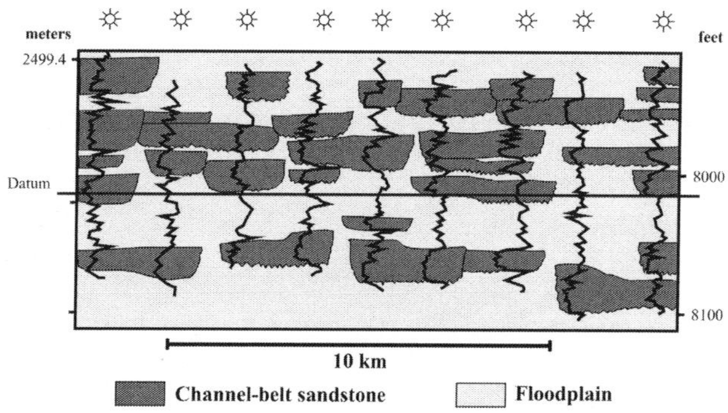

Figure 16.15. (a) Modern analog chosen for the depositional setting for the fluvially dominated deltas (Tye, 2004a). (b) Depositional architecture of the fluvial Travis Peak formation in the North Appleby field, Texas (Bridge and Tye, 2000). Using core and wireline-log data estimates of mean-bankfull channel depths and channel-belt widths, the model-driven depositional architecture of the reservoir section is interpreted as involving relatively small and relatively disconnected coarse sediment bodies of individual channel belts. (c) Example of the numerical simulation of the channel belt, crevasse splay, and floodplain facies associations distribution (left panel) using the predicted channel-belt dimensions for the fluvial Sterling formation in the Beluga River field, Alaska (Tye, 2004b). Lengths and widths of crevasse-splay deposits came from literature examples. Facies associations helped to guide the model-driven simulation of associated porosity distribution shown in the right panel. (d) Grain-size distribution model for the fluvial or friction-dominated delta (Wright, 1977). Note that the coarsest-grained and best-sorted sediment is deposited on the apex and frontal margins of the distributary mouth bar.

where A and B are constants and σ is the effective confining pressure. A and B can be determined from data distribution (Figure 16.11b). The fracture permeability can also be calculated from (Sigal, 1998, pers. comm.):

$$k^{1/3} = k_o^{1/3}\,(1 - \surd 2h/a_o * \log_e(\sigma/\sigma_o)), \qquad (16\text{-}5)$$

where k_o and a_o are the permeability and half-aperture of the fracture at a reference effective stress σ_o and h is the fracture surface roughness factor. Figure 16.11b indicates that there are relatively small permeability changes over a broad range of effective stresses, before the fracture permeability collapses because of pore fluid pressure dissipation below the value of the strength of the surrounding rock.

Permeability related to fracture and fault patterns located inside fault blocks will be described as developed either during the internal straining that precedes the development of the main boundary fault and/or extensional detachment or during this development.

During pre-faulting straining at differential stresses significantly smaller than those necessary for the main fault propagation, randomly oriented microcracks start to increase in number and form clusters, as rock mechanics tests show (e.g., Scholz, 1968; Krantz and Scholz, 1977; Lockner et al., 1992). Microcracks are generally accepted as a precursor to extensional or shear fractures (e.g., Brace et al., 1966; Kranz, 1983; Cox et al., 1991), developing at stresses from 50–95 percent of the fracture strength of the rock (Scholz, 1968). Meso-scale joints also progressively cluster as studies of dolomite reservoirs demonstrate (Antonellini and Mollema, 2000). In dolomite, which represents one of the most brittle reservoir rock types, small strains are accommodated by pervasive fracturing that affects a large reservoir volume. The appearance of the first small offset faults records more advanced stages of this pervasive fracturing. These

Figure 16.15 (*cont.*)

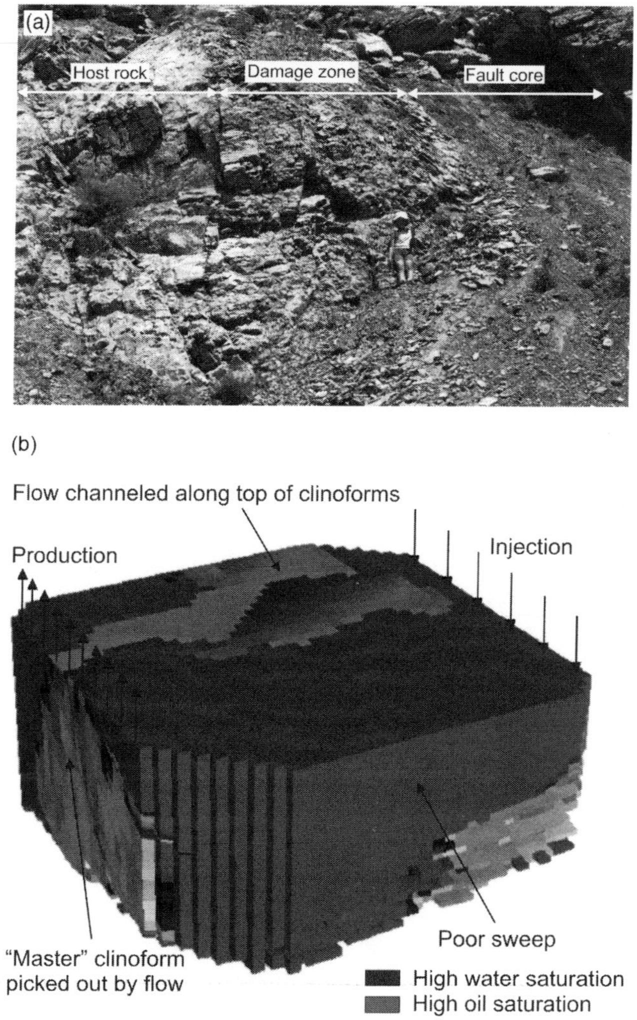

Figure 16.16. (a) Photo of fault zone architecture outcropping in Salt Lake City, Utah. Various types of fault core contain tectonic gouge, cataclasite, and mylonites (Nemčok et al., 2005a). Damage zones can be formed by small faults, fractures, cleavage, veins, and folds. (b) Fault zone permeability structures (Caine, 1999). A localized conduit is a feature of a fault zone that has slip localized along a single curviplanar surface or along discretely segmented planes. Both fault core and damage zone are absent or poorly developed. A distributed conduit is a feature of a fault zone that has slip accommodated along distributed surfaces. The fault core is absent or poorly developed. The damage zone has well-developed discrete slip surfaces and fracture networks. This fault zone architecture, known also as fault-fracture mesh (Sibson, 1996), is typical for modern accretionary prisms (Moore and Vrolijk, 1992). A localized barrier is a feature of a fault zone that has localized slip accommodated within a cataclastic zone. The fault core is well developed and the damage zone absent or poorly developed. The fault core behaves like an aquitard within the somewhat higher permeability aquifer of the host rock. Similar fault zones are reported as deformation bands in sandstones (Antonellini and Aydin, 1994). A combined conduit-barrier is a feature of a fault zone that has its deformation accommodated within a localized cataclastic zone and a distributed zone of subsidiary structures. The fault core behaves like an aquitard sandwiched between two aquifers in the damage zones.

faults have displacements of generally less than 30 mm and are characterized by en-echelon fracture patterns and breccia zones about 10 mm wide. Breccia zones form high-permeability paths, with permeabilities of 10^{-14} to $3*10^{-12}$ m^2, where fractures and high-porosity breccia are concentrated. Further fracture development results in medium-displacement faults. These faults have displacements of 1–10 m and are characterized by breccia zones 1–2 m wide and high-density fracture zones along their brecciated fault cores. The breccias may have porosities as high as 10 percent and represent zones of preferred fluid flow. The final stage of fracture development is represented by large-displacement faults with displacements greater than 10 m. These faults contain a wide zone of low-porosity breccia in their cores and form permeability barriers. Their fault cores are surrounded by high-permeability fracture zones, with permeabilities reaching values of 10^{-14} to $3*10^{-12}$ m^2. Similar fracture development can be expected in granitic basement rocks within thick-skin thrustbelt structures. These rocks behave in a similar very brittle fashion as the outcrop fracture studies of Martel and colleagues (1988) document.

Limestones are not as brittle as dolomites and granites and as a result differences appear in their fracture development. Although the overall process of pervasive fracturing, fracture nucleation, and fault development remains roughly similar, the fracture nucleation process proceeds through different deformation mechanisms (e.g., Willemse et al., 1997). Fracture development in limestone reservoirs also starts with en echelon fractures. However, instead of breaching the bridge between en echelon fractures by cross-joints as in dolomite, two symmetrically arranged zones of solution seams develop in the contractional arches of a bending bridge. The next stage is characterized by incipient pull-aparts, which form because of shear along solution seams. The continued shear along the first generation of solution seams leads to further development of these pull-aparts. Eventually a second-generation of solution seams develops, together with tail cracks, which form at the tips of sheared first-generation solution surfaces. The second generation of solution seams, oriented at a high angle to the fault zone, undergoes antithetic slip. Later, a third generation of solution seams, together with tail cracks, forms at the tips of the sheared second-generation solution seams. The result is a complex anastomosing network of discontinuities across the fault zone, which undergoes a block rotation. Solution seams and en-echelon fractures eventually develop in a contractional relay ramp between the two side-stepping fault segments. Ongoing syn- and antithetic slip along the solution seams and en-echelon fractures causes the formation of opening-mode tail cracks and solution seams, fragmenting the relay ramp. A through-going fault eventually develops, linking both fault segments.

Both pervasive fractures and fault systems can act as either fluid conduits or barriers. In either case, these structures can significantly alter the distribution of permeability in a reservoir.

Faults with low-permeability tectonic gouge in their cores and carbonate-filled fractures restrict hydrocarbon distribution, while discontinuous open fractures enhance reservoir permeability (Lindquist, 1983; Tillman, 1989). Studies of fractured tuff (Davidson et al., 1998) and fractured limestone (Nemčok et al., 2002) indicate that permeability enhancements of up to several orders of magnitude can occur due to fractures capable of focusing fluid flow. If the reservoir contains a system of permeability-enhancing faults, an understanding of their damage zone geometries is needed to predict the performance of the field and to map areas of enhanced permeability.

There is a significant difference in fracturing between extensional terrains, which incorporate mostly lithified rocks and terrains, which contain mostly unlithified sediments. The initial features that develop in mostly unlithified sediments are mud-filled veins, driven by fluid pressures exceeding the cohesion of the sediment undergoing compaction. The extensional geometry of these veins provides the evidence for their origin by disaggregation, dilation, and mineral growth. They contain structures indicative of fluid flow (Knipe et al., 1991) that reflect the abrupt enhancement of the reservoir permeability and record temporary fluid storage.

We can conclude from the previous text that the mechanical stratigraphy of the deforming strata controls fracturing.

In nature, even fine-scale layering exerts a significant impact on the fracture development in reservoirs.

The layer thickness effect can be illustrated using an example from the Bristol Channel, United Kingdom (Nemčok et al., 1995). Here the outcrop study allows us to interpret that: 1) the thicker the bed the stronger the tensile strength is, and 2) the stronger the bed the larger the fracture spacing is. These conclusions are valid for both sedimentary rocks located away from the large-scale fault and rocks located inside the damage zone of the fault. A large difference in fracture density, porosity, and permeability between reservoirs away from faults and those located inside damage zones of faults can be documented by fracture porosity measurements from the Bristol Channel (Nemčok et al., 1995). Although the study did not analyze which fractures were open at any particular time in the history of the reservoir, it indicates that the maximum possible fracture porosity decreases away from the damage zone of the normal fault.

Nemčok and colleagues (2005a) illustrated the importance of layer rheology on shear fracture development, showing a propagation of en echelon shear fractures in adjacent quartzite and conglomerate in the Proterozoic-Cambrian outcrop at Antelope Island, Utah. Because the number of grain-to-grain contacts defines the angle of internal friction, which is one of the factors controlling shear strength, fine-grained quartzite is more brittle than conglomerate. Therefore, the parts of the shear fractures that developed in quartzite are long and have wide apertures, whereas those developed in conglomerate are short and narrow. A similar effect, but for tensile fractures, is shown in Figure 16.17a, where fracture apertures are considerably smaller in the less brittle shale compared to the brittle limestone. The smaller apertures in the shale result in a 4 percent reduction in its fracture porosity, which in turn leads to a difference in permeability between the shale and limestone of three orders of magnitude (Figure 16.17b).

Another controlling factor in fracture enhancement of reservoir permeability is the timing of fracturing relative to diagenesis. While the early fractures can become cemented during subsequent events, late fractures have a much better chance of remaining open and improving the present-day permeabilities. This point introduces a discussion on mechanisms that maintain fractures open once their propagation ceases.

As numerous reservoir studies point out, both tensile and shear fractures can either remain open or become sealed. Several mechanisms inhibit the diagenetic sealing of fractures. One of them is the influx of hydrocarbons that prevent cement growth. Other inhibitors include clay minerals and increased concentrations of Al^{+3} in the fluid. As pointed out by fracture cementation studies in sandstone and carbonate reservoirs (e.g., Fisher and Knipe, 1998; Nemčok et al., 2002), even when fracture cementation occurs, the fracture is unlikely to become homogeneously sealed. Cement distributions are typically uneven, especially in shear fractures, where the tendency to precipitate minerals is greatest in the most dilatant segments of the fracture. This leaves some segments of the fracture permeable. The chance for a fracture to remain partly open increases with increasing aperture.

As discussed a bit earlier, the fracture permeability is controlled by interactions between stress and fluid pressure (e.g., Carlsson and Olsson, 1979; Sibson, 1990b; Fisher et al., 1996; Figure 16.11b). An increase in fluid pressure can inhibit fracture collapse. The maintenance of fluid pressure depends on the local draining of fluids or the ability of the surrounding reservoir to feed fluid to the fracture. Thermodynamic arguments further suggest that quartz and calcite are

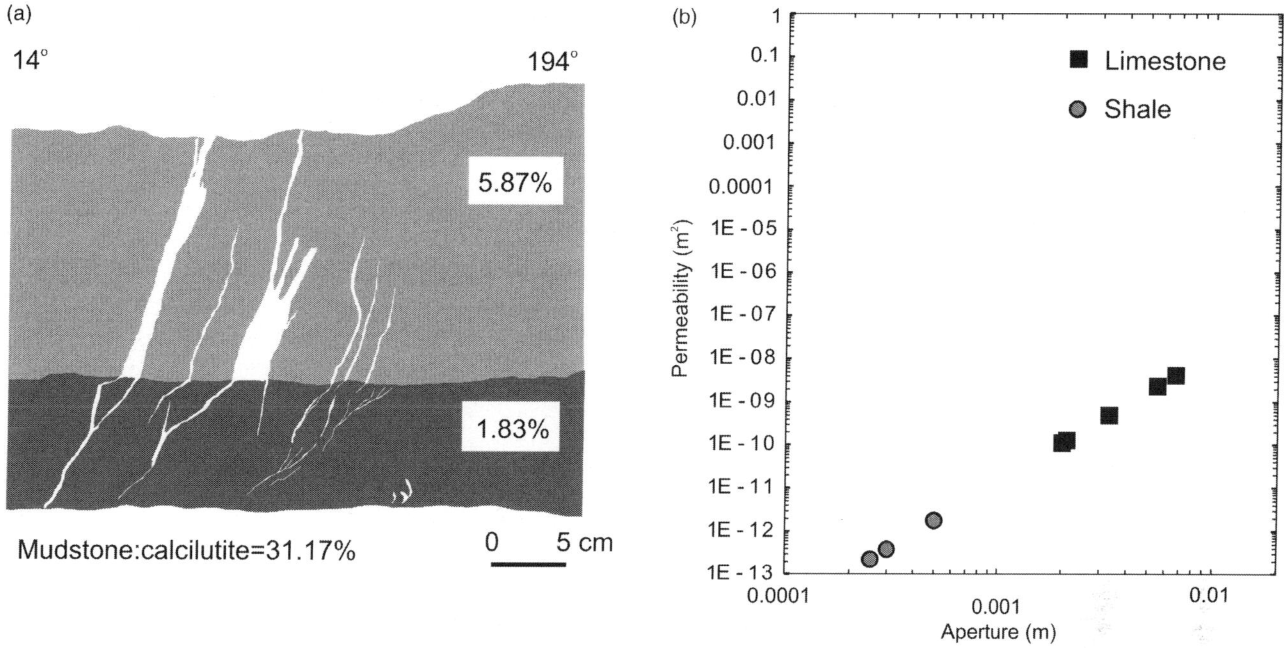

Figure 16.17. (a) Fracture aperture dependence on rock rheology as indicated by fracture porosity in Liassic sediments, St Donats, Bristol Channel (Nemčok et al., 2002). The light pattern represents bioclastic calcilutite and the dark pattern represents calcareous mudstone. Fracture porosity for each layer and the fracture porosity ratio are expressed as percentages. (b) Variation in permeability (expressed as the coefficient of hydraulic conductivity) with fracture aperture in fractures that propagate from mudstone to calcilutite in (a) (Nemčok et al., 2002). The calculation, which is based on Darcy's Law and the Navier-Stokes equation modified by Snow (e.g., Lee and Farmer, 1993), assumes a fluid density, r_f, of 1,000 kgm^{-3}, a dynamic fluid viscosity, η, of $1 * 10^{-3}$ $kgm^{-1}s^{-1}$ (Freeze and Cherry, 1979), a hydraulic gradient ratio, I, of 0.06 (Caine, 1999), and a fracture width of 0.173 m. Equations are: $Q = -a^3 r_f g/12\eta IW$ and $k = \eta Q/r_f gIA$, where k is the coefficient of hydraulic conductivity, Q is the volumetric flow rate in the direction of decreasing pressure, A is the cross-sectional area across which the perpendicular discharge Q flows, g is the acceleration of gravity, a is the fracture aperture, and W is the fracture width.

deposited along their flow paths along fractures with decreasing fluid pressure (Bruton and Helgeson, 1983; Capuano, 1990). Therefore, an increased fluid pressure would not only maintain fractures open, reducing the local stress, but also prevent cement precipitation along them (Roberts et al., 1996). This dynamic control has an important consequence for hydrocarbon production from fractured reservoirs, where production changes can affect fracture permeability by reducing the fluid pressure.

Additional factors, such as the presence of material that props the fracture open (propping material), cement bridges, and asperities, can prevent fractures from total closure. Cement bridges develop in cases of partial cementation and prevent total closure with increasing effective stress. The propping material can be formed by breccia, especially in very brittle reservoirs with a tendency to brecciate prior to fault development. Tensile fractures in the Triassic mudstones from the Bristol Channel, which were filled with fluids containing sand from isolated sand bodies, of those from the Kura Valley, provide an example of sand that acted as a natural propping material (Cosgrove, 2001; Nemčok et al., 2005a; Figures 18.13a–f). Asperities are common in shear fractures. They are formed by a juxtaposition of complex fracture wall geometries by strike-parallel displacement. This often results in features with positive relief keeping the fracture open at some residual permeability level.

Post-rift reservoirs in failed rift settings

Depositional environments of a post-rift time period of failed rifts vary among rift systems considerably. In order to illustrate several end-members, the following text will discuss:

1) reservoir deposition in the post-rift gulf open to marine conditions, the Cretaceous Qishn Formation of the Masila Basin, Yemen; and
2) the deposition of the Upper Cretaceous–Danian reservoirs in the Central Graben of the North Sea undergoing basin inversion.

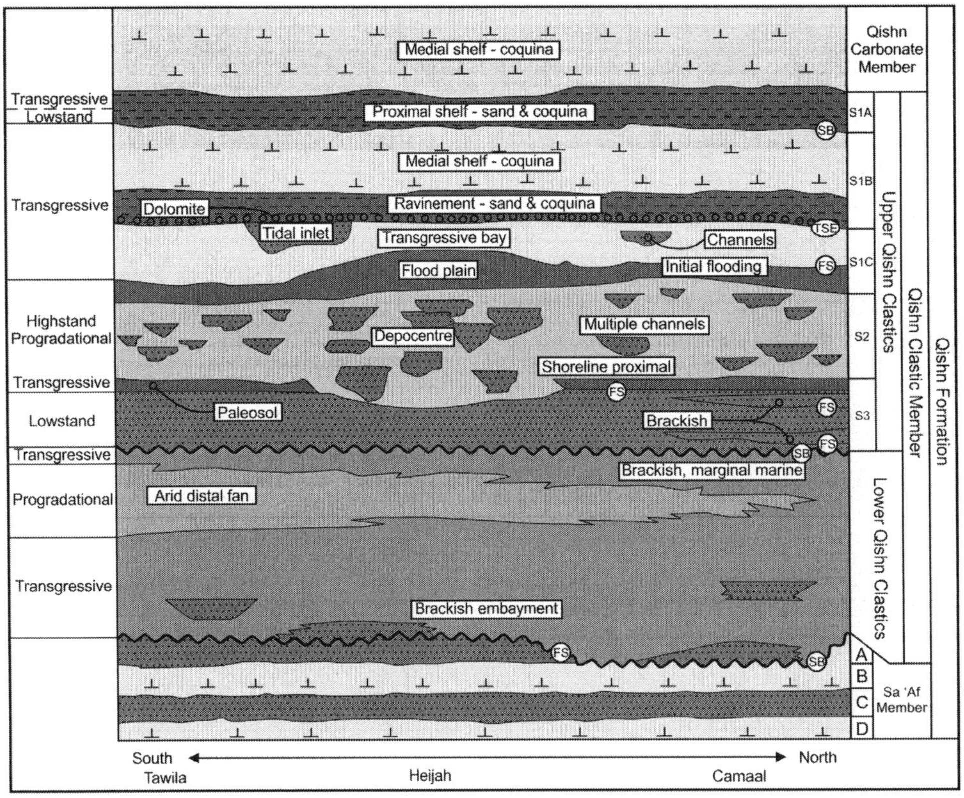

Figure 16.18. Sequence stratigraphic chart of the Qishn Formation of the Masila field, Republic of Yemen (Leckie and Rumpel, 2003). SB is the sequence boundary, TSE the transgressive surface of erosion, FS the flooding surface. Note that the chart has no scale included for confidentiality reasons.

Deposition in the post-rift gulf in Yemen allows us to discuss the post-rift deposition unaffected by any major tectonic events apart from the post-rift thermal subsidence. Here, the deposition of the Neocomian-Aptian Qishn Formation took place in an elongate gulf of the Say'un–al Masila Basin, which was open to a basin with marine conditions in the east, and which was undergoing carbonate deposition (Leckie and Rumpel, 2003). The tapering and constricting configuration of the gulf facing the paleo-Indian Ocean was an environment characterized by the development and amplification of tides. A nice present-day analog for such amplification is the Bristol Channel, where the difference between low and high tides can reach 14 m.

The Qishn Formation lapped in the eastward direction onto the preexisting Sa'af Member. The first sediment deposited above the regional unconformity is the Qishn clastic member, characterized by mixed carbonates and clastic sediments. It was deposited in a variety of environments, including brackish, tidal estuarine, open bay, and gulf (Figure 16.18). A subsequent deposition took place in environments such as arid nonmarine and coastal (Figure 16.18).

The basal unconformity separates the Qishn Formation from the underlying syn-rift sedimentary section associated with Late Jurassic–Early Cretaceous rifting. The westward transgression is apparent from the distribution of depositional environments, which are nonmarine in the west and progressively deeper marine in the east. Furthermore, the eastern side of the respective sedimentary section is characterized by carbonates intertonguing with fluvial deltaic to littoral sediments, which become fully siliciclastic toward the west.

The producing Qishn Formation reservoirs of the billion barrel Masila oil field are represented by sandstone bodies deposited in a variety of environments including:

1) sand and gravel braided rivers;
2) tidally influenced distributary channels;
3) sand flats;
4) sand shoals; and
5) tidal ridges.

The Lower Qishn clastic member (Figures 16.18) contains subarkose reservoirs. They have good to excellent quality. Porosity and permeability characterizing core samples are as high as 25 percent and more than 560 mD. Their controlling factors are:

1) content of carbonaceous-pyrite laminae;
2) degree of carbonate/silica cementation;
3) detrital clay content;
4) grain size;
5) sediment compaction; and
6) sorting.

The Upper Qishn Member (Figures 16.18) contains excellent sandstone reservoirs. Their improved quality in comparison with underlying subarkoses is due to a lack of inhibiting authigenic clays controlled by their quartz-dominated composition. Furthermore, their shallower burial controls a decreased degree of cementation. Porosity and permeability are as high as 20 percent and more than 1.5 D.

The Late Cretaceous–Danian deposition in the Central Graben of the North Sea allows us to discuss the effect of the basin inversion on the reservoir rock deposition. A coeval compression reactivated the syn-rift fault pattern of the Central Graben, which had north-northwest–south-southeast and northwest-southeast strikes, as reverse faults (Gowers, 1993). Chalk deposition dominated the syn-inversional deposition. While the outer margins of the Central Graben were formed by a stable platform and their chalk section is thin, the depocenter contains a thick chalk section.

A complexity to the distribution of sediment catchment areas during the basin inversion was added because of the effects of the Zechstein evaporite distribution, as one of the important salt mobilization events took place during the Tertiary period of the basin inversion. The effect of evaporites was complex due to their uneven initial thickness distribution. The evaporites were thick in the south, in the Netherlands, Denmark, Germany, and United Kingdom sectors. They were thin or absent in the north, in the Norwegian sector (e.g., Wride, 1994). The thick evaporites were extensively mobilized while the mobilization was missing in regions with thin or missing evaporites.

The basin inversion took place at different locations at different times. Seismic data tied to wells indicate that the Tomme-Litten, Edda, and Tor structures grew up during the deposition of the Campanian Magne chalk formation. The Ekofisk structure did not grow during this time period. However, the Tomme-Litten structure grew just a little during the deposition of the subsequent Campanian–Maastrichtian Tor chalk formation while the Edda, Tor, and Ekofisk structures underwent a significant growth. Later on, during the deposition of the Danian Ekofisk formation, there was no growth of the Tomme-Litten and Ekofisk structures while the Edda and Tor structures were still growing. The subsequent deposition of the Thanetian Rogaland group saw only a growth of the Tor structure and no growth of the Edda.

Furthermore, the biostratigraphic and paleoenvironmental data, such as those from the Valhall Field, document that some structures grew in cycles during relatively long time periods. Seismic images indicate that the Valhall and Hod Fields grew up during the deposition of Campanian Magne and the uppermost Campanian–Maastrichtian Tor chalk formations. The subsequent Danian Ekofisk Formation and Thanetian Rogaland Group are also part of the syn-tectonic strata but recording slower growth rates.

A temporally variable uplift of different structures controlled a (1) thickness reduction of various chalk reservoir horizons at a different time at different locations, and (2) uplift-driven gravity slumping of unconsolidated chalk material on slopes of growing structures.

Nice examples of the former come from the Valhall Field, where the Turonian-Campanian Hod Formation was partly eroded before an eastward thickening wedge of Campanian–Maastrichtian turbidites and debris flow covered the area. Even younger Tertiary clays drape the structure. As a result the entire chalk suite comprises numerous facies associations and many of them include reworked chalk material (see Joint Chalk Research, 1996; Sikora et al., 1999). Examples of the latter come from the Turonian-Campanian Hod and the Campanian–Maastrichtian Tor formations of the Lindesnes Ridge.

Mechanical properties are strongly porosity-dependent in chalk reservoirs (Joint Chalk Research, 1996). However, companion samples of similar porosity can behave differently in laboratory tests. This is because the mechanical variations are also a function of lithology, for example, due to silica variations. Some mechanical differences can be controlled by depositional mechanisms. A temperature effect on different lithologies is also important (Joint Chalk Research, 1996).

Production data from chalk fields also indicate a contribution of the fracture permeability to the total reservoir permeability. While the contribution of the fracture permeability and its competition with matrix permeability was discussed earlier in this chapter, the chalk fracture permeability is discussed here because of its special character.

The growth of structures induced fracture patterns of different scales into the deforming sedimentary strata, as can be seen at the Valhall Field. Although the gas cloud presence in its crest complicates the interpretation of large-scale faults from the seismic image, these faults can be interpreted from the Tor Formation isochors. Their development was controlled by dextral transpression, inhomogeneous shortening of the Valhall structure, and its crestal collapse. Many of them exert some degree of syn-depositional controls, such as:

1) hardground development on tops of crestal grabens;
2) the effect on different facies location; and
3) the effect on thickness distribution.

The meso-scale fault population is represented by groups including:

1) sealed fractures with primary calcite of small crystal size that attaches directly to the fracture wall and a second calcite generation represented by larger crystals filling the remaining space along the fracture;
2) open fractures with residual space not completely filled by cements, leaving additional open/vuggy space;
3) open fractures characterized by sharp fracture walls with a certain degree of brecciation; and
4) closed fractures represented by hairline traces in cores with fracture density as high as two to five fractures/cm, which occur in porous formations and originate as shear fractures with microscopic displacement.

Many cores from chalk reservoirs in this region contain intense micro-fracturing, usually treated as matrix porosity. While the matrix permeability in producing the facies of the Hod and Tor Formations can be as high as 0.5–14 mD, the fracture enhancement due to fracturing developed inside the fault block or fault damage zone can improve the total permeability to values as high as 10–130 mD.

Post-rift reservoirs in passive margin settings

Note that each passive continental margin setting can contain both pre-rift and syn-rift reservoirs buried underneath the passive margin sedimentary section. Also note that a significant difference exists between mature rifting in rifts, which eventually fail and rifts, which evolve all the way to the continental breakup. Accordingly, there is a difference between reservoir rocks deposited in these two scenarios.

Numerous successful applications of the uniform stretching model (McKenzie, 1978) to initial-stage rift terrains indicate that most of them formed by mechanisms dominated by passive rifting. Many studies of the more mature-stage rift terrains, however, required various modifications of the uniform stretching model to explain the observed data. This indicates that active rifting plays a progressively more important role in advanced rifting stages and rifting, which makes it to seafloor spreading. Studies of passive margins, structures that went beyond advanced rifting development, document that there are practically no passive margins that have been developed by homogeneous stretching.

Reflection seismic imaging at the Iberian margin (Reston et al., 1995, Reston, 1996; Manatschal et al., 2001; Wilson et al., 2001), the Gabon margin (Rosendahl et al., 2005; own observations), the Goban Spur margin (Davis and Kusznir, 2004), and the East Indian continental margin (Nemčok et al., 2010, 2012; Sinha et al., 2010) indicate that the continental breakup included a continental mantle unroofing. This mechanism has been also documented by studies drawing from deep oceanic well data (Boillot et al., 1980, 1987; Beslier et al., 1996; Whitmarsh et al., 2001), geochemical analyses (Abe, 2001; Hébert et al., 2001), West Alpine field analogs (Manatschal and Nievergelt, 1997; Manatschal and Bernoulli, 1998; Manatschal, 2004 and references therein), analog material modeling (Brun and Beslier, 1996; Brun, 1999), and numeric modeling (Huismans and Beaumont, 2005).

Seismic imagery of the East India margin, among others, further documents that the distal continental margin can consist of upper continental crust blocks resting on either significantly thinned lower continental crust or, relatively frequently, directly on continental lithospheric mantle (see Chapter 2). East Indian data further indicate that the lower crust withdrew from certain regions of the distal continental margin and inflated the lower crustal thickness in foots of prominent footwalls located more landward. In other words, the inhomogeneous crustal thinning in East India should be accommodated by a system of upper crustal fault blocks "floating" on top of the thinning ductile lower crust and, eventually, the less thinned upper crustal section should find itself resting directly on the mantle undergoing unroofing in the most distal margin.

These studies, which describe margins developed by crust-first–mantle-second breakup scenarios, indicate that rifting transitioning into continental breakup subsequently localizes in more distal portions of the future continental margin. As a result, proximal and distal margins have different crustal architecture (see Lemoine et al., 1986; Eberli, 1988; Manatschal and Bernoulli, 1998; Figures 2.8, 2.13, 2.14, 2.19b), which seems to be the case also for margins that evolved from the mantle-first–crust-second breakup scenarios

(Figures 2.6, 2.15). This difference has an important impact on the syn-tectonic deposition.

Proximal margins in East Coast India (Figure 16.19), Newfoundland, and Iberia are formed by systems of half-grabens and grabens bounded by relatively high-angle faults (e.g., Montenat et al., 1988; Driscoll et al., 1995; Pérez-Gussinyé et al., 2003). They are further characterized by prominent structural highs separating different extensional provinces (Figure 2.8, 2.13b–c, 2.14). The sediment accommodation space bounded by high-angle faults is predominantly controlled by vertical and steeply inclined shear (Figure 16.19a–b). Therefore, syn-rift fill including reservoir rocks in these accommodation spaces can develop large thicknesses and seismic sections would show divergent wedges, which are the result of sediment interaction with syn-sedimentary controlling faults.

Distal margins in published transects (e.g., Manatschal, 2004) also contain the structural grain of the proximal margins but as the older generation of structures. However, the initial subsidence patterns are later replaced by uplift during advanced rifting. The most prominent younger-generation rifting structure is a pattern of low-angle detachment faults (see Figure 16.19c). The sediment accommodation space bounded by low-angle detachment faults is predominantly controlled by a low-angle inclined shear. Therefore, syn-rift fill including reservoir rocks in these accommodation spaces develops only minimal thicknesses and seismic sections would not show any divergent wedges.

Further difference between reservoir rock controls in the already described rift settings and following passive margin settings is the post-breakup uplift of the continental margin, which controls the sediment input into the deep basin.

Uplifted flanks, or shoulders, are common features of rifts and post-breakup passive margins. The uplifted region has typically asymmetric geometry. It has a steep scarp facing the ocean and a gentle slope facing the land. Such uplifts have been explained by models based on transient and permanent uplift mechanisms. Transient uplift models are related to the thermal effects of rifting and include depth-dependent extension (Royden and Keen, 1980; Hellinger and Sclater, 1983; Watts and Thorne, 1984; Morgan et al., 1985), lateral heat flow (Steckler, 1981b; Cochran, 1983; Alvarez et al., 1984; Buck et al., 1988), and secondary convection under basin flanks (Keen, 1985a; Steckler, 1985; W. R. Buck, 1986). These mechanisms, which can create 500–1500 m topographic elevation, operate only during the elevated thermal regime of the lithosphere. Created positive topography will decay over the time period equivalent to the thermal time constant of the lithosphere, roughly 60 Ma. Permanent uplift models are related to the magmatic underplating (Cox, 1980; Ewart et al., 1980; McKenzie, 1984; White and McKenzie, 1988) and lithospheric unloading and/or plastic necking (Zuber and Parmentier, 1986; Parmentier, 1987; Issler et al., 1989; Braun and Beaumont, 1989; Weissel and Karner, 1989; Chery et al., 1992). Only relatively recently, processes of erosion and deposition started to be incorporated into existing models of the flank uplift (e.g., Van Balen et al., 1995; Van der Beek et al., 1995; Burov and Cloetingh, 1997; Figure 2.7).

The importance of the uplifted area is in:

1) preventing sediment transport from land to the oceanic basin apart from specific entry points; and
2) serving as a local sediment provenance, which can be unimportant in certain cases and important in other cases.

A nice case example of the preventing function comes from the Moroccan passive margin for the Middle Jurassic–Early Cretaceous time period (Nemčok et al., 2005b; Figure 7.21). The continental breakup timing for this region taking place during the later Early Jurassic can be derived either from the age of the breakup unconformity or magnetic anomalies of the oceanic crust. The breakup unconformity is as old as the Hettangian-Toarcian and Hettangian-Pliensbachian in the Georges Bank Basin and its southern Moroccan counterpart, the Tarfaya Basin, respectively (Davison et al., 2002). The age of the breakup unconformity in the Scotian Basin is Aalenian (Cloetingh et al., 1989; Tankard et al., 1989; Welsink et al., 1989), although Wade and McLean (1990) infer an older, Sinemurian-Toarcian age. The age of this unconformity in the northern Moroccan counterpart, the d'Abda Basin, is Toarcian-Aalenian, and becomes as young as the Callovian northward (Aajdour et al., 1999; Echarfaoui et al., 2002). Magnetic data for the northern Central Atlantic segment, which joined Nova Scotia and central and northwestern Morocco, indicate that the drifting of the North American and African continents started either before 191 Ma (Toarcian) or 175 Ma (Bajocian-Bathonian) (Klitgord and Schouten, 1986; Srivastava and Tapscott, 1986; Verhoef and Srivastava, 1989).

Figure 7.21a shows that following the continental breakup, the Middle–Late Jurassic marginal uplift prevents siliciclastic sediments from making it from land to the oceanic basin, with the exception of the entry point, located in the central segment of the margin, and controlled by the preexisting pull-apart basin lows. The rest of the margin was undergoing carbonate deposition in

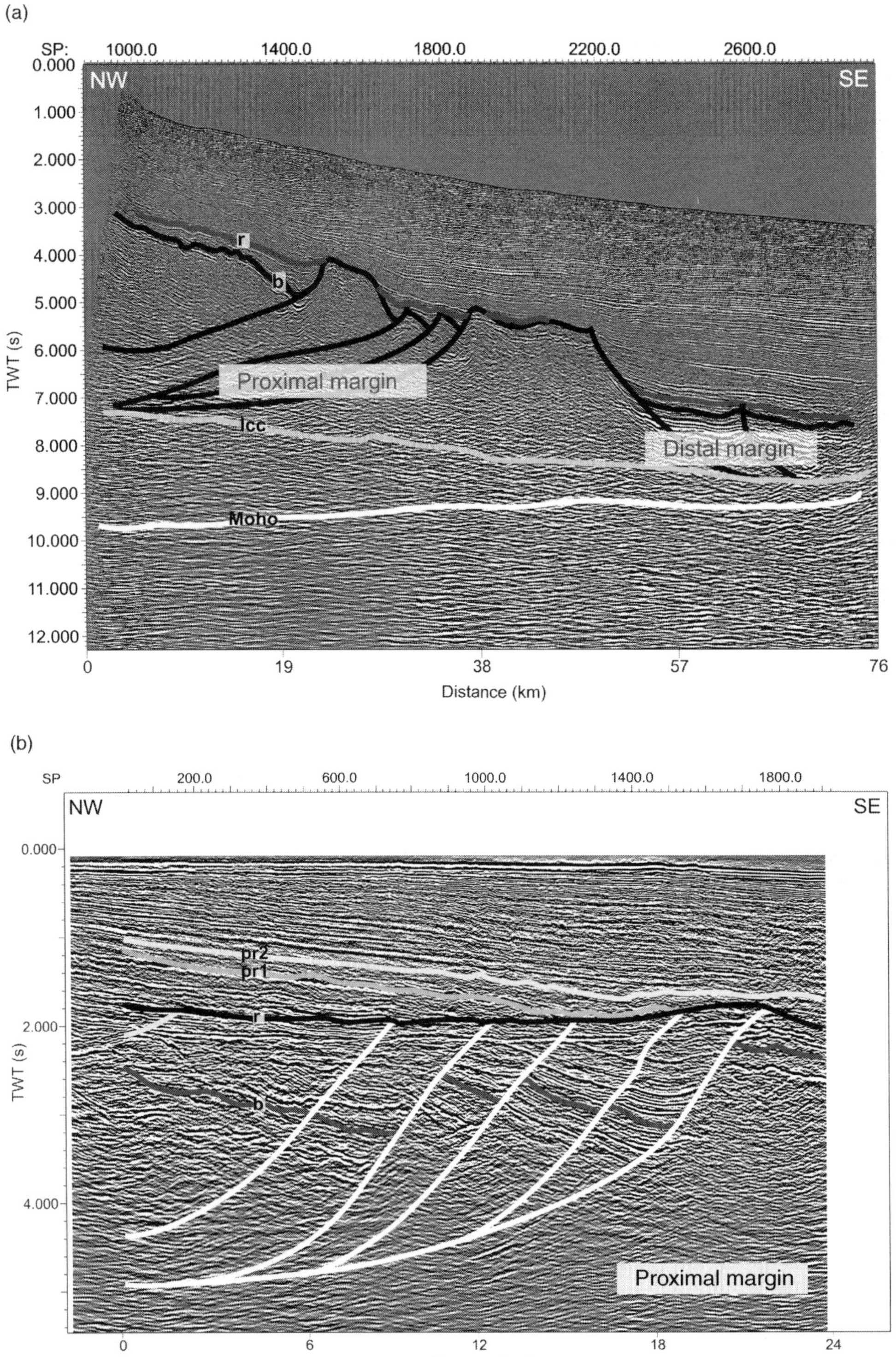

Figure 16.19. (a) Dip-oriented seismic profile through the East Indian continental margin imaging a large high dividing proximal and distal continental margins (Nemčok et al., 2007). Ucc – upper continental crust, lcc – lower continental crust, m – continental lithospheric mantle. (b) Dip-oriented section through the East Indian margin imaging proximal margin (Nemčok et al., 2007). (c) Dip-oriented section through the East Indian margin imaging distal margin (Nemčok et al., 2007).

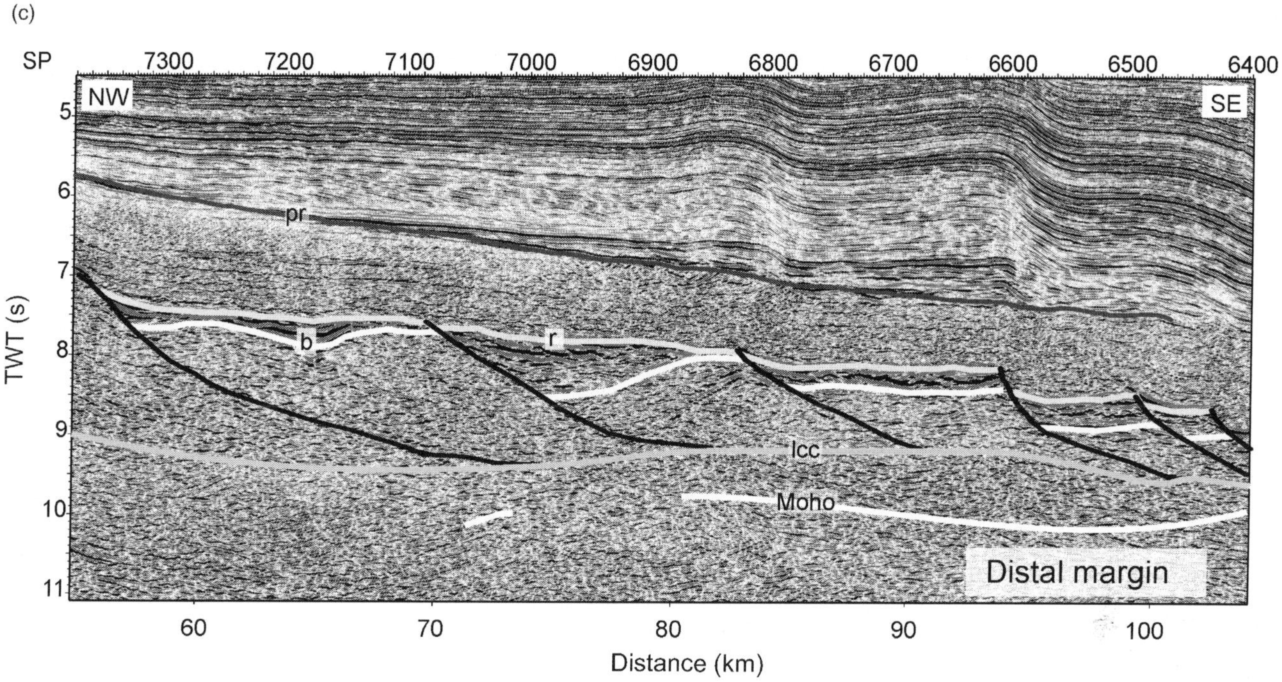

Figure 16.19 (*cont.*)

the uplifting zone where carbonate building was not disturbed by clastic sediment transport to any significant extent. Large bodies of turbidites were in front of the central margin segment, in front of wide zones of paralic sands and fanglomerates. The southern and northern segments had small turbiditic bodies occasionally in front of them, in front of an almost nonexistent paralic sand zone and a narrow fanglomerate zone.

It was only the Early Cretaceous when the uplifted margin was eroded off to the extent allowing for clastic sediment from land making it into the oceanic basin everywhere along the entire margin, replacing the carbonate-dominant deposition in the southern and northern segments (Figure 7.21b). This was the time when one could see broad paralic sand, a fanglomerate-to-delta plain, and fanglomerate zones in front of the entire margin and turbiditic bodies everywhere in front of them.

While the discussed Morocco example represents a passive margin end-member with marginal uplift practically following the continental breakup, the Santos segment of the Brazilian eastern margin is the opposite end-member, characterized by the marginal uplift significantly delayed. Its present-day morphology reaches the elevation of up to 2 km. Onshore apatite fission-track data (Gallagher et al., 1994) and offshore sedimentological data (e.g., de Almeida, 1976; Pereira et al., 1986; Pereira, 1990; Pereira and Macedo, 1990; Cainelli and Mohriak, 1999) indicate a rift shoulder uplift that started roughly from the Late Cretaceous. The present-day erosional remnant of the rift shoulder, the Serra do Mar Mountains, is located about 172–247 km landward from its initial Late Cretaceous position (Figure 4.12b).

Offshore sedimentological data from the Santos Basin (Figure 7.23) do not conform to the results of flexural models coupled with erosion and deposition. The characteristic offlap pattern (Figure 2.7b) did not follow the Neocomian-Aptian rifting. Instead of being formed during the Albian, the offlap pattern was formed during the Senonian, with a delay of about 24 Ma. This delay indicates that the rift shoulder uplift timing was affected by subsurface processes and yet another surface process on top of the erosion and deposition. Mohriak and colleagues (2008) argue that the additional process is represented by large-scale gravity gliding, which creates an extra see-saw effect by unloading the upper slope and loading the basin.

Note that the Santos and neighbor passive margins contain very prolific syn-rift strata, as proven by relatively recent discoveries of giant syn-rift oil fields such as Cariocas-Sugar Loaf, Tupi, and Jupiter in the Santos, Campos, and Espírito Santo Basins, which are ranked as the third, forty-third, and forty-fourth largest oil or condensate fields (see Berman, 2008).

The syn-rift fill (Figure 7.23) starts with the Neocomian Camboriú basalts (134.4–130.8 Ma). They are followed by the Barremian-Aptian alluvial fan/fan

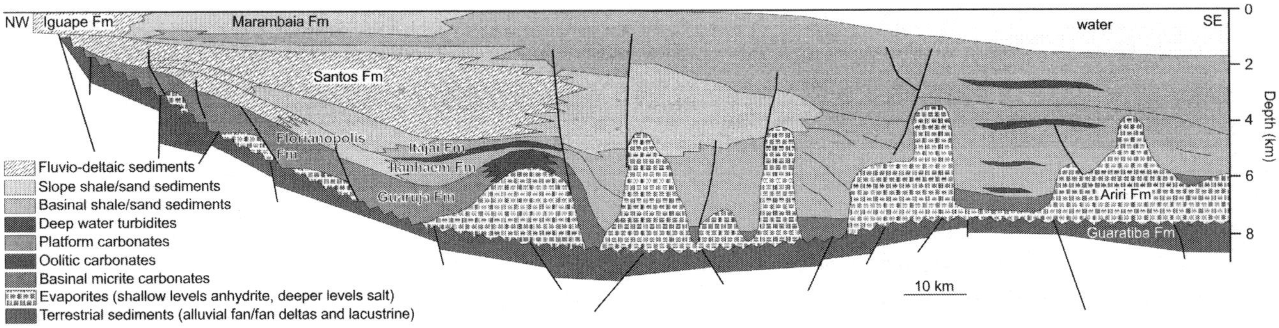

Figure 16.20. Simplified dip-oriented profile through the Santos Basin showing the main facies distribution (Nemčok, 2001).

delta and lacustrine sediments (130.8–119.2 Ma) of the Guaratiba Formation and the Aptian shallow-marine Ariri evaporites (119.2–112 Ma). Evaporites were deposited during the thermal subsidence that followed the rifting stage. Their deposition was followed by the erosion of about 104 m of sedimentary cover that took place during 120–119.2 Ma and created a breakup unconformity that indicates the timing of the continental breakup.

The syn-drift sediments start with the Albian Guarujá carbonates (112–105.2 Ma), represented by platform carbonates at the basin margin, oolitic limestone in a depth zone coincident with the occurrence of the first salt walls, and basinal micrite in the deep basin (Figure 16.20). Carbonates are overlaid by the regressive wedge (slope shale/sand) of the Albian–Cenomanian Itanhaém Formation (105.2–97.1 Ma), passing in the deeper parts into the basinal sand/shale sediments, and the Cenomanian-Turonian shale of the Itajaí Formation (96–90 Ma), representing a marine transgression onlapping an unconformity (97.1–96 Ma).

The top of the Itajaí Formation is formed by an unconformity (90–85 Ma), overlain by a regressive wedge of the Santos Formation (85–66.6 Ma) (Figure 16.20). Note that although the Santos Formation is simplified in Figure 16.20 into a single unit, it is represented by fluvio-deltaic sediments of the basin margin passing into shale/sand deposits on slope and shale/sand deposits in the basin. The deposition rate calculated for the maximum thickness of this unit gives the values comparable with values for planktonic sediment deposition in Persian Gulf or nannoplanktonic ooze deposition in the Mediterranean Sea. Some portions of the prograding wedge can also have lateral deepwater turbiditic equivalents (Fred Strawson, pers. comm., 2001).

There is an unconformity (66.6–65 Ma) at the top of the Santos Formation, followed by an onlapping Paleocene unit (65–60 Ma), represented by shale/sand sediments. Its top is formed by an unconformity (60–56.7 Ma) and overlaid by several regressive sedimentary packages. Sometimes possible distinction of the lowermost Eocene and Oligocene units, passing from prograding wedge into turbidite in deep basin (Gonzalo Enciso, Fred Strawson, pers. comm., 2001), can be used for the timing of tectonic events in the basin.

Marginal uplifts, regardless of whether they include post-breakup uplift or their driving forces are of different origin, are capable of changing drainage patterns completely, as the following Gulf of Guinea example illustrates (see Nemčok et al., 2004b, 2012).

The drifting in the Gulf of Guinea started in late Albian. The continental breakup was followed by marine transgression and the deposition rate decreased from Aptian-Albian rates down to one-tenth. The late Albian breakup was associated with regional erosion.

The following late Albian–early Tertiary period was still characterized by restricted water circulation, like the syn-rift period. Upwelling patterns were located on west- and south-facing coasts. It was only after the beginning of Tertiary when fracture zones and ridges of the continental crust among ancestral pull-apart basins ceased to act as topographic barriers and the effective water circulation started. From Santonian onward, the Dahomey margin started to behave as a typical passive margin experiencing thermal subsidence. The Upper Cretaceous lithostratigraphy here is complete on the lower slope and in the deep basin. It is reduced or eroded off on the shelf and upper slope.

During this time, the deep canyons were cut into the shelf during the sea-level falls. This concentrated the sediments and deposited them as larger turbidite fans on the slope. The canyons were also the place of shale deposition. North-south trending rivers characterized the drainage system during the Cenomanian–Turonian. There was one larger delta located at the Benin–Nigeria border. It deposited reservoir rocks occurring in the Seme Field, Benin, and the Aje Field, Nigeria. The Upper Cretaceous fans determined in offshore Benin are related to this delta.

The north-south trending rivers got cut from their drainage basins during the middle Turonian because of the east-west trending zone of margin uplifts located in the southern part of the West African block. The cut-off drainage basins joined the drainage basin of the Niger River, which was developing its delta on top of the Lower Cretaceous attenuated continental margin at the junction of two main strike-slip zones; the dextral Gulf of Guinea/Central African fault zone and the sinistral Sergipe-Alagoas/Benue/Trans-Saharan fault zone (Burke, 1972; Damuth, 1994).

The system of uplifts located in the southern West African block had positive topography until the Middle Eocene. A large Niger drainage basin contained Benue and Bida failed rift basins in its lower reach. The sediment input was more or less continuous since the mid-Turonian, although regressive sequences have been occasionally interrupted by episodic transgressions (Doust and Omatsola, 1990). The Middle Eocene end of active uplifts of Niger provenances apparently did not stop Niger from bringing sediment from northwest, north, and east to the delta.

Apart from a large-scale control on reservoir rock distribution, represented by the timing and geometry of the post-breakup marginal uplift, there are smaller-scale controls, represented by local tectonic and depositional settings of passive margins. Local settings are typically associated with extensional and contractional domains of gravity glide systems. Examples of the role of extensional domain on reservoir rock distributions come from the Niger Delta (see Doust and Omatsola, 1990; Van Heijst et al., 2002). An example of the role of contractional domain on reservoir rock distribution also comes from the Niger Delta (Hooper et al., 2002).

Reservoir distribution in the Niger Delta is best understood as a framework of depobelts separated by large seaward- and landward dipping fault systems (Figure 19.14; Knox and Omatsola, 1989; Doust and Omatsola, 1990; Cohen and McClay, 1996; Hooper et al., 2002). These depobelts were in their time a system of transient basinal areas accommodating the deposition of the delta prograding seaward (Figure 16.21a; Doust, 1990; Doust and Omatsola, 1990). Subsidence and deposition in each depobelt was facilitated by gravity gliding of the undercompacted and overpressured marine shales of the Upper Cretaceous–Eocene age under the loading by advancing paralic clastic wedge of the delta. The presence of depobelts indicates that, at a certain stage, each depobelt experienced no further accommodation of subsidence by the withdrawal of marine shale and focus of deposition shifted seaward to form a new depobelt. Each such seaward shift resulted in local shifts in tripartite stratigraphy (Doust and Omatsola, 1990; Tuttle et al., 1999) and the cease of the syn-depositional faulting in the abandoned depobelt. Particularly sudden shifts are recorded by the uppermost alluvial sequence by the escalator regressive style (Knox and Omatsola, 1989). It is controlled by a continuing sediment supply and cessation of subsidence in a depobelt, recording occasional rapid seaward advances of alluvial sands. These regressive pulses are so rapid that it is impossible to recognize any age difference across the width of many depobelts. Modern 2-D and 3-D data from the outer shelf and upper slope of the southern Niger Delta lobe indicate that depobelts initiated as a deposition behind a foldbelt, which was subsequently buried by the seaward-advancing delta wedge.

Extensional faults with the largest throws are located along the boundaries among depobelts. These boundaries are frequently a location of the highly deformed shale-cored structures (Damuth, 1994; Cohen and McClay, 1996). These normal faults initiate when contraction in a distal portion of the depobelt starts to decease (Hooper et al., 2002). Throws of the boundary normal faults are so big that a hanging wall paralic sequence is usually younger than a footwall paralic sequence. Extensional faults inside depobelts do not have such large throws. However, they are large enough to control a unique sand-shale distribution pattern in an associated basin. Each depobelt consists of its own family of temporally and spatially linked fault systems, consisting of normal faults, relay ramp faults, and linking strike-slip faults located behind a system of toe thrusts (Figure 16.21b; Hooper et al., 2002). Bounding and internal extensional faults of each depobelt have listric geometries that detach at various levels of the overpressured marine shale, depending on the position in the gravity glide. A detachment level progressively deepens seaward from the Tertiary levels typical for the rear portion of the glide to the levels inside the Upper Cretaceous shale typical for the center of the gravity glide and shallow again toward the frontal thrusts.

Distal portions of such a depobelt contain southward-dipping flanks with bedding tilted toward landward-dipping growth faults (Hooper et al., 2002). The boundary itself is formed by a seaward-dipping normal fault. Blocks among the most distal landward-dipping faults and boundary seaward-dipping faults are thought to be ridges formed by mobile shale withdrawn from beneath depobelts. Hooper and colleagues' (2002) detailed seismostratigraphic study indicates that the front of each depobelt was initiated by contraction, leaving a set of thrust faults in the sedimentary record. They deform pre-gliding sedimentary units and some of the subsequent syn-contractional units. They are later cut by seaward-dipping normal

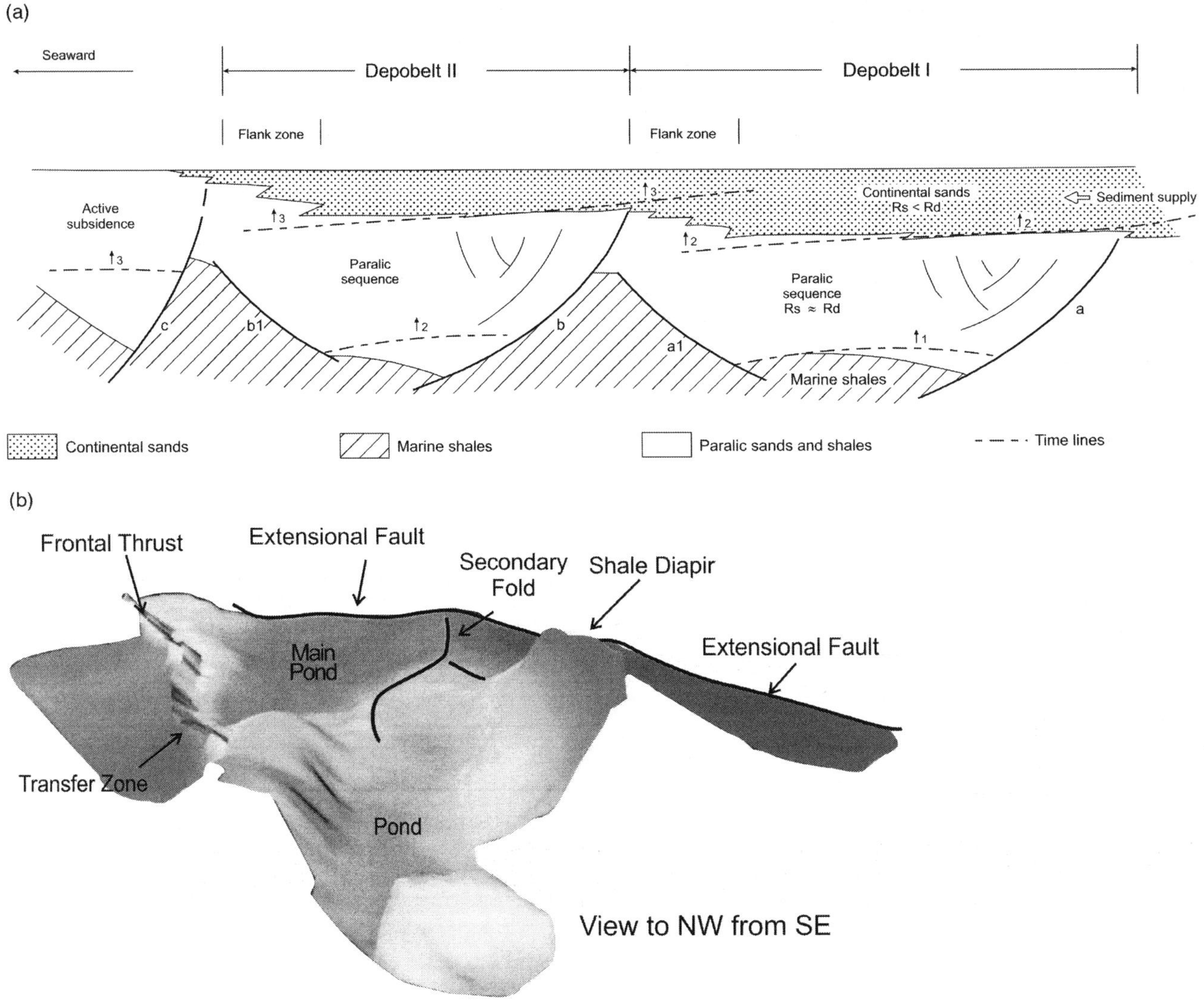

Figure 16.21. (a) Schematic diagram illustrating the successive development of depobelts (Doust and Omatsola, 1990). T1 is the time when subsidence and paralic deposition commence in Depobelt 1, mostly between faults a and a1. T2 is the time of subsidence cease in the proximal portion of Depobelt 1, the time of rapid advance of the alluvial wedge. During this time, the main paralic deposition occurs in Depobelt 2, mostly between faults b and b1. T3 is the time when subsidence ceases in the proximal part of Depobelt 2 and the area is filled by the rapidly advancing alluvial wedge. Paralic deposition shifts to a new depobelt seaward. Subsidence in distal flanks of depobelts continues longer than in proximal flanks. Paralic deposition lasts for about 5–10 Ma in each depobelt, reaching a thickness of about 5 km. (b) 3-D model of the base of the ponded section (Hooper et al., 2002). The structural control comprises a series of en echelon, thrust-cored folds and associated ponded slope basins, a set of nested, overlapping, extensional growth faults, and a shale-cored diapir. The crestal height of the folded structure is variable and is at minimum in a prominent transfer zone. The maximum height of the fold crest above the adjacent main pond is about 3 km. A second fold, with lower relief, is present behind the main fold and links into the shale diapir. The prominent main pond is behind the frontal fold and is separated by a flexure from a secondary pond adjacent to the shale diapir. The height of the diapir crest above the adjacent secondary pond is about 5 km.

faults related to the extension in a younger depobelt, located seaward from the one that underwent the stop of extension and shale withdrawal.

The internal structure of the Niger Delta consists of roughly eleven depobelts, which differ in age and have complex 3-D relationships according to temporal and spatial changes in advance directions of various delta portions (Figure 19.14, Table 16.2).

Diapirism becomes more important with the thickness of the gravitationally gliding sediments in depobelts, producing seaward and landward thickening wedges controlled by normal faults located seaward and

Table 16.2. Ages of recognized depobelts (modified from Doust and Omatsola, 1990; Hooper et al., 2002).

Depobelt	Age of paralic sequence	Age of alluvial sequence
Northern Delta	Late Eocene–Early Miocene	Early Miocene
Greater Ughelli	Oligocene–Early Miocene	Early Miocene
Central Swamp I	Early–Middle Miocene	Middle Miocene
Central Swamp II	Middle Miocene	Middle Miocene
Coastal Swamp I	Middle–Late Miocene	Late Miocene
Coastal Swamp II	Middle–Late Miocene	Late Miocene
Offshore	Late Miocene	Latest Miocene
Distal Belt	no stratigraphy available	no stratigraphy available
Slope Depocenters	no stratigraphy available	
Diapir Belt	no stratigraphy available	
Toe Thrust	no stratigraphy available	

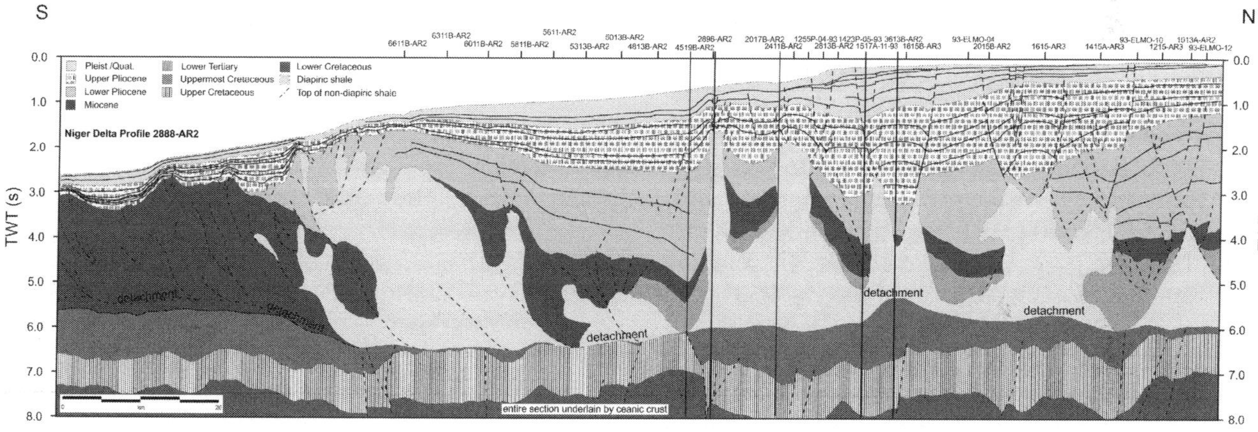

Figure 16.22. Out-of-sequence strength adjustments of the gravity-gliding wedge to make over basement steps formed by fault-bounded blocks of oceanic crust.

landward of the diapir. These wedges thin over the diapir and get deformed by collapse-related normal faults. Each depobelt preserves its system of frontal thrusts and linking strike-slip faults, which are, however, usually difficult to recognize in seismic images of older depobelts. Exceptions due to simpler geological settings and/or high-quality 3-D seismics are known from the outer shelf and upper slope of the southern Niger Delta lobe, in a belt of slope depocenters (see Hooper et al., 2002) and in frontal areas. Frontal present-day areas further indicate a complex relationship between the most external extensional depocenters and adjacent toe thrust. Some depocenters recorded activity of strike-slip faults linking various segments of the inhomogeneously shortened toe thrust system or recorded episodes of inversion preceding shortening events in the toe thrust. This inversion occurred as out-of-sequence adjustments of the gliding wedge to strengthen enough to overcome basement steps formed by fault-bounded blocks of oceanic crust (Figure 16.22; Nemčok et al., 2004b).

Each of the eleven depobelts contains seaward boundary thrust faults, and landward boundary-, seaward boundary-, internal seaward-dipping-, and local antithetic extensional faults, and crestal collapse faults.

Southern boundary thrust faults are preserved in the oldest pre-gliding and early syn-gliding sediments. They represent a record of a paleo-fold belt developed at early stages of the gravity gliding in each depobelt. At their time of growth, they were kinematically linked with both translational and extensional domains of the depobelt.

Landward boundary extensional faults represent crescentic segments of the landward boundary of the depobelt. They were propagated through the preexisting thrust faults of the older depobelt during the accelerating extension in the proximal portion of the younger depobelt. They are listric and seaward-dipping.

Seaward boundary normal faults represent crescentic segments of the seaward boundary of the depobelt, formed later after the deactivation of frontal contraction. They are listric and landward-dipping.

Internal seaward-dipping faults do not have large enough throw to perturb the seaward-dipping stratification of blocks bounded by major faults. They have arcuate geometries and play a role in local extensional accommodation.

Local antithetic normal faults are landward-dipping and related to major seaward-dipping faults forming their local accommodation structures. Crestal collapse normal faults are formed in thicker and younger depobelts. They are subparallel to axes of anticlines above the shale diapirs. They are usually listric but steeper than boundary faults of depobelts.

Extensional faults are the youngest fault system in depobelts and define their block-like structural style. Large-scale blocks frequently form roll-over anticlines but can be also represented by tilted flat blocks. Some of the roll-over anticlinal crests are dissected by crestal-collapse normal faults.

As discussed earlier, different combinations of extensional faults are characteristic for different depobelts. Depobelts close to the landward delta limit contain the oldest faults, progressively steeper seaward and relatively regularly spaced. Central depobelts are clearly defined and contain large-extension listric faults. Distal depobelts seem to be the most complex, affected by the growth of ridges and diapirs. It is these depobelts that contain crestal collapses, back-to back faults of opposite dips, and large seaward-dipping blocks cut by landward-dipping normal faults. It is also these depobelts that preserve the thrust fault systems of their paleo-thrustbelts best.

The Niger Delta also is the area where we can discuss the interaction of gravity-driven normal faults and deposition in detail. We can do it with the help of the analog material experiment originally done for the Miocene Imo River field (Van Heist et al., 2002), which simulated four sea-level change cycles, each of them showing a similar timing of the drainage evolution on the modeled shelf-margin delta.

All modeled highstand episodes produced delta progradation. During the subsequent sea-level fall, several canyon incisions developed in the exposed delta front, eroding toward the highstand delta apex. The deposition of the transgressive systems tract followed the retreating shoreline. At late rise to highstand, the retrogradation changed into aggradation, resulting in the restoration of the incised highstand delta on the footwall.

Despite numerous geological differences, the model and data from the Imo River field share common features, especially in the delta and fluvial architecture in both the footwall and hanging wall blocks. For example, the model shows incised valleys on the footwall block of the growth fault, which transported sediment to the hanging wall lowstand fan at times of the sea-level fall and lowstands. The seismic imaging in the field indicates a presence of similar valleys in the footwall in the middle-upper Miocene section (Van Heijst et al., 2002). The model also manages to reproduce the hanging wall preservation of all tracts, with coarse-grained deposits formed during the eustatic fall, lowstand, and early rise of the sea level. It also manages to reproduce the footwall succession, which consists of the stacked late transgressive and highstand sediments, which were repeatedly subject to erosion and sediment bypass as the shoreline moved seaward from the growth fault during lowstands.

The important outcome of modeling is the recognition of diachronous erosion and deposition along the stream profile as a consequence of the sea-level change (see also Van Heijst and Postma, 2001). The stream profile at maximum sea-level lowstand is an analog to the type 1 unconformity (sensu Vail and Todd, 1981), characterized by subaerial exposure and associated erosion from down-cutting stream. During the subsequent sea-level rise, a deposition develops strongly diachronically, commencing on the shelf and delta plain, while erosion continues in a landward direction. This diachroneity is further enhanced in growth fault scenarios (Figure 7.28, Table 7.1). For example, the upper delta and fluvial domain incisions continue during early sea-level rise and coincide with the deposition of the transgressive systems track in the hanging wall. Thus, the upstream part of the unconformity on the footwall does not record the lowstand hiatus but originates from early rise and becomes progressively younger up dip.

According to the interplay model of eustatic sea-level change and local subsidence (Posamentier and Allen, 1993), the interplay determines the type of unconformity and lowstand basin physiography, which would, in turn, control the morphology of the following transgressive systems tract. The interplay in the Imo River field resulted in three type 1 unconformities and one type 2 unconformity in the hanging wall and four type 1 unconformities in the footwall.

The type 1 unconformity is characterized by a few cross-shelf bypass valleys that act as point sources for lowstand fans (Figure 7.28a). These valleys transfer significant volumes of freshly eroded footwall sediment to the lowstand fans on the hanging wall, in addition to the sediment supplied by the fluvial valley. The type unconformity 2 is characterized by little or no incision of river valleys into the delta plain and shelf, so that the feeder system of a delta can potentially behave as a line source (Figure 7.28b).

17 Sealing characteristics

Introduction

All traps are transient (Sales, 1997). Given enough time, their seals undergo a certain amount of leakage. What makes seals different from the other lithostratigraphies in rift or passive margin settings are the capillary pressure and frictional resistivity force terms from Equation (17-1). They are higher than these terms in other lithostratigraphies.

The two forces that oppose migration can be expressed using a balance of migration driving and opposing forces (Mandl, 1988):

$$(\rho_w - \rho_o)g\sin\theta l + \Delta h = (P_{dr} - P_{imb}) + (u_o l\sin\theta\eta_o)/k_o, \qquad (17\text{-}1)$$

where ρ_w and ρ_o are the water and oil densities, g is the acceleration of gravity, θ is the angle between migration and horizontal directions, l is the oil column length, Δh is the hydraulic gradient, P_{dr} and P_{imb} are the drainage and imbibition capillary pressures, u_o is the Darcy velocity of the oil front, η_o is the oil viscosity, and k_o is the effective permeability of porous medium to oil. The terms on the left side express the buoyancy force and hydraulic gradient. The terms on the right side express the capillary pressure and frictional resistivity force.

Seals can fail under certain conditions, allowing the hydrocarbons to leak. This is indicated by the worldwide statistics on pooled hydrocarbons (Ulmishek and Klemme, 1990), which show that more than 80 percent of the world's oil and gas reserves were trapped since the Aptian, and almost 50 percent since the Oligocene, indicating that the older the trapping the better the seal required.

The following text will discuss conditions of such seal failure, and the different types of seals with respect to:

1) the medium providing the sealing;
2) seal geometry with respect to trap; and
3) sealing control,

and finish with a discussion of seal examples from rift and passive margin settings.

Sealing medium

The sealing media can be divided into lithological seals and fault seals.

The data on lithologic seals document that the most common hydraulic seals are formed by evaporites, shales, and sandstones (Hunt, 1990). Mineralized carbonates, such as the Cretaceous chalk in the North Sea province, are less common. In fact, mineralization further enhances the sealing properties of the respective lithology. Carbonate seals are frequently mineralized with silica. Shale seals are most typically mineralized with calcite (Hunt, 1990).

Statistical data on the top seals of apparently unfractured structural traps (Nederlof and Mohler, 1981) indicate that the original lithology is the most important factor in control of a good lithological seal. The overwhelming majority of known effective seals is formed by evaporites, fine-grained clastic sediments, and organic-rich sediments (Downey, 1994). This is because the other factors controlling a good seal, except suitable physical properties, are:

1) its lateral continuity;
2) its relative ductility; and
3) its areal extent within the basin.

Relative ductility is important because of the rock's deformational response to increased differential stresses. It varies with pressure and temperature. For example, evaporites form effective seals at depths below 1 km while they are relatively brittle and prone to fracturing at depths above 1 km. The majority of the largest hydrocarbon pools rely on an evaporite seal. The world's 176 giant gas fields almost all depend on evaporite or shale seals (Grunau, 1981).

Compaction and shear can enhance the effectiveness of the seal. Figure 17.1 shows how a sediment compaction in elastic and plastic deformation regions below the failure envelope decreases sediment porosity with increasing shear and mean effective stresses. This indicates that different tectonic scenarios decrease the matrix permeability of the seal by compaction due to progressive burial. The low-angle shear along detachments in rift settings and gravity glide systems of passive margins further modifies the seal texture.

Compaction modifies the permeability structure of the seal horizon, as indicated by vertical and horizontal hydraulic conductivity measurements in oedometrically consolidated clays (Basak, 1972; Tavenas et al., 1983; Leroueil et al., 1990). For example, kaolinite samples consolidated to 500kPa indicate permeability anisotropy of 2–3 (Al-Tabbaa and Wood, 1987) whereas natural silty clays indicate permeability anisotropies of 1.0–1.3. Both consolidation experiments indicate that the permeability structure is controlled by the resultant fabrics.

Another set of experiments on permeability structure changes are shear experiments made at effective stresses of about 4 MPa (Dewhurst et al., 1996a, 1996b). The results document distinct permeability anisotropy developed by shearing. It was added on permeability anisotropy developed by preexisting consolidation (Figure 17.2). The resultant shear-perpendicular hydraulic conductivity was distinctly and consistently lower than the shear-parallel conductivity. This conductivity difference between a horizontal and vertical direction was observed to increase with decreasing void ratio (Dewhurst et al., 1996a, 1996b).

Optical microscopy at low magnification allows us to see that consolidated clay is characterized by a subhorizontal alignment of silt particles (Figure 17.3a). We can occasionally see grain breakage in quartz particles. Higher magnification allows us to observe clay grains wrapping around silt particles and their tight packing in pore throats. In the pore throats and in the intergrain space are domains of strong alignment, mainly with subhorizontal trend. Uniaxial consolidation alone does not produce any measurable anisotropy in hydraulic conductivity in silty clay despite the anisotropic grain fabric developed by consolidation. The experiments indicate that pure clay has a maximum potential for fabric alignment. They show that the upper limit for the hydraulic conductivity anisotropy of pure clay is 2–4 (Al-Tabbaa and Wood, 1987; Leroueil, 1988; Leroueil et al., 1990).

Optical microscopy at low magnification shows that sheared clay is characterized by a thin detachment surface parallel to the shear zone boundary (Figure 17.3b). The vicinity of the shear zone contains intense grain alignment parallel to the slip surface. Strain localization has occurred at and near the slip surface, resulting in a highly deformed, chaotic fabric, in comparison to that of the host sediment. An alignment of clay plates and irregularly shaped quartz particles is visible at an

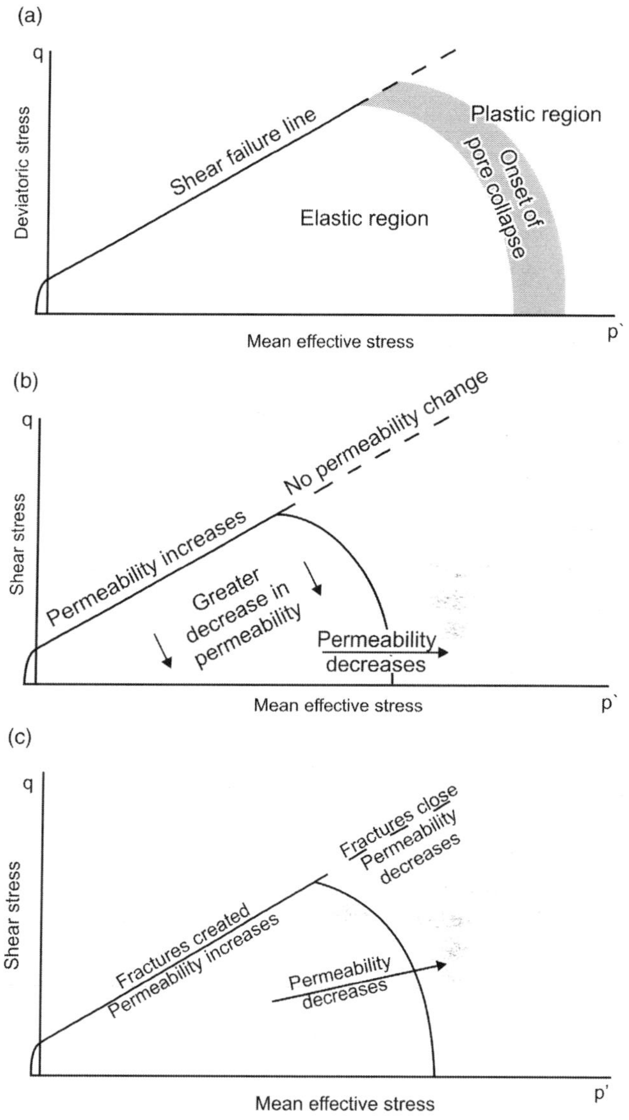

Figure 17.1. (a) The sediment (chalk) model (Joint Chalk Research, 1996). Vertical axis indicates deviatoric stress, $q = \sigma_1 - \sigma_3$, and horizontal axis indicates mean effective stress, $p' = (\sigma_1 + \sigma_2 + \sigma_3)/3$. Region where sediment behaves elastically is bounded by shear failure curve and yield surface. The elastic region along the negative mean effective stress line indicates tensile failure. The shear failure line is indicated with fracturing and involves a localized distortion and breakdown of structure along the failure plane. Yield surface is associated with pore collapse and involves pervasive breakdown of the structure, specifically destruction of cement bonding. The first yield stress is the point where the stress/strain curve departs from linearity. (b) Matrix permeability changes (Joint Chalk Research, 1996). Matrix permeability changes more for trajectories closer to the hydrostatic path, which runs from the origin along the p' axis. The permeability change when passing the shear failure line depends on whether the sample was in elastic or plastic region. (c) Fracture permeability changes (Joint Chalk Research, 1996).

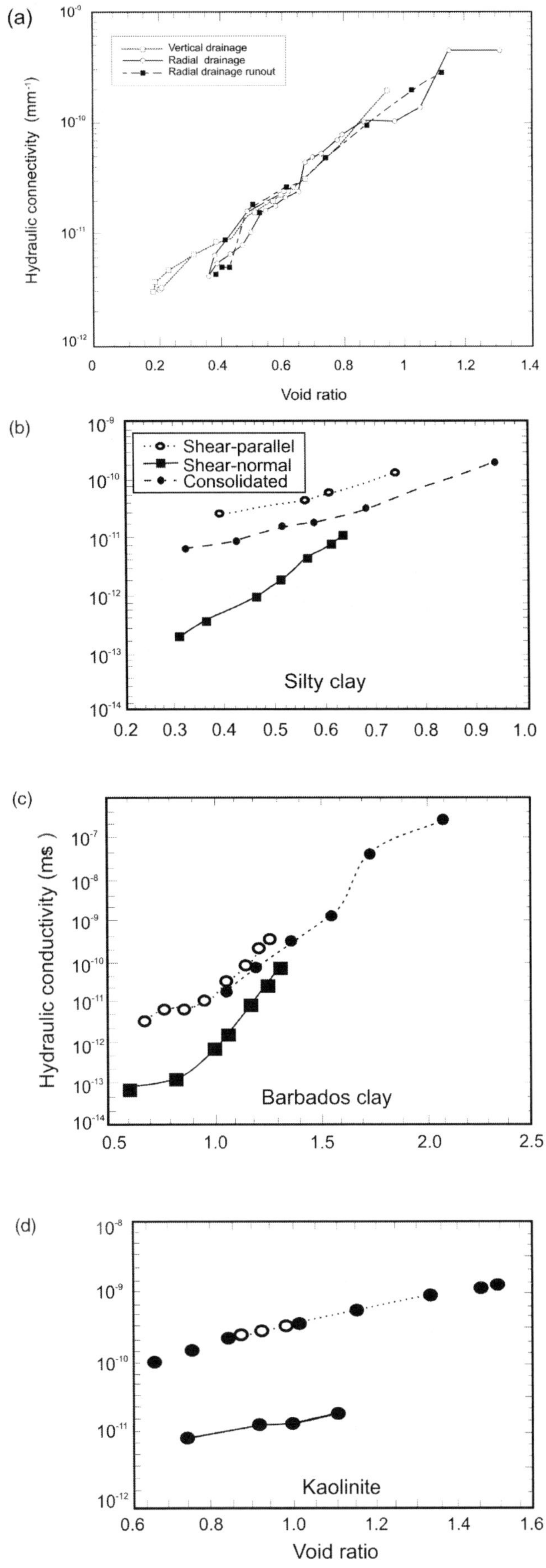

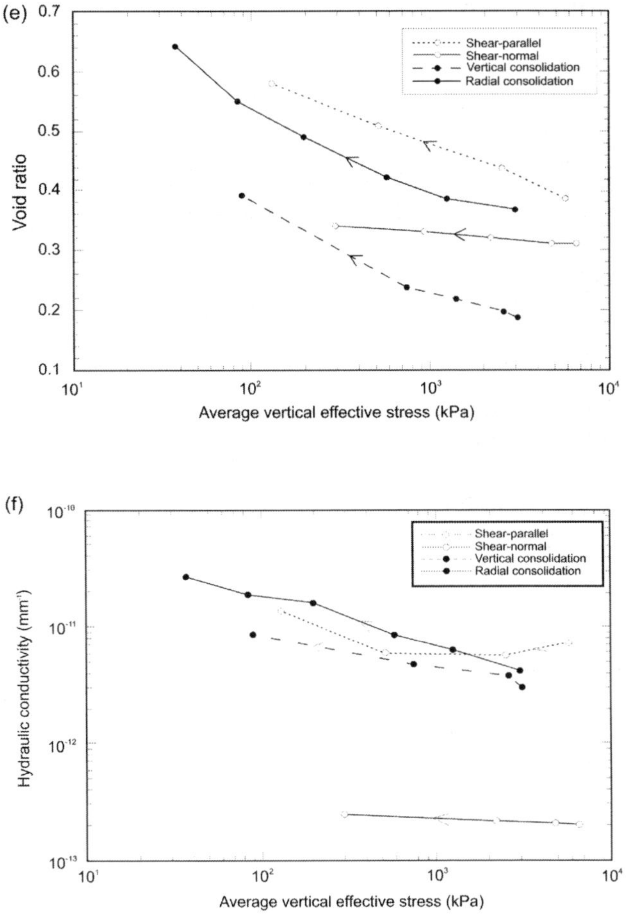

Figure 17.2. (a) Permeability anisotropy in consolidated and sheared clay (Dewhurst et al., 1996a). No discernible anisotropy is exhibited by the consolidated samples, but the sheared silty clay has an anisotropy that increases from three to sixteen with decreasing void ratio. (b) Hydraulic conductivity/void ratio for sheared and consolidated silty clay samples (Dewhurst et al., 1996a). Shear parallel/shear normal hydraulic conductivity ratio increases from three to sixteen with decreasing void ratio. (c) Hydraulic conductivity/void ratio for sheared and oedometrically consolidated clay (Dewhurst et al., 1996a). Shear parallel/shear normal hydraulic conductivity ratio increases from fifteen to thirty-five with decreasing void ratio. (d) Hydraulic conductivity/void ratio for sheared and consolidated kaolinite (Dewhurst et al., 1996a). Shear parallel/shear normal hydraulic conductivity ratio is approximately constant at twenty-five for all void ratios. (e) Decreasing effective stress conditions show no enhanced dilation of the sheared clay above that of the consolidated clay (Dewhurst et al., 1996a). (f) Sheared clays exhibiting no conductivity increase over that shown by consolidated clays under decreasing effective stress conditions (Dewhurst et al., 1996a).

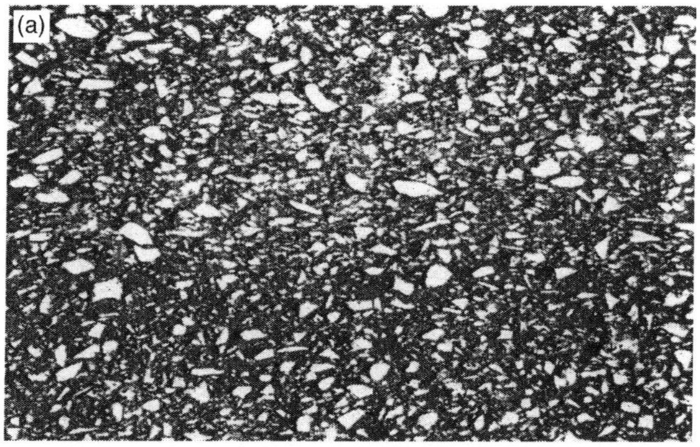

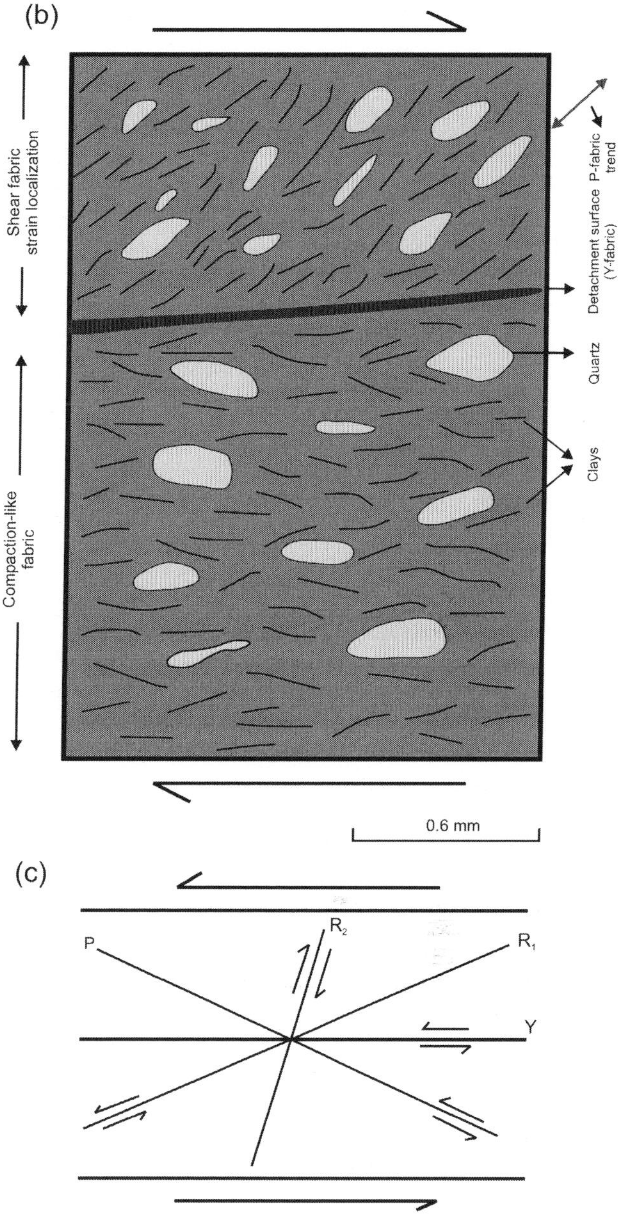

Figure 17.3. (a) Photomicrograph of a thin section of consolidated silty clay illustrating the alignment of the long axes of irregularly shaped silt particles normal to the direction of applied stress (Dewhurst et al., 1996b). (b) Line drawing of the thin section of sheared clay, illustrating the development of a detachment surface near the top of the sample (Dewhurst et al., 1996b). Above the detachment, the long axes of some silt and clay particles are aligned oblique and synthetic to the shear zone. Beneath the detachment, the long axes of the silt grains are oriented normal to the direction of applied stress. The sense of shear is dextral throughout the figure. (c) Principal fabric elements of shear zones (sensu Logan et al., 1979). Y-shear indicates late developing, through-going shears associated with slip at residual strength. P-fabrics are flattening fabrics associated with an inclined maximum principal compressive stress direction during shearing. R1 and R2 fabrics are Riedel shear orientations.

angle of 35–35° to the slip surface, antithetic to the shear zone. The rock in the vicinity of the shear zone consists of somewhat random fabric, apart from the small domains of intensely aligned clays and a subhorizontal orientation of quartz particles, being very similar to the fabric of consolidated clay. Secondary electron imaging shows that strain is spread through the rock, generating the synthetic Riedel fabric of streaked-out clay aggregates.

Similar shear fabric develops in quartz and calcite fault gouges (Logan et al., 1979). Alignment in these shear zones (Figures 17.2b–d) reduces hydraulic conductivity across them considerably (Dewhurst et al., 1996a; Figures 17.2f, 17.3b–c). Experiments show that an effective barrier to fluid migration can be produced in sheared clays even at high silt content (Dewhurst et al., 1996a, 1996b). Grain alignment near shear zones is unlikely to contribute to the overall anisotropy of the sheared silty clays (e.g., Arch and Maltman, 1990; Brown and Moore, 1993). The strain localization during residual shear produces a local volume loss and permeability decrease (Dewhurst et al., 1996a). The alignment of this thin barrier in a direction parallel to the shear and the strain localization associated with this slip surface explains the levels of absolute hydraulic conductivity and permeability anisotropy magnitudes of three to sixteen. In nature, the slip zone can develop either a parallel or an anastomosing array of similar surfaces, which together compose an effective barrier to fluid movement (e.g., Skempton, 1985).

The clay content and composition control the magnitude of the shear-parallel/shear-perpendicular hydraulic conductivities (Figures 17.2b–d). The smectite-rich clay attains the highest anisotropy ratio. A lower value than that characterizes the hydraulic conductivity anisotropy of kaolinite. The silty clay has the lowest anisotropy and clay content. The mentioned difference among hydraulic conductivity anisotropies of the three lithologies is controlled by different deformation fabrics that relate to different clay and silt contents (Lupini et al., 1981). Kaolinite and the smectite-rich clay have clay contents of 99.8 percent and less than 90 percent, respectively. Their fabrics are dominated by sliding shear elements. The silty clay contains 50 percent clay and 50 percent silt-sized quartz. Both sliding shear and turbulent processes are present.

Shale horizons are excellent candidates for seals because of the discussed reduction in their already low permeabilities by shearing. Such scenarios are known from the décollements of thrustbelts (Nemčok et al., 2005a). It would be logical to expect analogs from detachment faults of the large gravity glide systems at passive margins.

The fault seals are mostly generated by the late-stage changes in the fault zone porosity by various processes such as:

1) grain boundary sliding;
2) fracturing;
3) dissolution;
4) corrosion and alteration; and
5) cementation.

Displacement-controlled fault rock development during the pre- and post-seismic stages of the earthquake cycle involving aseismic creep, discussed by Nemčok and colleagues (2005a), is responsible for the reduction of the high permeability window. The effectiveness of various processes involved in this reduction determines the efficiency of fault sealing. Although many seals involve a combination of these processes, it is often possible to determine the dominant one. These processes of porosity reduction operate at different rates and, thus, each allows the high permeability window associated with faulting (Knipe, 1993) to remain open for different time periods. Therefore, the sealing properties of the fault zone can be understood as developing at different rates. In the development of clay smears, maintaining a dilation level or porosity in the grain aggregate high enough to allow fluid flow requires maintaining high fluid pressure. Reduction of this pressure tends to cause an abrupt permeability collapse. Similar reduction in fluid pressure in quartz aggregate tends to result in subtle permeability decrease, due to the higher strength of the grain aggregate. In this case, the permeability reduction will be more protracted because it is controlled by diffusion mass transfer processes.

Depending on controlling mechanisms, the fault seals can be divided into three broad categories (Knipe, 1992):

1) collapse seals;
2) cement seals; and
3) juxtaposition seals.

Collapse seals

Collapse seals develop by the pore volume decrease caused by grain reorganization due to grain boundary sliding, grain deformation, fracturing, and dissolution. They also include smear seals (Knipe, 1992).

Figure 17.4a shows an FMI image of the clay smear seal developed along the fault zone deforming eolian sandstone in the North Sea (Adams and Dart, 1998). The fault rock developed by linkage of propagating deformation bands, in accordance to well core observations from the same area. The damage zone sandwiching the fault core contains a series of deformation bands, which can be interpreted from increased density and image more resistive than that of the host sandstone. The fault core comprises clay, as can be implied from its high-conductivity character in an FMI image and open-hole response.

The analysis of anastomosing zones in the clay smears suggests that the associated fluid migration is episodic and related to the migration of dilation-displacement-collapse events along the fault zones (Knipe, 1986; Anderson et al., 1994a, 1994b; Losh et al., 1999, 2002). Permeability along such phyllosilicates-rich faults is associated with high dynamic porosity and permeability, which decays after the fault activity terminates (Knipe et al., 1991).

Natural examples from the North Sea document a grain size reduction along faults as a collapse seal mechanism (Figure 17.4b). Figure 17.4b shows an FMS image of the series of deformation bands developed in sediment prone to strain hardening. The

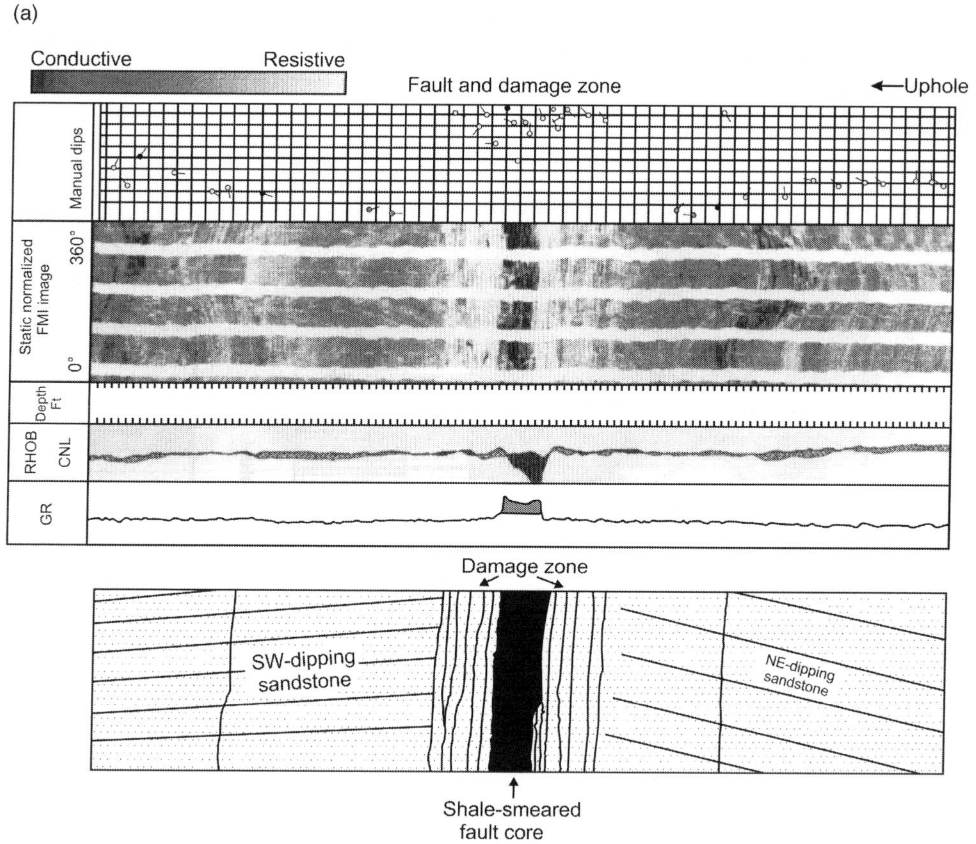

Figure 17.4. (a) Shale smear fault zone in eolian sandstone in horizontal well in FMI image (Adams and Dart, 1998). The central fault zone shows a fault core 1.2 m wide, interpreted as clay-smeared core on the basis of the high conductivity and open-hole response. The damage zone on either side of the fault core is more resistive than the surrounding sandstone, with elevated density log readings, and is interpreted to consist of a series of deformation bands, with possible associated cementation. Such features were seen in cored intervals of nearby wells. (b) Grain-size reduction fault seal in clastic rocks in FMS image (Adams and Dart, 1998). It comprises a zone of deformation bands formed by repeated strain causing widening of the deformation zone by strain hardening. The internal banded structure of the zone of deformation is its typical characteristic. Cement may be present, but the open hole logs do not show much change, suggesting that any decrease in porosity is primarily due to grain-size reduction.

strain hardening causes that to propagate a new band, which becomes energetically more favorable than maintaining the slip along already-existing bands relatively early in their development history (see Mandl, 1988). Although this example does not indicate any significant associated cementation, the late-stage collapse processes in cataclastic fault zones are frequently associated with dissolution and cementation (Knipe, 1993). The importance of such a diffusive mass transfer or pressure solution mechanism (Rutter, 1976, 1983; Gratier and Guiguet, 1986; Spiers and Schutjens, 1990) to the final stages of the fine-grained cataclastic zone development has been observed in several studies (e.g., Rutter and White, 1979; Stel, 1985; Knipe, 1989). This mechanism is one of the controls of the stick-slip behavior of faults (Angevine et al., 1982).

There are three scenarios where enhanced diffusion mass transfer develops with a decrease in the strain rate, including (Knipe, 1992):

1) quiescence intervals in faulting cycles;
2) fault activity pattern changes when the preexisting faults cease their activity and a new fault starts to accommodate the block movements during the fault array evolution; and
3) termination of the regional fault activity.

While the first scenario has a temporary character, the latter two scenarios tend to be permanent. After the relative quiescence, the compacted cataclasite is frequently fractured by subsequent fault events. The second and third scenarios are frequently characterized by progressive compaction, and associated compaction may be enhanced by sediment loading.

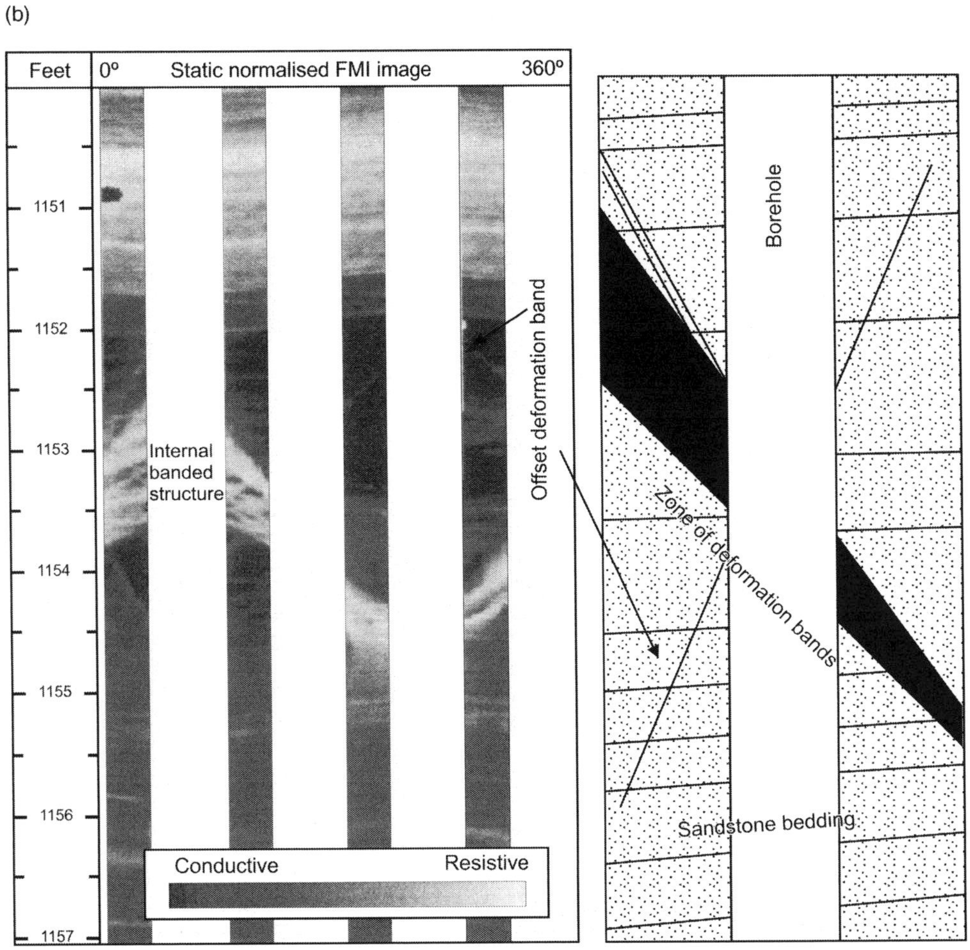

Figure 17.4. (*cont.*)

The permeability collapse during the slow or decreasing displacement rate represents the final stage of the high permeability decay associated with fault development history (Knipe, 1993). Note that a quantitative description of this fault development stage is difficult because of the interaction of numerous factors in control of the diffusive mass transfer rate, although rough estimates are possible (e.g., Gratier and Guiguet, 1986; Spiers and Schutjens, 1990). If the diffusion mass transfer creep rates in fine-grained cataclastic quartz aggregates at temperatures of 150–200°C are approximately 10^{-10} s^{-1} to 10^{-11} s^{-1} for a grain size of about 10 μm (Rutter, 1976) and shear strains of about 0.5 are needed to significantly reduce the porosity of the fractured aggregate, then about 100–1,000 years are needed for the associated fault permeability collapse (Knipe, 1993). This estimate indicates that the time period for fluids to use the permeability structure of the compacting cataclastic fault rock can be at least two orders of magnitude longer than the time period of the after-slip stage (Knipe, 1993). Analyses of aseismic creep events, which are about one to two hours long, indicate that after-slip and associated creep events are not caused by diffusion mass transfer, which will dominate with time (Wesson, 1988).

The interaction of the fault rock with migrating external fluids can have an important impact on its permeability structure as indicated by sandstone diagenesis studies in southern Utah (Bowen et al., 2007 and references therein). Also examples of widespread carbonate cement dissolution controlled by fault-hosted fluid flow creating secondary porosity are common in nature (Burley et al., 1989). When dissolution of various phases is focused along the fault zone, both permeability and strength of the fault rock can be affected. While the initial development of the secondary porosity increases the fault rock permeability, the associated large-scale dissolution may weaken the rock, causing the permeability structure collapse. The example of this mechanism comes from the Stillwater normal fault

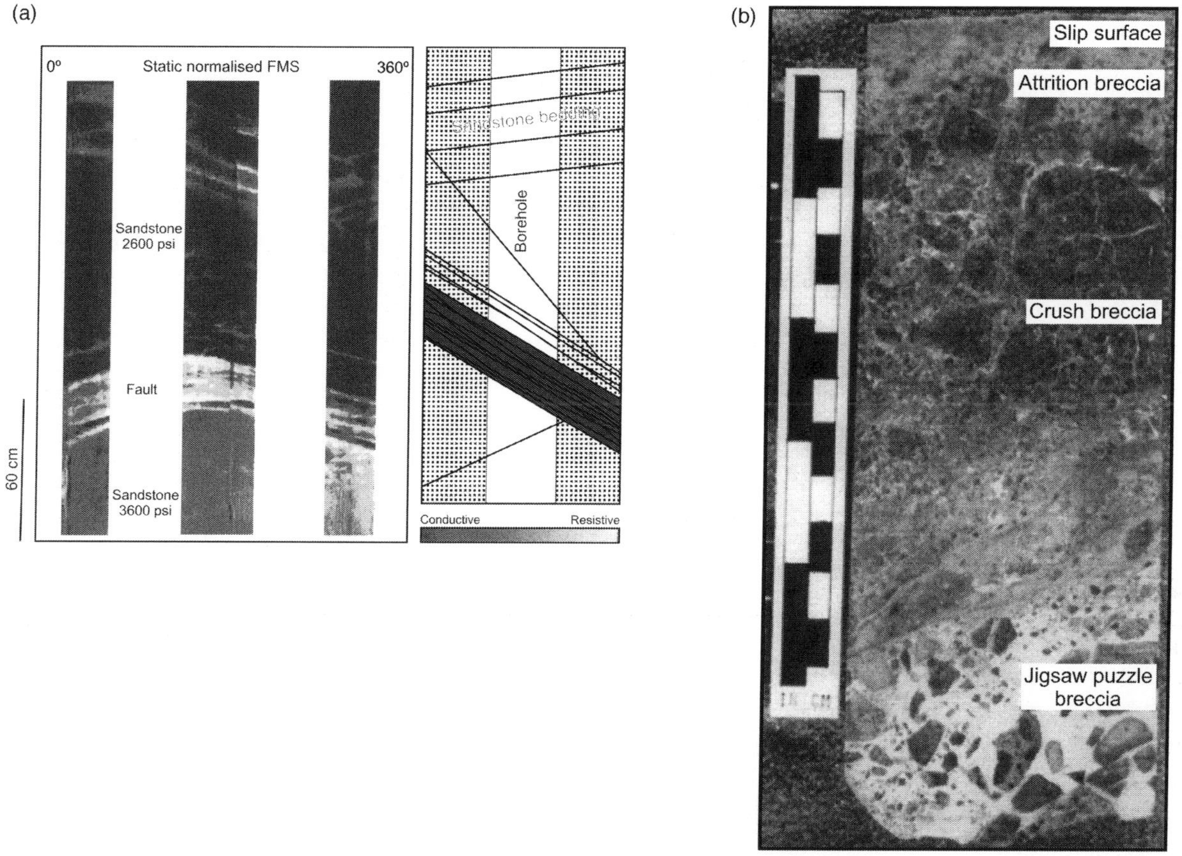

Figure 17.5. (a) Electrical FMS image across a cemented fault (Adams and Dart, 1998). This fault occurs in the moderate porosity sandstones of the Claymore Sandstone Member of the Kimmeridge Clay Formation, central North Sea. The open hole logs show a very strong response (high density, low neutron), indicating that the fault is cemented, and this is verified by the very strong resistivity of the fault zone. Formation pressure measurements indicate that this fault is responsible for a 1,000 psi pressure difference between the hanging wall and footwall. The darker color of the hanging wall suggests a water saturation higher than that in the footwall. (b) Photograph of a polished sample from the Mirrors locality fault core (Caine, 1999). The top of the sample is a slip surface. All three breccia types are exhibited. From top to bottom there is attrition breccia, crush breccia, and jigsaw puzzle breccia. The attrition breccia grades into crush breccia, indicating frictional wear of the crush breccia. The contact between the crush breccia and the jigsaw puzzle breccia is sharp, indicating the cross-cutting of the former by the latter.

zone of the Basin and Range province (Caine, 1999). The fault that formed in granodiorite can be characterized by the dissolution of feldspar and quartz and clay cementation in the fault core.

The feldspar dissolution in the fault core is particularly interesting for seal assessment because the produced clay phases may collapse to form smear seals during subsequent compaction. The sealing efficiency depends on the initial feldspar content and migration pathway of dissolving fluid inside the fault zone.

Cement seals

Cements seals form where the reduced flow across or along the fault is achieved by the precipitation of minerals in and around the fault zone (Knipe, 1992). The precipitation is linked to (Knipe, 1993):

1) fluid pressure changes associated with the faulting event;
2) temperature changes of the migrating fluid;
3) solubility changes arising from fault-induced mixing of fluids having different compositions; and
4) fluid-rock interactions involved in dissolution-precipitation processes of diffusive mass transfer.

These controlling factors are not confined to the late-stage evolution of the fault zone. They can be induced by the early parts of an earthquake cycle. When the first three factors operate, the fault zone behaves as an open system. The fourth factor can operate even when the fault behaves as a closed system without any distinct fluid flow.

Figure 17.5a shows an example of a cemented fault from the North Sea where the cement phase forms a significant portion of the fault zone volume. The fault is developed in sandstone, which tends to deform in a strain hardening fashion, especially when it has higher porosity. On the contrary, cemented faults developed in rocks, which undergo dilatancy, are characterized by cemented breccias. Such breccias involve a large amount of local dilation and may initiate where the wall-rock asperities and strengths cause or preserve dilation or where the fluid pressure magnitudes can maintain dilation.

The breccias can be formed by a variety of mechanisms, as documented from the Stillwater normal fault zone (Caine, 1999). Here the fault core consists of three different breccia types (Figure 17.5b):

1) attrition breccia;
2) crush breccia; and
3) jigsaw puzzle breccia.

Attrition breccia, located next to the principal displacement zone, is clast-supported, moderately sorted, and composed of angular and rounded clasts that result from frictional wear (Figure 17.5b). Most clasts are silicified and porphyroclastic, indicating multiple generations of deformation combined with fluid flow.

Further away from the principal displacement zone, the attrition breccia shows a gradational contact with crush breccia. The crush breccia is also clast-supported. It is poorly sorted, shows inter-clast and intra-clast veining, and is composed of angular and silicified host rock clasts (Figure 17.5b). It shows minimal evidence of cataclastic support, such as rounded grains or movement of grains with respect to each other. The presence of multiple veining generations indicates episodic deformation and fluid flow.

The jigsaw breccia, also known as implosion breccia (Sibson, 1986), forms pods with sharp boundaries located inside the volumes of the other two breccias (Figure 17.5b). These pods represent the last deformation and fluid flow event. The breccia texture is characterized by angular clasts that could be restored back together. The jigsaw breccia is matrix-supported, poorly sorted, highly silicified, and lacks veins and fractures. The ultra-fine grained matrix indicates that it may have precipitated rapidly. Porphyroclastic clasts in the jigsaw breccia indicate multiple episodes of brecciation and fluid flow.

Cementation events associated with fault activity are not necessarily confined to the fault zone itself. The host rock can also become cemented in the fault zone vicinity. In fact, a complex cement distribution in sandstone reservoirs has been recognized and assigned to fluid entries along faults (Flournoy and Ferrell, 1980; Burley et al., 1989; Bowen et al., 2007). Border zone seals may develop because of the damage zone in the case of a distributed and composite deformation zone (see Figures 1.25 and 16.16b) formed by linked fracture array surrounding the fault core. Border zone seals may also be related to preferential fluid penetration along high-permeability zones (Knipe, 1993). Border zone seals can increase in strength in comparison with the fault core and later fault reactivation may fail to breach the strengthened and cemented zone, which results in maintaining a seal (Knipe, 1992). An important consequence of the border seal's existence is its potential for channeling the fluid migration associated with later events along fault zones, which can result in fluids bypassing particular sealed horizons.

Juxtaposition seals

Juxtaposition seals are developed by fault displacement bringing reservoirs from one fault block against seals of the opposite fault block. Natural examples of juxtaposition seals come from the North Sea province (Figure 17.6). The most typical cases are juxtapositions of permeable limestone or sandstone against low-permeability mudstone.

The juxtaposition seal assessment requires detailed knowledge of the fault geometry and construction of fault-plane displacement and lithology juxtaposition maps from interpreted seismic data (Allan, 1989; Chapman and Meneilly, 1990). Figure 17.7a (Knipe, 1997) provides an example of such assessment. It is represented by normal fault with scoop-shaped geometry and displacement decreasing from the center toward the propagation tips. With increasing displacement, hanging wall layers are juxtaposed against progressively older footwall layers. For example, the offset along the cross-section FF' is large enough to juxtapose the youngest hanging wall layer A against the footwall layer D. Each juxtaposition detail along the fault plane, as Figure 17.7b suggests, is critical to sealing-fault versus open-fault behavior. The fault plane "lmno" is composed of areas that represent the parts of the fault plane where particular juxtapositions occur. For example, the triangle "uvw" represents the fault plane part where layer A is juxtaposed against itself across the fault. The parallelogram "vwxy" is the part of fault

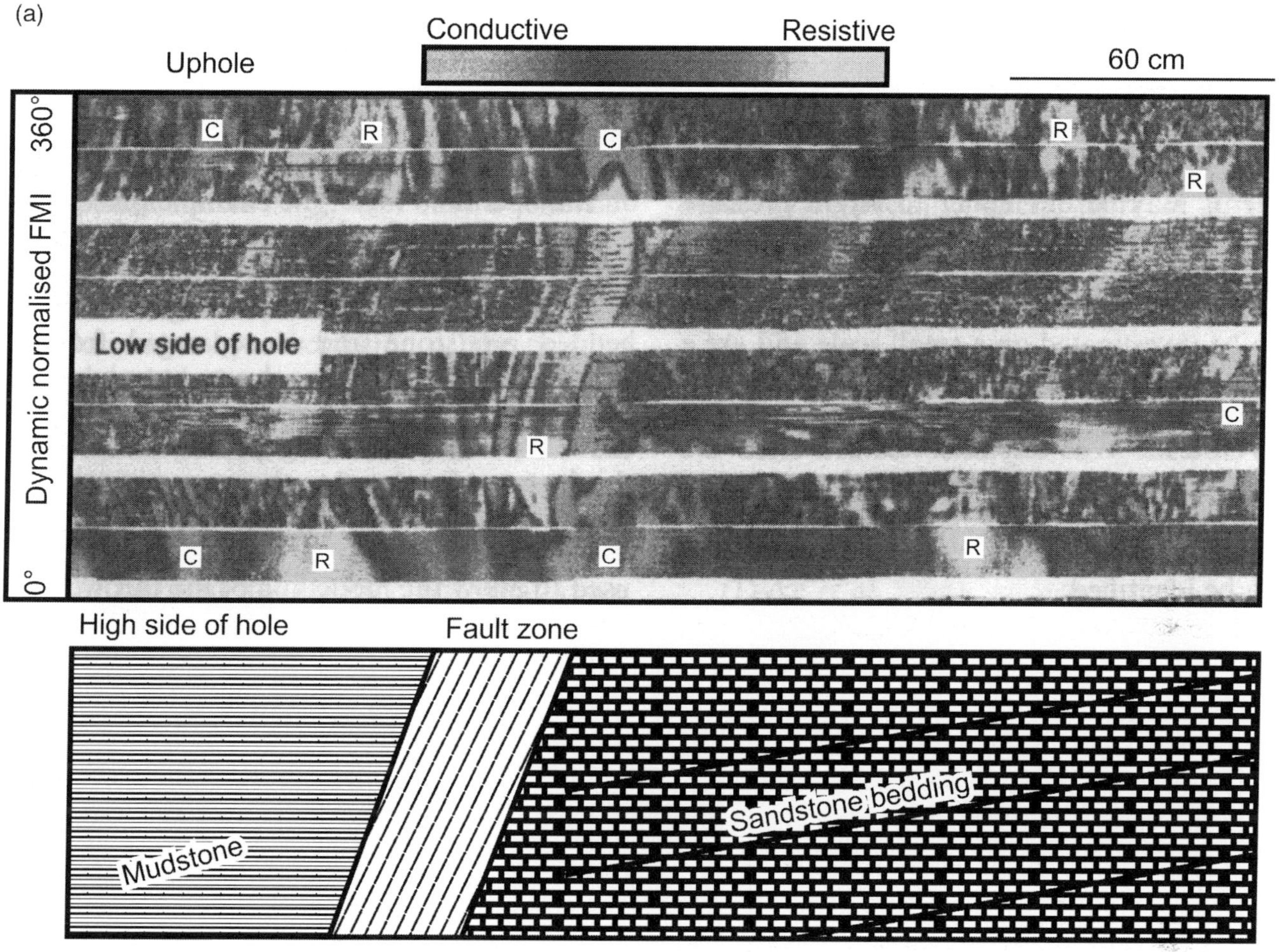

Figure 17.6. (a) Juxtaposition seal in chalk in a horizontal well in an FMI image (Adams and Dart, 1998). This fault juxtaposes impermeable mudstone with permeable limestone. The fault zone is approximately 60 cm wide, formed by anastomosing slip zones. Labels C and R show main conductive and resistive rock volumes. (b) Juxtaposition fault seal in clastics in UBI image (Adams and Dart, 1998). This fault juxtaposes mudstone in the footwall with sandstone in the hanging wall, resulting in an effective seal. The fault is a single plane. Its width is not resolvable from this image.

plane where hanging wall layer A is juxtaposed against footwall layer B.

At the end of the fault seal discussion, note that none of the sealing processes operates alone. For example, sealing by cementation may either be linked to a rapid pressure drop following the faulting event, inducing a rapid permeability collapse, or it may be controlled by a slower dissolution reaction rate.

Furthermore, the fault rock lithology and fault zone architecture are controlled by stress, strain, temperature, degree of lithification, the degree of fragmentation, fluid chemistry, and host rock lithology (e.g., Chester and Logan, 1986; Smith et al., 1990; Figure 17.8; Fisher and Knipe, 1998; Table 17.1; Caine, 1999; Figures 1.25, 16.16b). The higher the differential stress, the larger the damage to the deformed host rock is. In brittle rocks like dolomite, limestone, and granite, which are prone to dilatancy under stress relatively early, faults will have a character of either a distributed deformation zone or a composite deformation zone (Figures 1.25, 16.16b), depending on the amount of displacement along the fault. Faults developed in higher-porosity sandstones, which have a suppressed or no tendency to dilate during failure, have a character of either composite or localized deformation zones (Figures 1.25, 16.16b).

The fault development stage also controls how fault permeability structure evolves, as demonstrated by a faulting study in dolomite (Antonellini and Mollema, 2000). The study shows how the fault architecture in this very brittle rock evolves from a distributed deformation

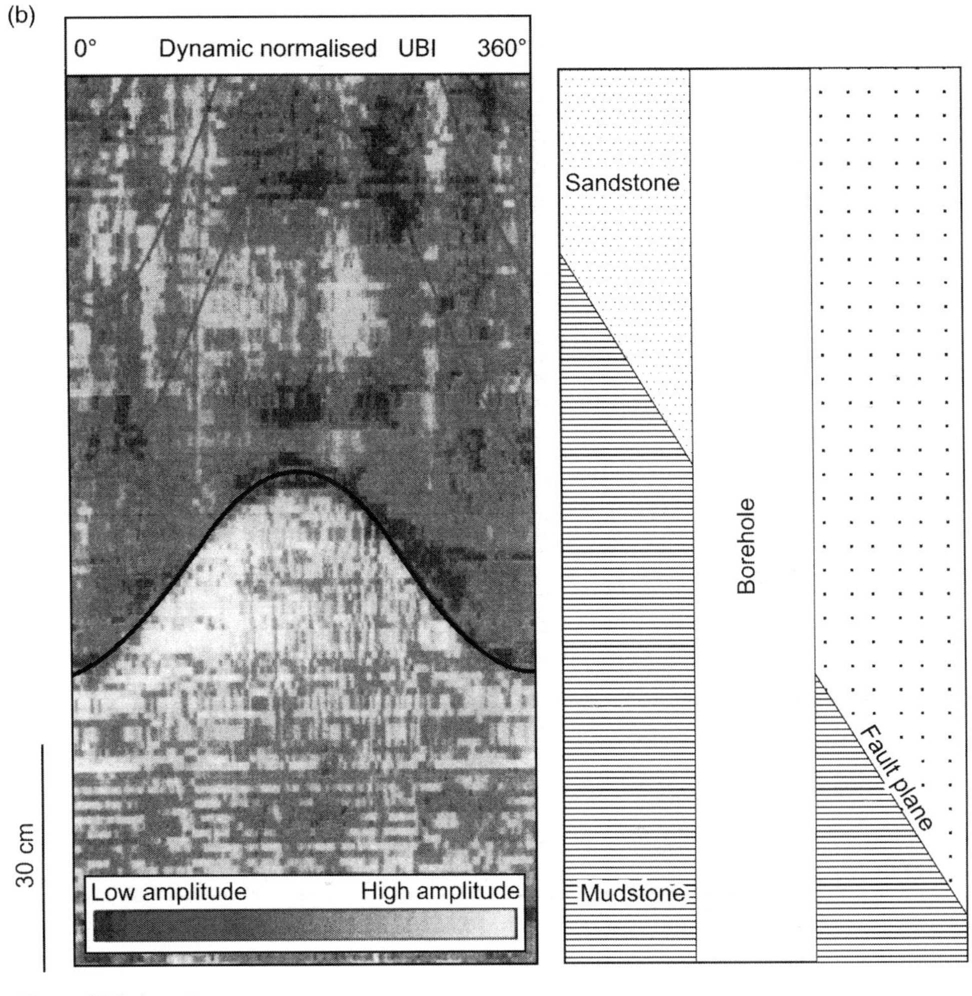

Figure 17.6. (*cont.*)

zone to a composite deformation zone, changing its permeability structure from a distributed conduit to a combined conduit/barrier. The fault development starts with an extensional fracturing stage, continues with a cross-joint and breccia development stage, and ends with a fault core development stage. The rock is characterized by pervasive fracturing accommodating already small strains. The fracturing develops in a broad area. The fault architecture stage with small-offset faults and a breccia zone is characterized by a permeability range of 10^{-14}–$3*10^{-12}$ m^2. The stage with medium-offset faults and initial brecciated fault core development is characterized by 10 percent fracture porosity and preferred fluid flow. The stage with large offset faults and full fault core development is characterized by a low-permeability barrier surrounded by high permeability zones of 10^{-14}–$3*10^{-12}$ m^2.

The lithology-controlled aspect of fault development can be highlighted by the comparison of fault development in dolomite and sandstone. A faulting example in sandstone comes from southern Utah (Antonellini and Aydin, 1994). Here, in porous sandstone, faulting had evolved from a single deformation band through deformation band zones to a fault plane (Aydin, 1977, 1978a, 1978b; Aydin and Johnson, 1978, 1983; Figures 17.9a–b). A deformation band is a thin fault-like planar feature, about 0.5–2 mm thick, which is usually characterized by distinct lateral continuity (Zhao and Johnson, 1991). Apart from cataclasis, the permeability structure of the bands may be reduced by cementation by iron oxides, calcite, or other diagenetic minerals. A deformation band accommodates a small amount of displacement and is characterized by localized cataclasis and volume changes (Antonellini and Aydin, 1994).

Volume changes are represented by distinct porosity changes between the deformation band and the host rock (Antonellini and Aydin, 1994). The porosity reduction of two or three times in the band is associated with three orders of magnitude permeability reduction

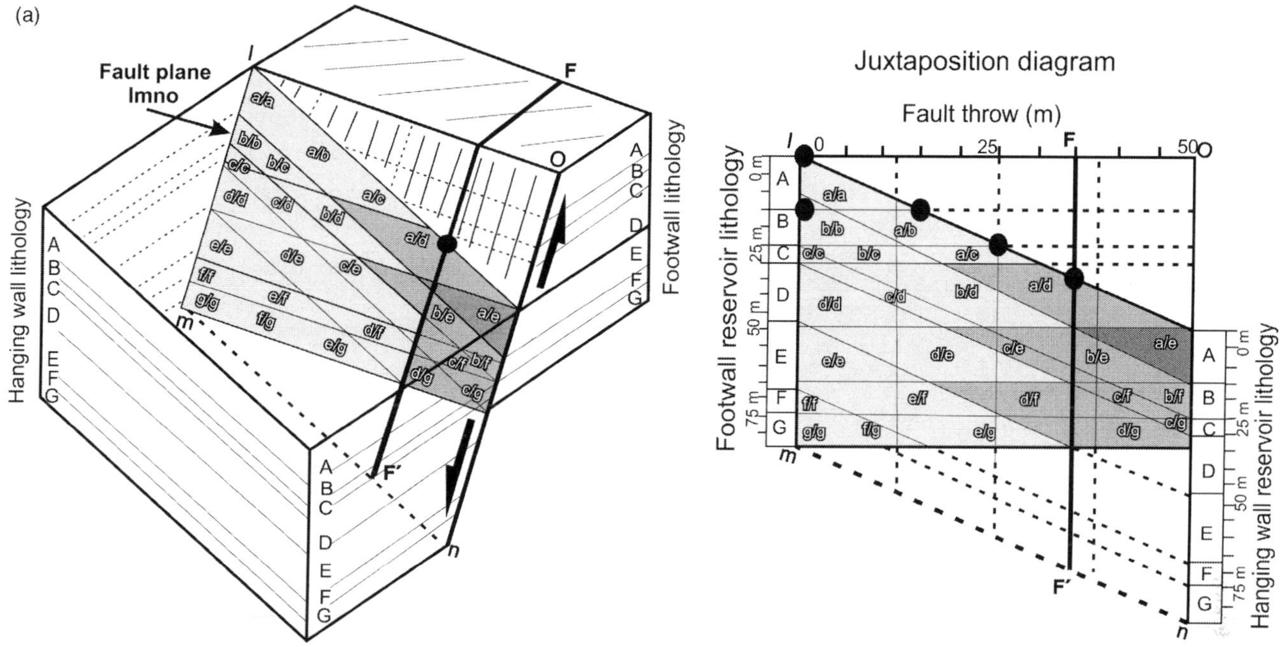

(b)

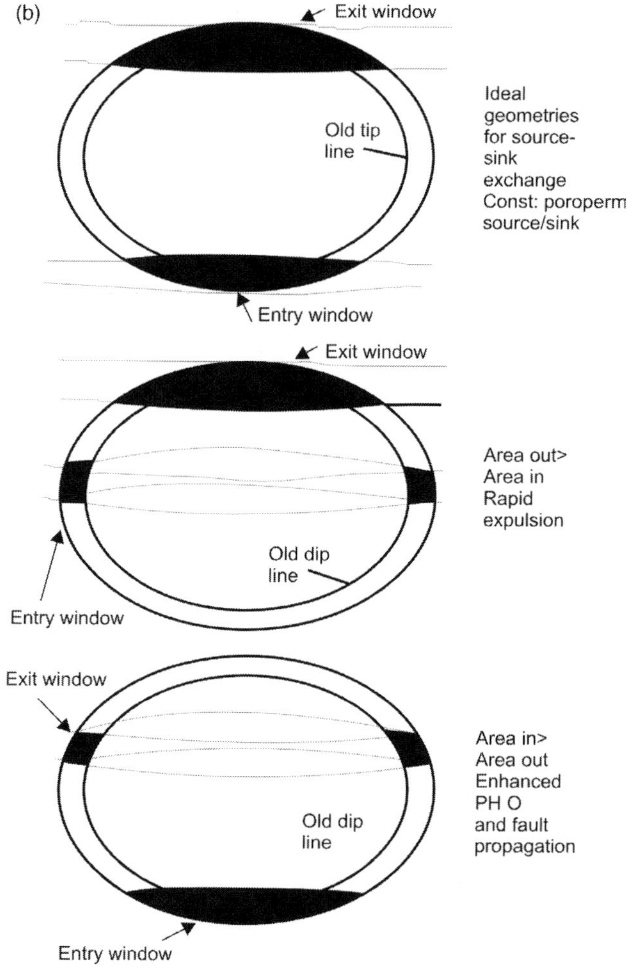

Figure 17.7. (a) Juxtaposition analysis along the fault (Knipe, 1997). a) Diagram illustrates a three-dimensional fault with varying displacement along its length. The juxtaposition analysis is based on extracting the fault plane (lmno) and its intersections with the stratigraphy in the hanging wall and footwall. The juxtaposition diagram can be understood as a "see-through" fault. Footwall stratigraphic units, like unit C, intersect with the fault (the footwall cutoffs) from behind the diagram plane. The hanging wall cutoffs intersect from in front of the diagram plane. Fault can be shown as fault traces (e.g., FF'). Juxtaposition can be assessed by reading a selected point on the fault trace (e.g., dot on fault FF'), looking first horizontally and left, back to the footwall stratigraphy (i.e., the upper unit D forms the footwall at this point), and then looking down and right to the hanging wall stratigraphy (top unit A is juxtaposed in the hanging wall). Each of the triangle areas (e.g., uvw) and trapezoidal areas (e.g., vwxy) on the diagram represents areas on the fault plane with different juxtapositions. For example, the triangle uvw represents the area where unit A in the hanging wall is juxtaposed against unit A in the footwall. Trapezoid vwxy represents the area where the juxtaposition of unit A in the hanging wall is against unit B in the footwall. (b) Fluid behavior and the exact location of the fault tip line with respect to the position of the entry and exit windows in the source and sink horizons involved in flow (Knipe, 1993). Each example assumes that the fault zone before the rupturation event does not connect juxtaposed lithologies, because of the border zone cement seals or juxtaposition with low-permeability horizons. The uppermost sketch illustrates a simple geometry where the entry and exit windows have similar areas, and the porosity/permeability characteristics of the source and sink are also similar. The middle sketch shows a situation where the entry window dimensions are much more restricted than the exit window. This may lead to rapid loss of fluid pressure and fluid expulsion from the fault zone. The lowermost sketch illustrates a case where the entry window allows rapid access to the fault zone but expulsion from the fault plane is restricted by the small exit window. This may lead to the rapid reestablishment of high fluid pressure in the fault zone.

Table 17.1. Diagram showing the types of fault rocks developed in the North Sea and their relationship to the composition of the host sediment and the extent of grain-size reduction and post-deformation lithification experienced (Fisher and Knipe, 1998).

	Fragmentation	Clean sandstone, silt	Impure sandstone, silt	Claystone, shale
Clay content		0–15%	15–40%	40–100%
	0–10%	Disaggregation zone (clay poor)	Disaggregation zone (intermediate clay content)	Disaggregation zone (clay-rich hydroplastic)
	10–50%	Proto-cataclasis (clay-poor)	Proto-cataclasis (intermediate clay content)	Proto-cataclasis (clay rich)
	50–90%	Cataclasite (clay poor)	Cataclasite (intermediate clay content)	Cataclasite (clay rich)
	90–100%	Ultra-cataclasite (clay-poor)	Ultra-cataclasite (intermediate clay content)	Ultra-cataclasite (clay rich)
		Quartz-rich Lithologies (deformation band) Fault rock series	Phyllosilicate framework fault rock series	Clay smear series

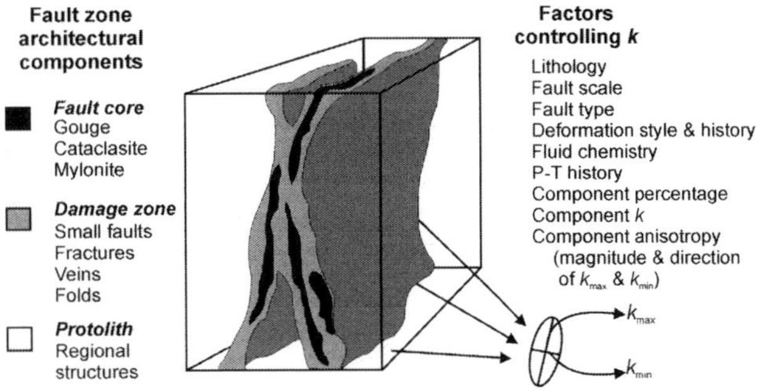

Figure 17.8. Conceptual model of fault zone (Chester and Logan, 1986; Smith et al., 1990). Ellipse represents relative magnitude and orientation of the bulk two-dimensional permeability tensor that might be associated with each distinct architectural component of the fault zone.

(Figure 17.9c). Thanks to this, the deformation band behaves as a localized deformation zone and a barrier to fluid flow. This behavior is further enhanced in the case of the deformation band zone and fault. The final result of the faulting process, the fault, represents large permeability anisotropy, as documented by permeability measurements (Antonellini and Aydin, 1994; Figure 17.9d).

Seal geometry

The seals can be further differentiated according to their position in relationship to the reservoir. They include top and lateral seals.

The top seal is a low-permeability layer that overlies the petroleum reservoirs and retards the upward migration of trapped hydrocarbons. The layer can be represented by primary lithology, lithology improved by cementation or by ice-rich permafrost. Top seals can exhibit both membrane and hydraulic sealing, discussed in the next section.

The lateral seal is a low-permeability body forming a lateral boundary of the petroleum reservoir, capable of retarding the lateral migration of trapped hydrocarbons. The body can be represented by a fault zone, shale body, and evaporite body. Lateral seals can exhibit both membrane and hydraulic sealing, discussed in the next section.

Sealing control

Based on differing modes of failure, the seals can be defined as membrane and hydraulic seals (Watts, 1987). The failure of the membrane seal is primarily controlled by the capillary entry threshold pressure, p_c, which is the pressure required for entry through the largest interconnected pore throat of the seal (Berg, 1975).

The current theory of trapping, which has a static nature, is usually derived by making a force balance between the buoyancy and capillary forces (Berg, 1975; Schowalter, 1979; England et al., 1987). The static theory draws from the setting filtration velocity of oil, u_o, as equal to total flux, u_t, set as equal to zero in the equation (Siddiqui and Lake, 1997):

$$u_o = fu_t + fk\lambda_{rw}\Delta\rho g - fk\lambda_{rw}\partial p_c \partial x, \quad (17\text{-}2)$$

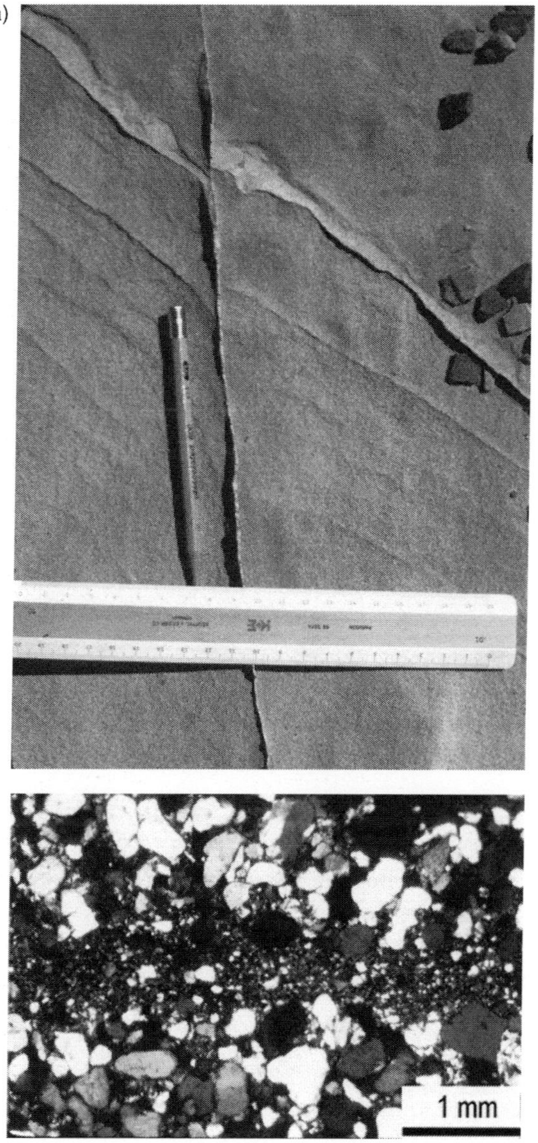

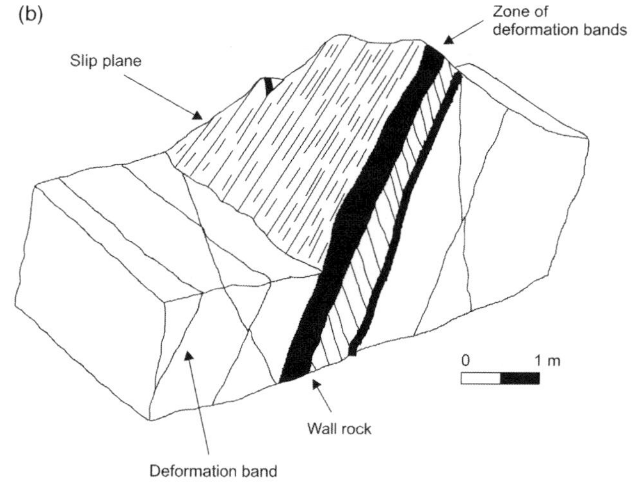

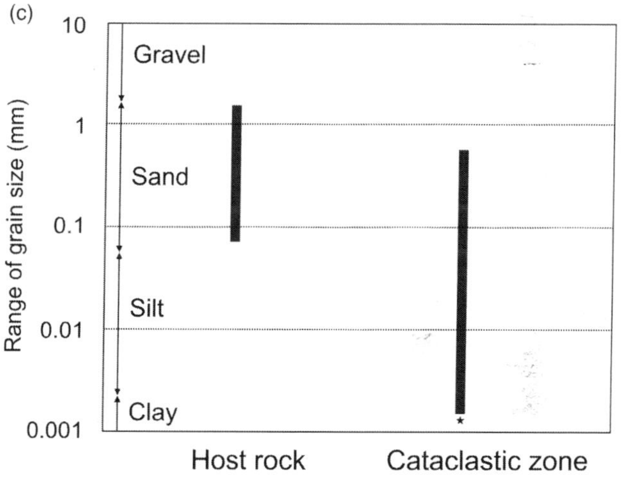

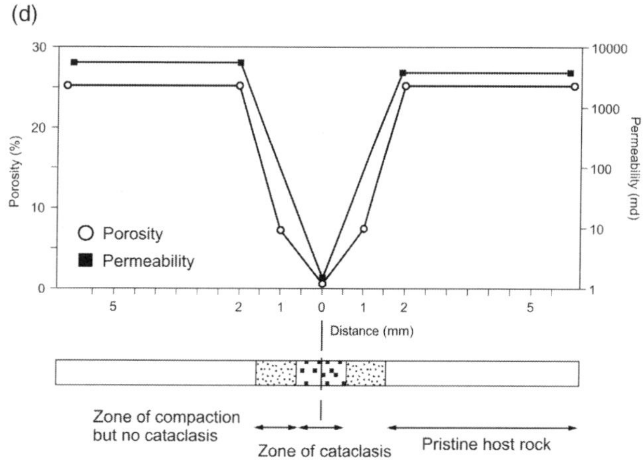

Figure 17.9. (a) Photomicrograph and outcrop photograph of shear bands formed in the Entrada Sandstone in the San Rafael Desert, Utah (Aydin et al., 2006). (b) Sketch of a fault plane in sandstone (Antonellini and Aydin, 1994). Note numerous bands forming the fault. (c) Grain size in host rock and in the cataclastic zone inside a deformation band in sandstone (Antonellini and Aydin, 1994). Grain sizes of gravel, sand, silt, and clay are given by arrows for comparison. The star indicates the limit of resolution using a petrographic microscope. b) Permeability and porosity profile through the deformation band developed in the Entrada Sandstone (Antonellini and Aydin, 1994). (d) Permeability and porosity profile through the deformation band developed in the Entrada Sandstone (Antonellini and Aydin, 1994).

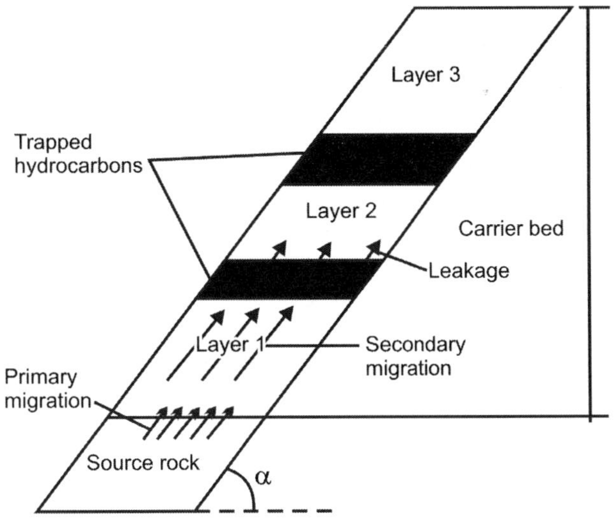

Figure 17.10. Different components of a hydrocarbon migration system in a multilayer carrier bed (Siddiqui and Lake, 1997).

which gives:

$$\partial p_c \partial x = \Delta\rho g, \quad (17\text{-}3)$$

where f is the fractional flow of oil, k is the intrinsic permeability, λ_{rw} is the relative mobility of water, $\Delta\rho$ is the density difference between oil and brine, g is the acceleration of gravity, p_c is the capillary pressure, and x is direction perpendicular to reservoir and seal horizons.

The fractional flow of oil is given by:

$$f = \lambda_{ro}/(\lambda_{ro} + \lambda_{rw}), \quad (17\text{-}4)$$

where λ_{ro} is the relative mobility of oil. Relative mobilities of oil and water are controlled by:

$$\lambda_{ro} = k_{ro}/\eta_o \text{ and } \lambda_{rw} = k_{rw}/\eta_w, \quad (17\text{-}5)$$

where k_{ro} and k_{rw} are relative permeabilities to oil and water, and η_w is the water viscosity.

The static theory is based on a permeability contrast between the reservoir and seal (Figure 17.10). Contrasting permeabilities control the displacement pressures (Siddiqui and Lake, 1997):

$$p_c = j(S_o) \cdot [(\gamma\cos\theta)/\sqrt{(k/n)}], \quad (17\text{-}6)$$

where γ is the interfacial tension, θ is the wettability, k is the permeability, n is the porosity, and $j(S_o)$ has the Brooks-Corey form of the Leverett j-function (Corey, 1986):

$$j(S_o) = (1-S_o)^{-npc}, \quad (17\text{-}7)$$

where S_o is the normalized oil saturation.

The resultant displacement pressure for the seal, p_{dS}, is normally greater than the displacement pressure for the reservoir, p_{dR}. As a consequence, a hydrocarbon slug that reached the reservoir–seal boundary cannot enter the seal. The condition for entering would be the increase of the hydrocarbon phase pressure until it equals the displacement pressure of the seal. Until it does, the hydrocarbons remain trapped below the reservoir–seal boundary.

As the hydrocarbon migration from the source rock continues to supply the reservoir, the length of the trapped hydrocarbon column progressively increases. If the capillary pressure, p_c, at the bottom of the hydrocarbon column in the reservoir, where $x = 0$, equals the displacement pressure for the reservoir, p_{dR}, the hydrocarbon column length, h_{static}, required to attain the displacement pressure for the seal, p_{dS}, at the reservoir–seal boundary can be found by integrating Equation (17-3) (Siddiqui and Lake, 1997);

$$\text{from } p_{dR} \text{ to } p_{dS}\, dp_c = \text{from } x = 0 \text{ to } h_{static}\, \Delta\rho g dx. \quad (17\text{-}8)$$

When the $\Delta\rho$ is set as a constant, Equation (17-8) changes into:

$$h_{static} = (p_{dS} - p_{dR})/\Delta\rho g, \quad (17\text{-}9)$$

which expresses the maximum hydrocarbon column length that can be trapped underneath a respective seal. Any further supply of hydrocarbons would initiate the failure of a membrane seal.

The static calculation serves as the first approximation of the hydrocarbon column trapped underneath a specific seal. The disadvantage of the static theory is its underestimation of the maximum length of the trapped hydrocarbon column (e.g., Berg, 1975). The error is caused by ignoring the other forces from the force balance expressed in Equation (17-1) (e.g., Berg, 1975; Tissot and Welte, 1984; Siddiqui and Lake, 1997).

The frictional resistivity term from Equation (17-1) controls the trapping behavior of a seal depending on whether the fluid supply from the source rock is smaller or greater than the seal capacity. As experiments demonstrate (Tokunaga et al., 2000), when the supply is greater than the capacity, the flow pattern is represented by stable displacement. In this case, the initial shock-front fluid migration through the reservoir reaches the reservoir–seal boundary where the lower seal capacity prevents the complete transmitting of the flowing oil into the seal. This results in a reflected shock wave (Siddiqui and Lake, 1997). Thus the frictional resistivity term controlled by forces due to viscosity and relative permeability adds a dynamic sealing on top of static sealing. When the hydrocarbon supply from the source rock stops, all dynamically trapped hydrocarbons leak off. The leakage time can be geologically significant.

Equations (17-3) to (17-9) indicate that there is a certain saturation, S*, that corresponds to the displacement pressure of the seal, p_{dS}. It defines a critical value of the hydrocarbon saturation at the top of the reservoir at which the trapped hydrocarbons start to invade the seal. Similarly, there is a certain saturation, S_A, that corresponds to the minimum saturation causing the mentioned reflected shock wave. This value sets the limit between static and dynamic seal behavior.

While the dynamic pressure gradients in reservoir and seal cannot be neglected in the dynamic seal case, they are negligible in the static seal case and the pressure behavior corresponds to the static pressure profile. Typical permeabilities for good hydrocarbon seals vary around 10^{-21} m^2. Such seals can trap hydrocarbon columns greater than static columns for geologically significant times (Siddiqui and Lake, 1997).

If the hydrocarbon supply from the source rock terminates before the hydrocarbon column reaches its h_{static} length, there is no fluid invasion into the seal. The termination in the static seal case has no effect on the hydrocarbon column in the reservoir. It can only result in no further leakage into the seal as no more hydrocarbons are available.

The difference between static and dynamic behavior of the seal defines a good trap. A good trap is charged by a high enough hydrocarbon supply capable of turning the static seal into a dynamic one. Dynamically trapped hydrocarbons can adjust to statically trapped ones once the supply stops. The traps never reach the stage of dynamic seal if they do not have a high enough hydrocarbon expulsion rate.

The resistance of a seal against fluid invasion is controlled by wettability. An increase in wettability to a certain fluid phase shifts the maximum oil flux to a higher saturation of that phase (Siddiqui, 1996). The increase results in slowing down the rate of the upward-moving fluid migration shocks.

Equation (17-6) shows a wettability control on capillary pressure. A combination of an oil-wet reservoir with an oil-wet seal results in a minor invasion of the seal, controlling a seal better than a water-wet seal of the same intrinsic permeability. A combination of a water-wet reservoir with oil-wet seal results in a transient trap incapable of trapping hydrocarbons permanently, contrary to a water-wet seal, which loses only a portion of its trapped hydrocarbons, retaining an oil column with a height defined by h_{static} (Siddiqui and Lake, 1997).

The failure of the hydraulic seal is controlled by an entry pressure exceeding the strength of the seal rock (Berg and Gangi, 1999). Such failure is represented by the hydraulic fracturing of the seal rock associated with overpressure venting through the propagated/reactivated fractures, as discussed in Chapter 18 (Figure 18.13).

Hydraulic seals can be recognized in wells based on the large fluid pressure differences across short depth intervals, which are caused by seals slowing down all fluid phase movements considerably on a distinct time scale (Hunt, 1990). Well pressure/depth profiles modified by seals distributed in the penetrated sedimentary section differ from profiles characterized by hydrostatic gradient, which has values of 9.79 kPam^{-1} and 11.9 kPam^{-1} for freshwater and saturated salt solution, respectively (Hunt, 1990; Figures 17.11a, b). Such profiles are characterized by the seal separating fluid compartments with different pore fluid pressure regimes.

The seal layer represents a low permeability layer that permits a transient fluid flow of all phases. The initial condition at the base of seal horizon is specified as (Deming, 1994c):

$$h(z, t = 0) = 0, \qquad (17\text{-}10)$$

where h is the hydraulic head, z is the seal layer thickness, and t is the time (Figure 17.11c). To illustrate the hydraulic seal behavior, we can instantaneously raise the head to value Δh at the base of seal layer, hold it constant, and assume a constant-density fluid. In such a case, the hydraulic head can be treated as a potential field, and the propagation of excess hydraulic head into the seal can be described by the diffusion equation (Deming, 1994c):

$$\delta h/\delta t = D(\delta^2 h/\delta z^2), \qquad (17\text{-}11)$$

where D is the hydraulic diffusivity. If we ignore a small change of head, which is caused by elevation changes, the solution of Equation (17-11) becomes (Deming, 1994c):

$$h(z, t) = \Delta h\{1 - \mathrm{erf}[z/(4Dt)^{1/2}]\}, \qquad (17\text{-}12)$$

where erf is the error function. Equation (17-12) allows us to calculate the time, t, which takes the transient disturbance at the base of the seal to propagate a distance z into the seal layer. The time, t_{ch}, which takes the potential disturbance at the base of seal to propagate a distance z into the seal layer is given by (Lachenbruch and Sass, 1977):

$$t_{ch} = z^2/4D. \qquad (17\text{-}13)$$

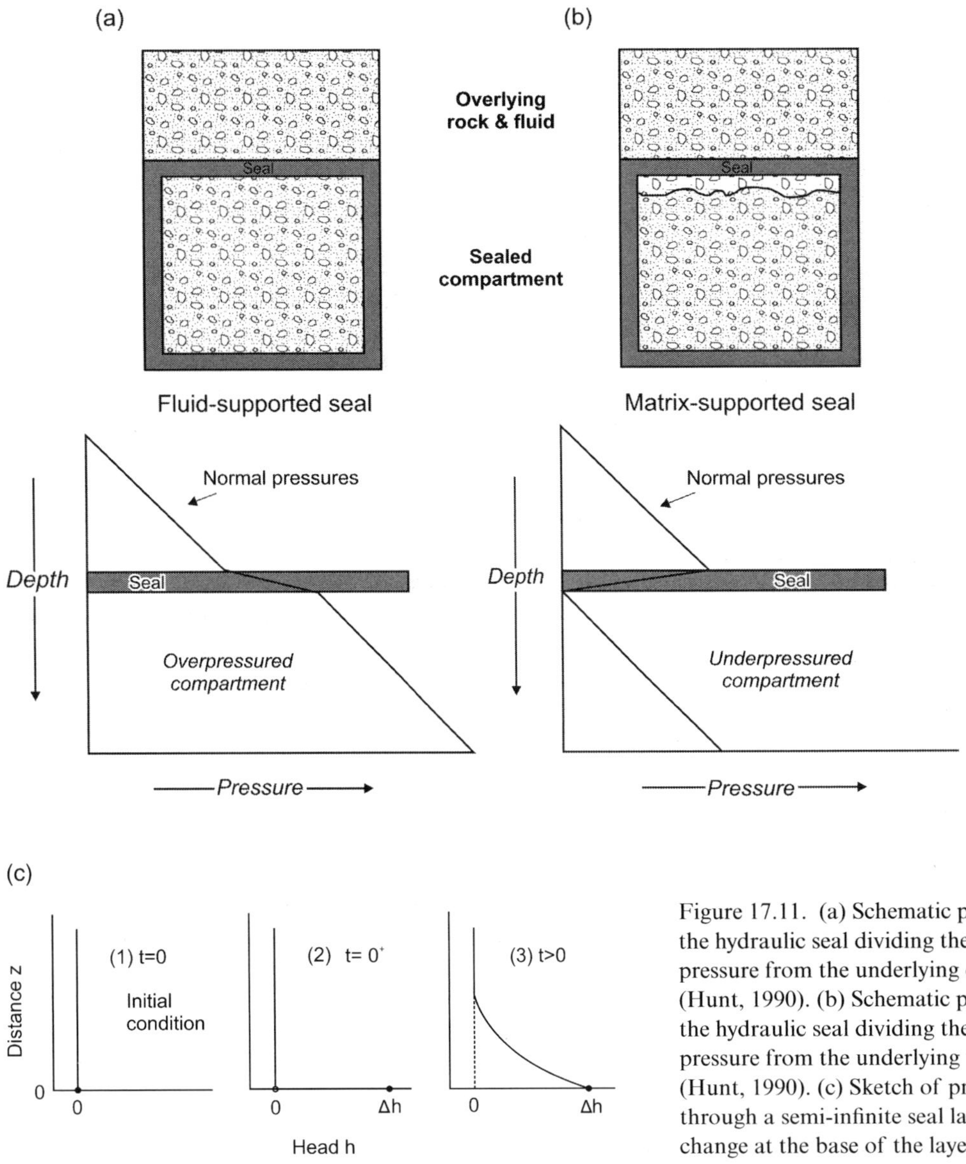

Figure 17.11. (a) Schematic pressure/depth profile showing the hydraulic seal dividing the compartment with normal pressure from the underlying overpressured compartment (Hunt, 1990). (b) Schematic pressure/depth profile showing the hydraulic seal dividing the compartment with normal pressure from the underlying underpressured compartment (Hunt, 1990). (c) Sketch of propagation of head transient through a semi-infinite seal layer following sudden head change at the base of the layer (Deming, 1994c).

The hydraulic diffusivity, D, is a function of hydraulic conductivity, K:

$$D = K/S_s, \qquad (17\text{-}14)$$

where S_s is the specific storage.

Equations (17-13) and (17-14) can be rephrased as (Ge and Garven, 1992):

$$t_d = z^2S_s/K, \qquad (17\text{-}15)$$

allowing us to calculate the fluid diffusion time, t_d, as controlled by the thickness of strata, z, and their hydraulic conductivity and specific storage. The specific storage refers to the physical ability of porous material to store fluid. It is largely determined by the compressibility of both fluid and porous material (Ge and Garven, 1992):

$$S_s = [\alpha_n/(1-n) + n\alpha_s + n\beta]Dp/Dt, \qquad (17\text{-}16)$$

where α_n and α_s are compressibilities of pore space and solid grains, β is the fluid compressibility, and n is the porosity. Compressibilities α_n and α_s are controlled by:

$$\alpha_n = -\partial/\sigma_e$$

and

$$\alpha_s = 1/\rho_s(\partial\rho_s/\sigma_e), \qquad (17\text{-}17)$$

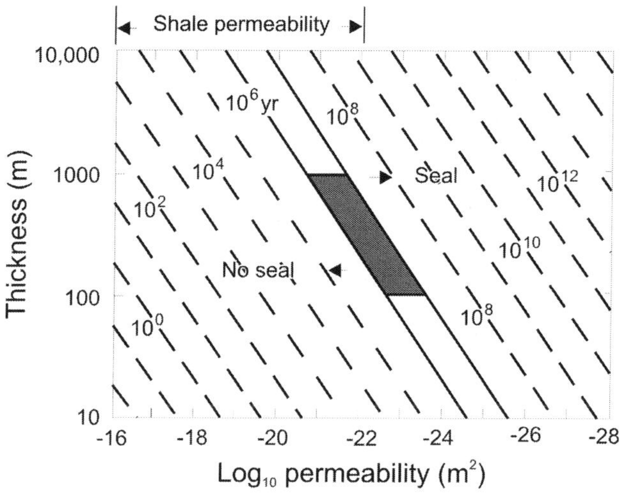

Figure 17.12. Maximum time over which a seal layer of given thickness and permeability may confine excess pressures (Deming, 1994c). The time is given in years. The shaded area indicates the approximate permeability required to sustain a seal layer 100–1,000 m thick over a time of about 1 Ma.

where ρ_s is the solid grain density and σ_e is the effective stress expressed as the total mean stress minus the pore fluid pressure ($\sigma_m - p$).

Hydraulic conductivity refers to the property of both the porous material of the seal layer and the fluid moving through it (Darcy, 1856; Hubert, 1969):

$$K = k\rho g/\eta, \tag{17-18}$$

where k is the intrinsic permeability, ρ is the fluid density, g is acceleration of gravity, and η is the dynamic fluid viscosity.

The time within which a seal layer may confine a sudden change in head at its base is proportional to the specific storage, S_s, and the square of its thickness, z, and inversely proportional to its permeability, k, as shown in combined Equations (17-13), (17-14), (17-15), and 17-16 (Deming, 1994c; Figure 17.12):

$$T = [z^2S_s/k][\eta/4\rho g]. \tag{17-19}$$

A compartment sealed by the seal layer can undergo several overpressure-generating mechanisms, such as:

1) chemical reactions leading from kerogen to hydrocarbons (Osborne and Swarbrick, 1997);
2) disequilibrium compaction (Maltman, 1994; Osborne and Swarbrick, 1997);
3) seismically induced liquefaction (Obermeier, 1996);
4) smectite-to-illite transition (Bekins et al., 1994; Osborne and Swarbrick, 1997);
5) gypsum-to-anhydrite transformation (Jowett et al., 1993);
6) diagenesis itself (Osborne and Swarbrick, 1997 and references therein); and
7) aquapressuring (Shi and Wang, 1986).

The slope of the pressure/depth curve within the fluid compartment would remain the same as that in the normally pressured sediments overlying the seal layer, assuming the density of the fluid above and below the seal is the same and that internal hydraulic communication exists within the sealed sediment volume (Figure 17.11a). Fluid below the seal layer supports the weight of the seal and portions of the overlying sedimentary sequence and fluid loads. The magnitude of the resulting pore fluid pressure change depends on how much of the weight of the overlying rock column is borne by the fluids of the sealed compartment in comparison to what is borne by the rock skeleton of the sealed compartment. Natural examples of this situation come from the Ernie Dome in the Central Transylvanian Basin, Romania (Powley, 1980; Figure 17.13a) and Cook Inlet in Alaska (Powley, 1980).

The Cook Inlet case, however, is special in that its seal is not a lithological seal. It does not follow any specific lithostratigraphy, which is a phenomenon more common in clastic sedimentary sections than carbonate ones (Hunt, 1990). The seal cuts indiscriminately across structures and is interpreted as a thermocline where bands of calcite precipitation have plugged the original porosity (Powley, 1980). This case also serves as an example of a composite seal formed by a system of permeable and impermeable layers.

A compartment sealed by the seal layer can also experience the underpressure generation, given an optimal combination of its permeability structure, uplift, and erosion. The latter two mechanisms result in the cooling of the sealed compartment and reduction of its pore fluid pressure (Figure 17.11b). The magnitude of the pressure reduction is controlled by how much of the overburden weight is shifted from being borne by pore fluid to being borne by the rock skeleton below the seal layer. A natural example of this case comes from the Keyes Dome in Oklahoma (Powley, 1980; Figure 17.13b).

The hydraulic seal can fail in shear tension, if the pore fluid pressure regime in the sealed compartment modifies the stress regime to cause the failure (Figures 17.14, 17.1). The failure results in fluid discharge along fractures (Figures 17.15, 17.16a, 18.13) and a related drop in fluid pressure.

Multiple cycles of pore fluid pressure buildup in sealed compartments followed by cycles of seal failure

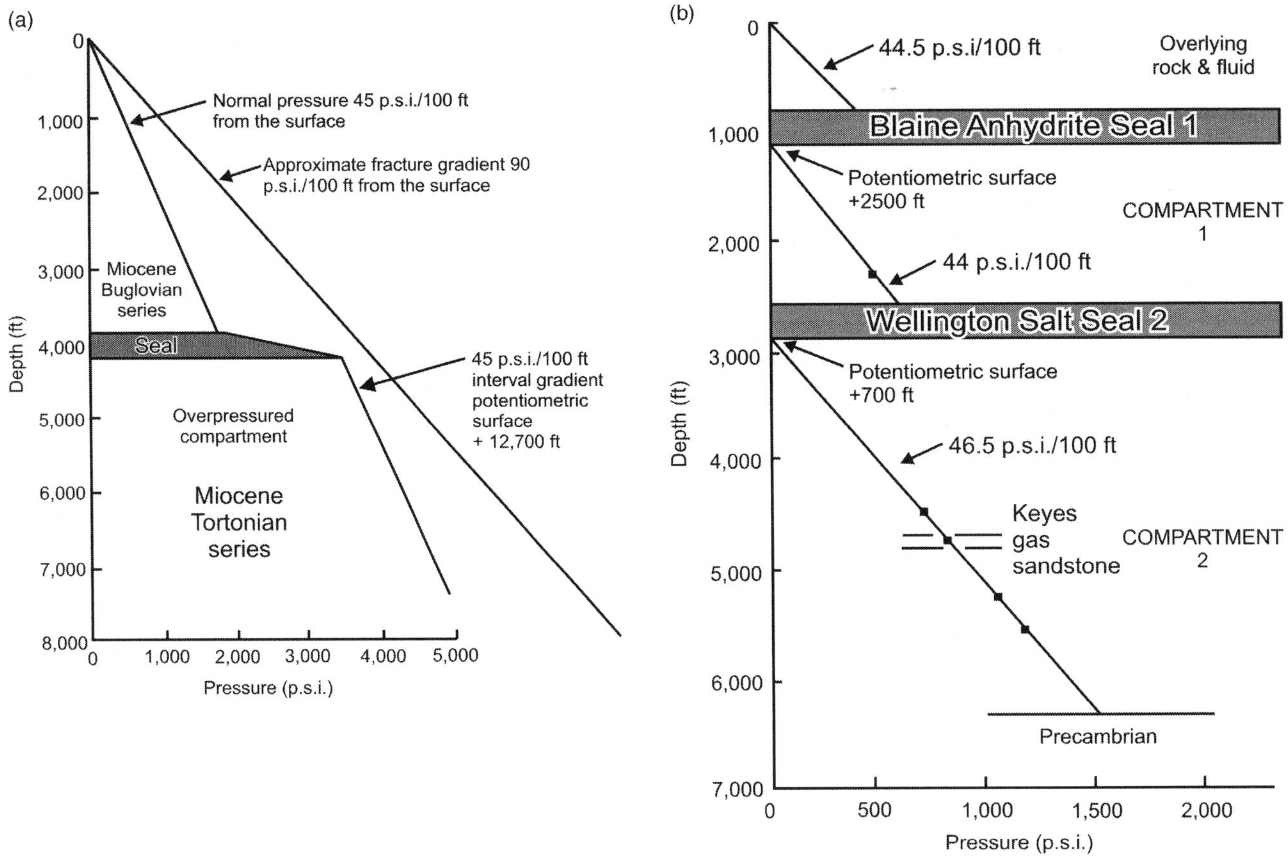

Figure 17.13. (a) Pressure/depth curve along the well in Ernie Dome in the central Transylvanian Basin, Romania (Powley, 1980). (b) Pressure/depth curve along the well in the Keyes Field, Oklahoma (Powley, 1980).

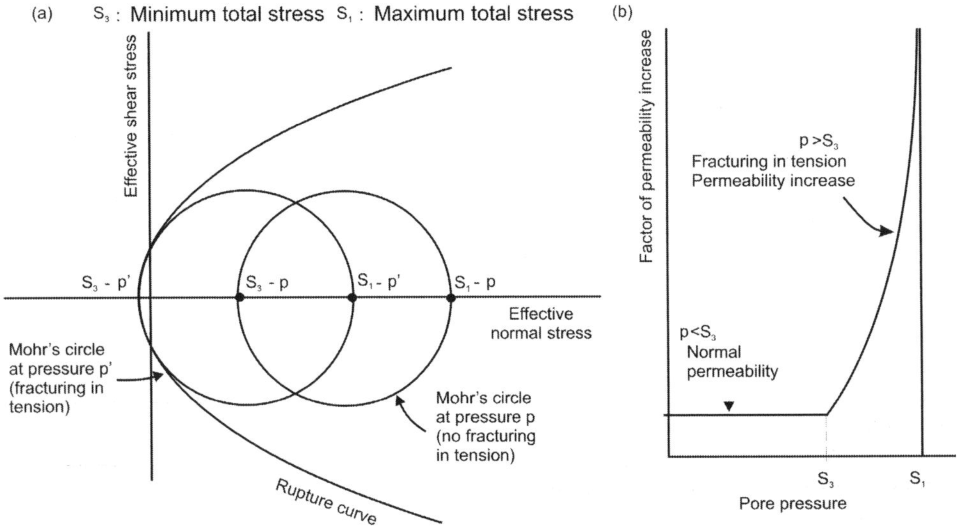

Figure 17.14. (a) Mohr's diagram of the dependence of effective stress on pore fluid pressure (Ungerer, 1990). Tensile fractures occur only if the lowest component of the effective stress tensor (S_3-p) reaches a negative value (Du Rouchet, 1981). (b) Typical permeability–pore fluid pressure relationship controlling the hydraulic fracturing. Tensile strength of the rock matrix is assumed as negligible and permeability increases as soon as minimum effective stress (S_3-p) becomes negative. The factor of the permeability increase approaches infinity as pore fluid pressure approaches the maximum total stress S_1. Pore fluid pressure is maintained lower than S_1 and maximum effective stress remains positive.

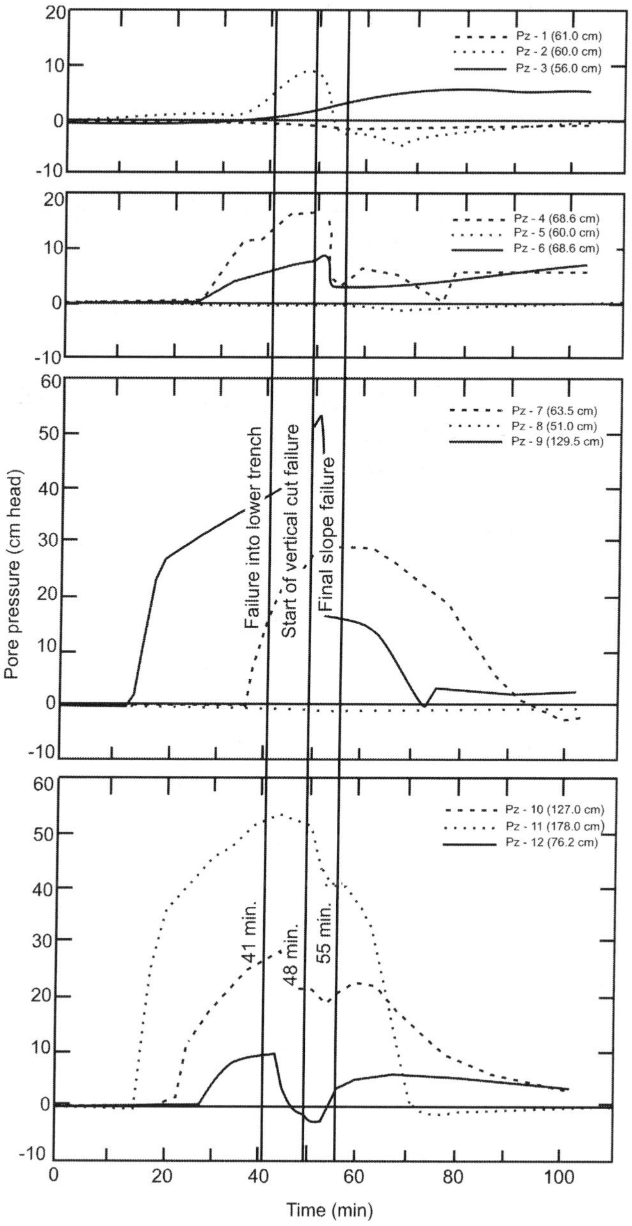

Figure 17.15. Artificial landslide experiment at Monroe Canyon, California (Harp et al., 1990). Diagram shows pore pressure history versus time during soil failure. Vertical lines across records denote times of specified events. The experiment started with pouring water into an artificial trench above the anticipated future landslide. All piezometers, except 1, 3, 6, and 8, had shown significant pore pressure rises within the first thirty-seven minutes. Piezometers 1, 3, and 6 were too shallow and too close to the top of the future slump to record any distinct curves. Piezometer 8 was installed outside of the experiment area. At fifty or five minutes before the final failure, well-placed piezometers showed an abrupt pore pressure drop related to the effects of subsurface removal of fine-grained material and dilatation of the landslide mass in its initial stages of failure. Dilation fracturing was documented as the development of smaller failures prior to the main slope failure and their progressive widening one to two seconds before failure, but a noticeable pore pressure drop started before visible cracks were produced. At fifty-five minutes, a purely translational failure of 1.3 m depth and 14 m^3 volume slid down the slope. Water poured rapidly from the failure and fractures distributed around it.

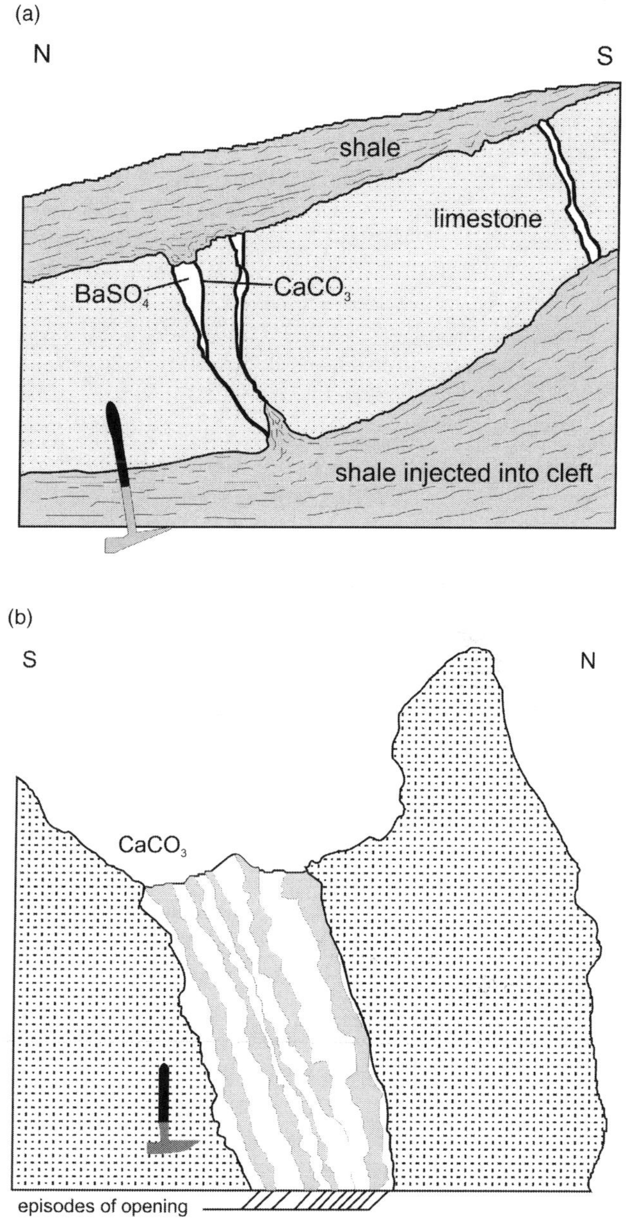

Figure 17.16. (a) Shale injections in the limestone layer of the Lower Jurassic limestone/mudstone sequence at St. Audrie's Bay, Bristol Channel, UK (Nemčok and Gayer, 1996). Line tracing shows the limestone layer separated by extensional veins. Internal parts of the veins are filled by barite, external parts by calcite. Clefts formed at the tips of veins are filled by injected mudstone. (b) Multiply reopened extensional calcite veins at Porthcawl, Bristol Channel (Nemčok and Gayer, 1996). Each episode of extensional fracturing was triggered by new pore fluid pressure buildup, which shifted the stress state toward failure.

by fracturation have been described from the syn-rift fill of the Bristol Channel (Figure 17.16b; Nemčok et al., 1995; Nemčok and Gayer, 1996). Here each new pore fluid pressure buildup cycle resulted in fracturing. The fracturing was followed by pore pressure release due to fluid discharge along neo-formed and reactivated fractures. The flow allowed mineral precipitation inside fractures. This led to fracture sealing and the onset of a new pore pressure buildup cycle in the sealed compartment.

Pore pressure studies in the Gulf Coast of the United States have also shown that when a seal fails and fracture-related high-permeability pathways are formed, the fluid pressure in previously sealed compartment drops to a pressure/depth dependent gradient of about 13.5 $kPam^{-1}$ but does not drop all the way to the hydrostatic pressure. The reason is that the fracture permeability gets reduced by fractures resealing already at moderate overpressures in the sealed-off compartments (e.g., Hunt, 1990).

Hydraulic or at least partially hydraulic seals are quite common, as indicated by a database of seals from about 180 basins published by Hunt (1990). Some of them have recognized sealed compartments with abnormal pore fluid pressure regimes. Only a small portion of seal data is associated with underpressured compartments. This is probably an artifact of the data collection. They are more common and have an important role in preventing water washing and biodegradation of trapped oils (Hunt, 1990).

If the seal is poor, prone to failure either as a membrane or hydraulic seal, one needs to rely on the active petroleum system capable of charging the leaky traps. Portions of the Taranaki Basin, the South Caspian Basin, the Bohai Bay Basin, the North Sea, and the northern Gulf of Mexico provinces would serve as examples of the recently active petroleum systems (e.g., Anderson et al., 1994a, 1994b; Roberts and Carney, 1997; Feyzullayev et al., 2001; Guliyev et al., 2001b; Losh et al., 2002; Kubala et al., 2003; Hao et al., 2009).

Seals in the rift settings

The discussion of seal examples in rift settings requires us to find a mature hydrocarbon province with a large database on pre-rift, syn-rift, and post-rift seal scenarios known not only from the exploration but also from the production standpoint. The classical North Sea province fulfills these criteria and provides a basin inversion-related modification of the preexisting pre-rift, syn-rift, and post-rift traps in several areas as an extra complexity.

The best examples of seals working with pre-rift reservoirs come from the plays in the Lower–Middle Jurassic section (Husmo et al., 2002). Most fields of this type are trapped in rotated fault blocks and rely on sealing from the younger shales. Depending on the depth of erosion at footwall edges, the seals are either syn-rift Upper Jurassic shales or younger syn-rift Lower Cretaceous shales, or even post-rift Upper Cretaceous shales. Although the intra-formational shales inside the Lower–Middle Jurassic Brent Group are not considered seals on a geological scale, they behave as seals at the production scale. Most of the traps also rely on lateral seals provided by faults juxtaposing the Lower–Middle Jurassic reservoirs against the Upper Jurassic shales. Examples come from the Gullfaks and Statfjord fields. Rarely, when permeable strata are juxtaposed against permeable strata, the trap relies on a shale smear sealing along the fault, such as in the Deveron field (Williams, 1991). About 25 percent of the fields with Lower–Middle Jurassic reservoirs contain several compartments separated by such shale-smear faults. Their individual compartments have varying initial pressures and hydrocarbon column heights. Examples come from the Gullfaks, Gullfaks Sor, Sleipner Vest, Oseberg Sor, Thistle, and Murchison fields. The quality of sealing depends on the shale content in the fault rock, as core studies in the Oseberg and Sleipner Vest fields have documented (see Fristad et al., 1997; Ottesen et al., 1998).

The Upper Jurassic section of the North Sea provides examples of the syn-rift reservoirs in both shallow-marine and deep-marine sandstones (Fraser et al., 2002). The deep-marine sandstones rely on sealing by the partially encasing organic-rich muds and lateral fault seals. A combination of structural and stratigraphic trapping is involved in practically all fields in deep-marine sandstones, especially in fields of the Viking Graben and Outer Moray Firth. Traps are represented by:

1) footwall-related traps;
2) hanging wall-related traps;
3) salt movement-related traps; and
4) basin inversion-related traps.

The first three categories represent 90 percent of the existing fields and discoveries.

Footwall traps are provided by the same scenario as the underlying Lower–Middle Jurassic traps described earlier. Examples having juxtaposition seal faults as lateral seals come from the Brage, Troll, Piper, Rob Roy, and Telford fields. The lateral seal faults relying on clay-smear sealing are controlling the Scott, Glamis, Ivanhoe, and Hamish fields. A comparison of fields such as Piper, Scott, Snorre, Statfjord North, Telford, and Saltire reveals a significant variation in the depth of erosion at the footwall edges of these traps. In an extreme case, an erosional truncation provides the top seal, with the controlling fault not directly involved in trapping, such as is known from the Magnus and Claymore fields.

Hanging wall traps comprise downthrown or low-side closures adjacent to rotated fault blocks. In the South Viking Graben, along the Brae trend, the faults seal by juxtaposing the Upper Jurassic reservoirs against the tight Devonian sandstones of the footwall. Hanging wall-related slope apron/basin floor fan reservoirs occur in the Outer Moray Firth Basin, including the Lowlander, Tartan, Galley North, Perth, and Saltire fields. Some of these fields rely on both structural and stratigraphic trapping, having reservoirs sealed partly by the conformably overlying syn-rift Kimmeridge Clay Formation and partly by the truncation followed by the onlap of Lower Cretaceous shales (Harker et al., 1991; Casey et al., 1993). The cases of the Lowlander, South Scott, and Chanter fields indicate that the most critical aspect of this type of trap is a lateral fault-seal effectiveness.

Salt-related traps are represented by closures controlled by the movement of the underlying Zechstein salt, which is superimposed on extensional architecture. An example comes from the Fulmar field, which has been interpreted as controlled by salt withdrawal, by the grounding and inversion of the Triassic primary rim synclines, and by the subsequent extensional faulting. Its top seal is formed by the conformably overlying syn-rift Kimmeridge Clay Formation to the southeast and by the unconformably overlying post-rift chalk section to the northeast. Traps such as those in the Kittiwake, Durward, and Dauntless fields are both dip closure- and reservoir pinch-out-dependent.

The inversion-related traps are not primarily related to rift setting but relatively common in the rift settings subsequently modified by younger contraction. Inversion processes in the North Sea province did not result in new traps but rather modified, or sometimes destroyed, the preexisting syn-rift traps. The Ula field serves as an example of this trap type.

A category of the North Sea syn-rift traps would not be complete without traps with Lower Cretaceous reservoirs (see Copestake et al., 2002). Many of them rely on seals represented by condensed claystones bounding the depositional sequences. In the cases of stacked sandstone reservoirs, such as in the Agat, Scapa, and Britannia fields, these condensed claystones may form intra-reservoir seals or fluid flow barriers.

A unique combination of source, reservoir, and seal is available in the Halibut Horst area where a series of discrete mass-flow sandstone units occurs inside the shale-dominant basin fill, where the condensed claystones bound depositional sequences. This unique depositional system was prone to the development of stratigraphic/structural traps developed by differential compaction or by draping over sandstone bodies (Law et al., 2000). One of the fields in the Halibut Horst area is the Captain field, which is a combination of a stratigraphic/structural trap formed by onlap and the drape of reservoirs over the westward plunging end of the horst. The top seal is formed by a thin claystone section, usually less than 10 m thick. The younger Chalk Group sequences provide extra sealing help.

The Agat and Britannia fields provide other examples of the traps with Lower Cretaceous reservoirs. The former relies on a top seal formed by flooding surface. The latter also relies on the lateral sealing provided by faults. An extra modification of the trap was provided by the Eocene basin inversion (Porter et al., 2000). The principal phase of sand deposition in the Britannia field is interpreted as terminated by flooding of the source terrain during the deposition of the condensed flooding claystone. This unit forms a top seal although the most effective top seal is probably higher up in the stratigraphy, which is the only one available in the areas affected by erosion removing the mentioned condensed flooding claystone.

The remaining category of traps with post-rift reservoirs starts with the Upper Cretaceous section (Surlyk et al., 2002). The seal for the fields is dominantly provided by the overlying Paleocene mudstones whose areal extent is one of the controlling factors of the chalk play in the North Sea province (Spencer et al., 1996). This seal has two major roles:

1) to keep the underlying chalk package sealed; and
2) to keep the underlying chalk package overpressured, which preserves its high porosity.

The sealing capability of the Paleocene mudstones relies on suitable facies being compromised in the sites of sandstone deposition.

The chalk fields, with the exception of the stratigraphically trapped Danish Halfdan field, are represented by structural traps developed because of either basin inversion or salt movement (e.g., D'Heur, 1984; Megson, 1992; Albrechtsen et al., 2001). Inversional examples from Norway come from the Edda, Eldfisk, Valhall, and Hod fields. Other inversion-related fields occur in the Danish sector, such as the Roar, Adda, Tyra, Sif, Halfdan, and Igor fields. The salt-related trap examples include a broad range of geometries, starting from a gentle swell, such as those in the Ekofisk and Dan fields, and ending with salt diapirs, which rupture the sub-chalk syn-rift intervals and strongly deform the post-rift chalk package.

The younger post-rift reservoirs reside in the Paleocene sedimentary section (Ahmadi et al., 2002). The seals are

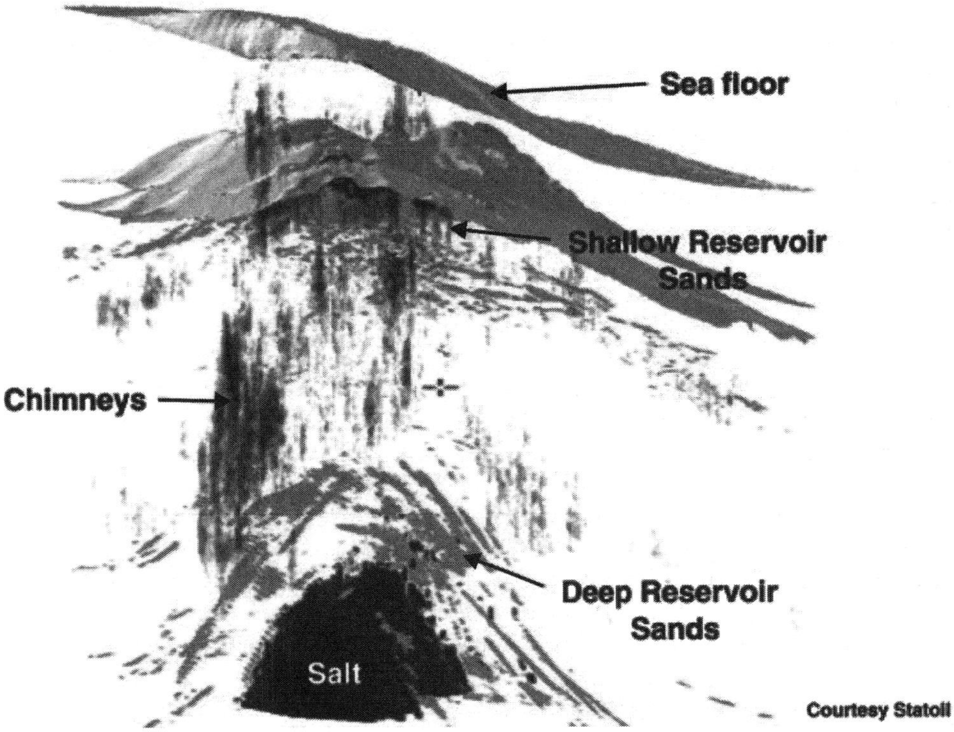

Figure 17.17. Seal failure associated with seismically imaged diffuse chimney above the trap from unspecified rift or passive margin hydrocarbon province (Aminzadeh and Connolly, 2002).

either formed by the intra-Paleocene shales like that in the Mariner discovery in the UK sector or reside in the younger sedimentary section as in the case of the Jotun field in the Norwegian sector, which is a trap composed of two four-way dip-closed structures and one stratigraphic trap sealed by mudstones overlying the Upper Paleocene Lista Formation. An extra example comes from the Pierce field in the UK sector, which relies also on a lateral seal, formed by a salt diapir.

The Eocene post-rift reservoirs of the North Sea province rely mainly on the regional seal formed by the claystone of the Horda Formation and partly on the intra-formational seals (Jones et al., 2002). An example comes from the Frigg field region where the reservoir sandstone has been deposited in a submarine fan system of the South Viking Graben and where closures are formed by a combination of the fault seal and differential compaction across the fan. This area is known to contain numerous gas chimneys in seismic images, indicating active gas migration through fractured shale horizons.

Seals in passive margin settings

Several passive margins provide us with examples emphasizing that all traps are inherently transient. For example, Aminzadeh and Connolly (2007) discussed the reasons for dry and sub-economic wells in deep gas plays in the Tertiary strata and concluded that about 6 percent of them are represented by seal and migration failures. Most typical seal failures observed in the North Sea and the Gulf of Mexico can be grouped into several categories (Brooke et al., 1995; Heggland, 1998; Bolchert et al., 2000; Cole et al., 2001):

1) recent faults reaching the surface and, connecting it with an active hydrocarbon kitchen through the seal horizon;
2) shallow salt bodies associated with faults, which connect the failed seal with the surface;
3) mud volcanoes and other localized seeps associated with mud mounds, authigenic carbonates, carbonate mounds, hydrate mounds, and chemosynthetic communities, having feeder systems penetrated through the seal horizon to the surface; and
4) localized and broadly distributed fracture clusters breaching the failed seal horizon, indicated by shallow subsurface anomalies such as carbonate cementation features, shallow gas accumulations, and wipeout zones from gas chimneys.

A robust observational database from the entire northern continental slope of the Gulf of Mexico, composed of 3-D seismic data, high resolution acoustic surveys, and submersible survey data, suggests that venting through such seal failure occurs on a variety of time scales and at various rates (Roberts and Carney, 1997). Most of the venting sites are controlled by faults. As the case histories of the GB 338 and GB 382 mounds document, the seepage rates vary as the seal failure evolves with time. The evolution is controlled by the interplay between:

1) sedimentary loading;
2) sea-level changes;
3) fault movements; and
4) salt/shale body adjustments.

Seal failures associated with seismically imaged diffuse chimneys above the trap in provinces such as the Gulf of Mexico, the Niger Delta, and the North Sea (Figure 17.17) observed above 10–20 percent of the producing fields indicate that fracture-controlled failure does not necessarily breach the seal to a significant extent (Heggland, 1998; Bolchert et al., 2000; Connolly et al., 2004; Connolly and Aminzadeh, 2006). On the contrary, no producing fields fall into a category of the failure of the sealing fault, as the database of Connolly and Aminzadeh (2006) records.

The broader Equatorial Atlantic region (see Nemčok et al., 2004b) provides a good natural laboratory for seals in the passive margin setting. The discussion of seals in this region would benefit from separating the margins into segments as based on the mechanism of the upper continental crust breakup that preceded their passive margin development (Figure 16.13). Such division yields extensional and transform passive margin segments. The reason for this division is the observation (Nemčok et al., 2012b) that gravity gliding starts at extensional margins relatively early while it starts rather late at transform margins, if it takes place at all.

The following text will discuss the sealing characteristics of both extensional and transform margin types, separating their sedimentary packages into the pre-rift, syn-rift, and post-rift portions.

Seals at extensional passive margin segments

For seals, which are most typical for extensional segments, we need to look at the Foz do Amazonas and Potiguar Basins, the Caete sub-basin of the Barreirinhas Basin, the Mundaú sub-basin of the Ceará Basin, and the Demerara Plateau on the South American side and offshore Gabon and Liberia Basins, Niger Delta, and the Guinea Plateau on the African side of the Equatorial Atlantic.

Syn-rift

Typical seals in the syn-rift section of the Caete sub-basin interpreted in exploration studies are Aptian-Albian shale horizons deposited in fluvial, deltaic, and lacustrine environments. A fairly commonly assumed top seal for the syn-rift traps is a middle Albian breakup unconformity and intra-formational shales of onlapping formations. All these seals are proven in the oil-producing Mundaú sub-basin where the existing fields are related to eroded edges of tilted fault blocks and roll-over anticlines (De Azevedo, 1991). The most typical top seals here are the Aptian-Albian shale horizons. Top sealing in the South Gabon Basin, where the syn-rift fields are located mostly on the platform, is provided by either intra-formational shales within the syn-rift section or by evaporite intervals of the rift-drift transitional section. Another example of the syn-rift seals comes from the Senegal Basin where sealing is provided by the intra-formational shales of the fluvio-deltaic and lacustrine facies.

Numerous syn-rift traps rely on sealing faults, as documented in the syn-rift section of the South Gabon Basin, which contains three fields located in syn-rift anticlines and three fields located in tilted and truncated extensional fault blocks. The same applies to the Senegal Basin, where the trap scenarios include syn-rift tilted block-related structural traps, which sometimes contain a stratigraphic component, and to the Potiguar Basin, which was the second most important producing basin in Brazil in 2000 (Filho et al., 2000).

Transitional

As the oil-producing Mundaú sub-basin proves, syn-rift traps also rely on seals deposited during the rift-drift transition period, represented by the upper Aptian, and sometimes even lower Albian, evaporitic horizons here (Regalli, 1989). They can be locally up to 100 m thick.

Traps in the sedimentary section deposited during the rift–drift transition typically rely on a lateral sealing by salt structures in regions that contain evaporites in the stratigraphic section of the rift–drift transition. Typical examples of this sealing come from the Congo and Angola Basins where numerous fields produce from the platform carbonate and siliciclastic reservoirs of the Albian Madiela Formation.

Post-rift

Typical seals of the Caete sub-basin assumed in exploration studies include the Cretaceous-Tertiary intra-formational shales capable of sealing channel fills. The Miocene intra-formational seals are proven by the Calcutta discovery in the Guyana Basin, which documents oil pools in Miocene sandstones (Davison et al., 2002). The older, Upper Cretaceous–Tertiary seals are proven to seal oil pools in both the Paleocene sandstones of the near-shore region and the Upper Cretaceous turbidites in the deep offshore Guyana Basin (Davison et al., 2002). The sealing capability of the Paleocene-Eocene intra-formational shales is proven by the Tambaredjo field in Suriname (Popescu, 1995). The field is mostly based on the stratigraphic trap in Paleocene stratigraphy. The producing facies was deposited over the truncated Cretaceous section in a beach environment. The top seal is a clay horizon 3–8 m thick. The bottom seal is formed by the aforementioned unconformity. While seals are not difficult to find in the Guyana Basin itself, characterized by the wealth of the Upper Cretaceous–Miocene marine shales formed by a background deposition at numerous stratigraphic levels, they are a problem in the shelf environments at basement highs such as the Demerara Plateau and Pomeroon High, bounding the basin from east and west, respectively (Popescu, 1995).

The seals of thick carbonate platforms rimming large regions of the northern Brazilian Equatorial Atlantic margins are known to be problematic. For example, top seals in the Caete sub-basin sealing porous platform carbonates deposited from the Albian to recent times are apparently not efficient (Davison, 1999).

Foz do Amazonas represents a basin at a passive margin segment affected by a large river entering a deep basin. Schenk and colleagues (2003) compiled a regional hydrocarbon assessment summary for this basin. It includes the Neogene hydrocarbon systems associated with a 10 km-thick sedimentary section deposited in the deltaic and slope environments off the Amazon shelf since the Miocene. The majority of the listed potential traps reside in both the extensional and contractional domains of the gravity glide system. Hanging wall closures in extensional domains rely on the sealing normal faults deforming layers dipping in the same direction as faults. Footwall closures rely on a combination of sealing normal fault with uplifted footwall. Traps contain a stratigraphic component including channel and pinching-out sediments developed at delta flanks.

Yet another example of the gravity glide system detached inside shale comes from the Niger Delta. While both exploration and production during the past decade are in decline, the province still ranks as one of the largest provinces, with proven ultimate recoverable reserves of about 26 billion bbl in 2000 (see Haack et al., 2000). Traps in the extensional domain consist of anticlinal, hanging wall, and footwall closures. The footwall closures are provided by a combination of sealing normal fault and uplifted footwall, with sediments and fault dipping in opposite directions. Some hanging wall closures are provided by a combination of sealing fault with strata dipping in the same direction. These producing traps are rare but do occur at two locations. The first of them is related to an internal seaward-dipping fault that forms a juxtaposition seal. The second location is related to an internal seaward-dipping fault that juxtaposes hanging wall to shale diapir.

Some Niger Delta traps contain a stratigraphic component, provided by the channel- and pinch-out-related features developed at delta flanks. Truncations against clay-filled channels are known from the Afam and Qua Iboe channels. Regional sand pinchouts are known from some regional flank areas where the sequence thins against the basement-controlled ridge structures.

Typical top seals in the Niger Delta are formed by multiple fine clastic intercalations in the Abgada Formation. Lateral seals provided by faults have been proven by numerous examples from producing fields. Growth faults are commonly sealing because of a juxtaposition of sand horizons against shale horizons. Some normal faults without growth character also seal because of juxtaposition or efficient clay smear along them. A rule of thumb for the Niger Delta (Weber and Daukoru, 1975) anticipates efficient clay smears when the shale/sand ratio in the hanging wall is 1:4, honoring the fact that one needs thick shale to provide a high smear potential.

While the Foz do Amazonas Basin and Niger Delta both contain gravity glide systems developed on top of shale, offshore Gabon provides examples of gliding on top of evaporites. Almost three-fourths of Gabon's oil reserves listed in 1990 was trapped in structures associated with salt tectonics deforming the Upper Cretaceous–Tertiary reservoirs sands. The structures typically rely on lateral sealing provided by salt structures. Top seals are provided by marine shales in the Upper Cretaceous and Tertiary sedimentary systems. While all post-rift structural traps of Gabon listed in 1990 were salt-related, there are post-rift stratigraphic traps developed independently of salt tectonics, associated with Miocene channel sandstones.

Seals at transform margin segments

For the most typical seals at transform margin segments one needs to look at the Para-Maranhao and Sergipe-Alagoas Basins, the Piauí-Camocim sub-basin of the Ceará Basin, and the Tutoia sub-basin of the Barreirinhas Basin on the South American side and the offshore Benin, Ghana, Côte d'Ivoire, Togo Basins and the Rio Muni Basin on the African side of the Equatorial Atlantic.

Pre-rift

Occurrences, such as that in the Saltpond field in offshore Ghana, document the importance of the pre-rift reservoirs in the passive margin province. The field contains Devonian reservoirs. Geologic profiles through the field indicate intense fracturing and strike-slip deformation (Best et al., 1985), active during rift-early drift time periods. The highly compartmentalized character of the field is proven by producing wells showing a rapid decrease in flow after days or weeks. Similar compartmentalized field behavior is known from the fields such as the Sao Joao, Espigao, and Oeste de Canoas in the opposite Brazilian margin, in the Barreirinhas Basin.

Various Accra-Keta prospects in offshore Ghana were also explored in the pre-rift Devonian sediments, bounded to the east by the Romanche Fracture Zone.

Syn-rift

Typical syn-rift reservoir seals in the Benin region are:

1) intra-formational shales;
2) Albian breakup unconformity;
3) Albian post-rift sediments; and
4) prominent regional Cenomanian-Turonian shale-dominant sedimentary package.

Albian breakup unconformity is a fairly common seal also in the Côte d'Ivoire province, typically sealing fault-block traps in the underlying syn-rift fill. Proof of fault-block traps occurring in the tilted syn-rift blocks comes from the Carmopolis field of the Sergipe-Alagoas Basin (Lana, 1993), which is currently the largest oil field in onshore Brazil, with reserves of up to 1.2. MMbbl. This field represents about 54 percent of the oil production in the Sergipe-Alagoas Basin. Successful syn-rift and rift-drift transition seals seal traps such as:

1) fault-bounded tilted syn-rift blocks;
2) fractured basement highs;
3) sand-prone pinching-out wedges in the syn-rift section;
4) wrench-related drag folds;
5) footwall overturned beds along certain strike-slip fault segments; and
6) fault scarp sediments created by mass wasting.

The best example is provided by the Aptian shales and evaporites functioning as a top seal in the Carmopolis field.

Additional top seal in the Sergipe-Alagoas province is provided by the Albian marine shales, marls, and limestones. The Lower Cretaceous lacustrine shales also form top seals in rare cases where the reservoir is formed by the fractured Precambrian basement located in footwall edges associated with rifting. Some of the discovered hydrocarbon pools in the syn-rift package exist because of the normal and strike-slip faults providing lateral seals, because of either favorable juxtaposition or a development of the tectonic clay.

Various Accra-Keta prospects were studied in the Albian sediments bounded to the east by the Romanche Fracture Zone. One of them is the Albian flower structure with trapping in tilted block.

The Lome-1 well in the Togo province discovered sub-commercial oil pooled in the Aptian–middle Albian fluvial channel sandstone. Oil has an API gravity of 44.5–50.0°. The average reservoir porosity is 12 percent. The fault-block structure here failed to have a closure where projected but stratigraphic traps are still possible. The planned prospect was a high block between the Keta Low pull-apart basin and a strike-slip fault along its south-southeastern boundary. Well-penetrated residual oil appears in the Lower Cretaceous strata, which were missing in the stratigraphically higher Lower Cretaceous sandstone horizons. The general knowledge from the Togo province allows us to conclude that Albian sediments, if developed below the Cenomanian/Turonian unconformity, contain seal horizons.

Transitional

The Seme field discovered in eastern Benin is formed by a complicated system of different reservoir units sealed by the Turonian shales (Best et al., 1985). The field is highly compartmentalized by the syn-rift strike-slip faults, some of which were reactivated during the early-drift period. The trap is an anticlinal structure truncated by unconformity to the northeast.

The Cenomanian shale is known to be widespread in the Togo shelf area, marking a prominent deepening of the depositional system after the continental breakup. Turonian sealing works in the case of the Aje field located at the Benin–Nigeria border, which contains gas and condensate reservoired in the Turonian sandstone.

The oil and gas occurrences located onshore and on the shelf, in the Deep Ivorian and Tano Basins, such as the Espoir, Foxtrot, Lion, and Panthere fields, are associated with Albian sandstone reservoirs. They are associated with onshore seeps. They have to rely on sealing either from the transitional or early-drift sedimentary packages. The Espoir field, the largest field in offshore Côte d'Ivoire, has reservoirs trapped in tilted fault blocks, which rely on the laterally sealing faults. Its top seal is represented by the post-Albian breakup unconformity. The North Tano field in Ghana proves the rift-drift transition-related seals, holding dominantly gas in the mid-upper Albian sandstone. Similar seals control the South Tano oil and condensate field.

Both fields are sealed by the breakup unconformity and rely on the sealing faults providing lateral seals.

Seismic images from western offshore Côte d'Ivoire highlight the Upper Cretaceous seals sealing the Albian prograded shelf sandstones occurring in fault blocks formed by the Upper Cretaceous gravity gliding detached on shale. They are located either in the rift–drift transition or syn-drift sedimentary packages. Seismic interpretations of Coterill and colleagues (2002) in eastern offshore Côte d'Ivoire indicate that the activity of the Saint Paul strike-slip fault zone ceased by the Cenomanian and the Albian steeper-dipping slope changed into the Cenomanian shallower-dipping slope, resulting in the development of the Cenomanian seals. They are formed by an increase in shale in siliciclastic deposits and the change of canyon formation to a formation of meandering valley systems.

Sealing from transitional or early-drift package is required for the exploration play of Albian raft traps in the Rio Muni Basin (see Dailly, 2000; Coterill et al., 2002). Existing wells have tested some of them. Their success depends on hydrocarbon migration from a syn-rift source kitchen, which should be located, based on the analogy to the North Gabon Basin, under the shelf. Note that the effects of salt tectonics, controlling the lateral sealing, change along the offshore from north to south. The amount of evaporites increases southward and it controls both the increase in size of gravity glides southward and the change from simple to complex salt structures, as Coterill and colleagues (2002) documented. While the northerly seismic sections image salt tongues, the southerly ones image salt canopies and sheets.

Yet another exploration play of the Rio Muni Basin is the Aptian clastic sediments on structural highs. Existing wells have tested some of them. Their success depends on hydrocarbon migration from a syn-rift source kitchen, which should be located under the shelf.

Post-rift

The most typical seals in the post-rift sedimentary package are the Upper Cretaceous–Tertiary intra-formational shale intervals, Miocene intra-formational shale intervals, such as those in offshore Benin and the Tutoia sub-basin of the Barreirinhas Basin. The early syn-drift section frequently needs lateral sealing by shale ridges and diapirs as can be seen in the Piaui-Camocim sub-basin of the Ceará Basin.

The Benin margin shows a rather complex set of traps and seals because the major strike-slip faulting did not stop at the time of continental breakup. It just changed from dextral transtensional faulting to transpressional faulting because of a minor change in drift vector orientation and a fact that the thinned and warm distal continental margin was highly prone to deformation (see Nemčok et al., 2004b, 2012b). As a result, several rather specific trap-seal scenarios are observed in this region in seismic images of exploration blocks. The first of them is represented by the main strike-slip-related anticlines, which have the Upper Cretaceous shales frequently eroded from their crestal areas so that the structural closure has to be sealed against the base-Tertiary unconformity and overlying Paleogene sediments. Some of the growing anticlines affected the location of channels in the Upper Cretaceous section where sand-prone channel fills are sealed laterally and above by shales within the Upper Cretaceous section. Growing anticlines were further responsible for the occurrence of several upslope pinching-out sand-prone wedges on the seaward limb of the main anticlines. These sediments are probably basin-floor fans uplifted into their current lower slope position by the growth of main strike-slip fault-related folds.

The trap-seal scenarios in offshore Benin, which are independent of the aforementioned strike-slip faulting, include the prograding Senonian wedges located inside the subhorizontally layered shelf sedimentary sections and underneath the Senonian unconformity, and the Paleogene wedges of sand-prone sediments located on the lower slope and pinching out upslope, sealed by the surrounding shale-prone sediments.

The eastern offshore Côte d'Ivoire does not seem to have the large transpression-related anticlines such as those in Benin. However, it contains the Upper Cretaceous ponded turbiditic sediments sealed by the Upper Cretaceous shales. The Upper Cretaceous shales serve as seals also in the Cape Three Points area, offshore Ghana, where they control stratigraphic traps. Dana's West oil discovery in the Tano Basin proves the existence of intra-formational Maastrichtian shale capable of sealing a channel fill (Craven, 2000). The Belier oil field in the Côte d'Ivoire produces from the submarine channel sandstones sealed within the Upper Cretaceous sedimentary package.

A similar sealing scenario is known from the Rio Muni Basin center, where Amerada Hess' exploration resulted in the discovery of the Ceiba, Oveng, and Okume fields producing from the Upper Cretaceous channelized and/or ponded turbidites. Similar trap-seal scenarios have been interpreted further south in the basin by Vanco.

Drilling in the Rio Muni basin resulted in proving the upper Aptian-Albian and Maastrichtian lithostratigraphies as behaving as efficient seals. Other evidence comes from the seismostratigraphic studies indicating numerous flooding surfaces in the deepwater region, having the potential to contain efficient sealing lithologies.

The early Tertiary sequence deposited in the upper slope fan system with numerous slumps and bypass canyons forms a potential top lateral seal as indicated in seismic images (Coterill et al., 2002). These canyons are filled by shale and oriented perpendicular to the Senonian amalgamated channels.

Proof of efficient Upper Cretaceous–lower Tertiary seal lithologies also comes from the offshore Sergipe-Alagoas Basin (see Bacoccoli, 1991). The example comes from the Guaricema field holding hydrocarbons in the Upper Cretaceous–lower Tertiary sandy channels and turbidites of the Calumbi Formation. Its oil pools are sealed by the deepwater shales and marls deposited during transgressive and regressive events in the syn-drift section.

The Oligocene unconformity in the Côte d'Ivoire province is also known to behave as a seal in several instances, sealing gravity gliding-related fault-block traps. The Tertiary shales form potential top seals.

18 Models of hydrocarbon migration

Introduction

Petroleum, which is generated within a volume of source rock in rift or passive margin settings, migrates out of the hydrocarbon generation kitchen and through the basin along a carrier system that contains carrier beds, faults, and fracture systems. Usually, this combination of carriers either disperses the petroleum to isolated volumes within the sedimentary section or transports it to a surface where it undergoes seepage (Macgregor, 1993). Less usually, the petroleum migrates into structural or stratigraphic traps that retard its further updip migration. However, only a subgroup of traps contains a commercially important volume of petroleum.

If we characterize each stage of the petroleum expulsion and migration from the petroleum generation kitchen as process leading to dispersion, seepage, and trapping (Figure 18.1), following Biteau and colleagues (2003), we can derive an efficiency ratio of each petroleum system, that is, petroleum system yield, PSY, as the ratio of hydrocarbons accumulated, HCA, to hydrocarbons generated, HCG.

PSY usually ranges between 0.1 percent and 30 percent, although the majority fits between 1 percent and a few percent (Biteau et al., 2003). Different petroleum systems in the same basin may have different petroleum system yields as is the case in the Jeanne d'Arc Basin of the Labrador-Newfoundland shelf, a passive margin setting where the Egret-Hibernia(!) and Egret-Avalon(!) petroleum systems have been assigned the petroleum system yields of 44 percent and 28 percent, respectively (Magoon et al., 2005). The high efficiencies of these systems are controlled by the expelled petroleum migrating updip to nearby faulted, anticlinal traps, where much of it migrated across faults and upsection.

Factors that contribute to an efficient petroleum system can be documented in the example of the Cretaceous and Tertiary West African petroleum systems. Being sourced by world-class source rocks such as the Barremian Bucomazi-Melania-Marnes Noires, Albian-Senonian Iabe, Turonian Azile, and other coeval strata, the respective systems further capitalize on the superposed or imbricated source rocks with plays. A similar combination of the world-class source rock and source rock/reservoir imbrications is known from the upper Oxfordian–lower Ryazanian sedimentary section of the North Sea province (e.g., Leythaeuser et al., 1988; Kubala et al., 2002).

Petroleum migration is divided into primary and secondary. Primary migration is the expulsion of generated petroleum molecules from the kerogen followed by their pressure-driven movement through the source rock matrix and transient microfractures (e.g., Leythaeuser et al., 1987; Littke et al., 1988; Mann, 1994; Mann et al., 1997). Secondary migration represents a long-distance movement through the reservoir rock matrix and fractures. It is driven mainly by buoyancy resulting from the density contrast of petroleum fluids with respect to coexisting formation water (e.g., Mann et al., 1997). The secondary migration is complex. It is controlled by factors such as the separate phase flow, diffusion, solution, and dissolution of gas in oil and water, and chemical cracking.

Migrating oil and gas can become entrained in an externally driven water flux, described in Chapter 8. Because oil and gas solubilities are very low, this is not counted as an important migration mechanism (McAuliffe, 1979). Water flow is more effective at dispersing the lighter, water-soluble components of petroleum, rather than concentrating them in traps. However, lighter oil components can migrate in a gaseous solution (Thompson, 1988; Leythaeuser and Poelchau, 1991). Yet another migration mechanism is the transport along concentration gradients by diffusion or osmosis. Its rates are several orders of magnitude lower than those of the buoyancy-driven flow (Krooss et al., 1991). Therefore, this mechanism is important only in primary migration. Primary and secondary migrations can be treated as multiphase fluid flow driven by petroleum fluid potential gradients (England et al., 1987), but on different scales.

The relationship of maturation and expulsion can be rather complex (Mann et al., 1997). As discussed in Chapter 15, the type of petroleum generated and its chemical composition depends on the type of kerogen

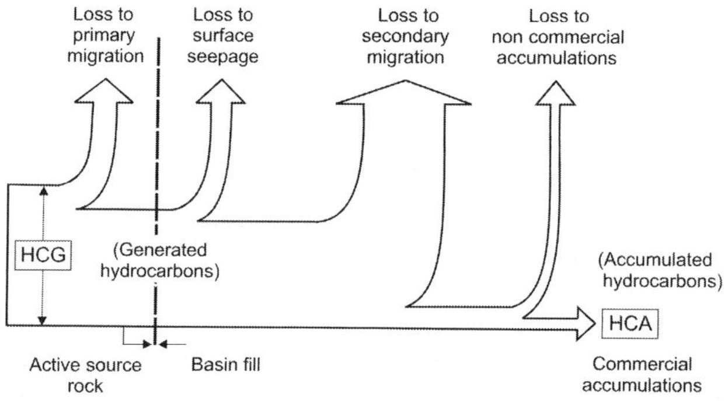

Figure 18.1. Hydrocarbon loss during migration from an active source rock (England, 1994). Only a small portion of the hydrocarbons generated (HCG) is expelled from the source rock (HCE) to find its way into commercial accumulations (HCA). Most of the expelled petroleum is dispersed in the secondary migration carrier system, lost to surface seeps or accumulated in sub-commercial pools.

within the source rock and the maturity stage of the source rock. The generated petroleum type and the rate at which it leaves the source rock depend on the expulsion behavior of the source rock, which in turn depends on the expulsion mechanism. The expulsion efficiency influences the maturity level of the hydrocarbon released, and thereby its composition. If the source rock continues to be thermally stressed, oil not expelled is ultimately transformed to a more easily expelled gas.

Further complexities for petroleum generation, expulsion, and migration are introduced by the following factors:

1) Rates of syn-tectonic burial and uplift are responding to tectonic loading, deposition, and isostatic rebound. Active regional tectonics, gravity gliding, and the location of large rivers have the effect of compartmentalizing the deformed basin, creating adjacent segments with markedly different burial and thermal histories.
2) The extensional stresses associated with the rift setting and an extensional domain of gravity gliding and compressional stresses associated with the contractional domain of gravity gliding may have more than the normal secondary influence on the generation of petroleum. In addition, they may contribute to the hydrofracturing of maturing source rocks. This results in the development of microfractures that aid in the migration of petroleum out of the source rock, increasing the expulsion efficiency.
3) The complex array of faults and fracture networks associated with regional extension and gravity gliding links deformed carrier beds and modifies the efficiency of the carrier beds to form an extremely complicated and ever-changing plumbing system for petroleum migration.
4) Topographically driven hydrodynamics can be unusually active in rift settings that have their portion subaerially exposed.

All of these factors make the investigation and modeling of petroleum generation, expulsion, and migration challenging.

Petroleum generation

Petroleum generation results from a set of chemical reactions transforming kerogen to liquid and gaseous products of lower molecular weight and to residues of an increasing degree of condensation (Schenk et al., 1997). With increasing maturity, crude oils undergo a progressive increase in hydrogen content, API gravity, and gas-oil ratio. Progressive increase in hydrogen content also characterizes gas expelled from kerogen. Eventually, oil generation gives way to condensate and, finally, dry gas generation. The calculated oil generation occurs in the 0.6 to 1.1 percent vitrinite reflectance (Ro) range, gas generation in the 0.5 to 2.2 percent Ro window, and oil cracking in the 1.6 to 3.3 percent Ro range (Sweeney and Burnham, 1990), although these limits differ among different kerogen types as discussed in detail in Chapter 15.

The majority of molecular transformations composing the thermal alteration of organic matter to petroleum during the catagenesis stage involve hemolytic bond breaking, that is, cracking. They are governed by the laws of chemical kinetics (Vandenbroucke et al., 1999). Most of them are quasi first-order reactions in which the concentration of the reacting molecule, c, decreases according the equation:

$$-dc/dt = kc, \tag{18-1}$$

where t is the time and k is the rate of kerogen transformation to petroleum.

The petroleum generation is controlled by time and temperature. The rate of kerogen transformation to petroleum is described by the semi-empirical Arrhenius equation:

$$k = Ae^{-E/RT}, \tag{18-2}$$

where A is the Arrhenius constant, E is the apparent activation energy, R is the gas constant, and T is the absolute temperature.

Equation (18-2) shows that a temperature represents an important control of kerogen conversion to petroleum. The conversion rate almost doubles with a 10°C increase in temperature. The factor RT represents a measure of the thermal energy of the system at a given temperature. If the factor RT is large in comparison to the activation energy, the petroleum generation is fast.

Modern methods of petroleum systems analysis simulate the petroleum generation in the basin with geologic time using models based on Equation (18-2). Its kinetic parameters A and E are routinely determined by programmed pyrolysis experiments on samples from immature source rocks of various kerogen types (e.g., Tissot and Espitalié, 1975). Despite the fact that the range of reactions and processes leading to the thousands of molecular products of the petroleum generation from kerogen is very large and complex, the pyrolysis data attempt to capture, using the bulk kinetic behavior of a single or limited number of components, the essence of the thousands of molecular transformations leading to crude oil and gas.

The philosophy behind numerical simulation of petroleum generation is the assumption that the rates of petroleum generation by programmed pyrolysis at temperatures of about 450°C are predicted by kinetic parameters the same as those predicting petroleum generation by nature at temperatures of about 100°C. Note that this assumption is not always appropriate because of:

1) errors in values of kinetic parameters determined by pyrolysis; and
2) potential non-universality of the underlying assumption.

Furthermore, the applicability of different pyrolysis methods, including open versus closed or hydrous versus anhydrous systems, is still a matter of debate (see Lewan, 1994, 1997; Pepper and Corvi, 1995; Pepper and Dodd, 1995; Schenk and Horsfield, 1998; Vandenbroucke et al., 1999; Henry and Lewan, 2001; Ruble et al., 2003). Further complexity can be brought into the petroleum generation by clay catalysis (Kissin, 1987) and transition metal catalysis (Mango, 1997), which has been proposed as an important factor in the generation of light hydrocarbons. A further controlling factor of molecular transformations is pressure, the effect of which is measurable, but relatively small in comparison with the effect of temperature (Schenk et al., 1997). Despite these complications, numerical kinetic modeling is currently the most standard tool for predicting the time, depth, and products of petroleum genesis under geologic conditions (Waples et al., 1992a, 1992b; Waples, 1994a), drawing from a variety of computer programs.

Two widely used tables of kinetic parameters for modeling hydrocarbon generation are those published by the Lawrence Livermore National Laboratory (e.g., Burnham et al., 1995; Table 18.1) and the Institute Français du Petrol (e.g., Tissot et al., 1987; Ungerer and Pelet, 1987). Kinetic parameters for source rocks with kerogens of types I, II, and III have been the results of numerous experiments (e.g., Tissot et al., 1987; Lewan, 1993; Tegelar and Noble, 1994; Burnham et al., 1995; Reynolds and Burnham, 1995). Their data document that marine type II and terrigenous type III kerogens generally have broad distributions of activation energies (Table 18.1, Figures 15.14a, 15.12a), while type I kerogens have very narrow distributions of activation energies with a slightly higher maximum (Figure 15.13a). Furthermore, the results indicate that distribution maxima of activation energies can be highly variable depending on sulfur content and the mineralogy of the source rock, among other variables (Ungerer, 1990). For example, data from the type II-S kerogens of the Monterey Formation in California have especially low distributions of activation energies because their C-S bonds are more readily hydrolyzed than C-C bonds (Tegelar and Noble, 1994; Waples, 1994a).

In rift and passive margin scenarios, where sedimentary and tectonic burial is followed in rapid succession by uplift, erosion, and possibly reburial, the differing kinetic behaviors of kerogen types are especially important to the generation of petroleum. For example, type II and type III kerogens mature over a wide temperature/depth range (Figures 15.14b, 15.12b) and are more likely to continue to generate during multiple episodes of maturation, while type I kerogens mature over a very narrow temperature range and expel rapidly (Figure 15.13b) because of their narrow activation energy distributions (Figure 15.13a). They are then more likely to exhaust their generative potential in a single burial event.

While all kerogen types can generate methane and other natural gases, the sapropelic source rocks with types I, II, and II-S kerogens can initially yield volumes of liquid hydrocarbons larger than those from type III humic kerogens (Figure 18.2). They are considered oil-prone, while humic kerogens are gas-prone. As pyrolysis experiments indicate, humic kerogens are associated with gas-oil ratios higher than those of sapropelic kerogens (Lewan and Henry, 2001).

Table 18.1. LLNL kerogen kinetic parameters (Burnham et al., 1995).

Cracking P-primary S-secondary	Fract. (0–1)	Activation Energy (kcal/mol)	Arrhenius Constant (1/my)	Reaction Constants
Type I				
P	0.07	49.0	$1.6*10^{27}$	Primary oil = 650
P	0.90	53.0	$1.6*10^{27}$	Primary gas = 70
P	0.03	54.0	$1.6*10^{27}$	
S	1.00	54.0	$3.2*10^{25}$	Secondary gas yield = 50%
Type II				
P	0.05	49.0	$9.5*10^{26}$	Primary oil = 350
P	0.20	50.0	$9.5*10^{26}$	Primary gas = 65
P	0.50	51.0	$9.5*10^{26}$	
P	0.20	52.0	$9.5*10^{26}$	
P	0.05	53.0	$9.5*10^{26}$	
S	1.00	54.0	$3.2*10^{25}$	Secondary gas yield = 50%
Type III				
P	0.04	48.0	$5.1*10^{26}$	Primary oil = 50
P	0.14	50.0	$5.1*10^{26}$	Primary gas = 110
P	0.32	52.0	$5.1*10^{26}$	
P	0.17	54.0	$5.1*10^{26}$	
P	0.13	56.0	$5.1*10^{26}$	
P	0.10	60.0	$5.1*10^{26}$	
P	0.07	64.0	$5.1*10^{26}$	
P	0.03	68.0	$5.1*10^{26}$	
S	1.00	54.0	$3.2*10^{25}$	Secondary gas yield = 50%

Typical ratios for sapropelic and humic kerogens are within the intervals 382–2,381 and 883–2,831 scf/bbl, respectively.

Because sapropelic kerogen has a higher elemental concentration of hydrogen at the outset, it will have a larger yield of both oil and methane than that of humic kerogen (Figure 18.2). In fact, sapropelic kerogens can generate twice as much hydrocarbon gas per gram of organic matter as humic kerogen (Lewan and Henry, 2001). To a large extent the methane and other hydrocarbon gases from sapropelic kerogen are generated from residual oil, called bitumen, which remains in the source rock as it is buried to deeper and hotter levels of the basin. Therefore, prolific natural gas provinces are not restricted to coal-bearing basins.

The temperature is the key parameter in petroleum generation (Quigley and Mackenzie, 1988). While good oil-prone source rocks reach peak maturity at temperatures of 100–150°C, lean oil-prone source rocks reach it at temperatures of 120–150°C. At temperatures in excess of 150°C, oil retained in the source rock cracks to gas.

A generation of liquid hydrocarbons from coals is a complex process. It is controlled by factors such as

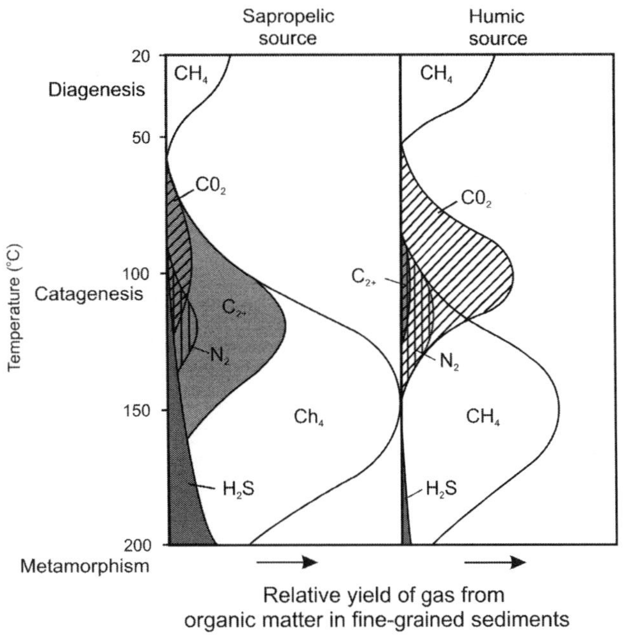

Figure 18.2. A generation of natural gases with a depth of burial (Hunt, 1979). C_{2+} represents hydrocarbons in the gas phase heavier than CH_4.

(Hunt, 1991; Katz et al., 1991; Thompson et al., 1994; Petersen et al., 1996):

1) the type of macerals composing the kerogen;
2) the microstructure of the coals;
3) petrophysical properties controlling involved constituent macerals; and
4) hydrocarbon saturation thresholds.

The pyrolytic products of coals depend on maceral content more than on age or geographic location (Katz et al., 1991). Vitrinite-rich coals generate aromatic hydrocarbons with minor amounts of n-paraffins. Algae-rich coals and some other exinites generate mainly paraffinic hydrocarbons. Resin-rich coals generate naphthenic and aromatic oils.

Two processes are responsible for methane generation: biogenic and thermogenic. The first process occurs in sediments at very shallow depths and involves the microbial breakdown of organic matter called methanogenesis. The microbes responsible for biogenic methane do not survive in temperature conditions greater than about 40–50°C (Hunt, 1979; Figure 18.2). A natural gas of microbial origin (Martini et al., 2003) is nearly pure methane with no other hydrocarbon gases, having low wetness value and various amounts of carbon dioxide (Figure 18.2). Because of the metabolic selectivity of the microbes, biogenic methane is isotopically lighter than thermogenic methane, having larger negative $\delta^{13}C$. The $\delta^{13}C$ of microbial methane is typically lighter than -55 ‰ (Mattavelli and Novelli, 1988; Whiticar, 1994).

The second process occurs at greater depths and involves thermal disaggregation of organic molecules containing hydrogen. The chemical processes leading to thermogenic methane become significant at temperatures greater than 135–175°C (Pepper and Corvi, 1995). There are two main sources of thermogenic gas: maturing source rocks and the in situ cracking of oil in reservoirs (Pepper and Dodd, 1995; Schenk and Horsfield, 1998). Low-maturity thermogenic gas produced from the primary cracking of kerogen is primarily methane (Seewald et al., 1998). In contrast, mature thermogenic gas associated with the onset of bitumen generation (Figure 18.2) consists of methane together with other hydrocarbon gases, resulting in a high wetness value. Its $\delta^{13}C$ values are only slightly lighter than those of the precursor organic matter. It is also associated with non-hydrocarbon gases, such as nitrogen, helium, and/or hydrogen sulfide. As the thermal maturity of the natural gas increases, its wetness value decreases. Thus, shallow microbial and deep, highly mature thermogenic methanes are both a dry gas and low on the fraction of other hydrocarbon gases and have a low wetness value. They can be distinguished only by their isotopic compositions.

Most C_1-C_4 generation occurs after the peak of oil generation. However, the CO_2 generation occurs before, during, and after the generation of liquid hydrocarbons (Seewald et al., 1998). Type I sapropelic kerogens tend to start gas generation at thermal maturity stages slightly higher than those of either Type II or Type III kerogens (Figures 15.12b, 15.13b, 15.15b). Furthermore, gas generation is not restricted to high thermal maturities representing the post-oil generation, above the Ro = 1.2% threshold. Significant amounts of thermogenic gas can be generated coevally with oil (Schenk and Horsfield, 1998). Approximately 75 percent of gas from the source rock can be, thus, generated by the end of oil generation, at Ro = 1.1%. However, bitumen and kerogen retained in a maturing source rock that has passed through the oil window will crack to natural gas with further thermal stress (Pepper and Dodd, 1995).

Trying to simplify kinetic modeling, Pepper and Corvi (1995) statistically calculated the entry and deadline temperatures for the oil and gas windows for a reference heating rate of 2.0°C/Ma for different organofacies (Table 18.2). The resultant temperatures vary with kerogen type and increase by about 5°C with every

Table 18.2. Reference oil and gas window thresholds and deadlines for representative organofacies (modified after Pepper and Corvi, 1995). The calculated window boundaries are for a reference heating rate of 2.0°C/Ma. A doubling of the heating rate elevates each of the boundaries by about 5°C.

Organofacies	Precursor biomass	Environment/age association	Sulphur incorporation	Kerogen type	Oil window	Gas window
A – Aquatic, marine, siliceous, or carbonate/ evaporite	Marine algae and bacteria	Marine, upwelling zones, clastic-starved basins of any age	High	Type II-S	95–135°C	105–165°C
B – Aquatic, marine, siliciclastic	Marine algae and bacteria	Marine, clastic basins of any age	Moderate	Type II	105–145°C	140–210°C
C – Aquatic, non-marine, lacustrine	Freshwater algae and bacteria	Tectonic non-marine basins and coastal plains (Phanerozoic)	Low	Type I	120–140°C	135–170°C
D+E – Terrigeneous, non-marine, waxy	Higher plant cuticule, lignin, bacteria, and resin (facies D)	Mesozoic and younger "ever-wet" coastal plains	Low	Type III	120–160°C	175–220°C
F – Terrigeneous, non-marine, wax-poor	Lignin	Late Paleozoic and younger coastal plains	Low	Type III/IV	145–175°C	175–220°C

doubling of heating rate. However, this table provides only a useful rule of thumb for predicting the levels of generative windows and does not form a substitute for rigorous numerical simulation.

Primary migration

Primary hydrocarbon migration in a source rock can be understood as occurring initially as diffusion through micropores (Figure 18.3) connecting with mesopores to macropores and fractures (Mann, 1994). Migration at this point leads to the development of the petroleum bulk phase, which migrates using the macropore and fracture network of the source rock. The volume expansion of newly generated petroleum controls the overpressure development (Osborne and Swarbrick, 1997 and references therein). The overpressure is widely accepted as the principal driving mechanism of expulsion (England et al., 1987; Durand, 1988), but other factors can also play a role.

Primary migration occurs after newly generated hydrocarbons fully saturate the available pore space. There is a causal link between expulsion efficiency and the initial petroleum potential of a source rock (Mackenzie and Quigley, 1988).

The richer the source rock, the larger the excess volume of petroleum, which is available for expulsion after the pore and fracture network in the source rock is fully saturated. In rich source rocks, which can generate more than 5 kg of hydrocarbons from a ton of rock, the expulsion efficiency is in the order of 60–80 percent. In lean source rocks, the efficiency does not exceed 40 percent.

An alternative understanding of expulsion (Pepper, 1991) argues that unexpelled petroleum is not retained by capillary effects in the pore network. Instead, it is adsorbed onto organic matter. Pepper (1991) proposes that in tight, clay-rich source rocks the relative permeability to oil is greater than that to water. This is due to bound water adsorbed on hydrophilic/hydrophobic clay particles, affecting relative permeability so as to favor the Darcy flow of oil. This explains why he observes a good correlation between expulsion efficiency of a source rock and its hydrogen index, why the correlations of expulsion efficiency/total organic carbon and expulsion efficiency/initial potential remain inconclusive.

The easiest and most efficient oil expulsion has been recorded from type I source rocks characterized by a hydrogen index of up to 900 mg/g C_{org}. Relatively efficient and early oil expulsion has been determined from type II source rocks with a hydrogen index of 500–700 mg/g C_{org}, and some sapropelic coals. Efficient expulsion

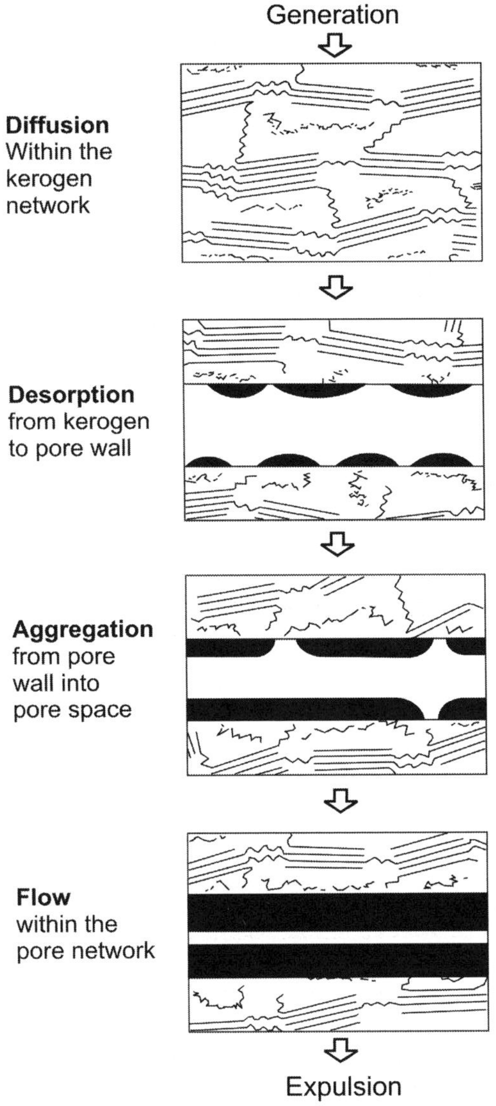

Figure 18.3. Primary migration mechanisms for hydrocarbons in source rocks (after Mann et al., 1997).

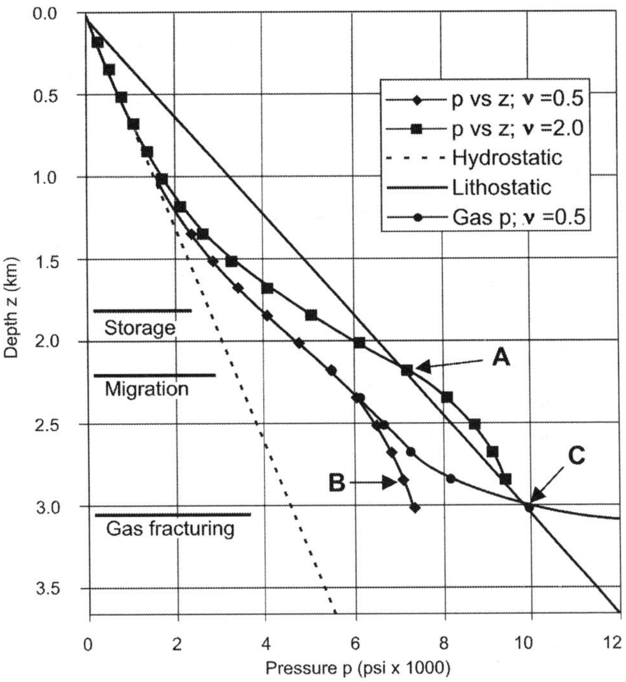

Figure 18.4. Modeled pressure buildup with depth during oil and gas generation in the Austin Chalk source rock (Berg and Gangi, 1999). Pressure curves are shown for a rich source rock with a TOC of about 2.7 wt% and a lean source rock with a TOC of 0.65 wt% having ratios of initial kerogen to initial water, v, of 2.0 and 0.5, respectively. Significant overpressures occur at a depth of 1,830 m, just below the top of the oil window and in a region in which generated oil is probably retained in the source rock. For the richer source rock, the fracture pressure is reached at a depth of 2,160 m (point A). The leaner source rock, however, does not reach fracture pressure through oil generation alone (B), but rather only after appreciable gas generation (point C) at a depth of 3,000 m.

of only gas/gas condensate at high levels of thermal maturity characterizes humic coals and type III source rocks. Only at high maturity can they expel more than 40 percent of their potential. Regardless of the source rock type, the gas expulsion efficiency is usually greater than 90 percent (Pepper, 1991).

The organic matter conversion to less dense oil and gas in very fine-grained source rocks with permeabilities of 10^{-19} to 10^{-20} m^2 commonly leads to pore pressure buildup (e.g., Momper, 1978; Ungerer et al., 1983; Osborne and Swarbrick, 1997 and references therein), which induces fracturing (e.g., Littke et al., 1988; Berg and Gangi, 1999). The fracturing enhances a rock permeability by up to six or seven orders of magnitude, up to a permeability of 10^{-13} m^2 (Capuano, 1993) and provides important expulsion conduits (Leythaeuser et al., 1988). Pressure-induced fracturing tends to occur in rich source rocks capable of generating large volumes of oil or at elevated levels of maturity characterized by gas formation (Littke et al., 1988; Berg and Gangi, 1999; Figure 18.4). The hydraulic fracturing is further controlled by factors such as high hydrocarbon generation rates and stress loading (Cosgrove, 2001).

Expulsion efficiency plays an important role in the oil-generating potential of sapropelic coal. Some coals can have high microporosity and absorption capacity, enhancing the entrapment of generated hydrocarbons in the source rock matrix, thus retarding the expulsion (Hunt, 1991). In order to undergo expulsion, these coals have to reach an oil saturation of about 30 mg

HC/g TOC (Snowdon, 1991). Such threshold value is difficult to reach for less oil-prone coals, which results in oil remaining in the coal to be cracked to gas at a further increased maturity level.

Secondary migration

Secondary migration of petroleum through a carrier bed or fault network from a source rock to a reservoir in the trap is driven dominantly by the buoyancy controlled by the density contrast between petroleum and formation water (Schowalter, 1979; England et al., 1987; England, 1994). The hydrodynamic flow, if present in the cases discussed in Chapter 8, may deflect migration pathways by acting either with buoyancy or against it (Hubbert, 1953; Dahlberg, 1995). The hydrodynamic flow can be further modified by the stress perturbations associated with regional tectonics. The main force opposing buoyancy is the capillary pressure. It is a function of the pore-throat radii of the carrier system, petroleum-water interface tension, and wettability (England, 1994). Another opposing force is the frictional resistivity force controlled by the fluid viscosity, its filtration velocity, and the permeability of the carrier bed.

Under hydrostatic conditions, the driving buoyancy force for a petroleum column or rivulet having a specific vertical height is:

$$P_b = gh(\rho_w - \rho_o), \tag{18-3}$$

where P_b is the driving buoyancy force, g is the hydrostatic gradient, h is the vertical height of a rivulet, and ρ_w and ρ_o are the densities of formation water and oil, respectively. In the case of a vertically rising petroleum rivulet being deflected along an inclined surface, the buoyancy force is reduced by the sinus of the dip angle.

The capillary pressure is controlled by forces acting both within fluids and between fluids and their hosting rocks (Vavra et al., 1992; Vandenbroucke, 1993). While forces between fluid and fluid are cohesive, forces between host rock and fluid are adhesive. If the adhesive forces overcome the cohesive forces, the fluid has a wetting character. In the opposite case, it is a non-wetting fluid. An example of wetting fluid is water, at least in most reservoir rocks.

The relative wettability of a fluid is characterized by the contact angle, θ, which is one of the controlling factors of the capillary pressure. It is the angle between the fluid–fluid interface as measured through the denser fluid. Remaining controlling factors of the capillary pressure are the interfacial tension between immiscible fluids, γ, and the pore throat size. The capillary pressure, P_c, can be calculated from the expression:

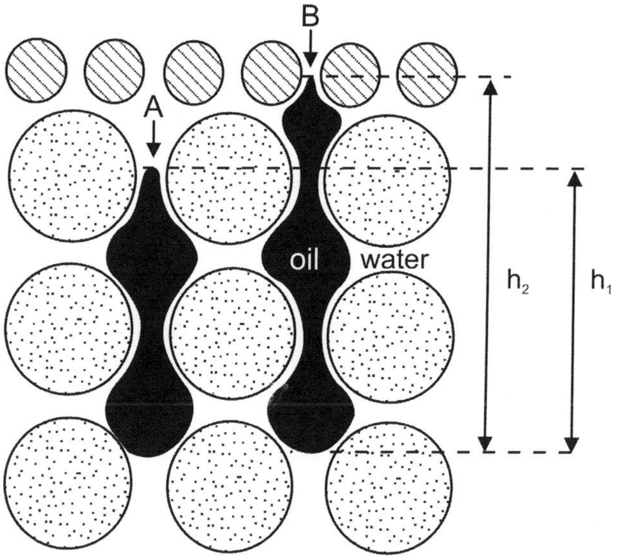

Figure 18.5. Relationship between buoyancy forces and capillary entry pressure in a porous medium (after Berg, 1975). Explanation in text. Copyright © 1974, CCC Republication.

$$P_c = 2\gamma\cos\theta/r, \tag{18-4}$$

where γ is the interfacial tension between oil and water, θ is the wettability, and r is the average radius of the interconnected pore space.

When the buoyancy pressure, P_b, is greater than the capillary entry pressure, P_c, or resistance of the water in the pore throats, A, oil or gas, which represents immiscible phases, migrates in rivulets or filaments upward through the pore system of the carrier bed (Berg, 1975; Figure 18.5). When the migrating fluid encounters a smaller pore throat size, B, the migration can continue only when the initial length of the rivulet, h_1, increases until an adequately long rivulet, h_2, is developed for the increased buoyancy, P_b, to force a breakthrough.

The migration is aided by relative permeability (Pepper, 1991; Matthews, 1999). When the pore network becomes filled by petroleum, its ability to transport water decreases. Its pore pressure increases, helping to push the petroleum phase through the capillary restrictions. Migration rates are controlled by the migration process (Thomas and Clouse, 1995; Matthews, 1999). Typical rates for buoyancy-driven migration are in the order of several mmy^{-1} for oil and up to a meter per day for gas. Typical rates for migration controlled by hydrodynamic flow range between 0.1 and 100 my^{-1}. Compaction- and diffusion-driven flow rates have

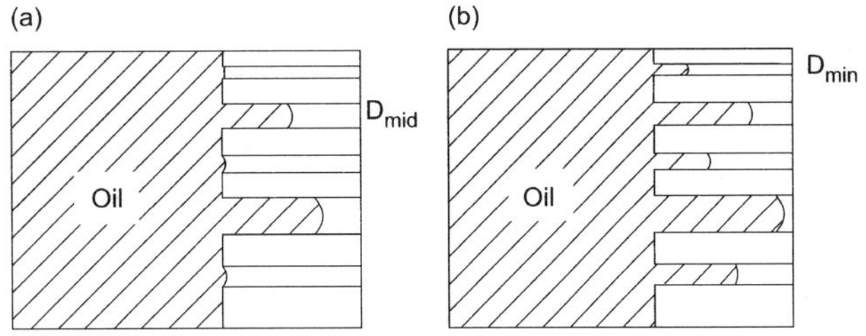

Figure 18.6. The effect of pore-throat size distribution on the difference of pressure jump with respect to the fluid injection rate (Tokunaga et al., 2000). (a) The case of low injection rate is characterized by excess pressure defined by a relatively large pore throat (d_{mid}) through an equation similar to Equation (18-4). (b) The case of high injection rate is associated with a non-wetting phase fluid having to intrude into the smaller pores, which results in increased excess pressure.

typical values of 0.001–1.0 my^{-1} and 1–10 m per 1 Ma, respectively.

Experimental results indicate that the vertical migration of the petroleum rivulet is faster than inclined migration (Dembicki and Anderson, 1989; Catalan et al., 1992; Thomas and Clouse, 1995). The general consensus was to understand it as a flow preferentially along pathways of maximum pore throat size representing maximum permeability. As oil migration experiments in glass cylinders packed with sorted glass beads (Tokunaga et al., 2000) have shown, however, this mechanism holds only under low injection rate conditions. Under a low injection rate, the excess pressure necessary for the non-wetting phase fluid to intrude into the porous medium becomes the same as the capillary pressure of the relatively large pore throats because non-wetting fluid tends to fill from the largest pore throat because of its low capillary pressure (Figure 18.6a). On the contrary, under a high injection rate, the non-wetting phase fluid has to fill the smaller pore throats to preserve the mass balance. Therefore, the excess pressure in this case is higher than that in the low injection rate (Figure 18.6b).

In nature, the secondary migration is further complicated by a 3-D permeability structure. This structure is made complex by the occurrence of dipping sedimentary textures involving grain size variations, various fault zone geometries, and fracture systems. They all tend to form more favorable pathways, those of highest permeability (Figure 18.7) (e.g., Losh et al., 2002; Boles et al., 2004; Tye, 2004b; Jonk et al., 2005; Magoon et al., 2005; Bowen et al., 2007; Jin et al., 2008). When inclined, these features induce a component of lateral migration. The most common geometry of migrating fluids in matrix-controlled carrier beds is rivulets. Many small rivulets rising from the interface between a source rock and carrier bed initially form an anastomosing network of filament-size flow paths (England, 1994). They soon amalgamate into fewer, highly efficient rivulets responsible for the bulk of petroleum migration. Well penetration of the fault acting as an oil and gas migration pathway in the Eugene Island South Addition Block 330 in a passive margin setting, offshore Louisiana, indicates that rivulets are also typical for fault-controlled migration, instead of sheets adopting the geometry of the fault zone (see Anderson et al., 1994a).

Migration paths are controlled by the morphologies of the dominating carrier beds adjacent to top seals in geologic scenarios such as:

1) the interbedded sandstones and shales (e.g., the South Caspian Basin – Bredehoeft et al., 1988; the Songliao Basin – Xie et al., 2003);
2) permeable fault surfaces (e.g., the Gulf of Mexico – Anderson et al., 1994a; Roberts and Nunn, 1995; Connolly et al., 2007; the Junggar Basin – Jin et al., 2008, the Kura Basin – own data; California – Boles et al., 2004; Perez and Boles, 2004; the Recôncavo Basin – Destro et al., 2003b); and
3) fracture networks (the Gulf of Mexico – Bolchert et al., 2000; the Niger Delta – Connolly et al., 2004; the North Sea – Heggland, 1998; Ligtenberg and Thomsen, 2003; Connolly et al., 2004; the West African margins – Connolly et al., 2002).

The carrier systems may act alone, in a series or in parallel, in complex geometric combinations in different parts of the basin (e.g., Allan, 1989; Alexander and Handschy, 1998). The areal extent of top seals plays an important role in vertical drainage localization by forcing petroleum to converge on faults and other types of migration windows through the seals and thus focusing the vertical drainage into chimneys through which vertical rise continues until another set of effective top seals is encountered (e.g., Cobbold et al., 2001; Nunn and Meulbroek, 2002).

In geologic time, the traps are not considered final end-points for migration, but rather a transient storage through which the petroleum will (Sales, 1997):

1) pass by breaching the top seal when the oil/gas column is sufficiently high, or
2) spill out through the bottom of the trap when it is full.

Top seal breaching favors the further migration of gas and condensate over oil (Schowalter, 1979), whereas the trap spilling favors the oil migration (Gussow, 1954).

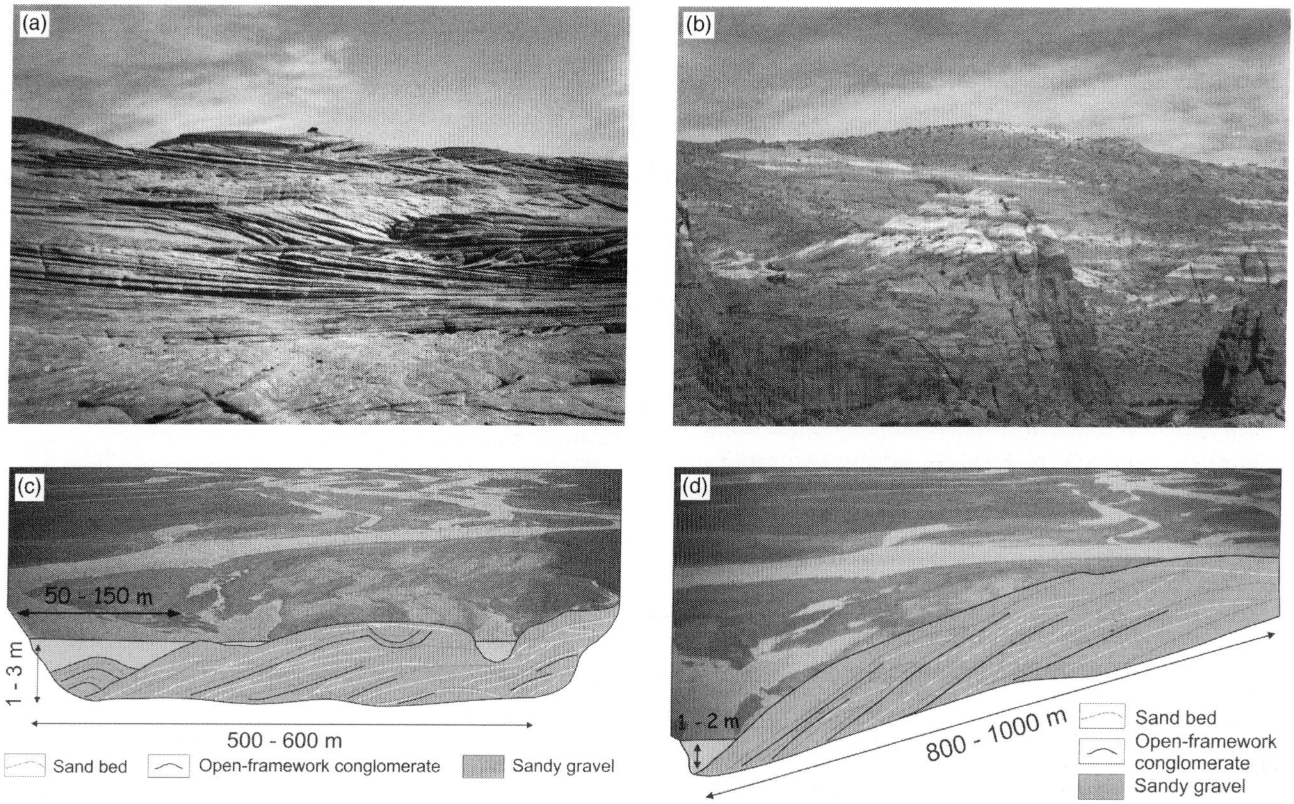

Figure 18.7. (a) Cross-bedded eolian Navaho sandstone, Snow Canyon, St. George, Utah. (b) Cross-bedded eolian Navaho sandstone, Snow Canyon, St. George, Utah, documenting preferential fluid flow pathways indicated by the distribution of bleached zones. The zones are parallel to depositional textures utilizing high permeability horizons and sets of horizons. (c–d) Present-day channel bar sediment analogs for the reservoir sweet spots in the Prudhoe Bay field, North Slope, Alaska (Tye, 2004b). Note that the dominant permeability is controlled by the distribution of open-framework conglomerates. (c) Sedimentary textures in an across-stream direction. Note that sand and sandy-gravel beds can be meters to hundreds of meters wide.

Hydrodynamic flow in a basin increases the possibility of migration pathway deflection from directions determined by morphology alone. The influence of basin hydrodynamics becomes important in geologic scenarios with lateral hydrocarbon migration. Readers interested in a mathematical description of migration pathways under combined buoyancy-hydrodynamic drives are referred to Hindle (1997). Examples of strong hydrodynamic flows in rift basins come from the Newark Basin (Steckler et al., 1993), the Rhine Graben (Person and Garven, 1992; Lampe and Person, 2002), and the Rio Grande Rift (Mailloux et al., 1999). The latter two are described in detail in Chapter 8. The first indications of hydrodynamic flow regimes at passive continental margins are emerging from the East Indian margin (Nemčok et al., 2008).

Experimental studies of petroleum migration (Dembicki and Anderson, 1989; Catalan et al., 1992; Thomas and Clouse, 1995) have simulated flow pathway configurations and flow rates. Thomas and Clouse (1995) have concluded that most of the carrier beds above the source rock would be contacted by oil, but at low saturations of about 5–10 percent. The vertical oil flow rate would reach a value of about 8 $mmyr^{-1}$. Once the vertically migrating oil reaches an inclined permeability barrier, it continues to move laterally through only a small portion of the carrier bed, just 0.5–1.0 m below the sealing surface. The facial control of the vertical versus lateral migration can be well understood from the study of the uppermost Miocene–Lower Pliocene fluvial and deltaic sediments of the paleo-Volga River entering the Caspian Sea (Reynolds et al., 1998). In these sediments, the character of inclined barriers varies:

1) from siltstone and mudstone layers of delta front facies association, which form important laterally extensive seal horizons, dramatically

reducing the vertical permeability of the system and making it prone to lateral migration (Figures 18.8a, b),

2) to siltstone and mudstone horizons of delta-plain and fluvial facies associations, which form local baffles and no baffles to flow, keeping the vertical permeability of the system at a reasonable level and making it prone to a reasonable amount of vertical migration (Figures 18.8c, d).

As discussed earlier, migration is accomplished through a network of thin petroleum rivulets that occupy very little of the pore space of the carrier bed, rather than a more pervasive sheet flow. Therefore, the rates of lateral migration are likely to be geologically very rapid and hydrocarbon losses within the migration network low.

The geometry of the carrier system controls whether migration is dispersive, resulting in the dilution of hydrocarbons away from the petroleum kitchen, or concentrating, resulting in the directing of hydrocarbons toward potential traps. Concave carrier systems tend to dilute whereas convex systems tend to concentrate.

An important control of trap filling is the distance of the potential trap from the petroleum kitchen (Hindle, 1997). Traps located vertically above a kitchen have the greatest potential for filling. Traps tens of kilometers away from the kitchen cannot receive more than a small fraction of the oil expelled, even if flow is focused along a convex carrier system. Both stratigraphic and structural factors control whether migration will be local in the close vicinity of the petroleum generative kitchen or will be long-range. Focused vertical migration is associated with a moderate to high degree of structural deformation resulting in selective breaching and local petroleum leakage through a laterally continuous seal. Lateral migration characterizes basins with a low degree of structural deformation unable to breach a laterally continuous seal. Such laterally drained petroleum systems have several features in common (Demaison and Huizinga, 1991):

1) Oil accumulations occur in immature strata at a considerable distance from the hydrocarbon kitchen.
2) A single reservoir succession underlying the most effective regional seal usually hosts most of the entrapped hydrocarbons in the petroleum system.
3) Faulting of the effective regional seal is minor or insignificant.

Both vertical drainage- and lateral drainage-dominated systems are known from rift and passive margin settings. The former is represented by examples such as the south Eugene Island in offshore Louisiana and the Penglai field in the Bohai Basin (Losh, 1998; Hao et al., 2009). The latter is represented by regions including the Pechelbronn and Landau oil fields in the Rhine Graben (Person and Garven, 1992; Lampe and Person, 2002) and the Central Sumatran Basin province (Williams and Eubank, 1995). While the maximum lateral migration is limited to the width of individual rift units in rift settings, it can be longer than that in the post-rift sections of passive margin settings. However, the length of migration in passive margin settings is frequently controlled by the width of individual contractional and extensional units of gravity glide systems, which are relatively common features of passive margins.

Preferential fluid flow through fault zones has been reasonably well documented, even in the deep crust (MacCaig, 1988; Barton et al., 1995). It is known that fault and fracture systems in sedimentary basins may serve as:

1) migration pathways or barriers to flow; or
2) both at different stages in their histories; or
3) both at the same development stage but different portions of the fault/fracture system.

During and shortly after their activity, the fault/fracture systems represent high-permeability conduits, which are favored by fluid flow (Hooper, 1991; G. P. Roberts 1991; Knipe, 1993; Sibson, 1994, 1996; Dholakia et al., 1998; Connolly and Cosgrove, 1999). They may exhibit fault valve behavior as they regulate the release of fluid overpressures (Sibson, 1990a, 2003; Grauls and Baleix, 1994; Smith and Wiltschko, 1996; Losh, 1998; Cosgrove, 2001), although this behavior is least frequent in rift and passive margin settings. Once the deformation ends, faults and fractures become merely part of the overall porous rock system, equally subject to the diagenetic cementation that also takes place in the matrix pore space (Laubach et al., 2000; Laubach, 2003).

A special category of deformation features in carrier beds are deformation bands. Antonellini and Aydin (1994) observed that the capillary pressure within deformation bands in sandstone produced by cataclasis is 10–100 times greater than that in the surrounding host rock. The deformation bands, instead of acting as subvertical conduits for fluid flow similarly to the previously mentioned faults and fractures, are highly effective barriers to flow.

In order to determine the control of faults, fractures, and deformation bands on fluid flow in a carrier bed, a suite of analytical methods such as stable isotope and fluid-inclusion analysis of vein and host rock cement minerals can be used. These methods have

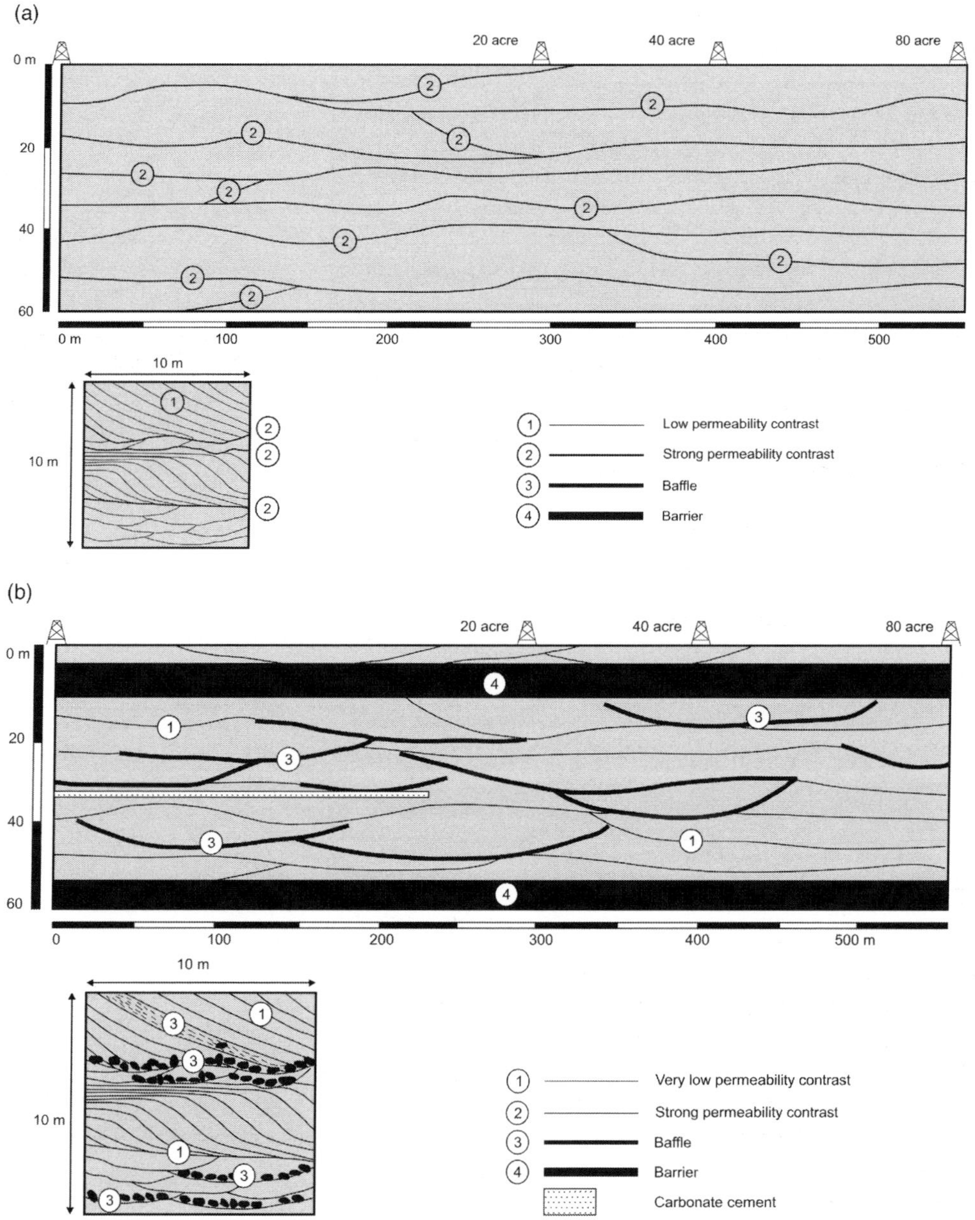

Figure 18.8. Reservoir models for four sedimentary facies of the fluvial-deltaic depositional system, South Caspian Basin (Reynolds et al., 1998). (a) A reservoir model based on outcrop observations of the fluvial facies association. There are no major barriers or baffles to flow, and only minor small-scale heterogeneities. Good reservoir communication is expected. (b) A reservoir model based on the delta-plain facies association. Delta-plain siltstones form laterally extensive seals that create a stacked series of reservoirs. These may themselves be compartmentalized by membrane-sealed faults. Within compartments, mudstone conglomerate bodies form baffles to flow. (c) A reservoir model based on the proximal delta-front facies association. The association is characterized by alternation between laterally extensive delta-front siltstones, and mouth-bar and channel sandstones. The result is a stacked succession of reservoirs vertically sealed by the delta-front siltstones. Internally, the sands display low heterogeneity. Coarsening-upward and fining-upward profiles occur and are likely to be reflected in systematic permeability trends. In addition, individual reservoirs are likely to be compartmentalized by sealing faults. (d) A reservoir model based on the distal delta-front facies association. The association is characterized by tabular sandstone beds interpreted as the product of hyperpycnal underflows separated by fine-grained siltstones and mudstones. This type of reservoir is likely to have zero effective vertical permeability and to have horizontal permeability strongly controlled by vertical faults.

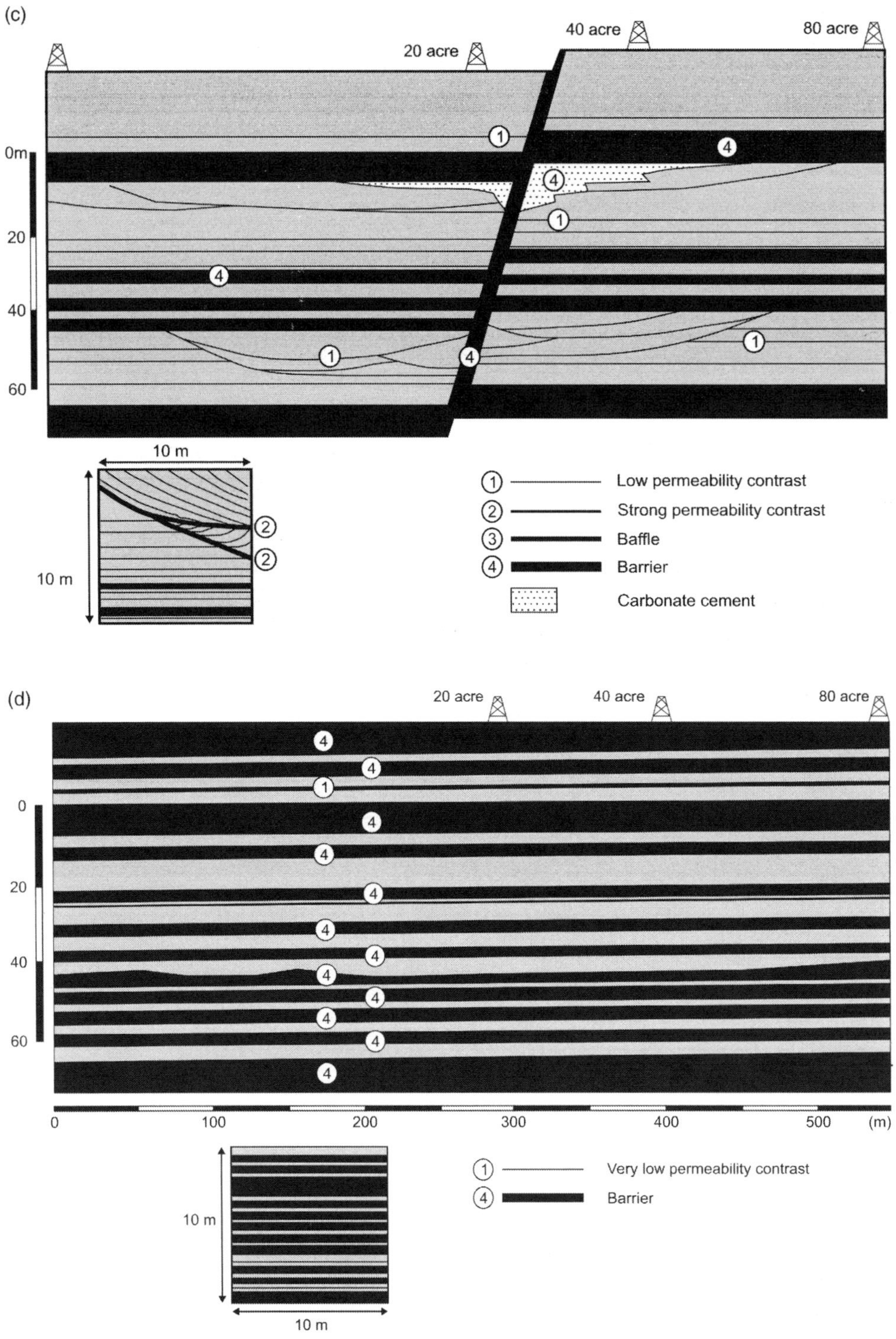

Figure 18.8 (*cont.*)

proven useful in constraining the history of fluid flow in deformed rocks, including the time of petroleum migration (Burruss et al., 1985; Earnshaw et al., 1993; Parris et al., 2003).

Potential hydrodynamic fluid flow regimes controlling hydrocarbon migration in rift and passive margin settings are related to compaction and topography-driven fluid flow mechanisms.

The first one draws from the development of the compaction-produced hydraulic heads increasing with depth in areas with fast deposition rates and fine-grained sediments (e.g., Bethke, 1985; Bethke and Corbet, 1985; Neuzil, 1986). The second one draws from the development of significant topographic gradients in subaerially exposed areas (e.g., Garven and Freeze, 1984a, 1984b). Both are, given a significant enough flow rate, capable of deflecting buoyancy-driven hydrocarbon migration (Hitchon, 1984; Garven, 1989; Ge and Garven, 1989; Montañez, 1994). One more hydrodynamic fluid flow regime potentially capable of controlling hydrocarbon migration is the buoyancy-driven flow, developed in a variety of geological scenarios described in Chapter 8.

Maturation, expulsion, and migration in rift and passive margin settings

The generation, expulsion, and migration of petroleum in the rift and passive margin settings are relatively simple and characteristic for these settings because of a syn-tectonic burial and thermal history that is not very complex and a relatively simple array of migration pathways presented by matrix- and fault/fracture-controlled permeable systems. Most of the known migration scenarios in both settings include vertical migration-dominated scenarios or scenarios with a lateral migration component less than 20 km (see Ahlbrandt, 2002; Connolly and Aminzadeh, 2006), represented by petroleum systems in provinces such as the Gulf of Mexico, the West African margins, the Niger Delta, and the North Sea.

Organic matter maturation and petroleum migration is controlled by the location of the petroleum system within the basin, as source rocks in rift and passive margin settings may generate laterally as well as vertically.

The maturation depends on the thermal history of the source rock (Deming and Chapman, 1989; Waples, 1992, 1994b). The history is controlled by the interplay of the burial history and changing heat flow regimes. The factors influencing hydrocarbon maturation in rift and passive margin-dominated settings can be understood by considering the following scenarios:

a) Rapid burial under relatively high heat flow regime during the extension in rift or pull-apart setting. The associated crustal thinning is responsible for the increased heat flows. The extra heat flow increase may be associated with magmatic activity, which can accompany crustal stretching as discussed in Chapters 1, 2, and 12.

b) Relatively slow burial under cooling heat flow regime during the sag stage of the rift basin as it subsides in response to crustal cooling.

c) Burial of varying rates under relatively high heat flow regime during the rift–drift transition. The degree of maturity the source rocks reach will depend on whether the proximal or distal part of the passive margin is involved, as can be implied from the proximal/distal margin discussion in Chapter 2.

d) Gradual burial under a normal or near-normal heat flow regime during the passive margin development that follows continental breakup.

e) Rapid burial under a normal or near-normal heat flow regime during the passive margin development influenced by a distinct delta entering the oceanic basin. Very high rates of deposition in this type of basin can cool the basin fill, which can be seen in the Amazon case discussed in Chapter 11.

f) Renewed rapid burial in the basin undergoing the younger event of accelerated subsidence due to either extension or encroaching orogens like the Balkans, the Greater Caucasus, and the Lesser Caucasus in the Black Sea case (Nemčok et al., 2009; Stuart et al., 2011). The former case is characterized by a relatively high background heat flow regime while the latter case is associated with a heat flow regime slightly cooler than normal. Very high rates of deposition in this type of basin can cool the basin fill.

g) Rapid burial under relatively low heat flow regime within short-lived foredeeps flanking the earliest basement inversions; scenario affecting the pre-existing passive margin setting, which controls the dominant characteristics of the respective portion of the petroleum province (see discussion of the Persian Gulf province in Chapter 14).

If the source rock has entered the generative kitchen because of the increasing burial, it is common for the hydrocarbons generated to migrate laterally updip along stratigraphic carriers or unconformities and vertically along fault and fracture systems either into traps or to the surface.

Portions of basins can differ from each other in the role of burial, uplift, and reburial events composing their development histories. In these histories, marine and terrestrial source rocks can expel hydrocarbons with each cycle of burial and uplift. Going through these cycles they generate oils that are progressively more mature and smaller in volume. The original kerogens become compositionally altered to partially depleted kerogen types, converging on the resemblance of Type III kerogen. However, not all kerogens types can expel hydrocarbons with multiple cycles. Because of the narrow range of activation energies (compare Figure 15.13a with Figures 15.14a and 15.12a), Type I kerogens of lacustrine source rocks are prone to exhausting their generative capacity in just a single burial-uplift cycle. Type II and Type III kerogens, having broader activation energy ranges (Figures 15.14a, 15.12a), can undergo a multiphase generation. In theory, source rocks, which are rapidly buried and then rapidly uplifted, can keep generating hydrocarbons at significantly decreased rates and volumes during uplift. In contrast, slowly buried source rocks lose their capacity to continue generating as they pass rapidly back up through the temperature gradient.

Most of the chemical reactions leading from kerogen to hydrocarbons, especially the cracking of organic molecules, involve volume increase. It contributes to overpressure generation controlled by other mechanisms such as disequilibrium compaction (Maltman, 1994; Osborne and Swarbrick, 1997), seismically induced liquefaction (Obermeier, 1996), smectite-to-illite transition (Bekins et al., 1994; Osborne and Swarbrick, 1997), gypsum-to-anhydrite transformation (Jowett et al., 1993), diagenesis itself (Osborne and Swarbrick, 1997 and references therein), and aquapressuring (Shi and Wang, 1986). Mentioned mechanisms have been interpreted as functioning to various degrees in the North Sea (Lonergan et al., 2000; Jolly and Lonergan, 2002; Duranti and Hurst, 2004; Huuse et al., 2004), the South Caspian Basin (Bredehoeft et al., 1988), the Songliao Basin (Xie et al., 2003), the Bristol Channel (Nemčok et al., 1995; Nemčok and Gayer, 1996), and the Gulf of Mexico (Bruce, 1984). The resultant overpressure has a tendency to retard a maturation of organic matter.

Organic matter maturation histories in various portions of the basin can also be modified by the fluid flow mechanisms discussed in Chapter 13. Because of its very high heat capacity, moving water is capable of significant modifications of the conductive geothermal regime of a basin (Nunn and Deming, 1991; Deming, 1994; Vasseur and Demongodin, 1995). This affects maturation histories of hydrocarbon kitchens in this basin. This situation can happen when the fluid flow crosses isotherms causing lateral fluid flow to effectively redistribute heat. Observed basin-scale thermal anomalies are consistent with such fluid-driven heat transport (e.g., Chapman et al., 1991; Jessop and Majorowitz, 1994).

One-dimensional modeling, using burial history curves alone, can provide us with relatively reliable estimates of the time of entry into petroleum maturation thresholds, especially if well tops and/or quality cross-sections are available. The method requires the knowledge of:

1) absolute ages of all involved stratigraphies;
2) duration of tectonic events;
3) exhumation history; and
4) burial history.

Thermochronology methods such as apatite and zircon fission-track and apatite (U-Th)/He thermochronometry methods, discussed in Chapter 4, can aid in constraining the thermal history of sedimentary basins (Laslett et al., 1982, 1987; Naeser et al., 1989; Vrolijk et al., 1992; Burtner et al., 1994; O'Sullivan et al., 1995; Ehlers and Farley, 2003). With knowledge of the thickness, richness, and kerogen type of the likely source rocks in the vertical profile, it is even possible to predict the petroleum yield as a function of time. Such one-dimensional modeling, despite its limitations (see discussion in Nemčok et al., 2005a), can be a relatively useful exercise for identifying the location and time of generation/expulsion of petroleum. The model results are normally displayed in a burial history diagram, but alternatively may be tabulated or mapped.

When migration starts, it is most efficient during active extension in the province, which keeps active faults open for flow along and across them during rupturing and for a certain time after rupturing. Cyclical reactivation of rift- or gravity-glide-related faults keeps fault conduits open for the expulsion of all fluids, including connate and meteoric fluids and hydrocarbons, as documented from provinces such as the Gulf of Mexico (Anderson et al., 1994a; Roberts et al., 1996; Roberts and Carney, 1997; Bolchert et al., 2000; Losh et al., 2002), the Bristol Channel (Nemčok et al., 1995), the Recôncavo Rift Zone (Destro et al., 2003b), southern Utah (Bowen et al., 2007), the Junggar Basin (Jin et al., 2008), the Santa Barbara Basin (Boles et al., 2004), the Bohai Bay Basin (Hao et al., 2009), the Grand Banks (Magoon et al., 2005), the Sea of Galilee (Gvirtzman et al., 1997), the Rio Grande Rift (Mailloux et al., 1999), and the Rhine Graben (Person and Garven, 1992; Lampe and Person, 2002). Thus, the ideal time for the operation of a rifting-related petroleum system is late in the development of the rift or pull-apart basin. At this stage, traps are in place or forming, good migration pathways are available, and hanging wall hydrocarbon kitchens are likely to be at peak maturities.

Numerous documented evidences exist of migration modified by fluid flow systems. For example, the Pechelbronn and Laudau oil fields in the Rhine Graben are located in the discharge regions of the hydrothermal fluid flow systems (Person and Garven, 1992). Evidences exist for the hydrodynamically assisted origin of petroleum in the Guayamas Basin in the Gulf of California and the Escanaba Trough of the northeastern Pacific Ocean (Kawka and Simoneit, 1987; Kvenvolden and Simoneit, 1990), the Dead Sea Rift (Gvirtzman and Stanislavsky, 2000), and the Grand Canyon and Bacon Flat oil fields in the Basin and Range province (Hulen et al., 1994). Therefore, in areas known to have undergone hydrodynamic regimes, a hydrodynamic analysis on the scale of a rift block or basin can be used to search for subtle migration pathways and traps (Davis, 1991).

One-dimensional petroleum maturation models, mentioned a bit earlier, cannot predict migration pathways that ultimately lead to the charging of potential traps and charge effectiveness. Such prediction requires a rigorous knowledge of the structural architecture in three dimensions and the fluid transmissivity properties of all involved carrier beds, faults, and fracture networks. It also requires knowledge of the time of movement on all potential carrier faults, and knowledge of the hydrodynamic history as these will affect heat flow, geothermal gradients, and petroleum migration.

Note that it is unusual to have this level of knowledge, even in an intensely explored and mature petroleum province. However, the two- and three-dimensional software tools that integrate maturation, thermal, and fluid flow modeling routines can be helpful for evaluating alternative scenarios for trap filling (Bishop et al., 1984) and thereby further constrain exploration risk assessments, if we refrain from modeling details and focus on differentiating main trends. This is a research area still in development and testing.

Migration case stories

Trap charge case stories from the rift and passive margin settings document hydrocarbon migration from single and multiple source rocks (e.g., Cornford, 1994; Magoon et al., 2005; Jin et al., 2008). They introduce a dynamic mixing concept for hydrocarbon mixing in trap (Peters et al., 2007). Apart from relatively simple secondary migration scenarios, some of them illustrate complex migration histories including remigration from earlier traps (e.g., Xie et al., 2003; Jin et al., 2008). A growing database of evidences about remigration or leaking trap (e.g., Brooke et al., 1995; Heggland, 1997, 1998) reinforces the concept of a trap representing a transient storage in geologic time (see Sales, 1997) introduced earlier.

Migration from a single source rock

A well-documented trap charge from a single source rock via carrier bed comes from the Junggar Basin, northwestern China, where only a single oil migration from the Lower Permian Fengcheng source rock mainly during the Late Triassic–Early Jurassic time period is observed in the slope area in the southern portion of the basin (Jin et al., 2008).

Migration from several source rocks

In contrast to the slope area of the Junggar Basin undergoing migration from a single source rock, the northern basin portion deformed by a basin-controlling fault system underwent fault-controlled migration from the Lower Permian Jiamuhe, Fengcheng, and Middle Permian lower Wuerhe formations (Jin et al., 2008). This migration fed multiple reservoirs of varying stratigraphies (Cao et al., 2006). The oil migration and mixing occurred during the Late Triassic–Early Jurassic and Cretaceous periods, followed by gas migration during the Tertiary, documenting a multistage migration.

An example of several source rocks but from the same stratigraphy comes from the Egret-Hibernia(!) petroleum system of the Grand Banks in Canada (Magoon et al., 2005), where geochemical data indicate that the Egret source rock contains various facies with different organic matter. The system of facies was deposited in an epeiric sea before the final rifting and Atlantic Ocean opening (e.g., Tankard and Welsink, 1987; Ziegler, 1989b). The western and thicker portion of the Jeanne d'Arc half-graben, adjacent to the controlling normal fault, contains two types of oil-prone source rock facies. The eastern and thinner portion of the basin, adjacent to the hinge zone, contains gas-prone facies (Von der Dick et al., 1989; Fowler and McAlpine, 1994; Huang et al., 1994). The hydrocarbon charge to traps across the entire basin resulted in trends such as a decrease in oil pool volume from west to east and a gas increase in the same direction (Magoon et al., 2005).

The Jeanne d'Arc half-graben is deformed by a system of faults perpendicular to the main boundary fault zone, and which result in various reservoir juxtapositions. These juxtapositions facilitated oil migration from stratigraphically older to younger reservoir rocks across the faults, controlling a lateral component of migration (Magoon et al., 2005). The vertical component was driven by overpressure cycles built in generating source rock, which resulted in fault reactivations and cycles of overpressure venting along faults. A portion of the vertical migration was matrix-controlled as there is no effective seal between the Jeanne d'Arc and Hibernia sandstone reservoirs.

Other examples of provinces with oil migrating from different facies of the same stratigraphic section come from the Kimmeridgian section of the North Sea (Cornford, 1994), the Canadian McKenzie Delta

(McCaffrey et al., 1994), and the Miocene section of the Mahakam Delta (Peters et al., 2000). This scenario is characterized by systematic variations in geochemistry of pooled oil across the province, which has been documented in the Los Angeles Basin (Jeffrey et al., 1991), the Beaufort Sea (McCaffrey et al., 1994), the Mahakam Delta, and the Makassar Slope (Peters et al., 2000).

Dynamic hydrocarbon mixing

The case of several source rock intervals providing hydrocarbons for the producing field comes from the Penglai 19-3 oil field in the Bohai Bay Basin in China (Hao et al., 2009), which is the largest Chinese offshore field. The area developed as a Mesozoic back-arc basin, which turned later into an intra-cratonic rifted basin (Wang and Qian, 1992; Ge and Chen, 1993; Huang and Pearson, 1999). The trap is formed by faulted anticline containing post-rift Miocene fluvial reservoir rocks. Seals are interbedded with reservoirs, while the source rocks are represented by three Eocene stratigraphies (Carroll and Bohacs, 2001; Harris et al., 2004; Hao et al., 2009):

1) the third member of the Shahejie Formation containing a hydrogen-rich, oil-prone organic matter;
2) the first member of the Shahejie Formation, also containing an oil-prone organic matter; and
3) the Dongyiong Formation containing mixed type I and III kerogens.

The indication about the pooled oil being a dynamic mixture of oils comes from correlations of six biomarker parameters such as C_{24} tetracyclic terpane/C_{26} tricyclic terpane, C_{19}/C_{23} tricyclic terpane, hopane/sterane, C_{23} tricyclic terpane/C_{30} hopane, gammacerane/$\alpha\beta C_{30}$ hopane, and 4-metyl steranes/ΣC_{29} steranes ratios (Figure 18.9a). In fact, virtually each well of the field can be characterized by a different biomarker assemblage (Hao et al., 2009). This indicates that pooled oil is a mixture of oils generated by different source rock intervals and different source kitchens (sensu Peters et al., 2007).

Hierarchical cluster analysis allows one to group the oil mixtures from various wells into three classes (Figure 18.9b):

1) Class I oils correlating with source rock No. 1;
2) Class II oils represented by mixtures sourced from source rocks Nos. 1 and 2; and
3) Class III oils correlated with dominant source No. 3 and a subordinate source No. 1.

The field is surrounded by four hydrocarbon kitchens containing all three source rock intervals; that is, the Miaoxi, Bodong, eastern Bozhong, and central Bozhong depressions. The migration directions into the field have been determined by directional variations in compositional geochemistry of pooled oils (sensu Leythaeuser and Ruckheim, 1989; Hao et al., 1998, 2000; Peters and Fowler, 2002) (Figure 18.10). The variations indicate that:

1) the Bodong and eastern Bozhong depressions provided Class III oils for the field from the north;
2) the Miaoxi depression provided Class I oils for the field from the east and southeast; and
3) the central Bozhong depression provided Class I and II oils for the field from the northwest.

The use of compositional variations is a reliable technique because the diffusive mixing required to homogenize the composition in large fields would take tens of millions of years (see Larter and Aplin, 1995).

Remigration

A nice remigration example comes from the Junggar Basin, northwestern China, where the pre-Upper Triassic reservoirs sealed by the Upper Triassic Baijiantan Formation were charged by the hydrocarbons from Permian source rocks during the Late Triassic–Early Jurassic (Jin et al., 2008). The migration pathways used available faults, fractures, and carrier beds (Cao et al., 2005). Subsequently, these reservoirs underwent cycles of overpressure and episodes of faulting. Episodic fault reactivation, which started during the Late Cretaceous and continues until now, breaches the Baijiantan top seal and pooled hydrocarbons remigrate along faults to the stratigraphically higher Jurassic–Cretaceous reservoirs.

A complex migration story comes from the Songliao Basin, one of the largest Mesozoic continental basins in China. It contains the fault-bounded Upper Jurassic–Lower Cretaceous rift fill overlain by the Upper Cretaceous fill of the subsequent sag basin, which underwent post-Cretaceous compression (Xie et al., 2003). Producing fields in the basin indicate a bizarre hydrocarbon distribution characterized by oil pools being distant from the source rock kitchen and located at flanks of the basin and gas pools being located right above the source rock kitchen in the basin center. A combination of the synthetic interpretation of thermal history, porosity-permeability, present-day fluid pressure, and fluid inclusion data with numerical simulation of the fluid pressure regimes during the basin

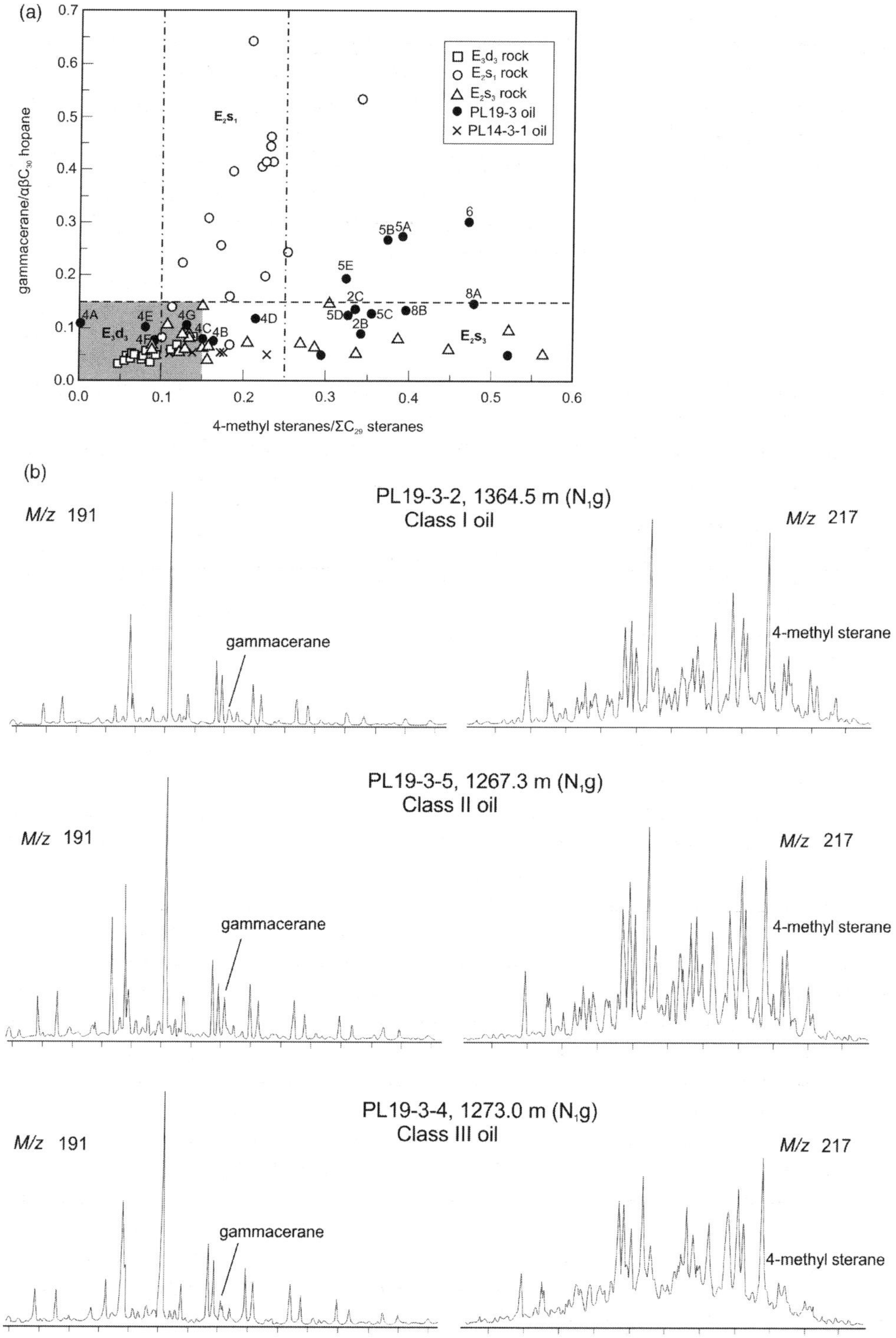

Figure 18.9. (a) Oil and source rock correlations using gammacerane/$\alpha\beta C_{30}$ hopane and 4-metyl steranes/ΣC_{29} steranes ratios (Hao et al., 2009). Oil is from the Penglai 19-3 field and source rocks are from the Bozhong sub-basin of the Bohai Bay Basin. (b) Mass chromatograms of hopane (m/z = 191) and sterane (m/z = 217) series for oil classes from the Penglai 19-3 field in the Bohai Bay Basin (Hao et al., 2009). Note the gammacerane and 4-methyl steranes contents varying among classes.

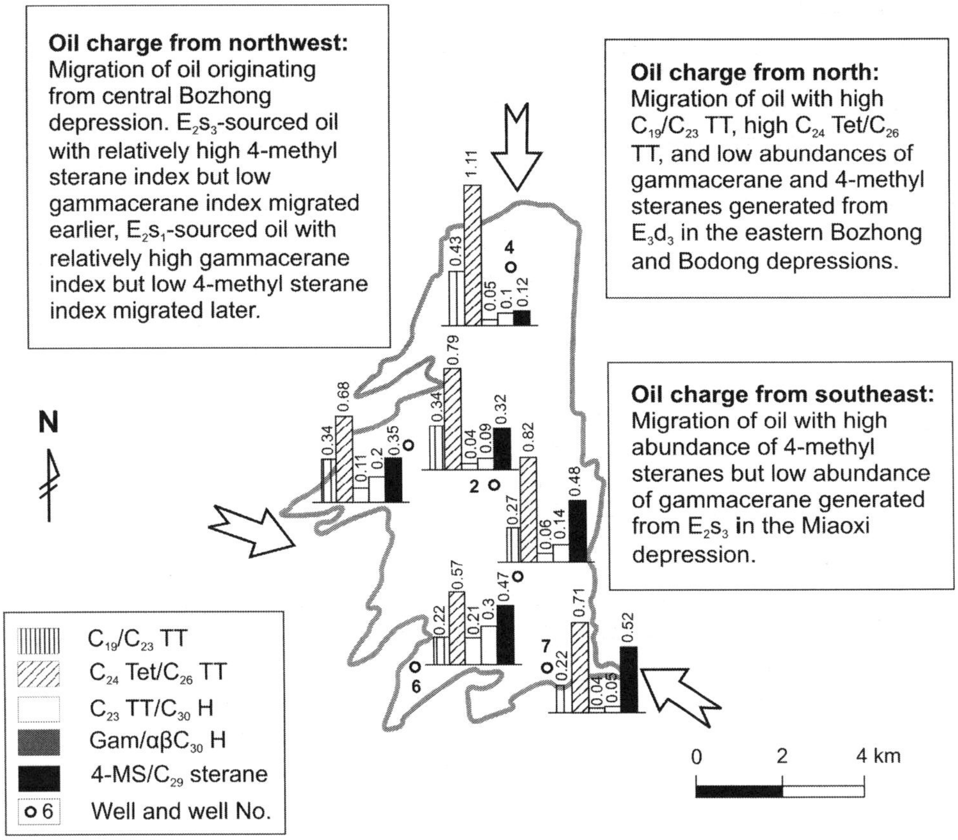

Figure 18.10. Directional variations in biomarker distributions in the Penglai 19-3 field of the Bohai Bay Basin indicating migration directions (Hao et al., 2009). TT – tricyclic terpane, Tet – tetracyclic terpane, C_{30} H – C_{30} hopane, Gam – gammacerane, 4-MS – 4-methyl steranes.

development indicates a following two-phase migration history. During the main rift stage, the initial rift sediments underwent the onset of oil generation contemporaneous with significant overpressure development. This, together with existing inclined layered sedimentary fill consisting of intercalated seal and carrier bed horizons, resulted in a lateral migration of hydrocarbons from the rift center to its flanks. Toward the end of the Cretaceous, the fluid pressure in the former rift center started to change from overpressure to underpressure due to the onset of a long-term tectonic uplift accompanied by exhumation and cooling. This basin development stage was characterized by:

1) the lateral fluid flow from former rift flanks to the underpressured former rift center;
2) some vertical migration from the source rock in the basin center; and
3) most of the source rocks in the basin reaching a gas window.

Apart from modeling and mentioned analytical data, the described migration history also explains the gas maturity trends characterized by gases in the center having higher maturity than gases at the flanks.

Very complex migration stories come from the gravity glide systems of passive margin segments affected by sediments brought by a large river, especially if gravity gliding takes place on top of evaporites and evaporites change the geometry of their autochthonous and allochthonous bodies with time. In complex migration scenarios like these from this setting, it is useful to include brine geochemistry studies in the exploration program. They can indicate the involvement of either connate or meteoric waters on top of hydrocarbon migration utilizing cycles of fault activity. Thinking about these two extra fluids involved in local fluid flow regimes, note that connate fluids ascend along faults only if they are overpressured in their source horizon.

An example of a very complex migration from the passive margin comes from the Eugene Island block 330 oil and gas field in the Gulf of Mexico. It is characterized by both lateral and vertical migration components and migration pathways evolving through time (Losh et al., 2002). The geology here is controlled by the interaction among salt tectonics, faulting, and rapid deltaic and slope deposition. Evolving pathways react to changes in geometries of evaporitic bodies in the section. Lateral migration components dominate where a very effective top seal is formed by allochthonous salt body. Vertical migration components dominate through windows in evaporitic seals due to the development of welds or fault/fracture systems. The total vertically migrated distance between the Mesozoic source rocks and the shallow Pliocene-Pleistocene reservoirs is about 14 km. Losh and colleagues' (2002) study focuses only on the youngest stage of migration history, when the salt weld formation triggered the onset of vertical migration into reservoirs deposited in a salt-withdrawal mini-basin (Figure 18.11a). Well data indicate that most of this vertical migration took place along faults. A combination of FMI logs with core data documents faults as wide, structurally complex zones containing high-strain tectonic gouge in the fault core and variably faulted/fractured sediment in the damage zone, which is oil charged. Oil and gas shows are typical for each well (see Anderson et al., 1994a, 1994b; Losh et al., 1999, 2002).

Isotopic compositions of strontium and concentrations of ^{129}I in the field brines indicate a migration from lithostratigraphies at least as old as the Oligocene (Hoefs, 1981). Based on their geochemistry, the brines can be divided into several types. They differ by the salinity and extent of interaction with hosting reservoir, which indicates their migration through a complex and temporally variable system of migration pathways (Losh et al., 2002).

The brine migration direction within each reservoir was from the south to the north. This was derived from brine geochemistry trends such as:

1) a northward increase in sodium deficiency indicating the amount of brine-sediment interaction along the increasing length of the flow path; and
2) a northward iodine concentration decrease indicating the dilution of deeply sourced brine high in iodine with local connate fluid low in iodine (Figures 18.11b–c).

The low extent of brine–sediment interaction at fault entries to reservoirs is consistent with a relatively fast deep brine ascent along the fault that prevented a significant brine–sediment interaction during the ascent (Losh et al., 2002). That different reservoir horizons contain different brines demonstrates a complex brine migration history controlled by fault-juxtaposition sealing windows evolving and changing with time.

The hydrocarbon migration scenario looks complex as well. The difference between gas and oil in source $\delta^{13}C$ and maturity indicates their separate expulsion and subsequent mixing at the later stages of migration history. The oil was generated from various facies of the underlying Jurassic–Lower Cretaceous marly section. The gas was generated by an oil-cracking in situ. On top of this, the upper reservoirs contain biodegraded oils due to the cut-off depth of bacterial activity (Whelan et al., 1994; Losh, 1998). The cut-off depth appears to be the depth with a temperature of 65°C (Losh et al., 2002).

Seepage and concept of transient trap

A growing pool of evidences about gas seepage (e.g., Roberts and Carney, 1997; Heggland, 1998; Bolchert et al., 2000; Connolly and Aminzadeh, 2003) illustrates the concept of trap being a transient feature (Sales, 1997) and vertical migration being the dominant migration in rift and important migration in passive margin settings (Perrodon, 1995). Lately, the seepages have been interpreted from vertical zones of disturbance in 3-D reflection seismic images, which are characterized by a reflector phase and amplitude distortion related or unrelated to faults. The seepage determination can be enhanced by techniques such as pattern recognition algorithms drawing from multidimensional, multi-attribute, and neural network modeling (e.g., Aminzadeh and Connolly, 2002, 2007).

Cases of gas chimneys above gas and oil fields (see Figure 17.26) and findings of shallow gas accumulations above seismically recognized gas chimneys have been cited from Danish and Norwegian sectors of the North Sea province (Brooke et al., 1995; Heggland, 1997, 1998). Apart from the present-day seepages, the gas seepages to a Pliocene seabed in the North Sea have been interpreted from a series of carbonate buildups at former Pliocene vents imaged by the reflection seismic data (Heggland, 1998). Similar buildups have been observed offshore western Ireland and offshore northwestern Australia (Hovland et al., 1994).

Migration controls

Fault- and fracture-controlled migration

Fault/fracture-controlled migration of hydrocarbons has been documented by a combination of fieldwork and geochemical studies of the exhumed Jurassic Navajo sandstone, which acted as a hydrocarbon reservoir associated with the Moab and Kaibab Faults in Utah (Garden et al., 2001; Bowen et al., 2007). The Kaibab study focuses on the fault, which bounds a structural high about 190 km long. The fault roots into the reactivated Proterozoic shear zone. The Navajo sandstone in the high contains extensive bleaching in the crestal area,

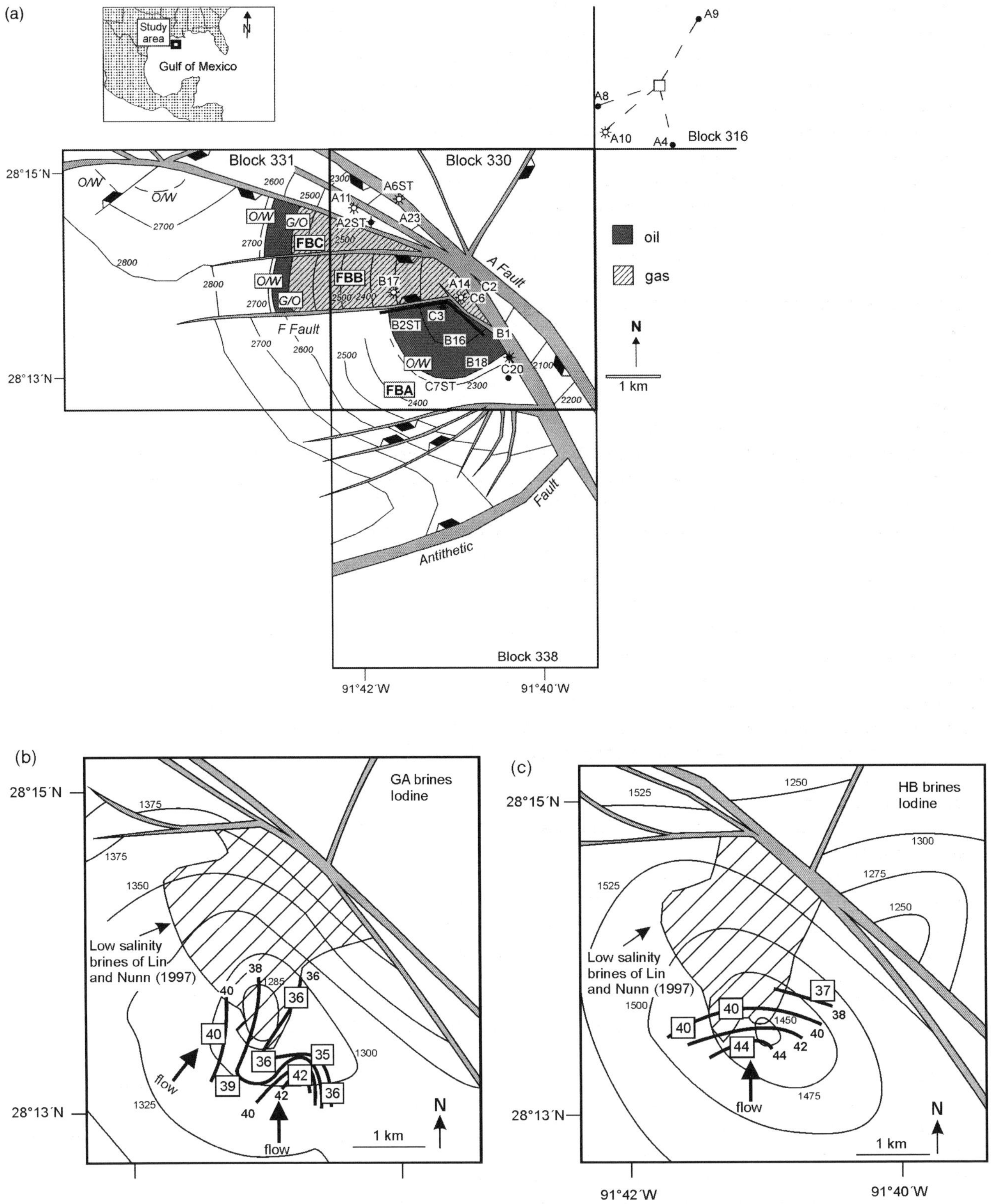

Figure 18.11. (a) Structural map on top of one of the Pliocene-Pleistocene reservoir horizons of the Eugene Island block 330 oil and gas field in the Gulf of Mexico, showing the fault system of the salt-withdrawal mini-basin and well locations (Losh et al., 2002). Contours are in meters subsea true vertical depth. (b–c) Structural map on top of one of the Pliocene-Pleistocene reservoir horizons of the Eugene Island block 330 oil and gas field in the Gulf of Mexico, showing an iodine concentration in mgl^{-1} decreasing from south to north (Losh et al., 2002).

containing no iron carbonate but a significant amount of kaolinite. The intersection of the Navajo reservoir with the Kaibab Fault contains an iron-rich zone post-dating the reservoir bleaching and caused by fluids migrating along the fault (Bowen et al., 2007).

A typical feature of the fault-controlled fluid migration is its episodic character (e.g., Sibson, 1981a, 1994; Hooper, 1991; Losh et al., 1999; Nemčok et al., 2002; Zeng and Jin, 2003; Nemčok et al., 2005a and references therein). Because diagenetic effects result from fluid–wall rock interaction (e.g., Surdam et al., 1989; Fisher and Boles, 1990; Caine 1999), banded veins are indicators of multiple fluid flow episodes (e.g., Suchy et al., 2000; Boles et al., 2004; Liu et al., 2004; Perez and Boles, 2004).

An example of the migration along the basin-bounding fault comes from the Refugio-Carneros Fault of the Santa Barbara Basin in California (Boles et al., 2004). The fault deforms Miocene and younger sediments. It has a throw of 120 m representing a normal faulting component along a sinistral oblique-slip fault. The U-Th dating of calcite indicates its Pleistocene age. The fault contains numerous veins with calcite fill bands, divided by bands of heavy minerals and quartz sand grains trapped at the growth faces along the bands, indicating multiple opening and fluid flow episodes. Detritus-rich and detritus-free bands have been interpreted as deposited during cycles of rapid and slow fluid flow, respectively. Micas in detritus-rich bands are oriented parallel to fracture walls as opposed to wall-perpendicular micas in the host rock, indicating mica erosion from wall rock by rapid flow and flow-parallel reorientation inside the fracture. Geochemical and crystallization temperature data from the fluid inclusion in calcite indicate that calcite precipitated from hot basinal fluids ascending along the fault. The fluids derived from the saline water mixed with meteoric one. Maintaining a relatively constant temperature and isotopic composition, the fluids must have had a large volume and a uniform source. The $\delta^{13}C$ isotopic compositions from -30 to -40‰ for most of the calcite crystals indicate carbon dioxide sourcing from thermogenic methane oxidation (see Hathaway and Degens, 1969; Donovan et al., 1974; Whiticar and Faber, 1986). Although a small volume of methane leakage could have produced relatively large calcite accumulations, they required a relatively large volume of aqueous solution because of low calcite solubility in water.

The recharge of meteoric water into the fluid system took place through the subaerially exposed Santa Ynez Mountains. The recharge decreased in a seaward direction as indicated by progressively more saline formation waters of offshore producing fields in the basin, indicating a progressively more prominent meteoric input in the present-day landward direction.

It is interesting to compare the discussed ancient fluid seeps along the Refugio-Carneros Fault, sourced by geothermal fluids and hydrocarbons (Boles et al., 2004), with modern seeps from the offshore Ellwood Fault, sourced dominantly by methane (Hornafius et al., 1999; Eichhubl et al., 2000; Boles et al., 2001). The comparison, checked by fluid flow modeling, indicates the following development history. The fault had acted as a mixing zone between the meteoric water from the north and the ascending hot formation fluid and hydrocarbons from the deep basin in the south. Cycles of overpressuring controlled by differential compaction and hydrocarbon expulsion have been released along the basin-bounding faults because of continued deformation in the basin. Over a time-scale of 10^5 years, the fluid pressure regime characterized by overpressure buildup cycles evolved into a hydrostatic regime that does not support episodic venting of hot basinal brines and hydrocarbons anymore (Boles et al., 2004). As a result, the modern seeps are characterized by buoyancy-driven seepage of hydrocarbons only.

A similar multistage precipitation story of cements along the fault zones comes from the Junggar Basin in China (Jin et al., 2008), the place where oil and gas fields can be correlated with locations of the basin-controlling fault pattern (Zhang, 2002; Cao et al., 2006). The cement samples for this study were taken from the feeder faults of the Hong-shanzui and Karamay oil and gas fields, and the Baijiantan, Wuerhe, Fengcheng, and Xiazijie oil fields. Calcium in sampled calcites has, at least partially, derived from the dissolution of plagioclase in reservoir sandstones. This was possibly caused by organic acids associated with migrating hydrocarbons. Both fluorescence colors of involved hydrocarbons in cements documenting varying oil maturity in different cement bands and the presence of hydrocarbon-bearing and hydrocarbon-free bands in veins indicate a multistage episodic fluid flow along the "field-feeding" faults. Varying oil maturities correspond to various mixtures of older biodegraded oil with younger nonbiodegraded oil, documenting a dynamic oil mixture from the remigration of primary hydrocarbon pools. The gas chromatography, gas chromatography-mass spectrometry, and Fourier transform infrared microspectrometry analyses document that the proportion of water to hydrocarbons varies significantly from band to band. This most likely indicates that, during the precipitation of calcite bands, there was a simultaneous movement and mixing of hydrocarbons and fluid to varying degrees, most likely controlled by the rupturing of seals among various compartments (see also Boles et al., 2004). Despite the complexity of dynamic fluid mixture, the following generalization can be made. The hydrocarbon-free bands, having high magnesium concentrations, can be characterized as precipitated dominantly from normal lacustrine formation water. The hydrocarbon-rich bands, containing various oil mixtures and increased Mn concentrations, can be characterized as precipitated from the mixture of hydrocarbon-bearing basinal fluids and meteoric water, the former containing various dynamic oil mixtures (Jin et al., 2008).

Matrix-controlled migration

A study of matrix-controlled fluid migration in the eolian sandstone in the Jurassic Navajo sandstone in southern Utah indicates that the present-day reservoir quality and permeability structure is dominantly controlled by fluid-related diagenetic processes (Bowen et al., 2007). The identification of alteration-related minerals and their spatial distribution mapping in this study was done by reflectance spectroscopy. The most common identified diagenetic chemical reactions include:

1) reduction of iron bleaching the sandstone: hematite + $3.75H^+$ + $0.25CH_4$

   $= 2.25H_2O + 2Fe^{2+} + 0.25HCO3^-$
2) alteration of K-feldspar to kaolinite and illite: K – feldspar + $0.5H_2O$ + H^+

   $= K^+$ + 0.5kaolinite + 2quartz
3) reaction: K – feldspar + $0.67H^+$ = $0.67K^+$ + 0.33illite + 2quartz
4) precipitation and dissolution of carbonate cement: calcite + H^+ = Ca^{2+} + $HCO3^-$
5) oxidationandprecipitationofironoxidefromsolution: $0.25O_2$ + Fe^{2+} + $1.5H_2O$ = goethite + 2H+
6) reaction: 2goethite = hematite + H_2O

Reactions 1 and 5 especially produce color changes easily recognizable at an outcrop (Figure 18.7a). Figure 18.7b shows a relatively unaltered Navajo sandstone having a reddish color due to a very thin hematite coating of sandstone grains bleached by migrating reducing fluids likely associated with hydrocarbons (see Chan et al., 2000; Beitler et al., 2003; Parry et al., 2004). Similar bleaching due to migrating hydrocarbons is known from:

1) the hydrocarbon reservoirs in this region, such as the White Rim sandstone in the Tar Sands triangle, the outcropping reservoirs at the Circle Cliffs uplift, and the reservoirs in Dubinky well (Chan et al., 2000; Beiter et al., 2003);
2) the laboratory experiments on sandstone bleaching (Surdam et al., 1993);
3) the geochemical modeling (Parry et al., 2004); and
4) the sandstone host rocks in the vicinity of bitumen veins in Moab, Utah (Chan et al., 2000).

The outcrop work by Bowen and colleagues (2007) indicates that subtle changes in diagenetic mineral content are correlable with sedimentary layering, which controlled the three-dimensional permeability structure available for fluids responsible for the diagenetic changes of reservoir. While Bowen and colleagues' (2007) observations suggest that mineral changes commonly occur in the zones subparallel to bedding, our photos of bleaching patterns in the Navajo sandstone from southern Utah (Figure 18.7b) suggest an addition of textural features to the control list, which is in accordance with:

1) observations from the present-day fluvial/deltaic sediments combined as analogs with core samples and production histories from the Prudhoe Bay field in the North Slope, Alaska (see Tye, 2004b; Figures 18.7c, d);
2) observations from the Pliocene fluvial/deltaic sediments in outcrops compared with coeval producing reservoirs deposited by the paleo-Volga River in the South Caspian Basin (Reynolds et al., 1998; Figure 18.8); and
3) observations from the Santonian marine shoreface-shelf sediments in outcrops in New Mexico compared as analogs with producing Toarcian reservoirs in the Wytch Farm Field, United Kingdom and Middle Jurassic reservoirs of the entire UK Brent province (Hampson et al., 2004; Sixsmith et al., 2004a, 2004b).

Vertical and lateral migration styles

Vertical migration in rift sections has several different scenarios, including:

1) migration inside the coarse-grained facies near the graben-controlling fault;
2) migration inside the cyclically opened fault-controlled pathway;
3) distributed hydrofracturing-assisted migration;
4) salt tectonic-assisted migration; and
5) migration inside injectite lithologies.

Fluid migration along coarse facies at the rift unit boundary was described in the case of brine migration in the Angolan Cretaceous Rift Basin in Chapter 13 (Person and Garven, 1994). Typical examples of these facies in the case of continental rift setting would include high-permeability alluvial fans and sand flats located along the boundary faults on the deeper side of a half-graben. A vertical migration in this setting can be further assisted by hydrodynamics controlled by either topography-driven or compaction-driven fluid flows.

Vertical migration along faults has been described as a charge-controlling scenario in numerous studies of rift and passive margin settings (e.g., the Gulf of Mexico – Anderson et al., 1994a; Roberts and Carney, 1997; Bolchert et al., 2000; Aminzadeh and Connolly, 2002; Losh et al., 2002; the Santa Barbara Basin – Boles et al., 2004; the Junggar Basin – Jin et al., 2008 and the Bohai Bay Basin – Hao et al., 2009).

An exceptionally good natural laboratory for the fluid migration study was the Pathfinder well penetrating a fault with currently migrating fluids, located in the Eugene Island South Addition Block 330 in the Gulf of Mexico (Anderson et al., 1994a). The analyses have demonstrated that fault-zone hydrocarbons were geochemically identical to those analyzed from producing reservoirs located within shallower splays of the fault zone. The drill-stem tests have shown that the migration rate does not reach economically viable values. Furthermore, the larger the drawn-down pressure applied across the perforations, the lower the permeability of the fault zone is.

A high amplitude connectivity mapping in the 3-D seismic volume indicates that migration conduits within the normal growth fault zone are convoluted and can be visualized more like rivulets rather than sheets paralleling the fault zone (see Anderson et al., 1994a). This interpretation is in accordance with observations of plethora of cement-free cavities inside fault cores and damage zones of fault zones in geothermal reservoirs such as those in the Karaha-Telaga Bodas in Java or The Geysers in California (e.g., Nemčok et al., 2006a).

The coring of the fault zone allowed us to see numerous slickensides having various dips (Anderson et al., 1994a). The oil that cathodoluminiscence identified was proven in some of them, lacking in others. The hanging wall was connected to the fault zone via a suite of hydraulic fractures.

The stress and production testing allowed us to identify the difference between σ_3 and fluid pressure in control of the hydraulic conductivity of the fault zone, which was gaining significant permeability during cycles of fluid pressure converging upon σ_3 (Anderson et al., 1994a). Direct measurements of the stress field in the fault zone and reservoirs immediately above it allowed us to determine that the fault zone would require about 500 psi increase in fluid pressure to undergo the reopening of its fracture system and fluid flow in this segment. Furthermore, the petroleum columns in the reservoirs just above this fault segment result in buoyant pressures of about 500 psi, indicating that these columns are most likely "topped off" in a state of dynamic equilibrium and that any new hydrocarbon volume arriving to this fault segment would be then able to migrate upward utilizing hydraulic fracturing. Anderson and colleagues (1994a) interpret vertical fractures as indicating this mechanism of fluid migration from fault to reservoir. Such a shortest path of venting of the increased fluid pressure using the steepest fractures has been documented as the choice from a wide range of fracture geometries during rapid venting in the Karaha-Telaga Bodas geothermal reservoir in Java (Nemčok et al., 2004a).

Results of fracture reopening cycles are the volumes of hydrocarbons released in transient bursts into a reservoir "fed" by a fault (Anderson et al., 1994a). Time periods among reopening cycles are characterized by fluid pressure buildup cycles in underlying turbiditic sediments, increasing until fracture reopening stresses are reached and faults "feed" the overlying reservoir with new volume of hydrocarbons. The fluid-pressure venting results in a decrease in the pressure level, which marks the fault closure and the initiation of the new pressure buildup cycle. The described mechanism is analogical to a mechanism known from the research on accretionary prisms and thrustbelts (see Nemčok et al., 2005a and references therein).

A knowledge of distributed hydrofracturing-assisted migration comes from fracture research in shale-dominated basins fills such as the Bristol Channel or the Kura Valley Basin, Georgia (e.g., Nemčok et al., 1995; Nemčok and Gayer, 1996; Cosgrove, 2001; Nemčok, 2012). Like in the case of faults, the time periods among fracture reopening cycles are controlled by the length of fluid pressure buildup cycles. However, uniaxial and triaxial compression tests indicate that microfracturing begins at stresses built up to about half the fracture strength of the rock, and continues until the stress level has reached about 95 percent of the fracture strength (Scholz, 1968). Similar relationships are valid for meso-scale fracture patterns associated with large faults (e.g., Martel et al., 1988; Willemse et al., 1997; Antonellini and Mollema, 2000). Therefore, fracturing-assisted migration operates at stresses smaller than those controlling fault-assisted migration.

One important observation from shale-dominated sedimentary sections is that fracturing is most commonly represented by micro-cracks, hydraulic extension fractures, and extensional shear fractures (see Sibson, 1996 and references therein). The micro-cracks developed under stress conditions related to grain impingements. The hydraulic extension fractures propagated under a stress criterion of $p_f = \sigma_3 + T$ and a stress condition of

$(\sigma_1 - \sigma_3) < 4T$. The extensional shear fractures developed under a criterion of $p_f = \sigma_n + [(4T^2 - \tau^2)/4T]$ and a condition of $4T < (\sigma_1 - \sigma_3) < 6T$. The fracturing in shale-dominated settings is only rarely represented by shear fractures. It is because, given a host rock rheology, the stress buildups terminate by the aforementioned three types of failure instead of reaching the differential stress high enough to control the shear fracturing.

The Lower Jurassic multilayer of the Bristol Channel has an interesting diagenetic history. It developed from a massive mudstone fill, which formed the limestone layers during its diagenesis thanks to the early calcite available from fossil-rich horizons. These limestone layers subsequently served as top and bottom seals controlling cycles of pore fluid pressure buildup in mudstone layers distributed among them. Each pressure-buildup cycle was terminated by extensional fracturing of the sealing layer and venting of the overpressured fluid from the sealed mudstone compartment into neighboring lower-pressure compartments (Figure 17.25a). This venting is recorded as taking part either upward or downward, depending on the local pressure gradient (Figure 17.25a). Extensional veins in sealing layers can contain multiple types of cements and can be monophase or polyphase features. Outcrops allow one to see that brittle seal horizons accumulated all fracture events while the mudstone horizons remained free of meso-scale fractures. The meso-scale fracturing is relatively well organized. It is characterized by both preferred spacing and geometry.

The migration of hydrocarbons along the zones of weakness related to salt mobilization has been recognized in an increasing number of cases thanks to the reflection seismic-based gas chimney indication techniques applied in the passive margin provinces such as the Gulf of Mexico (Aminzadeh and Connolly, 2002), offshore Nigeria (Graue, 2000), and North West Shelf, Australia (Cowley and O'Brien, 2000).

A typical character of this migration type is its episodic nature, as documented by the episodic fluid expulsion of geothermal fluids along faults near the diapiric structures of the Yinggehai Basin in the South China Sea (Xie et al., 2001). Both reflection seismic imaging (Figure 18.12a) and outcrops (Figure 18.12b) of diapir-sediment contacts from various rift and passive margin regions worldwide document episodes of faulting and/or hydraulic fracturing reacting to episodes of salt mobilization and pore fluid pressure changes.

Most robust data on structural controls of salt-controlled episodic petroleum venting come from the northern Gulf of Mexico (e.g., Roberts and Carney, 1997; Bolchert et al., 2000). Petroleum seeps from this data set have been identified by oil slicks, mud volcanoes, and authigenic carbonate buildups, each of them associated with a different fluid flow rate. Their relationship with salt mobilization events is documented by the fact that all studied seeps in the Green Canyon/Ewing Bank study area are located at boundaries of salt bodies with mini-basins (Beherens, 1988; Roberts et al., 1990; Sassen et al., 1993; MacDonald et al., 1996; Boettcher et al., 1998; Bolchert et al., 2000). Being an analog to other passive margins with salt tectonics, the petroleum here migrates vertically from the source rock until it reaches the base of the allochthonous salt body, which forces migration to change to the lateral underneath the body until the migrating fluid reaches:

1) the edge of the body or
2) salt weld or
3) any kind of window in salt distribution,

where the migration changes again to the vertical (e.g., the Gulf of Mexico – McBride et al., 1998a; Bolchert et al., 2000; the Santos Basin in Brazil – Cobbold et al., 2001; offshore Gabon – Nemčok and Rosendahl, 2006). Edges of the bodies can be associated with the variety of faults described in Chapter 19. Together with faults at the tops of the bodies, the fault types include extensional faults such as (Rowan et al., 1999):

1) symmetric faults
   a. peripheral faults
   b. crestal faults
   c. keystone faults
2) asymmetric faults
   a. basinward faults
      i. roller faults
      ii. ramp faults
      iii. shale-detachment faults
   b. landward faults
      i. counterregional faults
   c. variable faults
      i. flap faults (Figure 18.12a)
      ii. roll-over faults

and contractional faults such as:

1) toe-thrust faults
2) break-thrust faults

and lateral strike-slip faults. An intense and multi-event, meso-scale deformation associated with these fault types can be found in outcrops around salt bodies such as those located along Moab-Spanish Valley, Salt Valley, Castle Valley, and Onion Creek near Moab, southern Utah (Figure 18.12b).

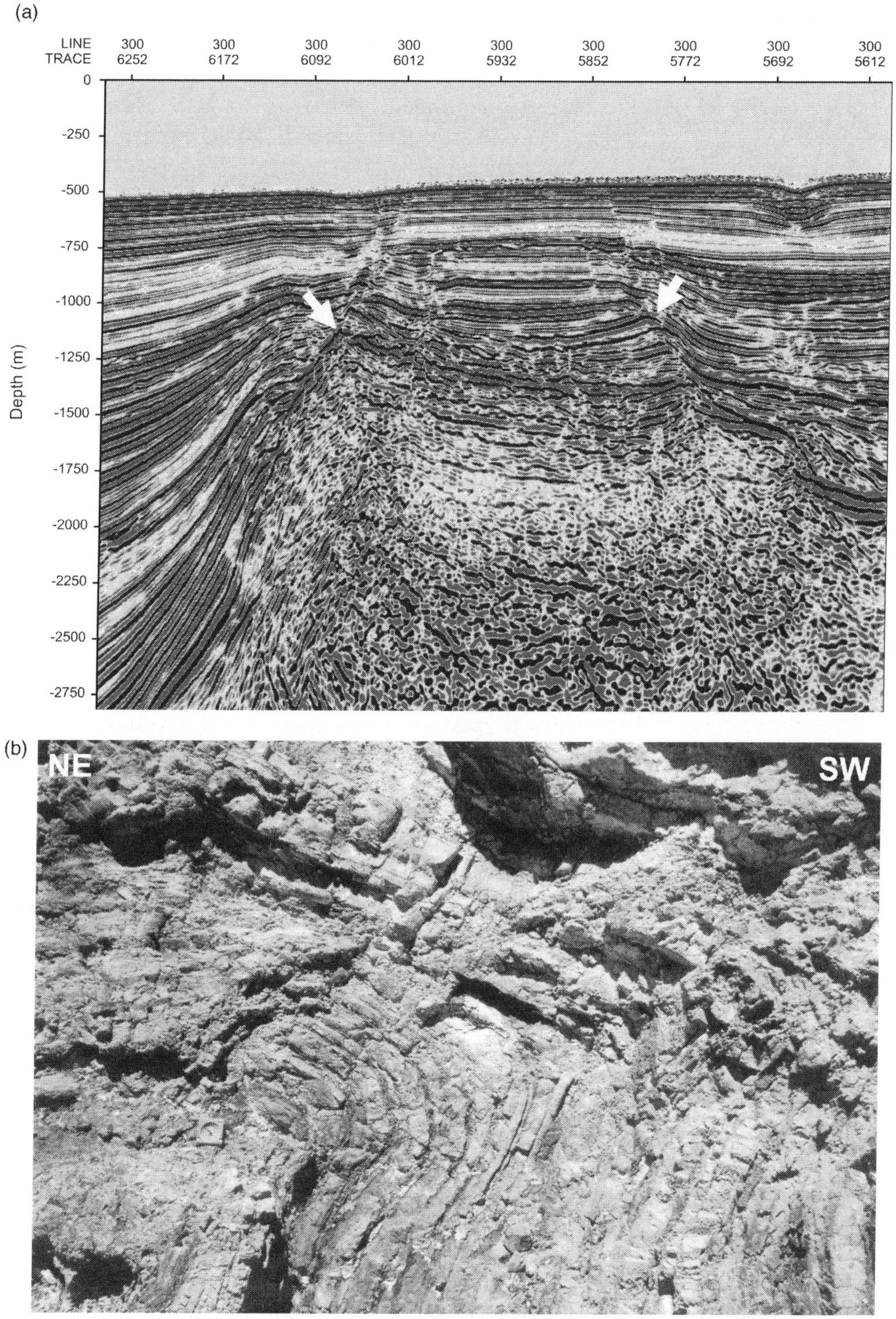

Figure 18.12. (a) Flap fault systems in GXT GabonSpan section through the Gabon passive margin (Nemčok and Rosendahl, 2006). (b) Detail of the salt body-host rock contact from Onion Creek Valley in Utah. Note the intense deformation in immediate cap rock of the diapir formed by the Pennsylvanian Paradox Formation. Note multiple faults and fractures deforming the folded strata.

Salt welds can form in association with salt evacuation of numerous salt body geometries such as (Rowan and Weimer, 1998):

1) welds associated with bulb-shaped salt stock evacuation including
    a. welds developed by ramp-fault rotation
    b. welds developed by counter-regional collapse
2) welds associated with salt sheet evacuation including
    a. welds developed by Roho system
    b. welds developed by stepped counterregional fault,

which are known to be associated with numerous petroleum seeps in the northern Gulf of Mexico (see Bolchert et al., 2000).

A relatively even distribution of weld types in control of these seeps indicates that all allochthonous salt systems create pathways for vertical petroleum migration. The timing of the weld formation marks the onset of vertical fluid migration (e.g., McBride, 1997, 1998a; Rowan and Weimer, 1998). Other possibilities for migration windows through the salt cover are represented by:

1) the welds associated with:
    a. thrust faults described in the Gulf of Mexico (e.g., Rowan et al., 1999)
    b. wrench faults described in the Santos Basin (Cobbold et al., 2001) such as those accommodating uneven gliding of different portions of the glide system in the southwestern Santos Basin (see Nemčok, 2001; Mohriak et al., 2008)
2) salt windows created by the salt horizon welding above a prominent pre-salt high, documented in offshore Gabon (Nemčok and Rosendahl, 2006).

Welds of various types take different time periods to develop as indicated by structural restorations made in an intra-slope setting of the Gulf of Mexico (McBride et al., 1998b). It takes about 0.3–0.8 Ma for ramp-fault systems, 1.4–5.5 Ma for counterregional systems, and more than 1.9 Ma for stepped-counter-regional systems.

A knowledge of the migration inside the injectite lithologies started to accumulate with an outcrop research of sand dykes in the United Kingdom (Waterston, 1950; Cosgrove, 2001; Figure 18.13a), the Sacramento Valley in California (Jolly et al., 1998), and the Kura Valley in Georgia (Nemčok, 2004, 2005; Figure 18.13b), their paleostress studies in southeastern France (Huang, 1988), Santa Cruz, California (Boehm and Moore, 2002), and northeastern Scotland (Jonk et al., 2003), and seismic, core, and well log-based reservoir studies in the North Sea province (e.g., the Balder and Jotun fields – Jenssen et al., 1993; Bergslien, 2002; the Gryphon field – Purvis et al., 2002; the Alba field in the Central Graben – Lonergan and Cartwright, 1999; Duranti and Hurst, 2004). The general consensus is that a sediment injection to the overlying or underlying sedimentary section requires a fluid overpressure (e.g., Nemčok and Gayer, 1996; Jolly and Lonergan, 2002; Duranti and Hurst, 2004). The overpressure is documented, for example at St Audrie's Bay, Bristol Channel, as forming V-shaped clefts that cut through preexisting calcite-filled veins (Nemčok and Gayer, 1996; Figure 17.25a). Reopened centers of the veins are filled by barite and calcite, in which the latter is clearly separated from the earlier calcite-fill event. Hydraulic fracturing by overpressured fluid is further indicated by some of the clefts filled by an untextured limestone/mudstone mixture derived from the underlying beds. The normally bed-parallel fabric of the underlying shale is tightly folded into the base of the clefts, suggesting its origin as a soft sediment injection structure. Similar hydraulic fractures in the Watchet region of the Bristol Channel have been formed in the Triassic Mercia mudstone, preserved in the mudstone section thanks to their satin spar fill or injection by unconsolidated sand sourced from small pockets of sand entrapped inside the thick mudstone section (Cosgrove, 2001; Figures 18.13c, d). Unfolding of the vertically contracted veins at Rhoose Point, Bristol Channel (Nemčok and Gayer, 1996), sand-cored hydraulic fractures at Watchet (Figure 18.13e), and sand-cored hydraulic fractures from the Eocene section of the Alba field (Hillier and Cosgrove, 2002) indicate that the overpressure release-driven hydraulic fracturing occurs relatively early in the compaction history of the basin fill. It can start as early as at the first hundred meters of burial when freshly deposited mud-supported sediment reaches a stable grain framework in the ongoing burial (see Moore, 1989). The Bristol Channel studies recognize a disequilibrium compaction, eventually assisted by early cementation responsible for a mud section turning into a section of intercalated mudstone and limestone layers, as the fluid overpressuring mechanism. The hydraulic fracturing in the Tertiary reservoirs of the North Sea was controlled to a various extent also by seismically induced liquefaction (Obermeier, 1996; Jolly and Lonergan, 2002; Jonk et al., 2005) and the addition of extra pore fluid volumes such as hydrocarbons (Lonergan et al., 2000; Mazzini et al., 2003; Jonk et al., 2005).

Outcrop evidence from mineralized and sand-propped veins in the Bristol Channel (Nemčok and Gayer, 1996;

Figure 18.13. (a) Hydraulic fracture filled by gypsum and sandstone at Watchet, Bristol Channel, sourced from a small pocket of sand encapsulated in the thick Mercia mudstone section. (b) Sand-propped hydraulic fracture in the shale of the Akchagil Formation at location M125 in Taribani, Kura Valley (Nemčok, 2005). (c) Hydraulic fractures filled by gypsum and sandstone at Watchet, Bristol Channel. (d) Typical size of the sand body trapped inside the Mercia mudstone sequence feeding sand-propped hydraulic fractures at Watchet, Bristol Channel. (e) Hydraulic fracture filled by gypsum and sand, formed at the early stages of burial and subsequently folded by ongoing compaction, Watchet, Bristol Channel. (f) Cemented hydraulic fracture with several episodes of opening indicating cycles of pore pressure buildup followed by failure in the shale of the Shiraki Formation at location M13 in Taribani, Kura Valley (Nemčok, 2005). Mineral fibers have a northwest–southeast trend.

Figure 18.13 (*cont.*)

Figure 18.13c) and the Kura Valley (Nemčok, 2005; Figure 18.13f) indicates that numerous hydraulic fractures were multiply reopened, serving as temporary vents for escaping fluids after each cycle of overpressure buildup by increasing burial in sealed-off mudstone horizons. Although the origin of sand-propped and cemented hydraulic fractures is the same, their fluid venting capability differs.

The multiply reopened and subsequently cemented hydraulic fractures vented the overpressured horizons through sandwiching seals in the Bristol Channel only while they were open (Nemčok et al., 1995, Nemčok and Gayer, 1996). On the contrary, sand-propped hydraulic fractures, in theory, form long-lasting fluid conduits (Jonk et al., 2005). In reality, their permeability in time is controlled by a diagenetic history following their development.

Such history was studied in the Eocene mudstone-dominant Balder and Frigg formations with sandstone bodies in 4 hydrocarbon fields of the South Viking Graben in the North Sea (Jonk et al., 2005). It drew from a chemical composition of diagenetic minerals, fluid inclusions, and analysis of stable isotopes of carbon and oxygen. The parent sandstone bodies here were deposited by gravity flows. They are characterized by original depositional structures such as parallel lamination, erosional bases, concordant tops, and mudstone clasts aligned parallel to bedding only in a few cases. The sandstone remobilization is more common either in the external portions of larger bodies or in the entire small bodies characterized by structureless sandstone. Injectites are branching off parent bodies; forming zones of sandstone dikes and sills up to 100 m thick. Uncemented dikes underwent ptygmatitic folding due to differential compaction such as that in Figure 18.13e. Cemented dikes underwent imbrication such as some near Watchet in the Bristol Channel, which documents the role of dike's mechanical stratigraphy on its subsequent deformation style. Injectites are associated with sand-supported mud-clast breccias developed from a mudstone section pervasively fractured and invaded by fluidized sand.

An interesting observation is that a mechanical compaction-controlled porosity reduction has been recorded by all types of observed lithologies while the porosity loss is dominated by cementation only in

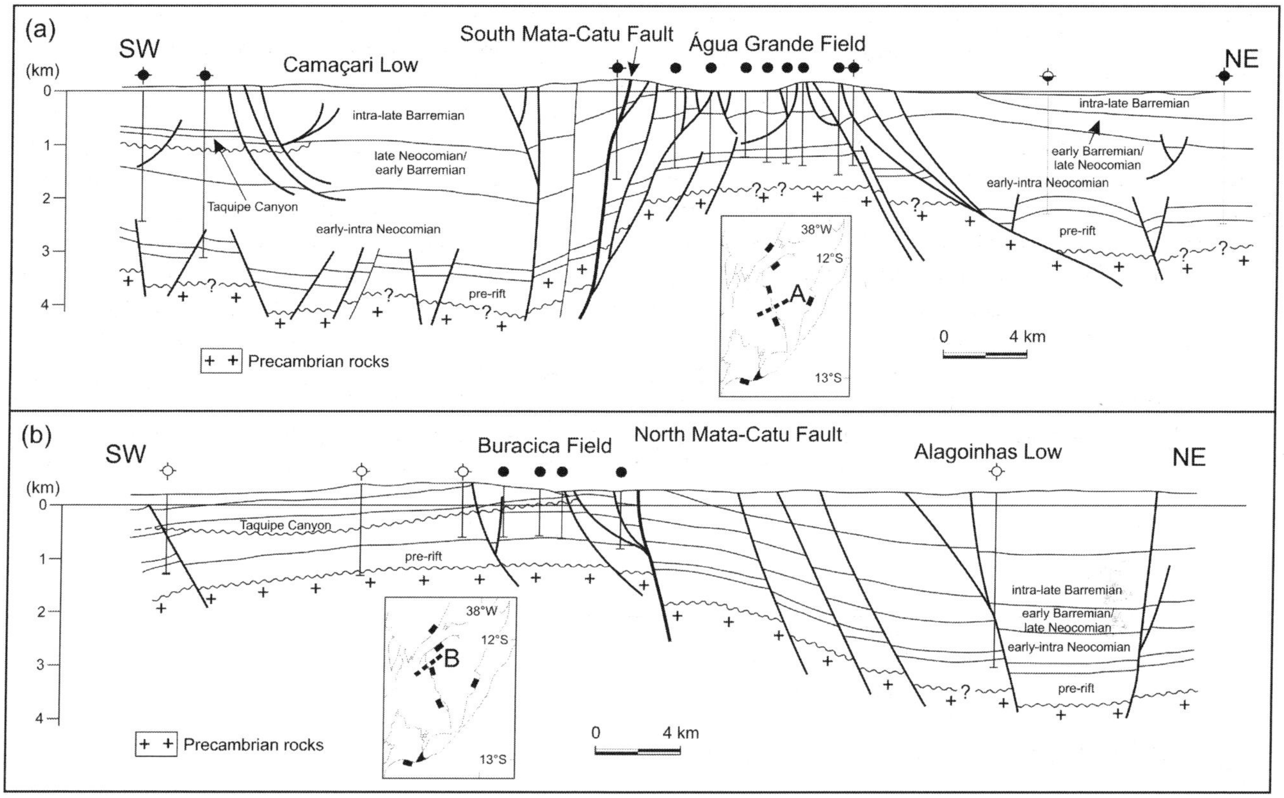

Figure 18.14. Geological cross-sections through the South (a) and North (b) Mata-Catu Faults in the Recôncavo Rift Zone showing the locations of release faults, oil fields in the footwalls of release faults, and oil-generating kitchens in the hanging walls of release faults (Destro et al., 2003b). Note that vertical and lateral migration components are comparable in the case of the Agua Grande Field while lateral migration dominates in the case of the Buracica Field.

injectites. The thinner the injectite body, the more dominant the cementation is. The study of various subsequent cementation phases and their relative timing based on the fluid inclusion data indicates that injectites continued to act as vents for trapped fluids for an extensive period of time after their development, although their cementation history started immediately after their emplacement (Jonk et al., 2005). Fluid inclusions in different cement generations further indicate that hydrocarbons must have been one of the overpressuring controls, as they were vented during the injection and subsequent fluid escape events responsible for younger cementation events.

The geochemical and stable isotope data further indicate that the injectites served as conduits for moderate-salinity basinal brines and hydrocarbons migrating upward and low-salinity mixed marine-meteoric fluids migrating downward (Jonk et al., 2005). The meteoric waters were recharging during the Paleogene, when the northern North Sea formed a narrow, deep-marine basin bordered by subaerially exposed highs in Norway and the East Shetland Platform in response to the North Atlantic rifting (Huuse, 2002; Fyfe et al., 2003; Jones et al., 2003). This exposure was further enhanced by a pronounced sea-level drop at the Eocene–Oligocene boundary (Lear et al., 2000; Huuse, 2002).

A cementation history of injectites is strongly dependent on the hydrocarbon charge timing with respect to the timing of injectite emplacement, because of the oil biodegradation by descending low-salinity fluids being an important bicarbonate source for the pervasive carbonate cementation (see Macaulay et al., 2000). The injectites emplaced coevally with oil migration tend to get significantly cemented while injectites significantly predating oil migration undergo a relatively moderate cementation.

An interesting combination of vertical and lateral migration in rift settings can occur controlled by hanging wall relief rising toward both ends of the graben-controlling normal fault due to bending stresses of the hanging wall block caused by a variation of displacement along the strike of normal fault. This relief can be further dissected, forming local footwall culminations, by release faults that accommodate differential down-dip movements of the hanging wall (Figure 1.24).

Examples of such migration scenarios come from the region along the South and North Mata-Catu release faults in the Recôncavo Rift Zone, where the oil fields and source rock kitchens are located in footwalls and hanging walls of release faults, respectively (see Destro et al., 2003b; Figure 18.14). The source rocks are represented by the syn-rift Candeias Formation of Lower Cretaceous age, while reservoirs have a pre-rift Jurassic age.

The vertical migration took place along the release faults, characterized by some fault segments leaking some sealing (see Magnavita, 2000). The lateral migration along carrier beds was on occasion as long as 40 km, as documented by the Buracica oil field. The release faults were apparently dilatant, as documented by patterns of accompanying release fractures, tensile fractures enhancing the surrounding reservoir permeability as it can be seen in the Candeias oil field (Destro et al., 2003b).

The examples of combined vertical and lateral migrations come from provinces with normal faults, which served as vertical migration pathways, and also as facilitators of juxtaposition of reservoirs of different stratigraphies, developing juxtaposed reservoir combinations serving as lateral migration pathways. An example of this situation comes from the Egret-Hibernia(!) petroleum system of the Grand Banks area, Canada (Magoon et al., 2005). The area underwent several rift and drift events, such as the Triassic-Jurassic rifting related to the Pangea breakup (e.g., Tankard et al., 1987; Kusznir and Egan, 1989; Ziegler, 1989b) and the Early Cretaceous rifting resulting in the North Atlantic ocean and Labrador Sea opening (Ziegler, 1989b; Balkwill and McMillan, 1990). The former rift fill contains the shallow marine Egret member, the richest source rock in the province. The later rift fill contains seals in the Nautilus shale and Whiterose shale sections. The main reservoir rocks are represented by the Hibernia and Avalon formations located in the later rift fill. There are six more known reservoirs in the Upper Jurassic–Paleocene section deposited during both rift events. The existence of pooled hydrocarbons in eight reservoirs in the province indicates a complex migration history.

19 Trapping Styles

Introduction

A petroleum trap can be characterized as a reservoir/seal contact that has closed contours and positive morphology, represented by a variety of geologic features (Biddle and Wielchowsky, 1994). Depending on their controlling mechanism, the traps include structural, stratigraphic, and a combination of structural/stratigraphic traps. If the trap undergoes the pore fluid flow in the reservoir driving the petroleum against the trapping surfaces, it represents a hydrodynamic trap (Hubbert, 1953; Goolsby et al., 1988). If not, it is a hydrostatic trap. A combination of sealing surfaces can include (Milton and Bertram, 1992):

1) a conformable bedding surface;
2) a facies change boundary;
3) an unconformity; and
4) a fault.

Traps that have a single sealing surface most commonly rely on specific lithostratigraphy, bedding, or an unconformity surface. It can be a detachment fault or a salt weld in gravity glide. Multi-seal traps are typically combinations of stratigraphic and fault surfaces.

The most common traps in failed rift provinces are syn-rift traps with pre-rift reservoirs and post-rift traps with post-rift reservoirs with trapping geometry created by basin inversion and diapirism (e.g., Zanella and Coward, 2003). Passive margins with large deltas typically have traps from the extensional domains of gravity glides (e.g., Doust and Omatsola, 1990).

The following text describes the traps of rifts and passive margins, how common specific traps are, and their advantages and problems. It is based on knowledge from the classical West Siberian, North Sea, and North African rift and the Persian Gulf, Gulf of Mexico, and offshore West African passive margin provinces, among others. Discussion of the probability of occurrence draws from our database on about 1,000 hydrocarbon fields worldwide (Tables 19.1, 19.2).

Traps in rift setting

The most common traps in rift terrains include (Morley et al., 1990; Destro et al., 2003b; Zanella and Coward, 2003):

1) traps related to release faults;
2) traps related to various kinds of accommodation zones;
3) traps related to simple rifting including hanging wall and footwall types;
4) traps related to the interference faulting due to structural complexity driven by subsequent extension events;
5) traps related to the existence of a lateral shear component on top of a normal faulting component in areas undergoing reactivation of preexisting anisotropy – that is, coexistence of extension and inversion in rotated fault blocks;
6) traps related to syn-rift halokinesis;
7) traps related to post-rift halokinesis;
8) traps related to syn-inversion halokinesis;
9) traps related to syn-rift gravity gliding;
10) traps related to post-rift gravity gliding;
11) traps related to the deformation of the preexisting sediment entry point containing the coarsest sediments (relay ramp, low-relief accommodation zone, linkage of neighbor growing faults cutting off preexisting sediment entry point, etc.);
12) traps controlled by listric faults related to mass wasting of the fault scarp of the main boundary fault;
13) stratigraphic traps related to formation pinch-outs;
14) inversion-related traps;
15) drape folds over diapirs; and
16) drape folds over structural highs.

Table 19.1 indicates that the distribution of traps in the combined West Siberia, North Sea, and North Africa provinces is similar to the worldwide one. Therefore, these provinces will serve as our key study areas. For example, their fields document that:

1) the post-rift basin inversion acts as an important trap enhancement factor in rift setting;
2) the existence of evaporites may or may not give a chance for rift architecture to play a significant role in trapping;
3) the existence of evaporites reduces the need for basin inversion improvement of the original rift architecture in order to enhance trapping; and

Table 19.1. Similarity of trap distribution in the worldwide rift provinces and combined North Africa, North Sea, and West Siberia provinces. Distribution is shown in % from total.

Province	Structural traps	Stratigraphic traps	Combined structural stratigraphic traps
North Africa	86	7	7
North Sea	64	13	23
West Siberia	62	34	4
above combined	*71*	*18*	*11*
worldwide	*71*	*20*	*9*
Province	**Inversion traps**	**Salt tectonics traps**	**Rift traps**
North Africa	76	0	24
North Sea	38	24	38
West Siberia	100	0	0
above combined	*71*	*8*	*21*
worldwide	*72*	*8*	*20*

Table 19.2. Similarity of trap distribution in the World's passive margin provinces and combined Persian Gulf, Gulf of Mexico, and West Africa provinces. Distribution is shown in % from total.

Province	Structural traps	Stratigraphic traps	Combined structural/ stratigraphic traps
Arabian Gulf	98	0	2
Gulf of Mexico	51	17	32
West Africa	66	22	12
above combined	*72*	*13*	*15*
worldwide	*74*	*12*	*14*
Province	**Inversion, drape, gravity-glide-anticline traps**	**Extensional salt tectonics traps**	**Rift traps**
Arabian Gulf	79	3	6
Gulf of Mexico	38	30	1
West Africa	45	12	4
above combined	*74*	*21*	*5*
worldwide	*73*	*15*	*12*

4) the increased importance of the primary rift architecture in trapping requires the help of stratigraphic trapping to structural trapping.

However, note that the database may be skewed by the fact that the current strategy of exploration results in discoveries of structural traps much more often than discoveries of stratigraphic traps.

Rifting-related structural traps

Simple up-thrown tilted fault and complex up-thrown tilted fault blocks

A simple up-thrown tilted fault block trap (Figure 19.1a) is not known from the West Siberia province but forms about 7 percent of the traps in the North Sea and North Africa provinces. The North Sea province is represented by seventeen fields. While most of these fields rely on the pre-rift reservoirs, Claymore has both pre-rift and syn-rift reservoirs and Magnus, Piper, and Troll have only syn-rift reservoirs. The North Africa province is represented by the Abu Attifel and Masrab fields of the Sirte Basin. This type of trap is also known from the back-arc setting, where the example comes from the Renqiu field of the Baoding Basin, China, and cratonic sag, where the example is represented by the Bodalla South field of the Eromanga Basin, Australia.

The complex up-thrown tilted fault block trap (Figure 19.1b) is not known from the West Siberia province but forms about 9 percent and 10 percent of traps in the North Sea and North Africa provinces, respectively. The difference between this trap and the simple up-thrown tilted fault block is in the degradation of the crest and/or the fault scarp by small-scale normal faults. The fields in the North Sea province include Brent, Cormorant, Fram, Gloa, Ninian, Statfjord (Figure 19.2), and Varg (Zanella and Coward, 2003). The fields in North Africa include the Amal and Gialo fields of the Sirte Basin and the Hassi Berki field of the Ghadames Basin (see Berman, 2008).

These two types of traps in the North Sea province are represented by rotated fault blocks containing pre-rift Lower-Middle Jurassic and syn-rift Upper Jurassic reservoirs. They rely on sealing from the younger shales, the stratigraphy of which depends on the depth of erosion at footwall edges. Consequently, the seals are either syn-rift Upper Jurassic–Lower Cretaceous shales or post-rift Upper Cretaceous shales. In both types of trap, reservoirs always predate fault-block rotation. The example comes from the Claymore field where the reservoir is formed by the Upper Jurassic Claymore sandstone member and the trap is developed by Early Cretaceous tilting (Zanella and Coward, 2003).

Most of the traps also rely on lateral seals provided by faults juxtaposing reservoirs against shale horizons. Examples of the Lower–Middle Jurassic reservoirs juxtaposed against the Upper Jurassic shales come from the Gullfaks and Statfjord fields. Rarely, when permeable strata are juxtaposed against permeable strata, the traps rely on the shale-smear sealing along the fault, such as that in the Deveron field (Williams, 1991). About 25 percent of the fields with Lower–Middle Jurassic reservoirs contain several compartments separated by such shale-smear faults.

Examples of juxtaposition seal faults as lateral seals for syn-rift reservoirs come from the Brage, Troll, Piper, Rob Roy, and Telford fields. The lateral seal faults relying on clay-smear sealing are controlling the Scott, Glamis, Ivanhoe, and Hamish fields. A comparison of fields such as Piper, Scott, Snorre, Statfjord North,

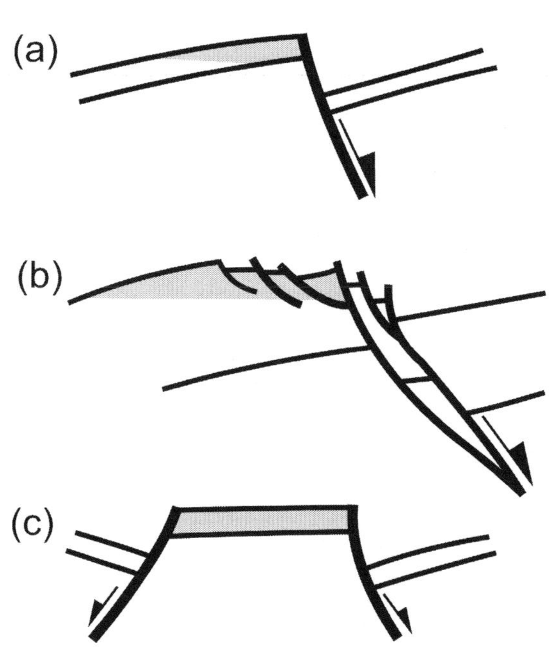

Figure 19.1. Rifting-related traps (Zanella and Coward, 2003) including a) simple up-thrown tilted fault block trap, b) complex up-thrown tilted fault block trap, and c) horst structure trap.

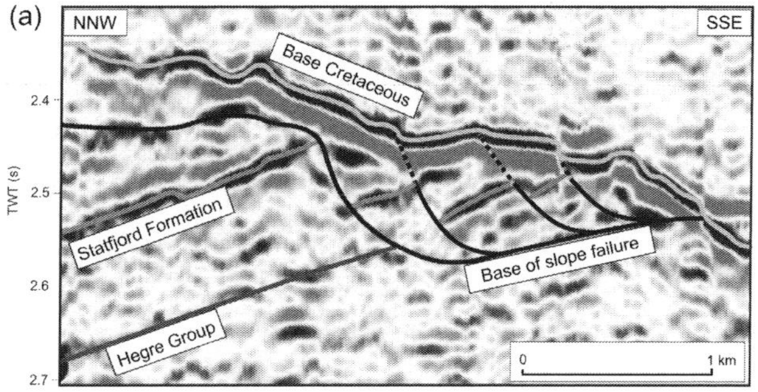

Figure 19.2. (a) Interpreted reflection seismic section through the eastern portion of the Statfjord Field (Hesthammer et al., 1999). Note the crest affected by listric faults driven by gravity collapse. (b) Tectonic development sketch of the Statfjord field (Hesthammer and Fossen, 1999). 1) Gravitational instability of the Brent group results in gravity gliding detached inside the shaly Ness Formation. 2–3) Continuous activity of the boundary fault results in exhumation of the Dunlin group and a new cycle of gravitational instability of the fault scarp. The detachment developed inside the shaly Amundsen Formation. 4–5) Continuing activity of the boundary fault results in the exhumation of the Statfjord Formation and gravity gliding detached along the shaly lower portion of the Statfjord Formation. 6) Eventually the boundary fault becomes inactive with its adjacent zones deformed by systems of gravity glides detached at various stratigraphic levels.

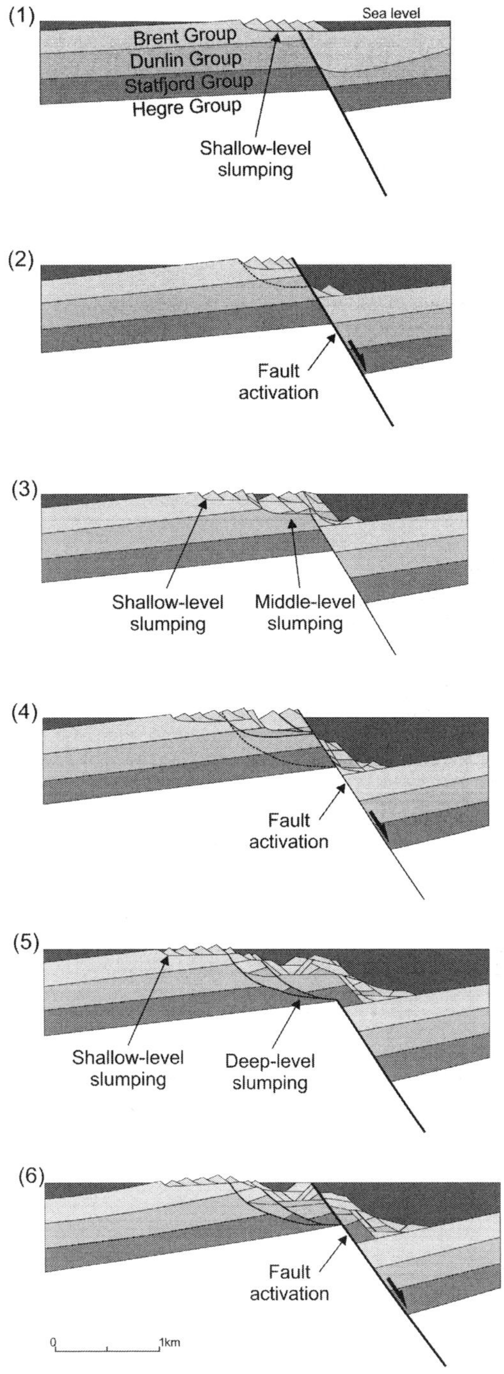

Figure 19.2 (*cont.*)

Telford, and Saltire reveals a significant variation in depth of erosion at the footwall edges of these traps. In extreme cases, the erosional truncation provides a top seal, with a controlling fault not directly involved in trapping, such as is known from the Magnus and Claymore fields.

An interaction of the two major boundary faults can lead to the development of a corner structure in the uplifted footwall. Examples come from the Gullfaks and Hutton fields of the North Sea province (Figure 19.3; Zanella and Coward, 2003).

Horst Structure

This trap (Figure 19.1c) is not very frequent in the key rift provinces, forming only 3 percent of the traps in the North African province, 2 percent of the traps in the North Sea province, and missing in the West Siberian province. An example from North Africa comes from the Sarir C field in the Sirte Basin. The North Sea province contains examples such as the Auk, Buchan, Beryl A, Hamilton, Hamilton N, Hutton, Judy, and Tern-Eider fields. Zanella and Coward (2003) describe fields apart from Hamilton and Hamilton N as having pre-rift reservoirs, which indicates a tendency of these highs to undergo either nondeposition or a combination of reduced deposition/subsequent erosion.

Horst structure traps are also known to occur in other extensional settings. The example from the back-arc setting comes from the Amposta field from the western Mediterranean region.

Down-Thrown Fault Block

Rift-related down-thrown fault block traps (Figure 19.4a) form the second largest population of structural traps in the North Sea province while they are missing in the West Siberian and North African provinces. A potential reason for such variation is the trap domination by inversion structures in North Africa (63%), and in West Siberia (50%), which must have formed on account of preexisting hanging wall traps. The ratio of rift-related down-thrown fault block traps to inversion-controlled traps in the North Sea is 11 percent to 29 percent, which, perhaps, indicates a less important modification of the rift-related hanging wall traps by post-rift basin inversion in comparison to the North Africa and West Siberia provinces.

Examples of the down-thrown traps from the North Sea include the Brae trend fields (Figures 19.4b–c); the Beryl A, Bruce, Oseberg, and Thistle fields from the Viking Graben; the Angus, Cook, Curlew-B, and Curlew-D fields from the Central Graben; the Chanter field from the Moray Firth Basin; and the Tern field from the East Shetland Basin, occurring in all three branches of the North Sea Rift System.

A comparison of 11 percent of hanging wall traps with 18 percent of footwall traps in the North Sea province indicates that hanging wall traps are less successful rift-related traps. It is due to the higher risk for hydrocarbon charge and seal efficiency (see Zanella and

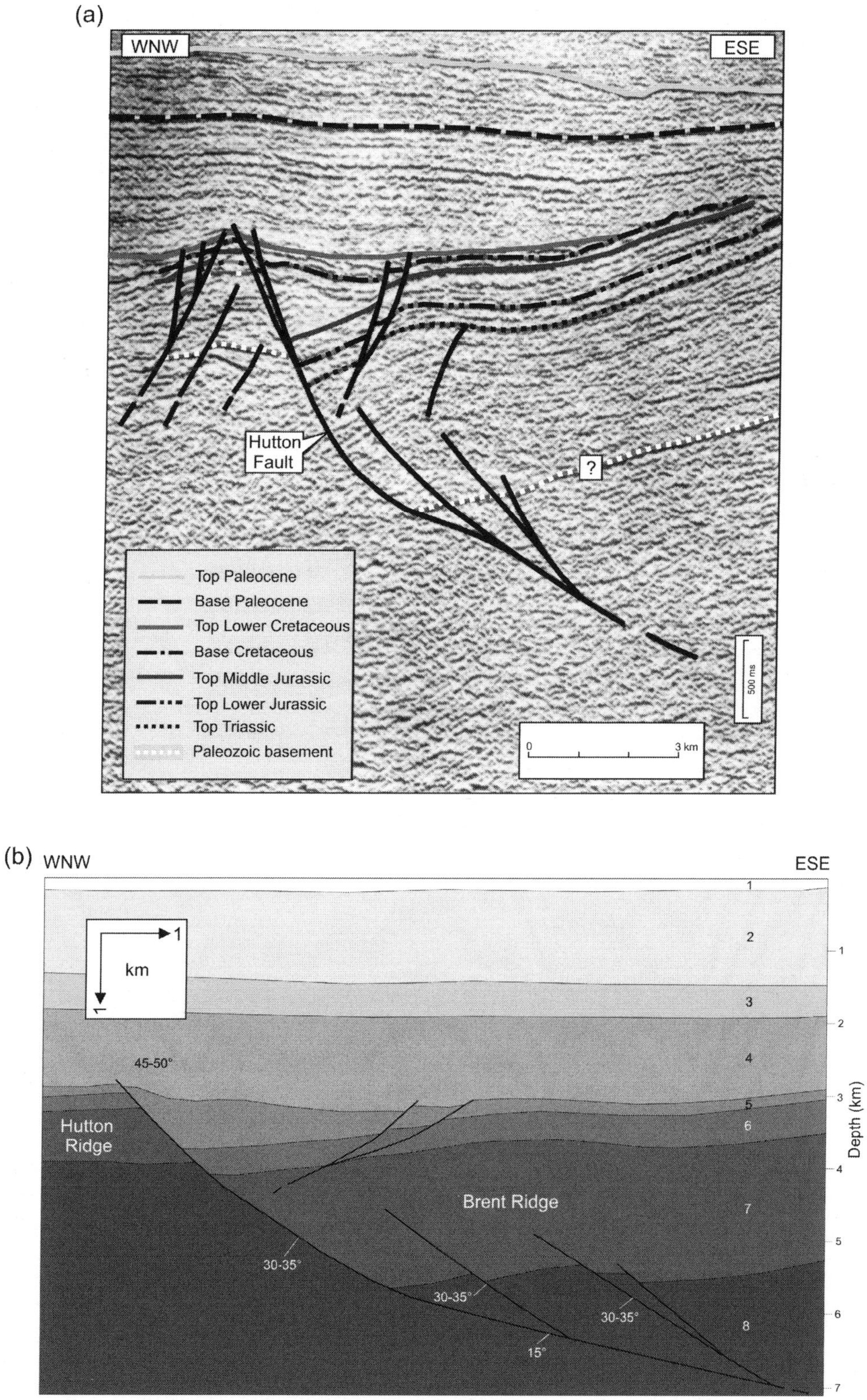

Figure 19.3. Time- (a) and depth-migrated (b) interpreted seismic sections through the Hutton Field (Thomas, 1996a, 1996b). The uppermost portion of the boundary fault is the steepest, having a dip of 45–50°; the lowermost portion has the shallowest dip of about 15°, documenting a listric geometry.

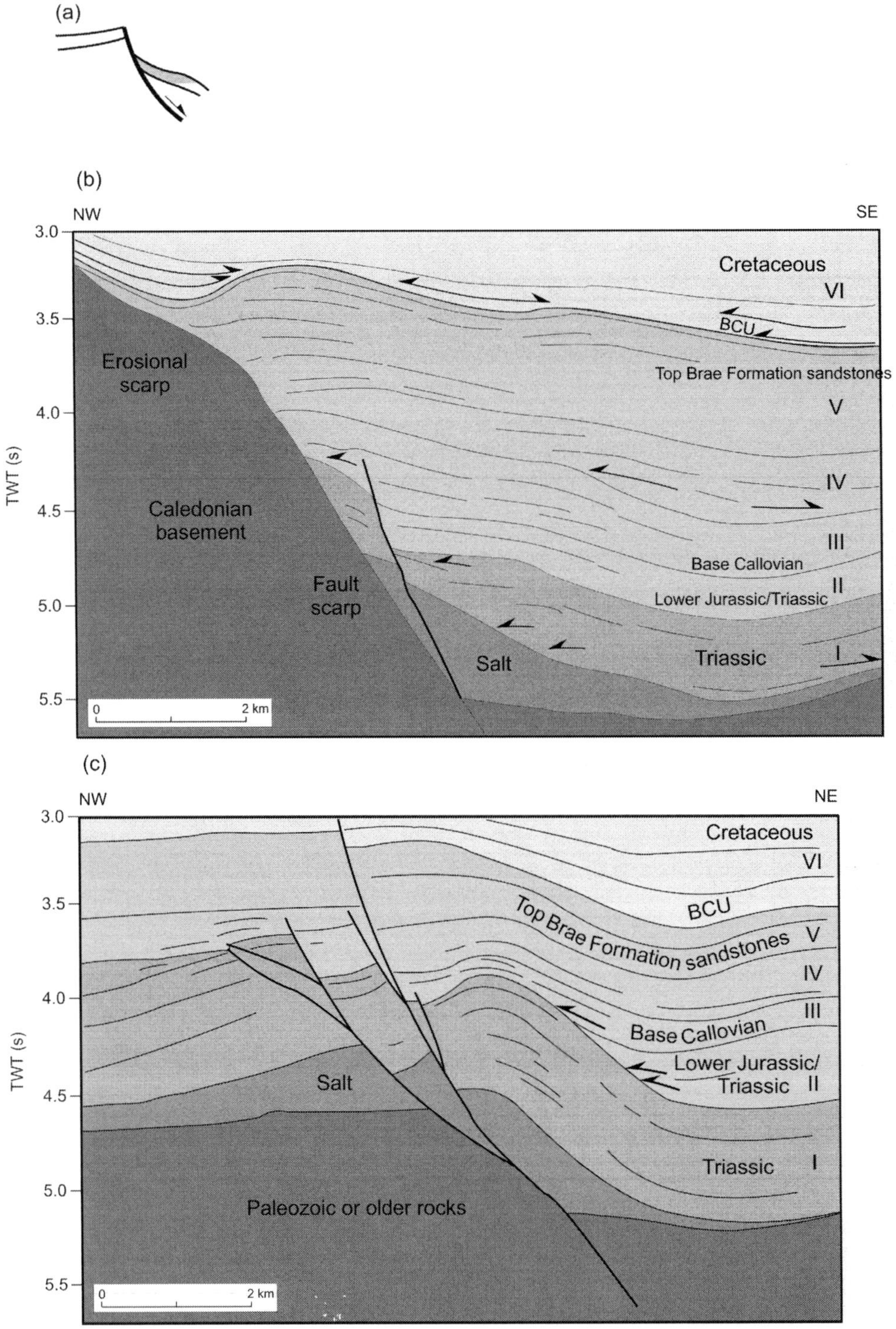

Figure 19.4. (a) Down-thrown fault block trap (Zanella and Coward, 2003). (b–c) Cross-sections through the South Viking Graben in the Brae area.

Coward, 2003). In fact, as documented in the Brae area of the South Viking Graben, several hanging wall traps are successful because of the subsequent basin inversion enhancement. Here the hydrocarbons are trapped in syn-rift conglomeratic reservoirs deposited near boundary faults during their Late Jurassic activity. The subsequent inversion turned preexisting hanging walls of boundary faults into fault-bend folds developed by

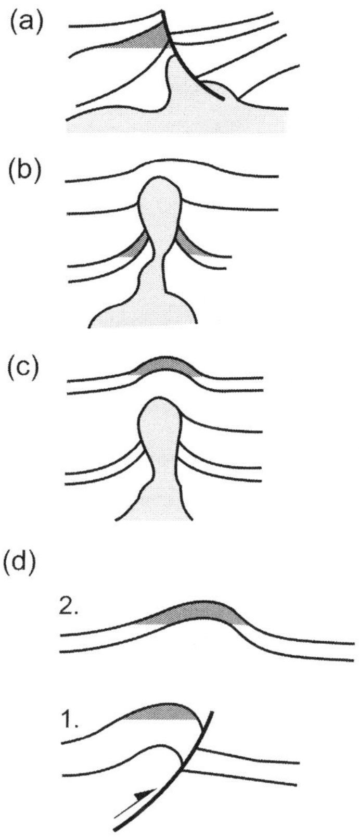

Figure 19.5. Salt-related traps, represented by (a) tilted fault block traps on salt pods, (b) diapir flank traps and (c) traps formed by anticlines related to salt diapir, and d) inversion-related traps, represented by a simple inversion anticline (Zanella and Coward, 2003).

oblique contraction. Such traps rely on sealing inverted fault and mudstone top seal.

Post-rift structural traps

This category contains two important groups: salt-related and basin inversion-related traps. Salt-related traps include (Figure 19.5):

1) gravity-gliding-related footwall traps;
2) gravity-gliding-related hanging wall traps;
3) structural traps located at flanks of diapirs;
4) structural/stratigraphic traps located at flanks of diapirs; and
5) structural traps associated with salt domes.

Inversion-related anticlinal traps can be divided into several varieties including:

1) simple anticlines;
2) faulted anticlines;
3) anticlines with drapes above it; and
4) eroded anticlines with unconformity.

Salt-related traps

Among the three key rift provinces, West Siberia, the North Sea, and North Africa, only the North Sea contains salt-related traps.

Gravity gliding-related traps in the North Sea province contain only footwall traps represented by tilted fault blocks located on salt detachment pods located above salt rollers (Figure 19.5a). Examples come from the Heron, Marnock, Puffin, and Yme fields, which rely on pre-rift sandstone reservoirs of the Triassic-Jurassic section with the exception of the Puffin field that contains both pre-rift and syn-rift reservoirs.

Structural traps at the flanks of a salt diapir (Figure 19.5b) form 1 percent of the North Sea traps in our database. They are formed around the piercing diapirs of the pre-rift Zechstein salt, characterized by upturned reservoir strata. The sealing is provided by evaporites and/or overlying shale horizons. An example comes from the Rev field, which contains the Upper Jurassic syn-rift sandstones around a salt structure.

Other examples of traps at diapir flanks, such as those at the Gannet, Guillemot, Lomond, Mungo, and Pierce (Figure 19.6) fields, are structural/stratigraphic, forming 8 percent of the North Sea traps in our database.

Reservoirs in salt flank-related fields such as those at Banff, Gannet B, Kyle, Machar, Monan, and Pierce contain post-rift reservoirs from the Paleocene-Eocene section.

Remaining salt-related traps in the North Sea province are structural traps associated with salt domes, represented by four-way dip-closure anticlines developed above salt diapirs (Figure 19.5c). This trap category represents about 9 percent of the North Sea traps in our database. Their geometry in map view is either circular or elliptical and their width can be as large as 6–8 km, as the Ekofisk field documents. Most of these traps, as documented by the Albuskjell, Cod, Ekofisk, Eldfisk, Maureen, Tommeliten Gamma, and West Ekofisk fields, are a combination of structural and stratigraphic traps. Fields such as Albuskjell, Andrew, Cod, Dan, Ekofisk, Eldfisk, Gorm, Lomond, Machar, and Tor document that the reservoirs come from the post-rift section.

Inversion-related traps

The basin inversion apparently acts as trap enhancer, as can be implied from its dominance in trap controls in all three key provinces, contributing to 100 percent of traps in the West Siberian province, 76 percent of traps in the North Africa province, where the inversion-related traps dominate, and 38 percent of traps in the North

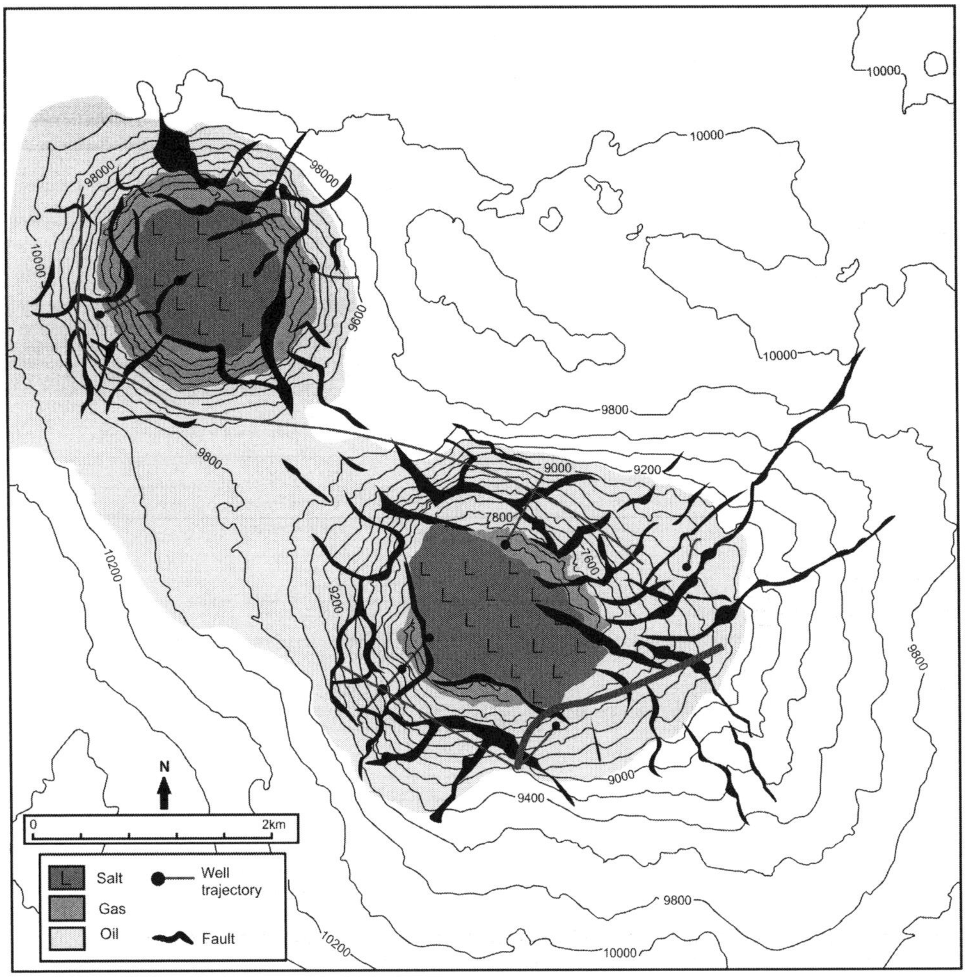

Figure 19.6. A depth map to the top Forties sandstone member in the Pierce Field, North Sea province (Ahmadi et al., 2003). Contours are in feet, drawn in 200-foot intervals. Typical structural feature are radial faults propagated from the diapir into the host sediments.

Sea province, where inversion traps are as important as rift-related traps. While the inversional traps in North Africa are all purely structural, some of them needed stratigraphic trapping enhancement in the North Sea and West Siberian provinces.

Simple inversional anticline

Simple inversional anticlines are dominant leaders in trap type in the salt-lacking provinces of West Siberia and North Africa, forming 44 percent and 56 percent of their traps, respectively. The salt tectonic involvement in trap forming in the North Sea province apparently correlates with a drop of the percentage of simple inversion anticlines down to 3 percent.

As we can see in the North Sea, simple inversional anticlines have compressional origin and offer a four-way dip closure, having their crests developed above the propagation tip of the reverse fault (Figure 19.5d). The simple inversion anticlines of the North Sea trap hydrocarbons in post-rift reservoirs of the Upper Cretaceous–Paleocene section, as documented by the Adda, Hod, Roar, Sleipner East, Tyra, Valdemar, and Valhall fields. Some of these anticlines can have a truncation trapping element due to localized erosion, such as that in the Fulmar field. As the Tyra field shows, it is sometimes difficult to determine whether a simple anticline is an inversion- or salt-related trap.

The North African province indicates that simple inversional traps, apart from simple anticlines, can include monoclinal folds. Fields such as the Alrar, In Amenas, and Zarzaitine from the Illizi Basin form

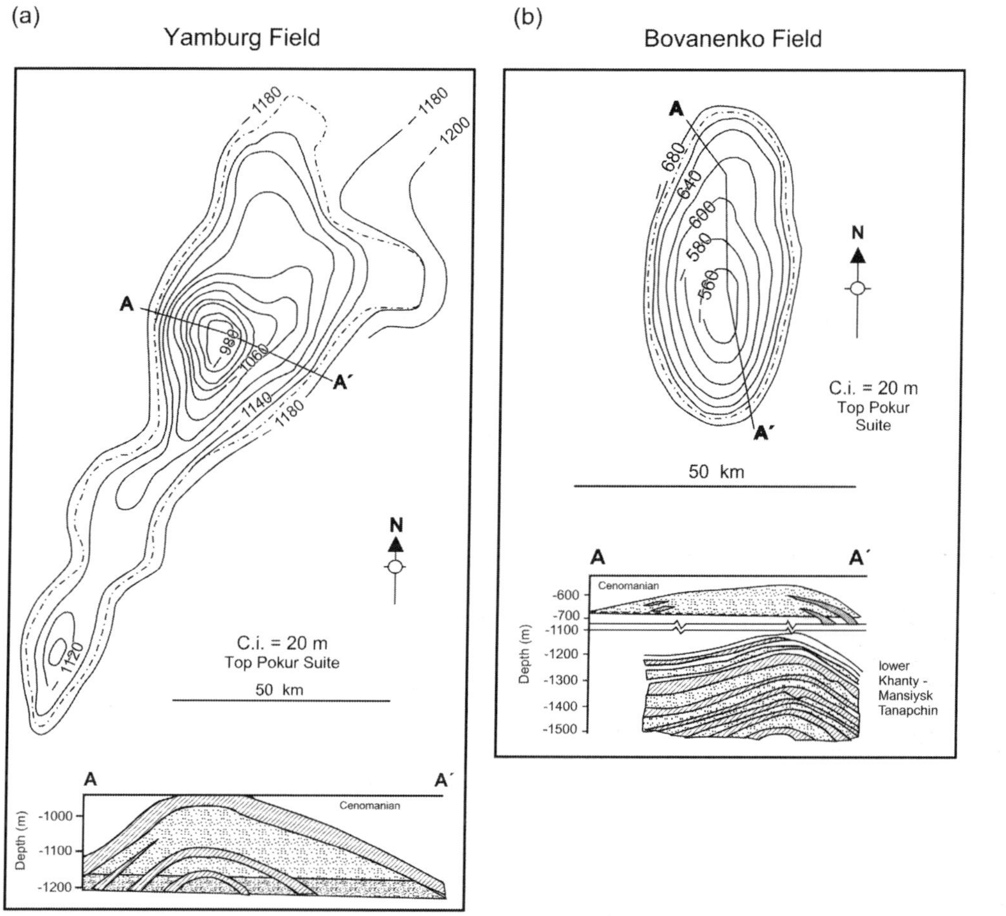

Figure 19.7. (a) Map of the Yamburg trap geometry and dip-oriented cross-section through the field (Grace and Hart, 1986). (b) Map of the Bovanenko trap geometry and strike-oriented cross-section through the field (Gustavson Associates, 1992).

about one-sixth of the simple inversional anticlines in the province.

Examples of this trap type from the West Siberian province include the supergiant Bovanenko, Urengoy, and Yamburg gas fields (see Meyerhoff, 1980a, 1980b, 1982a, 1982b; Grace and Hart, 1986; Gustavson Associates, 1992; Figure 19.7). Note that the leads located in inversional traps in the West Siberia province had about a 64 percent success rate by the year 1992. At that time, the province had about 1,100 structures identified, most of them associated with reverse faults related to basement uplift. Three hundred eight of them contained oil and gas fields, 174 were proven as nonproductive, and about 600 remained untested (see Meyerhoff, 1982a, 1982b; Gustavson Associates, 1992). Dry structures are usually located along basin flanks. Their problems are related to (Gustavson Associates, 1992):

1) flushing;
2) absence of the regional sealing provided by the Turonian Kuznetsov Formation or other seal horizon;
3) lack of good quality source rocks in basin margin areas;
4) lack of closure; and
5) a combination of these or other unspecified factors.

The typical characteristics of producing anticlines include enormous areal extent, low relief, and structural simplicity. An example of all three attributes comes from the Urengoy field, the world's largest gas field, which is about 200 km long and up to 30 km wide. Its low relief is demonstrated by a maximum closure of only about 200 m. The trap is a simple elongate anticline without translation by major faulting. Smaller-scale faults were discovered relatively recently from seismic imaging and geochemical data. The gas is pooled in multiple post-rift sandstone reservoirs in the Neocomian Megion and Vartov Groups, and the Cenomanian Pokur Group. Apart from gas, the condensate and oil occur in the post-rift Megion and clastic Upper Jurassic–Lower Cretaceous Vasyugan formations.

The Bovanenko field is another example of the simple anticlinal trap with low relief, trapping gas in Neocomian groups and the Cenomanian Pokur Group and being one of the largest gas fields of the West Siberian province.

The main factors behind the success of this province include (Peterson and Clarke, 1989):

1) a high content of disseminated and concentrated organic matter of humic composition matured at a sub-bituminous stage of catagenesis to form large quantities of gas;
2) large volumes of source and reservoir rocks with high migration capacity;
3) large structures with great closure located far from the borders of the basin, which have an increased risk of flushing;
4) presence of thick seals; and
5) relatively young timing of charge.

The importance of simple inversional anticline in trap forming during the post-rift period in rift settings is further highlighted by the occurrence of these traps in back-arc and cratonic sag settings. Simple inversional anticlines in the back-arc setting are represented by the Infantas and La Cira fields in the Middle Magdalena Valley Basin, Colombia, and the Matzen field in the Vienna Basin, Austria. Examples from the cratonic sag setting come from the Jackson, King Christian, Kristoffer, and Thor fields of the Sverdrup Basin, Canada, and the Lindsborg field of the Salina Basin, the United States.

Inversion-related faulted anticline traps and syn-inversion drape-related traps

While simple inversional anticlines are the leaders in trap type in salt-lacking provinces, drape folds and subordinately faulted anticlines combined are the leaders in trap type in salt-involved provinces. This trend can be documented by comparing 2 percent and 7 percent in the West Siberian and North African provinces with 25 percent in the North Sea province. The dominance of drape folds over faulted anticlines is documented by the fact that drape folds represent fifteen out of eighteen North Sea field examples in our database.

The only example of this trap type in our database from the West Siberian province, that is, the Samotlor field, and examples from the North Africa province, that is, the El Borma and Raguba fields, are represented by faulted anticlines.

From all of this we can interpret the occurrence of drape folds in basins as not only affected by inhomogeneous compaction of clastic and carbonate sediments but as significantly enhanced by the presence of an evaporite horizon. Its capability to accumulate and withdraw in different areas seems to play an important role in inhomogeneous compaction.

Examples of the drape fold from the North Sea province include the Arbroath, Arkwright, Cyrus, Montrose, and Nelson fields from the Central Graben; the Captain, Cromarty, and Hannay fields from the Moray Firth Basin; and the Frigg, Glitne, Moira, Nuggets N1, Nuggets N2, Nuggets N3, and Nuggets N4 fields from the Viking Graben. The remaining North Sea examples of this trap type are faulted anticlines represented by the Beinn, Sleipner Ost, and Sleipner Vest fields.

Inhomogeneous compaction-related drape structures, formed above deeper anticlines, horsts, and edges of tilted fault blocks can be characterized as low-relief anticlines with four-way dip closure. Examples of the drape folds over underlying structural highs and edges of tilted fault blocks come from the Forties-Montrose High of the Central Graben, represented by the Arbroath, Forties, Heimdal, Montrose, and Nelson (Figure 19.8) fields.

An example of this trap type from the West Siberian province comes from the supergiant Samotlor oil field (see Meyerhoff, 1980a, 1980b, 1982a, 1982b; Grace and Hart, 1986; Gustavson Associates, 1992). The field is currently ranked as the fiftieth largest oil field in the world (see Berman, 2008). Its trap has a domal geometry, being about 40 km wide. Its producing area is more than 800 km^2 and its maximum closure on the Aptian stratigraphic level is less than 100 m. The oil is pooled in ten sandstone reservoirs occurring in the Neocomian post-rift Megion and Vartov Formations. Fine-to-medium grained sandstone and siltstone reservoirs were deposited in a deltaic environment.

The importance of drape folds and faulted inversional anticlines in trap forming during the post-rift period in rift settings is further highlighted by the occurrence of these traps in different kinds of extensional settings, as documented by examples coming from:

1) the Prinos field in the Prinos Basin of the North Aegean Sea, which has a back-arc setting;
2) the Drake Point field in the Sverdrup Basin, which has a cratonic sag setting; and
3) the Kapuni field in the Taranaki Basin, New Zealand, which has an inter-arc setting.

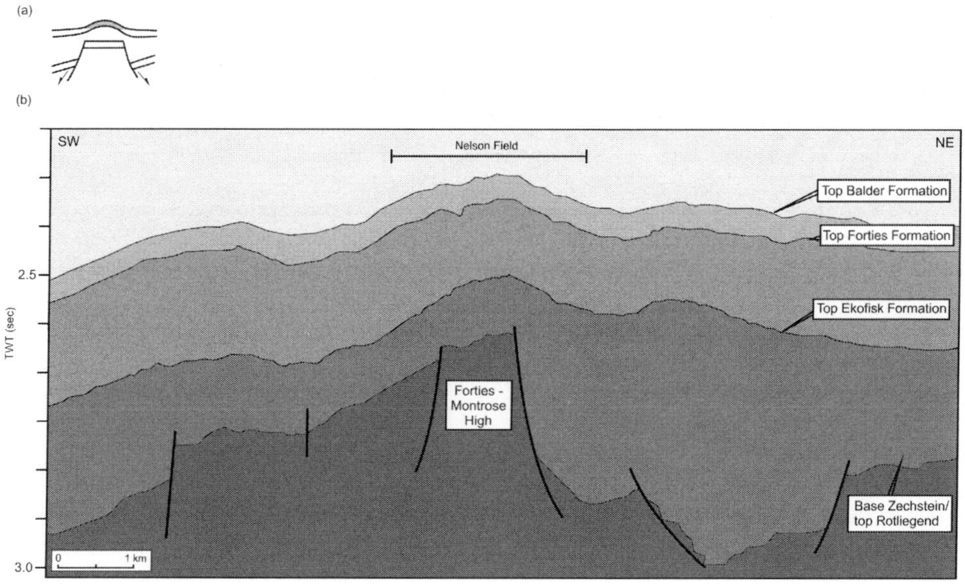

Figure 19.8. (a) Drape fold formed by differential compaction above the deeper-seated horst (Zanella and Coward, 2003). (b) Reflection seismic section through the Nelson field of the North Sea province field imaging the deep-seated horst controlling the differential compaction and development of the drape fold above it (Ahmadi et al., 2003).

Inversional anticline traps with a stratigraphic trapping component

The last type of inversion-related traps is a structural, anticlinal trap enhanced by a stratigraphic trapping component. This trap type represents 4 percent of the traps in the West Siberian province, 1 percent of the traps in the North Sea province. It is not present in the North African province in our database.

Examples from West Siberia include the Taz, Tyan, and Ust-Balyk fields (see Meyerhoff, 1980a, 1980b, 1982a, 1982b; Grace and Hart, 1986; Gustavson Associates, 1992; Figure 19.9). Oil in the Ust-Balyk field is trapped in an anticlinal structure serving as a four-way dip closure, which is 40 km long and 8 km wide. The structure is elongated and has gently dipping flanks. It has deep-seated fault control. Oil is pooled in the Valanginian-Hauterivian sandstone reservoirs of the Vartov Group, some of which pinch out inside shale-siltstone sediments, form lenses in it, or undergo lateral facies change. The only example of this trap type in the North Sea province in our database is the Beinn field in the South Viking Graben. The stratigraphic trapping component comes from the erosion of the inversional anticline and associated unconformity.

Structural traps related to release faults and accommodation zones

Examples of the traps related to release faults come from the Água Grande field next to the South Mata Catu cross-fault and the Buracica oil field next to the North Mata Catu fault in the Recôncavo Rift Zone (Destro et al., 2003b; Figure 18.14). All their compartments are located in footwalls of cross-faults and represented by simple up-thrown tilted fault block traps, complex up-thrown tilted fault block traps, and horst structures. The main petroleum system in this area is located in the Upper Jurassic pre-rift and Neocomian main syn-rift strata, with the pre-rift eolian-fluvial Sergi Formation serving as the main reservoir. What differentiates cross-fault-related footwall traps from those described earlier in this chapter is the fact that the erosion of the footwalls associated with cross faults is less significant than the erosion of isostatically uplifting footwalls adjacent to main boundary normal faults. Further north from the Mata Catu release faults lies the Itanagra-Aracas release fault. It also contains a footwall trap represented by the Miranga oil field but, in this case, is complicated by the presence of shale diapirs.

The most unique examples of the release fault-related fields from the Recôncavo Rift Zone come from the Brejinho and Canabrava oil fields located at the western flank of the zone. They are producing from the hanging wall traps represented by anticlinal structures associated with reverse displacement. While the South Cassarongongo Fault has a normal slip displacement, the North Cassarongongo Fault has a reverse displacement. This combination of displacements can be predicted from the release faulting model because the portions of a hanging wall block located at opposite

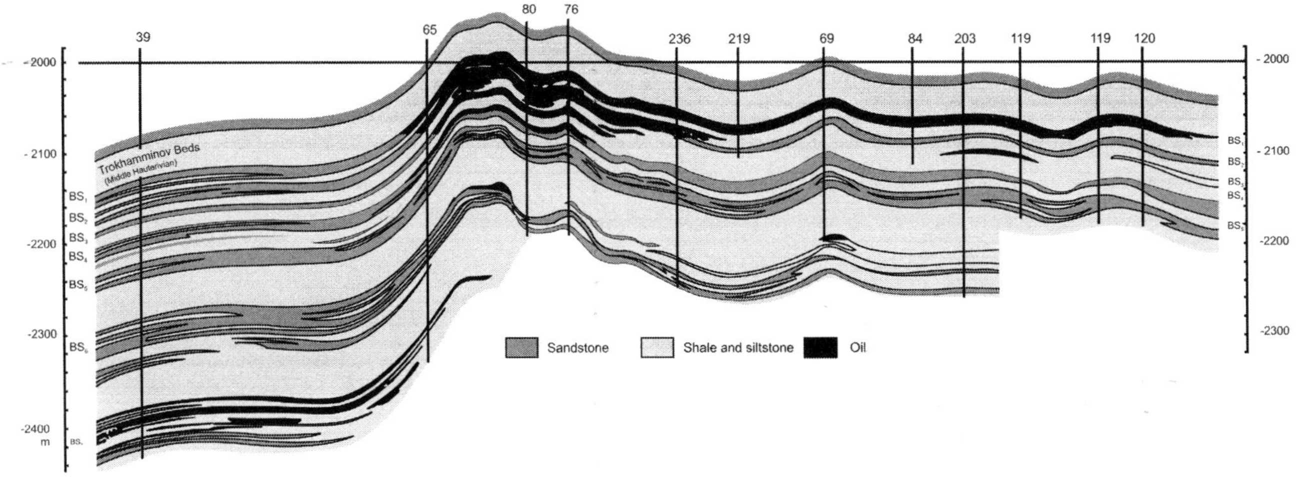

Figure 19.9. Strike-oriented cross-section through the Ust-Balyk Field of the West Siberian province with an indication of lithostratigraphies and pooled hydrocarbons (Meyerhoff, 1980a, 1980b).

sides of the line of neutral strike-slip movement moved toward each other. The South Cassarongongo fault, similarly to the Mata Catu Faults and the Itanagra-Aracas Fault, is associated with footwall traps as documented by the Cassarongongo oil field.

Examples of the traps related to accommodation zones come from the North Sea, Gulf of Suez, North African, and North West Australian provinces (Morley et al., 1990). Contrary to the previously discussed trap types, we cannot describe a trapping mechanism specific for accommodation zones because the trap location in this setting is simply a result of a specific reservoir, seal, trap, and migration combinations (Figure 19.10). Accordingly, we can observe an increased concentration of the North Sea fields along synthetic and conjugate transfer zones (Morley et al., 1990; Fraser et al., 2002; Husmo et al., 2002).

A similar field clustering can be interpreted for the Gulf of Suez, where Morley and colleagues (1990) observe fields such as El Morgan, located along the complex conjugate overlapping transfer zones dividing regions of different dips, or Belayim, October, and Ras Budran, located along synthetic overlapping transfer zones dividing rift units in the region of the same dip.

The footwalls of boundary faults and conjugate divergent, overlapping transfer zones tend to form drainage-dividing elevated topography. Examples come from Lake Tanganyika in East Africa and the Hateiba field of the Sirte Basin, North Africa (e.g., Harding, 1984; Rosendahl, 1987; Morley et al., 1990; Figure 19.10). In this case, the syn-rift reservoirs are most likely eroded off such topography. However, the Sirte Basin examples indicate a presence of structural/stratigraphic traps with syn-rift reservoirs located on the flanks of such horsts in hanging walls of the boundary faults (see Harding, 1984). More typical for highs associated with conjugate divergent, overlapping transfer zones is to rely on pre-rift reservoirs, post-rift top seals, and lateral syn-rift lithological seals as can be seen in the Bohai Basin, China (Desheng, 1980) and the Sirte Basin (Harding, 1984). The exception from these characteristics is the case of a post-rift reservoir trapped in a broad drape over a conjugate divergent, overlapping transfer zone, represented by the Forties field in the North Sea province.

Most of the other transfer zones tend to control broader and more subtle intra-basin highs. Examples come from the Zelten and Lahib fields of the Sirte Basin, the Argyll field in the North Sea, the Rankin Platform fields in North West Australia, and the Ramadan field in the Gulf of Suez (Brennand and Van Veen, 1975; Pennington, 1975; Harding, 1984; Veenstra, 1985; Morley et al., 1990). Apart from being structural traps, intra-basin highs can be the sites with turbiditic sandstones onlapping on their flanks, also providing a potential for stratigraphic traps.

Structural highs can be associated with drape folds above them, such as the fields in the Zelten Platform of the Sirte Basin, containing late syn-rift and post-rift reservoirs.

Stratigraphic traps

Growing evidence of stratigraphic and structural/stratigraphic traps from the rift settings, making 38 percent, 36 percent, and 14 percent of the West Siberian, North Sea, and North African traps, indicates their importance either as stand-alone traps or

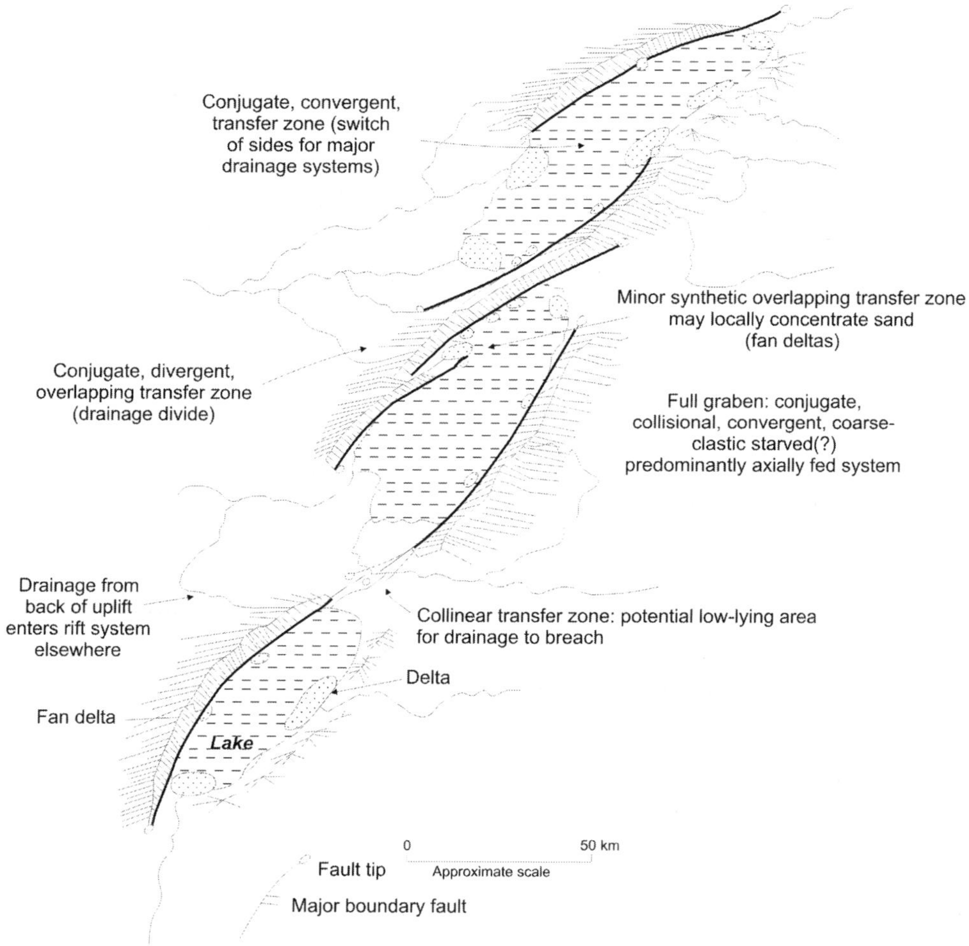

Figure 19.10. Simplified interaction of drainage, sand distribution, boundary faults, and isostatic uplift of footwalls that can influence the location of traps related to accommodation zones (Morley et al., 1990). The isostatically uplifting footwalls of the main boundary fault create barriers for sediment transport into rift depressions. As a consequence, any larger sediment transport systems can enter the depressions only through "gaps" between the boundary faults.

a structural traps-enhancing factor. Stratigraphic trap controlling component includes variety of scenarios such as:

1) organic buildup;
2) angular unconformity;
3) sealed-off depositional lens;
4) horizon pinch-out;
5) lateral facies change;
6) horizon draping; and
7) differential compaction over preexisting structure.

It can have pre-rift, syn-rift, and post-rift timing. Examples of the organic buildup-related traps from the three key provinces in our database come only from the North African province, represented by the Intisar (Ildris) "A" and Intisar (Ildris) "D" fields of the Sirte Basin. This small amount, perhaps, indicates that this trap type is not important for large fields, although this conclusion may be significantly biased by using only the three key provinces instead of numerous ones, which would likely indicate a climatic control over the location of this trap type.

Angular unconformity-controlled stratigraphic traps are not represented in the discussed key rift provinces, although the unconformity assistance to structural traps has been mentioned earlier in this chapter in the case of the Beinn field in the South Viking Graben, North Sea.

Although the stratigraphic traps due to a sealed-off depositional lens are not represented in our database on the North Sea and North Africa provinces, they achieve the second largest representation among the West Siberian trap types, represented by 22 percent. The largest amount of published data exists on the

Salym oil field (e.g., Dikenshteyn et al., 1982). Its reservoir is formed by fractured Upper Jurassic–Tithonian Bazhenov shale.

Horizon pinch-out-controlled stratigraphic traps form 7 percent and 1 percent of the West Siberian and North Sea traps. Examples from West Siberia include the Komsomol'sk, Myl'dzhino, Sugmut, Vachim, Var'yegan Severnyy, and Vengayakha fields. The Dauntless field in the UK Central Graben represents an example from the North Sea province.

Lateral facies change-controlled stratigraphic trapping is not very common, represented by 3 percent in both the West Siberian and North Sea provinces. However, two examples from the West Siberia province, Krasnoleninsk and Priobskoe, place fortieth and fifty-first in the list of the world's largest oil and condensate fields (see Berman, 2008). This trap type in the North Sea province is represented by the Buzzard, Grane, and Gungne fields.

Traps in passive margin settings

As in the rift settings, we will try to use the key passive margin provinces, that is, the Persian Gulf, the Gulf of Mexico, and West Africa, as a statistically representative subset of the world's passive margin provinces (Table 19.2). However, before we start, it would be interesting to compare the main trap categories to those of the rift setting (Table 19.1). The differences are partly due to the different evolution of passive margins from their initiators, rift settings, and partly due to the current state of exploration at offshore passive margins, which usually reside under the thicker water column. The dominance of structural traps over stratigraphic and combined structural/stratigraphic traps at passive margins is more pronounced than that in rift settings. This is most likely an artifact of the structural trap preference over stratigraphic trap in the strategy of the deepwater exploration. Furthermore, contrary to rift settings, combined structural/stratigraphic traps are more important than stratigraphic traps at passive margins. The reason is most likely geological, as passive margin provinces frequently contain numerous robust gravity gliding systems, which are not as common for rift settings. The changing geometry of structures in gravity glides interacts with ongoing deposition (e.g., Martin, 1978; Diegel et al., 1995; Rowan et al., 1999; Combellas-Biggot and Galloway, 2002; Mohriak et al., 2008), which causes the development of combined structural/stratigraphic traps at numerous margins.

A comparison of anticlinal, salt body-related, and rift-related traps (Tables 19.1, 19.2) indicates that anticlinal traps are even more dominant at passive margins. However, in rift settings, they are represented by inversion-related anticlines, while at passive margins they are a composition of drape folds, inversion-related anticlines, and gravity-glide-related extensional roll-over and contractional anticlines. While in the Persian Gulf and West Africa our database allows the unmixing of this composition, this is not possible in the case of the Gulf of Mexico. Nevertheless, the help from either a post-rift inversional event or post-rift gravity gliding seems to significantly enhance the number of post-rift structural traps. This is, perhaps, the reason why passive margins occupy the first place in the ranked list of settings based on the number of giant fields while rifts are second (Mann et al., 2004; Table 14.2). Contrary to rift settings, the other types of salt tectonics-related traps dominate over the rift-related traps. This should be partly due to the existence of robust salt-detached gravity glide systems on passive margins but also due to the fact that current exploration only relatively recently advances into the deepest, syn-rift stratigraphic levels, which are sometimes difficult to image under evaporites and, therefore, remain the last ones to find at passive margins.

This leads to the changing perception about evaporites' role at passive margins. While at the beginning of this century explorers sought migration windows through the autochthonous evaporites (e.g., Cobbold et al., 2001; Meisling et al., 2001), the discoveries of syn-rift oil fields such as Cariocas-Sugar Loaf, Tupi, and Jupiter in the Santos, Campos, and Espírito Santo Basins in Brazil, ranked as the third, forty-third, and forty-fourth largest oil or condensate fields in the world (see Berman, 2008), indicate that a lack of those windows can assist in the preservation of surprisingly important syn-rift petroleum systems.

Table 19.2 further documents that passive margin provinces can vary significantly with respect to distribution of trap types. This is because not all of them have undergone a post-rift inversion, being spared any subsequent orogenic effects. Therefore, one can find as important variations in the occurrence of the post-rift inversion as:

1) a dominance of inversion-related traps in the Persian Gulf province, expressed as 79 percent of all traps (Table 19.2);
2) a small contribution of inversion (transpression)-related traps in the Gulf of Mexico province, expressed as small percentage of traps located in southeastern Mexico and the United States (see Guzman and Marquez-Dominguez, 2001; Meneses-Rocha, 2001; Prost and Aranda, 2001; Ambrose et al., 2002); and

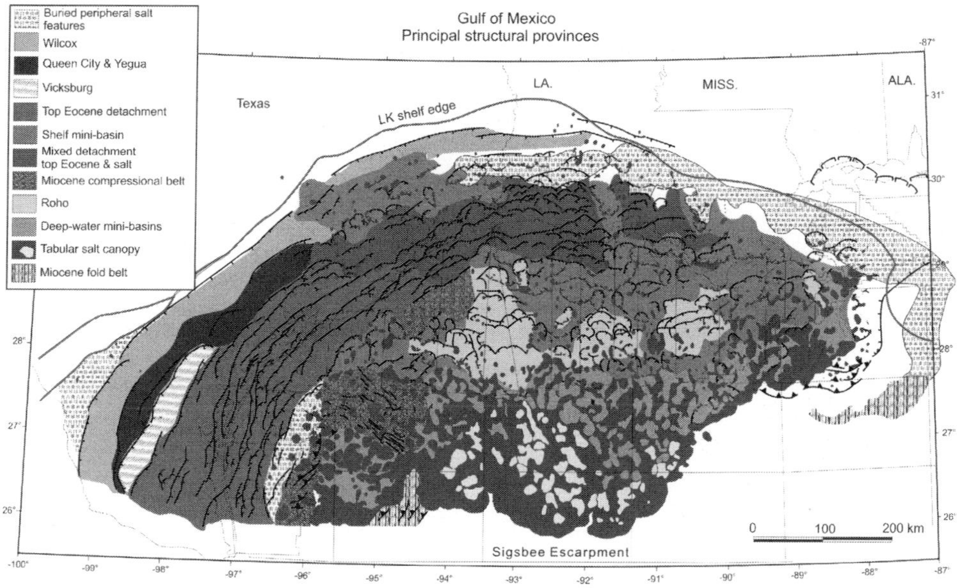

Figure 19.11. Map of tectono-stratigraphic provinces of the northern Gulf of Mexico (Diegel et al., 1995; Karlo and Shoup, 2000).

3) a very small contribution of inversion-related traps in the West African province, apart from several traps offshore between the Côte d'Ivoire and southern Gabon (see Nemčok et al., 2004b).

Further difference among these three provinces is represented by no gravity gliding in the Persian Gulf province, while the other two are typical for that (Figure 19.11). Actually, only certain segments of the West African province are typical for that while the others are not. As Nemčok and colleagues (2012b) pointed out, the occurrence versus lack of gravity gliding correlates with the type of passive margin segment (Figures 6.13a, b). The segments originated by strike-slip faulting start their development with a large topographic gradient that requires a very long time for deposition to turn it into a gently dipping slope that can subsequently react to uplift, earthquakes, and increasing sedimentary load by episodes of gravitational instability triggering gravity gliding. On the contrary, the segments originated by normal faulting have such a slope practically from the time of continental breakup and undergo gravity gliding from the early post-salt deposition period, as documented, for example, in the Rio Muni Basin, where the gliding started during the deposition of the first post-salt, Albian formation (see Dailly, 2000).

Further difference between transform and extensional margin segments is due to the fact that it is the strike-slip faulting-dominated tectonics at transform margins that control their pre-breakup structural trap development while it is normal faulting-dominated tectonics at extensional margins that control their pre-breakup structural trap development. Furthermore, as observed at the Ghana Ridge (Nemčok et al., 2012b), the faulting at transform margins stops only when the seafloor-spreading centers clear the contact with the continental margin. On the contrary, the faulting dies out already at the time of continental breakup at extensional margin segments. Therefore, the geological differences between strike-slip- and normal faulting-originated passive margins lead to differences in inventories of potential traps (compare Figures 19.12a and b).

Table 19.2 indicates that structural traps represent about 72 percent of the passive margin trap population, stratigraphic traps about 13 percent, and combined structural/stratigraphic traps about 15 percent. Structural traps of a known type can be divided into more than 26 percent of post-rift contraction/transpression-related traps, more than 15 percent of salt structure-related traps, more than 14 percent of gravity-gliding-related roll-over anticline traps above salt or shale detachment, and only about 4 percent of traps related to the original rift architecture. The strikingly low representation of traps related to original rift architecture is caused by the fact that:

1) subsequent contraction/transpression took them out from this category in the Persian Gulf;
2) they are too deep-seated to be found in any important quantity in the Gulf of Mexico yet (see Sartain and See, 1997; Biles et al., 2000; Prost and Aranda, 2001); and

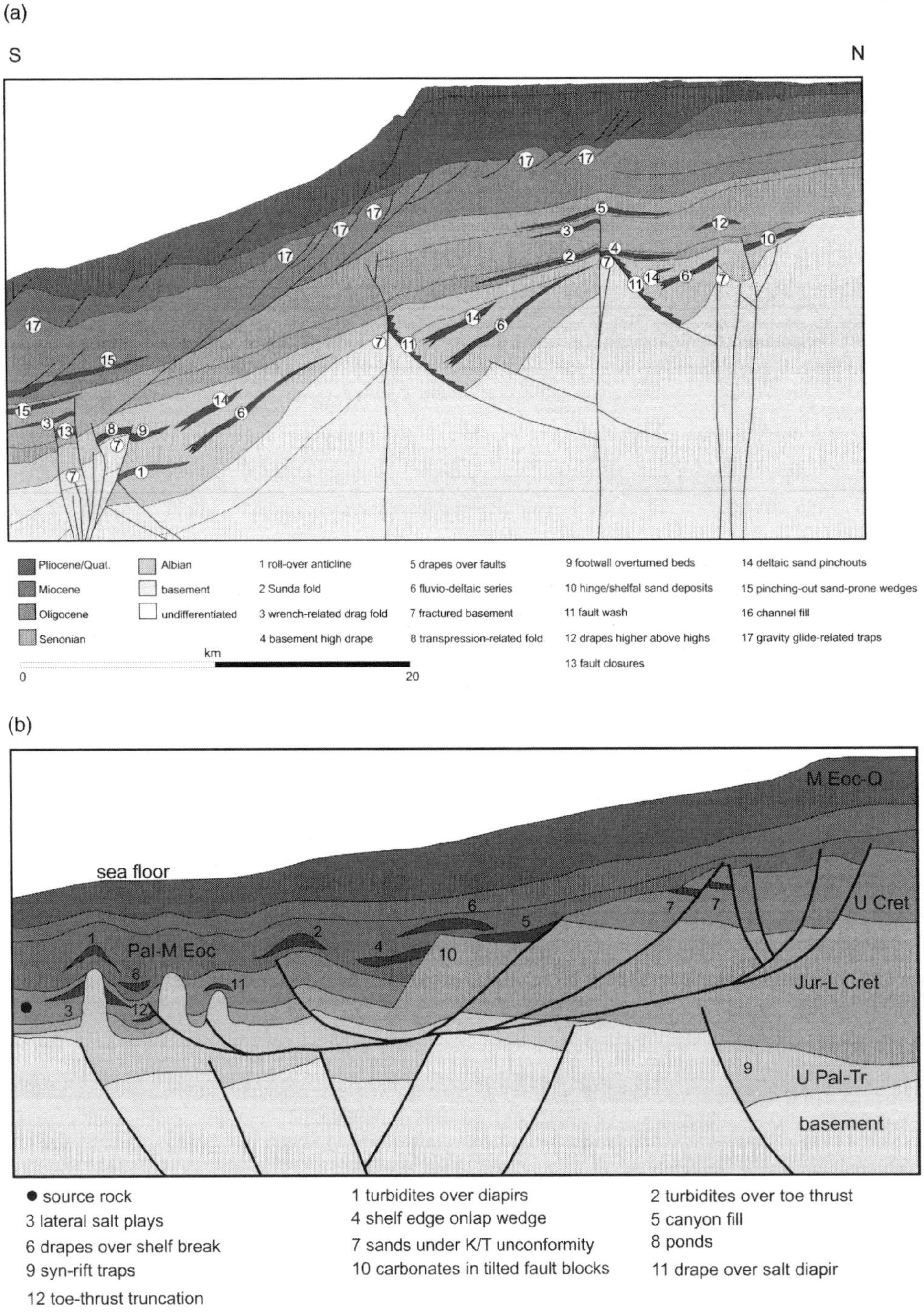

Figure 19.12. (a) A list of potential traps sketched on the portion of the interpreted reflection seismic profile through a typical portion of the strike-slip-originated passive margin in the Equatorial Atlantic region (Nemčok et al., 2004b). (b) A list of potential traps typical for an extensional margin segment sketched on a simplified section through the western side of the Guinea Plateau (Nemčok et al., 2004b).

3) they are too deep to be found in a more important quantity in extensional margin segments of the West African province and frequently modified by subsequent transpression in transform segments (see Nemčok et al., 2004b),

indicating that they are an important target for future passive margin exploration because our database does not prove them as not working. The proofs of this claim came from the syn-rift sections of the Santos, Campos, and Espírito Santo basins in offshore Brazil (see Izundu, 2007; Koike et al., 2007; Berman, 2008; Watkins, 2008).

Structural traps

Rifting-related and sub-salt traps

This group of traps is generally considered either finished with its development during the latest rifting or located underneath the layer of autochthonous evaporites, which may have syn-rift or rift-drift transition stratigraphy (compare syn-rift evaporites of Morocco and Nova Scotia with rift–drift transition ones of the northern South Atlantic – for example, Dailly, 2000; Nemčok et al., 2005b and references therein). The group contains:

1) simple up-thrown tilted fault block (Figures 7.8, 7.9, 7.10a);
2) complex up-thrown tilted fault block (Figures 7.10c, 7.13a, 7.14c); and
3) horst structure (Figure 7.10b).

The evaporite stratigraphy represents an important constraint of the regional extent of associated seals for mentioned traps, as the late syn-rift evaporites have a patchy distribution associated with individual grabens while the transitional evaporites tend to form areally extensive horizons (see comparison of the Central Atlantic versus South Atlantic evaporites in Tari et al., 2000, 2001, 2002). The examples come from the syn-rift Triassic–Lower Jurassic evaporites of Morocco and the post-rift Aptian evaporites of Cameroon/Gabon/Congo/Angola. In the former case, the extensional domain of the gravity glide may not be the ultimate driving force for the contractional domain of the gravity glide down-dip because of the lack of a basin-wide salt-located detachment efficiently linking all the areas of the gravity glide system. This difference appears to impact the distribution of salt-related structural traps significantly in these two cases areas (Tari et al., 2001).

Simple up-thrown tilted fault and complex up-thrown tilted fault blocks

Among the three key provinces, only the Persian Gulf and West Africa contain simple up-thrown tilted fault blocks, represented by the Dilam, Hawtah, Raghib, and Wadayhi fields in the former case and the Cabinda Onshore South Block fields in the Angolan portion of the Lower Congo Basin, representing 4 percent and 2 percent of their traps in our database, respectively. Out of forty-three producing fields in Gabon in 1990 (Teisserenc and Villemin, 1990), three were located in tilted and truncated extensional fault blocks and another three occurred in pre-salt anticlines associated with extensional tectonics.

Among the key provinces, only the Gulf of Mexico contains complex up-thrown tilted fault blocks, represented by the Sitio Grande field in the Villahermosa Uplift of Mexico.

Horst structure

Horst structures in our database on key provinces are relatively rare, documented by fields such as Ghawar, Farhah and Khurais of the Persian Gulf province and the Rabi-Kounga field of the Gabon Coastal Basin, West Africa province, representing 2 percent of fields in both cases.

Other pre-drift traps

Equatorial Atlantic margins being initiated by strike-slip faults and pull-apart basins, as opposed to normal faulting-initiated margin segments, contain extra traps. They are related to syn-rift faulting with the possibility of early-drift activity. They are represented by fault-block traps, relying on fault sealing. Examples come from the Espoir and Tano fields of the Côte d'Ivoire. These traps are expected to reside on the slope and in the strike-slip duplexes and ridges of the continental crust parallel to the Romanche Fracture Zone. This trap here has a high risk of being breached by the subsequent Cenomanian-Turonian (Santonian) erosion related to the post-breakup isostatic uplift of the continental margin. However, it is proven to work in the western deep-water offshore Côte d'Ivoire (St John, 2000). The Espoir field, for example, was developed and produces an oil of 31° API gravity from Albian fluvial sandstones. The post-Albian unconformity provides the sealing.

Post-rift traps

Post-rift structural traps contain salt structure-, gravity gliding-, post-rift draping-, and contraction/transpression-related groups of traps. Salt-related traps include:

1) structural traps located at flanks of diapirs (Figures 7.24, 18.12a, 19.13a);
2) structural/stratigraphic traps located at flanks of diapirs (Figure 7.24); and
3) structural traps associated with salt domes (Figure 7.24).

Structural and structural/stratigraphic traps located at flanks of diapirs

A structural subgroup of this type of trap is quite common, being represented by 6 percent and 1 percent of all traps in the West African and Persian Gulf provinces.

Examples come from the Iranian Siri D field of the Rub Al Khali Basin of the Persian Gulf and the Brame Marine, Mandaros Marine, M'Polunie, and Torpille North East Marine fields of the Gabon Coastal Basin.

The situation in offshore Gabon in 1990 (Teisserenc and Villemin, 1990) further documents the importance of this trap type for the extensional margin segments with gravity glides. Out of forty-three fields, three were located on crests and flanks of piercement domes on the platform.

Structural/stratigraphic derivation of this type comes from the Gulf of Mexico, representing 9 percent of the traps in this province in our database. It is represented by fields such as Bass Lite, Bullwinkle, Genesis, Lobster, Marquette, Ness, Seattle Slew, and Tahiti in the U.S. portion of the province.

Structural traps associated with salt domes

This trap type forms the second most prominent group of traps in the Gulf of Mexico, represented by 21 percent of traps in our database. Examples include the Bay Marchand and Timbalier Bay fields of the Gulf Cenozoic OCS; the Bastian Bay, Bayou Sale, and Thompson fields of the Western Gulf and Cactus; and the Iris and Samaria fields of the Villahermosa Uplift. A typical field associated with this trap has Miocene to Pleistocene reservoirs, having non-associated gas reservoired in progradational sand. An average trapped volume is below 8 BCF of reserves and an average trap area is less than about 2 km^2 (Kim, 2002).

This trap plays a distinct role in the West African province, being represented by Angola's Girassol and Congo's Emeraude Marin, Loango, and N'Kossa Marine fields, forming 6 percent of the traps in our database on West Africa. This trap type is also present in the Persian Gulf, representing 2 percent of its traps in our database. Examples come from the Pars North field of the Mesopotamian foredeep basin and the G 3 field of the Qatar Arch.

The situation in offshore Gabon in 1990 (Teisserenc and Villemin, 1990) further documents the importance of this trap type for the extensional segments with gravity glides. Out of forty-three producing fields, seventeen were located in association with non-piercement salt domes in the deep basin and turtle anticlines on the platform (Figure 19.13b).

An example of this type of trap complicated by normal faulting of the Paleocene–Lower Miocene cover of the salt body comes from the Chinguetti oil field in offshore Mauritania (Figure 19.13c).

Gravity gliding-related traps, apart from those listed under salt traps, include:

1) gravity gliding-related footwall traps (Figure 7.24); and
2) gravity gliding-related hanging wall traps (Figures 7.24, 19.13d).

Gravity gliding-related hanging wall and footwall traps of extensional domains

Figure 19.14 indicates that both these trap types dominate in gravity glide systems detached on shale. Examples come from the Niger Delta, the Foz do Amazonas Basin, the Krishna-Godavari and Mahanadi Basins of the East Indian offshore, and the northeast and west slopes of the Gulf of Mexico.

Forty-one percent of anticlinal traps in our database on the West African province are represented by roll-over anticlines of extensional domains of gravity glide systems while the additional 2 percent are formed by a combination of roll-over anticline and overlying drape fold. Examples of the former type include the Amenam/Kpono, Asasa, Apoi, Bisemi, Cawthorne Channel, Imo River, Kokori, Nembe Creek, Odidi, and Soku fields of the Niger Delta, the Congo's Sendji and Moho Marine fields, and the Angolan Malongo N, S, and W, and Takula fields. Examples of the latter type include the Rosa-Block 17 field of the Lower Congo Basin and the Dalia-Block 17 field of west-central coastal Angola. Another example of this trap type comes from the Banda gas field in offshore Mauritania (Fusion Oil and Gas, 2002; Figure 19.13c).

Hanging wall traps in the extensional domain of the Niger Delta gravity glide system form one of the simplest and most common producing traps in this region. Anticlinal closures here include unfaulted roll-over anticline (Figure 19.13d-1), faulted roll-over anticline with fault throws less than a minimum reservoir thickness, which is about 15 m (Figure 19-13d-2), and faulted roll-over anticline with non-sealing faults. Other traps are provided by a combination of a sealing fault with sedimentary layers dipping in the same direction (Figure 19.13d-3). These traps are relatively rare and occur in two settings. The first is controlled by the internal seaward-dipping fault that forms a juxtaposition seal. The second is controlled by the seaward-dipping fault that juxtaposed a hanging wall to shale diapir. The remaining trap type is an anticline deformed by collapse-driven normal faults.

The extensional domain of the Niger Delta also offers a variety of gravity-gliding-related footwall traps (Figure 19.14). The closures here are provided by a combination of sealing fault and an uplifted footwall characterized by fault and sedimentary layers dipping in opposite directions. They occur in three settings. The first includes a landward-dipping antithetic fault on the seaward flank of the major rollover anticline. The second is the seaward-dipping synthetic

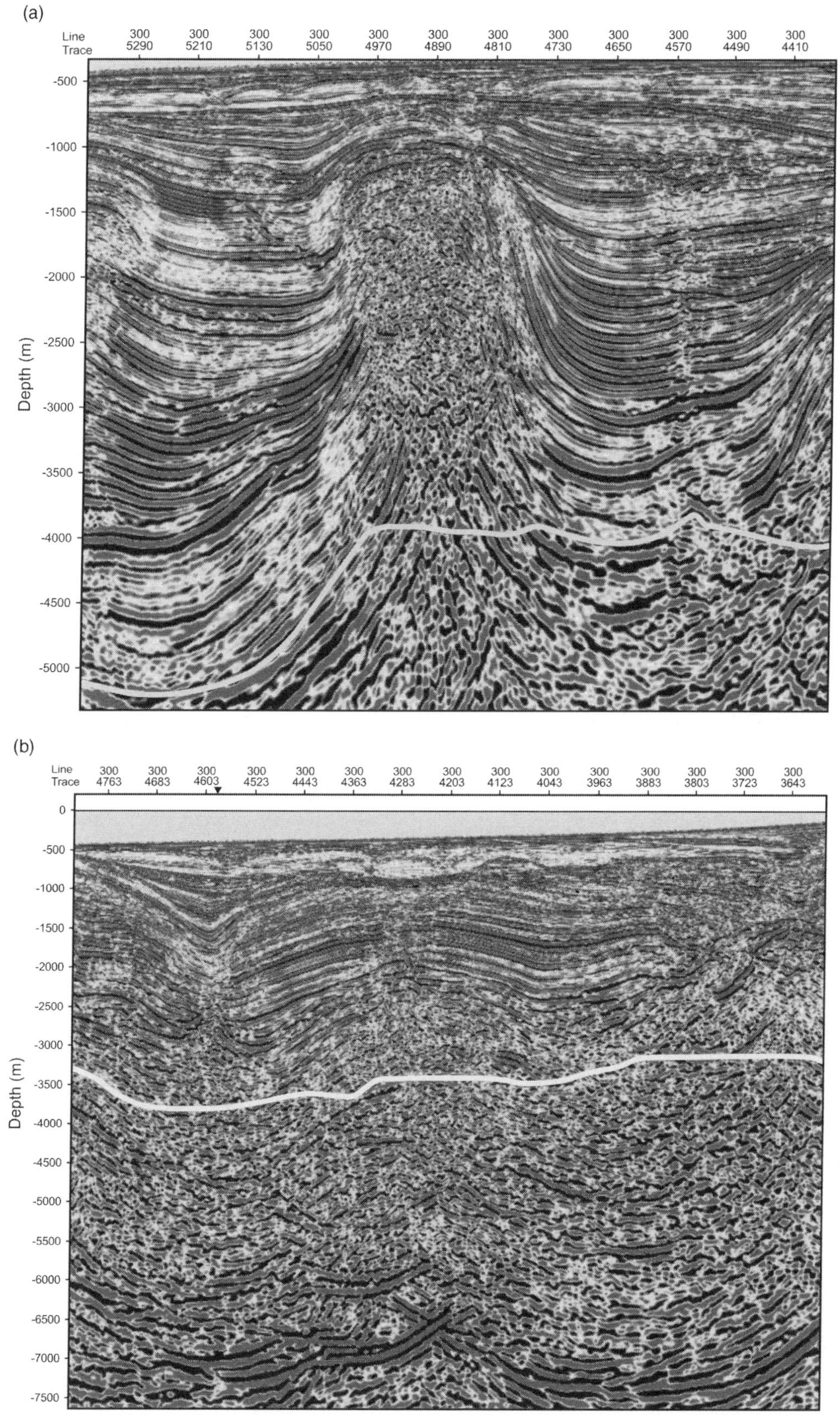

Figure 19.13. (a) Salt stock with small flap fault near the crest and steeply dipping to overturned strata at its flanks, offshore Gabon (Nemčok and Rosendahl, 2006). (b) Turtle anticline in offshore Gabon (Nemčok and Rosendahl, 2006). (c) Geologic cross-section through the Chinguetti oil and Banda gas fields, offshore Mauritania (Fusion Oil and Gas, 2002). (d) Principal producing trap types of the Niger Delta (Doust and Omatsola, 1990).

(c)

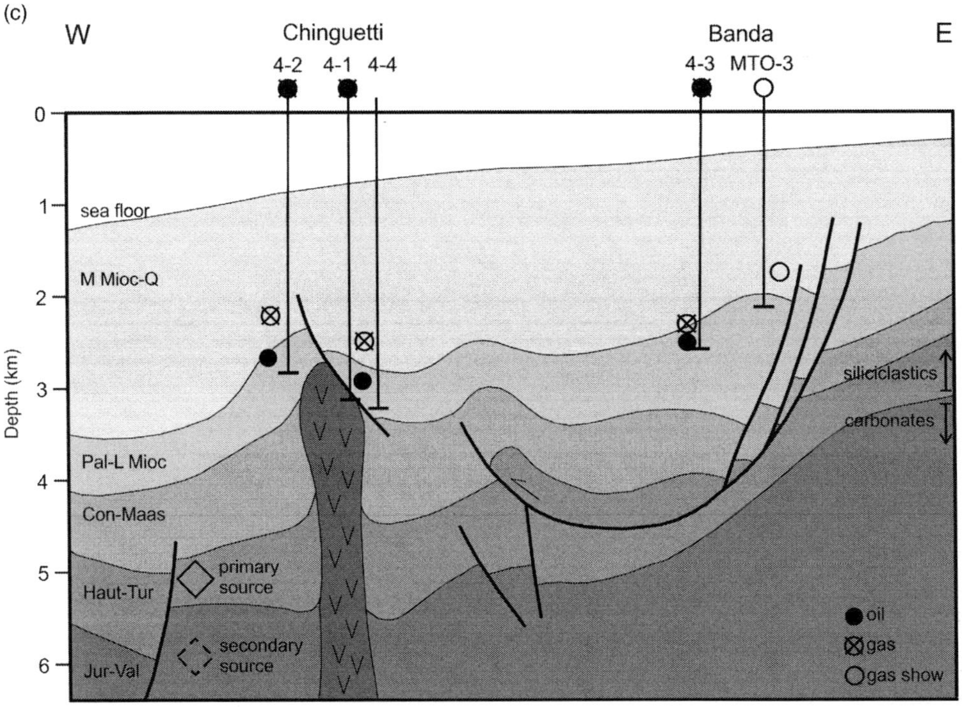

(d)

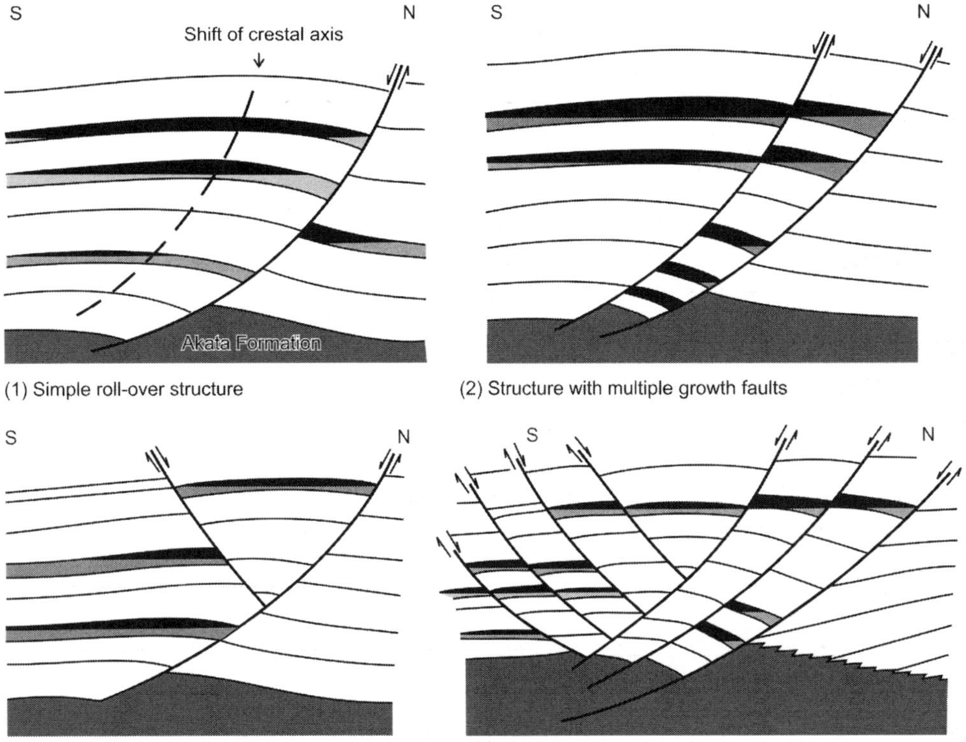

(1) Simple roll-over structure

(2) Structure with multiple growth faults

(3) Structure with antithetic fault

(4) Collapsed crest structure

Figure 19.13 (*cont.*)

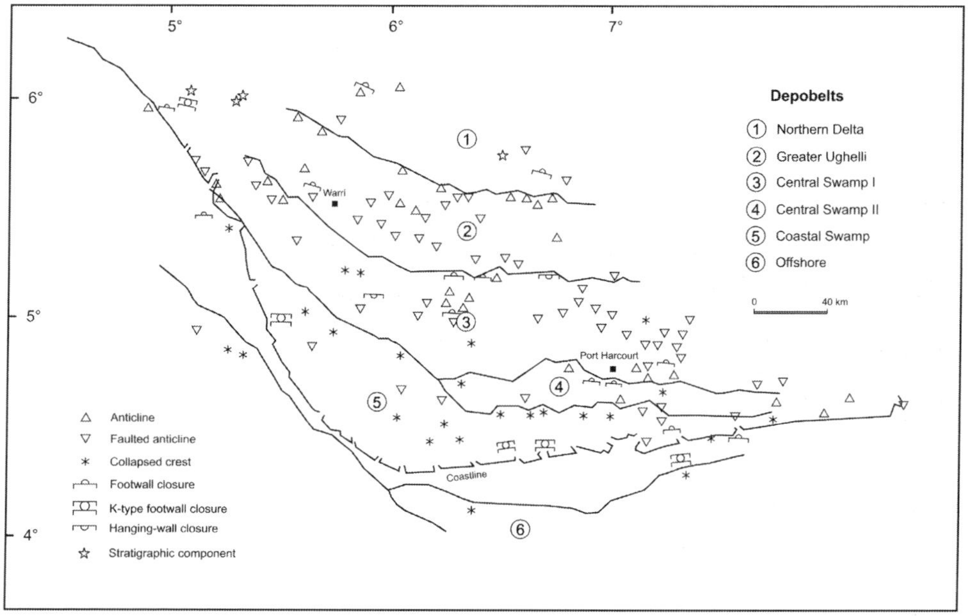

Figure 19.14. Distribution of structural types in the Shell fields of the Niger Delta (Doust and Omatsola, 1990). A preponderance of anticlines and faulted anticlines in the older depobelts gives way to more complex structures such as the collapsed crests and footwall closures related to cross-cutting normal faults in the distal part of the delta.

fault on the landward flank of the major roll-over anticline. The third is at the roll-over controlling normal fault. All three trap types form the second most important trap group in the extensional domain of the Niger Delta.

The Rio Muni Basin provides us with examples of Albian raft traps, formed in strata deposited right above the evaporites. Some of them were tested by wells (see Dailly, 2000). Their success depends on the charge from a syn-rift source rock kitchen, which should be located, based on an analogy with the North Gabon Basin, under the shelf.

While it is impossible to estimate the amount of roll-over anticlines represented in our database on anticlinal traps from the Gulf of Mexico because they are grouped together with drape and inversional folds, other databases, such as that of Kim (2002) (Table 19.3), list roll-over anticlines as comprising 13 percent of the Gulf of Mexico structures. Our grouped percentage is rather high, being 38 percent, which indirectly indicates the importance of roll-over anticlines, which form the largest subgroup of this group. The Gulf of Mexico traps of this type are associated with two kinds of gravity gliding, the first detached along evaporites, the second detached along shale. The former occupies the northwestern Gulf of Mexico and a corner between the Yucatan Peninsula and continental Mexico. The latter occurs in the northeastern and western Gulf of Mexico (e.g., Watkins et al., 1978).

Note that gravity-gliding-related hanging wall and footwall traps form large trap groups at extensional margin segments while they pale out at transform margins, and are practically missing at margins affected by subsequent inversion such as a portion of the Persian Gulf.

Drape folds

The drape folds are difficult to extract from our database on anticlinal traps from the Gulf of Mexico, where they are mixed with other types of anticlinal traps, contributing to 38 percent of traps represented by the anticlinal trap type. However, our West African data set documents that the drape folds act as contributors to various kinds of traps including:

1) syn-rift tilted footwalls and horst structures (Figure 4.7);
2) gravity gliding-related roll-over anticlines, represented by the already mentioned Angolan Rosa-Block 17 and Dalia-Block 17 fields; and
3) contraction/transpression anticlines (Figure 19.15).

Reflection seismic imaging of the West African province indicates that the remaining types of drape folds occur above highs of various origins, located on continental or even proto-oceanic crusts (Nemčok and Rosendahl, 2006).

Drape fold-related fields such as the Plutonio field in the Lower Congo Basin and the Pelican field in offshore Mauritania indicate that drape fold traps do not always enhance other trap types but can occur as stand-alone traps.

Contraction/transpression-related traps
This trap type has the potential to dominate in certain passive margin provinces such as the Persian Gulf, where it forms as much as 79 percent of all traps in our database. The simple anticlines dominate while the faulted anticlines are rare, represented by the Kuwaiti Greater Burgan and the Iraqi Nahr Umr fields of the Mesopotamian foredeep basin. They are ranked as the world's second and forty-second largest oil and condensate fields (see Berman, 2008). The examples of simple anticline inside the world's fifty largest oil and condensate fields are Rumaila North & South (4th), Safaniya (6th), Zakum (9th), Manifa (11th), Baghdad East (12th), Zuluf (17th), Majnoon (19th), Abqaiq (23rd), Berri (28th), Khafji (32nd), Zubair (34th), and Abu Sa'fah (35th). This documents post-rift contraction's important role in the determination of the size of traps in a passive margin province.

Table 19.3. Gulf of Mexico field characterization, based on respective share from total reserves (Kim, 2002).

Trap type	Share (%)
fault-sealed	21
salt, shale diapirs/domes	27
anticline	44
unknown	5
other	3
structure type	Share (%)
roll-over into growth fault	13
fault-sealed	22
anticline	27
salt-related	31
unknown	2
other	5

The development of this trap type depends on preexisting passive margin proximity to subsequent orogen and the degree of contractional deformation of the margin. While it was dominating in the Persian Gulf, it affected the Gulf of Mexico province only marginally and the West African province to a very small extent.

The development of contractional/transpressional traps in the Gulf of Mexico took part only in its western portion, controlled by Tertiary transpression driven by the subduction of the Farallon Plate underneath the North American and Caribbean Plates, associated with ongoing eastward migration of the triple junction, Oligocene development of the Sierra Madre Occidental, the Late Miocene development of the Sierra de Chiapas, and the activity of the sinistral wrench fault bounding both the Veracruz Basin and the Gulf of Mexico, and the accommodating movement of the Yucatan Peninsula located to the north of it (see, e.g., Pindell, 1993; Scotese, 1998;

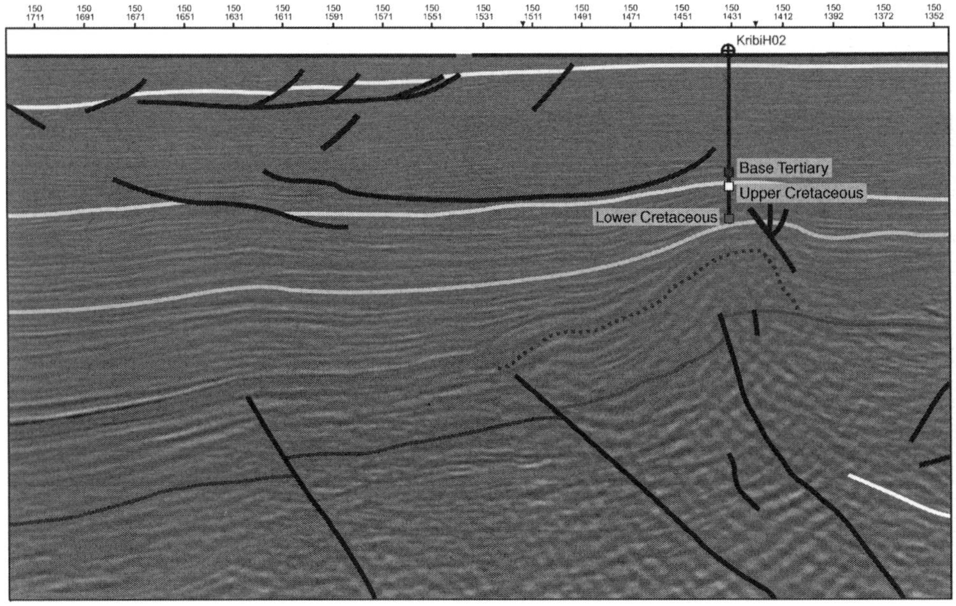

Figure 19.15. Two compression/transpression–driven structures of the continental margin, offshore Cameroon (Nemčok and Rosendahl, 2006). Both have drapes formed over them and the second structure is salt-cored.

Golonka, 2000; Prost and Aranda, 2001). The affected areas include, for example, the Chiapas-Tabasco area of the Sureste Basin of southeastern Mexico, where this trap type controls seventy-five fields producing light-medium density oil from the Upper Jurassic–middle Cretaceous carbonates, and the Veracruz Basin (see Meneses-Rocha, 2001; Ambrose et al., 2002). Contractional folds are also known from the Macuspana Basin located along the western flank of the Yucatan Peninsula.

Examples coming from the West African province frequently include a local controlling mechanism. For example, only a portion of the Gulf of Guinea contains the post-breakup contraction/transpression folds. Numeric simulation by Nemčok and colleagues (2012b), discussed in Chapter 12, indicates that some of them, as those in offshore Benin, were most likely related to drift vector change during the early drift, while continental margins were still relatively warm and weak. The simulation done for the Ghana Ridge proves that already a drift vector change of as little as 9° would be sufficient for triggering a transpression-related uplift of the continental margin zone adjacent to the oceanic crust and reactivating the preexisting steep-dipping faults from the syn-rift period. Such reactivation-related anticlines (Figure 19.16) have been formed by Coniacian–Maastrichtian and Danian–Thanetian events along the Romanche Transform Zone, which was approximately 80 km wide. They usually deform a syn-rift section and a younger section deposited before the time of the respective folding. Numerous structures have been already recognized associated with this mechanism, including inversion-related structural highs such as Belier, Lion, Foxtrot, Espoir, and Quebec. The Late Cretaceous Belier field produces an oil of 33° API gravity from the Upper Cretaceous submarine channel sandstones. The Lion field produces oil and gas from the Lower Cretaceous syn-rift sandstones and the Foxtrot field developed with prevailing gas and condensate had a discovery well proving a reservoir in the Albian channel sandstone. The Seme field from the northeastern

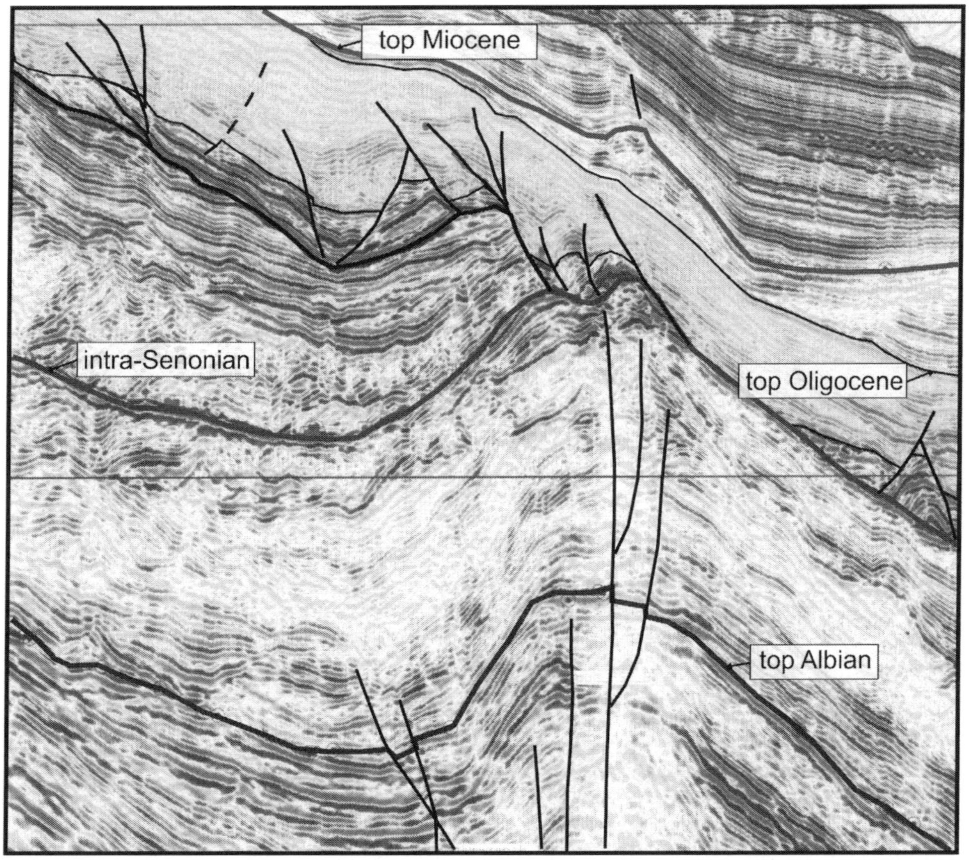

Figure 19.16. The early syn-drift deformation of the previously extended continental crust in offshore Benin (Nemčok et al., 2012b). The dextral-transpression-driven fold has chevron geometry. The Upper Cretaceous–lower Paleogene sediments thinning over its crest indicate its Senonian–early Paleogene growth.

portion of the offshore Benin also falls into this trap category. Its trap is an anticlinal structure truncated by unconformity, holding reserves exceeding 100 million barrels of oil. Additional reservoirs and horizons with oil shows in this region are in the Upper Cretaceous, Lower Cretaceous, and Devonian strata.

Another mechanism developing transpression-related anticlines is the contraction component at contractional quadrants located at terminations of the strike-slip fault zones (see Gamond, 1987; Mollema and Antonellini, 1996; Willemse et al., 1997; Nemčok et al., 2002). This mechanism has been noted in the western Côte d'Ivoire Basin (Nemčok et al., 2004b). The associated traps are not necessarily only anticlinal traps.

Fault block traps are also occurring in association with reactivated strike-slip faults, as interpreted prospects in offshore Benin (Nemčok et al., 2004b) document. Trapping in tilted fault block was also observed in offshore Ghana, in the Albian strike-slip fault zone, which is one of the various Accra-Keta prospects explored in Devonian pre-rift and Albian rift-drift transition sediments. Two more examples come from the Tano Basin, Ghana, having an earlier syn-rift timing component and subsequently an early syn-drift timing component, as the traps are associated with syn-rift faulting reactivated during the rift–drift transition period. The first example is the North Tano field, which is mainly a gas field reservoired in the middle-upper Albian sandstone. It has associated oil with an API gravity of 32°. The second example is the South Tano field, which hosts saturated oil and condensate in the upper Albian sandstone. The traps are a combination of structures associated with faulting and mid-Cretaceous unconformity.

Hanging wall anticlines of compressional domains of gravity glides

This trap type is important at fronts of large gravity glide systems at extensional passive margin segments. Examples come from the Niger Delta (Hooper et al., 2002; Nemčok et al., 2004b; Figure 19.17a), the Foz do Amazonas Basin, offshore West Africa (Nemčok and Rosendahl, 2006; Figure 19.17b), and the Gulf of Mexico (Diegel et al., 1995; Rowan et al., 1999, Karlo and Shoup, 2000; Peel et al., 2001; Stephens, 2001; Figure 19.18).

The contractional domain of the Niger Delta gravity glide system contains potential traps such as anticlinal closures, overturned footwall beds, drapes over diapirs and anticlines, and stratigraphic traps related to reservoir ponding between neighboring anticlines or an anticline/extensional fault combination. Similar traps from the Rio Muni Basin involve the Upper Cretaceous clastic sediments (Dailly, 2000). They are located on the lower slope and deep basin. Their success depends on the charge from post-rift Albian and Cenomanian-Turonian coastal plain and marine source facies.

Stratigraphic traps

The importance of purely stratigraphic traps varies among the previously discussed three key provinces. It varies from 0 percent in the Persian Gulf province, through 17 percent in the Gulf of Mexico province, to 22 percent in the West African province.

Stratigraphic traps can be divided into those related to:

1) organic buildup;
2) angular unconformity;
3) depositional lens;
4) pinch-out; and
5) lateral facies change.

Organic buildup-related traps are documented only from the Cerro Azul field of the Tampico-Misantla Basin in the Gulf of Mexico. Other examples are not fields but interpreted structures, such as the system of organic buildups developed on top of the continental margin in offshore Cameroon, which was uplifted during the Late Cretaceous (see Nemčok and Rosendahl, 2006; Figure 19.19).

An interesting play is present in the Campeche area of the Mexican portion of the broader Gulf of Mexico region, where twenty-four fields are producing from the Upper Cretaceous–Paleocene reef talus breccia (Meneses-Rocha, 2001). An example of this play comes from the Cantarel field, which practically belongs to a category of compressional/transpressional anticlinal traps and holds reserves of 12 billion bbls of oil with an API gravity of 24° (Guzman and Marquez-Dominguez, 2001). Further examples come from the adjacent Ku-Maloob-Zaap, Abkutum, Pol, Chuc, and Caan fields.

Traps associated with angular unconformity (Figure 19.20) are not represented in our field database on the three key provinces. However, examples of interpreted traps of this type come from the Equatorial Atlantic margins, where they are represented by syn-rift traps sealed by the breakup unconformity and subsequent condensed sedimentary section (Nemčok et al., 2004b). Figure 19.21 documents a large number of regional and local unconformities developed as controlled by different factors at Equatorial Atlantic margins, documenting the offshore Côte d'Ivoire case example. Other scenarios of angular unconformity-related traps include syn-gravity gliding traps sealed by local unconformities such as

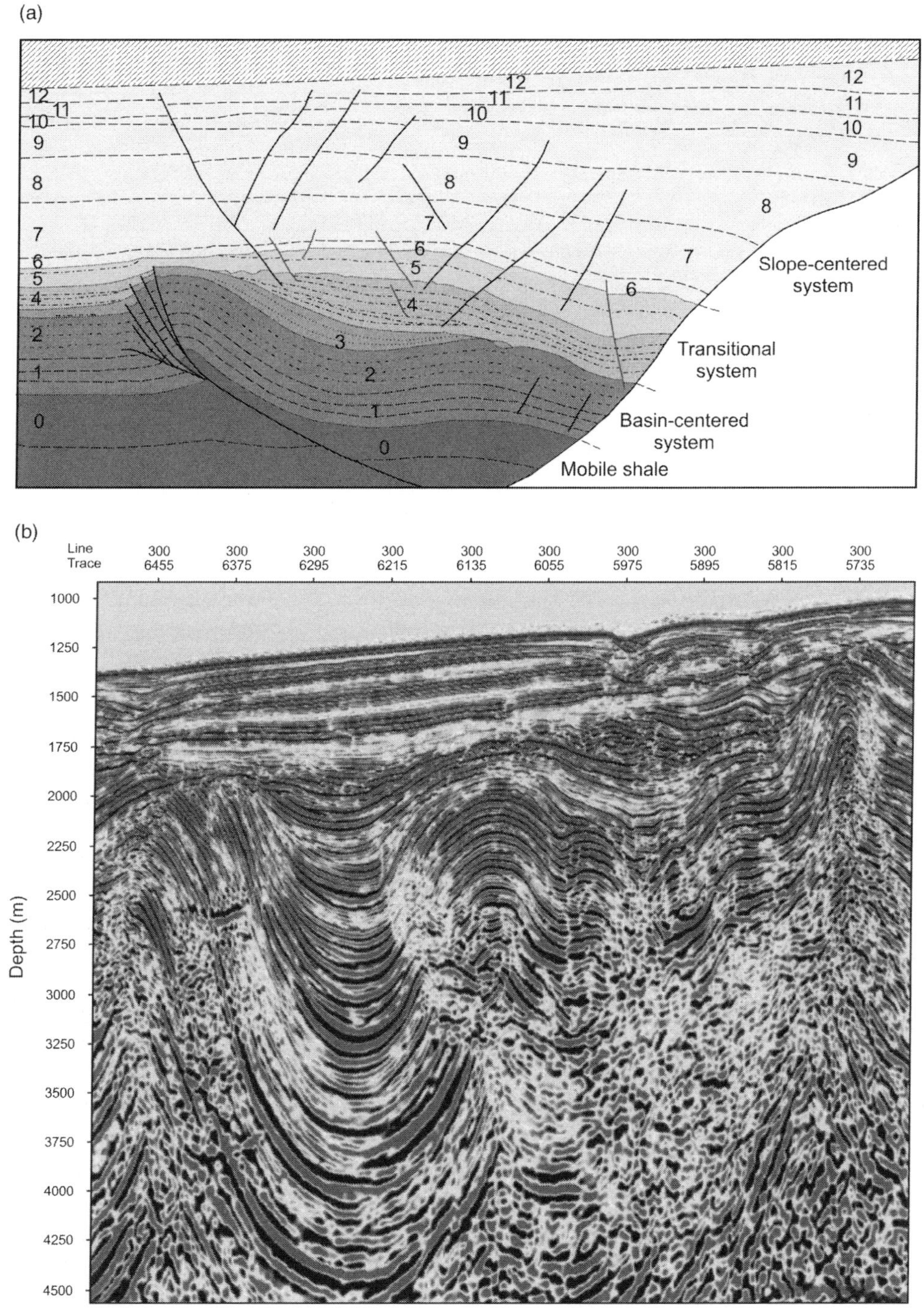

Figure 19.17. (a) Geoseismic section through the frontal fold of the paleo-fold belt that illustrates the form of the associated ponded slope-basin and extensional faults (Hooper et al., 2002). 0 represents a marine shale unit, 1–2 are pre-contractional units, 3–4 are syn-contractional early-ponding units, 5 is the last syn-contractional and ponding unit, 6–12 are syn-extensional units. Three major stratigraphic envelopes can be defined in the deltaic wedge, including basin-centered, transitional, and slope-centered systems. The ponded section 3–5 can be divided into several units confined between the pre-contractional (or pre-kinematic) strata below and syn-extensional strata above. The depositional axis of the pond migrates as the structure develops. Strata deposited early during the fold development are progressively rotated into the back limb of the fold. Note that the form of the whole ponded section has been modified by rotation associated with the regional extensional fault. (b) Contractional domain of the gravity glide system, offshore Gabon (Nemčok and Rosendahl, 2006).

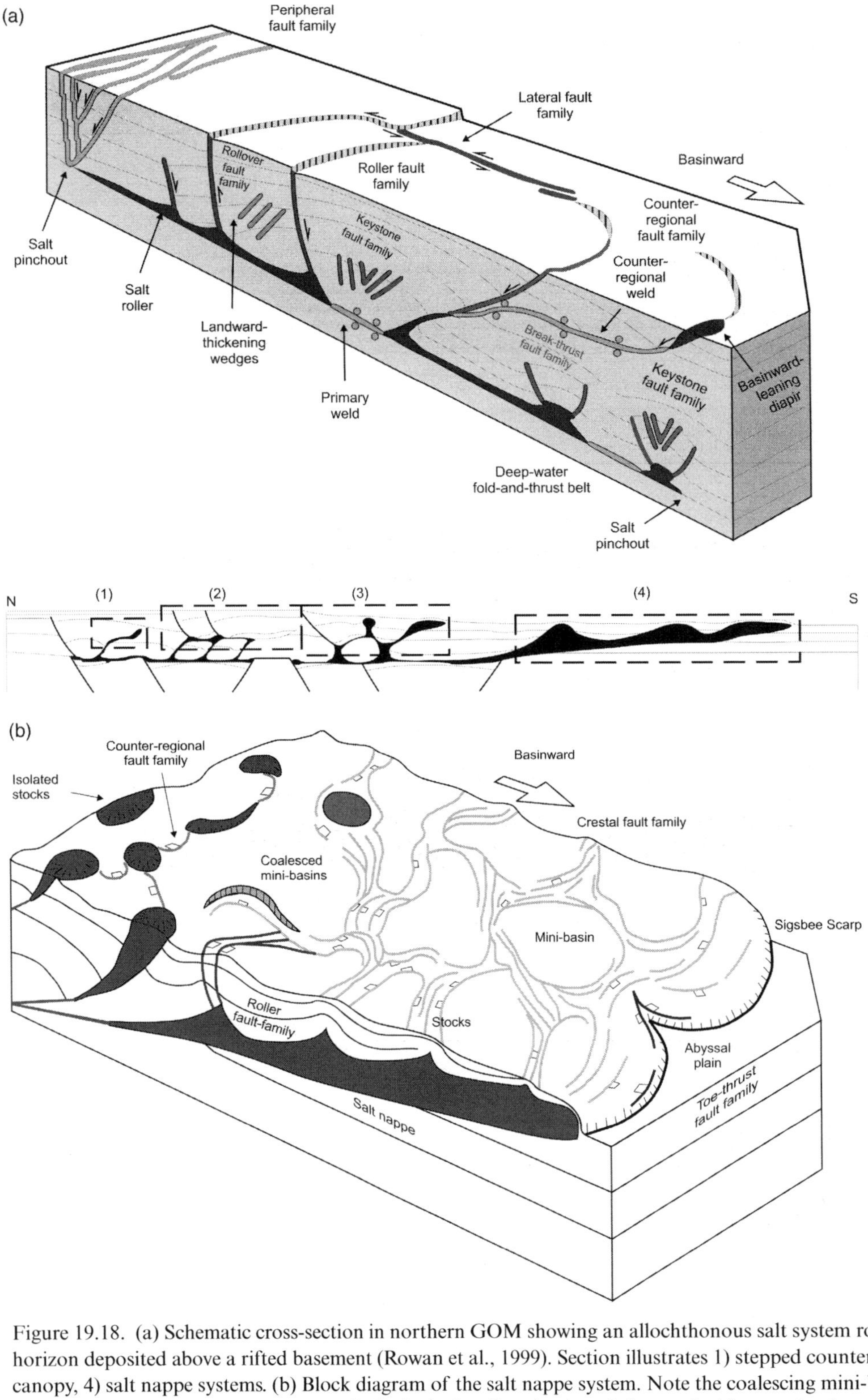

Figure 19.18. (a) Schematic cross-section in northern GOM showing an allochthonous salt system rooted in an autochthonous salt horizon deposited above a rifted basement (Rowan et al., 1999). Section illustrates 1) stepped counter-regional, 2) Roho, 3) salt-stock canopy, 4) salt nappe systems. (b) Block diagram of the salt nappe system. Note the coalescing mini-basins located above the inflated growing salt body, toe thrusts at down-dip limit, fault rollers developed above salt ridges, and polygonal pattern of roller faults.

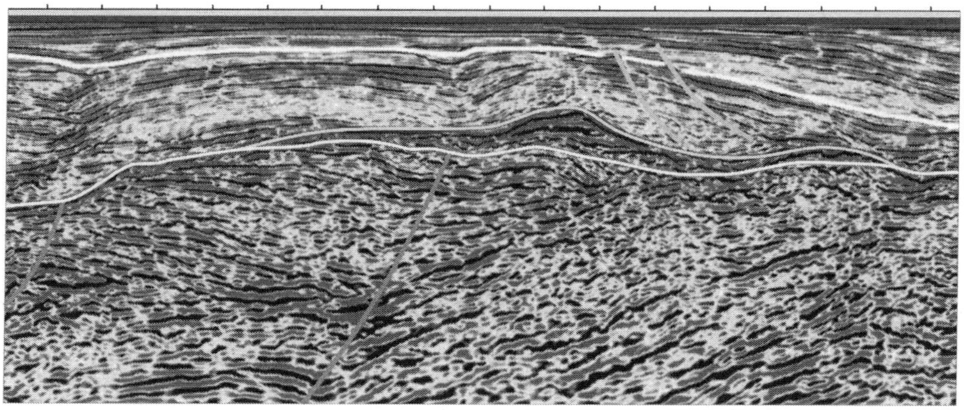

Figure 19.19. Interpreted reef carbonate body grown on top of the continental margin, which was uplifted and eroded during the Late Cretaceous, offshore Cameroon (Nemčok and Rosendahl, 2006).

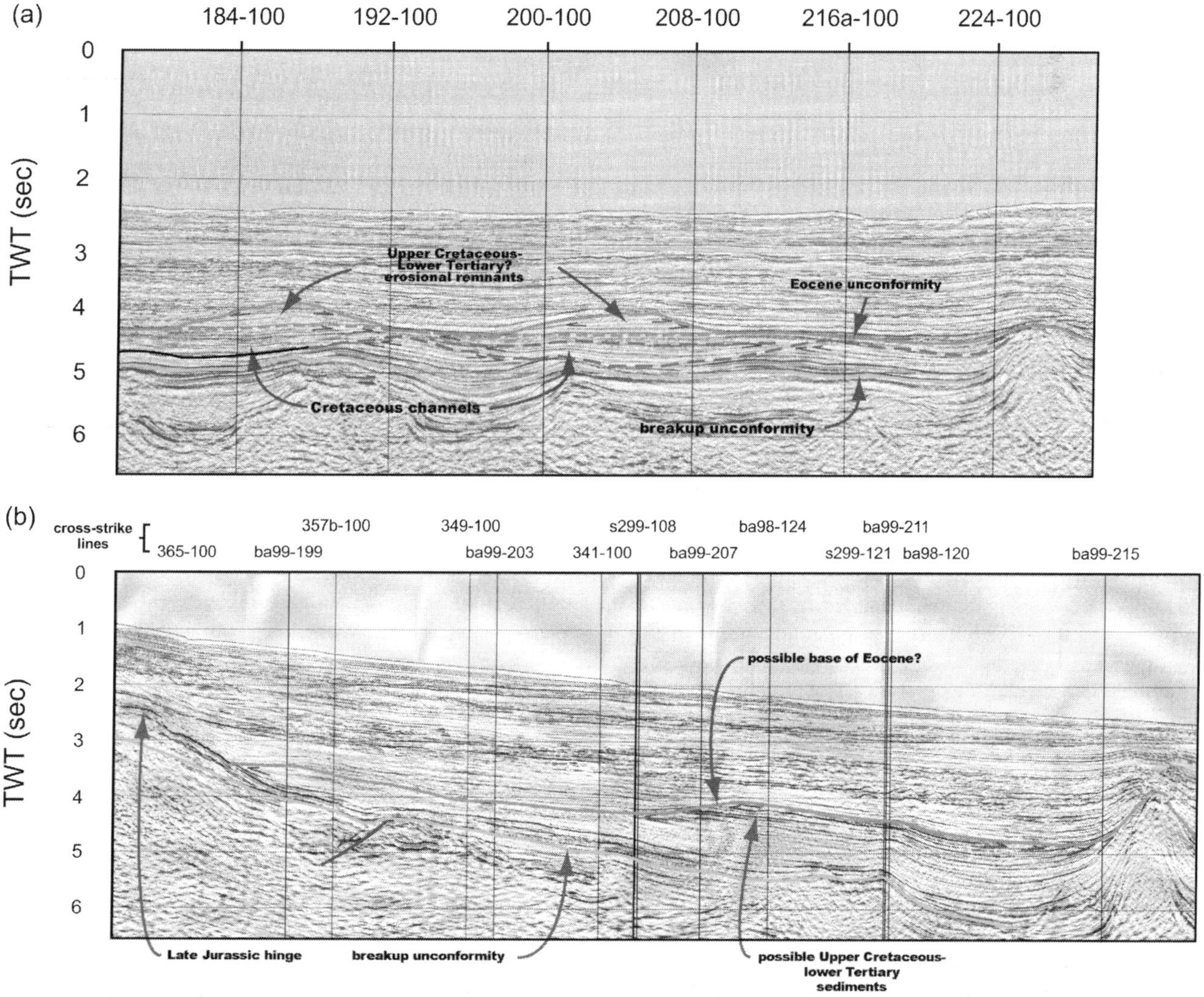

Figure 19.20. (a) Strike-oriented profile segment with submarine channels and unconformities in Upper Cretaceous and Lower Tertiary intervals (Nemčok et al., 2005b). (b) Dip-oriented profile segment showing basinward subsidence from a Late Jurassic hinge line, the breakup unconformity, and an onlap of syn-drift stratigraphic units.

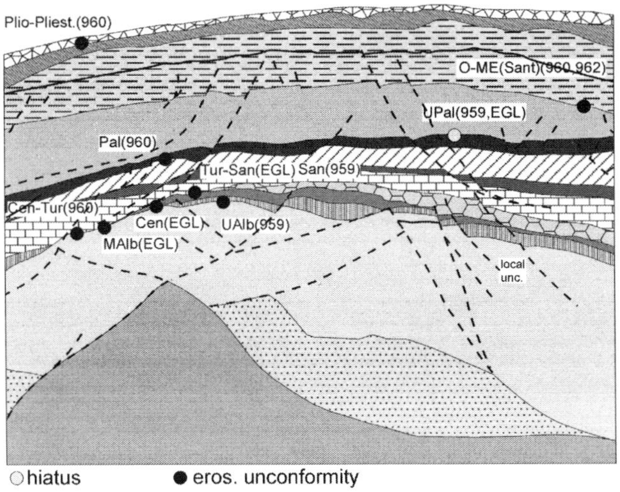

Figure 19.21. A list of erosional unconformities and hiatuses sketched on the portion of the interpreted profile, recognized by the biostratigraphic study in wells combined with seismic interpretation (Nemčok et al., 2004b). 959–962 are ODP site, EGL – East Grand Lahou well.

(a)

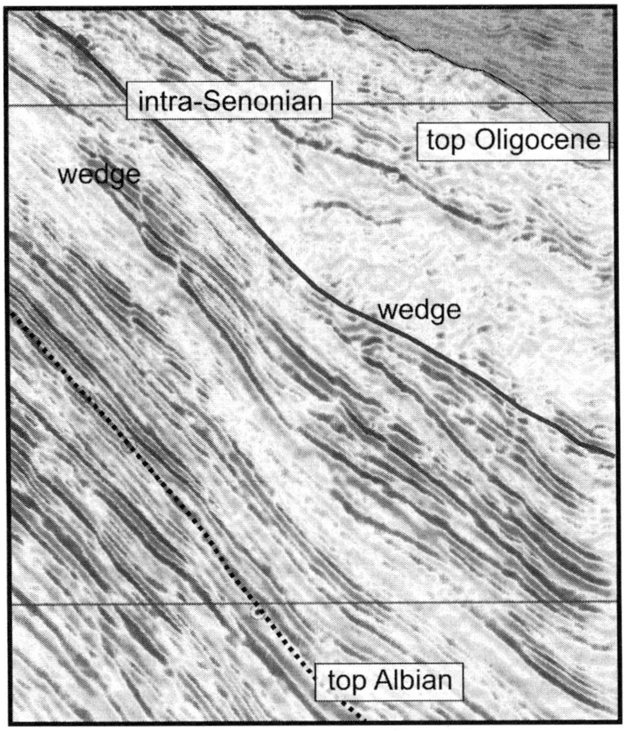

Figure 19.22. (a) Upper Cretaceous pinching-out wedges on the lower slope, offshore Benin (Nemčok et al., 2004b). Senonian lower slope sediments comprise upward pinching-out sand-prone wedges in seaward limbs of the main anticlines. Some of the associated reflectors brighten upward. Sediments of the basin floor fans were uplifted into their current lower slope position by the growth of main strike-slip fault–related folds. Their age is Cenomanian–Turonian. (b) Upper Cretaceous–lower Tertiary sediments pinching out against the continental margin in offshore Cameroon, which was uplifted and eroded during the Late Cretaceous (Nemčok and Rosendahl, 2006).

those from offshore Gabon (Nemčok and Rosendahl, 2006; Figure 19.17c). One more example comes from the Seme field in offshore Benin. The trap enhanced by unconformity in its northeastern part is an anticline developed by the transpressional reactivation of a syn-rift strike-slip fault during early drift. An unconformity trapping component also exists in the fault-related North Tano and South Tano traps in offshore Ghana, discussed earlier. The unconformity seals also exist in the case of the Espoir field offshore Côte d'Ivoire, which is listed under other pre-drift traps. It is the post-Albian unconformity developed by the erosion associated with post-breakup uplift.

Depositional lens-related traps (Figure 4.5b) are the second/third most important stratigraphic traps in the discussed key provinces. They form 1 percent and 6 percent of all traps in the Gulf of Mexico and West Africa, respectively. Examples come from the Chicontepec field of the Tampico-Misantla Basin, the Angolan Benguela and Kuito fields of the Lower Congo Basin, and the Agbami and Bonga fields of the Niger Delta. Apart from extensional domain, the Niger Delta also offers the lens-related traps from the contractional domain of the gravity glide system. They are related to reservoir ponding between neighboring anticlines or an anticline/extensional fault combination.

Depositional lens are common for gravity glide systems, where the relationship of growing structures and deposition constantly changes the relief of the sediment catchment region. This results in various ponding scenarios changing spatially and temporally. Examples of this interaction from gravity glides gliding on salt come from the Gulf of Mexico (e.g., Martin, 1978; Diegel et al., 1995; Karlo and Shoup, 2000; Rowan et al., 1999; Combellas-Biggot and Galloway, 2002; Holcombe et al., 2002) while the shale-detachment examples come from the Niger Delta (e.g., Doust and Omatsola, 1990; Hooper et al., 2002).

Other quite well-known examples of the lens-related traps come either from the Cenomanian channel fill system in the Rio Muni Basin, represented by the Ceiba, Oveng, and Okume fields, producing from channelized and/or ponded turbidites (Dailly, 2000; Dailly et al., 2002), or from the Upper Cretaceous channel prospects mapped in offshore Benin (Nemčok et al., 2004b). An example of the channel fill trap from Nigeria comes from the Aje field. There are also other Equatorial Atlantic

(b)

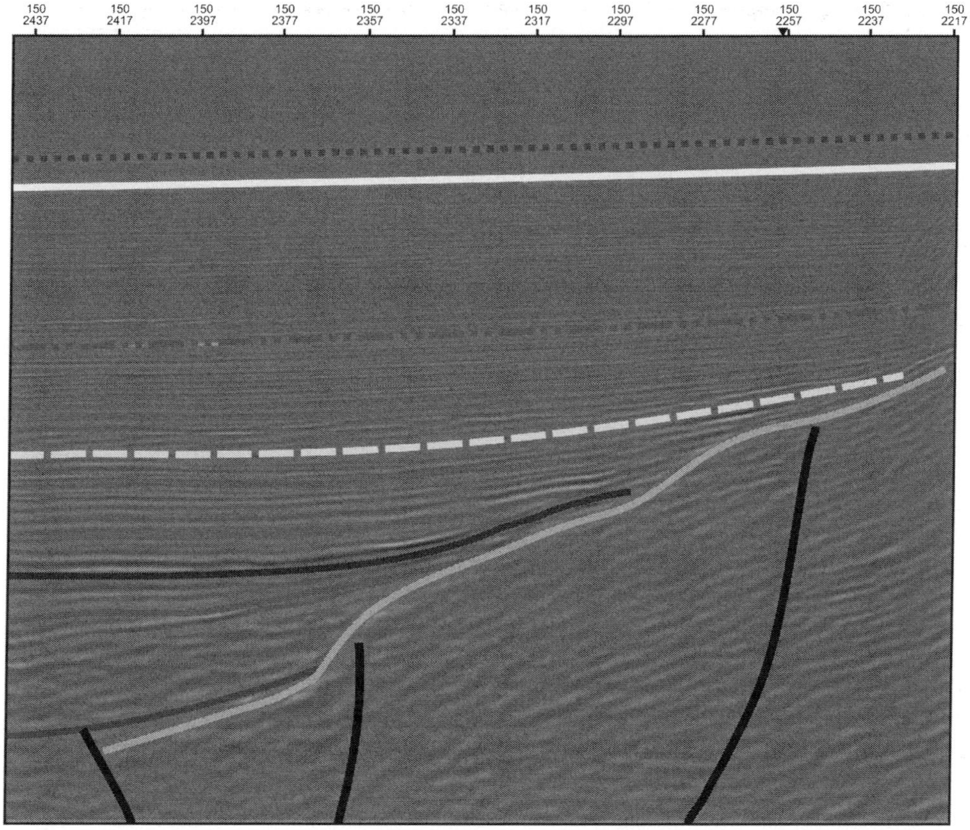

Figure 19.22 (*cont.*)

margins examples, including traps from offshore Côte d'Ivoire and Ghana, including sand-prone or turbiditic sediments ponded above former depressions remaining from the syn-rift period. Sediment sources were local north-to-south flowing rivers, including the largest of them, paleo-Volta. One of the examples is the Foxtrot field proving gas and condensate in the Albian channel sandstone. Offshore Togo also contains an example of the possible channel-related trap. The Lome-1 well discovered sub-commercial oil reservoired in the syn-rift Aptian-middle Albian fluvial channel sandstone. The oil has an API gravity of 44.5–50.0°. The average reservoir porosity is 12 percent. Although the assumed fault-block structure has failed to have a closure, stratigraphic traps are still possible.

Further south in the South Atlantic, out of the forty-three fields in Gabon producing in 1990 (Teisserenc and Villemin, 1990), only one was associated with a channel trap, which is located in the Miocene sedimentary section.

The most important stratigraphic traps in our key passive margin provinces are stratigraphic pinch-out-related traps (Figure 19.21). They are represented by 4 percent and 6 percent of all traps in the fields of the Gulf of Mexico and the West African provinces, respectively. Examples come from the Allegheny, Green Canyon 29, Oyster, and Popeye fields of the U.S. portion of the Gulf of Mexico, the Coco Marine field of the Central Douala Basin in Cameroon, the Landana field of the Lower Congo Basin in Angola, the Kudu field of the Orange River costal area in Namibia, and the West Cape Three Point Mahogany-1 field of west-central coastal Ghana. Any of the West African margin segments affected by significant post-breakup uplift contain pinch-outs at various stratigraphic levels. Typical examples come from offshore Benin (Figure 19.22a) and Cameroon (Figure 19.22b).

The offshore Côte d'Ivoire provides examples of pinch-out traps associated with structural highs inherited from the syn-rift time (Figures 4.7, 19.21). They are usually associated with Upper Cretaceous and Tertiary turbidites lapping against them. The Niger Delta provides some of the extensional traps with a stratigraphic component, represented by pinch-out- and channel-related features developed at delta flanks. Truncations against clay-filled channels are known from the Afam and Qua

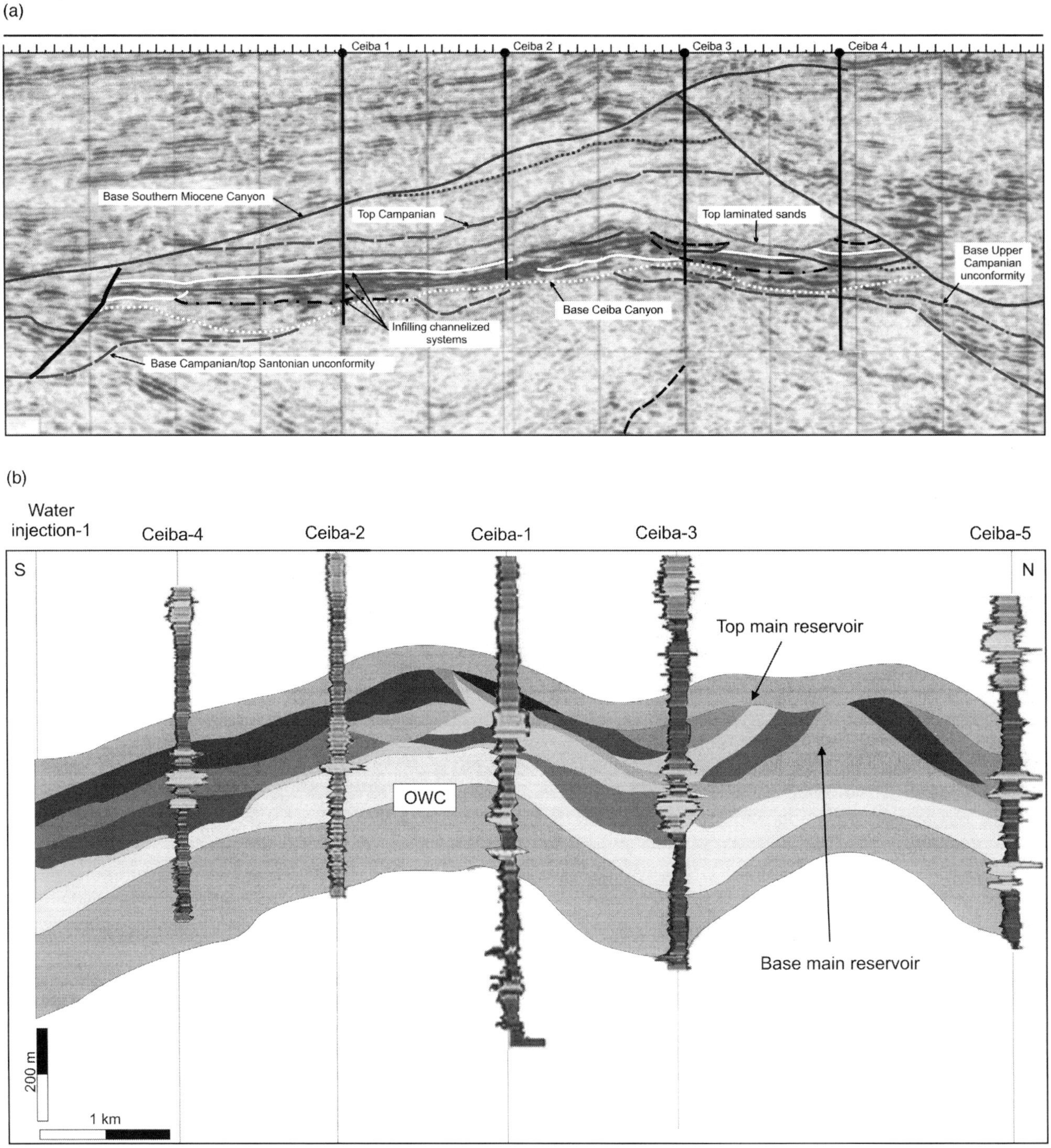

Figure 19.23. Details about the reservoir situation in the Ceiba Field (Dailly et al., 2002). (a) North–south cross-section along the axis of the main channelized reservoir fairway. (b) Log correlation for the Ceiba reservoir in wells Ceiba 1 to 5. The interbedded, channelized nature of the turbiditic reservoir sands is shown below the "Top of the Main Reservoir" correlation horizon. Ceiba 3 to 5 wells have been drilled along the axis of the main Ceiba channel complex and encountered similar pay sands to those seen previously in Ceiba 1 and 2. (c) An rms amplitude map showing the channelized nature of the reservoir and development/production well locations. Appraisal wells, drilled in 2000, confirmed the presence, connectivity, and quality of the reservoirs encountered in the discovery wells.

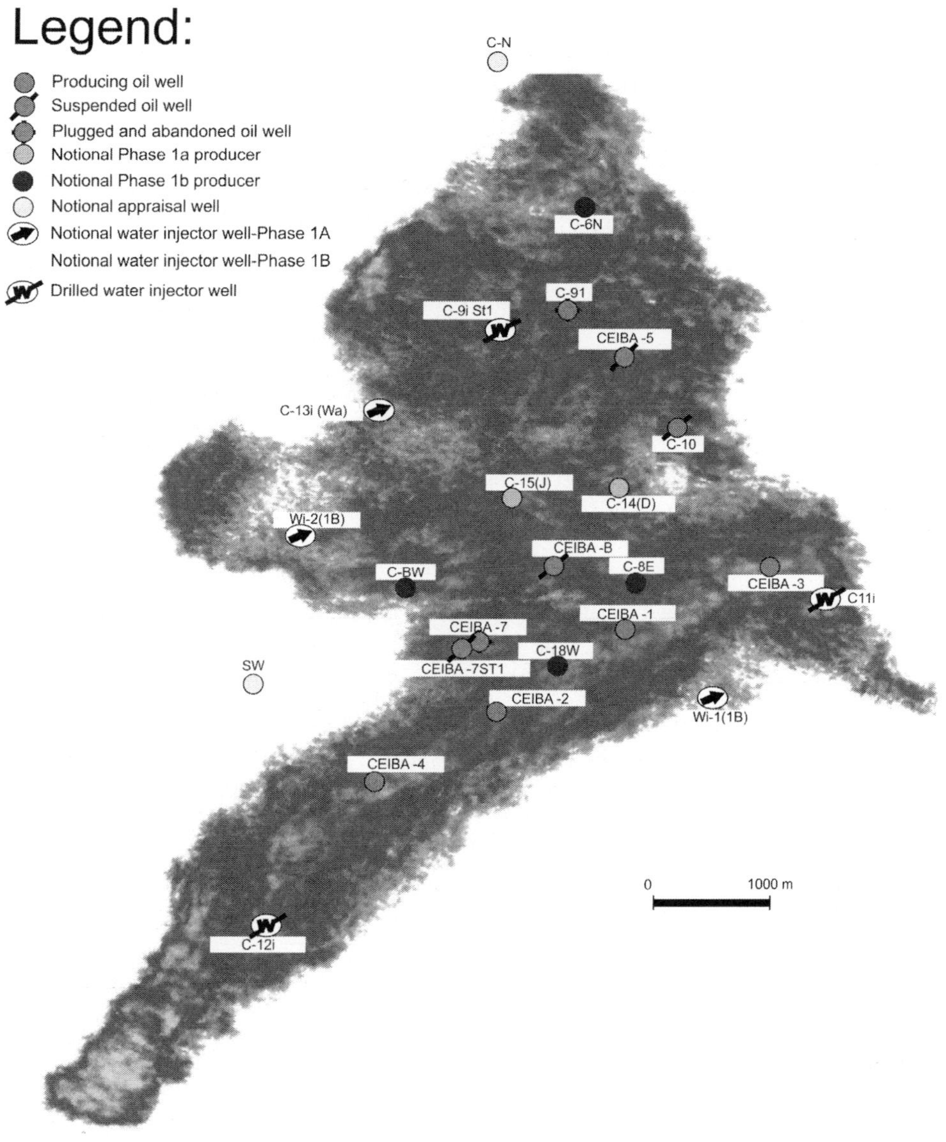

Figure 19.23 (*cont.*)

Iboe channels. Regional sand-pinch-outs are known from some regional flank areas where the sequence thins against basement-controlled ridge structures.

Among other provinces, pinch-outs are also known from the Central Atlantic, represented by shelf onlapping coarse wedges in Tertiary stratigraphies.

The lateral facies change-related traps (Figure 7.19b, c) represent the second/third most important traps in the Gulf of Mexico and the Western African provinces, represented by 1 percent and 6 percent of existing traps in our database. Examples come from the Shasta field of the U.S. portion of the Gulf of Mexico, the Angolan Gindungo and Lobito/Tomboco fields of the Lower Congo Basin, the Xikomba field of west-central coastal Angola, and the Ceiba field of offshore Equatorial Guinea (Figures 19.23a–c). This trap overlaps a bit with depositional lens-related traps because they are both characterized by abrupt lateral facies changes.

While potential traps offshore Côte d'Ivoire, Equatorial Atlantic are sealed by the Oligocene unconformity, this unconformity, together with the late Albian and Senonian unconformities, frequently forms a base for mounded and channelized sand-prone features. Mounds can be large. Some examples reach sizes of 80 km^2 and more. These stratigraphic traps have been reported from the Tano and Saltpond Basins. This trap relies on sealing intra-formational shale documented from several small traps in the Côte d'Ivoire Basin. Channel features over the Senonian unconformity are Maastrichtian channels known from both shallow-water and deepwater regions of the Côte d'Ivoire Basin. An oil discovery was made in some of them in the western Tano Basin and in the shallow-water Lion field, which produces from the Lower Cretaceous syn-rift sandstones containing oil and gas, just like the example of the Panthere field (Craven, 2000). Even larger features are present in the Grand Lahou Depression, one of the depressions inherited from the syn-rift period and maintained later. Also the Belier field, described earlier under contractional/transpressional traps, produces from submarine channel sediments deposited during the Late Cretaceous.

Among other provinces, canyon fills are also known from the Central Atlantic, occurring in the Upper Cretaceous and Tertiary stratigraphies of the Senegal Basin.

20 Hydrocarbon Preservation

Migrating petroleum undergoes alteration in composition and physical properties from the time it leaves the source rock. The alteration occurs during migration, during accumulation, and after entrapment (Blanc and Connan, 1993).

The main controlling factors of petroleum alteration before accumulation are:

1) source rock parameters, such as:
   a. kerogen type and
   b. maturation level;
2) expulsion efficiency; and
3) conditions of the secondary migration.

The principal factors controlling the composition of the oil accumulated in the reservoir are temperature and pressure. They affect the gas-oil ratio, GOR, and related physical properties during accumulation. API gravity and GOR of the unaltered oil and gas accumulated in the reservoir usually range from 25° to 40° and from 100 to 200 m^3/m^3, respectively. The temperature represents an important control over oil viscosity that is highly dependent on temperature. At normal reservoir temperatures, the most mature, non-waxy, and unaltered oils have viscosities of less than 100 mPas (Mann et al., 1997).

The alteration that takes place after entrapment includes:

1) thermal maturation;
2) biological and physical degradation;
3) de-asphalting; and
4) dysmigration.

Because of rapid rates of burial, uplift, and denudation of topography, rift flanks and proximal continental margins are susceptible to secondary alteration. The processes involved in petroleum alteration and degradation include:

1) temperature change related to changes in burial depth;
2) flushing by formation and meteoric waters; and
3) microbial activity in regions of freshwater circulation.

The temperature increase controlled by distinct burial and geothermal gradient perturbations may result in thermal maturation and the cracking of accumulated oils. In situ cracking, which takes part within the trap, controls:

1) API gravity increase;
2) depletion of polycyclic biomarkers;
3) formation of light hydrocarbons and gas;
4) GOR increase;
5) specific gravity decrease;
6) sulfur content decrease; and
7) viscosity decrease.

Oils accumulated in deep reservoirs tend to reach depletion in aromatics and, consequently, enrichment in n-alkanes (Vandenbroucke et al., 1999). The cracking reactions lead into a pyrobitumen residue formation. This conserves the hydrogen balance. The highly dehydrogenated residue has parameters that differentiate it from asphaltines precipitated in the reservoir. For example, it has vitrinite reflectance values of 1.5–2.5 percent and a very high T_{max}, characterized by values exceeding 460°C (Blanc and Connan, 1994).

The temperature interval associated with an onset of oil cracking depends on the oil composition (Schenk et al., 1997). However, there is an ongoing discussion about which parameters control this process and how. The alternative views include:

1) the process is controlled by chemical kinetics alone (e.g., Vandenbroucke et al., 1999);
2) the light hydrocarbons in oils are largely the products of transition-metal catalysis (Mango, 1997); and
3) a complex chemical segregation process occurs during secondary migration (Price and Schoell, 1995).

Having done chemical kinetic calculations, Pepper and Dodd (1995) describe that oils, which are n-alkane-rich, crack relatively slowly over a temperature range of 155–205°C. In contrast, oils, which are n-alkane-poor, crack easier, over a range of 115–145°C. Heating of the accumulated oil, which exceeds a temperature of 225°C, results in its thermal decomposition to just methane and pyrobitumen.

With regards to other alteration-controlling factors, it is usually rather difficult to separate the influences of

Table 20.1. Levels of biodegradation as indicated by organic compound classes removed by microbial oxidation or consequently enriched in the residual oil (Volkman et al., 1983).

Biodegradation level		Compounds removed or enriched
Minor	1	None
	2	Short n-alkanes absent
Moderate	3	>90% of n-alkanes removed
	4	Alkyleyelohexanes absent and isoprenoids reduced
	5	Isoprenoids absent
Extensive	6	Bicyclic alkanes absent
	7	>50% of regular steranes removed
Severe	8	Steranes altered and demethylated hopanes abundant
Extreme	9	Demethylated hopanes predominate, disteranes formed, steranes absent

water-washing and biodegradation. This is due to the fact that both processes commonly take place at shallow depths where oil is in contact with meteoric water and both of them shift the composition of the altered oil in the same direction.

Water-washing controls the preferential removal of the more water-soluble compounds. They include mostly the aromatics and n-alkanes lighter than C_{15}. This occurs without the alteration of naphthenes, which reflects their relative aqueous solubilities (Lafargue and Barker, 1988). Water-washing can be associated with waters of any salinity and can occur at any depth.

Biodegradation occurs at temperatures less than about 80°C and under a continuous flow of oxygen-charged fresh or brackish water (Hunt, 1979). Strong biodegradation changes both the chemical and physical properties of normal oil by (Connan, 1984, 1993; Wenger et al., 2002):

1) GOR decrease;
2) light hydrocarbon content decrease;
3) sulfur content increase; and
4) viscosity increase.

Microbial oxidation controls a selective removal of aromatics and alkanes. It also leads to a consequent enrichment of nitrogen-sulfur-oxygen-rich polar compounds. The end product can be represented by heavy oil with a gravity of 8–12° API and a viscosity of thousands of mPas. Such oil has a lower commercial quality. It is difficult to produce. The oxidation removes normal alkanes first. This is then followed by the removal of isoprenoids, regular steranes, disteranes, hopanes, and finally neohopanes (Volkman et al., 1983). Understanding the order of removal of compound classes allows one to identify the level of biodegradation (Table 20.1).

The addition of gas to accumulated oil changes the bulk chemistry of the oil. This is represented by its shift toward a lower average molecular weight. Furthermore, this change is also associated with the precipitation of a hydrogen-poor asphaltene residue, solid bitumen. The deasphalting results in a phase separation in the reservoir leading to oil that is substantially lighter than the original oil. Wet gas formed during the in situ thermal cracking of oil causes natural deasphalting in a process analogous to the propane deasphalting of residuum oil in a refinery (Hunt, 1979). In such cases, deasphalting joins forces with thermal maturation in developing lighter oil at greater depths.

Another way of adding gas or condensate into an accumulated oil is secondary migration (Blanc and Connan, 1994). This addition controls the deasphalting independently from the burial and temperature increases. It can occur even in relatively shallow oil pools.

Depending on its seal efficiency, a trap can preferentially leak light-fraction hydrocarbons represented by gas and condensate. This results in the evaporative fractionation involving the differential solution of light hydrocarbons from the oil into the gaseous solution (Thompson, 1987, 1988). The removal of light hydrocarbons from the accumulated oil and their addition to the gas phase controls an increase of light aromatic and naphthenic hydrocarbons in relationship to paraffin in the residual oil. The leaking of the gas phase may occur, for example, because of tectonically induced pressure changes in the gas-saturated oil reservoir (Leythaeuser and Poelchau, 1991). If the migrating leaked-off fraction becomes accumulated in a shallower trap at lower pressure and temperature than those in the original accumulation, it will condense as light oil with a gas cap by the process of dysmigration (Blanc and Connan, 1994). This process differs from gas flushing, in which the creation of a gas cap in a rising or cooling oil pool causes some portion of the pooled oil to spill out of the trap. In gas flushing, it is the oil fraction, made heavier by phase separation of gas and condensate, which migrates away from the original oil pool.

References

Aajdour, M., Hssaïda, T., Fedan, B., El Morabet, A., Oumalch, F., and Fakhi, S. (1999). Contribution à l'étude du Jurassique de la marge marocaine, Apports de l'analyse palynologique dans l'évolution géodynamique. *1st Colloquium on Moroccan Jurassic, Morocco*, pp. 1–2.

Abe, N. (2001). Petrochemistry of serpentinized peridotite from the Iberia abyssal plain (ODP Leg 173): its character intermediate between suboceanic and sub-continental upper mantle. *Geological Society of London Special Publications*, **187**, 143–59.

Abu El Karamat, S. and Fouda, H. G. (1990). Cross fault pattern and its impact on clysmic faults in the Gulf of Suez. *Egyptian 10th Exploration and Production Conference*, p. 14.

Achauer, U. and the KRISP Teleseismic Working Group (1994). New ideas on the Kenya Rift based on the inversion of the combined dataset of the 1985 and 1989/90 seismic tomography experiments. *Tectonophysics*, 236, 305–30.

Adám, Á. (1976). Results of deep electromagnetic investigations (in Pannonian Basin). Geoelectric and Geothermal Studies (East-Central Europe, Soviet Asia). In *KAPG Geophysical Monograph*, ed. Á. Adám and Akadémiai Kiadó, Budapest, pp. 547–60.

(1997). Mantle diapir – mantle plumes in the Pannonian Basin (investigations on the asthenosphere in Hungary from the sixties till the present). In *Physics and Geophysics with Special Historical Case Studies*, ed. W. Schröder. Bremen-Roennebeck, Germany, pp. 71–87.

Adams, J. T. and Dart, C. (1998). The appearance of potential sealing faults on borehole images. In *Faulting, Fault Sealing and Fluid Flow in Hydrocarbon Reservoirs*, ed. G. Jones, Q. J. Fisher, and R. J. Knipe. *Geological Society of London Special Publications*, **147**, 71–86.

Aguiar, L. A. M., Jinno, K., and Neto, J. B. M. (1986). Transpressional tectonics in the Foz do Amazonas Basin. Internal report, Petrobras (in Portuguese). In *Tectonic Evolution of Brazilian Equatorial Continental Margin Basins*, R. P. De Azevedo. (1991). PhD thesis, Imperial College, London, pp. 1–403.

Aguilera, R. and van Poollen, H. K. (1979). Occurrence of fractured reservoirs. *Oil and Gas Journal*, 77, 70–4.

Ahlbrandt, T. S. (2002). Assessment of global oil, gas, and NGL resources based on the total petroleum system concept. *American Association of Petroleum Geologists Bulletin*, 86, 1847.

Ahmadi, Z., Sawyers, M., Kenyon-Roberts, S., Stanworth, B., Kugler, K., Kristensen, J., and Fugelli, E. (2002). Chapter 14. Paleocene. In *The Millennium Atlas: Petroleum Geology of the Central and Northern North Sea*, ed. D. Evans, C. Graham, A. Armour, and P. Bathurst. Geological Society of London, pp. 549–97.

(2003). Chapter 14. Paleocene. In *The Millennium Atlas: Petroleum Geology of the Central and Northern North Sea*, ed. D. Evans, C. Graham, A. Armour, and P. Bathurst. Geological Society of London, pp. 235–59.

Ahnert, F. (1970). Functional relationships between denudation, relief, and uplift in large, mid-latitude drainage basins. *American Journal of Science*, 268, 243–63.

Aizenshtat, Z., Miloslavski, I., Ashengrau, D., and Oren, A. (1999). Hypersaline depositional environments and their relation to oil generation. In *Microbiology and Biogeochemistry of Hypersaline Environments: The Microbiology of Extreme and Unusual Environments*, ed. A. Oren. CRC Press, Boca Raton, FL, pp. 89–108.

Al-Tabaa, A. and Wood, D. M. (1987). Some measurements of permeability of kaolin. *Geotechnique*, 37, 499–503.

Albrechtsen, T., Andersen, S. J., Dons, T., Engstrom, F., Jorgensen, O., Sorensen, F. W., and Maersk Olie og Gas AS (2001). Developing non-structurally trapped oil in North Sea chalk. Conference paper. *Society of Petroleum Engineers Annual Technical Conference and Exhibition*, New Orleans, Louisiana, September 30–October 3, 2001.

Alexander, L. L. and Handschy, J. W. (1998). Fluid flow in a faulted reservoir system: fault trap analysis for the Block 330 field in Eugene Island, south addition, offshore Louisiana. *American Association of Petroleum Geologists Bulletin*, 82, 387–411.

Allan, U. S. (1989). A model for the migration and entrapment of hydrocarbons within faulted structures. *American Association of Petroleum Geologists Bulletin*, 75, 803–21.

Allemand, P. (1988). Approche experimentale de la mécanique du rifting continental. *Memoires et Documents du Centre Armoricain D'etude Structurale des Socles*, Rennes, 38, 1–192.

Allemand, P. and Brun, J. P. (1991). Width of continental rifts and rheological layering of the lithosphere. *Tectonophysics*, 188, 63–9.

Allemand, P., Brun, J. P., Davy, P., and van den Driessche, J. (1989). Symetrie et asymetrie des rifts et mecanismes d'amincissement de la lithosphere. Symmetry and asymmetry of rifts and mechanism of lithospheric thinning. *Bulletin de la Societe Geologique de France*, Huitieme Serie, 5, 445–51.

Allen, P. A. and Allen, J. R. (1990). *Basin Analysis: Principles and Applications*. Blackwell Scientific Publications, Oxford, p. 451.

Allen, P. A. and Densmore, A. L. (2000). Sediment flux from an uplifting fault block. *Basin Research*, 12, 367–80.

Allis, R. G., Armstrong, P. A., and Funnell, R. H. (1995). Implications of high heat flow anomaly around New Plymouth. *New Zealand Journal of Geology and Geophysics*, 38, 121–30.

Allix, P., Grosdidier, E., Jardine, S., Legoux, O., and Popoff, M. (1981). Decouverte d'Aptien superieur a Albien inferieur date par microfossiles dans la serie detritique cretacee du fosse de la Benoue (Nigeria). Upper Aptian and lower Albian deposits, dated from microfossils, in the Cretaceous detrital series of the Benue Valley, Nigeria. *Comptes-Rendus des Seances de l'Academie des Sciences, Serie 2: Mecanique-Physique, Chimie, Sciences de l'Univers, Sciences de la Terre*, 292, 1291–4.

Almeida, F. F. M., Dal Re Carneiro, C., Machado, D. D. L. Jr., and Dehira, L. K. (1988). Magmatismo Pos-Paleozoico no nordeste oriental do Brasil. Post-Paleozoic magmatism of the eastern part of northeastern Brazil. *Revista Brasileira de Geociencias*, 18, 451–62.

Al'Subbary, A. K., Nichols, G. N., Bosence, D. W. J., and Al-Kadasi, M. (1998). Pre-rift doming, peneplanation or subsidence in the southern Red Sea?: Evidence from the Medj-Zir Formation (Tawilah Group) of western Yemen. In *Sedimentation and Tectonics in Rift Basins: Red Sea–Gulf of Aden Case*, ed. B. Purser and D. Bosence. Chapman and Hall, London, pp. 119–34.

Alt, J. C. (1995). Subseafloor processes in mid-oceanic ridge hydrothermal systems. *American Geophysical Monograph*, 91, 85–114.

Alvarez, F., Virieux, J., and Le Pichon, X. (1984). Thermal consequences of lithosphere extension: the initial stretching phase. *Geophysical Journal of the Royal Astronomic Society*, 78, 389–411.

Ambrose, G. (2000). The geology and hydrocarbon habitat of the Sarir sandstone, SE Sirt Basin, Libya. *Journal of Petroleum Geology*, 23, 165–92.

Ambrose, W. A., Jones, R. H., Wawrzyniec, T. F., Fouad, K., Dutton, S. P., Jennette, D. C., Elshayeb, T., Sanchez-Barreda, L., Solis, H., Meneses-Rocha, J. L., Lugo, J. E., Aguilera, L., Berlanga, J., Miranda, L., Ruiz Morales, J., and Rojas, R. (2002). Upper Miocene and Pliocene shallow-marine and deepwater, gas-producing systems in the Macuspana Basin, southeastern Mexico. In *Gulf Coast Association of Geological Societies and Gulf Coast Section SEPM, Technical Papers and Abstracts*, ed. S. P. Dutton, S. C. Ruppel, and T. F. Hentz. *Transactions – Gulf Coast Association of Geological Societies*, 52, 3–12.

Aminzadeh, F. and Connolly, D. (2002). Looking for gas chimneys and faults. *American Association of Petroleum Geologists Explorer*, 23, 20–1.

(2007). Distinguishing hydrocarbon migration pathways in seismic data. *Bulletin of the South Texas Geological Society*, 47, 15–16.

Amrani, A., Lewan, M. D., and Aizenshtat, Z. (2005). Stable sulfur isotope partitioning during simulated petroleum formation as determined by hydrous pyrolysis of Ghareb Limestone, Israel. *Geochimica et Cosmochimica Acta*, 69, 5317–31.

Anderholm, S. K. (1983). Hydrogeology of the Socorro and La Jencia Basins, Socorro County, New Mexico. *Field Conference Guidebook, New Mexico Geological Society*, 34, 303–10.

(1987). Hydrogeology of the Socorro and La Jencia Basins, Socorro County, New Mexico. *U.S. Geological Survey Water Resources Investigation Report*, 84–4342.

Anders, D. E. and Gerrild, P. M. (1984). Hydrocarbon generation in lacustrine rocks of Tertiary age, Uinta Basin, Utah – organic carbon, pyrolysis yield, and light hydrocarbons. In *Hydrocarbon Source Rocks of the Greater Rocky Mountain Region*, ed. J. Woodward, F. F. Meissner, and J. L. Clayton. Rocky Mountain Association of Geologists, Denver, pp. 513–29.

Anders, D. E., Palacas, J. G., and Johnson, R. C. (1992). Thermal maturity of rocks and hydrocarbon deposits, Uinta Basin, Utah. In *Hydrocarbon and Mineral Resources of the Uinta Basin, Utah and Colorado*, ed. T. D. Fouch, V. F. Nuccio, and T. C. Chidsey. *Utah Geological Association Guidebook*, 20, 53–76.

Anders, M. H. and Schlische, R. W. (1994). Overlapping faults, intrabasin highs, and the growth of normal faults. *Journal of Geology*, 102, 165–79.

Anders, M. H., Spiegelman, M., Rodgers, D. W., and Hagstrum, J. T. (1993). The growth of fault-bounded tilt blocks. *Tectonics*, 12, 1451–9.

Anderson, D. L. (1994). The sublithospheric mantle as the source of continental flood basalts: the case against the continental lithosphere and plume head reservoirs. *Earth and Planetary Science Letters*, 123, 269–80.

(2000). The thermal state of the upper mantle: no role for mantle plumes. *Geophysical Research Letters*, 27, 3623–6.

Anderson, D. L., Zhang, Y. S., and Tanimoto, T. (1992). Plume heads, continental lithosphere, flood basalts and tomography. In *Magmatism and the Causes of Continental Break-up*, ed. B. C. Storey, T. Alabaster, and R. J. Pankhurst. *Geological Society of London Special Publications*, **68**, 99–124.

Anderson, R. N., Flemings, P. B., Losh, S., Austin, J., and Woodhams, R. (1994a). Gulf of Mexico growth fault drilled, seen as oil, gas migration pathway. *Oil and Gas Journal*, 94, 97–104.

Anderson, R. N., Flemings, P. B., Losh, S., Whelan, J., Billeaud, J., Austin, J., and Woodhams, R. (1994b). The Pathfinder drilling program into a major growth fault in Eugene Island 330, Gulf of Mexico: implications for behavior of hydrocarbon migration pathways.

CD-ROM prepared under Department of Energy contract DE-FC22-93BC14961.

Anderson, R. N., Hobart, M. A., and Langseth, M. G. (1979). Geothermal convection through oceanic crust and sediments in the Indian Ocean. *Science*, 204, 828–32.

Andrews-Speed, C. P., Oxburgh, E. R., and Cooper, B. A. (1984). Temperatures and depth-dependent heat flow in western North Sea. *American Association of Petroleum Geologists Bulletin*, 68, 1764–81.

Angelier, J. (1985). Extension and rifting: the Zeit region, Gulf of Suez. *Journal of Structural Geology*, 7, 605–12.

Angelier, J. and Colletta, B. (1983). Tension fractures and extensional tectonics. *Nature*, 301, 49–51.

Angevine, C. L., Turcotte, D. L., and Furnish, M. D. (1982). Pressure solution lithification as a mechanism for the stick-slip behavior of faults. *Tectonophysics*, 1, 151–60.

Anhaeusser, C. R., Mason, R., Vilhoen, M. J., and Viljoen, R. P. (1969). A reappraisal of some aspects of Precambrian shield geology. *Geological Society of America Bulletin*, 80, 2175–200.

Anisimov, P. D., Dixon, J. E., and Langmuir, C. H. (2004). A hydrous melting and fractionation model for mid-ocean ridge basalts: application to the Mid-Atlantic Ridge near Azores. *Geochemistry, Geophysics, Geosystems*, 5.

ANP (2013). http://www.anp.gov.br/ accessed on March 1, 2013. Lithostratigraphic chart of the Santos Basin accessed at: http://www.anp.gov.br/brasil-rounds/round4/round4/workshop/restrito/portugues/Santos_port.pdf. Lithostratigraphic chart of the Campos Basin accessed at: http://www.anp.gov.br/brasil-rounds/round4/round4/workshop/restrito/portugues/Campos_port.PDF.

Antonellini, M. and Aydin, A. (1994). Effect of faulting on fluid flow in porous sandstones: petrophysical properties. *American Association of Petroleum Geologists Bulletin*, 78, 355–77.

Antonellini, M. and Mollema, P. N. (2000). A natural analog for a fractured and faulted reservoir in dolomite: Triassic Sella Group, northern Italy. *American Association of Petroleum Geologists Bulletin*, 84, 314–44.

AOA Geophysics (2001). *Boujdour permit area, Morocco*. Phase 1 investigation AOA Geophysics, Houston.

Arabasz, W. J., Pechmann, J. C., and Brown, E. D. (1992). Observational seismology and the evaluation of earthquake hazards and risk in the Wasatch Front area, Utah. In *Assessment of Regional Earthquake Hazards and Risk along the Wasatch Front, Utah*, ed. P. L. Gori and W. W. Hays. *U.S. Geological Survey Professional Paper*, 1500-A-J, pp. 1–36.

Arad, A. and Bein, A. (1986). Saline versus freshwater contribution to the thermal waters of the northern Jordan Rift Valley. *Journal of Hydrology*, 83, 49–66.

Arch, J. and Maltman, A. J. (1990). Anisotropic permeability and tortuosity in deformed wet sediments. *Journal of Geophysical Research*, 95, 9035–47.

Arch, J., Maltman, A. J., and Knipe, R. J. (1988). Shear-zone geometries in experimentally deformed clays: the influence of water content, strain rate and primary fabric. *Journal of Structural Geology*, 10, 91–9.

Areshev, E. G., Dong, T. L., San, N. T., and Shnip, O. A. (1992). Reservoirs in fractured basement on the continental shelf of Southern Vietnam. *Journal of Petroleum Geology*, 15, 451–64.

Argent, J. D., Stewart, S. A., Green, P. F., and Underhill, J. R. (2002). Heterogeneous exhumation in the Inner Moray Firth, UK North Sea: constraints from new AFTA and seismic data. *Journal of the Geological Society of London*, 159, 715–29.

Argent, J. D., Stewart, S. A., and Underhill, J. R. (2000). Controls on the Lower Cretaceous Punt Sandstone member, a massive deep-water clastic deposystem, Inner Moray Firth, UK North Sea. *Petroleum Geoscience*, 6, 275–85.

Armijo, R., Tapponnier, P., Mercier, J. L., and Han, T. L. (1986). Quaternary extension in southern Tibet: field observations and tectonic implications. *Journal of Geophysical Research*, 91, 13803–72.

Armstrong, L. A., Ten Have, A., and Johnson, H. D. (1987). The geology of the Gannet fields, Central North Sea, UK Sector. In *Petroleum Geology of North West Europe*, ed. J. Brooks and K. W. Glennie. Graham and Trotman, London, pp. 533–48.

Armstrong, P. A. and Chapman, D. S. (1998). Beyond surface heat flow: an example from a tectonically active sedimentary basin. *Geology*, 26, 183–6.

Armstrong, P. A., Ehlers, T. A., Chapman, D. S., Farley, K. A., and Kamp, P. J. J. (2003). Exhumation of the central Wasatch Mountains, Utah: 1. Patterns and timing of exhumation deduced from low-temperature thermochronology data. *Journal of Geophysical Research*, 108.

Armstrong, P. A., Kamp, P. J. J., Allis, R. G., and Chapman, D. S. (1997). Thermal effects of intrusion below the Taranaki Basin (New Zealand): evidence from combined apatite fission track age and vitrinite reflectance data. *Basin Research*, 9, 151–69.

Arndt, N. T. and Christensen, U. (1992). The role of the lithospheric mantle in continental flood volcanism: thermal and geochemical constraints. *Journal of Geophysical Research*, 97, 10967–81.

Arrowsmith, J. R. (1995). *Coupled Tectonic Deformation and Geomorphic Degradation along the San Andreas Fault System*. Stanford University, Stanford, California, p. 347.

Artemieva, I. M. and Mooney, W. D. (2001). Thermal thickness and evolution of Precambrian lithosphere: a global study. *Journal of Geophysical Research*, 106, 16387–414.

(2002). On the relations between cratonic lithosphere thickness, plate motions, and basal drag. *Tectonophysics*, 358, 211–31.

Artemjev, M. E. and Artyushkov, E. V. (1971). Structure and isostasy of the Baikal Rift and the mechanism of rifting. *Journal of Geophysical Research*, 76, 1197–211.

Artyushkov, E. V., Morner, N. A., and Tarling, D. H. (2000). The cause of loss of lithospheric rigidity in areas far from plate tectonic activity. *Geophysical Journal International*, 143, 752–76.

Asimow, P. D., Dixon, J. E., and Langmuir, C. H. (2004). A hydrous melting and fractionation model for mid-ocean ridge basalts: application to the Mid-Atlantic Ridge near Azores. Geochem. *Geophys. Geosyst.*, 5.

Aslanian, D., Moulin, M., Olivet, J.-L., Unternehr, P., Matias, L., Bache, F., Rabineau, M., Nouzé, H., Klingelhoefer, F., Contrucii, I., and Labails, C. (2009). Brazilian and African passive margins of the Central Segment of the South Atlantic Ocean: kinematic constraints. *Tectonophysic*, 468, 98–112.

Athy, L. F. (1930). Density, porosity and compaction of sedimentary rocks. *American Association of Petroleum Geologists*, 14, 1–24.

Atwood, G. I. and Forman, M. J. (1959). Nature and growth of southern Louisiana salt domes and its effect on petroleum accumulation. *American Association of Petroleum Geologists Bulletin*, 43, 2595–622.

Audet, D. M. and McConell, J. D. C. (1992). Establishing resolution limits for tectonic subsidence curves by forward basin modeling. *Marine and Petroleum Geology*, 11, 400–11.

Aulbach, S., Rudnick, R. L., and McDonough, W. F. (2008). Li-Sr-Nd isotope signatures of the plume and cratonic lithospheric mantle beneath the margin of the rifted Tanzanian craton (Labait). *Contribution to Mineralogy and Petrology*, 155, 79–92.

Austin, J. A., Stoffa, P. L., Phillips, J. D., Oh, J., Sawyer, D. S., Purdy, G. M., Reiter, E., and Makris, J. (1990). Crustal structure of the southeast Georgia Embayment–Carolina Trough: preliminary results of a composite seismic image of a continental suture (?) and a volcanic passive margin. *Geology*, 18, 1023–7.

Avdulov, M. V. (1970). *Stroyeniye zemnoy kory Krymskogo poluostrova po rezul'tatam geofizicheskikh issledovaniy. Structure of the Earth's crust of the Crimea Peninsula according to geophysical survey results*. Kompleksnyye issledovaniya Chernomorskoy vpadiny. Akademia Nauk SSSR, Mezhduved. Geofiz. Kom., Rezul't. Issled. Mezhdunar. Geofiz. Proyekt., Moscow.

Aydin, A. (1977). Small faults formed as deformation bands in sandstone. In *Proceedings of Conference II: Experimental Studies of Rock Friction with Application to Earthquake Prediction*, ed. J. F. Evernden. U.S. Geological Survey, Office of Earthquake Studies, Menlo Park, California, pp. 617–53.

(1978a). Faulting in sandstone. Doctoral dissertation, Stanford University.

(1978b). Small faults formed as deformation bands in sandstone. *Pure and Applied Geophysics*, 116, 913–30.

Aydin, A., Borja, R. I., and Eichhubl, P. (2006). Geological and mathematical framework for failure modes in granular rock. *Journal of Structural Geology*, 28, 83–98.

Aydin, A. and Johnson, A. M. (1978). Development of faults as zones of deformation bands and as slip surfaces in sandstone. *Pure and Applied Geophysics*, 116, 931–42.

(1983). Analysis of faulting in porous sandstones. *Journal of Structural Geology*, 5, 19–35.

Aydin, A. and Nur, A. (1985). The types and role of stepovers in strike-slip tectonics. In *Strike-Slip Deformation, Basin Formation and Sedimentation*, ed. K. T. Biddle and N. Christie-Blick. *Special Publication – Society of Economic Paleontologists and Mineralogists*, pp. 35–44.

Babuška, V. and Plomerová, J. (1988). Subcrustal continental lithosphere: a model of its thickness and anisotropic structure. *Physics of the Earth and Planetary Interiors*, 51, 130–2.

Bachler, D., Kohl, T., and Rybach, L. (2003). Impact of graben-parallel faults on hydrothermal convection – Rhine Graben case study. *Physics and Chemistry of the Earth*, 28, 431–41.

Bachu, S. (1985). Influence of lithology and fluid flow on the temperature distribution in a sedimentary basin: a case study from the Cold Lake area, Alberta, Canada. *Tectonophysics*, 120, 257–84.

(1991). On the effective thermal and hydraulic conductivity of binary heterogeneous sediments. *Tectonophysics*, 190, 299–314.

Bachu, S., Ramon, J. C., Villegas, M. E., and Underschultz, J. R. (1995). Geothermal regime and thermal history of the Llanos Basin, Colombia. *American Association of Petroleum Geologists Bulletin*, 79, 116–29.

Bacoccoli, G. (1991). O futuro da exploração de petróleo na Bacia de Sergipe-Alagoas: Relatório Interno Petrobrás, Depex, Rio de Janeiro. Soil structure and its effects on hydraulic conductivity. *Soil Science*, 114, 417–22.

Bailey, C. C., Price, I., and Spencer, A. M. (1981). The Ula oil field, Block 7/12, Norway. In *The Sedimentation of the North Sea Reservoir Rocks*. Norwegian Symposium of Exploration, Report of the Norwegian Petroleum Society, Oslo, 18, pp. 1–26.

Bailey, N. J. L., Burwood, R., and Harriman, G. E. (1990). Application of pyrolysate carbon isotope and biomarker technology to organofacies definition and oil correlation problems in North Sea basins. *Organic Geochemistry*, 16, 1157–72.

Bailey, R. C. (1990). Trapping of aqueous fluids in the deep crust. *Geophysical Research Letters*, 17, 1129–32.

Baird, R. A. (1986). Maturation and source rock evaluation of Kimmeridge Clay, Norwegian North Sea. *American Association of Petroleum Geologists Bulletin*, 7, 1–11.

Baker, A. (1933). Geology and oil possibilities of the Moab district, Grand and San Juan Counties. *Geological Survey Bulletin*, Utah, p. 841.

Baker, J., Chazot, G., Menzies, M. A., and Thirlwall, M. (2002). Lithospheric mantle beneath Arabia; a

Pan-African protolith modified by the Afar and older plumes, rather than a source for continental flood volcanism? In *Volcanic Rift Margins*, ed. M. A. Menzies, S. L. Klemperer, C. Ebinger, and J. Baker. Geological Society of America Special Paper, 362, pp. 65–80.

Baker, J., MacPhearson, C. J., Menzies, M. A., Thirwall, M. F., Al-Kadasi, M., and Mattey, D. P. (2000). Resolving crustal and mantle contributions to continental flood volcanism, Yemen: constraints from mineral oxygen isotope data. *Journal of Petrology*, 41(12), 1805–20.

Baker, J., Snee, L., and Menzies, M. A. (1996a). A brief Oligocene period of flood volcanism in Yemen: implications for the duration and rate of continental flood volcanism at the Afro-Arabian triple junction. *Earth and Planetary Science Letters*, 138, 39–56.

Baker, J., Thirwall, M. F., and Menzies, M. A. (1996b). Sr-Nd_Pb isotopic and trace element evidence for crustal contamination of a mantle plume: Oligocene flood volcanism in western Yemen. *Geochemica et Cosmochimica Acta*, 60, 2559–81.

Bakun, W. H. and Lindh, A. G. (1985). The Parkfield, California earthquake prediction experiment. *Science*, 229, 619–24.

Balkwill, H. R. and McMillan, N. J. (1990). Geology of the Labrador Shelf, Baffin Bay and Davis Strait: Part 1. Mesozoic–Cenozoic geology of the Labrador Shelf. In *Geology of the Continental Margin of Eastern Canada*, ed. M. J. Keen and G. L. Williams. *Geological Survey of Canada*, 2, pp. 295–324.

Balling, N. (1985). Thermal structure of the lithosphere beneath the Norwegian-Danish Basin and the southern Baltic shield: a major transition zone. *Terra cognita*, 5, pp. 377–8.

Bally, A. W. (1982). Musings over sedimentary basin evolution. In *Evolution of Sedimentary Basins*, ed. P. Kent, M. H. P. Botts, D. P. McKenzie, and C. A. Williams. *Philosophical Transactions of the Royal Society of London, Mathematical and Physical Sciences*, 305, pp. 325–38.

Bally, A. W., Bernoulli, D., Davis, G. A., and Montadert, L. (1981). Listric normal faults. *Oceanologica Acta*, 4, 87–101.

Bally, A. W., Gordy, P. L., and Stewart, G. A. (1966). Structure, seismic data, and orogenic evolution of southern Canadian Rocky Mountains. *Bulletin of Canadian Petroleum Geology*, 14, 337–81.

Banda, E., Gallart, J., Garcia-Duenas, V., Danobeitia, J. J., and Makris, J. (1993). Lateral variation of the crust in the Iberian Peninsula: new evidence from the Betic Cordillera. *Tectonophysics*, 221, 53–66.

Bankey, V. A., Cevas, D., Daniels, A. A., Finn. I., Hermandez and Project Members (2002). Digital data grids for the magnetic anomaly map of North America: [Nova Scotia aeromagnetic field map]. *U.S. Geological Survey, Open-File Report*, 02-0414.

Barbarand, J., Carter, A., Wood, I., and Hurford, T. (2003). Compositional and structural control of fission-track annealing in apatite. *Chemical Geology*, 198, 107–37.

Barnard, P. C., Collins, A. G., and Cooper, B. S. (1981). Identification and distribution of kerogen facies in a source rock horizon: examples from the North Sea Basin. In *Organic Maturation Studies and Fossil Fuel Exploration*, ed. J. Brook. Academic Press, London, pp. 271–82.

Barnard, P. C. and Cooper, B. S. (1981). Oils and source rocks in the North Sea area. In *Petroleum Geology of the Continental Shelf of Northwest Europe*, ed. L. V. Illing and G. D. Hobson. Heyden and Son, London, pp. 169–75.

Barnouin-Jha, K., Parmentier, E. M., and Sparks, D. W. (1997). Buoyant mantle upwelling and crustal production at oceanic spreading centers: on-axis segmentation and off-axis melting. *Journal of Geophysical Research*, 102, 11979–89.

Barr, D. (1987). Lithospheric stretching, detached normal faulting and footwall uplift. *Geological Society Special Publications*, 28, 75–94.

Barroll, M. W. and Reiter, M. (1990). Analysis of the Socorro hydrogeothermal system: central New Mexico. *Journal of Geophysical Research*, 95, 21949–63.

Barton, A. J. and White, R. S. (1997a). Crustal structure of Edoras Bank continental margin and mantle thermal anomalies beneath the North Atlantic. *Journal of Geophysical Research*, 102, 3109–29.

(1997b). Volcanism on the Rockall continental margin. *Journal of the Geological Society*, 154, 531–6.

Barton, C. A. and Moss, D. (1988). Analysis of macroscopic fractures in the Cajon Pass scientific drillhole: over the interval 1829–2115 meters. *Geophysical Research Letters*, 15, 1013–16.

Barton, C. A. and Zoback, M. D. (1992). Self-similar distribution and properties of macroscopic fractures at depth in crystalline rock on the Cajon Pass scientific drill hole. *Journal of Geophysical Research*, 97, 5181–200.

Barton, C. A., Zoback, M. D., and Moos, D. (1995). Fluid flow along potentially active faults in crystalline rock. *Geology*, 23, 683–6.

Basak, P. (1972). Soil structure and its effects on hydraulic conductivity. *Soil Science*, 114, 417–22.

Basile, C., Mascle, J., Benkhelil, J., and Bouillin, J. P. (1998). Geodynamic evolution of the Côte d'Ivoire–Ghana transform margin: an overview of Leg 159 results. In *Proc. ODP, Sci. Results, 159, 101–110*, ed. J. Mascle, G. P. Lohmann and M. Moullade. Available at: http://www.odp.tamu.edu/publications/159_SR/CHAPTERS/CHAP_11.PDF accessed December 5, 2001.

Baskin, D. K. and Peters, K. E. (1992). Early generation characteristics of a sulfur-rich Monterey kerogen. *American Association of Petroleum Geologists Bulletin*, 76, 1–13.

Baskin, V. N., Dmitrieva, L. S., Makhov, G. T., Ponomareva, T. V., and Polyakov, V. A. (1979). Effect of the duration of storage of standard specimens on their chemical composition. *Industrial Laboratory (USSR)*, 45, 159–60.

Bassi, G., Keen, C. E., and Potter, P. (2003). Contrasting styles of rifting: models and examples from the eastern Canadian margin. *Tectonics*, 12, 639–55.

Bauer, K., Neben, S., Schrekenberger, B., Emmerman, R., Hinz, K., Jokat, W., Schulze, A., Trumbull, R. B., and Weber, K. (2000). Deep structure of the Namibia continental margin as derived from integrated geophysical studies. *Journal of Geophysical Research*, 105, 25829–53.

Baum, M. S., Withjack, M. O., and Schlische, R. W. (2003). Controls of structural geometries associated with rift-basin inversion. *2003 AAPG Annual Convention with SEPM, Extended Abstracts*, Tulsa, Oklahoma, 12, 10.

(2008). The ins and outs of buttress folds: examples from the Fundy rift basin, Nova Scotia and New Brunswick, Canada. In *Central Atlantic Conjugate Margins*, ed. D. E. Brown. Dalhousie University, Halifax, Nova Scotia, pp. 53–61.

Baxter, K., Cooper, G. T., Hill, K. C., and O'Brien, G. W. (1999). Late Jurassic subsidence and passive margin evolution in the Vulcan Sub-basin, north-west Australia: constraints from basin modelling. *Basin Research*, 11, 97–111.

Bear, J. (1972). *Dynamics of Fluids in Porous Media*. Elsevier, New York.

Beaumont, C., Ellis, S., and Pfiffner, A. (1999). Dynamics of sediment subduction-accretion at convergent margins: short-term modes, long-term deformation, and tectonic implications. *Journal of Geophysical Research*, 104, 17573–601.

Beaumont, C., Keen, C. E., and Boutilier, R. (1982). On the evolution of rifted continental margins: comparison of models and observations for the Nova Scotian margin. *Geophysical Journal of the Royal Astronomical Society*, 70, 667–715.

Beck, W. C. and Chapin, C. E. (1994). Structural and tectonic evolution of the Joyita Hills, central New Mexico: implications of basement control on Rio Grande Rift. *Special Paper – Geological Society of America*, 291, 187–205.

Becker, H., Horan, M. F., Walker, R. J., Gao, S., Lorand, J. P., and Rudnick, R. L. (2006). Highly siderophile element composition of the Earth's primitive upper mantle: constraints from new data on peridotite massifs and xenoliths. *Geochimica et Cosmochimica Acta*, 70, 4528–50.

Bederke, E. (1966). The development of European rifts. *Geological Survey of Canada*, 66, 213–19.

Bedrosian, P. A., Unsworth, M. J., and Egbert, G. D. (2002). Magnetotelluric imaging of the creeping segment of the San Andreas Fault near Hollister. *Geophysical Research Letters*, 29, 1-1-1-4.

Bedrosian, P. A., Unsworth, M. J., Egbert, G. D., and Thurber, C. H. (2004). Geophysical images of the creeping segment of the San Andreas Fault: implications for the role of crustal fluids in the earthquake process. *Tectonophysics*, 385, 137–58.

Beekman, F. (1994). Tectonic modeling of thick-skinned compressional intraplate deformation. PhD thesis, Vrije University, Amsterdam, p. 152.

Beer, J. A., Allmendinger, R. W., Figueroa, D. E., and Jordan, T. E. (1990). Seismic stratigraphy of a Neogene piggyback basin, Argentina. *American Association of Petroleum Geologists Bulletin*, 74, 1183–202.

Begg, G. C., Griffin, W. L., Natapov, L. M., O'Reilly, S. Y., Grand, S. P., O'Neill, C. J., Hronsky, J. M. A., Poudjom Djomani, Y., Swain, C. J., Deen, T., and Bowden, P. (2009). The lithospheric architecture of Africa: seismic tomography, mantle petrology, and tectonic evolution. *Geosphere*, 5, 23–50.

Beherens, W. E. (1988). Geology of a continental slope oil seep, Gulf of Mexico. *American Association of Petroleum Geologists Bulletin*, 72, 105–14.

Behn, M. D. and Lin, J. (2000). Segmentation in gravity and magnetic anomalies along the U.S. East Coast passive margin: implications for incipient structure of the oceanic lithosphere. *Journal of Geophysical Research*, 105, 25769–90.

Behrendt, J. C., Hamilton, R. M., Ackermann, H. D., and Henry, V. J. (1981). Cenozoic faulting in the vicinity of the Charleston, South Carolina, 1886 earthquake. *Geology*, 9, 117–22.

Beitler, B., Chan, M. A., and Parry, W. T. (2003). Bleaching of Jurassic Navajo Sandstone on Colorado Plateau Laramide highs: evidence of exhumed hydrocarbon supergiants? *Geology*, 31, 1041–4.

Bekins, B. A., McCaffrey, A. M., and Dreiss, S. J. (1994). The influence of kinetics on the smectite to illite transition in the Barbados accretionary prism. *Journal of Geophysical Research*, 99, 18145–58.

Belgasem, B. A. (1990). An evaluation of an oil-bearing granite reservoir from well logs. *The Log Analyst*, 32, 404.

Belgasem, B. A., Barlai, Z., and Rez, F. (1991). Interpretation of well logs in the basement oil-bearing reservoir, Nafoora-Augile Field, Libya. *The Log Analyst*, 32, 42–3.

Bellani, S., Brogi, A., Lazzarotto, A., Liotta, D., and Ranalli, G. (2004). Heat flow, deep temperatures and extensional structures in the Larderello geothermal field (Italy): constraints on geothermal fluid flow. *Journal of Volcanology and Geothermal Research*, 132, 15–29.

Belousov, V. V. (1988). Structure and evolution of the Earth's crust and upper mantle of the Black Sea. *Bolletino di Geofisica Teorica ed Applicata*, 3, 109–96.

Ben-Avraham, Z., Ginzburg, A., and Yuval, Z. (1981). Seismic reflection and refraction investigations of Lake Kinneret-central Jordan Valley, Israel. *Tectonophysics*, 80, 165–81.

Ben-Avraham, Z., Hanel, R., and Villinger, H. (1978). Heat flow through the Dead sea rift. *Marine Geology*, 28, 253–67.

Ben-Avraham, Z., Hartnady, C. J. H., and Kitchin, K. A. (1997). Structure and tectonics of the Agulhas-Falkland fracture zone. *Tectonophysics*, 282, 83–98.

Ben-Avraham, Z. and Zoback, M. D. (1992). Transform-normal extension and asymmetric basins; an alternative to pull-apart models. *Geology*, 20, 423–6.

Bender, A. A. (1987). *O comportamento termomecânico do Terciário da Bacia do Pará-Maranhão*. University Ouro Preto.

Benes, V. and Davy, P. (1996). Modes of continental lithospheric extension: experimental verification of strain localization processes. *Tectonophysics*, 254, 69–87.

Benkhelil, J. (1988). Structure et evolution geodynamique du bassin intracontinental de la Benoue (Nigeria). Structure and geodynamic evolution of the intracontinental Benue Basin, Nigeria. *Bulletin des Centres de Recherches Exploration-Production Elf-Aquitaine*, 12, 29–128.

Benkhelil, J., Guiraud, M., Mascle, J., Basile, C., Bouillin, J.-P., Mascle, G., and Cousin, M. (1996). Structural record of the Africa/Brazil sliding in the Cretaceous sediments of the Cote d'Ivoire-Ghana transform margin. *Comptes rendus de l'Academie des sciences, Série 2*, 323, 73–90.

Bennett, R. A., Wernicke, B. P., and Davis, J. L. (1998). Continuous GPS measurements of contemporary deformation across the northern Basin and Range Province. *Geophysical Research Letters*, 25, 563–6.

Benson, R. N. (2002). Age estimates of the seaward-dipping volcanic wedge, earliest oceanic crust, and earliest drift-stage sediments along the North American Atlantic continental margin. *Geophysical Monograph*, 136, 61–75.

Berckhemer, H., Baier, B., Bartelsen, H., Behle, A., Burkhart, H. et al. (1975). Deep seismic sounding in the Afar region and on the highland of Ethiopia. In *Afar Depression of Ethiopia*, vol. 1, ed. A. Pilger and A. Roesler. E. Schweizer Verlagsbuchhandl, Suttgart.

Berg, R. R. (1975). Capillary pressure in stratigraphic traps. *American Association of Petroleum Geologists Bulletin*, 59, 939–56.

Berg, R. R. and Gangi, A. F. (1999). Primary migration by oil-generation microfracturing in low-permeability source rocks: application to the Austin Chalk, Texas. *American Association of Petroleum Geologists Bulletin*, 83, 727–56.

Berger, M. and Roberts, A. M. (1999). The Zeta structure: a footwall degradation complex formed by gravity sliding on the western margin of the Tampen Spur, Northern North Sea. In *Petroleum Geology of Northwest Europe*, ed. A. J. Fleet and S. A. R. Boldy. *Proceedings of the 5th Conference, The Geological Society of London*, pp. 107–16.

Bergslien, D. (2002). Balder and Jotun – Two sides of the same coin? A comparison of two Tertiary oil fields in the Norwegian North Sea. *Petroleum Geoscience*, 8, 349–63.

Berkowitz, N. (1997). *Fossil Hydrocarbons: Chemistry and Technology*. Academic Press, San Diego, California.

Berman, A. (2008). Three super-giant fields discovered offshore Brazil. *World Oil Magazine*, 229. http://www.worldoil.com/Magazine/MAGAZINE_DETAIL.asp?ART_ID=3450&MONTH_YEAR=Feb-2008.

Bertotti, G. (1991). Early Mesozoic extension and Alpine shortening in the western Southern Alps: the geology of the area between Lugano and Menaggio (Lombardy, northern Italy). *Memorie di Scienze Geologiche di Padova*, 43, 17–123.

Bertotti, G., Gouiza, M., Foeken, J., and Andriessen, P. (2010). Late Jurassic–Early Cretaceous Tectonics and Exhumation Onshore Morocco: implications for Terrigenous Sand Reservoirs in the Offshore of NW-Africa. *Adapted from poster presentation at AAPG Annual Convention and Exhibition*, New Orleans, Louisiana, April 11–14, 2010.

Bertotti, G., Siletto, G. B., and Spalla, M. I. (1993). Deformation and metamorphism associated with crustal rifting: the Permian to Liassic evolution of the Lake Lugano–Lake Como area (Southern Alps). *Tectonophysics*, 226, 271–84.

Bertotti, G. and ter Voorde, M. (1994). Thermal effects of normal faulting during rifted basin formation, 2, The Lugano-Val Grande normal fault and the role of pre-existing thermal anomalies. *Tectonophysics*, 240, 145–57.

Bertram, G. T. and Milton, N. J. (1989). Reconstructing basin evolution from sedimentary thickness: the importance of palaeobathymetric control, with reference to the North Sea. *Basin Research*, 1, 247–57.

Bertrand, R. and Achab, A. (1990). Equivalents between the reflectance of vitrinite, zooclasts (chitinozoans, graptolites and scolenodonts) and the color alteration of palynomorphs (spores and acritarchs). *Palynology*, 15, 13–280.

Beslier, M. O., Cornen, G., Agrinier, P., Feraud, G., and Girardeau, J. (1996). Structure and evolution of a passive continental margin: main results of ODP Leg 149 in the ocean–continent transition of the Iberia abyssal plain. In *First Eurocolloquium of the Ocean Drilling Program*, ed. H. Ristedt. Oldenburg, Federal Republic of Germany, pp. 21–2.

Best, P. et al. (1985). Ghana Project – Mid Term Report. In *Tectonic Evolution of Brazilian Equatorial Continental Margin Basins*, ed. R. P. De Azevedo. PhD thesis, Imperial College, London, p. 455.

Bethke, C. M. (1985). A numerical model of compaction-driven groundwater flow and heat transfer and its application to the paleohydrology of intracratonic sedimentary basins. *Journal of Geophysical Research*, 90, 6817–28.

Bethke, C. M. and Corbet, T. F. (1985). Linear and nonlinear solutions for one-dimensional compaction flow in sedimentary basins. *Water Resources Research*, 24, 461–7.

Bethke, C. M. and Marshak, S. (1990). Brine migrations across North America – the plate tectonics of groundwater.

Annual Reviews of Earth and Planetary Sciences, 18, 287–315.

Betts, P. G., Giles, D., and Lister, G. S. (2003). Tectonic environment of shale-hosted massive sulfide Pb-Zn-Ag deposits of Proterozoic northeastern Australia. *Economic Geology*, 98, 557–76.

Beydoun, Z. R. (1997). Introduction to the revised Mesozoic stratigraphy and nomenclature for Yemen. *Marine and Petroleum Geology*, 14, 617–29.

Beydoun, Z. R., As-Saruri, M. A. L., El-Nakhal, H., Al-Ganad, I. N., Baraba, R. S., Nani, A. A. S. O., and Al-Aawah, M. H. (1998). *International Lexicon of Stratigraphy, v. III.* Asia, Republic of Yemen, International Union of Geological Sciences Publication, 34, 245.

Biddle, K. T. and Wielchowski, C. C. (1994). Hydrocarbon traps. In *The Petroleum System – From Source to Trap*, ed. L. B. Magoon and W. G. Dow. *American Association of Petroleum Geologists*, 60, 219–35.

Bielik, M., Blizkovský, M., Burda, M., Fusan, O., Huebner, M., Herrmann, H., Novotný, A., Suk, M., Tomek, C., and Vyskočil, V. (1994). Density models of the earth's crust along seismic profiles. In *Crustal Structure of the Bohemian Massif and the West Carpathians*, ed. M. Burda, M. Suk, J. Šefara, and G. Pliva. Academy of Sciences Czech Republic, Geophysical Institute, CZ-14131 Prague 4, Czechoslovakia CSK, pp. 177–87.

Bielik, M., Šefara, J., Bezák, V., and Kubeš, P. (1995). Deep-seated models of the Western Carpathians. *Proceedings of the 1st Slovak Geophysical Conference*, GISAS, pp. 7–11.

Biggar, N. E. and Adams, J. A. (1987). Dates derived from Quaternary strata in the vicinity of Canyonlands National Park. In *Geology of Cataract Canyon and Vicinity*, ed. J. A. Campbel. Four Corners Geological Society, 10th Field Conference Guidebook, pp. 127–36.

Bijwaard, H. and Spakman, W. (1999). Tomographic evidence for a narrow whole mantle plume below Iceland. *Earth and Planetary Science Letters*, 166, 121–6.

Biles, N. E., Hannan, A. E., Jamieson, G. A., and Bain, J. E. (2000). Geologic overview of the NE Mississippi fan and shelf to West Florida terrace region, offshore Gulf of Mexico. In *Transactions – Gulf Coast Association of Geological Societies*, ed. J. A. Ragsdale and N. C. Rosen. Gulf Coast Association of Geological Societies and Gulf Coast Section SEPM, 50th Annual Meeting, 50, 129–36.

Bill, M., O'Dogherty, L., Guex, J., Baumgartner, P. O., and Masson, H. (2001). Radiolarite ages in Alpine-Mediterranean ophiolites: constraints on the oceanic spreading and the Tethys-Atlantic connection. *Geological Society of America Bulletin*, 113, 129–43.

Billiris, H., Paradissis, D., Veis, G., England, P., Featherstone, W., Parsons, B., Cross, P., Rands, P., Rayson, M., Sellers, P., Ashkenazi, V., Davison, M., Jackson, J., and Ambraseys, N. (1991). Geodetic determination of tectonic deformation in central Greece from 1900 to 1988. *Nature*, 350, 124–9.

Binks, R. M. and Fairhead, J. D. (1992). A plate tectonic framework for the evolution of the Cretaceous rift basins in West and Central Africa. In *Geodynamics of Rifting. Volume II. Case history studies on rifts: North and South America and Africa*, ed. P. A. Ziegler. *Tectonophysics*, 213, 141–51.

Biot, M. A. (1941). General theory of three dimensional consolidation. *Journal of Applied Physics*, 12, 155–64.

(1956). Theory of propagation of elastic waves in a fluid-saturated porous solid: I. Low-frequency range. *Journal of the Acoustical Society of America*, 28, 168–78.

Birch, F. and Clark, H. (1940). The thermal conductivity of rocks and its dependence upon temperature and composition. *American Journal of Science*, 238, 529–58.

Bird, P. (1978a). Finite element modeling of lithosphere deformation: the Zagros collision orogeny. *Tectonophysics*, 50, 307–36.

(1978b). Initiation of intracontinental subduction in the Himalayas. *Journal of Geophysical Research*, 10, 4975–87.

Bischoff, M., Endrum, B., and Meier, T. (2006). Lower crustal anisotropy in Central Europe deduced from dispersion analysis of Love and Rayleigh waves. *Geophysical Research Abstracts*, 8.

Bishop, M. G. (2001). South Sumatra Basin province, Indonesia: the Lahat/Talang Akar-Cenozoic total petroleum system. *USGS Open-File Report*, 99-50-S, p. 22.

Bishop, R. S., Gehman, J. H. M., and Young, A. (1984). Concepts for estimating hydrocarbon accumulation and dispersion. In *Petroleum Geochemistry and Basin Evaluation*, ed. G. Demaison and R. J. Murris. *American Association of Petroleum Geologists*, 35, 41–52.

Bissada, K. K. (1982). Geochemical constraints on petroleum generation and migration – a review. *Proceeding of the Second ASCOPE Conference*, Manilla, October 1981, pp. 69–87.

Biteau, J. J., Choppin de Janvry, G., and Perrodon, A. (2003). How the petroleum system relates to the petroleum province. *Oil and Gas Journal*, 101, 46–9.

Bjorlykke, K., Ramm, M., and Saigal, G. C. (1989). Sandstone diagenesis and porosity modification during basin evolution. *Geologische Rundschau*, 78, 243–68.

Bjorlykke, M., Dypvik, H., and Finstad, K. G. (1975). The Kimmeridge shale, its composition and radioactivity. In *Proceedings from NPS Jurassic Northern North Sea Symposium*, ed. K. G. Finstad and R. C. Selley. Norwegian Petroleum Society, Bergen, pp. 12.1–12.20.

Blackwell, D. D. (1978). Heat flow and energy loss in the Western United States. In *Cenozoic Tectonics and Regional Geophysics of the Western Cordillera*, ed. G. P. Eaton and R. B. Smith. *Geological Society of America Memoir*, 152, 209–50.

Blackwell, D. D. and Richards, M. C. (2004). *Geothermal Map of North America. Scale 1:6,500,000.* AAPG, Tulsa, Oklahoma.

Blackwell, D. D. and Steele, J. L. (1989). Thermal conductivity of sedimentary rocks: measurement and significance. In *Thermal History of Sedimentary Basins – Methods and Case Histories*, ed. N. D. Naeser and T. H. McCulloh. Heidelberg, Springer, pp. 14–36.

Blackwell, D. D., Steele, J. L., and Brott, C. A. (1981). Heat flow in the Pacific Northwest. In *Physical Properties of Rocks and Minerals, Vol. II-2*, ed. Y. S. Toulokian, W. R. Judd, and R. F. Roy. McGraw-Hill, New York, pp. 1–510.

Blanc, P. and Connan, J. (1993). Crude oils in reservoirs: the factors influencing their composition. In *Applied Petroleum Geochemistry*, ed. M. L. Bordenave. Technip, Paris, pp. 151–74.

(1994). Preservation, degradation, and destruction of trapped oil. In *The Petroleum System – from Source to Trap*, ed. L. B. Magoon and W. G. Dow. *American Association of Petroleum Geologists Memoir*, 60, 237–47.

Blanpied, M. L., Lockner, D. A., and Byerlee, J. D. (1995). Frictional slip of granite at hydrothermal conditions. *Journal of Geophysical Research*, 100, 13045–64.

Bodell, J. M. and Chapman, D. S. (1982). Heat flow in the north-central Colorado Plateau. *Journal of Geophysical Research*, 87, 2869–84.

Bodenlos, A. J. (1970). Cap rock development and salt stock movement. In *The Geology and Technology of Gulf Coast Salt*, ed. D. H. Kupfer. Symposium Proceedings, Louisiana State University, Baton Rouge, p. 192.

Boehm, A. and Moore, J. C. (2002). Fluidized sandstone intrusions as an indicator of paleostress orientation, Santa Cruz, California. *Geofluids*, 2, 147–61.

Boettcher, S. S., James, A., Wenger, L., Gross, O., Harrison, S., and Hood, K. E. (1998). Integration of seismic and surface geochemistry data for evaluation of hydrocarbon systems in the Gulf of Mexico. *AAPG International Conference and Exhibition*, 82, p. 1893.

Bohacs, K. M. (1998). Contrasting expressions of depositional sequences in mudrocks from marine to nonmarine environs. In *Shales and Mudstones I*, ed. J. Schieber, W. Zimmerle, and P. Sethi. Schweizerbartische Verlangsbuchhandlung, Stuttgart, pp. 33–78.

Bohacs, K. M., Carroll, A. R., Neal, J. E., and Mankiewicz, P. J. (2000). Lake-basin type, source potential, and hydrocarbon character: an integrated sequence-stratigraphic-geochemical framework. In *Lake Basins through Space and Time*, ed. E. H. Gierlowski-Kordesh and K. R. Kelts. *American Association of Petroleum Geologists Studies in Geology*, 46, 3–34.

Bohacs, K. M. and Suter, J. (1997). Sequence stratigraphic distribution of coaly rocks: fundamental controls and paralic examples. *American Association of Petroleum Geologists Bulletin*, 81, 1612–39.

Bohannon, R. G., Naeser, C. W., Schmidt, D. L., and Zimmermann, R. A. (1989). The timing of uplift, volcanism and rifting peripheral to the Red Sea: a case for passive rifting? *Journal of Geophysical Research*, 94, 1683–701.

Boillot, G., Girardeau, J., and Kornprobst, J. (1989). Rifting of the West Galicia continental margin: a review. *Bulletin de la Societe Geologique de France*, 8, 393–400.

Boillot, G., Grimaud, S., Mauffret, A., Mougenot, D., Kornprobst, J., Mergoil-Daniel, J., and Torrent, G. (1980). Ocean–continent boundary off the Iberian margin: a serpentinite diapir west of the Galicia Bank. *Earth and Planetary Science Letters*, 48, 23–34.

Boillot, G., Recq, M., Winterer, E., Meyer, A. W., Applegate, J., Baltuck, M., Bergen, J. A., Comas, M. C., Davis, T. A., Dunham, K., Evans, C. A., Girardeau, J., Goldberg, D. G., Haggerty, J., Jansa, L. F., Johnson, J. A., Kasahara, J., Loreau, L.-P., Luna-Sierra, E., Moullade, M., Ogg, J., Sarti, M., Thurow, J., and Williamson, M. (1987). Tectonic denudation of the upper mantle along passive margins: a model based on drilling results (ODP Leg 103, western Galicia margin, Spain). *Tectonophysics*, 132, 335–42.

Bolchert, G., Weimer, P., and McBride, B. C. (2000). Structural and stratigraphic controls on petroleum seeps, Green Canyon and Ewing Bank, Northern Gulf of Mexico: implications for petroleum migration. *Gulf Coast Association of Geological Societies Transactions*, 50, 65–74.

Boles, J. R., Clark, J. F., Washburn, L., and Leifer, I. (2001). Temporal variation in natural methane seep rate due to tides and other factors, Coal Point area, California. *Journal of Geophysical Research*, 106, 27077–86.

Boles, J. R., Eichhubl, P., Garven, G., and Chen, J. (2004). Evolution of a hydrocarbon migration pathway along basin-bounding faults: evidence from fault cement. *American Association of Petroleum Geologists Bulletin*, 88, 947–70.

Bonatti, E. (1976). Serpentinite protrusions in the oceanic crust. *Earth and Planetary Science Letters*, 32, 107–13.

(1978). Vertical tectonism in oceanic fracture zones. *Earth and Planetary Science Letters*, 37, 369–79.

(1985). Punctiform initiation of seafloor spreading in the Red Sea during transition from a continental to an oceanic rift. *Nature*, 316, 33–7.

Bonatti, E., Colantoni, P., Della Vedova, B., and Taviani, M. (1984). Geology of the Red Sea transitional zone (22°N–25°N). *Oceanologica Acta*, 7, 385–98.

Boness, N. L. and Zoback, M. D. (2004). Multi-scale crustal seismic anisotropy in the region surrounding the San Andreas Fault near Parkfield California. *American Geophysical Union, Fall Meeting 2004, Abstract*, T11F-05.

Bos, B. (2000). Faults, fluids and friction: effect of pressure solution and phyllosilicates on fault slip behaviour, with implications for crustal rheology. PhD thesis, Universiteit Utrecht, Utrecht, p. 58.

Bos, B. and Spiers, C. J. (2002). Frictional-viscous flow of phyllosilicate-bearing fault rock: microphysical model and implications for crustal strength profiles. *Journal of Geophysical Research*, 107, 1–13.

Bosence, D. W. J. (1998). Stratigraphic and sedimentological models of rift basins. In *Sedimentation and Tectonics of Rift Basins: Red Sea-Gulf of Aden*, ed. B. H. Purser and D. W. J. Bosence. Chapman and Hall, London, pp. 9–25.

Bostick, N. H. (1979). *Maturation of Organic Matter and Generation of Petroleum in Tertiary Oil Basins*. Geological Survey Professional Paper, Washington, DC, p. 1150.

Bostick, N. H. and Alpern, B. (1977). Principles of sampling, preparation and constituent selection for microphotometry in measurement of maturation of sedimentary organic matter. *Journal of Microscopy*, 109, 41–7.

Bosworth, W. (1985). Geometry of propagating continental rifts. *Nature*, 316, 625–7.

(1992). Mesozoic and early Tertiary rift tectonics in East Africa. *Tectonophysics*, 209, 115–37.

(1995). A high-strain rift model for the southern Gulf of Suez (Egypt). *Geological Society Special Publications*, 80, 75–102.

Bott, M. H. P. and Dean, D. S. (1973). Stress diffusion from plate boundaries. *Nature*, 243, 339–41.

Bott, M. H. P. and Kusznir, N. J. (1984). The origin of tectonic stress in the lithosphere. *Tectonophysics*, 105, 1–13.

Bottero, C. J., Desneulin, J., Joubert, J. B., and Martine, J. P. (1989). Final geological report of (the) Matondo 1 well, Equatorial Guinea, Elf Aquitaine.

Bouillin, J. P., Basile, C., Labrin, E., and Mascle, J. (1998). Thermal constraints on the Cote D'Ivoire–Ghana transform margin: evidence from apatite fission track ages. *Proceedings of the Ocean Drilling Program, Scientific Results*, 159, 43–8.

Bowen, B. B., Martini, B. A., Chan, M. A., and Parry, W. T. (2007). Reflectance spectroscopic mapping of diagenetic heterogeneities and fluid-flow pathways in the Jurassic Navajo Sandstone. *American Association of Petroleum Geologists Bulletin*, 91, 173–90.

Bown, J. W. and White, R. S. (1994). Variation with spreading rate of oceanic crustal thickness and geochemistry. *Earth and Planetary Science Letters*, 121, 435–49.

(1995). Effect of finite extension rate on melt generation at rifted continental margins. *Journal of Geophysical Research*, 100, 18011–29.

Boyd, F. R. (1989). Compositional distinction between oceanic and cratonic lithosphere. *Earth and Planetary Science Letters*, 96, 15–26.

Brace, W. F. (1961). Dependence of fracture strength of rocks on grain size. *Bulletin of the Mineral Industries Experiment Station, Mining Engineering Series, Rock Mechanics*, 76, 99–103.

Brace, W. F. and Kohlstedt, D. L. (1980). Limits on lithospheric stress imposed by laboratory experiments. *Journal of Geophysical Research*, 85, 6248–52.

Brace, W. F., Paulding, B. W., and Scholz, C. (1966). Dilatancy in the fracture of crystalline rocks. *Journal of Geophysical Research*, 71, 3939–53.

Bradley, W. H. and Eugster, H. P. (1969). Geochemistry and paleolimnology of the trona deposits and associated authigenic minerals of the Green River Formation of Wyoming. *USGS Professional Paper*, 496, 1–86.

Brandt, J. L. (1965). *Stratigraphic clay-mineral distribution in the Cretaceous Colorado Group near Saskatoon (Saskatchewan)*. Master's thesis, University of Saskatchewan, Saskatoon.

Brantley, S. and Glicken, H. (1986). Volcanic debris avalanches. *Earthquakes and Volcanoes, U.S. Geological Survey Report*, 18, 195–206.

Braun, J. and Beaumont, C. (1987). Styles of continental rifting: results from dynamic models of lithosphere extension. In *Sedimentary Basins and Basin-Forming Mechanisms*, ed. C. Beaumont and A. J. Tankard. *Canadian Society of Petroleum Geologists*, 12, 241–58.

(1989). A physical explanation of the relation between flank uplifts and the breakup unconformity at rifted margins. *Geology*, 17, 760–4.

Brassell, S. C., Sheng, G., Fu, J., and Eglinton, G. (1988). Biological markers in lacustrine Chinese oil shales. In *Advances in Organic Geochemistry 1985*, ed. D. Leythaeuser and J. Rulkotter. Pergamon Press, Oxford, pp. 927–41.

Bredehoeft, J. D., Djevanshir, R. D., and Belitz, K. R. (1988). Lateral fluid flow in a compacting sand-shale sequence: South Caspian Basin. *American Association of Petroleum Geologists Bulletin*, 72, 416–24.

Bredehoeft, J. D. and Ingebritsen, S. E. (1990). Degassing of carbon dioxide as a possible source of high pore pressure in the crust. In *The Role of Fluids in Crustal Processes*, ed. J. D. Bredehoeft and D. L. Norton. National Academy Press, Washington, DC, pp. 158–64.

Bredehoeft, J. D. and Papadopulos, I. S. (1965). Rates of vertical groundwater movement estimated from the Earth's thermal profile. *Water Resources Research*, 1, 325–8.

Brekke, H. (2000). The tectonic evolution of the Norwegian Sea Continental Margin with emphasis on the Vøring and Møre Basins. In *Dynamics of Norwegian Margin*, ed. A. Nottvedt, B. T. Larsen, S. Olaussen, B. Tørudbakken, J. Skogseid, R. H. Gabnelson, H. Brekke, and O. Birkeland. *Geological Society of London Special Publications*, **167**, 327–8.

Brennand, T. P. and Van Veen, F. R. (1975). The Auk oil-field. In *Petroleum and the Continental Shelf of North-West Europe, v. 1, Geology*, ed. A. W. Woodland. Halstead Press, New York, pp. 275–83.

Briais, A., Patriat, P., and Tapponnier, P. (1993). Updated interpretation of magnetic anomalies and seafloor spreading stages in the South China Sea: implications for the Tertiary tectonics of Southeast Asia. *Journal of Geophysical Research*, 98, 6299–328.

Briais, A. and Rabinowicz, M. (2002). Temporal variations of the segmentation of slow to intermediate spreading mid-ocean ridges 1. Synoptic observations based on satellite altimetry data. *Journal of Geophysical Research*, 107, 3-1–3–17.

Bridge, J. S. (1999). Alluvial architecture of the Mississippi Valley: predictions using a 3D simulation model. In

Floodplains: Interdisciplinary Approaches, ed. S. B. Marriott and J. Alexander. *Geological Society Special Publications*, 163, 269–78.

(2003). *Rivers and floodplains: forms, processes, and sedimentary record.* Blackwell Publishing, Oxford, p. 491.

Bridge, J. S. and Tye, R. S. (2000). Interpreting the dimensions of ancient fluvial channel bars, channels, and channel belts from wireline-logs and cores. *American Association of Petroleum Geologists Bulletin*, 84, 1205–28.

Brigaud, F., Chapman, D. S., and Le Douaran, S. (1990). Estimating thermal conductivity in sedimentary basins using lithologic data and geophysical well logs. *American Association of Petroleum Geologists Bulletin*, 74, 1459–77.

Briggs, D. J. (1980). An investigation into variations in mineralogical and mechanical properties of coal measure strata. PhD thesis, University of Wales, Cardiff.

Briole, P., Rigo, A., Lyon-Caen, H., Ruegg, J. C., Papazissi, K., Mitsakaki, C., Balodimou, A., Veis, G., Hatzfeld, D., and Deschamps, A. (2000). Active deformation of the Corinth Rift, Greece: results from repeated global positioning system surveys between 1990 and 1995. *Journal of Geophysical Research*, 105, 25605–25.

Brocher, T. M. (2005). Empirical relations between elastic wavespeeds and density in the Earth's crust. *Bulletin of the Seismological Society of America*, 95, 2081–92.

Brodie, J. and White, N. (1994). Sedimentary basin inversion caused by igneous underplating: northwest European continental shelf. *Geology*, 22, 147–50.

(1995). The link between sedimentary basin inversion and igneous underplating. In *Basin Inversion*, ed. J. G. Buchanan and P. G. Buchanan. *Geological Society Special Publications*, 88, 21–38.

Bromann-Klausen, M. and Larsen, H. C. (2002). East-Greenland coast-parallel dike swarm and its role in continental break-up. In *Volcanic Rifted Margins*, ed. M. A. Menzies, S. L. Klemperer, C. J. Ebinger, and J. Baker. *GSA Special Papers*, 362, 133–58.

Brooke, C. M., Trimble, T. J., and Mackay, T. A. (1995). Mounded shallow gas sands from the Quaternary of the North Sea: analogues for the formation of sand mounds in deep water Tertiary sediments? *Geological Society London Special Publication*, 94, 95–101.

Brooks, C. K. and Nielsen, T. D. F. (1982). The E Greenland continental margin: a transition between oceanic and continental magmatism. *Journal of the Geological Society of London*, 136, 265–75.

Brooks, M., Hillier, B. V., and Miliorizos, M. (1993). New seismic evidence for a major geological boundary at shallow depth, N Devon. *Journal of the Geological Society of London*, 150, 131–5.

Brooks, M., Trayner, P. M., and Trimble, T. (1988). Mesozoic reactivation of Variscan thrusting in the Bristol Channel area, UK. *Journal of the Geological Society of London*, 145, 439–44.

Brophy, P. (2003). Geothermal exploration techniques. *Short course, GRC Annual Meeting 2003*, Morelia.

Brosse, E. and Huc, A. Y. (1986). Organic parameters as indicators of thermal evolution in the Viking Graben. In *Thermal Modelling in Sedimentary Basins*, ed. J. Burrus. Technip, Paris, pp. 173–95.

Brott, C. A., Blackwell, D. D., and Ziagos, J. P. (1981). Thermal and tectonic implications of heat flow in the eastern Snake River, Idaho. *Journal of Geophysical Research*, 86, 11709–34.

Brown, D. W. (1986). The Bay of Fundy: thin-skinned tectonics and resultant early Mesozoic sedimentation. *Basins of Eastern Canada and Worldwide Analogues, Atlantic Geoscience Society Programme, Abstracts*, Halifax, 26.

Brown, K. M. and Moore, J. C., 1993. Comment on "Anisotropic permeability and tortuosity in deformed wet sediments" by J. Arch and A. J. Maltman. *Journal of Geophysical Research*, 98, 17859–64.

Brown, R. W. (1991). Backstacking apatite fission-track "stratigraphy": a method for resolving the erosional and isostatic rebound components of tectonic uplift histories. *Geology*, 19, 74–7.

Brown, R. W., Gallagher, K., Gleadow, A. J. W., and Summerfield, M. A. (2000). Morphotectonic evolution of the South Atlantic margins of Africa and South America. In *Geomorphology and Global Tectonics*, ed. M. A. Summerfield. John Wiley and Sons, Chichester, UK, pp. 255–81.

Brown, R. W., Rust, D. J., Summerfield, M. A., Gleadow, A. J. W., and de Wit, M. C. J. (1990). An Early Cretaceous phase of accelerated erosion on the southwestern margin of Africa: evidence from apatite fission track analysis and the offshore sedimentary record. *Nuclear Tracks and Radiation Measurements*, 17, 339–50.

Brown, S. (1990). Jurassic. In *Introduction to the Petroleum Geology of the North Sea*, ed. K. W. Glennie. Blackwell Scientific Publications, Oxford, pp. 219–49.

Brownfield, M. E. and Charpentier, R. R. (2003). Assessment of the undiscovered oil and gas of the Senegal Province, Mauritania, Senegal, Gambia, and Guinea-Bissau, Northwest Africa. *U.S. Geological Survey Bulletin*, 2207, 1–26.

Brozena, J. M. and White, R. S. (1990). Ridge jumps and propagations in the South Atlantic Ocean. *Nature*, 348, 149–52.

Bruce, C. H. (1984). Smectite dehydration – its relation to structural development and hydrocarbon accumulation in northern Gulf of Mexico Basin. *American Association of Petroleum Geologists Bulletin*, 68, 673–83.

Bruhn, R. L., Yonkee, W., and Parry, W. (1990). Structural and fluid-chemical properties of seismogenic normal faults. *Tectonophysics*, 175, 139–57.

Brun, J. P. (1999). Narrow rifts versus wide rifts: inferences for the mechanics of rifting from laboratory experiments. *Philosophical Transactions – Royal Society, Mathematical, Physical and Engineering Sciences*, 357, 695–712.

Brun, J. P. and Beslier, M. O. (1996). Mantle exhumation at passive margins. *Earth and Planetary Science Letters*, 142, 161–73.

Brun, J. P. and Choukroune, P. (1983). Normal faulting, block tilting, and decollement in a stretched crust. *Tectonics*, 2, 345–56.

Brun, J. P., Gutscherr, M. A., and ECORS-DEKORP team (1992). Deep structure of the Rhine Graben from DEKORP-ECORS seismic reflection data, a summary. In *Geodynamics of Rifting, Volume I, Case History Studies on Rifts: Europe and Asia*, ed. P. A. Ziegler. *Tectonophysics*, 208, 139–47.

Brun, J. P., Sokoutis, D., and van den Driessche, J. (1994). Analogue modeling of detachment fault systems and core complexes. *Geology*, 22, 319–22.

Bruton, D. J. and Helgeson, H. C. (1983). Calculation of the chemical and thermodynamic consequences of differences between fluid and geostatic pressure in hydrothermal systems. *American Journal of Science*, 283, 540–88.

Bryant, B. (1990). *Geologic map of the Salt Lake City 30' x 60' quadrangle, north-central Utah, and Uinta County, Wyoming. U.S. Miscellaneous Investigations Series Map I-1944, scale 1:100,000.*

Buck, P. (1986). Sedimentology and micropalaeontology of gravity cores from the NE Atlantic Ocean (south west of Ireland). *Technical Report – Marine Geoscience Unit, Joint Geological Survey/University of Cape Town*, 16, 106–12.

Buck, W. R. (1985). When does small-scale convection begin beneath oceanic lithosphere? *Nature*, 313, 775–7.

(1986). Small scale convection induced by passive rifting: the cause for uplift of rift shoulders. *Earth Planetary Science Letters*, 77, 362–72.

(1991). Modes of continental lithospheric extension. *Journal of Geophysical Research*, 96, 20161–78.

(2004). Consequences of asthenospheric variability on continental rifting. In *Rheology and Deformation of the Lithosphere at Continental Margins*, ed. G. D. Karner, B. Taylor, N. W. Driscolland, and D. L. Kohlstedt. Columbia University Pages, pp. 1–31.

Buck, W. R., Lavier, L. L., and Poliakov, A. N. B. (1999). How to make a rift wide. *Mathematical, Physical and Engineering Sciences*, 357, 671–93.

(2005). Modes of faulting at mid-ocean ridges. *Nature*, 43, 719–23.

Buck, W. R., Martinez, F., Steckler, M. S., and Cochran, J. R. (1988). Thermal consequences of lithospheric extension: Pure and simple. *Tectonics*, 7, 213–34.

Buntebarth, G. (1984). *Geothermics*. Springer-Verlag, New York.

Buntebarth, G. and Rybach, L. (1981). Linear relationships between petrophysical properties and mineralogical constitution: preliminary results. *Tectonophysics*, 75, 41–6.

Burbank, D. W., Derry, L. A., and France-Lanord, C. (1993). Reduced Himalayan sediment production 8 Myr ago despite an intensified monsoon. *Nature*, 364, 48–50.

Burchfiel, B. C., Hodges, K. V., and Royden, L. H. (1987). Geology of Panamint Valley-Saline Valley pull-apart system, California: palinspastic evidence for low-angle geometry of a Neogene range-bounding fault. *Journal of Geophysical Research*, 92, 10422–6.

Burchfiel, B. C. and Stewart, J. H. (1966). "Pull-apart" origin of the central segment of Death Valley, California. *Geological Society of America Bulletin*, 77, 439–41.

Burford, R. O. and Harsh, P. W. (1980). Slip on the San Andreas Fault in central California from alignment array surveys. *Bulletin of the Seismological Society of America*, 70, 1233–61.

Burgess, C. F., Rosendahl, B. R., Sander, S., Burgess, C. A., Lambiase, J., Derksen, S., and Meader, N. (1988). The structural and stratigraphic evolution of Lake Tanganyika: a case study of continental rifting. In *Triassic-Jurassic Rifting: Continental Breakup and the Origin of the Atlantic and Passive Margins*, ed. W. Manspeizer. Elsevier, Amsterdam, pp. 859–81.

Burgess, P. M. and Hovius, N. (1998). Rates of delta progradation during highstands: consequences for timing of deposition in deep-marine systems. *Journal of the Geological Society of London*, 155, 217–22.

Burke, K. (1972). Longshore drift, submarine canyons and submarine fans in development of Niger delta. *American Association of Petroleum Geologists Bulletin*, 56, 1975–83.

Burley, S. D. (1984). Distribution and origin of authigenic minerals in the Triassic Sherwood Sandstone Group, UK. *Clay Minerals*, 19, 403–40.

Burley, S. D., Mullis, J., and Matter, A. (1989). Timing of diagenesis in the Tartan Reservoir (UK North Sea): constraints from combined cathodoluminescence microscopy and fluid inclusion studies. *Marine and Petroleum Geology*, 6, 98–120.

Burnham, A. K., Schmidt, B. J., and Braun, B. L. (1995). A test of the parallel reaction model using kinetic measurements on hydrous pyrolysis residues. *Organic Geochemistry*, 23, 931–9.

Burns, B. A., Heller, P. L., Marzo, M., and Paola, C. (1997). Fluvial response in a sequence stratigraphic framework: example from the Montserrat fan delta, Spain. *Journal of Sedimentary Research*, 67, 311–21.

Burov, E. and Cloetingh, S. (1997). Erosion and rift dynamics: new thermomechanical aspects of post-rift evolution of extensional basins. *Earth and Planetary Science Letters*, 150, 7–26.

Burov, E. B. and Diament, M. (1995). The effective elastic thickness (Te) of continental lithosphere: What does it really mean? *Journal of Geophysical Research*, 100, 3905–28.

Burov, E. and Poliakov, A. (2001). Erosion and rheology controls on synrift and postrift evolution: verifying old and new ideas using a fully coupled numerical model. *Journal of Geophysical Research*, 106, 16461–81.

Burov, E. B. and Watts, A. B. (2006). The long-term strength of continental lithosphere: "jelly sandwich" or "crème brulée"? *GSA Today*, 16.

Burr, T. N. and Currey, D. R. (1988). The Stockton Bar. In *In the Footsteps of G. K. Gilbert – Lake Bonneville and Neotectonics of the Eastern Basin and Range Province*, ed. M. N. Machettte. Utah Geological and Mineral Survey Miscellaneous Publication, 88, 66–73.

Burr, T. N. and Currey, D. R. (1992). *The Stockton Bar* Miscellaneous Publication – Utah Geological Survey, Salt Lake City.

Burrus, J., Schneider, F., and Wolf, S. (1994). Modeling overpressures in sedimentary basins: consequences for permeability and rheology of shales, and petroleum expulsion efficiency. *American Association of Petroleum Geologists Bulletin*, 78, 1137.

Burruss, R. C., Cercone, K. R., and Harris, P. M. (1985). Timing of hydrocarbon migration: evidence from fluid inclusions in calcite cements, tectonics and burial history. In *Carbonate Cements Revisited*, ed. P. M. Harris and N. M. Schneidermann. *SEPM Special Publication*, pp. 277–89.

Burtner, R. L., Nigrini, A., and Donelick, R. A. (1994). Thermochronology of Lower Cretaceous source rocks in the Idaho-Wyoming thrust belt. *American Association of Petroleum Geologists Bulletin*, 78, 1613–36.

Burton, K. W., Schiano, P., Birck, J. L., Allegre, C. J., Rehkamper, M., Halliday, A. N., and Dawson, J. B. (2000). The distribution and behavior of rhenium and osmium amongst mantle minerals and the age of the lithospheric mantle beneath Tanzania. *Earth and Planetary Science Letters*, 183, 93–106.

Burwood, R., De Witte, S. M., Mycke, B., and Paulet, J. (1995). Petroleum geochemical characterization of the Lower Congo coastal basin Bucomazi Formation. In *Petroleum Source Rocks*, ed. B. J. Katz. Springer-Verlag, Berlin, pp. 235–64.

Burwood, R., Mycke, B., Paulet, J., Jacobs, L., and Hall, D. (1997). The Central Graben Upper Jurassic source rock system: a sequence stratigraphic approach to oil provenance. In *Organic Geochemistry: Developments and Applications to Energy, Climate, Environment and Human History*, ed. J. O. Grimalt and C. Dorronsoro. Selected papers from 17th International Meeting on Organic Geochemistry, San Sebastian, Spain, 4–8 September 1995, pp. 249–52.

Byerlee, J. D. (1968). Brittle-ductile transition in rocks. *Journal of Geophysical Research*, 73, 4741–50.

Byerlee, J. (1978). Friction of rocks. *Pure and Applied Geophysics*, 116, 615–26.

Caine, J. S. (1999). *The Architecture and Permeability Structure of Brittle Fault Zones*. Salt Lake City: University of Utah.

Caine, J. S., Evans, J. P., and Forster, C. B. (1996). Fault zone architecture and permeability structure. *Geology*, 24, 1025–8.

Cainelli, C. and Mohriak, W. U. (1999). Some remarks on the evolution of sedimentary basins along the eastern Brazilian continental margin. *Episodes*, 22, 206–16.

Callot, J.-P. (2002). *Structure et developpement des marges volcaniques: l'examplе du Groenland*. PhD thesis, University of Paris, p. 584.

Callot, J.-P., Geoffroy, L., Aubourg, C., Pozzi, J.-P., and Mege, D. (2001). Magma flow directions of shallow dykes from the East-Greenland volcanic margin inferred from magnetic fabric studies. *Tectonophysics*, 335, 313–29.

Callot, J.-P., Geoffroy, L., and Brun, J. P. (2002). 3D analogue modelling of volcanic passive margins. *Tectonics*, 21.

Camelbeeck, T. and Iranga, M. D. (1996). Deep crustal earthquakes and active faults along the Rukwa Trough, eastern Africa. *Geophysical Journal International*, 124, 612–30.

Camp, W. K. (2000). Geologic model and reservoir description of the deepwater "p sand" at subsalt Mahogany Field, Gulf of Mexico. In *Volume Integration of Geologic Models for Understanding Risk in the Gulf of Mexico*, ed. R. Shoup, J. Watkins, J. Karlo, and D. Hall. Papers from Hedberg Conference, 1998, Galveston, AAPG/ Databases Discovery Series 1 (CD-ROM).

Campanha, V. A. (1987). *Análises bioestratigráficas do poço 2-AP-1-CE, ITP report, Relatório 24769*. Instituto De Pesquisas Técnicas, São Paulo, p. 33.

Campbell, I. H. and Griffiths, R. W. (1990). Implications of mantle plume structure for the evolution of flood basalts. *Earth and Planetary Science Letters*, 99, 79–93.

Canales, J. P., Detrick, R. S., Lin, J., Collins, J. A., and Toomey, D. R. (2000). Crustal and upper mantle seismic structure beneath the rift mountains and across a nontransform offset at the Mid-Atlantic Ridge. *Journal of Geophysical Research*, 105, 2699–720.

Cande, S. C., LaBrecque, J. L., and Haxby, W. F. (1988). Plate kinematics of the South Atlantic: Chron C34 to present. *Journal of Geophysical Research*, 93, 13479–92.

Cande, S. C. and Rabinowicz, P. D. (1979). *Magnetic Anomalies of the Continental Margin of Brazil*. American Association of Petroleum Geologists, Tulsa, Oklahoma.

Cann, J. R., Blackman, D. K., Smith, D. K., McAllister, E., Janssen, B., Mello, S., Avgerinos, E., Pascoe, A. R., and Escartín, J. (1997). Corrugated slip surfaces formed at ridge-transform intersections on the Mid-Atlantic Ridge. *Nature*, 385, 329–32.

Cannat, M., Rommevaux-Jestin, C., Sauter, D., Deplus, C. and Mendel, V. (1999). Formation of the axial relief at the very slow spreading Southwest Indian Ridge (49 degrees to 69 degrees E). *Journal of Geophysical Research*, 104.B10, 22825–22843.

Cao, J., Yao, S. P., Jin, Z. J., Hu, W. X., Zhang, Y. J., Wang, X. L., Zhang, Y. Q., and Tang, Y. (2006). Petroleum migration and mixing in NW Junggar Basin (NW China): constraints from oil-bearing fluid inclusion analyses. *Organic Geochemistry*, 37, 827–46.

Cao, J., Zhang, Y. J., Hu, W. X., Yao, S. P., Wang, X. L., Zhang, Y. Q., and Tang, Y. (2005). The Permian hybrid petroleum system in the northwest margin of the Junggar Basin. *Marine and Petroleum Geology*, 22, 331–49.

Cappetti, G., Celati, R., Cigni, U., Squarci, P., Stefani, G. C., and Taffi, L. (1985). Development of deep exploration in the geothermal areas of Tuscany, Italy. *International Symposium on Geothermal Energy*, Kailua-Kona, Hawaii, International Volume, pp. 303–9.

Capuano, R. M. (1990). Hydrochemical constraints on fluid-mineral equilibria during compaction diagenesis of kerogen-rich geopressured sediments. *Geochimica et Cosmochimica Acta*, 54, 1283–99.

(1993). Evidence of fluid flow in microfractures in geopressured shales. *American Association of Petroleum Geologists Bulletin*, 77, 1303–14.

Carey, S. W. (1958). The tectonic approach to continental drift. In *Continental Drift – A Symposium*, ed. S. W. Carey. University of Tasmania, Hobart, pp. 177–363.

Carlson, W. D., Donelick, R. A., and Ketcham, R. A. (1999). Variability of apatite fission-track annealing kinetics: I. Experimental results. *American Mineralogist*, 84, 1213–23.

Carlsson, A. and Olsson, T. (1979). Hydraulic conductivity and its stress dependence. *Proceedings: Workshop on Low-Flow, Low Permeability Measurements in Largely Impermeable Rocks*, Paris, pp. 249–59.

Carroll, A. R. and Bohacs, K. M. (1995). A stratigraphic classification of lake types and hydrocarbon source potential: balancing climatic and tectonic controls. *First International Limno-Geological Congress, Abstract volume*, pp. 18–19.

(1999). Stratigraphic classification of ancient lakes: balancing tectonic and climatic controls. *Geology*, 27, 99–102.

(2001). Lake-type controls on petroleum source rock potential in nonmarine basins. *American Association of Petroleum Geologists Bulletin*, 85, 1033–53.

Carroll, A. R., Brassell, S. C., and Graham, S. A. (1992). Upper Permian lacustrine oil shales, southern Junggar basin, northwest China. *American Association of Petroleum Geologists Bulletin*, 76, 1874–902.

Carruthers, A., McKie, T., Price, J., Dyer, R., Williams, G., and Watson, P. (1996). The application of sequence stratigraphy to the understanding of Late Jurassic turbidite plays in the Central north Sea, UKCS. In *Geology of the Humber Group: Central Graben and Moray Firth, UKCS*, ed. A. Hurst, H. D. Johnson, S. D. Burley, A. C. Canham, and D. S. Mackertich. *Special Publication of Geological Society of London*, 114, 29–45.

Carslaw, H. S. and Jaeger, J. C. (1959). *Conduction of Heat in Solids*. Oxford University Press, New York.

Carter, N. L. (1976). Steady-state flow in rocks. *Reviews of Geophysics and Space Physics*, 14, 301–60.

Carter, N. L. and Tsenn, M. C. (1987). Flow properties of continental lithosphere. *Tectonophysics*, 136, 27–63.

Cartwright, J. A. and Mansfield, C. S. (1998). Lateral displacement variation and lateral tip geometry of normal faults in the Canyonlands National Park, Utah. *Journal of Structural Geology*, 20, 3–19.

Cartwright, J. A., Mansfield, C. S., and Trudgill, B. D. (1996). The growth of normal faults by segment linkage. In *Modern Developments in Structural Interpretation, Validation, and Modelling*, ed. P. G. Buchanan and D. A. Nieuwland. *Geological Society Special Publications*, pp. 163–77.

Cartwright, J. A., Trudgill, B. D., and Mansfield, C. S. (1995). Fault growth by segment linkage: an explanation for scatter in maximum displacement and trace length data from the Canyonlands Grabens of SE Utah. *Journal of Structural Geology*, 17, 1319–26.

Casey, B. J., Romani, R. S., and Schmitt, R. H. (1993). Appraisal geology of the Saltire field, Witch Ground Graben, North Sea. In *Petroleum Geology of Northwest Europe*, ed. J. R. Parker. Proceedings of the 4th Conference, Geological Society of London, pp. 507–17.

Cassignol, C., Cornette, Y., David, B., and Gillot, P.-Y. (1978). Technologie potassium-argon. *Rapport CEA R-4802, C.E.N. Saclay Publication*, p. 37.

Catalan, L., Xiaowen, F., Chatzis, I., and Dullien, F. A. I. (1992). An experimental study of secondary oil migration. *American Association of Petroleum Geologists Bulletin*, 76, 638–50.

Cathles, L. M., Erendi, A. H. J., and Barrie, T. (1997). How long can a hydrothermal system be sustained by a single intrusive event? *Economic Geology*, 92, 766–71.

Catuneanu, O. (2002). Sequence stratigraphy of clastic systems: concepts, merits and pitfalls. *Journal of African Earth Sciences*, 35, 1–43.

Cayley, G. T. (1987). Hydrocarbon migration in the central North Sea. In *Petroleum Geology of North West Europe*, ed. J. Brooks and K. W. Glennie. Graham and Trotman, London, pp. 549–55.

Čermák, V. (1994). Results of heat flow studies in Czechoslovakia. In *Crustal Structure of the Bohemian Massif and the West Carpathians*, ed. V. Bucha and M. Blížkovský. Academia Praha, Springer-Verlag, pp. 85–120.

Čermák, V. and Bodri, L. (1992). Crustal thinning during rifting: a possible signature in radiogenic heat production. *Tectonophysics*, 209, 227–39.

Čermák, V., Bodri, L., and Rybach, L. (1991). Radioactive heat production in the continental crust and its depth dependence. In *Terrestrial Heat Flow and the Lithosphere Structure*, ed. V. Čermák and L. Rybach. Springer, Berlin, pp. 23–69.

Čermák, V. and Haenel, R. (1988). Geothermal maps. In *Handbook of Terrestrial Heat-Flow Density Determination: With Guidelines and Recommendations of the International Heat Flow Commission*, ed. R. Haenel, L. Rybach, and L. Stegena. Kluwer Academic Publishers, Dordrecht, pp. 261–300.

Čermák, V. and Hurtig, E. (1977). *Preliminary heat flow map of Europe, 1:5,000,000; explanatory text.* Potsdam: IASPEI, International Heat Flow Communication.

Čermák, V. and Rybach, L. (1982). *Thermal Conductivity and Specific Heat of Minerals and Rocks. Landolt-Bornstein – Group V Geophysics*, Springer-Verlag, pp. 305–43.

CGMW (1985–1990). *International Geological Map of Africa*. Commission for the Geological Map of the World and UNESCO.

CGMW (2000). *Geologic Map of South America*. Brazil Ministry of Mines and Energy, Commission for the Geological Map of the World.

Chan, M. A., Parry, W. T., and Bowman, J. R. (2000). Diagenetic hematite and manganese oxides and fault-related fluid flow in Jurassic sandstones southeastern Utah. *American Association of Petroleum Geologists Bulletin*, 84, 1281–310.

Chang, H. K., Kowsmann, R. O., and Figueiredo, A. M. F. (1988). New concepts on the development of East Brazilian marginal basins. *Episodes*, 11, 194–202.

Chang, H. K., Kowsmann, R. O., Figueiredo, A. M. F., and Bender, A. A. (1992). Tectonics and stratigraphy of the East Brazil Rift System: an overview. *Tectonophysics*, 213, 97–138.

Chapman, D. S. and Rybach, L. (1985). Heat flow anomalies and their interpretation. *Journal of Geodynamics*, 4, 3–37.

Chapman, D. S., Willett, S. D., and Clauser, C. (1991). Using thermal fields to estimate basin-scale permeabilities. In *Petroleum Migration*, ed. W. A. England and A. J. Fleet. *Geological Society of London Special Publications*, **59**, 123–5.

Chapman, T. J. and Meneilly, A. W. (1990). Fault displacement analysis in seismic exploration. *First Break*, 8, 11–12.

Charlou, J.-L., Bougault, H., Aprioui, P., Nielsen, T., and Rona, P. (1991). Different TDM/CH4 hydrothermal plume signatures: TAG site at 26o N and serpentinized ultrabasic diaper at 15o 05'N on the Mid-Atlantic Ridge. *Geochimica et Cosmochimica Acta*, 55, 3209–22.

Charlou, J.-L. and Donval, J.-P. (1993). Hydrothermal methane venting between 12 degrees N and 26 degrees N along the Mid-Atlantic Ridge. *Journal of Geophysical Research*, 98, 9625–42.

Chekunov, A. V. (1989). Dynamic model of the geotraverse through the Dnieper-Donets paleorift, Ukrainian Shield, southern Carpathians. *Geotectonics*, 23, 467–75.

Chen, W.-P. and Molnar, P. (1983). Focal depths of intracontinental and intraplate earthquakes and their implications for the thermal and mechanical properties of the lithosphere. *Journal of Geophysical Research*, 88, 4183–214.

Chen, Z., Zhou, G., and Alexander, R. (1994). A biomarker study of immature crude oils from the Shengli oilfield, People's Republic of China. *Chemical Geology*, 113, 117–32.

Cherry, S. T. J. (1993). The interaction of structure and sedimentary processes controlling deposition of the Upper Jurassic Brae Formation conglomerate, Block 16/17, North Sea. In *Petroleum Geology of Northwest Europe*, ed. J. R. Parker. *Proceedings of the 4th Conference, Geological Society of London*, pp. 387–400.

Chery, J., Lucazeau, M., Daignieres, M., and Vilotte, J. P. (1992). Large uplift of rift flanks: A genetic link with lithospheric rigidity? *Earth Science and Planetary Letters*, 112, 195–211.

Chery, J., Zoback, M. D., and Hassani, R. (2001). An integrated mechanical model of the San Andreas Fault in Central and Northern California. *Journal of Geophysical Research*, 106, 22051–66.

Chesley, J. T., Rudnick, R. L., and Lee, C. T. (1999). Re-Os systematic of mantle xenoliths from the East African Rift: age, structure, and history of the Tanzanian craton. *Geochimica et Cosmochimica Acta*, 63, 1203–17.

Chester, F. M. and Logan, J. M. (1986). Composite planar fabric of gouge from the Punchbowl Fault, California. *Journal of Structural Geology*, 9, 621–34.

Chian, D., Louden, K. E., Minshull, T. A., and Whitmarsh, R. B. (1999). Deep structure of the 600 ocean–continent transition in the southern Iberia Abyssal Plain from seismic refraction 601 profiles: 1. Ocean Drilling Program (Legs 149 and 173) transect. *Journal of Geophysical Research*, 104, 7443–62.

Chian, D., Louden, K. E., and Reid, I. (1995). Crustal structure of the Labrador Sea conjugate margin and implications for the formation of nonvolcanic continental margins. *Journal of Geophysical Research*, 100, 24239–53.

Childs, C., Watterson, J., and Walsh, J. J. (1995). Fault overlap zones within developing normal fault systems. *Journal of the Geological Society of London*, 152, 535–49.

Chilingarian, G. V. (1983). Compactional diagenesis. In *Sediment Diagenesis*, ed. A. Parker and B. W. Sellwood. Reidel Publishing Company, Dordrecht, pp. 57–168.

Chiozzi, P., Pasquale, M., and Verdoya, M. (2006). Seimicity and rheological modelling in an extensional-compressional tectonic realm. *Geophysical Research Abstracts*, 8, 02185.

Chorowicz, J. (1989). Transfer and transform fault zones in continental rifts: examples in the Afro-Arabian rift system, implications of crust breaking. *Journal of African Earth Sciences*, 8, 203–14.

Chorowicz, J. and Mukonki, M. N. B. (1980). Lineaments anciens, zones transformantes recentes et geotectonique des fosses de l'Est Africain, d'apres la teledetection et la microtectonique. Ancient lineaments, recent transform zones, and East African geotectonic trenches according to remote sensing and microtectonics. *Rapport Annuel – Musee Royal de l'Afrique Centrale*, Departement de Geologie et de Mineralogie, pp. 143–67.

Choubert, G. and Faure-Muret, A. (1985). *International Geological Map of Africa*. Commission for the Geological Map of the World and UNESCO, Paris.

Choubert, G. et al. (1988). International geological map of Africa. Commission for the Geological Map of the World and UNESCO, Paris.

Choudhuri, M., Guha, D., Dutta, A., Sinha, S., and Sinha, N. (2006). Spatio-temporal variations and kinematics of shale mobility in the Krishna-Godavari Basin, India. *Poster presentation at AAPG Hedberg Conference*, Port of Spain, Trinidad and Tobago, June 5–7, 2006.

Christensen, C., Nemčok, M., McCulloch, J., and Moore, J. (2002). The characteristics of productive zones in the Karaha–Telaga Bodas geothermal system. *Geothermal Resources Council Transactions*, 26, 623–6.

Christensen, N. I. (1989). Reflectivity and seismic properties of the deep continental crust. *Journal of Geophysical Research*, 94, 17793–804.

Christensen, N. I. and Mooney, W. D. (1995). Seismic velocity structure and composition of the continental crust. *Journal of Geophysical Research*, 100, 9761–88.

Christiansen, L. B. and Garven, G. (2003). A theoretical comparison of buoyancy-driven and compaction-driven fluid flow in oceanic sedimentary basins. *Journal of Geophysical Research*, 108.

(2004a). Transient hydrogeologic models for submarine flow in volcanic seamounts: 1. The Hawaiian Islands. *Journal of Geophysical Research*, 109.

(2004b). Transient hydrogeologic models for submarine flow in volcanic seamounts: 2. Comparison of the Hawaiian, Canary, and Marquesas Islands. *Journal of Geophysical Research*, 109.

Christie-Blick, N. and Biddle, K. T. (1985). Deformation and basin formation along strike-slip faults. *Special Publication – Society of Economic Paleontologists and Mineralogists*, 37, 1–34.

Chung, H. M., Wingert, W. S., and Claypool, G. E. (1992). Geochemistry of oil from the northern Viking Graben. In *Giant Oil and Gas Fields of the Decade 1978–1988*, ed. M. T. Halbouty. AAPG Memoir, 54, pp. 277–86.

Chung, W.-Y. and Kanamori, H. (1980). Variation of seismic source parameters and stress drops within a descending slab and its implications in plate mechanics. *Physics of the Earth and Planetary Interiors*, 23, 134–59.

Clark, J. A., Stewart, S. A., and Cartwright, J. A. (1998). Evolution of the NW margin of the North Permian Basin, UK North Sea. *Journal of the Geological Society of London*, 155, 663–76.

Clark, Jr., S. P. (1966). *Handbook of Physical Constants*. Geological Society of America Memoir, New York, 97.

Clarke, B. J. (2002). *Early Cenozoic Denudation of the British Isles: A Quantitative Stratigraphic Approach*. University of Cambridge, Cambridge.

Clausen, S., Nemčok, M., Moore, J., Hulen, J., and Bartley, J. (2006). Mapping fractures in the Medicine Lake geothermal system. *Geothermal Resources Council Transactions*, 30, 383–6.

Clauser, C. and Villinger, H. (1990). Analysis of conductive and convective heat transfer in a sedimentary basin, demonstrated for the Rheingraben. *Geophysical Journal International*, 100, 393–414.

Clayton, J. L., Yang, J., King, J. D., Lillis, P. G., and Warden, A. (1997). Geochemistry of oils from the Junggar basin, northwest China. *American Association of Petroleum Geologists Bulletin*, 81, 1926–44.

Clem, K. (1985). Oil and gas production summary of the Uinta basin. In *Geology and Energy Resources, Uinta Basin of Utah*, ed. M. D. Picard. Utah Geological Association, Salt Lake City, pp. 159–67.

Clemetz, D. M., Demaison, G. J., and Daly, A. R. (1979). Well site geochemistry by programmed pyrolysis. *Proceedings of the 11th Annual Offshore Technology Conference*, pp. 465–70.

Clemson, J., Cartwright, J., and Swart, J. (1999). The Namib Rift: a rift system of possible Karoo age, offshore Namibia. In *The Oil and Gas Habitats of the South Atlantic: Geological Society*, ed. N. R. Cameron, R. H. Bate, and V. S. Clure. [London] Special Publication 153, pp. 381–402.

Clift, P. D. (1997). The thermal impact of Paleocene magmatic underplating in the Faeroe-Shetland-Rockall region. In *Petroleum Geology of Northwest Europe*, ed. A. J. Fleet and S. A. R. Boldy. *Proceedings of the 5th Conference*, London, UK, pp. 585–93.

Clift, P. D. and Lorenzo, M. (1999). Flexural unloading and uplift along the Cote d'Ivoire Ghana transform margin, Equatorial Atlantic. *Journal of Geophysical Research*, 104, 25257–74.

Clift, P. D. and Turner, J. (1998). Paleogene igneous underplating and subsidence anomalies in the Rockall-Faeroe-Shetland area. *Marine and Petroleum Geology*, 15, 223–43.

Cloetingh, S. (1986). Intraplate stresses: a new tectonic mechanism for fluctuations of relative sea level. *Geology*, 14, 617–20.

Cloetingh, S. and Banda, E. (1992). Europe's lithosphere: physical properties, mechanical structure. In *A Continent Revealed – the European Geotraverse*, ed. D. Blundell, R. Freeman, and S. Mueller. Cambridge University Press, Cambridge, pp. 80–91.

Cloetingh, S., Kooi, H., and Groenewoud, W. (1989). Intraplate stresses and sedimentary basin evolution. *Geophysical Monograph*, 48, 1–16.

Cloetingh, S., Tankard, A. J., Welsink, H. J., and Jenkins, W. A. M. (1989). Vail's coastal onlap curves and their correlation with tectonic events, offshore eastern Canada. In *Extensional Tectonics and Stratigraphy of the North Atlantic Margins*, ed. A. J. Tankard and H. R. Balkwill. *American Association of Petroleum Geologists Memoir*, 46, 282–93.

Cloetingh, S., Zoetemeijer, R., and van Wees, J. D. (1995). Tectonics I. Tectonics and basin formation in convergent settings: thermo-mechanical evolution of the lithosphere and basin evolution in compressive tectonic regimes. *Short Course*, Vrije University, Amsterdam.

Coakley, B. J. and Watts, A. B. (1991). Tectonic controls on the development of unconformities: the North Slope, Alaska. *Tectonics*, 10, 101–30.

Cobbold, P. R., Meisling, K. E., and Mount, V. S. (2001). Reactivation of an obliquely rifted margin, Campos and Santos Basins, southeastern Brazil. *American Association of Petroleum Geologists Bulletin*, 85, 1925–44.

Cobbold, P. R., Mourgues, R., and Boyd, K. (2004). Mechanism of thin-skinned detachment in the Amazon fan: assessing the importance of fluid overpressure and hydrocarbon generation. *Marine and Petroleum Geology*, 21, 1013–25.

Coblentz, D. D., Richardson, R. M., and Sandiford, M. (1994). On the gravitational potential of the Earth's lithosphere. *Tectonics*, 13, 929–45.

Cochran, E. S., Li, Y.-G., and Vidale, J. E. (2006). Anisotropy in the shallow crust observed around the San Andreas

Fault before and after the 2004 M 6.0 Parkfield earthquake. *Bulletin of the Seismological Society of America*, 96, 364–75.

Cochran, J. R. (1973). Gravity and magnetic investigations in the Guiana Basin, Western Equatorial Atlantic. *Geological Society of America Bulletin*, 84, 3249–68.

(1982). The magnetic quiet zone in the eastern Gulf of Aden: implications for the early development of the continental margin. *Geophysical Journal of the Royal Astronomical Society*, 68, 171–201.

(1983). Effects of finite extension times on the development of sedimentary basins. *Earth and Planetary Science Letters*, 66, 289–302.

(2005). Northern Red Sea: nucleation of an oceanic spreading center within a continental rift. *Geochemistry, Geophysics, Geosystems*, 6.

Cochran, J. R. and Karner, G. D. (2007). Constraints on the deformation and rupturing of continental lithosphere of the Red Sea: the transition from rifting to drifting. In *Imaging, Mapping and Modelling Continental Lithosphere Extension and Breakup*, ed. G. D. Karner, G. Manatschal, and L. M. Pinheiro. *Special Publication of the Geological Society of London*, 282, 265–89.

Cochran, J. R. and Martinez, F. (1988). Evidence from the northern Red Sea on the transition from continental to oceanic rifting. *Tectonophysics*, 153, 25–53.

Cockings, J. H., Kessler, L. G., Mazza, T. A., and Riley, L. A. (1992). Bathonian to mid-Oxfordian sequence stratigraphy of the South Viking Graben, North Sea. In *Exploration Britain: Geological Insights into the Next Decade*, ed. R. F. P. Hardman. *Special Publication of the Geological Society of London*, 67, 65–105.

Cogne, N., Gallagher, K., and Cobbold, P. R. (2011). Post-rift reactivation of the onshore margin of southwest Brazil: evidence from apatite (U-Th)/He and fission-track data. *Earth and Planetary Science Letters*, 309, 118–30.

Cohee, G. V., Applin, P. L., Bass, N. W., Bell, A. H., Billings, M. P., DeFord, R. K., Dobbin, C. E., Donnell, J. R., Forrester, J. D., Gilluly, J., Hoots, H. W., Kind, P. B., Longwell, C. R., Lyons, P. L., Murray, G. E., and Waters, A. C. (1962). *Tectonic Map of the United States Exclusive of Alaska and Hawaii*. AAPG and USGS, Tulsa, Oklahoma.

Cohen, H. A. and McClay, K. R. (1996). Niger Delta shale tectonics. *Marine and Petroleum Geology*, 13, 313–28.

Cole, G., Yu, A., Peel, F., Requejo, R., DeVay, J., Brooks, J., Bernard, B., Zumberge, J., and Brown, S. (2001). Constraining source and charge risk in deepwater areas. *World Oil Magazine*, 222, 69–77.

Colella, A. (1988a). Pliocene-Holocene fan deltas and braid deltas in the Crati Basin, southern Italy: a consequence of varying tectonic conditions. In *Fan Deltas*, ed. W. Nemec and R. J. Steel. Blackie, London, pp. 50–74.

(1988b). Marine fault-controlled Gilbert-type fan deltas. *American Association of Petroleum Geologists Bulletin*, 72, 995.

Colletta, B., Le Quellec, P., Letouzey, J., and Moretti, I. (1988). Longitudinal evolution of the Suez rift structure (Egypt). *Tectonophysics*, 153, 221–33.

Collier, J. S., Sansom, V., Ishizuka, O., Taylor, R. N., Minshull, T. A., and Whitmarsh, R. B. (2008). Age of Seychelles–India break-up. *Earth and Planetary Science Letters*, 272, 264–77.

Collister, J. W., Simmons, R. E., Lichtfouse, E., and Hayes, J. M. (1992). An isotopic biochemical study of the Green River oil shale. *Organic Geochemistry*, 19, 265–76.

Colwell, J. B., Symonds, P. A., and Crawford, A. J. (1994). The nature of the Walaby (Curvier) Plateau and other igneous provinces of the West Australian Margin. In *Geology of the Outer North West Shelf, Australia: Australian Geological Survey Organization*, ed. N. Exon, *AGSO Journal of Australian Geology and Geophysics*, 15(1), 137–56.

Combarnous, M. H. and Bories, S. A. (1975). Hydrothermal convection in saturated porous media. *Advances in Hydrosciences*, 10, 231–307.

Combellas-Biggot, R. I. and Galloway, W. E. (2002). Depositional history and genetic sequence stratigraphic framework of the middle Miocene depositional episode south Louisiana. *Annual Meeting Expanded Abstracts – American Association of Petroleum Geologists*, p. 33.

Conceicao, J. C. de J., Zalán, P. V., and Wolff, S. (1988). Mecanismo, evolucao e cronologia do rift Sul-Atlantico. Mechanism, evolution and chronology of South Atlantic rifting. *Boletim de Geociencias da PETROBRAS*, 2, 255–65.

Coney, D., Fyfe, T. B., Retail, P., and Smith, P. J. (1993). Clair appraisal: the benefits of a co-operative approach. In *Petroleum Geology of Northwest Europe: Proceedings of the 4th Conference*, ed. J. R. Parker. The Geological Society of London, pp. 1409–20.

Coney, P. J. (1980). Cordilleran metamorphic core complexes: an overview. *Geological Society of America Memoir*, 153, 7–31.

Connan, J. (1984). Biodegradation of crude oils in reservoirs. In *Advances in Petroleum Geochemistry, 1*, ed. J. Brooks and D. Welte. London, Academic Press, pp. 299–335.

(1993). Origin of severely biodegraded oils: a new approach using biomarker pattern of asphaltene pyrolysates. In *Applied Petroleum Geochemistry*, ed. M. L. Bordenave. Technip, Paris, pp. 457–63.

Connolly, D. L. and Aminzadeh, F. (2003). Exploring for deep gas in the Gulf of Mexico shelf and deepwater using gas chimney processing. *Gulf Coast Association of Geological Societies Transactions*, 53, 135–42.

(2006). Prospect risking using 3-D seismic derived gas chimney volumes. *Abstracts, Annual Meeting – American Association of Petroleum Geologists*, 15, 21.

Connolly, D. L., Aminzadeh, F., and Ligtenberg, H. (2004). Reducing seal and charge risk by detecting fluid migration pathways using seismic data. *Abstracts,*

Annual Meeting – American Association of Petroleum Geologists, 13, 27.

Connolly, D. L., Fraser, B., and Aminzadeh, F. (2007). Distinguishing static and dynamic faults: application for Gulf of Mexico deep gas play. *Abstracts, Annual Meeting – American Association of Petroleum Geologists*, 27.

Connolly, D. L., Sawyer, R. K., Aminzadeh, F., de Groot, P. F. M., and Ligtenberg, H. J. (2002). Gas chimney processing as a new exploration tool: a West Africa example. *Abstracts, Annual Meeting – American Association of Petroleum Geologists*, 34.

Connolly, J. A. D. and Podladchikov, Y. Y. (2000). Temperature-dependent viscoelastic compaction and compartmentalization in sedimentary basins. *Tectonophysics*, 324, 137–68.

Connolly, P. and Cosgrove, J. (1999). Prediction of fracture-induced permeability and fluid flow in the crust using experimental stress data. *American Association of Petroleum Geologists Bulletin*, 83, 757–77.

Constenius, K. N. (1996). Late Paleogene extensional collapse of the Cordilleran foreland fold and thrust belt. *Geological Society of America Bulletin*, 108, 20–39.

Contrucci, I., Matias, L., Moulin, M., Geli, L., Klingelhofer, F., Nouze, H., Aslanian, D., Olivet, J. L., Rehault, J. P., and Sibuet, J. C. (2004). Deep structure of the West African continental margin (Congo, Zaire, Angola), between 5u S and 8u S, from reflection/refraction seismics and gravity data. *Geophysical Journal International*, 158, 529–53.

Cook, D. R. (1987). The Goban Spur: exploration in a deep water frontier basin. In *Petroleum Geology of North West Europe*, ed. J. Brooks and K. W. Glennie. Graham and Trotman, London.

Coolbaugh, M., Blackwell, D., and Richards, M. (2005). Temperature gradient map of the Great Basin. Great Basin Center for Geothermal Energy, University of Nevada, Reno. www. unr.edu/Geothermal/.

Cooper, B. S. and Barnard, P. C. (1984). Source rocks and oils of the central and northern North Sea. In *Petroleum Geochemistry and Basin Evaluation*, ed. G. Demaison and R. J. Murris. *American Association of Petroleum Geologists Memoir*, 35, 303–14.

Cooper Jr. H. H. (1966). The equation of groundwater flow in fixed and deforming coordinates. *Journal of Geophysical Research*, 71, 4785–90.

Copestake, P., Sims, A., Crittenden, S., Hamar, G., Ineson, J., Rose, P., and Tringham, M. (2002). Chapter 12. Lower Cretaceous. In *The Millennium Atlas: Petroleum Geology of the Central and Northern North Sea*, ed. D. Evans, C. Graham, A. Armour, and P. Bathurst. Geological Society of London, pp. 441–88.

(2003). Chapter 12. Lower Cretaceous. In *The Millennium Atlas: Petroleum Geology of the Central and Northern North Sea*, ed. D. Evans, C. Graham, A. Armour, and P. Bathurst. Geological Society of London, pp. 191–211.

Corey, A. T. (1986). *Mechanics of Immiscible Fluids in Porous Media*. Water Resources Publications, Littleton.

Cornea, I., Rădulescu, F. L., and Sova, A. (1981). Deep seismic sounding in Romania. *Pure and Applied Geophysics*, 119, 1144–56.

Corner, B., Cartwright, J., and Swart, R. (2002). Volcanic passive margin of Namibia: a potential fields perspective. In *Volcanic Rift Margins*, ed. M. A. Menzies, S. L. Klemperer, C. Ebinger, and J. Baker. Geological Society of America Special Paper, 362, pp. 203–20.

Cornford, C. (1994). The Mandal-Ekofisk(!) petroleum system in the central graben of the North Sea. In *The Petroleum System – From Source to Trap*, ed. L. B. Magoon and W. G. Dow. *American Association of Petroleum Geologists Memoir*, 60, 537–71.

(1998). Source rock and hydrocarbons of the North Sea. In *Petroleum Geology of the North Sea, Basic Concepts and Recent Advances (fourth edition)*, ed. K. W. Glennie. Blackwell Scientific Publications, Oxford, pp. 376–462.

Cornford, C., Bray, R., and Kieft, R. (2005). Deepwater source rocks – accumulation and efficiencies. *The 4th HGS/PESGB International Conference on African E7P*, Houston Program, 4.

Cornford, C., Needham, C. E. J., and de Walque, L. (1986). Geochemical habitat of North Sea oils and gases. In *Habitat of Hydrocarbons on the Norwegian Continental Shelf*, ed. A. M. Spencer. Graham and Trotman, London, pp. 39–54.

Corti, G., Bonini, M., Conticelli, S., Innocenti, F., Manetti, P., and Sokoutis, D. (2003). Analogue modelling of continental extension: a review focused on the relations between the patterns of deformation and the presence of magma. *Earth-Science Reviews*, 63, 169–247.

Cosgrove, J. W. (2001). Hydraulic fracturing during the formation and deformation of a basin: a factor in the dewatering of low-permeability sediments. *American Association of Petroleum Geologists Bulletin*, 85, 737–48.

Coterill, K., Tari, G., Molnar, J., and Ashton, P. R. (2002). Comparison of depositional sequences and tectonic styles among the West African deepwater frontiers of western Ivory Coast, southern Equatorial Guinea, and northern Namibia. *The Leading Edge*, 1103–11.

Courtillot, V., Devaille, A., Besse, J., and Stock, J. (2003). Three distinct types of hotspots in the earth's mantle. *Earth and Planetary Science Letters*, 205, 295–308.

Coutts, S. D., Larsson, S. Y., and Rosman, R. (1996). Development of the slumped crestal area of the Brent reservoir, Brent Field: an integrated approach. *Petroleum Geoscience*, 2, 219–29.

Coward, R. N., Clark, N. M., and Pinnock, S. J. (1991). The Tartan field, Block 15/16, UK North Sea. In *United Kingdom Oil and Gas Fields, 25 Years Commemorative Volume*, ed. I. L. Abbotts. *Memoir of the Geological Society of London*, 14, 377–84.

Cowie, P. A. (1998). A healing-reloading feedback control on the growth rate of seismogenic faults. *Journal of Structural Geology*, 20, 1075–87.

Cowie, P. A., Gupta, S., and Dawers, N. H. (2000). Implications of fault array evolution for synrift depocentre development: insights from a numerical fault growth model. *Basin Research*, 12, 241–61.

Cowie, P. A. and Scholz, C. H. (1992). Physical explanation for the displacement-length relationship of faults using a post-yield fracture mechanics model. *Journal of Structural Geology*, 14, 1133–48.

Cowley, R. and O'Brien, G. W. (2000). Identification and interpretation of leaking hydrocarbons using seismic data: a comparative montage of examples from the major fields in Australia's North West Shelf and Gippsland Basin. *APPEA Journal*, 40, 121–50.

Cox, K. G. (1980). A model for flood basalt volcanism. *Journal of Petrology*, 21, 629–50.

Cox, S. J. D., Meredith, P. G., and Stuart, C. E. (1991). Microfracturing during brittle rock failure: a model for the Kaiser effect including sub-critical crack growth. *Seventh International Congress on Rock Mechanics*, pp. 703–7.

Coyle, B. J. and Zoback, M. D. (1988). In situ permeability and fluid pressure measurements at ~2km depth in the Cajon Pass research well. *Geophysical Research Letters*, 15, 1029–32.

Crampin, S. (1990). The scattering of shear waves in the crust. *Pure and Applied Geophysics*, 132, 67–91.

Craven, J. (2000). Petroleum system of the Ivorian/Tano Basin in the W Tano Contract area, offshore west Ghana. *Abstracts Volume, Petroleum Systems and Developing Technologies in African Exploration and Production, Geol. Soc./PESGB Conference*, May, 2000.

Creaney, S. and Allan, J. (1992). Petroleum systems in the foreland basin of western Canada. In *Foreland Basin and Fold Belts*, ed. R. W. Macqueen and D. A. Leckie. *American Association of Petroleum Geologists Memoir*, 55, 279–308.

Creaney, S. and Passey, Q. R. (1993). Recurring patterns of total organic carbon and source rock quality within a sequence stratigraphic framework. *American Association of Petroleum Geologists Bulletin*, 77, 386–401.

Crider, J. G. and Pollard, D. D. (1998). Fault linkage: three-dimensional mechanical interaction between echelon normal faults. *Journal of Geophysical Research*, 103, 24373–91.

Criss, R. E. and Hofmeister, A. M. (1991). Application of fluid dynamics principles in tilted permeable media to terrestrial hydrothermal systems. *Geophysical Research Letters*, 18, 199–202.

Crittenden, M. D., Coney, P. J., and Davis, G. H. (1980). Cordilleran metamorphic core complexes. *Geological Society of America Memoir*, 153, 490.

Crossley, R. (1984). Controls of sedimentation in the Malawi Rift Valley, central Africa. *Sedimentary Geology*, 40, 33–50.

Crowell, J. C. (1974a). Sedimentation along the San Andreas Fault, California. In *Modern and Ancient Geosynclinal Sedimentation*, ed. R. H. Dott and R. H. Shaver. *Society of Economic Sedimentologists and Mineralogists Special Publications*, 19, 292–303.

(1974b). Origin of late Cenozoic basins in southern California. *Society of Economic Sedimentologists and Mineralogists Special Publications*, 22, 190–204.

(1982). The tectonics of the Ridge Basin, southern California. In *Geologic History of the Ridge Basin*, ed. J. C. Crowell and M. H. Link. *Pacific Section of the Society of Economic Geologists, Paleontologists and Mineralogists*, pp. 25–41.

Crowell, J. C. and Link, M. H. (1982). *Geologic History of Ridge Basin, southern California*. Los Angeles, Pacific Section, Society of Economic Paleontologists and Mineralogists, p. 304.

Crowley, K. D., Cameron, M., and Schaefer, R. L. (1991). Experimental studies of annealing of etched fission tracks in fluorapatite. *Geochimica et Cosmochimica Acta*, 55, 1449–65.

Curiale, J. A. (1994). Correlation of oils and source rocks: a conceptual and historical perspective. In *The Petroleum System – From Source to Trap*, ed. L. B. Magoon and W. G. Dow. *American Association of Petroleum Geologists Memoir*, 60, 251–60.

Currey, D. R., Oviatt, C. G., and Czarnomski, J. E. (1984). Late Quaternary geology of Lake Bonneville and Lake Waring. In *Geology of Northwest Utah, Southern Idaho and Northeast Nevada*, ed. G. J. Kerns and R. L. Kerns Jr. *Utah Geological Association Publication*, 13, 227–37.

Currie, S. (1996). The development of the Ivanhoe, Rob Roy and Hamish Fields, Block 15/21a, UK North Sea. In *Geology of the Humber Group: Central Graben and Moray Firth, UKCS*, ed. A. Hurst, H. D. Johnson, S. D. Burley, A. C. Canham, and D. S. Mackertich. *Special Publication of Geological Society of London*, 114, 329–41.

Currie, S., Gowland, S., Taylor, A., and Woodward, M. (1999). The reservoir development of the Fife Field. In *Petroleum Geology of Northwest Europe*, ed. A. J. Fleet and S. A. Boldy. Proceedings of the 5th Conference, Geological Society, London, pp. 1135–46.

Curtis, C. D. (1983). The link between aluminum mobility and destruction of secondary porosity. *American Association of Petroleum Geologists Bulletin*, 67, 380–4.

Curtis, J. B., Kotarba, M. J., Lewan, M. D., and Wieclaw, D. (2004). Oil/source rock correlations in the Polish Flysch Carpathians and Mesozoic basement and organic facies of the Oligocene Menilite Shales: insights from hydrous pyrolysis experiments. *Organic Geochemistry*, 35, 1573–96.

d'Acremont, E., Leroy, S., and Burov, E. B. (2003). Numerical modelling of a mantle plume: the plume head-lithosphere interaction in the formation of an oceanic large igneous province. *Earth and Planetary Science Letters*, 206, 379–96.

D'Agostino, N., Giuliani, R., Mattone, M., and Bonci, L. (2001). Active crustal extension in the Central

Apennines (Italy) inferred from GPS measurements in the interval 1994–1999. *Geophysical Research Letters*, 28, 2121–4.

Dahl, J. E. P., Moldowan, J. M., Teerman, S. C., McCaffery, M. A., Sundararaman, P., and Stelting, C. E. (1994). Source rock quality determination from oil biomarkers I: A new geochemical technique. *American Association of Petroleum Geologists Bulletin*, 78, 1507–26.

Dahl, N. and Solli, T. (1993). The structural evolution of the Snorre Field and surrounding areas. In *Petroleum Geology of Northwest Europe*, ed. J. R. Parker. Proceedings of the 4th Conference, Geological Society of London, pp. 1159–66.

Dahlberg, E. C. (1995). *Applied Hydrodynamics in Petroleum Exploration*. Springer-Verlag, New York.

Dailly, P. (2000). Tectonic and stratigraphic development of the Rio Muni Basin, Equatorial Guinea: the role of transform zones in Atlantic Basin evolution. *Geophysical Monograph*, 115, 105–28.

Dailly, P., Lowry, P., Goh, K., and Monson, G. (2002). Exploration and development of Ceiba Field, Rio Muni Basin, southern Equatorial Guinea. *The Leading Edge*, 1140–6.

Daly, R. A., Manger, G. E., and Clark, S. P. Jr. (1966). Density of rocks. *Handbook of Physical Constants*, 4, 19–26.

Damtoft, K., Nielsen, L. H., Johannessen, P. N., Thomsen, E., and Andersen, P. R. (1992). Hydrocarbon plays of the Danish Central Trough. In *Generation, Accumulation and Production of Europe's Hydrocarbons I.*, ed. A. M. Spencer. *Special Publication of the European Association of Petroleum Geoscientists*, 2, 35–58.

Damuth, J. E. (1994). Neogene gravity tectonics and depositional processes on the deep Niger Delta continental margin. *Marine and Petroleum Geology*, 11, 320–46.

Darcy, J. (1856). *Les Fontaines Publiques de la Ville de Dijon*. Dalmont, Paris.

Dart, C. J., Collier, R. E. L., Gawthorpe, R. L., Keller, J. V. A., and Nichols, G. (1994). Sequence stratigraphy of (?) Pliocene-Quaternary synrift, Gilbert-type fan deltas, northern Peloponnesos, Greece. *Marine and Petroleum Geology*, 11, 545–60.

Davidson, G. R., Bassett, R. L., Hardin, E. L., and Thompson, D. L. (1998). Geochemical evidence of preferential flow of water through fractures in unsaturated tuff, Apache Leap, Arizona. *Applied Geochemistry*, 13, 185–95.

Davies, G. F. (1994). Thermomechanical erosion of the lithosphere by mantle plumes, *J. Geophys. Res.*, 99(B8), 15709–22.

Davies, P., Williams, A. T., and Bomboe, P. (1991). Numerical modelling of Lower Lias rock failures in the coastal cliffs of South Wales. In *Coastal Sediments '91: Proceedings of a Special Conference on Quantitative Approaches to Coastal Processes*, ed. N. C. Kraus, K. J. Gingerich, and D. L. Kriebel. American Society of Civil Engineers, New York, pp. 1599–612.

Davies, R., England, P., Parsons, B., Billiris, H., Paradissis, D., and Veis, G. (1997). Geodetic strain of Greece in the interval 1892–1992. *Journal of Geophysical Research*, 102, 24571–88.

Davies, S. J., Dawers, N. H., McLeod, A. E., and Underhill, J. R. (2000). The structural and sedimentological evolution of early synrift successions: the Middle Jurassic Tarbert Formation, North Sea. *Basin Research*, 12, 343–65.

Davis, G. A. (1980). *Problems of Intraplate Extensional Tectonics, Western United States*. National Academy of Sciences, Washington, DC.

Davis, G. A., Anderson, J. L., Frost, E. G., and Shackelford, T. J. (1980). Mylonitization and detachment faulting in the Whipple-Buckskin-Rawhide Mountains terrane, southeastern California and western Arizona. *Geological Society of America*, 153, p. 79.

Davis, G. H. (1983). Shear-zone model for the origin of metamorphic core complexes. *Geology*, 11, 342–7.

Davis, M. and Kusznir, N. (2002). Are buoyancy forces important during the formation of rifted margins? *Geophysical Journal International*, 149, 524–33.

(2004). Depth-dependent lithospheric stretching at rifted continental margins. In *Rheology and Deformation of the Lithosphere at Continental Margins MARGINS Theoretical and Experimental Earth Science Series*, ed. G. D. Karner, B. Taylor, N. W. Driscoll, and D. L. Kohlstedt. Columbia University Press, New York, pp. 92–137.

Davis, P. J., McKenzie, J. A., Palmer-Julson, A., and colleagues (1991). Explanatory notes. *Proceedings of the Ocean Drilling Program, Initial Reports*, 133, 31–58.

Davis, R. W. (1991). Integration of geological data into hydrodynamic analysis of hydrocarbon movement. In *Petroleum Migration*, ed. W. A. England and A. J. Fleet. *Geological Society of London Special Publications*, **59**, 127–35.

Davison, I. (1989). Extensional domino fault tectonics: kinematics and geometrical constraints. *Annales Tectonicae*, 3, 12–24.

(1999). Tectonics and hydrocarbon distribution along the Brazilian South Atlantic margin. In *The Oil and Gas Habitats of the South Atlantic*, ed. N. R. Cameron, R. H. Bate, and V. S. Clure. *Geological Society of London Special Publications*, **153**, 133–51.

(2005). Central Atlantic margin basins of northwest Africa: geology and hydrocarbon potential (Morocco to Guinea). *Journal of African Earth Sciences*, 43, 254–74.

Davison, I., Al-Kadasi, M., Al-Khirbash, S., Al-Subbary, A., Baker, J., Blakely, S., Bosence, D., Dart, C., Owen, L., Menzies, M., McClay, K., Nichols, G., and Yelland, A. (1994). Geological evolution of the Southern red Sea rift margin-republic Yemen. *Geological Society of American Bulletin*, 106, 1474–93.

Davison, I., Taylor, M., and Longacre, M. (2002). *Northwest Africa – Iberia – USA – Canada. Central Atlantic reconstruction, Scale 1: 5 000 000, Earthmoves, Ltd.* Richmond International Geoscience, MBL, Inc.

Davy, P. (1986). *Modélisation thermo-mécanique de la collision continentale*. University of Rennes, Rennes.

Davy, P. and Cobbold, P. R. (1991). Experiments on shortening of a 4-layer model of the continental lithosphere. *Tectonophysics*, 1–25.

Daw, G. P., Howell, F. T., and Woodward, F. A. (1974). The effect of applied stress upon the permeability of some Permian and Triassic sandstones of northern England: advances in rock mechanics. *Proceedings of the Third International Congress of Rock Mechanics*, II, 537–42.

Dawers, N. H. and Anders, M. H. (1995). Displacement-length scaling and fault linkage. *Journal of Structural Geology*, 17, 607–14.

Dawers, N. H., Berge, A. M., Haeger, K. O., Puigdefabregas, C., and Underhill, J. R. (1999). Controls on Late Jurassic, subtle sand distribution in the Tampen Spur area, northern North Sea. In *Petroleum Geology of Northwest Europe*, ed. A. J. Fleet and S. A. R. Boldy. *Proceedings of 5th conference*, London, UK, pp. 827–38.

Dawers, N. H. and Underhill, J. R. (2000). The role of fault interaction and linkage in controlling synrift stratigraphic sequences: Late Jurassic, Statfjord East area, northern North Sea. *American Association of Petroleum Geologists Bulletin*, 84, 45–64.

de Almeida, F. F. M. (1976). The system of continental rifts bordering the Santos Basin, Brazil. *Anais da Academia Brasileira de Ciencias*, 48, 15–26.

De'Ath, N. G. and Schuyleman, S. F. (1981). The geology of the Magnus oilfield. In *Petroleum Geology of the Continental Shelf of North-West Europe*, ed. L. V. Illing and G. D. Hobson. Heyden and Son, London, pp. 342–51.

de Azevedo, R. P. (1991). *Tectonic evolution of Brazilian equatorial continental margins basins*. Thesis, Imperial College, London, p. 455.

de Boer, J. Z. (1992). Stress configurations during and following emplacement of ENA basalts in the northern Appalachians. *Special Paper – Geological Society of America*, 268, 361–78.

de Boer, J. Z. and Clifton, A. E. (1988). Mesozoic tectogenesis: development and deformation of "Newark" rift zones in the Appalachians (with special emphasis on the Hartford Basin, Connecticut). In *Triassic–Jurassic Rifting, Continental Breakup and the Origin of the Atlantic Ocean Passive Margins, Part A*, ed. W. Manspeizer. Elsevier, New York, pp. 275–306.

De Bremaeker, J. C. (1983). Temperature, subsidence and hydrocarbon maturation in extensional basins: a finite element model. *American Association of Petroleum Geologists Bulletin*, 67, 1410–14.

de Charpal, O., Guennoc, P., Montadert, L., and Roberts, D. G. (1978). Rifting, crustal attenuation and subsidence in the Bay of Biscay. *Nature*, 275, 706–11.

de Graciansky, P. C., Dardeau, G., Lemoine, M., and Tricart, P. (1989). The inverted margin of the French Alps and foreland basin inversion. In *Inversion Tectonics*, ed. M. A. Cooper and G. D. Williams. *Geological Society of London Special Publications*, **44**, 87–104.

de Graciansky, P. C., Poag, C. W., and Foss, G. (1985). Drilling on the Goban Spur: objectives, regional geological setting, and operational summary. *Initial Reports of the Deep Sea Drilling Project*, 80, 5–13.

de Graciansky, P. C., Poag, C. W., Cunningham, R. and colleagues (1985). Site 550. *Initial Reports of the Deep Sea Drilling Project*, 80, 251–356.

De-Hua, H. and Batzle, M. (2006). Velocities of deepwater reservoir sands. *The Leading Edge*, pp. 260–6.

De Marsily, G. (1981). *Hydrogeologie quantitative*. Paris, Mason.

de Matos, R. M. D. (1987). Sistema de rifts Cretáceos do Nordeste Brasiliero, Anais, Tectos. *Petrobrás-Depex*, Rio de Janeiro, Brazil, pp. 126–59.

(1992). The Northeast Brazilian rift system. *Tectonics*, 11, 766–91.

de Matos, R. M. D., de Lima Neto, F. F., Alves, A. C., and Waick, R. N. (1987). O rift Potiguar, gênese, preenchimento e acumulações de hidrocarbonetos, Anais, Rifts intracontinentais. *Petrobrás-Depex*, Rio de Janeiro, Brazil, pp. 160–77.

Dean, S. M., Minshull, T. A., Whitmarsh, R. B., and Louden, K. E. (2000). Deep structure of the ocean–continent transition in the southern Iberia abyssal plain from seismic refraction profiles: the IAM-9 transect at 40 degrees 20'N. *Journal of Geophysical Research*, 105, 5859–85.

DeCarlo, E. H., McMurtry, G. M., and Yeh, H.-W. (1983). Geochemistry of hydrothermal deposits from Loihi submarine volcano, Hawaii. *Earth and Planetary Science Letters*, 66, 438–49.

DeCelles, P. G., Gray, M. B., Ridgway, K. D., Cole, R. B., Srivastava, P., Pequera, N., and Pivnik, D. A. (1991). Kinematic history of a foreland uplift from Paleocene synorogenic conglomerate, Beartooth Range, Wyoming and Montana. *Geological Society of America Bulletin*, 103, 1458–75.

Decker, K., Peresson, H., and Hinsch, R. (2005). Active tectonics and Quaternary basin formation along the Vienna Basin transform fault. *Quaternary Science Reviews*, 24, 305–20.

Deenen, M. H. L., Ruhl, M., Boris, N. L., Krijgsman, W., Kuerschner, W. M., Reitsma, M., and van Bergen, M. J. (2010). A new chronology for the End-Triassic Mass Extinction. *Earth and Planetary Science Letters*, 291, 113–25.

Dehler, S. A. (2010). Initial rifting and break-up between Nova Scotia and Morocco: an examination of new geophysical data and models. In *Conjugate Margins II*, Lisbon 2010, Metedo Directo, VIII, 79–82. http://metedodirecto.pt/CM2010.

Dehler, S. A. and Welford, J. K. (2012). Variations in rifting style and structure of the Scotian margin, Atlantic Canada, from 3D gravity inversion. In *Conjugate Divergent Margins*, ed. W. U. Mohriak, A. Danforth,

P. J. Post, D. E. Brown, G. M Tari, M. Nemčok, and S. T. Sinha. *Geological Society of London Special Publications*, **369**, 289–300.

Demaison, G. J., Holck, A. J. J., Jones, R. W., and Moore, G. T. (1983). Predictive source bed stratigraphy: a guide to regional petroleum occurrences. *Proceedings of the 11th World Petroleum Congress*, London, pp. 1–13.

Demaison, G. J. and Huizinga, B. J. (1991). Genetic classification of petroleum systems. *American Association of Petroleum Geologists Bulletin*, 75, 1626–43.

Dembicki, Jr. H. and Anderson, M. L. (1989). Secondary migration of oil experiments supporting efficient movement of separate, buoyant oil phase along limited conduits. *American Association of Petroleum Geologists Bulletin*, 73, 1018–21.

Demercian, L. S. (1996). *A halocinese na evolucao do Sul da Bacia de Santos do Aptiano ao Cretaceo Superior*. MS thesis, Universidade Federal do Rio Grande do Sul, Porto Alegre, Brazil, p. 201.

Demercian, S., Szatmari, P., and Cobbold, P. R. (1993). Style and pattern of salt diapirs due to thin-skinned gravitational gliding, Campos and Santos basins, offshore Brazil. *Tectonophysics*, 228, 393–433.

Demetrescu, C. and Veliciu, S. (1991). Heat flow and lithosphere structure in Romania. In *Terrestrial Heat Flow and Lithospheric Structure*, ed. V. Čermák and L. Rybach. Springer-Verlag, Berlin, pp. 187–205.

Deming, D. (1994a). Fluid flow and heat transport in the upper continental crust. In *Geofluids: Origin, Migration and Evolution of Fluids in Sedimentary Basins*, ed. J. Parnell. *Geological Society of London Special Publications*, **78**, 27–42.

(1994b). Overburden rock, temperature, and heat flow. In *The Petroleum System – From Source to Trap*, ed. L. B. Magoon and W. G. Dow. *American Association of Petroleum Geologists Memoir*, 60, 165–86.

(1994c). Factors necessary to define a pressure seal. *Bulletin of the American Association of Petroleum Geologists*, 78, 1005–9.

Deming, D. and Chapman, D. S. (1989). Thermal histories and hydrocarbon generation: example from Utah-Wyoming thrust belt. *American Association of Petroleum Geologists Bulletin*, 73, 1455–71.

Deming, D. and Nunn, J. A. (1991). Numerical simulations of brine migration by topographically driven recharge. *Journal of Geophysical Research*, 96, 2485–99.

Deming, D., Nunn, J. A., and Evans, D. G. (1990a). Thermal effects of compaction-driven groundwater flow from overthrust belts. *Journal of Geophysical Research*, 95, 6669–83.

Deming, D., Nunn, J. A., Jones, S., and Chapman, D. S. (1990b). Some problems in thermal history studies. In *Applications of Thermal Maturity Studies to Energy Exploration*, ed. F. Nuccio and C. E. Barker. Rocky Mountain Section – Society of Exploration Paleontologists and Mineralogists, 61–80.

Densmore, A. L., Ellis, M. A., and Anderson, R. S. (1998). Landsliding and the evolution of normal fault-bounded mountains. *Journal of Geophysical Research*, 103, 15203–19.

Deptuck, M. E. (2011). Proximal to distal postrift structural provinces of the western Scotian Margin, offshore Eastern Canada. *CNSOPB Geoscience Open File Report*, 2011-001MF, p. 42.

Desheng, L. (1980). Geological structure and hydrocarbon occurrence of the Bohai Gulf oil and gas basin (China). In *Petroleum Geology in China*, ed. J. F. Mason. Penn Well Books, Tulsa, pp. 180–92.

Desmurs, L., Manatschal, G., and Bernoulli, D. (2001). The Steinmann Trinity revisited: mantle exhumation and magmatism along an ocean–continent transition: the Platta Nappe, eastern Switzerland. *Geological Society of Special Publications*, 187, 235–66.

Destro, N., Alkmim, F. F., Magnavita, L. P., and Szatmari, P. (2003a). The Jeremoabo transpressional transfer fault, Reconcavo-Tucano Rift, NE Brazil. *Journal of Structural Geology*, 25, 1263–79.

Destro, N., Szatmari, P., Alkmim, F. F., and Magnavita, L. P. (2003b). Release faults, associated structures, and their control on petroleum trends in the Reconcavo rift, northeast Brazil. *American Association of Petroleum Geologists Bulletin*, 87, 1123–44.

Devey, C. W. and Shipboard Scientific Party (2002). *Hydrothermal Studies of Grimsey Field, Volcanic Studies of Kolbeinsey Ridge*. University of Bremen.

Devlin, W. J., Cogswell, J. M., Gaskins, G. M., Isaksen, G. H., Pitcher, D. M., Puls, D. P., Stanley, K. O., and Wall, G. R. T. (1999). South Caspian Basin: young, cool, and full of promise. *GSA Today*, 9, 1–9.

Dewhurst, D. N., Brown, K. M., Clennell, M. B., and Westbrook, G. K. (1996a). A comparison of the fabric and permeability anisotropy of consolidated and sheared silty clay. *Engineering Geology*, 42, 253–67.

Dewhurst, D. N., Clennell, M. B., Brown, K. M., and Westbrook, G. K. (1996b). Fabric and hydraulic conductivity of sheared clays. *Geotechnique*, 46, 761–8.

D'Heur, M. (1984). Porosity and hydrocarbon distribution in the North Sea chalk reservoirs. *Marine and Petroleum Geology*, 1, 211–38.

Dholakia, S. K., Aydin, A., Pollard, D. D., and Zoback, M. D. (1998). Fault-controlled hydrocarbon pathways in the Monterey Formation, California. *American Association of Petroleum Geologists Bulletin*, 82, 1551–74.

Dias-Brito, D. (1987). A Bacia de Campos no Mesocretaceo: uma contribuicao a paleoceanografia do Atlantico sul primitivo. The Campos Basin of the Middle Cretaceous: a contribution to paleo-oceanography of the ancient South Atlantic. *Revista Brasileira de Geociências*, 17, 162–7.

Dibblee, T. R. (1980). Geology along the San Andreas fault from Gilroy to Parkfield. In *Studies of the San Andreas Fault Zone in Northern California*, ed. R. Streitz and R.

Sherburne. *California Division of Mines and Geology, Sacramento, Special Report*, 140, 3–18.

Dickinson, G. (1953). Geological aspects of abnormal reservoir pressures in Gulf Coast, Louisiana. *American Association of Petroleum Geologists*, 37, 410–32.

Diegel, F. A., Karlo, J. F., Schuster, D. C., Shoup, R. C., and Tauvers, P. R. (1995). Cenozoic structural evolution and tectono-stratigraphic framework of the northern Gulf Coast continental margin. In *Salt Tectonics: A Global Perspective*, ed. M. P. A. Jackson, D. G. Roberts, and S. Snelson. *American Association of Petroleum Geologists Memoir*, 65, 109–51.

Dikenshteyn, G. K., Maksimov, S. P., and Ivanova, T. D. (1982). *Tectonics of Oil-Gas Provinces and Regions of the U.S.S.R. (in Russian)*. Nedra, Moscow, p. 223.

Dinter, D. A. and Royden, L. (1993). Late Cenozoic extension in northeastern Greece; Strymon Valley detachment system and Rhodope metamorphic core complex. *Geology*, 21, 45–8.

Discovery 215 Working Group (1998). Deep structure in the vicinity of the ocean-continent 618 transition zone under the southern Iberia Abyssal Plain. *Geology*, 26, 743–6.

Dmitriyevskiy, A. N., Kireyev, F. A., Bochko, R. A., and Fedorova, T. A. (1993). Hydrothermal origin of oil and gas reservoirs in basement rock of the South Vietnam continental shelf. *International Geology Review*, 35, 621–30.

Dodson, M. H. (1973). Closure temperature in cooling geochronological and petrological systems. *Contributions to Mineralogy and Petrology*, 40, 259–74.

Doglioni, C. (1993). Some remarks on the origin of foredeeps. In *Crustal Controls on the Internal Architecture of Sedimentary Basins*, ed. R. A. Stephenson. Elsevier, Amsterdam, 228, 1–20.

Doligez, B., Bessi, F., Burrus, J., Ungerer, P., and Chenet, P. Y. (1986). Integrated numerical simulation of the sedimentation heat transfer, hydrocarbon formation and fluid migration in a sedimentary basin: the Themis model. In *Thermal Modeling in Sedimentary Basins*, ed. J. Burrus. Technip, Paris, pp. 173–95.

Donath, F. A. (1961). Experimental study of shear failure in anisotropic rocks. *Geological Society of America Bulletin*, 72, 985–9.

Donelick, R. A. (1994). Selected Poland samples. Apatite fission track data. *Donelick Analytical Report*, 76, 1–114.

Donelick, R. A., O'Sullivan, P. B., and Ketcham, R. A. (2005). Apatite fission-track analysis. In *Low-Temperature Thermochronology: Techniques, Interpretations and Applications*, ed. P. W. Reiners and T. A. Ehlers. *Reviews in Mineralogy and Geochemistry*, 58, 49–94.

Donovan, T. J., Friedman, I., and Gleason, J. D. (1974). Recognition of petroleum-bearing traps by unusual isotopic composition of carbonate-cemented surface rocks. *Geology*, 2, 351–4.

Doré, A. G., Lundin, E. R., Birkeland, O., Eliassen, P. E., and Jensen, L. N. (1997). The NE Atlantic Margin: implications of late Mesozoic and Cenozoic events for hydrocarbon prospectivity. *Petroleum Geosciences*, 3, 117–31.

Dorsey, R. J., Stone, K. A., and Umhoefer, P. J. (1997a). Stratigraphy, sedimentology, and tectonic development of the southeastern Pliocene Loreto Basin, Baja California Sur, Mexico. In *Pliocene Carbonates and Related Facies Flanking the Gulf of California, Baja California, Mexico*, ed. M. E. Johnson and J. Ledesma-Vazquez. *Special Paper – Geological Society of America*, pp. 83–109.

Dorsey, R. J. and Umhoefer, P. J. (2000). Tectonic and eustatic controls on sequence stratigraphy of the Pliocene Loreto Basin, Baja California Sur, Mexico. *Geological Society of America Bulletin*, 112, 177–99.

Dorsey, R. J., Umhoefer, P. J., and Falk, P. (1997b). Earthquake clustering and stacked Gilbert-type fan deltas in the Pliocene Loreto Basin, Baja California Sur, Mexico. *Geology*, 25, 679–82.

Dorsey, R. J., Umhoefer, P. J., and Renne, P. R. (1995). Rapid subsidence and stacked Gilbert-type fan deltas, Pliocene Loreto Basin, Baja California Sur, Mexico. *Sedimentary Geology*, 98, 181–204.

Doubre, C. (2004). *Structure et mécanisme des segments de rift volcanotectoniques*. PhD thesis, Université du Maine, Angers, France, p. 423.

Doubre, C. and Geoffroy, L. (2003). Model for stress inversions and rift-zone development around a hot-spot-related magma centre on the Isle of Skye, Scotland. *Terra Nova*, 15, 230–7.

Doust, H. (1990). Petroleum geology of the Niger Delta. In *Classic Petroleum Provinces*, ed. J. Brooks. *Geological Society Special Publication*, 50, 365.

Doust, H. and Omatsola, E. (1990). Niger Delta. In *Divergent/Passive Margin Basins*, ed. J. D. Edwards and P. A. Santogrossi. *American Association of Petroleum Geologists Memoir*, 48, 201–38.

Dow, W. (1977). Kerogen studies and geological interpretations. *Journal of Geochemical Exploration*, 7, 79–99.

Downes, H. and Vaselli, O. (1995). The lithospheric mantle beneath the Carpathian-Pannonian region: a review of trace element and isotopic evidence from ultramafic xenoliths. *Acta Vulcanologica*, 7, 219–29.

Downey, M. W. (1994). Hydrocarbon seal rocks. In *The Petroleum System – From Source to Trap*, ed. L. B. Magoon and W. G. Dow. *American Association of Petroleum Geologists Memoir*, 60, 159–64.

Driscoll, N. W., Hogg, J. R., Christie-Blink, N., and Karner, G. D. (1995). Extensional tectonics in the Jeanne d'Arc Basin, offshore Newfoundland: implications for the timing of break-up between Grand Banks and Iberia. In *The Tectonics, Sedimentation and Paleoceanography of the North Atlantic Region*, ed. R. A. Scrutton. *Special Publications*, pp. 1–28.

Driscoll, N. W. and Karner, G. D. (1998). Lower crustal extension across the northern Carnarvon Basin, Australia: evidence for an eastward dipping detachment. *Journal of Geophysical Research*, 103, 4975–91.

Droz, L., Marsset, T., Onreas, H., Lopez, M., Savoye, B., and Spy-Anderson, F.-L. (2003). Architecture of an active mud-rich turbidite system: the Zaire Fan (Congo-Angola margin Southeast Atlantic): results from ZaiAngo 1 and 2 cruises. *American Association of Petroleum Geologists Bulletin*, 87, 1145–68.

Drucker, D. C. and Prager, W. (1952). Soil mechanics and plastics analysis for limited design. *Quarterly of Applied Mathematics*, 10, 157–65.

Drummond, B. J. (1988). A review of crust/upper mantle structure in the Precambrian areas of Australia and implications for Precambrian crustal evolution. *Precambrian Research*, 40–41, 101–16.

Du, K. E., Pai, S., Brown, J., Moore, R. M., and Simmons, M. (2000). Optimising the development of Blake field under tough economic and environmental conditions. *Society of Petroleum Engineers*, SPE 64714.

Du Rouchet, J. (1981). Stress fields. A key to oil migration. *American Association of Petroleum Geologists Bulletin*, 65, 74–85.

Duan, Z., Moller, N., and Weare, J. H. (1996). A general equation of state for supercritical fluid mixtures and molecular dynamics simulation of mixture PVTX properties. *Geochimica et Cosmochimica Acta*, 60, 1209–16.

Duddy, I. R., Green, P. F., Bray, R. J., and Hegarty, K. A. (1994). Recognition of the thermal effects of fluid flow in sedimentary basins. In *Origin, Migration and Evolution of Fluids in Sedimentary Basins*, ed. J. Parnell. *Geological Society Special Publication*, 78, 325–45.

Duddy, I. R. and Kelly, P. R. (1999). Uranium in mineral sands: measurement and uses. *Australian Institute of Geoscientists Bulletin*, 26, 4.

Duffy, C. J. and Al-Hassan, S. (1988). Groundwater circulation in a closed desert basin: topographic scaling and climatic forcing. *Water Resources Research*, 24, 1675–88.

Dumitru, T. A., Hill, K. C., Coyle, D. A., Duddy, I. R., Foster, D. A., Gleadow, A. J. W., Green, P. F., Kohn, B. P., Laslett, G. M., and O'Sullivan, A. J. (1991). Fission track thermochronology: application to continental rifting of south-eastern Australia. *APEA Journal*, 31, 131–42.

Dupre, S., Bertotti, S., and Cloetingh, S. A. P. L. (2007). Tectonic history along the South Gabon Basin: anomalous early post-rift subsidence. *Marine Petroleum Geology*, 24, 151–72.

Durand, B. (1988). Understanding of HC migration in sedimentary basins (present state of knowledge). *Organic Geochemistry*, 13, 445–59.

Duranti, D. and Hurst, A. (2004). Fluidisation and injection in the deep-water sandstones of the Eocene Alba Formation (UK North Sea). *Sedimentology*, 51, 503–29.

Durney, D. W. and Ramsay, J. G. (1973). Incremental strains measured by syntectonic crystal growths. In *Gravity and Tectonics*, ed. K. A. de Jong and R. Scholten. Wiley, New York, 67–96.

Durrheim, R. J. and Mooney, W. D. (1991). Archean and Proterozoic crustal evolution: evidence from crustal seismology. *Geology*, 19, 606–9.

(1992). Archean and Proterozoic crustal evolution: evidence from crustal seismology. *Geology*, 20, 665–6.

(1994). Evolution of the Precambrian lithosphere: seismological and geochemical constraints. *Journal of Geophysical Research*, 99, 15359–74.

Dydik, B. M., Simoneit, B. R. T., Brassell, S. C., and Eglinton, G. (1978). Organic geochemical indicators of palaeoenvironmental conditions of sedimentation. *Nature*, 272, 216–22.

Dziak, R. P., Fox, C. G., Embley, R. W., Lupton, J. E., Johnson, G. C., Chadwick, W. W., and Koski, R. A. (1996). Detection of and response to a probable volcanogenic T-wave event swarm on the western Blanco Transform Fault Zone. *Geophysical Research Letters*, 23, 873–6.

Earnshaw, J. P., Hogg, A. J. C., Oxtoby, N. H., and Cawley, S. J. (1993). Petrographic and fluid inclusion evidence for the timing of diagenesis and petroleum entrapment in the Papuan Basin. In *Petroleum Exploration, Development and Production in Papua New Guinea: Proceedings of the Second PNG Petroleum Convention 31st May-2nd June*, ed. G. J. Carman and Z. Carman. Port Moresby, Petroleum Exploration and Development, pp. 459–75.

Ebdon, C. C., Granger, P. J., Johnson, H. D., and Evans, A. M. (1995). Early Tertiary evolution and sequence stratigraphy of the Faeroe-Shetland Basin; implications for hydrocarbon prospectivity. In *The Tectonics, Sedimentation, and Palaeoceanography of the North Atlantic Region*, ed. R. A. Scrutton, M. S. Stoker, G. B. Shimmield, and A. W. Tudhope. *Geological Society Special Publications*, pp. 51–69.

Eberli, G. P. (1988). The evolution of the southern continental margin of the Jurassic Tethys ocean as recorded in the Allgaeu Formation of the Austroalpine Nappes of Graubuenden (Switzerland). *Eclogae Geologicae Helvetiae*, 81, 175–214.

Ebinger, C. J. and Casey, M. (2001). Continental breakup in magmatic provinces: an Ethiopian example. *Geology*, 29, 527–30.

Ebinger, C. J., Jackson, J. A., Foster, A. N., and Hayward, N. J. (1999). Extensional basin geometry and the elastic lithosphere. *Philosophical Transactions of the Royal Society*, 357, 741–65.

Ebner, F. and Sachsenhofer, R. F. (1995). Palaeogeography, subsidence and thermal history of the Neogene Styrian Basin (Pannonian Basin system, Austria): tectonics of the Alpine-Carpathian-Pannonian region. *ALCAPA Meeting on Geological Evolution of the Internal Zones in the Alps, the Carpathians and of the Pannonian Basin*, Graz, Austria, July 1–3, 242, 133–50.

Echarfaoui, H., Hafid, M., Ait Salem, A. A., and Ait Fora, A. (2002). Seismo-stratigraphic analysis of the Abda Basin, western Morocco: a case of inverse structures during Atlantic rifting. *Comptes Rendus Geoscience (Academie des Sciences)*, 334, 371–7.

Eckstein, Y. and Simmons, G. (1978). Measurements and interpretation of terrestrial heat flow in Israel. *Geothermics*, 6, 117–42.

Edwards, M. H., Kurras, G. J., Tolstoy, M., Bohnenstiehl, M. Coakley, B. J. and Cochran, J. R. (2001). Evidence of recent volcanic activity on the ultraslow-spreading Gakkel ridge. *Nature*, 409.6822, 808–812.

Egan, S. S. and Meredith, D. J. (2005). The structural and geodynamic evolution of the Black Sea Basin. *InterMARGINS Workshop: Modeling the Extensional Deformation of the Lithosphere. (IMEDL 2004), Abstract*, Pontresina, Switzerland.

Egloff, F., Rihm, R., Makris, J., Izzeldin, Y. A., Bobsien, M., Meier, K., Junge, I., Noman, T., and Warsi, W. (1991). Contrasting structural styles of the eastern and western margins of the southern Red Sea: the 1988 SONNE experiment. *Tectonophysics*, 198, 329–53.

Ehlers, T. A., Armstrong, P. A., and Chapman, D. S. (2001). Normal fault thermal regimes and interpretation: low-temperature thermochronometer data. *Physics of the Earth and Planetary Interiors*, 126, 179–94.

Ehlers, T. A. and Chapman, D. S. (1999). Normal fault thermal regimes: conductive and hydrothermal heat transfer surrounding the Wasatch fault, Utah. *Tectonophysics*, 312, 217–34.

Ehlers, T. A. and Farley, K. A. (2003). Apatite (U-Th)/He'thermochronometry, methods, and applications to problems in tectonic and surface processes. *Earth and Planetary Science Letters*, 206, 1–14.

Ehlers, T. A., Willett, S. D., Armstrong, P. A., and Chapman, D. S. (2003). Exhumation of the central Wasatch Mountains, Utah; 2, Thermokinematic model of exhumation, erosion, and thermochronometer interpretation. *Journal of Geophysical Research*, 108, 18.

Eichhubl, P., Greene, H. G., Naehr, T., and Maher, N. (2000). Structural control of fluid flow: offshore seepage in the Santa Barbara Basin, California. *Journal of Geochemical Exploration*, 69–70, 545–9.

Elders, W. A., Rex, R. W., Meidav, T., Robinson, P. T., and Biehler, S. (1972). Crustal spreading in southern California. *Science*, 178, 15–24.

Eldholm, O. and Grue, K. (1994). North Atlantic volcanic margins: dimensions and production rates. *Journal of Geophysical Research*, 99, 2955–68.

Eldholm, O., Skogseid, J., Planke, S., and Gladczenko, P. (1995). Volcanic margin concept. In *Rifted Ocean–Continent Boundaries*, ed. E. Banda. Kluwer Academic Publishers, p. 16.

Eldholm, O., Thiede, J., and Taylor, E. (1989). Evolution of the Vøring volcanic margin. In *Proceedings of Ocean Drilling Program (ODP)*, ed. O. Eldholm, J. Thiede, and E. Taylor. *Scientific Results*, 104, 1033–65.

Eliet, P. P. and Gawthorpe, R. L. (1995). Drainage development and sediment supply within rifts, examples from the Sperchios Basin, central Greece. *Journal of the Geological Society of London*, 152, 883–93.

Eliuk, L. and Wach, G. D. (2010). Large scale mixed carbonate-siliciclastic clinoform systems: three types from the Mesozoic North American Atlantic offshore. AAPG Annual Conference, New Orleans, April 11–14.

Ellevset, S., Ottesen, D., Knipe, R. J., Olsen, T. S., Fisher, Q. J., and Jones, G. (1998). Fault controlled communication in the Sleipner Vest Field, Norwegian continental shelf: detailed, quantitative input for reservoir simulation and well planning. In *Faulting, Fault Sealing and Fluid Flow in Hydrocarbon Reservoirs*, ed. G. Jones, Q. J. Fisher, and R. J. Knipe. *Geological Society of London Special Publications*, **147**, 283–97.

Ellis, S. and Beaumont, C. (1999). Models of convergent boundary tectonics: implications for the interpretation of Lithoprobe data. *Canadian Journal of Earth Sciences*, 36, 1711–41.

Embry, A. F. (1990). A tectonic origin for third-order depositional sequences in extensional basins; implications for basin modeling. In *Quantitative dynamic stratigraphy*, ed. T. A. Cross. Prentice Hall, Englewood Cliffs, New Jersey.

Enescu, D., Danchiv, D., and Bálá, A. (1992). Lithosphere structure in Romania II: thickness of the Earth's crust, depth dependent propagation velocity curves for P and S waves. *Studii Cercetri Geofizica*, 30, 3–19.

England, P. C. (1983). Constraints on extension of continental lithosphere. *Journal of Geophysical Research*, 88, 1145–52.

England, P. C. and Molnar, P. (1990). Surface uplift, uplift of rocks, and exhumation of rocks. *Geology*, 18, 1173–7.

England, P. C. and Richardson, S. W. (1977). The influence of erosion upon mineral facies of rocks from different metamorphic environments. *Journal of the Geological Society of London*, 134, 201–13.

England, W. A. (1994). Secondary migration and accumulation of hydrocarbons. In *The Petroleum System – From Source to Trap*, ed. L. B. Magoon and W. G. Dow. *American Association of Petroleum Geologists Memoir*, 60, 211–17.

England, W. A., Mackenzie, A. S., Mann, D. M., and Quigley, T. M. (1987). The movement and entrapment of petroleum fluids in the subsurface. *Journal of the Geological Society of London*, 144, 327–47.

Erratt, D. (1993). Relationships between basement faulting, salt withdrawal and Late Jurassic rifting, UK Central North Sea. In *Petroleum Geology of Northwest Europe, Proceedings of the 4th Conference*, ed. J. R. Parker. The Geological Society of London, 1211–19.

Escalona, A. (2003). *Regional tectonics, sequence stratigraphy and reservoir properties of Eocene clastic sedimentation, Maracaibo Bain, Venezuela*. PhD thesis, University of Texas, Austin, p. 222.

Escartín, J., Mevel, C., MacLeod, C. J., and McCaig, A. M. (2003). Constraints on deformation conditions and the origin of oceanic detachments: the Mid-Atlantic Ridge core complex at 15 degrees 45'N. *Geochemistry, Geophysics, Geosystems*, 4.

Espitalié, J., Madec, M., Tissot, B., Menning, J. J., and Leplat, P. (1977a). Source rock characterization method for petroleum exploration. *Ninth Annual Offshore Technology Conference*, 439–48.

Espitalié, J., Laporte, J. L., Madec, M., Marquis, F., Leplat, P., Paulet, J. and Boutefeu, A. (1977b). Rapid method for source rock characterization, and for determination of their petroleum potential and degree of evolution. *Revue de l'Institut Francais du Petrole et Annales des Combustibles Liquides*, 32, 23–42.

Esteoule-Choux, J. (1983). Kaolinitic weathering profiles in Britany: genesis and economic importance. *Contributions to Mineralogy and Petrology*, 75, 129–52.

Eugster, H. P. and Hardie, L. A. (1975). Sedimentation in an ancient playa-lake complex: the Wilkins Peak Member of the Green River Formation of Wyoming. *Geological Society of America Bulletin*, 86, 319–34.

Evans, A. C. and Parkinson, D. N. (1983). A half-graben and tilted fault block structure in the northern North Sea. In *Seismic Expression of Structural Styles: A Picture and Work Atlas*; Volume 2, ed. A. W. Bally. *American Association of Petroleum Geologists Studies in Geology*, 15, 2.2.2-7 – 2.2.2-11.

Evans, D. G. and Nunn, J. A. (1989). Free thermohaline convection in sediments surrounding a salt column. *Journal of Geophysical Research*, 94, 12413–22.

Evans, D. G., Nunn, J. A., and Hanor, J. S. (1991). Mechanisms driving groundwater flow near salt domes. In *Crustal-Scale Fluid Transport: Magnitude and Mechanisms*, ed. T. Torgensen. Geophysical Research Letters, 18, 927–30.

Evans, K. F., Burford, R. O., and King, G. C. P. (1981). Propagating episodic creep and the aseismic slip behavior of the Calaveras Fault north of Hollister, California. *Journal of Geophysical Research*, 86, 3721–35.

Ewart, A., Baxter, K., and Ross, J. A. (1980). The petrology and petrogenesis of the Tertiary anorogenic mafic lavas of southern and central Queensland, Australia – possible implications for crustal thickening. *Contributions to Mineralogy and Petrology*, 75, 129–52.

Exon, N. and Colwell, J. B. (1994). Geological history of the outer North West Shelf of Australia: a synthesis. In *Geology of the Outer North West Shelf, Australia: Australian Geological Survey Organization*, ed. N. Exon. *AGOSO Journal of Australian Geology and Geophysics*, 15(1), 177–90.

Faccenna, C., Davy, P., Brun, J. P., Funiciello, R., Giardini, D., Mattei, M., and Nalpas, T. (1996). The dynamics of back-arc extension: an experimental approach to the opening of the Tyrrhenian Sea. *Geophysical Journal International*, 126, 781–95.

Fachmann, S. (2001). *Geologische Entwicklung im Umfeld des Mahanadi-Riftes (Indien)*. Universitat Freiberg.

Faerseth, R. B., Knudsen, B. E., Liljedahl, T., Midboe, P. S., and Soderstrom, B. (1997). Oblique rifting and sequential faulting in the Jurassic development of the northern North Sea. *Journal of Structural Geology*, 19, 1285–302.

Fainstein, R., Milliman, J. D., and Jost, H. (1975). Magnetic character of the Brazilian continental shelf and upper slope. *Revista Brasileira de Geociencias*, 5, 198–211.

Fairhead, J. D. (1988a). Mesozoic plate tectonic reconstructions of the central South Atlantic Ocean: the role of the West and Central African rift system. *Tectonophysics*, 155, 181–91.

(1988b). Late Mesozoic rifting in Africa. In *Triassic-Jurassic Rifting: Continental Breakup and the Origin of the Atlantic Ocean and Passive Margins*, ed. W. Manspeizer. *Developments in Geotectonics*, 22, 821–31.

Fairhead, J. D. and Binks, R. M. (1991). Differential opening of the Central and South Atlantic Oceans and the opening of the West African rift system. *Tectonophysics*, 187, 191–203.

Fairhead, J. D. and Okereke, C. S. (1987). A regional gravity study of the West African rift system in Nigeria and Cameroon and its tectonic interpretation. *Tectonophysics*, 143, 141–59.

Faleide, J. I., Kyrkjebo, R., Kjennerud, T., Gabrielsen, R., Jordt, J., Fanavoll, S., and Bjerke, M. D. (2002). Tectonic impact on sedimentary processes during Cenozoic evolution of the northern North Sea and surrounding areas. In *Exhumation of the North Atlantic Margin: Timing, Mechanisms and Implications for Petroleum Exploration*, ed. A. G. Dore, J. A. Cartwright, M. S. Stoker, J. P. Turner, and N. J. White. *Geological Society Special Publications*, 196, 235–69.

Falt, U., Guerin, G., Retail, P., and Evans, M. (1992). Clair discovery: evaluation of natural fracturation in a horizontal well drilled in the basement and producing from overlying sediments. *Proceedings: European Petroleum Conference: Moving the Frontiers, Sharing Solutions, Society of Petroleum Engineers*, 2, 11–21.

Farley, K. A. (2000). Helium diffusion from apatite: general behavior as illustrated by Durango fluorapa. *Journal of Geophysical Research*, 105, 2000.

(2002). (U-Th)/He dating: techniques, calibrations, and applications. *Reviews in Mineralogy and Geochemistry*, 47, 819–43.

Farley, K. A., Kohn, B. P., and Pillans, B. (2002). The effects of secular disequilibrium on (U-Th)/He systematics and dating of Quaternary volcanic zircon and apatite. *Earth and Planetary Science Letters*, 201, 114–25.

Farley, K. A., Wolf, R. A., and Silver, L. T. (1996). The effects of long alpha-stopping distances on (U-Th)/He ages. *Geochimica et Cosmochimica Acta*, 60, 4223–9.

Fassoulas, C., Kilias, A., and Mountrakis, D. (1994). Postnappe stacking extension and exhumation of high-pressure/low-temperature rocks in the Island of Crete, Greece. *Tectonics*, 13, 125–38.

Faugère, E. and Brun, J. P. (1984). Modelisation experimentale de la distention continentale. Experimental models of a stretched continental crust. *Compte Rendu Academie des Sciences*, 299, 365–70.

Faugère, E., Brun, J. P., and van den Driessche, J. (1986). Bassins asymetriques en extension pure et en decrochement: modeles experimentaux. Asymmetric extension and unhooked basins: experimental models. *Bulletin des Centres de Recherches Exploration-Production Elf-Aquitaine*, 10, 13–21.

Faulds, J. E., Geissman, J. W., and Mawer, C. K. (1990). Structural development of a major extensional accommodation zone in the Basin and Range Province, northwestern Arizona and southern Nevada: implications for kinematic models of continental extension. *Memoir – Geological Society of America*, 176, 37–76.

Faure, G. and Mensing, T. M. (2005). *Isotopes. Principles and Applications*. 3rd edition, John Wiley and Sons, p. 897.

Fechtig, H. and Kalbitzer, S. (1966). *The Diffusion of Argon in Potassium-Bearing Solids*. Springer, pp. 68–107.

Feighner, M. A., Kellogg, L. H., and Travis, B. J. (1995). Numerical modeling of chemically buoyant mantle plumes at spreading ridges. *Geophysical Research Letters*, 22, 715–18.

Feighner, M. A. and Richards, M. A. (1995). The fluid dynamics of plume-ridge and plume-plate interactions: an experimental investigation. *Earth and Planetary Science Letters*, 129, 171–82.

Feraud, G., Hofmann, C., and Alric, V. I. (1996). Successful (super 40) Ar/ (super 39) Ar laser probe dating of whole rock basalts affected by 39Ar recoil. *Eos Transactions American Geophysical Union*, 77, 93–4.

Ferentinos, G., Papatheodorou, G., and Collins, M. B. (1988). Sediment transport processes on an active submarine fault escarpment: Gulf of Corinth, Greece. *Marine Geology*, 83, 43–61.

Ferguson, I. J., Westbrook, G. K., Langseth, M. G., and Thomas, G. P. (1993). Heat flow and thermal models of the Barbados Ridge accretionary complex. *Journal of Geophysical Research*, 98, 4121–42.

Fernàndez, M., Ayala, C., Torne, M., Vergés, J., Gómez, M., and Karpuz, R. (2005). Lithospheric structure of the Mid-Norwegian Margin: comparison between the Møre and Vøring margins. *Journal of Geological Society of London*, 162, 1005–12.

Ferrara, G. C., Palmerini, G. C., and Scappini, U. (1985). Update report on geothermal development in Italy. *International Symposium on Geothermal Energy, International Volume(GRCT)*, 95–105.

Ferre, E. C., Bordarier, C., and Marsh, J. S. (2002). Magma flow inferred from AMS fabrics in a layered mafic sill, Insizwa, South Africa. *Tectonophysics*, 354, 1–23.

Ferriday, I. and Hall, K. (1993). Sourcing of oils on the eastern flank of the Northern Viking Graben. *Proceedings of the 16th International Meeting on Organic Geochemistry*, Stavanger, Norway, September 20–4, pp. 43–5.

Feyzullayev, A. A., Guliyev, I. S., and Tagiyev, M. F. (2001). Source potential of the Mesozoic-Cenozoic rocks in the South Caspian basin and their role in forming the oil accumulations in the lower Pliocene reservoirs. *Petroleum Geoscience*, 7(4), 409–17.

Field, J. D. (1985). Organic geochemistry in exploration of the northern North Sea. In *Petroleum Geochemistry in Exploration of the Norwegian Shelf*, ed. B. M. Thomas, A. G. Dore, S. S. Eggen, P. C. Home, and L. R. Mange. Graham and Trotman, London, pp. 39–57.

Figueiredo, A. M. F. 1985. In De Azevedo, R. P. (1991). Tectonic evolution of Brazilian equatorial continental margin basins. PhD thesis, Imperial College, London, pp. 1–403.

Filho, J. D. S., Correa, A. C. F., Neto, E. V. S., and Trinidade, L. A. F. (2000). Alagamar-Acu petroleum system, onshore Potiguar Basin, Brazil: a numerical approach for secondary migration. In *Petroleum Systems of South Atlantic Margins*, ed. M. R. Mello and B. J. Katz. *American Association of Petroleum Geologists Memoir*, 73, 151–8.

Filjak, R., Pletikapic, Z., Nikolic, D., and Aksin, V. (1969). Geology of petroleum and natural gas from the Neogene complex and its basement in the southern part of the Pannonian basin, Yugoslavia. *Conference of Institute of Petroleum and AAPG, Brighton, Proceedings*, 113–30.

Filmer, P. E., McNutt, M. K., Webb, H. F., and Dixon, D. J. (1994). Volcanism and archipelagic aprons in the Marquesas and Hawaiian Islands. *Marine Geophysical Research*, 16, 385–406.

Finlayson, D. M., Collins, C. D. N., Lukaszyk, I., and Chudyk, E. C. (1998). A transect across Australia's southern margin in the Otway Basin region: crustal architecture and the nature of rifting from wide-angle seismic profiling. *Tectonophysics*, 288, 177–89.

Fisher, A. T. and Becker, K. (1995). Correlation between seafloor heat flow and basement relief: observational and numerical examples and implications for upper crustal permeability. *Journal of Geophysical Research*, 100, 12641–57.

Fisher, A. T., Zwart, G., Shipley, T., Ogawa, Y., Ashi, J., Blum, P., Brueckmann, W., Filice, F., Goldberg, D., Henry, P., Housen, B., Jurado, M. J., Kastner, M., Labaume, P., Laier, T., Leitch, E., Maltman, A., Meyer, A., Moore, J. C., Moore, G., Peacock, S., Rabaute, A., Steiger, T., and To, H. (1996). Relation between permeability and effective stress along a plate-boundary fault, Barbados accretionary complex. *Geology*, 24, 307–10.

Fisher, J. B. and Boles, J. R. (1990). Water-rock interaction in Tertiary sandstones, San Joaquin Basin, California, U.S.A.: Diagenetic controls on water composition. *Chemical Geology*, 82, 83–101.

Fisher, M. J. and Miles, J. A. (1983). Kerogen type, organic maturation and hydrocarbon occurrences in the Moray Firth and South Viking Graben, North Sea Basin. In *Petroleum Geochemistry and Exploration of Europe (second edition)*, ed. J. Brooks. *Geological Society of London Special Publications*, **12**, 195–201.

Fisher, Q. J. and Knipe, R. J. (1998). Fault sealing processes in siliciclastic rocks. In *Faulting, Fault Sealing and Fluid Flow in Hydrocarbon Reservoirs*, ed. G. Jones, Q. J. Fisher, and R. J. Knipe. *Geological Society of London Special Publications*, **147**, 117–34.

Fishwick, S., Heintz, M., Kennett, B. L. N., Reading, A. M., and Yoshizawa, K. (2008). Steps in lithospheric thickness within eastern Australia: evidence from surface wave tomography. *Tectonics*, 27.

Fitzgerald, P. G. (1992). The Transantarctic Mountains of southern Victoria Land: the application of apatite fission track analysis to a rift shoulder uplift. *Tectonics*, 11, 634–62.

Fjellanger, E., Olsen, T. R., and Rubino, R. L. (1996). Sequence stratigraphy and paleogeography of the Middle Jurassic Brent and Vestland deltaic systems, Northern North Sea. *Norsk Geologist Tiddskrift*, 76, 75–106.

Fleitout, L., Froidevaux, C., and Yuen, D. (1986). Active lithosphere thinning. *Tectonophysics*, 132, 271–8.

Fleitout, L. and Yuen, D. A. (1984a). Secondary convection and the growth of the oceanic lithosphere. *Physics of the Earth and Planetary Interiors*, 36, 181–212.

(1984b). Steady state, secondary convection beneath lithospheric plates with temperature- and pressure-dependent viscosity. *Journal of Geophysical Research*, 89, 9227–44.

Florensov, N. A. (1966). *The Baikal Rift Zone*. Nauka, Moscow.

Flournoy, L. A. and Ferrell, R. E. (1980). Geopressure and diagenetic modifications of porosity in the Lirette field area, Terrebonne Parish, Louisiana. *Gulf Coast Geological Association Transactions*, 30, 341–5.

Flovenz, O. G. and Saemundsson, K. (1993). Heat flow and geothermal processes in Iceland. *Tectonophysics*, 225, 123–38.

Forbes, P. L., Ungerer, P. M., Kuhfuss, A. B., Riis, F., and Eggen, S. (1991). Compositional modeling of petroleum generation and expulsion: trial application to a local mass balance in Smorbuk Sor field, Haltenbanken area, Norway. *American Association of Petroleum Geologists Bulletin*, 75, 873–93.

Fournier, R. O. (1991). The transition from hydrostatic to greater than hydrostatic fluid pressure in presently active continental hydrothermal systems in crystalline rock. *Geophysical Research Letters*, 18, 955–8.

(1999). Hydrothermal processes related to movement of fluid from plastic into brittle rock in the magmatic – epithermal environment. *Economic Geology*, 94, 1193–212.

Fournier, R. O., White, D. E., and Truesdell, A. H. (1974). Geochemical indicators of subsurface temperatures: Part 1. Basic assumptions. *Journal of Geophysical Research*, 101, 25499–509.

Foster, D. A. and Gleadow, A. J. W. (1994). Tertiary structural framework of basement flanking the Kenya Rift: constraints from an apatite fission-track thermal-stratigraphy. *Abstracts – Geological Society of Australia*, 37, 115.

Fowler, M. G. and McAlpine, K. D. (1994). The Egret Member: a prolific Kimmeridgian source rock from offshore eastern Canada. In *Petroleum Source Rocks*, ed. B. J. Katz. Springer-Verlag, New York, pp. 111–30.

Francis, T. J. G., Davies, D., and Hill, M. N. (1966). Crustal structure between Kenya and the Seychelles. *Royal Society of London Philosophical Transactions*, 259, 240–61.

Francis, T. J. G. and Shor, G. G. (1966). Seismic refraction measurement in the northwest Indian Ocean. *Journal of Geophysical Research*, 71, 427–49.

Frape, S. K. and Fritz, P. (1987). Geochemical trends for groundwaters from the Canadian Shield. In *Saline Water and Gases in Crystalline Rocks*, ed. P. Fritz and S. K. Frape. *Geological Association of Canada Special Papers*, 33, 19–38.

Fraser, A. R. and Tonkin, P. C. (1991). The Glamis Field, Block 16/21a, UK North Sea. In *United Kingdom Oil and Gas Fields, 25 Years Commemorative Volume*, ed. I. L. Abbotts. *Memoir of the Geological Society of London*, 14, 317–22.

Fraser, S., Robinson, A. Q., Johnson, H., Underhill, J., Kadolsky, D., Connell, R., Johannessen, P., and Ravnas, R. (2002). Chapter 11. Upper Jurassic. In *The Millennium Atlas: Petroleum Geology of the Central and Northern North Sea*, ed. D. Evans, C. Graham, A. Armour, and P. Bathurst. *Geological Society of London*, pp. 157–89.

Frederiksen, S., Nielsen, S. B., and Balling, N. (2001). Post-Permian evolution of the central North Sea: a numerical model. *Tectonophysics*, 343, 185–203.

Freeze, R. A. and Cherry, J. A. (1979). *Groundwater*. Prentice Hall, Englewood Cliffs, New Jersey, p. 604.

Frey, F. A., McNaughton, N. J., Nelson, D. R., de Laeter, J. R., and Duncan, R. T. A. (1996). Petrogenesis of the Bunbury Basalt, Western Australia: between the Kerguelen plume and Gondwana lithosphere? *Earth and Planetary Science Letters*, 144(1–2), 163–83.

Friedman, G. M. (1999). Thermal anomalies associated with forced and free ground-water convection in the Dead Sea Rift Valley: Discussion. *GSA Bulletin*, 111, 1098–101.

Fristad, T., Gorth, A., Yielding, G., and Freeman, B. (1997). Quantitative fault seal prediction – a case study from Oseberg Syd. In *Hydrocarbon Seals – Importance for Exploration and Production*, ed. P. Moller-Pederson and A. G. Koestler. *Special Publication of the Norwegian Petroleum Society*, 7, 107–24.

Frost, B. R. and Bucher, K. (1994). Is water responsible for geophysical anomalies in the deep continental crust? A petrological perspective. *Tectonophysics*, 231, 293–309.

Frostick, L. and Reid, I. (1987). Is structure the main control of river drainage and sedimentation in rifts? *Journal of African Earth Sciences*, 8, 165–82.

Fu, J., Shen, G., Peng, P., Brassell, S. C., Elington, G., and Liang, J. (1986). Peculiarities of salt lake sediments as potential source rocks in China. *Organic Geochemistry*, 10, 119–26.

Fuis, G. S. and Kohler, W. M. (1984). Crustal structure and tectonics of the Imperial Valley region, California. *Field Trip Guidebook – Pacific Section: Society of Economic Paleontologists and Mineralogists*, 40, 1–13.

Funck, T., Hopper, J. R., Larsen, H. C., Louden, K. E., Tucholke, B. E., and Holbrook, W. S. (2003). Crustal structure of the ocean–continent transition at Flemish Cap: seismic refraction results. *Journal of Geophysical Research*, 108, 20.

Funck, T. and Schmincke, H. U. (1998). Growth and destruction of Gran Canaria deduced from seismic reflection and bathymetric data. *Journal of Geophysical Research*, 103, 15393–407.

Fusion Oil and Gas (2002). *Deepwater Northwest Africa: A Geological Tour*. Presentation in Cape Town.

Fyfe, J. A., Gregersen, U., Jordt, H., Rundberg, Y., Eidvin, T., Evans, D., Stewart, D., Hovland, M., and Andresen, P. (2003). Oligocene to Holocene. In *The Millennium Atlas: Petroleum Geology of the Central and Northern North Sea*, ed. D. Evans, C. Graham, A. Armour, and P. Bathurst. *The Geological Society of London*, pp. 279–87.

Gaarenstrom, L., Tromp, R. A. J., Jong, M. C., and Brandenberg, A. M. (1993). Overpressures in the Central North Sea: implications for trap integrity and drilling safety. In *Petroleum Geology of Northwest Europe*, ed. J. R. Parker. *Proceedings of the 4th Conference, The Geological Society of London*, 1305–13.

Gaina, C., Muller, R. D., Brown, B., and Ishihara, T. (2003). Micro-continent formation around Australia. In *The Evolution and Dynamics of the Australian Plate*, ed. R. Hillis and R. D. Muller. *Joint Geological Association of Australia and Geological Society of America Special Paper*, 22, 399–410.

Galanis, Jr. S. P., Sass, J. H., Munroe, R. J., and Abu-Ajamieh, M. (1986). Heat flow at Zerka-Ma'in and Zara and a geothermal reconnaissance of Jordan. *US Geological Survey Open-File Report*, 86, 110.

Galbraith, R. F. (1988). Graphical display of estimates having differing standard errors. *Technometrics*, 30, 271–81.

(1992). Statistical models for mixed ages. *International Workshop on Fission Track Thermochronology*, Philadelphia.

Gallagher, K. and Brown, R. (1999). The Mesozoic denudation history of the Atlantic margins of southern Africa and southeast Brazil and the relationship to offshore sedimentation. In *The Oil and Gas Habitats of the South Atlantic*, ed. N. R. Cameron, R. H. Bate, and V. S. Clure. *Geologic Society of London Special Publications*, 153, 41–53.

Gallagher, K., Hawkesworth, C. J., and Mantovani, M. S. M. (1994). The denudation history of the onshore continental margin of SE Brazil inferred from apatite fission track data. *Journal of Geophysical Research*, 99, 18117–45.

Galloway, W. E. (1989a). Depositional framework and hydrocarbon resources of the early Miocene (Fleming) episode, Northwest Gulf Coast Basin. *Marine Geology*, 90, 19–29.

(1989b). Genetic stratigraphic sequences in basin analysis; I, Architecture and genesis of flooding-surface bounded depositional units. *American Association of Petroleum Geologists Bulletin*, 73, 125–42.

Gamond, J. F. (1987). Bridge structures as sense of displacement criteria in brittle fault zones. *Journal of Structural Geology*, 9, 609–20.

Gans, P. B. (1987). An open-system, two-layer crustal stretching model for the eastern Great Basin. *Tectonics*, 6, 1–12.

Garcia, A. J. V., Morad, S., De Ros, L. F., and Al-Aasm, I. S. (1998). Palaeogeographical, palaeoclimatic and burial history controls on the diagenetic evolution of reservoir sandstones: evidence from the Lower Cretaceous Serraria sandstones in the Sergipe-Alagoas Basin, NE Brazil. *Special Publication of the International Association of Sedimentologists*, 26, 107–40.

Garden, I. R., Guscott, S. C., Burley, S. D., Foxford, K. A., Walsh, J. J., and Marshall, J. (2001). An exhumed palaeo-hydrocarbon migration fairway in a faulted carrier system, Entrada Sandstone of SE Utah, U.S.A. *Geofluids*, 1, 195–213.

Gardner, G. H. F., Gardner, L. W., and Gregory, A. R. (1974). Formation velocity and density – the diagnostic basics for stratigraphic traps. *Geophysics*, 39, 770–80.

Garfunkel, Z. B. Y (1977). The tectonics of the Suez Rift. *Bulletin – Geological Survey of Israel*, 71, 44.

Garland, C. R. (1993). Miller Field: reservoir stratigraphy and its impact on development. In *Petroleum Geology of Northwest Europe. Proceedings of the 4th Conference*, ed. J. R. Parker. *Geological Society of London*, 401–14.

Garven, G. (1989). A hydrogeologic model for the formation of the giant oil sand deposits of the western Canada sedimentary basin. *American Journal of Science*, 289, 105–66.

Garven, G. and Freeze, R. A. (1984a). Theoretical analysis of the role of groundwater flow in the genesis of stratabound ore deposits: 1. Mathematical and numerical model. *American Journal of Science*, 284, 1085–124.

(1984b). Theoretical analysis of the role of groundwater flow in the genesis of stratabound ore deposits: 2. Quantitative results. *American Journal of Science*, 284, 1124–56.

Gaulier, J. M., Le Pichon, X., Lyberis, N., Avedik, F., Gely, I., Moretti, I., Deschamps, A., and Hafez, S. (1988). Seismic study of the crustal thickness, northern Red Sea and Gulf of Suez. *Tectonophysics*, 153, 55–88.

Gawthorpe, R. L., Jackson, C. A. L., Young, M. J., Sharp, I. R., Moustafa, A. R., and Leppard, C. W. (2003). Normal fault growth, displacement localisation and the evolution of normal fault populations: the Hammam Faraun fault block, Suez Rift, Egypt. *Journal of Structural Geology*, 25, 883–95.

Gawthorpe, R. L., Fraser, A. J., and Collier, R. E. L. (1994). Sequence stratigraphy in active extensional basins: implications for the interpretation of ancient basin-fills. *Marine and Petroleum Geology*, 11, 642–58.

Gawthorpe, R. L. and Hurst, J. M. (1993). Transfer zones in extensional basins: their structural style and influence on drainage development and stratigraphy. *Journal of the Geological Society of London*, 150, 1137–52.

Gawthorpe, R. L., Hurst, J. M., and Sladen, C. P. (1990). Evolution of Miocene footwall-derived coarse-grained deltas, Gulf of Suez, Egypt: implications for exploration. *American Association of Petroleum Geologists Bulletin*, 74, 1077–86.

Gawthorpe, R. L. and Leeder, M. R. (2000). Tectono-sedimentary evolution of active extensional basins. *Basin Research*, 12, 195–218.

Gawthorpe, R. L., Sharp, I. R., Underhill, J. R., and Gupta, S. (1997). Linked sequence stratigraphic and structural evolution of propagating normal faults. *Geology*, 25, 795–8.

Ge, S. and Garven, G. (1989). Tectonically induced transient groundwater flow in foreland basins. In *Origin and Evolution of Sedimentary Basins and Their Energy and Mineral Resources*, ed. R. A. Price. American Geophysical Union, Geophysical Monograph, 48, 145–57.

(1992). Hydromechanical modeling of tectonically driven groundwater flow with application to the Arkoma foreland basin. *Journal of Geophysical Research*, 97, 9119–44.

Ge, T. and Chen, Y. (1993). *Liaohe Oil Field: Petroleum Geology of China III (in Chinese)*. Petroleum Industry Press, Beijing, p. 39.

Gebauer, D. (1993). The pre-Alpine evolution of the continental crust of the Central Alps – an overview. In *Pre Mesozoic Geology in the Alps*, ed. J. F. Von Raumer and F. Neubauer. Springer, Berlin, pp. 93–117.

Gee, M., Watts, A. B., Masson, D. G., and Mitchell, N. C. (2001). Landslides and the evolution of the El Hierro in the Canary Island. *Marine Geology*, 177, 271–93.

Geiser, P. (1974). Cleavage in some sedimentary rocks of the central Valley and Ridge province, Maryland. *Geological Society of America Bulletin*, 85, 1399–412.

Genik, G. J. (1993). Petroleum geology of the Cretaceous-Tertiary rift basins in Niger, Chad and Central African Republic. *American Association of Petroleum Geologists Bulletin*, 77, 1405–34.

Geoffroy, L. (1994). *Contraintes et volcanisme: Le domaine Nord-Est Atlantique au Tertiaire inferieur*. PhD thesis, University of Paris, p. 426.

(2001). The structure of volcanic margins: some problematics from the North Atlantic/Baffin Bay system. *Marine and Petroleum Geology*, 18, 463–9.

(2005). Volcanic passive margins. *Comptes Rendus Geosciences*, 337, 1395–408.

Geoffroy, L., Callot, J. P., Caillet, S., Skuce, A. S., Gelard, J. P., Ravilly, M., Angelier, J., Bonin, B., Cayet, C., Perrot-Galmiche, K., and Lepvrier, C. (2001). Southeast Baffin volcanic margin and the North American–Greenland plate separation. *Tectonics*, 20, 566–84.

George, R., Rogers, N., and Kelley, S. (1998). Earliest magmatism in Ethiopia: evidence for two mantle plumes in one flood basalt providence: *Geology*, 26, 923–6.

Geotrack-International (1999a). *(U-Th)/He dating of apatite. Improved resolution at low temperatures. Geotrack Information Sheet 99/8*. Geotrack International Pty Ltd, Brunswick West.

Geotrack-International (1999b). *Thermal history interpretation of AFTA data. Geotrack Information Sheet 99/2*. Geotrack International Pty Ltd, Brunswick West.

Gernigon, L. (2002). *Extension et magmatisme en context de marge passive volcanique: deformation et satructure crustale de la marge norvegienne externe*. PhD thesis. University of Eastern Bretagne, p. 301.

Geyh, M. A. and Schleicher, H. (1990). *Absolute Age Determination: Physical and Chemical Dating Methods and Their Application*. Springer, Berlin.

Ghebreab, W. and Talbot, C. J. (2000). Red Sea extension influenced by pan-African tectonic grain in eastern Eritrea. *Journal of Structural Geology*, 22, 931–46.

Ghignone, J. I. (1979). Geologia dos sedimentos fanerozoicos do estado da Bahia. Geology of Phanerozoic sediments in Bahia. In *Quantificacao dos recursos hidricos subterraneos do aquifero reconcavo na bacia do rio Capivara*, ed. S. S. Cavalcanti. Tese de Doutorado em Geofísica. Instituto de Geociencias, Universidade Federal da Bahia. Salvador – Bahia, p. 121.

(1986). Estratigrafia, estrutura e possibilidades de petróleo das bacias do Araripe, Iguatu e Rio do Peixe. Anais do XXXIV Congresso Brasileiro de Geologia, Goiânia, **1**, 271–85.

Giambalvo, E. R., Steefel, C. I., Fisher, A. T., Rosenberg, N. D., and Wheat, C. G. (2002). Effects of fluid-sediment reaction on hydrothermal fluxes of major elements, eastern flank of the Juan de Fuca Ridge. *Geochimica et Cosmochimica Acta*, 66, 1739–57.

Gibbs, A. D. (1984). Structural evolution of extensional basin margins. *Journal of the Geological Society of London*, 141, 609–20.

Gibbson, I. L. and Love, D. (1989). A listric fault model for the formation of the dipping reflectors penetrated during the drilling of hole 642E, ODP Leg 104. In *ODP Proceedings*, ed. O. Eldholm, J. Thiede, E. Taylor and colleagues. *Scientific Results*, 104, 979–83.

Gilchrist, A. R. and Summerfield, M. A. (1990). Differential denudation and flexural isostasy in formation of rifted-margin upwarps. *Nature*, 346, 739–42.

Gillot, P.-Y. and Cornette, Y. (1986). The Cassignol technique for potassium-argon dating, precision and accuracy: examples from the late Pleistocene to Recent volcanics from southern Italy. *Chemical Geology: Isotope Geoscience Section*, 59, 205–22.

Gilpin, B. and Lee, T. C. (1978). A microearthquake study in the Salton Sea geothermal area, California. *Bulletin of the Seismological Society of America*, 68, 441–50.

Girdler, R. W. (1970). A review of Red Sea heat flow. *Philosophical Transactions of the Royal Society of London*, 267, 191–203.

Gladczenko, T., Hinz, K., Eldholm, O., Meyer, H., Neben, S., and Skogseid, J. (1997). South Atlantic volcanic margins. *Journal of the Geological Society of London*, 154, 465–70.

Glasby, G. P., Stuben, D., Jeschke, G., Stoffers, P., and Garbe-Schonberg, C.-D. (1997). A model for the formation of the hydrothermal manganese crusts from the Pitcairn Island Hotspot. *Geochimica et Cosmochimica Acta*, 61, 4583–97.

Glasmann, J. R., Lundegard, P. D., Clark, R. A., Penny, B. K., and Collins, I. D. (1989). Geochemical evidence for the history of diagenesis and fluid migration: Brent Sandstone, Heather Field, North Sea. *Clay Minerals*, 24, 255–84.

Glasspool, I. J., Edwards, D., and Axe, L. (2004). Charcoal in the Silurian as evidence for the earliest wildfire. *Geology*, 32, 381–3.

Gleadow, A. J. W. and Fitzgerald, P. G. (1987). Uplift history and structure of the Transantarctic Mountains: new evidence from fission track dating of basement apatites in the Dry Valleys area, southern Victoria Land. *Earth and Planetary Science Letters*, 82, 1–14.

Gleason, G. C. and Tullis, J. (1995). A flow law for dislocation creep of quartz aggregates determined with the molten salt cell. *Tectonophysics*, 247, 1–23.

Glennie, K. W. and Armstrong, L. A. (1991). The Kittiwake Field, Block 21/18, UK North Sea. In *United Kingdom Oil and Gas Fields, 25 Years Commemorative Volume*, ed. I. L. Abbotts. *Memoir of the Geological Society of London*, 14, 339–45.

Goddard, J. V. and Evans, J. P. (1995). Chemical changes and fluid-rock interaction in faults of crystalline thrust sheets, northwestern Wyoming, U.S.A. *Journal of Structural Geology*, 17, 533–47.

Godfrey, N. J., Beaudoin, B. C., Klemperthe, S. L., Levander, A., Luetgert, J., Meltzer, A., Mooney, W., and Trehu A. (1997). Ophiolitic basement to the Great Valley forearc basin, California, from seismic and gravity data: implications for crustal growth at the North American continental margin. *Geological Society of America Bulletin*, 109, 1536–62.

Goetze, C. and Evans, B. (1979). Stress and temperature in the bending lithosphere as constrained by experimental rock mechanics. *Geophysical Journal of the Royal Astronomic Society*, 59, 463–78.

Goff, F. (1980). Geology of the Geysers-Clear Lake geothermal regime, Northern California. In *Continental Scientific Drilling Program Thermal Regimes: Comparative Site Assessment Geology of Five Magma-Hydrothermal Systems*, ed. F. Goff and A. C. Waters. *Los Alamos Scientific Laboratory*, Report LA-8550-OBES, pp. 7–28.

Goff, J. C. (1983). Hydrocarbon generation and migration from Jurassic source rocks in the E Shetland Basin and Viking Graben of the northern North Sea. *Journal of the Geological Society of London*, 140, 445–74.

Goldflam, P., Hinz, K., Weigel, W., and Wissmann, G. (1980). Some features of the northwest African margin and magnetic quiet zone. *Philosophical Transactions of the Royal Society of London, Series A: Mathematical and Physical Sciences*, 294, 87–96.

Golonka, J. (2000). *Cambrian-Neogene Plate Tectonic Maps*. Wydawnictwo Uniwersytetu Jagiellonskiego, Krakow, p. 125.

Gomes, J. R. C., Gatto, C. M. P. P., Souza, G. M. C., Luz, D. S., Pires, J. L., and Teixeira, W. (1981). Projeto Radambrasil, Folhas SB 24–25, Jaguaride – Natal. *Geologia, geomorfologia, pedologia, vegetação e uso potencial da terra*, 27, 741.

Gomes, P. O., Gomes, B. S., Palma, J. J. C., Jinno, K., and de Souza, J. M. (2000). Ocean-continent transition and tectonic framework of the oceanic crust at the continental margin of NE Brazil: results of LEPLAC Project. *Geophysical Monograph*, 115, 261–91.

Goolsby, S. M., Druyff, L., and Fryt, M. S. (1988). Trapping mechanisms and petrophysical properties of the Permian Kaibab Formation, south-central Utah. In *Properties of Carbonate Reservoirs in the Rocky Mountain Region Occurrence and Petrophysical*, ed. S. M. Goolsby and M. W. Longman. Denver, Rocky Mountain Association of Geologists, 1988 Guidebook, pp. 193–210.

Goth, K., de Leeuw, J. W., Puttmann, W., and Tegelaar, E. W. (1988). Origin of Messel oil shale kerogen. *Nature*, 336, 759–61.

Govers, R. and Wortel, M. J. R. (1995). Extension of stable continental lithosphere and the initiation of lithospheric scale faults. *Tectonics*, 14, 1041–55.

Gowers, M. B., Holtar, E., and Swensson, E. (1993). The structure of the Norwegian Central Trough (Central Graben area). In *Petroleum Geology of Northwest Europe*, ed. J. R. Parker. *Proceedings of the 4th Conference, The Geological Society of London*, 1245–54.

Grace, J. D. and Hart, G. F. (1986). Giant fields of northern West Siberia. *American Association of Petroleum Geologists Bulletin*, 70, 830–52.

Gradijan, S. J. and Wiik, M. (1987). Statfjord Nord. In *Geology of the Norwegian Oil and Gas Fields*, ed. A. M. Spencer, C. J. Campbell, S. H. Hanslien, E. Holter, P. H. H. Nelson, E. Nysaether, and G. Ormaasen. Graham and Trotman, London, pp. 341–50.

Gradstein, F. M., Ogg, J. G., Smith, A. G., Bleeker, W., and Lourens, L. J. (2004). A new geologic time scale, with special reference to Precambrian and Neogene. *Episodes*, 27(2), 83–100.

Graham, S. A. (1976). *Tertiary sedimentary tectonics of the central Salinian Block of California*. PhD thesis, Stanford University, California, p. 510.

Graham, S. A. and Williams, L. A. (1985). Tectonic, depositional, and diagenetic history of Monterey Formation (Miocene), central San Joaquin Basin, California. *American Association of Petroleum Geologists Bulletin*, 69, 385–411.

Grandt, C. (2005). http://www.christoph-grandt.com.

Gras, R. and Thusu, B. (1998). Trap architecture of the Early Cretaceous Sarir sandstone in the eastern Sirt Basin,

Libya. In *Petroleum Geology of North Africa*, ed. D. S. MacGregor, R. T. J. Moody, and D. D. Clark-Lowes. *Geological Society of London Special Publications*, **132**, 317–34.

Grasemann, B. and Mancktelow, N. S. (1993). Two dimensional thermal modeling of normal faulting: the Simplon Fault Zone, central Alps, Switzerland. *Tectonophysics*, 225, 155–65.

Gratier, J. P. and Guiguet, R. (1986). Experimental pressure solution-deposition on quartz grains: the crucial effect of the nature of the fluid. *Journal of Structural Geology*, 8, 845–56.

Graue, K. (2000). Mud volcanoes in deepwater Nigeria. *Marine and Petroleum Geology*, 17, 959–74.

Grauls, D. J. and Baleix, J. M. (1994). Role of overpressures and in situ stresses in fault-controlled hydrocarbon migration: a case study. *Marine and Petroleum Geology*, 11, 734–42.

Gray, D. I. (1987). Troll. In *Geology of the Norwegian Oil and Gas Fields*, ed. A. M. Spencer, C. J. Campbell, S. H. Hanslien, E. Holter, P. H. H. Nelson, E. Nysaether, and G. Ormaasen. Graham and Trotman, London, pp. 389–401.

Green, P. F. (1981). A new look at statistics in fission-track dating. *Nuclear Tracks*, 5, 77–86.

Green, P. F., Crowhurst, P. V., and Duddy, I. R. (2004). Integration of AFTA and (U-Th)/He thermochronology to enhance the resolution and precision of the thermal history reconstruction in the Anglesea-1 well, Otway Basin, SE Australia. PESA Eastern Australian Basins Symposium II. Adelaide, September 19–22, 2004, pp. 117–31.

Green, P. F., Duddy, I. R., Gleadow, A. J. W., Tingate, P. R., and Laslett, G. M. (1986a). Fission track annealing in apatite: track length measurements and the form of the Arrhenius plot. *Nuclear Tracks*, 10, 323–8.

(1986b). Thermal annealing of fission tracks in apatite 1. A qualitative description. *Chemical Geology*, 59, 237–53.

Green, P. F., Duddy, I. R., and Hegarty, K. A. (2002). Quantifying exhumation from apatite fission-track analysis and vitrinite reflectance data: precision, accuracy and latest results from the Atlantic margin of NW Europe. In *Exhumation of the North Atlantic Margin: Timing, Mechanisms and Implications for Petroleum Exploration*, ed. A. G. Dore, J. Cartwright, M. S. Stoker, J. P. Turner, and N. White. *Geological Society of London Special Publications*, **196**, 331–54.

Green, P. F., Duddy, I. R., Hegarty, K. A., Gleadow, A. J. W., and Lovering, J. F. (1989). Thermal annealing of fission tracks in apatite 4. Quantitative modeling techniques and extention to geological timescales. *Chemical Geology*, 79, 155–82.

Green, W. V., Achauer, U., and Meyer, R. P. (1991). A three-dimensional seismic image of the crust and upper mantle beneath the Kenya Rift. *Nature*, 354, 199–203.

Gregory, A. R. (1977). Aspects of rock physics from laboratory and log data that are important to seismic interpretation. In *Seismic Stratigraphy, Applications to Hydrocarbon Exploration*, ed. C. E. Payton. *American Association of Petroleum Geologists Memoir*, 26, 15–46.

Gretener, P. E. (1981). Geothermics: using temperature in hydrocarbon exploration. *American Association of Petroleum Geologists Short Course Notes*, 17, 1–156.

Griffin, W. L., O'Reilly, S. Y., Alonso, J. C., and Begg, G. C. (2009). The composition and evolution of lithospheric mantle: a re-evaluation and its tectonic implications. *Journal of Petrology*, 50, 1185–204.

Griffin,W. L., O'Reilly, S. Y., and Ryan, C. G. (1999a). The composition and origin of subcontinental lithospheric mantle. In *Mantle Petrology: Field Observations and High Pressure Experimentation: A Tribute to Francis F. (Joe) Boyd*, ed. Y. Fei, C. M. Bertka, and B. O. Mysen. The Geochemical Society, pp. 13–45.

Griffin, W. L., Shee, S. R., Ryan, C. G., Win, T. T., Wyatt, B. A. (1999b). Harzburgite to lherzolite and back again: metasomatic processes in ultramafic xenoliths from the Wesselton kimberlite, Kimberley, South Africa. *Contributions to Mineralogy and Petrology*, 134, 232–50.

Griffiths, R. W. and Campbell, I. H. (1991). Interaction of mantle plume heads with the Earth's surface and onset of small-scale convection. *Journal of Geophysical Research*, 96, 18295–310.

Grist, A. and Zentilli, M. (2000). Apatite fission track constraints on a widespread Early Cretaceous heating episode in the Canadian Atlantic margin: added effect of a climatically perturbed gradient? http://earthsciences.dal.ca/people/zentilli/zenabs13.htm.

Groschel-Becker, H. M. (1996). *Formational processes of oceanic crust at sedimented spreading centers; perspectives from the West African continental margin and Middle Valley, Juan de Fuca Ridge*. PhD thesis, University of Miami (Florida), Coral Gables.

Gross, G. W. and Wilcox, R. (1983). Groundwater circulation in the Socorro geothermal area. *Field Conference Guidebook*, New Mexico Geological Society, 34, 311–18.

Grove, T. L. and Parman, S. W. (2004). Thermal evolution of the Earth as recorded by komatiites. *Earth and Planetary Science Letters*, 219, 173–87.

Grunau, H. R. (1981). Worldwide review of seals for major accumulation of natural gas. *American Association of Petroleum Geologists Bulletin*, 65, 933.

Guardado, L. R., Gamboa, L. A. P., and Lucchesi, C. F. (1990). Petroleum geology of the Campos Basin, Brazil: a model for a producing Atlantic-type basin. In Divergent/Passive Margins, ed. J. D. Edwards and P. A. Santogrossi. *American Association of Petroleum Geologists Memoir*, 48, 3–79.

Guillou, H., Carracedo, J. C., Perez Torrado, F., and Rodriguez Badiola, E. (1996a). K-Ar ages and magnetic stratigraphy of ahotspot-induced, fast grown

oceanic island: El Hiero, Canary Islands. *Journal of Volvanology and Geothermal Research*, 73, 141–55.
Guillou, H., Turpin, L. and Garcia, M. O. (1996b). Unspiked K-Ar dating of young submarine volcanic rocks from Loihi and Mauna Loa. *Eos, Transactions, American Geophysical Union*, 77, 812.
Guillou, H., Turpin, L., Garnier, F., Charbit, S., and Thomas, D. M. (1997). Unspiked K-Ar dating of Pleistocene tholeiitic basalts from the deep core SOH-4, Kilauea, Hawaii. *Chemical Geology*, 140, 81–8.
Guillou-Frottier, L., Burov, E., Nehlig, P., and Wyns, R. (2007). Deciphering plume–lithosphere interactions beneath Europe from topographic signatures, *Global and Planetary Change*, 58, 119–40.
Guiraud, R. and Bosworth, W. (1997). Senonian basin inversion and rejuvenation of rifting in Africa and Arabia: synthesis and implications to plate-scale tectonics. *Tectonophysics*, 282, 39–82.
Guiraud, R. and Maurin, J. C. (1991). Le Rifting en Afrique au Cretace inferieur: synthese structurale, mise en evidence de deux etapes dans la genese des bassins, relations avec les ouvertures oceaniques peri-africaines. Lower Cretaceous rifting in Africa: structural synthesis, evidence for two-stage basin formation, relationship to the opening of the peri-African oceans. *Bulletin de la Societe Geologique de France, Huitieme Serie*, 162, 811–23.
(1992). Early Cretaceous rifts of Western and Central Africa: an overview. *Tectonophysics*, 213, 153–68.
Gulbrandsen, A. (1987). Agat. In *Habitat of Hydrocarbons on the Norwegian Continental Shelf*, ed. A. M. Spencer and 6 others. Graham and Trotman, London, pp. 363–70.
Guliyev, I. S., Feyzullayev, A. A., and Tagiyev, M. F. (2001a). Source potential of the Mesozoic-Cenozoic rocks in the South Caspian Basin and their role in forming the oil accumulations in the Lower Pliocene reservoirs. *Petroleum Geoscience*, 7, 409–17.
Guliyev, I. S., Tagiyev, M. F., and Feyzullayev, A. A. (2001b). Geochemical characteristics of organic matter from Maykop rocks of eastern Azerbaijan. *Lithology and Mineral Resources*, 36, 280–5.
Gunnell, Y., Gallagher, K., Carter, A., Widdowson, M., and Hurford, A. J. (2003). Denudation history of the continental margin of western peninsular India since the early Mesozoic – reconciling apatite fission-track data with geomorphology. *Earth and Planetary Science Letters*, 215, 187–201.
Gunnell, Y. and Radhakrishna, B. P. (2001). Shyadri, the Great Escarpment of the Indian subcontinent. Patterns of landscape development in the Western Ghats. *Geological Society of India Memoir*, 47, 1054.
Gupta, S., Cowie, P. A., Dawers, N. H., and Underhill, J. R. (1998). A mechanism to explain rift-basin subsidence and stratigraphic patterns through fault-array evolution. *Geology*, 26, 595–8.
Gupta, S., Underhill, J. R., Sharp, I. R., and Gawthorpe, R. L. (1999). Role of fault interactions in controlling synrift sediment dispersal patterns: Miocene, Abu Alaqa Group, Suez Rift, Sinai, Egypt. *Basin Research*, 11, 167–89.
Gussow, W. C. (1954). Differential entrapment of gas and oil: a fundamental principle. *American Association of Petroleum Geologists Bulletin*, 38, 816–53.
Gustavson Associates, Inc. (1992). *Petroleum Geology and Exploration Potential in the Former Soviet Republics*. Gustavson Associates, Inc., Boulder.
Guterch, A., Grad, M., Janik, T., Materzok, R., Luosto, U., Yliniemi, J., Lück, E., Schulze, A., and Förste, K. (1994). Crustal structure of the transition zone between Precambrian and Variscan Europe from new seismic data along LT-7 profile (NW Poland and eastern Germany). *Comptes Rendus de l'Academie des Sciences, Serie II, Sciences de la Terre et des Planetes*, 319, 1489–96.
Guterch, A., Kowalski, T., Materzok, R., and Toporkiewicz, S. (1976). Seismic refraction study of the Earth's crust in the Teisseyre-Tornquist line zone in Poland along the regional profile LT-2. *Publications of the Institute of Geophysics, Series A: Physics of the Earth Interior*, 2, 15–23.
Guzman, A. and Marquez-Dominguez, B. (2001). The Gulf of Mexico Basin south of the border: the petroleum province of the twenty-first century. In *Petroleum Provinces of the Twenty-first Century*, ed. M. W. Downey, J. C. Threet and W. A. Morgan. *American Association of Petroleum Geologists Memoir*, 74, 337–51.
Gvirtzman, H., Garven, G., and Gvirtzman, G. (1997). Thermal anomalies associated with forced and free ground-water convection in the Dead Sea rift valley. *Geological Society of America Bulletin*, 109, 1167–76.
Gvirtzman, H. and Stanislavsky, E. (2000). Paleohydrology of hydrocarbon maturation, migration and accumulation in the Dead Sea rift. *Basin Research*, 12, 79–93.
GX TECHNOLOGY (2005a). EquatorSPAN gravity and magnetic maps. Archive of GX Technology, Houston.
GX TECHNOLOGY (2005b). EquatorSPAN in Gabon. Archive of GX Technology, Houston.
GX TECHNOLOGY (2005c). CongoSPAN in Congo. Archive of GX Technology, Houston.
Haack, R. C., Sundararaman, P., Diedjomahor, J. O., Xiao, H., Gant, N. J., May, E. D., and Kelsch, K. (2000). Niger Delta petroleum systems, Nigeria. In *Petroleum Systems of South Atlantic Margins*, ed. M. R. Mello and B. J. Katz. *American Association of Petroleum Geologists Memoir*, 73, 213–31.
Haberland, C., Agnon, A., El-Kelani, R., Naercklin, N., Qabbani, I., Rumpker, G., Ryberg, T., Scherbaum, F., and Weber, M. (2003). Modeling of seismic guided waves at the Dead Sea Transform. *Journal of Geophysical Research*, 108.
Habermehl, G. and Hundrieser, H. J. (1983). Fossile Relikte der "Wasserblute" im Messeler Olschiefer. *Naturwissenschaften*, 70, 566–7.
Habib, D. (1982). Sediment supply origin of Cretaceous black shales. In *Nature and Origin of Cretaceous Carbon-rich*

Facies, ed. S. O. Schlanger and M. B. Cita. London, Academic Press, pp. 113–27.

Haenel, R., Rybach, L., and Stegena, L. (1988). *Handbook of Terrestrial Heat-Flow Density Determination*. Kluwer Academic Publishers, Dordrecht.

Hall, S. A. (1992). The Angus Field, a subtle trap. In *Exploration Britain: Geological Insights into the Next Decade*, ed. R. F. P. Hardman. *Special Publication of Geological Society of London*, 67, 151–85.

Hallam, A. (1984). Pre-Quaternary sea-level changes. *Annual Review of Earth and Planetary Science*, 12, 205–43.

Halliday, A. N., Dickin, A. P., Fallick, A. E., and Fitton, J. G. (1988). Mantle dynamics – a Nd, Sr, Pb and O isotopic study of the Cameroon Line volcanic chain. *Journal of Petroleum*, 29, 181–211.

Hamblin, W. K. (1965). Origin of "reverse drag" on the downthrown side of normal faults. *Geological Society of America Bulletin*, 76, 1145–64.

(1976). Patterns of displacement along the Wasatch Fault. *Geology*, 4, 619–22.

Hamdani, Y., Mareschal, J.-C., and Arkani-Hamed, J. (1994). Phase change and thermal subsidence of the Williston Basin. *Geophysical Journal International*, 116, 585–97.

Hames, W. E., Renne, P. R., and Ruppel, C. (2000). New evidence for geologically-instantaneous emplacement of earliest Jurassic Central Atlantic magmatic province basalts on the North American margin. *Geology*, 28, 859–62.

Hamilton, R. M., Behrendt, J. C., and Ackermann, H. D. (1983). Land multichannel seismic-reflection evidence for tectonic features near Charleston, South Carolina. In *Studies Related to the Charleston, South Carolina, Earthquake of 1886 – Tectonics and Seismicity*, ed. G. S. Gohn. *Geological Survey Professional Paper*, 1313, 1–18.

Hamilton, W. B. (1987). Crustal extension in the Basin and Range Province, Southwestern United States. *Geological Society Special Publications*, 28, 155–76.

Hammond, D. E., Leslie, B. W., and Ku, T.-L. (1988). 222Rn concentrations in deep formation waters and the geohydrology of the Cajon Pass borehole. *Geophysical Research Letters*, 15, 1045–8.

Hammond, W. C. and Thatcher, W. (2005). Nortwest Basin and Range tectonic deformation observed with the Global Positioning System, 1999–2003. *Journal of Geophysical Research*, 110 (B10405).

Hampson, G. J. (2004). Facies architecture and stratigraphy of: stray "shelf" sandstones, Late Cretaceous Mancos Shale, northern Utah and Colorado. *AAPG Annual Meeting Expanded Abstracts*, 13, 57058.

Hampson, G. J., Rodriguez, A. B., Storms, J. E. A., Johnson, H. D., and Meyer, C. T. (2008). Geomorphology and high-resolution stratigraphy of progradational wave-dominated shoreline deposits: impact on reservoir-scale facies. In *Recent Advances in Models of Siliciclastic Shallow Marine Stratigraphy*, ed. G. J. Hampson, R. J. Steel, P. M. Burgess, and R. W. Dalrymple. *Special Publication of Society for Sedimentary Geology*, 90, 117–42.

Hampson, G. J., Sixsmith, P. J., and Johnson, H. D. (2004). A sedimentological approach to refining reservoir architecture in a mature hydrocarbon province: the Brent Province, UK North Sea. *Marine and Petroleum Geology*, 21, 457–84.

Hancock, N. J. and Fisher, M. J. (1981). Middle Jurassic North Sea deltas with particular reference to Yorkshire. In *Petroleum Geology of the Continental Shelf of North-West Europe*, ed. L. V. Illing and G. D. Hobson. Heyden and Son, London, pp. 186–95.

Handy, M. R. and Stuenitz, H. (2002). Strain localization by fracturing and reaction weakening: a mechanism for initiating exhumation of subcontinental mantle beneath rifted margins. *Geological Society Special Publications*, 200, 387–407.

Handy, M. R. and Zingg, A. (1991). The tectonic and rheological evolution of an attenuated cross section of the continental crust: Ivrea crustal section, southern Alps, northwestern Italy and southern Switzerland. *Geological Society of America Bulletin*, 103, 236–53.

Hannington, M., Herzig, P., Stoffers, P., Scholten, J., Botz, R., Garbe-Schonberg, D., Jonasson, I. R., Roest, W., and Shipboard Scientific Party (2001). First observations of high-temperature submarine hydrothermal vents and massive anhydrite deposits off the north coast of Iceland. *Marine Geology*, 177, 199–220.

Hanor, J. S. (1979). Sedimentary genesis of hydrothermal fluids. In *Geochemistry of Hydrothermal Ore Deposits*, ed. H. L. Barnes. John Wiley, New York, pp. 137–68.

(1987). Kilometer-scale thermohaline overturn of pore fluid in the Louisiana Gulf Coast. *Nature*, 327, 501–3.

Hansen, D. L. and Nielsen, S. B. (2002). Does thermal weakening explain basin inversion? Stochastic modelling of the thermal structure beneath sedimentary basins. *Earth and Planetary Science Letters*, 198, 113–27.

Hao, F., Li, S. T., Sun, Y. C., and Zhang, Q. M. (1998). Geology, compositional heterogeneities and geochemical origin of the Yacheng gas field in the Qiongdongnan Basin, South China Sea. *American Association of Petroleum Geologists Bulletin*, 82, 1372–84.

Hao, F., Li, S. T., Gong, Z. S., and Yang, J. M. (2000). Thermal regime, inter-reservoir compositional heterogeneities, and reservoir-filling history of the Dongfang gas field, Yinggehai Basin, South China: Evidence for episodic fluid injections in overpressured basins? *American Association of Petroleum Geologists Bulletin*, 84, 607–26.

Hao, F., Zhou, X., Zhu, Y., Bao, X., and Yang, Y. (2009). Charging of the Neogene Penglai 19-3 field, Bohai Bay Basin, China: oil accumulation in a young trap in an active fault zone. *American Association of Petroleum Geologists Bulletin*, 93, 155–79.

Haq, B. U., Hardenbol, J., and Vail, P. R (1987). Chronology of fluctuating sea levels since the Triassic. *Science*, 235, 1156–67.

Hardarson, B. S., Fitton, J. G., Ellam, R. M., and Pringle, M. S. (1997). Rift relocation – a geochemical and geochronological investigation of a paleo-rift in northwest Iceland. *Earth and Planetary Science Letters*, 153, 181–96.

Harding, T. P. (1984). Graben hydrocarbon occurrences and structural styles. *American Association of Petroleum Geologists Bulletin*, 68, 333–62.

Hardy, S. and Gawthorpe, R. L. (1998). Effects of variations in fault slip rate on sequence stratigraphy in fan deltas; insights from numerical modeling. *Geology* , 26, 911–14.

Harker, S. D. (1998). The palingenesy of the Piper oil field, UK North Sea. *Petroleum Geoscience*, 4, 271–86.

Harker, S. D. and Chermak, A. (1992). Detection and prediction of Lower Cretaceous sandstone distribution in the Scapa field, North Sea. In *Exploration Britain: Geological Insights into the Next Decade*, ed. R. F. P. Hardman. *Geological Society of London Special Publications*, **67**, 221–46.

Harker, S. D., Green, S. C. H., and Romani, R. S. (1991). The Claymore field, Block 14/19, UK North Sea. In *United Kingdom Oil and Gas Fields, 25 Years Commemorative Volume*, ed. I. L. Abbotts. *Memoir of the Geological Society of London*, 14, 269–78.

Harp, E. L., Wells, W. G., and Sarmiento, J. G. (1990). Pore pressure response during failure in soils. *Geological Society of America Bulletin*, 102, 428–38.

Harris, N. B., Freeman, K. H., Pancost, R. D., White, T. S., and Mitchell, G. D. (2004). The character and origin of lacustrine source rocks in the Lower Cretaceous synrift section, Congo Basin, West Africa. *American Association of Petroleum Geologists Bulletin*, 88, 1163–84.

Harris, R. N., Von Herzen R. P., McNutt, M. K., Garven, G., and Jordahl, K. (2000). Submarine hydrogeology of the Hawaiian archipelagic apron 1. Heat flow patterns north of Oahu and Maro Reef. *Journal of Geophysical Research*, 105, 21353–69.

Harrison, T. M., Copeland, P., Hall, S. A., Quade, J., Burner, S., Ojha, T. P., and Kidd, W. S. F. (1993). Isotopic preservation of Himalayan/Tibetan uplift, denudation, and climatic histories of two molasse deposits. *Journal of Geology*, 101, 157–75.

Harry, D. L. and Sawyer, D. S. (1992). Basaltic volcanism, mantle plumes, and the mechanics of rifting: the Paraná flood basalt province of South America. *Geology*, 20, 207–10.

Hasebe, N., Barbarand, J., Jarvis, K., Carter, A., and Hurford, A. J. (2004). Apatite fission-track chronometry using laser ablation ICP-MS. *Chemical Geology*, 207, 135–45.

Haszeldine, R. S., Wilkinson, M., Darby, D., Macaulay, C. I., Couples, G. D., Fallick, A. E., Fleming, C. G., Stewart, R. N. T., and McAulay, G. (1999). Diagenetic porosity creation in an overpressured graben. *The Geological Society of London*, 5, 1339–50.

Hatcher, R. D., Osberg, P. H., and Drake Jr. A. A. (1986). *Tectonic Map of the U.S.* Appalachians, DNAG Appalachians – Ouachitas Volume, Plate 1.

Hathaway, J. C. and Degens, E. T. (1969). Methane-derived marine carbonates of Pleistocene age. *Science*, 165, 609–92.

Haughton, P. D. W., McCaffrey, W. D., Davis, C., and Barker, S. P. (2008). Sediment gravity flow deposits and bed-scale heterogeneity: lessons from North Sea fields. In *Answering the Challenges of Production from Deep-water Reservoirs: Analogues and Case Histories to Aid a New Generation*, ed. K. Schofield, N. C. Rosen, and D. Pfeiffer. *Papers presented at the Gulf Coast Section, Society of Economic Paleontologists and Mineralogists Foundation Annual Bob F. Perkins Research Conference*, 28, 407.

Hawkesworth, C. J., Gallagher, K., Kelley, S., Mantovani, M. S. M., Peate, D. W., Regelous, M., and Rogers, N. W. (1992). Parana magmatism and the opening of the South Atlantic. In *Magmatism and the Causes of Continental Break-up*, ed. B. Storey, A. Alabaster, and R. Pankhurst. Geological Society [London] Special Publication 68, pp. 221–40.

Haworth, R. T., Daniels, D. L., Williams, H., and Zietz, I. (1980). *Bouguer gravity anomaly map of the Appalachian Orogen – Map No. 3*. Memorial University of Newfoundland, St John's, Newfoundland.

Haxby, W. F. (1985). *Gravity Field of the World's Oceans. Lamont-Doherty Geological Observatory*, Columbia University, Palisades, New York. (http://www.ngdc.noaa.gov).

Hayward, N. J. and Ebinger, C. J. (1996). Variations in the along-axis segmentation of the Afar Rift system. *Tectonics*, 15, 244–57.

He, B., Xu, Y.-G., Chung, S.-L., Xiao, L., and Wang, Y. (2003). Sedimentary evidence for a rapid doming prior to the eruption of the Emeishan flood basalts. *Earth and Planetary Science Letters*, 213, 389–403.

Hébert, H., Deplus, C., Huchon, P., Khanbari, K., and Audin, L. (2001). Lithospheric structure of a nascent spreading ridge inferred from gravity data: the western Gulf of Aden. *Journal of Geophysical Research*, 106, 26345–63.

Hecker, S. (1993). Quaternary tectonics of Utah with emphasis on earthquake hazard characterization. *Utah Geological Survey Bulletin*, 127, 1–157.

Hedenquist, J. W., Arribas, A. Jr., and Reynolds, T. J. (1998). Evolution of an intrusion – centered hydrothermal system: Far Southeast – Lepanto porphyry and epithermal Cu-Au deposits, Philippines. *Economic Geology*, 93, 373–404.

Heggland, R. (1997). Detection of gas migration from a deep source by the use of exploration 3D seismic data. *Marine Geology*, 137, 41–7.

(1998). Gas seepage as an indicator of deeper prospective reservoirs: a study based on exploration 3-D seismic data. *Marine and Petroleum Geology*, 15, 1–9.

Hein, J. R., Koski, R. A., Embley, R. W., Reid, J., and Chang, S.-W. (1999). Diffuse-flow hydrothermal field in an oceanic fracture zone setting, northeast Pacific: deposit composition. *Exploration and Mining Geology*, 8, 299–322.

Hekinian, R., Hoffert, M., Larque, P., Cheminee, J. L., Stoffers, P., and Bideau, D. (1993). Hydrothermal Fe and Si oxyhydroxide deposits from the South Pacific intraplate volcanoes and East Pacific Rise axial and off-axial regions. *Economic Geology*, 88, 2099–121.

Helgeson, D. E. (1999). Structural development and trap formation in the central North Sea HP/HT play. *The Geological Society of London*, 5, 1029–34.

Heling, D. and Teichmuller, M. (1974). The transition zone between montmorillonite and mixed-layer minerals and its relation to coalification in the Graue Beds of the Oligocene in the Upper Rhine Graben. *Fortschritte in der Geologie von Rheinland und Westfalen, Inkohlung und Erdoel*, 24, 113–28.

Helland-Hansen, W., Ashton, M., Lomo, L., and Steel, R. (1992). Advance and retreat of the Brent delta: recent contributions to the depositional model. In *Geology of the Brent Group*, ed. A. C. Morton, R. S. Haszeldine, M. R. Giles, and S. Brown. *Geological Society of London Special Publications*, **61**, 109–28.

Hellinger, S. J. and Sclater, J. G. (1983). Some remarks on two-layer extensional models or the evolution of sedimentary basins. *Journal of Geophysical Research*, 88, 8251–69.

Hempton, M. R. and Neher, K. (1986). Experimental fracture, strain and subsidence patterns over en echelon strike-slip faults: implications for the structural evolution of pull-apart basins. *Journal of Structural Geology*, 8, 597–605.

Hendrie, D. B., Kusznir, N. J., Morley, C. K., and Ebinger, C. J. (1994). Cenozoic extension in northern Kenya: a quantitative model of rift basin development in the Turkana region. *Tectonophysics*, 236, 409–38.

Henk, A. (2006). *Impact of crustal rheology on subsidence history and basin geometry- numerical experiments of half-graben formation*. EGU General Assembly. University of Freiburg, Germany, Wien.

Henry, A. A. and Lewan, M. D. (2001). Comparison of kinetic-model prediction of deep gas generation. In *Geologic Studies of Deep Natural Gas Resources*, ed. T. S. Dyman and V. A. Kuuskraa. *U.S. Geological Survey Digital Data Series*, **67**, 1–25.

Henry, S., Danforth, A., and Venkatraman, S. (2005). New insights into petroleum systems and plays in Angola, the Congo and Gabon from PDSM sub-salt imaging. In *Petroleum Systems of Divergent Continental Margin Basin*, ed. P. J. Post, N. Rosen, D. L. Olson, S. L. Palmes, K. T. Lyons, and G. B. Newton. *GCSSEPM25th Annual Bob F. Perkins Research Conference*, SEPM Foundation, Houston, **25**, 49–55.

Hermanrud, C. (1986). On the importance to the petroleum generation of heating effects from compaction-derived water: an example from the northern North Sea. In *1st IFP Exploration Research Conference*, ed. J. Burrus. Technip, Paris.

Herzig, C. T., Mehegan, J. M., and Stelting, C. E. (1988). Lithostratigraphy of the State 2–14 Borehole; Salton Sea Scientific Drilling Project. *Journal of Geophysical Research*, 93, 12969–80.

Hesthammer, J. and Fossen, H. (1999). Evolution and geometries of gravitational collapse structures with examples from the Statfjord Field, northern North Sea. *Marine and Petroleum Geology*, 16, 259–81.

Hesthammer, J., Jourdan, C. A., Nielsen, P. E., Ekern, T. E., and Gibbons, K. A. (1999). A tectonostratigraphic framework for the Statfjord Field, northern North Sea. *Petroleum Geoscience*, 5, 241–56.

Hickman, S., Sibson, R. H., and Bruhn, R. (1995). Introduction to special section: mechanical involvement of fluids in faulting. *Journal of Geophysical Research*, 100, 12831–40.

Hill, D. P., Eaton, J. P., and Jones, L. M. (1990). *Seismicity 1980–86*. US Geological Survey.

Hillier, R. D. and Cosgrove, J. W. (2002). Core and seismic observations of overpressure-related deformation within Eocene sediments of the Outer Moray Firth, UKCS. *Petroleum Geoscience*, 8, 141–9.

Hindle, A. D. (1997). Petroleum migration pathways and charge concentration: a three-dimensional model. *American Association of Petroleum Geologists Bulletin*, 81, 1451–81.

Hinz, K., Neben, S., Schreckenberger, B., Roeser, H. A., Block, M. et al. (1999). The Argentine continental margin north of 48 degrees S: sedimentary successions, volcanic activity during breakup. *Marine and Petroleum Geology*, 16, 1–25.

Hinz, K. and Weber, J. (1976). Zum geologischen Augbau des Norwegischen Kontinental Randes und der Barents Sea nach Reflections seismichen Messungen. *Erdol und Kohle, Erdgas, Petrochemie*, 75–6, 3–29.

Hirth, G. and Kohlstedt, D. L. (1996). Water in the oceanic upper mantle: implications for rheology, melt extraction and the evolution of the lithosphere. *Earth and Planetary Science Letters*, 144, 93–108.

(2004). Rheology of the upper mantle and the mantle wedge: a view from the experimentalists. In *Inside the Subduction Factory*, ed. J. Eiler. *AGU Monograph*, **138**, 83–105.

Hitchon, B. (1984). Geothermal gradients, hydrodynamics, and hydrocarbon occurrences, Alberta, Canada. *American Association of Petroleum Geologists Bulletin*, 68, 713–43.

Hite, R. J. and Anders, D. E. (1991). Petroleum and evaporates. In *Evaporites, Petroleum and Mineral Resources*, ed. J. L. Melvin. Amsterdam: Elsevier, pp. 349–411.

Hochstein, M. P. and Browne, P. R. L. (2000). Surface manifestations of geothermal systems with volcanic heat sources. In *Encyclopedia of Volcanoes*, ed. H. Sigurdsson, B. F. Houghton, S. R. McNutt, H. Rymer, and J. Stix. Academic Press, San Diego, pp. 835–55.

Hodgson, N. A., Farnsworth, J., and Fraser, A. J. (1992). Salt-related tectonics, sedimentation and hydrocarbon plays in the Central Graben, North Sea, UKCS. 31–63. In *Exploration Britain: Geological Insights into the Next Decade*, R. F. P. Hardman. *Special Publication of the Geological Society of London*, 67.

Hodkinson, R. A., Stoffers, P., Scholten, J., Cronan, D. S., Jeschke, G., and Roger, T. D. S. (1994). Geochemistry of hydrothermal manganese deposits from the Pitcairn Island hotspot, southeastern Pacific. *Geochimica et Cosmochimica Acta*, 58, 5011–29.

Hoefs, J. (1981). Isotopic composition of the ocean-atmospheric system in the geologic past. In *Evolution of the Earth*, ed. R. J. O'Connell and W. S. Fyfe. *Geodynamic Series*, **5**, 110–19.

Hofmann, C., Courtillot, V., Feraud, F., Rochette, P., Yirgu, G., Ketefo, E., and Pik, R. (1997). Timing of the Ethiopian flood basalt event: implications for plume birth and global change. *Nature*, 389, 838–40.

Hohndorf, A., Haenel, R., and Giesel, W. (1975). Geothermal models of the Ivrea zone. *Journal of Geophysics*, 41, 179–87.

Hoholick, J. D., Metarko, T., and Potter, P. E. (1984). Regional variation of porosity and cement: St. Peter and Mount Simon sandstones in Illinois Basin. *American Association of Petroleum Geologists Bulletin*, 68, 753–64.

Holbrook, W. S. and Kelemen, P. B. (1993). Large igneous province on the US Atlantic margin and implications for magmatism during continental break-up. *Nature*, 364, 433–6.

Holbrook, W. S., Larsen, H. C., Korenaga, J., Dahl-Jensen, T., Reid, L. D., Kelemen, P. B., Hopper, J. R., Kent, G. M., Lizarralde, D., Bernstein, S., and Detrick, R. S. (2001). Mantle thermal structure and active upwelling during continental breakup in the North Atlantic. *Earth and Planetary Science Letters*, 190, 251–66.

Holbrook, W. S., Mooney, W. D., and Christensen, N. I. (1992). The seismic velocity structure of the deep continental crust. In *Continental Lower Crust*, ed. D. M. Fountain, R. Arculus, and R. W. Kay. *Developments in Geotectonics*, Elsevier, pp. 1–44.

Holcombe, T. L., Bryant, W. R., Bouma, A. H., Taylor, L. A., and Liu, J. Y. (2002). Northern Gulf of Mexico bathymetry and feature names. *Gulf Coast Association of Geological Societies, Technical Papers and Abstracts*, 52, 397–405.

Hold, I. M., Brussee, N. J., Schouten, S., and Sinninghe Damste, J. S. (1998). Changes in the molecular structure of a Type II-S kerogen (Monterey Formation, U.S.A.) during sequential chemical degradation. *Organic Geochemistry*, 29, 1403–17.

Holmes, M. A. (1998). Thermal diagenesis of Cretaceous sediment recovered at the Côte d'Ivoire Ghana transform margin. In *Proceedings of the Ocean Drilling Program*, ed. J. Mascle, G. P. Lohmann, and M. Moullade. *Scientific Results*, 159, 53–70.

Holt, W. E. and Stern, T. A. (1991). Sediment loading on the western platform of the New Zealand continent: implication for the strength of a continental margin. *Earth and Planetary Science Letters*, 107, 523–38.

Homberg, C., Hu, J. C., Angelier, J., Bergerat, F., and Lacombe, O. (1997). Characterization of stress perturbations near major fault zones: insights from 2-D distinct-element numerical modelling and field studies (Jura Mountains). *Journal of Structural Geology*, 19, 703–18.

Hooper, E. C. D. (1991). Fluid migration along growth faults in compacting sediments. *Journal of Petroleum Geology*, 14, 161–80.

Hooper, R. J., Fitzsimmons, R. J., Grant, N., and Vendeville, B. C. (2002). The role of deformation in controlling depositional patterns in the south-central Niger Delta, West Africa. *Journal of Structural Geology*, 24, 847–59.

Hopper, J. R. and Buck, W. R. (1996). The effect of lower crustal flow on continental extension and passive margin formation. *Journal of Geophysical Research*, 101, 20175–94.

(1998). Styles of extensional decoupling. *Geology*, 26, 699–702.

Hopper, J. R., Funck, T., Tucholke, B. E., Larsen, H. C., Holbrook, W. S., Louden, K. E., Shillington, D., and Lau, H. (2004). Continental breakup and the onset of ultraslow seafloor spreading off Flemish Cap on the Newfoundland rifted margin. *Geology*, 32, 93–6.

Hopper, J. R., Mutter, J. C., Larson, R. L., and Mutter, C. Z. (1992). Magmatism and rift margin evolution: evidence from northwest Australia. *Geology*, 20, 853–7.

Horn, M. K. (2003). Giant fields 1868–2003 (CD-ROM). In *Giant Oil and Gas Fields of the Decade 1990–1999*, ed. M. K. Halbouty. *American Association of Petroleum Geologists Bulletin*, **78**, 340.

(2004). Giant fields 1868–2004 (CD-ROM). *AAPG/Datapages Miscellaneous Data Series*, version 1.2, revision of Horn, 2003.

Horn, P., Muller-Sohnius, D., and Schult, A. (1988). Potassium-argon ages on a Mesozoic tholeiitic dike swarm in Rio Grande do Norte, Brazil. *Rev. Bras. De Geociences*, 18, 50–3.

Hornafius, J. S., Quigley, D., and Luyendyk, B. P. (1999). The world's most spectacular marine hydrocarbon seeps (Coal Oil Point, Santa Barbara Channel, California): quantification of emissions. *Journal of Geophysical Research*, 104, 20703–11.

Horsfield, B., Curry, D. J., Bohacz, K., Littke, R., Rullkötter, J., Schenk, H. J., Radke, M., Schaefer, R. G., Carroll, A. R., Isaksen, G., and Witte, E. G. (1994). Organic geochemistry of freshwater and alkaline lacustrine environments, Green River Formation, Wyoming. *Organic Geochemistry*, 22, 415–40.

Horsfield, W. T. (1980). Contemporaneous movement along crossing conjugate normal faults. *Journal of Structural Geology*, 2, 305–10.

Horváth, F. (1993). Towards a mechanical model for the formation of the Pannonian Basin. *Tectonophysics*, 226, 333–57.

House, M. A., Kohn, B. P., Farley, K. A., and Raza, A. (2000). (U-Th)/He thermochronometry in southeastern Australia; confirmation of laboratory diffusion experiments and insights into the Cenozoic thermal history of the Otway Basin. *Abstracts – Geological Society of Australia*, 58, 167–71.

House, M. A., Wernicke, B. P., Farley, K. A., and Dumitru, T. A. (1997). Cenozoic thermal evolution of the central Sierra Nevada, California, from (U-Th)/He thermochronometry. *Earth and Planetary Science Letters*, 151, 167–79.

Houseman, G. and England, P. (1986). A dynamical model of lithosphere extension and sedimentary basin formation. *Journal of Geophysical Research*, 91, 719–29.

Houseman, G. A., McKenzie, D. P., and Molnar, P. (1982). Convective instability of a thickened boundary layer and its relevance for the thermal evolution of continental convergent belts. *Journal of Geophysical Research*, 86, 6115–32.

Hovland, M., Croker, P. F., and Martin, M. (1994). Fault-associated seabed mounds (carbonate knolls?) off western Ireland and north-west Australia. *Marine and Petroleum Geology*, 11, 232–46.

Howell, D. G., Crouch, J. K., Greene, H. G., McCulloch, D. S., and Vedder, J. G. (1980). Basin development along the late Mesozoic and Cainozoic California margin: a plate tectonic margin of subduction, oblique subduction and transform tectonics. *Special Publication of the International Association of Sedimentologists*, 4, 43–62.

Hoye, T., Damsleth, E., and Hollund, K. (1994). Stochastic modeling of Troll West, with special emphasis on the thin oil zone. In *Stochastic Modeling and Geostatistics: Principles, Methods and Case Studies*, ed. J. M. Yarus and R. L. Chambers. *AAPG Computer Applications in Geology*, **3**, 217–39.

Huang, D., Li, J., Zhang, D., Huang, X., and Zhou, Z. (1991). Maturation sequence of Tertiary crude oils in the Qaidam basin and its significance in petroleum resource assessment. *Journal of Southeast Asian Earth Sciences*, 5, 359–66.

Huang, H. P. and Pearson, J. M. (1999). Source rock paleoenvironments and controls on the distribution of dibenzothiophenes in lacustrine crude oils, Bohai Bay Basin, eastern China. *Organic Geochemistry*, 30, 1455–70.

Huang, Q. (1988). Geometry and tectonic significance of Albian sedimentary dykes in the Sisteron area, SE France. *Journal of Structural Geology*, 10, 453–62.

Huang, Z., Williamson, M. A., Fowler, M. G., and McAlpine, K. D. (1994). Predicted and measured petrophysical and geochemical characteristics of the Egret Member oil source rock, Jeanne d'Arc Basin, offshore eastern Canada. *Marine and Petroleum Geology*, 11, 294–306.

Hubbert, M. K. (1940). The theory of ground-water motion. *Journal of Geology*, 48, 785–944.

(1953). Entrapment of petroleum under hydrodynamic conditions. *American Association of Petroleum Geologists Bulletin*, 37, 1954–2026.

(1969). *The Theory of Ground-Water Motion and Related Papers*. Hafner Publication, Co New York-London.

Hubbert, M. K. and Willis, D. G. (1955). Important fractured reservoirs in the United States. *Proceedings of the World Petroleum Congress, Actes et Documents, Congres Mondial du Petrole*, pp. 57–81.

Huc, A. Y. (1988). Sedimentology of organic matter. In *Humic Substances and Their Role in the Environment: Report of the Dahlem Workshop*, ed. F. H. Frimmel and R. F. Christman. *Life Sciences Research Reports*, **41**, 215–43.

Hudec, M. R. and Jackson, M. P. A. (2006). Advance of allochthonous salt sheets in passive margins and orogens. *American Association of Petroleum Geologists Bulletin*, 90, 1535–64.

Hughes, W. B., Holba, A. G., and Dzou, L. I. (1995). The ratios of dibnzothiophene to phenathrene and pristine to phytane as indicators of depositional environment and lithology of petroleum source rocks. *Geochimica et Cosmochimica Acta*, 59, 3581–98.

Huismans, R. S. (1998). *Dynamic Modelling of the Transition from Passive to Active Rifting*. Vrije University, Amsterdam, p. 182.

(1999). *Dynamic modeling of the transition from passive to active rifting. Application to the Pannonian basin*. PhD thesis, Institute of Earth Sciences, Vrije Universiteit Amsterdam, p. 196.

Huismans, R. S. and Beaumont, C. (2005). Effect of lithospheric stratification on extensional styles and rift basin geometry. In *Petroleum Systems of Divergent Continental Margin Basins*, ed. P. J. Post, N. C. Rosen, D. L. Olson, S. L. Palmes, K. T. Lyons, and G. B. Newton. *25th Annual GCSSEPM Foundation Bob F. Perkins Research Conference*, GCSSEPM, Houston, 25, 12–55.

(2007). Roles of lithospheric strain softening and heterogeneity in determining the geometry of rifts and continental margins. *Geological Society of London Special Publications*, **282**, 111–38.

(2008). Complex rifted continental margins explained by dynamical models of depth-dependent lithospheric extension. *Geology*, 36, 163–6.

(2011). Depth-dependent extension, two-stage breakup and cratonic underplating at rifted margins. *Nature*, 473, 74–9.

Huismans, R. S., Buiter, S. J. H., and Beaumont, C. (2005). Effect of plastic-viscous layering and strain softening on mode selection during lithospheric extension. *Journal of Geophysical Research*, 110, 17.

Huismans, R. S., Podladchikov, Y. Y., and Cloetingh, S. A. P. L. (2001). Transition from passive to active rifting: relative importance of asthenospheric doming and passive extension of the lithosphere. *Journal of Geophysical Research*, 106, 11271–91.

Hulen, J. B., Goff, F., Ross, J. R., Bortz, L. C., and Bereskin, S. R. (1994). Geology and geothermal origin of Grant Canyon and Bacon Flat oil fields, Railroad Valley, Nevada. *American Association of Petroleum Geologists Bulletin*, 78, 596–623.

Hulen, J. B., Norton, D., Kaspereit, D., Murray, L., van de Putte, T., and Wright, M. (2003). Geology and working conceptual model of the Obsidian Butte (Unit 6) sector of the Salton Sea geothermal field, California. *Geothermal Resources Council Transactions*, 27, 227–40.

Hulen, J. B. and Pulka, F. S. (2001). Newly-discovered, ancient extrusive rhyolite in the Salton Sea geothermal field, Imperial Valley, California. *Twenty-Sixth Workshop on Geothermal Reservoir Engineering*, Stanford, California, January 29–31, 2001, pp. 1–8.

Hunstad, I., Selvaggi, G., D'Agostino, N., England, P., Clarke, P., and Pierozzi, M. (2003). Geodetic strain in peninsular Italy between 1875 and 2001. *Geophysical Research Letters*, 30, 1181.

Hunt, C. B. and Mabey, D. R. (1966). General geology of Death Valley, California: stratigraphy and structure. *U.S. Geological Survey Professional Paper*, 0494-A.

Hunt, J. M. (1979). *Petroleum Geochemistry and Geology*. W. H. Freeman and Company, San Francisco, California.

(1990). Generation and migration of petroleum from abnormally pressured fluid compartments. *American Association of Petroleum Geologists Bulletin*, 74, 1–12.

(1991). Generation of gas and oil from coal and other terrestrial organic matter. *Organic Geochemistry*, 17, 673–80.

Huntoon, P. W. (1982). The Meander Anticline, Canyonlands, Utah: an unloading structure resulting from horizontal gliding on salt. *Geological Society of America Bulletin*, 93, 941–50.

Hurford, A. J. and Green, P. F. (1983). The zeta age calibration of fission-track dating. *Chemical Geology*, 41, 285–317.

Hurlbut, C. S. and Klein, C. (1977). *Manual of mineralogy (after James D. Dana)*. John Wiley and Sons, New York, p. 532.

Hurlow, H. A., Snoke, A. W., and Hodges, K. V. (1991). Temperature and pressure of mylonitization in a Tertiary extensional shear zone, Ruby Mountains–East Humboldt Range, Nevada: tectonic implications. *Geology*, 19, 82–6.

Husmo, T., Hamar, G., Hoiland, O., Johannessen, E. P., Romuld, A., Spencer, A., and Titterton, R. (2002). Chapter 10. Lower and Middle Jurassic. In *The Millennium Atlas: Petroleum Geology of the Central and Northern North Sea*, ed. D. Evans, C. Graham, A. Armour, and P. Bathurst. *Geological Society of London*, pp. 129–55.

Hutchinson, I. (1985). The effects of sedimentation and compaction on oceanic heat flow. *Geophysical Journal of the Royal Astronomic Society*, 82, 439–59.

Hutton, A. C. and Cook, A. C. (1980). Influence of alginite on the reflectance of vitrinite from Joadia, NSW, and some other coals and oils shales containing alginite. *Fuel*, 59, 711–14.

Hutton, A. C., Kantsler, A. J., Cook, A. C., and McKirdy, D. M. (1980). Organic matter in oil shales. *The APEA Journal*, 20, 44–67.

Huuse, M. (2002). Cenozoic uplift and denudation of southern Norway: insights from the North Sea Basin. In *Exhumation of the North Atlantic Margin: Timing, Mechanisms and Implications for Petroleum Exploration*, ed. A. G. Dore, J. A. Cartwright, M. S. Stoker, J. P. Turner, and N. White. *Geological Society of London Special Publications*, **196**, 209–33.

Huuse, M., Duranti, D., Steinsland, N., Guargena, C. G., Prat, P., Holm, K., Cartwright, J. A., and Hurst, A. (2004). Seismic characteristics of large-scale sandstone intrusions in the Paleogene of the south Viking Graben, UK and Norwegian North Sea. In *3D Seismic Data: Application to the Exploration of Sedimentary Basins*, ed. R. J. Davies, J. A. Cartwright, S. A. Stewart, J. R. Underhill, and M. Lappin. *Geological Society of London Memoir*, **29**, 257–71.

Ibrahim, A. E., Ebinger, C. J., and Fairhead, J. D. (1991). Interpretation of the Central African Rift system of Sudan based on new gravity and aeromagnetic data. *EOS*, 72, 462.

Ibrmajer, J., Suk, M., and 35 others (1989). *Geophysical Picture of Czechoslovak Socialistic Republic (in Czech)*. ÚÚG Praha, p. 354.

IES (1993) In H. S. Poelchau, D. R. Baker, T. Hantschel, B. Horsfield, and B. Wygrala (1997). Basin simulation and the design of the conceptual basin model. In *Petroleum and Basin Evolution: Insights from Petroleum Geochemistry, Geology and Basin Modeling*, ed. D. H. Welte, B. Horsfield, and D. R. Baker. Springer, Berlin, pp. 3–70.

Illies, J. H. (1981). Mechanism of graben formation. *Tectonophysics*, 73, 249–66.

Ingebritsen, S. E. and Manning, C. E. (1999). Geological implications of a permeability-depth curve for the continental crust. *Geology*, 27, 1107–10.

International Heat Flow Commission (1991). *A New Global Heat Flow Compilation Assembled by Pollack, H. N., Hurter, S. J. and Johnson, J. R.* Department of Geological Sciences, University of Michigan, Ann Arbor.

Inyang, M. I., Ekweozor, C. M., and Pratt, L. M. (1995). Mid-Cretaceous anoxic events in southeastern Nigeria sedimentary basins – geochemical signatures and petroleum potential implications. *Nigerian Association of Petroleum Explorationists, Official Programme*, p. 34.

ION/GX Technology 2007. *IndiaSpan East, Phase 1, Interpretation Report*. Archive of Reliance, Mumbai, 39.

Irwin, W. P. and Barnes, I. (1975). Effect of geologic structure and metamorphic fluids on seismic behavior of the San Andreas fault system in Central and Northern California. *Geology*, 3, 713–16.

(1980). Tectonic relations of carbon dioxide discharges and earthquakes. *Journal of Geophysical Research*, 85, 3115–21.

Isaacs, C. M. and Garrison, R. E. (1983). *Petroleum Generation and Occurrence in the Miocene Monterey Formation, California*. Guidebook, Los Angeles, SEPM, Pacific Section.

Issler, D. H., McQueen, H., and Beaumont, C. (1989). Thermal and isostatic consequences of simple shear extension of the continental lithosphere. *Earth and Planetary Science Letters*, 91, 341–58.

Issler, D. R. and Beaumont, C. (1989). A finite-element model of the subsidence and thermal evolution of extensional basins: application to the Labrador continental margin. In *Thermal History of Sedimentary Basins, Methods and Case Histories*, ed. N. D. Naeser and T. H. McCulloch. Springer-Verlag, Berlin, pp. 239–67.

Ito, G., Lin, J., and Gable, C. W. (1996). Dynamics of mantle flow and melting at a ridge-centered hotspot: Iceland and the Mid-Atlantic Ridge. *Earth and Planetary Science Letters*, 144, 53–74.

Ito, G., Lin, J., and Gable, C. (1997). Interaction of mantle plumes and migrating mid-ocean ridge systems: implications for the Galapagos plume-ridge system. *Journal of Geophysical Research*, 102, 15403–17.

Ivanov, A. V., Boven, A. A., Brandt, I. S., and Rasskazov, S. V. (2002). Achievements and Limitations of the K-Ar and 40Ar/39AR Methods: What's in it for dating the Quaternary Sedimentary Deposits? *International Symposium – Speciation of Ancient Lakes, Berliner Palaobiologische Abhandlungen*, Irkutsk, pp. 65–75.

Izundu, U. (2007). Petrobras to start Tupi oil development by 2011. *Oil and Gas Journal*, p. 2.

Jackson, J. (1994). Active tectonics of the Aegean region. *Annual Review of Earth and Planetary Sciences*, 22, 239–71.

Jackson, J. A., Gagnepain, J., Houseman, G., King, G. C. P., Papadimitriou, P., Soufleris, C., and Virieux, J. (1982). Seismicity, normal faulting, and the geomorphological development of the Gulf of Corinth (Greece): the Corinth earthquakes of February and March 1981. *Earth and Planetary Science Letters*, 57, 377–97.

Jackson, J. A. and Leeder, M. (1994). Drainage systems and the development of normal faults: an example from Pleasant Valley, Nevada. *Journal of Structural Geology*, 16, 1041–59.

Jackson, J. A. and McKenzie, D. (1983). The geometrical evolution of normal fault systems. *Journal of Structural Geology*, 5, 471–82.

Jackson, J. A., White, N. J., Garfunkel, Z., and Anderson, H. (1988). Relations between normal-fault geometry, tilting and vertical motions in extensional terrains: an example from the southern Gulf of Suez. *Journal of Structural Geology*, 10, 155–70.

Jackson, M. P. A., Cramez, C., and Fonck, J.-M. (2000). Role of subaerial volcanic rocks and mantle plumes in creation of South Atlantic margins: implications for salt tectonics and source rocks. *Marine and Petroleum Geology*, 17, 477–98.

Jaeger, J. C. (1960). Shear failure of anisotropic rocks. *Geological Magazine*, 97, 65–72.

James, E. W. and Silver, L. T. (1988). Implications of zeolites and their zonation in the Cajon Pass deep drillhole. *Geophysical Research Letters*, 15, 973–6.

Janecky, D. R. and Seyfried, W. E. Jr. (1986). Hydrothermal serpentinization of peridotite within the oceanic crust: experimental investigations of mineralogy and major element chemistry. *Geochimica et Cosmochimica Acta*, 50, 1357–78.

Jansa, L. F. and Pe-Piper, G. (1985). Early Cretaceous volcanism on the northeastern American margin and implications for plate tectonics. *GSA Bulletin*, 96, 83–91.

Jardim de Sá, E. F. (1984). A evolução Proterozoica da Província Borborema. *Atas do 11th simposio de geologia do nordeste*, Natal, Brazil, pp. 297–315.

Jarrige, J.-J., Ott d'Estevou, P., Burollet, P. F., Montenat, C., Prat, P., Richert, J.-P., and Thiriet, J. P. (1990). The multistage tectonic evolution of the Gulf of Suez and northern Red Sea continental rift from field observations. *Tectonics*, 9, 441–65.

Jeffrey, A. W. A., Alimi, H. M., and Jenden, P. D. (1991). Geochemistry of Los Angeles basin oil and gas systems. In *Active Margin Basins*, ed. K. T. Biddle. *American Association of Petroleum Geologists Memoir*, 52, 197–219.

Jensenius, J. and Munksgaard, N. C. (1989). Large scale hot water migration systems around salt diapirs in the Danish central trough and their impact on diagenesis of chalk reservoirs. *Geochimica et Cosmochimica Acta*, 53, 79–88.

Jenssen, A. I., Bergslien, D., Rye-Larsen, M., and Lindholm, R. M. (1993). Origin of complex mound geometry of Palaeocene submarine-fan sandstone reservoirs, Balder field, Norway. In *Petroleum Geology of Northwest Europe*, ed. J. R. Parker. *Proceedings of the 4th Conference, Geological Society of London*, pp. 135–43.

Jeremiah, J. M. and Nicholson, P. H. (1999). Middle Oxfordian to Volgian sequence stratigraphy of the Greater Shearwater area. In *Petroleum Geology of Northwest Europe*, ed. A. J. Fleet and S. A. R. Boldy. *Proceedings of the 5th Conference, The Geological Society of London*, pp. 153–70.

Jerram, D. A., Mountney, N., Holzforster, F., and Stollhofen, H. (1999). Internal stratigraphic relationships in the Etendeka Group in the Huab Basin, NW Namibia: understanding the onset of flood volcanism. *Journal of Geodynamics*, 28, 393–418.

Jessop, A. M. and Majorowicz, J. A. (1994). Fluid flow and heat transfer in sedimentary basins. In *Geofluids: Origin, Migration and Evolution of Fluids in Sedimentary Basins*, ed. J. Parnell. *Geological Society of London Special Publications*, **78**, 43–54.

Jin, Z., Cao, J., Hu, W., Zhang, Y., Yao, S., Wang, X., Zhang, Y., Tang, Y., and Shi, X. (2008). Episodic petroleum fluid migration in fault zones of the northwestern Junggar Basin (northwest China): evidence from hydrocarbon-bearing zoned calcite cement. *American Association of Petroleum Geologists Bulletin*, 92, 1225–43.

Jiracek, G. R. (1995). Geoelectromagnetics charges on. U.S. Quadrennial Report to IUGG. *Reviews on Geophysics*, 33, 169–76.

Johannessen, P. N. and Andsbjerg, J. (1993). Middle to Late Jurassic basin evolution and sandstone distribution in the Danish Central Trough. In *Petroleum Geology of Northwest Europe*, ed. J. R. Parker. *Proceedings of the 4th Conference, Geological Society of London*, pp. 271–83.

Johannessen, P. N., Dybkjaer, K., and Rasmussen, E. S. (1996). Sequence stratigraphy of Upper Jurassic reservoir sandstones in the northern part of the Danish Central Trough, North Sea. *Marine and Petroleum Geology*, 13, 755–70.

Johannssen, E. P., Mjøs, R., Renshaw, D., Dalland, A., and Jacobsen, T. (1995). Northern limit of the "Brent delta" at the Tampen Spur – a sequence stratigraphic approach for sandstone prediction. In *Sequence Stratigraphy on the Northwest European Margin*, ed. R. J. Steel, V. L. Felt, E. P. Johannssen, and C. Mathieu. *Special Publication of the Norwegian Petroleum Society*, Amsterdam, Elsevier, **5**, 213–56.

John, B. E. (1984). Primary corrugations in Tertiary low-angle normal faults, SE California: porpoising mullion structures? *Geological Society of America Abstracts with Programs*, 16, 291.

Johnson, A. and Eyssautier, M. (1987). Alwyn North Field and its regional geological context. In *Petroleum Geology of North West Europe*, ed. J. Brooks and K. W. Glennie. Graham and Trotman, London, pp. 963–77.

Johnson, H. D. and Fisher, M. J. (1998). North Sea plays: geological controls on hydrocarbon distribution. In *Petroleum Geology of the North Sea, Basic Concepts and Recent Advances (fourth edition)*, ed. K. W. Glennie. Blackwell Scientific Publications, Oxford, pp. 463–547.

Johnson, H. D., Mackay, T. A., and Stewart, D. J. (1986). The Fulmar oil field (Central north Sea): geological aspects of its discovery, appraisal and development. *Marine and Petroleum Geology*, 3, 99–125.

Johnson, N. M., Officer, C. B., Opdyke, N. D., Woodard, G. D., Zeitler, P. K, and Lindsay, H. (1983). Rates of late Cenozoic tectonism in the Vallecito-Fish Creek basin, western Imperial Valley, California. *Geology*, 11, 664–7.

Joint Chalk Research (1996). *Geology, rock mechanics, rock properties and improved oil recovery in chalk of the Danish and Norwegian sectors of the North Sea*. Joint Chalk Research, Phase IV Monograph.

Jolley, D. W. (1997). Palaeosurface palynofloras of the Skye lava field and the age of British Tertiary volcanic province. In *Palaesurfaces: Recognition, Reconstruction and Palaeo-environmental Interpretation*, ed. M. Widdowson. Geological Society [London] Special Publication 120, pp. 67–94.

Jolly, R. J. H., Cosgrove, J. W., and Dewhurst, D. N. (1998). Thickness and spatial distributions of clastic dykes, northwest Sacramento Valley, California. *Journal of Structural Geology*, 20, 1663–72.

Jolly, R. J. H. and Lonergan, L. (2002). Mechanisms and controls on the formation of sand intrusions. *Journal of Geological Society*, 159, 605–17.

Jones, A. G. (1992). Electrical conductivity of the continental lower crust. In *Continental Lower Crust*, ed. D. M. Fountain, R. J. Arculus, and R. W. Kay. Elsevier, Amsterdam, pp. 81–143.

Jones, C. H., Unruh, J. R., and Sonder, L. J. (1996). The role of gravitational potential energy in active deformation in the Southwestern United States. *Nature*, 381, 37–41.

Jones, E., Jones, R., Ebdon, C., Ewen, D., Milner, P., Plunkert, J., Hudson, G., and Slater, P. (2003). Eocene. In *The Millennium Atlas: Petroleum Geology of the Central and Northern North Sea*, ed. D. Evans, C. Graham, A. Armour, and P. Bathurst. *The Geological Society of London*, pp. 261–77.

Jones, L. S., Garrett, S. W., Macleod, M., Guy, M., Condon, P. J., and Notman, L. (1999). Britannia field, UK Central North Sea: modeling heterogeneity in unusual deep-water deposits. In *Petroleum Geology of Northwest Europe. Proceedings of the 5th Conference*, ed. A. J. Fleet and S. A. R. Boldy. *The Geological Society of London*, pp. 1115–24.

Jones, S. M., White, N., Clarke, B. J., Rowley, E., and Gallagher, K. (2002). Present and past influence of the Iceland Plume on sedimentation. In *Exhumation of the North Atlantic Margin: Timing, Mechanisms, and Implications for Petroleum Exploration*, ed. A. G. Dore, J. A. Cartwright, M. S. Stoker, J. P. Turner, and N. White. *Geological Society of London Special Publications*, pp. 13–25.

Jones, S. M., White, N., and Lovell, B. (2001). Cenozoic and Cretaceous transient uplift in the Porcupine Basin and its relationship to a mantle plume. In *The Petroleum Exploration of Ireland's Offshore Basins*, ed. P. M. Shannon, P. D. W. Haughton, and D. Corcoran. *Geological Society of London Special Publications*, **188**, 345–60.

Jongmans, D. and Malin, P. E. (1995). Microearthquake S-wave observations from 0 to 1 km in the Varian well at Parkfield, California. *Bulletin of the Seismological Society of America*, 85, 1805–21.

Jonk, R., Duranti, D., Parnell, J., Hurst, A., and Fallick, A. E. (2003). The structural and diagenetic evolution of injected sandstones: examples from the Kimmeridgian of NE Scotland. *Journal of the Geological Society of London*, 160, 881–94.

Jonk, R., Hurst, A., Duranti, D., Parnell, J., Mazzini, A., and Fallick, A. E. (2005). Origin and timing of sand injection, petroleum migration, and diagenesis in Tertiary reservoirs, south Viking Graben, North Sea. *American Association of Petroleum Geologists Bulletin*, 89, 329–57.

Joppen, M. and White, R. S. (1990). The structure and subsidence of Rockall Trough from two-ship seismic experiments. *Journal of Geophysical Research*, 95, 19821–37.

Jordan, T. E. and Flemings, P. B. (1991). Large-scale stratigraphic architecture, eustatic variation, and unsteady tectonism: a theoretical evaluation. *Journal of Geophysical Research*, 96, 6681–99.

Jordan, T. H. (1978). Composition and development of the continental tectosphere. *Nature*, 274, 544–8.

Jowett, E. C., Cathles, L. M., and Davis, B. W. (1993). Predicting depths of gypsum dehydration in evaporitic sedimentary basins. *American Association of Petroleum Geologists Bulletin*, 77, 402–13.

Juhlin, C. (1990). Interpretation of the reflections in the Siljan Ring area based on results from the Gravberg-1 borehole. *Tectonophysics*, 173, 345–60.

Jurine, D., Jaupart, C., Brandeis, G., and Tackley, P. J. (2005). Penetration of mantle plumes through depleted lithosphere. *Journal of Geophysical Research*, 110, B10.

Kadolsky, D., Johansen, S. J., and Duxbury, S. (1999). Sequence stratigraphy and sedimentary history of the Humber Group (Late Jurassic – Ryazanian) in the Outer Moray Firth (UKCS, North Sea). In *Petroleum Geology of Northwest Europe*, ed. A. J. Fleet and S. A. R. Boldy. *Proceedings of the 5th Conference, Geological Society of London*, pp. 839–60.

Kaila, K. L., Krishna, V. G., and Mall, D. M. (1981). Crustal structure along Mehmadabad-Billimora profile in Cambay basin, India from deep seismic soundings. *Tectonophysics*, 76, 99–130.

Kaila, K. L. and Sain, K. (1997). Variation of crustal velocity structure in India as determined from DSS studies and their implications as regional tectonics. *Journal of the Geological Society of India*, 49, 395–407.

Kaila, K. L., Tewari, H. C., Krishna, V. G., Dixit, M. M., Sarkar, D., and Reddy, M. S. (1990). Deep Seismic Sounding studies in the north Cambay and Sanchor basins, India. *Geophysical Journal International*, 103, 621–37.

Kappelmeyer, O. and Haenel, R. (1974). *Geothermics with Special Reference to Application*. Gebrueder Borntraeger, Berlin.

Karato, S.-I. and Wu, P. (1993). Rheology of the upper mantle: a synthesis. *Science*, 260, 771–8.

Karig, D. E. (1986). Physical properties and mechanical state of accreted sediments in the Nankai Trough, Southwest Japan Arc. In *Structural Fabric in Deep Sea Drilling Project Cores from Forearcs*, ed. J. C. Moore. *Geological Society of America Memoir*, p. 117–33.

Karim, A., Pe-Piper, G., Piper, D. J. W., and Hanley, J. J. (2010). Thermal history and the relationship between diagenesis and the reservoir connectivity: Venture field, offshore Nova Scotia, eastern Canada. *Open File Report – Geological Survey of Canada*, 6557, 1–256.

Karl, D. M., McMurtry G. M., Malahoff, A., and Garcia, M. O. (1988). Loihi Seamount, Hawaii: a mid-plate volcano with a distinctive hydrothermal system. *Nature*, 335, 532–5.

Karlo, J. F. and Shoup, R. C. (2000). Classification of syndepositional: structural systems, northern Gulf of Mexico. *The Bulletin of the Houston Geological Society*, 42, 11–12.

Karner, G. D. (2000). Rifts of the Campos and Santos basins, southeastern Brazil: distribution and timing. *American Association of Petroleum Geologists Memoir*, 73, 301–15.

(2008). Depth-dependent extension and mantle exhumation: an extreme passive margin endmember or a new paradigm? *Central Atlantic Conjugate Margins Conference*, August 13–15, 2008, Halifax, Program & Extended Abstracts, 10–16. http://www.conjugatemargins.com/abstracts/by_presenter/jp, accessed on January 27, 2012.

Karner, G. D. and Dewey, J. F. (1986). Rifting: lithospheric versus crustal extension as applied to the Ridge Basin of southern California. *American Association of Petroleum Geologists Memoir*, A131, 317–37.

Karner, G. D. and Driscoll, N. W. (1999). Style, timing and distribution of tectonic deformation across the Exmouth Plateau, northwest Australia, determined from stratal architecture and quantitative basin modelling. *Geological Society of London Special Publications*, **164**, 271–311.

(1999). Tectonic and stratigraphic development of the West African and eastern Brazilian margins: insights from quantitative basin modelling. *Geological Society of London Special Publications*, **153**, 11–40.

Karner, G. D., Driscoll, N. W., and Barker, D. H. N. (2003). Syn-rift region subsidence across the West African continental margin: the role of lower plate ductile extension. In *Petroleum Geology of Africa: New Themes and Developing Technologies*, ed. T. J. Arthur, D. S. Macgregor, and N. Cameron. *Geological Society of London Special Publications*, **207**, 105–29.

Karner, G. D. and Gamboa, L. A. P. (2007). Timing and origin of the South Atlantic pre-salt basins and their capping evaporites. *Geological Society of London Special Publications*, **285**, 15–35.

Karner, G. D., Neal, W., McGinnis, J. P., Brumbaugh, W. D., and Cameron, N. R. (1997). Tectonic significance of syn-rift sediment packages across the Gabon-Cabinda continental margin. *Marine and Petroleum Geology*, 14, 973–1000.

Karner, G. D. and Watts, A. B. (1982). On isostasy at Atlantic-type continental margins. *Journal of Geophysical Research*, 87, 2923–48.

Karson, J. A. and Brooks, C. K. (1999). Structural and magmatic segmentation of the Tertiary east Greenland volcanic rifted margin. *Geological Society of London Special Publications*, **164**, 313–18.

Katili, J. A. and Sudradjat, A. (1984). *Galunggung: the 1982–83 eruption*. Volcanological Survey of Indonesia, Bandung, p. 102.

Katz, B. J. (1983). Limitations of "Rock-Eval" pyrolysis for typing organic matter. *Organic Geochemistry*, 4, 195–9.

(1988). Clastic and carbonate lacustrine systems: an organic geochemical comparison (Green River Formation and East African lake sediments). *Geological Society of London Special Publications*, **40**, 81–90.

(1990). Control son distribution of lacustrine source rocks through time and space. In *Lacustrine Basin Exploration – Case Studies and Modern Analogues*, ed. B. J. Katz. *American Association of Petroleum Geologists Memoir*, **50**, 61–76.

(1995a). Petroleum source rocks: an introductory overview. In *Petroleum Source Rocks*, ed. B. J. Katz. Springer-Verlag, Berlin, pp. 1–8.

(1995b). A survey of rift basin source rocks. In *Hydrocarbon Habitat in Rift Basins*, ed. J. J. Lambiase. *Geological Society of London Special Publications*, **80**, 213–42.

Katz, B. J., Kelley, P. A., Royle, R. A., and Jorjorian, T. (1991). Hydrocarbon products of coals as revealed by pyrolysis-gas chromatography. *Organic Geochemistry*, 17, 711–22.

Katz, B. J. and Mello, M. R. (2000). Petroleum systems of South Atlantic marginal basins – an overview. In *Petroleum Systems of South Atlantic Margins*, ed. M. R. Mello and B. J. Katz. *American Association of Petroleum Geologists Memoir*, **73**, 1–13.

Kaus, B. J. P., Connolly, J. A. D., Podladchikov, Y. Y., and Schmalholz, S. M. (2005). Effect of mineral phase transitions on sedimentary basin subsidence and uplift. *Earth and Planetary Science Letters*, 233, 213–28.

Kawka, O. E. and Simoneit, B. R. T. (1987). Survey of hydrothermally-generated petroleums from the Guaymas Basin spreading center. *Organic Geochemistry*, 11, 311–28.

Keaton, J. R., Currey, D. R., and Olig, S. S. (1993). Paleoseismicity and earthquake hazards evaluation of the West Valley fault zone, Salt Lake City urban area, Utah. *Contract Report*, 93–8, 1–55.

Keefer, D. K. (1984). Landslides caused by earthquakes: source characteristics. *Bulletin of the Geological Society of America*, 95, 406–21.

Keen, C. E. (1985a). The dynamics of rifting: deformation of the lithosphere by active and passive driving forces. *Geophysical Journal of the Royal Astronomical Society*, 80, 95–120.

(1985b). Evolution of rifted continental margins. *Paper – Geological Survey of Canada*, 85, 8–10.

Keen, C. E. and Boutilier, R. R. (1995). Lithosphere-asthenosphere interaction below rifts. *Rifted Ocean-Continent Boundaries*, 463, 17–30.

Keen, C. E., Boutilier, R., de Voogd, B., Mudford, B., and Enachescu, M. E. (1987). Crustal geometry and extensional models for the Grand Banks, Eastern Canada: constraints from deep seismic reflection data. *Atlantic Geoscience Society Special Publication*, 5, 101–15.

Keen, C. E., Kay, W. A., and Roest, W. R. (1990). Crustal anatomy of a transform continental margin. *Tectonophysics*, 173, 527–44.

Keen, C. E., Kay, W. A., Keppie, J. D., Marillier, F., Pe-Piper, G., and Waldron, J. W. F. (1991). Deep seismic reflection data from the Bay of Fundy and Gulf of Maine: tectonic implications for the Northern Appalachians. *Canadian Journal of Earth Sciences*, 28, 1096–111.

Keen, C. E. and Potter, D. P. (1995). Formation and evolution of the Nova Scotia rifted margin: evidence from deep seismic reflection data. *Tectonics*, 14, 918–32.

Keith, J. D. and Shanks, W. C. (1988). Chemical evolution and volatile fugacities of the Pine Grove porphyry molybdenum and ash-flow tuff system, southwestern Utah. In *Recent Advances in the Geology of Granite-Related Mineral Deposits*, ed. R. P. Taylor and D. F. Strong. *Canadian Institute of Mining and Metallurgy Special Volume*, **39**, 402–23.

Keith, J. F., Nemčok, M., Fleischmann, K. H., Nemčok, J., Gross, P., Rudinec, R., and Kanes, W. H. (1991). Sedimentary basins of Slovakia, Part III, Hydrocarbon potential of the Central Carpathian Paleogene Basin. University of South Carolina, Columbia, *Earth Sciences and Resources Institute Technical Report*, 90-07-306, pp. 1–302.

Kelemen, P. B. and Holbrook, W. S. (1995). Origin of thick, high-velocity igneous crust along the U.S. East Coast Margin. *Journal of Geophysical Research*, 100, 10077–94.

Keller, G. R., Braile, L. W., and Morgan, P. (1979). Crustal structure, geophysical models and contemporary tectonism of the Colorado Plateau. *Tectonophysics*, 61, 131–47.

Kelley, P. A., Mertani, B., and Williams, H. H. (1995). Brown Shale Formation: Paleocene lacustrine source rocks of central Sumatra. In *Petroleum Source Rocks*, ed. B. J. Katz. Springer-Verlag, Berlin, pp. 283–308.

Kendall, C. (2008). Sequence stratigraphy provides a basic framework to conceptual models used to interpret depositional systems: the key to simplification of the complex terminology of sequence stratigraphy is to use simple depositional models. *International Congress, Abstracts – Congres Geologique International, Resumes*, 33, 1255505.

Kendall, C. and Haughton, P. (2008). Sequence stratigraphy of deepwater fans and mass transport debris. In SEPM Sequence Stratigraphy Web. http://strata.geol.sc.edu/Deepwater/DeepwaterCLasticSeqStrat.ht.

Kendall, J. M., Stuart, G. W., Ebinger, C. J., Bastow, I. D., and Keir, D. (2005). Magma-assisted rifting in Ethiopia. *Nature*, 433, 146–8.

Kennedy, B. M., Kharaka, Y. K., Evans, W. C., Ellwood, A., DePaolo, D. J., Thordsen, J., Ambats, G., and Mariner, R. H. (1997). Mantle fluids in the San Andreas fault system, California. *Science*, 278, 1278–81.

Kent, R. W., Saunders, A. D., Kempton, P. D., and Ghose, N. C. (1997). Rajmahal basalts, eastern India: mantle sources and melt distribution at a volcanic rifted margin. *Large Igneous Provinces: Continental Oceanic and Planetary Flood Volcanism*, ed. J. Mahoney and M. F. Coffin. American Geophysical Union Geophysical Monograph 100, 145–82.

Keppie, J. D. (1982). Tectonic map of Nova Scotia. *Abstracts with Programs – Geological Society of America*, 14, 30.

Kerr, A. C. (1994). Lithospheric thinning during the evolution of continental large igneous provinces: a case study from the North Atlantic Tertiary Province. *Geology*, 22, 1027–30.

Kerr, D. R., Pappajohn, S., and Bell, P. J. (1979). Neogene continental rifting and sedimentation in the western Salton Trough, California. *Abstracts with Programs – Geological Society of America*, 11, 457.

Khalil, S. M. (1998). *Tectonic and structural development of the eastern Gulf of Suez margin*. PhD thesis, Royal Holloway University of London, p. 349.

Kharaka, Y. K., Ambats, G., Evans, W. C., and White, A. F. (1988b). Geochemistry of water at Cajon Pass, California: preliminary results. *Geophysical Research Letters*, 15, 1037–40.

Kharaka, Y. K., Thordsen, J. J., Evans, W. C., and Kennedy, B. M. (1999). Geochemistry and hydromechanical interactions of fluids associated with the San Andreas fault system, California. In *Faults and Subsurface Fluid Flow in the Shallow Crust*, ed. W. C. Haneberg, P. S. Mozley, J. C. Moore, and L. B. Goodwin. *American Geophysical Union*, pp. 129–48.

Kharaka, Y. K., White, L. D., Ambats, G., and White, A. F. (1988a). Origin of subsurface water at Cajon Pass, California. *Geophysical Research Letters*, 15, 1049–52.

Kim, E. M. (2002). Natural gas resource characterization of the Federal Offshore Gulf of Mexico: data analysis of key sand-reservoir attributes for future exploration and development opportunities. *Gulf Coast Association of Geological Societies and Gulf Coast Transactions*, 52, 527–36.

Kincaid, C., Sparks, D. W., and Detrick, R. S. (1996). The relative importance of plate-driven and buoyancy-driven flow at mid-ocean ridges. *Journal of Geophysical Research*, 101, 16177–93.

King, S. D. (2005). Archean cratons and mantle dynamics. *Earth Planetary Science Letters*, 234, 1–14.

King, S. D. and Anderson, D. L. (1995). An alternative mechanism of flood basalt formation. *Earth and Planetary Science Letters*, 136, 269–79.

(1998). Edge-driven convection. *Earth and Planetary Science Letters*, 160, 289–96.

Kincaid, C., Ito, G., and Gable, C. (1995a). Laboratory investigation of the interaction of off-axis mantle plumes and spreading centres. *Nature*, 376, 758–61.

Kincaid, C., Schilling, J.-G., and Gable, C. (1995b). The dynamics of off-axis plume-ridge interaction in the uppermost mantle. *Earth and Planetary Science Letters*, 137, 29–43.

Kirby, S. H. (1983). Rheology of the lithosphere. *Reviews of Geophysics*, 21, 1458–87.

(1985). Rock mechanics observations pertinent to the rheology of the continental lithosphere and the localization of strain along shear zones. *Tectonophysics*, 119, 1–27.

Kirby, S. H. and Kronenberg, A. K. (1987). Rheology of the lithosphere: selected topics. *Reviews of Geophysics*, 25, 1219–44.

Kirby, S. H. and McCormick, J. W. (1984). Inelastic properties of rocks and minerals; strength and rheology. In *Handbook of Physical Properties of Rocks*, ed. R. S. Carmichael. CRC Press, Boca Raton, **3**, 140–280.

Kiss, B. and Toth, J. (1985). Well log interpretation of metamorphic hydrocarbon-bearing formations. *First Break*, 3, 24–31.

Kissin, Y. V. (1987). Catagenesis and composition of petroleum: origin of n-alkanes and isoalkanes in petroleum crudes. *Geochimica et Cosmochimica Acta*, 51, 2445–57.

Kissling, E., Deichmann, N., Husen, S., and Zappone, A. (2006). Lower crustal seismic velocities and seismicity in the northern Alpine foreland: II. Geodynamical interpretation. *EGU General Assembly*, 2–7 April, 2006, Vienna, p. 562.

Klausen, B. M. and Larsen, H. C. (2002). East Greenland coast-parallel dike swarm and its role in continental breakup. In *Volcanic Rift Margins*, ed. M. A. Menzies, S. L. Klemperer, C. Ebinger, and J. Baker. Geological Society of America Special Paper, 362, pp. 133–58.

Klemme, H. D. (1994). Petroleum systems in the world that involve Upper Jurassic source rocks. *AAPG Memoir*, 60, 51–72.

Klemme, H. D. and Ulmishek, G. F. (1991). Effective petroleum source rocks of the world: stratigraphic distribution and controlling depositional factors. *American Association of Petroleum Geologists Bulletin*, 75, 1809–51.

Klemperer, S. and Hobbs, R. (1991). *The BIRPS Atlas: Deep Seismic Reflection Profiles around the British Isles*. Cambridge University Press.

Klitgord, K. D. and Schouten, H. (1986). Plate kinematics of the central Atlantic. In *The Western North Atlantic Region: The Geology of North America*, ed. P. R. Vogt and B. E. Tucholke. Geological Society of America, Boulder, pp. 351–78.

Kluegel, A., Hansteen, T. H., and Schmincke, H. U. (1997). Rates of magma ascent and depths of magma reservoir beneath La Palma, Canary Islands. In *International Workshop on Volcanism and Volcanic Hazards in Immature Intraplate Oceanic Islands. Estanción Volcanológica de Canarias, La Palma, Canary Islands, Spain*, pp. 29–30.

Knipe, R. J. (1986). Faulting mechanisms in slope sediments: examples from deep sea drilling project cores. In *Structural Fabric in Deep Sea Drilling Project Cores from Forearcs*, ed. J. C. Moore. *Geological Society of America Memoir*, 166, 45–54.

(1989). Deformation mechanisms: recognition from natural tectonites. *Journal of Structural Geology*, 11, 127–46.

(1992). Faulting processes and fault seal. In *Structural and Tectonic Modeling and Its Application to Petroleum Geology*, ed. R. M. Larsen, H. Brekke, B. T. Larsen, and E. Talleraas. *Proceedings of the Norwegian Petroleum Society Special Publication*, **1**, 325–42.

(1993). The influence of fault zone processes and diagenesis on fluid flow. In *Diagenesis and Basin Development*, ed. A. D. Horbury and A. Robinson. *American Association of Petroleum Geologists Studies in Geology*, 36, 135–51.

(1997). Juxtaposition and seal diagrams to help analyze fault seals in hydrocarbon reservoirs. *American Association of Petroleum Geologists Bulletin*, 81, 187–95.

Knipe, R. J., Agar, S. M., and Prior, D. J. (1991). The microstructural evolution of fluid flow paths in semi-lithified sediments from subduction complex. In *The Behaviour and Influence of Fluids in Subduction Zones*, ed. J. Tarney, K. T. Pickering, R. J. Knipe, and J. F. Dewey. *Philosophical Transactions of the Royal Society of London*, **335**, 261–73.

Knox, G. J. and Omatsola, E. M. (1989). Development of the Cenozoic Niger delta in terms of the "Escalator Regression" model and impact on hydrocarbon distribution. *Proceedings KNGMG Symposium "Coastal Lowlands, Geology and Geotechnology*," Dordrecht, pp. 181–202.

Knox, R. W. O. B. (1996). Correlation of the early Paleogene in Northwest Europe: an overview. In *Correlation of the Early Paleogene in Northwest Europe*, ed. R. W. O. B. Knox, R. M. Corfield, and R. E. Dunay. *Geological Society Special Publications*, pp. 1–11.

Knutson, C. A. and Munro, I. C. (1991). The Beryl Field, block 9/ 13, UK North Sea. 33–42 in *United Kingdom Oil and Gas Fields. Memoir of the Geological Society of London*, 14, 33–42.

Kohl, T., Bachler, D., and Rybach, L. (2000). Step towards a comprehensive thermo-hydraulic analysis of the HDR test site Soultz-Sous-Forets. *Proceedings of the World Geothermal Congress 2000*, Kyushu-Tohoku, May 28–June 10, 2671–6.

Kohl, T., Signorelli, S., and Rybach, L. (2001). Three-dimensional (3-D) thermal investigation below high Alpine topography. *Physics of the Earth and Planetary Interiors*, 126, 195–210.

Koike, L., Kiang, C. H., and Reis, F. A. M. (2007). Organic geochemistry of Santos basin oil, Brazil. *Abstracts of Reports – International Congress on Organic Geochemistry, The 23rd International Meeting on Organic Geochemistry*, Torquay, UK, **23**, 921–2.

Koning, T. and Karsani, A. (1985). Petroleum geology of the Ombilin intermontane basin, West Sumatra. *Proceedings of the Annual Convention – Indonesian Oetroleum Association*, 14, 117–37.

Kooi, H., Cloetingh, S. A. P. L., and Burrus, J. (1992). Lithospheric necking and regional isostasy at extensional basins 1. Subsidence and gravity modeling with an application to the Gulf of Lions margin (SE France). *Journal of Geophysical Research*, 97, 17553–71.

Kopietz, J. and Jung, R. (1978). *Geothermal in situ Experiments in the Asse Salt-mine. Seminar on in situ Heating Experiments in Geological Formations*. OECD Publications, Paris, France.

Kopylova, M. G., Russell, J. K., and Cookenboo, H. (1999). Petrology of peridotite and pyroxenite xenoliths from Jericho Kimberlite: implications for the thermal state of the mantle beneath the Slave Craton, northern Canada. *Journal of Petrology*, 40, 79–104.

Korenaga, J., Holbrook, S., Kent, G., Kelemen, P., Detrick, R., Larsen, H.-C., Hopper, J., and Dahl-Jensen, T. (2000). Crustal structure of the southeast Greenland margin from joint refraction and reflection seismic tomography. *Journal of Geophysical Research*, 105, 21591–614.

Kotarba, M. J. and Koltun, Y. V. (2006). Origin and habitat of hydrocarbons of the Polish and Ukrainian parts of the Carpathian province. In *The Carpathians: Geology and Hydrocarbon Resources*, ed. J. Golonka and F. Picha. *American Association of Petroleum Geologists Memoir*, **84**, 395–443.

Koukovelas, I., Mpresiakas, A., Sokos, E., and Doutsus, T. (1996). The tectonic setting and earthquake ground hazard of the 1993 Pyrogos earthquake, Peloponnese, Greece. *Journal of the Geological Society of London*, 153, 1137–49.

Kováč, M., Baráth, I., and Nemčok, M. (1992). Bericht 1991 ueber geologische Aufnahmen im Quartaer und Tertiaer im Suedoestlichen Teil des Wiener Beckens auf Blatt 77 Eisenstadt. *Jahrbuch der Geologischen Bundensanstalt*, 135, 701–3.

Koyi, H., Jenyon, M. K., and Petersen, K. (1993). The effect of basement faulting on diapirism. *Journal of Petroleum Geology*, 16, 285–311.

Kranz, R. L. (1983). Microcracks in rocks: a review. In *Continental Tectonics: Structure, Kinematics and Dynamics*, ed. M. Friedman and M. N. Toksoez. Tectonophysics, **100**, 449–80.

Kranz, R. L. and Scholz, C. H. (1977). Critical dilatant volume of rocks at the onset of tertiary creep. *Journal of Geophysical Research*, 82, 4893–8.

Krawczyk, C. M., Reston, T. J., Beslier, M.-O., and Boillot, G. (1996). Evidence for detachment tectonics on the Iberia Abyssal Plain rifted margin. *Proceedings of Ocean Drill Program Scientific Results*, 149, 603–15.

Krebs, W. N., Wescott, W. A., Nummedal, D., Gaafar, I., Azazi, G., and Karamat, S. (1997). Graphic correlation and sequence stratigraphy of Neogene rocks in the Gulf of Suez. *Bulletin de la Societe Geologique de France*, 168, 63–71.

Kreemer, C., Holt, W. E., and Haines A. J. (2003). An integrated global model of present-day plate motions and plate boundary deformation. *Geophysics Journal International*, 154, 8–34.

Krishna, K. S., Gopala Rao, D., and Sar, D. (2006). Nature of the crust in the Laxmi Basin (14° – 20°N), western continental margin of India. *Tectonics*, 25, 198.

Krooss, B. M., Brothers, L., and Engel, M. H. (1991). Geochromatography in petroleum migration: a review. In *Petroleum Migration*, ed. W. A. England and A. J. Fleet. *Geological Society of London Special Publications*, **59**, 149–63.

Kruijs, E. and Barron, E. (1990). Climatic model prediction of paleoproductivity and potential source-rock distribution. In *Deposition of Organic Facies*, ed. A. Y. Huc. *American Association of Petroleum Geologists Studies in Geology*, **30**, 195–216.

Krumbein, W. C. and Monk, G. D. (1943). Permeability as a function of the size parameters of unconsolidated sand. *Trans. AIME*, 151, 153.

Krus, S. and Šutora, A. (1986). *Geophysical-geological Atlas of the Alpine-Carpathian Mountain System*. Report, Archív Geofyziky, Brno.

Ksiazkiewicz, M. (1962). *Geological Atlas of Poland, Stratigraphic and Facial*. Instytut Geologiczny, Warszawa.

Kubala, M., Bastow, M., Thompson, S., Scotchman, I., and Kjell, O. (2002).Chapter 17. Geothermal regime, petroleum generation and migration. In *The Millennium Atlas: Petroleum Geology of the Central and Northern North Sea*, ed. D. Evans, C. Graham, A. Armour, and P. Bathurst. *Geological Society of London*, 298–315.

Kuhn, T., Herzig, P. M., Hannington, M. D., Garbe-Schonberg, D., and Stoffers, P. (2003). Origin of fluids and anhydrite precipitation in the sediment-hosted Grimsey hydrothermal field north of Iceland. *Chemical Geology*, 202, 5–21.

Kuhnt, W. (1992). An early Campanian paleoceanographic event in the Northern Atlantic and Western Tethys? *Fourth International Conference on Paleoceanography, ICP IV*, 57, 171.

Kuhnt, W. and Wiedmann, J. (1995). Cenomanian-Turonian source rocks: paleobiogeographic and paleoenvironmental aspects. *American Association of Petroleum Geologists Studies in Geology*, 40, 213–31.

Kukal, Z. (1990). The rate of geological processes. *Earth-Science Reviews*, 28, 1–284.

Kusznir, N. J. (1991). The distribution of stress with depth in the lithosphere. Thermorheological and geodynamic constraints. *Philosophical Transactions of the Royal Society of London*, 337, 95–110.

(2009). South Australia – Antarctica conjugate rifted margins: mapping crustal thickness and lithosphere thinning using satellite gravity inversion. A report for Geoscience Australia. https://www.ga.au/products/servlet/controller?event=GEOCAT_DETAILS&catno=68655 (accessed on August 5, 2010).

Kusznir, N. J., Davis, M. J., Roberts, A. M., and Baxter, K. (2000). Depth-dependent lithosphere stretching, mantle exhumation and the upper plate paradox at rifted continental margins. *Eos, Transactions, American Geophysical Union*, 81, 1112.

Kusznir, N. J. and Egan, S. S. (1989). Simple-shear and pure-shear models of extensional sedimentary basin formation: application to the Jeanne d'Arc basin, Grand Banks of Newfoundland. In *Extensional Tectonics and Stratigraphy of the North Atlantic Margins*, ed. A. J. Tankard and H. R. Balkwill. *American Association of Petroleum Geologists Memoir*, **46**, 305–22.

Kusznir, N. J. and Karner, G. D., 2007. Continental lithospheric thinning and breakup in response to upwelling divergent mantle flow: application to the Woodlark, Newfoundland and Iberia margins. *Geological Society of London Special Publications*, **282**, 389–419.

Kusznir, N. J. and Park, R. G. (1987). The extensional strength of the continental lithosphere: its dependence on geothermal gradient, and crustal composition and thickness. *Geological Society Special Publications*, 28, 35–52.

Kusznir, N. J. and Ziegler, P. A. (1992). The mechanics of continental extension and sedimentary basin formation: a simple-shear/pure-shear flexural cantilever model. *Tectonophysics*, 215, 117–31.

Kvenvolden, K. A. and Simoneit, B. R. T. (1990). Hydrothermally derived petroleum: examples from Guaymas Basin, Gulf of California, and Escanaba Trough, Northeast Pacific Ocean. *American Association of Petroleum Geologists Bulletin*, 74, 223–37.

Lachenbruch, A. H. (1968). Preliminary geothermal model of the Sierra Nevada. *Journal of Geophysical Research*, 73, 6977–89.

(1970). Crustal temperature and heat production: implications of the linear heat-flow relation. *Journal of Geophysical Research*, 75, 3291–300.

Lachenbruch, A. H. and Sass, J. H. (1977). Heat flow in the United States and the thermal regime of the crust. In *The Earth's Crust: Its Nature and Physical Properties*, ed. J. G. Heacock. *American Geophysical Union, Geophysical Monograph*, **20**, 626–75.

(1988). The stress heat-flow paradox and thermal results from Cajon Pass. *Geophysical Research Letters*, 15, 981–4.

Lafargue, E. and Barker, C. (1988). Effect of water washing on crude oil compositions. *American Association of Petroleum Geologists Bulletin*, 72, 263–76.

LakeNet (2005). Global Lake Database. http://worldlakes.org/searchlakes.asp.

Lama, R. D. and Vutukuri, V. S. (1978). *Handbook on Mechanical Properties of Rocks*. Transtech Publications, Bay Village.

Lambe, T. W. and Whitman, R. V. (1969). *Soil Mechanics*. John Wiley and Sons, New York.

Lampe, C., Noth, S., and Ricken, W. (1999). Burial and thermal history of Cenozoic sediments in the northern Rhinegraben: a 1D simulation based on well Nordheim-1. *Zentralblatt fur Geologie und Palaontologie Teil 1*, 10–12, 1375–89.

Lampe, C. and Person, M. (2002). Advective cooling within sedimentary rift basins – application to the Upper Rhinegraben (Germany). *Marine and Petroleum Geology*, 19, 361–75.

Lan, C. Y., Lee, T., and Lee, C. W. (1990). The Rb-Sr isotopic record in Taiwan gneisses and its tectonic implication. *Tectonophysics*, 183, 129–43.

Lana, M. C. (1993). Potencial petrolífero e exploração em agues profundas na Bacia de Sergipe- Alagoas. *Relatório Interno Petrobrás*, DEPEX / DINORD / SESEA, p. 36.

Landes, K. K. (1959). *Petroleum Geology*. John Wiley and Sons, New York, p. 443.

Landes, K. K., Amoruso, J. J., Charlesworth, L. J., Heany, F. Jr., and Lesperho ance, P.-J. (1960). Petroleum resources in basement rocks. *American Association of Petroleum Geologists Bulletin*, 44, 1682–91.

Lankreijer, A. C. (1998). *Rheology and basement control on extensional basin evolution in Central and Eastern Europe: Variscan and Alpine-Carpathian-Pannonian tectonics*. PhD thesis, Vrije University, Amsterdam, p. 158.

Lankreijer, A., Mocanu, V., and Cloetingh, S. (1997). Lateral variations in lithosphere strength in the Romanian Carpathians: constraints on basin evolution. *Tectonophysics*, 272, 269–90.

Larsen, H.-C. and Saunders, A. (1998). *Scientific Results, Ocean Drilling Program, Leg 152: Tectonism and Volcanism at the Southeast Greenland Rifted Margin: A Record of Plume Impact and Later Continental Rupture*. College Station, Texas, Ocean Drilling program, pp. 503–33.

Larsen, P.-H. (1988). Relay structures in a Lower Permian basement-involved extension system, East Greenland. *Journal of Structural Geology*, 10, 3–8.

Larsen, R. M. and Jarvik, L. J. (1980). The geology of the Sleipner field complex. *Paper 15 in The sedimentation of the North Sea reservoir rocks*, Geilo, May 11–14.

Larsen, S. C. and Reilinger, R. (1991). Age constraints for the present fault configuration in the Imperial Valley, California: evidence for northwestward propagation of the Gulf of California rift system. *Journal of Geophysical Research*, 96, 10339–46.

Larter, S. R. and Aplin, A. C. (1995). Reservoir geochemistry: methods, applications and opportunities. In *The Geochemistry of Reservoirs*, ed. J. M. Cubitt and W. A. England. *Geological Society Special Publication*, **86**, 5–32.

Laslett, G. M., Green, P. F., Duddy, I. R., and Gleadow, A. J. W. (1987). Thermal annealing of fission tracks in apatite, 2: a quantitative analysis. *Chemical Geology*, 65, 1–13.

Laslett, G. M., Kendall, W. S., Gleadow, A. J. W., and Duddy, I. R. (1982). Bias in measurement of fission-track length distributions. *Nuclear Tracks and Radiation Measurements*, 6, 79–85.

Latter, J. H. (1987). Volcanoes and volcanic risks in the circum-Pacific. *Pacific Rim Congress*, 87, 745–52.

Laubach, S. E. (2003). Practical approaches to identifying sealed and open fractures. *American Association of Petroleum Geologists Bulletin*, 87, 561–79.

Laubach, S. E., Marrett, R., and Olson, J. (2000). New directions in fracture characterization. *The Leading Edge*, 19, 704–11.

Lavier, L. and Manatschal, G. (2006). A mechanism to thin the continental lithosphere at magma-poor margins. *Nature*, 440, 324–8.

Lavier, L., Buck, W. R., and Poliakov, A. N. B. (1999). Self-consistent rolling-hinge model for the evolution of large-offset lowangle normal faults. *Geology*, 27, 1127–30.

Laville, E. and Petit, J.-P. (1984). Role of synsedimentary strike-slip faults in the formation of Moroccan Triassic basins. *Geology*, 12, 424–7.

Law, A., Raymond, A., White, G., Atkinson, A., Clifton, M., Atherton, T., Dawes, I. A. N., Robertson, E., Melvin, A., and Brayley, S. (2000). The Kopevik fairway, Moray Firth, UK. *Petroleum Geoscience*, 6, 256–74.

Lawrence, S. R. (1990). Aspects of the petroleum geology of the Junggar basin, Northwest China. In *Classic Petroleum Provinces*, ed. J. Brooks. *Geological Society of London Special Publications*, **50**, 545–57.

Lay, T. and Wallace, T. C. (1995). *Modern Global Seismology*. Academic Press, San Diego.

Le Pichon, X. and Alvarez, F. (1984). From stretching to subduction in back-arc regions: dynamic considerations. *Tectonophysics*, 102, 343–57.

Le Roy, P. and Piqué, A. (2001). Triassic-Liassic Western Moroccan synrift basins in relation to the central Atlantic opening. *Marine Geology*, 172, 359–81.

Lear, C. H., Elderfield, H., and Wilson, P. A. (2000). Cenozoic deep-sea temperatures and global ice volumes from Mg/Ca in benthic foraminiferal calcite. *Science*, 287, 269–72.

Leckie, D. A. and Rumpel, T. (2003). Tide-influenced sedimentation in a rift basin – Cretaceous Qishn Formation, Masila, Yemen – a billion barrel oil field. *American Association of Petroleum Geologists Bulletin*, 87, 987–1013.

Lee, C. H. and Farmer, I. (1993). *Fluid Flow in Discontinuous Rocks*. Chapman & Hall, London.

Lee, C. T. A., Lenardic, A., Cooper, C. M., Niu, F., and Levander, A. (2005). The role of chemical boundary layers in regulating the thickness of continental and oceanic thermal boundary layers. *Earth and Planetary Science Letters*, 230, 379–95.

Lee, C. T. and Rudnick, R. L. (1999). Compositionally stratified cratonic lithosphere petrology and geochemistry of peridotite xenoliths from the Labait tuff cone, Tanzania. In *Proceedings of the VIIth International Kimberlite Conference*, ed. J. J. Gurney, M. D. Gurney, S. Pascoe, and S. R. Richardson. Red Roof Design cc, Cape Town, pp. 503–21.

Lee, M. J. and Hwang, Y. J. (1993). Tectonic evolution and structural styles of the East Shetland Basin. In *Petroleum geology of Northwest Europe; 4th conference*, ed. J. R. Parker. Shell UK Exploration and Production, London, pp. 1137–49.

Leeder, M. R. and Gawthorpe, R. L. (1987). Sedimentary models for extensional tilt-block/half-graben basins, extensional tectonics: regional-scale processes. *Geological Society of London Special Publications*, **28**, 139–52.

Leeder, M. R. and Jackson, J. A. (1993). The interaction between normal faulting and drainage in active extensional basins, with examples from the western United States and central Greece. *Basin Research*, 5, 79–102.

Leeder, M. R., Ord, D. M., and Collier, R. (1988). Development of alluvial fans and fan deltas in neotectonic extensional settings: implications for the interpretation of basin-fills. In *Fan Deltas: Sedimentology and Tectonic Settings*, ed. W. Nemec and R. J. Steel. Blackie and Son, Glasgow, pp. 173–85.

Lees, C. H. (1910). On the shape of the isotherms under mountain ranges in radio-active districts. *Proceedings of the Royal Society of London*, 83, 339–46.

Lemoine, M., Bas, T., Arnaud Vanneau, A., Arnaud, H., Dumont, T., Gidon, T., Bourbon, M., de Graciansky, P. C., Rudkiewicz, J. L, Mégard-Galli, J., and Tricart, P. (1986). The continental margin of the Mesozoic Tethys in the Western Alps. *Marine and Petroleum Geology*, 3, 179–99.

Lenkey, L. (1999). *Geothermics of the Pannonian Basin and Its Bearing on the Tectonics of Basin Evolution*. Vrije University, Amsterdam, p. 215.

Leroueil, S. (1988). Tenth Canadian Geotechnical Colloquium: Recent developments in consolidation of natural clays. *Canadian Geotechnical Journal*, 25, 85–107.

Leroueil, S., Bouclin, G., Tavenas, F., Bergeron, I., and La Rochelle, P. (1990). Permeability anisotropy of natural clays as a function of strain. *Canadian Geotechnical Journal*, 27, 568–79.

Leslie, S. C., Moore, G. F., Morgan, J. K., and Hills, D. J. (2002). Seismic stratigraphy of the Frontal Moat: implications for sedimentary processes at the leading edge of an oceanic hotspot trace. *Marine Geology*, 184, 143–62.

Lespinasse, M. C., Leroy, J. L., Pironon, J., and Boiron, M.-C. (1998). The paleofluids from the marginal ridge of the Côte d'Ivoire-Ghana transform margin (Hole 960A) as thermal indicators. In *Proceedings of Ocean Drilling Programme*, ed. J. Mascle, G. P. Lohmann, and M. Moullade. *Scientific Results*, **159**, 49–52.

Levitte, D., Maurath, G., and Eckstein, Y. (1984). Terrestrial heat flow in a 3.5 km deep borehole in the Jordan-Dead Sea rift valley. *Geological Society of America Abstracts with Programs*, 16, 575.

Lewan, M. D. (1993). Laboratory simulation of petroleum formation: hydrous pyrolysis. In *Organic Geochemistry*, ed. M. H. Engel and S. A. Macko. Plenum Press, pp. 419–42.

(1994). Assessing natural oil expulsion from source rocks by laboratory pyrolysis. In *The Petroleum System – From Source to Trap*, ed. L. B. Magoon and W. G. Dow. *American Association of Petroleum Geologists Memoir*, **60**, 201–10.

(1997). Experiments on the role of water in petroleum formation. *Geochimica et Cosmochimica Acta*, 61, 3691–723.

Lewan, M. D. and Henry, A. A. (2001). Gas: oil ratios for source rocks containing type-I, -II, -IIS, and -III kerogens as determined by hydrous pyrolysis. In *Geologic Studies of Deep Natural Gas Resources*, ed. T. S. Dyman and V. A. Kuuskraa. *U.S. Geological Survey Digital Data Series*, **67**, 1–9.

Lewan, M. D., Kotarba, M. J., Curtis, J. B., Wieclaw, D., and Kosakowski, P. (2006). Oil-generation kinetics for organic facies with Type-II and -IIS kerogen in the Menilite Shales of the Polish Carpathians. *Geochimica et cosmochimica acta*, 70, 3351–68.

Lewis, C. R. and Rose, S. C. (1970). A theory relating high temperatures and overpressures. *Journal of Petroleum Technology*, 22, 11–16.

Lewis, R. Q. Sr and Campbell, R. H. (1965). Geology and uranium deposits of Elk Ridge and vicinity, San Juan County, Utah. *Publications of the U. S. Geological Survey, Reston*, P 0474-B.

Leythaeuser, D., Little, R., Radke, M., and Schaefer, R. G. (1987). Geochemical effects of petroleum migration and expulsion from Toarcian source rocks in the Hils syncline area, NW-Germany. *Organic Geochemistry*, 13, 489–502.

Leythaeuser, D. and Poelchau, H. S. (1991). Expulsion of petroleum from type III kerogen source rocks in gaseous solution: modeling of solubility fractionation. In *Petroleum Migration*, ed. W. A. England and A. J. Fleet. *Geological Society of London Special Publications*, **59**, 33–46.

Leythaeuser, D. and Ruckheim, J. (1989). Heterogeneity of oil composition within a reservoir as a reflection of accumulation history. *Geochimica et Cosmochimica Acta*, 53, 2119–23.

Leythaeuser, D., Schaefer, R. G., and Radke, M. (1988). Geochemical effects of primary migration of petroleum in Kimmeridge source rocks from Brae Field area, North Sea; I, Gross composition of C_{15}-soluble organic matter and molecular composition of C_{15}-saturated hydrocarbons. *Geochimica et Cosmochimica Acta*, 52, 701–13.

Lezzar, K. E., Tiercelin, J.-J., Turdu, C. L., Cohen, A. S., Reynolds, D. J., Le Gall, B., and Scholz, C. A. (2002). Control of normal fault interaction on the distribution of major Neogene sedimentary depocenters, Lake Tanganyika, East African Rift. *American Association of Petroleum Geologists Bulletin*, 86, 1027–59.

Li, R., Wang, Z., Lin, D., and Xin, M. (1988). Organic geochemical analysis of environments of Dongpu basin. *Chinese Journal of Geochemistry*, 7, 193–206.

Li, R., Wu, H., Lin, D., Wang, Z., Chen, R., Tian, X., and Zhang, R. (1992). Biomarker features of the Chinese Meso-Cenozoic saline lake deposits. *Advances in Geoscience*, 2, 275–92.

Li, Y.-G., Chen, P., Cochran, E. S., Vidale, J. E., and Burdette, T. (2006). Seismic evidence for rock damage and healing on the San Andreas Fault associated with the 2004 M 6.0 Parkfield earthquake. *Bulletin of the Seismological Society of America*, 96, 349–63.

Li, Y.-G., Ellsworth, W. L., Thurber, C. H., Malin, P. E., and Aki, K. (1997). Fault-zone guided waves from explosions in the San Andreas Fault at Parkfield and Cienga Valley, California. *Bulletin of the Seismological Society of America*, 87, 210–21.

Li, Y.-G., Vidale, J. E., and Cochran, E. S. (2004). Low-velocity damaged structure of the San Andreas Fault at

Parkfield from fault trapped waves. *Geophysical Research Letters*, 31, L12S06.

Li, Y. H. (1976). Denudation of Taiwan island since the Pliocene epoch. *Geology*, 4, 105–7.

Ligi, M., Bonatti, E., Bortoluzzi, G., Brunelli, D., Caratori Tontini, F., Cipriani, A., Cocchi, I., Cuffaro, M., Ferrante, V., Khalil, S., Mitschell, N. C., Rasul, N., and Schetino, A. (2008). Sea-floor spreading initiation: constraints from geophysical data of the Thetis Deep, northern Red Sea. *Abstract T33F-08, EOS Transactions*, 89, 53.

Ligi, M., Bonatti, E., Tontini, F. C., Cipriani, A., Cocchi, L., Schettino, A., Bortoluzzi, G., Ferrante, V., Khalil, S., Mitchell, N. C., and Rasul, N. (2011). Initial burst of oceanic crust accretion in the Red Sea due to edge-driven mantle convection. *Geology*, 39, 1019–22.

Ligtenberg, J. H. and Thomsen, R. O. (2003). Fluid migration path detection and its application to basin modeling. *EAGE 65th Conference, Stavanger, Norway, Extended abstracts*, C27.

Lillie, R. J. (1999). *Whole Earth Geophysics. An Introductory Textbook for Geologists and Geophysicists*. Prentice Hall, Upper Saddle River, New Jersey, p. 361.

Lin, G., Nunn, J. A., and Deming, D. (2000). Thermal buffering of sedimentary basins by basement rocks: implications arising from numerical simulations. *Petroleum Geoscience*, 6, 299–307.

Lindquist, S. J. (1983). Nugget formation reservoir characteristics affecting production in the overthrust belt of southwestern Wyoming. *Journal of Petroleum Technology*, 35, 1355–65.

Liotta, D. and Ranalli, G. (1999). Correlation between seismic reflectivity and rheology in extended lithosphere: southern Tuscany, inner Northern Apennines, Italy. *Tectonophysics*, 315, 109–22.

Lippman, P. W., Logatchev, N. A., and Zorin, Y. A. (1989). Intra-continental rift comparisons. *EOS Transactions*, 70, 578–88.

Lippolt, H. J., Leitz, M., Wernicke, R. S., and Hagedorn, B. (1994). (Uranium+thorium)/helium dating of apatite: experience with samples from different geochemical environments. *Chemical Geology*, 112, 179–91.

Liro, L. M. and Coen, R. (1995). Salt deformation history and postsalt structural trends, offshore southern Gabon, West Africa. *American Association of Petroleum Geologists Memoir*, 65, 323–31.

Liro, L. M. and Pardus, Y. C. (1990). Seismic facies analysis of fluvial-deltaic lacustrine systems – upper Fort Union Formation (Paleocene), Wind River basin, Wyoming. In *Lacustrine Basin Exploration – Case Studies and Modern Analogs*, ed. B. J. Katz. *American Association of Petroleum Geologists Memoir*, **50**, 225–42.

Lisker, F. and Fachmann, S. (2001). Phanerozoic history of the Mahanadi region, India. *Journal of Geophysical Research*, 106, 22027–50.

Lister, G. S. and Davis, G. A. (1989). The origin of metamorphic core complexes and detachment faults formed during Tertiary continental extension in the northern Colorado River region, U.S.A. *Journal of Structural Geology*, 11, 65–94.

Lister, G. S., Etheridge, M. A., and Symonds, P. A. (1986). Detachment faulting and the evolution of passive continental margins. *Geology*, 14, 246–50.

Lister, J.-R. and Kerr, R.-C. (1991). Fluid-mechanical models of crack propagation and their application to magma transport in dykes. *Journal of Geophysical Research*, 96, 10049–77.

Littke, R. (1994). *Deposition, Diagenesis, and Weathering of Organic Matter-Rich Sediments*. Springer-Verlag, Heidelberg.

Littke, R., Baker, D. R., and Leythaeuser, D. (1988). Microscopic and sedimentologic evidence for the generation and migration of hydrocarbons in Toarcian source rocks of different maturities. *Organic Geochemistry*, 13, 549–59.

Littke, R., Baker, D. R., and Rullkötter, J. (1997). Deposition of petroleum source rocks. In *Petroleum and Basin Evolution: Insights from Petroleum Geochemistry, Geology and Basin Modeling*, ed. D. H. Welte, B. Horsfield, and D. R. Baker. Springer-Verlag, Berlin, pp. 273–333.

Littke, R., Kroos, B., Idiz, E., and Frielingsdorf, J. (1995). Molecular nitrogen in natural gas accumulations: generation from sedimentary organic matter at high temperatures. *American Association of Petroleum Geologists Bulletin*, 9, 410–30.

Liu, L., Sun, X. M., Dong, F. X., Ma, F., and Ma, Y. P. (2004). Geochemical characteristics and fluid inclusion in calcite veins of lower part of member 1 of Shahejie Formation, offshore area, Dagang oil field: a case study of well Gangshen 67 (in Chinese). *Journal of Jilin University, Earth Science Edition*, 34, 49–54.

Livera, S. E. and Gdula, J. E. (1990). Brent oil field. *AAPG Treatise of Petroleum Geology, Atlas of Oil and Gas Fields*, A-017, 21–63.

Lizarralde, D. and Holbrook, W. S. (1997). U.S. mid-Atlantic Margin structure and early thermal evolution. *Journal of Geophysical Research*, 102, 22855–75.

Lockner, D. A., Moore, D. E., and Reches, Z. (1992). Microcrack interaction leading to shear fracture. *33rd Symposium on Rock Mechanics*, pp. 807–16.

Logan, J. M., Friedman, M., Higgs, M., Dengo, C., and Shimamoto, T. (1979). Experimental studies of simulated gouge and their application to studies of natural fault zones. *Proceedings of VIII Conference on Analysis of Actual Fault Zones in Bedrock*, US Geological Survey, Menlo Park, pp. 305–43.

Lomando, A. J. and Engelder, T. (1984). Strain indicated by calcite twinning: implications for deformation of the early Mesozoic northern Newark basin, New York. *Northeastern Geology*, 6, 192–5.

Lombardo, B., Rubatto, D., and Castelli, D. (2002). Ion microprobe U-Pb dating of zircon from a Monviso metaplagiogranite: implications for the evolution of the Piedmont-Liguria Tethys in the Western Alps. *Ofioliti*, 27, 108–17.

Lonergan, L. and Cartwright, J. A. (1999). Polygonal faults and their influence on deep-water sandstone reservoir geometries, Alba field, United Kingdom central North Sea. *American Association of Petroleum Geologists Bulletin*, 83, 410–32.

Lonergan, L., Lee, N., Johnson, H. D., Cartwright, J. A., and Jolly, R. J. H. (2000). Remobilization and injection in deepwater depositional systems: implications for reservoir architecture and prediction. In *Deep Water Reservoirs of the World*, ed. P. Weimer, R. M. Slatt, J. Coleman, N. C. Rosen, H. Nelson, A. H. Bouma, M. J. Styzen, and D. T. Lawrence. Gulf Coast Section SEPM Foundation, 20th annual conference, Houston, pp. 515–32.

Lonsdale, P. (1989). Geology and tectonic history of the Gulf of California. In *The Eastern Pacific Ocean and Hawaii*, ed. E. L. Winterer, D. M. Hussong, and R. W. Decker. Geological Society of America, Denver, pp. 499–522.

(1991). Structural patterns of the Pacific floor offshore of peninsular California. *American Association of Petroleum Geologists Memoir*, 47, 87–125.

Lopatin, N. V. (1971). Temperature and geologic time as factors in coalifications (in Russian). *Akademiya Nauk SSR Izvestiya Seriya Geologicheskaya*, 3, 95–106.

Losh, S. (1998). Oil migration in a major growth fault: structural analysis of the Pathfinder core, South Eugene Island Block 330, offshore Louisiana. *American Association of Petroleum Geologists Bulletin*, 82, 1694–710.

Losh, S., Eglinton, L., Shoell, M., and Wood, J. (1999). Vertical and lateral fluid flow related to a large growth fault, South Eugene Island Block 330 field, offshore Louisiana. *American Association of Petroleum Geologists Bulletin*, 83, 244–76.

Losh, S., Walter, L., Meulbroek, P., Martini, A., Cathles, L., and Whelan, J. (2002). Reservoir fluids and their migration into the South Eugene Island Block 330 reservoirs, offshore Louisiana. *American Association of Petroleum Geologists Bulletin*, 86, 1463–88.

Louvel, V., Dyment, J., and Sibuet, J.-C. (1997). Thinning of the Goban Spur continental margin and formation of early oceanic crust: constraints from forward modelling and inversion of marine magnetic anomalies. *Geophysical Journal International*, 128, 188–96.

Lowell, J. D. (1995). Mechanics of basin inversion from worldwide examples. In *Basin Inversion*, ed. J. G. Buchanan and P. G. Buchanan. *Geological Society of London Special Publications*, **88**, 39–57.

Lowell, J. D. and Genik, G. J. (1972). Sea-floor spreading and structural evolution of the southern Red Sea. *American Association of Petroleum Geologists Bulletin*, 56, 247–59.

Lowell, R. P. (1975). Circulation in fractures, hot springs, and convective heat transport on mid-ocean ridge crests. *Geophysical Journal of the Royal Astronomical Society*, 40, 351–65.

Lowell, R. P. and Germanovich, L. N. (1994). On the thermal evolution of high-temperature hydrothermal systems at ocean ridge crests. *Journal of Geophysical Research*, 99, 565–75.

Lowell, R. P. and Rona, P. A. (1985). Hydrothermal models of the generation of massive sulfide ore deposits. *Journal of Geophysical Research*, 90, 8769–83.

Lucas, M., Hull, J., and Manspeizer, W. (1988). A foreland-type fold and related structures in the Newark rift basin. In *Triassic–Jurassic Rifting, Continental Breakup and the Origin of the Atlantic Ocean Passive Margins*, part A, ed. W. Manspeizer. New York, Elsevier, pp. 307–32.

Ludwig, W. J., Nafe, J. E., and Drake, C. L. (1970). Seismic refraction. In *The Sea*, ed. A. E. Maxwell. Willey-Interscience, New York, pp. 53–84.

Lundegard, P. D. (1992). Sandstone porosity loss – a "big picture" view of the importance of compaction. *Journal of Sedimentary Petrology*, 62, 250–60.

Lüning, S., Craig, J., Loydell, D. K., Storch, P., and Fitches, B. (2000). Lower Silurian "hot shales" in North Africa and Arabia: regional distribution and depositional model. *Earth-Science Reviews*, 49, 121–200.

Lupini, J. F., Skinner, A. E., and Vaugham, P. R. (1981). The drained residual strength of cohesive soils. *Geotechnique*, 31, 181–213.

Lyberis, N. (1988). Tectonic evolution of the Gulf of Suez and the Gulf of Aqaba. *Tectonophysics*, 153, 209–20.

Macaulay, C. I., Fallick, A. E., Haszeldine, R. S., and Macaulay, G. E. (2000). Oil migration makes the difference: regional distribution of carbonate cement $\delta^{13}C$ in northern North Sea Tertiary sandstones. *Clay Minerals*, 35, 69–76.

MacCaig, A. M. (1988). Deep fluid circulation in fault zones. *Geology*, 16, 867–70.

MacCready, T., Snoke, A. W., Wright, J. E., and Howard, K. A. (1997). Mid-crustal flow during Tertiary extension in the Ruby Mountains core complex, Nevada. *Geological Society of America Bulletin*, 109, 1576–94.

MacDonald, I. R., Reilly, J. F., Best, S. E., Venkataramaiah, R., Sassen, R., Guinasso, N. L. Jr., and Amos, J. (1996). Remote sensing inventory of active oil seeps and chemosynthetic communities in the northern Gulf of Mexico. In *Hydrocarbon Migration and Its Near-Surface Expression*, ed. D. Schumacher and M. A. Abrams. *American Association of Petroleum Geologists Memoir*, **66**, 27–37.

Macdonald, K. C. (1982). Mid-ocean ridges; fine scale tectonic, volcanic and hydrothermal processes within the plate boundary zone. *Annual Review of Earth and Planetary Sciences*, 10, 155–90.

Macedo, J. and Marshak, S. (1999). Controls on the geometry of fold-thrust belt salients. *Geological Society of America Bulletin*, 111, 1808–22.

Macgregor, D. S. (1993). Relationships between seepage, tectonics and subsurface petroleum reserves. *Marine and Petroleum Geology*, 10, 606–19.

(1995). Hydrocarbon habitat and classification of inverted rift basins. In *Basin Inversion*, ed. J. G. Buchanan and

P. G. Buchanan. *Geological Society of London Special Publications*, **88**, 83–93.

Machette, M. N., Personius, S. F., and Nelson, A. R. (1991a). Paleoseismology of the Wasatch fault zone: a summary of recent investigations, interpretations, and conclusions. *U.S. Geological Survey Professional Paper*, 1500.

Machette, M. N., Personius, S. F., Nelson, A. R., Schwartz, D. P., and Lund, W. R. (1991b). The Wasatch fault zone, Utah: segmentation and history of Holocene earthquakes. *Journal of Structural Geology*, 13, 137–49.

Mackenzie, A. S., Leythaeuser, D., Altebaeumer, F.-J., Disko, U., and Rullkoetter, J. (1988). Molecular measurements of maturity for Lias delta shales in N.W. Germany. *Geochimica et Cosmochimica Acta*, 52, 1145–54.

Mackenzie, A. S. and Quigley, T. M. (1988). Principles of geochemical prospect appraisal. *American Association of Petroleum Geologists Bulletin*, 72, 399–415.

MacLeod, C. J., Escartín, J., Banerji, D., Banks, G. J., Gleeson, M., Irving, D. H. B., Lilly, R. M., McCaig, A. M., Niu, Y., Allerton, S., and Smith, D. K. (2002). Direct geological evidence for oceanic detachment faulting: the Mid-Atlantic Ridge, 15 degrees 45'N. *Geology*, 30, 879–82.

Magara, K. (1976). Water expulsion from elastic sediments during compaction: directions and volumes. *American Association of Petroleum Geologists Bulletin*, 60, 543–53.

Maggi, A., Jackson, J. A., McKenzie, D., and Priestley, K. (2000a). Earthquake focal depths, effective elastic thickness, and the strength of the continental lithosphere. *Geology*, 28, 495–8.

Maggi, A., Jackson, J. A., Priestley, K., and Baker, C. (2000b). A re-assessment of focal depth distributions in southern Iran, the Tien Shan and northern India: Do earthquakes really occur in the continental mantle? *Geophysical Journal International*, 143, 629–61.

Magnavita, L. P. (1992). *Geometry and kinematics of the Reconcavo-Tucano-Jatoba rift, NE Brazil*. PhD thesis, University of Oxford, England.

(2000). Deformation mechanisms in porous sandstones: implications for the development of fault seal and migration paths in the Reconcavo basin, Brazil. In *Petroleum Systems of South Atlantic Margins*, ed. M. P. Mello and B. J. Katz. *American Association of Petroleum Geologists Memoir*, **73**, 195–212.

Magoon, L. B. and Dow, W. G. (1994). The petroleum system. In *The Petroleum System – From Source to Trap*, ed. L. B. Magoon and W. G. Dow. *American Association of Petroleum Geologists Memoir*, **60**, 3–24.

Magoon, L. B., Hudson, T. L., and Peters, K. E. (2005). Egret-Hibernia(!), a significant petroleum system, northern Grand Banks area, offshore eastern Canada. *American Association of Petroleum Geologists Bulletin*, 89, 1203–37.

Maher, C. E. (1980). The Piper Oilfield. In *Giant Oil and Gas Fields of the Decade: 1968–1978*. *American Association of Petroleum Geologists Memoir*, **30**, 131–72.

(1981). The Piper Oilfield. In *Petroleum Geology of the Continental Shelf of North-West Europe*, ed. L. V. Illing and G. D. Hobson. Heyden and Son, London, pp. 358–70.

Maher, C. E. and Harker, S. D. (1987). The Claymore oil field. In *Petroleum Geology of North West Europe*, ed. J. Brooks and K. W. Glennie. Graham and Trotman, London, pp. 835–45.

Mahoney, J. and Coffin, M. F. (1997). Large igneous provinces: continental, oceanic and planetary flood volcanism. *AGU Geophysical Monograph*, 100, 438.

Mailloux, B. J., Person, M., Kelley, S., Dunbar, N., Cather, S., Strayer, L., and Hudleston, P. (1999). Tectonic controls on the hydrogeology of the Rio Grande rift, New Mexico. *Water Resources Research*, 35, 2641–59.

Majer, E. L. and McEvilly, T. V. (1979). Seismological investigations at The Geysers geothermal field. *Geophysics*, 44, 246–69.

Makhous, M. and Galushkin, Y. I. (2005). *Basin Analysis and Modeling of the Burial, Thermal and Maturation Histories in Sedimentary Basins*. Technip, Paris, p. 380.

Malahoff, A., McMurty, G. M., Wiltshire, J. C., and Yeh, H.-W. (1982). Geology and geochemistry of hydrothermal deposits from active submarine volcano Loihi, Hawaii. *Nature*, 298, 234–9.

Maltman, A. J. (1994). Prelithification deformation. In *Continental Deformation*, ed. P. L. Hancock. Pergamon Press, Tarrytown, pp. 143–58.

Mammerickx, J. and Sandwell, D. T. (1986). Rifting of old oceanic lithosphere. *Journal of Geophysical Research*, **91**, 1975–88.

Manatschal, G. (2004). New models for evolution of magma-poor rifted margins based on a review of data and concepts from West Iberia and the Alps. *International Journal of Earth Sciences*, 93, 432–66.

Manatschal, G. and Bernoulli, D. (1998). Rifting and early evolution of ancient ocean basins: the record of the Mesozoic Tethys and the Galicia-Newfoundland margins. *Marine Geophysical Researches*, 20, 371–81.

(1999). Architecture and tectonic evolution of nonvolcanic margins; present-day Galicia and ancient Adria. *Tectonics*, 18, 1099–119.

Manatschal, G., Froitzheim, N., Rubenach, M., and Turrin, B. D. (2001). The role of detachment faulting in the formation of an ocean-continent transition: insights from the Iberia Abyssal Plain. In *Non-volcanic Rifting of Continental Margins: A Comparison of Evidence from Land and Sea*, ed. R. C. L. Wilson, R. B Whitmarsh, B. Taylor, and N. Froitzheim. *Geological Society of London Special Publications*, **187**, 405–28.

Manatschal, G. and Niegervelt, P. (1997). A continent-ocean transition recorded in the Err and Platta nappes (eastern Switzerland). *Eclogae Geologicae Helvetiae*, 90, 3–27.

Mancktelow, N. S. and Grasemann, B. (1997). Time-dependent effects of heat advection and topography on cooling histories during erosion. *Tectonophysics*, 270, 167–95.

Mandl, G. (1987). Tectonic deformation by rotating parallel faults: the "bookshelf" mechanism. *Tectonophysics*, 141, 277–316.

(1988). *Mechanics of Tectonic Faulting: Models and Basic Concepts. Developments in Structural Geology*. Elsevier Science Publications, Amsterdam, Netherlands (NLD).

Mango, F. D. (1997). The light hydrocarbons in petroleum: a critical review. *Organic Geochemistry*, 26, 417–40.

Mann, D. M. and Mackenzie, A. S. (1990). Prediction of pore fluid pressures in sedimentary basins. *Marine and Petroleum Geology*, 7, 55–65.

Mann, P., Gahagan, L., and Gordon, M. B. (2004). Tectonic setting of the world's giant oil and gas fields. In *Giant Oil and Gas Fields of the Decade 1990–1999*, ed. T. Halbouty. American Association of Petroleum Geologists Memoir, **78**, 15–105.

Mann, U. (1994). An integrated approach to the study of primary petroleum migration. In *Geofluids: Origin, Migration and Evolution of Fluids in Sedimentary Basins*, ed. J. Parnell. *Geological Society of London Special Publications*, **78**, 233–60.

Mann, U., Hantschel, T., Schaefer, R. G., Krooss, B., Leythaeuser, D., Littke, R., and Sachsenhofer, R. F. (1997). Petroleum migration: mechanisms, pathways, efficiencies and numerical simulations. In *Petroleum and Basin Evolution: Insights from Petroleum Geochemistry, Geology and Basin Modeling*, ed. D. H. Welte, B. Horsfield, and D. R. Baker. Springer, Berlin, pp. 405–520.

Marcano, M., Lohman, K. C., and Pickett, E. A. (1998). Geochemistry of pore-filling and fracture-vein carbonates: Cote d'Ivoire–Ghana Marginal Ridge. *Proceedings of the Ocean Drilling Program, Scientific Results*, 159, 71–80.

Marks, K. M. and Stock, J. M. (1994). Variations in ridge morphology and depth-age relationships on the Pacific-Antarctic Ridge. *Journal of Geophysical Research*, 99.B1, 531–541.

Mart, Y. and Dauteuil, O. (2000). Analogue experiments of propagation of oblique rifts. *Tectonophysics*, 316, 121–32.

Martel, S. J., Pollard, D. D., and Segall, P. (1988). Development of simple strike-slip fault zones, Mount Abbot Quadrangle, Sierra Nevada, California. *Geological Society of America Bulletin*, 100, 1451–65.

Martin, M. A. and Pollard, J. E. (1996). The role of trace fossil (ichnofabric) analysis in the development of depositional models for the Upper Jurassic Fulmar Formation of the Kittiwake Field (Quadrant 21, UKCS). In *Geology of the Humber Group: Central Graben and Moray Firth, UKCS*, ed. A. Hurst, H. D. Johnson, S. D. Burley, A. C. Canham, and D. S. Mackertich. *Special Publication of the Geological Society of London*, **114**, 163–83.

Martin, R. G. (1978). Northern and eastern Gulf of Mexico continental margin; stratigraphic and structural framework. In *Framework, Facies, and Oil-Trapping Characteristics of the Upper Continental Margin*, ed. A. H. Bouma, G. T. Moore, and J. M. Coleman. *American Association of Petroleum Geologists Studies in Geology*, **7**, 21–42.

Martínez-Martínez, J. M., Soto, J. I., and Balanyá, J. C. (2004). Domes in extended orogens: a mode of mountain rift in the Betics, southeast Spain. *Special Paper – Geological Society of America*, 380, 243–65.

Martini, A. M., Walter, L. M., Ku, T. C. W., Budai, J. M., McIntosh, J. C., and Schoell, M. (2003). Microbial production and modification of gases in sedimentary basins: a geochemical case study from a Devonian shale gas play, Michigan basin. *American Association of Petroleum Geologists Bulletin*, 87, 1355–75.

Marton, L. G., Tari, G. C., and Lehmann, C. T. (2000). Evolution of the Angolan passive margin, West Africa, with emphasis on post-salt structural styles. *Geophysical Monograph*, 115, 129–49.

Marzoli, A., Bertrand, H., Knight, K. B., Cirilli, S., Buratti, N., Vérati, C., Nomade, S., Renne, P. R., Youbi, N., Martini, R., Allenbach, K., Neuwerth, R., Rapaille, C., Zaninetti, L., and Bellieni, G. (2004). Synchrony of the Central Atlantic Magmatic Province and the Triassic-Jurassic boundary climatic and biotic crisis. *Geology*, 32, 973–6.

Mascle, J., Lohmann, G.P., Clift, P.D. et al. (1996). Proc. ODP Init. Repts., 159, College Station, TX (Ocean Drilling Program).

Mascle, J., Lohmann, G.P., and Clift, P. (1997). Development of a passive transform margin: Cote d'Ivoire-Ghana transform margin – ODP Leg 159 preliminary results. Geo-Marine Letters, 17, 4–11.

Massari, F. and Colella, A. (1988). Evolution and types of fan-delta systems in some major tectonic settings. In *Fan Deltas: Sedimentology and Tectonic Settings*, ed. W. Nemec and R. J. Steel. Blackie and Son, Glasgow, pp. 103–22.

Masson, D. G. (1996). Catastrophic collapse of the volcanic island of Hierro 15 ka ago and the history of landslides in the Canary Islands. *Geology*, 24, 231–4.

Mattavelli, L. and Novelli, L. (1988). Geochemistry and habitat of natural gases in Italy. *Organic Geochemistry*, 13, 1–13.

Matthews, M. D. (1999). Migration of petroleum. In *Exploring for Oil and Gas Traps*, ed. E. A. Beaumont and N. H. Foster. *American Association of Petroleum Geologists Treatise of Petroleum Geology*, **7**, 31–8.

Matyas, J. (1996). How detrical mica affects compactional porosity loss. *Terra Nova*, 8, 425–9.

Mavko, G. M. and Nur, A. (1979). Wave attenuation in partially saturated rocks. *Geophysics*, 44, 161–78.

May, S. R., Ehman, K. D., Gray, G. G., and Crowell, J. C. (1993). A new angle on the tectonic evolution of the Ridge Basin, a "strike-slip" basin in Southern California. *Geological Society of America Bulletin*, 105, 1357–72.

Mazzini, A., Duranti, D., Jonk, R., Parnell, J., Cronin, B., Hurst, A., and Quine, M. (2003). Palaeo-carbonate seep structures above an oil reservoir, Gryphon field, Tertiary, North Sea. *Geomarine Letters*, 23, 323–39.

McAfee, A. (2005). Hinge zone control on reservoir quality in the West African syn-rift succession. *The 4th HGS/PESGB International Conference on African E7P*, Houston Program, 9.

McAuliffe, C. D. (1979). Oil and gas migration: chemical and physical constraints. *American Association of Petroleum Geologists Bulletin*, 73, 1455–71.

McBirney, A. R. (1995). Mechanisms of differentiation in the Skaergaard intrusion. *Journal of Geological Society of London*, 152, 421–35.

McBride, B. C. (1997). *The geometry and evolution of allochthonous salt and its impact on petroleum systems, northern Gulf of Mexico basin: studies in three and four-dimensional analysis*. PhD thesis, University of Colorado, Boulder, p. 276.

McBride, B. C., Rowan, M. G., and Weimer, P. (1998a). The evolution of allochthonous salt systems, Northern Green Canyon and Ewing Bank (Offshore Louisiana), Northern Gulf of Mexico. *American Association of Petroleum Geologists Bulletin*, 82, 1013–36.

McBride, B. C., Weimer, P., and Rowan, M. G. (1998b). The effect of allochthonous salt on petroleum systems of Northern Green Canyon and Ewing Bank (Offshore Louisiana), Northern Gulf of Mexico. *American Association of Petroleum Geologists Bulletin*, 82, 1083–112.

McBride, J. H. (1991). Constraints on the structure and tectonic development of the Early Mesozoic South Georgia rift, southeastern United States: seismic reflection data processing and interpretation. *Tectonics*, 10, 1065–83.

McBride, J. H. and Brown, L. D. (1986). Reanalysis of the COCORP deep seismic reflection profile across the San Andreas fault, Parkfield, California. *Bulletin of the Seismological Society of America*, 76, 1668–86.

McCabe, P. J. and Parrish, J. T. (1992). Tectonic and climatic controls on the distribution and quality of Cretaceous coals. In *Controls on the Distribution and Quality of Cretaceous Coals*, ed. P. J. McCabe and J. T. Parrish. *Geological Society of America Special Paper*, 267, 1–15.

McCaffrey, M. A., Dahl, J. E., Sundararaman, P., Moldowan, J. M., and Schoell, M. (1994). Source quality determination from oil biomarkers II – a case study using Tertiary-reservoired Beaufort Sea oils. *American Association of Petroleum Geologists Bulletin*, 10, 1527–40.

McClay, K. R. (1995). The geometries and kinematics of inverted fault systems: a review of analogue model studies. In *Basin Inversion*, ed. J. G. Buchanan and P. G. Buchanan. *Geological Society of London Special Publications*, **88**, 97–118.

McClay, K. R., Dooley, T., Whitehouse, P., and Mills, M. (2002). 4-D evolution of rift systems: insights from scaled physical models. *American Association of Petroleum Geologists Bulletin*, 86, 935–59.

McClay, K. R. and Khalil, S. (1998). Extensional hard linkages, eastern Gulf of Suez, Egypt. *Geology*, 26, 563–6.

McClay, K. R., Nichols, G. J., Khalil, S. M., Darwish, M., and Bosworth, W. (1998). Extensional tectonics and sedimentation, eastern Gulf of Suez, Egypt. In *Sedimentation and Tectonics of Rift Basins: Red Sea Gulf of Aden*, ed. B. H. Purser and D. W. J. Bosence. Chapman and Hall, London, pp. 223–38.

McClure, N. M. and Brown, A. A. (1992). Miller Field: a subtle Upper Jurassic submarine fan trap in the South Viking Graben, U.K. Sector, North Sea. In *Giant Oil and Gas Fields of the Decade: 1978–1988*, ed. M. T. Halbouty. *American Association of Petroleum Geologists Memoir*, **54**, 1519.

McDougall, I. and Harrison, T. M. (1988). Geochronology and thermochronology by the (super 40) Ar/ (super 39) Ar method. *Oxford Monographs on Geology and Geophysics*, 9, 212.

McGann, G. J., Green, S. C. H., Harker, S. D., and Romani, R. S. (1991). The Scapa Field, Block 14/19, UK North Sea. In *United Kingdom Oil and Gas Fields, 25 Years Commemorative Volume*, ed. I. L. Abbotts. *Memoir of the Geological Society of London*, **14**, 39–376.

McGee, K. A. and Gerlach, T. M. (1998). Annual cycle of magmatic CO_2 in a tree-kill soil at Mammoth Mountain, California: implications for soil acidification. *Geology*, 26, 463–6.

McGill, G. E. and Stromquist, A. W. (1975). Origin of graben in the Needles District, Canyonlands National Park, Utah. *Field Symposium – Guidebook of the Four Corners Geological Society*, 8, 235–43.

(1979). The grabens of Canyonlands National Park, Utah: geometry, mechanics, and kinematics. *Journal of Geophysical Research*, 84, 4547–63.

McHone, J. G. (1996). Broad-terrane Jurassic flood basalts across northeastern North America. *Geology*, 24, 319–22.

McHone, J. G. (2000). Non-plume magmatism and rifting during the opening of the central Atlantic Ocean. *Tectonophysics*, 316, 287–96.

McHone, J. G. and Puffer, J. H. (2003). *Flood Basalt Provinces of the Pangean Atlantic Rift: Regional Extent and Environmental Significance*. Columbia University Press, New York.

McInnes, B. I. A., Farley, K. A., Sillitoe, R. H., and Kohn, B. P. (1999). Application of apatite (U-Th)/He thermochronometry to the determination of the sense and amount of vertical fault displacement at the Chuquicamata porphyry copper deposit, Chile. *Economic Geology and the Bulletin of the Society of Economic Geologists*, 94, 937–47.

McKenzie, D. (1978). Some remarks on the development of sedimentary basins. *Earth and Planetary Science Letters*, 40, 25–32.

McKenzie, D. and Bickle, M. J. (1988). The volume and composition of melt generated by extension of the lithosphere. *Journal of Petrology*, 29, 625–79.

McKenzie, D. and Jackson, J. (1986). A block model of distributed deformation by faulting. *Journal of the Geological Society of London*, 143, 349–53.

McKenzie, D. and Priestley, K. (2008). The influence of lithospheric thickness variations on continental evolution. *Lithos*, 102, 1–11.

McKenzie, D. P. (1984). A possible mechanism for epeirogenetic uplift. *Nature*, 307, 616–19.

McKibben, M. A. and Hardie, L. A. (1997). Ore-forming brines in active continental rifts. In *Geochemistry of Hydrothermal Ore Deposits*, ed. H. L. Barnes. John Wiley and Sons, New York, pp. 877–935.

McLeod, A. E., Dawers, N. H., and Underhill, J. R. (2000). The propagation and linkage of normal faults: insights from the Strathspey-Brent-Statfjord fault array, northern North Sea. *Basin Research*, 12, 263–84.

McLeod, A. E. and Underhill, J. R. (1999). Processes and products of footwall degradation, northern Brent Field, northern North Sea. In *Petroleum Geology of Northwest Europe: Proceedings of the 5th Conference*, ed. A. J. Fleet and S. A. R. Boldy. American Geological Institute, London, pp. 91–106.

McLeod, A. E., Underhill, J. R., Davies, S. J., and Dawers, N. H. (2002). The influence of fault array evolution on synrift sedimentation patterns: controls on deposition in the Strathspey-Brent-Statfjord half graben, northern North Sea. *American Association of Petroleum Geologists Bulletin*, 86, 1061–93.

McNutt, M. (1987). Lithospheric stress and deformation. *Reviews of Geophysics*, 25, 1245–53.

Mechie, J. and Brooks, M. (1984). A seismic study of deep geological structure in the Bristol Channel area, SW Britain. *Geophysical Journal of the Royal Astronomical Society*, 78, 661–89.

Mechie, J., Keller, G. R., Prodehl, C., Gaciri, S., Braile, L. W., Mooney, W. D., Gajewski, D., and Sandmeler, K. J. (1994). Crustal structure beneath the Kenya Rift from axial profile data. *Tectonophysics*, 236, 179–99.

Meesters, A. G. C. A. and Dunai, T. J., (2002a). Solving the production-diffusion equation for finite diffusion domains of various shapes: Part I, Implications for low-temperature (U-Th)/He thermochronology. *Chemical Geology*, 186, 333–44.

Meesters, A. G. C. A. and Dunai, T. J. (2002b). Erratum: solving the production-diffusion equation for finite diffusion domains of various shapes: Part II, Application to cases with alpha -ejection and nonhomogeneous distribution of the source [modified]. *Chemical Geology*, 186, 347–63.

Megson, J. B. (1992). The North Sea chalk play: examples from the Danish Central Graben. In *Exploration in Britain: Geological Insights into the Next Decade*, ed. R. F. P. Hardman. *Special Publication of the Geological Society of London*, **67**, 247–82.

Meisling, K. E., Cobbold, P. R., and Mount, V. S. (2001). Segmentation of an obliquely rifted margin, Campos and Santos basins, southeastern Brazil. *American Association of Petroleum Geologists Bulletin*, 85, 1903–24.

Meissner, R. (1986). *The Continental Crust: A Geophysical Approach*. Academic Press, San Diego.

Meissner, R. and Strehlau, J. (1982). Limits of stresses in continental crusts and their relation to the depth-frequency distribution of shallow earthquakes. *Tectonics*, 1, 73–89.

Melinte, M. C. (2005). Oligocene palaeoenvironmental changes in the Romanian Carpathians, revealed by calcareous nannofossils. *Studia Geologica Polonica*, 124, 341–52.

Mello, M. R. and Maxwell, J. R. (1990). Organic geochemical and biological marker characterization of source rocks and oils derived from lacustrine environments in the Brazilian continental margin. In *Lacustrine Basin Exploration – Case Studies and Modern Analogs*, ed. B. J. Katz. *American Association of Petroleum Geologists Memoir*, **50**, 77–97.

Mello, M. R., Koutsoukos, E. A. M., and Santos Neto, E. V. (1991). Stratigraphy and depositional settings of Cretaceous organic-rich sequences in the Brazilian marginal basins. *American Association of Petroleum Geologists Bulletin*, 75, 633–4.

Mello, M. R., Telnaes, N., Gaglianone, P. C., Chicarelli, M. I., Brassell, S. C., and Maxwell, J. R. (1988). Organic geochemical characterisation of depositional palaeoenvironments of source rocks and oils in Brazilian marginal basins. *Organic Geochemistry*, 43, 31–45.

Mello, U. T. and Bender, A. A. (1988). On isostasy at the equatorial margin of Brazil. *Revista Brasileira de Geociencias*, 18, 237–46.

Menard, H. W. (1956). Archipelagic aprons (Pacific Ocean). *American Association of Petroleum Geologists Bulletin*, 40, 2195–210.

Meneses-Rocha, J. J. (2001). Tectonic evolution of the Ixtapa Graben, an example of a strike-slip basin of a southeastern Mexico: implications for regional petroleum systems. In *The Western Gulf of Mexico Basin: Tectonics, Sedimentary Basins, and Petroleum Systems*, ed. C. Bartolini, R. T. Buffler, and A. Cantu-Chapa. *American Association of Petroleum Geologists Memoir*, **75**, 183–216.

Menzies, M. A., Baker, J., Chazot, G., and Al'Kadasi, M. (1997b). Evolution of the Red Sea volcanic margin, western Yemen. In *Large Igneous Provinces: Continental Oceanic and Planetary Flood Colcanism*, ed. J. Mahoney and M. F. Coffin. American Geophysical Union Geophysical Monograph 100, pp. 29–43.

Menzies, M. A., Gallagher, K., Hurford, A., and Yelland, A. (1997a). Red Sea volcanic and the Gulf of Aden non-volcanic margins, Yemen: denudational histories and margin evolution: *Geochimica et Cosmochimica Acta*, 61, 2511–28.

Menzies, M. A., Klemperer, S. L., Ebinger, C. J., and Baker, J. (2002). Characteristics of volcanic rifted margins. In *Volcanic Rifted Margins*, ed. M. A. Menzies, S. L. Klemperer, C. J. Ebinger, and J. Baker. *GSA Special Paper*, **362**, 1–14.

Meredith, D. J. and Egan, S. S. (2002). The geological and geodynamic evolution of the eastern Black Sea basin: insights from 2-D and 3-D tectonic modelling. *Tectonophysics*, 350, 157–79.

Mero, W. E., Thurston, S. P., and Kropschot, R. E. (1992). The Point Arguello field. In *Giant Oil and Gas Fields of the Decade 1978–1988*, ed. M. T. Halbouty. *American Association of Petroleum Geologists Memoir*, **5**, 3–25.

Merrihue, C. and Turner, G. (1966). Potassium-argon dating by activation with fast neutrons. *Journal of Geophysical Research*, 71, 2852–9.

Meyerhoff, A. A. (1980a). Geology and petroleum fields in Proterozoic and Lower Cambrian strata, Lena-Tunguska petroleum province, Eastern Siberia, U.S.S.R. In *Giant Oil and Gas Fields of the Decade 1968–1978*, ed. M. T. Halbouty. *American Association of Petroleum Geologists Memoir*, **30**, 225–52.

(1980b). Petroleum basins of the Soviet Arctic. *Geological Magazine*, 117, 101–86.

(1982a). Hydrocarbon resources in Arctic and sub-Arctic regions. In *Arctic Geology and Geophysics*, ed. A. F. Embry and H. R. Balkwill. *Canadian Society of Petroleum Geologists Memoir*, **8**, 451–552.

(1982b). Petroleum basins of the Union of Socialist Soviet Republics and the People's Republic of China and the politics of petroleum. *Petroleum Exploration Society of Australia, Distinguished Lecture Series, Course Notes*, p. 341.

Meyers, J. B. (1995). *Rifted continental margin architecture off West Africa, as revealed by deep-penetrating multi-channel seismic reflection and potential field data*. PhD thesis, University of Miami (Florida), Coral Gables.

Meyers, J. B. and Rosendahl, B. R. (1991). Seismic reflection character of the Cameroon volcanic line: evidence for uplifted oceanic crust. *Geology*, 19, 1072–6.

Meyers, J. B., Rosendahl, B. R., Groschel-Becker, H., Austin, J. A. Jr., and Rona, P. A. (1996). Deep penetrating MCS imaging of the rift-to-drift transition, offshore Douala and North Gabon basins, West Africa. *Marine and Petroleum Geology*, 13, 791–835.

Meyers, J. B., Rosendahl, B. R., Harrison, C. G. A., and Ding, Z.-D. (1998). Deep-imaging seismic and gravity results from the offshore Cameroon volcanic line, and speculation of African hotlines. *Tectonophysics*, 284, 31–63.

Miall, A. D. (1985). Stratigraphic and structural predictions from a plate-tectonic model of an oblique-slip orogen: the Eureka Sound Formation (Campanian-Oligocene), Northeast Canadian Arctic Islands. *Special Publication – Society of Economic Paleontologists and Mineralogists*, 37, 361–74.

(1996). *The Geology of Fluvial Deposits: Sedimentary Facies, Basin Analysis, and Petroleum Geology*. Springer-Verlag, Berlin, p. 582.

Michon, L. and Merle, O. (2003). Mode of lithospheric extension: conceptual models from analogue modeling. *Tectonics*, 22, 1028.

Middleton, M. F. (1980). A model of intracratonic basin formation, entailing deep crustal metamorphism. *Geophysical Journal of the Royal Astronomical Society*, 62, 1–14.

(1982). Tectonic history from vitrinite reflectance. *Geophysical Journal of the Royal Astronomical Society*, 68, 121–32.

Miles, P. R., Munschy, M., and Segoufin, J. (1998). Structure and evolution of the Arabian Sea and the eastern Somali basin. *Geophysical Journal International*, 134, 876–88.

Miller, D. J. and Christensen, N. I. (1997). Seismic velocities of lower crustal and upper mantle rocks from the slow-spreading Mid-Atlantic Ridge, south of the Kane transform zone (MARK). *Proceedings of the Ocean Drilling Program, Scientific Results*, 153, 437–54.

Miller, D. S. and Duddy, I. R. (1989). Early Cretaceous uplift and erosion of the northern Appalachian Basin, New York, based on apatite fission track analysis. *Earth and Planetary Science Letters*, 93, 35–49.

Miller, R. G. (1990). A paleoceanographic approach to Kimmeridge Clay Formation. In *Deposition of Organic Facies*, ed. A. Y. Huc. *American Association of Petroleum Geologists Studies in Geology*, **30**, 13–26.

(1992). The global oil system: the relationship between oil generation, loss, half-life, and the world crude oil resource. *American Association of Petroleum Geologists Bulletin*, 76, 489–500.

Mills, R. A., Teagle, D. A. H., and Tivey, M. K. (1998). Fluid mixing and anhydrite precipitation within the TAG mound. In *Proceedings from the Oceanic Drilling Program Scientific Results*, ed. P. M. Herzig, S. E. Humphris, D. J. Miller, and R. A. Zierenberg. College Station, Texas, **158**, 119–27.

Mills, R. A. and Tivey, M. K. (1999). Sea water entrainment and fluid evolution within the TAG hydrothermal mound: evidence from analyses of anhydrite. In *Dynamics of Processes Associated with New Ocean Crust*, ed. J. H. Cann and A. Laughton. Cambridge University Press, Cambridge, pp. 225–63.

Milner, P. S. and Olsen, T. (1998). Predicted distribution of the Hugin Formation reservoir interval in the Sleipner Ostfield, South Viking Graben: the testing of a three-dimensional sequence stratigraphy model. In *Sequence Stratigraphy: Concepts and Applications*, ed. F. M. Gradstein, K. O. Sandvik, and N. J. Milton. *Special Publication of the Norwegian Petroleum Society*, **8**, 337–54.

Milner, S. C., Duncan, A. R., Whittigham, A. M., and Ewart, A. (1995). Trans-Atlantic correlation of eruptive sequences and individual silicic volcanic units within the Parana-Etendeka igneous province. *Journal of Volcanology and Geothermal Research*, 69(3–4), 137–57.

Milton, N. J. and Bertram, G. T. (1992). Trap styles: a new classification based on sealing surfaces. *American Association of Petroleum Geologists Bulletin*, 76, 983–99.

Milton, N. J., Bertram, G. T., and Vann, I. R. (1990). Early Paleogene tectonics and sedimentation in the central North Sea. In *Tectonic Events Responsible for*

Britain's Oil and Gas Reserves, ed. R. F. P. Hardman and J. Brooks. *Geological Society of London Special Publications*, pp. 339–51.

Min, D.-H., Bullister, J. L., and Weiss, R. F. (2000). Constant ventilation age of thermocline water in the eastern subtropical North Pacific Ocean from chlorofluorocarbon measurements over a 12-year period. *Geophysical Research Letters*, 27, 3909–12.

Min, K., Farley, K. A., Renne, P. R., and Marti, K. (2003). Single grain (U-Th)/He ages from phosphates in Acapulco Meteorite and implications for thermal history. *Earth and Planetary Science Letters*, 209, 323–36.

Minshull, T. A., Dean, S. M., White, R. S., and Whitmarsh, R. B. (2001). Anomalous melt production after continental breakup in the southern Iberia Abyssal Plain. In *Non-Volcanic Rifting of Continental Margins: Evidence from Land and Sea*, ed. R. C. L. Wilson, R. B. Whitmarsh, B. Taylor, and N. Froitzheim. *Geological Society of London Special Publications*, **187**, 537–50.

Minshull, T. A., White, R. S., Mutter, J. C., Buhl, P., Detrick, R. S., Williams, C. A., and Morris, E. (1991). Crustal structure at the Blake Spur fracture zone from expanding spread profiles. *Journal of Geophysical Research*, 96, 9955–84.

Mitra, S. (1987). Regional variations in deformation mechanisms and structural styles in the central Appalachian orogenic belt. *Geological Society of America Bulletin*, 98, 569–90.

Mittelstaedt, E. and Ito, G. (2005). Plume–ridge interaction, lithospheric stresses, and the origin of near-ridge volcanic lineaments. *Geochemistry Geophysics* Geosystems, **6**.

Mittelstaedt, E., Ito, G., and Behn, M. D. (2008). Mid-ocean ridge jumps associated with hotspot magmatism. *Earth and Planetary Science Letters*, 266, 256–70.

MLWG (1997). MONA LISA: deep seismic investigations of the lithosphere in the southeastern North Sea. *Tectonophysics*, 269, 1–19.

Mocanu, V. I. and Rădulescu, F. (1994). Geophysical features of the Romanian territory. *Romanian Journal of Tectonics and Regional Geology*, 75, 17–36.

Mohriak, W. U., Bassetto, M., and Vieira, I. S. (1998). Crustal architecture and tectonic evolution of the Sergipe-Alagoas and Jacuipe basins, offshore northeastern Brazil. *Tectonophysics*, 288, 199–220.

Mohriak, W. U., Nemčok, M., and Enciso, G. (2008). South Atlantic divergent margin evolution: rift-border uplift and salt tectonics in the basins of SE Brazil. In *West Gondwana: Pre-Cenozoic Correlations across the South Atlantic Region*, ed. R. J. Pankhurst, R. A. J. Trouw, B. B. de Brito Neves, and M. J. De Witt. *Geological Society of London Special Publications*, **294**, 365–98.

Mohriak, W. U., Rosendahl, B. R., Turner, J. P., and Valente, S. C. (2002). Crustal architecture of South Atlantic volcanic margins. In *Volcanic Rift Margins*, ed. M. A. Menzies, S. L. Klemperer, C. Ebinger, and J. Baker. Geological Society of America Special Paper, 362, pp. 159–202.

Moldowan, J. M., Albrecht, P., and Philp, R. P. (1993). *Biological Markers in Sediments and Petroleum*. Prentice Hall, Englewood Cliffs, New Jersey.

Molina, P. (1988). Correlation geologique Afrique-Amerique du Sud et provinces uraniferes. African-South American geologic correlations and uraniferous provinces. *Journal of African Earth Sciences*, 7, 489–97.

Mollema, P. N. and Antonellini, M. A. (1996). Compaction bands: a structural analog for anti-mode I cracks in aeolian sandstone. *Tectonophysics*, 267, 209–28.

Molnar et al. (2005). In *Paleogeography and maturation from Congo Span PSDM data*, ed. S. G. Henry, A. Danforth, A. G. Requejo, G. F. Schiefelbein, and S. Venkatraman. *The 4th HGS/PESGB International Conference on African E7P*, Houston Program, 3.

Momper, J. A. (1978). Oil migration limitations suggested by geological and geochemical considerations. *American Association of Petroleum Geologists Bulletin Continuing Education Course Notes*, 8, 1–60.

Mongelli, F. (1985). Heat flow along the Italian segment of the EGT. *Terra cognita*, 5, 385.

Mongelli, F., Loddo, M., and Tramacere, A. (1982). Thermal conductivity, diffusivity and specific heat variation of some Travale field (Tuscany) rocks versus temperature. *Tectonophysics*, 83, 33–43.

Montadert, L., de Charpel, O., Roberts, D., Guennoc, P., and Sibuet, J. C. (1979). Northeast Atlantic passive continental margins; rifting and subsidence processes. In *Deep Drilling Results in the Atlantic Ocean: Continental Margins and Paleoenvironment*, ed. M. H. Talwani and W. B. F. Ryan. Maurice Ewing Series, Harriman, New York, pp. 154–86.

Montañez, I. P. (1994). Late diagenetic dolomitization of Lower Ordovician, Upper Knox carbonates: a record of the hydrodynamic evolution of the Southern Appalachian Basin. *American Association of Petroleum Geologists Bulletin*, 78, 1210–39.

Montenat, C., Guery, F., and Jamet, M. (1988). Mesozoic evolution of the Lusitanian Basin: comparison with the adjacent margin. *Proceedings of the Ocean Drilling Program, Scientific Results*, 103, 757–75.

Moore, C. H. (1989). *Carbonate Diagenesis and Porosity*. Elsevier, Amsterdam.

Moore, D. G. and Curray, J. R. (1982). Geologic and tectonic history of the Gulf of California. *Initial Reports of the Deep Sea Drilling Project*, 64, 1279–94.

Moore, D. M. and Reynolds, R. C. Jr. (1997). *X-ray Diffraction and the Identification and Analysis of Clay Minerals*. Oxford University Press, Oxford, United Kingdom.

Moore, J. C., Orange, D. L., and Kulm, L. D. (1990). Interrelationship of fluid venting and structural evolution, Oregon margin. *Journal of Geophysical Research*, 95, 8795–808.

Moore, J. C. and Vrolijk, P. (1992). Fluids in accretionary prisms. *Reviews of Geophysics*, 30, 113–35.

Moore, J. G., Normark, W. R., and Holcomb, R. T. (1994). Giant Hawaiian landslides. *Annual Review of Earth and Planetary Sciences*, 22, 119–44.

Moore, J. N. and Adams, M. C. (1988). Evolution of the thermal cap in two wells from the Salton Sea geothermal system, California. *Geothermics*, 17, 695–710.

Moore, M. E., Gleadow, A. J. W., and Lowering, J. F. (1986). Thermal evolution of rifted continental margins: new evidence from fission tracks in basement apatites from southeastern Australia. *Earth and Planetary Science Letters*, 78, 255–70.

Mordecai, M. and Morris, L. H. (1971). An investigation into the changes of permeability occurring in a sandstone when failed under triaxial stress condition. In *Dynamic Rock Mechanics*, ed. D. B. Clark. *Proceedings of the 12th Symposium of Rock Mechanics*, American Institute of Mining, Metallurgical and Petroleum Engineers, pp. 221–40.

Morency, C., Doin, M.-P., and Dumoulin, C. (2002). Convective destabilization of thickened continental lithosphere. *Earth and Planetary Science Letters*, 202, 303–20.

Morgan, J. K. and Clague, D. A. (2003). Volcanic spreading on Mauna Loa volcano, Hawaii: evidence from accretion, alteration, and exhumation of volcanoclastic sediments. *Geology*, 31, 411–14.

Morgan, P. (1984). The thermal structure and thermal evolution of the continental lithosphere. In *Structure and Evolution of the Continental Lithosphere*, ed. H. N. Pollack and V. R. Murthy. *Physics and Chemistry of the Earth*, **15**, 107–93.

Morgan, P., Boulos, F. K., Hennin, S. F., El-Sherif, A. A., El-Sayed, A. A., Basta, N. Z., and Melek, Y. S. (1985). Heat flow in eastern Egypt: the thermal signature of a continental breakup. *Journal of Geodynamics*, 4, 107–31.

Morgan, P., Harder, V., Swanberg, C. A., and Daggett, P. H. (1981). A groundwater convection model for the Rio Grande rift geothermal resources. *Geothermal Research Council Transactions*, 5, 193–6.

Morgan, W. J. (1978). Rodriguez, Darwin, Amsterdam, …, a second type of hotspot island. *Journal of Geophysical Research*, 83, 5355–60.

Morin, R. and Silva, A. J. (1984). The effects of high pressure and high temperature on physical properties of ocean sediments. *Journal of Geophysical Research*, 89, 511–26.

Morley, C. K. (1999). Marked along-strike variations in dip of normal faults: the Lokichar Fault, N. Kenya Rift: a possible cause for metamorphic core complexes. *Journal of Structural Geology*, 21(5), 479–92.

Morley, C. K., Nelson, R. A., Patton, T. L., and Munn, S. G. (1990). Transfer zones in the East African Rift system and their relevance to hydrocarbon exploration in rifts. *American Association of Petroleum Geologists Bulletin*, 74, 1234–53.

Morley, C. K. and Wescott, W. A. (1999). *Sedimentary environments and geometry of sedimentary bodies determined from subsurface studies in East Africa*. University of Brunei, Department of Petroleum Geoscience, Darusalam, Brunei.

Morley, C. K., Wescott, W. A., Stone, D. M., Harper, R. M., Wigger, S. T., and Karanja, F. M. (1992). Tectonic evolution of the northern Kenyan Rift. *Journal of the Geological Society of London*, 149, 333–48.

Morley, C. K., Wonganan, N., Sankumarn, N., Hoon, T. B., Alief, A., and Simmons, M. (2001). Late Oligocene–recent stress evolution in rift basins of northern and central Thailand: implications for escape tectonics. *Tectonophysics*, 334, 115–50.

Morris, J. E., Hampson, G. J., and Johnson, H. D. (2006). A sequence stratigraphic model for an intensely bioturbated shallow-marine sandstone: the Bridport Sand Formation, Wessex Basin, UK. *Sedimentology*, 53, 1229–63.

Morrow, C. and Byerlee, J. (1988). Permeability of rock samples from Cajon Pass, California. *Geophysical Research Letters*, 15, 1033–6.

Morse, S. A. (1986). Thermal structure of crystallizing magma with two-phase convection. *Geological Magazine*, 123, 205–14.

Morton, A. C. and Taylor, P. N. (1987). Lead isotope evidence for the structure of the Rockall dipping-reflector passive margin. *Nature*, 326, 381–3.

Moser, F. and Frisch, W. (1996). Tertiary deformation in the Southern Carpathians – structural analysis of a brittle deformation. *TSK 6 Symposium Erweitere Kurzfassungen*, Vienna, Facultas-Universitatsverlag, pp. 280–2.

Moss, B., Barson, D., Rakhit, K., Dennis, H., and Swarbrick, R., 2002. Formation pore pressures and formation waters. In *The Millennium Atlas: Petroleum Geology of the Central and Northern North Sea*, ed. D. Evans, C. Graham, A. Armour, and P. Bathurst. *Geological Society of London*, pp. 317–29.

Mottana, A., Nicoletti, M., Petrucciani, C., Liborio, G., De Capitani, L., and Bocchio, R. (1985). Pre-Alpine and Alpine evolution of the South-alpine basement of the Southern Alps. *Geologische Rundschau*, 74, 353–66.

Moulin, M., Aslanian, D., Olivet, J.-L., Contrucci, I., Matias, L., Geli, L., Klingelhoefer, F., Nouze, H., Rehault, J.-P., and Unternehr, P. (2005). Geological constraints on the evolution of the Angolan margin based on reflection and refraction seismic data (ZaiAngo Project). *Geophysical Journal International*, 162, 793–810.

Moustafa, A. R. (1992). Structural setting of the Sidri-Feiran area, eastern side of the Suez Rift. *Earth Science Series*, 6, 44–54.

(1996). Internal structure and deformation of an accommodation zone in the northern part of the Suez Rift. *Journal of Structural Geology*, 18, 93–107.

(1997). Controls on the development and evolution of transfer zones: the influence of basement structure and sedimentary thickness in the Suez Rift and Red Sea. *Journal of Structural Geology*, 19, 755–68.

(2002). Controls on the geometry of transfer zones in the Suez Rift and Northwest Red Sea: implications for the structural geometry of rift systems. *American Association of Petroleum Geologists Bulletin*, 86, 979–1002.

Mueller, S. (1983). The EGT project. *EOS Transactions*, 64, 458.

Muentener, O., Desmurs, L., Pettke, T., and Schaltegger, U. (2002). Melting and melt/rock reaction in extending mantle lithosphere; trace element and isotopic constraints from passive margin peridotites. *Geochimica et Cosmochimica Acta*, 66, 526.

Muentener, O. and Hermann, J. (2001). The role of lower crust and continental upper mantle during formation of non-volcanic passive margins: evidence from the Alps. *Geological Society of London Special Publications*, **187**, 267–88.

Muentener, O. and Piccardo, G. B. (2003). Melt migration in ophiolitic peridotites: the message from Alpine-Apennine peridotites and implications for embryonic ocean basins. *Geological Society of Special Publications*, 218, 69–89.

Muffler, L. J. P. and White, D. E. (1968). Origin of CO2 in the Salton Sea geothermal system, southeastern California, U.S.A. *Proceedings of International Geological Congress*, 97, 185–94.

Mukhopadhyay, P. K. (1994). Vitrinite reflectance as a maturity parameter: petrographic and molecular characterization and its applications to basin modeling. In *Vitrinite Reflectance as a Maturity Parameter: Applications and Limitations*, ed. P. K. Mukhopadhyay and W. G. Dow. American Chemical Society Symposium Series, Washington, DC, **570**, 1–24.

(2002). Evaluation of petroleum systems for five dummy wells from various seismic sections of the Scotian slope, offshore Nova Scotia, based on one-dimensional numerical modelling and using geochemical concepts. *Confidential Internal Report*, Canada-Nova Scotia Offshore Petroleum Board.

Muller, M. R., Robinson, C. J., Minshull, T. A., White, R. S., and Bickle, M. J. (1997). Thin crust beneath Ocean Drilling Program Borehole 735B at the Southwest Indian Ridge? *Earth and Planetary Science Letters*, 148, 93–107.

Müller, R. D., Roest, W. R., Royer, J.-Y., Gahagan, L. M., and Sclater, J. G. (1993). A digital age map of the ocean floor. SIO Reference Series 93–30. Scripps Institution of Oceanography, San Diego. http://www.ngdc.noaa.gov/mgg/fliers/96mgg04.html.

Müller, R. D., Roest, W. R., Royer, J. Y., Gahagan, L. M., and Sclater, J. G. (1997). Digital isochrons of the world's ocean floor. *Journal of Geophysical Research*, 102, 3211–14.

Mutter, J. C., Buck, W. R., and Zehnder, C. M. (1988). Convective partial melting. 1. A model for the formation of thick basaltic sequences during the initiation of spreading. *Journal of Geophysical Research*, 93, 1031–48.

Mutter, J., Talwani, M., and Stofa, P. L. (1982). Origin of seaward-dipping reflectors in oceanic crust off the Norwegian margin by "subaerial sea-floor spreading." *Geology*, 10, 353–7.

Nadeau, R. M. and McEvilly, T. V. (1997). Seismological studies at Parkfield: V, Characteristic microearthquake sequences as fault-zone drilling targets. *Bulletin of the Seismological Society of America*, 87, 1463–72.

Nadin, P. A. and Kusznir, N. J. (1995). Palaeocene uplift and Eocene subsidence in the northern North Sea basin from 2D forward and reverse stratigraphic modelling. *Journal of the Geological Society of London*, 152, 833–48.

Nadin, P. A., Kusznir, N. J., and Cheadle, M. J. (1997). Early Tertiary plume uplift of the North Sea and Faeroe-Shetland basins. *Earth and Planetary Science Letters*, 148, 109–27.

Nadin, P. A., Kusznir, N. J., and Toth, J. (1995). Transient regional uplift in the early Tertiary of the northern North Sea and the development of the Iceland Plume. *Journal of the Geological Society of London*, 152, 953–8.

Naeser, N. D., Naeser, C. W., and McCulloh, T. H. (1989). The application of fission-track dating to depositional and thermal history of rocks in sedimentary basins. In *Thermal History of Sedimentary Basins*, ed. N. D. Naeser and T. H. McCulloh. Springer-Verlag, New York, pp. 157–80.

Naylor, M. A. and Spring, L. Y. (2002). Exploration strategy development and performance management: a portfolio-based approach. *The Leading Edge*, 21, 159–67.

Naylor, P. H., Bell, B. R., Jolley, D. W., Durnall, P., and Fredsted, R. (1999). Palaeogene magmatism in the Faeroe-Shetland Basin: influences on uplift history and sedimentation. In *Petroleum Geology of Northwest Europe*, ed. A. J. Fleet and S. A. R. Boldy. *Proceedings of the 5th Conference, Geological Society of London*, pp. 545–58.

Neculae, P. and Stanescu, V. (2001). *Oil Fields in the External Carpathian Flysch*. Vergiliu, Bucharest, p. 209.

Nederlof, M. H. and Mohler, H. P. (1981). Quantitative investigation of trapping effect of unfaulted caprock. *American Association of Petroleum Geologists Bulletin*, 65, 964.

Nelson, R. A. (2001). *Geologic Analysis of Naturally Fractured Reservoirs*. Gulf Professional Publishing, Boston.

Nemčok, M. (2001). Preliminary Petroleum Systems and Source Rock Risk Evaluation: Santos Basin, Brazil – Phase II. *EGI Report*, 01-00059-5000-50500773.

(2004). Niko-1 well project for Frontera Resources. *EGI Report*, 01-00059-5000-50501264.

(2005). Combined structural/sedimentological research in Block 12 for Frontera Resources during year 2005. *EGI Report*, 01-00059-5000-50501492.

(2012). Structural aspects of tight shales. In *Evolution of the Mental Picture of Tight Shales*, ed. J. McLennan. Energy and Geoscience Institute, Salt Lake City, pp. 41–58.

Nemčok, M. and Allen, R. (2001). Fault propagation timing in the northern portion of the Gulf of Guinea, Africa. *EGI report*, 50500854, p. 51.

Nemčok, M. and Gayer, R. A. (1996). Modelling palaeostress magnitude and age in extensional basins: a case study from the Mesozoic Bristol Channel Basin, UK. *Journal of Structural Geology*, 18, 1301–14.

Nemčok, M., Gayer, R. A., and Miliorizos, M. (1995). Structural analysis of the inverted Bristol Channel Basin: implications for the geometry and timing of the fracture porosity. Basin inversion. *Geological Society of London Special Publications*, **82**, 355–92.

Nemčok, M., Henk, A., Allen, R., Sikora, P. J., and Stuart, C. (2012b). Continental break-up along strike-slip fault zones: observations from Equatorial Atlantic. In *Conjugate Divergent Margins*, ed. W. U. Mohriak, A. Danforth, P. J. Post, D. E. Brown, G. M. Tari, M. Nemčok, and S. T. Sinha. *Geological Society of London Special Publications*, **369**.

Nemčok, M., Henk, A., Gayer, R. A., Vandycke, S., and Hathaway, T. M. (2002). Strike-slip fault bridge fluid pumping mechanism: Insights from field-based palaeostress analysis and numerical modeling. *Journal of Structural Geology*, 24, 1885–901.

Nemčok, M. and Kantor, J. (1990). Movement study in the selected area of the Velký Bok Unit (in Slovak). *Regionálna Geológia Západných Karpát*, 75–83.

Nemčok, M., Konecny, P., and Lexa, A. (2000). Calculations of tectonic, magmatic and residual stress in the Stiavnica stratovolcano, Western Carpathians: implications for mineral precipitation paths. *Geologica Carpathica*, 51, 19–36.

Nemčok, M. and Lexa, J. (1990). Evolution of the basin and range structure around Ziar mountain range. *Geologicky Sbornik*, 41, 229–58.

Nemčok, M., Moore, J. N., Allis, R., and McCulloch, J. (2004a). Fracture development within a stratovolcano: the Karaha-Telaga Bodas geothermal field, Java volcanic arc. In *The Initiation, Propagation, and Arrest of Joints and other Fractures: A Field Workshop Dedicated to the Memory of Paul Hancock*, ed. J. Cosgrove and T. Engelder. *Geological Society of London Special Publications*, **231**, 223–42.

Nemčok, M., Moore, J. N., Christensen, C., Allis, R., Powell, T., Murray, B., and Nash, G. (2006a). Controls on the Karaha-Telaga Bodas geothermal reservoir, Indonesia. *Geothermics*, 36, 9–46.

Nemčok, M., Pogacsás, G., and Pospíšil, L. (2006b). Activity timing of the main tectonic systems in the Carpathian-Pannonian region in relation to the roll-back destruction of the lithosphere. The Carpathians and their foreland: geology and hydrocarbon resources. *American Association of Petroleum Geologists Memoir*, 84, 743–66.

Nemčok, M., Pospíšil, L., Lexa, J., and Donelick, R. A. (1998). Tertiary subduction and slab break-off model of the Carpathian-Pannonian region. *Tectonophysics*, 295, 307–40.

Nemčok, M. and Rosendahl, B. R. (2006). CameroonSpan and GabonSpan interpretation. *EGI Report*, 01-00059-5000-50501399.

Nemčok, M., Rosendahl, B. R., Segall, M., Silva, C., Stuart, C., Allen, R., Sikora, P., Christensen, C., Francu, J., and Abrams, M. (2004b). Equatorial Atlantic margins basins project. A joint EGI-Industry evaluation of the evolution, development and prospectivity of basins along the eastern and western Equatorial Atlantic continental margins. *EGI Report*, 01-0059-5000-50500936.

Nemčok, M., Schamel, S., and Gayer, R. (2005a). *Thrustbelts: Structural Architecture, Thermal Regimes and Petroleum Systems*. Cambridge University Press, Cambridge, United Kingdom, p. 541.

Nemčok, M., Sheya, C., Welker, C., Smith, S., and Rybár, S. (2010). Crustal types, structural architecture, and plate configurations study of the Reliance West Coast region, Phase 1 project. *EGI Report*, 01-00059-5000-50501853 for Reliance (confidential).

Nemčok, M., Sinha, S. T., Stuart, C., Welker, C., Choudhuri, M., Sharma, S. P., Misra, A. A., Sinha, N., and Venkatraman, S. (2012c). East Indian margin evolution and crustal architecture: integration of deep reflection seismic interpretation and gravity modeling. In *Conjugate Divergent Margins*, ed. W. U. Mohriak, A. Danforth, P. J. Post, D. E. Brown, G. M. Tari, M. Nemčok, and S. T. Sinha. *Geological Society of London Special Publications*, **369**.

Nemčok, M., Stuart, C., Odegard, M., Sheya, C., Welker, C., Dvorakova, V., Meissner, A., Bubniak, I., Bubniak, A., Vangelov, D., and Baraboshkin, E. (2009). Tectonic development of the Black Sea region. Phase 2 of the circum-Black Sea project. *EGI Report*, 01-00059-5000-50501563.

Nemčok, M., Stuart, C., Moore, J. N., Smith, S., Welker, C., and Sheya, C. (2008). Crustal types, structural architecture, and plate configurations study of the Reliance East Coast region. Phase 1-extension study. *EGI report*, 01-00059-5000-50501784-2, p. 37.

Nemčok, M., Stuart, C. J., Rosendahl, B. R., Welker, C., Smith, S., Sheya, C., Sinha, S. T., Choudhuri, M., Allen, R., Reeves, C., Sharma, S. P., Venkatraman S., and Sinha, N. (2012a). Continental break-up mechanism: lessons from intermediate- and fast-extension settings. In *Conjugate Divergent Margins*, ed. W. U. Mohriak, A. Danforth, P. J. Post, D. E. Brown, G. M. Tari, M. Nemčok, and S. T. Sinha. *Geological Society of London Special Publications*, **369**.

Nemčok, M., Stuart, C., Segall, M. P., Allen, R. B., Christensen, C., Hermeston, S. A., and Davison, I. (2005b). Structural development of South Morocco: interaction of tectonics and deposition. In *Petroleum Systems of Divergent Continental Margin Basins*, ed. P. J. Post, N. Rosen, D. L. Olson, S. L. Palmes, K. T. Lyons, and G. B. Newton. 25th Annual GCSSEPM Foundation Bob F. Perkins Research Conference, Session I, Crustal architecture of divergent margins, Houston, pp. 151–202.

Nemčok, M., Stuart, C., Welker, C., Smith, S., Yalamanchili, R., Srivastava, D. C., Sinha, S., and Choudhuri, M. (2007). Crustal types, structural architecture, and plate configurations study of the Reliance East Coast region, Phase 1 project. *EGI Report*, 01-00059-5000-50501546 for Reliance (confidential).

Nemčok, M., Vangelov, D., Stuart, C., Francu, J., Christensen, C., Abrams, M., Moore, J., Jones, C., Clausen, S., Sahm, E., and Jones, P. (2004c). Bulgaria Bourgas Block Project Phase 2. *EGI Report*, 50501139.

Nemec, W. (1990). Aspects of sediment movement on steep delta slopes. In *Coarse Grained Deltas*, ed. A. Colella and D. B. Prior. *Special Publication of the International Association of Sedimentologists*, pp. 29–73.

Netto, A. S. T. and de Oliveira, J. J. (1985). O preenchimento do rift-valley na Bacia do Reconcavo. The filling of the rift valley in the Reconcavo Basin. *Revista Brasileira de Geociencias*, 15, 97–102.

Neuzil, C. E. (1986). Groundwater flow in low-permeability environments. *Water Resources Research*, 22, 1163–95.

Newmark, R. L., Kasameyer, P. W., and Younker, L. W. (1988). Shallow drilling in the Salton Sea region: the thermal anomaly. *Journal of Geophysical Research*, 93, 13005–23.

Nilsen, T. H. and McLaughlin, R. J. (1985). Comparison of tectonic framework and depositional patterns of the Hornelen strike-slip basin of Norway and the Ridge and Little Sulphur Creek strike-slip basins of California. *Special Publication – Society of Economic Paleontologists and Mineralogists*, 37, 79–103.

Noguti, I. and Santos, J. F. D. (1972). Zoneamento preliminar por foraminiferos planctonicos do aptiano ao mioceano na plataforma continental do Brasil. Preliminary zoning of planktonic foraminifera of the Aptian to Miocene rocks of the continental shelf of Brazil. *Boletim Tecnico da PETROBRAS*, 15, 265–83.

Noll, C. A. and Hall, M. (2006). Normal fault growth and its function on the control of sedimentation during basin formation: a case study from field exposures of the Upper Cambrian Owen Conglomerate, West Coast Range, western Tasmania, Australia. *American Association of Petroleum Geologists Bulletin*, 90, 1609–30.

Norris, R. J. and Carter, R. M. (1980). Offshore sedimentary basins at the southern end of the Alpine Fault, New Zealand. *Special Publication of the International Association of Sedimentologists*, 4, 237–65.

Norton, D. and Knight, J. (1977). Transport phenomena in hydrothermal systems: cooling plutons. *American Journal of Science*, 277, 937–81.

Nuckolls, H. M. and McCulley, B. L. (1987). Origin of saline springs in Cataract Canyon, Utah. *Field Symposium – Guidebook of the Four Corners Geological Society*, 10, 193–200.

Nunn, J. A. (1985). State of stress in the northern Gulf Coast. *Geology*, 13, 429–32.

Nunn, J. A. and Deming, D. (1991). Thermal constraints on basin-scale flow systems. In *Crustal-Scale Fluid Transport – Magnitude and Mechanisms*, ed. T. Torgensen. *Geophysical Research Letters*, 18, 967–70.

Nunn, J. A. and Meulbroek, P. (2002). Kilometer-scale upward migration of hydrocarbons in geopressured sediments by buoyancy-driven propagation of methane-filled fractures. *American Association of Petroleum Geologists Bulletin*, 86, 907–18.

Nunn, J. A. and Sassen, R. (1986). The framework of hydrocarbon generation and migration, Gulf of Mexico continental slope. *Gulf Coast Association of Geological Societies Transactions*, 35, 257–62.

Obermeier, S. (1996). Use of liquefaction-induced features for paleoseismic analysis – an overview of how seismic liquefaction features can be distinguished from other features and how their regional response distribution and properties of source sediment can be used to infer the location and strength of Holocene paleo-earthquakes. *Engineering Geology*, 44, 1–76.

O'Connell, R. J. and Budiansky, B. (1974). Seismic velocities in dry and saturated cracked solids. *Journal of Geophysical Research*, 79, 5412–26.

O'Connell, R. J. and Wasserburg, G. J. (1972). Dynamics of submergence and uplift of a sedimentary basin underlain by a phase-change boundary. *Reviews of Geophysics and Space Physics*, 10, 335–68.

Offshore Energy Technical Research (OETR) Association (2011). Play Fairway Analysis. http://www.offshoreenergyresearch.ca/OETR/OETRPlayFairwayProgramMainPage/tabid/402/Default.aspx.

Oh, J., Austin, J. A., Phillips, J. D., Coffin, M. F., and Stoffa, P. L. (1995). Seaward-dipping reflectors offshore the southeastern United States: seismic evidence for extensive volcanism accompanying sequential formation of the Carolina Trough and Blake Plateau basin. *Geology*, 23, 9–12.

O'Hara, I., Hower, J. G., and Rimmer, S. M. (1990). Constraints on the emplacement and uplift history of the Pine Mountain thrust sheet, eastern Kentucky: evidence from coal rank trends. *Journal of Geology*, 98, 43–51.

Ohlmacher, G. C. and Aydin, A. (1997). Mechanics of vein, fault and solution surface formation in the Appalachian Valley and Ridge, northeastern Tennessee, U.S.A.: implications for fault friction, state of stress and fluid pressure. *Journal of Structural Geology*, 19, 927–44.

Ojeda, H. A. O. (1982). Structural framework, stratigraphy, and evolution of Brazilian marginal basins. *American Association of Petroleum Geologists Bulletin*, 66, 732–49.

Oliver, J. (1986). Fluids expelled tectonically from orogenic belts: their role in hydrocarbon migration and other geologic phenomena. *Geology*, 14, 99–102.

Olsen, K. H. and Morgan, P. (1995). Introduction: progress in understanding continental rifts. *Developments in Geotectonics*, 25, 3–26.

Olsen, P. E. (1990). Tectonic, climatic and biotic modulation of lacustrine ecosystems – examples from Newark

Supergroup of eastern North Americas. In *Lacustrine Basin Exploration – Case Studies and Modern Analogs*, ed. B. J. Katz. *American Association of Petroleum Geologists Memoir*, 50, 209–24.

Olsen, P. E. and Schlische, R. W. (1990). Transtensional arm of the early Mesozoic Fundy rift basin: penecontemporaneous faulting and sedimentation. *Geology*, 18, 695–8.

Olsen, P. E., Schlische, R.W., and Gore, P. J. W. (1990). Tectonic, depositional, and paleoecological history of early Mesozoic rift basins, eastern North America. *Geology*, 18, 174.

Ord, A. and Hobbs, B. E. (1989). The strength of the continental crust, detachment zones and the development of plastic instabilities. *Tectonophysics*, 158, 269–89.

Orr, W. L. (1986). Kerogen/asphaltene/sulfur relationships in sulfur-rich Monterey oils. *Organic Geochemistry*, 10, 499–506.

(1993). Guidelines for Type II-S Kerogens in basin modeling: organic sulfur content of kerogens and crude oils. *American Association of Petroleum Geologists Bulletin*, 77, 1652.

Ortoleva, P., Merino, E., Moore, C., and Chadam, J. (1987). Geochemical self-organization, I. Reaction-transport feedbacks and modeling approach. *American Journal of Science*, 287, 979–1007.

Osborne, M. J. and Swarbrick, R. E. (1997). Mechanisms for generating overpressures in sedimentary basins: a reevaluation. *American Association of Petroleum Geologists Bulletin*, 81, 1023–41.

Ostermeier, R. M. (2001). Compaction effects on porosity ad permeability: deepwater Gulf of Mexico turbidites. *Journal of Petroleum Technology*, 53, 68–74.

O'Sullivan, P. B., Hanks, C. I., Wallace, W. K., and Green, P. F. (1995). Multiple episodes of Cenozoic denudation in the northeastern Brooks Range: fission-track data from the Okpilak batholith, Alaska. *Canadian Journal of Earth Sciences*, 32, 1106–18.

Ottesen, E. O., Knipe, R. J., Olsen, T. S., Fisher, Q. J., and Jones, G. (1998). Fault controlled communication in the Sleipner vest Field, Norwegian Continental Shelf: detailed, quantitative input for reservoir simulation and well planning. In *Faulting, Fault Sealing and Fluid Flow in Hydrocarbon Reservoirs*, ed. G. Jones, Q. L. Fisher, and R. J. Knipe. *Geological Society of London Special Publications*, **147**, 269–82.

Pajdušák, P., Plomerová, J., and Babuška, V. (1988). Lithosphere deep structure of the West Carpathians determined on the basis of P – residue. In *Proceedings of the Conference on Deep Structure Investigation of Czechoslovakia*. VEDA, Bratislava, Slovak Academy of Sciences, pp. 151–8.

Palciauskas, V. V. (1986). Models for thermal conductivity and permeability in normally compacting basins. In *Thermal Modeling of Sedimentary Basins*, ed. J. Burrus. Technip, Paris, pp. 323–36.

Palciauskas, V. V. and Domenico, P. A. (1982). Characterization of drained and undrained response of thermally loaded repository rocks. *Water Resources Research*, 18, 281–90.

P'An, C.-H. (1982). Petroleum in basement rocks. *American Association of Petroleum Geologists Bulletin*, 66, 1597–643.

Pandey, O. P., Agarwal, P. K., and Negi, J. G. (1995). Lithospheric structure beneath Laxmi Ridge and late Cretaceous geodynamic events. *Geo-Marine Letters*, 15, 85–91.

Parker, R. H. (1991). The Ivanhoe and Rob Roy Fields, Blocks 15/21a-b, UK North Sea. *Memoir of the Geological Society of London*, 14, 331–8.

Parmentier, E. M. (1987). Dynamic topography in rift zones: implications for lithospheric heating. *Royal Society of London Philosophical Transactions*, 321, 23–5.

Parnell, J. (1987). Secondary porosity in hydrocarbon-bearing transgressive sandstones on an unstable lower Paleozoic continental shelf, Welsh Borderland. In *Diagenesis of Sedimentary Sequences*, ed. J. D. Marshall. Queen's University of Belfast, pp. 297–312.

(2002). Diagenesis and fluid flow in response to uplift and exhumation. In *Diagenesis and Fluid Flow in Response to Uplift and Exhumation*, A. G. Dore, J. A. Cartwright, M. S. T. Stoker, N. J. Jonathan, and P. White. *Geological Society of London Special Publications*, pp. 433–46.

Parris, T. M., Burruss, R. C., and O'Sullivan, P. B. (2003). Deformation and the timing of gas generation and migration in the eastern Brooks Range foothills, Arctic National Wildlife Refuge, Alaska. *American Association of Petroleum Geologists Bulletin*, 87, 1823–46.

Parrish, J. T. (1982). Upwelling and petroleum source beds with reference to the Paleozoic. *American Association of Petroleum Geologists Bulletin*, 66, 750–74.

Parrish, R. R. (1983). Cenozoic thermal evolution and tectonics of the Coast Mountains of British Columbia, 1. Fission-track dating, apparent uplift rates, and patterns of uplift. *Tectonics*, 2, 601–31.

Parry, W. T. and Bruhn, R. L. (1987). Fluid inclusion evidence for minimum 11 km vertical offset on the Wasatch fault, Utah. *Geology*, 15, 67–70.

(1990). Fluid pressure transients on seismogenic normal faults. *Tectonophysics*, 179, 335–44.

Parry, W. T., Chan, M. A., and Beitler, B. (2004). Chemical bleaching indicates episodes of fluid flow in deformation bands in sandstone. *American Association of Petroleum Geologists Bulletin*, 88, 1–17.

Parry, W. T., Wilson, P., and Bruhn, R. L. (1988). Pore fluid chemistry and chemical reactions on the Wasatch normal fault, Utah. *Geochimica et Cosmochimica Acta*, 52, 2053–63.

Parsons, B. and Sclater, J. G. (1977). An analysis of the variation of ocean floor bathymetry and heat flow with age. *Journal of Geophysical Research*, 82, 803–27.

Partington, M. A., Copestake, P., Mitchener, B. C., and Underhill, J. R. (1993). Biostratigraphic calibration of genetic stratigraphic sequences in the Jurassic – lowermost Cretaceous (Hettangian – Ryazanian) of the

North Sea and adjacent areas. In *Petroleum Geology of the Northwest Europe*, ed. J. R. Parker. *Proceedings of the 4th Conference, Geological Society of London*, pp. 371–86.

Pasquale, V. (2006). Effects of crustal heat source redistribution on the strength envelopes. *Geophysical Research Abstracts*, **8**.

Patton, T. L. (1984). Surface studies of normal-fault geometries in the pre-Miocene stratigraphy of west central Sinai Peninsula. *Egyptian General Petroleum Corporation, Sixth Exploration Seminar*, pp. 437–52.

Patton, T. L., Moustafa, A. R., Nelson, R. A., and Abdine, A. S. (1994). Tectonic evolution and structural setting of the Suez Rift. *American Association of Petroleum Geologists Memoir*, 59, 9–55.

Peacock, D. C. P. and Sanderson, D. J. (1991). Displacements, segment linkage and relay ramps in normal fault zones. *Journal of Structural Geology*, 13, 721–33.

Peakall, J. (1998). Axial river evolution in response to half-graben faulting: Carson River, Nevada, U.S.A. *Journal of Sedimentary Research*, 68, 788–99.

Peate, D. (1997). The Paraná-Etendeka Province. In *Large Igneous Provinces: Continental Oceanic and Planetary Flood Volcanism*, ed. J. Mahoney and M. F. Coffin. American Geophysical Union Geophysical Monograph 100, pp. 217–45.

Pecova, J., Petr, V., Praus, O., Babuška, V., and Plancar, J. (1979). Internal electrical conductivity distribution on Czechoslovak territory. In *Geodynamical Investigations in Czechoslovakia, Bratislava*, ed. J. Vanek. Veda-SAV, pp. 119–27.

Pécskay, Z., Lexa, J., Szakács, A., Balogh, K., Seghedi, I., Konecný, V., Kovács, M., Márton, E., Kaliciak, M., Széky-Fux, V., Póka, T., Gyarmati, P., Edelstein, O., Rosu, E., and Zec, B. (1995). Space and time distribution of Neogene-Quaternary volcanism in the Carpatho-Pannonian region. *Acta Vulcanologica*, 7, 15–28.

Peddy, C., Pinet, B., Masson, D., Scrutton, R., Sibuet, J.-C., Warner, M. R., Leford, J.-P., and Shroeder, I. J. (1989). Crustal structure of the Goban Spur continental margin, Northeast Atlantic, from deep seismic reflection profiling. *Journal of the Geological Society of London*, **146**, 427–37.

Pedersen, T. F. and Calvert, S. E. (1990). Anoxia vs. productivity: What controls the formation of organic-carbon-rich sediments and sedimentary rocks? *American Association of Petroleum Geologists Bulletin*, 74, 454–66.

Pedersen, T. and Ro, H. E. (1992). Finite duration extension and decompression melting. *Earth and Planetary Science Letters*, 113, 15–22.

Pedersen, T. and Skogseid, J. (1989). Vøring Plateau volcanic margin: extension, melting and uplift. *Proceedings of the Ocean Drilling Program, Scientific Results*, 104, 985–91.

Peel, F. J., Scott, E., and Evans, N. (2001). Abnormal processes as major influences on deepwater seafloor morphology and deposition. *Annual Meeting Expanded Abstracts, American Association of Petroleum Geologists*, p. 155.

Peltier, W. R., Jarvis, G. T., Forte, A. M., and Solheim, L. P. (1989). The radial structure of the mantle general circulation. *The Fluid Mechanics of Astrophysics and Geophysics*, 4, 765–815.

Peng, Z. and Ben-Zion, Y. (2005). Temporal changes of shallow seismic velocity around the Daradere-Duzce branch of the north Anatolian fault and strong ground motion. *Pure and Applied Geophysics*, 163, 567–99.

Pennington, J. J. (1975). The geology of the Argyll field. In *Petroleum and the Continental Shelf of North-west Europe*, ed. A. W. Woodland. Applied Science Publishers, Barking, pp. 285–91.

PennWell (1982). *Oil and Gas Fields of the United States*. PennWell Publishing Company, Tulsa, Scale 1: 3 380 952.

Pepper, A. S. (1991). Estimating the petroleum expulsion behavior of source rocks: a novel quantitative approach. In *Petroleum Migration*, ed. W. A. England and A. J. Fleet. *Special Publication of Geological Society of London*, **59**, 9–31.

Pepper, A. S. and Corvi, P. J. (1995). Simple kinetic models of petroleum formation. Part I: Oil and gas generation from kerogen. *Marine and Petroleum Geology*, 12, 291–319.

Pepper, A. S. and Dodd, T. A. (1995). Simple kinetic models of petroleum formation. Part II: Oil-gas cracking. *Marine and Petroleum Geology*, 12, 321–40.

Pe-Piper, G. and Reynolds, P. H. (2000). Early Mesozoic alkaline mafic dykes, southwestern Nova Scotia. *The Canadian Mineralogist*, 38, 217–32.

Pereira, E. B., Hamza, V. M., Furtado, V. V., and Adams, J. A. (1986). U, Th, and K content, heat production and thermal conductivity of Sao Paulo, Brazil, continental shelf sediments: a reconnaissance work. *Chemical Geology*, 58, 217–26.

Pereira, M. J. (1990). Revised stratigraphic column of the Santos Basin (in Portuguese). Parameters controlling porosity and permeability in clastic reservoirs of the Merluza deep field, Santos Basin, Brazil. *Boletim de Geociencias da PETROBRAS*, 4, 451–66.

Pereira, M. J. and Macedo, J. M. (1990). A Bacia de Santos; perspectivas de uma nova provincia petrolifera na plataforma continental Sudeste Brasileira. *Boletim de Geociencias da PETROBRAS*, 4, 3–11.

Perez, R. J. and Boles, J. R. (2004). Mineralization, fluid flow, and sealing properties associated with an active thrust fault: San Joaquin Basin, California. *American Association of Petroleum Geologists Bulletin*, 88, 1295–314.

Pérez-Gussinyé, M., Ranero, C. R., Reston, T. J., and Sawyer, D. (2003). Mechanisms of extension at nonvolcanic margins: evidence from the Galicia interior basin, west of Iberia. *Journal of Geophysical Research*, 108.

Pérez-Gussinyé, M. and Reston, T. J. (2001). Rheological evolution during extension at nonvolcanic rifted margins: onset of serpentinization and development of detachments leading to continental breakup. *Journal of Geophysical Research*, 106, 3961–76.

Perón-Pinvidic, G. and Manatschal, G. (2010). From microcontinents to extensional allochthons: witnesses of how continents rift and break apart? *Petroleum Geoscience*, 16, 189–97.

Perón-Pinvidic, G., Manatschal, G., Minshull, T. A., and Sawyer, D. S. (2007). Tectonosedimentary evolution of the deep Iberia-Newfoundland margins: evidence for a complex breakup history. *Tectonics*, 26.

Perrodon, A. (1980). *Géodynamique pétrolière. Genése et répartition des gisements d'hydrocarbures*. Paris, Masson Elf-Aquitaine Pierce.

(1995). Petroleum systems and global tectonics. *Journal of Petroleum Geology*, 18, 471–6.

Perrot, K., Geoffroy, L., and Dyment, J. (2003). *Magnetic structure of the Greenland Volcanic Passive margin*. Abstracts, AGU-EUG-EGS meeting, Nice. April 6–11.

Person, M. (1990). *Hydrologic constraints on the thermal evolution of continental rift basins: implications for petroleum maturation*. PhD thesis, John Hopkins University, Baltimore, Maryland, p. 271.

Person, M. and Garven, G. (1992). Hydrologic constraints on petroleum generation within continental rift basins: theory and application to the Rhine Graben. *American Association of Petroleum Geologists Bulletin*, 76, 468–88.

(1994). A sensitivity study of the driving forces on fluid flow during continental-rift basin evolution. *Geological Society of America Bulletin*, 106, 461–75.

Peters, K. E. (1986). Guidelines for evaluating petroleum source rock using programmed pyrolysis. *American Association of Petroleum Geologists Bulletin*, 70, 318–29.

Peters, K. E. and Cassa, M. R. (1994). Applied source rock geochemistry. In *The Petroleum System – From Source to Trap*, ed. L. B. Magoon and W. G. Dow. *American Association of Petroleum Geologists Memoir*, **60**, 93–120.

Peters, K. E., Cunningham, A. E., Walters, C. C., Jigang, J., and Zhaoan, F. (1996). Petroleum systems in the Jiangling-Dangyang area, Jianghan basin, China. *Organic Geochemistry*, 24, 1035–60.

Peters, K. E. and Fowler, M. G. (2002). Applications of petroleum geochemistry to exploration and reservoir management. *Organic Geochemistry*, 33, 5–36.

Peters, K. E. and Moldowan, J. M. (1993). *The Biomarker Guide: Interpreting Molecular Fossils in Petroleum and Ancient Sediments*. Prentice Hall, Englewood Cliffs, New Jersey, p. 363.

Peters, K. E., Moldowan, J. M., Driscole, A. R., and Demaison, G. J. (1989). Origin of Beatrice oil by co-sourcing from Devonian and Middle Jurassic source rocks, Inner Morray Firth, United Kingdom. *AAPG Bulletin*, 73, 454–71.

Peters, K. E., Pytte, M. H., Elam, T. D., and Sundararaman, P. (1994). Identification of petroleum systems adjacent to the San Andreas fault, California. In *The Petroleum System – From Source to Trap*, ed. L. B. Magoon and W. G. Dow. *American Association of Petroleum Geologists Memoir*, **60**, 423–36.

Peters, K. E., Ramos, L. S., Zumberge, J. E., Valin, Z. C., Scotese, C. R., and Gautier, D. L. (2007). Circum-Artic petroleum systems identified using decision-tree chemometrics. *American Association of Petroleum Geologists Bulletin*, 91, 877–913.

Peters, K. E., Snedden, J. W., Sulaeman, A., Sarg, J. F., and Enrico, R. J. (2000). A new geochemical-stratigraphic model for the Mahakam delta and Makassar Slope, Kalimantan, Indonesia. *American Association of Petroleum Geologists Bulletin*, 84, 12–44.

Peters, T. and Stettler, A. (1987). Radiometric age, thermobarometry, and mode of emplacement of the Totalp peridotite in the Eastern Swiss Alps. *Schweizerische Mineralogische und Petrographische Mitteilungen = Bulletin Suisse de Mineralogie et Petrographie*, 67, 285–94.

Petersen, H. I., Bojesen-Koefoed, J. A., and Thomsen, E. (1995). Cola-derived petroleum in the Middle Jurassic Bryne Formation in the Danish North Sea – a new type of play. In *Organic Geochemistry: Developments and Applications to Energy, Climate, Environment and Human History*, ed. J. O. Grimalt and C. Dorronsoro. Proceedings of the 17th International Meeting on Organic Geochemistry, San Sebastian, Spain, September 4–8, pp. 473–5.

Petersen, H. I., Rosenberg, P., and Andsbjerg, J. (1996). Organic geochemistry in relation to the depositional environments of Middle Jurassic coal seams, Danish Central Graben, and implications for hydrocarbon generative potential. *American Association of Petroleum Geologists Bulletin*, 80, 47–62.

Peterson, J. A. and Clarke, J. W. (1989). West Siberian oil-gas province. United States Geological Survey, *Open File Report*, 89–192, 142.

Petit, J. P. (1987). Criteria for the sense of movement on fault surfaces in brittle rocks. *Journal of Structural Geology*, 9, 597–608.

Philip, H., Rogozhin, E., Cisternas, A., Bosquet, J. C., Borisov, B., and Karakhanian, A. (1992). The Armenian Earthquake of 1988 December 7: Faulting and folding, neotectonics and palaeoseismicity. *Geophysical Journal International*, 110, 141–58.

Phillips, O. M. (1991). *Flow and Reactions in Permeable Rocks*. Cambridge University Press, New York.

Philp, R. P. (1994). Geochemical characteristics of oils derived predominantly from terrigenous source materials. In *Coal and Coal-bearing Strata as Oil-prone Source Rocks*, ed. A. C. Scott and A. J. Fleet. *Geological Society of London Special Publications*, **77**, 71–91.

Philp, R. P. and Fan, Z. (1987). Geochemical investigation of oils and source rocks from Qianjiang Depression of Jianghan basin, a terrigenous saline basin, China. *Organic Geochemistry*, 11, 549–62.

Philp, R. P. and Lewis, C. A. (1987). Organic geochemistry of biomarkers. *Annual Review of Earth and Planetary Science*, 15, 363–95.

Pickup, S. L. B., Whitmarsh, R. B., Fowler, C. M. R., and Reston, T. J. (1996). Insight into the nature of the ocean-continent transition off West Iberia from a deep multichannel seismic reflection profile. *Geology*, 24, 1079–82.

Pierce, C., Whitmarsh, R. B., Scrutton, R. A., Pontoisse, B., Sage, R., and Mascle, J. (1996). Cote d'Ivoire-Ghana margin: seismic imaging of passive rifted crust adjacent to a transform continental margin. *Geophysics Journal International*, 125, 781–95.

Pierce, W. H. and Khalifa bin Mbarak, A.-H. (1993). Southern Arabian Basin oil habitat: seals and gathering areas. *Society of Petroleum Engineers of the American Institute of Mining, Metallurgical and Petroleum Engineers*, 8, 103–11.

Pindell, J. (1993). Mesozoic-Cenozoic paleogeographic evolution of northern South America. *American Association of Petroleum Geologists Bulletin*, 77, 340.

Pinet, P. and Souriau, M. (1988). Continental erosion and large-scale relief. *Tectonics*, 7, 563–82.

Pinnock, S. J., Clitheroe, A. R., and Rose, P. T. S. (2002). Appraisal and development of a viscous oil field. In *United Kingdom Oil and Gas Fields, Commemorative Millennium Volume*, ed. J. G. Gluyas and H. M. Hichens. *Geological Society of London Memoir*, **20**, 431–41.

Piper, D. Z. and Link, P. K. (2002). An upwelling model for the Phosphoria sea: a Permian, ocean-margin sea in the northwest United States. *American Association of Petroleum Geologists Bulletin*, 86, 1217–35.

Piqué, A., Jeannette, D., and Michard, A. (1980). The Western Meseta shear zone, a major and permanent feature of the Hercynian Belt in Morocco. *Journal of Structural Geology*, 2, 55–61.

Piqué, A., Le Roy, P., and Amrhar, M. (1998). Transtensive synsedimentary tectonics associated with ocean opening: the Essaouira-Agadir segment of the Moroccan Atlantic margin. *Journal of the Geological Society of London*, 155, 913–28.

Pitman, W. C. and Andrews, J. A. (1985). Subsidence and thermal history of small pull-apart basins. In *Strike-Slip Deformation, Basin Formation and Sedimentation*, ed. K. T. Biddle and N. Christie-Blick. *Special Publication – Society of Economic Paleontologists and Mineralogists*, San Antonio, Texas, pp. 45–119.

Planert, M. and Williams, J. S. (1995). *Groundwater Atlas of the United States, Segment, California, Nevada*. HA-0730-B, US Geological Survey.

Planke, S., Symonds, P. A., Alvestad, E., and Skogseid, J. (2000). Seismic volcanostratigraphy of large-volume basaltic extrusive complexes on rifted margins. *Journal of Geophysical Research*, 105, 19335–52.

Platte River Associates, 1995. BasinMod 1-D for WindowsTM. Basin Modeling System. Document Version: 5.0.

Plothner, D. (1988). Entwicklung und Anwending einer erdolgeologisch-geochemischen Explorationsmethode unter besonderer Berucksichtigung der Hydraulik im Pechbronner Gebiet-Fachbericht: das Pechelbronner Feld. *Bundensanstalt fur Geowissenschaft und Rohstoffe*, BMFT-Forschungsvorhaben, 032 6476, 103 464, 1–8.

Poag, C. W. (1979). Stratigraphy and depositional environments of Baltimore Canyon Trough. *AAPG Bulletin*, 63, 1452–66.

Podladchikov, Y. Y., Poliakov, A. N. B., and Yuen, D. A. (1994). The effect of lithospheric phase transitions on subsidence of extending continental lithosphere. *Earth and Planetary Science Letters*, 124, 95–103.

Poelchau, H. S., Baker, D. R., Hantschel, T., Horsfield, B., and Wygrala, B. (1997). Basin simulation and the design of the conceptual basin model. In *Petroleum and Basin Evolution: Insights from Petroleum Geochemistry, Geology and Basin Modeling*, ed. D. H. Welte, B. Horsfield, and D. R. Baker. Springer, Berlin, p. 3–70.

Pollack, H. N. (1986). Cratonization and thermal evolution of the mantle. *Earth and Planetary Science Letters*, 80, 175–82.

Pollack, H. N. and Chapman, D. S. (1977). On the regional variation of heat flow, geotherms, and lithospheric thickness. *Tectonophysics*, 38, 279–96.

Popescu, B. M. (1995). The Guyanas (Guyana, Suriname, French Guyana). In *Regional Petroleum Geology of the World*, ed. H. Kulke. **22**, 603–12.

Popoff, M. (1988). South Atlantic Gondwana: connections of the Benue Trench with northeastern Brazilian basins up to the opening of the Gulf of Guinea in the Lower Cretaceous. *Journal of African Earth Sciences*, pp. 409–31.

Porath, H. (1971). Magnetic variation anomalies and seismic low-velocity zone in the western United States. *Journal of Geophysical Research*, 76, 2643–8.

Porter, J. R., Knipe, R. J., Fisher, Q. J., Farmer, A. B., Allin, N. S., Jones, L. S., Palfrey, A. J., Garrett, S. W., and Lewis, G. (2000). Deformation processes in Britannia fields, UKCS. *Petroleum Geoscience*, 6, 241–54.

Posamentier, H. W. and Allen, G. P. (1993). Variability of the sequence stratigraphic model; effects of local basin factors. *Sedimentary Geology*, 86, 91–109.

(1999). Siliciclastic sequence stratigraphy: concepts and applications. *SEPM Concepts in Sedimentology and Paleontology*, 7, 210.

Posamentier, H. W., Jervey, M. T., and Vail, P. R. (1988). Eustatic controls on clastic deposition; I, Conceptual framework. *Special Publication – Society of Economic Paleontologists and Mineralogists*, 42, 109–24.

Posamentier, H. W. and Kolla, V. (2003). Seismic geomorphology and stratigraphy of depositional elements in deep-water settings. *Journal of Sedimentary Research*, 73, 367–88.

Posamentier, H. W. and Vail, P. R. (1988). Eustatic controls on clastic deposition: II, Sequence and systems tract models. In *Sea-level Changes: An Integrated Approach*, ed. C. K. Wilgus, B. S. Hastings, C. A. Ross, H. W. Posamentier, J. Van Wagoner, and C. Kendall. *Special Publication – Society of Economic Paleontologists and Mineralogists*, **42**, 125–54.

Posgay, K., Albu, I., Mayerová, M., Nakládalová, Z., Ibrmayer, I., Blizkovsky, M., Aric, K., and Gutdeutsch, R. (1991). Contour map of the Mohorovicic discontinuity beneath central Europe. *Geophysical Transactions*, 36, 7–13.

Posgay, K., Bodoky, T., Hajnal, Z., Tóth, T. M., Fancsik, T., Hegedüs, E., Kovács, A. C., and Takács, E. (1996). International deep reflection survey along the Hungarian geotraverse. *Geophysical Transactions*, 40, 1–44.

Pospíšil, L., Ádám, A., Bimka, J., Bodlak, P., Bodoky, T., Dövényi, P., Granser, H., Hegedüs, E., Joó, I., Kendzera, A., Lenkey, L., Nemčok, M., Posgay, K., Pylypyshyn, B., Sedlák, J., Stanley, W. D., Starodub, G., Szalaiová, V., Šály, B., Šutora, A., Várga, G., and Zsíros, T. (2006). Regional geophysical data on the Carpathian-Pannonian lithosphere. In *The Carpathians and Their Foreland: Geology and Hydrocarbon Resources*, ed. F. Pícha and J. Golonka. *American Association of Petroleum Geologists Memoir*, pp. 651–97.

Postma, G. (1984). Mass-flow conglomerate in a submarine canyon: Abrioja fan delta, Pliocene, southwest Spain. In *Sedimentology of Gravels and Conglomerates*, ed. E. H. Koster and R. J. Steel. *Canadian Society of Petroleum Geologists Memoir*, **10**, 237–8.

(1995). Sea-level-related architectural trends in coarse-grained delta complexes. *Sedimentary Geology*, 98, 3–12.

Postma, G. and Roep, T. B. (1985). Resedimented conglomerates in the bottomsets of Gilbert-type gravel deltas. *Journal of Sedimentary Petrology*, 55, 874–85.

Poudjom-Djomani, Y. H., O'Reilly, S. Y., Griffin, W. L., and Morgan, P. (2001). The density structure of subcontinental lithosphere through time. *Earth and Planetary Science Letters*, 184, 605–21.

Powell, T. G. (1986). Petroleum geochemistry and deposition setting of lacustrine source rocks. *Marine and Petroleum Geology*, 3, 200–19.

Powell, T. G. and Boreham, C. J. (1994). Terrestrially sources oils: Where do they exist and what are our limits of knowledge? – A geochemical perspective. In *Coal and Coal-bearing Strata as Oil-prone Source Rocks*, ed. A. C. Scott and A. J. Fleet. *Geological Society of London Special Publications*, **77**, 11–29.

Powell, W. G. (1997). *Thermal state of the lithosphere in the Colorado Plateau-Basin and Range transition zone, Utah*. PhD thesis, University of Utah, p. 232.

Powley, D. E. (1980). *Pressures, normal and abnormal*. American Association of Petroleum Geologists Advanced Exploration Schools Unpublished Lecture Notes.

Prat, P., Montenat, C., Ott d'Estevou, P., and Bolze, J. (1986). The western margin of the Gulf of Suez from the study of Gebel Zeit and Gebel Mellaha. *Documents et Travaux Institut Géologique Albert de Lapparent*, Paris, 10, 45–74.

Price, L. C. and Barker, C. E. (1985). Suppression of vitrinite reflectance in amorphous rich kerogen: a major unrecognised problem. *Journal of Petroleum Geology*, 8, 59–84.

Price, L. C. and Schoell, M. (1995). Constraints on the origins of hydrocarbon gas from compositions of gases at their site of origin. *Nature*, 378, 368–71.

Price, L. C. and Wenger, L. M. (1992). The influence of pressure on petroleum generation and maturation as suggested by aqueous pyrolysis. *Organic Geochemistry*, 19, 141–59.

Priest, S. S., Sass, J. H., Ellsworth, B., Farrar, C. D., Sorey, M. L., Hill, D. P., Bailey, R., Jacobson, R. D., Finger, J. T., McConnell, V. S., and Zoback, M. (1998). Scientific drilling in Long Valley, California – what will we learn? *USGS Fact Sheet*, pp. 77–97.

Prince, C. M. (1999). Textural and diagenetic controls on sandstone permeability. *Gulf Coast Association of Geological Societies Transactions*, 49, 42–53.

Prosser, S. (1993). Rift-related linked depositional systems and their seismic expression. In *Tectonics and Seismic Sequence Stratigraphy*, ed. G. D. Williams and A. Dobbs. *Geological Society Special Publications*, **71**, 35–66.

Prost, G. and Aranda, M. (2001). Tectonics and hydrocarbon systems of the Veracruz Bain, Mexico. In *The Western Gulf of Mexico Basin: Tectonics, Sedimentary Basins, and Petroleum Systems*, ed. C. Bartolini, R. T. Buffler, and A. Cantu-Chapa. *American Association of Petroleum Geologists Memoir*, **75**, 271–91.

Purser, B. H. and Bosence, D. W. J. (1998). *Sedimentation and Tectonics in Rift Basins: Red 354 Sea – Gulf of Aden*. Chapman and Hall, London, p. 663.

Purvis, K., Kao, J., Flanagan, K., Henderson, J., and Duranti, D. (2002). Complex reservoir geometries in a deep water clastic sequence, Gryphon field, UKCS: Injection structures, geological modelling and reservoir simulation. *Marine and Petroleum Geology*, 19, 161–79.

Puteanus, D., Glasby, G. P., Stoffers, P., and Dunzendorf, H. (1991). Hydrothermal iron-rich deposits from the Teahitia-Mehitia and Mackonalk hot spot areas, Southwest Pacific. *Marine Geology*, 98, 389–409.

Quigley, T. M. and Mackenzie, A. S. (1988). The temperatures of oil and gas formation in the subsurface. *Nature*, 333, 549–52.

Raab, M. J., Brown, R. W., Gallagher, K., and Weber, K. (2001). The geomorphic response of the passive continental margin of Namibia to post break-up tectonics. In *International Conference and Annual Meeting of the Deutsche Geologische Gesellschaft and Geologische Vereinigung*, ed. S. Roth and A. Rueggeberg. Schriftenreihe der Deutschen Geologischen Gesellschaft, Kiel, Federal Republic of Germany, p. 157.

Rabinowicz, M., Dandurand, J. L., Jakubowski, M., Schott, J., and Casson, J. P. (1985). Convection in a North Sea reservoir: influences on diagenesis and hydrocarbon migration. *Earth and Planetary Science Letters*, 74, 387–404.

Radhakrishna, M., Verma, R. K., and Purushotham, A. K. (2002). Lithospheric structure below the eastern Arabian Sea and adjoining west coast of India based on integrated analysis of gravity and seismic data. *Marine Geophysical Researchers*, 23, 25–42.

Radney, B. and Byerlee, J. D. (1988). Laboratory studies of the shear strength of montmorillonite and illite under crustal conditions. *Eos, Transactions, American Geophysical Union*, 69, 1463.

Rădulescu, D. P., Sandulescu, M., and Veliciu, S. (1983). A geodynamic model of the East Carpathians and the thermal field in the lithosphere. *Anuarul Institutului de Geologie si Geofizica – Annuaire de l'Institut de Geologie et de Geophysique*, 63, 135–44.

Rădulescu, F. (1988). Seismic models of the crustal structure in Romania. *Revue Roumaine de Géologie, Géophysique et Géographie*, 32, 13–19.

Raffensperger, J. P. and Garven, G. (1995). The formation of unconformity-type uranium ore deposits. 1. Coupled groundwater flow and heat transport modeling. *American Journal of Science*, 295, 581–636.

Raharjo, I., Wannamaker, P., Moore, J. N., Allis, R., Chapman, D., et al., 2002. Magmatic chimney beneath Telaga Bodas revealed by magnetotellurics profiling: a case study at the Karaha Bodas geothermal system, Indonesia. *Eos, Transactions, American Geophysical Union*, 83, 1423.

Rahe, B., Ferrill, D., Morris, A., and Alan P. (1998). Physical analog modeling of pull-apart basin evolution. *Tectonophysics*, 285, 21–40.

Răileanu, V., Diaconescu, C., and Radulescu, F. (1994). Characteristics of Romanian lithosphere from deep seismic reflection profiling. *Tectonophysics*, 239, 165–85.

Rampone, E. and Piccardo, G. B. (2000). The ophiolite-oceanic lithosphere analogue: new insights from the Northern Apennines (Italy). *Geological Society of America*, 349, 21–34.

Ranalli, G. (1995). *Rheology of the Earth*. Chapman and Hall, London, p. 413.

Ranalli, G. and Murphy, D. C. (1987). Rheological stratification of the lithosphere. *Tectonophysics*, 132, 281–95.

Ranalli, G. and Rybach, L. (2005). Heat flow, heat transfer and lithosphere rheology in geothermal areas: features and examples. *Journal of Volcanology and Geothermal Research*, 148, 3–19.

Ranganathan, V. and Hanor, J. S. (1987). A numerical model for the formation of saline waters due to diffusion of dissolved NaCl in subsiding sedimentary basins with evaporates. *Journal of Hydrology*, 92, 97–102.

Rankenburg, K., Lassiter, J. C., and Brey, G. (2005). The role of continental crust and lithospheric mantle in the genesis of Cameroon Volcanic Line lavas: constraints from isotopic variations in lavas and megacrysts from the Biu and Jos plateaux. *Journal of Petroleum*, 46, 169–90.

Rasskazov, S. V., Boven, A., Ivanov, A. V., and Semenova, V. G. (2000). Middle Quanternary volcanic impulse in the Olekma-Stanovoy mobile system: 40Ar-39Ar dating of volcanics from the Tokinsky Stanovik. *Geology of the Pacific Ocean*, 19, 19–28.

Ratcliffe, N. M. and Burton, W. C. (1985). Fault reactivation models for origin of the Newark Basin and studies related to eastern U.S. seismicity. *U. S. Geological Survey Circular*, 946, 36–45.

Rattey, R. P. and Hayward, A. P. (1993). Sequence stratigraphy of a failed rift system: the Middle Jurassic to Early Cretaceous basin evolution of the Central and Northern North Sea. In *Petroleum Geology of Northwest Europe*, ed. J. R. Parker. *Proceedings of the 4th Conference*. The Geological Society of London, pp. 215–49.

Ravnas, R. and Steel, R. J. (1998). Architecture of marine rift-basin successions. *American Association of Petroleum Geologists Bulletin*, 82, 100–46.

Ray, D. K. (1963). *Tectonic Map of India (1:2,000,000)*. Geological Society of India, Calcutta.

Raymo, M. E., Hodell, D., and Jansen, E. (1992). Response of deep ocean circulation to initiation of northern hemisphere glaciation (3–2 Ma). *Paleoceanography*, 7, 645–72.

Reading, H. G. (1991). The classification of deep-sea depositional systems by sediment calibre and feeder system. *Journal of the Geological Society of London*, 148, 427–30.

Reddy, P. R., Venkateswarulu, N., Koteswara Rao, P., and Prasad, A. S. S. S. R. S. (1999). Crustal structure of peninsular shield India from DSS studies. *Current Science*, 77, 1606–11.

Rees, B. A., Detrick, R. S., and Coakley, B. J. (1993). Seismic stratigraphy of the Hawaiian flexural moat. *Geological Society of America Bulletin*, 105, 189–205.

Regalli, M. S. P. (1989). A idade dos evaporitos da plataforma continental do Ceara, Brasil, e sua relacao com os outros evaporitos das bacias nordestinas. Boletin de Instituto de Geociencias, Universidade de Sao Paul, Publicacao Especial, **7**, 139–43.

Regelous, M., Niu, Y. L., Abouchanmi, W., and Castillo, P. R. (2009). Shallow origin for South Atlantic Dupal Anomaly from lower continental crust: geochemical evidence from the Mid-Atlantic Ridge at 26 degrees S. *Lithos*, 112, 57–72.

Rehrig, W. A. and Reynolds, S. J. (1980). Geologic and geochronologic reconnaissance of a northwest-trending zone of metamorphic core complexes in southern and western Arizona. In *Cordilleran Metamorphic Core Complexes*, ed. M. D. Jr. Crittenden, P. J. Coney, and G. H. Davis. *Geological Society of America Memoir*, 153, 131–57.

Reid, M. E., Sisson, T. W., and Brien, D. L. (2001). Volcano collapse promoted by hydrothermal alteration and edifice shape, Mount Rainier, Washington. *Geology*, 29, 779–82.

Reid, S. A. and McIntyre, J. L. (2001). Monterey Formation porcelanite reservoirs of the Elk Hills field, Kern County, California. *American Association of Petroleum Geologists Bulletin*, 85, 169–89.

Reiners, P. W., Campbell, I. S., Nicolescu, S., Allen, C., Garver, J. I., and Palin, J. M. (2002). Single crystal helium-lead dating of detrital zircon. *Abstracts with Programs – Geological Society of America*, 34, 484.

Reisser, K. (2000). Petroleum geology of the western Ust-Yurt Basin, Republic of Kazakhstan. *The Bulletin of the Houston Geological Society*, 43, 20–1.

Reiter, M., Edwards, C. L., Hartman, H., and Weidman, C. (1975). Terrestrial heat flow along the Rio Grande rift, New Mexico and southern Colorado. *Geological Society of America Bulletin*, 86, 811–18.

Reiter, M. A., Mansure, A. J., and Shearer, C. (1979). Geothermal characteristics of the Colorado Plateau. *Tectonophysics*, 61, 183–95.

Ren, S., Faleide, J. I., Eldholm, O., Skogseid, J., and Gradstein, F. (2003). Late Cretaceous-Paleocene tectonic development of the NW Vøring Basin. *Marine and Petroleum Geology*, 20, 177–206.

Renne, P., Deckart, K., Ernesto, M., Feraud, G., and Piccirillo, E. (1996a). Age of the Ponta Grossa dyke swarm (Brazil), and implications to the Parana flood volcanism. *Earth and Planetary Science Letters*, 144, 199–211.

Renne, P., Ernesto, M., Pacca, I. G., Coe, R. S., Glen, J. M., Prevot, M., and Perrin, M. (1992). The age of Paraná flood volcanism, rifting of Gondwanaland and the Jurassic-Cretaceous boundary: *Science*, 258, 975–9.

Renne, P., Glen, J. M., Milner, S. C., and Duncan, A. R. (1996b) Age of Etendeka flood volcanism and associated intrusions in south-western Africa. *Geology*, 24(7), 659–62.

Reston, T. J. (1996). The S reflector west of Galicia: the seismic signature of a detachment fault. *Geophysical Journal International*, 127, 230–44.

Reston, T. J., Krawczyk, C. M., and Hoffmann, H. J. (1995). Detachment tectonics during Atlantic rifting: analysis and interpretation of the S reflection, the West Galicia margin. *Geological Society Special Publications*, 90, 93–109.

Reston, T. J., Krawczyk, C. M., and Klaeschen, D. (1996). The S reflector west of Galicia 727 (Spain): evidence from prestack depth migration for detachment faulting during 728 continental breakup. *Journal of Geophysical Research*, 101, 8075–91.

Reston T. J., Pennell, J., Stubenrauch, A., Walker, I., and Perez-Gussinye, M. (2001). Detachment faulting, mantle serpentinization, and serpentinite-mud volcanism beneath the Porcupine Basin, southwest of Ireland. *Geology*, 29, 587–90.

Reynolds, A. D., Simmons, M. D., Bowman, M. B. J., Henton, J., Brayshaw, A. C., Ali-Zade, A. A., Guliyev, I. S., Syleymanova, S. F., Ateava, E. Z., Mamedova, D. N., and Koshkarly, R. O. (1998). Implications of outcrop geology for reservoirs in the Neogene productive series: Apsheron Peninsula, Azerbaijan. *American Association of Petroleum Geologists Bulletin*, 82, 25–49.

Reynolds, D. J. (1984). *Structural and dimensional repetition in continental rifting*. Master's thesis, Duke University, Durham, North Carolina, p. 175.

Reynolds, J. G. and Burnham, A. K. (1995). Comparison of kinetic analysis of source rocks and kerogen concentrates. *Organic Geochemistry*, 23, 11–19.

Reynolds, S. J. (1985). *Geology of the South Mountains, central Arizona*. Bulletin-State of Arizona, Bureau of Geology and Mineral Technology, Geological Survey Branch, p. 61.

Reynolds, T. (1994). Quantitative analysis of submarine fans in the Tertiary of the North Sea Basin. *Marine and Petroleum Geology*, 11, 202–7.

Ribe, N. M. (1996). The dynamics of plume-ridge interaction, 2. Off- ridge plumes. *Journal of Geophysical Research*, 101, 16195–204.

Ribe, N. M., Christensen, U. R., and Theiβing, J. (1995). The dynamics of plume-ridge interaction, 1: Ridge-centered plumes. *Earth and Planetary Science Letters*, 134, 155–68.

Ribe, N. M. and Delattre, W. L. (1998). The dynamics of plume-ridge interaction, III: The effects of ridge migration. *Geophysical Journal International*, 133, 511–18.

Rice, D. D., Fouch, T. D., and Johnson, R. C. (1992). Influence of source rock type, thermal maturity, and migration on composition and distribution of natural gases, Uinta basin, Utah. In *Hydrocarbon and Mineral Resources of the Uinta Basin, Utah and Colorado*, ed. T. D. Fouch, V. F. Nuccio, and T. C. Chidsey. *Utah Geological Association Guidebook*, **20**, 95–109.

Rice, J. R. (1992). Fault stress states, pore pressure distributions, and the weakness of the San Andreas fault. In *Fault Mechanics and Transport Properties of Rocks*, ed. B. Evans and T.-F. Wong. Academic Press, San Diego, California, pp. 475–503.

Richard, P. and Krantz, R. W. (1991). Experiments on fault reactivation in strike-slip mode: experimental and numerical modelling of continental deformation. *5th meeting of the European Union of Geosciences, Symposium on Experimental and Numerical Modelling of Continental Deformation*, Strasbourg, France, March 20–23, 1989, **188**, 117–31.

Richard, P. D., Naylor, M. A., and Koopman, A. (1995). Experimental models of strike-slip tectonics. *Petroleum Geoscience*, 1, 71–80.

Richards, M. A., Duncan, R. A., and Courtillot, V. E. (1989). Flood basalts and hot-spot tracks – Plume heads and tails. *Science*, 246, 105–7.

Richardson, R. M. and Coblentz, D. D. (1994). Stress modeling in the Andes: constraints on the South American intraplate stress magnitudes. *Journal of Geophysical Research*, 99, 22015–26.

Richardson, S. W. and Oxburg, E. R. (1979). The heat flow field in mainland UK. *Nature*, 282, 565–7.

Ridley, J. (1993). The relations between mean rock stress and fluid flow in the crust; with reference to vein- and lode-style gold deposits. *Ore Geology Reviews*, 8, 23–37.

Ritchie, J. D., Gatliff, R. W., and Richards, P. C. (1999). Early Tertiary magmatism in the offshore NW UK margin and surrounds. In *Petroleum Geology of Northwest Europe*, ed. A. J. Fleet and S. A. R. Boldy. *Proceedings of the 5th Conference*, London, pp. 573–84.

Ritsema, J. and van Heist, H. (2000). New seismic model of the upper mantle beneath Africa. *Geology*, 28, 63–6.

Ritter, O., Ryberg, T., Weckmann, U., Hoffmann-Rothe, A., Abueladas, A., Garfunkel, Z., and DESERT Research Group (2003). Geophysical images of the Dead Sea Transform in Jordan reveal an impermeable barrier for fluid flow. *Geophysical Research Letters*, 30.

Robert, P. (1988). *Organic Metamorphism and Geothermal History*. Elf-Aquitaine and Reidel Publishing, Dordrecht.

Roberts, A. M., Lundin, E. R., and Kusznir, N. J. (1997). Subsidence of the Vøring Basin and the influence of the Atlantic continental margin. *Journal of the Geological Society of London*, 154, 551–7.

Roberts, A. M., Yielding, G., and Badley, M. E. (1993). Tectonic and bathymetric controls on stratigraphic sequences within evolving half-graben. In *Tectonics and Seismic Sequence Stratigraphy*, ed. G. D. Williams and A. Dobb. *Geological Society of London Special Publications*, **71**, 87–121.

Roberts, D. G., Montadert, L., and Searle, R. C. (1979). The western Rockall Plateau: stratigraphy and structural evolution. *Initial Reports of the Deep Sea Drilling Project*, 48, 1061–88.

Roberts, G. P. (1991). Structural controls on fluid migration through the Rencurel thrust zone, Vercors, French Sub-Alpine chains. In *Petroleum Migration*, ed. W. A. England and A. J. Fleet. *Geological Society of London Special Publications*, **59**, 245–62.

(2007). Fault orientation variations along the strike of active normal fault systems in Italy and Greece: implications for predicting the orientations of subseismic-resolution faults in hydrocarbon reservoirs. *American Association of Petroleum Geologists Bulletin*, 91, 1–20.

Roberts, G. P. and Ganas, A. (2000). Fault-slip directions in central and southern Greece measured from striated and corrugated fault planes: comparison with focal mechanism and geodetic data. *Journal of Geophysical Research*, 105, 23443–62.

Roberts, H. H., Aharon, P., Carney, R., and Sassen, R. (1990). See floor responses to hydrocarbon seeps, Louisiana continental slope. *Geo-Marine Letters*, 10, 232–43.

Roberts, H. H. and Carney, R. S. (1997). Evidence of episodic fluid, gas and sediment venting on the Northern Gulf of Mexico continental slope. *Economic Geology*, 92, 863–79.

Roberts, J. A. and Cramp, A. (1996). Sediment stability on the western flanks of the Canary Islands. *Marine Geology*, 134, 13–30.

Roberts, M. J. (1991). The South Brae Field, Block 16/07a, U.K. North Sea. In *United Kingdom Oil and Gas Fields, 25 Years Commemorative Volume*, ed. I. L. Abbotts. *Memoir of the Geological Society of London*, **14**, 55–62.

Roberts, S. J. and Nunn, J. A. (1995). Episodic fluid expulsion from geopressured sediments. *Marine and Petroleum Geology*, 12, 195–204.

Roberts, S. J., Nunn, J. A., Cathles, L., and Cipriani, F. D. (1996). Expulsion of abnormally pressured fluids along faults. *Journal of Geophysical Research*, 101, 28231–52.

Robinson, D. and Santana Zamora, A. (1999). The smectite to chlorite transition in the Chipilapa geothermal system, El Salvador. *American Mineralogist*, 84, 607–19.

Robinson, N., Eglington, G., Brassel, S. C., and Cranwell, P. A. (1984). Dinoflagellate origin for sedimentary 4-Alpha-methylsteroids and 5-Alpha (H)-stanols. *Nature*, 308, 439–42.

Robinson, V. D. and Engel, M. H. (1993). Characterization of the source horizons within the Late Cretaceous transgressive sequence of Egypt. In *Source Rocks in a Sequence Stratigraphic Framework*, ed. B. J. Katz and L. M. Pratt. *American Association of Petroleum Geologists Studies in Geology*, **37**, 101–17.

Robson, D. (1991). The Argyll, Duncan and Innes fields, Blocks 30/24 and 30/25a, UK North Sea. In *United Kingdom Oil and Gas Fields, 25 Years Commemorative Volume*, ed. I. L. Abbotts. *Memoir of the Geological Society of London*, **14**, 219–25.

Rockwell, T. K., Keller, E. A., and Dembroff, G. R. (1988). Quaternary rate of folding of the Ventura anticline, western Transverse Ranges, southern California. *American Geological Society Bulletin*, 100, 850–8.

Roden-Tice, M. K. and Wintsch, R. P. (2002). Early Cretaceous normal faulting in southern New England: evidence from apatite and zircon fission-track ages. *Journal of Geology*, 110, 159–78.

Rodrigues, R., Santos, A. S., and Costa, L. A. (1984). Avaliacao geoquimica da Bacia de Barreirinhas. *Internal report Petrobras*, Siex 130–4017.

Roehler, H. W. (1992). Correlation, composition, areal distribution, and thickness of Eocene stratigraphic units, greater Green River basin, Wyoming, Utah, and Colorado. *USGS Professional Paper*, 1506, 1–49.

Rona, P. A., Bougault, H., Charlou, J. L., Appriou, P., Nelsen, T. A., Trefy, J. H., Eberhart, G. L., Barone, A., and Needham, H. D. (1992). Hydrothermal circulation, serpentinization, and degassing at a rift valley-fracture zone intersection: Mid-Atlantic Ridge near 15 degrees N, 45 degrees W. *Geology*, 20, 783–6.

Rook, S. H. and Williams, G. C. (1942). Imperial carbon dioxide gas field: summary of operations, California oil fields. *Annual Report of the State Oil and Gas Supervisor*, 28, 12–33.

Rooksby, S. K. (1991). The Miller Field, Blocks 16/7B, 16/8B, UK North Sea. In *United Kingdom oil and gas fields, 25 Years Commemorative Volume*, ed. I. L. Abbotts. *Memoir of the Geological Society of London*, **14**, 159–64.

Rose, P. R. (2001). Risk analysis and management of petroleum exploration ventures. *American Association of Petroleum Geologists Methods in Exploration Series*, 12, 178.

Rosendahl, B. R. (1987). Architecture of continental rifts with special reference to East Africa. *Annual Review of Earth and Planetary Sciences*, 15, 445–503.

Rosendahl, B. R. and Groschel-Becker, H. (1999). Deep seismic structure of the continental margin in the Gulf of Guinea: a summary report. *Geological Society Special Publications*, 153, 75–83.

Rosendahl, B. R., Groschel-Becker, H., Meyers, J., and Kaczmarick, K. (1991). Deep seismic reflection study of a passive margin, southeastern Gulf of Guinea. *Geology*, 19, 291–5.

Rosendahl, B. R., Mohriak, W. U., Nemčok, M., Odegard, M. E., Turner, J. P., and Dickson, W. G. (2005). West African and Brazilian conjugate margins: crustal types, architecture, and plate configurations. *The 4th HGS/PESGB International Conference on African E7P*, Houston Program, **3**.

Rosendahl, B. R., Reynolds, D. J., Lorber, P. M., Burgess, C. F., McGill, J., Scott, D., Lambiase, J. J., and Derksen, S. J. (1986). Structural expressions of rifting: lessons from Lake Tanganyika, Africa. In *Sedimentation in the African Rifts*, ed. L. E. Frostrick. *Geological Society of London Special Publications*, **25**, 29–43.

Rowan, M. G., Hart, B. S., Nelson, S., Flemings, P. B., and Trudgill, B. D. (1998). Three-dimensional geometry and evolution of a salt-related growth-fault array: Eugene Island 330 Field, offshore Louisiana, Gulf of Mexico. *Marine and Petroleum Geology*, 15, 309–28.

Rowan, M. G., Jackson, M. P. A., and Trudgill, B. D. (1999). Salt-related fault families and fault welds in the northern Gulf of Mexico. *American Association of Petroleum Geologists Bulletin*, 83, 1454–84.

Rowan, M. G. and Weimer, P. (1998). Salt–sediment interaction, northern Green Canyon and Ewing Bank (offshore Louisiana), northern Gulf of Mexico. *American Association of Petroleum Geologists Bulletin*, 82, 1055–82.

Roy, R. F., Beck, A. E., and Toulokian, Y. S. (1981). Thermophysical properties of rocks. In *Physical Properties of Rocks and Minerals*, ed. Y. S. Toulokian, W. R. Judd, and R. F. Roy. McGraw-Hill, New York, pp. 409–88.

Royden, L. H. (1985). The Vienna Basin: a thin-skinned pull-apart basin. Strike-slip deformation, basin formation, and sedimentation. *SEPM, Special Publications*, 37, 319–38.

Royden, L. and Keen, C. E. (1980). Rifting process and thermal evolution of the continental margin of eastern Canada determined from subsidence curves. *Earth and Planetary Science Letters*, 51, 343–61.

Rubey, W. W. and Hubbert, M. K. (1960). Role of fluid pressure in mechanics of overthrust faulting, II: Overthrust belt in geosynclinal area of western Wyoming in light of fluid pressure hypothesis. *Geological Society of America Bulletin*, 60, 167–205.

Ruble, T. E., Lewan, M. D., and Philp, R. P. (2001). New insights on the Green River petroleum system in the Uinta Basin from hydrous pyrolysis experiments. *American Association of Petroleum Geologists Bulletin*, 85, 1333–71.

(2003). New insights on the Green River petroleum system in the Uinta basin from hydrous pyrolysis experiments. Reply. *American Association of Petroleum Geologists Bulletin*, 87, 1535–41.

Ruble, T. E. and Philp, R. P. (1998). Stratigraphy, depositional environments and organic geochemistry of source-rocks in the Green River petroleum system, Uinta basin, Utah. In *Modern and Ancient Lake Systems: New Problems and Perspectives*, ed. J. K. Pitman and A. R. Carroll. *Geological Association Guidebook*, Salt Lake City, Utah, **26**, 289–328.

Rudnick, R. L., Ireland, T. R., Gehrels, G., Irving, A. J., Chesley, J. T., and Hanchar, J. M. (1999). Dating mantle metasomatism: U-Pb geochronology of zircons in cratyonic mantle xenoliths from Montana and Tanzania. In *The Nixon Volume*, ed. J. J. Gurney, M. D. Gurney, S. Pascoe and S. R. Richardson. *Proceedings of the VIIth International Kimberlite Conference*, pp. 503–21.

Rudnick, R. L., McDonough, W. F., and O'Connell, R. J. (1998). Thermal structure, thickness and composition of continental lithosphere. *Chemical Geology*, 145, 395–411.

Ruppel, C. (1995). Extensional processes in continental lithosphere. *Journal of Geophysical Research*, 100, 24187–215.

Russell, S. M. and Whitmarsh, R. B. (2003). Magmatism at the west Iberia non-volcanic rifted continental margin: evidence from analyses of magnetic anomalies. *Geophysical Journal International*, 154, 706–30.

Rust, D. J. and Summerfeld, M. A. (1990). Isopach and borehole data as indicators of rifted margin evolution in southwestern Africa. *Marine and Petroleum Geology*, 7, 277–87.

Rutter, E. H. (1976). The kinetics of rock deformation by pressure solution. *Philosophical Transactions of the Royal Society of London*, 283, 203–20.

(1983). Pressure solution in nature, theory and experiment. *Journal of Geological Society of London*, 140, 725–40.

Rutter, E. H. and Brodie, K. H. (1988). Experimental approaches to the study of deformation metamorphism relationships. *Mineralogical Magazine*, 52, 35–42.

Rutter, E. H. and White, S. H. (1979). The microstructures and rheology of fault gouges produced experimentally under wet and dry conditions at temperatures up to 400 degrees centigrade. *Bulletin de Mineralogie*, 102, 102–9.

Rybach, L. (1976). Radioactive heat production: a physical property determined by the chemistry of rocks. In *The Physics and Chemistry of Minerals and Rocks*, ed. R. G. J. Strens. Wiley and Sons, London, pp. 309–18.

(1985). The interpretation of heat flow measurements in terms of crustal and lithospheric structure. *Terra cognita*, 5, 33–4.

(1986a). Radioactive heat production in rocks and its relation to other petrophysical parameters. *Pure and Applied Geophysics*, 114, 309–17.

(1986b). Amount and significance of radioactive heat sources in sediments. In *Thermal Modeling in Sedimentary Basins*, ed. J. Burrus. Technip, Paris, pp. 311–22.

Rybach, L. and Buntebarth, G. (1984). The variation of heat generation, density and seismic velocity with rock type in the continental lithosphere. *Tectonophysics*, 103, 335–44.

Sage, F., Basile, C., Mascle, J., Pontoise, B., and Whitmarsh, R. B. (2000). Crustal structure of the continent–ocean transition off the Cote d'Ivoire-Ghana transform margin: implications for thermal exchanges across the palaeotransform boundary. *Geophysics Journal International*, 143, 62–678.

Sakai, H. et al. (1997). Paleomagnetic study with (super 40) Ar/ (super 39) Ar dating of Rajmahal Hills and Mahanadi Graben in India: reconstruction of Gondwanaland. Chishitsugaku Zasshi. *Journal of the Geological Society of Japan*, 103, 192–202.

Sales, J. K. (1997). Seal strength vs. trap closure: a fundamental control on the distribution of oil and gas. In *Seals, Traps and the Petroleum System*, ed. R. C. Surdam. *American Association of Petroleum Geologists Memoir*, **67**, 57–83.

Salmon, P., Taruno, R. J., and Sadirsan, W. S. (1996). Hydrocarbon exploration in Indonesia in the first half of 1990s decade. *International Geological Congress, Abstracts – Congres Geologique Internationale, Resumes*, 30, 565.

Sandwell, D. T. and Smith, W. H. F. (1997). Marine gravity anomaly from Geosat and ERS 1 satellite altimetry. *Journal of Geophysical Research*, 102, 10039–54.

Sartain, S. M. and See, B. E. (1997). The South Georgia Basin: an integration of Landsat, gravity, magnetics and seismic data to delineate basement structure and rift basin geometry. In *1997 Gulf Coast Association of Societies: Geology across the Gulf, New Offshore Technologies, Keys to Onshore Revitalization*, ed. W. W. Craig and B. Kohl. *Transactions – Gulf Coast Association of Geological Societies*, **47**, 493–7.

Sassen, R., Roberts, H. H., Aharon, P., Larkin, J., Chinn, E. W., and Carney, R. (1993). Chemosynthetic bacterial mats at cold hydrocarbon seeps, Gulf of Mexico continental slope. *Organic Geochemistry*, 20, 77–89.

Saunders, A. D., Fitton, J. G., Kerr, A. C., Norry, M. J., and Kent, R. W. (1997). The North Atlantic igneous province. In *Large Igneous Provinces: Continental, Oceanic and Planetary Flood Volcanism*, ed. J. Mahoney and M. F. Coffin. *AGU Geophysical Monograph*, **100**, 45–94.

Saunders, I. and Young, A. (1983). Rates of surface processes on slopes, slope retreat and denudation. *Earth Surface Processes and Landforms*, 8, 473–501.

Sawyer, D. S., Swift, A., Sclater, J. G., and Toksoz, N. (1982). Extensional model for the subsidence of the northern United States Atlantic continental margin, *Geology*, 10, 134–40, 3/82.

Sawyer, M. J. and Keegan, J. B. (1996). Use of palynofacies characterization in sand-dominated sequences, Brent Group, Ninian Field, UK North Sea. *Petroleum Geoscience*, 2, 289–97.

Schaltegger, U., Gebauer, D., and von Quadt, A. (2002). The mafic-ultramafic rock association of Loderio-Biasca (lower Pennine nappes, Ticino, Switzerland): Cambrian oceanic magmatism and its bearing on early Paleozoic paleogeography. *Chemical Geology*, 186, 265–79.

Schamel, S. (1993). Hydrocarbon Prospects in the Pechora Basin. Earth Science & Resources Institute, University of South Carolina, *ESRI Technical Report*, 92-09-373(1).

Schamel, S. and Ressetar, R. (1998). Tectonics and sedimentation of the Great Salt Lake and Uinta Basins, Utah: analogues for nonmarine basins in China. *EGI Field Seminar*, I00743, 170.

Schardt, C., Large, R., and Yang, J. (2006). Controls on heat flow, fluid migration, and massive sulfide formation of an off-axis hydrothermal system – the Lau Basin perspective. *American Journal of Science*, 306, 103–34.

Scheck, M. and Bayer, U. (1999). Evolution of the Northeast German Basin: inferences from a 3D structural model and subsidence analysis. *Tectonophysics*, 313, 145–69.

Schenk, C. J., Higley, D. K., and Magoon, L. B. (2003). Region 6 Assessment Summary – Central and South America. *U.S. Geological Survey Digital Data Series*, 60.

Schenk, H. J. and Horsfield, B. (1998). Using natural maturation series to evaluate the utility of parallel reaction kinetics models: an investigation of Toarcian shales and Carboniferous coals, Germany. *Organic Geochemistry*, 29, 137–4.

Schenk, H. J., Horsfield, B., Krooss, B., Schaefer, R. G., and Schwochau, K. (1997). Kinetics of petroleum formation and cracking. In *Petroleum and Basin Evolution: Insights from Petroleum Geochemistry, Geology and Basin Modeling*, ed. D. H. Welte, B. Horsfield, and D. R. Baker. Springer-Verlag, Berlin, pp. 233–69.

Schilling, J.-G. (1973). Iceland mantle plume: geochemical evidence along Reykjanes Ridge. *Nature*, 242, 565–71.

(1985). Upper mantle heterogeneities and dynamics. *Nature*, 314, 62–7.

Schilling, J.-G., Kingsley, R., Fontignie, D., Pordea, R., and Xue, S. (1999). Dispersion of the Jan Mayen and Iceland mantle plumes in the Arctic: a He-Pb-Nd-Sr isotope tracer study of basalts from the Kolbeinsey, Mohns, and Knipovich Ridges. *Journal of Geophysical Research*, 104, 10543–69.

Schleicher, A. M., van der Pluijm, B. A., Solum, J. G., and Warr, L. N. (2006). Origin and significance of clay-coated fractures in mudrock fragments of the SAFOD borehole (Parkfield, California). *Geophysical Research Letters*, 33, 16313.

Schlindwein, V. and Jokat, W. (1999). Structure and evolution of the continental crust of northern east Greenland from integrated geophysical studies. *Journal of Geophysical Research*, 104, 15227–45.

Schlische, R. W. (1991). Half-graben basin filling models: new constraints on continental extensional basin development. *Basin Research*, 3, 123–41.

(1992). Structural and stratigraphic development of the Newark extension basin, eastern North America: implications for the growth of the basin and its bounding structures. *Bulletin of Geological Society of America*, 104, 1246–63.

(1993). Anatomy and evolution of the Triassic-Jurassic continental rift system, eastern North America. *Tectonics*, 12, 1026–42.

(1995). Geometry and origin of fault-related folds in extensional settings. *American Association of Petroleum Geologists Bulletin*, 79, 1661–78.

(2003). Progress in understanding the structural geology, basin evolution, and tectonic history of the eastern North American Rift System. In *Aspects of Triassic-Jurassic Rift Basin Geoscience*, ed. P. M. LeTorneau and P. E. Olsen. Columbia University Press, New York.

Schlische, R. W. and Anders, M. H. (1996). Stratigraphic effects and tectonic implications of the growth of normal faults and extensional basins. In *Reconstructing the History of Basin and Range Extension Using Sedimentology and Stratigraphy*, K. K. Beratan. *Special Paper – Geological Society of America*, pp. 183–203.

Schmitt, H. R. H. (1991). The Chanter Field, Block 15/17, UK North Sea. In *United Kingdom Oil and Gas Fields, 25 Years Commemorative Volume*, ed. I. L. Abbotts. *Memoir of the Geological Society of London*, **14**, 261–8.

Schmitt, H. R. H. and Gordon, A. F. (1991). The Piper Field, Block 15/17, UK North Sea. In *United Kingdom Oil and Gas Fields, 25 Years Commemorative Volume*, ed. I. L. Abbotts. *Memoir of the Geological Society of London*, **14**, 361–8.

Schmucker, U. (1969). Conductivity anomalies, with special reference to the Andes. In *The Application of Modern Physics to the Earth and Planetary Interiors*, ed. S. K. Runcorn. Wiley-Interscience, pp. 125–38.

Schoell, M. (1984). Recent advances in petroleum isotope geochemistry. *Organic Geochemistry*, 6, 645–63.

Schoell, M., Hwang, R. J., Carlson, R. M. K., and Welton, J. E. (1994). Carbon isotopic composition of individual biomarkers in gilsonites. *Organic Geochemistry*, 21, 673–83.

Schöellkopf, N. B. and Patterson, B. A. (1997). Petroleum systems of Cabinda, Angola. *AAPG/ ABGP Hedberg Research Symposium, Extended Abstracts Volume*, Rio de Janeiro, Brazil.

Scholl, D. W., Christensen, M. N., Von Huene, R., and Marlow, M. S. (1970). Peru-Chile trench sediments and sea-floor spreading. *Geological Society of America Bulletin*, 81, 1339–60.

Scholten, J. C., Scott, S. D., Garbe-Schoenberg, D., Fietzke, J., Thomas, B., and Kennedy, C. B. (2004). Hydrothermal iron and manganese crusts from the Pitcairn Hotspot region. In *Oceanic Hotspots: Intraplate Submarine Magmatism and Tectonism*, ed. R. Hekinian, P. Stoffers, and J.-L. Cheminee. Springer, Berlin.

Scholz, C. H. (1968). Microfracturing and the inelastic deformation of rock in compression. *Journal of Geophysical Research*, 73, 1417–32.

Schon, J. (1983). *Petrophysik: Physikalische Eigenschaften von Gesteinen und Mineralen*. Ferdinand Enke Verlag, Stuttgart.

Schou, L., Eggen, S., and Schoell, M. (1985). Oil-oil and oil-source rocks correlation, northern North Sea. In *Petroleum Geochemistry in Exploration of the Norwegian Shelf*, ed. B. M. Thomas, A. G. Dore, S. S. Eggen, P. C. Home, and R. Mange Larsen. Graham and Trotman, London, pp. 101–17.

Schowalter, T. T. (1979). Mechanics of secondary hydrocarbon migration and entrapment. *American Association of Petroleum Geologists Bulletin*, 63, 723–60.

Schubert, C. (1980). Late-Cenozoic pull-apart basins, Bocono fault zone, Venezuelan Andes. *Journal of Structural Geology*, 2, 463–8.

Schueller, S., Gueydan, F., and Davy, P. (2005). Brittle-ductile coupling: role of ductile viscosity on brittle fracturing. *Geophysical Research Letters*, 32, 4.

Schulte, W. M., Van Rossem, P. A. H., and Van de Vover, W. (1994). Current challenges in the Brent Field. *Journal of Petroleum Technology*, 46, 1073–9.

Schultz, R. A. and Moore, J. M. (1996). New observations of grabens from the Needles district, Canyonlands National Park, Utah. In *Geology and Resources of the Paradox Basin*, ed. A. C. Huffman, W. R. Lund, and L. H. Godwin. Utah Geological Association, pp. 295–302.

Sclater, J. G. and Christie, P. A. F. (1980). Continental stretching: an explanation of the post mid-Cretaceous subsidence of the central North Sea basin. *Journal of Geophysical Research*, 85, 3711–39.

Sclater, J. G., Royden, L., Horváth, F., Burchfield, C., Sempken, S., and Stegena, S. (1980). The formation of the intra-Carpathian basins as determined from subsidence data. *Earth and Planetary Science Letters*, 51, 137–62.

Scotese, C. R. (1998). Computer software to produce plate tectonic reconstructions. In *Gondwana 10: Event Stratigraphy of Gondwana*, ed. J. Almond, J. Anderson, P. Booth, A. Chinsamy-Turan, D. Cole, M. J. De Wit, B. Rubridge, R. Smith, J. van Bever Donker, and B. C. Storey. *Journal of African Earth Sciences*, **27**, 171–2.

Scrutton, R. A. (1984). Modeling of mnagmatic and gravity anomalies at Goiban Spur, northeast Atlantic. In *Initial Reports of the Deep Sea Drilling Project 80*, ed. P. C. de Graciansky and C. W. Poag. U.S. Government Printing Office, Washington, DC.

Sebai, A., Stutzmann, E., Montager, J. P., Sicilia, D., and Beucler, E. (2006). Anisotropic structure of the African upper mantle from Rayleigh and Love wave tomography. *Physics of the Earth and Planetary Interiors*, 155, 48–62.

Sedwick, P. N., McMurtry, G. M., and Macdougall, J. D. (1992). Chemistry of hydrothermal solutions from Pele's Vents, Loihi Seamount, Hawaii. *Geochimica et Cosmochimica Acta*, 56, 3643–67.

Seewald, J. S., Benitez-Nelson, B. C., and Whelan, J. K. (1998). Laboratory and theoretical constraints on the generation and composition of natural gas. *Geochimica et Cosmochimica Acta*, 62, 1599–617.

Segall, M. P., Nash, G., Umbriaco, J., Kessler, C., Dudley-Murphy, E., and Sawyer, R. K. (2003). BHP-Billiton – EGI DSDP-ODP digital database project. Phase I – Northwest Africa. *EGI Report*, 50500985-03-30-03, p. 20, 32 enclosures, 3 CDs.

Segall, M. P. and Pollard, D. D. (1980). Mechanics of discontinuous faults. *Journal of Geophysical Research*, 85, 4337–50.

Seiglie G. and Baker, M. (1984). Relative sea level changes during the Middle and Late Cretaceous from Zaire to Cameroon (central West Africa). In *Interregional Unconformities and Hydrocarbon Accumulation*, ed. J. S. Schlee. AAPG Memoir 36, pp. 81–8.

Self, S., Thordarson, T., and Keszthelyi, L. (1997). Emplacement of continental flood basalt lava flows. In *Large Igneous Provinces: Continental, Oceanic and Planetary Flood Volcanism*, ed. J. J. Mahoney and M. F. Coffin. *AGU Geophysical Monograph*, **100**, 381–410.

Sengör, A. M. C. (1976). Collision of irregular continental margins: implications for foreland deformation of Alpine-type orogeny. *Geology*, 4, 779–85.

(2001). Elevation as indicator of mantle-plume activity. In *Mantle Plumes: Their Identification through Time*, ed. R. E. Ernst and K. L. Buchan. *Geological Society of America Special Paper*, **352**, 183–225.

Sengör, A. M. C. and Burke, K. (1978). Relative timing of rifting and volcanism on Earth and its tectonic implications. *Geophysical Research Letters*, 5, 419–21.

Shaler, N. S. and Woodworth, J. B. (1899). Geology of the Richmond Basin, Virginia. *U.S. Geological Survey Annual Report*, 19, 1246–63.

Sharp, I. R., Gawthorpe, R. L., Armstrong, B., and Underhill, J. R. (2000). Propagation history and passive rotation of mesoscale normal faults: implications for synrift stratigraphic development. *Basin Research*, 12, 285–305.

Sharp, J. M. (1978). Energy and momentum transport model of the Ouachita basin and its possible impact on formation of economic mineral deposits. *Economic Geology*, 73, 1057–68.

Jr. (1983). Permeability controls on aquathermal pressuring. *American Association of Petroleum Geologists Bulletin*, 67, 2057–61.

Shaw, W. J. and Lin, J. (1996). Models of ocean ridge lithospheric deformation: dependence on crustal thickness, spreading rate, and segmentation. *Journal of Geophysical Research*, 101, 17977–93.

Shearman, D. J., Mossop, G., Dunsmore, H., and Martin, M. (1972). Origin of gypsum veins by hydraulic fracture. *Institution of Mining and Metallurgy, Transactions, Applied Earth Science*, 81, 149–55.

Sheng, G., Fu, J., Brassell, S. C., Gowar, A. P., Eglinton, G., Damste, J. S. S., de Leeuw, J. W., and Schenk, P. A. (1987). Sulfur-containing compounds in sulfur-rich crude oils from hypersaline lake sediments and their geochemical implications. *Geochemistry*, 6, 115–26.

Shepherd, M. (1991). The Magnus Field, Blocks 211/7a, 12a, UK North Sea. In *United Kingdom Oil and Gas Fields, 25 Years Commemorative Volume*, ed. I. L. Abbotts. *Memoir of the Geological Society of London*, **14**, 153–7.

Sheridan, R. E., Musser, D. L., Glover, L., Talwani, M., Ewing, J. I., Holbrook, W. S., Purdy, G. M., Hawman, R., and Smithson, S. (1993). Deep seismic reflection data of EDGE U.S. Atlantic continent margin experiment: implications for Appalachian sutures and Mesozoic rifting and magmatic underplating. *Geology*, 21, 563–7.

Sheth, H. C. (2005). From Deccan to Réunion: no trace of a mantle plume. *Geological Society of America Special Paper*, 388, 477–501.

Shevenell, L. (2005a). *Crustal Thickness Map*. Great Basin Center, University of Nevada, Reno.

(2005b). *Shear Strain Distribution*. Great Basin Center, University of Nevada, Reno.

Shi, J. Y., Mackenzie, A. S., Alexander, R., Eglinton, G., Gowar, A. P., Wolff, G., and Maxwell, J. R. (1982). A biological marker investigation of petroleums and shales from the Shengli oilfield, the People's Republic of China. *Chemical Geology*, 35, 1–31.

Shi, Y. and Wang, C. Y. (1986). Pore pressure generation in sedimentary basins: overloading versus aquathermal. *Journal of Geophysical Research*, 91, 2153–62.

Shikazono, N. and Holland, H. D. (1983). The partitioning of strontium between anhydrite and aqueous solutions from 150 degrees to 250 degrees C. *Economic Geology Monograph*, 5, 320–8.

Shillington, D. J., Holbrook, W. S., Tucholke, B. E., Hopper, J. R., Louden, K. E., Larsen, H. C., Van Avendonk, H. J. A., Deemer, S., and Hall, J. (2004). Data report: marine geophysical data on the Newfoundland nonvolcanic rifted margin around SCREECH Transect 2. *Proceedings of the Ocean Drilling Program, Initial Report*, 210.

Shinohara, H., Kazahaya, K., and Lowenstern, J. B. (1995). Volatile transport in a convecting magma column: implications for porphyry Mo mineralization. *Geology*, 23, 1091–4.

Shirley, D. N. (1986). Compaction of igneous cumulates. *Journal of Geology*, 94, 795–809.

(1987). Differentiation and compaction in the Palisades Sill. *Journal of Petrology*, 28, 835–65.

Sial, A. N. (1976). The post-Paleozoic volcanism of Northeast Brazil and its tectonic significance. *Anais da Academia Brasileira de Ciencias*, 48, 299–311.

Sial, A. N., Long, L. E., Pessoa, D. A. R., and Kawashita, K. (1981). Potassium-argon ages and strontium isotope geochemistry of Mesozoic and Tertiary basaltic rocks, northeastern Brasil. *Anais da Academia Brasileira de Ciencias*, 53, 115–22.

Sibson, R. H. (1974). Frictional constraints on thrust, wrench and normal faults. *Nature*, 249, 542–4.

(1981a). Controls on low-stress hydro-fracture dilatancy in thrust, wrench and normal fault terrains. *Nature*, 289, 665–7.

(1981b). Fluid flow accompanying faulting: field evidence and models. In *Earthquake Prediction: An International Review*, ed. D. W. Simpson and P. G. Richards. *American Geophysical Union, Maurice Ewing Series*, 4, 593–603.

(1982). Fault zone models, heat flow, and the depth distribution of seismicity in the continental crust of the United States. *Seismological Society of America Bulletin*, 72, 151–63.

(1986). Brecciation processes in fault zones: inferences from earthquake rupturing. *Pure and Applied Geophysics*, 124, 159–75.

(1990a). Conditions for fault-valve behaviour. In *Deformation Mechanisms, Rheology and Tectonics*, ed. R. J. Knipe and E. H. Rutter. *Geological Society of London Special Publications*, pp. 15–28.

(1990b). Faulting and Fluid Flow. In *Fluids in Tectonically Active Regimes of the Continental Crust*, ed. B. E. Nesbitt. *Mineralogical Association of Canada Short Course*, 18, 93–132.

(1994). Hill fault/fracture meshes as migration conduits for overpressured fluids. In *Proceedings of Workshop LXIII, US Geological Survey Red-Book Conference on the Mechanical Involvement of Fluids in Faulting*, ed. H. Hickman, R. H. Sibson, R. L. Bruhn, and M. L. Jacobson. Open-*File Report*, 94–0228, pp. 224–30.

(1996). Structural permeability of fluid-driven fault-fracture meshes. *Journal of Structural Geology*, 18, 1031–42.

(2003). Brittle-failure controls on maximum sustainable overpressure in different tectonic regimes. *American Association of Petroleum Geologists Bulletin*, 87, 901–8.

Sibson, R. H., Robert, F., and Poulsen, K. H. (1988). High-angle reverse faults, fluid-pressure cycling, and mesothermal gold-quartz deposits. *Geology*, 16, 551–5.

Siddiqui, F. I. (1996). *A dynamic theory of hydrocarbon migration and trapping*. PhD thesis, University of Texas, Austin.

Siddiqui, F. I. and Lake, L. W. (1997). A comprehensive dynamic theory of hydrocarbon migration and trapping. *Proceedings of the SPE 72nd Annual Technical Conference*, San Antonio, October 5–8, 1997, pp. 395–410.

Silver, L. T., James, E. W., and Chappell, B. W. (1988). Petrological and geochemical investigations at the Cajon Pass deep drillhole. *Geophysical Research Letters*, 15, 961–4.

Sikora, P. J., Bergen, J. A., and Farmer, C. A. (1999). Chalk palaeoenvironments and depositional model, Valhall-Hod fields, southern Norwegian North Sea. In *Biostratigraphy in Production and Development Geology*, ed. R. W. Jones and M. D. Simmons. *Geological Society of London Special Publications*, **152**, 113–38.

Simancas, J. F., Carbonnell, R., Gonzalez Lodeiro, F., Perez Estaun, A., Juhlin, C., Ayarza, P., Kashubin, A., Azor, A., Martinez Poyatos, D., Almodovar, G. R., Pascula, E., Saez, R., and Exposito, I. (2003). Crustal structure of the transpression Variscan orogen of SW Iberia: SW Iberia deep seismic reflection profile (IBERSEIS). *Tectonics*, 22, 19.

Simon, N. S. C. and Podladchikov, Y. Y. (2006). The effect of mantle petrology on lithosphere dynamics during extension. *EGU General Assembly Vienna, Scientific Programme*, 279.

Simoneit, B. R. T. and Stuermer, D. H. (1982). Organic geochemical indicators for sources of organic matter and paleoenvironmental conditions in Cretaceous oceans. In *Nature and Origin of Cretaceous Carbon-rich Facies*, ed. S. O. Schlanger and M. B. Cita. Academic Press, London, pp. 145–63.

Simons, F. J., Zielhuis, A., and van der Hilst, R. D. (1999). The deep structure of the Australian continent from surface wave tomography. *Lithos*, 48, 17–43.

Sims, D., Ferrill, D. A., and Stamatakos, J. A. (1999). Role of a ductile decollement in the development of pull-apart basins; experimental results and natural examples. *Journal of Structural Geology*, 21, 533–54.

Sinclair, H. D. (1997). Tectonostratigraphic model for underfilled peripheral foreland basins: an Alpine perspective. *Geological Society of America Bulletin*, 109, 324–46.

Sinclair, H. D., Coakley, B. J., Allen, P. A., and Watts, A. B. (1991). Simulation of foreland basin stratigraphy using a diffusion model of mountain belt uplift and erosion: an example from the Central Alps, Switzerland. *Tectonics*, 10, 599–620.

Singh, A. P. (1999). The deep crustal accretion beneath the Laxmi Ridge in the northeastern Arabian Sea: the plume model again. *Journal of Geodynamics*, 27, 609–22.

(2002). Impact of Deccan volcanism on deep crustal structure along western part of Indian mainland and adjoining Arabian Sea. *Current Science*, 82, 316–25.

Singh, A. P. and Mall, D. M. (1998). Crustal accretion beneath Koyana coastal region (India) and late Cretaceous geodynamics. *Tectonophysics*, 290, 285–97.

Singh, R. N. (1976). *Measurement and analysis of strata deformation around mining excavations*. PhD thesis, University of Wales, Cardiff.

Sinha, S. T. (2013). *Factors controlling a micro-continent development during continental break up: the Elan Bank case study*. PhD thesis, Masaryk University, Brno.

Sinha, S. T., Nemčok, M., Choudhuri, M., Misra, A. A., Sharma, S. P., Sinha, N., and Sujata Venkatraman, S. (2010). The crustal architecture and continental break up of East India Passive margin: an integrated study of deep reflection seismic interpretation and gravity modeling, AAPG Annual Convention & Exhibition, New Orleans, USA, Search and Discovery Article #40611.

Sinninghe Damste, J. S., Kenig, F., Koopmans, M. P., Koster, J., Schouten, S., Haynes, J. M., and de Leeuw, J. W. (1995). Evidence for gammacerane as an indicator of water column stratification. *Geochimica et Cosmochimica Acta*, 59, 1895–900.

Sixsmith, P. J., Hampson, G. J., Gupta, S., and Johnson, H. D. (2004a). Sedimentology and facies architecture of a transgressive sandstone: the Cretaceous Hosta Sandstone, New Mexico, USA. *American Association of Petroleum Geologists Annual Meeting Expanded Abstracts*, 13, 129.

Sixsmith, P. J., Hampson, G. J., Gupta, S., Johnson, H. D. and Fofana, J. F. (2004b). Facies architecture of a net transgressive sandstone reservoir analog: the Cretaceous Hosta Tongue, New Mexico. *AAPG Bulletin*, 92(4), 513–47.

Sixsmith, P. J., Hampson, G. J., Gupta, S., Johnson, H. D., and Fofana, J. F. (2008). Facies architecture of a net transgressive sandstone reservoir analog: the Cretaceous Hosta Tongue, New Mexico. *American Association of Petroleum Geologists Bulletin*, 92, 513–47.

Skempton, A. W. (1985). Residual strength of clays in landslides, folded strata and the laboratory. *Geotechnique*, 35, 3–18.

Skjerven, J., Rijs, F., and Kalheim, J. E. (1983). Late Paleozoic to early Cenozoic structural development of the south-southeastern Norwegian North Sea. Geologie en Mijnbouw. *Netherland Journal of Geosciences*, 62, 35–45.

Skogseid, J. (2001). Volcanic margins: geodynamic and exploration aspects. *Marine and Petroleum Geology*, 18, 457–61.

Skogseid, J., Pedersen, T., Eldholm, O., and Larsen, B. T. (1992). Tectonism and magmatism during NE Atlantic continental break-up: the Vøring Margin. *Geological Society of London Special Publications*, **68**, 305–20.

Skogseid, J., Planke, S., Faleide, J. I., Pedersen, T., Eldholm, O., and Neverdal, F. (2000). NE Atlantic continental rifting and volcanic margin formation. In *Dynamics of Norwegian Margin*, ed. A. Nottvedt, B. T. Larsen, S. Olaussen, B. Tørudbakken, J. Skogseid, R. H. Gabnelson, H. Brekke, and O. Birkeland. *Geological Society of London Special Publications*, **167**, 295–326.

Sleep, N. H. (1990). Hotspots and mantle plumes: some phenomenology. *Journal of Geophysical Research*, 95, 6715–36.

(2005). Evolution of the continental lithosphere. *Annual Reviews of the Earth and Planetary Sciences*, 33, 369–93.

Sloss, L. L. (1979). Global sea level change: a view from the craton. *American Association of Petroleum Geologists Memoir*, 29, 461–7.

(1988). Forty years of sequence stratigraphy. *Geological Society of America Bulletin*, 100, 1661–5.

Sluijk, D. and Nederlof, M. H. (1984). Worldwide geological experience as a systematic basis for prospect appraisal. In *Petroleum Geochemistry and Basin Evaluation*, ed. G. Demaison and R. J. Murris. *American Association of Petroleum Geologists Memoir*, **35**, 15–26.

Smallwood, J. R., Staples, R. K., Richardson, K. R., and White, R. S. (1999). Crust generated above the Iceland mantle plume: from continental rift to oceanic spreading center. *Journal of Geophysical Research*, 104, 22885–902.

Smallwood, J. R. and White, R. S. (2002). Ridge-plume interaction in the North Atlantic and its influence on continental breakup and seafloor spreading. In *The North Atlantic Igneous Province: Stratigraphy, Tectonic, Volcanic and Magmatic Processes*, ed. D. W. Jolley and B. R. Bell. *Geological Society of London Special Publications*, **197**, 15–37.

Smith, A. D. and Lewis, C. (1999). The planet beyond the plume hypothesis. *Earth Science Reviews*, 48, 135–82.

Smith, J. T. (1994). The petroleum system logic as an exploration tool in a frontier setting. In *The Petroleum System – From Source to Trap*, ed. L. B. Magoon and W. G. Dow. *American Association of Petroleum Geologists Memoir*, **60**, 25–49.

Smith, K. and Ritchie, J. D. (1993). Jurassic volcanic centres in the Central North Sea. In *Petroleum Geology of Northwest Europe: Proceedings of the 4th Conference*, ed. J. R. Parker. *The Geological Society of London*, pp. 519–31.

Smith, L. and Chapman, D. S. (1983). On the thermal effects of groundwater flow, 1: Regional scale systems. *Journal of Geophysical Research*, 88, 593–608.

Smith, L., Forster, C. B., and Evans, J. P. (1990). Interaction of fault zones, fluid flow, and heat transfer at the basin scale. *International Association of Hydrogeologists, Selected Papers*, 2, 41–67.

Smith, R. B., Nagy, Walter C., Julander, Kelsey A., Viveiros, John J., Barker, Craig A., Gants, Donald G. (1989). Geophysical and tectonic framework of the eastern Basin and Range-Colorado Plateau-Rocky Mountain transition. *Memoir – Geological Society of America*, 172, 205–33.

Smith, R. E. and Wiltschko, D. V. (1996). Generation and maintenance of abnormal fluid pressures beneath a ramping thrust sheet: isotropic permeability experiments. *Journal of Structural Geology*, 18, 951–70.

Smith, R. I., Hodgson, N., and Fulton, M. (1993). Salt control on Triassic reservoir distribution, UKCS Central North Sea. In *Petroleum Geology of Northwest Europe: Proceedings of the 4th Conference*, ed. J. R. Parker. Geological Society of London, Bath, pp. 547–57.

Smith, R. L. (1987). The structural development of the Clyde Field. In *Petroleum Geology of North West Europe*, ed. J. Brooks and K. W. Glennie. Graham and Trotman, London, pp. 523–31.

Smith, W. H. F. and Sandwell, D. T. (1997). Seafloor topography from satellite altimetry and ship depth soundings. *Science*, 277, 1956–62.

Smoot, J. P. (1983). Depositional subenvironments in an arid closed basin: the Wilkins Peak Member of the Green River Formation (Eocene), Wyoming, U.S.A. *Sedimentology*, 30, 801–27.

Snoke, A. W., Howard, K., and Lush, A. P. (1984). Polyphase Mesozoic-Cenozoic deformational history of the northern Ruby Mountains-East Humboldt Range, Nevada. In *Field Trip Guide Book*, ed. J. Lintz, Jr. University of Nevada.

Snowdon, L. R. (1991). Oil from type III organic matter: resinite revisited. *Organic Geochemistry*, 17, 743–7.

Soderstrom, B., Forsberg, A., Holtar, E., and Rasmussen, B. A. (1991). The Mjolner Field, a deep Upper Jurassic oil field in the Central North Sea. *First Break*, 9, 156–71.

Sofer, Z. (1984). Stable carbon isotope composition of crude oils: applications to source depositional environments and petroleum alteration. *AAPG Bulletin*, 68, 31–49.

Solli, T. (1995). Upper Jurassic play concept – an integrated study in Block 34/7, Norway. *First Break*, 13, 21–30.

Sommer, F. (1978). Diagenesis of Jurassic sandstones in the Viking Graben. *Journal of the Geological Society of London*, 135, 63–7.

Sorel, D. (2000). A Pleistocene and still-active detachment fault and the origin of the Corinth-Patras Rift, Greece. *Geology*, 28, 83–6.

Sorel, D., Keraudren, B., Muller, C., Bahain, J.-C., and Falgueres, C. (1997). Holocene uplift and palaeoseismicity on the Eliki fault, western Gulf of Corinth, Greece. *Annales Geofisicae*, 39, 575–88.

Sorey, M., Evans, B., Kennedy, M., Rogie, J., and Cook, A. (1999). Magmatic gas emissions from Mammoth Mountain, Mono County, California. *California Geology*, 52, 4–16.

Souza, J. M. (1982). *Transmission of seismic energy through the Brazilian Parana Basin layered basalt stack*. Master's thesis, University of Texas, Austin.

Spaak, P., Almond, J., Salahudin, S., Mohd Salleh, Z., and Tosun, O. (1999). Fulmar: a mature field revisited. In *Petroleum Geology of Northwest Europe*, ed. A. J. Fleet and S. A. R. Boldy. *Proceedings of the 5th Conference, The Geological Society of London*, pp. 1089–100.

Spadini, G., Robinson, A. G., and Cloetingh, S. A. P. L. (1996). Western versus eastern Black Sea tectonic evolution: pre-rift lithospheric controls on basin formation. *Tectonophysics*, 266, 139–54.

(1997). Thermomechanical modeling of Black Sea basin formation, subsidence, and sedimentation. *American Association of Petroleum Geologists Memoir*, 68, 19–38.

Sparks, R. S. J., Huppert, H. E., Kerr, R. C., McKenzie, D. P., and Tait, S. R. (1985). Postcumulus processes in layered intrusions. *Geological Magazine*, 122, 555–68.

Spencer, A. M., Home, P. C., and Wiik, V. (1986). Habitat of hydrocarbons in the Jurassic Ula trend, Central Graben, Norway. In *Habitat of Hydrocarbons on the Norwegian Continental Shelf*, ed. A. M. Spencer, C. J. Campbell, S. H. Hanslein, P. H. Nelson, E. Nysaether, and E. G. Ormaasen. Graham and Trotman, London, pp. 111–28.

Spencer, A. M., Leckie, G. G., and Chew, K. J. (1996). North Sea hydrocarbon plays and their resources. *First Break*, 14, 345–57.

Spiegelman, M. and McKenzie, D. (1987). Simple 2-D models for melt extraction at mid-ocean ridges and island arcs. *Earth and Planetary Science Letters*, 83, 137–52.

Spiegelman, M. and Reynolds, J. R. (1999). Combined dynamic and geochemical evidence for convergent melt flow beneath the East Pacific Rise. *Nature*, 402, 282–5.

Spiers, C. J. and Schutjens, P. M. T. (1990). Densification of crystalline aggregates by fluid phase diffusional creep. In *Deformation Processes in Minerals, Ceramics and Rocks*, ed. D. J. Barber and P. G. Meredith. *Mineralogical Society Series*, **1**, 334–52.

Srivastava, S. P., Sibuet, J.-C., Cande, S., Roest, W. R., and Reid, I. D. (2000). Magnetic evidence for slow spreading during the formation of the Newfoundland and Iberian margins. *Earth and Planetary Science Letters*, 182, 61–76.

Srivastava, S. P. and Tapscott, C. R. (1986). Plate kinematics of the North Atlantic. In *The Western North Atlantic Region: The Geology of North America*, ed. P. R. Vogt and B. E. Tucholke. *Geological Society of America, Boulder*, pp. 379–404.

Stacey, F. D. (1992). *Physics of the Earth*. Brookfield Press, Brisbane, p. 513.

Stach, E., Mackowsky, M. T., Teichmueller, M., Taylor, G. H., Chandra, D., and Teichmueller, R. (1982). *Stach's Textbook of Coal Petrology, Third Revised and Enlarged Edition*. E. Schweizerbart'sche Verlagsbuchhandlung, Stuttgart, p. 569.

Stahle, V., Frenzel, G., Kobe, B., Michard, A., Puchelt, H., and Schneider, W. (1990). Zircon syenite pegmatites in the Finero peridotite (Ivrea Zone): evidence for a syenite from mantle source. *Earth and Planetary Science Letters*, 101, 196–205.

Steckler, M. S. (1981a). *The thermal and mechanical evolution of Atlantic – type continental margins*. PhD thesis, New York, Columbia University, p. 261.

Steckler, M.S. (1981b). A thermomechanical model of the subsidence of continental margins. *Eos, Transactions, American Geophysical Union*, 62, 390.

Steckler, M. S. (1985). Uplift and extension of the Gulf of Suez: indications of induced mantle convection. *Nature*, 317, 135–9.

Steckler, M. S., Berthelot, F., Lyberis, N., and Le Pichon, X. (1988). Subsidence in the Gulf of Suez: implications for rifting and plate kinematics. *Tectonophysics*, 153, 249–70.

Steckler, M. S., Lavier, L., Feinstein, S., Kohn, B. P., and Eyal, M. (1994). Pattern of mantle thinning from subsidence and heat flow measurements in the Gulf of Suez: a case for along-strike flow from the Red Sea. *Eos, Transactions, American Geophysical Union*, 75, 589.

Steckler, M. S., Omar, G. I., Karner, G. D., and Kohn, B. P. (1993). Pattern of hydrothermal circulation within the Newark basin from fission-track analysis. *Geology*, 21, 735–8.

Steel, R., Gjelberg, J., Helland-Hansen, W., Kleinspehn, K., Nottvedt, A., and Rye-Larsen, M. (1985). Tertiary strike-slip basins and orogenic belt of Spitsbergen. *Special Publication – Society of Economic Paleontologists and Mineralogists*, 37, 339–59.

Stel, H. (1985). Crystal growth in cataclasites: diagnostic microstructures and implications. *Tectonophysics*, 78, 585–600.

Stephens, B. P. (2001). Basement controls on hydrocarbon systems, depositional pathways, and exploration plays beyond the Sigsbee Escarpment in the central Gulf of Mexico. In *Petroleum Systems of Deep-water Basins: Global and Gulf of Mexico Experience*, ed. R. H. Fillon, N. C. Rosen, P. Weimer, A. Lowrie, H. Pettingill, R. L. Phair, H. H. Roberts, and B. van Hoorn. *Program and Abstracts – Society of Economic Paleontologists*, Gulf Coast Section, Research Conference, **21**, 129–58.

Stephenson, L. P. (1977). Porosity dependence on temperature: limits on maximum possible effect. *American Association of Petroleum Geologists Bulletin*, 61, 407–15.

Stephenson, M. A. (1991). The North Brae Field, Block 16/07a, U.K. North Sea. In *United Kingdom Oil and Gas Fields, 25 Years Commemorative Volume*, ed. I. L. Abbotts. *Memoir of the Geological Society of London*, **14**, 43–8.

Stephenson, R. and Lambeck, K. (1985). Erosion-isostatic rebound models for uplift: an application to south-eastern Australia. *Geophysical Journal of the Royal Astronomical Society*, 82, 31–55.

Stevens, D. A. and Wallis, R. J. (1991). The Clyde Field, Block 30/17b, UK North Sea. In *United Kingdom Oil and Gas Fields, 25 Years Commemorative Volume*, ed. I. L. Abbotts. *Memoir of the Geological Society of London*, **14**, 339–45.

Stewart, D. M. and Faulkner, A. J. G. (1991). The Emerald Field, Blocks 2/10a, 2/15a, 3/11b, U.K. North Sea. In *United Kingdom Oil and Gas Fields, 25 Years Commemorative Volume*, ed. I. L. Abbotts. *Memoir of the Geological Society of London*, **14**, 111–16.

Stewart, I. J. (1993). Structural controls on the Late Jurassic age shelf system, Ula trend, Norwegian North Sea. In *Petroleum Geology of Northwest Europe. Proceedings of the 4th Conference*, ed. J. R. Parker. *Geological Society of London*, pp. 469–83.

Stewart, I. S. and Hancock, P. L. (1988). Normal fault zone evolution and fault scarp degradation in the Aegean region. *Basin Research*, 1, 139–53.

Stewart, J. H. (1983). Extensional tectonics in the Death Valley area, California: transport of the Panamint Range structural block 80 km northwestward. *Geology*, 11, 153–7.

Stewart, K., Turner, S., Kelley, S., Hawkesworth, C.J., Kirstein, L., and Mantovani, M. (1996). 3-D, $^{40}Ar/^{39}Ar$ geochronology in the Parana continental flood basalt province. *Earth and Planetary Science Letters*, 143, 95–109.

Stewart, S. A. (1996). Tertiary extensional fault systems on the western margin of the North Sea Basin. *Petroleum Geoscience*, 2, 167–76.

Stewart, S. A. and Clark, J. A. (1999). Impact of salt on the structure of the Central North Sea hydrocarbon fairways. In *Petroleum Geology of Northwest Europe*, ed. A. J. Fleet and S. A. R. Boldy. *Proceedings of the 5th Conference, The Geological Society of London*, pp. 179–200.

Stewart, S. A., Fraser, S. I., Cartwright, J. A., Clark, J. A., and Johnson, H. D. (1999). Controls on Upper Jurassic sediment distribution in the Durward-Dauntless area, Blocks 21/11, 21/16. In *Petroleum Geology of Northwest Europe*, ed. A. J. Fleet and S. A. R. Boldy. *Proceedings of the 5th Conference, The Geological Society of London*, pp. 879–96.

Stewart, S. A. and Reeds, A. (2003). Geomorphology of kilometer-scale extensional fault scarps: factors that impact seismic interpretation. *American Association of Petroleum Geologists Bulletin*, 87, 251–72.

Stierman, D. J. and Williams, A. E. (1984). Hydrologic and geochemical properties of the San Andreas Fault at the Stone Canyon well. *Pure and Applied Geophysics*, 122, 403–24.

St John, W. (2000). The role of transform faulting in formation of hydrocarbon traps in the Gulf of Guinea, West Africa. *Offshore West Africa Conference*, Abidjan, March 21–23, 2000.

Stockbridge, C. P. and Gray, D. I. (1991). The Fulmar Field, Blocks 30/16 and 30/11b, UK North Sea. In *Petroleum Geology of Northwest Europe*, ed. A. J. Fleet and S. A. R. Boldy. *Proceedings of the 5th Conference, The Geological Society of London*, 14, 309–16.

Stockli, D. F., Dumitru, T. A., McWilliams, M. O., and Farley, K. A. (2003). Cenzoic tectonic evolution of the White Mountains, California and Nevada. *Geological Society of America Bulletin*, 115, 788–816.

Stoffers, P., Glasby, G. P., Stuben, D., Renner, R. M., Pierre, T. G., Webb, J., and Cardile, C. M. (1993). Comparative mineralogy and geochemistry of hydrothermal iron-rich crusts from the Pitcairn, Teahitia-Migetia and Macdonald hot spot areas of the S. W. Pacific. *Marine Georesources and Geotechnology*, 11, 45–86.

Stoll, R. D. and Bryan, G. M. (1979). Physical properties of sediments containing gas hydrates. *Journal of Geophysical Research*, 84, 1629–34.

Storey, M., Kent, R. W., Saunders, A. D., Hergt, J., Salters, V. J. M., Whitechurch, H., Sevigny, J. H., Thirlwall, M. F., Leat, P., Ghose, N. C., and Gifford, M. (1992). Scientific Results, Ocean Driling Program, Leg 120: Lower Cretaceous volcanic rocks on continental margins and their relationship to the Kerguelen Plateau: College Station, Texas, Ocean Drilling Program, pp. 33–53.

Stow, D. A. V. and Atkin, B. P. (1987). Sediment facies and geochemistry of Upper Jurassic mudrocks in the Central North Sea area. In *Petroleum Geology of North West Europe*, ed. J. Brooks and K. W. Glennie. Graham and Trotman, London, pp. 797–808.

Strehlau, J. (2006). Earthquakes and nonvolcanic tremor in the lower continental crust: open questions. *EGU annual meeting*, Vienna, April 2–7, 2006, Scientific Programme, p. 562.

Strehlau, J. and Stange, S. (2006). Earthquakes in the lower crust under the Northern Alpine foreland basin: seismological detection of active metamorphism? *EGU annual meeting*, Vienna, April 2–7, 2006, Scientific Programme, p. 562.

Streit, J. E. (1997). Low frictional strength of upper crustal faults: a model. *Journal of Geophysical Research*, 102, 24619–26.

Strong, G. E. and Milodowski, A. E. (1987). Aspects of the diagenesis of the Sherwood sandstones of the Wessex Basin and their influence on reservoir characteristics. In *Diagenesis of Sedimentary Sequences*, ed. J. D. Marshall. *Geological Society Special Publications*, pp. 325–37.

Stuart, C. J., Nemčok, M., Vangelov, D., Higgins, E. R., Welker, C., and Meaux, D. (2011). Deposition/tectonic interaction in the active and subsequently extensionally dismembered thrustbelt; Eastern Balkans, Bulgaria. *American Association of Petroleum Geologists Bulletin*, 95, 649–73.

Stuiver, M., Reimer, P. J., and Braziunas, T. F. (1998). High-precision radiocarbon age calibration for terrestrial and marine samples. *Radiocarbon*, 40, 1127–51.

Stüwe, K. L., White, L., and Brown, R. (1994). The influence of eroding topography on steady-state isotherms. Applications to fission track analysis. *Earth and Planetary Science Letters*, 124, 63–74.

Suardy, A., Simbolon, B., and Taruno, P. J. (1987). Two decades of hydrocarbon exploration activities in Indonesia. *Transactions of the Circum-Pacific Energy and Mineral Resources Conference*, 4, 243–61.

Subrahmanyam, C., Thakur, N. K., Gangadhara Rao, T., Ramesh, K., Ramana, M. V., and Subrahmanyam, V. (1999). Tectonics of the Bay of Bengal: new insights from satellite-gravity and ship-borne geophysical data. *Earth and Planetary Science Letters*, 171, 237–51.

Suchy, V., Heijlen, W., Sykorova, I., Muchez, P., Dobes, P., Hladikova, J., Jackova, I., Safanda, J., and Zeman, A. (2000). Geochemical study of calcite veins in the Silurian and Devonian of the Barrandien Basin (Czech Republic): evidence for widespread post-Variscan fluid flow in the central part of the Bohemian Massif. *Sedimentary Geology*, 131, 201–19.

Suppe, J. (1985). *Principles of Structural Geology*. Prentice Hall, Englewood Cliffs, New Jersey.

Surdam, R. C., Crossey, L. J., Hagen, E. S., and Heasler, H. P. (1989). Organic-inorganic and sandstone diagenesis. *American Association of Petroleum Geologists Bulletin*, 73, 1–23.

Surdam, R. C., Jiao, Z. S., and MacGowan, D. B. (1993). Redox reactions involving hydrocarbon and mineral oxidants: a mechanism for significant porosity enhancement in sandstones. *American Association of Petroleum Geologists Bulletin*, 77, 1509–18.

Surlyk, F., Dons, T., Clausen, C. K., and Higham, J. (2002). Chapter 13. Upper Cretaceous. In *The Millennium Atlas: Petroleum Geology of the Central and Northern North Sea*, ed. D. Evans, C. Graham, A. Armour, and P. Bathurst. *Geological Society of London*, pp. 490–547.

Swanberg, C. A. (1972). Vertical distribution of heat generation in the Idaho batholith. *Journal of Geophysical Research*, 77, 2508–13.

Sweeney, J. J. and Burnham, A. K. (1990). Evaluation of a simple model of vitrinite reflectance based on chemical kinetics. *American Association of Petroleum Geologists Bulletin*, 74, 1559–70.

Sweeney, J. J., Burnham, A. K., and Braun, R. L. (1987). A model of hydrocarbon generation from type I kerogen: application to Uinta Basin, Utah. *American Association of Petroleum Geologists Bulletin*, 71, 967–85.

Sylvester, A. G. (1988). Strike-slip faults. *Geological Society of America Bulletin*, 100, 1666–703.

Syms, R. M., Savory, D. F., Ward, C. J., Ebdon, C. C., and Griffin, A. (1999). Integrating sequence stratigraphy in field development and reservoir management – the Telford Field. In *Petroleum Geology of Northwest Europe*, ed. A. J. Fleet and S. A. R. Boldy. *Proceedings of the 5th Conference, The Geological Society of London*, pp. 1075–87.

Szafián, P. (1999). *Gravity and Tectonics: A Case Study in the Pannonian Basin and the Surrounding Mountain Belt*. Vrije University, Amsterdam, p. 154.

Tagami, T. and O'Sullivan, P. B. (2005). Fundamentals of fission-track thermochronology. In *Low-Temperature Thermochronology: Techniques, Interpretations and Applications*, ed. P. W. Reinersand and T. A. Ehlers. *Reviews in Mineralogy and Geochemistry*, **58**, 19–48.

Tait, S. R., Huppert, H. E., and Sparks, R. S. J. (1984). The role of compositional convection in the formation of accumulate rocks. *Lithos*, 17, 139–46.

Tait, S. R. and Kerr, R. C. (1987). Experimental modelling of interstitial melt convection in cumulus piles. In *Origins of Igneous Layering*, ed. I. Parson. *NATO ASI Series C*, **196**, 569–87.

Talbot, C. J. and Ghebreab, W. (1997). Red Sea detachment and basement core complexes in Eritrea. *Geology*, 25, 655–8.

Talbot, M. R. (1988). The origins of lacustrine oil source rocks: evidence from the lakes of tropical Africa. In *Lacustrine Petroleum Source Rocks*, ed. A. J. Fleet, K. Kelts, and M. R. Talbot. *Geological Society Special Publication*, **40**, 29–43.

Talwani, M., Windisch, C. C., and Langseth, M. G. Jr. (1971). Reykjanes Ridge crest: a detailed geophysical study. *Journal of Geophysical Research*, 76, 473–517.

Tankard, A. J. and Welsink, H. J. (1987). Extensional tectonics and stratigraphy of Hibernia oil field, Grand Banks, Newfoundland. *American Association of Petroleum Geologists Bulletin*, 71, 1210–32.

(1989). Mesozoic extension and styles of basin formation in Atlantic Canada. *American Association of Petroleum Geologists Memoir*, 46, 175–95.

Tankard, A. J., Welsink, H. J., and Jenkins, W. A. M. (1989). Structural styles and stratigraphy of the Jeanne d'Arc Basin, Grand Banks of Newfoundland. *American Association of Petroleum Geologists Memoir*, 46, 265–82.

Tard, F., Masse, P., Walgenwitz, F., and Gruneisen, P. (1991). The volcanic passive margin in the vicinity of Aden, Yemen. *Bulletin des centres de recherche et d'exploration-production Elf-Aquitaine*, 15, 1–9.

Tari, G., Ashton, P., Coterill, K., Molnar, J., Sorgenfrei, M., Thompson, W. A. P., Valasek, D., and Fox, J. (2002). Are West Africa deepwater salt tectonics analogous to the Gulf of Mexico? *Oil and Gas Journal*, 4, 1–16.

Tari, G. and Horváth, F. (2006). Alpine evolution and hydrocarbon geology of the Pannonian Basin: an overview. In *The Carpathians and Their Foreland: Geology and Hydrocarbon Resources*, ed. J. Golonka and F. J. Pícha. *American Association of Petroleum Geologists Memoir*, **84**, 605–18.

Tari, G., Horváth, F., and Rumpler, J. (1992). Styles of extension in the Pannonian Basin. *Tectonophysics*, 208, 203–19.

Tari, G., Molnar, J., Ashton, P., and Hedley, R. (2000). Salt tectonics in the Atlantic margin of Morocco. *Leading Edge*, 19, 1074–76.

(2001). Exploration in syn-rift versus post-rift salt basins of West Africa: Are there significant differences? *Annual Meeting Expanded Abstracts – American Association of Petroleum Geologists*, p. 198.

Tavenas, F., Jean, P., Leblond, P., and Leroueil, S. (1983). The permeability of natural soft clays. Part 2: Permeability characteristics. *Canadian Geotechnical Journal*, 20, 645–60.

Tayebi, M. (1989). *Le segment hercynien du haut-Atlas occidental dans les Ait Chaib, Maroc. Stratigraphie, tectonique et role de la zone failee ouest altasique*. PhD thesis, University of d'Aix-Marseille III, p. 234.

Taylor, A. M. and Gawthorpe, R. L. (1993). Application of sequence stratigraphy and trace fossil analysis to reservoir description: examples from the Jurassic of the North Sea. In *Petroleum Geology of Northwest Europe*, ed. J. R. Parker. *Proceedings of the 4th Conference, The Geological Society of London*, pp. 317–35.

Taylor, B., Crook, K., and Sinton, J. (1994). Extensional transform zones and oblique spreading centers. *Journal of Geophysical Research*, 99, 19707–18.

Taylor, G. H., Teichmueller, M., Davis, A., Diessel, C. F. K., Littke, R., Robert, P., Glick, D. C., Smyth, M., Swaine, D. J., Vandebroucke, M., and Espitalie, J. (1998). *Organic Petrology*. Gebrueder Borntraeger, Berlin, p. 704.

Tegelaar, E. W., Matthezing, R. M., Jansen, J. B. H., Horsfield, B., and de Leeuw, J. W. (1989). Possible origin of n-alkanes in high-wax crude oils. *Nature*, 342, 529–31.

Tegelaar, E. W. and Noble, T. A. (1994). Kinetics of hydrocarbon generation as a function of the molecular structure of kerogen as revealed by pyrolysis-gas chromatography. *Organic Geochemistry*, 22, 543–74.

Teisserenc, P. and Villemin, J. (1990). Sedimentary basin of Gabon-geology and oil systems. In *Divergent/Passive Margin Basins*, ed. J. D. Edwards and P. A. Santogrossi. *American Association of Petroleum Geologists Memoir*, **48**, 117–99.

Ten Brink, U. S. and Ben-Avraham, Z. (1989). The anatomy of a pull-apart basin: seismic reflection observations of the Dead Sea basin. *Tectonics*, 8, 333–50.

Ten Brink, U. and Stern, T. (1992). Rift flank uplifts and hinterland basins: comparison of the Transatlantic Mountains with the Great Escarpment of southern Africa. *Journal of Geophysical Research*, 97, 569–85.

Ter Voorde, M. and Bertotti, G. (1994). Thermal effects of normal faulting during rifted basin formation: 1, A finite difference model. *Tectonophysics*, 240, 133–44.

TGS-NOPEC (2003). Problems in exploration – Part 1. Need to find faults? *American Association of Petroleum Geologists Explorer*, 11.

Thatcher, W. R., Foulger, G. R., Julian, B. R., Svarc, J. L., and Quilty, E. G. (1998). Present day deformation across the Basin and Range Province, Western United States. *Eos, Transactions, American Geophysical Union*, 79, 559.

Theil, K. (1995). *Wlasciwosci fizyko-mechaniczne i modele masywow skalnych polskich Karpat fliszowych*. Gdansk, Institut Budownictwa Wodnego, Polska Akademia Nauk.

Thomas, D. W. (1996a). *Tectonic evolution and kinematics of the Mesozoic rift system, UK Northern North Sea*. PhD thesis, University of London.

(1996b). Mesozoic regional tectonics and South Viking Graben formation: evidence for localised thin-skinned detachments during rift development and inversion. *Marine and Petroleum Geology*, 13, 149–77.

Thomas, M. M. and Clouse, J. A. (1995). Scaled physical model of secondary oil migration. *American Association of Petroleum Geologists Bulletin*, 79, 19–29.

Thompson, A. H., Katz, A. J., and Krohn, C. E. (1987). The microgeometry and transport properties of sedimentary rock. *Advances in Physics*, 36, 625–94.

Thompson, G. and Melson, W. G. (1972). The petrology of oceanic crust across fracture zones in the Atlantic Ocean: evidence of a new kind of sea-floor spreading. *Journal of Geology*, 80, 526–38.

Thompson, K. F. M. (1987). Fractionated aromatic petroleums and the generation of gas-condensates. *Organic Geochemistry*, 18, 573–90.

(1988). Gas-condensate migration and oil fractionation in deltaic systems. *Marine and Petroleum Geology*, 5, 237–46.

Thompson, R. N. (1974). Primary basalts and magma genesis; I, Skye, North-West Scotland. *Contributions to Mineralogy and Petrology*, 45, 317–41.

Thompson, R. N. and Morrison, M. A. (1988). Asthenospheric and lower-lithospheric mantle contributions to continental extensional magmatism: an example from the British Tertiary Province. *Chemical Geology*, 68, 1–15.

Thompson, S., Cooper, B. S., and Barnard, P. C. (1994). Some examples and possible explanations for oil generation from coals and coaly sequences. In *Coal and Coal-bearing Strata as Oil-prone Source Rocks*, ed. A. C. Scott and A. J. Fleet. *Geological Society of London Special Publications*, **77**, 119–37.

Thompson, V. (2009). *Potential-field and 2D seismic analysis of a volcanic rifted margin: implications for crustal architecture and petroleum maturation off the west coast of South Africa*. Master's thesis, University of Utah, Salt Lake City, p. 142.

Thurber, C., Roecker, S., Ellsworth, W., Chen, Y., and Lutter, W. (1997). Two-dimensional seismic image of the San Andreas Fault in the Northern Gabilan Range, central California: Evidence from fluids in the fault zone. *Geophysical Research Letters*, 24, 1591–4.

Thurber, C., Roecker, S., Roberts, K., Gold, M., Powell, L., and Rittger, K. (2003). Earthquake locations and three-dimensional fault zone structure along the creeping section of the San Andreas fault near Parkfield, CA: preparing for SAFOD. *Geophysical Research Letters*, 30.

Thurber, C., Roecker, S., Zhang, H., Baher, S., and Ellsworth, W. L. (2004). Fine-scale structure of the San Andreas fault zone and location of the SAFOD target earthquakes. *Geophysical Research Letters*, 31.

Tiercelin, J. J. and Faure, H. (1978). Rates of sedimentation and vertical subsidence in neorifts and paleorifts. In *Tectonics and Geophysics of Continental Rifts*, ed. I. B. Ramberg and E. R. Neumann. Reidel Publishing Company, Dordrecht, pp. 41–7.

Tigre, C. A., Schaller, H., Del Luchese, C. Jr., and Possato, S. (1983). Pampo, Linguado, and Badejo fields: their discoveries, appraisals, and early production systems. *Proceedings of the Offshore Technology Conference*, 15, 407–14.

Tillman, L. E. (1989). Sedimentary facies and reservoir characteristics of the Nugget Sandstone (Jurassic), Painter Reservoir Field, Uinta County, Wyoming. In *Petrogenesis and Petrophysics of Selected Sandstone Reservoirs of the Rocky Mountain Region*, ed. E. B. Coalson, S. S. Kaplan, C. W. Keighin, C. A. Oglesby, and J. W. Robinson. *Rocky Mountain Association of Geologists*, pp. 97–108.

Tissot, B. P., Durand, B., Espitalié, J., and Combaz, A. (1974). Influence of nature and diagenesis of organic matter in formation of petroleum. *American Association of Petroleum Geologists Bulletin*, 58, 499–506.

Tissot, B. P. and Espitalié, J. (1975). The thermal evolution of organic matter in sediments: application of a mathematical model simulation. *Revue de l'Institut Français du Pétrol*, 30, 743–77.

Tissot, B. P., Pelet, R., and Ungerer, P. (1987). Thermal history of sedimentary basins, maturation indices, and kinetics of oil and gas generation. *American Association of Petroleum Geologists Bulletin*, 71, 1445–66.

Tissot, B. P. and Welte, D. H. (1984). *Petroleum Formation and Occurrence, a New Approach to Oil and Gas Exploration*. Springer-Verlag, New York.

Toksoz, M. N., Cheng, C. H., and Timur, A. (1976). Velocities of seismic waves in porous rocks. *Geophysics*, 41, 621–43.

Tokunaga, T., Mogi, K., Matsubara, O., Tosaka, H., and Kojima, K. (2000). Buoyancy and interfacial force effects on two-phase displacement patterns: an experimental study. *American Association of Petroleum Geologists Bulletin*, 84, 65–74.

Tomek, Č., Dvořáková, L., and Ibrmajer, I. (1987). Crustal profiles of active continental collisional belt: Czechoslovak deep seismic profiling in the West Carpathians. *Geophysical Journal of the Royal Astronomical Society*, 89, 383–8.

Tomek, C., Ibrmajer, I., and Korab, T. (1989). Kôrové štruktúry Západních Karpát na hlubinném reflexním seizmickém profilu. Crustal structure of the Western Carpathians as determined by deep seismic sounding. *Mineralia Slovaca*, 21, 3–26.

Tomic, J., Behar, F., Vandenbroucke, M., and Tang, Y. (1995). Artificial maturation of Monterey kerogen (Type II-S) in a closed system and comparison with Type II kerogen: implications on the fate of sulfur. *Organic Geochemistry*, 23, 647–60.

Torgensen, T. and Clarke, W. B. (1992). Geochemical constraints on formation fluid ages, hydrothermal heat flux, and crustal mass transport mechanisms at Cajon Pass. *Journal of Geophysical Research*, 97, 5031–8.

Toulokian, Y. S., Judd, W. R., and Roy, R. F. (1981). *Physical properties of rocks and minerals*. McGraw-Hill, New York, p. 548.

Trewin, N. H. and Bramwell, M. G. (1991). The Auk field, Block 30/16 UK North Sea. In *United Kingdom Oil and Gas Fields, 25 Years Commemorative Volume*, ed. I. L. Abbotts. *Memoir of the Geological Society of London*, **14**, 227–36.

Trinidade, L. A. F., Dias, J. L., and Mello, M. R. (1995). Sedimentological and geochemical characterization of the Lagoa Feia Formation, rift phase of the Campos Basin. In *Petroleum Source Rocks*, ed. B. J. Katz. Springer-Verlag, Berlin, pp. 149–65.

Tron, V. and Brun, J.-P. (1991). Experiments on oblique rifting in brittle-ductile systems. *Tectonophysics*, 188, 71–84.

Trudgill, B. D. (2002). Structural controls on drainage development in the Canyonlands grabens of Southeast Utah. *American Association of Petroleum Geologists Bulletin*, 86, 1095–112.

Trudgill, B. and Cartwright, J. (1994). Relay-ramp forms and normal-fault linkages, Canyonlands National Park, Utah. *Geological Society of America Bulletin*, 106, 1143–211.

Trumbull, R. B., Sobolev, S. V., and Bauer, K. (2002). Petrophysical modeling of high seismic velocity crust at the Namibian volcanic margin. In *Volcanic Rift Margins*, ed. M. A. Menzies, S. L. Klemperer, C, Ebinger, and J. Baker. Geological Society of America Special Paper, 362, pp. 221–30.

Tsenn, M. C. and Carter, N. L. (1987). Upper limits of power law creep of rocks. *Tectonophysics*, 136, 1–26.

Tucholke, B. E., Lin, J., and Kleinrock, M. C. (1998). Megamullions and mullion structure defining oceanic metamorphic core complexes on the Mid-Atlantic Ridge. *Journal of Geophysical Research*, 103, 9857–66.

Tucker, G. E. and Slingerland, R. (1996). Predicting sediment flux from fold and thrust belts. *Basin Research*, 8, 329–49.

Turcotte, D. L. and Emerman, S. H. (1983). Mechanisms of active and passive rifting. *Tectonophysics*, 94, 39–50.

Turcotte, D. L. and Schubert, G. (1982). *Geodynamics: Applications of Continuum Physics to Geological Problems*. John Wiley and Sons, New York, p. 450.

(2002). *Geodynamics, Second edition*. Cambridge University Press, Cambridge, p. 456.

Turner, C. C. and Allen, P. J. (1991). The Central Brae Field, Block 16/07a, U.K. North Sea. In *United Kingdom Oil and Gas Fields, 25 Years Commemorative Volume*, ed. I. L. Abbotts. *Memoir of the Geological Society of London*, **14**, 49–54.

Turner, C. C., Cohen, J. M., Connell, E. R., and Cooper, D. M. (1987). A depositional model for the South Brae oilfield. *Proceedings of the Third Canadian/American Conference on Hydrogeology*, June 22–6, 1986, Bannf, Alberta, pp. 853–64.

Turner, C. C. and Connell, E. R. (1991). Stratigraphic relationships between Upper Jurassic submarine fan sequences in the Braere, U.K. North Sea: the implications for reservoir distribution. *23rd Annual Offshore Technology Conference*, Dallas, pp. 83–91.

Turner, F. J. and Weiss, L. E. (1963). *Structural Analysis of Metamorphic Tectonites*. New York, McGraw-Hill, p. 545.

Turner, P. J. (1993). Clyde: reappraisal of a producing field. In *Petroleum Geology of Northwest Europe*, ed. J. R. Parker. *Proceedings of the 4th Conference, Geological Society of London*, pp. 1503–12.

Turner, S., Regelous, M., Kelley, S., Hawkesworth, C., and Mantovani, M. (1994). Magmatism and continental break-up in the South Atlantic: high precision ^{40}Ar-^{39}Ar geochronology. *Earth and Planetary Science Letters*, 121, 333–48.

Tuttle, M. L. W., Charpentier, R. R., and Brownfield, M. E. (1999). The Niger Delta petroleum system: Niger Delta province, Nigeria, Cameroon, and Equatorial Guinea, Africa. *USGS Open-File Report*, 99-50-H, p. 65.

Tye, R. S. (1991). Fluvial-sandstone reservoirs of the Travis Peak Formation, East Texas Basin. *Concepts in Sedimentology and Paleontology*, 3, 172–88.

(2004a). Geomorphology; an approach to determining subsurface reservoir dimensions. *American Association of Petroleum Geologists Bulletin*, 88, 1123–47.

(2004b). Reservoir description and unique horizontal-well designs boost. Primary and EOR production from the fluvio-deltaic Prudhoe Bay field, Alaska. *AAPG Distinguished Lecture Funded by the AAPG Foundation in honor of Roy M. Huffington.*

Tye, R. S. and Coleman, J. M. (1989a). Depositional processes and stratigraphy of fluvially dominated lacustrine deltas: Mississippi delta plain. *Journal of Sedimentary Petrology*, 59, 973–96.

(1989b). Evolution of Atchafalaya lacustrine deltas, south-central Louisiana. *Sedimentary Geology*, 65, 95–112.

Uenzelmann Neben, G., Siess, V., and Bleil, U. (1997). A seismic reconnaissance survey of the northern Congo fan. *Marine Geology*, 140, 283–306.

Ukstins, I., Renne, P., Wolfenden, E., Baker, J., Ayalew, D., and Menzies, M. A. (2002). Matching conjugate volcanic rifted margins: $^{40}Ar/^{39}Ar$ chronostratigraphy of pre- and syn-rift bimodal flood volcanism in Ethiopia and Yemen. *Earth and Planetary Science Letters*, 198, 289–306.

Ulmishek, G. F. and Klemme, H. D. (1990). Depositional controls, distribution, and effectiveness of world's petroleum source rocks. *U. S. Geological Survey Bulletin*, 1931.

Underhill, J. R. (1991a). Implications of Mesozoic – recent basin development in the Inner Moray Firth, UK. *Marine and Petroleum Geology*, 8, 359–69.

(1991b). Controls on Late Jurassic seismic sequences, Inner Moray Firth: a critical test of Exxon's original sea-level chart. *Basin Research*, 3, 79–98.

(1994). Discussion on palaeoecology and sedimentology across a Jurassic fault scarp, NE Scotland. *Journal of the Geological Society of London*, 151, 729–31.

Underhill, J. R. and Partington, M. A. (1993). Jurassic thermal doming and deflation in the North Sea: implication of the sequence stratigraphic evidence. In *Petroleum Geology of Northwest Europe*, ed. J. R. Parker. *Proceedings of the 4th Conference, The Geological Society of London*, pp. 337–46.

Underhill, J. R., Sawyer, M. J., Hodgson, P., Shallcross, M. D., and Gawthorpe, R. L. (1997). Implications of fault scarp degradation for Brent Group prospectivity, Ninian Field, northern North Sea. *American Association of Petroleum Geologists Bulletin*, 81, 999–1022.

Underwood, M. B. and Howell, D. G. (1987). Thermal maturity of the Cambria Slab, an inferred trench-slope basin in Central California. *Geology*, 15, 216–19.

Ungerer, P. (1990). State of the art of research in kinetic modeling of oil formation and expulsion. *Organic Geochemistry*, 16, 1–25.

Ungerer, R., Behar, F., and Discamp, D. (1983). Tentative calculation of the overall volume expansion of organic matter during hydrocarbon genesis from geochemistry data. Implications for primary migration. In *Advances in Organic Geochemistry 1981*, ed. M. Bjoroy, P. Albrecht, C. Cornford, K. de Groot, G. Elington, and E. Galimov. John Wiley, Chichester, pp. 129–35.

Ungerer, P., Burrus, J., Doligez, B., Chenet, P. Y., and Bessis, F. (1990). Basin evaluation by integrated two-dimensional modeling of heat transfer, fluid flow, hydrocarbon generation, and migration. *American Association of Petroleum Geologists Bulletin*, 74, 309–35.

Ungerer, P. and Pelet, R. (1987). Extrapolation of the kinetics of oil and gas formation from laboratory to sedimentary basin. *Nature*, 327, 52–4.

United States Geological Survey (USGS) (1996). Global 30-Arc-Second Elevation Data Set. United States Geological Survey, Sioux Falls, South Dakota.

United States Geological Survey (USGS) (2006). Quaternary Fault and Fold Database of the United States. http://earthquake.usgs.gov/regional/qfaults.

Unsworth, M. and Bedrosian, P. A. (2004). Electrical resistivity structure at the SAFOD site from magnetotelluric exploration. *Geophysical Research Letters*, 31.

Unsworth, M., Bedrosian, P. A., Eisel, M., Egbert, G., and Siripunvaraporn, W. (2000). Along strike variations in the electrical structure of the San Andreas Fault at Parkfield, California. *Geophysical Research Letters*, 27, 3021–4.

Unsworth, M., Egbert, G., and Booker, J. (1999). High-resolution electromagnetic imaging of the San Andreas fault in Central California. *Journal of Geophysical Research*, 104, 1131–50.

Unsworth, M., Malin, P. E., Egert, G. D., and Booker, J. R. (1997). Internal structure of the San Andreas fault at Parkfield, California. *Geology*, 25, 359–62.

Unternehr, P., Curie, D., Olivet, J. L., Goslin, J., and Beuzart, P. (1988). South Atlantic fits and intraplate boundaries in Africa and South America. *Tectonophysics*, 155, 169–79.

Urgeles, R., Canals, M., Baraza, J., and Alonso, B. (1998). Seismostratigraphy of the western flanks of El Hierro and La Palma (Canary Islands): record of Canary Island volcanism. *Marine Geology*, 146, 225–41.

Urgeles, R., Canals, M., Baraza, J., Alonso, B., and Masson, D. (1997). The most recent megalandslides of the Canary Islands: El Golfo debris avalanche and Canary debris flow, West El Hierro Island. *Journal of Geophysical Research*, 102, 20305–23.

Urgeles, R., Canals, M., Roberts, J., and the SNV "Las Palmas" Shipboard party (2000). Fluid flow from pore pressure measurements off La Palma, Canary Islands. *Journal of Volcanology and Geothermal Research*, 101, 253–71.

Urgeles et al. (1999). Fluid flow from pore pressure measurements off La Palma, Canary Islands. *Journal of Volcanology and Geothermal Research*, 94(1–4), 305–21.

Vagnes, E., Boavida, J., Jeronimo, P., de Brito, M., and Peliganga, J. M. (2005). Crustal architecture of West African rift basins in the deep water province. In *Petroleum Systems of Divergent Continental Margin Basins*, ed. P. Post. *25th Gulf Coast Section, Bob F. Perkins Research Conference*, 4–7 December 2005, CD-ROM.

Vail, P. R. (1987). Seismic stratigraphy interpretation using sequence stratigraphy: Part 1, Seismic stratigraphy interpretation procedure. *American Association of Petroleum Geologists Studies in Geology*, 27, 1–10.

Vail, P. R., Mitchum, R. M. Jr., and Thompson, S. (1977). Seismic stratigraphy and global changes of sea level: Part 3, Relative changes of sea level from coastal onlap. Seismic stratigraphy: applications to hydrocarbon exploration. *American Association of Petroleum Geologists*, pp. 63–81.

Vail, P. R. and Todd, R. G. (1981). Cause of northern North Sea Jurassic unconformities. *American Association of Petroleum Geologists Bulletin*, 65, 1003.

Vail, P. R., Todd, R. G., and Sangree, J. B. (1977). Seismic stratigraphy and global changes of sea level: Part 5. Chronostratigraphic significance of seismic reflections: Section 2. *Application of Seismic Reflection Configuration to Stratigraphic Interpretation Memoir*, 26, 99–116.

Van Avendonk, H. J. A., Holbrook, W. S., Nunes, G. T., Shillington, D. J., Tucholke, B. E., Louden, K. E., Larsen, H. C., and Hopper, J. R. (2006). Seismic velocity structure of the rifted margin of the eastern Grand Banks of Newfoundland, Canada. *Journal of Geophysical Research*, 111, B11404.

Van Avendonk, H. J. A., Lavier, L. L., Shillington, D. J., and Manatschal, G. (2009). Extension of continental crust at the margin of the eastern Grand Banks, Newfoundland. *Tectonophysics*, 468, 131–48.

Van Balen, R. T., Van der Beek, P. A., and Cloetingh, S. A. P. L. (1995). The effect of rift shoulder erosion on stratal patterns at passive margins: implications for sequence stratigraphy. *Earth and Planetary Science Letters*, 134, 527–44.

Van der Beek, P., Andriessen, P., and Cloetingh, S. (1995). Morphotectonic evolution of rifted continental margins: inferences from a coupled tectonic-surface processes model and fission track thermochronology. *Tectonics*, 14, 406–21.

Van der Beek, P., Cloetingh, S., and Andriessen, P. (1994). Mechanisms of extensional basin formation and vertical motions at rift flanks: constraints from tectonic modelling and fission-track thermochronology. *Earth and Planetary Science Letters*, 121, 417–33.

Van den Beukel, P. J. (1990). Thermal and mechanical modelling of convergent plate margins. *Geologica Ultraiectina*, 62, 126.

Van der Helm, A. A., Gray, D. I., Cook, M. A., and Schulte, A. M. (1990). Fulmar, the development of a large North Sea field. In *North Sea Oil and Gas Reservoirs, Vol. 2*, ed. A. T. Buller et al. *Proceedings of the North Sea Oil and Gas Reservoirs Conference*, Graham and Trotman, London, pp. 25–45.

Van der Kamp, G. and Bachu, S. (1989). Use of dimensional analysis in the study of thermal effects of various hydrogeological regimes. In *Hydrogeological Regimes and Their Subsurface Thermal Effects*, ed. A. E. Beck, G. Garven, and L. Stegena. *Geophysical Monograph*, **47**, 23–8.

Van Heijst, M. W. I. M. and Postma, G. (2001). Fluvial response to sea-level changes: a quantitative analogue, experimental approach. *Basin Research*, 13, 269–92.

Van Heijst, M. W. I. M., Postma, G., van Kesteren, W. P., and de Jongh, R. G. (2002). Control of syndepositional faulting on systems tract evolution across growth-faulted shelf margins: an analog experimental model of the Miocene Imo River Field, Nigeria. *American Association of Petroleum Geologists Bulletin*, 86, 1335–66.

Van Schmus, W. R. (1984). Radioactivity properties of minerals and rocks. In *Handbook of Physical Properties of Rocks*, ed. R. S. Carmichael. CRC Press, Boca Raton, pp. 281–93.

Van Wagoner, J. C., Posamentier, H. W., Mitchum, R. M., Vail, P. R., Sarg, J. F., Loutit, T. S., and Hardenbol, J. (1988). An overview of the fundamentals of sequence stratigraphy and key definitions. In *Sea-Level Changes: An Integrated Approach*, ed. C. K. Wilgus, B. S. Hastings, C. G. C. St. Kendall, H. W. Posamentier, C. A. Ross, and J. C. Van Wagoner. *Special Publication – Society of Economic Paleontologists and Mineralogists*, pp. 39–45.

Van Wijk, J. W. and Cloetingh, S. A. P. L. (2002). Basin migration caused by slow lithospheric extension. *Earth and Planetary Science Letters*, 198, 275–88.

Van Wijk, J. W., Huismans, R. S., ter Voorde, M., and Cloetingh, S. A. P. L. (2001). Melt generation at volcanic continental margins: no need for a mantle plume? *Geophysical Research Letters*, 28, 3995–8.

Vandenbroucke, M. (1993). Migration of hydrocarbons. In *Applied Petroleum Geochemistry*, ed. M. L. Bordenave. Technip, Paris, pp. 125–48.

Vandenbroucke, M., Behar, F., and Rudkiewicz, J. L. (1999). Kinetic modeling of petroleum formation and cracking: implication from the high pressure/high temperature Elgin Field (UK, North Sea). *Organic Geochemistry*, 30, 1105–25.

Vandervoort, D. S. (1997). Stratigraphic response to saline lake-level fluctuations and the origin of cyclic nonmarine evaporite deposits: the Pleistocene Blanca Lila Formation, northwest Argentina. *Geological Society of America Bulletin*, 109, 210–24.

Vasseur, G. and Burrus, J. (1990). Contraintes hydrodynamiques et thermiques sur la genese des gisements stratiformes a plomb-zinc. In *Mobilite et concentration des metaux de base dans les couvertures sedimentaires, manifestations, mecanismes, prospection; actes du colloque international*, ed. H. Pelissonnier and J. F. Sureau. *Documents – B. R. G. M.*, **183**, 305–54.

Vasseur, G. and Demongodin, L. (1995). Convective and conductive heat transfer in sedimentary basins. *Basin Research*, 7, 67–79.

Vauchez, A., Dineur, F., and Rudnick, R. L. (2005). Microstructure, texture and seismic anisotropy of the lithospheric mantle above a mantle plume: insights from the Labait volcano xenoliths (Tanzania). *Earth and Planetary Science Letters*, 232, 295–314.

Vavra, C. L., Kaldi, J. G., and Sneider, R. M. (1992). Geological applications of capillary pressure: a review. *American Association of Petroleum Geologists Bulletin*, 76, 840–50.

Veenstra, E. (1985). Rift and drift in the Dampier sub-basin, a seismic and structural interpretation. *The APEA Journal*, 25, 177–89.

Veldkamp, J. J., Gailllard, M. G., Jonkers, H. A., and Levell, B. K. (1996). A Kimmeridgian time-slice through the Humber Group of the central North Sea: a test of sequence stratigraphic methods. In *Geology of the Humber Group: Central Graben and Moray Firth, UKCS*, ed. A. Hurst, H. D. Johnson, S. D. Burley, A. C. Canham, and D. S. Mackertich. *Special Publication of the Geological Society of London*, **114**, 1–28

Vendeville, B., Cobbold, P. R., Davy, P., Brun, J. P., and Choukroune, P. (1987). Physical models of extensional tectonics at various scales. *Geological Society of London Special Publications*, **28**, 95–107.

Vendeville, B. C. and Jackson, M. P. A. (1992). The fall of diapirs during thinskinned extension. *Marine and Petroleum Geology*, 9, 354–71.

Venkatakrishnan, R. and Lutz, R. (1988). A kinematic model for the evolution of Richmond Triassic basin, Virginia. *Developments in Geotectonics*, 22, 445–62.

Verhoef, J. and Srivastava, S. P. (1989). Correlation of sedimentary basins across the North Atlantic as obtained from gravity and magnetic data, and its relation to the early evolution of the North Atlantic. *American Association of Petroleum Geologists Memoir*, 46, 131–47.

Versfelt, J. A. and Rosendahl, B. R. (1989). Relationships between pre-rift structure and rift architecture in Lakes Tanganyika and Malawi, East Africa. *Nature*, 337, 354–7.

Villemin, T., Alvarez, F., and Angelier, J. (1986). The Rhinegraben: extension, subsidence and shoulder uplift. *Tectonophysics*, 128(1–2), 47–59.

Villemin, T., Angelier, J., and Sunwoo, C. (1995). Fractal distribution of fault length and offsets: implications of brittle deformation evaluation: the Lorraine coal basin. In *Fractals in the Earth Sciences*, ed. C. Barton and P. R. La Pointe. Plenum Press, New York, pp. 205–26.

Vincent, M. W. and Ehlig, P. L. (1988). Laumontite mineralization in rocks exposed north of San Andreas Fault at Cajon Pass, southern California. *Geophysical Research Letters*, 15, 977–80.

Volkman, J. K. (1988). Biological marker compounds as indicators of the depositional environments of petroleum source rocks. In *Lacustrine Petroleum Source Rocks*, ed. A. J. Kelts and M. R. Talbot. *Geological Society of London Special Publications*, **40**, 103–22.

Volkman, J. K., Alexander, R., Kagi, P. I., and Woodhouse, G. W. (1983). Demethylated hopanes in crude oils and their applications in petroleum geochemistry. *Geochimica et Cosmochimica Acta*, 47, 785–94.

Vollset, J. and Dore, A. G. (1984). A revised Triassic and Jurassic lithostratigraphic nomenclature for the Norwegian North Sea. *Norwegian Petroleum Directorate Bulletin*, 3.

Von der Dick, H., Meloche, J. D., Dwyer, J., and Gunther, P. (1989). Source-rock geochemistry and hydrocarbon generation in the Jeanne d'Arc Basin, Grand Banks, offshore eastern Canada. *Journal of Petroleum Geology*, 12, 51–68.

Von Rad, U. and Thurow, J. (1992). Bentonitic clay as indicators of early Neocomian post-breakup volcanism off Northwest Australia: proceedings of the Ocean Drilling Program, Exmouth Plateau, covering Leg 122 of the cruises of the drilling vessel JOIDES Resolution, Singapore, Republic of Singapore, sites 759–764: College Station, Texas, Ocean Drilling Program, pp. 213–32.

Vrolijk, P., Donelick, R. A., Queng, J., and Cloos, M. (1992). Testing models of fission track annealing in apatitie in a simple thermal setting: Site 800. Leg 129. In *Proceedings of the Ocean Drilling Program Scientific Results*, ed. R. I. Larson, Y. Lancelot, and others. *Ocean Drilling Program*, College Station, **129**, 169–76.

Vutukuri, V. S., Lama, R. D., and Saluja, S. S. (1974). *Handbook on Mechanical Properties of Rocks: Testing Techniques and Results, 1*. Trans Tech Publications, Bay Village.

Waddams, P. and Clark, N. M. (1991). The Petronella Field, Block 14/20b, UK North Sea. In *United Kingdom Oil and Gas Fields, 25 Years Commemorative Volume*, ed. I. L. Abbotts. *Memoir of the Geological Society of London*, **14**, 353–60.

Wade, D. N., Lawrence, D. A., and Riley, L. A. (1995). The Rowan Sandstone Member (Upper Jurassic to Lower Cretaceous): stratigraphic definition and implications for North Sea exploration. *Journal of Petroleum Geology*, 18, 223–30.

Wade, J. A. and MacLean, B. C. (1990). The geology of the southeastern margin of Canada. In *Geology of the Continental Margin of Eastern Canada: Geology of North America*, ed. M. J. Keen and G. L. Williams. *Geological Survey of Canada*, **2**, 190–238.

Wade, S. C. and Reiter, M. (1994). Hydrothermal estimation of vertical ground-water flow, Canutillo, Texas. *Groundwater*, 325, 735–42.

Wakefield, L. L., Droste, H., Giles, M. R., and Janssen, R. (1993). Late Jurassic plays along the western margin of the Central Graben. In *Petroleum Geology of Northwest Europe*, ed. J. R. Parker. *Proceedings of the 4th Conference, Geological Society of London*, pp. 1167–78.

Walker, I. M., Berry, K. A., Bruce, J. R., Bystol, L., and Snow, J. H. (1997). Structural modelling of regional depth profiles in the Vøring Basin: implications for the structural and stratigraphic development of the Norwegian passive margin. *Journal of the Geological Society of London*, 154, 537–44.

Walsh, J. B. (1981). Effect of pore pressure and confining pressure on fracture permeability. *International Journal of Rock Mechanics and Mining Sciences*, 18, 429–35.

Walsh, J. J. and Watterson, J. (1988). Analysis of the relationship between displacements and dimensions of faults. *Journal of Structural Geology*, 10, 239–47.

Walter, M. J. (1999). Melting residues of fertile peridotite and the origin of cratonic lithosphere. In *Mantle Petrology: Field Observations and High Pressure Experimentation: A Tribute to Francis R. (Joe) Boyd*, ed. Y. Fei, C. M. Bertka and B. O. Mysen. *Geochemical Society Special Publication*, **6**, 225–39.

Walters, M. and Combs, J. (1989). Heat flow regime in the Geysers-Clear Lake region of northern California. *Geothermal Research Council Transactions*, 13, 491–502.

Walters, R. F. (1946). Burier pre-Cambrian hills ion northeastern Barton County, central Kansas. *American Association of Petroleum Geologists Bulletin*, 30, 660–710.

(1953). Oil production from fractured pre-Cambrian basement rocks in central Kansas. *American Association of Petroleum Geologists Bulletin*, 37, 300–13.

Walters, R. F. and Price, A. S. (1948). Kraft-Prusa oil field, Barton County, Kansas. *Oil and Gas Journal*, 45, 249–80.

Walton, N. R. G. (1981). A detailed hydrogeochemical study of groundwaters from the Triassic sandstone aquifer of south-west England. *Natural Environment Research Council, Institute of Geological Sciences*, p. 43.

Wang, B. and Qian, K. (1992). *Geology Research and Exploration Practice in the Shengli Petroleum Province (in Chinese)*. Petroleum University Press, Dongying, p. 239.

Wang, C.-Y. (1984). On the constitution of the San Andreas fault zone in Southern California. *Journal of Geophysical Research*, 89, 5858.

Wang, C.-Y., Rui, F., Zengsheng, Y., and Xingjue, S. (1986). Gravity anomaly and density structure of the San Andreas fault zone. *Pure and Applied Geophysics*, 124, 127–40.

Wannamaker, P. E., Doerner, W. M., Stodt, J. A., and Johnston, J. M. (1997). Subdued state of tectonism of the Great Basin interior relative to its eastern margin based on deep resistivity structure. *Earth and Planetary Science Letters*, 150, 41–53.

Wannamaker, P. E. and Hohman, G. W. (1991). Electromagnetic induction studies. *U.S. National Report to the IUGG, invited paper, Review of Geophysics*, pp. 405–15.

Wannesson, J., Icart, J.-C., and Ravat, J. (1991). Structure and evolution of two adjoining segments of the West African margin from deep seismic profiling. In *Continental Lithosphere: Deep Seismic Reflections*, ed. R. Meissner, L. Brown, H.-J. Duerbaum, W. Franke, K. Fuchs, and F. Seifert. *American Geophysical Union Geodynamics Series*, 22, 275–89.

Waples, D. W. (1980). Time and temperature in petroleum formation: application of Lopatin's method to petroleum exploration. *American Association of Petroleum Geologists Bulletin*, 64, 916–26.

Waples, D. W. (1985). *Geochemistry in Petroleum Exploration*. Reidel Publishing Company, Boston, p. 232.

(1992). Future developments in basin modeling. *29th International Geological Congress, Abstracts – Congres Geologique Internationale, Resumes*, **29**, 812.

(1994a). Maturity modeling: thermal indicators, hydrocarbon generation, and oil cracking. In *The Petroleum System – From Source to Trap*, ed. L. B. Magoon and W. G. Dow. *American Association of Petroleum Geologists Memoir*, **60**, 285–306.

(1994b). Modeling of sedimentary basins and petroleum systems. In *The Petroleum System – From Source to Trap*, ed. L. B. Magoon and W. G. Dow. *American Association of Petroleum Geologists Memoir*, **60**, 307–22.

Waples, D. W., Kamata, H., and Suizu, M. (1992a). The art of maturity modeling, Part 1: Finding a satisfactory geologic model. *American Association of Petroleum Geologists Bulletin*, 76, 31–46.

(1992b). The art of maturity modeling, Part 2: Alternative models and sensitivity analysis. *American Association of Petroleum Geologists Bulletin*, 76, 47–66.

Waples, D. W. and Machihara, T. (1991). Biomarkers for geologists. *American Association of Petroleum Geologists Methods in Exploration Series*, 9.

Warnock, A. C., Zeitler, P. K., Wolf, R. A., and Bergman, S. C. (1997). An evaluation of low-temperature apatite U-Th/He thermochronometry. *Geochimica et Cosmochimica Acta*, 61, 5371–7.

Warrington, G., Cope, J. C. W., and Ivimaycook, H. C. (1994). St. Audries Bay, Somerset, England: a candidate global stratotype section and point for the base of the Jurassic System. *Geological Magazine*, pp. 191–200.

Waschbusch, P. and Beaumont, C. (1996). Effect of slab retreat on crustal deformation in simple regions of plate convergence. *Journal of Geophysical Research*, 101, 28133–48.

Waterston, C. D. (1950). Note on the sandstone injections of Eathie Haven, Cromarty. *Geological Magazine*, 87, 133–9.

Watkeys, M. K. (2002). Development of the Lebombo rifted volcanic margin of Southeast Africa. In *Volcanic Rift Margins*, ed. M. A. Menzies, S. L. Klemperer, C. Ebinger, and J. Baker. Geological Society of America Special Paper, 362, pp. 27–46.

Watkins, E. (2008). Tupi area presalt oil discoveries likely contiguous. *Oil and Gas Journal*, 106, 32–4.

Watkins, J. S., Ladd, J. W., Buffler, R. T., Shaub, F. J., Houston, M. H., and Worzel, J. L. (1978). Occurrence and evolution of salt in deep Gulf of Mexico. In *Framework, Facies, and Oil-Trapping Characteristics of the Upper Continental Margin*, ed. A. H. Bouma, G. T. Moore, and J. M. Coleman. *American Association of Petroleum Geologists Studies in Geology*, **7**, 43–65.

Watson, E. B., Wark, D. A., and Thomas, J. B. (2006). Crystallization thermometers for zircon and rutile. *Contributions to Mineralogy and Petrology*, 151, 413–33.

Watts, A. B. (1978). An analysis of isostasy in the world's oceans: 1. Hawaiian-Emperor seamount chain. *Journal of Geophysical Research*, 83, 5989–6004.

(2001). *Isostasy and Flexure of the Lithosphere*. Cambridge University Press, Cambridge, p. 458.

Watts, A. B., Karner, G. D., and Steckler, M. (1982). Lithospheric flexure and the evolution of sedimentary basins. *Philosophical Transactions of the Royal Society of London*, 305, 249–81.

Watts, A. B., Peirce, C., Collier, J., Dalwood, R., Canales, J. P., and Henstock, T. J. (1997). A seismic study of lithospheric flexure in the vicinity of Tenerife, Canary Islands. *Earth and Planetary Science Letters*, 146, 431–47.

Watts, A. B. and Stewart, J. (1998). Gravity anomalies and segmentation of the continental margin offshore West Africa. *Earth and Planetary Science Letters*, 156, 239–52.

Watts, A. B. and Thorne, J. (1984). Tectonics, global changes in sea level and their relationship to stratigraphic sequences at the U.S. Atlantic continental margin. *Marine and Petroleum Geology*, 1, 319–39.

Watts, A. B. and Zhong, S. (2000). Observations of flexure and the rheology of oceanic lithosphere. *Geophysical Journal International*, 142, 855–75.

Watts, N. L. (1987). Theoretical aspects of cap-rock and fault seals for single- and two-phase hydrocarbon columns. *Marine and Petroleum Geology*, 4, 274–307.

Wdowinski, S. and Axen, G. J. (1992). Isostatic rebound due to tectonic denudation; a viscous flow model of a layered lithosphere. *Tectonics*, 11, 303–15.

Weast, R. C. (1974). *CRC Handbook of Chemistry and Physics*. CRC Press, Cleveland.

Weber, K. J. and Daukoru, E. (1975). Petroleum geology of the Niger delta. *9th World Petroleum Congress Proceedings*, Tokyo, 2, 209–21.

Weeks, L. G. (1952). Factors of sedimentary basin development that control oil occurrence. *American Association of Petroleum Geologists Bulletin*, 36, 2071–124.
Weimer, P., Varnai, P., Budhijanto, F. M., Acosta, Z. M., Martinez, R. E., Navarro, A. F., Rowan, M. G., McBride, B. C., Villamil, T., Arango, C., Crews, J. R., and Pulham, A. J. (1998). Sequence stratigraphy of Pliocene and Pleistocene turbidite systems, northern Green Canyon and Ewing Bank (offshore Louisiana), northern Gulf of Mexico. *American Association of Petroleum Geologists Bulletin*, 82, 918–60.
Weinberg, R. F. and Podladchikov, Y. Y. (1994). Diapiric ascent of magmas through power law crust and mantle. *Journal of Geophysical Research*, 99, 9543–59.
(1995). The rise of solid-state diapirs. *Journal of Structural Geology*, 17, 1183–95.
Weissel, J. K. and Karner, G. D. (1989). Flexural uplift of rift flanks due to mechanical unloading of the lithosphere during extension. *Journal of Geophysical Research*, 94, 13919–50.
Welsink, H. J., Dwyer, J. D., and Knight, R. J. (1989). Tectono-stratigraphy of the passive margin off Nova Scotia. *American Association of Petroleum Geologists Memoir*, 46, 215–31.
Welte, D. H., Hantschel, T., and Wygrala, B. (1997). Petroleum systems and the role of multi-dimensional basin modeling. *Proceedings of an International Conference on Petroleum Systems of SE Asia and Australasia*, pp. 23–4.
Wenger, L. M., Davis, C. L., and Isaksen, G. H. (2002). Multiple controls on petroleum biodegradation and impact on oil quality. *Society of Petroleum Engineers Reservoir Formation Evaluation and Engineering*, 5, 375–83.
Wernicke, B. (1985). Uniform-sense simple shear of the continental lithosphere. *Canadian Journal of Earth Sciences*, 22, 108–25.
Wernicke, B. P., Axen, G. J., and Snow, J. K. (1988). Basin and Range extensional tectonics at the latitude of Las Vegas, Nevada. *Geological Society of America Bulletin*, 100, 1738–57.
Wernicke, B. and Burchfiel, B. C. (1982). Modes of extensional tectonics. *Journal of Structural Geology*, 4, 105–15.
Wesson, R. L. (1988). Dynamics of fault creep. *Journal of Geophysical Research*, 93, 8929–51.
Wheatley, T. J., Biggins, D., Buckingham, J., and Holloway, N. H. (1987). The geology and exploration of the transitional shelf, an area to the west of the Viking Graben. In *Petroleum Geology of North West Europe*, ed. J. Brooks and K. W. Glennie. Graham and Trotman, London, pp. 979–89.
Whelan, J., Kennicut, M., Brooks, J., Schumacher, D., and Eglington, L. (1994). Organic geochemical indicators of dynamic fluid flow processes in petroleum basins. *Organic Geochemistry*, 22, 587–615.
White, D. A. (1993). Geologic risking guide for prospects and plays. *American Association of Petroleum Geologists Bulletin*, 77, 2048–64.
White, N. and Lovell, B. (1997). Measuring the pulse of a plume with the sedimentary record. *Nature*, 387, 888–91.
White, N. and McKenzie, D. (1988). Formation of the "steer's head" geometry of sedimentary basins by differential stretching of the crust and mantle. *Geology*, 16, 250–3.
White, R. S. (1992). Crustal structure and magmatism of North Atlantic continental margins. *Journal of the Geological Society of London*, 149, 841–54.
(1997). Rift-plume interaction in the North Atlantic. Series A. *Philosophical Transactions of the Royal Society of London*, 355, 319–39.
White, R. S. and McKenzie, D. (1989). Magmatism at rift zones: the generation of volcanic continental margins and flood basalts. *Journal of Geophysical Research*, 94, 7685–729.
White, R. S., McKenzie, D., and O'Nions, R. K. (1992). Oceanic crustal thickness from seismic measurements and rare earth element inversions. *Journal of Geophysical Research*, 97, 19683–715.
White, R. S., Spence, G. D., Fowler, S. R., McKenzie, D. P., Westbrook, G. K., and Bowen, A. N. (1987). Magmatism at rifted continental margins: *Nature*, 330, 439–44.
Whitehead, M. and Pinnock, S. J. (1991). The Highlander Field, Block 14/20b, UK North Sea. In *United Kingdom Oil and Gas Fields, 25 Years Commemorative Volume*, ed. I. L. Abbotts. *Memoir of the Geological Society of London*, **14**, 323–9.
Whiticar, M. J. (1994). Correlation of natural gases with their sources. In *The Petroleum System – From Source to Trap*, ed. L. B. Magoon and W. G. Dow. *American Association of Petroleum Geologists Memoir*, **60**, 261–83.
Whiticar, M. J. and Faber, E. (1986). Methane oxidation in sediment and water column environments – isotopic evidence. In *Advances in Organic Geochemistry*, ed. D. Leythaeuser and J. Rulkotter. Pergamon Press, New York, **10**, 759–68.
Whitmarsh, R. B., Dean, S. M., Minshull, T. A., and Tompkins, M. (2000). Tectonic implications of exposure of lower continental crust beneath the Iberia abyssal plain, Northeast Atlantic Ocean: geophysical evidence. *Tectonics*, 19, 919–42.
Whitmarsh, R. B., Manatschal, G., and Minshull, T. A. (2001). Evolution of magma-poor continental margins from rifting to seafloor spreading. *Nature*, 413, 150–4.
Whitmarsh, R. B. and Miles, P. R. (1995). Models of the development of the West Iberia rifted continental margin at 40°30'N deduced from surface and deep-tow magnetic anomalies. *Journal of Geophysical Research*, 100, 3789–806.
Whitmarsh, R. B., Miles, P. R., and Mauffret, A. (1990). The ocean-continent boundary off western continental margin of Iberia – I. Crustal structure at 40°30'N. *Geophysical Journal International*, 103, 509–31.
Whitmarsh, R. B., White, R. S., Horsefield, S. J., Sibuet, J.-C., Recq, M., and Louvel, V. (1996). The ocean-continent boundary off the western continental margin of Iberia: crustal structure west of Galicia Bank. *Journal of Geophysical Research*, 101, 28291–314.
Wignall, P. B. and Maynard, J. R. (1993). The sequence stratigraphy of transgressive black shales. In *Source Rocks*

in a Sequence Stratigraphic Framework, ed. B. J. Katz and L. M. Pratt. *American Association of Petroleum Geologists Studies in Geology*, **37**, 35–47.

Wikipedia (2005). http://en.wikipedia.org.

Wilhelm, B. and Somerton, W. H. (1967). Simultaneous measurement of pore and elastic properties of rocks under triaxial stress condition. *Journal of the Society of Petroleum Engineers*, 7, 283–94.

Willemse, E. J. M., Peacock, D. C. P., and Aydin, A. (1997). Nucleation and growth of strike-slip faults in limestones from Somerset, UK. *Journal of Structural Geology*, 19, 1461–77.

Williams, H. H. and Eubank, R. T. (1995). Hydrocarbon habitat in the rift graben of the central Sumatra Basin, Indonesia. In *Hydrocarbon Habitat in Rift Basins*, ed. J. J. Lambiase. *Geological Society of London Special Publications*, **80**, 331–71.

Williams, J. J., Conner, D. C., and Peterson, K. E. (1975). Piper oilfield, North Sea: fault-block structure with Upper Jurassic beach/bar reservoir sands. *American Association of Petroleum Geologists Bulletin*, 59, 1581–601.

Williams, R. R. (1991). The Deveron field, Block 211/18a, UK North Sea. In *United Kingdom Oil and Gas Fields, 25 Years Commemorative Volume*, ed. I. L. Abbotts. *Memoir of the Geological Society of London*, **14**, 83–7.

Wilson, D. L. (2002). Planning for success: Profit vs. reserves. *Oil and Gas Journal*, 100, 30–1.

Wilson, D. S., Teagle, D. A. H., Alt, J., Banerjee, N., Umino, S., Miyashita, S., and Acton, G. D. (2005). Drilling a complete in situ section of upper oceanic crust formed at superfast spreading rate: Hole 1256D. *EGU General Assembly*, Vienna, p. 374.

Wilson, M. (1997). Thermal evolution of the Central Atlantic passive margins: continental break-up above a Mesozoic super-plume. *Journal of the Geological Society*, 154, 491–5.

Wilson, P. G., Turner, J. P., and Westbrook, G. K. (2003). Structural architecture of the ocean-continent boundary at an oblique transform margin through deep-imaging seismic interpretation and gravity modelling: Equatorial Guinea, West Africa. *Tectonophysics*, 374, 19–40.

Wilson, R. C. L., Manatschal, G., and Wise, S. (2001). Rifting along non-volcanic passive margins: stratigraphic and seismic evidence from the Mesozoic successions of the Alps and western Iberia. *Geological Society Special Publications*, 187, 429–52.

Wilson, W. R. (1980). *Thermal studies in a geothermal area*. PhD thesis, University of Utah, Salt Lake City, p. 145.

Winterer, E. L. (1989). Sediment thickness map of the Northeast Pacific. In *The Eastern Pacific Ocean and Hawaii. The Geology of North America*, ed. E. L. H. Winterer, D. M. Decker, and W. Robert. *Geological Society of America*, Denver, Colorado.

Wise, D. U. (1992). Dip domain method applied to the Mesozoic Connecticut Valley rift basins. *Tectonics*, 11, 1357–68.

Witcher, J. C. (1988). Geothermal resources of southwestern New Mexico and southeastern Arizona. *Field Conference Guidebook, New Mexico Geological Society*, 39, 191–8.

Withjack, M. O., Meisling, K. E., and Russell, L. R. (1988). Forced folding and basement-detached normal faulting in the Haltenbanken area, offshore Norway. *American Association of Petroleum Geologists Bulletin*, 72, 259.

Withjack, M. O., Olsen, P. E., and Schlische, R. W. (1995). Tectonic evolution of the Fundy rift basin, Canada: evidence of extension and shortening during passive margin Development. *Tectonics*, 14, 390–405.

Withjack, M. O., Olson, J., and Peterson, E. (1990). Experimental models of extensional forced folds. *American Association of Petroleum Geologists Bulletin*, 74, 1038–54.

Withjack, M. O., Schlische, R. W., and Baum, M. S. (2009). Extensional development of the Fundy rift basin, southeastern Canada, *Geological Journal*, 44 (6), 631–51.

Withjack, M. O., Schlische, R. W., and Olsen, P. E. (1998): Diachronous rifting, drifting, and inversion on the passive margin of Central Eastern North America: an analog for other passive margins. *American Association of Petroleum Geologists Bulletin*, 82, 817–35.

Witkind, I. J., Myers, W. B., Hadley, J. B., Hamilton, W., and Fraser, G. D. (1962). Geologic features of the earthquake at Hebgen Lake, Montana, August 17, 1959. *Bulletin of the Seismological Society of America*, 52, 163–80.

Witlox, H. W. M. (1986). Finite element simulation of basal extension faulting within a sediment overburden. *Numerical Methods in Geomechanics – European Conference*, Stuttgard, 2, 765.

Wojtal, S. and Mitra, G. (1986). Strain hardening and strain softening in fault zones from foreland thrusts. *Geological Society of America Bulletin*, 97, 674–87.

Wolf, R. A., Farley, K. A., and Silver, L. T. (1996). Helium diffusion and low-temperature thermochronometry of apatite. *Geochimica et Cosmochimica Acta*, 60, 4231–40.

Wolfe, C. J., McNutt, M. K., and Detrick, R. S. (1994). The Marquesas archipelagic apron: seismic stratigraphy and implications for volcano growth, mass wasting, and crustal underplating. *Journal of Geophysical Research*, 99, 13591–608.

Wolff, G. A., Lamb, N. A., and Maxwell, J. R. (1986a). The origin and fate of 4-methyl steroid hydrocarbons. I. Diagenesis of 4-methyl steranes. *Geochimica et Cosmochimica Acta*, 50, 335–42.

(1986b). The origin and fate of 4-methyl steroid hydrocarbons. II. Dehydration of steroids and occurrence of C_{30} 4-methylsteranes. *Geochimica et Cosmochimica Acta*, 50, 965–74.

Wollard, G. P. (1959). Crustal structure from gravity and seismic measurements. *Journal of Geophysical Research*, 64, 1521–44.

Wood, D. (2003). More aspects of E&P asset and portfolio risk analysis. *Oil and Gas Journal*, 101, 28–32.

Wood, J. A. and Hewett, T. A. (1982). Fluid convection and mass transfer in porous sandstones – a theoretical approach. *Geochimica et Cosmochimica Acta*, 46, 1701–13.

(1984). Reservoir diagenesis and convective fluid flow. In *Clastic Diagenesis*, ed. D. A. McDonald and R. C. Surdam. *American Association of Petroleum Geologists Memoir*, **37**, 99–110.

Woodbury, H. O., Murray, I. B., Pickford, P. J., and Akers, W. H. (1973). Pliocene and Pleistocene depocenters, outer continental shelf of Louisiana and Texas. *American Association of Petroleum Geologists Bulletin*, 57, 2428–39.

Woodside, W. and Messmer, J. H. (1961). Thermal conductivity of porous media I: Unconsolidated sand. *Journal of Applied Geophysics*, 32, 1688–706.

Workman, R. K. and Hart, S. R. (2005). Major and trace element composition of the depleted MORB mantle (DMM). *Earth and Planetary Science Letters*, 231, 53–72.

Wride, V. C. (1994). Structural features and structural styles from the five countries area of the North Sea Central Graben. *First Break*, 13, 395–407.

Wright, L. A. (1974). *Fault Map of the Region of Central and Southern Death Valley, Eastern California and Western Nevada, Guidebook; Death Valley Region, California and Nevada*. Death Valley Publication Company.

Wright, L. D. (1977). Sediment transport and deposition at river mouths: a synthesis. *American Association of Petroleum Geologists Bulletin*, 88, 857–68.

Wright, T. L. (1991). Structural geology and tectonic evolution of the Los Angeles basin, California. In *Active Margin Basins*, ed. K. T. Biddle. *American Association of Petroleum Geologists Memoir*, **52**, 35–135.

Wygrala, B. P. (1989). *Integrated Study of an Oil Field in the Southern Po Basin, Northern Italy*. University of Koln.

Xie, X., Jiao, J. J., Tang, Z., and Zheng, C. (2003). Evolution of abnormally low pressure and its implications for the hydrocarbon system in the southeast uplift zone of Songliao Basin, China. *American Association of Petroleum Geologists Bulletin*, 87, 99–119.

Xie, X. N., Li, S. T., Dong, W. L., and Hu, Z. L. (2001). Evidence for episodic expulsion of hot fluids along faults near diapiric structures of the Yinggehai Basin, South China Sea. *Marine and Petroleum geology*, 18, 715–28.

Xiong, Y., Wang, Y., and Wang, Y. (2007). Selective chemical degradation of kerogen from Nenjiang Formation of the southern Songliao Basin. *Science in China, Series D, Earth Series*, 50, 1504–9.

Xue, L. and Galloway, W. E. (1993). Genetic sequence stratigraphic framework, depositional style, and hydrocarbon occurrence of the Upper Cretaceous QYN formations in the Songliao lacustrine basin, northeastern China. *American Association of Petroleum Geologists Bulletin*, 77, 1792–808.

Yalcin, M. N., Littke, R., and Sachsenhofer, R. F. (1997). Thermal history of sedimentary basins. In *Petroleum and Basin Evolution: Insights from Petroleum Geochemistry, Geology and Basin Modeling*, ed. D. H. Welte, B. Horsfield, and D. R. Baker. Springer-Verlag, Berlin, pp. 71–167.

Yaliz, A. 1991. The Crawford Field, block 9/28a, UK North Sea. In *United Kingdom Oil and Gas Fields, 25 Years Commemorative Volume*, ed. I. L. Abbotts. *Memoir of the Geological Society of London*, **14**, 287–93.

Yamasaki, T. and Nakada, M. (1997). The effects of the spinel-garnet phase transition on the formation of rifted sedimentary basins. *Geophysical Journal International*, 130, 681–92.

Yang, J., Large, R. R., Bull, S., and Scott, D. L. (2006). Basin-scale numerical modeling to test the role of buoyancy-driven fluid flow and heat transfer in the formation of stratiform Zn-Pb-Ag deposits in the northern Mount Isa Basin. *Economic Geology*, 101, 1275–92.

Yang, J., Latychev, K., and Edwards, R. N. (1998). Numerical computation of hydrothermal fluid circulation in fractured earth structures. *Geophysical Journal International*, 135, 627–49.

Yardley, B. W. D. (1986). Is there water in the deep continental crust? *Nature*, 323, 111.

Ye, S., Canales, J. P., Rihm, R., Danobeitia, J. J., and Gallart, J. (1999). A crustal transect through the northern and northeastern part of the volcanic edifice of Gran Canaria, Canary Islands. *Journal of Geodynamics*, 28, 3–26.

Yiang, W., Li, Y., and Gao, R. (1985). Formation and evolution of nonmarine petroleum in Songliao Basin, China. *American Association of Petroleum Geologists Bulletin*, 69, 1112–22.

Yielding, G., Badley, M. E., and Roberts, A. M. (1992). The structural evolution of the Brent Province. In *Geology of the Brent Group*, ed. A. C. Morton, R. S. Haszeldine, M. R. Giles, and S. Brown. *Geological Society Special Publications*, pp. 27–43.

Yielding, G., Jackson, J. A., King, G. C. P., Sinval, H., Vita-Finzi, C., and Wood, R. M. (1981). Relations between surface deformation, fault geometry, seismicity and rupture characteristics during the El Asnam (Algeria) earthquake of 10 October 1980. *Earth and Planetary Science Letters*, 56, 287–304.

Younes, A. I. (1996). *Fracture distribution on faulted basement blocks, Gulf of Suez, Egypt: reservoir characterization and tectonic implications*. PhD. thesis, Pennsylvania State University at University Park, University Park, p. 170.

Younes, A. I. and McClay, K. R. (2002). Development of accommodation zones in the Gulf of Suez–Red Sea Rift, Egypt. *American Association of Petroleum Geologists Bulletin*, 86, 1003–26.

Yuen, D. A. and Fleitout, L. (1985). Thinning of the lithosphere by small-scale convective destabilization. *Nature*, 313, 125–8.

Yukler, M. A. and Erdogan, T. L. (1996). Effects of geological parameters on temperature histories of basins. *Annual Meeting Abstracts – American Association of Petroleum Geologists and Society of Economic Paleontologists and Mineralogists*, 5, 157–8.

Zalán, P. V., Nelson, E. P., Warme, J. E., and Davis, T. (1985). The Piauí Basin, rifting and wrenching in an equatorial Atlantic transform basin. In *Strike-slip Deformation, Basin Formation, and Sedimentation*, ed. K. T. Biddle and N. Christie-Blick. Society of Economic Paleontologists and Mineralogists. Tulsa, OK, Special Publication 37, pp. 177–92.

Zalán, P. V., Severino, M. C. G., Rigoti, C. A., Magnavita, L. P., Oliveira, J. A. B., and Vianna, A. R. (2011). An entirely new 3-D view of the crustal and mantle structure of a South Atlantic passive margin – Santos, Campos and Espírito Santo Basins, Brazil. *Search and Discovery Article No. 30177, Adapted from expanded abstract presentation at AAPG Annual Convention and Exhibition*, Houston, Texas, April 10–13, 2011.

Zanella, E. and Coward, M. P. (2003). Chapter 4. Structural framework. In *The Millennium Atlas: Petroleum Geology of the Central and Northern North Sea*, ed. D. Evans, C. Graham, A. Armour, and P. Bathurst. *Geological Society of London*, pp. 45–59.

Zappone, A., Burlini, L., Connolly, J., Husen, S., and Kissling, E. (2006). Lower crustal seismic velocities and seismicity in the northern Alpine foreland: I. Observations and petrophysical experiments. *Geophysical Research Abstracts*, 8.

Zeitler, P. K., Herczeg, A. L., McDougall, I., and Honda, M. (1987). U-Th-He dating of apatite: a potential thermochronometer. *Geochimica et Cosmochimica Acta*, 51, 2865–8.

Zeng, J. H. and Jin, Z. J. (2003). Experimental investigation of episodic oil migration along fault systems. *Journal of Geochemical Exploration*, 78–79, 493–8.

Zhang, Y. J. (2002). *Petroleum migration and formation in fault-controlled Junggar Basin (in Chinese)*. PhD thesis, University of Petroleum, Beijing, p. 232.

Zhao, G. and Johnson, A. M. (1991). Sequential and incremental formation of conjugate faults. *Journal of Structural Geology*, 13, 887–896.

Ziegler, P. A. (1989a). Geodynamic model for Alpine intra-plate compressional deformation in Western and Central Europe. In *Inversion Tectonics Meeting*, ed. M. A. Cooper and G. D. Williams. *Geological Society of London Special Publications*, **44**, 63–85.

(1989b). Evolution of the North Atlantic – an overview. In *Extensional Tectonics and Stratigraphy of the North Atlantic Margins*, ed. A. J. Tankard and H. R. Balkwill. *American Association of Petroleum Geologists Memoir*, 46, 111–29.

(1990). *Geological Atlas of Western and Central Europe (second edition)*. Shell Internationale Petroleum Maatschappij and Geological Society, London, p. 239.

(1990). Tectonic and palaeogeographic development of the North Sea rift system. In *Tectonic Evolution of the North Sea rifts*, ed. D. J. Blundell and A. D. Gibbs. Clarendon Press, Oxford, p. 36.

(1994). Geodynamic processes governing development of rifted basins. In *Geodynamic Evolution of Sedimentary Basins, the International Symposium Held in Moscow, May 18–23, 1992*, ed. F. Roure, N. Ellouz, V. S. Shein, and I. I. Skvortsov. Editions Technip, Paris, pp. 19–37.

(1995). Cenozoic rift system of Western and Central Europe: an overview. *Geologie en Mijnbouw*, 73, 99–127.

Zijerveld, L., Stephenson, R., Cloetingh, S., Duin, E., and van den Berg, M. W. (1992). Subsidence analysis and modelling of the Roer Valley Graben (SE Netherlands): Geodynamics of rifting: Volume I, Case history studies on rifts; Europe and Asia. *Geodynamics of rifting symposium*, Glion-sur-Montreux, Switzerland, 208, 159–71.

Zimmerman, R. W., Somerton, W. H., and King, M. S. (1986). Compressibility of porous rocks. *Journal of Geophysical Research*, 91, 12765–77.

Zoback, M. L. (1983). Structure and Cenozoic tectonism along the Wasatch fault zone, Utah. In *Tectonics and Stratigraphy of the Eastern Great Basin*, ed. D. M. Miller, V. R. Todd, and K. A. Howard. *Geological Society of America Memoir*, 157, 3–37.

Zoetemeijer, R. (1993). *Tectonic modeling of foreland basins: thin-skinned thrusting, syntectonic sedimentation and lithospheric flexure*. PhD thesis,Vrije University, Amsterdam, p. 148.

Zoth, G. (1979). Temperaturmessungen in der Bohrung Hanigsen sowie Warmeleitfahigeitbestimmungen von Gesteinproben. *Report BGR/NLfB Hannover*, 81, 828.

Zoth, G., Haenel, R., Rybach, L and Stegena, L. (1988). Appendix. In *Handbook of Terrestrial Heat-flow Density Determination; with Guidelines and Recommendations of the International Heat Flow Commission*, ed. R. Haenel. Kluwer Academic Publishers, Dordrecht.

Zuber, M. T. and Parmentier, E. M. (1986). Lithospheric necking: a dynamic model for rift morphology. *Earth and Planetary Science Letters*, 77, 373–83.

Zwach, C., Poelchau, H. S., Hantschel, T., and Welte, D. H. (1994). Simulation with contrasting pore fluids: Can we afford to neglect hydrocarbon saturation in basin modeling? In *Basin modeling – What Have We learned?*, ed. S. Dueppenbecker and J. Iliffe. *Proceedings of Basin Modeling Conference*, British Geological Survey, p. 4.

Index